D0079244

Applied Circuit Analysis

Matthew N. O. Sadiku

Prairie View A&M University

Sarhan M. Musa

Prairie View A&M University

Charles K. Alexander

Cleveland State University

The McGraw-Hill Companies

APPLIED CIRCUIT ANALYSIS

Published by McGraw-Hill, a business unit of The McGraw-Hill Companies, Inc., 1221 Avenue of the Americas, New York, NY 10020. Copyright © 2013 by The McGraw-Hill Companies, Inc. All rights reserved. Printed in the United States of America. No part of this publication may be reproduced or distributed in any form or by any means, or stored in a database or retrieval system, without the prior written consent of The McGraw-Hill Companies, Inc., including, but not limited to, in any network or other electronic storage or transmission, or broadcast for distance learning.

Some ancillaries, including electronic and print components, may not be available to customers outside the United States.

This book is printed on acid-free paper

1 2 3 4 5 6 7 8 9 0 RJE/RJE 1 0 9 8 7 6 5 4 3 2

ISBN 978-0-07-802807-6
MHID 0-07-802807-8

Vice President & Editor-in-Chief: *Marty Lange*
Vice President & Director Specialized Publishing: *Janice M. Roerig-Blong*
Editorial Director: *Michael Lange*
Publisher: *Raghothaman Srinivasan*
Developmental Editor: *Darlene M. Schueller*
Senior Marketing Manager: *Curt Reynolds*
Senior Project Manager: *Lisa A. Bruflodt*
Design Coordinator: *Margarite Reynolds*
Cover Designer: *Studio Montage, St. Louis, Missouri*
Cover Image: *NASA/MSFC*
Buyer: *Susan K. Culbertson*
Media Project Manager: *Balaji Sundararam*
Compositor: *MPS Limited, a Macmillan Company*
Typeface: *10/12 Times Roman*
Printer: *R.R. Donnelley*

All credits appearing on page or at the end of the book are considered to be an extension of the copyright page.

Library of Congress Cataloging-in-Publication Data

Sadiku, Matthew N. O.
Applied circuit analysis / Matthew N. O. Sadiku, Sarhan M. Musa, Charles K. Alexander. — 1st ed.
p. cm. — (Applied circuit analysis)
ISBN 978-0-07-802807-6 (alk. paper)
1. Electric circuit analysis. I. Musa, Sarhan M. II. Alexander, Charles K. III. Title.

TK454.S3134 2013
621.319'2—dc23

2011037835

www.mhhe.com

Dedication

Matthew Sadiku dedicates this book to:

Loving memory of his father, Solomon, and mother, Ayisat

Sarhan Musa dedicates this book to:

His father, Mahmoud, mother, Fatmeh, and wife, Lama

Charles Alexander dedicates this book to:

His wife, Hannah

Table of Contents

Preface

This book is intended to present circuit analysis to engineering technology students in a manner that is clearer, more interesting, and easier to understand than other texts. This objective is achieved in the following ways:

- A course in circuit analysis is perhaps the first exposure students have to electrical engineering technology. We have included several features to help students feel at home with the subject. Each chapter opens with a historical profile or career talk. This is followed by an introduction that links the chapter with the previous chapters and states the chapter objectives. The chapter ends with a summary of key points and formulas.
- All principles are presented in a lucid, logical, step-by-step manner. As much as possible, we avoid wordiness and too much detail that could hide concepts and impede overall understanding of the material.
- Important formulas are boxed as a means of helping students sort out what is essential from what is not. Also, to ensure that students clearly get the gist of the matter, key terms are defined and highlighted.
- Thoroughly worked examples are liberally given at the end of every section. The examples are regarded as a part of the text and are clearly explained without asking the reader to fill in missing steps. Thoroughly worked examples give students a good understanding of the solution and the confidence to solve problems themselves. Some of the problems are solved in two or three ways to facilitate an understanding and comparison of different approaches.
- To give students practice opportunity, each illustrative example is immediately followed by a practice problem with the answer. The students can follow the example step-by-step to solve the practice problem without flipping pages or looking at end of the book for answers. The practice problem is also intended to test if students understand the preceding example. It will reinforce their grasp of the material before they move on to the next section.
- The last section in each chapter is devoted to application aspects of the concepts covered in the chapter. The material covered in the chapter is applied to at least one practical problem or device. This helps the students see how the concepts are applied to real-life situations.
- Ten review questions in multiple-choice form are provided at the end of each chapter, with answers. The review questions are intended to cover the little "tricks" that the examples and end-of-chapter problems may not cover. They serve as a self-test device and help students determine how well they have mastered the chapter.

- In recognition of the requirements by the ABET (Accreditation Board for Engineering and Technology) on integrating computer tools, the use of PSpice® and NI Multisim™ is encouraged in a student-friendly manner. Appendix C serves as a tutorial on PSpice for Windows, while Appendix D provides an introduction to Multisim. The latest versions of these software packages are used in the book. We also encourage the use of a TI-89 Titanium calculator and MATLAB® for number crunching.

Organization

This book was written for a two-semester or three-quarter course in linear circuit analysis. The book may also be used for a one-semester course by a proper selection of chapters and sections by the instructor. It is broadly divided into two parts. Part 1, consisting of Chapters 1 to 10, is devoted to dc circuits. Part 2, which contains Chapter 11 to 19, deals with ac circuits. The material in two parts is more than sufficient for a two-semester course, so the instructor must select which chapters or sections to cover. Sections preceded with the dagger sign may be skipped, explained briefly, or assigned as homework.

Prerequisites

As with most introductory circuit courses, the main prerequisite for a course using the text is physics. Although familiarity with complex numbers is helpful in the later part of the book, it is not required.

Online Resources

A website to accompany this text is available at www.mhhe.com/sadiku. The site includes a password-protected solutions manual, worked solutions in PSpice and Multisim, and an image library for instructors. Instructors can also obtain access to COSMOS for this text. COSMOS is a Complete Online Solutions Manual Organization System instructors can use to create exams and assignments, create custom content, and edit supplied problems and solutions.

McGraw-Hill Create™

Craft your teaching resources to match the way you teach! With McGraw-Hill Create, www.mcgrawhillcreate.com, you can easily rearrange chapters, combine material from other content sources, and quickly upload content you have written, such as your course syllabus or teaching notes. Find the content you need in Create by searching through thousands of leading McGraw-Hill textbooks. Arrange your book to fit your teaching style. Create even allows you to personalize your book's appearance by selecting the cover and adding your name, school, and course information. Order a Create book and you'll receive a complimentary print review copy in 3 to 5 business days or a complimentary electronic review copy (eComp) via e-mail in minutes. Go to www.mcgrawhillcreate.com today and register to experience how McGraw-Hill Create empowers you to teach *your* students *your* way.

Electronic Textbook Option

This text is offered through CourseSmart for both instructors and students. CourseSmart is an online resource where students can purchase the complete text online at almost half the cost of a traditional text. Purchasing the eTextbook allows students to take advantage of CourseSmart's web tools for learning, which include full text search, notes and highlighting, and email tools for sharing notes between classmates. To learn more about CourseSmart options, contact your sales representative or visit www.CourseSmart.com.

Acknowledgments

Special thanks are due to Robert Prather and Dr. Warsame Ali for their help with Multisim. We appreciate the support received from Dr. Kendall Harris, dean of the college of engineering at Prairie View A&M University. We would like to thank Dr. John Attia for his support and understanding. We extend our appreciation to Dr. Karl J. Huehne for thoroughly going through the text and the solutions of the problems making sure they are accurate. The insights and cooperation received from the McGraw-Hill team (Raghu Srinivasan, Darlene Schueller, Lora Neyens, Curt Reynolds, Lisa Bruflodt, Margarite Reynolds, LouAnn Wilson, Ruma Khurana, and Dheeraj Chahal) is very much appreciated.

We would like to thank the following reviewers for their comments:

Ryan Beasley, *Texas A&M University*
Michael E. Brumbach, *York Technical College*
Thomas Cleaver, *University of Louisville*
Walter O. Craig III, *Southern University*
Chad Davis, *University of Oklahoma*
Mark Dvorak, *Minnesota State University–Mankato*
Karl Huehne, *Indiana University–Purdue University Indianapolis*
Rajiv Kapadia, *Minnesota State University–Mankato*
Mequanint Moges, *University of Houston*
Jerry Newman, *University of Memphis*
Brian Norton, *Oklahoma State University*
Norali Pernalete, *California State Polytechnic University–Pomona*
John Ray, *Louisiana Tech University*
Barry Sherlock, *University of North Carolina–Charlotte*
Ralph Tanner, *Western Michigan University*
Wei Zhan, *Texas A&M University*

M. N. O. Sadiku, S. M. Musa and C. K. Alexander

Notes to Students

This may be one of your first courses in electrical engineering technology. Although electrical engineering technology is an exciting and challenging discipline, the course may intimate you. This book was written to prevent that. A good textbook and a good professor are an advantage-but you are the one who does the learning. If you keep the following ideas in mind, you will do very well in this course:

- This course is the foundation on which most other courses in the electrical engineering technology curriculum rest. For this reason, put in as much effort as you can. Study the course regularly.
- Problem solving is an essential part of learning process. Solve as many problems as you can. Begin by solving the practice problems following each example and then proceed to the end-of-chapter problems. The best way to learn is to solve a lot of problems. An asterisk in front of a problem indicates a challenging problem.
- Spice, a computer circuit analysis program, is used throughout the textbook. PSpice, the personal computer version of Spice, is the popular standard circuit analysis program at most universities. PSpice for Windows is described in Appendix C. Make an effort to learn PSpice, because you can check any circuit problem with PSpice and be sure you are handing in a correct problem solution.
- Multisim is another tool that helps you simulate what would otherwise be a real electronics workbench, complete with drawings, parts, and instruments. A quick introduction to Multisim is provided in Appendix D.
- MATLAB is a software package that is very useful in circuit analysis and other courses you will be taking. A brief tutorial on MATLAB is given in Appendix E to get you started. The best way to learn MATLAB is to start with it once you know a few commands.
- Each chapter ends with a section on how the material covered in the chapter can be applied to real-life situations. The concepts in this section may be new and advanced to you. No doubt, you will learn more of the details in other courses. We are mainly interested in gaining a general familiarity with these ideas.
- Attempt the review questions at the end of each chapter. They will help you discover some "tricks" not revealed in class or in the textbook.

A short review of finding determinants is covered in Appendix A, complex numbers in Appendix B, PSpice for Windows in Appendix C, Multisim in Appendix D, MATLAB in Appendix E, and the TI-89 Titanium calculator in Appendix F. Answers to odd-numbered problems are given in Appendix G.

Have fun!

PART ONE

DC Circuits

chapter

1

Basic Concepts

Technology feeds on itself. Technology makes more technology possible.

—Alvin Toffler

Historical Profiles

Alessandro Volta Alessandro Volta (1745–1827), an Italian physicist, invented the electric battery, which provided the first continuous flow of electricity, and the capacitor.

Born into a noble family in Como, Italy, Volta started performing electrical experiments at age 18. The invention of the battery by Volta in 1796 revolutionized the use of electricity. The publication of his work in 1800 marked the beginning of electric circuit theory. Volta received many honors during his lifetime, and the unit of voltage or potential difference, the *volt,* was named in his honor.

Alessandro Volta
© The Huntington Library, Burndy Library, San Marino, California

Andre-Marie Ampere Andre-Marie Ampere (1775–1836), a French mathematician and physicist, laid the foundation of electrodynamics (now called electromagnetism). It was during the 1820s that he defined electric current and developed a method to measure it.

Born in Lyons, France, Ampere mastered Latin rapidly because he was intensely interested in mathematics, and many of the best mathematical works at that time were in Latin. He was a brilliant scientist and a prolific writer. He invented the electromagnet and the ammeter and formulated the laws of electromagnetics. The unit of electric current, the *ampere,* was named after him.

Andre-Marie Ampere
© Pixtal/age Fotostock RF

1.1 Introduction

Electric circuit theory is basic to electrical engineering technology. Many branches of electrical engineering technology—such as power, electric machines, feedback and control systems, electronics, computers, communications, and instrumentation—are based on electric circuit theory. Circuit theory is the starting point for a beginning student in electrical engineering technology education and is the single most important course you will take. Circuit theory is also valuable to students specializing in other branches of the physical sciences, because circuits are a good model for the study of energy systems in general and because of the applied mathematics, physics, and topology involved. Virtually everything that plugs into a wall outlet or uses batteries or in some form uses electricity can be analyzed using the techniques described in this book.

In electrical engineering technology, we are often interested in communicating or transferring energy from one point to another. Doing this requires an interconnection of electrical devices. Such interconnection is referred to as an *electric circuit* and each component of the circuit is known as an *element.* Thus,

> An **electric circuit** is an interconnection of electrical elements.

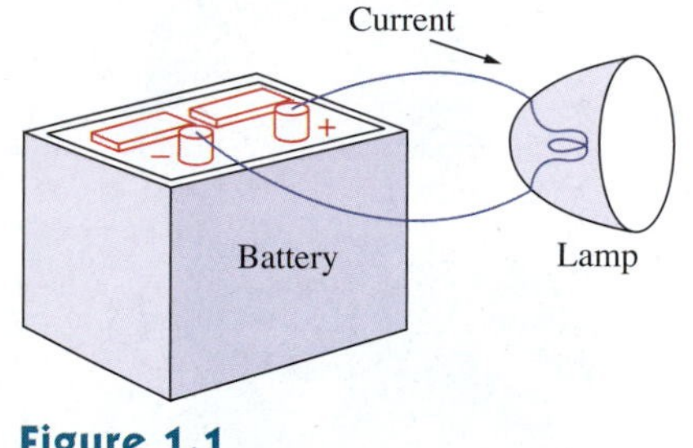

Figure 1.1
A simple electric circuit.

A simple electric circuit is shown in Fig. 1.1. It consists of three basic components: a battery, a bulb, and connecting wires. This simple circuit is used to power flashlights and searchlights, among other devices.

A complicated electric circuit is displayed in Fig. 1.2. It represents the schematic diagram for a radio transmitter. Although it seems complicated, it can be analyzed by using the techniques covered in this book. One of the goals of this course is for you to learn various analytical techniques and computer software for describing the behavior of circuits like this.

Electric circuits are used in numerous electrical systems to accomplish different tasks. Our objective in this book is not the study of the various uses and applications of circuits. Rather, our major concern is the *analysis* of electrical circuits. By this we mean a study of the

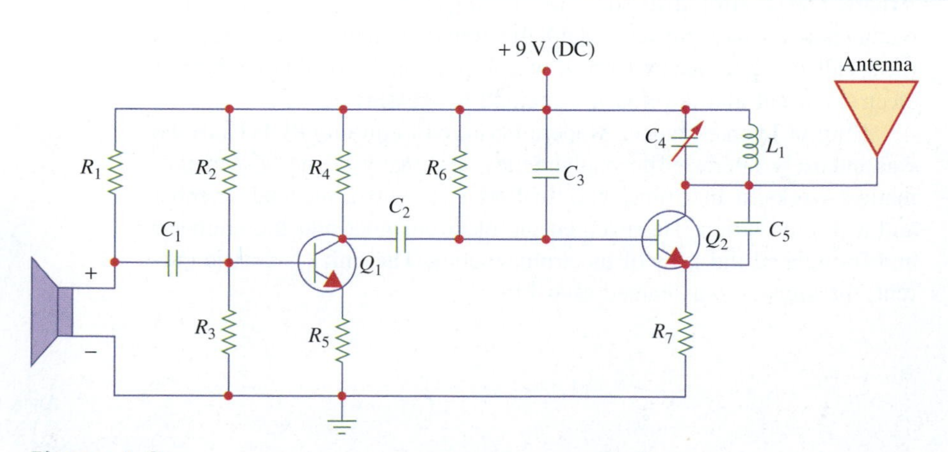

Figure 1.2
Electric circuit of a radio transmitter.

behavior of the circuit: How does it respond to a given input? How do the interconnected elements and devices in the circuit interact?

We begin our study by defining some basic concepts. Such concepts include charge, current, voltage, circuit elements, power, and energy. Before defining these concepts, we must first establish a system of units that will be used throughout the text.

1.2 International Systems of Units

As engineering technologists, we deal with measurable quantities. Our measurement, however, must be communicated in a standard language that all professionals can understand irrespective of the country in which the measurement is conducted. Such an international measurement language is the International System (SI) of Units, adopted by the General Conference on Weights and Measures in 1960. In this system, there are seven principal units from which the units of all other physical quantities can be derived. Table 1.1 shows the six units and one derived unit that are relevant to this text.

Although the SI units have been officially adopted by Institute of Electrical and Electronics Engineers (IEEE) and are used throughout this text, certain English units (non-SI units) are commonly used in practice in the United States. This is because the United States has only recognized the SI system and not officially mandated it. For example, distances are still specified in feet and miles, while electric motors are rated in horsepower. Thus, you will occasionally need to convert non-SI units to SI units using Table 1.2.

Example 1.1

Convert 42 inches to meters.

Solution:
From Table 1.2, 1 inch = 0.0254 meter. Hence

$$42 \text{ inches} = 42 \times 0.0254 \text{ meter} = 1.0668 \text{ meters}$$

Practice Problem 1.1

Convert 36 miles to kilometers.

Answer: 57.924 kilometers

TABLE 1.1

Six basic SI units and one derived unit relevant to this text.

Physical quality	Basic unit	Symbol
Length	meter	m
Mass	kilogram	kg
Time	second	s
Electric current	ampere	A
Thermodynamic temperature	kelvin	K
Luminous intensity	candela	cd
Charge	coulomb	C

TABLE 1.2

Conversion factors.

To convert from	To	Multiply by
Length		
inches(in)	meters (m)	0.0254
feet (ft)	meters (m)	0.3048
yards (yd)	meters (m)	0.9144
miles (mi)	kilometers (km)	1.609
mils (mil)	millimeters (mm)	0.0254
Volume		
gallons (gal)(U.S.)	liters (L)	3.785
cubic feet (ft^3)	cubic meter (m^3)	0.0283
Mass/weight		
pounds (lb)	kilogram (kg)	0.4536
Time		
hours (h)	seconds (s)	3600
Force		
pounds (lb)	newtons (N)	4.448
Power		
horsepower (hp)	watts (W)	746
Energy		
foot-pounds (ft-lb)	joules (J)	1.356
kilowatt-hours (kWh)	joules (J)	3.6×10^6

Example 1.2

An electric motor is rated at 900 watts. Express this in horsepower.

Solution:

From Table 1.2, 1 horsepower = 746 watts. Therefore, 1 watt = 1/746 horsepower. Thus,

$$900 \text{ watts} = 900 \times 1/746 = 1.206 \text{ horsepower}$$

Practice Problem 1.2

A force of 50 newtons is applied to a certain object. Express the force in pounds.

Answer: 11.241 pounds

1.3 Scientific and Engineering Notation

In science and engineering technology, we often encounter very small and very large numbers. These very small or large numbers can be expressed using of one of the following widely used notations:

- Scientific notation.
- Engineering notation.

Scientific notation uses the power of 10. In scientific notation, a number is usually expressed as $X.YZ \times 10^n$.

> In **scientific notation,** we express a number in powers of 10 with a single nonzero digit to the left of decimal point.

To write a number in scientific notation, we write it as a *coefficient* times 10 raised to an *exponent*. For example, we convert the number 0.000578 to scientific notation by shifting the decimal point four places to the right—that is, 5.78×10^{-4}. Similarly, in scientific notation, the number 423.56 becomes 4.2356×10^2, which is obtained by shifting the decimal point two places to the left. Thus, we note that 3.276×10^6 is in scientific notation, while 32.76×10^5 is not.

In **engineering notation**, we express a number by using certain powers of 10, as shown in Table 1.3.

In engineering technology in general and in circuit theory in particular, we are generally more interested in engineering notation. That is, engineering notation refers to the application of scientific notation in which the powers of 10 are limited to multiples of three. In fact, one great advantage of the SI units is that they use prefixes based on the power of 10 to relate larger and smaller units to the basic unit. Such SI prefixes and their symbols are shown in Table 1.3. Note that the prefixes are arranged in increments of three powers of 10. In engineering notation, a number can have from one to three digits to the left of the decimal point. For example, 0.0006 s is expressed in engineering notation as 600 μs.

We have moved the decimal point six places to the right, and we see that the prefix representing 10^{-6} is expressed as μ (micro). Similarly, 145,300 m is the same as 145.3 km in engineering notation. In this case, we have moved the decimal point three places to the left, and we see that the prefix representing 10^{+3} is expressed as k (kilo).

In electrical engineering technology, it is better to use the engineering notation than power of 10 and scientific notation. Certainly, you will find engineering notation used in all textbooks, technical manuals, and other technical material that you will be expected to read and use. Although this may seem difficult at first, it will become natural to you as you work with these prefixes.

TABLE 1.3

The SI prefixes.

Power of 10	Prefix	Symbol
10^{24}	yotta	Y
10^{21}	zetta	Z
10^{18}	exa	E
10^{15}	peta	P
10^{12}	tera	T
10^{9}	giga	G
10^{6}	mega	M
10^{3}	kilo	k
10^{-3}	milli	m
10^{-6}	micro	μ
10^{-9}	nano	n
10^{-12}	pico	p
10^{-15}	femto	f
10^{-18}	atto	a
10^{-21}	zepto	z
10^{-24}	yocto	y

Example 1.3

Express each of the following numbers in scientific notation:

(a) 621,409 (b) 0.00000548

Solution:

(a) The decimal point (not shown) is after 9—that is, 621409.0. If we shift the decimal point to five places to the left, we obtain

$$621409.0 = 6.21409 \times 10^5$$

which is in scientific notation.

(b) If we shift the decimal point six places to the right, we get

$$0.00000548 = 5.48 \times 10^{-6}$$

which is in scientific notation.

Practice Problem 1.3

Express the following numbers in scientific notation:

(a) 46,013,000 (b) 0.000245

Answer: (a) 4.6013×10^7 (b) 2.45×10^{-4}

Example 1.4

Express the following numbers in scientific notation:
(a) 2,563 m (b) 23.6 μs

Solution:

(a) 2563 m $= 2.563 \times 10^3$ m
(b) 23.6 μs $= 23.6 \times 10^{-6}$ s $= 2.36 \times 10^{-5}$ s

Practice Problem 1.4

Write the following numbers in scientific notation:
(a) 0.921 s (b) 145.6 km

Answer: (a) 9.21×10^{-1} s (b) 1.456×10^5 m

Example 1.5

Use engineering notation to represent the following numbers:
(a) 451,000,000 m (b) 0.0000782 s

Solution:

(a) 451,000,000 m $= 451 \times 10^6$ m $=$ 451 Mm (or 451,000 km)
(b) 0.0000782 s $= 78.2 \times 10^{-6}$ s $= 78.2$ μs

Practice Problem 1.5

Express the following numbers in engineering notation:
(a) 34,700,000,000 m (b) 0.0032 s

Answer: (a) 34.7 Gm (b) 3.2 ms

1.4 †Scientific Calculators[1]

Figure 1.3
Texas instruments TI-89 Titanium calculator.
© Sarhan M. Musa

Because circuit analysis involves a large number of calculations, you must master the use of an electronic hand calculator. The speed and accuracy of a handheld calculator are well worth the investment. The simplicity of a hand calculator is preferable to the sophistication of a personal computer. From time to time, calculator techniques will be given. Throughout the book, the calculator examples will be based on the TI-89 Titanium calculator, shown in Fig. 1.3. In fact, what we have done in Sections 1.2 and 1.3 (on conversions, scientific notation, and engineering notations) can easily be done using a TI-89 Titanium calculator. A brief review on how to use the calculator is provided in Appendix F. In case you do not have that type of calculator, make sure your calculator can at least perform the following operations: arithmetic ($+$, $-$, $\times$, $\div$), square root, sine, cosine, tangent, log (base 10), ln (base e), x^y (power), and exponential (e) and that it can convert rectangular to polar coordinates and vice versa. It does not need to have upgraded programmable features.

Most calculators can display numbers with 8 or 10 digits. It is not reasonable to show all the digits that we see on the display. In practice, numbers are usually rounded to three or four significant digits. For

[1]The dagger sign preceding a section heading indicates a section that your instructor may choose to skip, explain only briefly, or assign to you as homework.

example, the number 1.648247143 is rounded and recorded as 1.65, while the number 0.007543128 is recorded as 0.00754.

Example 1.6

Express the number 23,600 in engineering notation using the calculator.

Solution:
Using the TI-89 calculator, we take the following steps:

1. Press [MODE] to be in the Mode menu, where you specify how you want numbers.
2. Use the down-moving button to Exponential Format and select [Engineering].
3. Press [ENTER] to exit the mode screen.
4. Enter the number [2] [3] [6] [0] [0] [EE] [0] and press [ENTER].

The result will be displayed as

`23.6E3`

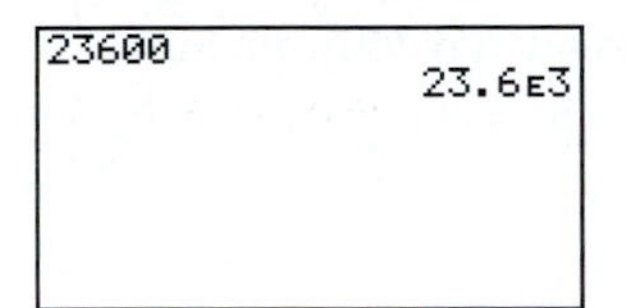

Practice Problem 1.6

Use a calculator to express 124,700 in engineering notation.

Answer: `124.7E3`

Example 1.7

Use a calculator to evaluate

$$\sqrt{\frac{45 - 7}{2}}$$

Solution:
We can do this in many ways. Perhaps it is easier to figure out the term under the square root sign and then take the square root of that.

1. Enter (45 − 7)/2 and press [ENTER]. The result is 19.
2. Press [2nd] and then [√]; enter [1] [9] [)] and press [♦] [ENTER].

The result is displayed as 4.3589.

```
(45-7)/2
                19
√19
     4.35889894354
■
```

Practice Problem 1.7

Using a calculator, find $\frac{125\pi}{\sqrt{36 + 17}}$.

Answer: 53.941

1.5 Charge and Current

Now that we are done with SI prefixes and the scientific calculator, we are ready to start our journey into electric circuit analysis. The most basic quantity in an electric circuit is *electric charge*. We all have experienced the effect of electric charge when we try to remove our wool sweater and have it stick to our body.

Charges of the same polarity (or sign) repel each other, while charges of opposite polarity attract each other. This means that all electric phenomena are manifestations of electric charge. Charges have polarity; they are either positive (+) or negative (−).

> **Electric charge** (Q) is an electrical property of matter responsible for electric phenomena, measured in coulombs (C).

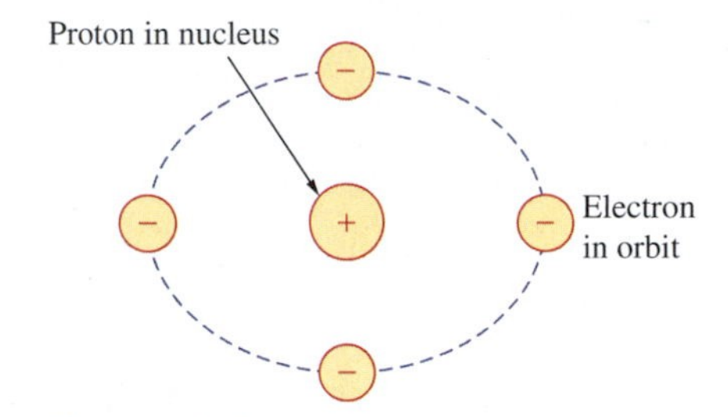

Figure 1.4
Atomic structure illustrating the nucleus and electrons.

We know from elementary physics that all matter is made of fundamental building blocks known as atoms and that each atom consists of electrons, protons, and neutrons as shown typically in Fig. 1.4. Because electrons carry negative charges, the charge carried by an electron is

$$e = -1.60 \times 10^{-19}\ \text{C} \tag{1.1}$$

A proton carries the same amount of charge but with positive polarity. The presence of equal number of protons and electrons leaves an atom electrically neutral. Charge comes in multiples of the charge on the electron or the proton.

Any material or body that has excess electrons is negatively charged, while any material or body with excess of protons (or deficiency of electrons) is positively charged. As shown in Fig. 1.5, unlike charges attract each other, while like charges repel each other.

The following points should be noted about electric charge:

1. The coulomb is a large unit for charges. In 1 C of charge, there are $1/(1.602 \times 10^{19}) = 6.24 \times 10^{18}$ electrons. Thus, realistic or laboratory values of charges are in the order of pC, nC, or μC.
2. According to experimental observations, the only charges that occur in nature are integral multiples of the electronic charge $e = 1.602 \times 10^{-19}$ C; that is, charge $Q = Ne$, where N is an integer.
3. The *law of conservation of charge* states that charge can be neither created nor destroyed; it can only be transferred. Thus, the algebraic sum of the electric charges in a closed system does not change.

We now consider the flow of electric charges. When a conducting wire consisting of trillions of atoms is connected to a battery (a source of electromotive force), the charges are compelled to move; positive

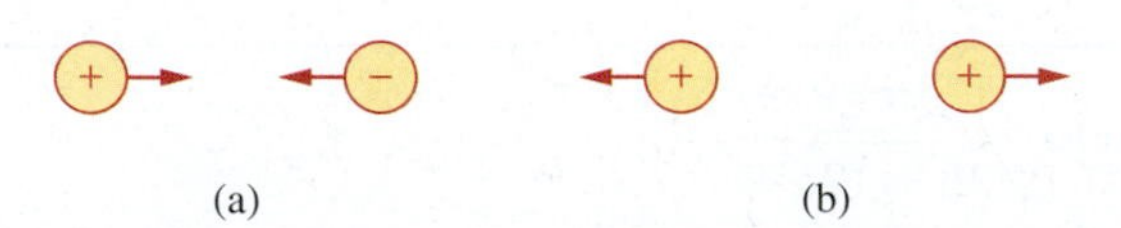

Figure 1.5
(a) Unlike charges attract; (b) like charges repel.

charges move in one direction while negative charges move in the opposite direction. This motion of charges creates electric current. It is conventional to take the direction of current as the movement of positive charges (i.e., opposite to the flow of negative charges), as illustrated in Fig. 1.6. (Note also that the positive side of the battery is the symbol with the longer bar.) This convention was introduced by Benjamin Franklin (1706–1790), an American scientist and inventor.[2] Although we now know that current in metallic conductors is due to negatively charged electrons, we will follow the universally accepted convention that current is the net flow of positive charges. Thus,

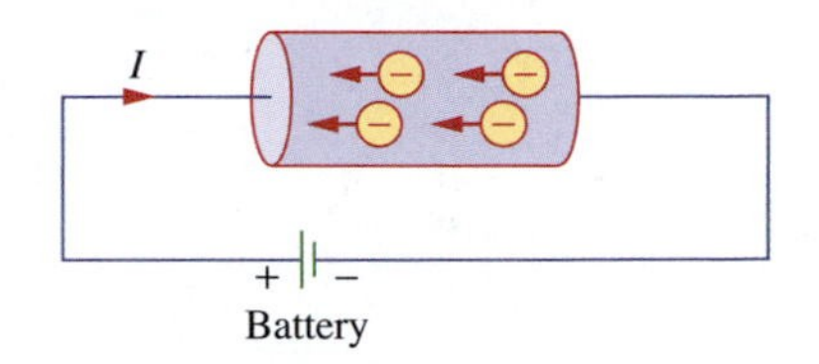

Figure 1.6
Electric current due to flow of electronic charge in a conductor.

> **Electric current** (I) is the time rate of flow of charge, measured in amperes (A).

Mathematically, the relationship between current I, charge Q, and time t is

$$I = \frac{Q}{t} \tag{1.2}$$

where current is measured in amperes (A)—that is, 1 ampere is 1 coulomb per second. There are several types of current; charge can vary with time in several ways, which are represented by different kinds of mathematical functions. If the current does not change with time, we call it *direct current* (dc). This is the current created by a battery. The symbol I is used to represent constant current.

> **Direct current** (dc) is current that remains constant with time.

A time-varying current is represented by the symbol i. A common form of time-varying current is the sinusoidal current or *alternating current* (ac). Alternating current is found in your household, used in running the heater, air conditioner, refrigerator, washing machine, and other electric appliances. Figure 1.7 graphs magnitude of current over time for direct current and alternating current, the two most common types of current. In general, alternating currents are currents that periodically reverse the direction of current flow. The sinusoidal current is clearly the most common and important type. Other types of electric current will be considered later in the book.

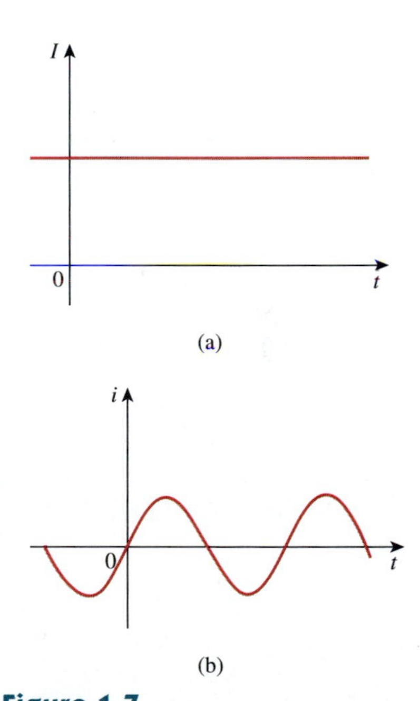

Figure 1.7
Two common types of current: (a) direct current (dc); (b) alternating current (ac).

> **Alternating current** (ac) is current that varies periodically with time.

Once we define current as the movement of charge, we expect current to have an associated direction of flow. As mentioned earlier, the direction of current is conventionally taken as the direction of positive charge movement. Based on this convention, a current of 5 A may be represented positively or negatively as shown in Fig. 1.8. In other words, a negative current of -5 A flowing in one direction as shown in Fig. 1.8(b) is the same as a current of $+5$ A flowing in the opposite direction as shown in Fig. 1.8(a).

It is expedient to consider different materials we may encounter. Generally speaking, materials may be divided into three categories,

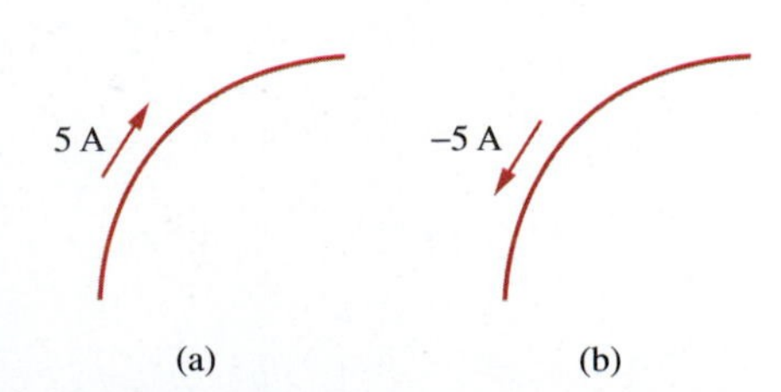

Figure 1.8
Conventional current: (a) positive current; (b) negative current.

[2] We will be using IEEE conventions throughout this book. A convention is a standard way of describing something so that others in the trade can understand what we mean.

depending on how easily they will allow charge to flow through:

- Conductors (e.g., copper, gold, silver).
- Semiconductors (e.g., silicon, germanium).
- Insulators (e.g., rubber, wood, plastic).

(Each class of materials is based on the number of valence electrons in the outer shell of the material.) Most materials are either conductors or insulators. In conductors, the electrons are so loosely bound to their atoms that they are free to move around. In other words, a conductor is a material containing free electrons capable of moving from one atom to another. In insulators, on the other hand, the electrons are much more tightly bound to the atoms and are not free to flow. Semiconductors are materials whose behavior falls between that of a conductor and that of an insulator.

Example 1.8

How much charge is represented by 4,600 electrons?

Solution:

Each electron has -1.602×10^{-19} C. Hence 4,600 electrons will have

$$-1.602 \times 10^{-19} \text{ C/electron} \times 4{,}600 \text{ electrons} = -7.3692 \times 10^{-16} \text{ C}$$

Practice Problem 1.8

Calculate the amount of charge represented by 2 million protons.

Answer: $+3.204 \times 10^{-13}$ C

Example 1.9

A charge of 4.5 C flows through an element for 0.2 second; determine the amount of current through the element.

Solution:

$$I = \frac{Q}{t} = \frac{4.5}{0.2} = 22.5 \text{ A}$$

Practice Problem 1.9

The current through a certain element is measured to be 8.6 A. How much time will it take for 2 mC of charge to flow through the element?

Answer: 0.2326 ms

1.6 Voltage

To move the electron in a conductor in a particular direction requires some work or energy transfer. This work is performed by an external electromotive force (emf), typically represented by the battery in Fig. 1.6. This emf is also known as *voltage* or *potential difference*. The voltage V_{ab} between two points a and b in an electric circuit is the energy (or work) W needed to move a charge Q from a to b divided by the charge. Mathematically,

$$V_{ab} = \frac{W}{Q} \tag{1.3a}$$

where W is energy in joules (J) and Q is charge in coulombs (C). We can manipulate Eq. (1.3a) to get

$$W = QV_{ab} \tag{1.3b}$$

$$Q = \frac{W}{V_{ab}} \tag{1.3c}$$

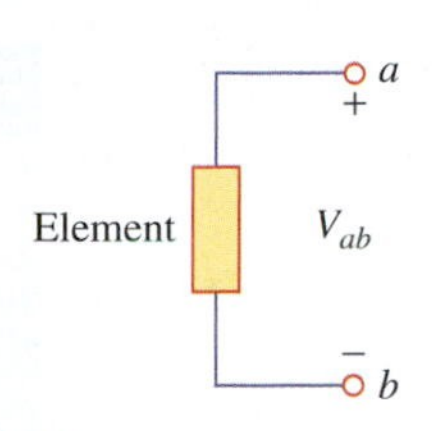

Figure 1.9
Polarity of voltage V_{ab}.

The voltage V_{ab}, or simply V, is measured in volts (V) named in honor of the Italian physicist Alessandro Antonio Volta. Thus,

> **Voltage** (or potential difference) is the energy required to move 1 coulomb of charge through an element, measured in volts (V).

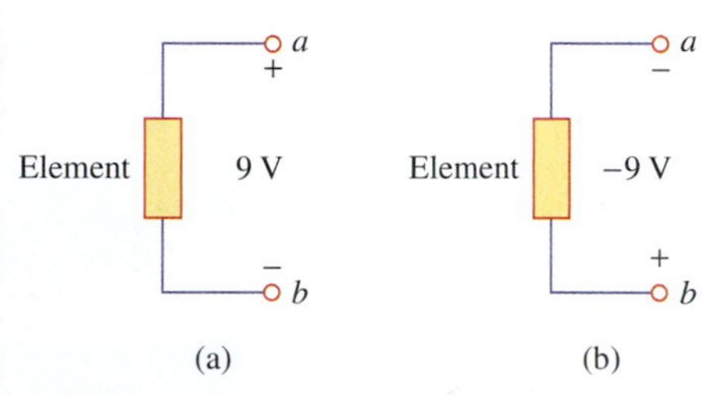

Figure 1.10
Two equivalent representations of the same voltage V_{ab}: (a) point a is 9 V above point b; (b) point b is −9 V above point a.

The voltage across an element (represented by a rectangular block) connected to points a and b is shown in Fig. 1.9. The plus (+) and minus (−) signs are used to define reference direction or voltage polarity. The voltage V_{ab} can be interpreted in two ways: (1) point a is at a potential of V_{ab} volts higher than point b, or (2) the potential at point a with respect to point b is V_{ab}. It follows logically that, in general,

$$\boxed{V_{ab} = -V_{ba}} \tag{1.4}$$

For example, in Fig. 1.10, we have two representations of the same voltage. In Fig. 1.10(a), point a is +9 V above point b; in Fig. 1.10(b), point b is −9 V above point a. We may say that in Fig. 1.10(a), there is a 9-V *voltage drop* from a to b or, equivalently, a 9-V *voltage rise* from b to a. In other words, a voltage drop from a to b is equivalent to a voltage rise from b to a.

Current and voltage are the two basic variables in electric circuits. Like electric current, a constant voltage is called a dc voltage, whereas a sinusoidally time-varying voltage is known as an ac voltage. A dc voltage is commonly produced by a battery such as are shown in Fig. 1.11; ac voltage is produced by an electric generator as shown in Fig. 1.12.

Figure 1.11
Electric car.
© VisionsofAmerica/Joe Sohm/Photodisc/Getty RF

Figure 1.12
Two photos of ac generators in hydroelectric power plants.
© Corbis RF

Example 1.10

If 20 J of energy is required to move 5 mC of charge through an element, what is the voltage across the element?

Solution:

$$V = \frac{W}{Q} = \frac{20}{5 \times 10^{-3}} = 4 \times 10^3 \text{ V} = 4 \text{ kV}$$

Practice Problem 1.10

Determine how much energy a 12-V battery uses to move 4.25 C.

Answer: 51 J

Example 1.11

How much work is performed by a 3-V battery in 8 seconds if the current in the conductor is 5 mA?

Solution:
The total charge moved is given by

$$Q = It = 5 \text{ mA} \times 8 \text{ s} = 40 \text{ mAs} = 40 \text{ mC}$$

The work done is given by

$$W = VQ = 3 \text{ V} \times 40 \text{ mC} = 120 \text{ m-VC} = 120 \text{ mJ}$$

Practice Problem 1.11

A current of 0.2 A flows through an element and releases 9 J of energy in 3 s. Calculate the voltage across the element.

Answer: 15 V

1.7 Power and Energy

Although current and voltage are the two basic variables in an electric circuit, the input and output of the circuit can be expressed in terms of power or energy. For practical purposes, we need to know how much power an electric device can handle. We all know from experience that a 100-watt bulb gives more light than a 60-watt bulb. We also know that when we pay our bills to the electric utility companies, we are paying for the electric energy consumed over a certain period of time. Thus, power and energy are important concepts in circuit analysis.

To relate power and energy to voltage and current, we recall from physics that:

> **Power** is the time rate of expending or absorbing energy, measured in watts (W).

Thus

$$P = \frac{W}{t} \tag{1.5}$$

where P is power in watts (W), W is energy in joules (J), and t is time in seconds (s). From Eqs. (1.2), (1.3), and (1.5), it follows that

$$P = \frac{W}{t} = \frac{W}{Q}\frac{Q}{t} = VI \tag{1.6}$$

or

$$P = VI \tag{1.7}$$

Thus, the power absorbed or supplied by an element is the product of the voltage across the element and the current through it. If the power has a + sign, power is being delivered to or absorbed by the element. If, on the other hand, the power has a − sign, power is being supplied by the element. But how do we know when the power has a negative or positive sign?

Current direction and voltage polarity play a major role in determining the sign of power. A load or element may be absorbing or delivering power. As shown in Fig. 1.13(a), when current flows into the element from the higher potential point (+), the element is absorbing power. On the other hand, if current is flowing into the element from the lower potential point (−), the element is delivering power. The *law of conservation of energy* must be obeyed in any electric circuit. For this reason, the algebraic sum of power in a circuit, at any instant of time, must be zero; that is,

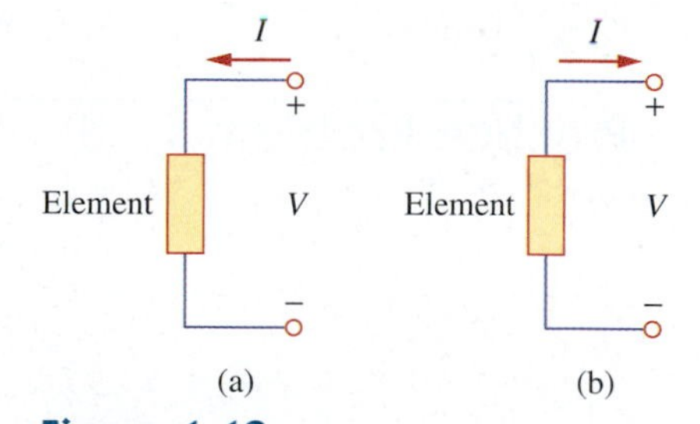

Figure 1.13
(a) Element absorbing power; (b) element delivering power.

$$\sum P = 0 \tag{1.8}$$

This again confirms the fact that the total power supplied to the circuit must balance the total power absorbed.

Energy is the capacity to do work, measured in joules (J).

From Eq. (1.5), the energy absorbed or supplied by an element for time t is

$$W = Pt = VIt \tag{1.9}$$

The electric power utility companies measure energy in watt-hours (Wh), where

$$1 \text{ Wh} = 3{,}600 \text{ J}.$$

Example 1.12

A 24-V source delivers 3 A from its positive terminal. How much power is delivered by the source?

Solution:

$$P = VI = 24 \times 3 = 72 \text{ W}$$

Practice Problem 1.12

A 30-W lamp is connected to a 120-V source. How much current through the lamp?

Answer: 0.25 A

Example 1.13

How much energy does a 100-W electric bulb consume in 2 hours?

Solution:

$$W = Pt = 100 \text{ W} \times 2 \text{ h} = 200 \text{ Wh}$$

This is the same as

$$\begin{aligned} W = Pt &= 100 \text{ (W)} \times 2 \text{ (h)} \times 60 \text{ (min/h)} \times 60 \text{ (s/min)} \\ &= 720{,}000 \text{ J} = 720 \text{ kJ} \end{aligned}$$

Practice Problem 1.13

A stove element draws 15 A when connected to a 120-V line. How long does it take to consume 30 kJ?

Answer: 16.67 s

1.8 †Applications

In this section, we will consider two practical applications of the concepts developed in this chapter. The first one deals with the TV picture tube and the other is concerned with how electric utility companies determine your electric bill.

1.8.1 TV Picture Tube

One important application of the motion of electrons is found in both the transmission and reception of TV signals. At the transmission end, a TV camera reduces a scene from an optical image to an electrical signal. Scanning is accomplished with a thin beam of electrons in an iconoscope camera tube.

At the receiving end, the image is reconstructed by using a cathode-ray tube (CRT) located in the TV receiver.[3] The CRT is depicted in Fig. 1.14. Unlike the iconoscope tube, which produces an electron beam of constant intensity, the CRT beam varies in intensity according to the incoming signal. The electron gun, maintained at a high potential, fires the electron beam. The beam passes through two sets of plates for vertical and horizontal deflections so that the spot on the screen where the beam strikes can move right and left, up and down. When the electron beam strikes the fluorescent screen, it gives off light at that spot. Thus, the beam can be made to "paint" a picture on the TV screen. Although the TV tube illustrates what we did in this chapter, a more modern device would be the charge-coupled device (CCD) camera. In fact, the liquid-crystal display televisions (LCD TVs) are more superior than CRT-based televisions.

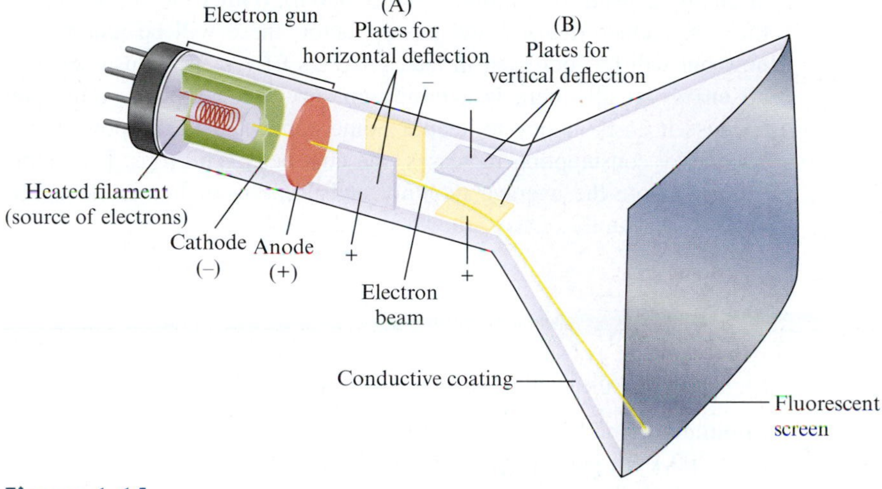

Figure 1.14
Cathode-ray tube.

Example 1.14

The electron beam in a TV picture tube carries 10^{15} electrons/second. As a design technologist, determine the voltage V_o needed to accelerate the electron beam to achieve 4 W.

Solution:

The charge on an electron is

$$e = -1.6 \times 10^{-19}\text{ C}$$

If the number of electrons is N, then $Q = Ne$ and

$$I = \frac{Q}{t} = e\frac{N}{t} = (-1.6 \times 10^{-19})(10^{15}) = -1.6 \times 10^{-4}\text{ A}$$

where the negative sign shows that the electrons flow in a direction opposite to electron flow as shown in Fig. 1.15, which is a simplified diagram of the CRT for the case when the vertical deflection plates carry no charge. The beam power is

$$P = V_o I \qquad \text{or} \qquad V_o = \frac{P}{I} = \frac{4}{1.6 \times 10^{-4}} = 25{,}000\text{ V}$$

Thus, the required voltage is 25 kV.

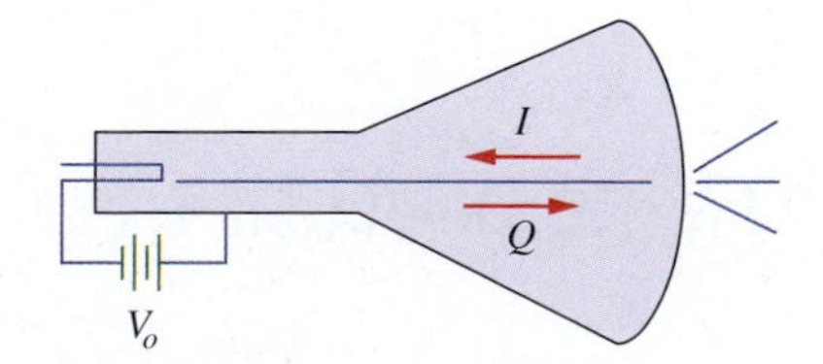

Figure 1.15
A simplified diagram of the cathode-ray tube; for Example 1.14.

[3] In practice, modern TV tubes use a different technology. Most cameras use some form of CCD to receive incoming light.

Practice Problem 1.14

If an electron beam in a TV picture tube carries 10^{13} electrons/second and is passing through plates maintained at a potential difference of 30 kV, calculate the power in the beam.

Answer: 48 mW

TABLE 1.4

Typical average monthly consumption of household appliances.

Appliance	kWh consumed
Water heater	500
Washing machine	120
Refrigerator/freezer	100
Lighting	100
Stove	100
Dryer	80
Dishwasher	35
Microwave oven	25
Electric iron	15
Personal computer	12
TV	10
Radio	8
Toaster	4
Clock	2

1.8.2 Electricity Bills

The second application deals with how electric utility companies charge their customers. The cost of electricity depends upon the amount of energy consumed in kilowatt-hours (kWh). (Other factors that affect the cost include demand and power factor; these will be ignored for now but will be addressed in later chapters.) Even if a consumer uses no energy at all, there is a minimum service charge one must pay because it costs money to keep a home connected to the power line. As energy consumption increases, the cost per kWh drops. It is interesting to note the average monthly consumption of household appliances for a family of five, shown in Table 1.4.

Example 1.15

A homeowner consumed 3,300 kWh in January. Determine his electricity bill for the month using the following residential rate schedule:

Minimum monthly charge = $12.00
First 100 kWh per month at 16 cents/kWh
Next 200 kWh per month at 10 cents/kWh
More than 300 kWh per month at 6 cents/kWh

Solution: The electricity bill is calculated as follows.

Minimum monthly charge = $12.00
First 100 kWh @ $0.16/kWh = $16.00
Next 200 kWh @ $0.10/kWh = $20.00

Remaining energy consumed

(3,300 − 300) = 3000 kWh @ $0.06/kWh = $180.00
Total bill = $228.00
Average cost = $228/3300 = 6.9 cents/kWh

Practice Problem 1.15

Using the residential rate schedule in Example 1.15, calculate the average cost per kWh if only 400 kWh are consumed in July when the family is on vacation most of the time.

Answer: 13.5 cents/kWh

1.9 Summary

1. An electric circuit consists of electrical elements connected together.
2. The International System (SI) of Units is the international measurement language, which enables engineers to communicate their

results. From the seven principal units, the units of other physical quantities can be derived.

3. Very large and very small quantities can be expressed in the power of 10 notation, in scientific notation, or engineering notation.
4. The scientific calculator is an important tool that a student must master.
5. Current is the rate of charge flow past a given point in a given direction.

$$I = \frac{Q}{t}$$

where I is current (in amperes), Q is charge (in coulombs), and t is time (in seconds)

6. Voltage is the energy required to move 1 C of charge through an element.

$$V = \frac{W}{Q}$$

where V is voltage (in volts), W is energy or work done (in joules), and Q is charge (in coulombs).

7. Power is the energy supplied or absorbed per unit time. It is also the product of voltage and current.

$$P = \frac{W}{t} = VI$$

where P is power (in watts), W is energy (in joules), t is time (in seconds), V is voltage (in volts), and I is current (in amperes).

8. The energy supplied or absorbed by an element for time t is

$$W = Pt = VIt$$

9. The law of conservation of energy must be obeyed by any electric circuit. Therefore, the algebraic sum of the power in a circuit at any instant of time must be zero.

$$\sum P = 0$$

10. Two areas of applications of the concepts covered in this chapter are the TV picture tube and electricity billing procedure.

Review Questions

1.1 One millivolt is one millionth of a volt.

(a) True (b) False

1.2 The prefix *micro* stands for:

(a) 10^6 (b) 10^3
(c) 10^{-3} (d) 10^{-6}

1.3 The voltage 2,000,000 V can be expressed in engineering notation as:

(a) 2 mV (b) 2 kV
(c) 2 MV (d) 2 GV

1.4 A charge of 2 C flowing past a given point each second is a current of 2 A.

(a) True (b) False

1.5 A 4-A current charging a certain material will accumulate a charge of 24 C after 6 s.

(a) True (b) False

1.6 The unit of current is:

(a) coulombs (b) amperes
(c) volts (d) joules

1.7 Voltage is measured in:

(a) watts (b) amperes

(c) volts (d) joules per second

1.8 The voltage across a 1.1-kW toaster that produces a current of 10 A is:

(a) 11 kV (b) 1100 V

(c) 110 V (d) 11 V

1.9 Watt is the unit of:

(a) charge (b) current

(c) voltage (d) power (e) energy

1.10 Which of these is not an electrical quantity?

(a) charge (b) time

(c) voltage (d) current (e) power

Answers: 1.1b, 1.2d, 1.3c, 1.4a, 1.5a, 1.6b, 1.7c, 1.8c, 1.9d, 1.10b

Problems

Section 1.2 International Systems of Units

1.1 Convert the following lengths to meters.

(a) 45 feet (b) 4 yards

(c) 3.2 miles (d) 420 mils

1.2 Express the following in joules:

(a) 28 foot-pounds (b) 4.6 kilowatt-hours

1.3 Express 32 horsepower in terms of watts.

1.4 Convert 124 miles to kilometers.

Section 1.3 Scientific and Engineering Notation

1.5 Express the following numbers in engineering notation.

(a) 0.004500 (b) 0.00926

(c) 7,421 (d) 26,356,000

1.6 Express the following numbers in scientific notation.

(a) 0.0023 (b) 6,400 (c) 4,300,000

1.7 Express the following numbers in scientific notation.

(a) 0.000126 (b) 98,000 (c) $\dfrac{1}{2{,}000{,}000}$

1.8 Express the following numbers in engineering notation.

(a) 160×10^{-7} s (b) 30×10^{4} V

(c) 1.3×10^{-3} J (d) 0.5×10^{-9} W

1.9 Perform the following operations, and express your answer in scientific notation.

(a) $(2 \times 10^{4})(6 \times 10^{5})$

(b) $(3.2 \times 10^{-3})(7 \times 10^{-6})$

(c) $\dfrac{(30{,}000)^2}{(0.04)^2}$ (d) $\dfrac{(500)^3(100)^2}{10^7}$

1.10 Perform the following operations, and express your answer in scientific notation.

(a) $0.003 + 542.8 + 641 \times 10^{-3}$

(b) $(25 \times 10^{3})(0.04)^2$

(c) $\dfrac{(40 + 10)^{-3}(6000)}{(3 \times 10^{-2})^2}$

(d) $\dfrac{(0.002)^2(100)^4}{10^6}$

Section 1.4 Scientific Calculators

1.11 Perform the following operations using a calculator.

(a) $12(8 - 6)$ (b) $\cos^{-1}\dfrac{2}{3}$

1.12 Use the calculator to perform the following operations.

(a) $\sqrt{\dfrac{120}{3^2 + 4^2}}$ (b) $\dfrac{\pi}{6^2 + 8^2}$

1.13 Perform the following operations using a calculator.

(a) 825×0.0012

(b) $(42.8 \times 11.5)/(12.6 + 7.04)$

1.14 Evaluate the following expressions using a calculator.

(a) $(3.6 \times 10^{3})^2$ (b) $(8.1 \times 10^{4})^{1/2}$ (c) $\dfrac{2 \times 10^2}{5 \times 10^{-1}}$

Section 1.5 Charge and Current

1.15 How many coulombs are represented by the following numbers of electrons?

(a) 6.482×10^{17} (b) 1.24×10^{18}

(c) 2.46×10^{19} (d) 1.628×10^{20}

1.16 If a 2-mA current passes through a wire for 2 min and 4 s, determine how many electrons have passed.

1.17 How much charge is 10^{20} electrons?

1.18 If 6×10^{22} electrons pass a wire in 42 s, what is the resulting current?

1.19 If 36 C of charge passes through a wire in 10 s, find the current through the wire.

1.20 How many electrons does a charge 2.4 μC represent?

1.21 How long does it take a charge of 50 μC to pass a point if the current is 24 mA?

1.22 If a charge of 700 C flows past a point in 5 min, what is the resulting value of current?

1.23 If the current through a certain circuit is 4 A, how long will it take to transfer 0.65 C of charge?

1.24 If a charge of 8.5 mC passes through a point in a conductor in 120 ms, determine the current.

Sections 1.6 and 1.7 Voltage and Power and Energy

1.25 Find the voltages V_{ab}, V_{bc}, V_{ac}, and V_{ba} in the circuit in Fig. 1.16.

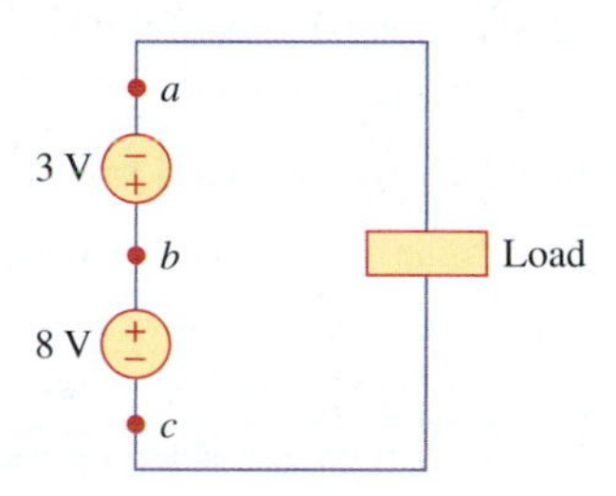

Figure 1.16
For Problem 1.25.

1.26 The potential difference between points a and b is 32 V. How much work is necessary to move a 2-C charge from a to b?

1.27 A 40-W motor operates from a 120-V source. How much current will the motor draw?

1.28 A power transmission line is rated at 1.04×10^6 V. What is the voltage in kV?

1.29 Energy of 450 J is necessary to move 2.6×10^{20} electrons from one point to another. Determine the potential difference between the two points.

1.30 Find the power consumed by the lamp in Fig. 1.17.

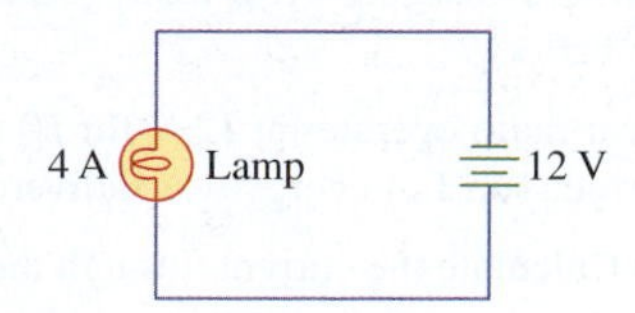

Figure 1.17
For Problem 1.30.

1.31 How much current is flowing through a 40-W, 120-V lamp?

1.32 A battery provides 120 J of energy to 18 C of charge. Find the terminal voltage of the battery.

1.33 What is the current in the heating element of a 45-W soldering iron if operated on 120 V?

1.34 An energy source forces a constant current of 2 A for 10 s to flow through a lightbulb. If 2.3 kJ is given off in the form of light and heat energy, calculate the voltage drop across the bulb.

1.35 To move charge Q from point a to point b requires 30 J. Find the voltage drop if

(a) $Q = 2$ C (b) $Q = -6$ C

1.36 If an automobile headlight takes 2.5 A at 8 V, how much power is consumed by the headlight?

1.37 An electric oven uses 38 A and 120 V. Calculate the power used.

1.38 An electric toaster is rated at 800 W and draws 7 A. Determine the operating voltage of the toaster.

1.39 The voltage between point a and point b is 120 mV. Find the charge Q if 360 μJ of work is required to move charge Q from point a to point b.

1.40 If a lamp with a 70-W lightbulb is plugged into a 120-V source, how much current flows through the circuit?

1.41 A 12-V battery can supply 100 A for 5 s. Determine the work done by the battery.

Section 1.8 Applications

1.42 It takes 8 photons to strike the surface of a photodetector in order to emit 1 electron. If 4×10^{11} photons/second strike the surface of the photodetector, calculate the amount of current.

1.43 Find the power rating of the following electrical appliances in your household:

(a) Lightbulb (b) Radio set
(c) TV set (d) Refrigerator
(e) Personal computer (f) PC printer
(g) Microwave oven (h) Blender

1.44 A 1.5-kW electric heater is connected to a 120-V source.

(a) How much current does the heater draw?

(b) If the heater is on for 45 min, how much energy is consumed in kilowatt-hours (kWh)?

(c) Calculate the cost of operating the heater for 45 min if the energy cost is 10 cents/kWh.

1.45 A 1.2-kW toaster takes roughly 4 min to heat four slices of bread. Find the cost of operating the toaster once per day for one month (30 days). Assume an energy cost of 9 cents/kWh.

1.46 A flashlight battery has a rating of 0.8 ampere-hour (Ah) and a lifetime of 10 hours (h).

(a) How much current can it deliver?

(b) How much power can it give if its terminal voltage is 1.5 V?

(c) How much energy is stored in the battery in kilowatt-hours (kWh)?

1.47 A 20-W incandescent lamp is connected to a 120-V source and is left burning continuously in an otherwise dark staircase. Determine (a) the current through the lamp and (b) the cost of operating the light for one non-leap year if the electricity cost is 12 cent per kWh.

1.48 An electric stove with four burners and an oven is used in preparing a meal as follows:

Burner # 1	20 min
Burner # 2	40 min
Burner # 3	15 min
Burner # 4	45 min
Oven	30 min

If each burner is rated at 1.2 kW and the oven at 1.8 kW, and if the electricity costs is 12 cents/kWh, calculate the cost of electricity used in preparing the meal.

1.49 PECO (the electric company in Philadelphia) charged a consumer $34.24 in a month for using 215 kWh. If the basic service charge is $5.10, how much did PECO charge per kWh?

1.50 A 600-W TV receiver is turned on for 4 h with nobody watching it. If electricity costs 10 cents/kWh, how much money is wasted?

1.51 A utility company charges 8.5 cents/kWh. If the consumer operates a 40-W lightbulb continuously for one day, how much is the consumer charged?

1.52 Reliant Energy (the electric company in Houston, Texas) charges customers as follows:

Monthly charge = $6

First 250 kWh @ $0.02/kWh

All additional kWh @ $0.07/kWh

If a customer uses 1218 kWh in one month, how much will Reliant Energy charge?

1.53 An elevator in a building can lift 7,000 lb to a height of 60 ft in 30 s. What is the required horsepower (hp) to operate the elevator? Assume that 1 hp = 550 ft-lb/s.

1.54 The following household appliances were running in a home:

A 3.2-kW air conditioner for 9 h

Eight 60-W lamps for 7 h

A 400-W TV receiver for 3 h

(a) Determine the total energy consumed in kWh.

(b) What is the total cost for running these appliances if the utility company charges $0.06/kWh?

1.55 (a) If a 70-W lightbulb is left on for 8 h every night, determine the energy used in one week.

(b) How much is the cost per month (assume 30 days) if the electricity cost is $0.09/kWh?

Comprehensive Problems

1.56 A telephone wire has a current of 20 μA flowing through it. How long does it take for a charge of 15 C to pass through the wire?

1.57 A lightning bolt carried a current of 2 kA and lasted for 3 ms. How many coulombs of charge was contained in the lightning bolt?

1.58 A battery may be rated in ampere-hours (Ah). An alkaline battery is rated at 160 Ah.

(a) What is the maximum current it can supply for 40 h?

(b) How many days will it last if it is discharged at 1 mA?

1.59 How much work is done by a 12-V automobile battery in moving 5×10^{20} electrons from the positive terminal to the negative terminal?

1.60 How much energy does a 10-hp motor deliver in 30 min? Assume that 1 hp = 746 W.

1.61 A 2-kW electric iron is connected to a 120-V line. Calculate the current drawn by the iron.

1.62 A lightning bolt with 30 kA strikes an airplane for 2 ms. How many coulombs of charge are deposited on the plane?

1.63 A 12-V battery requires a total charge of 40 Ah during recharging. How many joules are supplied to the battery?

1.64 A car radio operates at 12 V for 20 min. During that period, 108 J of energy was delivered to the radio.

(a) Calculate the current through the radio.

(b) How many coulombs of charge are used during the period?

1.65 A pocket calculator has a 4-V battery pack that generates 0.2 mA in 45 minutes.

(a) Calculate the charge that flows in the circuit.

(b) Find the energy that the battery delivers to the calculator circuit.

chapter

2

Resistance

No pain, no palm; no thorns, no throne; no gall, no glory; no cross, no crown.

—William Penn

Historical Profiles

Georg Simon Ohm (1787–1854), a German physicist, in 1826 experimentally determined the most basic law relating voltage and current for a resistor. Ohm's work was initially denied by critics.

Born of humble beginnings in Erlangen, Bavaria, Ohm threw himself into electrical research. Ohm's major interest was current electricity, which had recently been advanced by Alessandro Volta's invention of the battery. Using the results of his experiments, Ohm was able to define the fundamental relationship among voltage, current, and resistance. This resulted in his famous law—Ohm's law—which will be covered in this chapter. He was awarded the Copley Medal in 1841 by the Royal Society of London. He was also given the Professor of Physics chair in 1849 by the University of Munich. To honor him, the unit of resistance is named the *ohm*.

Georg Simon Ohm
© SSPL via Getty Images

Ernst Werner von Siemens (1816–1892) was a German electrical engineer and industrialist who played an important role in the development of the telegraph.

Siemens was born at Lenthe in Hanover, Germany, the oldest of four brothers—all of whom were distinguished engineers and industrialists. After attending grammar school at Lübeck, Siemens joined the Prussian artillery at age 17 for the training in engineering that his father could not afford. Looking at an early model of an electric telegraph, invented by Charles Wheatstone in 1837, Siemens realized its possibilities for making improvements and for international communication. He invented a telegraph that used a needle to point to the right letter, instead of using Morse code. He laid the first telegraph line in Germany with his brothers, William Siemens and Carl von Siemens. The unit of *conductance* is named in his honor.

Ernst Werner von Siemens
© Hulton Archive/Getty

2.1 Introduction

In the last chapter, we introduced some basic concepts such as current, voltage, and power in an electric circuit. To actually determine the values of these variables in a given circuit requires that we understand some fundamental laws that govern electric circuits. These laws—known as Ohm's law and Kirchhoff's laws—form the foundation upon which electric circuit analysis is built. Ohm's law will be covered in this chapter, while Kirchhoff's laws will be covered in Chapters 4 and 5.

We begin the chapter by first discussing resistance—its nature and characteristics. We then cover Ohm's law, conductance, and circular wires. We present color coding for physically small resistors. We will finally apply the concepts covered in this chapter to dc measurements.

2.2 Resistance

Materials in general have a characteristic behavior of opposing the flow of electric charge. This opposition is due to the collisions between electrons that make up the materials. This physical property, or ability to resist current, is known as *resistance* and is represented by the symbol R. Resistance is expressed in ohms (after Georg Simon Ohm), which is symbolized by the capital Greek letter omega (Ω). The schematic symbol for resistance or resistor is shown in Fig. 2.1, where R stands for the resistance of the resistor.

Figure 2.1
Circuit symbol for resistance.

> The **resistance** R of an element denotes its ability to resist the flow of electric current; it is measured in ohms (Ω).

The resistance of any material is dictated by four factors:

1. Material property—each material will oppose the flow of current differently.
2. Length—the longer the length ℓ, the more is the probability of collisions and, hence, the larger the resistance.
3. Cross-sectional area—the larger the area A, the easier it becomes for electrons to flow and, hence, the lower the resistance.
4. Temperature—typically, for metals, as temperature increases, the resistance increases.

Thus, the resistance R of any material with a uniform cross-sectional area A and length ℓ (as shown in Fig. 2.2) is directly proportional to the length and inversely proportional to its cross-sectional area. In mathematical form,

$$R = \rho \frac{\ell}{A} \tag{2.1}$$

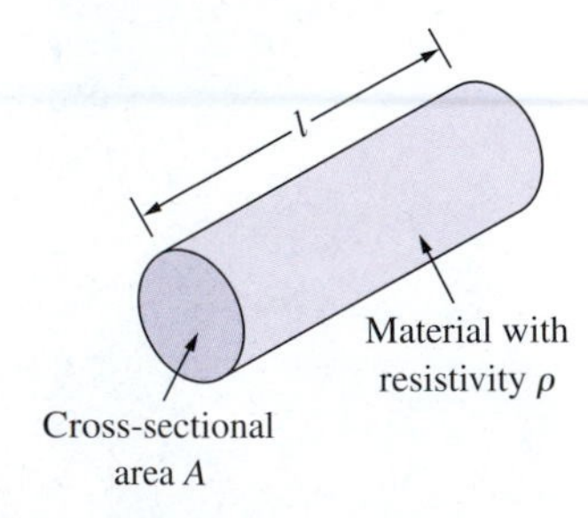

Figure 2.2
A conductor with uniform cross section.

where the Greek letter rho ρ is known as the *resistivity* of the material. Resistivity is a physical property of the material and is measured in ohm-meters (Ω-m).

The cross section of an element can be circular, square, rectangular, and so on. Because most conductors are circular in cross-section, the cross-sectional area may be determined in terms of the radius r or diameter d of the conductor as

$$A = \pi r^2 = \pi\left(\frac{d}{2}\right)^2 = \frac{\pi d^2}{4} \tag{2.2}$$

TABLE 2.1

Resistivities of common materials.

Material	Resistivity (Ω-m)	Usage
Silver	1.64×10^{-8}	Conductor
Copper	1.72×10^{-8}	Conductor
Aluminum	2.8×10^{-8}	Conductor
Gold	2.45×10^{-8}	Conductor
Iron	1.23×10^{-7}	Conductor
Lead	2.2×10^{-7}	Conductor
Germanium	4.7×10^{-1}	Semiconductor
Silicon	6.4×10^{2}	Semiconductor
Paper	10^{10}	Insulator
Mica	5×10^{11}	Insulator
Glass	10^{12}	Insulator
Teflon	3×10^{12}	Insulator

The resisitivity ρ varies with temperature and is often specified for room temperature.

Table 2.1 presents the values of ρ for some common materials at room temperature (20°C). The table also shows that materials can be classified into three groups according to their usage: conductors, insulators, and semiconductors. Good conductors, such as copper and aluminum, have low resistivities. Of those materials shown in Table 2.1, silver is the best conductor. However, a lot of wires are made of copper because copper is almost as good and is much cheaper. In general, the resistance of a conductor increases with a rise in temperature. Insulators, such as mica and paper, have high resistivities. They are used in forming the insulating coating of copper wires. Semiconductors, such as germanium and silicon, have resistivities that are neither high nor low. They are used in making transistors and integrated circuits. There is even a considerable range within the conductor group. Nichrome (an alloy of nickel, chrome, and iron) has resistivity roughly 58 times greater than that of copper. For this reason, Nichrome is used in making resistors and heating elements.

The circuit element used to model the current-resisting behavior of a material is the *resistor.* For the purpose of constructing circuits, resistors shown in Fig. 2.3 are usually made from metallic alloys and carbon compounds. The resistor is the simplest passive element.

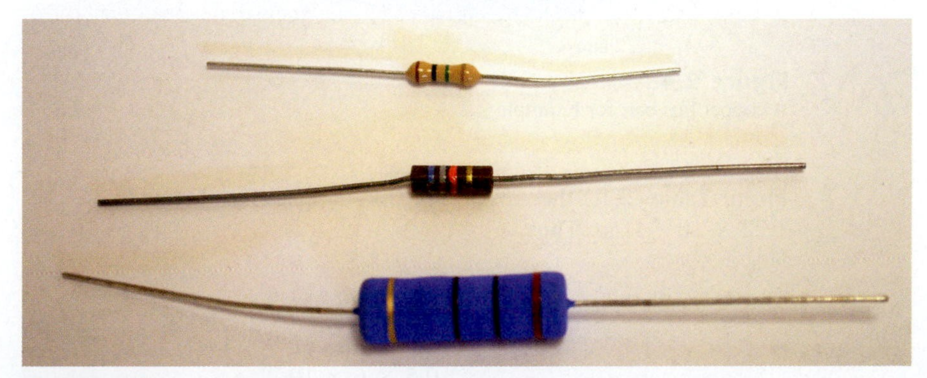

Figure 2.3
From top to bottom $\frac{1}{4}$-W, $\frac{1}{2}$-W, and 1-W resistors.
© Sarhan M. Musa

Example 2.1

Calculate the resistance of an aluminum wire that is 2 m long and of circular cross section with a diameter of 1.5 mm.

Solution:
We first calculate the cross-sectional area:

$$A = \frac{\pi d^2}{4} = \frac{\pi(1.5 \times 10^{-3})^2}{4} = 1.767 \times 10^{-6}\ \text{m}^2$$

From Table 2.1, we obtain the resistivity of aluminum as $\rho = 2.8 \times 10^{-8}\ \Omega\text{-m}$. Thus,

$$R = \frac{\rho \ell}{A} = \frac{2.8 \times 10^{-8} \times 2}{1.767 \times 10^{-6}}$$
$$= 31.69\ \text{m}\Omega$$

Practice Problem 2.1

Determine the resistance of an iron wire having a diameter of 2 mm and a length of 30 m.

Answer: 1.174 Ω

Example 2.2

A copper bus bar is shown in Fig. 2.4. Calculate the length of the bar that will produce a resistance of 0.5 Ω.

Solution:
The bus bar has a uniform cross section so that Eq. (2.1) applies. But the cross section is rectangular so that the cross-sectional area is

$$A = \text{Width} \times \text{Breadth} = (2 \times 10^{-3}) \times (3 \times 10^{-3})$$
$$= 6 \times 10^{-6}\ \text{m}^2 = 6\ \mu\text{m}^2$$

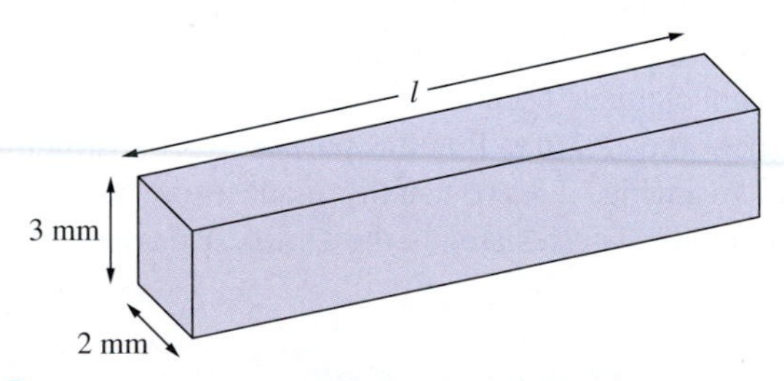

Figure 2.4
A copper bus bar; for Example 2.2.

From Table 2.1, the resistivity of copper is obtained as $\rho = 1.72 \times 10^{-8}\ \Omega\text{-m}$. Thus,

$$R = \rho\frac{\ell}{A} \longrightarrow \ell = \frac{RA}{\rho}$$

$$\ell = \frac{0.5 \times 6 \times 10^{-6}}{1.72 \times 10^{-8}} = 174.4\ \text{m}$$

Practice Problem 2.2

A conducting bar with triangular cross section is shown in Fig. 2.5. If the bar is made of lead, determine the length of the bar that will produce a resistance of 1.25 mΩ.

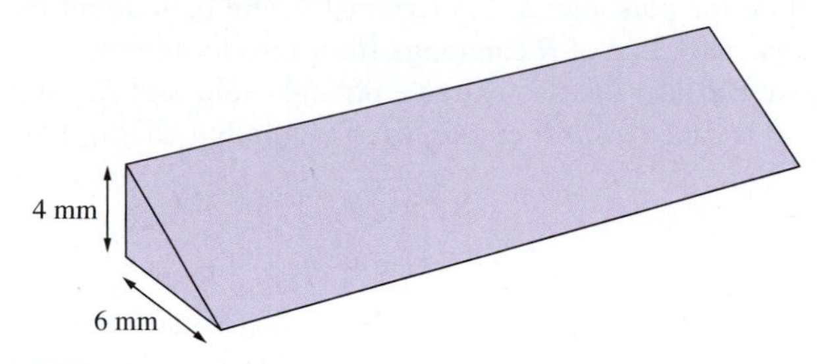

Figure 2.5
For Practice Problem 2.2.

Answer: 6.82 cm

2.3 Ohm's Law

Georg Simon Ohm (1787–1854), a German physicist, is credited with finding the relationship between current and voltage for a resistor. This relationship is known as *Ohm's law*. That is,

$$V \propto I \tag{2.3}$$

Ohm's law states that the voltage V across a resistor is directly proportional to the current I flowing through the resistor.

Ohm defined the constant of proportionality for a resistor to be the resistance R. (The resistance is a material property that could change if the internal or external conditions of the element were altered, e.g., if there were changes in the temperature.) Thus, Eq. (2.3) becomes

$$V = IR \tag{2.4}$$

which is the mathematical form of Ohm's law. In Eq. (2.4), we recall that the voltage V is measured in volts, the current I is measured in amperes, and the resistance R is measured in ohms. We may deduce from Eq. (2.4) that

$$R = \frac{V}{I} \tag{2.5}$$

so that

$$1\ \Omega = 1\ V/1\ A \tag{2.6}$$

We may also deduce from Eq. (2.4) that

$$I = \frac{V}{R} \tag{2.7}$$

Thus, Ohm's law can be stated in three different ways, as in Eqs. (2.4), (2.5), and (2.7).

To apply Ohm's law as stated in Eq. (2.4), for example, we must pay careful attention to the current direction and voltage polarity. The direction of current I and the polarity of voltage V must conform with the convention shown in Fig. 2.6. This implies that current flows from

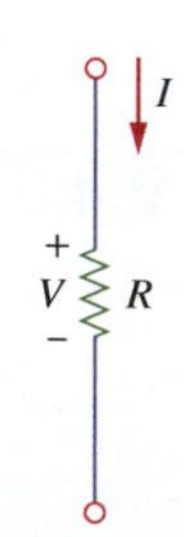

Figure 2.6
Direction of current I and polarity of voltage V across a resistor R.

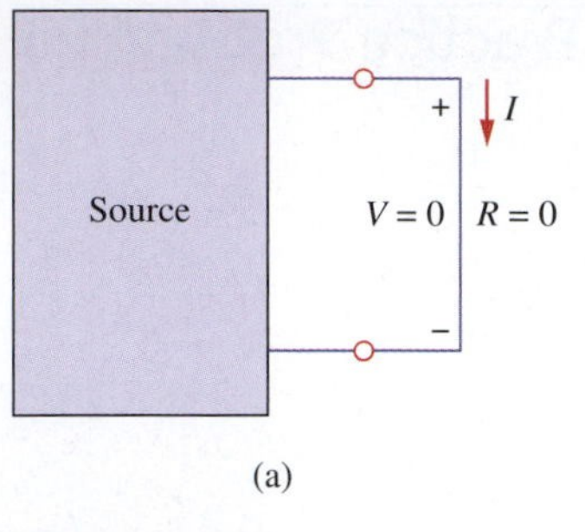

(a)

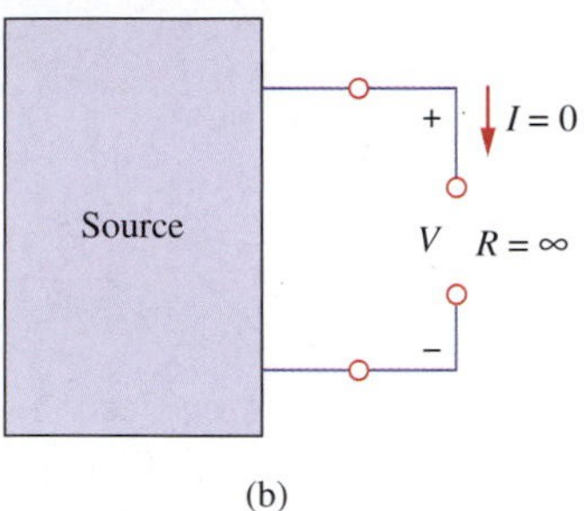

(b)

Figure 2.7
(a) Short circuit ($R = 0$); (b) open circuit ($R = \infty$).

a higher potential to a lower potential in order for $V = IR$. If current flows from a lower potential to a higher potential, then $V = -IR$. (When the polarity of the voltage across the resistor is not specified, always place the plus sign at the terminal where the current enters.)

Because the value of R can range from zero to infinity, it is important that we consider the two extreme possible values of R. An element with $R = 0$ is called a *short circuit*, as shown in Fig. 2.7(a). For a short circuit,

$$V = IR = 0 \tag{2.8}$$

showing that the voltage is zero but the current could be anything. In practice, a short circuit is usually a connecting wire assumed to be a perfect conductor. Thus

> A **short circuit** is a circuit element with resistance approaching zero.

Similarly, an element with $R = \infty$ is known as an *open circuit*, as shown in Fig. 2.7(b). For an open circuit,

$$I = \frac{V}{R} = \frac{V}{\infty} = 0 \tag{2.9}$$

indicating that the current is zero though the voltage could be anything. Thus,

> An **open circuit** is a circuit element with resistance approaching infinity.

Example 2.3

An electric iron draws 2 A at 120 V. Find its resistance.

Solution:
From Ohm's law,

$$R = \frac{V}{I} = \frac{120}{2} = 60\ \Omega$$

Practice Problem 2.3

The essential component of a toaster is an electrical element (a resistor) that converts electrical energy to heat energy. How much current is drawn by a toaster with resistance of 12 Ω at 110 V?

Answer: 9.17 A

Example 2.4

In the circuit shown in Fig. 2.8, calculate the current I.

Figure 2.8
For Example 2.4.

Solution:
The voltage across the resistor is the same as the source voltage (30 V) because the resistor and the voltage source are connected to the same pair of terminals. Hence,

$$I = \frac{V}{R} = \frac{30}{5 \times 10^3} = 6\text{ mA}$$

Practice Problem 2.4

If $I = 8$ mA in the circuit shown in Fig. 2.9, determine the value of resistance R.

Answer: 1.5 kΩ

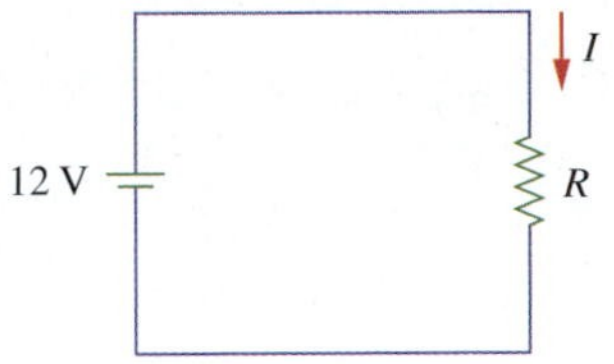

Figure 2.9
For Practice Problem 2.4.

2.4 Conductance

A useful quantity in circuit analysis is the reciprocal of resistance R, known as conductance and denoted by G:

$$G = \frac{1}{R} = \frac{I}{V} \tag{2.10}$$

The conductance is a measure of how well an element will conduct electric current. The old unit of conductance is the *mho* (ohm spelled backward) with symbol $\mho$, the inverted omega. Although engineers still use mhos, in this book we will prefer to use the SI unit of conductance, the siemens (S), in honor of Werner von Siemens:

$$1\ S = 1\ \mho = 1\ A/1\ V \tag{2.11}$$

Thus,

> **Conductance** is the ability of an element to conduct electric current; it is measured in siemens (S).

[We should not confuse S for siemens with s (seconds) for time.] The same resistance can be expressed in ohms or siemens. For example, 10 Ω is the same as 0.1 S. From Eqs. (2.1) and (2.10), we may write

$$G = \frac{A}{\rho \ell} = \frac{\sigma A}{\ell} \tag{2.12}$$

where the Greek letter sigma $\sigma = 1/\rho$ = conductivity of the material (in S/m).

Example 2.5

Find the conductance of the following resistors: (a) 125 Ω (b) 42 kΩ

Solution:

(a) $G = 1/R = 1/(125\ \Omega) = 8$ mS
(b) $G = 1/R = 1/(42 \times 10^3\ \Omega) = 23.8\ \mu$S

Practice Problem 2.5

Determine the conductance of the following resistors:

(a) 120 Ω
(b) 25 MΩ

Answers: (a) 8.33 mS (b) 40 nS

2.5 Circular Wires

Circular wires are commonly used in several applications. We use wires to connect elements, but those wires have resistance and a maximum allowable current. So we need to choose the right size. Wires are arranged in standard gauge numbers, known as AWG (American Wire Gauge). This designation of cables and wires is in the English system. In the English system,

$$1{,}000 \text{ mils} = 1 \text{ in} \tag{2.13a}$$

or

$$1 \text{ mil} = \frac{1}{1000} \text{ in} = 0.001 \text{ in} \tag{2.13b}$$

A unit of cross-sectional area used for wires is the *circular mil* (CM), which is the area of a circle having diameter of 1 mil. From Eq. (2.2),

$$A = \frac{\pi d^2}{4} = \frac{\pi (1 \text{ mil})^2}{4} = \frac{\pi}{4} \text{ sq mil} \tag{2.14}$$

Thus,

$$1 \text{ CM} = \frac{\pi}{4} \text{ sq mil} \tag{2.15a}$$

or

$$1 \text{ sq mil} = \frac{4}{\pi} \text{ CM} \tag{2.15b}$$

If the diameter of a circular wire is in mils, the area in circular mils is

$$\boxed{A_{\text{CM}} = d_{\text{mil}}^2} \tag{2.16}$$

A listing of data for standard bare copper wires is provided in Table 2.2, where d is the diameter and R is the resistance for 1000 ft. (Notice the wire diameter decreases as the gauge number increases.) As you might guess, the maximum allowable currents are just a rule of thumb. The steel industry uses a different numbering system for their wire thickness gages (e.g., U.S. Steel Wire Gauge) so that the data in Table 2.2 do not apply to steel wire. See Fig. 2.10 for different sizes of wires. Typical household wiring is AWG number 12 or 14. Telephone wire is usually 22, 24, or 26 gauge. The following examples will illustrate how to use the table.

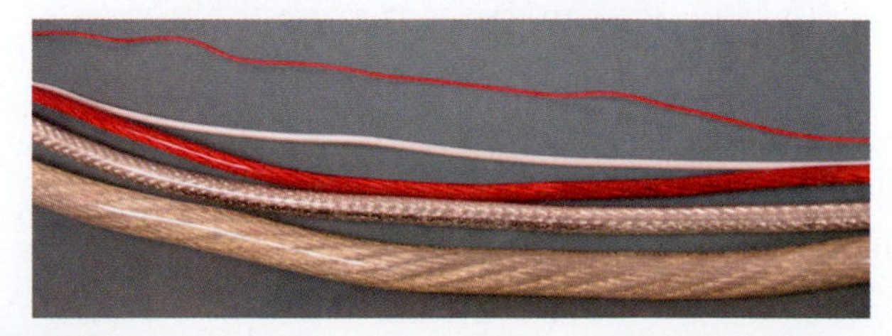

Figure 2.10
Insulated wires of different gauges.
© Sarhan M. Musa

TABLE 2.2

American wire gauge (AWG) sizes at 20°C.

AWG #	*d*(mil)	Area (CM)	*R* (Ω/1000 ft)	Maximum allowable current (A)
0000	460	211,600	0.0490	230
000	409.6	167,810	0.0618	200
00	364.8	133,080	0.0780	175
0	324.9	105,530	0.0983	150
1	289.3	83,694	0.1240	130
2	257.8	66,373	0.1563	115
3	229.4	52,634	0.1970	100
4	204.3	41,740	0.2485	85
5	181.9	33,102	0.3133	—
6	162	26,250	0.3951	65
7	144	20,820	0.4982	—
8	128.5	16,510	0.6282	50
9	114.4	13,090	0.7921	—
10	101.9	10,381	0.9989	30
11	90.74	8,234	1.260	—
12	80.81	6,530	1.588	20
13	71.96	5,178	2.003	—
14	64.08	4,107	2.525	15
15	57.07	3,257	3.184	
16	50.82	2,583	4.016	
17	45.26	2,048	5.064	
18	40.30	1,624	6.385	
19	35.89	1,288	8.051	
20	31.96	1,022	10.15	
21	28.46	810.10	12.80	
22	25.3	642.40	16.14	
23	22.6	509.5	20.36	
24	20.1	404.01	25.67	
25	17.9	320.40	32.37	
26	15.94	254.10	40.81	
27	14.2	201.50	51.57	
28	12.6	159.79	64.90	
29	11.26	126.72	81.83	
30	10.03	100.50	103.2	
31	8.928	79.70	130.1	
32	7.95	63.21	164.1	
33	7.08	50.13	206.9	
34	6.305	39.75	260.9	
35	5.6	31.52	329.0	
36	5	25	414.8	
37	4.5	19.83	523.1	
38	3.965	15.72	659.6	
39	3.531	12.47	831.8	
40	3.145	9.89	1049	

Example 2.6

Calculate the resistance of 840 ft of AWG #6 copper wire.

Solution:

From Table 2.2, the resistance of 1000 ft of AWG #6 is 0.3951 Ω. Hence, for a length of 840 ft,

$$R = 840 \text{ ft}\left(\frac{0.3951 \ \Omega}{1000 \text{ ft}}\right) = 0.3319 \ \Omega$$

Practice Problem 2.6

Find the resistance of 1200 ft of AWG #10 copper wire.

Answer: 199 Ω

Example 2.7

Find the cross-sectional area of a AWG #9 having a diameter of 114.4 mil.

$$A_{\text{CM}} = (114.4)^2 = 13{,}087 \text{ CM}$$

Practice Problem 2.7

What is the cross-sectional area in CM of a wire with a diameter of 0.0036 in.?

Answer: 12.96 CM

2.6 Types of Resistors

Different types of resistors have been created to meet different requirements. Some resistors are shown in Fig. 2.11. The primary functions of resistors are to limit current, divide voltage, and dissipate heat.

A resistor is either fixed or variable. Most resistors are of the fixed type; that is, their resistance remains constant. The two common types

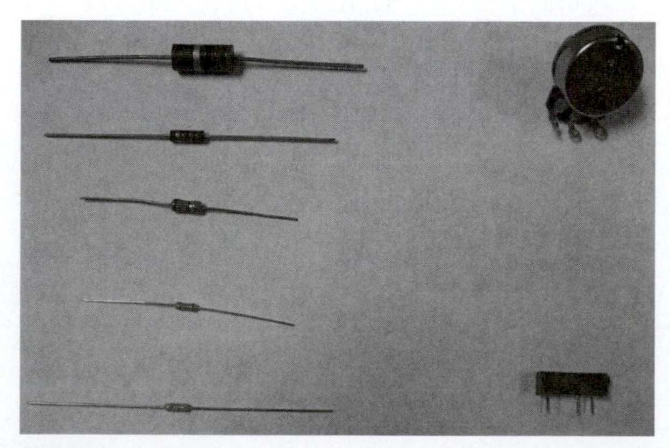

Figure 2.11
Different types of resistors.
© Sarhan M. Musa

of fixed resistors (wirewound and composition) are shown in Fig. 2.12. Wirewound resistors are used when there is a need to dissipate a large amount of heat, while the composition resistors are used when large resistance is needed. The circuit symbol in Fig. 2.1 is for a fixed resistor. Variable resistors have adjustable resistance. The symbol for a variable resistor is shown in Fig. 2.13. There are two main types of variable resistors: potentiometer and rheostat. The *potentiometer* or *pot* for short, is a three-terminal element with a sliding contact or wiper. By sliding the wiper, the resistances between the wiper terminal and the fixed terminals vary. The potentiometer is used to adjust the amount of voltage provided to a circuit, as typically shown in Fig. 2.14. A potentiometer with its adjuster is shown in Fig. 2.15. The *rheostat* is a two- or three-terminal device that is used to control the amount of current within a circuit, as typically shown in Fig. 2.16. As the rheostat is adjusted for more resistance and less current flow, and the motor slows down and vice versa. It is possible to use the same variable resistor as a potentiometer or a rheostat, depending on how it is connected. Like fixed resistors, variable resistors can either be of wirewound or composition type, as shown in Fig. 2.17. Although fixed resistors shown in Fig. 2.12 are used in circuit designs, today, most circuit components (including resistors) are either surface mounted or integrated, as typically shown in Fig. 2.18. Surface mount technology (SMT) is being used to implement both digital and analog circuits. An SMT resistor is shown in Fig. 2.19.

It should be pointed out that not all resistors obey Ohm's law. A resistor that obeys Ohm's law is known as a *linear* resistor. It has a constant resistance, and thus its voltage-current characteristic is as illustrated in Fig. 2.20(a); that is, its *V-I* graph is a straight line passing through the origin. A *nonlinear* resistor does not obey Ohm's law. Its resistance varies with current and its *V-I* characteristic is typically shown

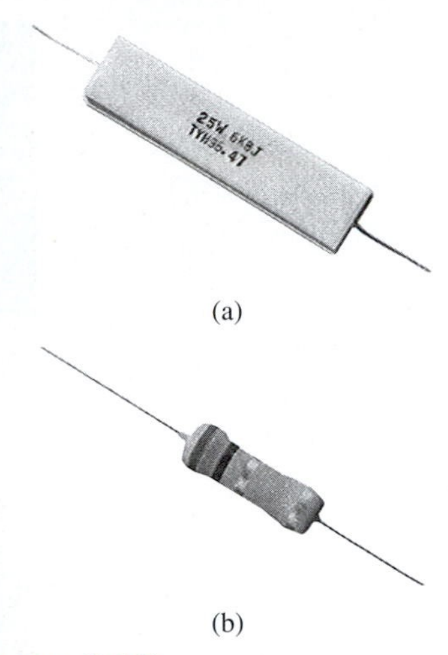

Figure 2.12
Fixed resistors: (a) wirewound type; (b) carbon film type.
Courtesy of Tech America

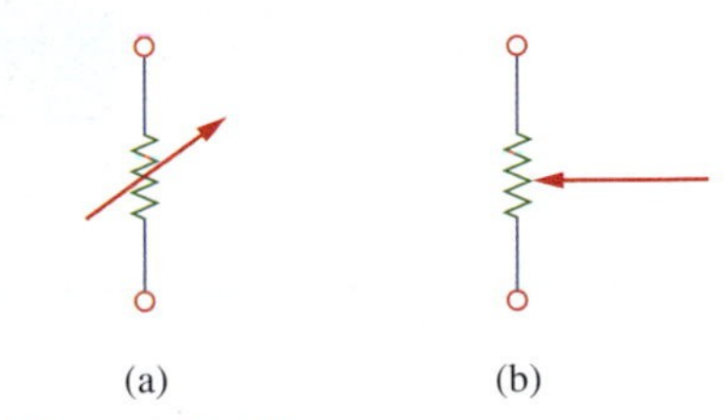

Figure 2.13
Circuit symbols for a variable resistor.

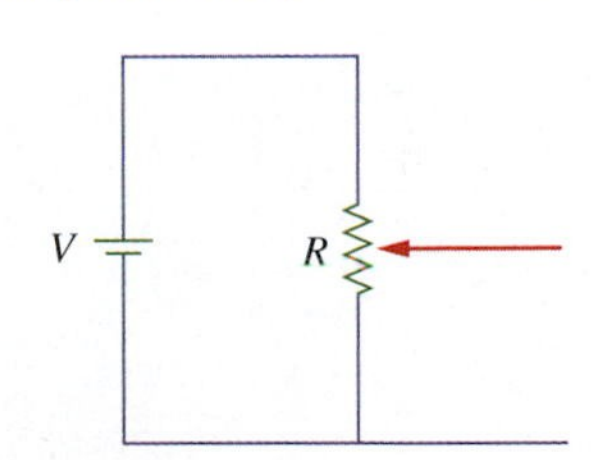

Figure 2.14
Variable resistor used as a potentiometer.

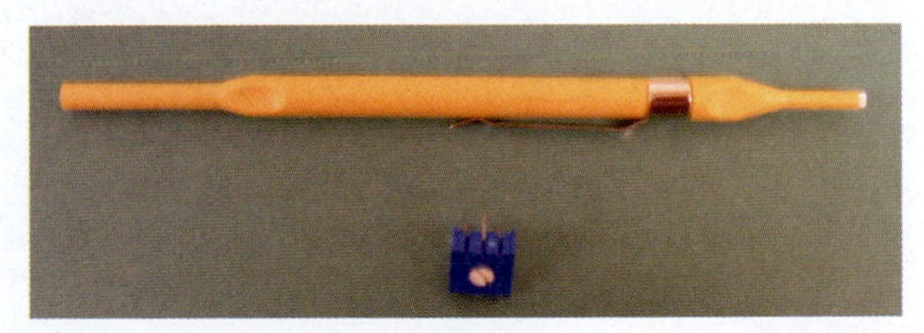

Figure 2.15
Potentiometers with their adjusters.
© Sarhan M. Musa

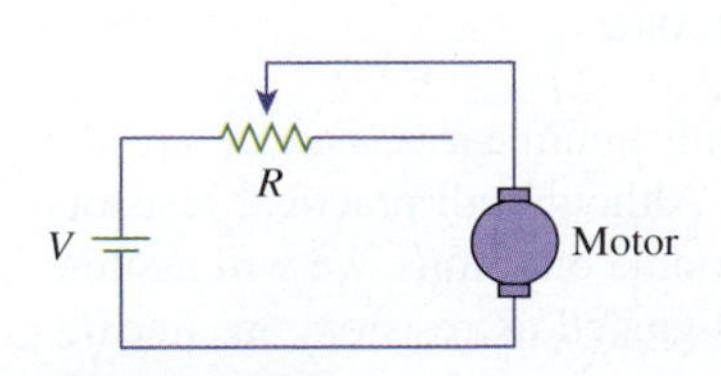

Figure 2.16
Variable resistor used as a rheostat.

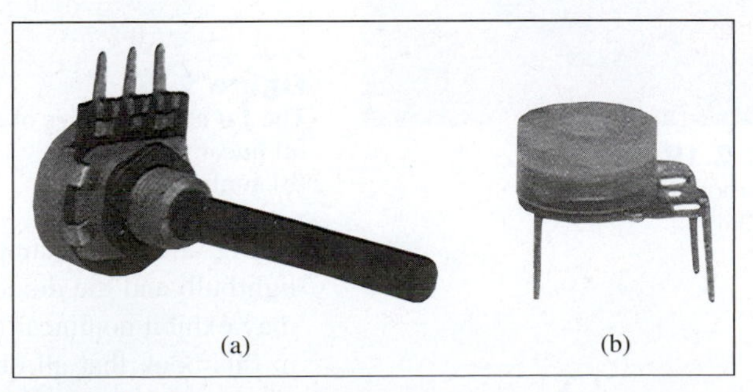

Figure 2.17
Variable resistors: (a) composition type; (b) slider pot.
Courtesy of Tech America

Figure 2.18
Resistors in an integrated circuit board.
© Eric Tomey/Alamy RF

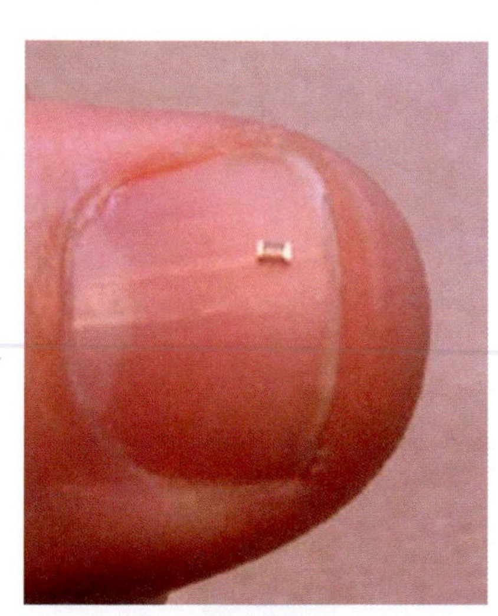

Figure 2.19
Surface mount resistor.
© Greg Ordy

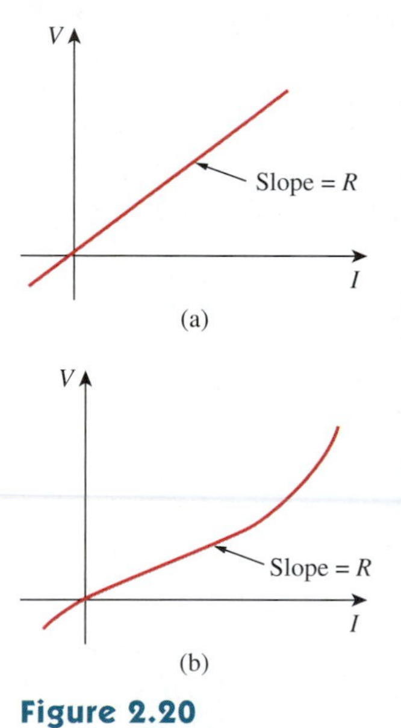

Figure 2.20
The *V*-*I* characteristics of a
(a) linear resistor;
(b) nonlinear resistor.

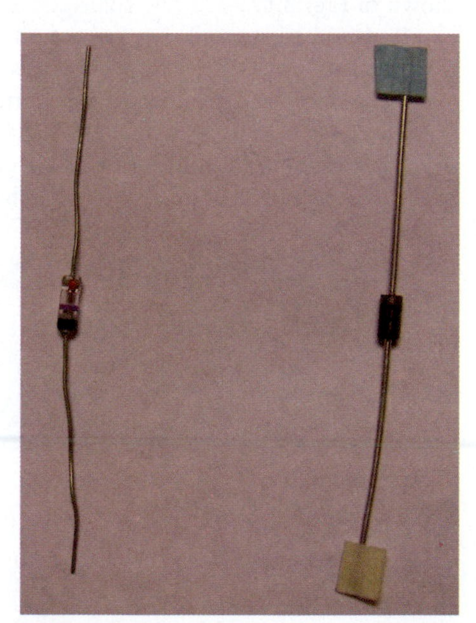

Figure 2.21
Diodes.
© Sarhan M. Musa

in Fig. 2.20(b). Examples of devices with nonlinear resistance are the lightbulb and the diode[1] (see Fig. 2.21). Although all practical resistors may exhibit nonlinear behavior under certain conditions, we will assume in this book that all objects actually designated as resistors are linear.

[1] A diode is a semiconductor device that acts like a switch; it allows charge/current to flow in only one direction.

2.7 Resistor Color Code

Some resistors are physically large enough to have their values printed on them. Other resistors are too small to have their values printed on them. For such small resistors, color coding provides a way of determining the value of resistance. As shown in Fig. 2.22, the color coding consists of three, four, or five bands of color around the resistor. The bands are illustrated in Table 2.3 and explained as follows:

Figure 2.22
Resistor color codes.

A = First significant figure of resistance value
B = Second significant figure of resistance value
C = Multiplier of resistance for resistance value
D = Tolerance rating (in %)
E = Reliability factor (in %)

**We read the bands from left to right.*

0		Black	Big
1		Brown	Boys
2		Red	Race
3		Orange	Our
4		Yellow	Young
5		Green	Girls
6		Blue	But
7		Violet	Violet
8		Gray	Generally
9		White	Wins

Figure 2.23
Memory aid for color codes.

The first three bands (A, B, and C) specify the value of the resistance. Bands A and B represent the first and second digits of the resistance value. Band C is usually given as a power of 10 as in Table 2.3. If present, the fourth band (D) indicates the tolerance percentage. For example, a 5 percent tolerance indicates that the actual value of the resistance is within ± 5 of the color-coded value. When the fourth band is absent, the tolerance is taken by default to be ± 20 percent. The fifth band (E), if present, is used to indicate a reliability factor, which is a statistical indication of the expected number of components that will fail to have the indicated resistance after working for 1,000 hours. As shown in Fig. 2.23, the statement "Big Boys Race Our Young Girls, But Violet Generally Wins" can serve as a memory aid in remembering the color code.

TABLE 2.3

Resistor color code.

Color	Band A significant figure	Band B significant figure	Band C multiplier	Band D tolerance	Band E reliability
Black	N/A	0	10^0		
Brown	1	1	10^1		1%
Red	2	2	10^2		0.1%
Orange	3	3	10^3		0.01%
Yellow	4	4	10^4		0.001%
Green	5	5	10^5		
Blue	6	6	10^6		
Violet	7	7	10^7		
Gray	8	8	10^8		
White	9	9	10^9		
Gold			0.1	5%	
Silver			0.01	10%	
No color				20%	

Example 2.8

Figure 2.24
For Example 2.8.

Determine the resistance value of the color-coded resistor shown in Fig. 2.24.

Solution:
Band A is blue (6); band B is red (2); band C is orange (3); band D is gold (5%); and band E is red (0.1%). Hence,

$$R = 62 \times 10^3\ \Omega \pm 5\% \text{ tolerance with a reliability of } 0.1\%$$
$$= 62\ \text{k}\Omega \pm 3.1\ \text{k}\Omega \text{ with a reliability of } 0.1\%$$

This means that the actual resistance of the color-coded resistor will fall between 58.9 kΩ (62 − 3.1) kΩ and 65.1 kΩ (62 + 3.1) kΩ. The reliability of 0.1% indicates that 1 out of 1,000 will fail to fall within the tolerance range after 1,000 hours of service.

Practice Problem 2.8

What is the resistance value, tolerance, and reliability of the color-coded resistor shown in Fig. 2.25?

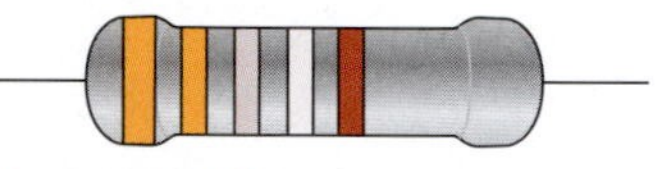

Figure 2.25
For Practice Problem 2.8.

Answer: 3.3 MΩ ± 10% with a reliability of 1%

Example 2.9

A resistor has three bands only—in order green, black, and silver. Find the resistance value and tolerance of the resistor.

Solution:
Band A is green (5); band B is black (0); and band C is silver (0.01). Hence

$$R = 50 \times 0.01 = 0.5\ \Omega$$

Because the fourth band is absent, the tolerance is, by default, 20 percent.

Practice Problem 2.9

What is the resistance value and tolerance of a resistor having bands colored in the order yellow, violet, white, and gold?

Answer: 47 GΩ ± 5%

Example 2.10

A company manufactures resistors of 5.4 kΩ with a tolerance of 10 percent. Determine the color code of the resistor.

Solution:

$$R = 5.4 \times 10^3 = 54 \times 10^2$$

From Table 2.3, green represents 5; yellow stands for 4; while red stands for10^2. The tolerance of 10 percent corresponds to silver. Hence, the color code of the resistor is:

Green, yellow, red, silver

Practice Problem 2.10

If the company in Example 2.10 also produces resistors of 7.2 MΩ with a tolerance of 5 percent and reliability of 1 percent, what will be the color codes on the resistor?

Answer: Violet, red, green, gold, brown

2.8 Standard Resistor Values

One would expect resistor values are commercially available in all values. For practical reasons, this would not make sense. Only a limited number of resistor values are commercially available at reasonable cost. The list of standard values of commercially available resistors is presented in Table 2.4. These are the standard values that have been agreed to for carbon composition resistors. Notice that the values range from 0.1 Ω to 22 MΩ. While 10 percent tolerance resistors are available only for those values in bold type at reasonable cost, 5 percent tolerance resistors are available in all values. For example, a 330-Ω resistor could be available either as a 5 or 10 percent tolerance component, while a 110-kΩ resistor is available only as 5 percent tolerance component.

When designing a circuit, the calculated values are seldom standard. One may select the nearest standard values or combine the standard values. In most cases, selecting the nearest standard value may

TABLE 2.4

Standard values of commercially available resistors.

Ohms (Ω)					Kilohms (kΩ)		Megohms (MΩ)	
0.10	**1.0**	**10**	**100**	**1000**	**10**	**100**	**1.0**	**10.0**
0.11	1.1	11	110	1100	11	110	1.1	11.0
0.12	**1.2**	**12**	**120**	**1200**	**12**	**120**	**1.2**	**12.0**
0.13	1.3	13	130	1300	13	130	1.3	13.0
0.15	**1.5**	**15**	**150**	**1500**	**15**	**150**	**1.5**	**15.0**
0.16	1.6	16	160	1600	16	160	1.6	16.0
0.18	**1.8**	**18**	**180**	**1800**	**18**	**180**	**1.8**	**18.0**
0.20	2.0	20	200	2000	20	200	2.0	20.0
0.22	**2.2**	**22**	**220**	**2200**	**22**	**220**	**2.2**	**22.0**
0.24	2.4	24	240	2400	24	240	2.4	
0.27	**2.7**	**27**	**270**	**2700**	**27**	**270**	**2.7**	
0.30	3.0	30	300	3000	30	300	3.0	
0.33	**3.3**	**33**	**330**	**3300**	**33**	**330**	**3.3**	
0.36	3.6	36	360	3600	36	360	3.6	
0.39	**3.9**	**39**	**390**	**3900**	**39**	**390**	**3.9**	
0.43	4.3	43	430	4300	43	430	4.3	
0.47	**4.7**	**47**	**470**	**4700**	**47**	**470**	**4.7**	
0.51	5.1	51	510	5100	51	510	5.1	
0.56	**5.6**	**56**	**560**	**5600**	**56**	**560**	**5.6**	
0.62	6.2	62	620	6200	62	620	6.2	
0.68	**6.8**	**68**	**680**	**6800**	**68**	**680**	**6.8**	
0.75	7.5	75	750	7500	75	750	7.5	
0.82	**8.2**	**82**	**820**	**8200**	**82**	**820**	**8.2**	
0.91	9.1	92	910	9100	91	910	9.1	

provide adequate performance. To ease calculations, most of the resistor values used in this book are nonstandard.

2.9 Applications: Measurements

Resistors are often used to model devices that convert electrical energy into heat or other forms of energy. Such devices include conducting wires, lightbulbs, electric heaters, stoves, ovens, and loudspeakers. Also, by their nature, resistors are used to control the flow of current. We take advantage of this property in several applications such as in potentiometers and meters. In this section, we will consider meters—the ammeter, voltmeter, and ohmmeter, which measure current, voltage, and resistance, respectively. Being able to measure current I, voltage V, and resistance R is very important.

The **voltmeter** is the instrument used to measure voltage; the **ammeter** is the instrument used to measure current; and the **ohmmeter** is the instrument used to measure resistance.

It is common these days to have the three instruments combined into one instrument known as a *multimeter,* which may be analog or digital. An analog meter is one that uses a needle and calibrated meter to display the measured value; that is, the measured value is indicated by the pointer of the meter. A digital meter is one in which the measured valued is shown in form of a digital display. The digital meters are more commonly used today. Because both analog and digital meters are used in the industry, one should be familiar with both. Figure 2.26 illustrates a typical analog multimeter (combining voltmeter, ammeter, and ohmmeter) and a typical digital multimeter. The digital multimeter (DMM) is the most widely used instrument. Its analog counterpart is the volt-ohm-milliammeter (VOM).

To measure voltage, we connect the voltmeter/multimeter across the element for which the voltage is desired, as shown in Fig. 2.27. The voltmeter measures the voltage across the load and is therefore connected in parallel[2] with the element.

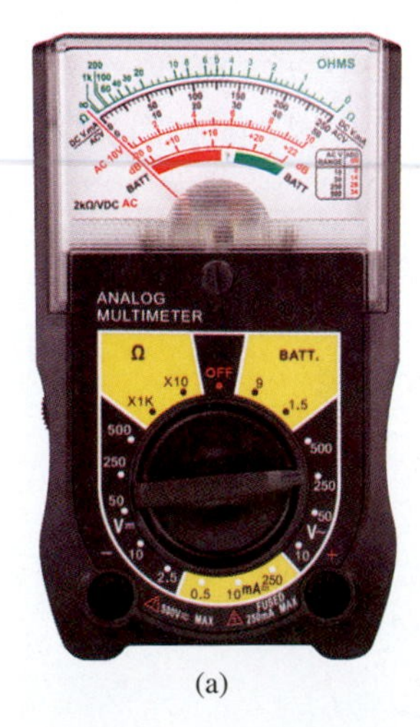

(a)

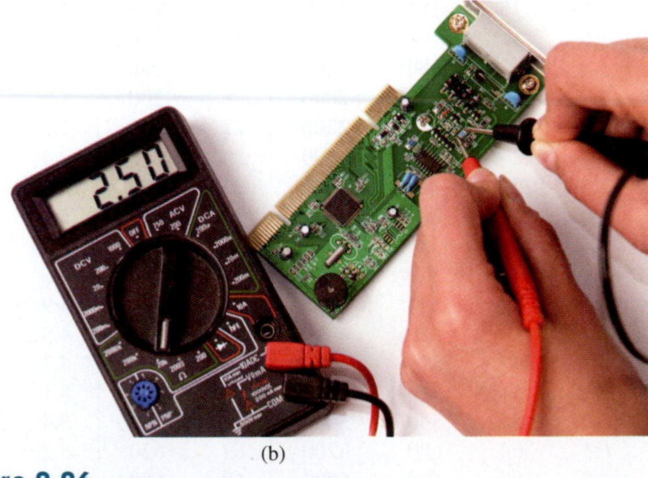
(b)

Figure 2.26
(a) Analog multimeter; (b) digital multimeter.
(a) © iStock; (b) © Oleksy Maksymenko/Alamy RF

[2] Two elements are in parallel if they are connected to the same two points.

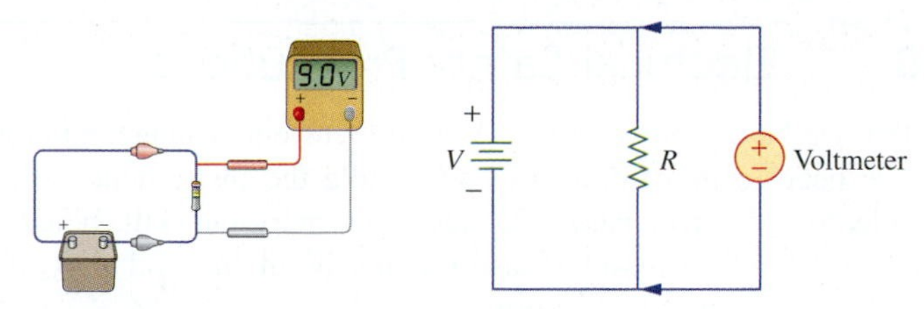

Figure 2.27
Measuring voltage.

To measure current, we connect the ammeter/multimeter in series[3] with the element under test, as shown in Fig. 2.28. The meter must be connected such that the current enters through the positive terminal to get a positive reading. The circuit must be "broken"; that is, the current path must be interrupted so that the current must flow through the ammeter. (The ampclamp is another device for measuring ac current.)

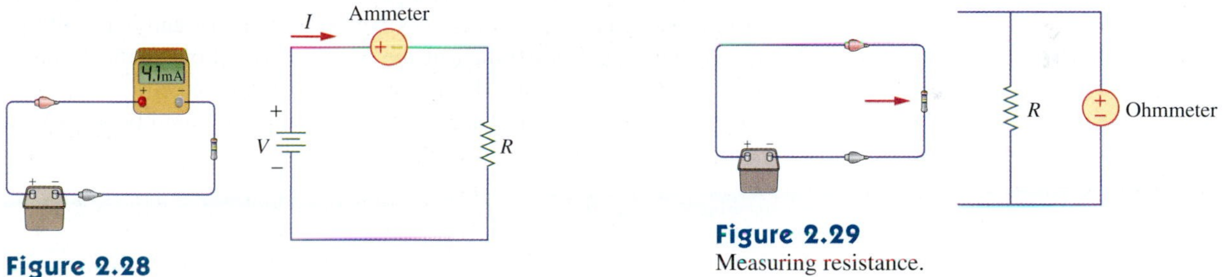

Figure 2.28
Measuring current.

Figure 2.29
Measuring resistance.

To measure resistance of an element, connect the ohmmeter/multimeter across it, as shown in Fig. 2.29. If the element is connected to a circuit, one end of the element must first be disconnected from the circuit before we measure its resistance. Because the resistance of a wire with no breaks is zero, the ohmmeter can be used to test for continuity. If the wire has a break, the ohmmeter connected across it will read infinity. Thus, the ohmmeter can be used to detect a short circuit (low resistance) and an open circuit (high resistance).

When working with any of the meters mentioned in this section, it is good practice to observe the following:

1. If possible, turn the circuit power off before connecting the meter.
2. To avoid damaging the instrument, it is best to always set the meter on the highest range and then move down to the appropriate range. (Most DMMs are auto-ranging.)
3. When measuring dc current or voltage, observe proper polarity.
4. When using a multimeter, make sure you set the meter in the correct mode (ac, dc, V, A, Ω), including moving the test idea to the appropriate jacks.
5. When the measurement is completed, turn off the meter to avoid draining the internal battery of the meter.

This leads to the issue of safety in electrical measurement.

[3] Two elements are in series if they are cascaded or connected sequentially.

2.10 Electrical Safety Precautions

Now that we have learned how to measure current, voltage, and resistance, we need to be careful how we handle the instruments so as to avoid electric shock or harm. Because electricity can kill, being able to make safe and accurate measurements is an integral part of the knowledge that you must acquire.

2.10.1 Electric Shock

When working on electric circuits, there is the possibility of receiving an electric shock. The shock is due to the passage of current through your body. An electric shock can startle you and cause you to fall down or be thrown down. It may cause severe, rigid contractions of the muscles, which in turn may result in fractures, dislocations, and loss of consciousness. The respiratory system may be paralyzed and the heart may beat irregularly or even stop beating altogether. Electrical burns may be present on the skin and extend into deeper tissue. High current may cause death of tissues between the entry and exit point of the current. Massive swelling of the tissues may follow as the blood in the veins coagulates and the muscles swell. Thus, electric shock can cause muscle spasms, weakness, shallow breathing, rapid pulse, severe burns, unconsciousness, or death.

Electric shock is an injury caused by an electrical current passing through the body.

The human body has resistance that depends on several factors such as body mass, skin moisture, and points of contact of the body with the electric appliance. The effects of various amounts of current in milliamperes (mA) is shown in Table 2.5.

2.10.2 Precautions

Working with electricity can be dangerous unless you adhere strictly to certain rules. The following safety rules should be followed whenever you are working with electricity:

- Always make sure that the circuit is actually dead before you begin working on it.
- Always unplug any appliance or lamp before repairing it.
- Always tape over the main switch, empty fuse socket, or circuit breaker when you're working. Leave a note there so no one will accidentally turn on the electricity. Keep any fuses you've removed in your pocket.

TABLE 2.5

Electric shock

Electric Current	Physiological effect
Less than 1mA	No sensation or feeling
1 mA	Tingling sensation
5–20 mA	Involuntary muscle contraction
20–100 mA	Loss of breathing, fatal if continued

- Handle tools properly and make sure that the insulation on metal tools is in good condition.
- If measuring V or I, turn on the power and record reading. If measuring R, do not turn on power.
- Refrain from wearing loose clothing. Loose clothes can get caught in an operating appliance.
- Always wear long-legged and long-sleeved clothes and shoes and keep them dry.
- Do not stand on a metal or wet floor. (Electricity and water do not mix.)
- Make sure there is adequate illumination around the work area.
- Do not work while wearing rings, watches, bracelets, or other jewelry.
- Do not work by yourself.
- Discharge any capacitor that may retain high voltage.
- Work with only one hand a time in areas where voltage may be high.

Protecting yourself from injury and harm is absolutely imperative. If we follow these safety rules, we can avoid shock and related accidents. Thus, our rule should always be "safety first."

2.11 Summary

1. A resistor is an element in which the voltage, V, across it is directly proportional to the current, I, through it. That is, a resistor is an element that obeys Ohm's law.

$$V = IR$$

   where R is the resistance of the resistor.

2. The resistance R of an object with uniform cross-sectional area A is evaluated as resistivity ρ times length ℓ divided by the cross-section area A, that is,

$$R = \frac{\rho\ell}{A}$$

3. A short circuit is a resistor (a perfectly conducting wire) with zero resistance ($R = 0$). An open circuit is a resistor with infinite resistance ($R = \infty$).
4. The conductance G of a resistor is the reciprocal of its resistance R:

$$G = \frac{1}{R}$$

5. For a circular wire, the cross-sectional area is measured in circular mils (CM). The diameter in mils is related to the area in CM as

$$A_{\text{CM}} = d_{\text{mil}}^2$$

6. American Wire Gauge is a standard system for designating the diameter of wires.
7. There are different types of resistors: fixed or variable, linear or nonlinear. Potentiometer and rheostat are variable resistors that are used to adjust voltage and current, respectively. Common types of

resistors include carbon or composition resistors, wirewound resistors, chip resistors, film resistors, and power resistors.

8. A resistor is usually color coded when it is not physically large enough to print the numerical value of the resistor on it. The statement "Big Boys Race Our Young Girls, But Violet Generally Wins" is a memory aid for the color code: black, brown, red, orange, yellow, green, blue, violet, gray, and white.
9. For carbon composition resistors, standard values are commercially available in the range of 0.1 Ω to 22 MΩ.
10. Voltage, current, and resistance are measured using a voltmeter, ammeter, and ohmmeter, respectively. The three are measured using a multimeter such as a digital multimeter (DMM) or a volt-ohm-milliammeter (VOM).
11. Safety is all about preventing accidents. If we follow some safety precautions, we should have no problems working on electric circuits.

Review Questions

2.1 Which of the following materials is not a conductor?

(a) Copper (b) Silver (c) Mica
(d) Gold (e) Lead

2.2 The main purpose of a resistor in a circuit is to:

(a) resist change in current
(b) produce heat
(c) increase current
(d) limit current

2.3 An element draws 10 A from a 120-V line. The resistance of the element is:

(a) 1200 Ω (b) 120 Ω
(c) 12 Ω (d) 1.2 Ω

2.4 The reciprocal of resistance is:

(a) voltage (b) current
(c) conductance (d) power

2.5 Which of these is not the unit of conductance?

(a) ohm (b) Siemen
(c) mho (d) ℧

2.6 The conductance of a 10-mΩ resistor is:

(a) 0.1 mS (b) 0.1 S
(c) 10 S (d) 100 S

2.7 Potentiometers are types of:

(a) fixed resistors (b) variable resistors
(c) meters (d) voltage regulators

2.8 What is the area in circular mils of a wire that has a diameter of 0.03 in.?

(a) 0.0009 (b) 9
(c) 90 (d) 900

2.9 All resistors are color coded.

(a) True (b) False

2.10 Digital multimeters (DMM) are the most widely used type of electronic measuring instrument.

(a) True (b) False

Answers: 2.1c, 2.2d, 2.3c, 2.4c, 2.5a, 2.6d, 2.7b, 2.8d, 2.9b, 2.10a

Problems

Section 2.2 Resistance

2.1 A 250-m-long copper wire has a diameter of 2.2 mm. Calculate the resistance of wire.

2.2 Find the length of a copper wire that has a resistance of 0.5 Ω and a diameter of 2 mm.

2.3 A 2-in. × 2-in. square bar of copper is 4 ft long. Find its resistance.

2.4 If an electrical hotplate has a power rating of 1200 W and draws a current of 6 A, determine the resistance of the hotplate.

2.5 A Nichrome ($\rho = 100 \times 10^{-8}$ Ωm) wire is used to construct heating elements. What length of a 2-mm-diameter wire will produce a resistance of 1.2 Ω ?

2.6 An aluminum wire of radius 3 mm has a resistance of 6 Ω. How long is the wire?

2.7 A graphite cylinder with a diameter of 0.4 mm and a length of 4 cm has resistance of 2.1 Ω. Determine the resistivity of the cylinder.

2.8 A certain circular wire of length 50 m and diameter 0.5 m has a resistance of 410 Ω at room temperature. Determine the material the wire is made of.

2.9 If we shorten the length of a conductor, why does the conductor decrease in resistance?

2.10 Two wires are made of the same material. The first wire has a resistance of 0.2 Ω. The second wire is twice as long as the first wire and has a radius that is half of the first wire. Determine the resistance of the second wire.

2.11 Two wires have the same resistance and length. The first wire is made of copper, while the second wire is made of aluminum. Find the ratio of the cross-sectional area of the copper wire to that of the aluminum wire.

2.12 High-voltage power lines are used in transmitting large amounts of power over long distances. Aluminum cable is preferred over copper cable due to low cost. Assume that the aluminum wire used for high-voltage power lines has a cross-sectional area of 4.7×10^{-4} m^2. Find the resistance of 20 km of this wire.

Section 2.3 Ohm's Law

2.13 Which of the graphs in Fig. 2.30 represent Ohm's law?

2.14 When the voltage across a resistor is 60 V, the current through it is 50 mA. Determine its resistance.

2.15 The voltage across a 5-kΩ resistor is 16 V. Find the current through the resistor.

2.16 A resistor is connected to a 12-V battery. Calculate the current if the resistor is:

(a) 2 kΩ (b) 6.2 kΩ

2.17 An air-conditioning compressor has resistance 6 Ω. When the compressor is connected to a 240-V source, determine the current through the circuit.

2.18 A source of 12 V is connected to a purely resistive lamp and draws 3 A. What is the resistance of the lamp?

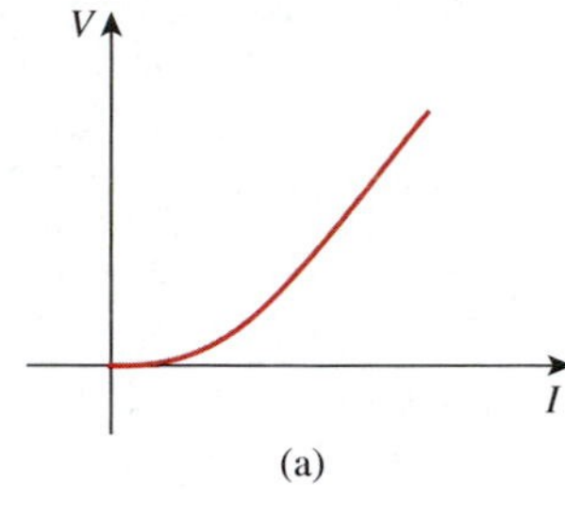

(a)

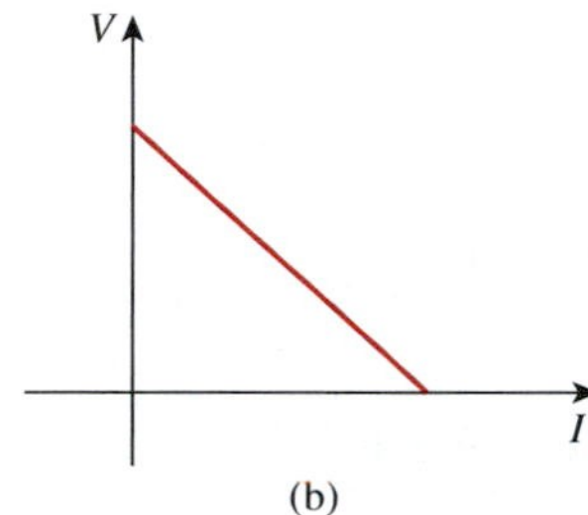

(b)

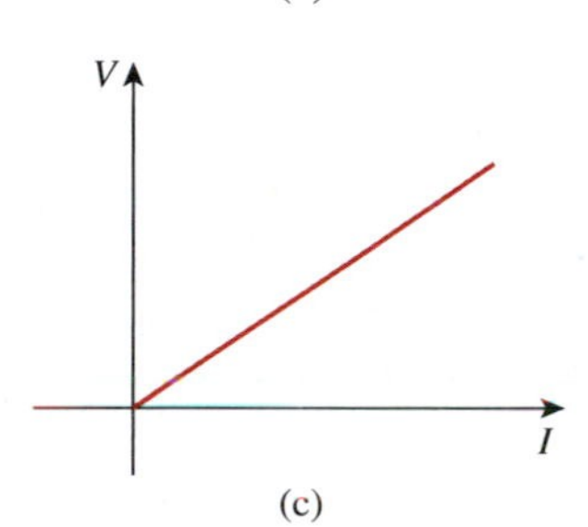

(c)

Figure 2.30
For Problem 2.13.

2.19 If a current of 30 μA flows through a 5.4-MΩ resistor, what is the voltage?

2.20 A current of 2 mA flows through a 25-Ω resistor. Find the voltage drop across it.

2.21 An element allows 28 mA of current to flow through it when a 12-V battery is connected to its terminals. Calculate the resistance of the element.

2.22 Find the voltage of a source which produces a current of 10 mA in a 50-Ω resistor.

2.23 A nonlinear resistor has $I = 4 \times 10^{-2} V^2$. Find I for $V = 10$, 20, and 50 V.

2.24 Determine the magnitude and direction of the current associated with the resistor in each of the circuits in Fig. 2.31.

2.25 Determine the magnitude and polarity of the voltage across the resistor in each of the circuits in Fig. 2.32.

2.26 A flashlight uses two 3-V batteries in series to provide a current of 0.7 A in the filament. (a) Find the potential difference across the flashlight bulb. (b) Calculate the resistance of the filament.

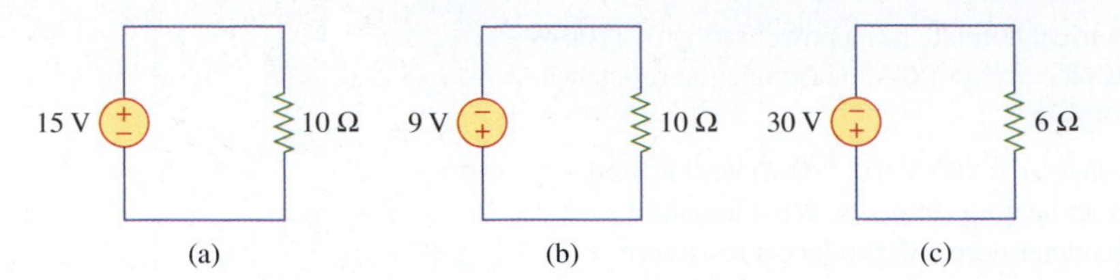

Figure 2.31
For Problem 2.24.

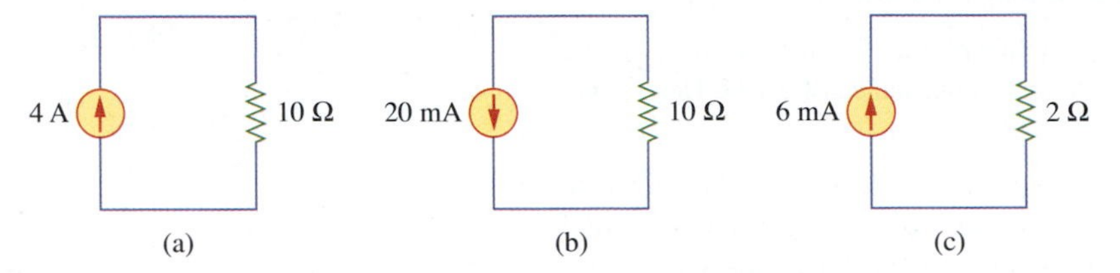

Figure 2.32
For Problem 2.25.

Section 2.4 Conductance

2.27 Determine the conductance of each of the following resistances:

(a) 2.5 Ω (b) 40 kΩ (c) 12 MΩ

2.28 Find the resistance for each of the following conductances:

(a) 10 mS (b) 0.25 S (c) 50 S

2.29 When the voltage across a resistor is 120 V, the current through it is 2.5 mA. Calculate its conductance.

2.30 A copper rod has a length of 4 cm and a conductance of 500 mS. Find its diameter.

2.31 Determine the battery voltage V in the circuit shown in Fig. 2.33.

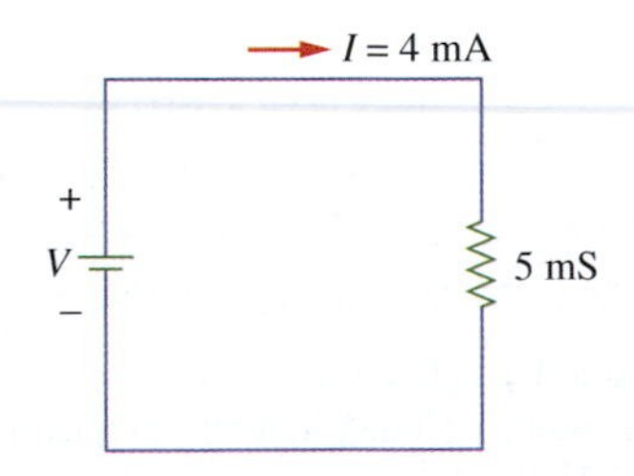

Figure 2.33
For Problem 2.31.

Section 2.5 Circular Wires

2.32 Using Table 2.2, determine the resistance of 600 ft of #10 and #16 AWG copper.

2.33 The resistance of a copper transmission line cannot exceed 0.001 Ω, and the maximum current drawn by the load is 120 A. What gauge wire is appropriate? Assume a length of 10 ft.

2.34 Find the diameter in inches for wires having the following cross-sectional areas:

(a) 420 CM (b) 980 CM

2.35 Calculate the area in circular mils of the following conductors:

(a) circular wire with diameter 0.012 in.

(b) rectangular bus bar with dimensions 0.2 in. × 0.5 in.

2.36 How much current will flow in a #16 copper wire 1 mi long connected to a 1.5-V battery?

Section 2.7 Resistor Color Code

2.37 Find the resistance value having the following color codes:

(a) blue, red, violet, silver

(b) green, black, orange, gold

2.38 Determine the range (in ohms) in which a resistor having the following bands must exist.

	Band A	Band B	Band C	Band D
(a)	Brown	Violet	Green	Silver
(b)	Red	Black	Orange	Gold
(c)	White	Red	Gray	—

2.39 Determine the color codes of the following resistors with 5 percent tolerance.

(a) 52 Ω (b) 320 Ω

(c) 6.8 kΩ (d) 3.2 MΩ

2.40 Find the color codes of the following resistors:

(a) 240 Ω (b) 45 kΩ (c) 5.6 MΩ

2.41 For each of the resistors in Problem 2.37, find the minimum and maximum resistance within the tolerance limits.

2.42 Give the color coding for the following resistors:

(a) 10 Ω, 10 percent tolerance

(b) 7.4 kΩ, 5 percent tolerance

(c) 12 MΩ, 20 percent tolerance

Section 2.9 Applications: Measurements

2.43 How much voltage is the multimeter in Fig. 2.34 reading?

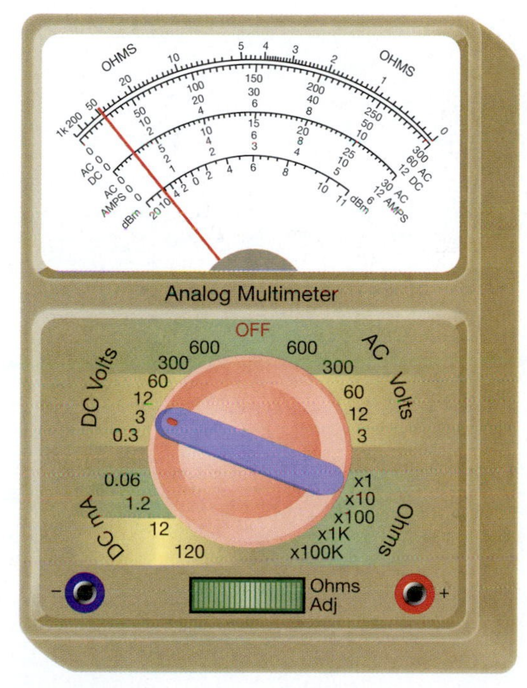

Figure 2.34
For Problem 2.43.

2.44 Determine the voltage reading for the multimeter in Fig. 2.35.

2.45 You are supposed to check a lightbulb to see whether is burned out or not. Using an ohmmeter, how would you do this?

2.46 What is wrong with the measuring scheme in Fig. 2.36?

2.47 Show how you would place a voltmeter to measure the voltage across resistor R_1 in Fig. 2.37.

2.48 Show how you would place an ammeter to measure the current through resistor R_2 in Fig. 2.37.

2.49 Explain how you would connect an ohmmeter to measure the resistance R_2 in Fig. 2.37.

2.50 How would you use an ohmmeter to determine the on and off states of a switch?

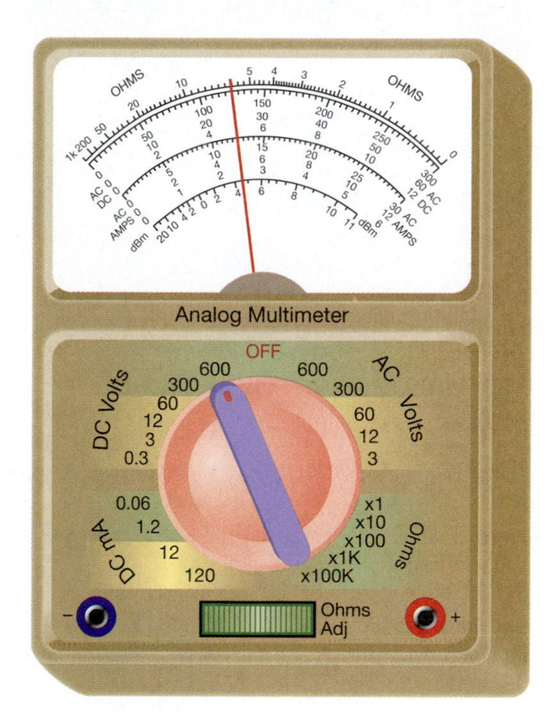

Figure 2.35
For Problem 2.44.

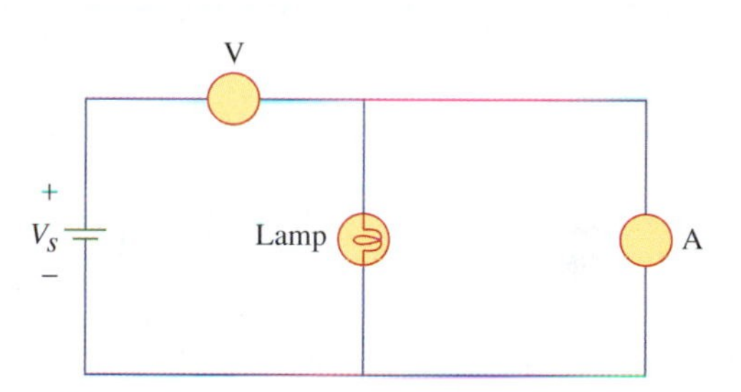

Figure 2.36
For Problem 2.46.

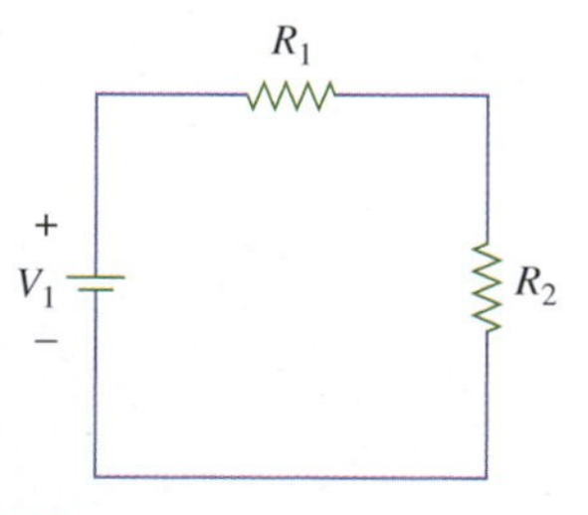

Figure 2.37
For Problems 2.47, 2.48, and 2.49.

Section 2.10 Electrical Safety Precautions

2.51 What causes electric shock?

2.52 Mention at least four safety precautions you would take when taking measurements.

chapter

3

Power and Energy

No one can make you feel inferior without your consent.

—Eleanor Roosevelt

Historical Profiles

James Watt (1736–1819), a Scottish inventor and mechanical engineer, is renowned for his improvements to the steam engine.

Born in Greenock, Scotland, Watt had little formal education due to poor health. He went to London to study instrument-making for a year, and then to Glasgow to set up shop, but the Guild denied permit because he hadn't apprenticed for seven years. He was given a workshop under university auspices. There, he established his own instrument-making business and, later, developed his interest in steam engines. Watt was also a renowned civil engineer, making several surveys of canal routes. In 1785, James Watt was elected Fellow of the Royal Society. The unit of power, the *watt,* is named in his honor.

James Watt
Courtesy University of Texas Library

James Prescott Joule (1818–1889), a British physicist, established that the various forms of energy—mechanical, electrical, and heat—are basically the same and can be changed from one form into another.

Born in Manchester, England, to a prosperous brewery owner, Joule was educated at home. He was sent to Cambridge at the age of 16 to study under the eminent British chemist John Dalton. Hoping to replace steam engines by electric motors, his first research sought to improve electric motor efficiency. He established the equivalence between amounts of heat and mechanical work. This led to the law of conservation of energy (the first law of thermodynamics), which states that energy used up in one form reappears in another and is never lost. His experiments led to Joule's law, which describes the amount of heat produced in a wire by an electrical current. The international unit of energy, the *joule,* is named in his honor.

James Prescott Joule
© National Bureau of Standards Archives, courtesy AIP Emilio Segre Visual Archives, E. Scott Barr

3.1 Introduction

In the previous chapter, our major concern was finding the resistance of an element. We introduced Ohm's law to that end. In this chapter, our major concern is in calculating power and energy (introduced in Chapter 1) and relating them to electric circuits.

Energy is a quantity that can be converted to different forms, including heat energy, kinetic energy, potential energy, and electromagnetic energy. It is the capacity for doing work. Most of the world's convertible energy comes from fossil fuels that are burned to produce heat that is then used as a transfer medium to mechanical or other forms in order to accomplish tasks. Power is the rate of flow of energy, or the rate at which work is done.

Power is the most important quantity in electric utilities and electronic systems, because such systems involve transmission of power from one point to another. Also, every industrial and household electrical device—every fan, motor, lamp, iron, TV, personal computer—has a power rating that indicates how much power the equipment requires; exceeding the power rating can do permanent damage to an electrical device.

We begin the chapter by defining energy and power and how to calculate them in a given electric circuit. We introduce the passive sign convention for determining the sign of power. We discuss the power or wattage rating of resistors and the efficiency of energy-conversion devices. We finally consider two applications: measurement of power using the wattmeter and measurement of energy using the watt-hour meter.

3.2 Power and Energy

Energy is the ability to do work, while *power* is the rate of expending energy. As mentioned in Section 1.7, power P and energy (or work) W are related as

$$\boxed{P = \frac{W}{t}} \tag{3.1}$$

or

$$W = Pt \tag{3.2}$$

where t is time in seconds. Power is measured in watts (in honor of James Watt), while energy is measured in joules (in honor of James Joule).

James Watt introduced the horsepower as a unit of mechanical power. Although horsepower is an old unit of power, it is still in use today. Electrical power in watts (W) is related to mechanical power in horsepower (hp) as

$$1 \text{ hp} = 746 \text{ W} \tag{3.3}$$

that is, 1 horsepower is approximately equal to 0.75 kW.

The power companies charge customers based on the amount of energy consumed. Because the power companies deal with large amounts of energy, instead of using joules (or watt-seconds) as the unit of energy, they prefer to use watt-hours (Wh) or kilowatt-hours (kWh). From Eq. (3.2),

$$\text{Energy (watt-hours)} = \text{Power (watts)} \times \text{Time (hours)} \tag{3.4}$$

Example 3.1

An electric motor delivers 30 kJ of energy in 2 min. What is the power in watts?

Solution:

$$P = \frac{W}{t} = \frac{30 \times 10^3 \text{ J}}{2 \times 60 \text{ s}} = 250 \text{ W}$$

Practice Problem 3.1

Determine the power consumed when 600 J of energy is expended in 5 min.

Answer: 2 W

Example 3.2

How many kilowatt-hours are consumed by a 250-W bulb left on for 16 h? How much will it cost to operate the bulb for that long if electricity costs 6.5 cents/kWh?

Solution:

$$W = Pt = 250 \times 16 = 4000 \text{ Wh} = 4 \text{ kWh}$$
$$\text{Cost} = 4 \times 6.5 = 26 \text{ cents}$$

Practice Problem 3.2

If 480 watts are being used for 8 h, find the energy consumption in kilowatt-hours and the cost if electricity costs 8 cents/kWh.

Answer: 3.84 kWh, 31 cents

3.3 Power in Electric Circuits

We always deal with power in electric circuits. In terms of current I and voltage V, Eq. (3.1) can be written as

$$P = \frac{W}{t} = \frac{WQ}{Qt} = VI$$

or

$$P = VI \tag{3.5}$$

If we now incorporate Ohm's law ($V = IR$), we can express Eq. (3.5) in terms of other circuit quantities. Substituting $V = IR$ in Eq. (3.5) yields

$$P = VI = (IR)I$$

or

$$P = I^2R \tag{3.6}$$

Substituting $I = V/R$ in Eq. (3.5) results in

$$P = VI = V\left(\frac{V}{R}\right)$$

that is,

$$P = \frac{V^2}{R} \tag{3.7}$$

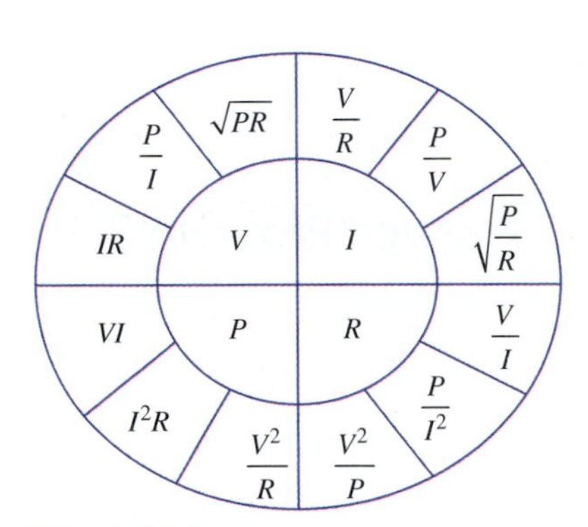

Figure 3.1
Relationships among V, I, P, and R.

Note that the formulas in Eqs. (3.5), (3.6), and (3.7) are known as Watt's law and are equivalent expressions for finding the power dissipated in a resistor. Which formula is used depends on the information you have. The four quantities I, V, P, and R are related as shown in Fig. 3.1. The term in each quadrant of the inner circle is equal to the each of the three formulas in that quadrant's outer circle. Hence V, for example, can be found by $V = IR$, $V = P/I$, or $V = \sqrt{PR}$.

Power can be delivered or absorbed depending on the polarity of the voltage and direction of current. All power delivered to a resistor is absorbed and dissipated in the form of heat.

Example 3.3

Determine the power in each of the circuits shown in Fig. 3.2.

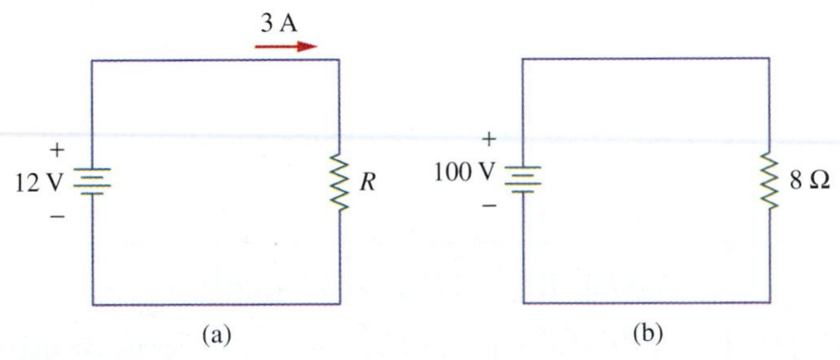

Figure 3.2
For Example 3.3.

Solution:

(a) For the circuit in Fig. 3.2(a),

$$P = VI = (12)(3) = 36 \text{ W}$$

(b) For the circuit in Fig. 3.2(b),

$$P = \frac{V^2}{R} = \frac{(100)^2}{8} = 1.25 \text{ kW}$$

Practice Problem 3.3

Calculate the power in each of the circuits shown in Fig. 3.3.

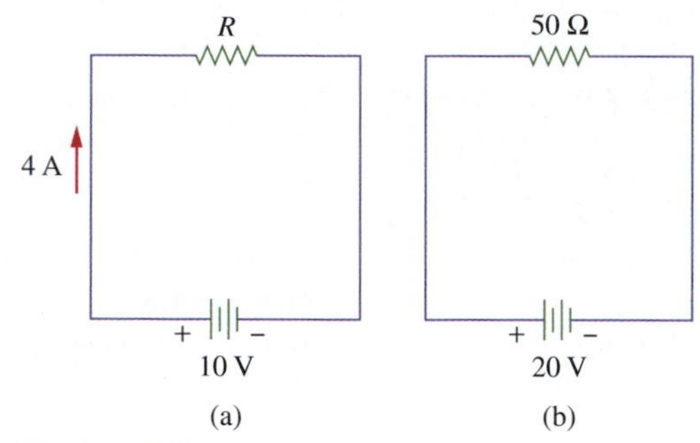

Figure 3.3
For Practice Problem 3.3.

Answer: (a) 40 W (b) 8 W

Example 3.4

Find the maximum allowable current that can flow through a 6-kΩ, 4-W resistor without exceeding its rating.

Solution:
We can rewrite Eq. (3.6) as

$$P = I^2R \quad \longrightarrow \quad I = \sqrt{\frac{P}{R}}$$

The maximum current is

$$I = \sqrt{\frac{4}{6 \times 10^3}} = 25.82 \text{ mA}$$

Practice Problem 3.4

In an electric circuit, an 10-mA current flows through in a 40-Ω resistor. Find the power absorbed by the resistor.

Answer: 4 mW

3.4 Power Sign Convention

Current direction and voltage polarity play a major role in determining the sign of power. Therefore, it is important that we pay attention to the relationship between current I and voltage V in Fig. 3.4(a). The voltage polarity and current direction must conform to those shown in Fig. 3.4(a) for the power to have a positive sign. This is known as the *passive sign convention*. By the passive sign convention, current enters through the positive polarity of the voltage. In this case, $P = +VI$ or $VI > 0$ implies that the element is absorbing power. However, if $P = -VI$ or $VI < 0$, as in Fig. 3.4(b), the element is releasing or supplying power. Thus,

> **Passive sign convention** is satisfied when the current enters through the positive terminal of an element and $P = +VI$. Otherwise, $P = -VI$.

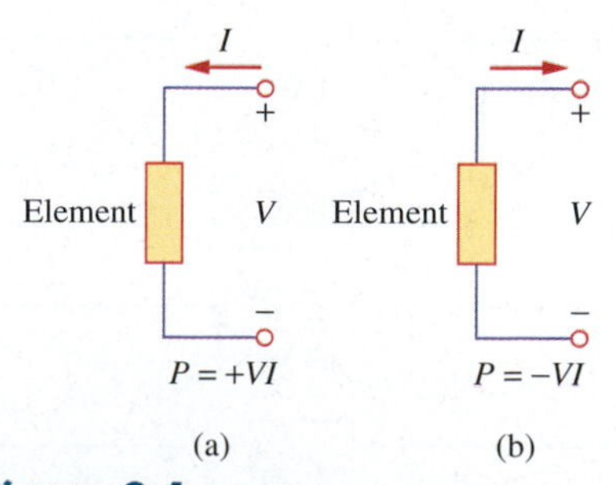

Figure 3.4
Reference polarities for power using the passive sign convention: (a) absorbing power; (b) supplying power.

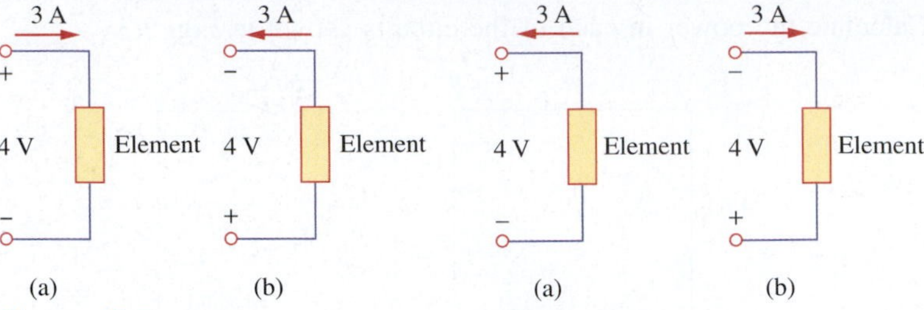

Figure 3.5
Two cases of an element with an absorbing power of 12 W:
(a) $P = 4 \times 3 = 12$ W;
(b) $P = (-4) \times (-3) = 12$ W.

Figure 3.6
Two cases of an element with a supplying power of 12 W:
(a) $P = 4 \times (-3) = -12$W;
(b) $P = 4 \times (-3) = -12$ W.

Unless otherwise stated, the passive sign convention will be followed throughout this book.

For example, the element in both circuits of Fig. 3.5 has an absorbing power of +12 W because a positive current enters the positive terminal in both cases. In Fig. 3.6, however, the element is supplying power of −12 W because a positive current enters the negative terminal. Of course, an absorbing power of +12 W is equivalent to a supplying power of −12 W. In general,

$$\text{Power absorbed} = -\text{Power supplied} \tag{3.8}$$

3.5 Resistor Power Ratings

In addition to specifying the value of resistance of a resistor, its power rating is usually specified.[1] For this reason, resistors are rated in watts, specifying the power or wattage rating.

> The **power rating** of a resistor is the maximum power that it can dissipate without it becoming too hot or risking damage to it.

A resistor must have a wattage or power rating high enough to dissipate the power produced by the current flowing through it without becoming too hot. The power rating of a resistor does not depend on its resistance but on its physical size. The larger the physical size of any resistor, the higher the wattage rating. This is true because a larger surface area of material radiates a greater amount of heat more easily. Resistors of the same resistance value are available in different wattage values. Carbon resistors, for example, are commonly made in wattage ratings of $\frac{1}{8}$, $\frac{1}{4}$, $\frac{1}{2}$, 1, and 2 W. The relative sizes of the resistors for different wattage ratings are shown in Fig. 3.7. As evident in Fig. 3.7, a larger physical size indicates a higher power rating. Also, higher-wattage resistors can operate at higher temperatures when dissipating less power than its rating. If resistors with wattage ratings greater than 5 W are needed, wirewound resistors are often used. Wirewound resistors are made with power ratings between 5 and 200 W.

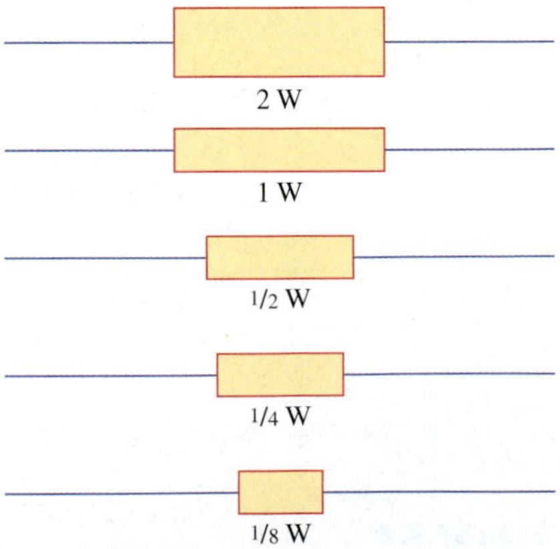

2 W
1 W
1/2 W
1/4 W
1/8 W

Figure 3.7
Metal-film resistors with standard power ratings of $\frac{1}{8}$ W, $\frac{1}{4}$ W, $\frac{1}{2}$ W, 1 W, and 2 W.

[1] In general, electrical components are given a power rating.

Example 3.5

A 0.2-Ω resistor is rated at 6 W. Is this resistor safe when conducting a current of 8 A?

Solution:
Using the power equation,

$$P = I^2R = (8)^2(0.2) = 12.8 \text{ W}$$

Because the calculated power is greater than the rated power of 6 W, the resistor will overheat and most likely be damaged.

Practice Problem 3.5

Calculate the maximum safe current flow in a 54-Ω, 4-W resistor.

Answer: 272 mA

3.6 Efficiency

The efficiency of a device or circuit is a means of comparing its useful output to its input. The law of conservation of energy states that energy can neither be created nor destroyed but can be converted from one form to another. Examples of this law are found in the conversion of electrical energy into thermal (or heat) energy by a resistor and the conversion of electrical energy into mechanical energy by an electric motor. In the process of energy conversion, part of the energy is converted into a form that is not useful; we refer to that as "lost" energy. This reduces the efficiency of the system.

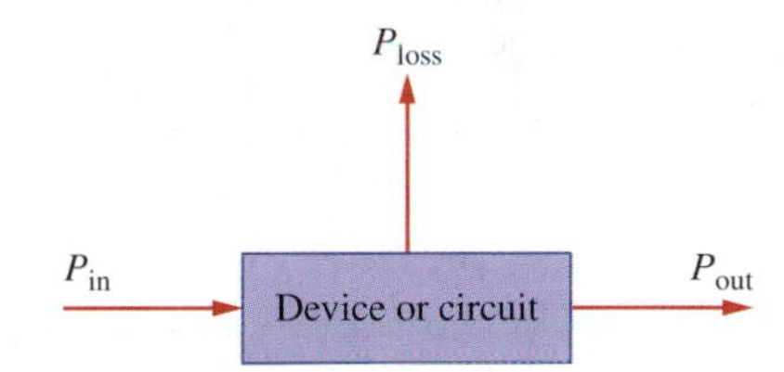

Figure 3.8
Loss of power due to energy conversion.

Efficiency is the ratio of useful output power to total input power.

The amount of input energy that an energy-converting device can convert into useful energy is its *efficiency*, denoted by the Greek letter eta (η). In the process of energy-conversion, part of the energy or power is lost, as shown in Fig. 3.8.

$$P_{in} = P_{out} + P_{loss} \tag{3.9}$$

The efficiency of a device or circuit is the ratio of the output power P_{out} to the input power P_{in}.

Efficiency η can be expressed in terms of power or energy. (It is always less than 1 or less than 100 percent.) In terms of power,

$$\eta = \frac{P_{out}}{P_{in}} \times 100\% \tag{3.10}$$

In terms of energy,

$$\eta = \frac{W_{out}}{W_{in}} \times 100\% \tag{3.11}$$

We notice that the two relationships in Eqs. (3.10) and (3.11) are the same, because $W = Pt$.

Example 3.6

Determine the efficiency of a 110-V motor that draws 15 A and develops an output power of 1.8 hp. How much power is lost? (1 hp = 746 W)

Solution:

$$P_{\text{in}} = VI = 110 \times 15 = 1650 \text{ W}$$

$$P_{\text{out}} = 1.8 \times 746 = 1342.8 \text{ W}$$

$$\eta = \frac{P_{\text{out}}}{P_{\text{in}}} \times 100\% = \frac{1342.8}{1650} \times 100\% = 81.38\%$$

$$P_{\text{loss}} = P_{\text{in}} - P_{\text{out}} = 1650 - 1342.8 = 307.2 \text{ W}$$

Practice Problem 3.6

The input power of a motor is 3260 W, while its useful output power is 2450 W. Calculate the efficiency of the motor.

Answer: 75.15%

Example 3.7

An electric motor develops a mechanical power of 20 hp with 88 percent efficiency. Finds its electric input power.

Solution:

$$P_{\text{out}} = 20 \times 746 = 14{,}920 \text{ W}$$

$$P_{\text{in}} = \frac{P_{\text{out}}}{\eta} = \frac{14{,}920}{0.88} = 16{,}955 \text{ W} = 16.955 \text{ kW}$$

Practice Problem 3.7

Determine the output power of an electric motor that draws 8 A from a 220-V source and is 85 percent efficient. How much power is lost?

Answer: 1.496 kW, 264 W

3.7 Fuses, Circuit Breakers, and GFCIs

As we know, the power dissipated in a resistance R varies as the square of the current I, that is, $P = I^2R$. Doubling the current causes the power to increase to four times. This increase in power is accompanied by an increase in the temperature of the resistor. For the same reason, the wires in a building can get hot enough to overheat or even ignite the structural materials and cause smoke or fire. Thus, some protective devices are needed to protect circuits from overcurrent. The protective devices ensure that the current through the line does not exceed the rated value.

Figure 3.9
Electrical fuses.
© Steve Cole/Getty RF

> A **fuse** is an electrical device that can interrupt the flow of electrical current when its rating is exceeded.

Fuses protect some of the electrical and electronic devices in your home or car. Figure 3.9 shows some electrical fuses. In case of a power surge, a fuse will "blow" so that the extra electricity does not reach the device. A good fuse is nearly a short circuit and reads 0 Ω, or a small fraction of an ohm, while a blown fuse is an open circuit and reads infinity on the ohmmeter. Figure 3.10 shows how a fuse protects a circuit. During normal operation of the circuit, as in Fig. 3.10(a), the current through the fuse is not high enough to blow it. When the load is short-circuited, as shown in Fig. 3.10(b), the high current through the fuse causes it to blow or melt. Consequently, no current passes through the load and the load is protected, as shown in Fig. 3.10(c). Fuses are rated according to the amount of current they can handle. The thinner the wire element in the fuse, the smaller the current rating. For example, automobile fuses are generally rated 10 to 30 A.

In recent years, household fuses have been replaced by circuit breakers. The function of a circuit breaker is, like a fuse, to break a circuit path when a predetermined amount of current is passed through it. The major difference between a fuse and a circuit breaker is this: when a fuse is blown or melted, the fault that caused it must be corrected and the fuse must be replaced by another fuse of the same rating; whereas when a circuit breaker opens, it can be simply reset and used again. A typical household electrical service panel containing several circuit breakers is shown in Fig. 3.11. A circuit breaker has a spring that expands with heat and trips open the circuit when the current exceeds the rating. A tripped circuit breaker is reset by pushing its switch. There are even automatic circuit breakers that reset themselves after they cool down.

> A **ground fault circuit interrupter** (GFCI) is a fast-acting, circuit breaker–like device that shuts off the associated circuit if there is an electrical leakage to ground.

An electrical ground is a common return path for electric current, or a direct physical connection to the earth. As shown in Fig. 3.12, every electrical appliance must be grounded for safety. A ground fault circuit interrupter (GFCI) is an electrical device that protects people by detecting potentially hazardous ground faults and quickly disconnecting power from the circuit. Fuses and circuit breakers protect a device or circuit from becoming overloaded. They do not protect a person from receiving a shock. The GFCI, on the other hand, can protect a person from getting shocked or electrocuted. GFCIs are found in newer homes, usually in the kitchen, bathroom, laundry room, garage, and outside outlets, where the risk of electric shock is greatest. Sometimes a GFCI is installed on the main breaker box, thereby protecting the entire building from ground faults. There are many types of GFCIs; a

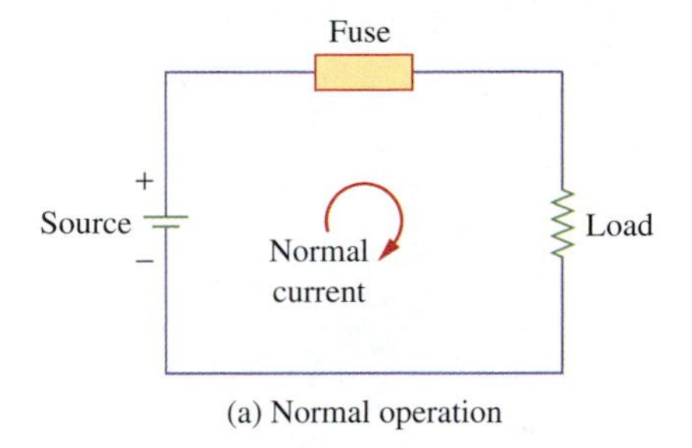

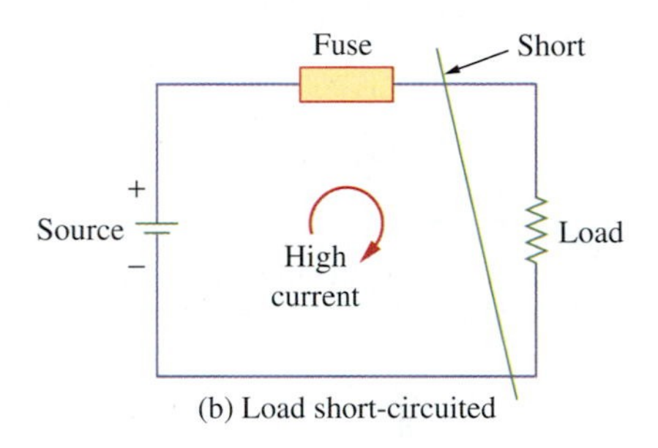

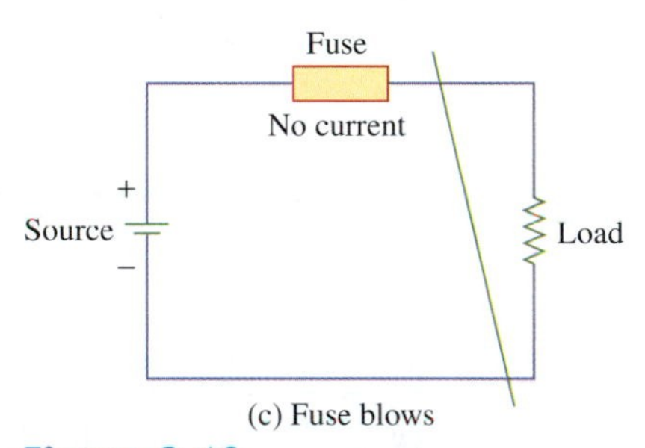

Figure 3.10
Using a fuse to protect the load.

Figure 3.11
An electrical panel of circuit breakers.
© Tetra Images/Getty RF

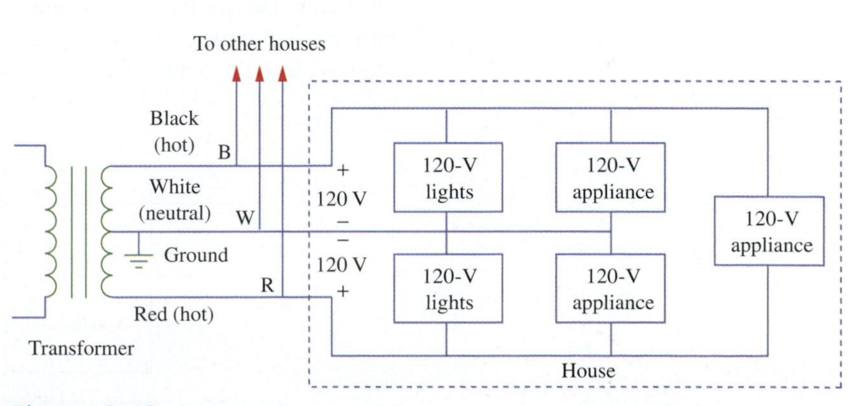

Figure 3.12
Three-wire residential wiring.

Figure 3.13
A typical ground fault circuit interrupter.
© TRBfoto/Getty RF

typical one is shown in Fig. 3.13. A GFCI works by comparing the amount of electrical current coming into a circuit (on the black wire) with the amount leaving (on the neutral or red wire). If more current enters the circuit through the red wire than leaves through the neutral wire, there is a current leak or ground fault. GFCI is able to detect a leak as little as a 5 mA, and it can shut down a circuit within 0.025 s (25 ms), helping to prevent serious electrical shocks. You should seriously consider adding GFCIs to any and all circuits that may have a ground hazard. GFCI protection should be provided at every location where someone might be harmed and the environment might be moist or damp. A person can inadvertently complete a circuit to ground by coming in contact with the red or black wire in an appliance while in contact with ground, including standing in wet or damp places.

3.8 †Applications: Wattmeter and Watt-hour Meter

In this section, we consider two important applications: how power is measured and how the energy you consume is measured by utility companies.

3.8.1 Wattmeter

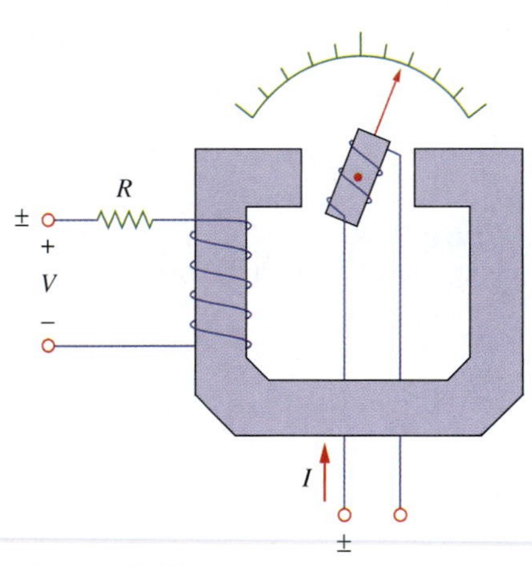

Figure 3.14
An analog wattmeter.

Electric power is measured by means of a wattmeter, shown in Fig. 3.14. The wattmeter basically consists of two coils: the current coil and the voltage coil. Because we define power as the product of voltage and current, any meter designed to measure power must account for both voltage and current. Wattmeters are often designed around dynamometer meter movements, which employ both voltage and current coils to move a needle. As illustrated in Fig. 3.15, the top (horizontal) coil of wire measures load current, while the bottom (vertical) coil measures load voltage. The activating force of a wattmeter comes from the field of its current coil and the field of its voltage coil. The force acting on the movable coil at any instant (tending to turn it) is proportional to the product of the instantaneous values of line current and voltage. Although there are digital wattmeters, most wattmeters in common use are analog. Recent emphasis on energy conservation has resulted in the availability of a small digital wattmeter, which is plugged into the wall outlet, and the appliance that you wish to measure is then plugged into the wattmeter.

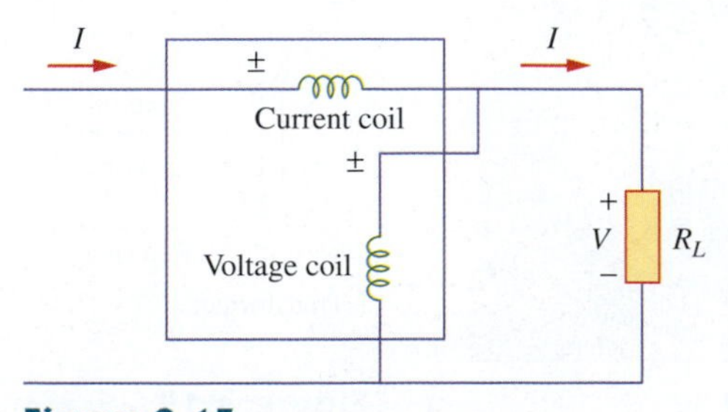

Figure 3.15
The wattmeter connected to a load.

Figure 3.16
A typical watt-hour meter.
© Comstock Images/Jupiterimages RF

3.8.2 Watt-hour Meter

The watt-hour meter is an instrument for measuring energy. Because energy is the product of power and time, the watt-hour meter must take into consideration both power and time. This meter is designed to measure accumulated kilowatt-hours on any electrical system. It consists of a motor whose torque is proportional to the current flowing through it and a register to count the number of revolutions the motor makes. A typical watt-hour meter is shown in Fig. 3.16. In this case, the meter is a four-dial type. The pointer on the right-hand dial registers 1 kWh (or 1,000 watt-hours) for each division of the dial. A complete revolution of the hand on this dial will move the hand of the second dial one division and register 10 kWh. A complete revolution of the hand of the second dial will move the third hand one division and register 100 kWh, and so on. The utility company charges based on the difference between the past and current readings of the watt-hour meter. Many modern watt-hour meters have digital displays.

In these days of advanced communication systems, smart meters are used to read power consumption. A smart meter is generally an electrical meter that identifies consumption in more detail (such as time-of-use) than a conventional meter and communicates that information via some network back to the local utility company for monitoring and billing purposes. Smart meters are communication systems that can capture and transmit information about energy use as it happens and allow consumers to keep much better track of their power usage.

3.9 Summary

1. Power is the rate of expending energy.

$$P = \frac{W}{t}$$

2. The power absorbed by a resistor is

$$P = VI$$

or

$$P = I^2R$$

or

$$P = \frac{V^2}{R}$$

3. The horsepower (hp) is a unit still in use today.

$$1 \text{ hp} = 746 \text{ W}$$

4. According to the passive sign convention, power assumes a positive sign when the current enters the positive polarity of the voltage across an element.
5. The wattage or power rating of a resistor indicates the maximum power the resistor is designed to dissipate rather than its normal operating power.
6. The efficiency η of a device is the ratio of its useful output power to its input power.

$$\eta = \frac{P_{\text{out}}}{P_{\text{in}}} \times 100\%$$

7. Fuses, circuit breakers, and GFCIs are protective devices that are deliberately used to create an open circuit when the current through them exceeds a predetermined value due to a malfunction in a circuit. While fuses and circuit breakers protect only the devices, the GFCI also protects the users.
8. Power is measured using the wattmeter, while energy is measured using the watt-hour meter.

Review Questions

3.1 What quantity is defined as rate at which energy is used?

(a) heat (b) voltage
(c) current (d) power

3.2 The power in a resistor is the product of the current and the voltage.

(a) True (b) False

3.3 An electric heater draws 2 A from a 110-V source, the power absorbed by the heater is:

(a) 220 W (b) 55 V
(c) 27.5 W (d) 18.18 mW

3.4 One horsepower equals approximately $\frac{3}{4}$ kW.

(a) True (b) False

3.5 Which of the following is not a unit of energy?

(a) joule (b) watt
(c) watt-second (d) kilowatt-hour

3.6 When current flows through a resistor, electrical energy is converted to heat energy.

(a) True (b) False

3.7 What is the energy used by a 60-W lamp in 10 h?

(a) 6 J (b) 600 J
(c) 0.6 kWh (d) 6 kWh

3.8 A particular motor develops 2 hp of mechanical output for an input of 3 hp. What is the efficiency of the motor?

(a) 33.33% (b) 50%
(c) 66.67% (d) 120%

3.9 It is possible to achieve an efficiency of 120%.

(a) True (b) False

3.10 An instrument designed to measure energy is called a:

(a) voltmeter (b) energy-meter
(c) wattmeter (d) watt-hour meter

Answers: 3.1d, 3.2a, 3.3a, 3.4a, 3.5b, 3.6a, 3.7c, 3.8c, 3.9b, 3.10d

Problems

Section 3.2 Power and Energy

3.1 How much power is consumed when 2,600 J of energy is expended in 2 h?

3.2 If a certain device absorbs 560 J of energy in 8 min, determine the power absorbed by the device.

3.3 If a resistor dissipates 7 W, how long does it take the resistor to consume an energy of 280 J?

3.4 A 40-W radio operates for 5 h. How much does it cost to operate the radio at 8 cents/kWh?

3.5 An electrical appliance uses 420 W. If it runs for 6 days, calculate how many kilowatt-hours are consumed.

3.6 How long does it take for a 120-W soldering iron to dissipate 1.5 kJ?

3.7 Convert the following to kilowatt-hours:

(a) 200 W for 56 s

(b) 180 W for 2 h

(c) 40,000 W for 4 h

3.8 A particular motor delivers 4.2 hp to a load. How many watts is this?

3.9 When a resistor is connected to an 8-V supply, a current of 4 A flows in it. Determine the time it takes for the resistor to dissipate 600 J.

3.10 A 3-kW room heater is connected to a 120-V source. Determine the resistance of the heater.

3.11 A 110-V television is rated 185 W. Find its rated current.

3.12 A 12-V battery is charged for 6 h at the rate of 2 A. Calculate the amount of energy consumed.

3.13 A 65-W soldering iron draws a current of 0.56 A. At what voltage must the iron operate?

3.14 Complete the following table.

R (Ω)	*V* (V)	*I* (A)	*P* (W)
—	120	0.04	—
60	—	—	0.8
—	24	—	2.2
42	—	0.1	—

3.15 A certain battery provides 2.4 A of current for 36 h. Calculate its ampere-hour rating.

3.16 How long will it take a 40-W radio to use 0.2 kWh of energy?

3.17 The resistance of a 120-V percolator heating element at the hot position is 4 Ω. Find: (a) the current drawn by the element, (b) the power consumed by the element, (c) the heat power in Btu/min if all the consumed electrical power is converted to heat. Assume that 1 W = 0.0569 Btu/min.

3.18 A vacuum cleaner operates with a voltage of 120 V. How much power is consumed by the vacuum cleaner if its current rating is 5 A?

3.19 An electrical designer decided to build a 30-A fuse. The fuse would melt in 3 s when its energy dissipation is greater than 30 J. How much power is required to melt the fuse?

3.20 A sliced bread toaster has a resistance of 12 Ω. If the toaster is connected to a 120-V outlet for 1 min, what is the energy delivered to the toaster?

3.21 A motor is rated at $\frac{1}{4}$ hp. If the motor operates from a 120-V source, how much current is supplied to the motor?

Section 3.3 Power in Electric Circuits

3.22 The voltage drop across a 10-Ω resistor is 12 mV. Calculate the power absorbed by the resistor.

3.23 Find the power delivered to each resistor in Fig. 3.17.

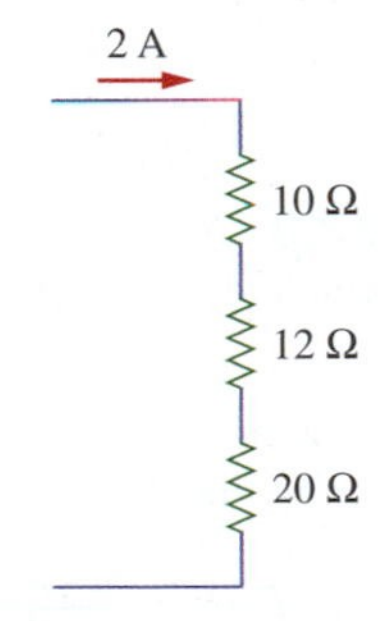

Figure 3.17
For Problem 3.23.

3.24 Determine the power absorbed by each resistor in Fig. 3.18.

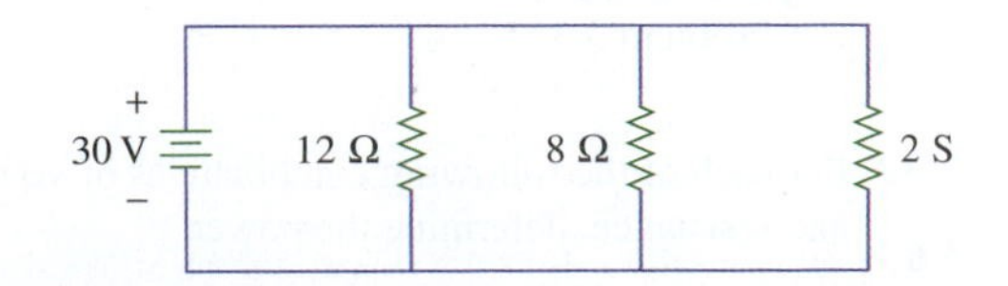

Figure 3.18
For Problem 3.24.

3.25 If the power dissipated in a 50-kΩ resistor is 400 mW, calculate the current through it and the voltage across it.

3.26 The heater of a cathode-ray tube is rated at 8.2 V, 0.8 A. Determine the rate at which energy is converted into heat.

3.27 Calculate the energy used by a 30-Ω heater operating from a 110-V source for 4 h.

3.28 For the circuit in Fig. 3.19, find the amount of energy that the circuit would consume in 2 h.

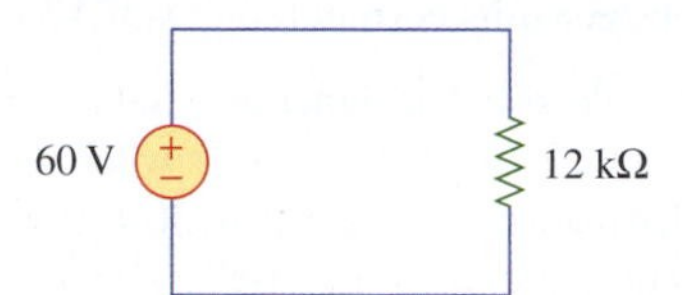

Figure 3.19
For Problem 3.28.

3.29 In a certain circuit, the applied voltage has doubled, while the resistance has decreased to one-half of its original value. What must have happened to the power dissipation?

3.30 Calculate the power delivered to resistors R_1 and R_2 in the circuit in Fig. 3.20.

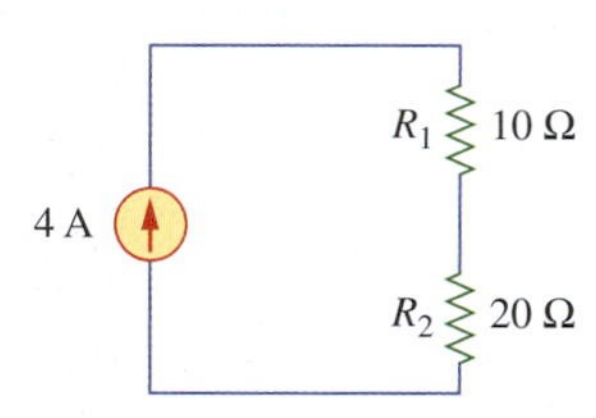

Figure 3.20
For Problem 3.30.

3.31 Determine the power delivered to resistors R_1 and R_2 in the circuit in Fig. 3.21.

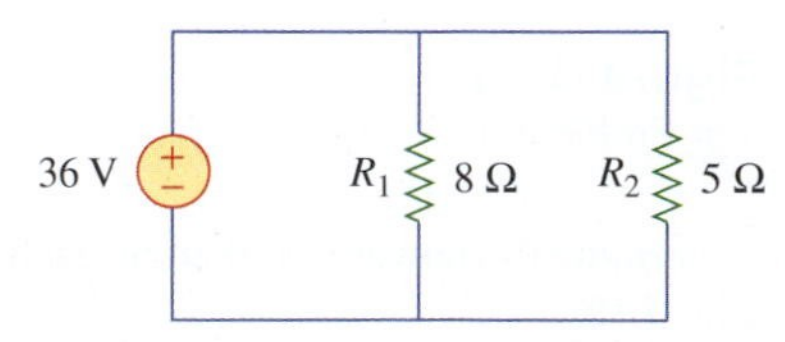

Figure 3.21
For Problem 3.31.

3.32 For each of the following combinations of voltage and resistance, determine the power.

(a) $V = 12$ V, $R = 3.3$ kΩ

(b) $V = 8.8$ V, $R = 150$ Ω

(c) $V = 120$ V, $R = 820$ Ω

Section 3.4 Power Sign Convention

3.33 For each element in Fig. 3.22, determine the power.

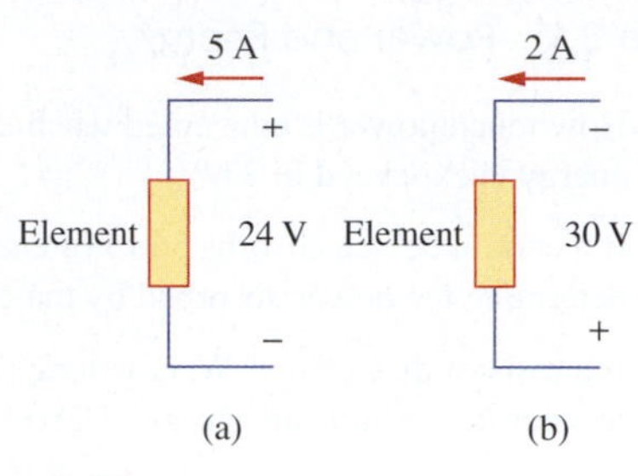

Figure 3.22
For Problem 3.33.

3.34 Each element in Fig. 3.23 may be a load or a source. For each element, find the power and state whether it is supplied or absorbed.

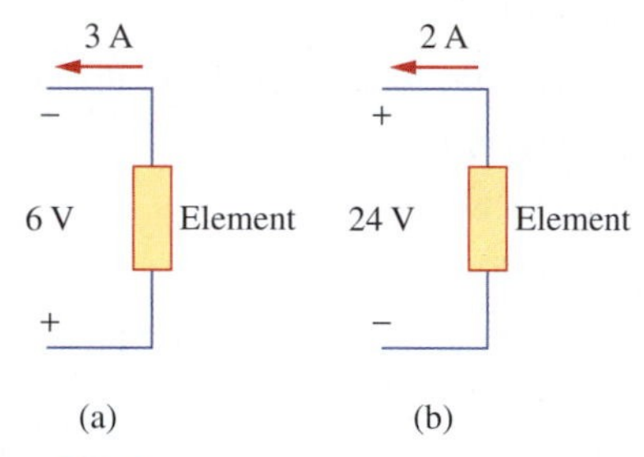

Figure 3.23
For Problem 3.34.

Section 3.5 Resistor Power Ratings

3.35 A 1-W resistor has a resistance of 2 kΩ. Determine the maximum current it can safely handle.

3.36 Is it safe to apply 50 V across an 8-kΩ, $\frac{1}{4}$-W resistor?

3.37 Determine which of the following resistors (if any) can be damaged by overheating.

(a) 1 W, 850 Ω with 110 V across it

(b) $\frac{1}{2}$ W, 8 Ω with 1 mA through it

(c) 10 W, 2 Ω with 4 A through it

3.38 A current of 5 mA through a resistor causes a voltage drop of 60 V across it. Determine the minimum wattage rating of the resistor.

3.39 Determine the maximum voltage that can be safely applied across a 5.6-kΩ, 2-W resistor.

3.40 If a 0.3-Ω resistor is rated at 8 W, determine how safe the resistor is when conducting a current of 15 A.

3.41 An air conditioner is rated at 1.5 kW and is run for 10 h/day. If the cost of electricity is $0.085/kWh, how much will it cost to run the air conditioner for 30 days?

Section 3.6 Efficiency

3.42 A power amplifier delivers 260 W to its speaker, while the power loss is 320 W. What is the efficiency of the system?

3.43 When in use, an automobile starter motor draws 70 A from a 12-V source. If the motor is rated at 1.2 hp, calculate its efficiency.

3.44 Calculate the efficiency of a motor that has an input of 500 W with an output of 0.6 hp.

3.45 A certain stereo system draws 3.2 A at 110 V when its output is 200 W. Calculate its efficiency and the power lost.

3.46 A certain system operates at efficiency of 92 percent. If losses are 12 kW, calculate P_{in} and P_{out}.

3.47 A motor is 84 percent efficient. If it takes 1.6 kW from an energy source, calculate the mechanical output in horsepower.

3.48 A system consists of two identical devices in cascade. If each device operates at an efficiency of 70 percent and the input energy is 40 J, calculate the output energy.

3.49 An electric lightbulb provides 10 W of useful output for 75 W of input. Determine its efficiency.

3.50 An electric motor develops mechanical energy at the rate of 2 hp with 85 percent efficiency. What current will the motor draw from a 220-V source?

3.51 A radio station with an efficiency of 60 percent transmits 32 kW for 24 h/day. Determine the cost of operating the station for 1 day at 8 cents/kWh.

3.52 A 5-hp motor runs 20 percent of the time over a 7-day period with an efficiency of 80 percent. If electricity costs 6 cents/kWh, how much will it cost the user?

3.53 Calculate the efficiency of a motor that has an input power of 800 W and an output power of 0.6 hp. Assume that 1 hp = 746 W.

3.54 A 70-W lamp uses 60 W to generate light and the other 10 W is dissipated as heat. Calculate the efficiency of the lamp.

3.55 A motor develops 24 kJ of mechanical output for an input of 30 kJ. (a) Determine the efficiency of the motor. (b) Determine the energy that the motor fails to convert to useful output. (c) What happens to the lost energy?

3.56 A certain 5-hp motor operates at an efficiency of 82 percent. If the input current is 8.2 A, calculate the input voltage. What is the input power?

3.57 A 200-V dc motor draws 15 A and develops an output power of 2.1 hp. (a) Determine the efficiency of the motor. (b) Calculate the lost power.

3.58 Two transducers in cascade provide an output power of 40 mW. If the first transducer has an efficiency of 80 percent and the second an efficiency of 95 percent, determine the input power.

3.59 A dc supply of 120 V draws 20 A and develops an output power of 1.5 hp when it is connected to an electrical motor. Assume 1 hp = 746 W. Find: (a) the input power P_{in} of the motor, (b) the output power P_{out} of the motor, (c) the motor efficiency, (d) the motor loss, (e) the heat power in Btu/min that is dissipated by the motor if all losses are converted to heat. (1 W = 0.0569 Btu/min)

Section 3.8 Applications: Wattmeter and Watt-hour Meter

3.60 You have a two-slice, 1200-W toaster that takes 1 min and 30 s to toast two slices. If there are 20 slices in a loaf of bread and energy costs \$0.08/kWh, calculate the cost of toasting the entire loaf.

chapter

4

Series Circuits

The stupid neither forgive or nor forget, the naïve forgive and forget; the wise forgive but do not forget.

—Thomas Szasz

Historical Profile

Gustav Robert Kirchhoff (1824–1887) was a German physicist, who contributed to the fundamental understanding of electrical circuits, spectroscopy, and the emission of blackbody radiation by heated objects.

Born the son of a lawyer in Konigsberg, East Prussia, Kirchhoff entered the University of Konigsberg at age 18 and later became a lecturer in Berlin. His collaborative work in spectroscopy with German chemist Robert Bunsen led to the discovery of cesium in 1860 and rubidium in 1861. He stated two basic laws in 1847 concerning the relationship between the currents and voltages in an electrical network. Kirchhoff's laws, along with Ohm's law, form the basis of circuit theory. Kirchhoff was also credited with the Kirchhoff law of radiation. Thus, Kirchhoff is famous among engineers, chemists, and physicists.

Gustav Robert Kirchhoff
© Pixtal/age Fotostock RF

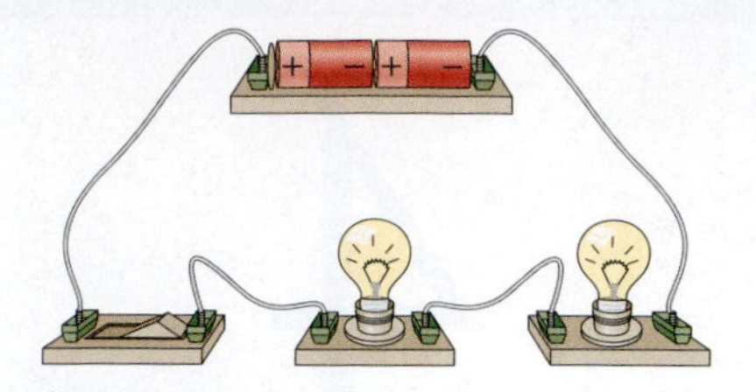

Figure 4.1
An example of a series circuit; two lamps in series with a dimmer.

4.1 Introduction

In previous chapters, we have, for the most part, limited circuits to having a single resistor. From now on, we will consider circuits with more than one resistor. Such resistive circuits can either be in series or in parallel or neither. A simple example of a series circuit is shown in Fig. 4.1. We will consider how to analyze series circuits in this chapter, while we deal with parallel circuits in the next chapter. Circuits that are neither exclusively series nor parallel will be covered in later chapters.

We begin the chapter by first introducing the basic concepts of nodes, branches, loops, series, and parallel. We then learn what series circuits are. Next, we introduce Kirchhoff's voltage law, which, along with Ohm's law, is very useful in analyzing circuits. We discuss voltage sources in series, voltage dividers, and power in series circuits. We learn how to analyze series circuits using PSpice and Multisim. We finally consider two simple applications of series circuits—using a resistor as a current limiter and electrical lighting systems.

4.2 Nodes, Branches, and Loops

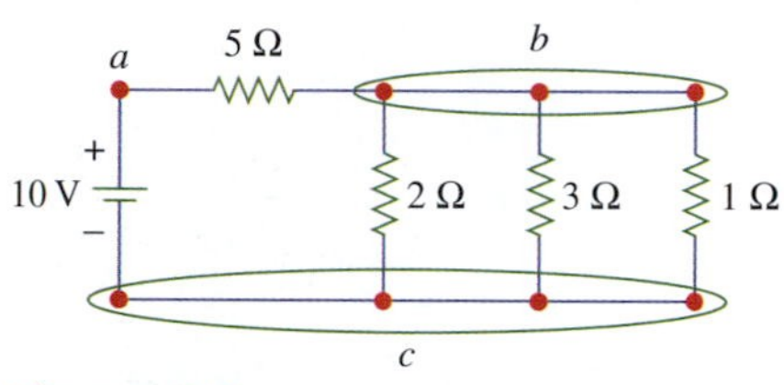

Figure 4.2
Nodes, branches, and loops.

There are two types of elements found in electric circuits: *passive* elements and *active* elements. An active element is capable of generating energy while a passive element is not. Examples of passive elements are resistors, capacitors, and inductors. Typical active elements include generators, batteries, and amplifiers.

Because the elements of an electric circuit can be interconnected in several ways, we need to understand some basic concepts of network topology. By *network topology*, we mean the properties relating to the placement of elements in the network and the geometric configuration of the network. Such properties include branch, node, and loop.

A branch is any two-terminal element. The circuit in Fig. 4.2 has five branches, namely, the 10-V battery and four resistors.

> A **branch** is a single element such as a voltage source or a resistor.

A node is usually indicated by a dot in a circuit, although we do not follow this convention in this book. If a short circuit (a connecting wire) connects two nodes, the two nodes constitute a single node. The circuit in Fig. 4.2 has three nodes *a*, *b*, and *c*. Notice that the three points that form node *b* are connected by perfectly conducting wires and therefore constitute a single point. The same is true of the four points forming node *c*. We demonstrate that the circuit in Fig. 4.2 has only three nodes by redrawing the circuit in Fig. 4.3. The two circuits in Figs. 4.2 and 4.3 are identical. However, for the sake of clarity, nodes *b* and *c* are spread out with perfect conductors as in Fig. 4.2.

> A **node** is the point of connection between two or more branches.

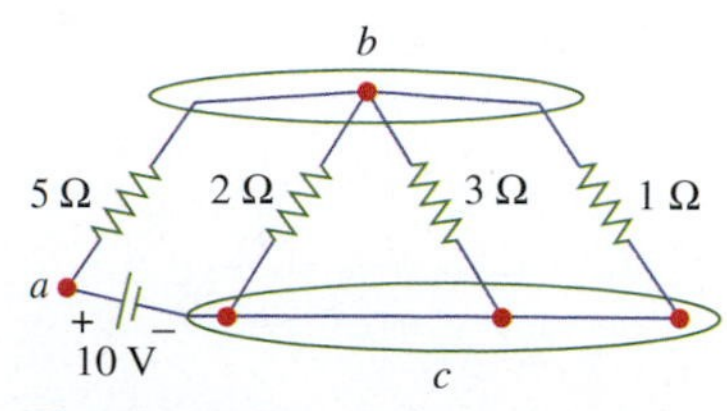

Figure 4.3
The three-node circuit in Fig. 4.2 is redrawn.

For example, the closed path *abca* containing the 2-Ω resistor in Fig. 4.3 is a loop. Another loop is the closed path *bcb* containing the 3-Ω resistor and the 1-Ω resistor. Although one can identify six loops in Fig. 4.3, only three of them are independent. A loop is said to be independent if it contains at least one branch that is not part of any other independent loop. Independent loops or paths result in an independent set of equations.

> A **loop** is any closed path in a circuit.

A network with b branches, n nodes, and ℓ independent loops will satisfy the fundamental theorem of network topology:

$$b = \ell + n - 1 \tag{4.1}$$

As the next two definitions show, circuit topology is of great value to the study of voltages and currents in an electric circuit:

> Two or more elements are in **series** if they are cascaded or connected sequentially and consequently carry the same current.

> Two or more elements are in **parallel** if they are connected to the same two nodes and consequently have the same voltage

Elements are in series when they are chain connected or connected sequentially, end to end. For example, two elements are in series if they share one common node and no other element is connected to that common node. Elements in parallel are connected to the same pair of terminals. Elements may be connected in a way that they are neither in series nor in parallel. In the circuit shown in Fig. 4.2, the battery and the 5-Ω resistor are in series because the same current will flow through them. The 2-Ω resistor, the 3-Ω resistor, and the 1-Ω resistor are in parallel because they are connected to the same two nodes (b and c) and consequently have the same voltage across them. The 5-Ω and 2-Ω resistors are neither in series nor in parallel.

Example 4.1

Determine the number of branches and nodes in the circuit shown in Fig. 4.4. Identify which elements are in series and which are in parallel.

Solution:

Because there are four elements in the circuit, the circuit has four branches; namely, 10 V, 5 Ω, 6 Ω, and 12 V. The circuit has three nodes as identified in Fig. 4.5. The 5-Ω resistor is in series with the 10-V voltage source because the same current would flow in both. The 6-Ω resistor is in parallel with the 12-V voltage source because both are connected to the same nodes 2 and 3.

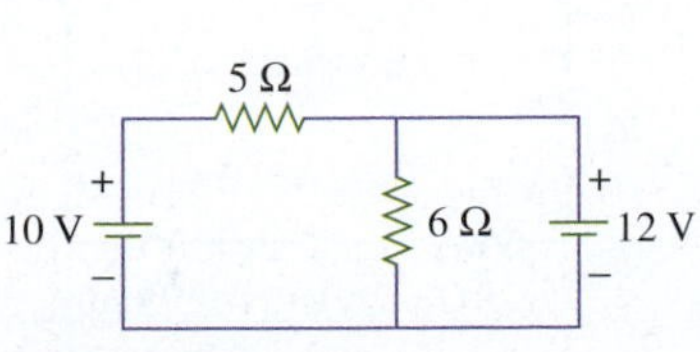

Figure 4.4
For Example 4.1.

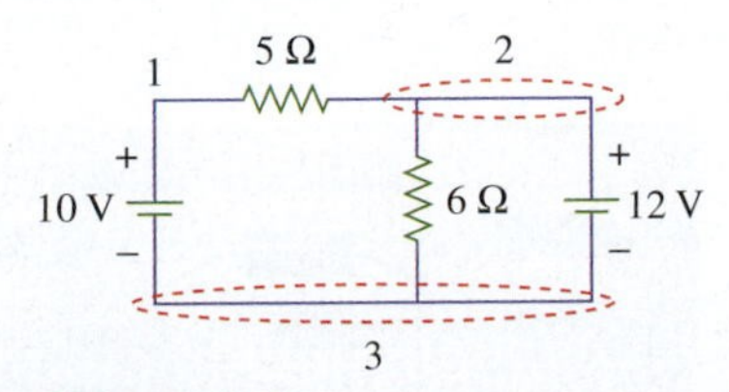

Figure 4.5
The three nodes in the circuit of Fig. 4.4.

Practice Problem 4.1

How many branches and nodes does the circuit in Fig. 4.6 have? Identify which elements are in series and in parallel.

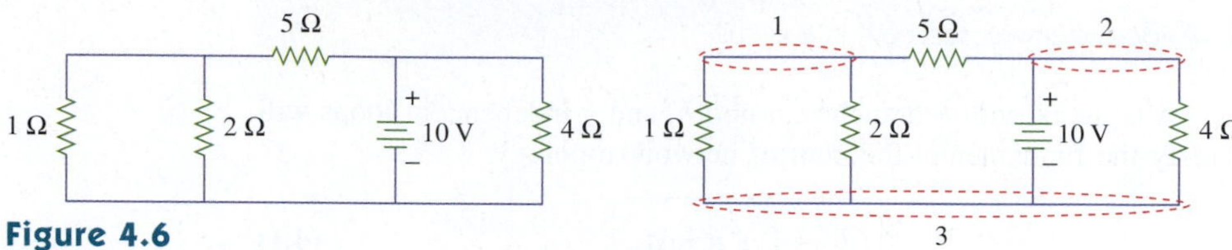

Figure 4.6
For Practice Problem 4.1.

Figure 4.7
Answer to Practice Problem 4.1.

Answer: Five branches and three nodes are identified in Fig. 4.7; the 1-Ω and 2-Ω resistors are in parallel; the 4-Ω resistor and 10-V source are also in parallel.

4.3 Resistors in Series

The need to combine resistors in series (or in parallel) occurs so frequently that it warrants special attention. Figure 4.8 shows resistors connected in series on a breadboard. When two or more resistors are connected end-to-end (or in tandem), the resistors are said to be connected in *series*. Because there is only one path for current, the same current flows through the resistors in series. Examples of series circuits are shown in Fig. 4.9.

> A **series circuit** is one in which resistors are connected in tandem and the same current flows through the resistors.

In Fig. 4.9(a), the total resistance $R_T = R_1 + R_2 + R_3$. In Fig. 4.9(b), the total resistance is $R_T = R_1 + R_2 + R_3 + \cdots + R_9$. In general,

> The **total resistance** of any number of resistors connected in series is the sum of the individual resistances.

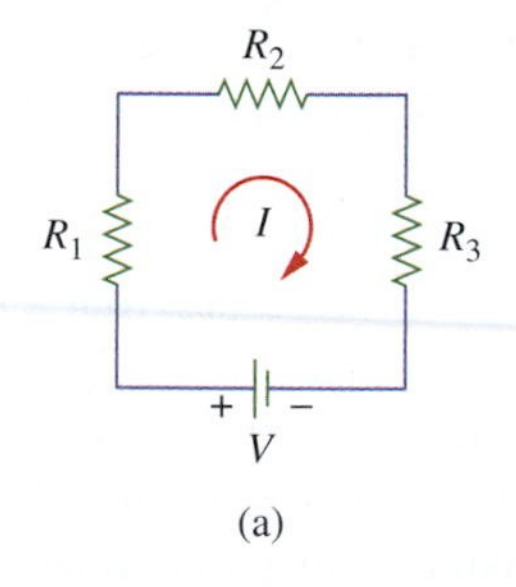

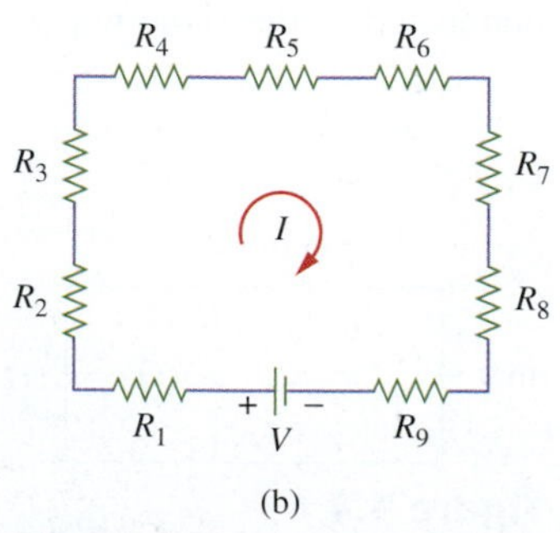

Figure 4.9
Examples of series circuits.

Figure 4.8
Resistors in series on a breadboard.
© Sarhan M. Musa

For N resistors in series, this is expressed as

$$R_T = R_1 + R_2 + R_3 + \cdots + R_N \tag{4.2}$$

Because the same current I flows through each resistor, we can calculate the voltage across each resistor (using Ohm's law) and the power absorbed by it as:

$$\begin{aligned} V_1 &= IR_1, \quad P_1 = IV_1 = I^2R_1 \\ V_2 &= IR_2, \quad P_2 = IV_2 = I^2R_2 \\ &\vdots \\ V_N &= IR_N, \quad P_N = IV_N = I^2R_N \end{aligned} \tag{4.3}$$

This indicates that voltage drop by each resistor in a series circuit depends on its resistance. The total power delivered to the series circuit is

$$\begin{aligned} P_T &= P_1 + P_2 + P_3 + \cdots + P_N \\ &= I^2(R_1 + R_2 + R_3 + \cdots + R_N) = I^2R_T \end{aligned} \tag{4.4}$$

Example 4.2

Consider the series circuit in Fig. 4.10. Find:

(a) the total resistance
(b) the current I
(c) the voltage across R_1, R_2, and R_3
(d) the power absorbed by R_1, R_2, and R_3
(e) the power delivered by the source

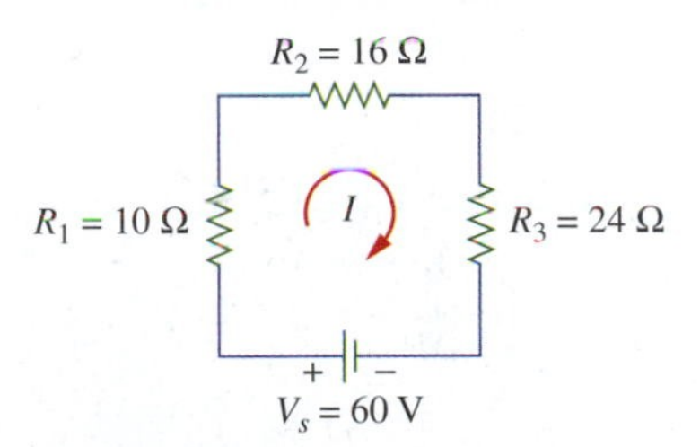

Figure 4.10
For Example 4.2.

Solution:

(a) The total resistance is

$$R_T = R_1 + R_2 + R_3 = 10 + 16 + 24 = 50\ \Omega$$

(b) Using Ohm's law,

$$I = \frac{V_s}{R_T} = \frac{60}{50} = 1.2\ \text{A}$$

(c) The voltages across the resistors are

$$\begin{aligned} V_1 &= IR_1 = 1.2 \times 10 = 12\ \text{V} \\ V_2 &= IR_2 = 1.2 \times 16 = 19.2\ \text{V} \\ V_3 &= IR_3 = 1.2 \times 24 = 28.8\ \text{V} \end{aligned}$$

indicating that source voltage is shared by the three resistors in proportion to their resistances.

(d) The powers absorbed by the resistors are:

$$\begin{aligned} P_1 &= IV_1 = 1.2 \times 12 = 14.4\ \text{W} \\ P_2 &= IV_2 = 1.2 \times 19.2 = 23.04\ \text{W} \\ P_3 &= IV_3 = 1.2 \times 28.8 = 34.56\ \text{W} \end{aligned}$$

The total power absorbed by the resistors is

$$P_T = P_1 + P_2 + P_3 = 14.4 + 23.04 + 34.56 = 72\ \text{W}$$

(e) The power supplied or delivered by the source is

$$P_d = V_s I = 60 \times 1.2 = 72 \text{ W}$$

which is the same as the total absorbed by the resistors.

Practice Problem 4.2

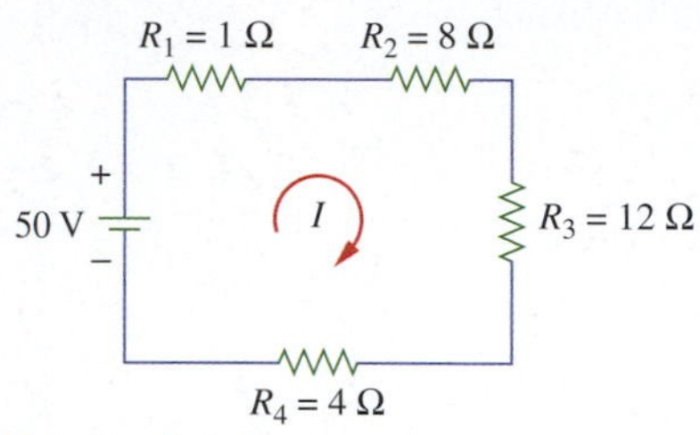

Figure 4.11
For Practice Problem 4.2.

Consider the series circuit in Fig. 4.11. Find:

(a) the total resistance
(b) the current I
(c) the voltages across R_1, R_2, R_3, and R_4
(d) the power absorbed by R_1, R_2, R_3, and R_4
(e) the power delivered by the source

Answer: (a) 25 Ω (b) 2A (c) 2, 16, 24, 8 V (d) 4, 32, 48, 16 W (e) 100 W

4.4 Kirchhoff's Voltage Law

Ohm's law by itself is not sufficient to analyze circuits. However, when it is coupled with Kirchhoff's two laws, we can analyze a large variety of electric circuits. Kirchhoff's laws were first introduced in 1847 by the German physicist Gustav Robert Kirchhoff. These laws are formally known as Kirchhoff's voltage law (KVL), which will covered in this chapter, and Kirchhoff's current law (KCL), which will be covered in the next chapter.

> **Kirchhoff's voltage law** (KVL) states that the algebraic sum of all voltages around a closed path (or loop) is zero.

Kirchhoff's voltage law is based on the principle of conservation of energy in electrical circuits. (Keep in mind that the electrical potential or voltage is the energy per unit charge.) The principle of conservation of energy implies that the algebraic sum of the electrical potential differences around a circuit must be zero.

Expressed mathematically, if there are N voltages in a loop or closed path,

$$V_1 + V_2 + V_3 + \cdots + V_N = 0 \qquad \textbf{(4.5)}$$

Expressed in symbolic form, where $\sum$ represents summation, KVL can be stated as

$$\boxed{\sum_{i=1}^{N} V_i = 0} \qquad \textbf{(4.6)}$$

where N is the number of voltages in the loop and V_i is the ith voltage.

To illustrate KVL, consider the circuit in Fig. 4.12. The sign on each voltage is the polarity of the terminal encountered first as we travel around the loop. (Keep in mind the relationship between the direction of current and the polarity of voltage across a resistor, as shown in Fig. 2.6.) We can start with any branch and go around the loop either clockwise or counterclockwise.[1] Suppose we start with the

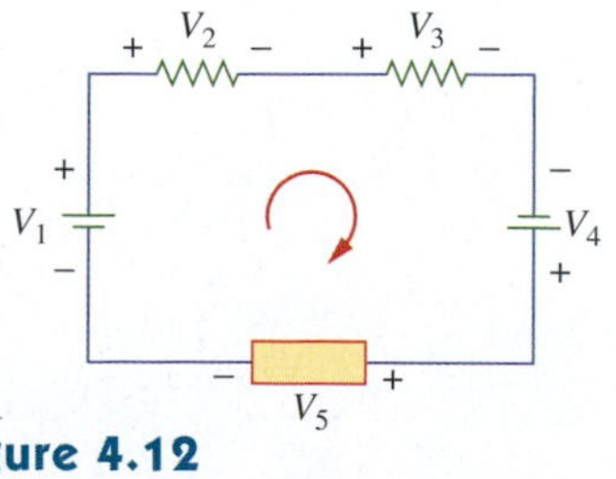

Figure 4.12
A single-loop circuit illustrating KVL.

[1] KVL can be applied in two ways: by taking a clockwise or counterclockwise trip around the loop. Either way, the algebraic sum of voltages around the loop is zero.

voltage source V_1 and go clockwise around the loop as shown; then voltages would be $-V_1$, $+V_2$, $+V_3$, $-V_4$ and $+V_1$ in that order. For example, as we reach branch 3, the positive terminal is met first; hence we have $+V_3$. For branch 4, we reach the negative terminal first; hence $-V_4$. Thus, KVL yields

$$-V_1 + V_2 + V_3 - V_4 + V_5 = 0 \tag{4.7}$$

Rearranging terms gives

$$V_2 + V_3 + V_5 = V_1 + V_4 \tag{4.8}$$

which may be interpreted as

$$\sum \text{voltage drops} = \sum \text{voltage rises} \tag{4.9}$$

This is an alternative form of KVL. Notice that if we had traveled counterclockwise starting with V_1, the result would have been $+V_1$, $-V_5$, $+V_4$, $-V_3$ and $-V_2$, which is the same as obtained before except that the signs are reversed. Hence, Eqs. (4.7) and (4.8) remain the same. Also, notice that a voltage rise occurs when we travel from $-$ to $+$ through an element, while a voltage drops occurs when we travel from $+$ to $-$ through an element. It is worth mentioning that voltage rise ($+$ relative to $-$) takes place in an active element, while voltage drop ($-$ relative to $+$) takes place in a passive element.

Example 4.3

Determine the unknown voltage V_x in the circuit in Fig. 4.13.

Solution:
By going around the loop as shown by the arrow, we apply Kirchhoff's voltage law and get

$$-24 + V_x + 30 = 0$$

or

$$V_x = 24 - 30 = -6 \text{ V}$$

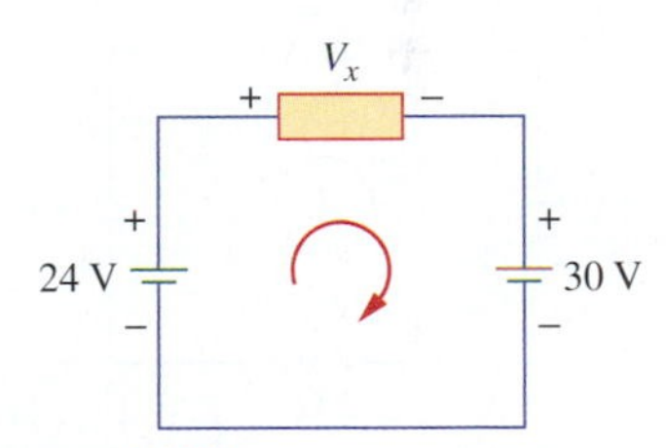

Figure 4.13
For Example 4.3.

Practice Problem 4.3

Find the unknown voltage V_x in the circuit in Fig. 4.14.

Answer: 8 V

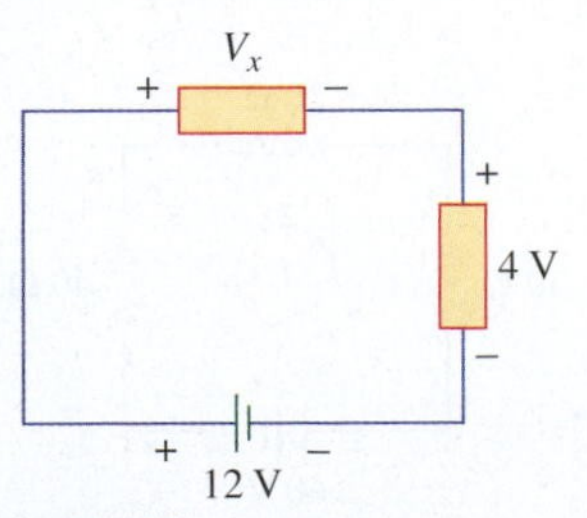

Figure 4.14
For Practice Problem 4.3.

Example 4.4

For the circuit in Fig. 4.15, find voltages V_1 and V_2.

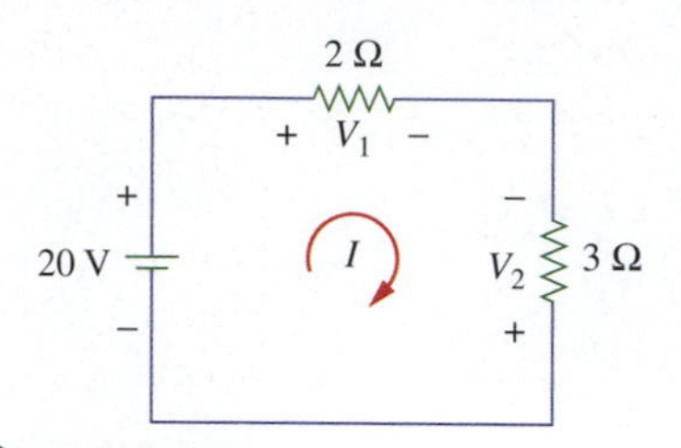

Figure 4.15
For Example 4.4.

Solution:
To find V_1 and V_2, we apply Ohm's law and Kirchhoff's voltage law. Assume that current I flows through the loop as shown in Fig. 4.15. From Ohm's law,

$$V_1 = 2I, \quad V_2 = -3I \tag{4.4.1}$$

Applying KVL around the loop gives

$$-20 + V_1 - V_2 = 0 \tag{4.4.2}$$

Substituting Eq. (4.4.1) into Eq. (4.4.2), we obtain

$$-20 + 2I + 3I = 0$$

or

$$5I = 20 \longrightarrow I = \frac{20}{5} = 4 \text{ A}$$

Substituting for I in Eq. (4.4.1) finally gives

$$V_1 = 2I = 8 \text{ V}, \qquad V_2 = -3I = -12 \text{ V}$$

The fact that the polarity of V_2 is negative indicates that its polarity should be reversed in Fig. 4.15. That is, we expect the arrow in Fig. 4.15 to hit the resistor on the + side first and then $V_2 = 12$ V.

Practice Problem 4.4

Find V_1 and V_2 in the circuit of Fig. 4.16.

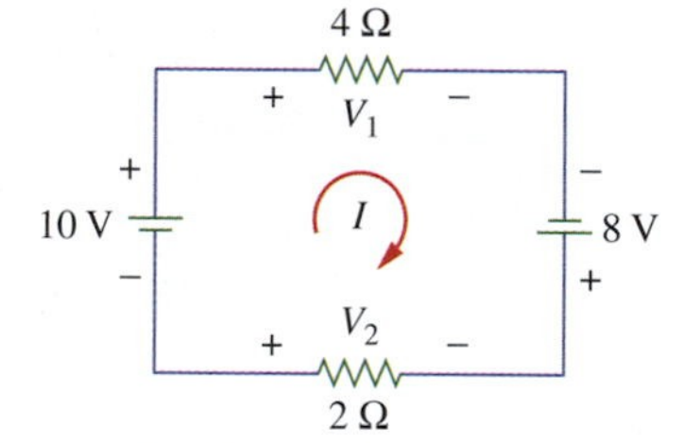

Figure 4.16
For Practice Problem 4.4.

Answer: 12 V; −6 V

Example 4.5

Using Kirchhoff's voltage law, determine the current I in the circuit of Fig. 4.17.

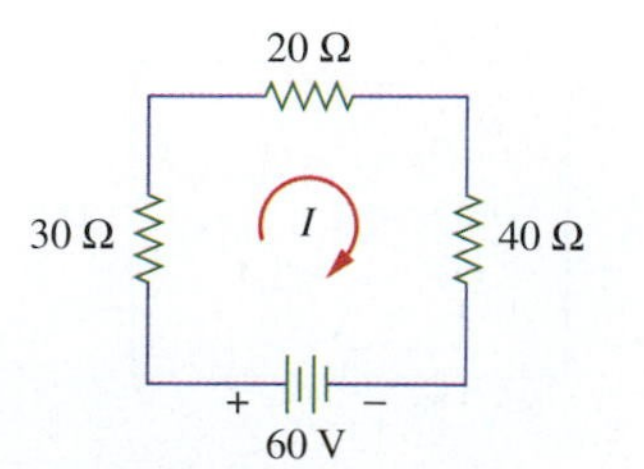

Figure 4.17
For Example 4.5.

Solution:
This can be solved in two ways.

■ **METHOD 1** The resistors are in series so that

$$R_T = 30 + 20 + 40 = 90\ \Omega$$

Using Ohm's law,

$$I = \frac{V_s}{R_T} = \frac{60}{90} = 0.667 \text{ A}$$

METHOD 2 Applying KVL to the loop gives

$$-60 + 30I + 20I + 40I = 0$$

or

$$90I = 60 \quad \text{i.e.,} \quad I = 60/90 = 0.667 \text{ A}$$

as obtained previously.

Practice Problem 4.5

Apply KVL to the circuit in Fig. 4.18 and find current I.

Answer: 0.6 A

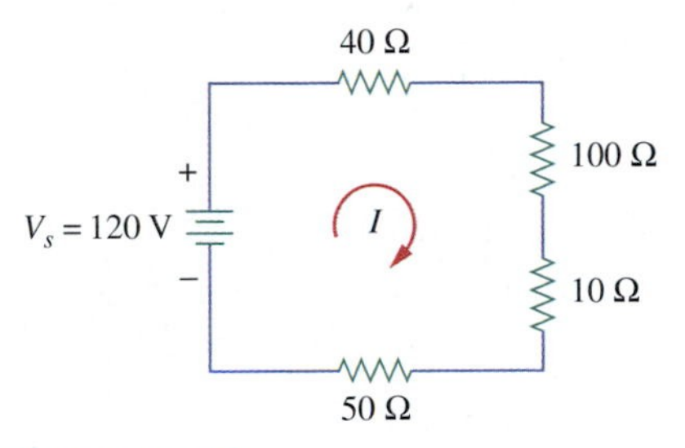

Figure 4.18
For Practice Problem 4.5.

4.5 Voltage Sources in Series

Two important applications of KVL deal with voltage sources in series and the concept of voltage division, which will be considered in the next section. When voltage sources are connected in series, KVL can be applied to obtain the total voltage. The combined voltage is the algebraic sum of the voltages of the individual sources. For example, for the voltage sources shown in Fig. 4.19(a), the combined or equivalent voltage source in Fig. 4.19(b) is obtained by applying KVL to Fig. 4.19(a).

$$-V_{ab} + V_1 + V_2 - V_3 = 0 \tag{4.10a}$$

or

$$V_{ab} = V_1 + V_2 - V_3 \tag{4.10b}$$

We should mention that the situation in Fig. 4.19(a) is theoretical; it would be counterproductive to actually connect the sources (such as batteries) backwards. Some elements such as diodes or LEDs (light-emitting diodes) are modeled as batteries connected backward.

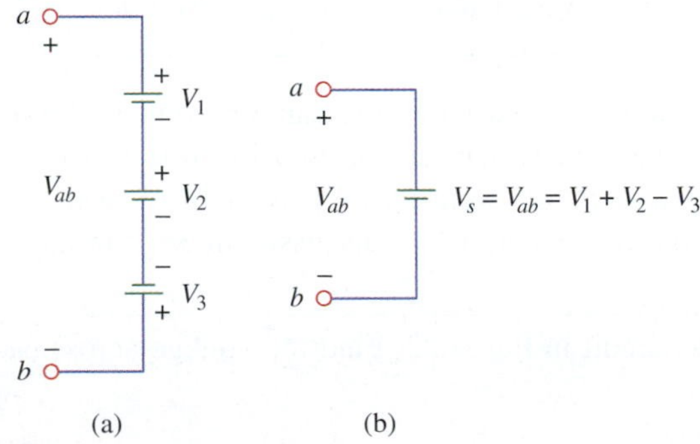

Figure 4.19
Voltage sources in series: (a) original circuit; (b) equivalent circuit.

4.6 Voltage Dividers

Series resistors are often used to provide voltage division. To determine the voltage across each resistor, we consider the circuit in Fig. 4.20. If we apply Ohm's law to each resistor, we get

$$V_1 = IR_1, \quad V_2 = IR_3, \quad \cdots \quad V_n = IR_n \tag{4.11}$$

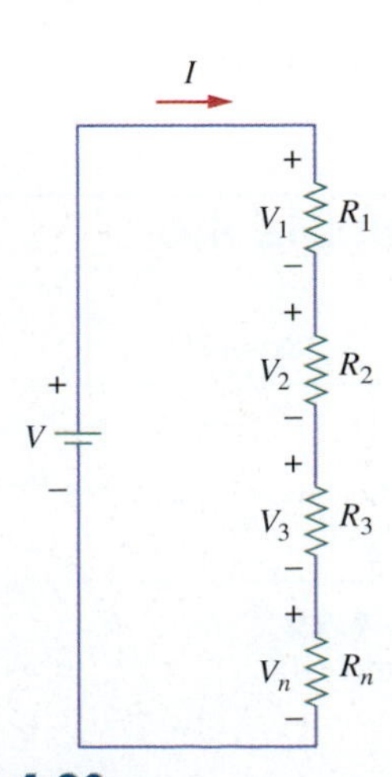

Figure 4.20
General voltage division.

Because the resistors are in series, the total or equivalent resistance is

$$R_{eq} = R_1 + R_2 + \cdots + R_n \tag{4.12}$$

The current I that flows through the resistors is

$$I = \frac{V}{R_{eq}} \tag{4.13}$$

Substituting Eq. (4.13) into Eq. (4.11) produces

$$V_1 = \left(\frac{R_1}{R_{eq}}\right)V, \qquad V_2 = \left(\frac{R_2}{R_{eq}}\right)V, \qquad \cdots \qquad V_n = \left(\frac{R_n}{R_{eq}}\right)V \tag{4.14}$$

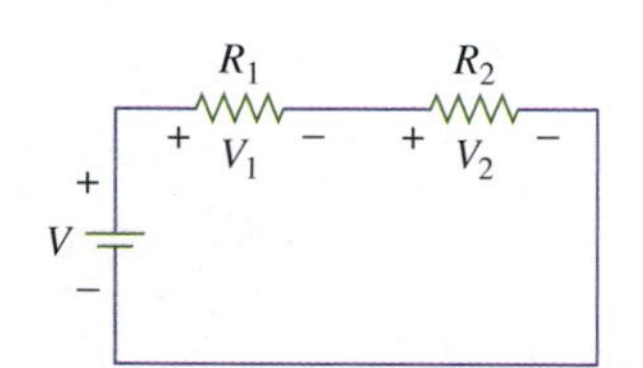

Figure 4.21
Voltage division for two resistors.

where R_n is the resistor the voltage drop is across, R_{eq} is the total resistance of the resistors in series, and V is the voltage across the resistors in series. Notice from Eq. (4.14) that the source voltage V is divided among the resistors in direct proportion to their resistances; the larger the resistance, the larger the voltage drop. This is referred to as the *voltage divider rule* (VDR) and the circuit in Fig. 4.20 is called a *voltage divider.*

In a **voltage divider,** the voltage drop across any resistor is proportional to the magnitude of the resistor.

We now consider the case where there are only two resistors connected in series, as shown in Fig. 4.21. In this case, the equivalent resistance is

$$R_{eq} = R_1 + R_2 \tag{4.15}$$

and Eq. (4.14) becomes

$$\boxed{V_1 = \left(\frac{R_1}{R_1 + R_2}\right)V, \qquad V_2 = \left(\frac{R_2}{R_1 + R_2}\right)V} \tag{4.16}$$

The VDR should be used when the same current is through the two resistors. As we showed earlier, the division rule can be extended to more than two resistors. Note that a potentiometer can be used as an adjustable voltage divider, as we discussed in Section 2.6.

Example 4.6

Consider the circuit in Fig. 4.22. Find the voltage across each resistor.

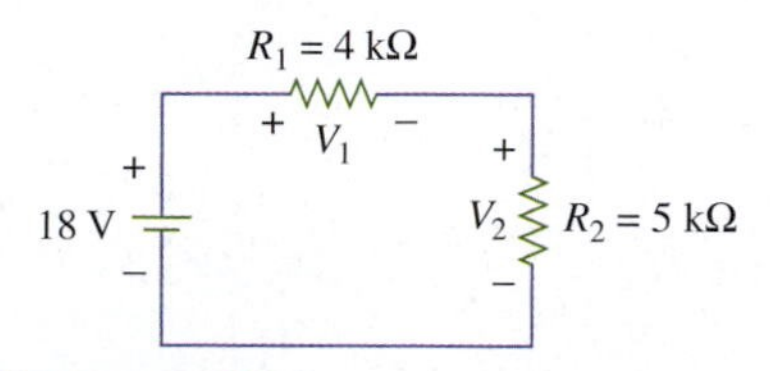

Figure 4.22
For Example 4.6.

Solution:

Because there are only two resistors, we use Eq. (4.16). $R_{eq} = 4 + 5 = 9\ \text{k}\Omega$.

$$V_1 = \frac{R_1}{R_{eq}}V = \frac{4}{9}(18) = 8\ \text{V}$$

$$V_2 = \frac{R_2}{R_{eq}}V = \frac{5}{9}(18) = 10\ \text{V}$$

Note that V_1 and V_2 add up to source voltage of 18 V.

Practice Problem 4.6

Determine V_1 and V_2 in Fig. 4.23 using voltage division.

Answer: $V_1 = 60$ V; $V_2 = 40$ V

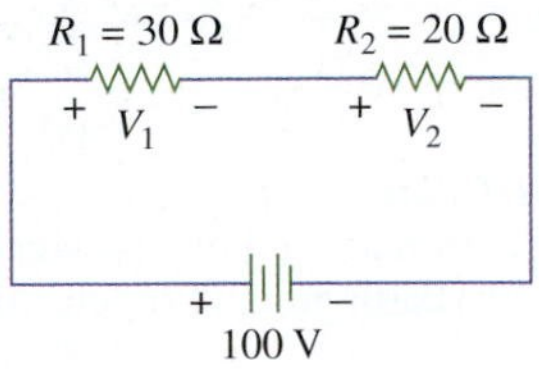

Figure 4.23
For Practice Problem 4.6.

Example 4.7

Use the voltage division rule to find V_1, V_2, and V_3 in the circuit of Fig. 4.24.

Solution:
In this case, we use Eq. (4.14). $R_{eq} = 10 + 12 + 8 = 30\ \Omega$. Hence

$$V_1 = \frac{R_1}{R_{eq}}V = \frac{10}{30}(60) = 20 \text{ V}$$

$$V_2 = \frac{R_2}{R_{eq}}V = \frac{12}{30}(60) = 24 \text{ V}$$

$$V_3 = \frac{R_3}{R_{eq}}V = \frac{8}{30}(60) = 16 \text{ V}$$

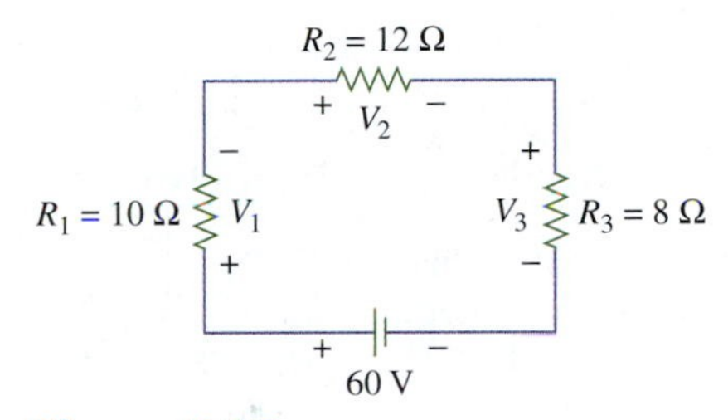

Figure 4.24
For Example 4.7.

Practice Problem 4.7

Determine V_1, V_2, and V_3 in the circuit of Fig. 4.25.

Answer: $V_1 = 25$ V; $V_2 = 10$ V; $V_3 = 15$ V

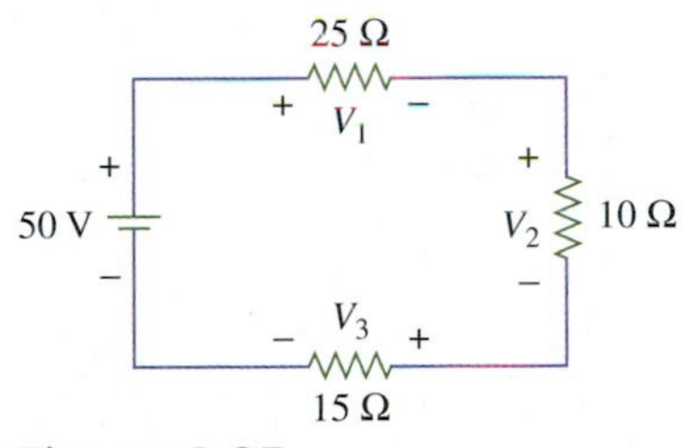

Figure 4.25
For Practice Problem 4.7.

4.7 Ground Connections

Like measuring distance, electric potential (voltage) must always be measured from a reference point. The most frequently used reference point is the Earth or, more specifically, the ground that your building is embedded in. This reference potential is declared to be zero volts and is commonly referred to as *ground*. Electrical equipment connected to the ground is said to be *grounded* or *earthed*. Part of the electrical wiring of every building is a wire that is connected to a large metal rod driven deep into the ground, thereby ensuring a good connection to ground.

A **ground** is an electrical connection to Earth.

Proper grounding is vital for making electrical equipment safe to use. (An exception to this would be a cordless phone or a cell phone.) Many devices (e.g., scopes) have a floating ground. Consider an ungrounded electrical device sitting on a wooden table. The table is an insulator. Assume the device is damaged so that charge starts to build on the frame of the device. People are conductors (albeit very poor ones) and are roughly at earth potential because they are in contact with floors, walls, and so forth. A person touching the device, or even approaching closely, provides a path for the accumulated charge to flow because of the potential difference between the device and the person.

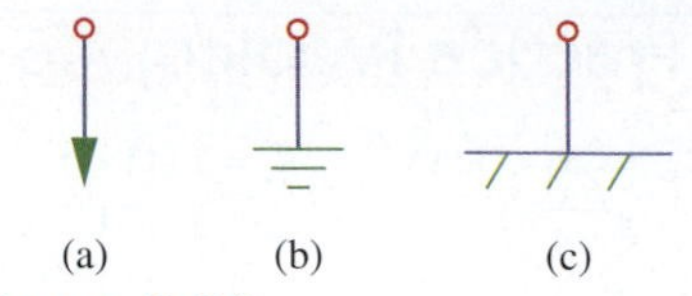

Figure 4.26
Common ground symbols: (a) signal ground; (b) earth ground; (c) chassis ground.

Large enough voltages and currents can burn and kill. Properly grounded devices would be at the same potential as the surroundings, and there would be no sparks or undesired currents.

The grounding of electrical systems in older homes was originally done by connecting a heavy gauge copper ground wire, known as a bond wire, to the main water supply pipe. In those days, a main water line was comprised of galvanized steel, an excellent electrical conductor. And because this pipe extended a considerable distance below ground, it served as an adequate basis for grounding the entire electrical system. Problems occurred, however, as these old waterlines became rusted and finally needed replacement. Some of the plumbers who replaced these deteriorated lines were focused primarily on plumbing concerns, with lesser attention to the status of the electrical system. Thus, many of these waterlines were replaced with PVC (polyvinyl chloride) plastic pipe. And because plastic has no capacity for conducting electricity, grounding for those homes was effectively eliminated.

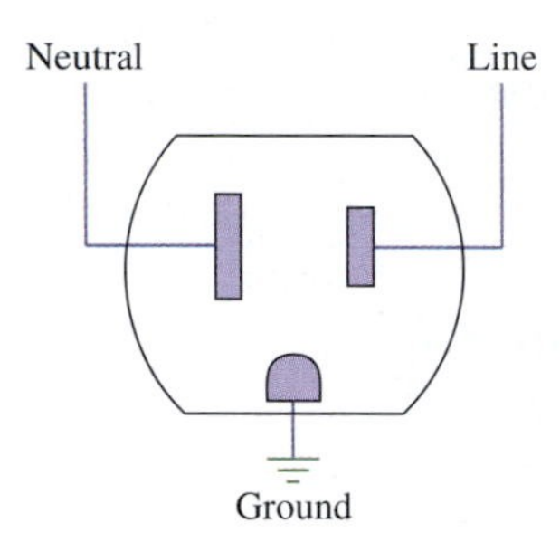

Figure 4.27
Ground connection in a 120-V outlet.

A ground is a point of reference or a common point in a circuit. Other points in the circuit take it as the reference potential to which the value of 0 V is attached. It is commonly indicated by any of the three symbols in Fig. 4.26. The type of ground in Fig. 4.26(c) is called a *chassis* ground and is used in devices where the case, enclosure, or chassis acts as a reference point for all circuits. When the potential of the earth is used as reference, we use the earth ground in Fig. 4.26(a) or (b). We shall always use the symbol in Fig. 4.26(b). As shown in Fig. 4.27, the rounded terminal of the 120-V electrical outlet is the ground terminal, which provides a common point for circuits connected to it. Although the electric circuits would function the same way with or without the ground, the ground is necessary as a safety precaution.

4.8 Computer Analysis

We will be using two computer packages in this text. One is PSpice from Cadence and the other is Multisim from Electronics Workbench, a group within National Instruments. Both are very useful for simulating electrical circuits before they are built. They also permit one to analyze the final design. A short tutorial on PSpice is given in Appendix C, while Appendix D provides a short tutorial on Multisim. We will soon notice that PSpice and Multisim are basically similar.

4.8.1 PSpice

PSpice is a computer software for circuit analysis that we will gradually learn to use in this textbook. This section illustrates how to use PSpice for Windows to analyze the dc circuits we have studied so far. The reader is expected to review Sections C.1 through C.3 of Appendix C before proceeding with this section. It should be noted that PSpice is only helpful in determining branch voltages and currents when the numerical values of all the circuit components are known.

Example 4.8

Use PSpice to find the node voltages in the circuit of Fig. 4.28. Determine current I.

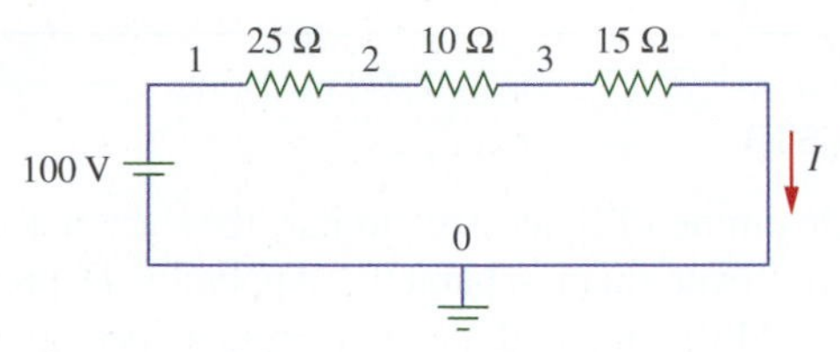

Figure 4.28
For Example 4.8.

Solution:

The first step is to draw the given circuit using Schematics. Following the instructions given in Appendix C, Sections C.2 and C.3, the schematic in Fig. 4.29 is produced. Because this is a dc analysis, we use voltage source VDC. Once the circuit is drawn and saved as exam48.dsn, we simulate the circuit by selecting **PSpice/New Simulation Profile**. This leads to the New Simulation dialog box. Type "exam48" as the name of the file and click Create. This leads to the Simulation Settings dialog box. This dialog box is important for transient and ac analyses. Because this is dc analysis, just click OK. Then select **PSpice/Run.** The circuit is simulated and some of the results are displayed on the schematic as in Fig. 4.29. Other results are in the output file. To see the output file, select **PSpice/View Output File.** The output file includes the following:

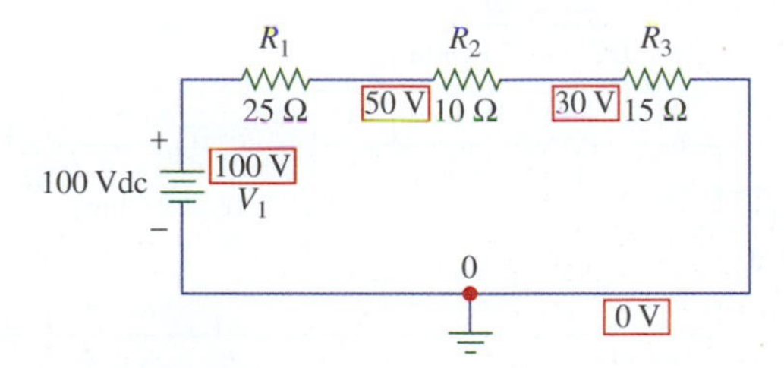

Figure 4.29
For Example 4.8; the schematic of the circuit in Fig. 4.28.

```
NODE       VOLTAGE NODE       VOLTAGE NODE       VOLTAGE
(N00127)   30.0000 (N00131)   50.0000 (N00135)   100.0000

VOLTAGE SOURCE CURRENTS

NAME        CURRENT
V_V1        -2.000E+00
```

indicating that $V_1 = 100$ V, $V_2 = 50$ V, and $V_3 = 30$ V. The current through the voltage source is -2 A, which implies that $I = 2$ A. Note that V_2 is the voltage between node N00131 and the ground; it is not the voltage across R_2. This same can be said about V_3.

Practice Problem 4.8

For the circuit in Fig. 4.30, use PSpice to find the node voltages.

Answer: $V_1 = 40$ V; $V_2 = 24$ V; $V_3 = 14$ V

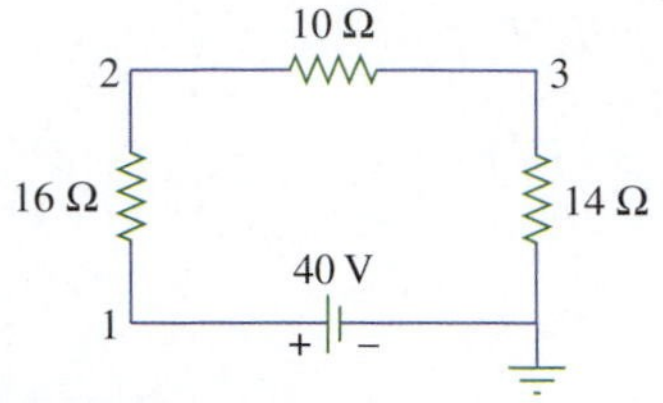

Figure 4.30
For Practice Problem 4.8.

4.8.2 Multisim

Although Multisim and PSpice are similar, they are not the same. For this reason, we cover them separately. Appendix D provides a brief introduction to Multisim, and the reader is expected to have read Appendix D (especially Sections D.1 and D.2) before proceeding with this section.

Example 4.9

Use Multisim to determine V_o and I_o in the circuit of Fig. 4.31.

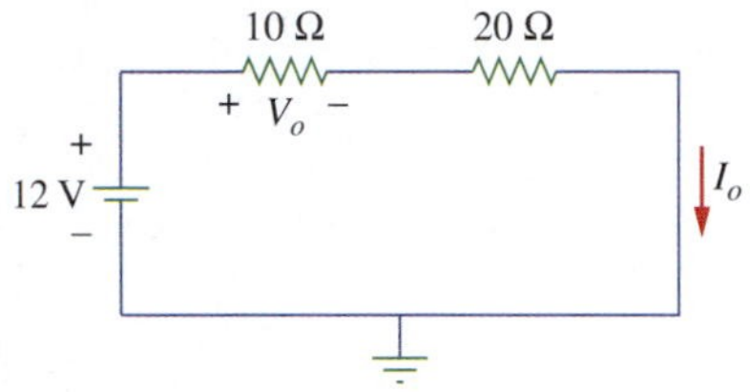

Figure 4.31
For Example 4.9.

Solution:
We draw the circuit following the instructions given in Appendix D, Section D.2. The result is found in Fig. 4.32. We add a voltmeter to measure V_o and an ammeter to measure I_o. Note that the voltmeter is connected in parallel with the 10-Ω resistor, while the ammeter is connected in series with the resistor. We need to add the ground for the circuit to simulate. We simulate the circuit by pressing the power switch or by selecting **Simulate/Run**. This produces the voltmeter and ammeter readings shown in Fig. 4.32, that is,

$$V_o = 4 \text{ V}, \qquad I_o = 0.4 \text{ A}$$

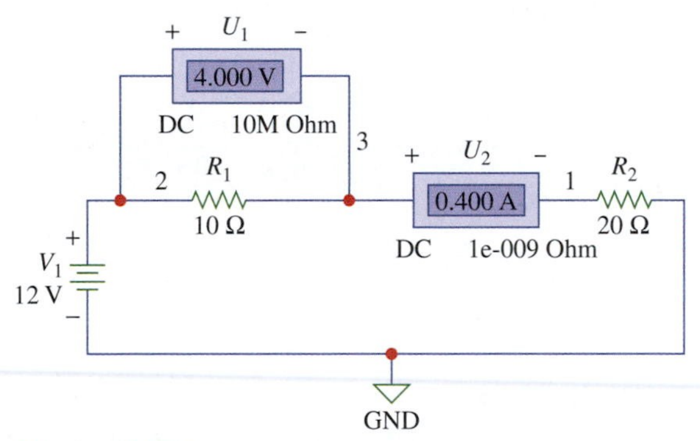

Figure 4.32
Multisim simulation of the circuit in Fig. 4.31.

Practice Problem 4.9

Use Multisim to find V_x and I_x in the circuit of Fig. 4.33.

Answer: 12 V; 2 A

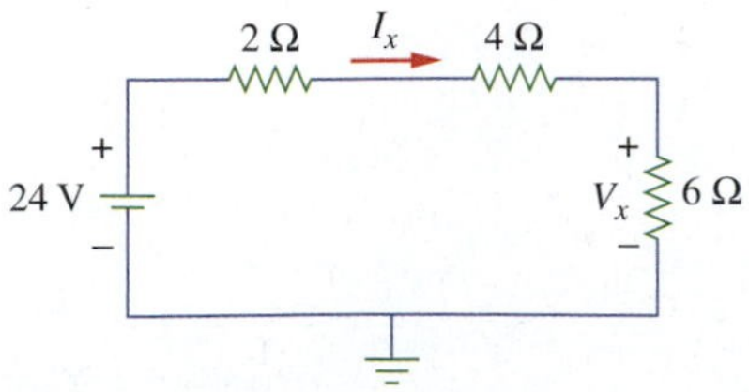

Figure 4.33
For Practice Problem 4.9.

4.9 Applications

There are several applications of series circuits considered in this chapter. One simple example is a light-emitting diode (LED) circuit, shown in Fig. 4.34. This circuit is designed to limit the current flowing through the diode to a certain amount. Without the series resistor R in place, the current through the diode would be too much, and the diode would be damaged. Thus, current limiting is a common application of series resistors.

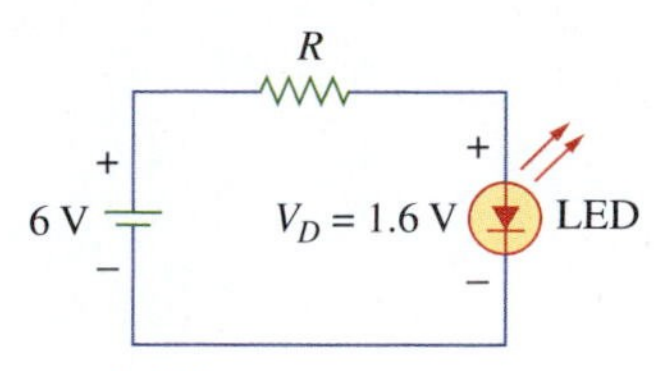

Figure 4.34
Using series resistor to limit current flow.

Another example is the lighting system, such as in a house or Christmas trees; it often consists of N lamps connected either in series or in parallel (to be discussed in the next chapter), as shown in Fig. 4.35. Each lamp is modeled as a resistor. Assuming that all the lamps are identical and V_o is the power-line voltage, the voltage across each lamp is V_o/N for the series connection. The series connection is easy to manufacture but is seldom used in practice for at least two reasons. First, it is less reliable because when a lamp fails, all the lamps go out. Second, it is harder to maintain because when a lamp is bad, one must test all the lamps one by one until the faulty one is found. Nevertheless, modern Christmas lights usually are series to avoid having 120 V across each bulb.

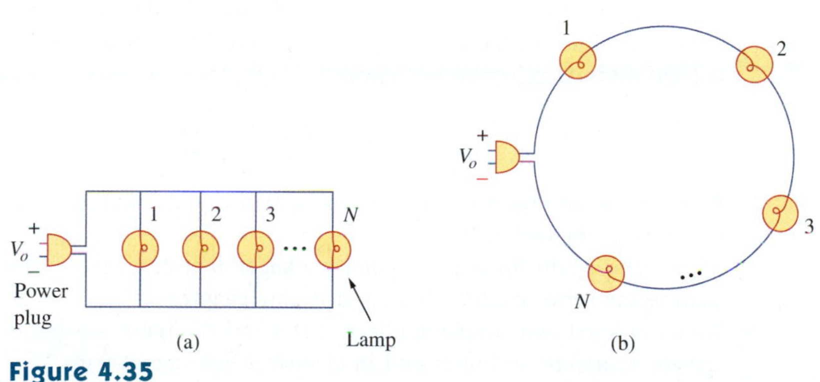

Figure 4.35
(a) Parallel connection of lightbulbs; (b) series connection of lightbulbs.

Example 4.10

If a current of 12 mA passes through the diode in Fig. 4.34, find R.

Solution:
By applying Kirchhoff's voltage law around the loop, we get

$$-6 + IR + 1.6 = 0$$

Hence,

$$R = \frac{6 - 1.6}{I} = \frac{4.4}{12 \text{ mA}} = 366.7\ \Omega$$

Practice Problem 4.10

If the resistance in the circuit of Fig. 4.34 is 270 Ω, find the diode current.

Answer: 16.3 mA

4.10 Summary

1. A branch is a single, two-terminal element in an electric circuit. A node is the point of connection between two or more branches. A loop is a closed path in a circuit. The number of branches b, the number of nodes n, and the number of independent loops ℓ in a network are related as:

$$b = \ell + n - 1$$

2. Two elements are in series when they are connected sequentially, end-to-end. When elements are in series, the same current flows through them.
3. The total resistance (or equivalent resistance) in a series circuit is equal to the sum of the individual resistances, that is,

$$R_T = R_1 + R_2 + R_3 + \cdots + R_N$$

4. Kirchhoff's voltage law (KVL) states that the voltages around a closed path algebraically sum to zero. In other words, the sum of voltage rises equals the sum of voltage drops.
5. When voltage sources are connected in series, the total voltage is the algebraic sum of the individual voltages.
6. In a voltage divider, the voltage across resistors divides according to the magnitude of the resistances. For two resistors in series, the voltage division principle becomes

$$V_1 = \left(\frac{R_1}{R_1 + R_2}\right)V, \quad V_2 = \left(\frac{R_2}{R_1 + R_2}\right)V$$

7. A ground is an electrical connection to Earth, and it serves as a reference point with 0 V.
8. PSpice and Multisim are computer packages that can be used to analyze the series circuits discussed in this chapter.
9. We considered two simple applications of series circuits—using a resistor as a current limiter and in electrical lighting systems.

Review Questions

4.1 A network has 12 branches and 8 independent loops. How many nodes are there in the network?

(a) 19 (b) 17
(c) 5 (d) 4

4.2 The electrical quantity common to all resistors in series is:

(a) power (b) energy
(c) voltage (d) current

4.3 A series circuit has four resistors with values 40 Ω, 50 Ω, 120 Ω, and 160 Ω. The total resistance is:

(a) 90 Ω (b) 280 Ω
(c) 370 Ω (d) 740 Ω

4.4 The current is zero in a series circuit with one or more elements open.

(a) True (b) False

4.5 In a series circuit, you can physically change the positions of the resistors without affecting the current or the total resistance.

(a) True (b) False

4.6 According to Kirchhoff's voltage law (KVL), the algebraic sum of the voltage drops in a circuit must be equal to the algebraic sum of the source voltages.

(a) True (b) False

4.7 A series circuit consists of a 10-V battery and two resistors (12 Ω and 8 Ω). How much current flows through the circuit?

(a) 1.25 A (b) 0.5 A
(c) 2 A (d) 200 A

4.8 In the circuit in Fig. 4.36, V_x is:

(a) 30 V (b) 14 V
(c) 10 V (d) 6 V

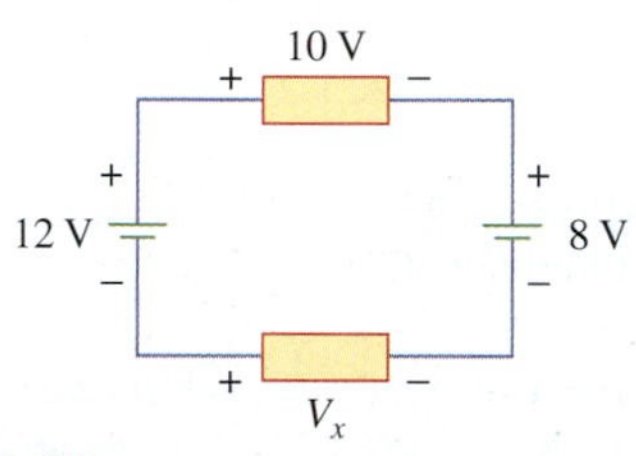

Figure 4.36
For Review Question 4.8.

4.9 In the circuit in Fig. 4.37, the terminal voltage V_{ab} is:

(a) 9 V (b) 7 V (c) 6 V (d) 2 V

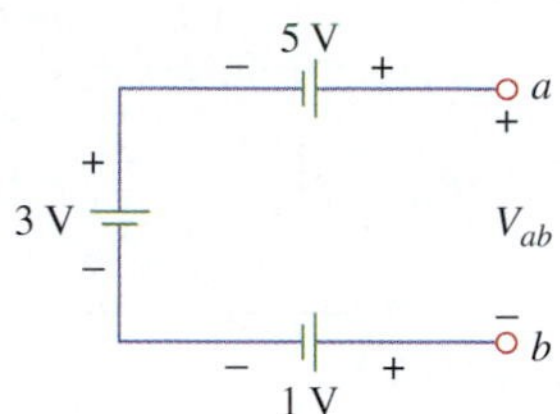

Figure 4.37
For Review Question 4.9.

4.10 If a series circuit has three resistors (12 kΩ, 20 kΩ, and 50 kΩ), the resistor that has the least voltage is:

(a) 12 kΩ (b) 20 kΩ
(c) 50 kΩ (d) cannot be determined from the given information

Answers: 4.1c, 4.2d, 4.3c, 4.4a, 4.5a, 4.6a, 4.7b, 4.8d, 4.9b, 4.10a

Problems

Section 4.2 Nodes, Branches, and Loops

4.1 Determine the numbers of nodes, loops, and branches in the circuit in Fig. 4.38.

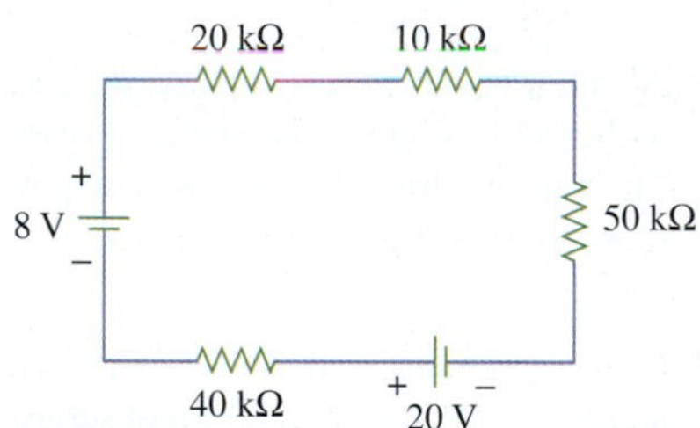

Figure 4.38
For Problem 4.1.

4.2 For the network shown in Fig. 4.39, determine the numbers of nodes, branches, and loops.

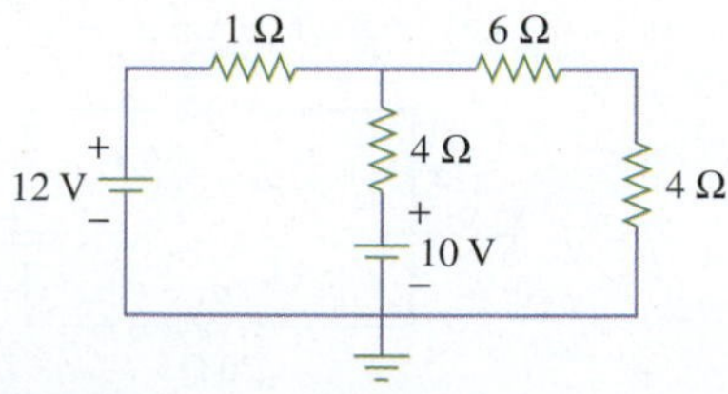

Figure 4.39
For Problems 4.2 and 4.48.

4.3 For the network graph in Fig. 4.40 (where each line represents an element), find the number of nodes, branches, and loops.

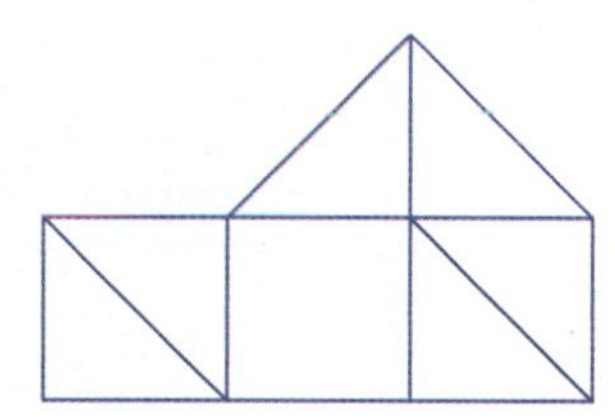

Figure 4.40
For Problem 4.3.

Section 4.3 Resistors in Series

4.4 Given $R_1 = 5.6$ kΩ, $R_2 = 47$ kΩ, $R_3 = 22$ kΩ, and $R_4 = 12$ kΩ, find the total resistance when the resistors are connected in series.

4.5 Find the total resistance in the circuit of Fig. 4.41.

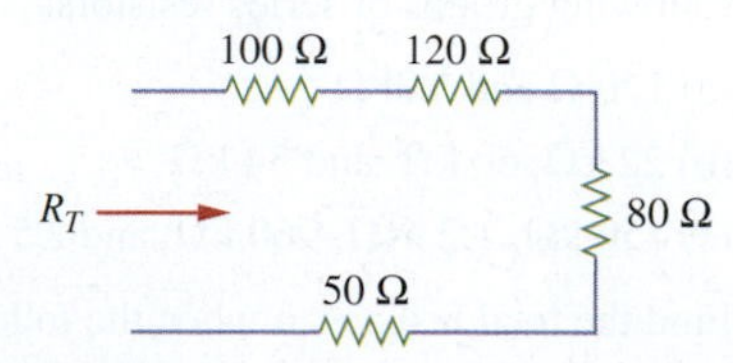

Figure 4.41
For Problem 4.5.

4.6 Determine R_{ab} in the circuit of Fig. 4.42.

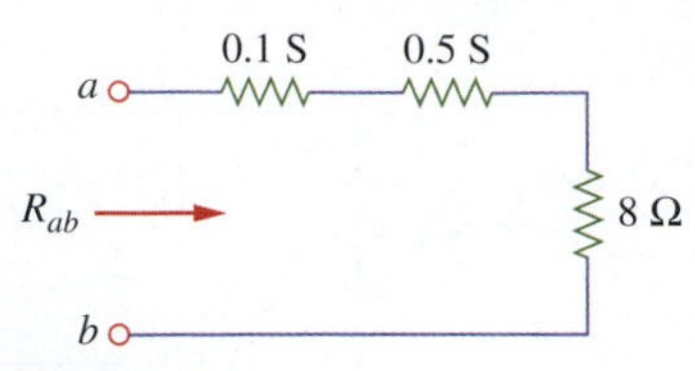

Figure 4.42
For Problem 4.6.

4.7 Find R_T in the circuit of Fig. 4.43.

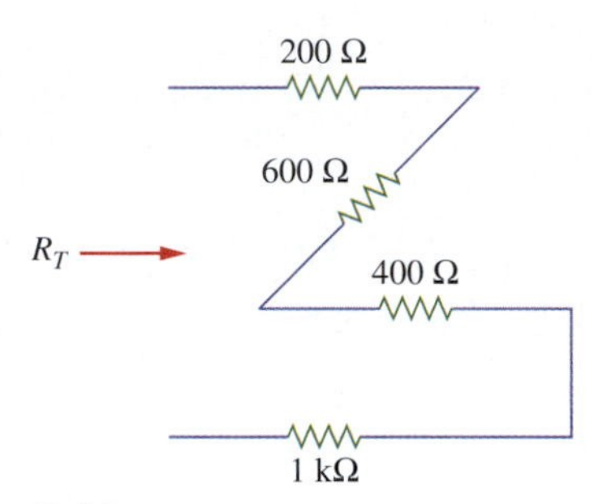

Figure 4.43
For Problem 4.7.

4.8 A string of four resistors is connected as shown in Fig. 4.44. Calculate the total resistance.

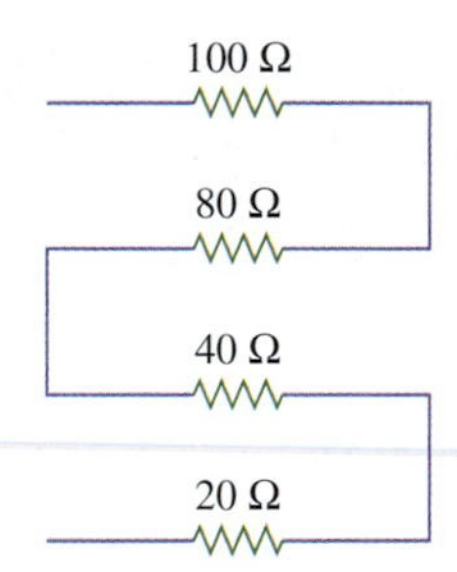

Figure 4.44
For Problem 4.8.

4.9 What is the total resistance of a circuit containing five 47-kΩ resistors in series?

4.10 Calculate the equivalent resistance of each of the following groups of series resistors:

(a) 120 Ω and 560 Ω

(b) 22 kΩ, 60 kΩ, and 34 kΩ

(c) 450 kΩ, 1.2 MΩ, 960 kΩ, and 2.5 MΩ

4.11 Find the total resistance when the following resistors are connected in series:

2.4 MΩ 480 kΩ 56 kΩ 4.2 MΩ

4.12 Calculate the resistance of a resistor to be connected in series with an 80-Ω resistor such that the 80-Ω resistor dissipates 20 W when the series combination is connected to a 110-V source.

4.13 Determine the values of R_1 and R_2 in the circuit in Fig. 4.45.

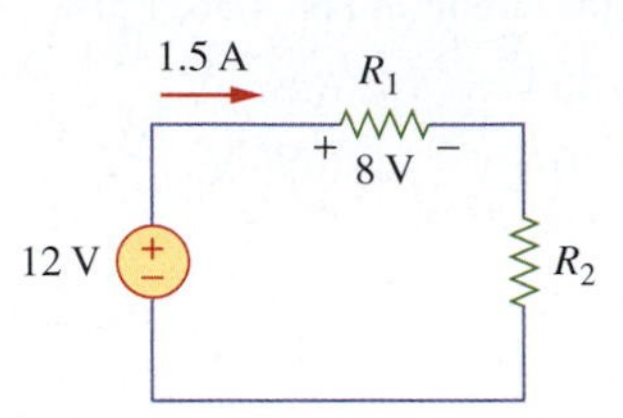

Figure 4.45
For Problem 4.13.

4.14 Three resistors R_1, R_2, and R_3 are connected in series to a 120-V source. The voltage drop across R_1 and R_2 is 90 V, and the voltage drop across R_2 and R_3 is 80 V. If the total resistance is 12 Ω, what is the resistance of the individual resistors?

4.15* Refer to the circuit in Fig. 4.46. Let $V_s = 120$ V, $R_1 = 8\ \Omega$, and $P_2 = 400$ W. Calculate the value of R_2.

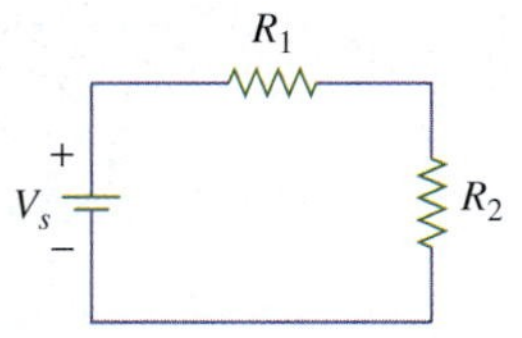

Figure 4.46
For Problem 4.15.

4.16 A circuit has three resistors R_1, R_2, and R_3 in series with a 42-V source. The voltage across R_1 is 10 V. The current through R_2 is 2 A. The power dissipated in R_3 is 40 W. Calculate the values of R_1, R_2, and R_3.

4.17 Four resistors $R_1 = 80\ \Omega$, $R_2 = 120\ \Omega$, $R_3 = 160\ \Omega$, and $R_4 = 40\ \Omega$ are connected in series along with a 6-V battery. Determine the current that flows through the resistors.

Section 4.4 Kirchhoff's Voltage Law

4.18 Find the current I in the circuit of Fig. 4.47.

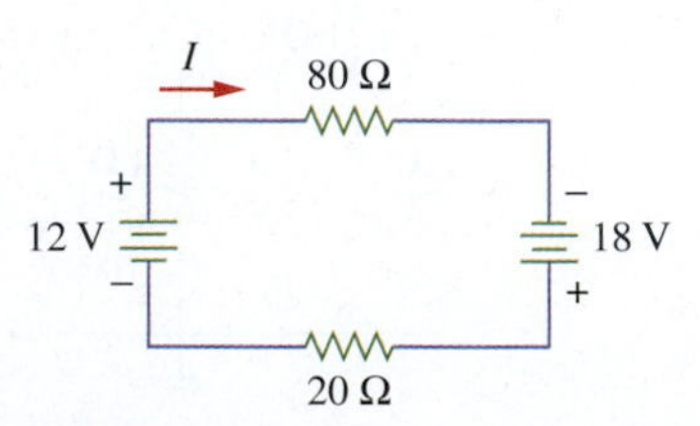

Figure 4.47
For Problem 4.18.

4.19 Determine the current I_x in the circuit of Fig. 4.48. Find the power absorbed by each resistor.

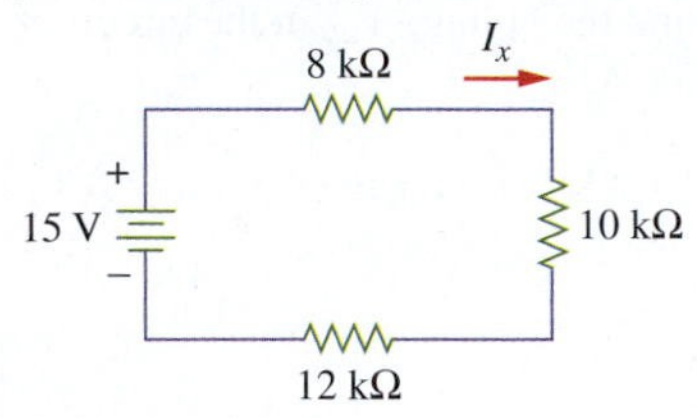

Figure 4.48
For Problem 4.19.

4.20 Use Kirchhoff's voltage law to determine the current I in the circuit of Fig. 4.49.

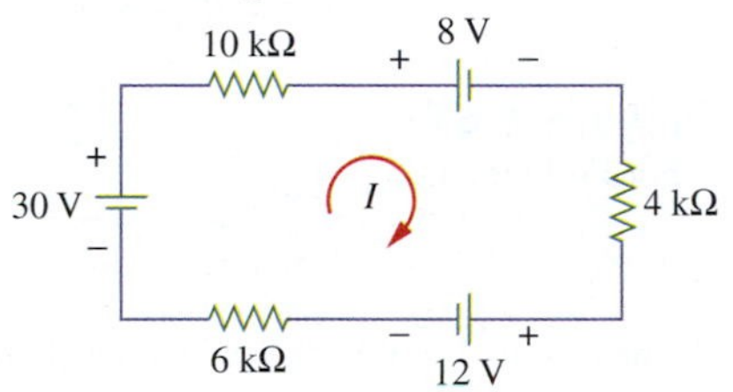

Figure 4.49
For Problem 4.20.

4.21 Determine the voltage V_x in the circuit of Fig. 4.50.

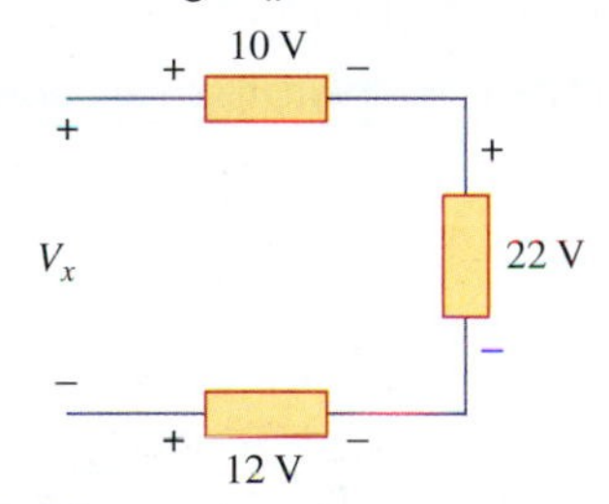

Figure 4.50
For Problem 4.21.

4.22 Find R_x in the circuit of Fig. 4.51.

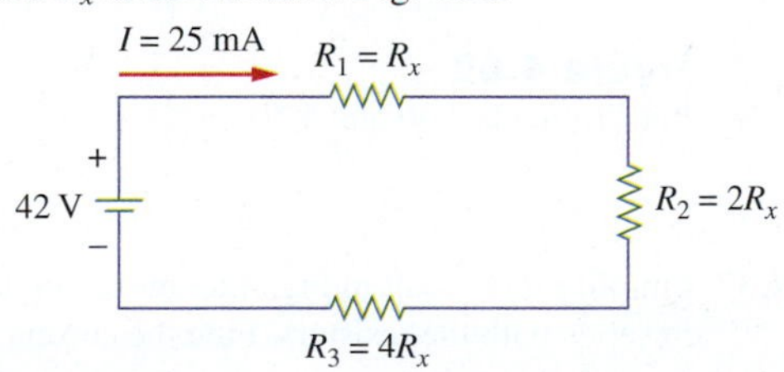

Figure 4.51
For Problem 4.22.

4.23 For the circuit in Fig. 4.52, find V and R.

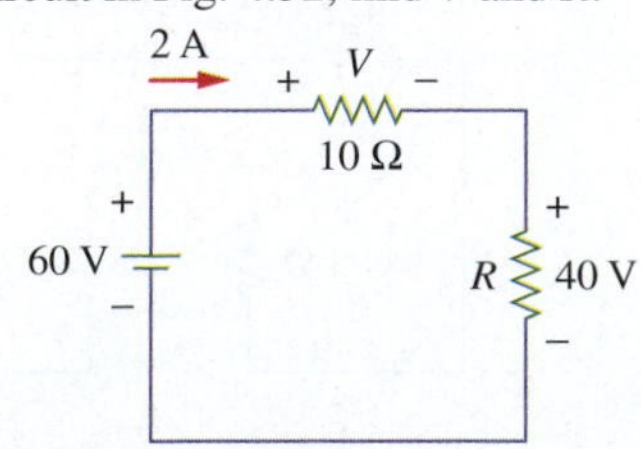

Figure 4.52
For Problem 4.23.

4.24 Find V_x and I_x in the circuit in Fig. 4.53.

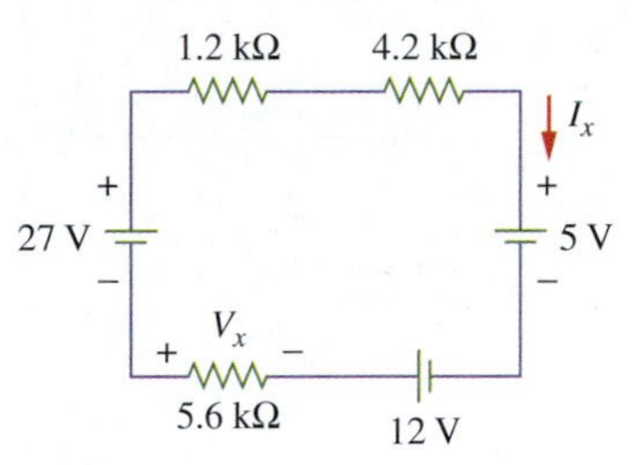

Figure 4.53
For Problem 4.24.

4.25 Find I and V in the circuit shown in Fig. 4.54.

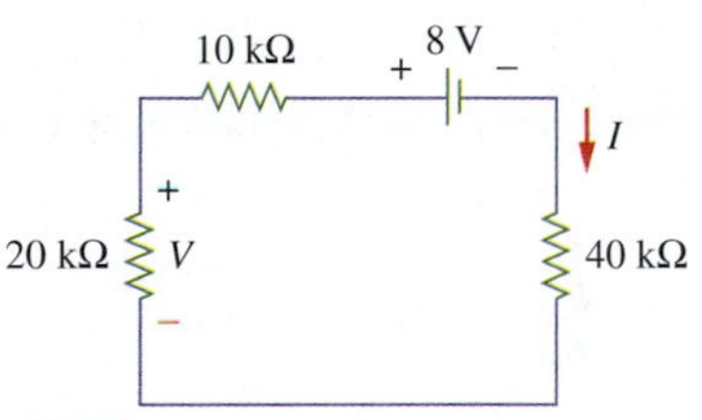

Figure 4.54
For Problem 4.25.

4.26 Determine V_1 and V_2 in the circuit of Fig. 4.55.

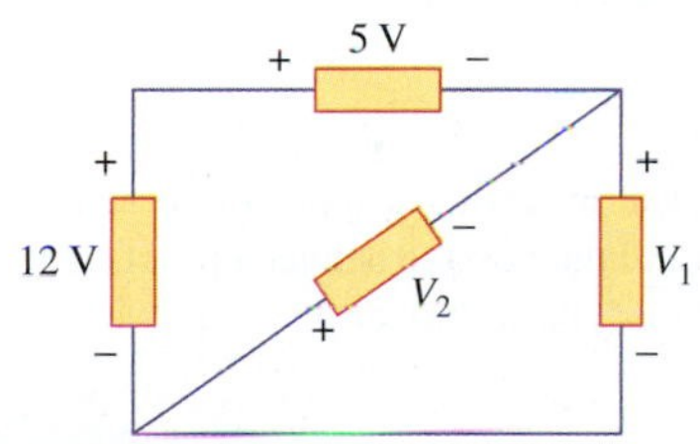

Figure 4.55
For Problem 4.26.

4.27 Determine I_x in the circuit of Fig. 4.56.

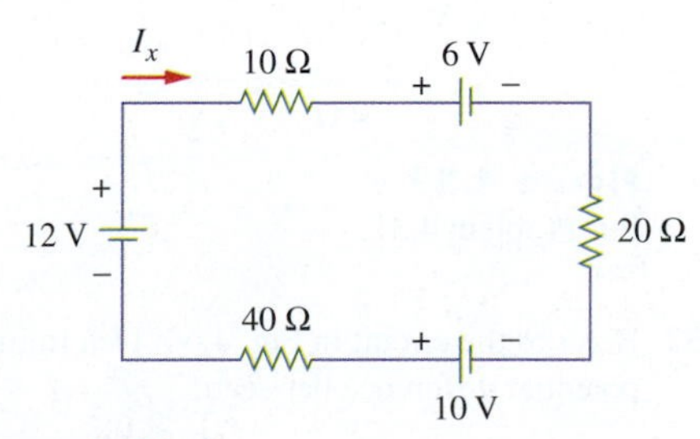

Figure 4.56
For Problem 4.27.

4.28 A series circuit contains components dropping 12 V, 16 V, 24 V, and 32 V, respectively.

(a) What is the applied voltage V_s?

(b) What is the algebraic sum of voltages around the loop including the source?

4.29 Find I and V_{ab} in the circuit of Fig. 4.57.

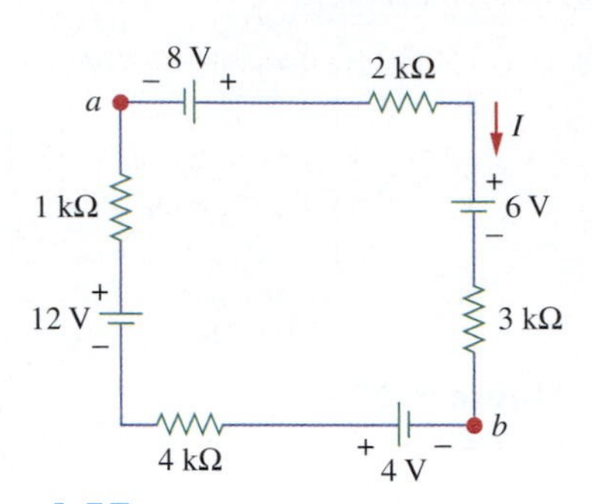

Figure 4.57
For Problem 4.29.

4.30 Determine I and V_{ab} in the circuit of Fig. 4.58.

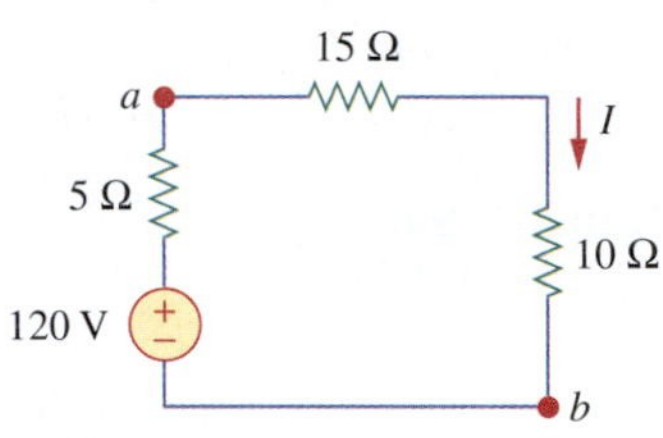

Figure 4.58
For Problem 4.30.

4.31 Determine the current I in the circuit in Fig. 4.59. Find the voltage between points x and y. Which point (x or y) is at higher potential?

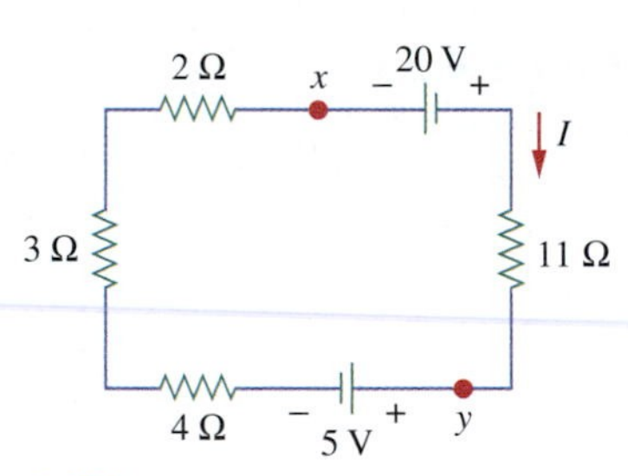

Figure 4.59
For Problem 4.31.

4.32 Refer to the circuit in Fig. 4.60. Determine the potential difference between:

(a) x and y

(b) x and z

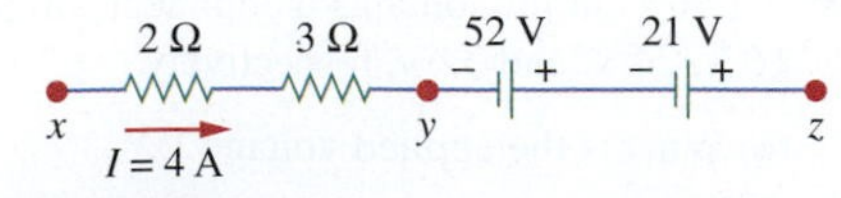

Figure 4.60
For Problem 4.32.

Section 4.5 Voltage Sources in Series

4.33 Find the voltage V_{ab} in the circuit of Fig. 4.61.

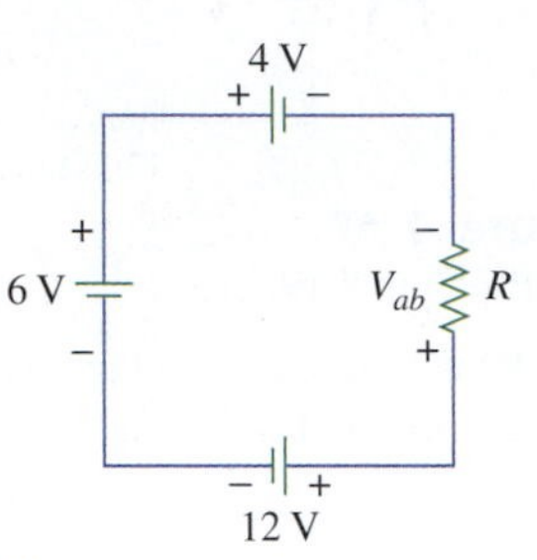

Figure 4.61
For Problem 4.33.

4.34 One of four 1.2-V batteries in a flashlight is put in backwards. Determine the voltage across the bulb.

4.35 How many 1.5-V batteries must be connected in series to produce 12 V?

4.36 Find the current I in the circuit of Fig. 4.62.

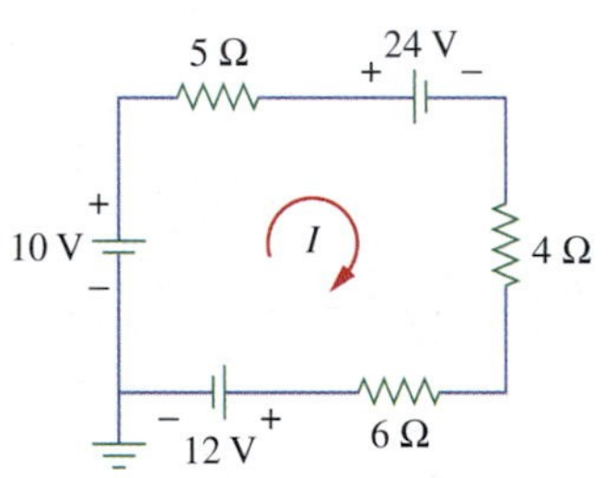

Figure 4.62
For Problems 4.36 and 4.50.

4.37 Simplify the circuit in Fig. 4.63 into a single source in series with the resistors. Find the current passing through the resistors.

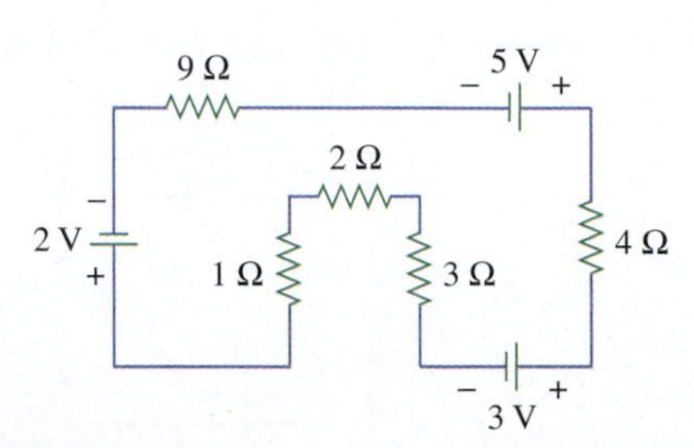

Figure 4.63
For Problem 4.37.

Section 4.6 Voltage Dividers

4.38 Find the voltage between points a and b in the circuit shown in Fig. 4.64.

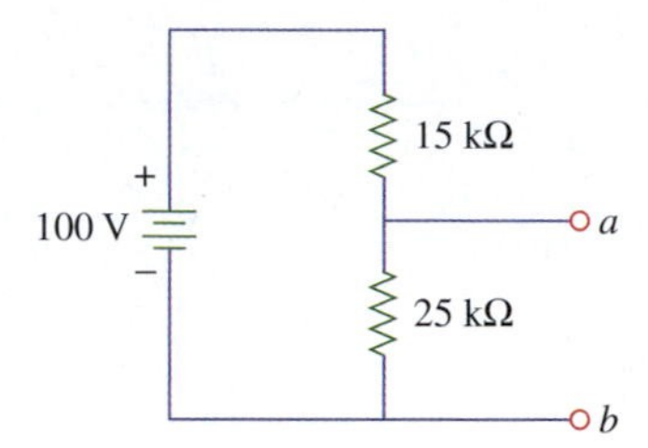

Figure 4.64
For Problems 4.38 and 4.51.

4.39 Determine the voltage V_{ab} in the circuit of Fig. 4.65.

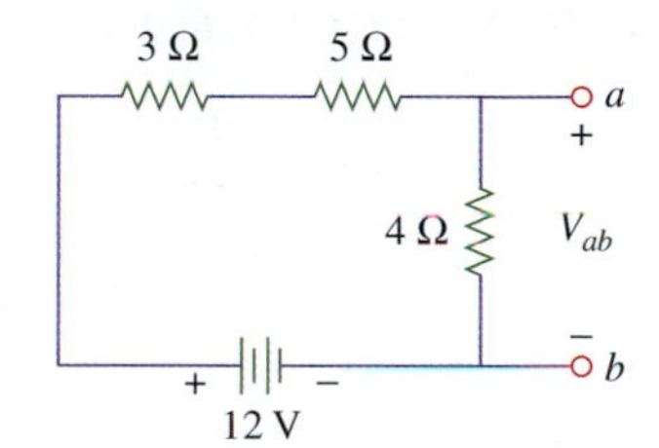

Figure 4.65
For Problems 4.39 and 4.53.

4.40 Apply the voltage divider rule to find the voltage across each resistor in the circuit of Fig. 4.66.

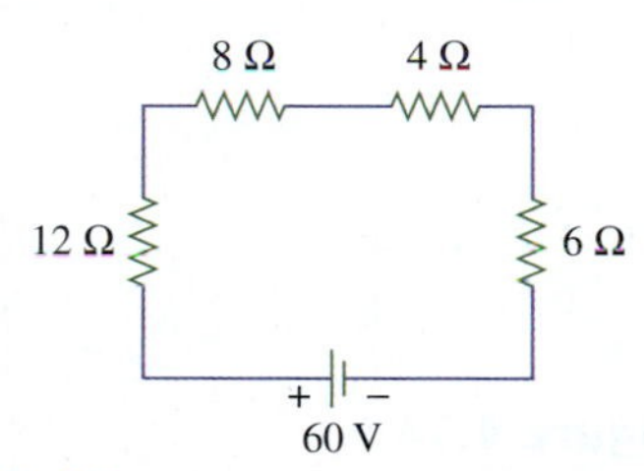

Figure 4.66
For Problem 4.40.

4.41 A voltage divider consists of resistors R_1 and R_2 and a 12-V battery. $R_1 = 5.6$ kΩ and the voltage across it is 4.5 V. Calculate the resistance of R_2.

4.42 Refer to the circuit in Fig. 4.67. If $V_s = 24$ V and $R_s = 100$ Ω, calculate R_L such that $V_L = V_s/2$. What powers are dissipated in R_L and R_s under this condition?

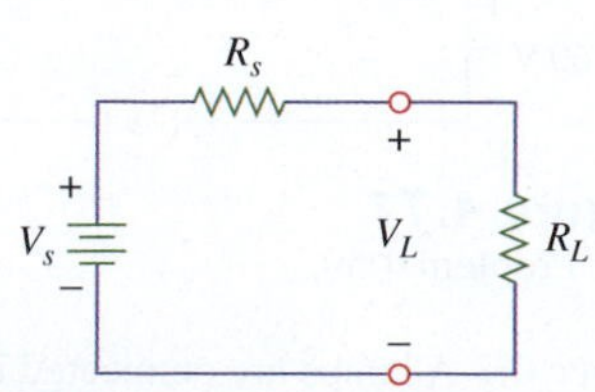

Figure 4.67
For Problem 4.42.

4.43 In the circuit of Fig. 4.68, R_2 is a 5-kΩ potentiometer (variable resistor).

(a) Calculate the maximum and minimum values of V_x.

(b) Calculate the maximum and minimum value of V_x if R_3 becomes short-circuited.

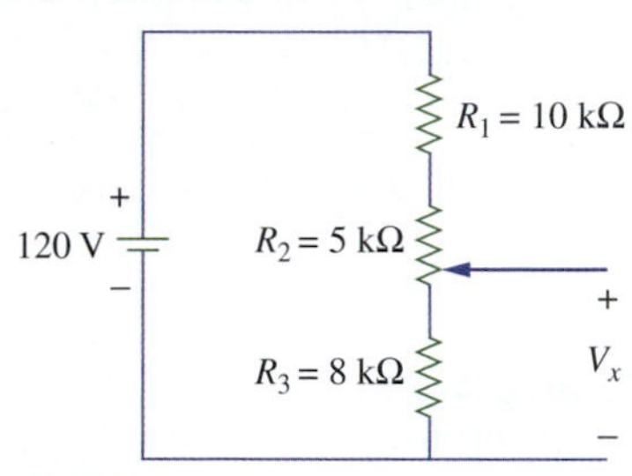

Figure 4.68
For Problem 4.43.

4.44 A 30-kΩ resistor and a 50-kΩ resistor are connected in series with a 120-V supply. Find (a) the voltage drop across the 30-kΩ resistor, and (b) the power dissipated by the 50-kΩ resistor.

4.45 Find the voltage at point A in Fig. 4.69.

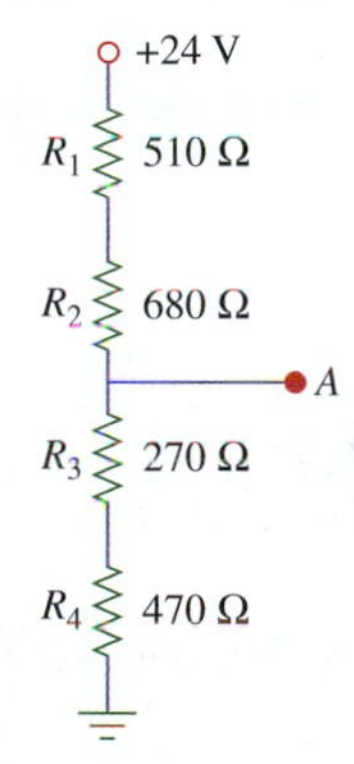

Figure 4.69
For Problem 4.45.

4.46 A 28-V voltage source has an internal resistance of 250 mΩ. Determine the terminal voltage when the source is connected to a load of 7 Ω.

Section 4.8 Computer Analysis

4.47 Refer to the circuit in Fig. 4.70. Use PSpice to find the voltages V_1, V_2, and V_3.

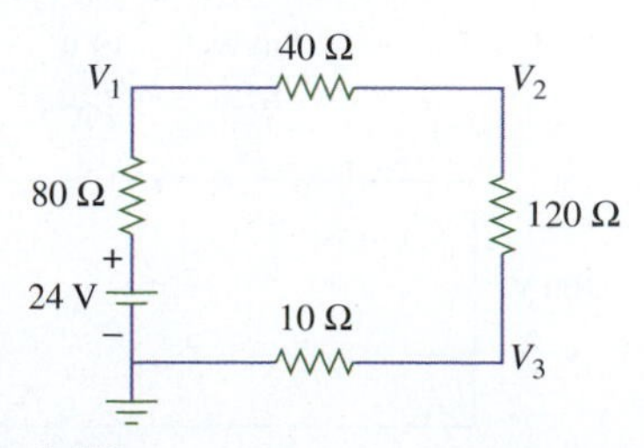

Figure 4.70
For Problem 4.47.

4.48 Use PSpice to determine the voltage across each resistor in the circuit of Fig. 4.39.

4.49 Use PSpice to find the node voltages V_1 and V_2 in the circuit of Fig. 4.71.

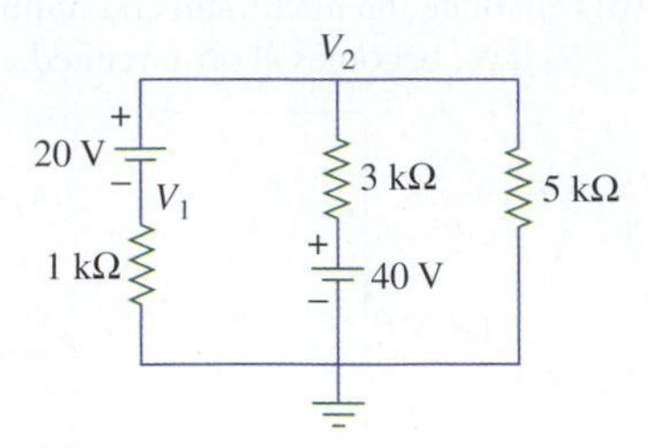

Figure 4.71
For Problem 4.49.

4.50 Use PSpice to determine current I in Fig. 4.62.

4.51 Use Multisim to find V_{ab} in the circuit of Fig. 4.64.

4.52 Given the circuit in Fig. 4.72, use Multisim to find I_o and the voltage across each resistor.

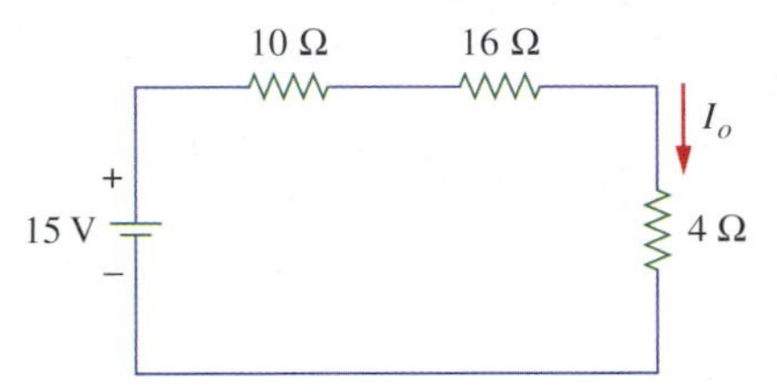

Figure 4.72
For Problem 4.52.

4.53 Refer to Fig. 4.65, and use Multisim to find V_{ab}.

Section 4.9 Applications

4.54 A Christmas tree light set consists of ten 8-W lamps in series. If the light set is connected to a 120-V source, calculate the "hot" resistance of each lamp.

4.55 Three lightbulbs are connected in series to a 100-V battery as shown in Fig. 4.73. Find the current I through the bulbs.

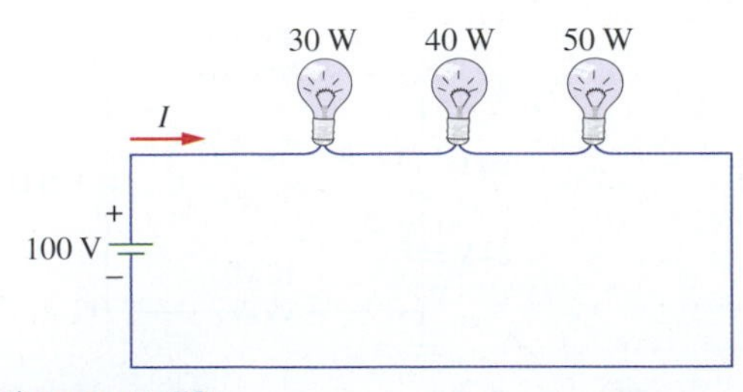

Figure 4.73
For Problem 4.55.

4.56 The potentiometer (adjustable resistor) R_x in Fig. 4.74 is to be designed to adjust current I_x from 1 A to 10 A. Calculate the values of R and R_x needed to achieve this.

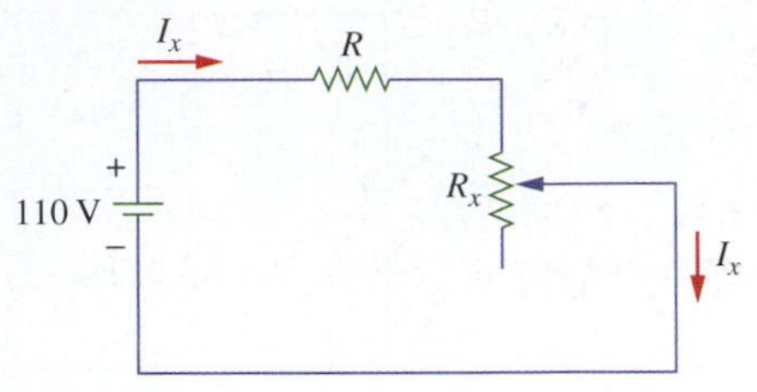

Figure 4.74
For Problem 4.56.

4.57 Figure 4.75 represents a model of a solar photovoltaic panel. Given that $V_s = 30$ V, $R_1 = 20\ \Omega$, and $I = 1$ A, find R_L.

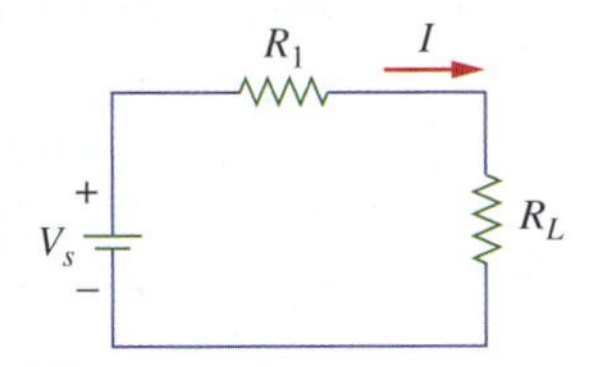

Figure 4.75
For Problem 4.57.

4.58 Determine the value of R in Fig. 4.76 that will limit the current flow to 2 mA. Assume that the LED drops 1.6 V.

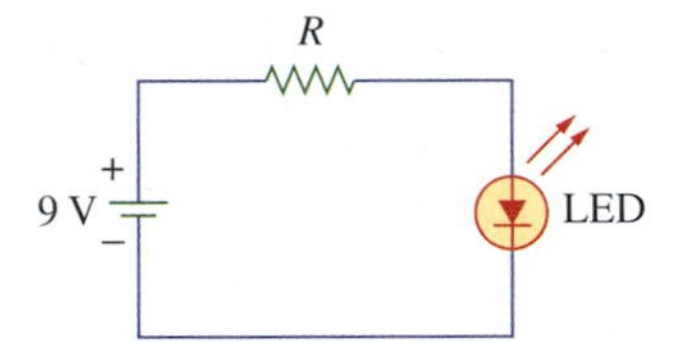

Figure 4.76
For Problem 4.58.

4.59 Find the current measured by the ammeter in Fig. 4.77 for each switch position.

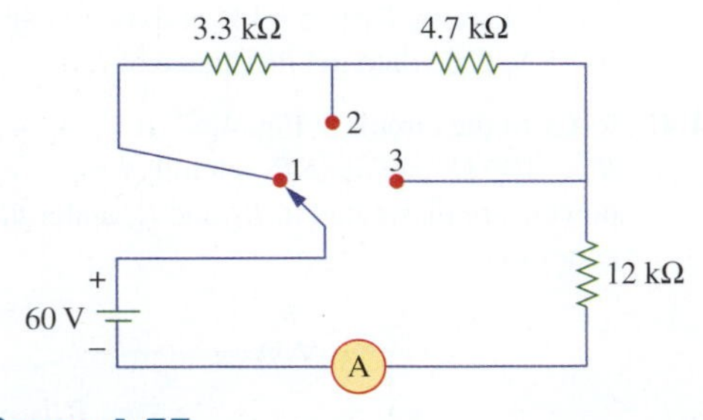

Figure 4.77
For Problem 4.59.

4.60 Three 0.7-A lamps are connected in series and each lamp drops 120 V. Determine (a) the total current, (b) the total voltage, and (c) the total power used.

4.61 A Christmas tree has eight lamps which are connected in series. If each lamp requires 14 V and draws 0.2 A, calculate (a) the total current, (b) the total voltage, and (c) the total power used.

4.62 Three 12-V batteries are connected in series with a load of 2 Ω. Assume that the batteries have internal resistances of 1 Ω, 2 Ω, and 3 Ω. Determine (a) the current through the load, (b) the voltage across the load, (c) the power dissipated by the load, and (d) the power supplied by the batteries.

4.63 Figure 4.78 shows a circuit consisting of a lamp with a resistance of 11.7 Ω and a battery with internal resistance $R_i = 0.3\ \Omega$. Find: (a) the current drawn from the battery, (b) the voltage drop across the internal resistance R_i, (c) the terminal voltage of the battery, (d) the power dissipated internally by the battery, (e) the power delivered to the load, and (f) the efficiency of the battery.

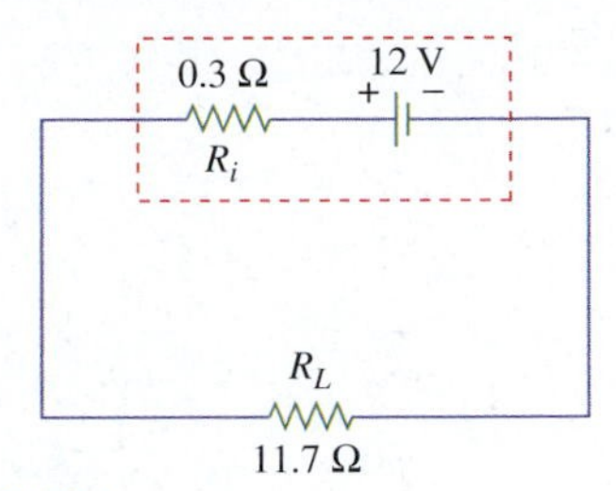

Figure 4.78
For Problem 4.63.

4.64 A flashlight has two batteries; each battery has an open circuit terminal voltage of 3 V and an internal resistance of 0.2 Ω. Determine (a) the total terminal voltage of the batteries for no-load condition, (b) the total internal resistance, and (c) the current that flows through the batteries when they are short-circuited.

chapter

5

Parallel Circuits

Learn to reason forward and backward on both sides of a question.

—Abraham Lincoln

Enhancing Your Career

Asking Questions

Students raising hands in computer class
© Photodisc/Getty RF

In more than 30 years of teaching, I have struggled with determining how best to help students learn. Regardless of how much time students spend in studying for a course, the most helpful activity for students is learning how to ask questions in class and then asking those questions. (Students can still ask questions when courses are offered online.) The student, by asking, becomes actively involved in the learning process and is no longer merely a passive receptor of information. I think this active involvement contributes so much to the learning process that it is probably the single most important aspect to the development of education. In fact, asking questions is the basis of science and technology. As Charles P. Steinmetz rightly said, "No man really becomes a fool until he stops asking questions."

It seems very straightforward and quite easy to ask questions. Have we not been doing that all of our lives? The truth is, however, that to ask questions in an appropriate manner and to maximize the learning process takes thought and preparation.

I am sure that there are several models you could effectively use. Let me share what has worked for me. The most important thing to keep in mind is that you do not have to form a perfect question. The question-and-answer format allows the question to easily be refined as you go. I frequently tell students that they are most welcome to read their questions in class.

Here are three things you should keep in mind when asking questions. First, prepare your question. If you are like many students who are either shy or have not learned to ask questions in class, you may wish to start with a question you have written down outside of class. Second, wait for an appropriate time to ask questions. Use your judgment on that; for example, you would not want to ask a question on circuit breakers in the middle of a discussion of efficiency. Third, be prepared to clarify your question by paraphrasing it or saying it in a different way in case you are asked to repeat the question.

One last comment: Not all professors like students to ask questions in class even though they may say they do. You need to find out which professors like classroom questions. Good luck in enhancing one of your most important skills as a technologist.

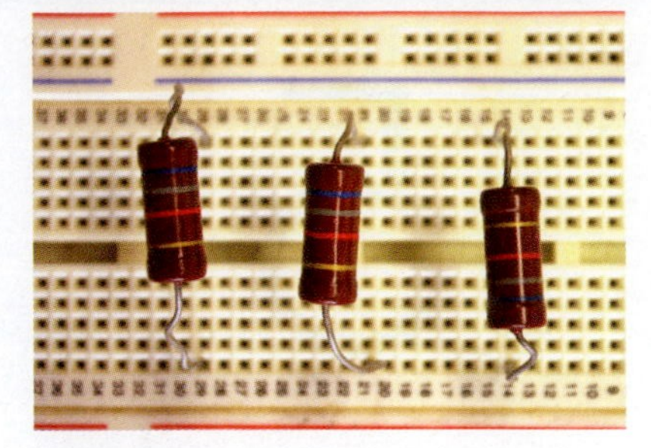

Figure 5.1
Resistors in parallel on a breadboard.
© Sarhan M. Musa

5.1 Introduction

In the previous chapter, we considered the series circuit and applied Kirchhoff's voltage law (KVL) to analyze it. In this chapter, we will learn about parallel circuits and apply Kirchhoff's current law (KCL) to analyze such circuits. Although complex circuits contain portions that are series or parallel circuits, some circuits contain only parallel circuit topology. A simple example of parallel circuits is shown in Fig. 5.1. For example, many automobile accessories are connected to the battery using parallel circuits. Also, parallel circuits are found in most household electrical wiring. This is done so that lights do not stop working just because you turned off your TV.

In this chapter, we begin by looking at the characteristics of parallel circuits. We will find that parallel circuits have some properties that are analogous to, but opposite of, those of series circuits. We present Kirchhoff's current law (KCL). We apply KCL in combining current sources in parallel, current dividers, and resistors in parallel. We will also consider some applications of parallel circuits. We will finally consider how to use PSpice and Multisim to analyze parallel circuits.

5.2 Parallel Circuits

A **parallel circuit** consists of resistors (or elements) that have two nodes in common; consequently, the same voltage appears across each resistor (or element).

Two resistors are in parallel when they are connected to the same two terminals or nodes. In general, when two or more resistors are connected to the two nodes, they are said to be in parallel.

Examples of parallel-circuits are shown in Fig. 5.2. From these examples, the features of a parallel circuit emerge:

1. Every element is connected to two nodes like other elements.
2. There are two or more paths for current flow.
3. The voltage across each parallel element is the same.

One advantage of a parallel circuit over a series circuit is that when one element becomes an open circuit, the other elements are not affected.

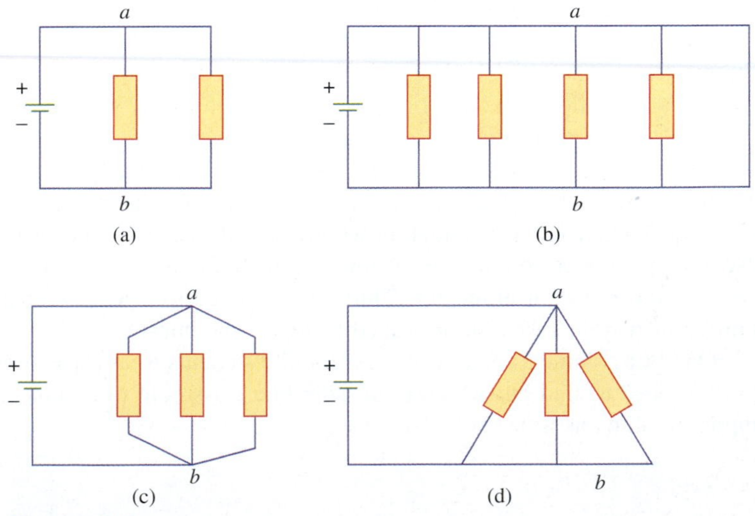

Figure 5.2
Examples of parallel circuits.

Example 5.1

For the circuit in Fig. 5.3, find the current through each resistor and the power absorbed by each resistor.

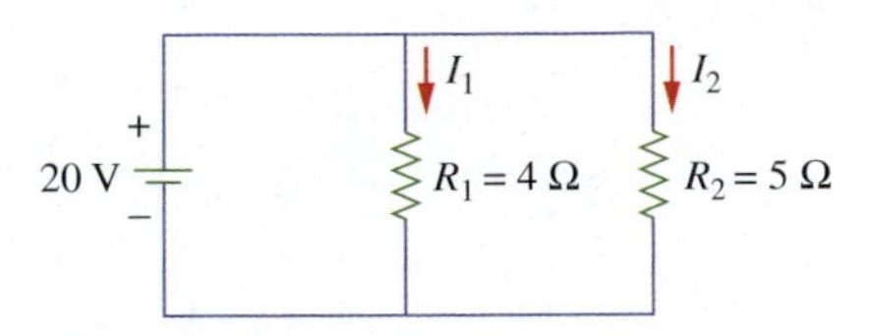

Figure 5.3
For Example 5.1 and Problem 5.51.

Solution:
Because the resistors are in parallel, they must have the same voltage of 20 V across them. Using Ohm's law,

$$I_1 = \frac{V}{R_1} = \frac{20}{4} = 5\ \text{A}$$

$$I_2 = \frac{V}{R_2} = \frac{20}{5} = 4\ \text{A}$$

The power absorbed by R_1 is

$$P_1 = VI_1 = 20 \times 5 = 100\ \text{W}$$

and the power absorbed by R_2 is

$$P_2 = VI_2 = 20 \times 4 = 80\ \text{W}$$

Practice Problem 5.1

Calculate the power absorbed by R_1, R_2, and R_3 in the circuit of Fig. 5.4.

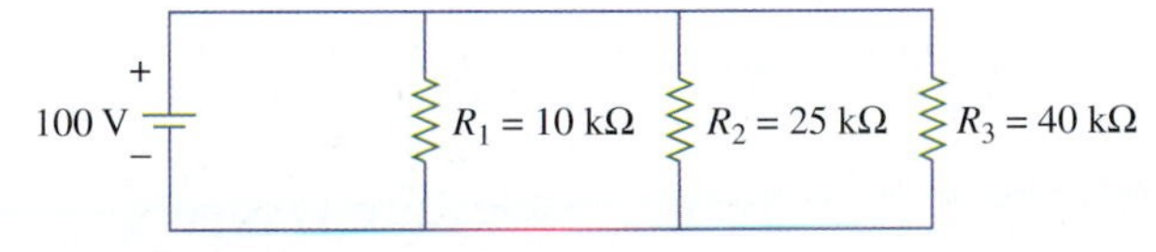

Figure 5.4
For Practice Problem 5.1.

Answer: 1 W; 0.4 W; 0.25 W

5.3 Kirchhoff's Current Law

Kirchhoff's current law is based on the law of conservation of charge, which requires that the algebraic sum of charges within a system cannot change.

> **Kirchhoff's current law** (KCL) states that the algebraic sum of currents entering a node is zero.

Mathematically, KCL implies that

$$I_1 + I_2 + I_3 + \cdots + I_N = 0 \tag{5.1}$$

where N is the number of branches connected to the node and I_n is the nth current entering (or leaving) the node, $n = 1, 2, 3, \ldots, N$. By this law, currents entering a node may be regarded as positive, while currents leaving the node may be taken as negative or vice versa.

Consider the node in Fig. 5.5. Applying KCL gives

$$+I_1 - I_2 + I_3 + I_4 - I_5 = 0 \tag{5.2}$$

because currents I_1, I_3, and I_4 are entering the node, while currents I_2 and I_5 are leaving it. By rearranging the terms, we get

$$I_1 + I_3 + I_4 = I_2 + I_5 \tag{5.3}$$

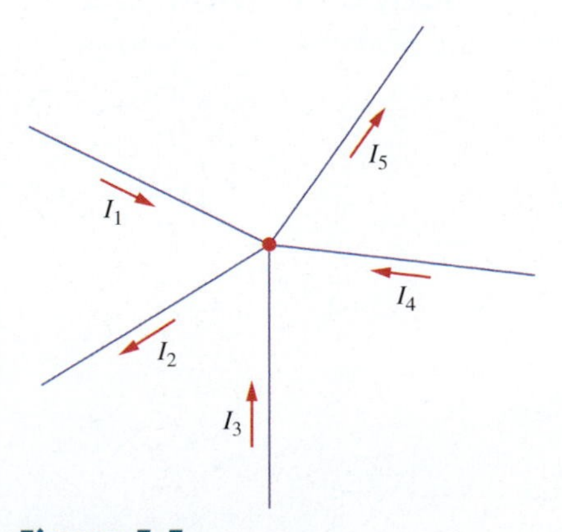

Figure 5.5
Currents at a node illustrating KCL.

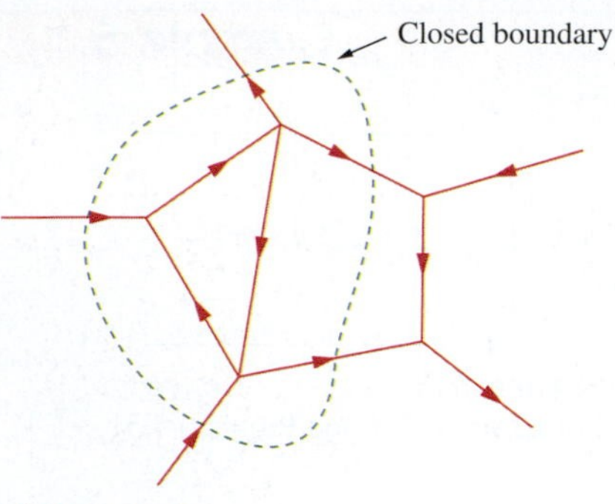

Figure 5.6
Applying KCL to a closed boundary.

Equation (5.3) is an alternative form of KCL:

> The sum of the currents entering a node is equal to the sum of the currents leaving the node.

Note that KCL also applies to a closed boundary. This may be regarded as a generalized case because a node may be regarded as a closed surface shrunk to a point. In two dimensions, a closed boundary is the same as a closed path. As illustrated in Fig. 5.6, the total current entering the closed surface is equal to the total current leaving the surface.

Example 5.2

Determine current I_3 in the circuit of Fig. 5.7.

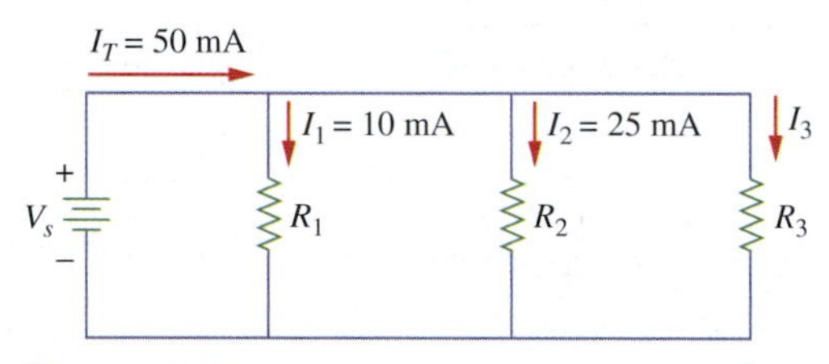

Figure 5.7
For Example 5.2.

Solution:
We apply KCL to the upper node and obtain

$$I_T = I_1 + I_2 + I_3$$

or

$$I_3 = I_T - I_1 - I_2 = 50 \text{ mA} - 10 \text{ mA} - 25 \text{ mA} = 15 \text{ mA}$$

Practice Problem 5.2

Find the current I_2 in the circuit of Fig. 5.8.

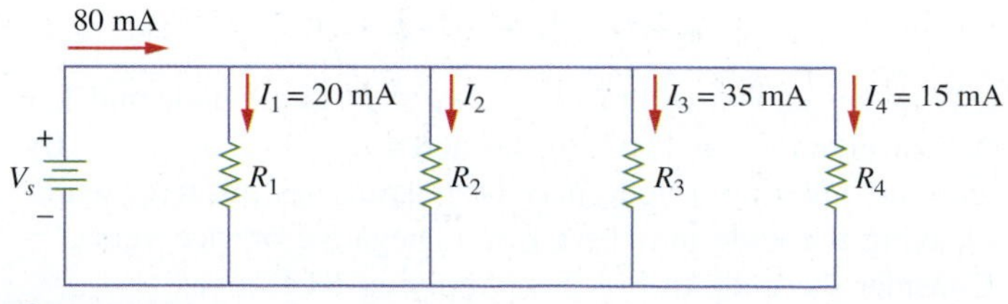

Figure 5.8
For Practice Problem 5.2.

Answer: 10 mA

Example 5.3

Use KCL to obtain I_1, I_2, and I_3 in the circuit of Fig. 5.9.

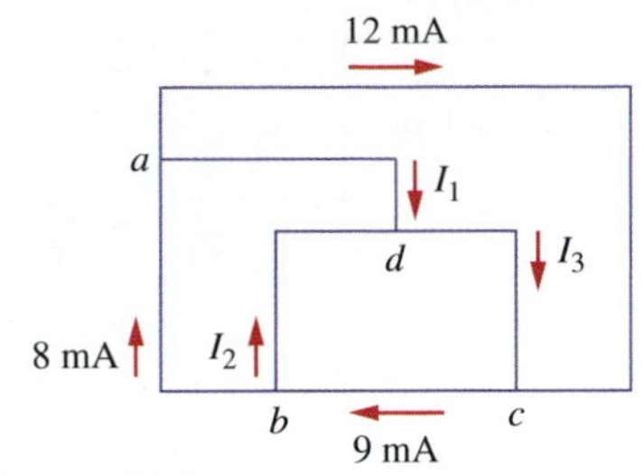

Figure 5.9
For Example 5.3.

Solution:
At node a,

$$8 \text{ mA} = 12 \text{ mA} + I_1$$

or

$$I_1 = 8 \text{ mA} - 12 \text{ mA} = -4 \text{ mA}$$

The negative sign shows that the actual flow of current is opposite to the assumed direction.

At node b,

$$9 \text{ mA} = 8 \text{ mA} + I_2$$

or

$$I_2 = 9 \text{ mA} - 8 \text{ mA} = 1 \text{ mA}$$

At node c,

$$9 \text{ mA} = 12 \text{ mA} + I_3$$

or

$$I_3 = 9 \text{ mA} - 12 \text{ mA} = -3 \text{ mA}$$

To check the values of the currents, we apply KCL at node d.

$$I_3 = I_1 + I_2 \quad \Rightarrow \quad -3 \text{ mA} = -4 \text{ mA} + 1 \text{ mA}$$

which checks.

Practice Problem 5.3

Find I_1, I_2, and I_3 in the circuit of Fig. 5.10.

Answer: 11 A; 4 A; 1 A

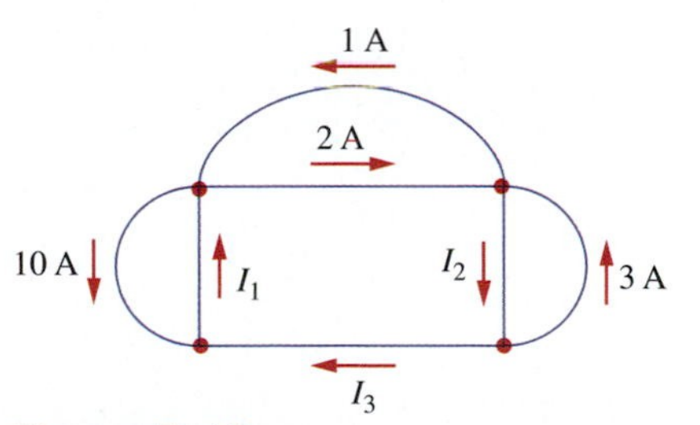

Figure 5.10
For Practice Problem 5.3.

5.4 Current Sources in Parallel

A current source is the dual of a voltage source. It is an energy source that provides a constant current to a load connected to it. Current sources are useful in applications where a constant current is required. The circuit symbol for a current source is shown in Fig. 5.11(a). A realistic or practical current source has some source resistance R_s, as shown in Fig. 5.11(b). An ideal current source has infinite source resistance ($R_s \rightarrow \infty\ \Omega$), as shown in Fig. 5.11(a). A practical, constant-current source uses an LM2576 regulator integrated circuit as shown in Fig. 5.12; its full discussion is beyond the scope of this book. Although we may not know how exactly LM2576 works, we know that it produces a constant-current source.

A simple application of KCL is combining current sources in parallel. The combined current is the algebraic sum of the current supplied by the individual sources. For example, the current sources shown in

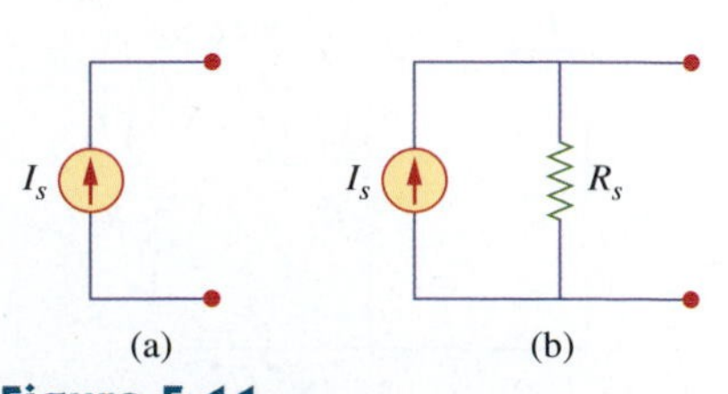

Figure 5.11
Current sources: (a) ideal current source; (b) practical current source

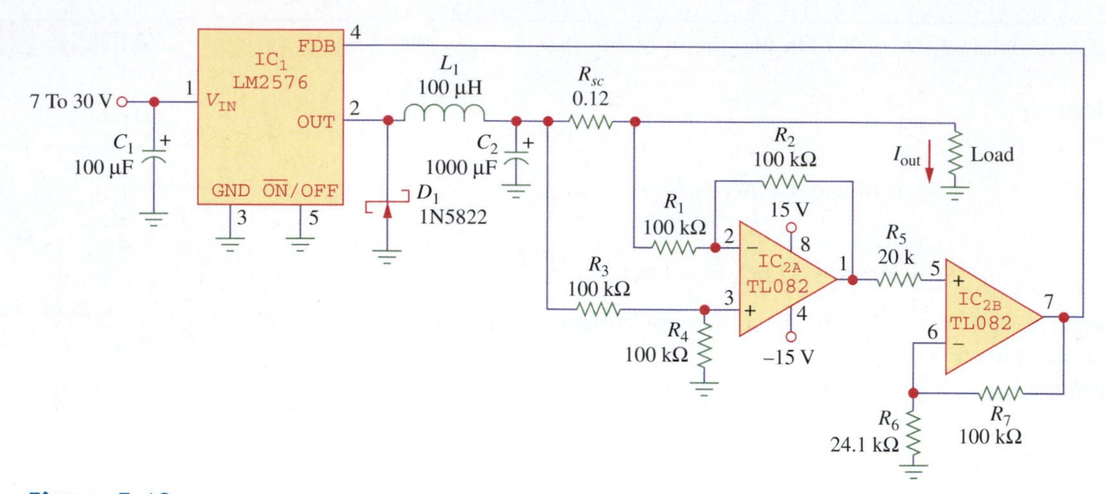

Figure 5.12
A constant-current source.
Courtesy of EDN

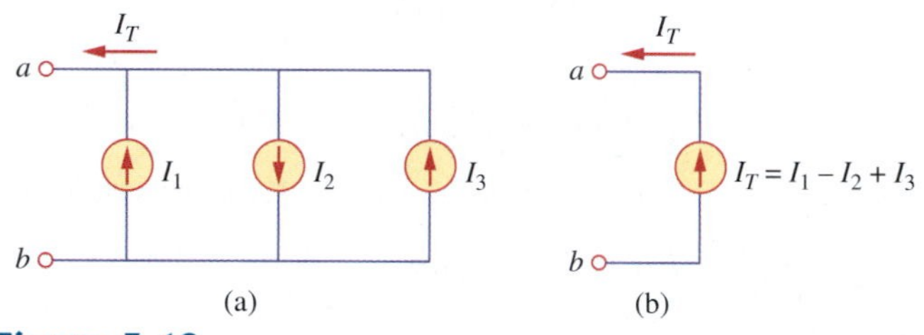

Figure 5.13
Current sources in parallel: (a) original circuit; (b) equivalent circuit.

Fig. 5.13(a) can be combined as in Fig. 5.13(b). The circuit in Fig. 5.13(b) is said to be the *equivalent circuit* of that in Fig. 5.13(a). Two circuits are said to be equivalent if they have the same current-voltage (*I-V*) characteristics. The combined or equivalent current source can be found by applying KCL to node *a*.

$$I_T + I_2 = I_1 + I_3$$

or

$$I_T = I_1 - I_2 + I_3 \tag{5.4}$$

A circuit cannot contain two different current sources I_1 and I_2 in series unless $I_1 = I_2$; otherwise KCL will be violated.

5.5 Resistors in Parallel

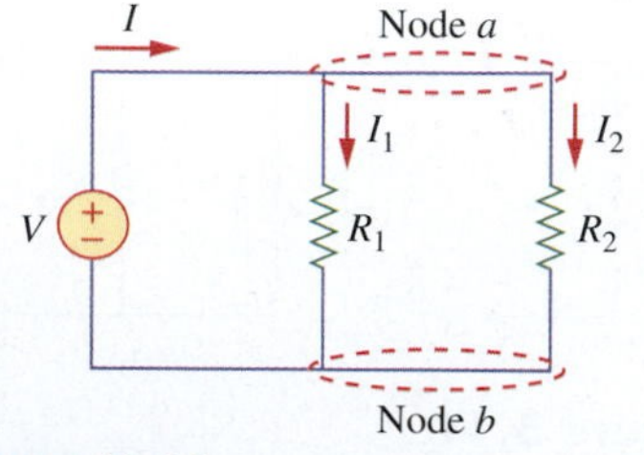

Figure 5.14
Two resistors in parallel.

Consider the circuit in Fig. 5.14, where two resistors are connected in parallel and therefore have the same voltage across them. (The two resistors "see" the same voltage drop, which is why they have the same voltage.) From Ohm's law,

$$V = I_1R_1 = I_2R_2$$

or

$$I_1 = \frac{V}{R_1}, \qquad I_2 = \frac{V}{R_2} \tag{5.5}$$

Applying KCL at node a gives the total current I as

$$I = I_1 + I_2 \tag{5.6}$$

Substituting Eq. (5.5) into Eq. (5.6), we get

$$I = \frac{V}{R_1} + \frac{V}{R_2} = V\left(\frac{1}{R_1} + \frac{1}{R_2}\right) = \frac{V}{R_{eq}} \tag{5.7}$$

where R_{eq} is the equivalent resistance of the resistors in parallel:

$$\frac{1}{R_{eq}} = \frac{1}{R_1} + \frac{1}{R_2} \tag{5.8}$$

or

$$\frac{1}{R_{eq}} = \frac{R_1 + R_2}{R_1 R_2}$$

or

$$\boxed{R_{eq} = \frac{R_1 R_2}{R_1 + R_2}} \tag{5.9}$$

This is known as the *product-over-sum equation*. Thus,

The **equivalent resistance** of two resistors in parallel is equal to the product of their resistances divided by the sum of their resistances.

It must be emphasized that this applies only to two resistors in parallel. From Eq. (5.9), if $R_1 = R_2$, then $R_{eq} = R_1/2$.

We can extend the result in Eq. (5.8) to the general case of a circuit with N resistors in parallel. The equivalent resistance is

$$\boxed{\frac{1}{R_{eq}} = \frac{1}{R_1} + \frac{1}{R_2} + \frac{1}{R_3} + \cdots + \frac{1}{R_N}} \tag{5.10}$$

Note that R_{eq} is always smaller than the resistance of the smallest resistor in the parallel combination. If $R_1 = R_2 = \cdots = R_N = R$, then

$$R_{eq} = R/N \tag{5.11}$$

For example, if four 100-Ω resistors are connected in parallel, their equivalent resistance is 25 Ω.

It is often more convenient to use conductance rather than resistance when dealing with resistors in parallel. From Eq. (5.10), the equivalent conductance for N resistors in parallel[1] is

$$\boxed{G_{eq} = G_1 + G_2 + G_3 + \cdots + G_N} \tag{5.12}$$

[1] Conductances in parallel behave as a single conductance whose value is equal to the sum of the individual conductances.

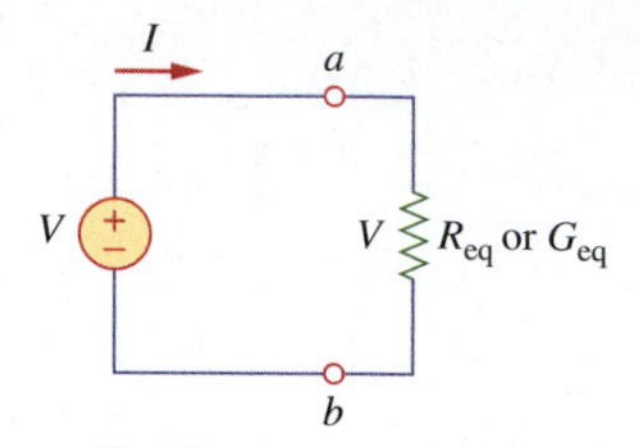

Figure 5.15
Equivalent circuit to Fig. 5.14.

where $G_{eq} = 1/R_{eq}, G_1 = 1/R_2, G_2 = 1/R_2, G_3 = 1/R_3, \ldots, G_N = 1/R_N$. Equation (5.12) states:

> The **equivalent conductance** of resistors connected in parallel is the sum of their individual conductances.

This means that we may replace the circuit in Fig. 5.14 with that in Fig. 5.15. In other words, the circuit in Fig. 5.15 is the equivalent circuit of that in Fig. 5.14. Notice the similarity between Eqs. (4.2) and (5.12)—that is, the equivalent conductance of resistors in parallel is obtained the same way as the equivalent resistance of resistors in series. In the same manner, the equivalent conductance of resistors in series is obtained just the same way as the equivalent resistance of resistors in parallel.

Example 5.4

Determine the equivalent resistance R_{eq} and the current I_T in the circuit of Fig. 5.16. Find the power supplied by the voltage source.

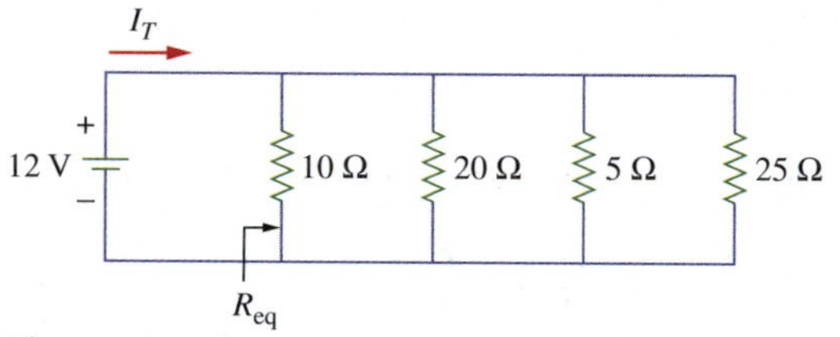

Figure 5.16
For Example 5.4.

Solution:

$$\frac{1}{R_{eq}} = \frac{1}{R_1} + \frac{1}{R_2} + \frac{1}{R_3} + \frac{1}{R_4} = \frac{1}{10} + \frac{1}{20} + \frac{1}{5} + \frac{1}{25} = 0.39 \text{ S} \qquad \textbf{(5.4.1)}$$

To obtain this on a TI-89 Titanium calculator, follow these steps.
Enter 1/10 + 1/20 + 1/5 + 1/25 and then press [♦] [Enter]
Display shows 0.39

$$R_{eq} = 1/0.39 = 2.564 \ \Omega$$

where each forward slash (/) is generated by the division symbol (÷). Notice that the equivalent resistance R_{eq} is smaller than the smallest resistance in Fig. 5.16.

Using Ohm's law,

$$I_T = \frac{V}{R_{eq}} = \frac{12}{2.564} = 4.68 \text{ A}$$

The power supplied is

$$P = VI_T = 12 \times 4.68 = 56.16 \text{ W}$$

Practice Problem 5.4

Find the equivalent resistance R_{eq} in the circuit of Fig. 5.17. Determine the power supplied by the voltage source.

Answer: 571.43 Ω; 1.008 W

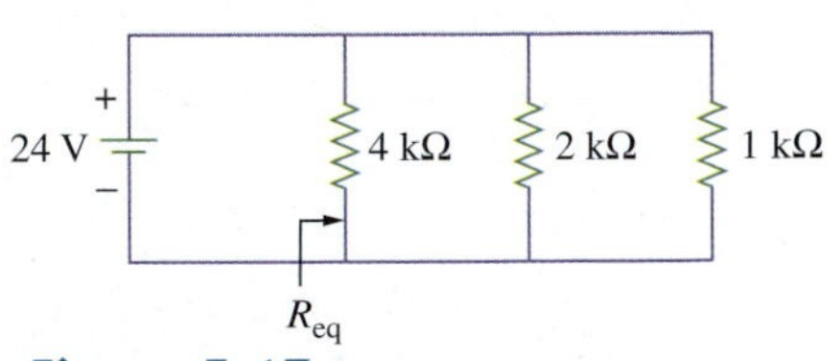

Figure 5.17
For Practice Problem 5.4.

Example 5.5

Determine the voltage V_{ab} in the circuit of Fig. 5.18 and find I_1 and I_2.

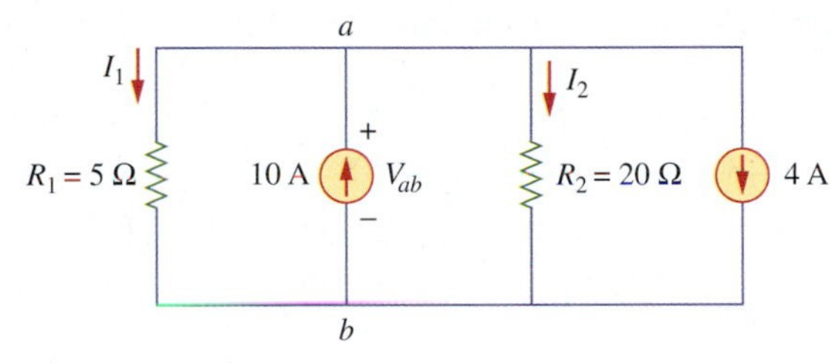

Figure 5.18
For Examples 5.5 and 5.8.

Solution:
V_{ab} is the potential at a with respect to b. It can be found in two ways.

■ **METHOD 1** We find the equivalent resistance as

$$R_T = \frac{R_1 R_2}{R_1 + R_2} = \frac{5 \times 20}{5 + 25} = 4\ \Omega \tag{5.5.1}$$

and the net current source entering node a as

$$I_T = 10 - 4 = 6\text{ A}$$

Thus, we can replace the circuit in Fig. 5.18 by its equivalent circuit in Fig. 5.19. Applying Ohm's law gives

$$V_{ab} = I_T R_T = 6 \times 4 = 24\text{ V} \tag{5.5.2}$$

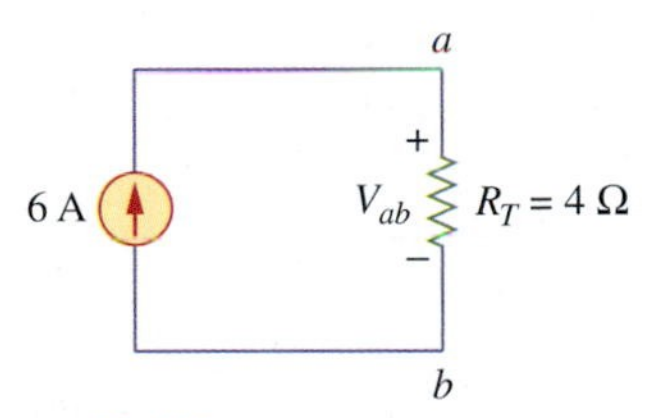

Figure 5.19
Equivalent circuit to that of Fig. 5.18.

■ **METHOD 2** We apply KCL at node a.

$$10 = 4 + I_1 + I_2 \tag{5.5.3}$$

But

$$I_1 = \frac{V_{ab}}{R_1} = \frac{V_{ab}}{5}, \qquad I_2 = \frac{V_{ab}}{R_2} = \frac{V_{ab}}{20}$$

Substituting these into Eq. (5.5.3) gives

$$6 = \frac{V_{ab}}{5} + \frac{V_{ab}}{20}$$

Multiplying through by 20 yields

$$120 = 4V_{ab} + V_{ab} = 5V_{ab}$$

or

$$V_{ab} = 120/5 = 24 \text{ V}$$

as obtained previously. Once we get V_{ab}, we can apply Ohm's law to get currents I_1 and I_2.

$$I_1 = \frac{V_{ab}}{R_1} = \frac{24}{5} = 4.8 \text{ A}$$

$$I_2 = \frac{V_{ab}}{R_2} = \frac{24}{20} = 1.2 \text{ A}$$

Practice Problem 5.5

Calculate the voltage V_{ab} in the circuit of Fig. 5.20 and find I_1, I_2, and I_3.

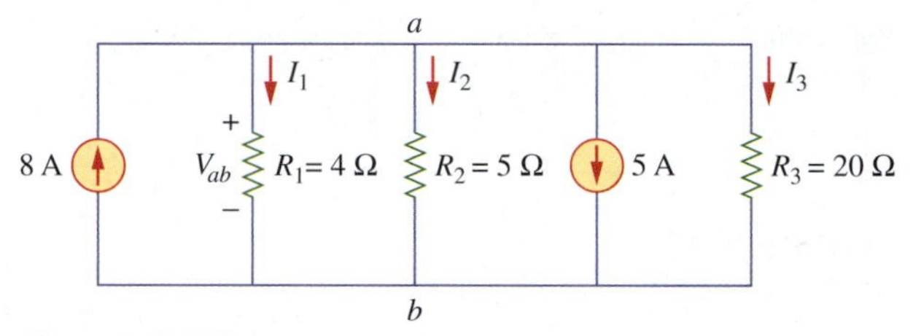

Figure 5.20
For Practice Problems 5.5 and 5.8.

Answer: 6 V; 1.5 A; 1.2 A; 0.3 A

5.6 Current Dividers

Given the total current I_T entering node a in Fig. 5.21, how do we obtain current I_1 and I_2? We know that the equivalent resistor has the same voltage, or

$$V = I_T R_{\text{eq}} = \frac{I_T R_1 R_2}{R_1 + R_2} \tag{5.13}$$

Combining Eqs. (5.5) and (5.13) results in

$$I_1 = \frac{R_2}{R_1 + R_2} I_T, \qquad I_2 = \frac{R_1}{R_1 + R_2} I_T \tag{5.14}$$

or

$$I_1 = \frac{R_{\text{eq}}}{R_1} I_T, \qquad I_2 = \frac{R_{\text{eq}}}{R_2} I_T \tag{5.15}$$

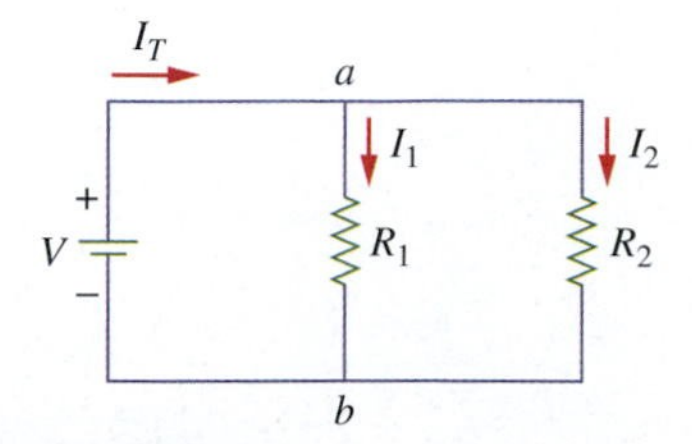

Figure 5.21
Parallel resistors divide current.

which shows that the total current I is shared by the resistors in inverse proportion to their resistances. This is known as the *principle of current division* or the *current divider rule* (CDR), and the circuit in Fig. 5.21 is referred to as a *current divider*. Notice that the larger current flows through the smaller resistance and that $I_T = I_1 + I_2$, satisfying KCL.

As an extreme case, suppose one of the resistors in Fig. 5.21 is zero, say, $R_2 = 0\ \Omega$; that is, R_2 is a short circuit as shown in Fig. 5.22(a). From Eq. (5.14), $R_2 = 0\ \Omega$ implies that $I_1 = 0$ A, $I_2 = I$. This means that the entire current I bypasses R_1 and flows through the short circuit $R_2 = 0\ \Omega$, the path of least resistance. Thus, when a circuit is short-circuited, as shown in Fig. 5.22(a), two things should be kept in mind:

1. The equivalent resistance $R_{eq} = 0\ \Omega$. [See what happens when $R_2 = 0\ \Omega$ in Eq. (5.9).]
2. The entire current flows through the short circuit.

As another extreme case, suppose $R_2 = \infty\ \Omega$; that is, R_2 is an open circuit, as shown in Fig. 5.22(b). The current flows through the path of least resistance R_1. By taking the limit of Eq. (5.9) as R_2 becomes infinite, we obtain $R_{eq} = R_1$ in this case.

The current division rule can be extended to a situation where you have more than two resistors. Consider the network in Fig. 5.23. Because the resistors are in parallel, they have the same voltage V across them. If R_{eq} is the equivalent resistance,

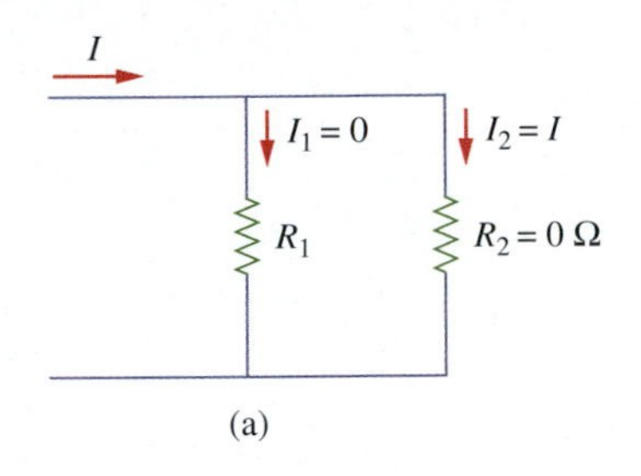

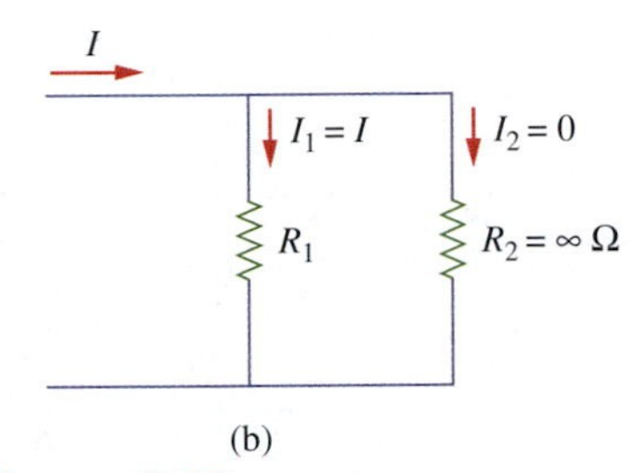

Figure 5.22
Shorts and opens: (a) a shorted circuit; (b) an open circuit.

$$V = I_T R_{eq} \quad \Rightarrow \quad I_T = \frac{V}{R_{eq}} \tag{5.16}$$

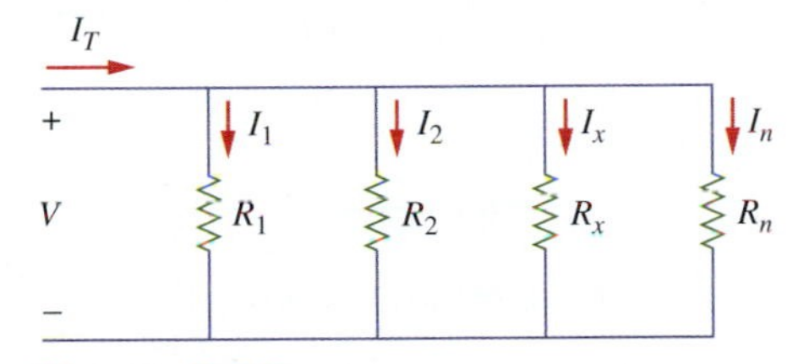

Figure 5.23
Current divider rule for n branches.

Similarly,

$$I_x = \frac{V}{R_x} \tag{5.17}$$

Substituting Eq. (5.16) into Eq. (5.17), we obtain

$$\boxed{I_x = \frac{R_{eq}}{R_x} I_T} \tag{5.18}$$

The current division rule can also be expressed in terms of the conductances. If we divide both the numerator and denominator by $R_1 R_2$, Eq. (5.14) becomes

$$I_1 = \frac{G_1}{G_1 + G_2} I_T, \qquad I_2 = \frac{G_2}{G_1 + G_2} I_T \tag{5.19}$$

Thus, in general, if a current divider has N conductors ($G_1, G_2, \ldots, G_N$) in parallel with the source current I_T, the kth conductor (G_N) will have current

$$I_k = \frac{G_k}{G_1 + G_2 + \cdots + G_N} I_T \tag{5.20}$$

or

$$I_k = \frac{G_k}{G_{eq}} I_T \tag{5.21}$$

Example 5.6

Refer to the circuit in Fig. 5.24. (a) Find R_2 such that the equivalent resistance is 4 kΩ. (b) Find the currents I_1 and I_2.

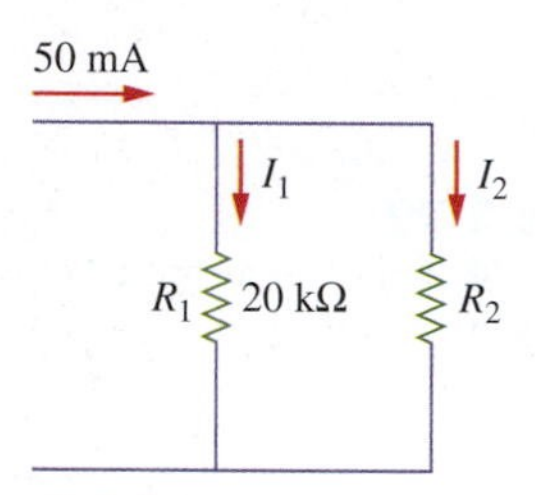

Figure 5.24
For Example 5.6.

Solution:
Because two resistors are connected in parallel, assuming R is in kΩ, we obtain

$$4 = \frac{20 \times R_2}{20 + R_2}$$

or

$$80 + 4R_2 = 20R_2$$

or

$$16R_2 = 80$$

$$R_2 = 80/16 = 5 \text{ k}\Omega$$

We find the currents by applying the current division principle

$$I_1 = \frac{R_2}{R_1 + R_2}(50 \text{ mA}) = \frac{5}{20 + 5}(50 \text{ mA}) = 10 \text{ mA}$$

$$I_2 = \frac{R_1}{R_1 + R_2}(50 \text{ mA}) = \frac{20}{20 + 5}(50 \text{ mA}) = 40 \text{ mA}$$

Practice Problem 5.6

Use current division principle to find currents I_1 and I_2 in the circuit of Fig. 5.25.

Answer: 37.5 mA; 12.5 mA

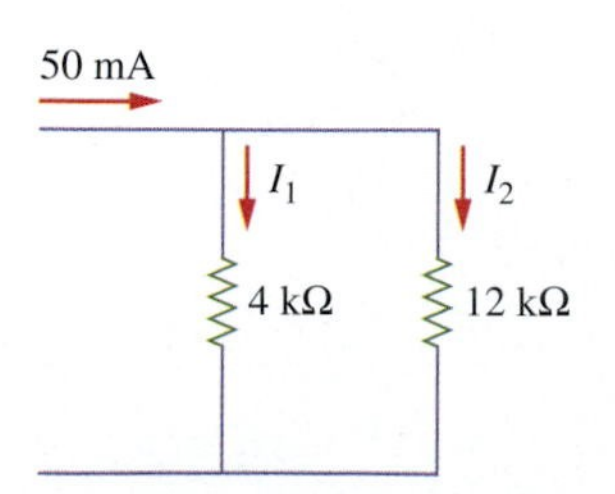

Figure 5.25
For Practice Problem 5.6.

Example 5.7

For the circuit in Fig. 5.26, find the currents I_1, I_2, and I_3.

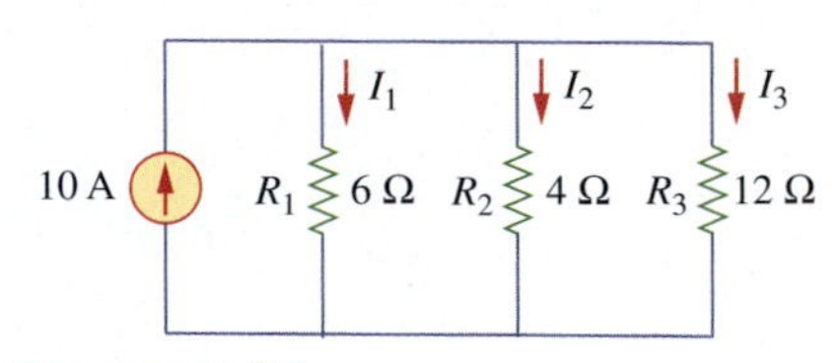

Figure 5.26
For Example 5.7.

Solution:
This can be solved in two ways.

■ **METHOD 1** This involves using conductances and applying Eq. (5.21).

$$G_{eq} = \frac{1}{6} + \frac{1}{4} + \frac{1}{12} = 0.5 \text{ S}$$

Hence,

$$I_1 = \frac{G_1}{G_{eq}} I = \frac{1/6}{0.5}(10) = 3.333 \text{ A}$$

$$I_2 = \frac{G_2}{G_{eq}} I = \frac{1/4}{0.5}(10) = 5 \text{ A}$$

$$I_3 = \frac{G_3}{G_{eq}} I = \frac{1/12}{0.5}(10) = 1.667 \text{ A}$$

■ **METHOD 2** This involves using resistances and applying Eq. (5.14). Here, we have to consider two resistors at a time. To that end, we redraw the circuit as shown in Fig. 5.27. The parallel combination of 4 Ω and 12 Ω is

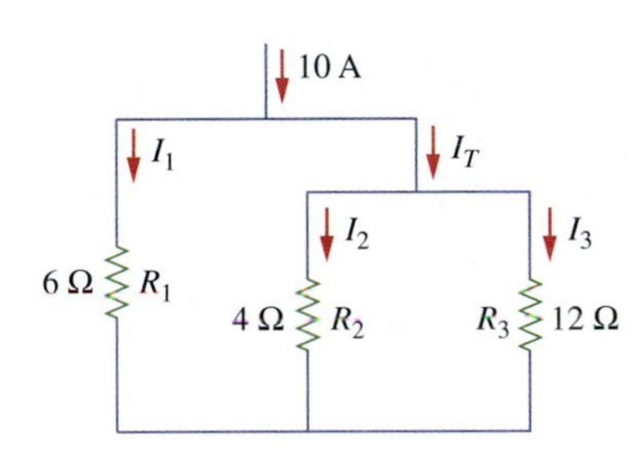

Figure 5.27
For Example 5.7.

$$R_T = 4 \parallel 12 = \frac{4 \times 12}{4 + 12} = 3 \, \Omega$$

We now divide the 10 A from the current source between the 6-Ω resistor and R_T using Eq. (5.14).

$$I_1 = \frac{R_T}{R_1 + R_T} I = \frac{3}{6+3}(10) = 3.333 \text{ A}$$

$$I_T = \frac{R_1}{R_1 + R_T} I = \frac{6}{6+3}(10) = 6.667 \text{ A}$$

The current I_T is now shared between 4-Ω and 12-Ω resistors.

$$I_2 = \frac{R_3}{R_2 + R_3} I_T = \frac{12}{4+12}(6.667) = 5 \text{ A}$$

$$I_3 = \frac{R_2}{R_2 + R_3} I_T = \frac{4}{4+12}(6.667) = 1.667 \text{ A}$$

which is what we got previously. Notice the smallest resistor gets the largest current, and the largest resistor gets the least current.

Practice Problem 5.7

Find the currents I_1, I_2, and I_3 in the circuit shown in Fig. 5.28.

Answer: 6 A; 4.8 A; 1.2 A

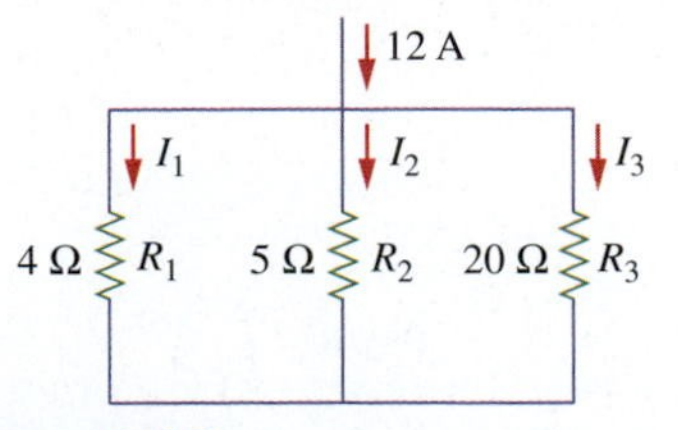

Figure 5.28
For Practice Problem 5.7.

5.7 Computer Analysis

5.7.1 PSpice

PSpice is a useful tool in analyzing parallel circuits. The following example illustrates how to use PSpice for handling such circuits. It is assumed that the reader has reviewed Sections C.1 through C.3 of Appendix C before proceeding with this section.

Example 5.8

Use PSpice to solve Example 5.5. (See Fig. 5.18.)

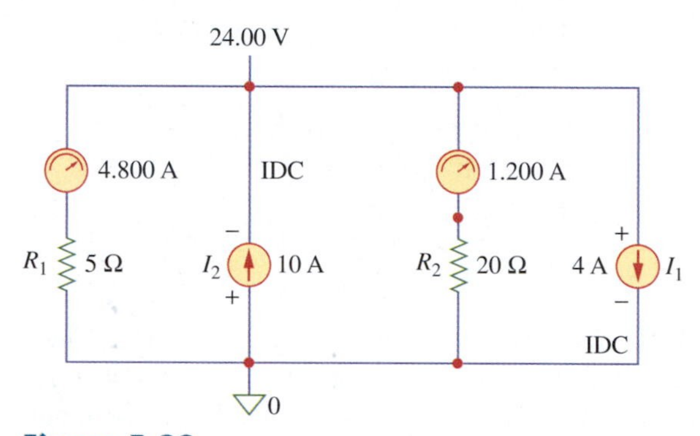

Figure 5.29
The PSpice schematic for Example 5.8.

Solution:
The schematic is shown in Fig. 5.29. Because this is a dc analysis, we use current source IDC. Once the circuit is drawn and saved as exam58.dsn, we simulate the circuit by selecting **PSpice/New Simulation Profile**. This leads to the New Simulation dialog box. Type "exam58" as the name of the file and click Create. This leads to the Simulation Settings dialog box. Because this is dc analysis, just click OK. Then select **PSpice/Run**. The circuit is simulated, and some of the results are displayed on the circuits as in Fig. 5.29. Other results are in the output file. To see the output file, select **PSpice/View Output File**. The results are:

$$V_{ab} = 24 \text{ V}, \quad I_1 = 4.8 \text{ A}, \quad I_2 = 1.2 \text{ A}$$

as obtained in Example 5.5.

Practice Problem 5.8

Rework Practice Problem 5.5 using PSpice.

Answer: 6 V; 1.5 A; 1.2 A; 0.3 A

5.7.2 Multisim

Parallel circuits are easily simulated using Multisim. The way Multisim handles such circuits is similar to the way series circuits are handled. The following example illustrates how Multisim is used in simulating parallel circuits.

Example 5.9

Use Multisim to determine the currents I_T, I_1, and I_2 in the circuit shown in Fig. 5.30.

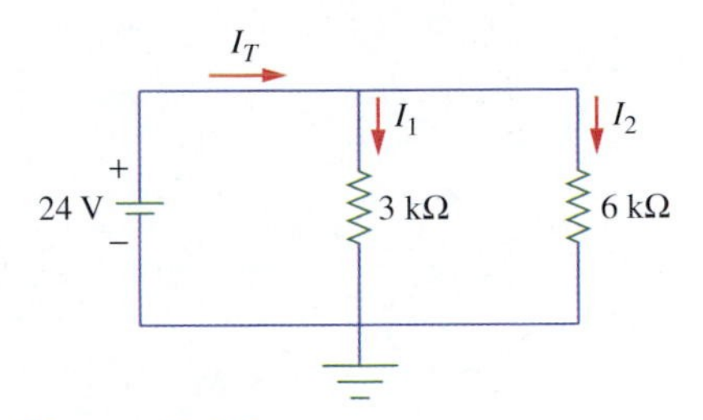

Figure 5.30
For Example 5.9.

Solution:
We draw the circuit as shown in Fig. 5.31. Insert ammeters to measure currents I_T, I_1, and I_2. (Instead of using ammeters, we could use multimeters, which are used to measure current, voltage, and resistance.) After adding the ground and saving the circuit, we simulate it by pressing the power switch or selecting **Simulate/Run**. The results are displayed in Fig. 5.31; that is,

$$I_T = 12 \text{ mA}, \quad I_1 = 8 \text{ mA}, \quad I_2 = 4 \text{ mA}$$

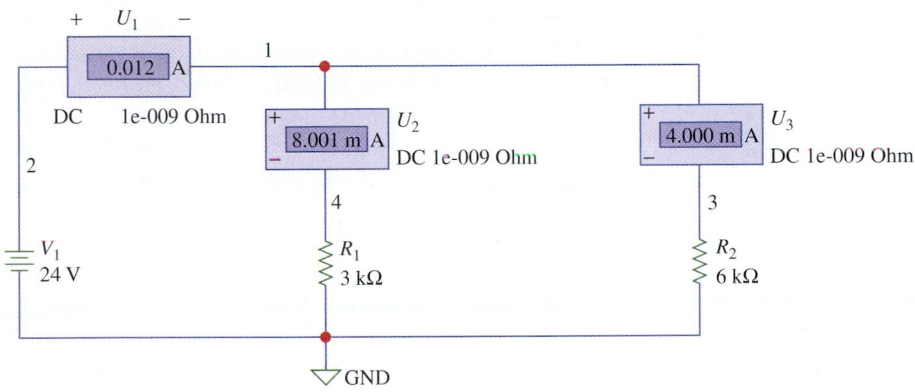

Figure 5.31
Multisim simulation of the circuit in Fig. 5.30.

Practice Problem 5.9

Refer to the circuit in Fig. 5.32. Use Multisim to find I_T and I_2.

Answer: $I_T = 14$ mA; $I_2 = 2$ mA

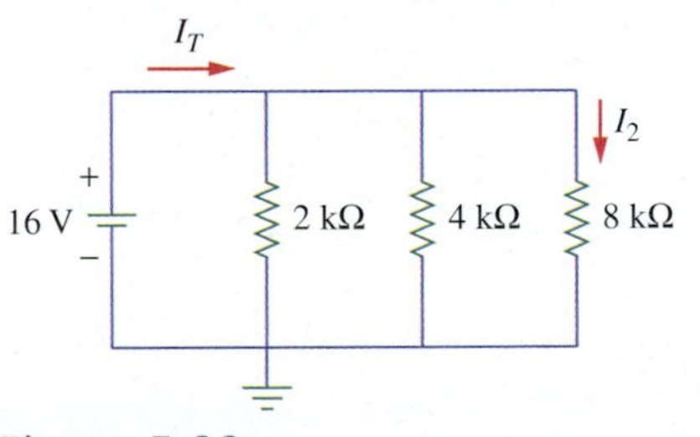

Figure 5.32
For Practice Problem 5.9.

5.8 Troubleshooting

A technologist or technician may be asked to troubleshoot a component, device, or system—from transistors to audio receivers to personal computers and beyond. Troubleshooting of any kind requires experience, thinking, sound deductive reasoning, and tracing from cause to effect.

> **Troubleshooting** is the process by which knowledge and experience are used to diagnose a defective circuit.

A circuit may be defective or may not operate properly for many reasons: the power supply may not be connected properly, a connection may be loose or open, an element may be short circuited or damaged, or a fuse may be blown. In order to troubleshoot such a defective circuit, one must have a good understanding of how the circuit is meant to work in the first place. With experience and knowledge of the basic laws of electric circuits, one can locate the cause of a defect in a given circuit. Some rules of thumb to follow include:

- Check the connections.
- Follow every line, from V{supply} to ground to −V{supply} to ground and back.
- Make sure every leg of every component is connected to the others that it is supposed to be connected to and not connected to anything to which it is not supposed to be connected.

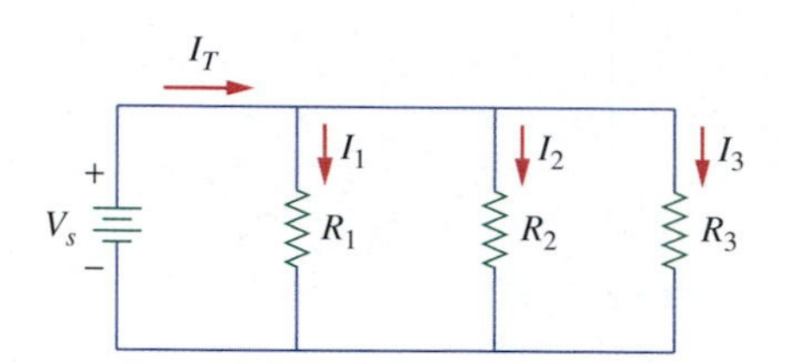

Figure 5.33
An open in a parallel circuit.

We begin with troubleshooting at a single component level. Our job is to identify the branch or element causing the trouble and replace it. There are several strategies that may be used to locate the problem in the circuit. The two easiest problems to find are opens and shorts. To check these conditions, remove the power and use the ohmmeter.

When a component in a series or parallel circuit develops an infinite resistance, this type of fault is called an *open*. The component is "burned out." One or more elements of circuit may develop an open. Consider the circuit in Fig. 5.33, which is a parallel circuit with an open because R_2 is burned out. We observe the following symptoms:

1. Because R_2 is open, $R_2 = \infty\Omega$, $I_2 = 0$ A.
2. The total current I_T will be less than normal.
3. The voltage would be normal.
4. If R_1 and R_3 are lamps, they will be on.
5. The defective device R_2 will not function. If R_2 is a lamp, it will be off.

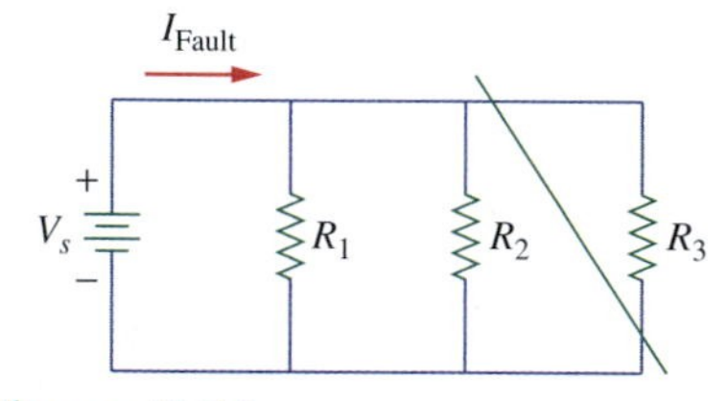

Figure 5.34
A short in a parallel circuit.

A circuit may develop another type of fault known as a *short*. Consider the parallel circuit in Fig. 5.34. There is a short that approaches 0 Ω. We observe the following symptoms on the circuit:

1. All current flows through the short, and no current flows through the other branches because current flows through the path of least resistance.
2. The total current flow will be greater than normal.
3. None of the loads will operate.

For example, for residential electrical problems, you can follow this approach. If one thing is not working, turn it on or replace it. If two or more things are not working, reset breakers and GFCIs. If they still do not work, locate the bad connection by improving connections at the interface between the live and dead portions of that circuit (if you know the circuit) or between the dead portions and any nearby live portions (if you don't know the circuit). If they still do not work, you can replace the breaker and check the neutral connections at the panel.

Although it is often easy to tell when you have a short or overloaded circuit—the lights go dead when you plug in the faulty toaster oven—it is not always simple to tell where in the system the fault has occurred. Start by turning off all wall switches and unplugging all lights and appliances. Then reset the circuit breaker or replace the blown fuse. When you have tried all the techniques that you know and yet the fault is not located, consult an experienced person.

Example 5.10

Refer to the circuit in Fig. 5.33. Calculate the currents and power during normal operation and abnormal situation. Take $V_s = 60$ V, $R_1 = 20\ \Omega$, $R_2 = 10\ \Omega$, and $R_3 = 5\ \Omega$.

Solution:
Under normal operations,

$$I_1 = \frac{V_s}{R_1} = \frac{60}{20} = 3 \text{ A}$$

$$I_2 = \frac{V_s}{R_2} = \frac{60}{10} = 6 \text{ A}$$

$$I_3 = \frac{V_s}{R_3} = \frac{60}{5} = 12 \text{ A}$$

The total current is

$$I_T = 3 + 6 + 12 = 21 \text{ A}$$

The total power is

$$P_T = V_1 I_1 + V_2 I_2 + V_3 I_3 = 60 \times 3 + 60 \times 6 + 60 \times 12 = 1260 \text{ W}$$

Under abnormal condition, R_2 becomes an open so that $I_2 = 0$ and $R = \infty$.

$$I_1 = \frac{V_s}{R_1} = \frac{60}{20} = 3 \text{ A}$$

$$I_2 = \frac{V_s}{\infty} = 0 \text{ A}$$

$$I_3 = \frac{V_s}{R_3} = \frac{60}{5} = 12 \text{ A}$$

The total current is not

$$I_T = 3 + 0 + 12 = 15 \text{ A}$$

The total power is

$$P_T = V_1 I_1 + V_2 I_2 + V_3 I_3 = 60 \times 3 + 0 + 60 \times 12 = 900 \text{ W}$$

showing the power is reduced because of the fault.

Practice Problem 5.10

Refer to Fig. 5.35. Find the current I during normal and abnormal operations. Let $V_s = 120$ V, $R_1 = 10\ \Omega$, $R_2 = 20\ \Omega$, and $R_3 = 5\ \Omega$.

Answer: normal: 3.429 A; abnormal: 4.8 A.

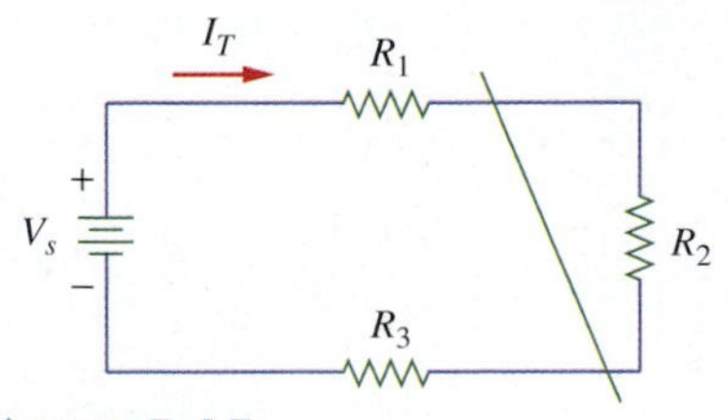

Figure 5.35
For Practice Problem 5.10.

5.9 †Applications

Parallel circuits have several applications in real life. Such applications include Christmas lighting (as we saw in Section 4.9), the lighting system of an automobile, and residential wiring. We consider two of these here.

The parallel circuit is used in power distribution, particularly in residential electrical systems. All appliances and lightbulbs are wired in parallel. Consider Fig. 5.36, indicating that electric outlets are wired

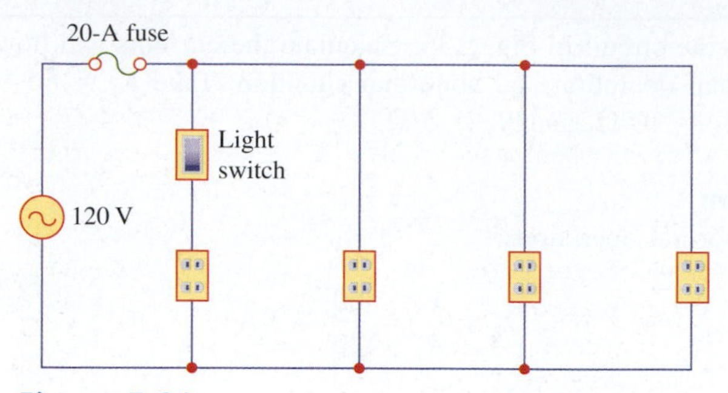

Figure 5.36
Electric outlets are wired in parallel.

in parallel. That means that the same voltage is available at all the outlets. The total current drain is the sum of the currents to all the parallel branches. Because of the dangers of electricity, house wiring is carefully regulated by a code drawn by local ordinances and by the National Electrical Code (NEC). To avoid trouble, insulation, grounding, fuses, and circuit breakers are used. Modern wiring codes require a third wire for a separate ground. The ground wire does not carry power like the neutral wire but enables appliances to have a separate ground connection.

Another common use of parallel circuits is found in the lighting system of an automobile. A simplified version of the system is shown in Fig. 5.37. Because of the parallel arrangement, when one headlight goes out (opens), it does not affect other lights. The electrical system of an automobile is run as a dc system. The 12-V charged battery is expected to provide the necessary dc for the entire system. The electrical system is basically a parallel system. Each component is connected to the battery and to the ground or chassis.

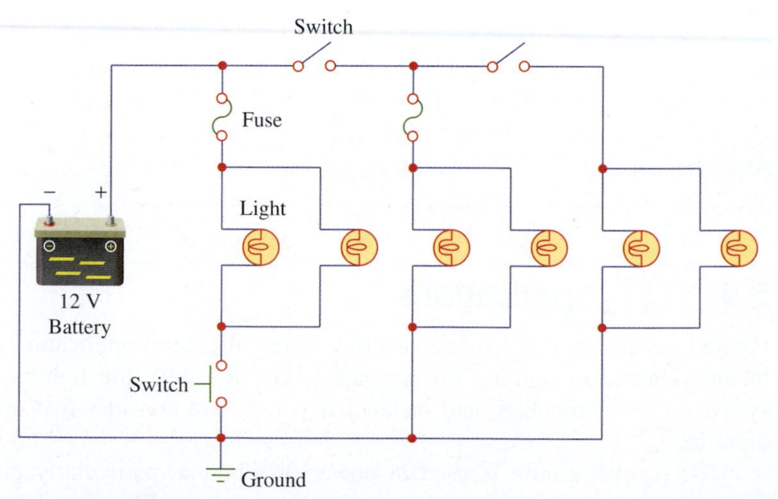

Figure 5.37
Simplified form of an automobile lighting system.

5.10 Summary

1. Two elements are in parallel if they are connected to the same two nodes. Elements in parallel always have the same voltage across them.
2. Kirchhoff's current law (KCL) states that currents at any node algebraically sum to zero. In other words, the sum of the currents entering a node equals the sum of the currents leaving the node.
3. Current sources in parallel add algebraically.
4. When two resistors R_1 $(=1/G_1)$ and R_2 $(=1/G_2)$ are in parallel, their equivalent resistance R_{eq} and equivalent conductance G_{eq} are

$$R_{\text{eq}} = \frac{R_1 R_2}{R_1 + R_2}, \qquad G_{\text{eq}} = G_1 + G_2$$

5. The current divider rule for two resistors in parallel is

$$I_1 = \frac{R_2}{R_1 + R_2} I, \qquad I_2 = \frac{R_1}{R_1 + R_2} I$$

   If a current divider has N conductors $(G_1, G_2, \ldots, G_n)$ in parallel with the source current I, the kth conductor (G_k) will have current

$$I_k = \frac{G_k}{G_{\text{eq}}} I$$

6. Troubleshooting is the process of locating a fault in a defective circuit.
7. Two applications of parallel circuits considered in this chapter are residential wiring system and automobile lighting system.

Review Questions

5.1 When resistors are connected in parallel, they have the same:

(a) current (b) voltage
(c) power (d) resistance

5.2 Voltage sources of different voltages can be placed in parallel.

(a) True (b) False

5.3 The value of R_{eq} is always less than the smallest resistor in a parallel circuit.

(a) True (b) False

5.4 Three resistors 240 Ω, 560 Ω, and 100 Ω are connected in parallel. The equivalent resistance is approximately:

(a) 900 Ω (b) 63 Ω
(c) 56 Ω (d) 22 Ω

5.5 The resistors R_1 and R_2 in Fig. 5.38 are in parallel.

(a) True (b) False

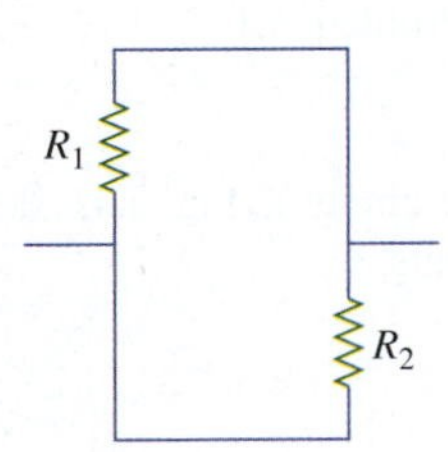

Figure 5.38
For Review Question 5.5.

5.6 When one of the resistors in parallel is short-circuited, the total resistance:

(a) doubles (b) increases
(c) increases (d) is zero

5.7 The current I_o in Fig. 5.39 is:

(a) −4 A (b) −2 A
(c) 4 A (d) 16 A

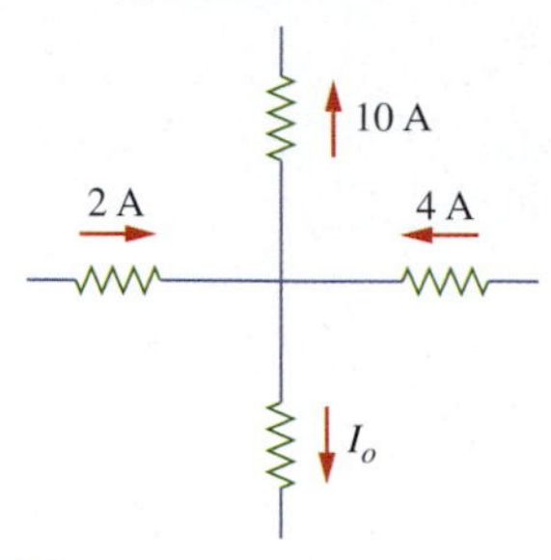

Figure 5.39
For Review Question 5.7.

5.8 The current divider principle is used when resistors are in parallel.

(a) True (b) False

5.9 If resistors 40 Ω, 60 Ω, 80 Ω, and 100 Ω are connected in parallel to a current source, the resistor that has the least current through it is:

(a) 40 Ω (b) 60 Ω
(c) 80 Ω (d) 100 Ω

5.10 Two resistors 40 Ω and 60 Ω are connected in parallel to a current source of 10 mA. The current that flows through 40 Ω is:

(a) 10 mA (b) 6 mA
(c) 4 mA (d) 0 mA

Answers: 5.1b, 5.2b, 5.3a, 5.4b, 5.5a, 5.6d, 5.7a, 5.8a, 5.9d, 5.10b

Problems

Section 5.3 Kirchhoff's Current Law

5.1 In Fig. 5.40, let $I_1 = -12$ A, $I_2 = 3$ A, and $I_4 = 5$ A. Find I_3.

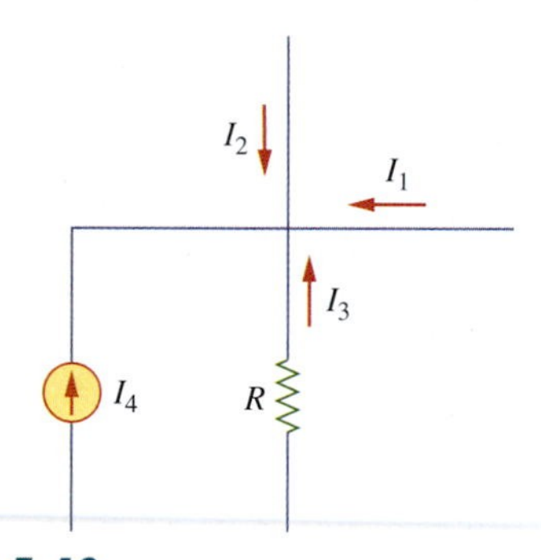

Figure 5.40
For Problem 5.1.

5.2 In the circuit in Fig. 5.41, determine the unknown currents.

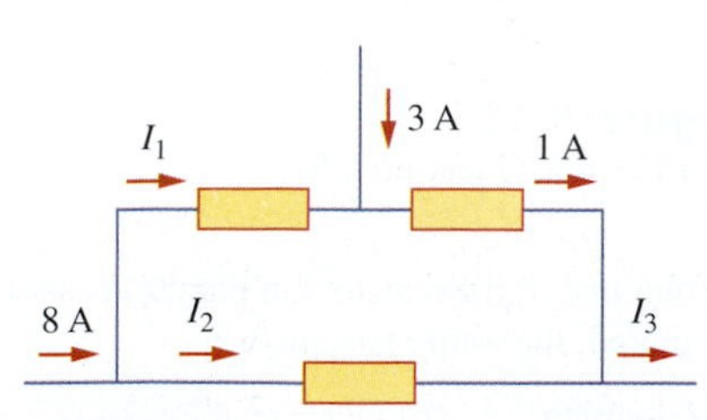

Figure 5.41
For Problem 5.2.

5.3 Use KCL to find the unknown currents I_1, I_2, and I_3 in the circuit of Fig. 5.42.

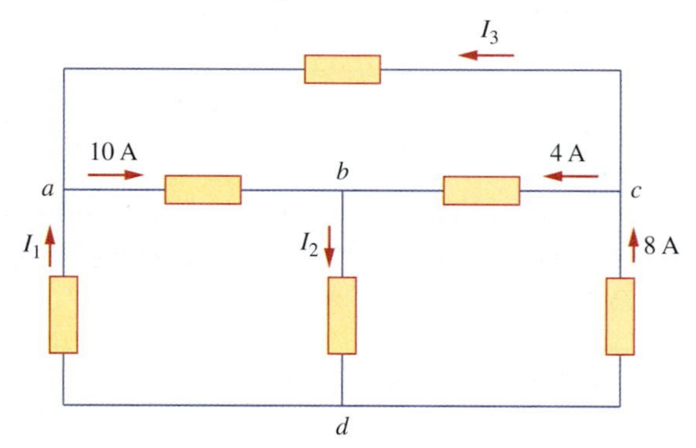

Figure 5.42
For Problem 5.3.

5.4 Find I_3 and I_4 in the circuit of Fig. 5.43.

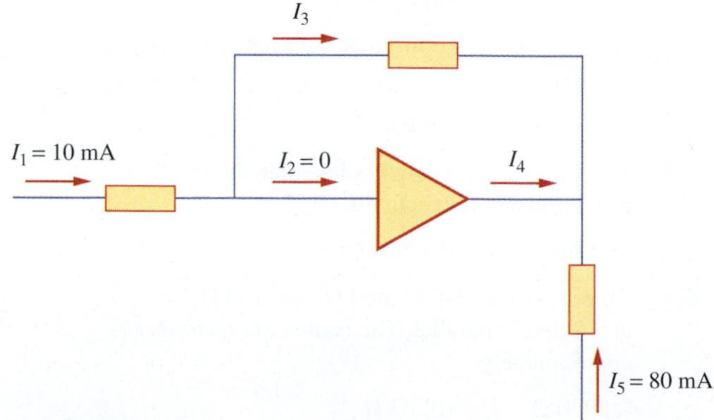

Figure 5.43
For Problem 5.4.

5.5 Determine I_1 and I_2 in the circuit of Fig. 5.44.

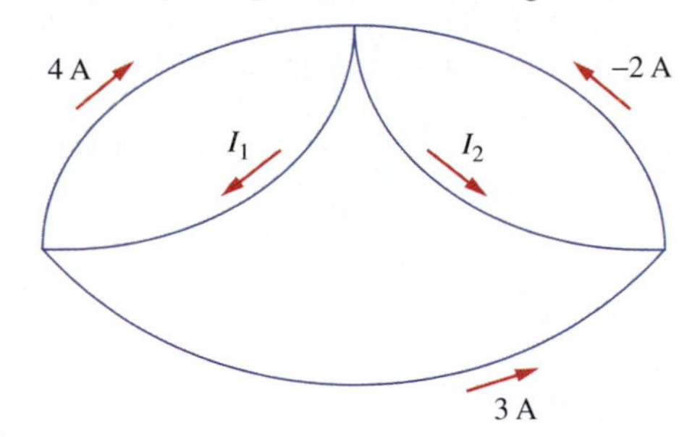

Figure 5.44
For Problem 5.5.

5.6 Find I_o in Fig. 5.45.

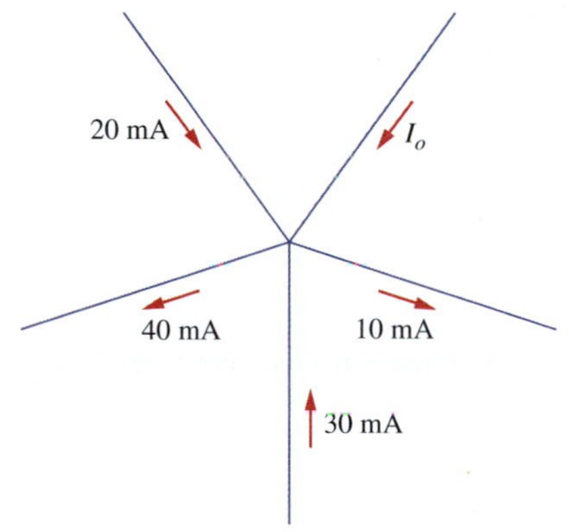

Figure 5.45
For Problem 5.6.

5.7 Find the current I_1 in the circuit of Fig. 5.46.

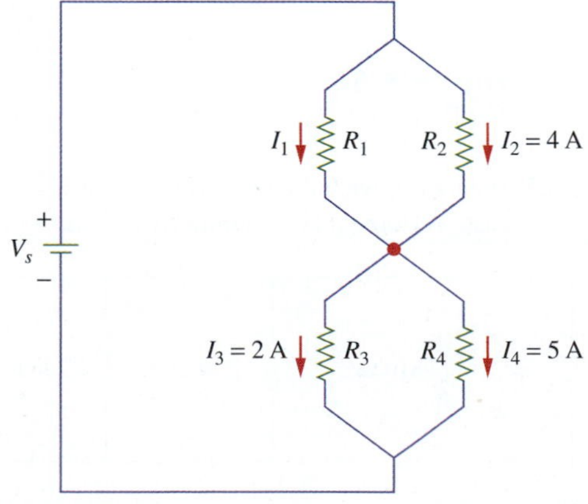

Figure 5.46
For Problem 5.7.

5.8 In Fig. 5.47, let $I_1 = 4$ A, $I_2 = 3$ A, and $I_3 = -12$ A. Find I_4, I_5, and V.

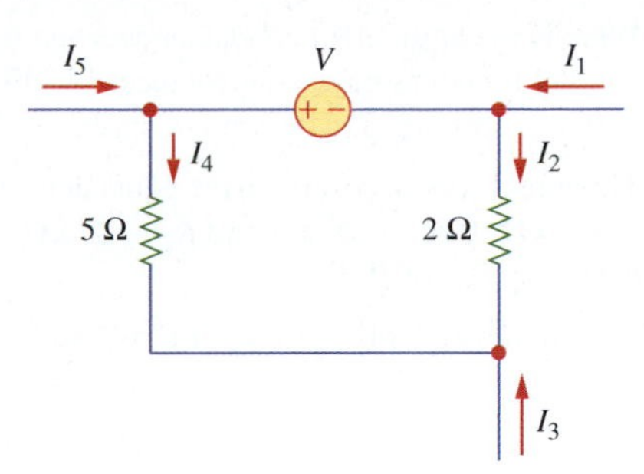

Figure 5.47
For Problem 5.8.

Section 5.4 Current Sources in Parallel

5.9 Find the current through the resistor in Fig. 5.48.

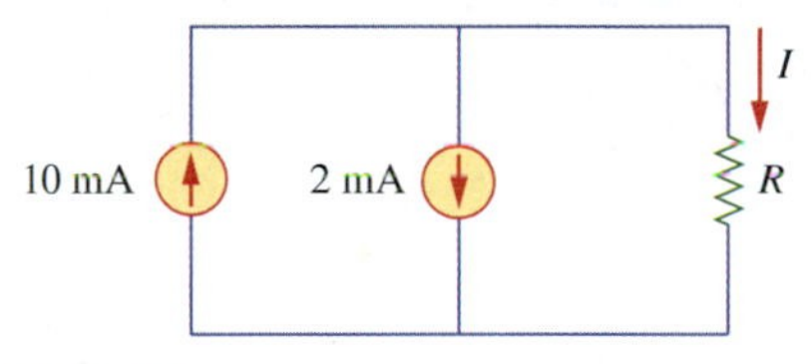

Figure 5.48
For Problem 5.9.

5.10 Refer to Fig. 5.49. Find the current through the resistor.

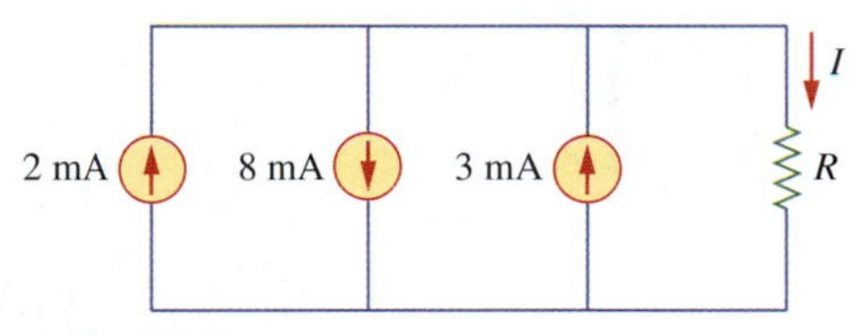

Figure 5.49
For Problem 5.10.

Section 5.5 Resistors in Parallel

5.11 Three lamps each having a resistance of 30 Ω are connected in parallel. Find the total resistance of the lamps.

5.12 Two 50-Ω resistors are connected in parallel. Find the total resistance.

5.13 Find the total resistance of each of the following groups of parallel resistors.

(a) 56 Ω and 82 Ω

(b) 12 kΩ, 36 kΩ, and 75 kΩ

(c) 1.2 MΩ, 5.6 MΩ, and 680 kΩ

5.14 What are the total resistance and total conductance of ten 24-kΩ resistors in parallel?

5.15 What is the equivalent conductance when four parallel branches have conductances of 750 mS, 640 mS, 480 mS, and 300 mS, respectively?

5.16 Three resistors in parallel have equivalent resistance of 4.2 kΩ. If $R_1 = 20$ kΩ and $R_2 = 25$ kΩ, what is the resistance of R_3?

5.17 Determine R_T for the circuit of Fig. 5.50.

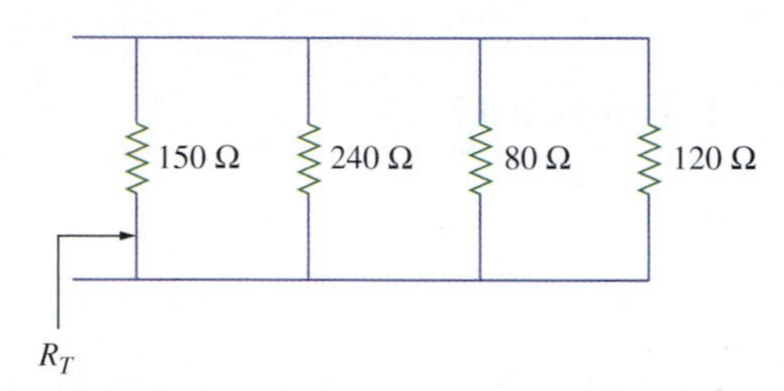

Figure 5.50
For Problem 5.17.

5.18 Find G_{eq} for the circuit of Fig. 5.51.

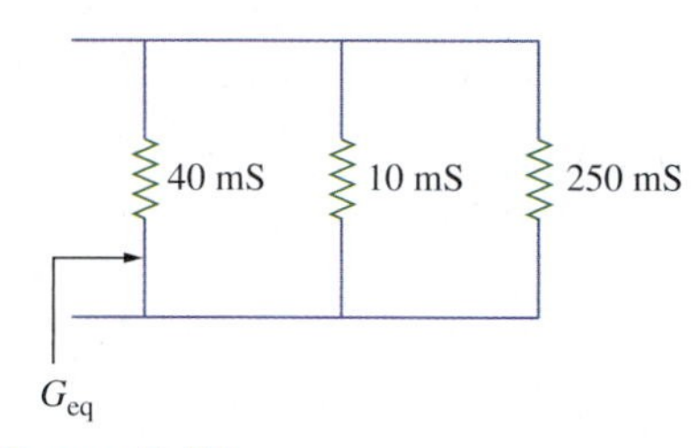

Figure 5.51
For Problem 5.18.

5.19 Find the resistance R_x in the circuit of Fig. 5.52.

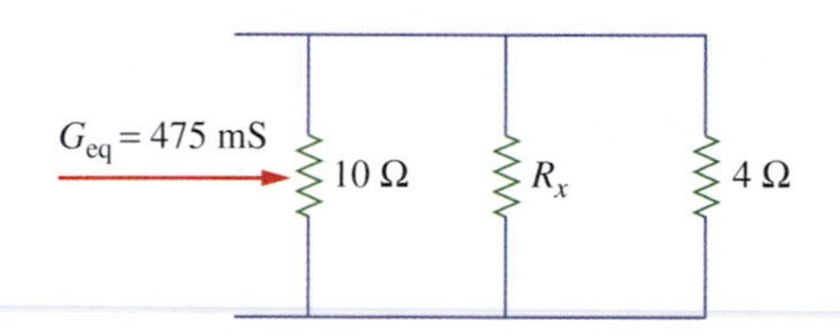

Figure 5.52
For Problem 5.19.

5.20 Calculate R_{eq} and G_{eq} for the circuit shown in Fig. 5.53.

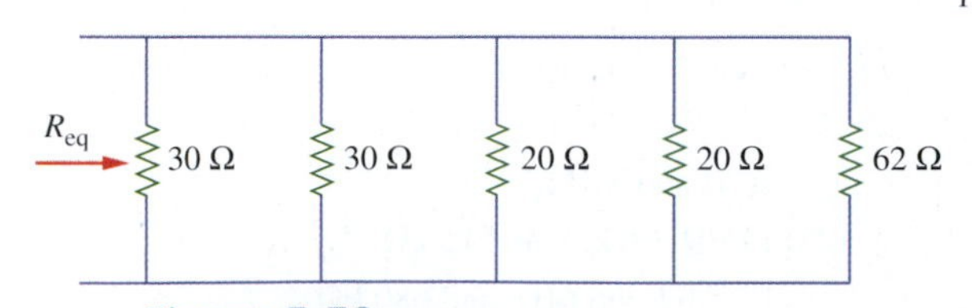

Figure 5.53
For Problem 5.20.

5.21 Find R_T in the circuit of Fig. 5.54.

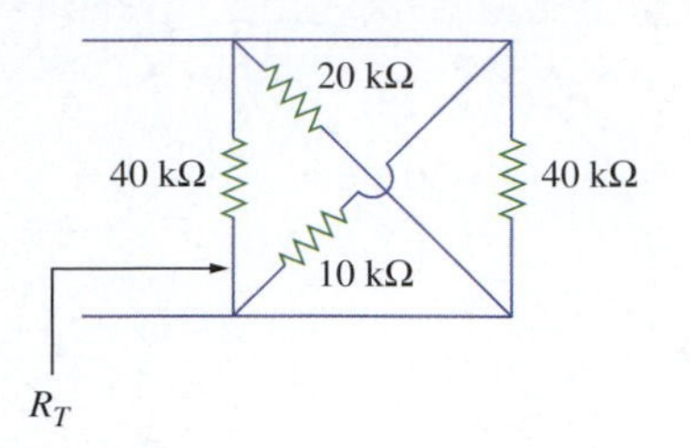

Figure 5.54
For Problem 5.21.

5.22 Determine the equivalent resistance of the circuit in Fig. 5.55.

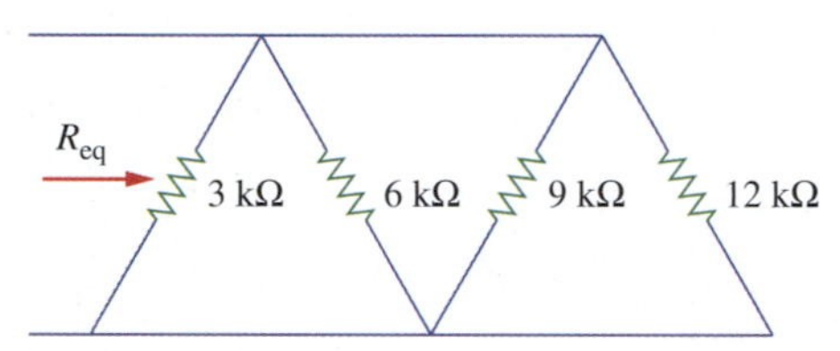

Figure 5.55
For Problem 5.22.

5.23 Four 1.2-kΩ resistors and two 300-Ω resistor are connected in parallel. Find the equivalent resistance.

5.24 Determine the value of R_L in Fig. 5.56 that will (a) reduce the terminal voltage of the current source to 8V and (b) draw one-quarter of the source current.

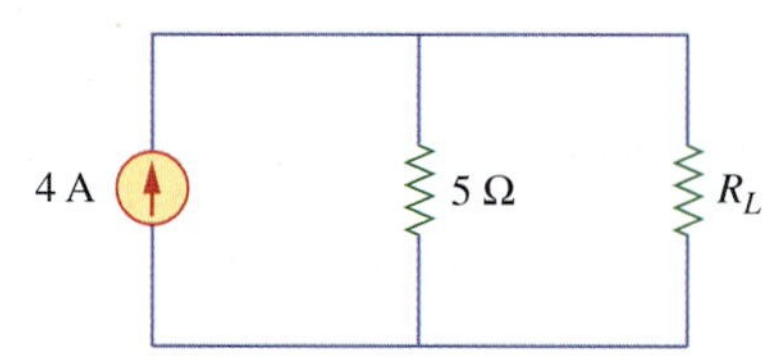

Figure 5.56
For Problem 5.24.

5.25 Figure out what the two ammeters in Fig. 5.57 will read. Assume that the ammeter resistances are zero.

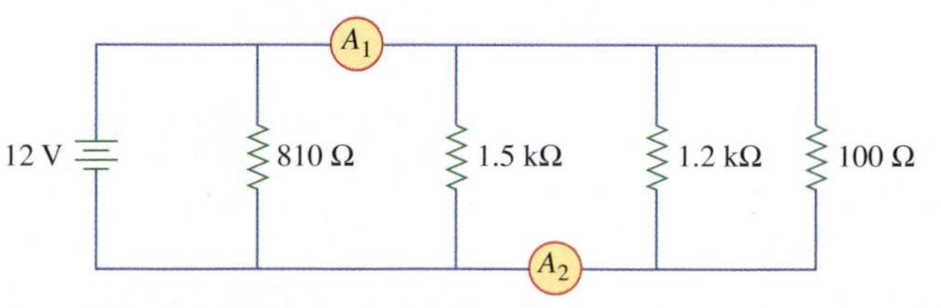

Figure 5.57
For Problem 5.25.

5.26 Two conductances $G_1 = 750$ μS and $G_2 = 500$ μS are connected in parallel. The parallel combination is connected in parallel with a 20-Ω resistor. Determine the total equivalent resistance of the parallel circuit.

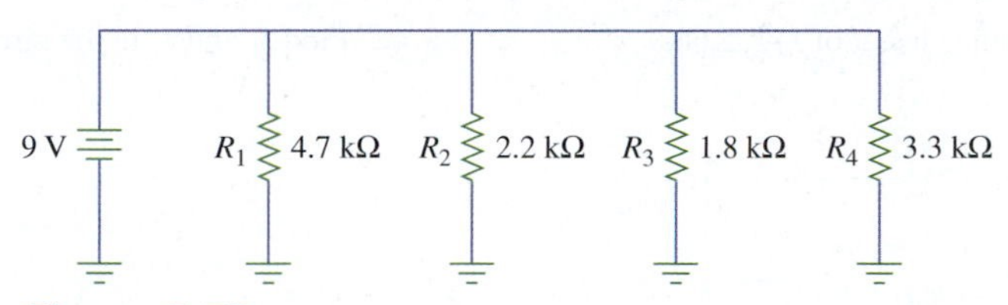

Figure 5.58
For Problem 5.27.

5.27 Determine the branch current and the power absorbed by each resistor in Fig. 5.58.

5.28 Determine the current through each resistor in Fig. 5.59 and the total battery current.

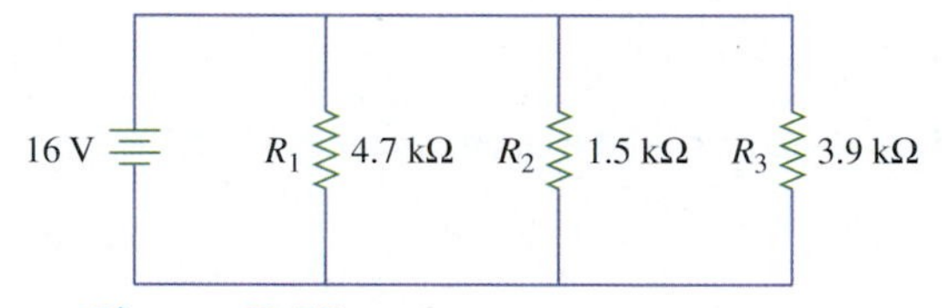

Figure 5.59
For Problem 5.28.

5.29 Three 0.7-A lamps are connected in parallel across a 120-V source. Determine (a) the total current, (b) the total voltage, and (c) the total power consumed by the lamps.

5.30 Find R_T in the circuit of Fig. 5.60.

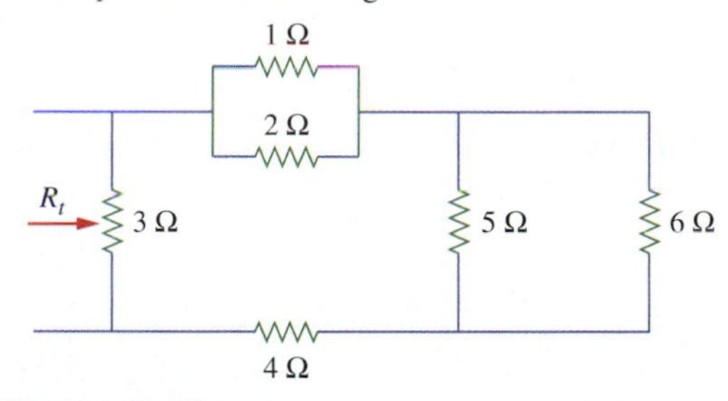

Figure 5.60
For Problem 5.30.

5.31 An ideal voltmeter is connected to a parallel circuit as shown in Fig. 5.61. Find (a) the voltage measured by the voltmeter, (b) the power absorbed by each of the two resistors, and (c) the power supplied by the current source.

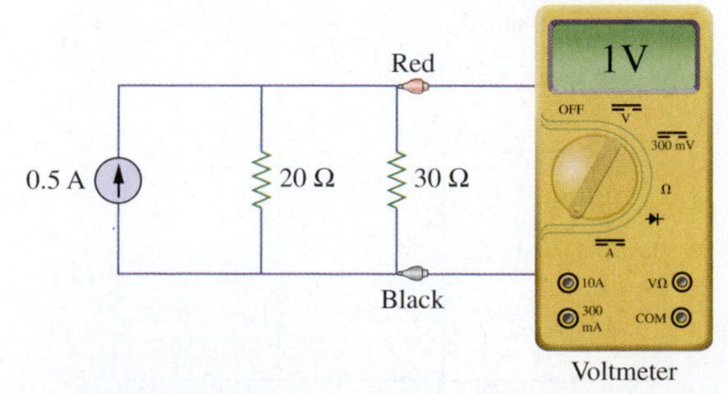

Figure 5.61
For Problem 5.31.

5.32 For the circuit in Fig. 5.62, (a) determine the resistance R when $I_R = 2$ A and (b) determine R when it absorbs a power of 700 W.

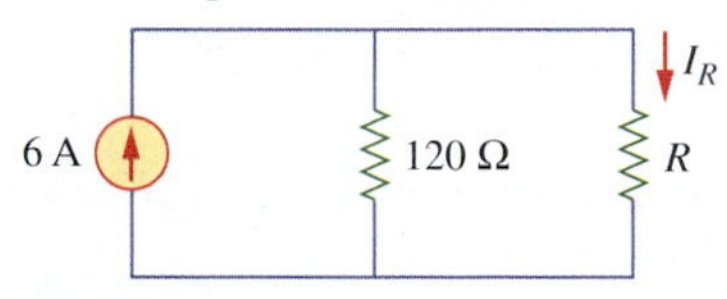

Figure 5.62
For Problem 5.32.

5.33 In the parallel circuit of Fig. 5.63, determine the value of R such that $I = 3$ A. Also, determine the terminal voltage V.

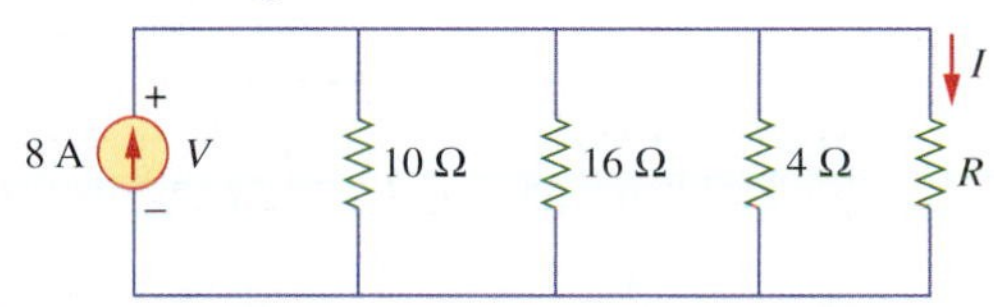

Figure 5.63
For Problem 5.33.

5.34 Determine the resistance that must be connected in parallel with a 20-Ω resistor to provide a total resistance of 8 Ω.

Section 5.6 Current Dividers

5.35 Find I_1 and I_2 in the circuit of Fig. 5.64.

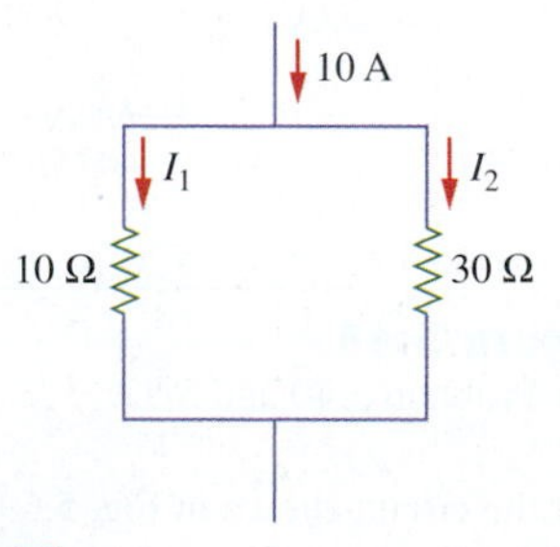

Figure 5.64
For Problem 5.35.

5.36 Determine I_1, I_2, and I_3 in the circuit of Fig. 5.65.

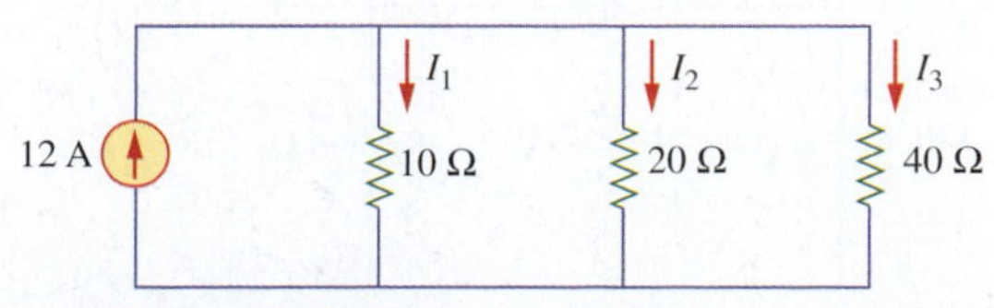

Figure 5.65
For Problems 5.36 and 5.47.

5.37 Find I_1, I_2, and I_3 in the circuit of Fig. 5.66.

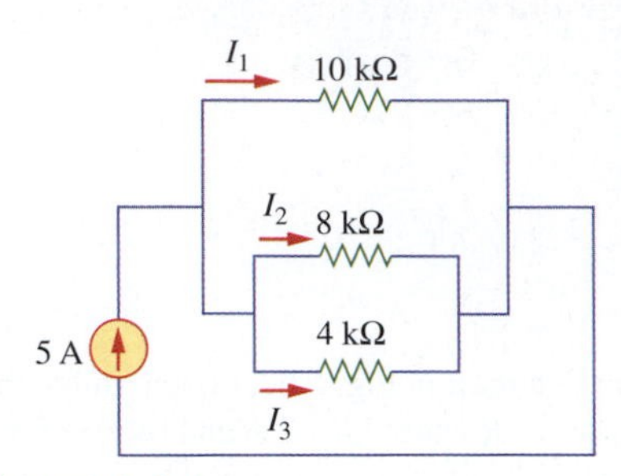

Figure 5.66
For Problems 5.37 and 5.48.

5.38 Determine the current through each resistor in the circuit of Fig. 5.67.

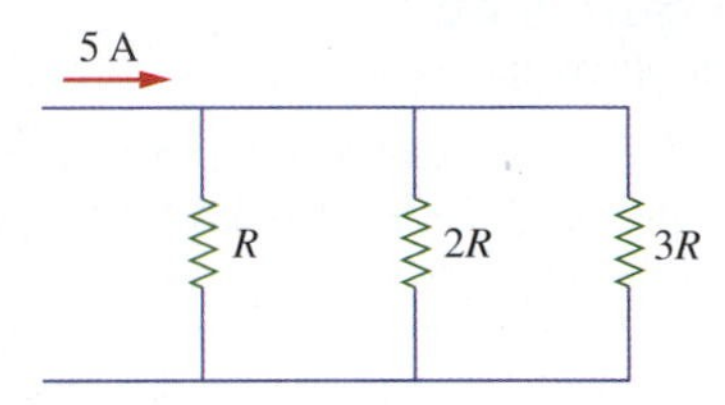

Figure 5.67
For Problem 5.38.

5.39 Three conductances 100 mS, 300 mS, and 600 mS connected in parallel draw a total current of 250 mA. Calculate the current through each conductance.

5.40 How much current is indicated by two ammeters inserted in the circuit in Fig. 5.68?

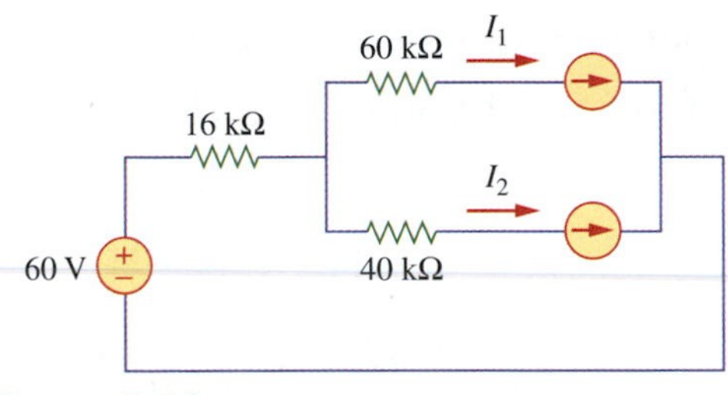

Figure 5.68
For Problem 5.40 and 5.52.

5.41 For the circuit shown in Fig. 5.69, determine (a) G_T and (b) R_T, I_T, and P_3.

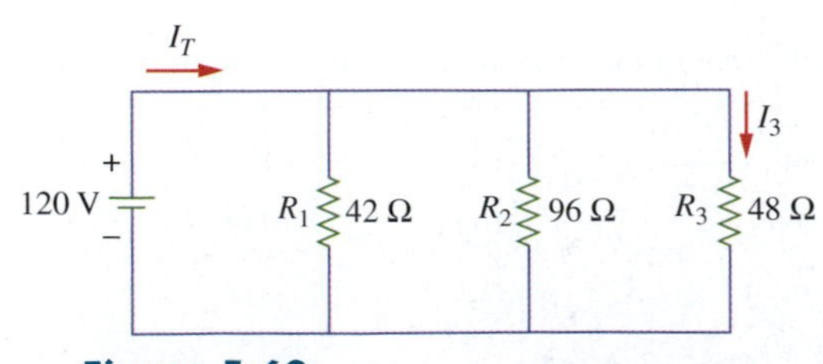

Figure 5.69
For Problem 5.41.

5.42 Find I_1 and I_2 in the circuit shown in Fig. 5.70.

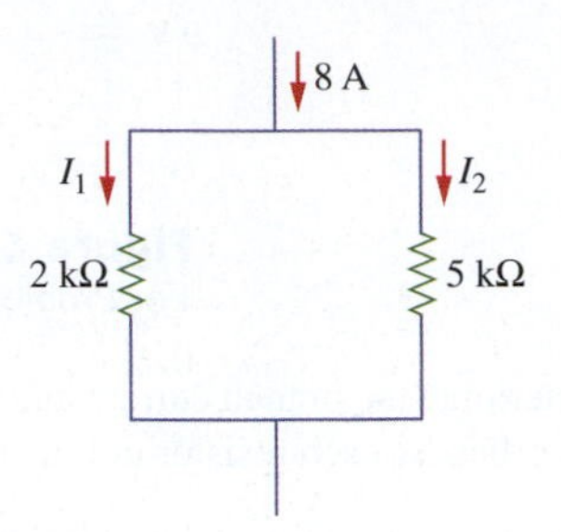

Figure 5.70
For Problem 5.42.

5.43 Find I_1, I_2, and I_3 in the circuit of Fig. 5.71.

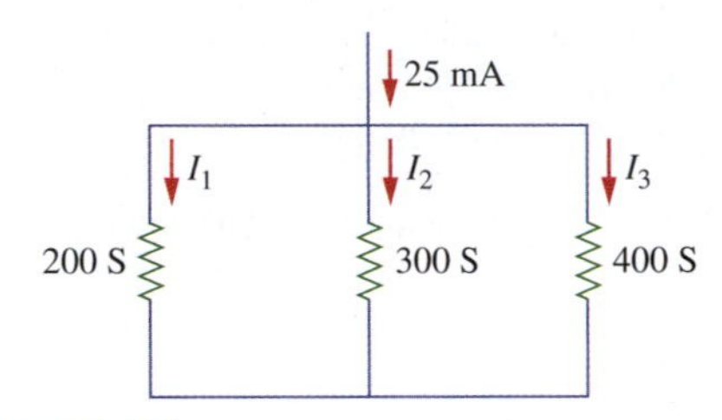

Figure 5.71
For Problem 5.43.

5.44 Given the circuit in Fig. 5.72, find (a) V_s and (b) I_x.

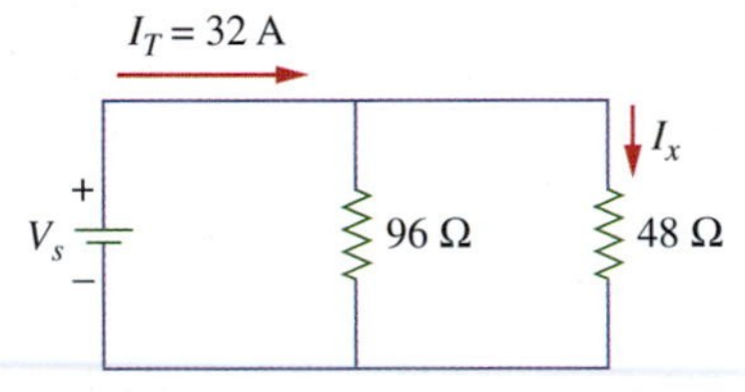

Figure 5.72
For Problem 5.44.

5.45 Consider the circuit in Fig. 5.73; find (a) R_T, (b) G_T, and (c) I_T and I_x.

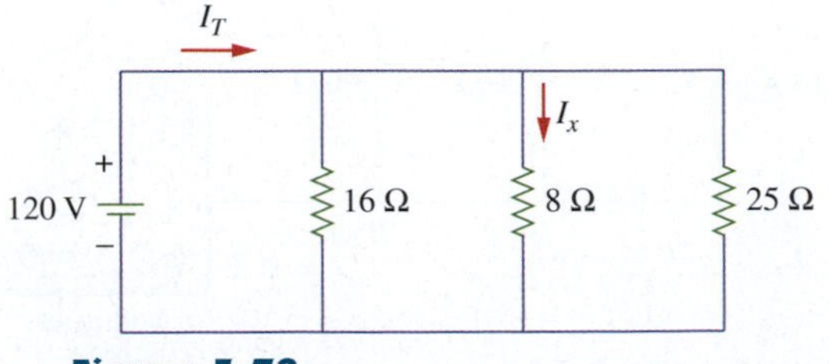

Figure 5.73
For Problem 5.45.

5.46 Find the current each ammeter in Fig. 5.74 will indicate.

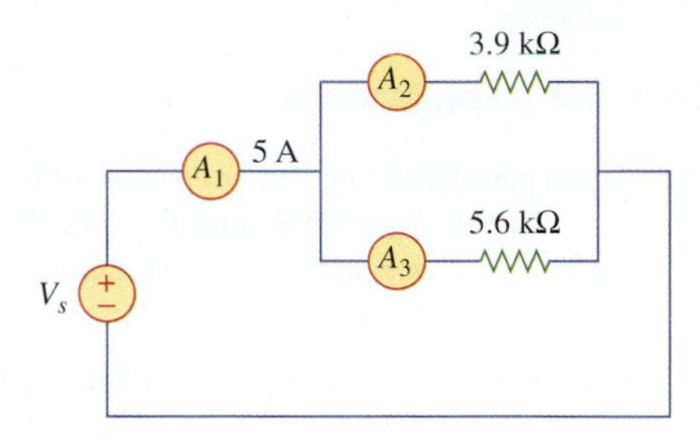

Figure 5.74
For Problem 5.46.

Section 5.7 Computer Analysis

5.47 Use PSpice to find the currents I_1 to I_3 in Fig. 5.65.

5.48 Rework Problem 5.37 using PSpice.

5.49 Use Multisim to find current I_T in the circuit of Fig. 5.75.

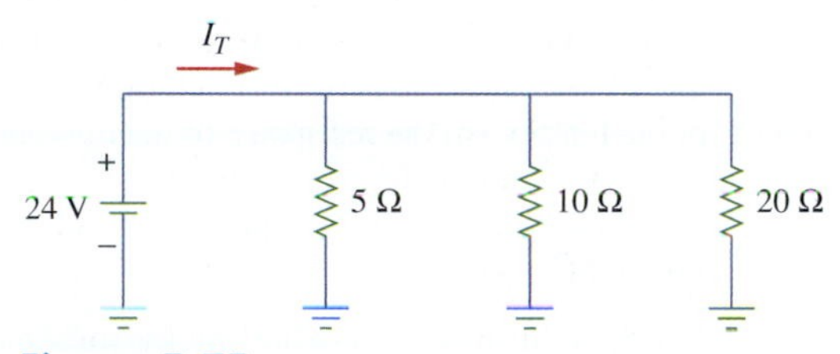

Figure 5.75
For Problem 5.49.

5.50 Find I_1 through I_5 in the circuit of Fig. 5.76 using Multisim.

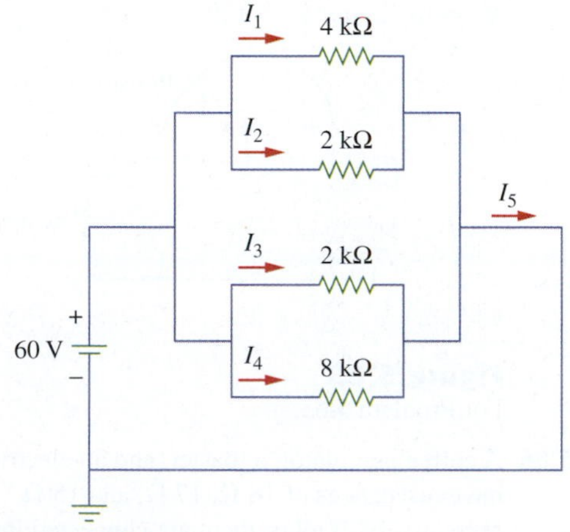

Figure 5.76
For Problem 5.50.

5.51 Use Multisim to find I_1 and I_2 in Fig. 5.3 (see Example 5.1).

5.52 Find I_1 and I_2 in the circuit of Fig. 5.68 using Multisim.

5.53 Refer to the circuit in Fig. 5.77. Use Multisim to determine the current through each resistor.

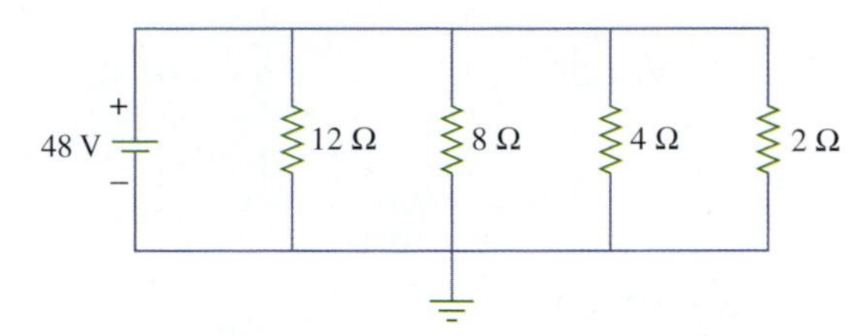

Figure 5.77
For Problem 5.53.

Section 5.8 Troubleshooting

5.54 In Fig. 5.78, assume that R_3 is open. What are the values of V_1, V_2, V_3, and V_4?

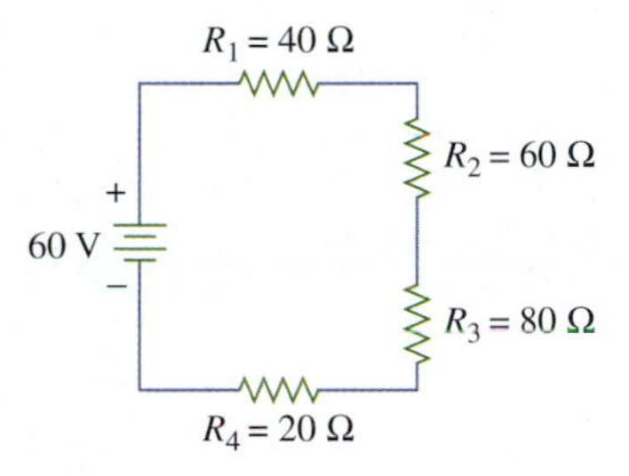

Figure 5.78
For Problems 5.54 and 5.55.

5.55 Repeat Problem 5.54 for the case in which R_3 is shorted.

5.56 Find I_o and V_{ab} in the circuit of Fig. 5.79.

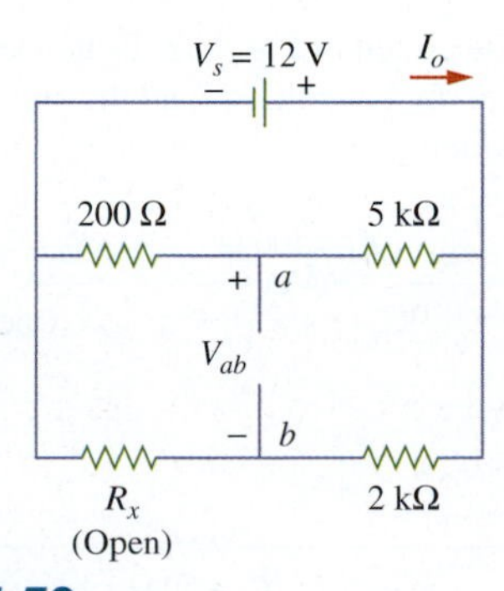

Figure 5.79
For Problem 5.56.

5.57 Voltage measurements are taken on the circuit in Fig. 5.80. The following table shows the measurements taken. For each row, identify which component is defective and the type of defect.

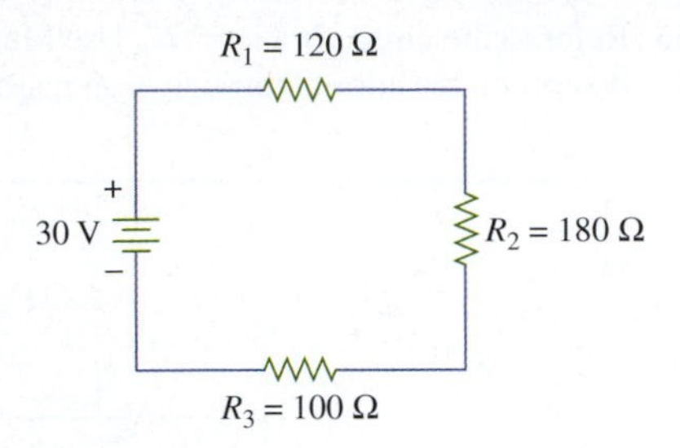

Figure 5.80
For Problem 5.57.

	V_1	V_2	V_3
Nominal	**9**	**13.5**	**7.5**
Trouble 1	0	30	0
Trouble 2	30	0	0
Trouble 3	0	0	30

5.58 Consider the circuit in Fig. 5.81. Find the current I under the normal condition and abnormal condition when R_2 is shorted.

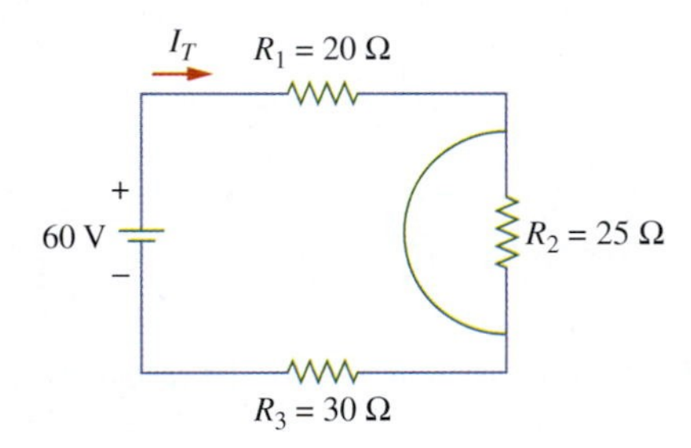

Figure 5.81
For Problem 5.58.

5.59 In the circuit of Fig 5.82, R_2 is open. Calculate I_T for the normal conditions and the abnormal (open R_2) situations.

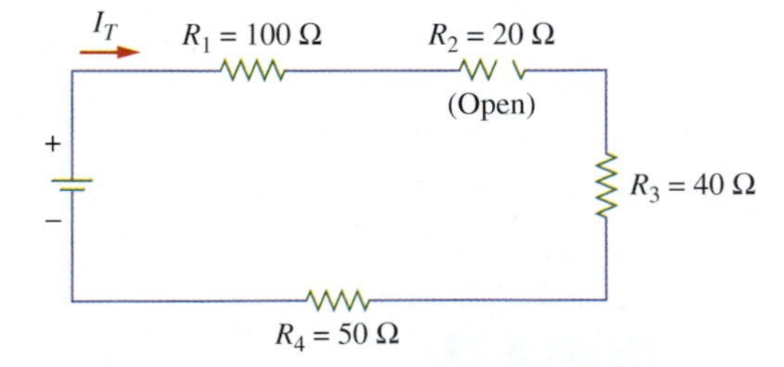

Figure 5.82
For Problem 5.59.

5.60 What technique is necessary in troubleshooting a defective series circuit that has a number of series branches?

Section 5.9 Applications

5.61 Three parallel-connected lamps use a 110-V supply: $P_1 = 120$ W, $P_2 = 80$ W, and $P_3 = 45$ W. Determine the power of one additional lamp that may be connected if the supply current must not exceed 4 A.

5.62 Four lamps are connected in parallel to a 110-V supply. The lamps are rated $P_1 = 120$ W, $P_2 = 80$ W, $P_3 = 60$ W, and $P_4 = 40$ W. Find the current through each lamp.

5.63 An electrical designer designs a flashlight by connecting three bulbs of resistances 7 Ω, 6 Ω, and 5 Ω in parallel with a 9-V voltage source. Determine the current through each of the lamps. Find the total current.

5.64 A wall receptacle of a house supplies 120 V. A toaster and a lightbulb are connected in parallel with the voltage supply using the wall receptacle. Assume the rated power of the toaster is 640 W and that of the light bulb is 62 W. Find: (a) the resistance of the toaster, (b) the resistance of the lightbulb, (c) the total resistance as seen by the voltage source, and (d) the current through the toaster and the lightbulb.

5.65 Fig. 5.83 illustrates a cooking unit for some electric grills. The cooking unit consists of two resistive heating elements and a special switch that connects them to the voltage source individually, in series, or in parallel. Determine the total maximum and minimum heating power when $R_1 = 10\ \Omega$, $R_2 = 14\ \Omega$, and $V_s = 120$ V.

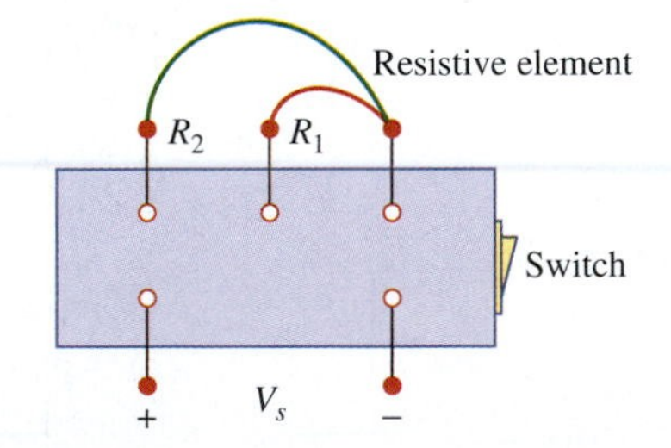

Figure 5.83
For Problem 5.65.

5.66 A coffee percolator, a toaster, and an electric iron have resistances of 18 Ω, 17 Ω, and 15 Ω, respectively. If all of them are connected in parallel across a 120-V line, determine (a) the total resistance and (b) the total current.

chapter 6

Series-Parallel Circuits

A wise man makes his own decisions, an ignorant man follows the public opinion.

—Chinese Proverb

Enhancing Your Career

Career in Electronics

An important area where electric circuits analysis is applied is electronics. The term *electronics* was originally used to distinguish circuits of very low current levels. This distinction no longer holds because power semiconductor devices operate at high levels of current. Today, electronics is regarded as the study of the behavior and effects of electrons in useful devices. It involves the motion of charges in a gas, vacuum, or semiconductors. Modern electronics involves transistors and transistor circuits. Electronic circuits used to be assembled from components. Many electronic circuits are now produced as integrated circuits, fabricated in a semiconductor substrate or chip.

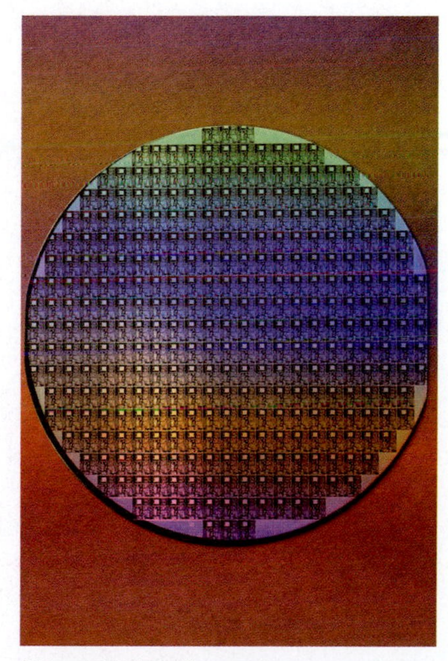

300-mm (12-in.) silicon wafer
© Corbis RF

Electronic circuits find applications in many areas such as automation, control, broadcasting, computers, and instrumentation. The range of devices that use electronic circuits is enormous and is limited only by our imagination. Such devices include radio, television, computers, and stereo systems to mention but a few.

An electrical engineering technologist usually performs diverse functions and is likely to use, design, or construct systems that incorporate some form of electronic circuits. Therefore, an understanding of the operation and analysis of electronics is essential to the electrical engineering technologist. Electronics has become a specialty distinct from other disciplines within electrical engineering technology. Because the field of electronics is ever-advancing, an electronics engineering technologist must update his/her knowledge from time to time through additional schooling, seminars, and by taking online classes. Another way to achieve this is by being a member of a professional organization such as Institute of Electrical and Electronics Engineers (IEEE), American Society of Certified Engineering Technicians (ASCET), the National Institute for Certification in Engineering Technologies (NICET), the Applied Science Technologists and Technicians of British Columbia (ASTTBC), and the Instrumentation, Systems, and Automation (ISA) Society. A member benefits immensely from the numerous magazines, journals, transactions, and conference/symposium proceedings published by such organizations. You should consider becoming a member of at least one of these organizations.

6.1 Introduction

Having mastered the series circuit and the parallel circuit, we can move on to series-parallel circuits, which are generally more complicated.

> A **series-parallel circuit** is a circuit that contains both series and parallel circuit topologies.

A series-parallel circuit is one that combines some of the attributes of both series and parallel circuit configurations. A typical example of a series-parallel circuit is shown in Fig. 6.1. You will notice that R_1 and R_2 are in series and so are R_3 and R_4, while R_5 and R_6 are in parallel. Although series-parallel circuits are generally more complicated than the series and parallel circuits, the same principles apply. We apply Ohm's law along with KVL and KCL to analyze series-parallel circuits.

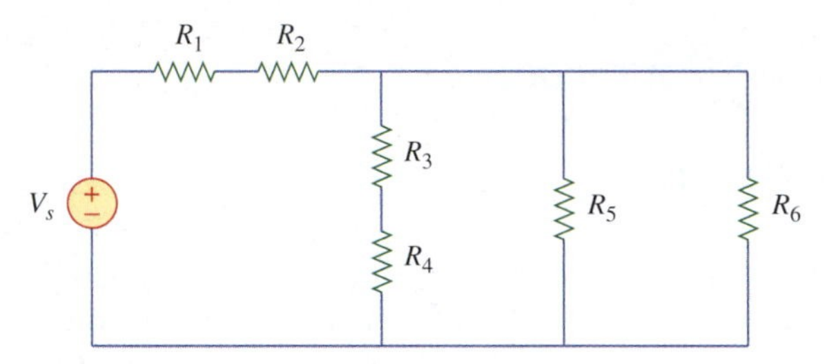

Figure 6.1
A typical series-parallel circuit.

We begin the chapter by looking at typical series-parallel circuits. We then discuss the ladder network and the Wheatstone bridge. Finally, we learn how to use PSpice and Multisim to analyze series-parallel circuits using a computer.

6.2 Series-Parallel Circuits

To analyze a series-parallel circuit, we resolve each purely series group into its equivalent resistance and resolve each parallel group into its equivalent resistance. We apply this process as many times as necessary. In other words, we substitute the equivalent resistance for various series or parallel portions of the circuit until the original circuit is reduced to a simple series or parallel circuit. Thus, the analysis of a series-parallel circuit involves using a combination of the following principles:

- Combination of resistors in series.
- Combination of resistors in parallel.
- Kirchhoff's voltage law (KVL).
- Kirchhoff's current law (KCL).
- Ohm's law.
- Voltage division principle.
- Current division principle.

We illustrate this with examples.

Example 6.1

Find I_1, I_2, and I_3 in the circuit in Fig. 6.2. Calculate the power absorbed by each resistor.

Solution:
First notice that R_2 and R_3 are in parallel because they are connected to the same two points. Their equivalent resistance is

$$R_{eq} = R_2 \| R_3 = \frac{R_2 R_3}{R_2 + R_3} = \frac{36 \times 72}{36 + 72} = 24\ \Omega$$

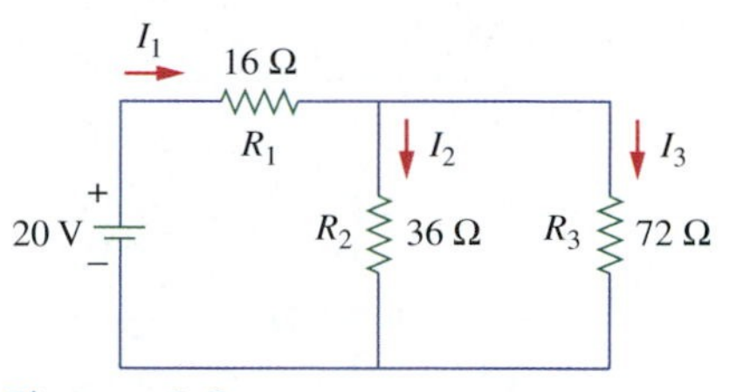

Figure 6.2
For Example 6.1.

(The symbol $\|$ is used to indicate a parallel combination.). By replacing R_2 and R_3 with R_{eq}, the equivalent circuit becomes that shown in Fig. 6.3. This is now a series circuit. We can readily apply KVL to the circuit.

$$-20 + I_1(16 + 24) = 0 \quad \Rightarrow \quad I_1 = 20/40 = 0.5\ \text{A}$$

Once we obtain I_1, we apply current division principle to obtain I_2 and I_3. Looking at the parallel portion of Fig. 6.2,

$$I_2 = \frac{R_3}{R_2 + R_3} I_1 = \frac{72}{36 + 72}(0.5) = 333.3\ \text{mA}$$

$$I_3 = \frac{R_2}{R_2 + R_3} I_1 = \frac{36}{36 + 72}(0.5) = 166.7\ \text{mA}$$

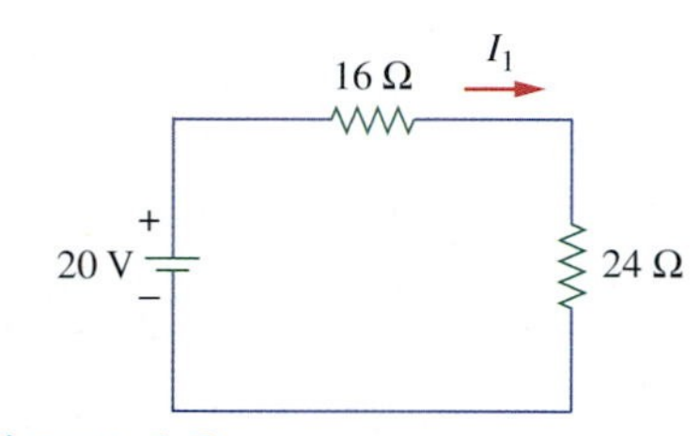

Figure 6.3
For Example 6.1.

Once we know the currents through the branches, we can find the power absorbed by them.

$$P_1 = I_1^2 R_1 = (0.5)^2 16 = 4\ \text{W}$$
$$P_2 = I_2^2 R_2 = (0.3333)^2 36 = 4\ \text{W}$$
$$P_3 = I_3^2 R_3 = (0.1667)^2 72 = 2\ \text{W}$$

Hence, the total absorbed power is 10 W. To check this result, we can calculate the total power supplied as

$$P_s = V_s I_1 = 20 \times 0.5 = 10\ \text{W}$$

Practice Problem 6.1

For the circuit in Fig. 6.4, find I_1, I_2, and I_3 and calculate the power absorbed by each resistor.

Answer: 1.2 A; 0.8 A; 2 A; 57.6 W; 38.4 W; 104 W

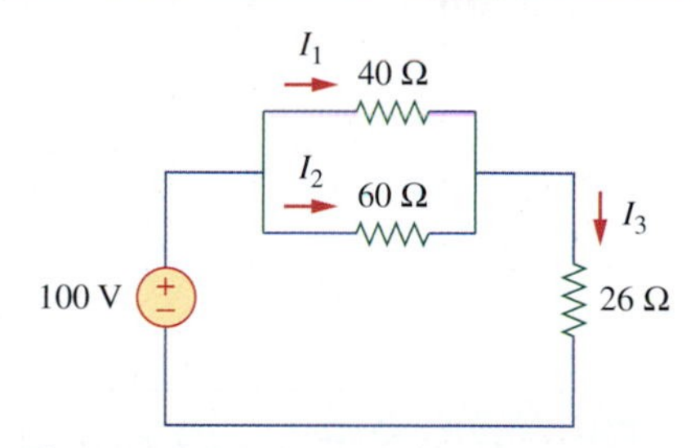

Figure 6.4
For Practice Problem 6.1.

Example 6.2

Find R_{eq} for the circuit shown in Fig. 6.5.

Solution:
To get R_{eq}, we combine resistors in series and in parallel. The 6-Ω and 3-Ω resistors are in parallel because they are connected to the same nodes. Their equivalent resistance is

$$6 \| 3 = \frac{6 \times 3}{6 + 3} = 2\ \Omega$$

Also, the 1-Ω and 5-Ω resistors are in series because the same current would flow through them; hence, their equivalent resistance is

$$1 + 5 = 6\ \Omega$$

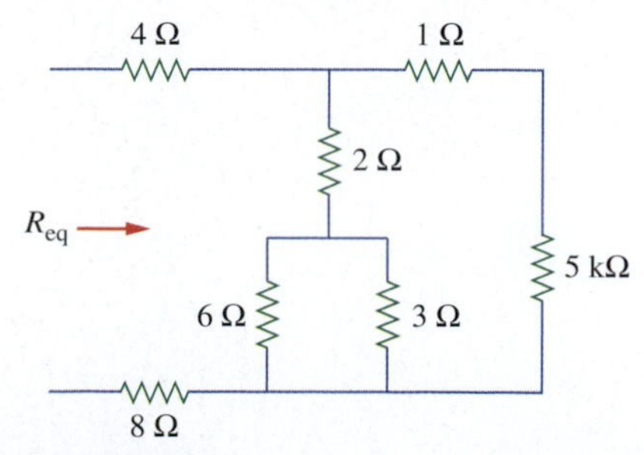

Figure 6.5
For Example 6.2.

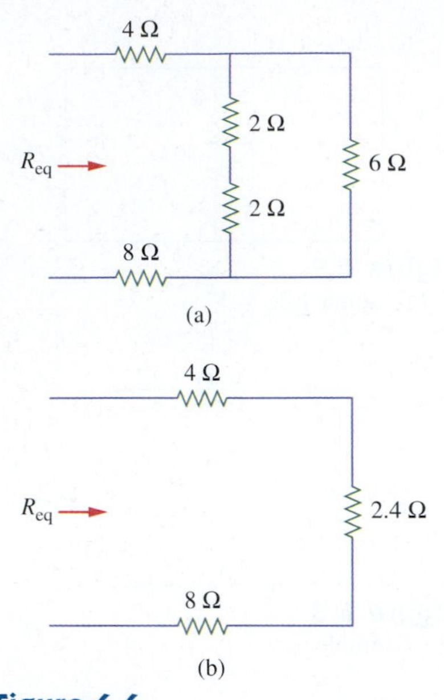

Figure 6.6
Equivalent circuits for Example 6.2.

Thus, the circuit in Fig. 6.5 is reduced to that in Fig. 6.6(a). In Fig. 6.6(a), we notice that the two 2-Ω resistors are in series, so their equivalent resistance is

$$2 + 2 = 4\ \Omega$$

The 4-Ω resistor is now in parallel with the 6-Ω resistor in Fig. 6.6(a); the equivalent resistance is

$$4 \parallel 6 = \frac{4 \times 6}{4 + 6} = 2.4\ \Omega$$

The circuit in Fig. 6.6(a) is now replaced with that in Fig. 6.6(b). In Fig. 6.6(b), the three resistors are in series. Hence, the equivalent resistance for the circuit is

$$R_{eq} = 4 + 2.4 + 8 = 14.4\ \Omega$$

Practice Problem 6.2

By combining resistors in Fig. 6.7, find R_{eq}.

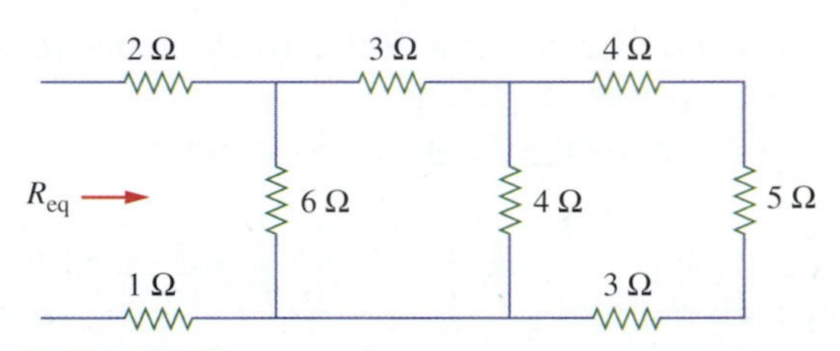

Figure 6.7
For Practice Problem 6.2.

Answer: 6 Ω

Example 6.3

Calculate the equivalent resistance R_{ab} in the circuit in Fig. 6.8.

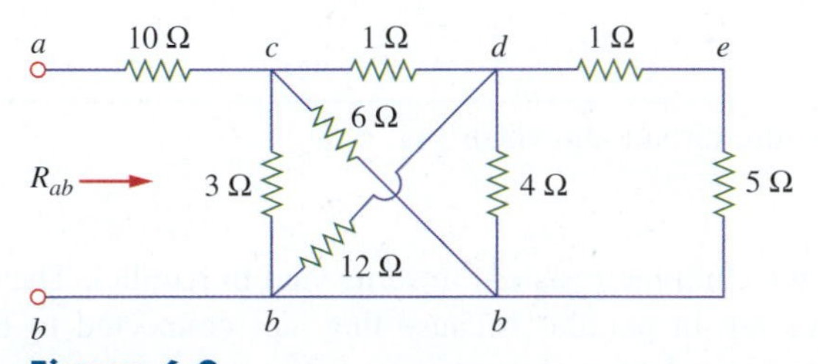

Figure 6.8
For Example 6.3.

Solution:
The 3-Ω and 6-Ω resistors are in parallel because they are connected to the same two nodes c and b. Their combined resistance is

$$3 \parallel 6 = \frac{3 \times 6}{3 + 6} = 2\ \Omega \tag{6.3.1}$$

Similarly, the 12-Ω and 4-Ω resistors are in parallel because they are connected to the same two nodes d and b. Hence,

$$12 \parallel 4 = \frac{12 \times 4}{12 + 4} = 3\ \Omega \qquad \textbf{(6.3.2)}$$

Also, the 1-Ω and 5-Ω resistors are in series; hence, their equivalent resistance is

$$1 + 5 = 6\ \Omega \qquad \textbf{(6.3.3)}$$

With these three combinations, we can replace the circuit in Fig. 6.8 with that in Fig. 6.9(a). In Fig. 6.9(a), 3 Ω in parallel with 6 Ω gives 2 Ω as calculated in Eq. (6.3.1). This 2-Ω equivalent resistance is now in series with the 1-Ω resistance to give a combined resistance of 1 + 2 = 3 Ω. Thus, we replace the circuit in Fig. 6.9(a) with that in Fig. 6.9(b). In Fig. 6.9(b), we combine the 2-Ω and 3-Ω resistors in parallel to get

$$2 \parallel 3 = \frac{2 \times 3}{2 + 3} = 1.2\ \Omega$$

This 1.2-Ω resistor is in series with the 10-Ω resistor so that

$$R_{ab} = 10 + 1.2 = 11.2\ \Omega$$

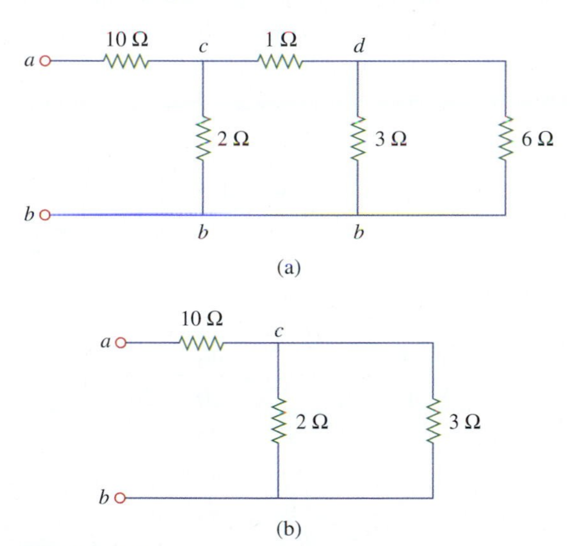

Figure 6.9
Equivalent circuits for Example 6.3.

Practice Problem 6.3

Find R_{ab} for the circuit in Fig. 6.10.

Answer: 11 Ω

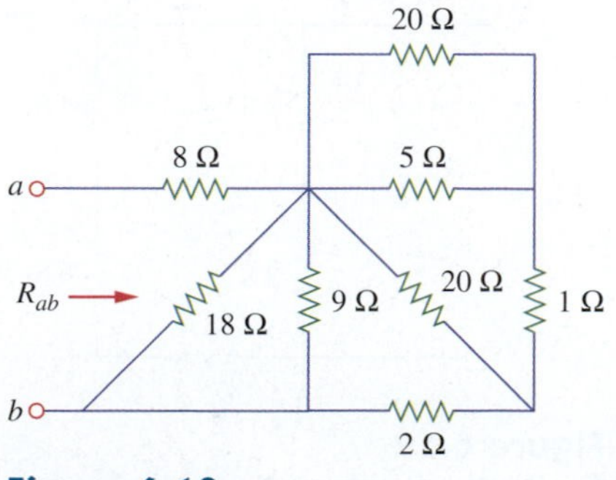

Figure 6.10
For Practice Problem 6.3.

Example 6.4

Find the equivalent conductance G_{eq} for the circuit in Fig. 6.11(a).

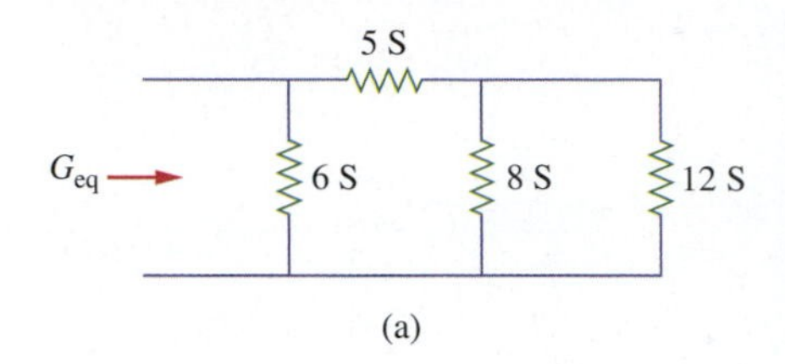

(a)

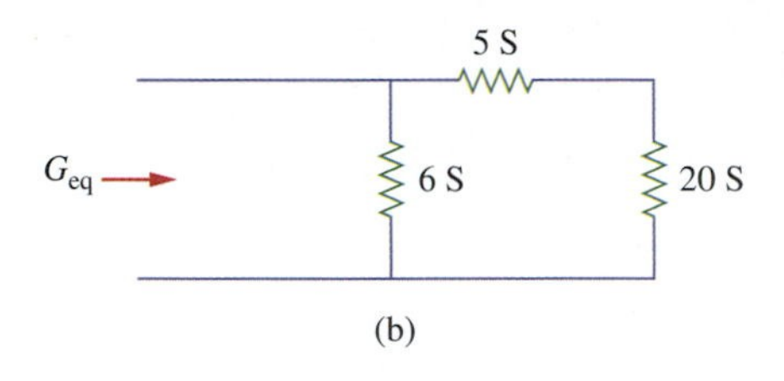

(b)

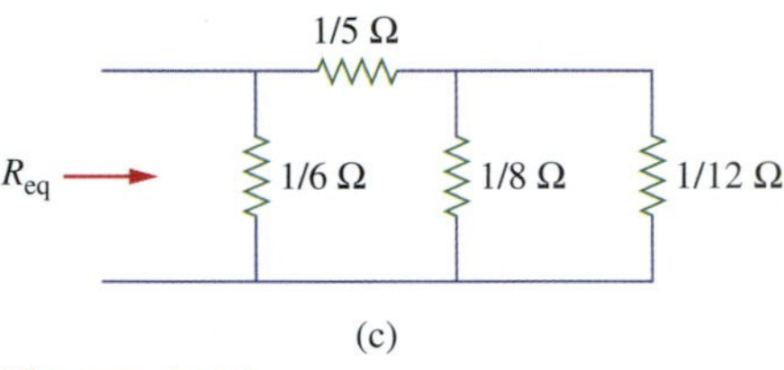

(c)

Figure 6.11
For Example 6.4: (a) original circuit; (b) its equivalent circuit; (c) same circuit as in (a), but resistors are expressed in ohms.

Solution:
We can solve this in two ways—working with conductances or with resistances.

■ **METHOD 1** The 8-S and 12-S resistors are in parallel so that their conductance is

$$8 + 12 = 20 \text{ S}$$

This 20-S resistor is now in series with 5 S as shown in Fig. 6.11(b) so that the combined conductance is

$$\frac{20 \times 5}{20 + 5} = 4 \text{ S}$$

This is in parallel with the 6-S resistor. Hence,

$$G_{eq} = 6 + 4 = 10 \text{ S}$$

■ **METHOD 2** We should note that the circuit in Fig. 6.11(c) is the same as that in Fig. 6.11(a). While the resistors in Fig. 6.11(a) are expressed in siemens, they are expressed in ohms in Fig. 6.11(c). To show that the circuits are the same, we find R_{eq} for the circuit in Fig. 6.11(c).

$$R_{eq} = \frac{1}{6} \left\| \left(\frac{1}{5} + \frac{1}{8} \right\| \frac{1}{12} \right) = \frac{1}{6} \left\| \left(\frac{1}{5} + \frac{1}{20} \right) = \frac{1}{6} \right\| \frac{1}{4} = \frac{\frac{1}{6} \times \frac{1}{4}}{\frac{1}{6} + \frac{1}{4}} = \frac{1}{10} \ \Omega$$

where we have used

$$\frac{1}{8} \left\| \frac{1}{12} = \frac{\frac{1}{8} \times \frac{1}{12}}{\frac{1}{8} + \frac{1}{12}} = \frac{\frac{1}{96}}{\frac{20}{96}} = \frac{1}{20} \right.$$

Hence,

$$G_{eq} = \frac{1}{R_{eq}} = 10 \text{ S}$$

This is the same answer that we obtained using method 1.

Practice Problem 6.4

Calculate G_{eq} in the circuit of Fig. 6.12.

Answer: 4 S

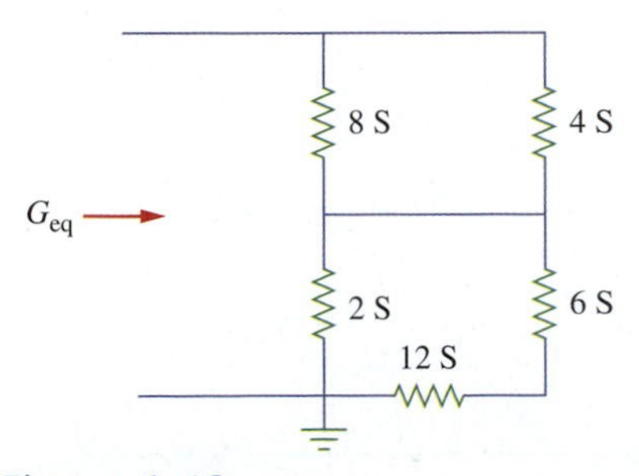

Figure 6.12
For Practice Problem 6.4.

Example 6.5

Find the voltage across each resistor in the series-parallel circuit of Fig. 6.13.

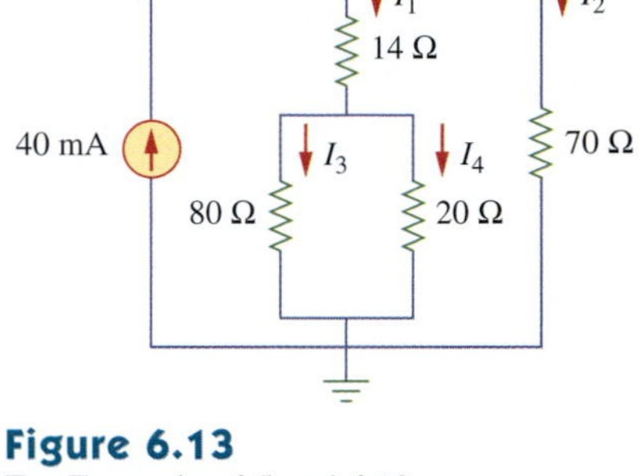

Figure 6.13
For Examples 6.5 and 6.12.

Solution:
The 80-Ω and 20-Ω resistors are in parallel. These can be combined to get

$$80 \parallel 20 = \frac{80 \times 20}{80 + 20} = 16\ \Omega$$

This 16-Ω resistor is in series with the 14-Ω resistor, giving

$$16 + 14 = 30\ \Omega$$

We can apply the current divider rule to find current I_1 and I_2.

$$I_1 = \frac{70}{70 + 30}(40\text{ mA}) = 28\text{ mA}$$

$$I_2 = \frac{30}{70 + 30}(40\text{ mA}) = 12\text{ mA}$$

We can apply the current divider rule again to share I_1 between I_3 and I_4.

$$I_3 = \frac{20}{20 + 80}(28\text{ mA}) = 5.6\text{ mA}$$

$$I_4 = \frac{80}{20 + 80}(28\text{ mA}) = 22.4\text{ mA}$$

Once we determine the branch currents, we can apply Ohm's law to determine the voltage drop across each resistor. For the 70-Ω resistor,

$$V_{70} = 70I_2 = 70 \times 12 \times 10^{-3} = 0.84\text{ V}$$

For the 14-Ω resistor,

$$V_{14} = 14I_1 = 14 \times 28 \times 10^{-3} = 0.392\text{ V}$$

Because the 80-Ω and 20-Ω resistors are in parallel, they have the same voltage.

$$V_{20} = V_{80} = 80I_3 = 80 \times 5.6 \times 10^{-3} = 0.448\text{ V}$$

Practice Problem 6.5

Determine the voltage drop across each resistor in the series-parallel circuit of Fig. 6.14.

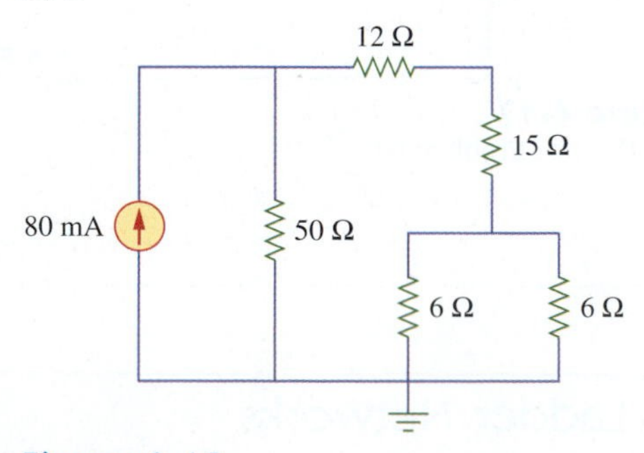

Figure 6.14
For Practice Problems 6.5 and 6.12.

Answer: $V_{50} = 1.5$ V; $V_{12} = 0.6$ V; $V_{15} = 0.75$ V; $V_6 = 0.15$ V

Example 6.6

Find V_{ab} in the circuit of Fig. 6.15.

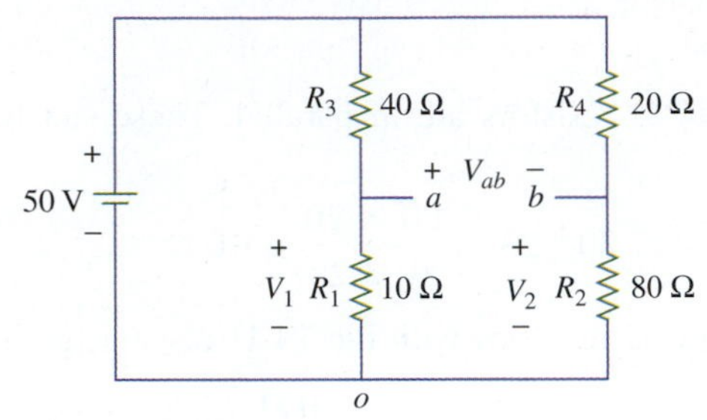

Figure 6.15
For Example 6.6.

Solution:
Because the 50-V voltage source is in parallel with the series combination of the 40-Ω and 10-Ω resistors and the series combination of the 20-Ω and 80-Ω resistors, we can use the voltage divider rule to find V_1 and V_2.

$$V_1 = \frac{10}{10 + 40}(50) = 10 \text{ V}$$

$$V_2 = \frac{80}{80 + 20}(50) = 40 \text{ V}$$

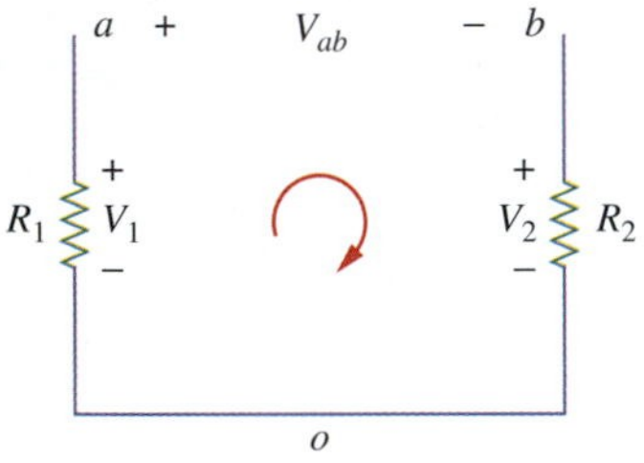

Figure 6.16
For Example 6.6.

We now apply KVL around loop *oabo* as shown in Fig. 6.16.

$$-V_1 + V_{ab} + V_2 = 0$$

or

$$V_{ab} = V_1 - V_2 = 10 - 40 = -30 \text{ V}$$

Practice Problem 6.6

Determine I in the circuit of Fig. 6.17.

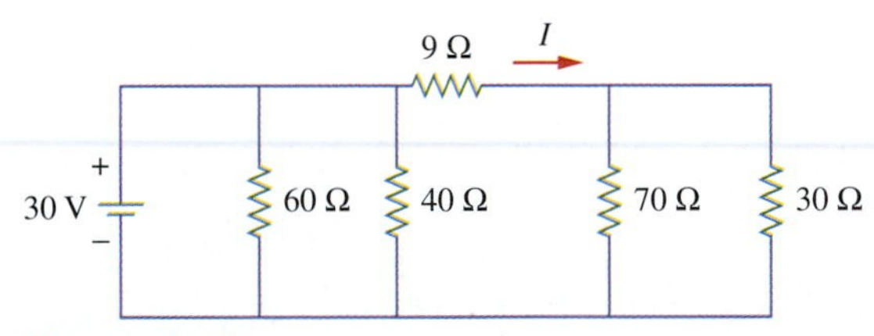

Figure 6.17
For Practice Problem 6.6.

Answer: 1 A

6.3 Ladder Networks

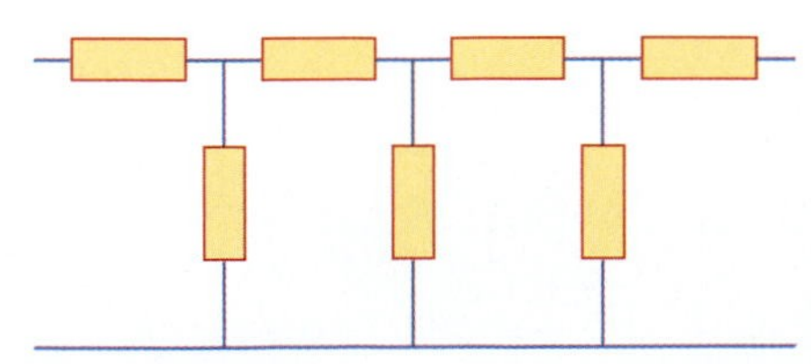

Figure 6.18
A ladder circuit.

A ladder network is a special series-parallel circuit. A typical ladder network is shown in Fig. 6.18. It consists of a set of series and parallel resistors. This representation is sometimes pictured as a "ladder" because of its ladderlike configuration.

Ladder networks are used in digital-to-analog converters (DACs) to provide reference voltages that are $\frac{1}{2}$, $\frac{1}{4}$, $\frac{1}{8}$, and so on of a source voltage. A special ladder used in DAC is called the $R/2R$ ladder, shown in Fig. 6.19 (LSB = least significant bit, MSB = most significant bit). Digital information is presented to the ladder as individual bits of a digital word switched between a reference voltage and ground.

A **ladder circuit** is a series-parallel circuit with a topology that resembles a ladder.

A ladder network is analyzed just like any other series-parallel circuit. But the analysis usually starts by working backward. This is best illustrated with an example.

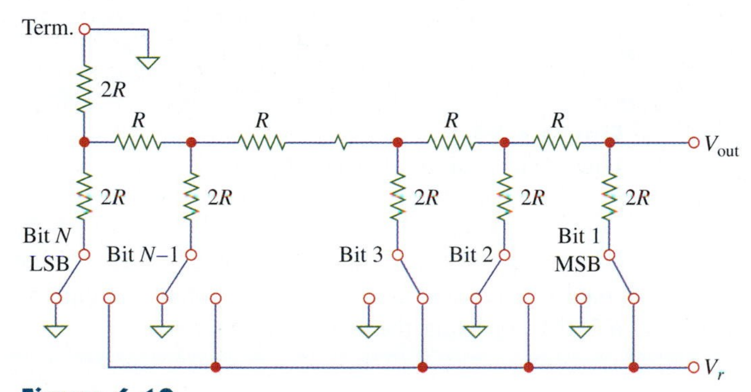

Figure 6.19
$R/2R$ ladder of N bits used for a digital-to-analog converter.
Courtesy of TT Electronics IRC

Example 6.7

Find V_o in the ladder network of Fig. 6.20.

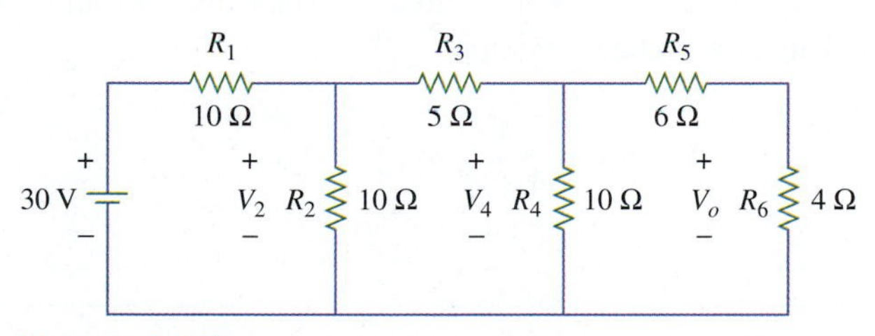

Figure 6.20
For Example 6.7.

Solution:
We notice from Fig. 6.20 that R_5 and R_6 are in series so that their combined resistance is

$$6 + 4 = 10\ \Omega$$

This 10 Ω is in parallel with $R_4 = 10\ \Omega$ so that their combined resistance is

$$10 \parallel 10 = 5\ \Omega$$

At this point, the equivalent circuit is shown in Fig. 6.21(a). The 5-Ω resistor is in series with $R_3 = 5\ \Omega$ and their combined resistance amounts to

$$5 + 5 = 10\ \Omega$$

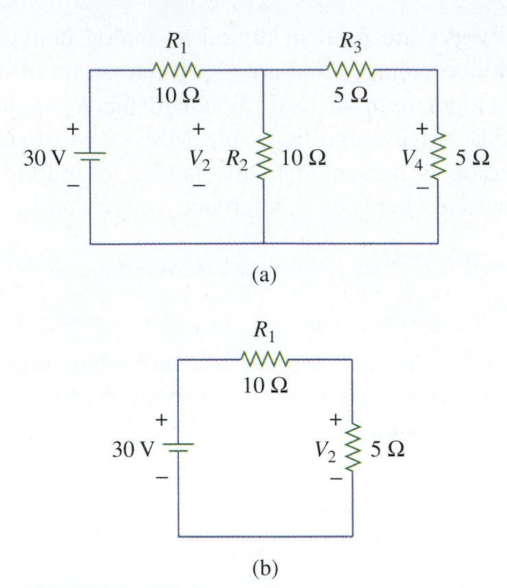

Figure 6.21
Equivalent circuits for Example 6.7.

Again, this 10-Ω resistor is in parallel with another 10-Ω resistor so that their combined resistance is 5 Ω. Thus, we obtain the equivalent circuit in Fig. 6.21(b). By using the voltage divider rule,

$$V_2 = \frac{5}{5 + 10}(30 \text{ V}) = 10 \text{ V}$$

We now can work forward to get V_o. The voltage $V_2 = 10$ V is shared equally by the two 5-Ω resistors in Fig. 6.21(a); that is, $V_4 = 10/2 = 5$ V. This voltage, $V_4 = 5$ V, is shared between the 6-Ω and the 4-Ω resistors. Thus, by voltage division,

$$V_o = \frac{4}{4 + 6}(5 \text{ V}) = 2 \text{ V}$$

Practice Problem 6.7

The circuit in Fig. 6.22 is called an $R/2R$ ladder. Let $R = 10$ kΩ and find I_x.

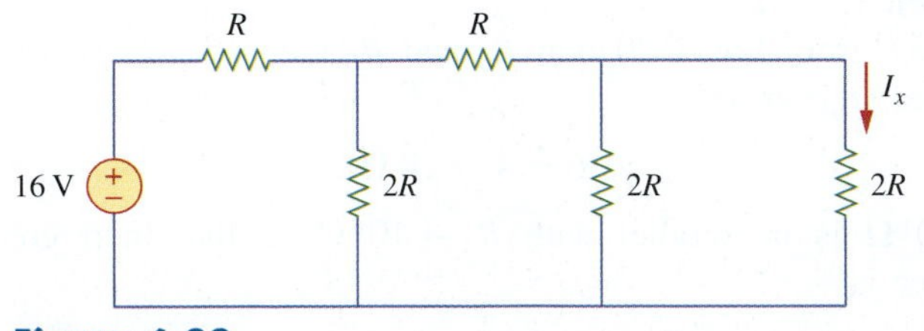

Figure 6.22
For Practice Problem 6.7.

Answer: 0.2 mA

6.4 Dependent Sources

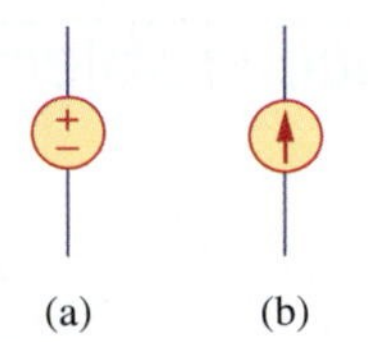

Figure 6.23
(a) Independent voltage source; (b) independent current source.

So far, we have been analyzing circuits with independent sources. The symbol for an independent voltage source is shown in Fig. 6.23(a), while that for an independent current source is shown in Fig. 6.23(b). Independent sources produce voltage or current unaffected by what is going on in the remainder of the circuit.

Dependent sources are important because they are used to model electronic circuit elements. They are usually designated by diamond-shaped symbols, as shown in Fig. 6.24. Because the control of the dependent source is achieved by a voltage or current of some element in the circuit, and source can be voltage or current, it follows that there are four possible types of dependent sources:

1. A voltage-controlled voltage source (VCVS); see Fig. 6.24(a).
2. A voltage-controlled current source (VCCS); see Fig. 6.24(b).
3. A current-controlled voltage source (CCVS); see Fig. 6.24(c).
4. A current-controlled current source (CCCS); see Fig. 6.24(d).

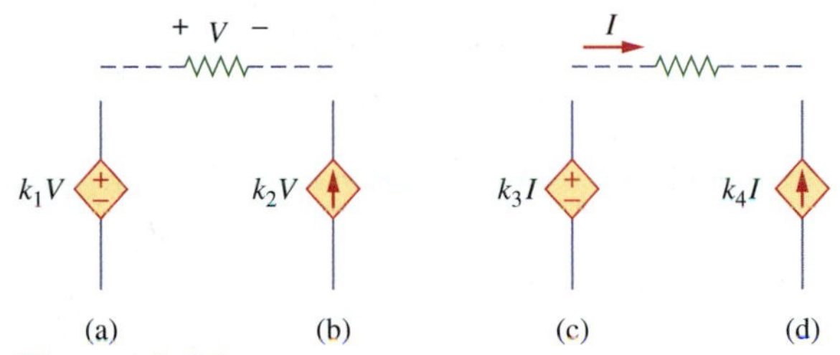

Figure 6.24
(a) voltage-controlled voltage source; (b) voltage-controlled current source; (c) current-controlled voltage source; (d) current-controlled current source.

Dependent sources do not serve as inputs to a circuit like independent sources. They are used to model active electronic circuit elements. For example, to describe an operational amplifier (op amp), we need a voltage-controlled voltage source.

> A **dependent source** is either a voltage or current source whose value is proportional to some other voltage or current in the circuit.

Example 6.8

Determine V_x in the circuit shown in Fig. 6.25.

Solution:
We apply KVL around the loop and get

$$-12 + 5I + 2V_x = 0$$

But $5I = V_x$,

$$-12 + V_x + 2V_x = 0 \quad \Rightarrow \quad 3V_x = 12$$

$$V_x = 4\ \text{V}$$

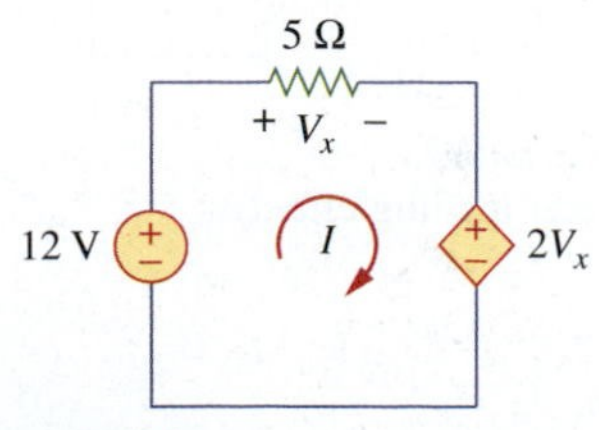

Figure 6.25
For Example 6.8.

Practice Problem 6.8

Find V_x in the circuit shown in Fig. 6.26.

Answer: 20 V

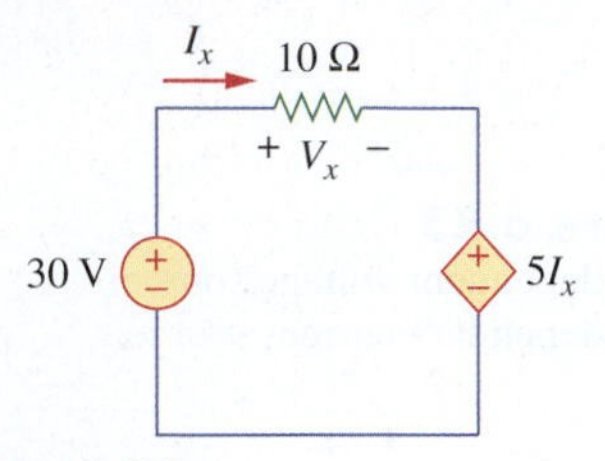

Figure 6.26
For Practice Problem 6.8.

Example 6.9

Calculate V_x in the circuit shown in Fig. 6.27.

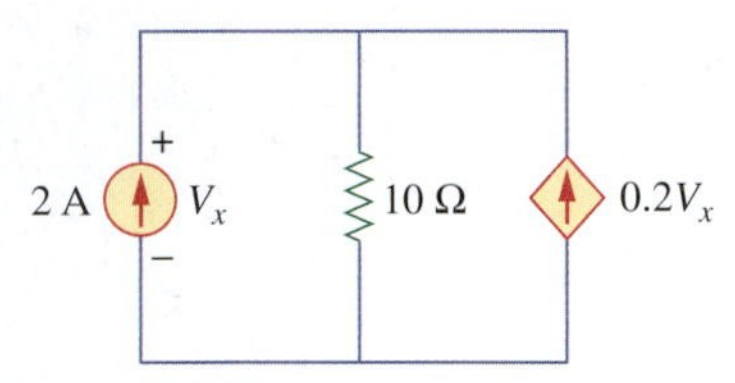

Figure 6.27
For Example 6.9.

Solution:
Let I be the current through the 10-Ω resistor and assume the current is flowing down the resistor.

$$I = 2 + 0.2V_x$$

$$V_x = 10I = 20 + 2V_x$$

We now solve for V_x.

$$V_x = -20 \text{ V}$$

Practice Problem 6.9

Find V_x in the circuit shown in Fig. 6.28.

Answer: –5 V

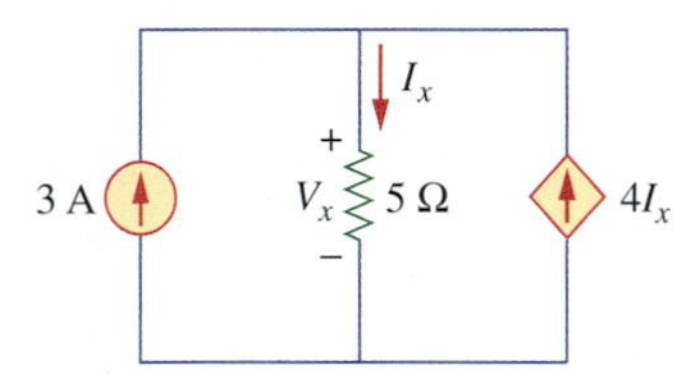

Figure 6.28
For Practice Problem 6.9.

6.5 Loading Effects of Instruments

When an instrument such as a voltmeter or an ammeter is connected to a circuit, its presence affects the reading due to its internal resistances and the fact that the instrument absorbs some energy from the circuit. This effect is called *loading*. The loading effect is important, especially when accuracy is of primary concern.

When a voltmeter is connected in parallel with a branch as shown in Fig. 6.29, it would be ideal if the voltmeter had an infinite resistance. This would mean that the circuit operating condition is not altered by the presence of the voltmeter. But a practical voltmeter has a finite resistance R_V. The voltmeter resistance should be as large as possible to minimize loading effect. The higher the value of R_V, the smaller the loading effect.

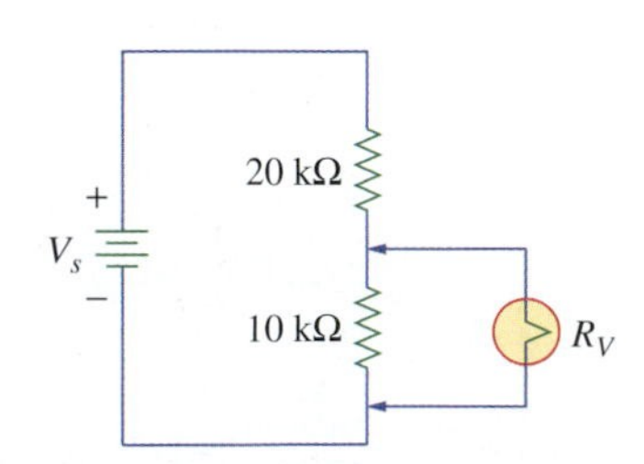

Figure 6.29
Voltmeter loading effect.

Similarly, when an ammeter is connected in series with a branch, typically as shown in Fig. 6.30, the resistance of the ammeter should ideally be zero so that the resistance of the circuit remains the same. However, a practical ammeter has a finite but small resistance R_A. If the ammeter is connected in series with a large resistance, R_A can be ignored. But if the ammeter is connected in series with a relatively small resistance, R_A cannot be ignored.

The percentage error introduced by the instrument is calculated as

$$\text{Error (\%)} = \frac{\text{Ideal value} - \text{Measured value}}{\text{Ideal value}} \times 100 \qquad \textbf{(6.1)}$$

where Ideal value is what we get when the instrument is either absent or ideal, while the Measured value is what we get with the instrument in place. The allowable error depends on the situation. A good rule of thumb is to ignore the loading effect if the error is less than 5 percent.

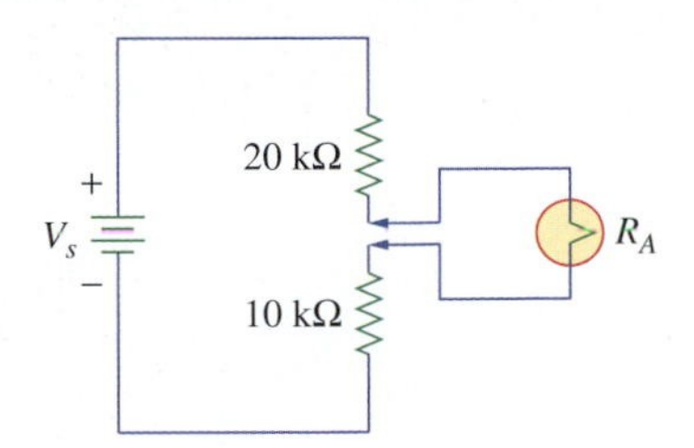

Figure 6.30
Ammeter loading effect.

Example 6.10

Consider the circuit shown in Fig. 6.31. If a voltmeter with an internal resistance of 10 MΩ is used to measure V_1 and V_2, calculate the loading effects.

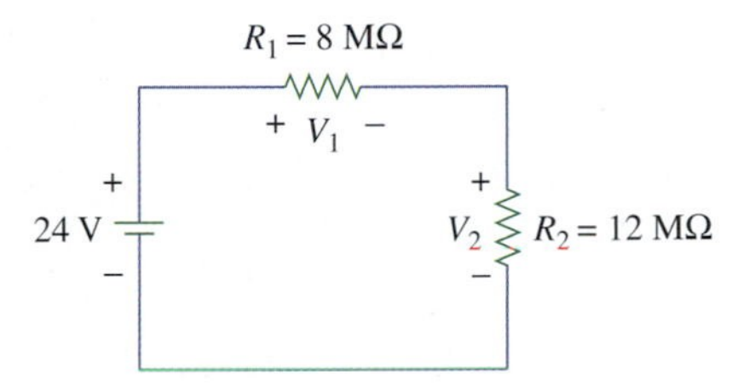

Figure 6.31
For Example 6.10.

Solution:
We need to calculate both the unloaded and loaded voltages. The unloaded (or ideal) voltages are those values of V_1 and V_2 when the voltmeter is not present. Using the voltage divider rule,

$$V_1 = \frac{R_1}{R_1 + R_2} V_s = \frac{8 \text{ M}\Omega}{8 \text{ M}\Omega + 12 \text{ M}\Omega}(24 \text{ V}) = 9.6 \text{ V}$$

$$V_2 = \frac{R_2}{R_1 + R_2} V_s = \frac{12 \text{ M}\Omega}{8 \text{ M}\Omega + 12 \text{ M}\Omega}(24 \text{ V}) = 14.4 \text{ V}$$

The loaded (or measured) voltages are the values of V_1 and V_2 when the voltmeter is present. For V_1, the voltmeter is connected in parallel with R_1 as shown in Fig. 6.32(a). The parallel combination of R_1 and R_V is

$$R_{T1} = R_1 \parallel R_V = \frac{R_1 R_V}{R_1 + R_V} = \frac{8 \times 10}{8 + 10} \text{ M}\Omega = 4.444 \text{ M}\Omega$$

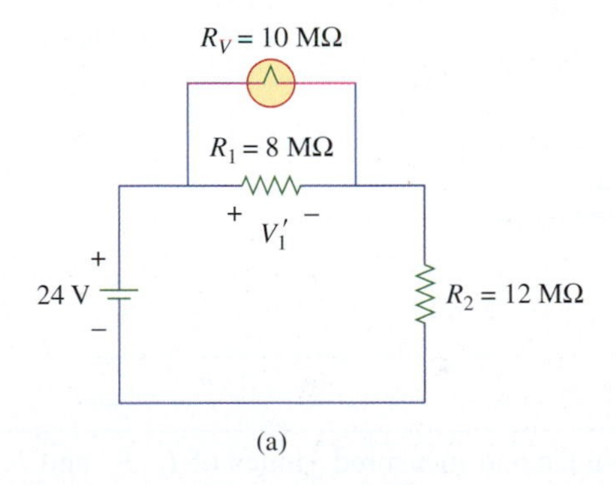

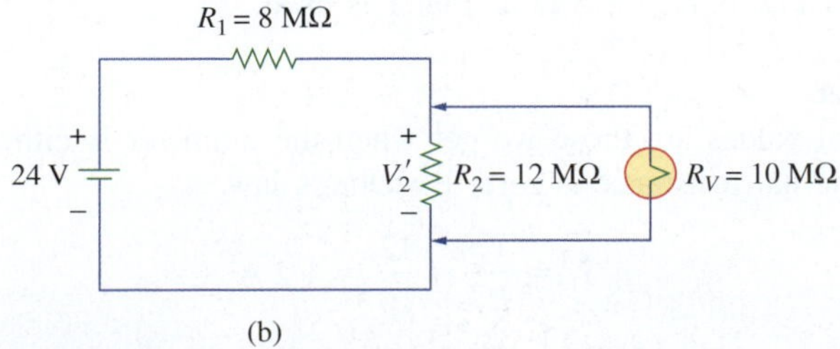

Figure 6.32
Reading effects: (a) measuring V_1; (b) measuring V_2.

Hence,

$$V_1' = \frac{R_{T1}}{R_{T1} + R_2}V_s = \frac{4.444\text{ M}\Omega}{4.444\text{ M}\Omega + 12\text{ M}\Omega}(24\text{ V}) = 6.486\text{ V}$$

For V_2, the voltmeter is connected in parallel with R_2 as shown in Fig. 32(b). The parallel combination of R_2 and R_V is

$$R_{T2} = R_2 \parallel R_V = \frac{R_2R_V}{R_2 + R_V} = \frac{12 \times 10}{12 + 10}\text{M}\Omega = 5.455\text{ M}\Omega$$

Therefore,

$$V_2' = \frac{R_{T2}}{R_{T2} + R_1}V_s = \frac{5.455\text{ M}\Omega}{5.455\text{ M}\Omega + 8\text{ M}\Omega}(24\text{ V}) = 9.730\text{ V}$$

The error introduced by the voltmeter in measuring V_1 is

$$\text{Error (\%)} = \frac{V_1 - V_1'}{V_1} \times 100 = \frac{9.6 - 6.486}{9.6} \times 100 = 32.44\%$$

while the error introduced in measuring V_2 is

$$\text{Error (\%)} = \frac{V_2 - V_2'}{V_2} \times 100 = \frac{14.4 - 9.730}{14.4} \times 100 = 32.43\%$$

We notice that the error percentages are relatively high. This is due to the fact that R_V is comparable to the values of R_1 and R_2. We can reduce the errors by increasing R_V.

Practice Problem 6.10

For the circuit shown in Fig. 6.33, calculate the percentage error introduced by measuring V_3. Assume that the internal resistance of the voltmeter is 12 MΩ.

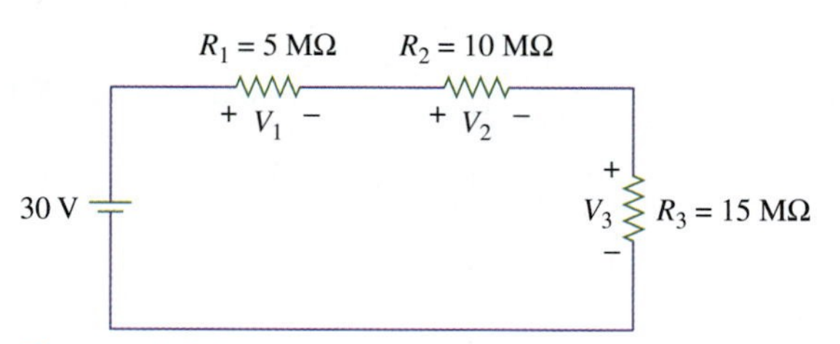

Figure 6.33
For Practice Problem 6.10.

Answer: 38.46%

Example 6.11

Determine the ideal values and measured values of I_1, I_2, and I_T in the circuit of Fig. 6.34 if a 5-Ω ammeter is used.

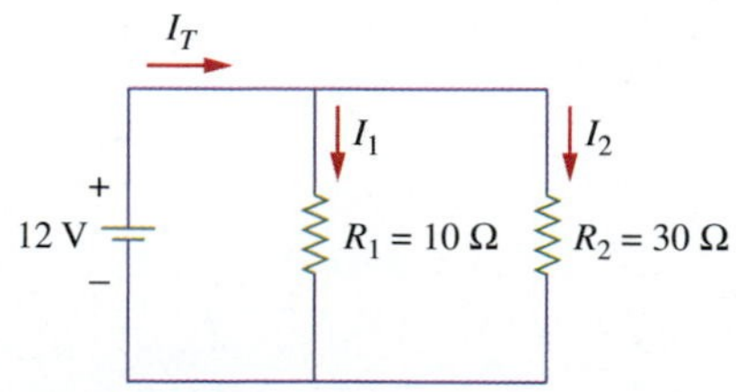

Figure 6.34
For Example 6.11.

Solution:
The ideal values are those we get when the ammeter is either absent or its internal resistance is zero. By Ohm's law,

$$I_1 = \frac{V_s}{R_1} = \frac{12}{10} = 1.2\text{ A}$$

$$I_2 = \frac{V_s}{R_2} = \frac{12}{30} = 0.4\text{ A}$$

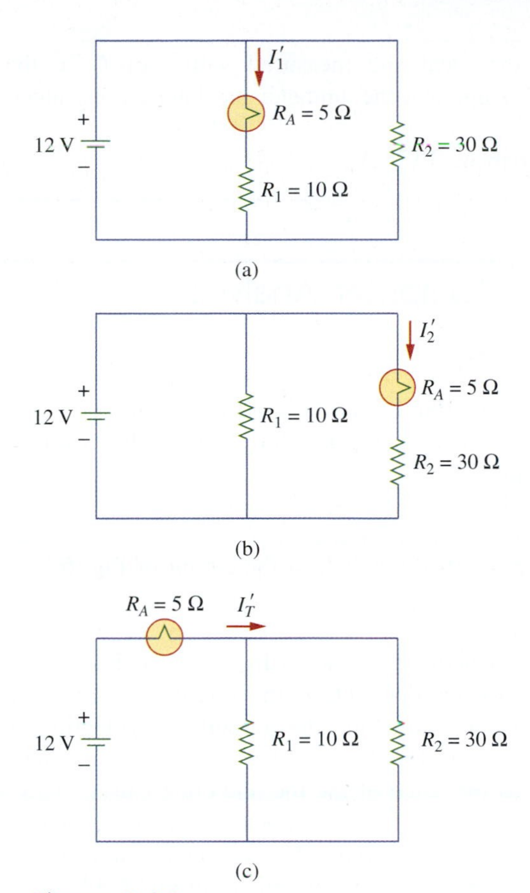

Figure 6.35
Measuring: (a) I_1; (b) I_2; (c) I_T.

Using Kirchhoff's current law,

$$I_T = I_1 + I_2 = 1.2 + 0.4 = 1.6 \text{ A}$$

We determine the measured values by inserting the ammeter in series with each of the branches. To measure I_1, consider the circuit in Fig. 6.35(a). The current through the ammeter is

$$I_1' = \frac{V_s}{R_1 + R_A} = \frac{12}{10 + 5} = 0.8 \text{ A}$$

To measure I_2, consider the circuit in Fig. 6.35(b).

$$I_2' = \frac{V_s}{R_2 + R_A} = \frac{12}{30 + 5} = 0.343 \text{ A}$$

To measure I_T, refer to the circuit in Fig. 6.29(c). The total resistance is

$$R_T = R_A + R_1 \parallel R_2 = 5 + \frac{10 \times 30}{10 + 30} = 5 + 7.5 = 12.5\ \Omega$$

By Ohm's law,

$$I_T' = \frac{12}{12.5} = 0.96 \text{ A}$$

Notice that

$$I_T' \neq I_1' + I_2'$$

Practice Problem 6.11

Determine the ideal and measured values of I_T in the circuit of Fig. 6.36. Assume that the ammeter has internal resistance of 2 Ω.

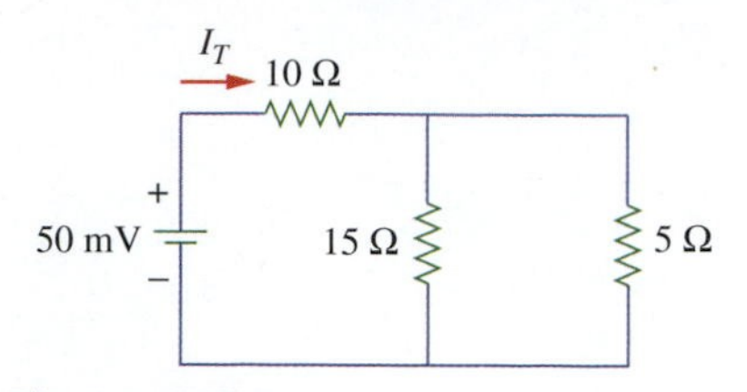

Figure 6.36
For Practice Problem 6.11.

Answer: 3.636 A; 3.175 A

6.6 Computer Analysis

6.6.1 PSpice

With little effort, PSpice can be used to find the branch currents and node voltages of a series-parallel circuit. The following example illustrates this.

Example 6.12

Find the currents I_1 through I_4 in the circuit of Fig. 6.13.

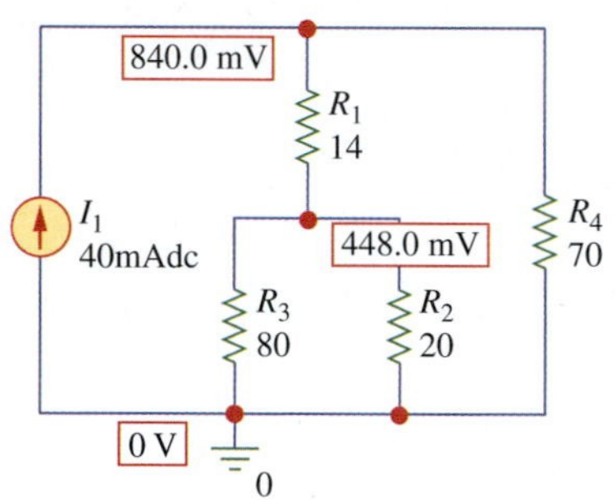

Figure 6.37
PSpice schematic for Example 6.12.

Solution:
We draw the schematic as shown in Fig. 6.37. For a dc analysis, we use current source IDC. Once the circuit is drawn and saved as exam612.dsn, we simulate the circuit by selecting **PSpice/New Simulation Profile**. This leads to the New Simulation dialog box. Type "exam612" as the name of the file and click Create. This leads to the Simulation Settings dialog box. Because this is dc analysis, just click OK. Then select **PSpice/Run**. The circuit is simulated and the node voltages are displayed on the circuits as in Fig. 6.37. We can calculate the currents by hand:

$$I_1 = \frac{840 - 448}{14} = 28 \text{ mA}$$

$$I_2 = \frac{840}{70} = 12 \text{ mA}$$

$$I_3 = \frac{448}{80} = 5.6 \text{ mA}$$

$$I_4 = \frac{448}{20} = 22.4 \text{ mA}$$

which agrees with our results in Example 6.5.

Practice Problem 6.12

Use PSpice to determine the voltage drop across each resistor in the series-parallel circuit shown earlier in Fig. 6.14.

Answer: $V_{50} = 1.5$ V; $V_{12} = 0.6$ V; $V_{15} = 0.75$ V; $V_6 = 0.15$ V

6.6.2 Multisim

Multisim can be used to analyze series-parallel circuits just as we used it to analyze series and parallel circuits in previous chapters. Multisim also provides a multimeter that can be used to measure resistance and a wattmeter to measure power. The following example illustrates how we use Multisim to analyze a series-parallel circuit.

Example 6.13

Refer to the circuit in Fig. 6.38. (a) Use Multisim to find R_{eq}. (b) Use Multisim to find V_o.

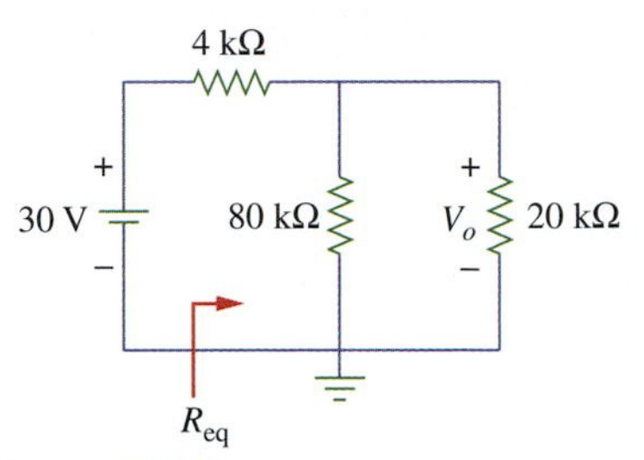

Figure 6.38
For Example 6.13.

Solution:

(a) The circuit is wired as shown in Fig. 6.39(a). The voltage source is replaced by a multimeter, which can measure voltage, current, or resistance. Here, we use it to measure resistance R_{eq}. We double-click the multimeter and select the ohmmeter (Ω) function. After saving and simulating the circuit, we obtain

$$R_{eq} = 20\text{ k}\Omega$$

Unlike voltmeters and ammeters, which display the measured results, the multimeter does not display the results directly. After simulation, you double-click the multimeter and the result is displayed.

(b) The circuit is wired as shown in Fig. 6.39(b). Although we can determine V_o using a voltmeter, we choose to use a multimeter instead so that we can become familiar with it. Here, V_o is measured by connecting the multimeter in parallel with the 20-kΩ resistor and selecting the dc voltmeter function of the multimeter. Once the circuit is saved and simulated, we double-click on the multimeter and get

$$V_o = 24\text{ V}$$

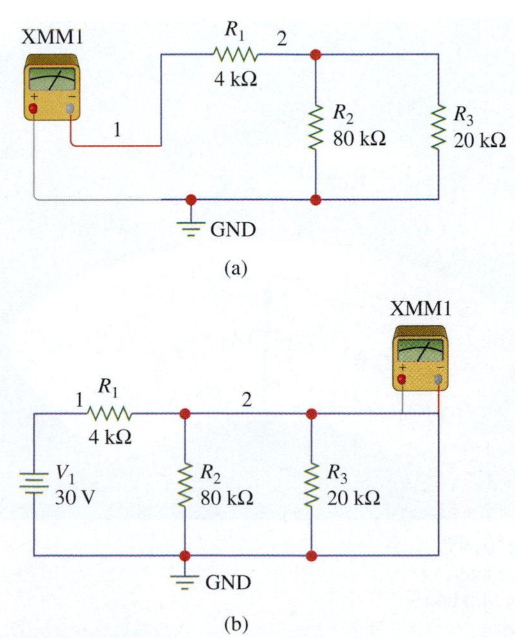

Figure 6.39
For Example 6.13: (a) measuring R_{eq}; (b) measuring V_o.

Practice Problem 6.13

Use Multisim to determine branch current I_o in the circuit of Fig. 6.40.

Answer: 0.5 mA

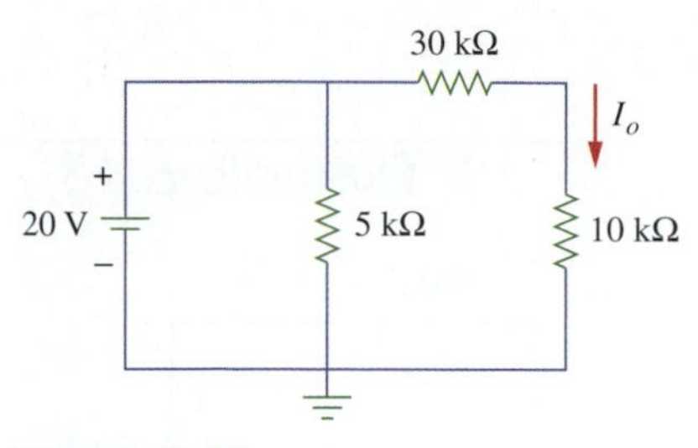

Figure 6.40
For Practice Problem 6.13.

6.7 †Application: Wheatstone Bridge

Besides the ladder circuits, there are several other applications of series-parallel circuits. Most electronic devices such as radios, televisions, and computers contain series-parallel circuits. In this section, we consider one important application—the Wheatstone bridge circuit.

> The **Wheatstone bridge** is an electrical circuit that is used to determine an unknown resistance by adjusting a known resistance so that the measured current is zero.

Although the ohmmeter method provides the simplest way to measure resistance, more accurate measurement may be obtained by using the Wheatstone bridge. While ohmmeters are designed to measure resistance in low, middle, or high range, a Wheatstone bridge is used in measuring resistance in the mid-range, say, between 1 Ω and 1 MΩ. Very low values of resistance are measured with a *milliohmmeter,* while very high values are measured with a *Megger tester.*

The Wheatstone bridge[1] (or resistance bridge) circuit is used in a number of applications. Here, we will use it to measure an unknown resistance. The unknown resistance R_x is connected to the bridge as shown in Fig. 6.41. The variable resistance is adjusted until no current flows through the galvanometer (see Fig. 6.42), which is essentially a d'Arsonval movement operating as a sensitive current-indicating device. (A d'Arsonval movement is a dc coil-type movement in which an electromagnetic core is suspended between the poles of a permanent magnet). A galvanometer is the historical name given to a moving-coil electric

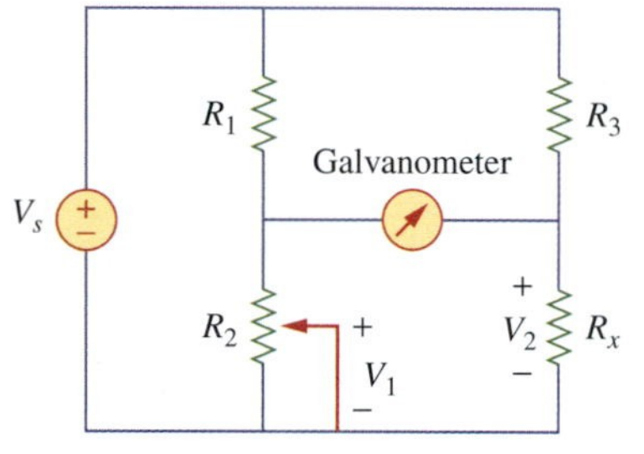

Figure 6.41
Wheatstone bridge; for Example 6.14 and Problem 6.63.

Figure 6.42
Galvanometer.
© Sarhan M. Musa

[1] HISTORICAL NOTE: The bridge was invented by Charles Wheatstone (1802–1875), a British professor who also invented the telegraph, as Samuel Morse independently did in the United States.

current detector. It is similar to an analog ammeter. When a current is passed through a coil in a magnetic field, the coil experiences a torque proportional to the current. When no current flows through the galvanometer, $V_1 = V_2$ and the bridge is said to be *balanced*. Because no current flows through the galvanometer, R_1 and R_2 behave as though they were in series; so do R_3 and R_x. Applying the voltage division principle,

$$V_1 = \frac{R_2}{R_1 + R_2}V = V_2 = \frac{R_x}{R_3 + R_x}V \tag{6.2}$$

Hence, no current flows through the galvanometer when

$$\frac{R_2}{R_1 + R_2} = \frac{R_x}{R_3 + R_x} \quad \Rightarrow \quad R_2R_3 = R_1R_x$$

or

$$\boxed{R_x = \frac{R_3}{R_1}R_2} \tag{6.3}$$

If $R_1 = R_3$, and R_2 is adjusted until no current flows through the galvanometer, then the value of $R_x = R_2$. In addition to measuring resistance, the Wheatstone bridge is used in measuring capacitance and inductance, which will be considered in later chapters.

A Wheatstone bridge can be used in both balanced and unbalanced modes. How do we find the current through the galvanometer when the Wheatstone bridge is *unbalanced*? In fact, we handled a similar situation in Example 6.6. We simply apply KVL. The unbalanced Wheatstone bridge is useful in measuring several types of physical quantities such as strain, temperature, and pressure. The value of the quantity being measured can be determined by the degree to which the bridge is unbalanced.

Example 6.14

In Fig. 6.41, $R_1 = 500\ \Omega$, $R_3 = 200\ \Omega$. The bridge is balanced when R_2 is adjusted to be $125\ \Omega$. Determine the unknown resistance R_x.

Solution:

Using Eq. (6.3),

$$R_x = \frac{R_3}{R_1}R_2 = \frac{200}{500}125 = 50\ \Omega$$

Practice Problem 6.14

A Wheatstone bridge has $R_1 = R_3 = 1\ \text{k}\Omega$. Here, R_2 is adjusted until no current flows through the galvanometer. At that point, $R_2 = 3.2\ \text{k}\Omega$. What is the value of the unknown resistance?

Answer: $3.2\ \text{k}\Omega$

6.8 Summary

1. A series-parallel circuit combines the features of series and parallel circuits.
2. To find the total resistance of a series-parallel circuit, we combine series and parallel resistances.
3. To find the branch current and branch voltages, we apply KVL, KCL, Ohm's law, and the voltage divider and current divider rules.

4. A ladder network is a series-parallel circuit with a schematic that resembles a ladder.
5. Dependent sources are used to model active electronic circuit elements. The value of a dependent source is proportional to some other voltage or current in the circuit.
6. Due to the loading effect, error is introduced into measurements made by a voltmeter or an ammeter.
7. Series-parallel circuits can be analyzed using PSpice and Multisim.
8. The Wheatstone bridge is an electrical circuit for the precise measurement of resistances. The bridge is balanced when its output voltage is zero. This condition is met when the ratio of resistances on one side of the bridge is equal to the ratio on the other side. Then,

$$R_x = \frac{R_3}{R_1} R_2$$

Review Questions

6.1 The parallel combination of a 60-Ω and a 40-Ω resistor is in series with a series combination of a 10-Ω and a 30-Ω resistance. The total resistance is:

(a) 140 Ω (b) 64 Ω
(c) 31.5 Ω (d) 7.5 Ω

6.2 The current I_x in the circuit of Fig. 6.43 is:

(a) 12 A (b) 7 A
(c) 5 A (d) 2 A

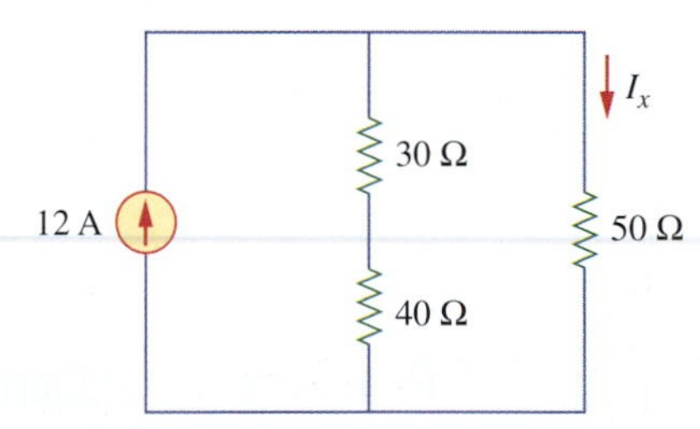

Figure 6.43
For Review Questions 6.2 through 6.4.

6.3 If the 30-Ω resistor in Fig. 6.43 is short-circuited, I_x becomes:

(a) 12 A (b) 6.777 A
(c) 5.333 (d) 0 A

6.4 If the 40-Ω resistor in Fig. 6.43 is open-circuited, I_x becomes:

(a) 12 A (b) 7.5 A
(c) 4.5 A (d) 0 A

6.5 What type of circuit is shown in Fig. 6.44?

(a) series (b) parallel
(c) ladder (d) Wheatstone bridge

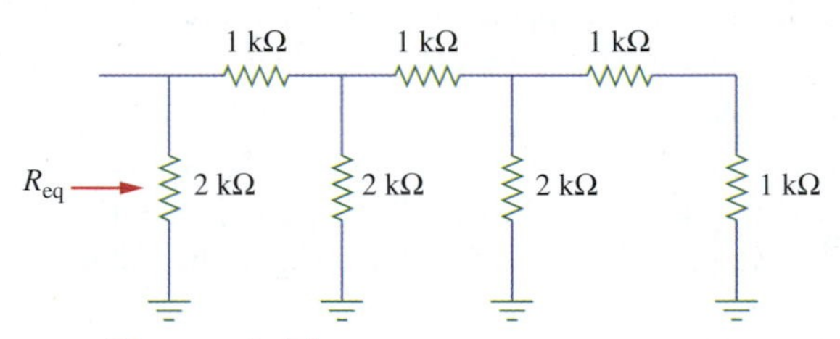

Figure 6.44
For Review Questions 6.5 and 6.6.

6.6 The equivalent resistance R_{eq} of the circuit in Fig. 6.44 is:

(a) 1 kΩ (b) 2 kΩ
(c) 3 kΩ (d) 10 kΩ

6.7 In a ladder circuit, simplifying the circuit should begin at:

(a) the source
(b) the center
(c) the resistor closest to the source
(d) the resistor farthest from the source

6.8 The Wheatstone bridge is a useful tool in measuring very small changes in resistance.

(a) True (b) False

6.9 The Wheatstone bridge shown in Fig. 6.45 is balanced.

(a) True (b) False

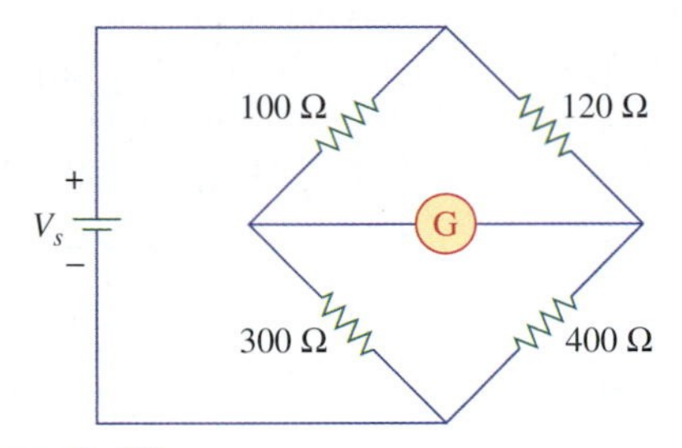

Figure 6.45
For Review Question 6.9.

6.10 Which of the following quantities cannot be measured with a Wheatstone bridge?

(a) resistance (b) inductance
(c) temperature (d) power

Answers: 6.1b, 6.2b, 6.3c, 6.4a, 6.5c, 6.6a, 6.7d, 6.8a, 6.9b, 6.10d

Problems

Section 6.2 Series-Parallel Circuits

6.1 For the circuit of Fig. 6.46, identify the series and parallel relationships of the resistors.

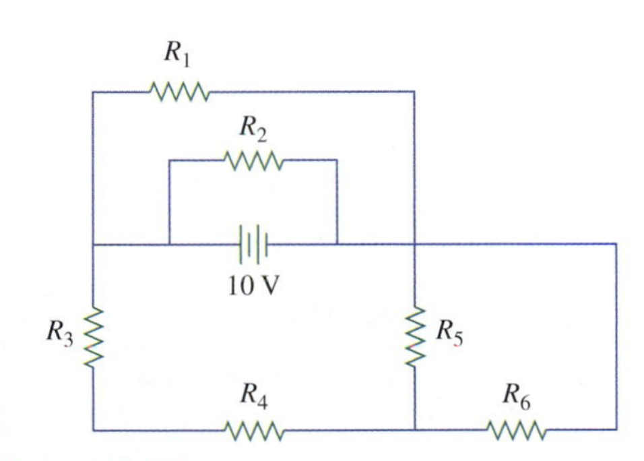

Figure 6.46
For Problem 6.1.

6.2 Determine R_{ab} in the circuit in Fig. 6.47.

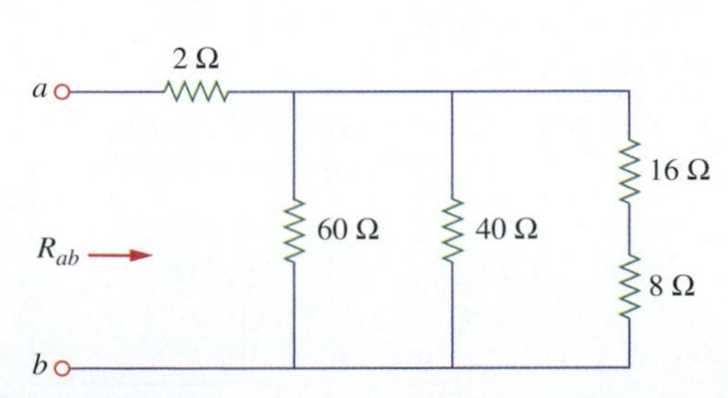

Figure 6.47
For Problem 6.2.

6.3 Find R_T in the circuit in Fig. 6.48.

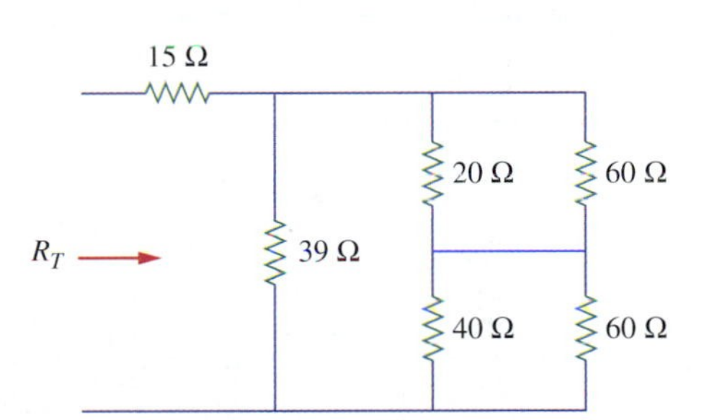

Figure 6.48
For Problem 6.3.

6.4 Consider the circuit of Fig. 6.49. Find the equivalent resistance between terminals *a-b*.

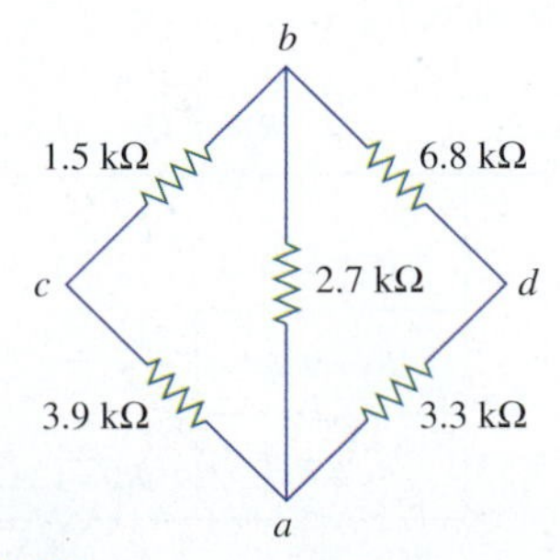

Figure 6.49
For Problem 6.4.

6.5 Find the equivalent resistance of the circuit in Fig. 6.50.

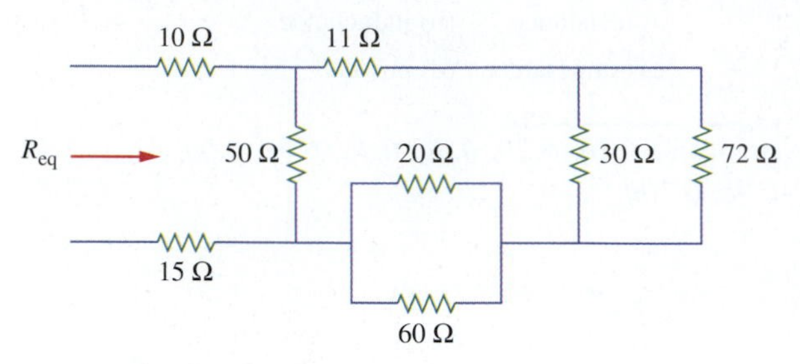

Figure 6.50
For Problem 6.5.

6.6 Find R_{eq} in the circuit in Fig. 6.51. Let $R = 5$ kΩ.

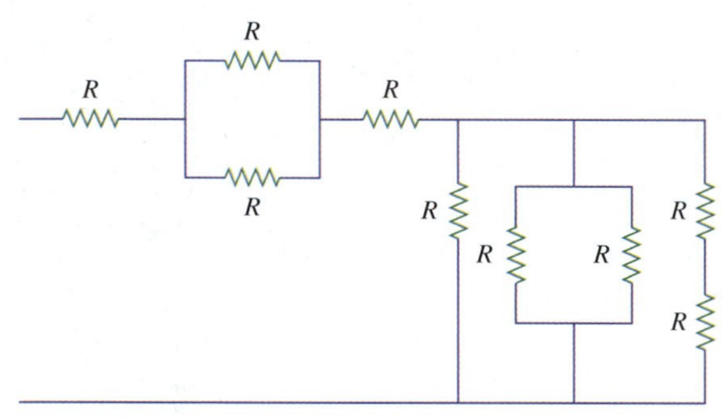

Figure 6.51
For Problem 6.6.

6.7 Determine R_{ab} in the circuit in Fig. 6.52.

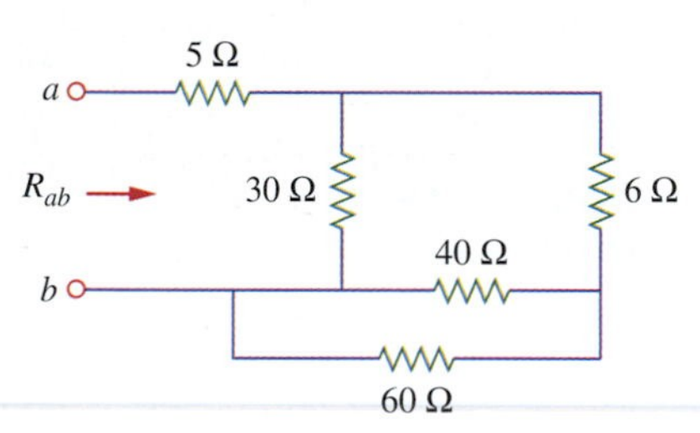

Figure 6.52
For Problem 6.7.

6.8 In the circuit in Fig. 6.53, determine the equivalent resistance R_{ab}.

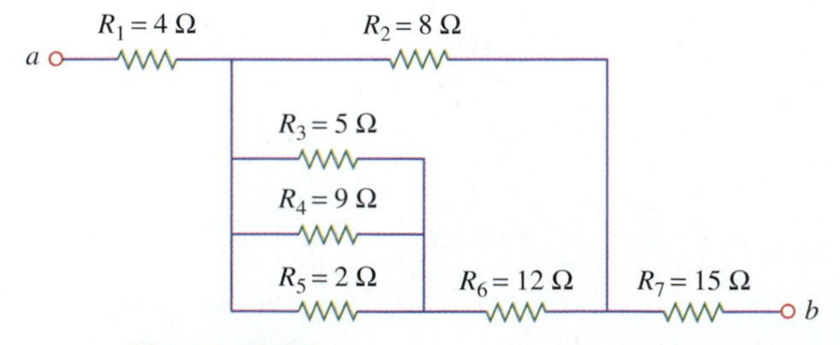

Figure 6.53
For Problem 6.8.

6.9 Determine the total resistance R_T in the circuit of Fig. 6.54.

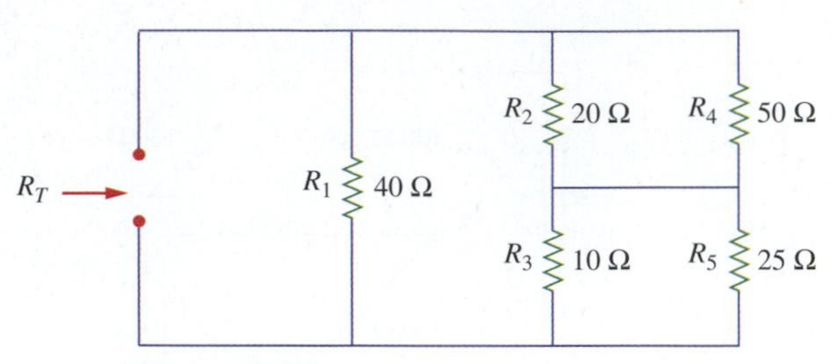

Figure 6.54
For Problem 6.9.

6.10 For the circuit in Fig. 6.55, find the equivalent resistance R_{eq} and the current I_T.

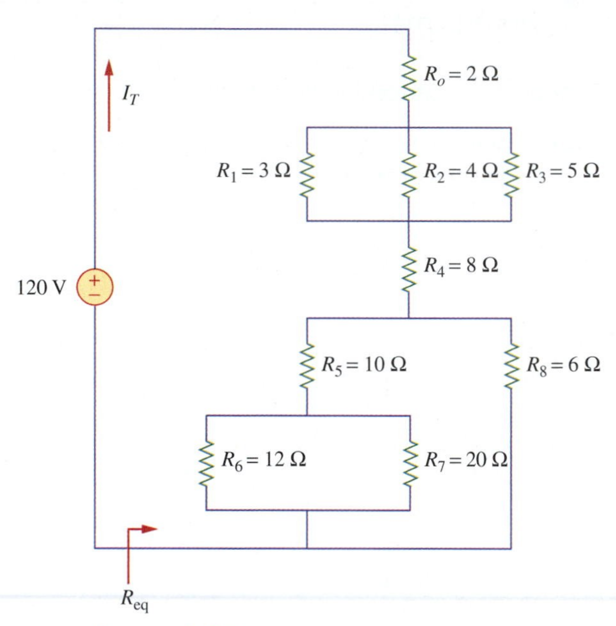

Figure 6.55
For Problem 6.10.

6.11 Calculate I_o in the circuit in Fig. 6.56.

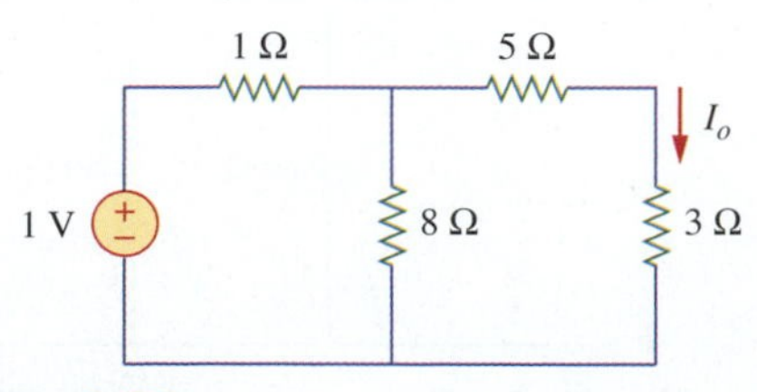

Figure 6.56
For Problem 6.11.

6.12 Determine the voltage drop across each resistor in Fig. 6.57.

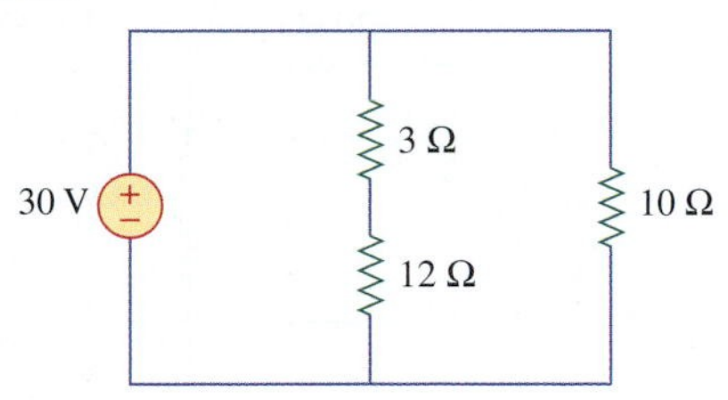

Figure 6.57
For Problem 6.12.

6.13 Find the current I in the circuit in Fig. 6.58.

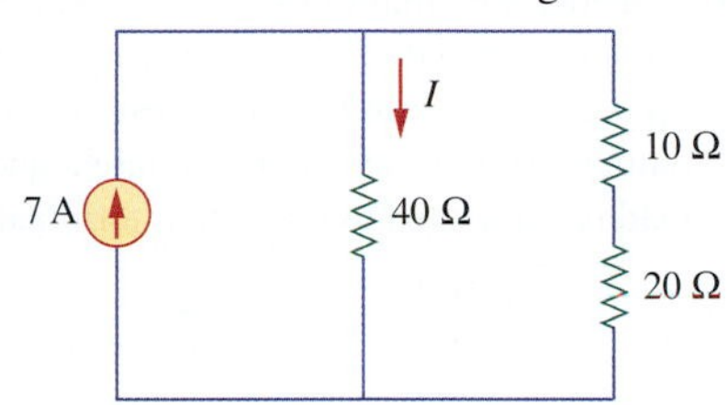

Figure 6.58
For Problem 6.13.

6.14 Find the voltage V_{ab} in the circuit in Fig. 6.59.

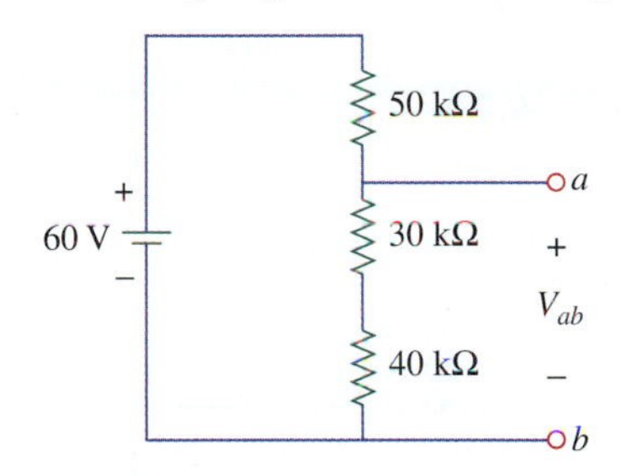

Figure 6.59
For Problem 6.14.

6.15 Find V_o in the circuit in Fig. 6.60.

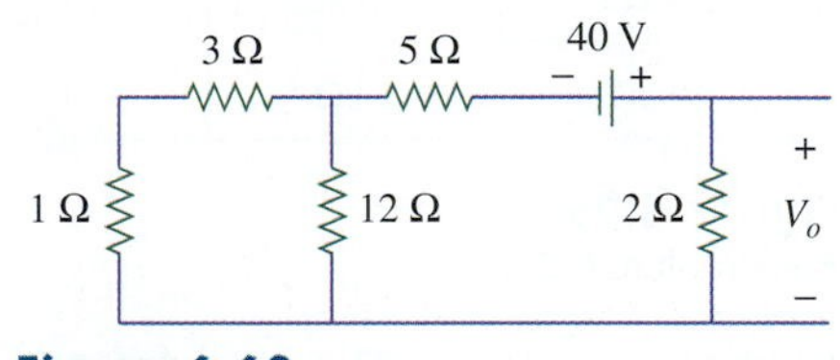

Figure 6.60
For Problem 6.15.

6.16 Determine I_o in the circuit in Fig. 6.61.

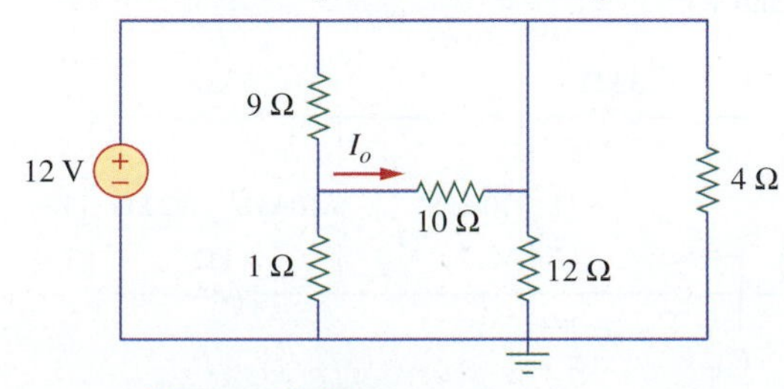

Figure 6.61
For Problems 6.16 and 6.51.

6.17 Calculate the voltage drop across each resistor in Fig. 6.62.

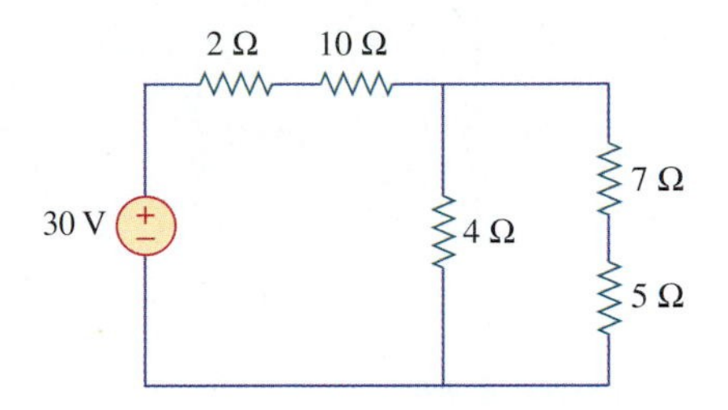

Figure 6.62
For Problem 6.17.

6.18 Find current I_x in the circuit in Fig. 6.63.

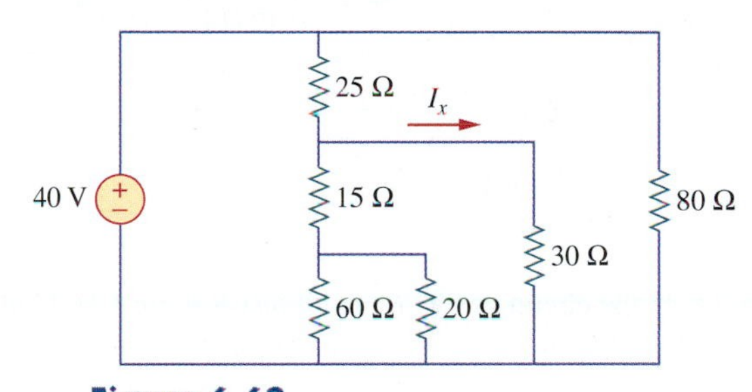

Figure 6.63
For Problem 6.18.

6.19 Determine the voltage drop across each resistor in Fig. 6.64.

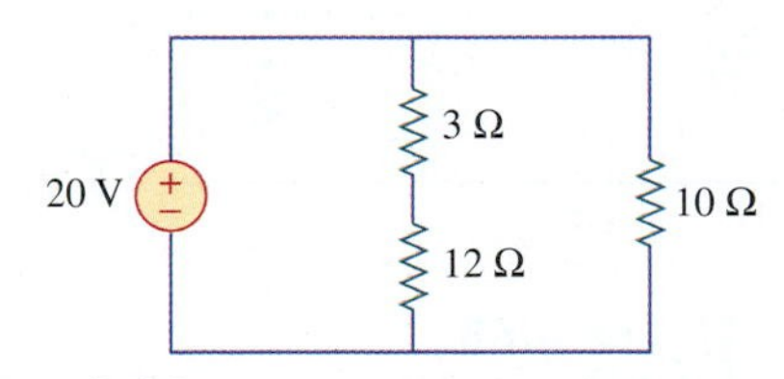

Figure 6.64
For Problem 6.19.

6.20 For the circuit in Fig. 6.65, find the node voltages V_1 and V_2.

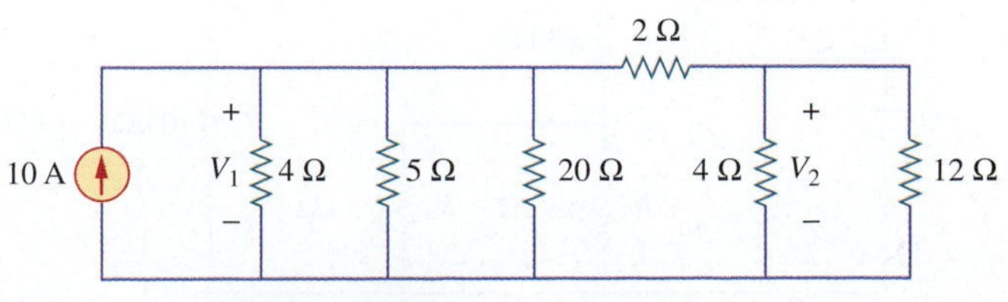

Figure 6.65
For Problem 6.20.

6.21 In the circuit shown in Fig. 6.66, find I_T.

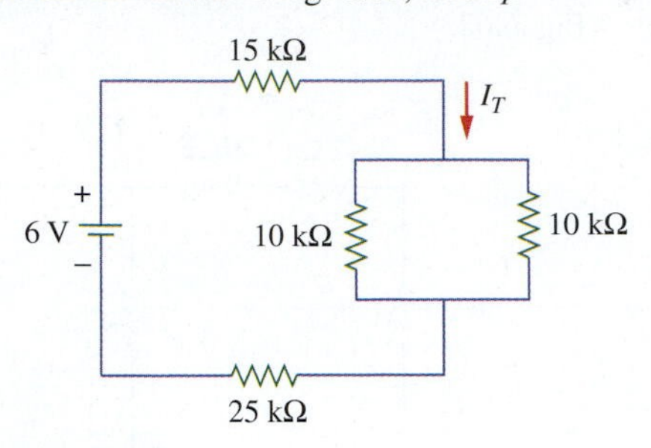

Figure 6.66
For Problem 6.21.

6.22 Find V_s and V_o in the circuit in Fig. 6.67.

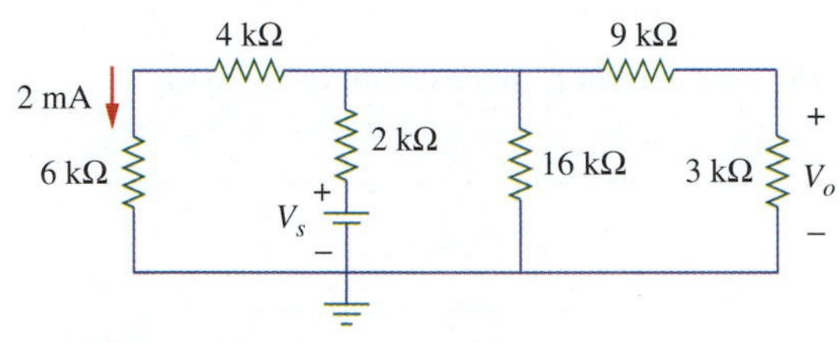

Figure 6.67
For Problem 6.22.

6.23 In the circuit of Fig. 6.68, all resistors are 1 Ω. Find R_T at (a) terminals *a–b* and (b) terminals *c–d*.

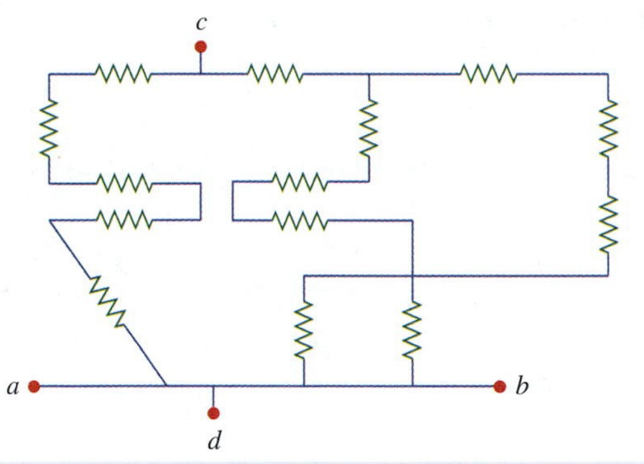

Figure 6.68
For Problem 6.23.

6.24 (a) For the circuit in Fig. 6.69, determine V_o.
(b) Repeat part (a) with R_4 short-circuited.
(c) Repeat part (a) with R_5 open-circuited.

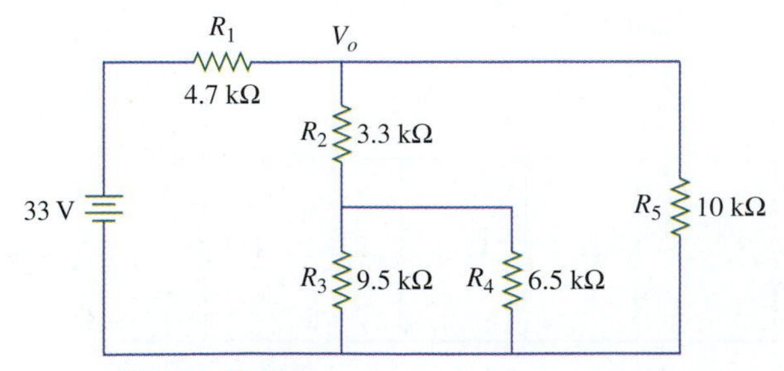

Figure 6.69
For Problem 6.24.

6.25 If *AB* in Fig. 6.70 is open-circuited, calculate V_{AB}.

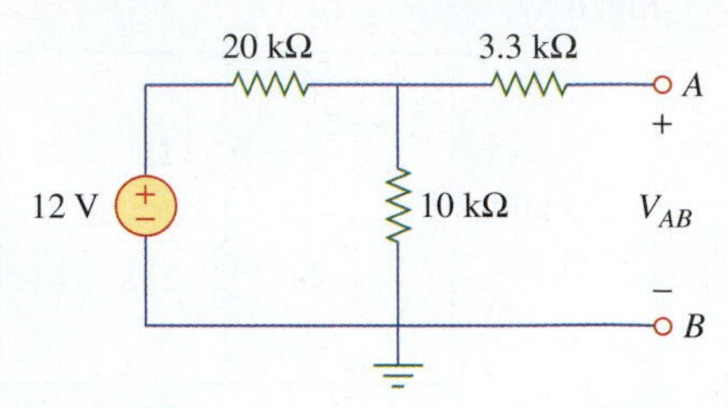

Figure 6.70
For Problem 6.25.

6.26 Consider the circuit in Fig. 6.71. Determine the resistance R_{AB} under the following conditions: (a) the output terminals are open-circuited, (b) the output terminals are short-circuited, and (c) a 200-Ω load is connected to the output terminals.

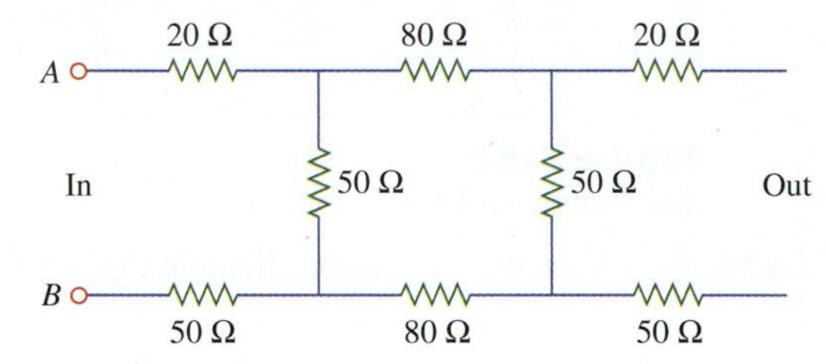

Figure 6.71
For Problem 6.26.

6.27 Find the equivalent resistance R_{AB} in the circuit of Fig. 6.72.

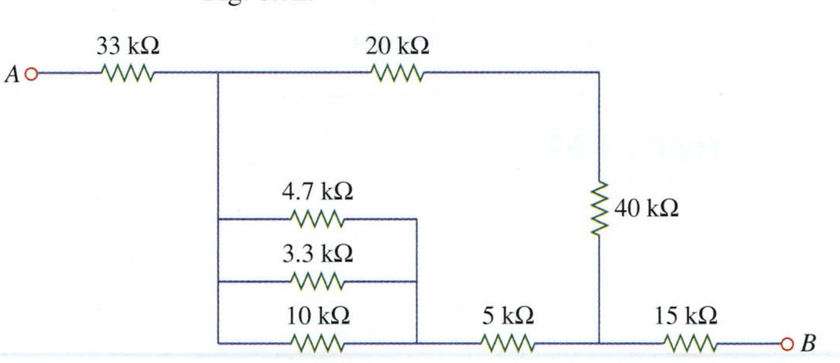

Figure 6.72
For Problem 6.27.

6.28 For the circuit of Fig. 6.73, find (a) the equivalent resistance R_{eq}, (b) the current *I*, (c) the total dissipated power, and (d) the voltages V_1 and V_2.

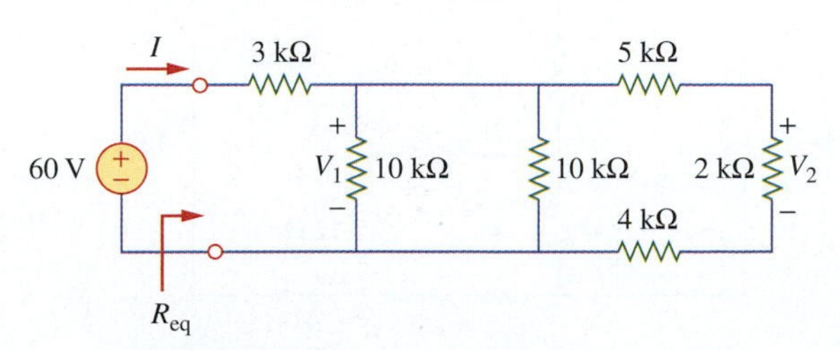

Figure 6.73
For Problem 6.28.

6.29 For the circuit shown in Fig. 6.74, find R_{eq} and I.

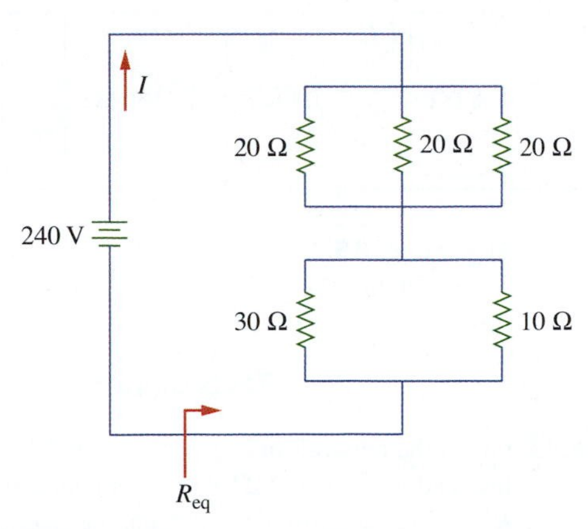

Figure 6.74
For Problem 6.29.

6.30 Refer to the circuit in Fig. 6.75. Determine the equivalent resistance R_{eq} and the current I_t.

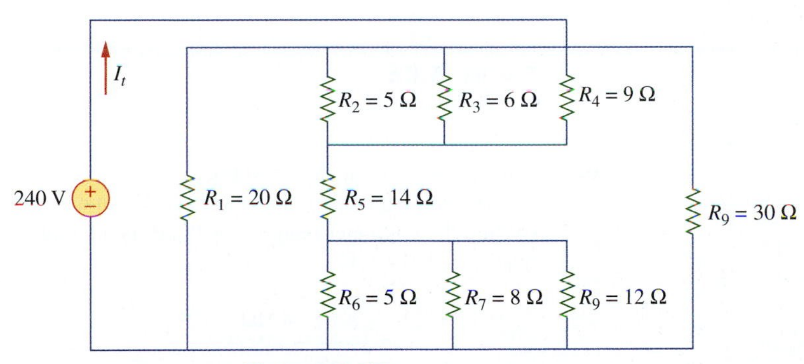

Figure 6.75
For Problem 6.30.

Section 6.3 Ladder Networks

6.31 Calculate the current through each resistor in Fig. 6.76.

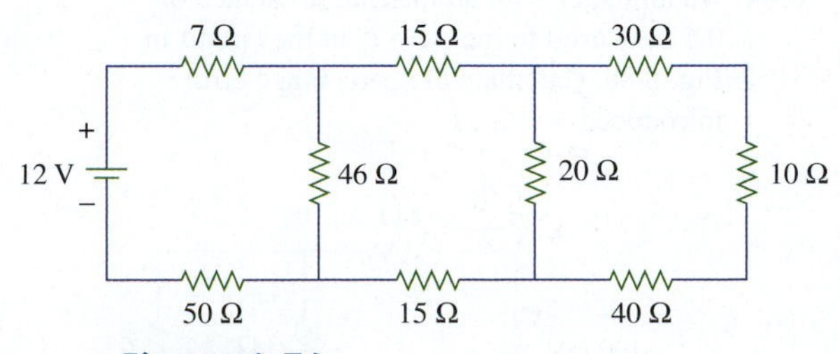

Figure 6.76
For Problems 6.31 and 6.32.

6.32 For the circuit in Fig. 6.76, find the voltage drop across each resistor.

6.33 Find V_o in the circuit in Fig. 6.77.

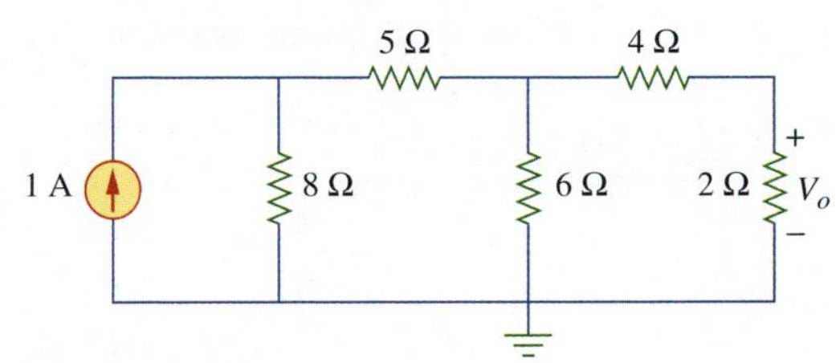

Figure 6.77
For Problems 6.33 and 6.52.

6.34 Find V_x in the circuit in Fig. 6.78.

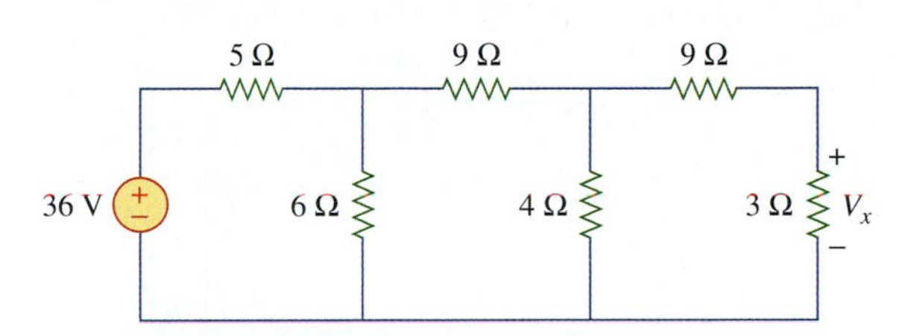

Figure 6.78
For Problem 6.34.

6.35 In the circuit in Fig. 6.79, express currents I_1 through I_6 in terms of I_o.

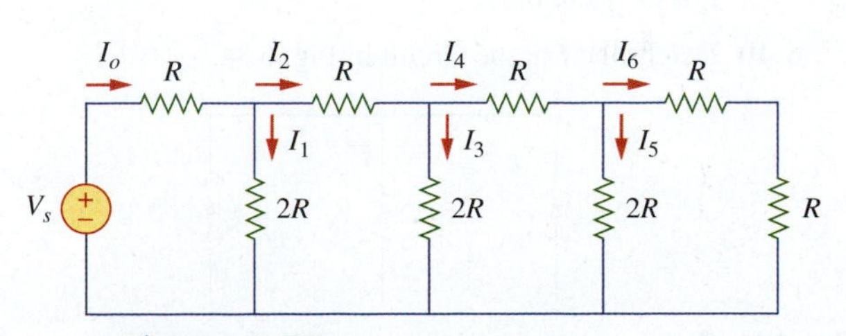

Figure 6.79
For Problem 6.35.

Section 6.4 Dependent Sources

6.36 Figure 6.80 shows a voltage-controlled voltage source. Find I and V_o.

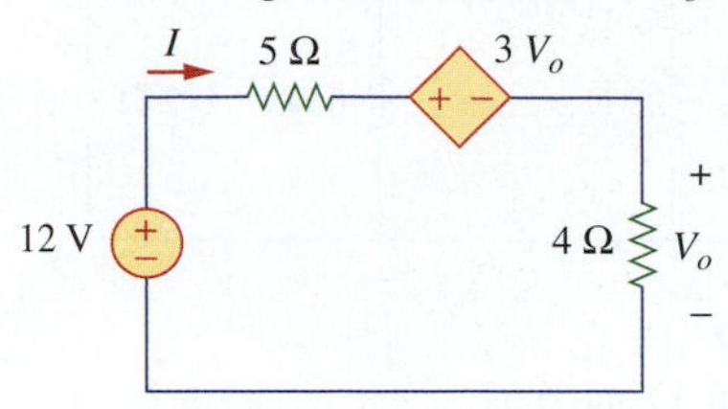

Figure 6.80
For Problem 6.36.

6.37 Calculate R and V in the circuit in Fig. 6.81.

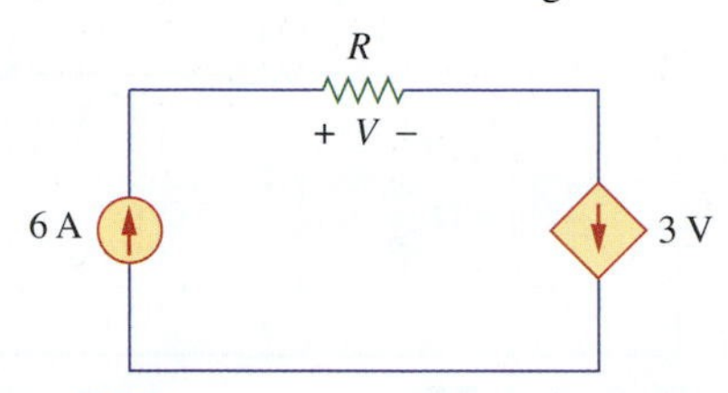

Figure 6.81
For Problem 6.37.

6.38 Find the value of R in the circuit in Fig. 6.82 if the current I_o is 2 A.

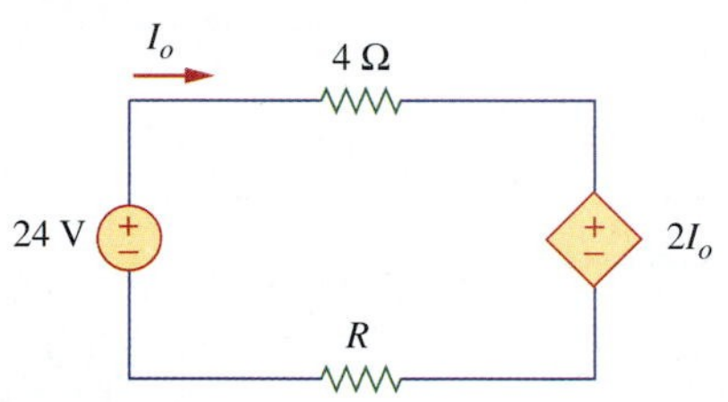

Figure 6.82
For Problem 6.38.

6.39 Find the current I in the circuit shown in Fig. 6.83.

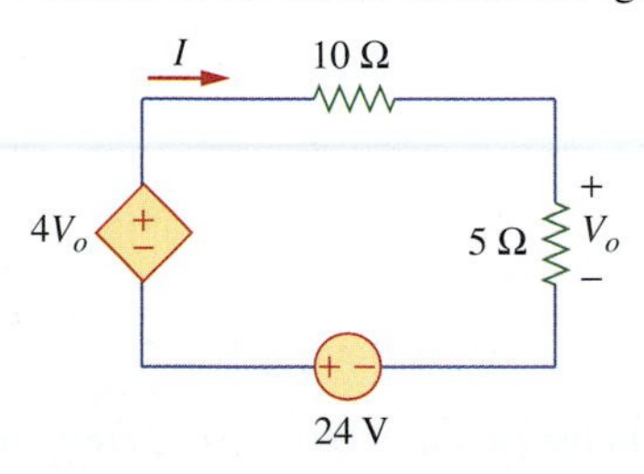

Figure 6.83
For Problem 6.39.

6.40 Determine I in the circuit in Fig. 6.84.

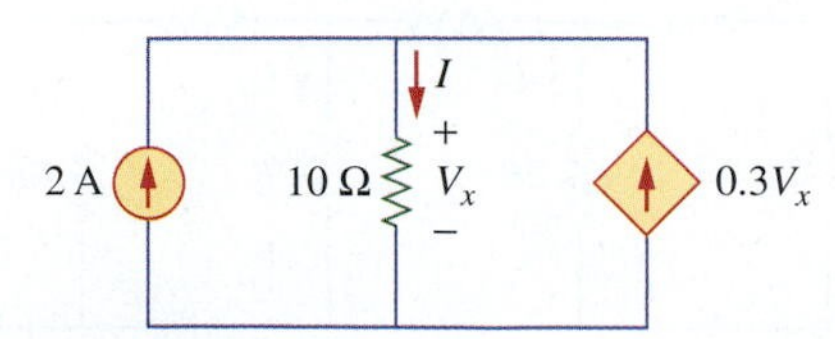

Figure 6.84
For Problem 6.40.

6.41 Determine V_o in the circuit in Fig. 6.85.

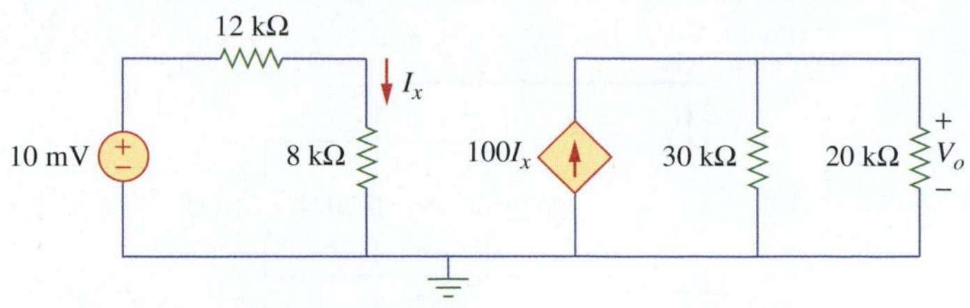

Figure 6.85
For Problem 6.41.

Section 6.5 Loading Effects of Instruments

6.42 Given the network in Fig. 6.86: (a) find V_1, (b) find the reading of a 10-MΩ voltmeter connected across R_1, and (c) determine the percentage error introduced.

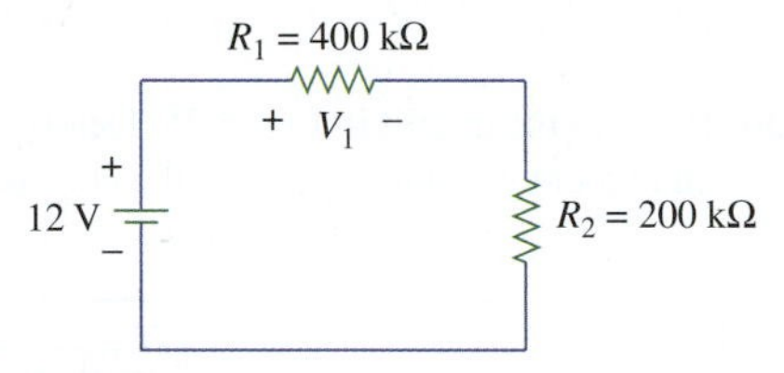

Figure 6.86
For Problem 6.42.

6.43 Consider the circuit shown in Fig. 6.87.
(a) Determine the open-circuit voltage V_{ab} and
(b) find the voltmeter reading for V_{ab} if its internal resistance is 12 MΩ.

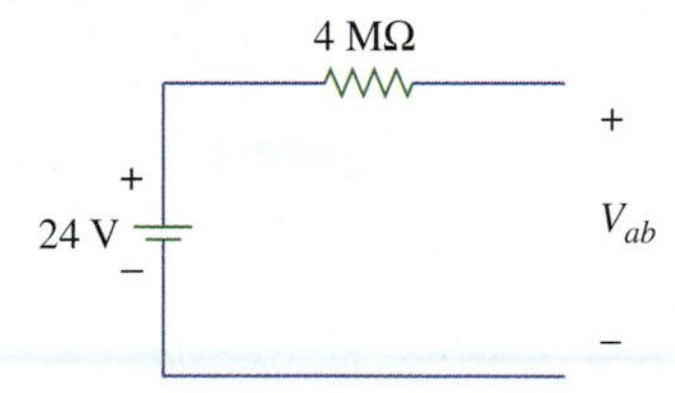

Figure 6.87
For Problem 6.43.

6.44 An ammeter with an internal resistance of 0.5 Ω is used to measure I_T in the circuit in Fig. 6.88. Calculate the percentage error introduced.

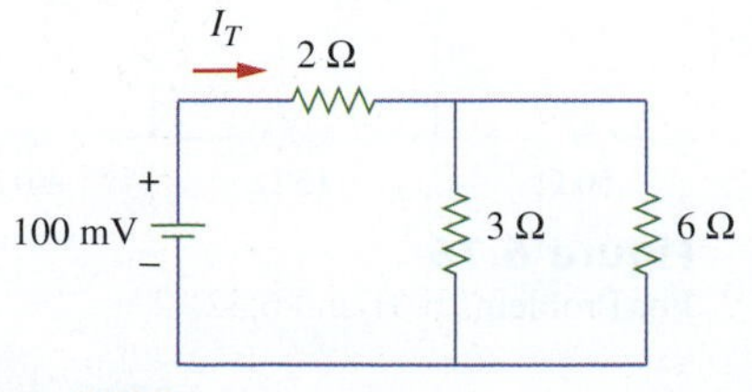

Figure 6.88
For Problem 6.44.

6.45 A 200-kΩ voltmeter is connected across the 40-kΩ resistor as shown in Fig. 6.89. (a) What would the voltmeter read? (b) What is the ideal value of V_1? (c) Calculate the percentage error.

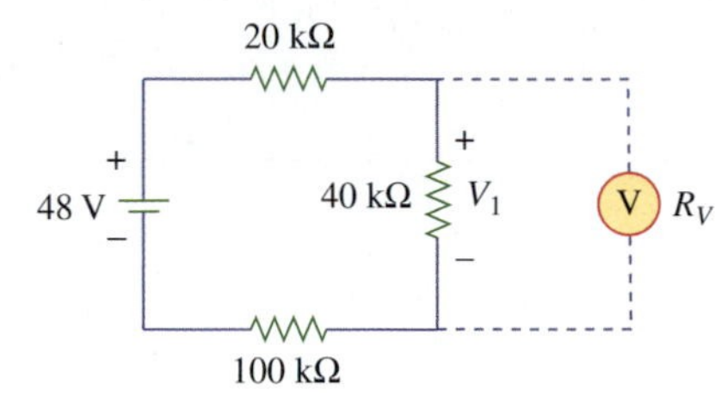

Figure 6.89
For Problem 6.45.

6.46 A voltmeter is used to measure the voltage across resistor R_3 in the circuit shown in Fig. 6.90. How much less is the voltage measured than the ideal or actual voltage? Assume $R_v = 20$ kΩ.

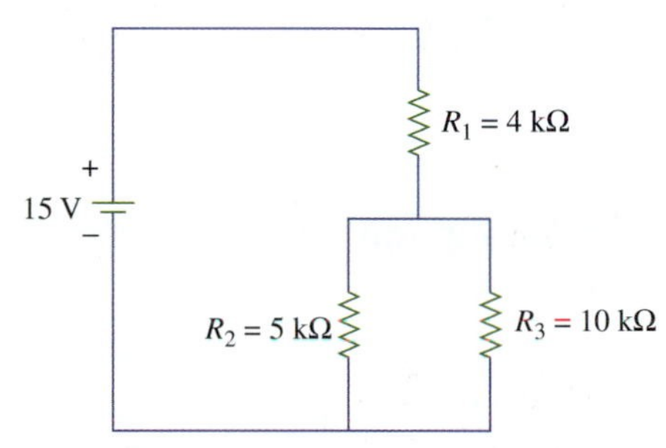

Figure 6.90
For Problem 6.46.

6.47 A digital multimeter (DMM) with internal resistance of 10 MΩ is used to measure the voltage across the 2-MΩ resistor as shown in Fig. 6.91. If the meter shows 20 V, calculate the value of the supply voltage V_s.

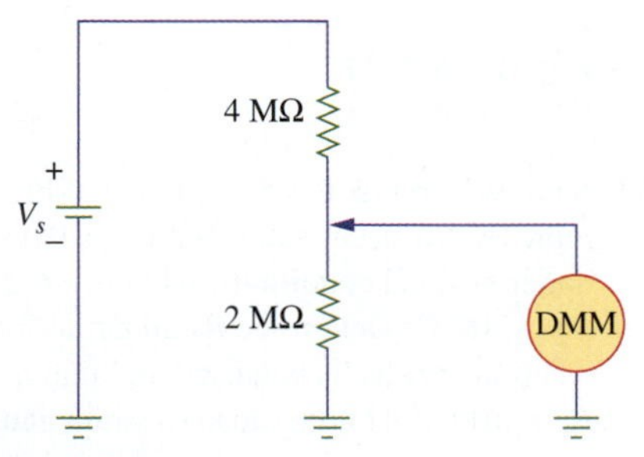

Figure 6.91
For Problem 6.47.

6.48 Two 850-kΩ resistors are connected in series with a 30-V voltage source. If a DMM with internal resistance of 10 MΩ is used to measure the voltage across either resistor, how much will the meter indicate? What is the percentage error?

Section 6.6 Computer Analysis

6.49 Find V_1 through V_3 in the ladder circuit of Fig. 6.92 using PSpice.

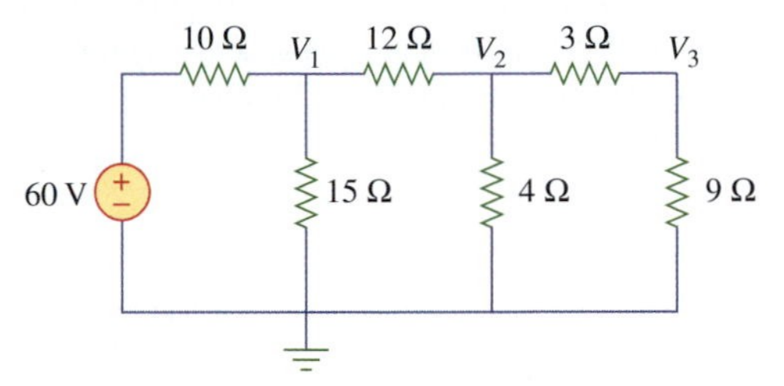

Figure 6.92
For Problem 6.49.

6.50 Use PSpice to simulate the circuit in Fig. 6.93 and obtain the node voltages V_1 through V_3.

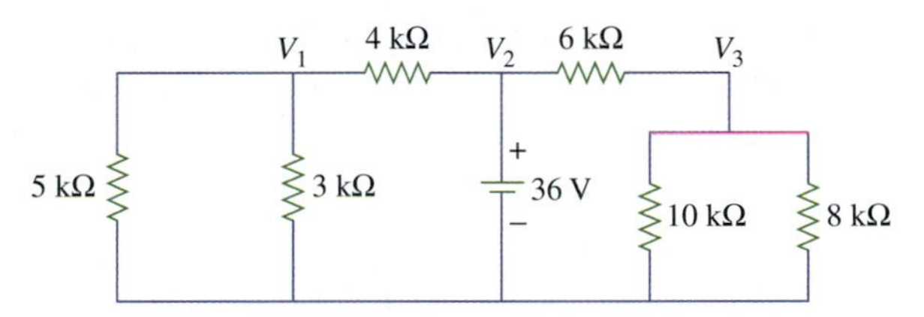

Figure 6.93
For Problems 6.50 and 6.56.

6.51 Use PSpice to find I_o in the circuit in Fig. 6.61.

6.52 Use PSpice to find V_o in the circuit in Fig. 6.77.

6.53 Determine V_x in the circuit in Fig. 6.94 using PSpice.

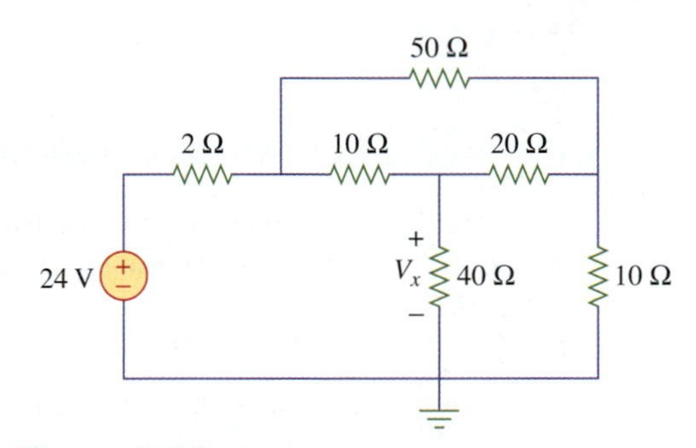

Figure 6.94
For Problems 6.53 and 6.55.

6.54 Use Multisim to find the resistance R_o in the circuit in Fig. 6.95.

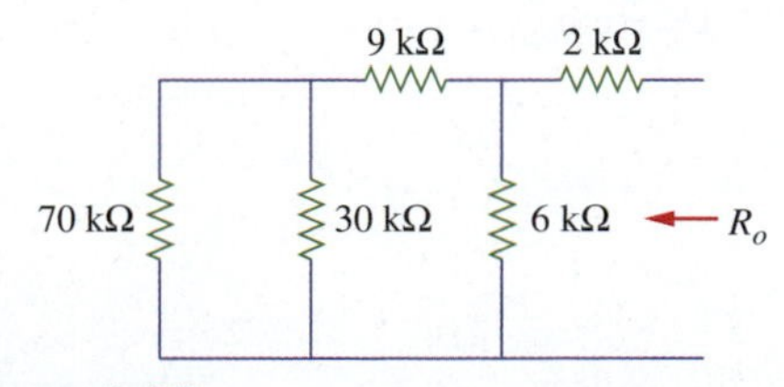

Figure 6.95
For Problem 6.54.

6.55 Find V_x in the circuit of Fig. 6.94 using Multisim.

6.56 Use Multisim to find V_1 through V_3 in the circuit in Fig. 6.93.

6.57 Refer to the circuit in Fig. 6.96. (a) Use Multisim to find the equivalent resistance R. (b) Determine currents I_1 and I_2 using Multisim.

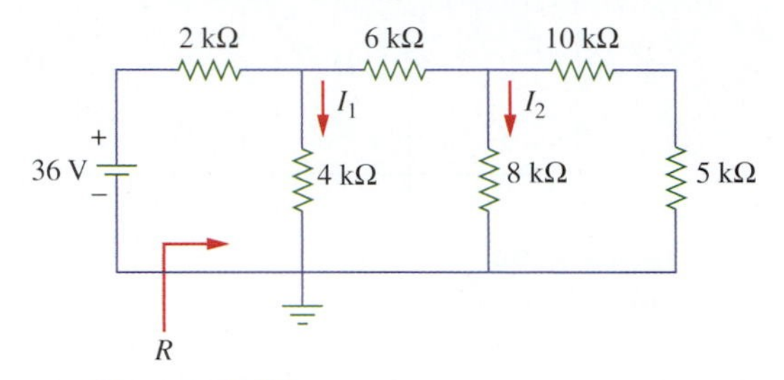

Figure 6.96
For Problem 6.57.

6.58 For the circuit shown in Fig. 6.97, use Multisim or PSpice to find the voltage at each node with respect to ground.

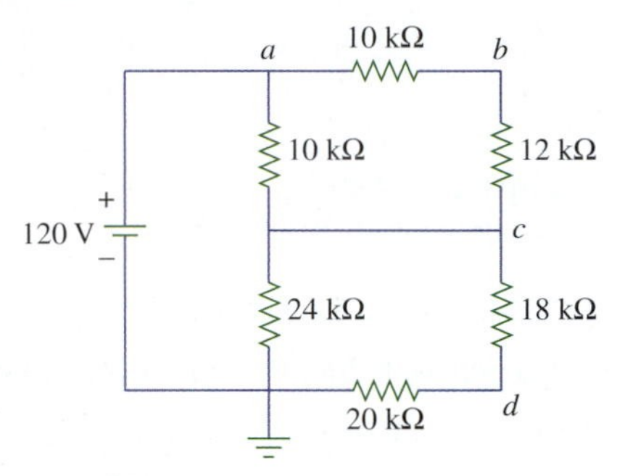

Figure 6.97
For Problem 6.58.

Section 6.7 Application: Wheatstone Bridge

6.59 The bridge circuit shown in Fig. 6.98 is balanced when $R_1 = 120\ \Omega$, $R_2 = 800\ \Omega$, and $R_3 = 300\ \Omega$. What is R_x?

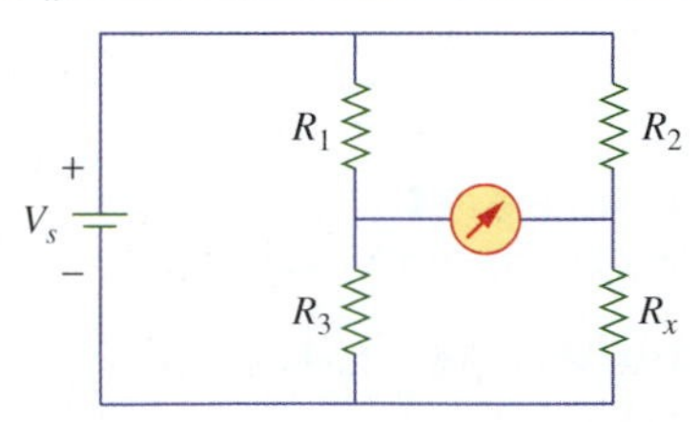

Figure 6.98
For Problem 6.59.

6.60 The Wheatstone bridge in Fig. 6.99 operates under a balanced condition. Determine R_x if $R_1 = 50\ \text{k}\Omega$, $R_2 = 30\ \text{k}\Omega$, and $R_3 = 100\ \Omega$.

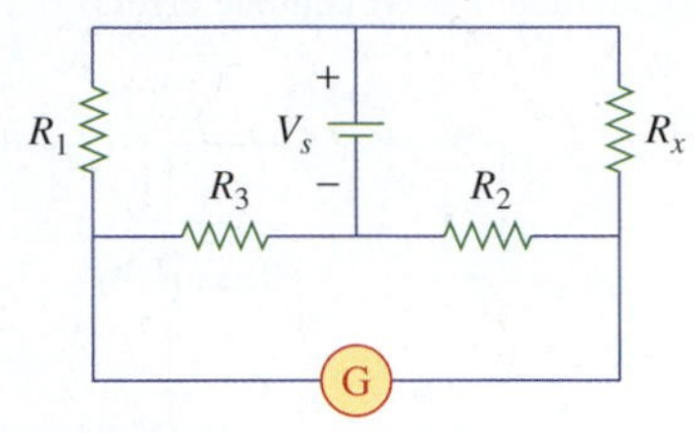

Figure 6.99
For Problem 6.60.

6.61 In the Wheatstone bridge shown in Fig. 6.100, find V_{ab} under a balanced condition.

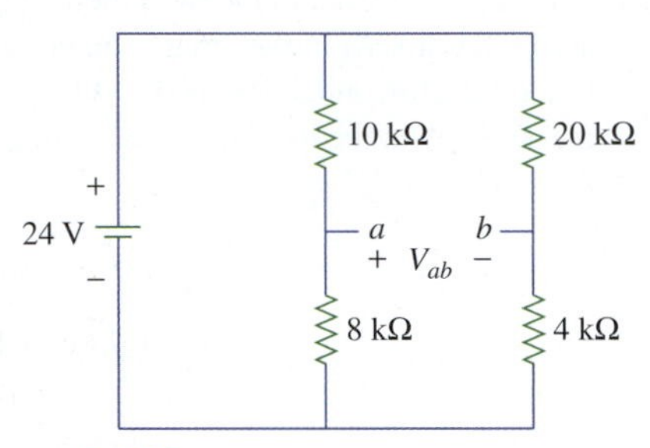

Figure 6.100
For Problem 6.61.

6.62 Find V_{ab} in the circuit in Fig. 6.101 under a balanced condition.

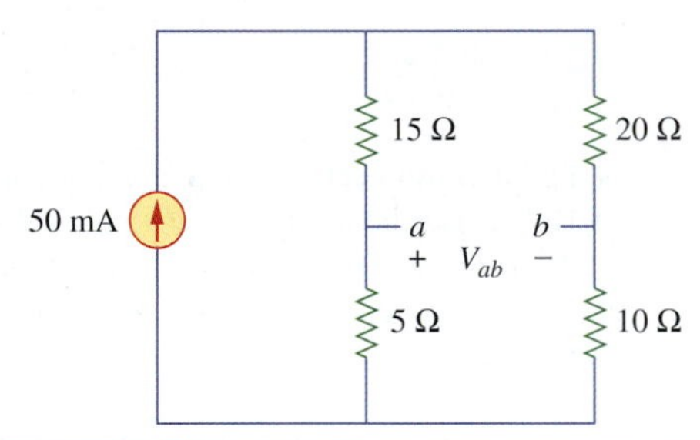

Figure 6.101
For Problem 6.62.

6.63 Suppose the unknown resistance in Fig. 6.41 represents a strain gauge with resistance $R_x = 100\ \Omega$ under no-load conditions, while $R_1 = 250\ \Omega$ and $R_2 = 300\ \Omega$. Determine R_3 under no-load conditions when the bridge is balanced and find its value when $R_x = 100.25\ \Omega$ under load (a strain gauge is a device used to measure the mechanical strain of an object).

chapter

7

Methods of Analysis

Always do right. This will gratify some people, and astonish the rest.

—Mark Twain

Enhancing Your Career

Career in Electronic Instrumentation

Instruments for measurement.
© Sarhan M. Musa

Engineering technology involves applying physical principles to design devices for the benefit of humanity. But physical principles cannot be understood without measurement. In fact, physicists often say that physics is the science that measures reality. Just as measurements are tools for understanding the physical world, instruments are tools for measurement.

Electronic instruments are used in all fields of science and engineering technology. They have proliferated in science and technology to the extent that it would be ridiculous to have a scientific or technical education without exposure to electronic instruments. For example, physicists, physiologists, chemists, and biologists must learn to use electronic instruments. For electrical engineering technology students in particular, the skill in operating digital and analog electronic instruments is crucial. Such instruments include ammeters, voltmeters, ohmmeters, oscilloscopes, spectrum analyzers, breadboards, and signal generators, some of which are shown in Figs. 7.1 through 7.7.

Beyond developing the skill for operating the instruments, some electrical engineering technologists specialize in designing and constructing electronic instruments. These technologists derive pleasure in building their own instruments. Many of them invent new circuits and patent them. Specialists in electronic instruments find employment in medical schools, hospitals, research laboratories, aircraft industries, and thousands of other industries where electronic instruments are routinely used.

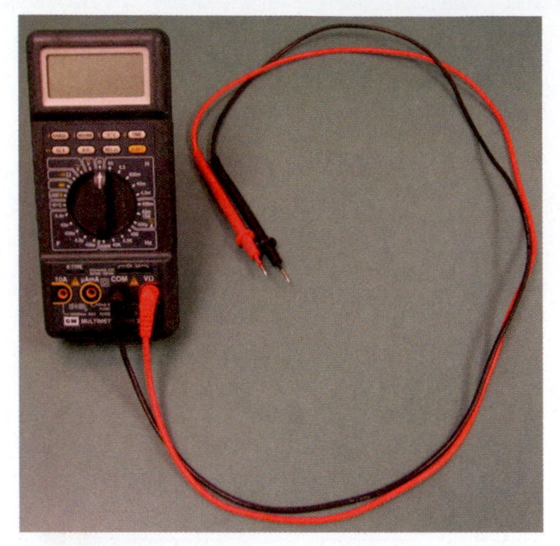

Figure 7.1
Digital multimeter with the test leads.
© Sarhan M. Musa

Figure 7.2
Analog microammeters.
© Sarhan M. Musa

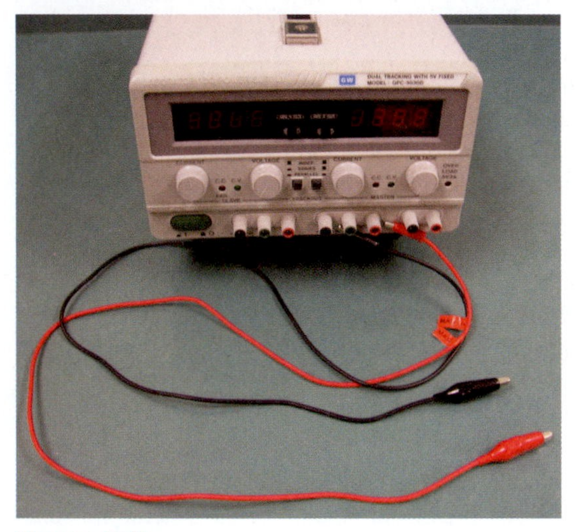

Figure 7.3
DC power supply with test leads.
© Sarhan M. Musa

Figure 7.4
Breadboard electronic trainer.
© Sarhan M. Musa

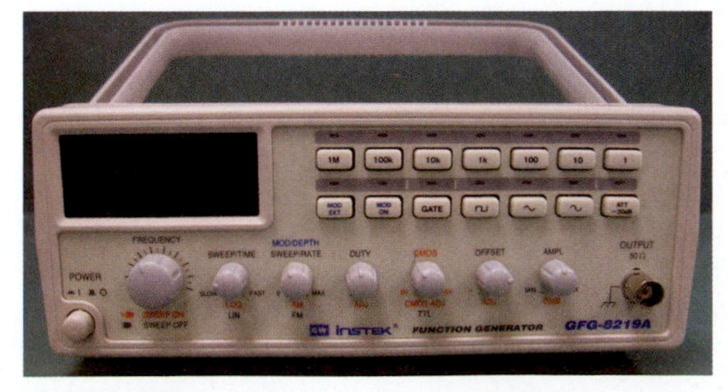

Figure 7.5
Function generator (ac).
© Sarhan M. Musa

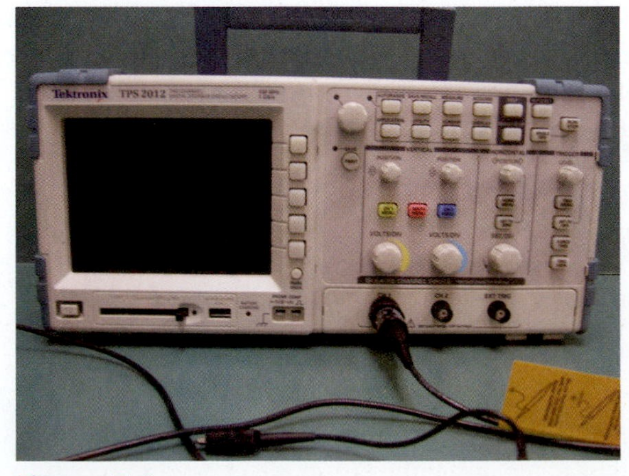

Figure 7.6
Two-channel digital oscilloscope with test leads.
© Sarhan M. Musa

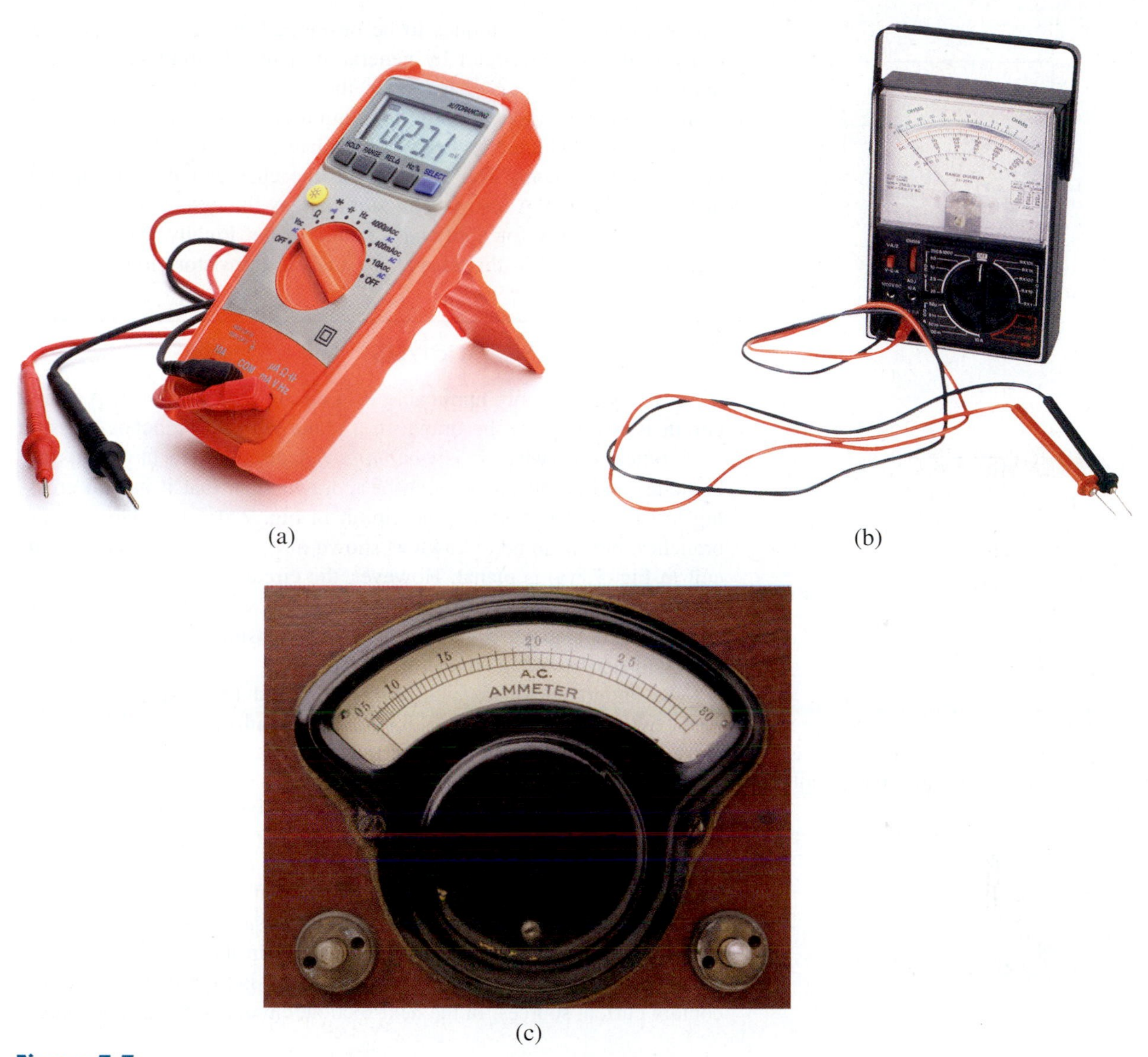

(a) (b) (c)

Figure 7.7
(a) Ohmmeter; (b) voltmeter; (c) ammeter.
(a) © iStock; (b) © Comstock/Jupiter RF; (c) © Sarhan M. Musa

7.1 Introduction

Having understood the fundamental laws of circuit theory (Ohm's law and Kirchhoff's laws), we are now prepared to apply these laws to develop two powerful techniques for circuit analysis: *mesh analysis,* which is based on a systematic application of Kirchhoff's voltage law (KVL) and *nodal analysis,* which is based on a systematic application of Kirchhoff's current law (KCL). (Thus, this chapter is just a formalized way of using what we learned in the previous chapters.) The two techniques are so important that this chapter should be regarded as the most important in this book. Students are therefore encouraged to pay careful attention.

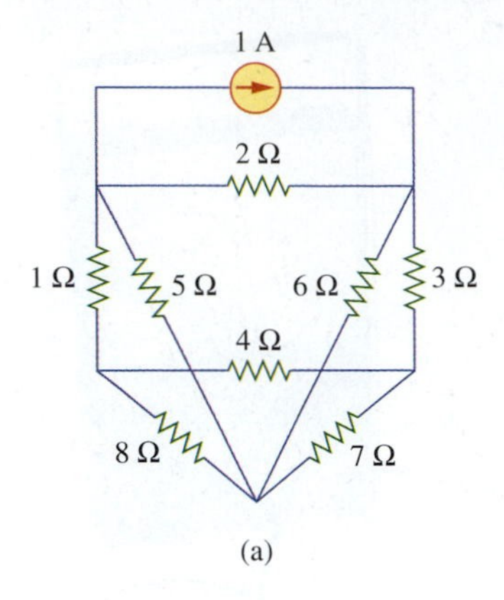

(a)

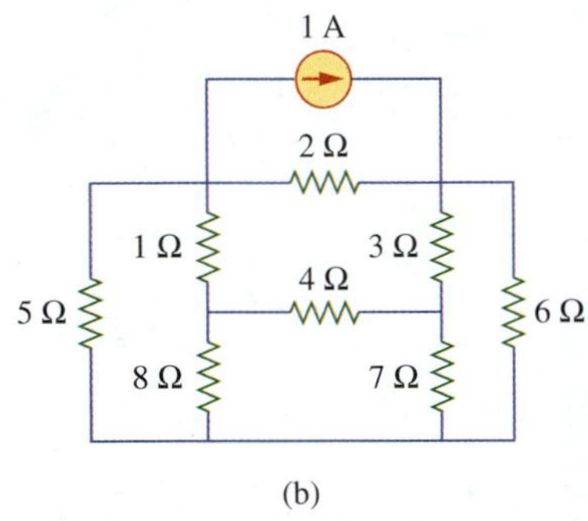

(b)

Figure 7.8
Circuit topology: (a) a planar circuit with crossing branches; (b) the same circuit redrawn with no crossing branches.

With the two techniques to be developed in this chapter, we can analyze almost any circuit by generating a set of simultaneous equations that are then solved to obtain the required values of current or voltage. One method of solving simultaneous equations involves Cramer's rule, which allows us to calculate circuit variables as a quotient of determinants. The examples in the chapter will illustrate this method; Appendix A also briefly summarizes the essentials the reader will need to know for applying Cramer's rule. Finally, we apply the techniques learned in this chapter to analyze transistor circuits.

7.2 Mesh Analysis

Mesh analysis[1] is only applicable to a circuit that is *planar*. A planar circuit is one that can be drawn in a plane with no branches crossing each other; otherwise, it is *nonplanar*. A circuit may have crossing branches and be planar if it can be redrawn such that it has no crossing branches. For example, the circuit in Fig. 7.8(a) has two crossing branches, but it can be redrawn as shown in Fig. 7.8(b). Hence, the circuit in Fig. 7.8(a) is planar. However, the circuit in Fig. 7.9 is nonplanar because there is no way to redraw it and keep the branches from crossing. Nonplanar circuits can be handled using nodal analysis, but they will not be considered in this text.

To understand mesh analysis, we should first explain what we mean by a mesh. In Fig. 7.10, for example, paths *abefa* and *bcdeb* are meshes but path *abcdefa* is not a mesh.[2]

A **mesh** is a loop that does not contain any other loops within it.

The current in a mesh is known as *mesh current*. In mesh analysis, we are interested in applying KVL to find the mesh currents in a given circuit. A mesh current does not necessarily correspond to any measurable physical current actually flowing in the circuit.

In this section, we will apply mesh analysis to circuits that do not contain current sources. In the next section, circuits with current sources

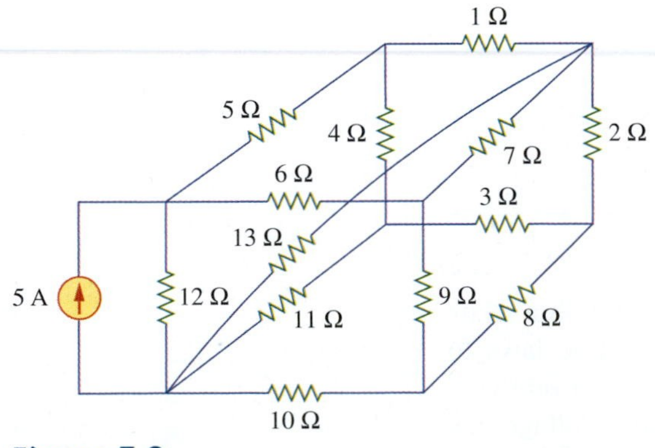

Figure 7.9
A nonplanar circuit.

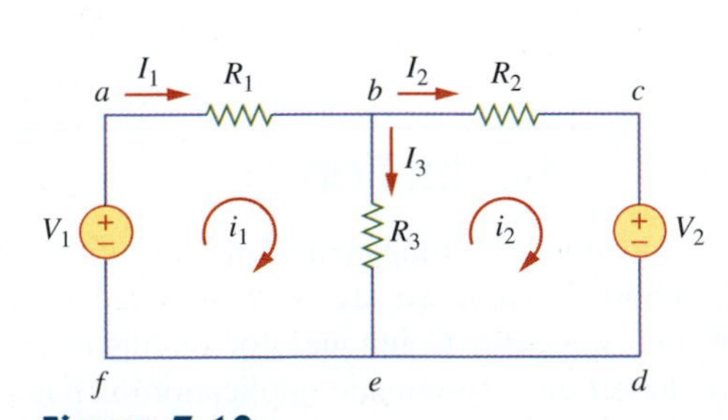

Figure 7.10
A circuit with two meshes.

[1] Note: The mesh analysis is also known as *loop analysis* or the *mesh-current method.*

[2] Note: Although path *abcdefa* is a loop and not a mesh, KVL still holds. This is the reason for loosely using the terms *loop analysis* and *mesh analysis* to mean the same thing.

will be considered. In the mesh analysis of a circuit with n meshes, we take the following three steps:

1. Assign mesh currents $i_1, i_2, \ldots, i_n$ to the n meshes.
2. Apply KVL to each of the n meshes. Use Ohm's law to express the voltages in terms of the mesh currents.
3. Solve the resulting n simultaneous equations to get the mesh currents.

To illustrate the steps, consider the circuit in Fig. 7.10. The first step requires that mesh currents i_1 and i_2 are assigned to meshes 1 and 2. Although a mesh current may be assigned to each mesh in an arbitrary direction, it is conventional to assume that each mesh current flows clockwise. If we analyze the same circuit in Fig. 7.10 by assuming the mesh currents flow counterclockwise, we would get the same result.

As the second step, we apply KVL to each mesh. Applying KVL to mesh 1, we obtain

$$-V_1 + R_1 i_1 + R_3(i_1 - i_2) = 0$$

Note that $-i_2 R_3$ is negative because its contribution is opposite the clockwise movement of i_1. Thus,

$$(R_1 + R_3)i_1 - R_3 i_2 = V_1 \tag{7.1}$$

For mesh 2, applying KVL gives

$$R_2 i_2 + V_2 + R_3(i_2 - i_1) = 0$$

or

$$-R_3 i_1 + (R_2 + R_3)i_2 = -V_2 \tag{7.2}$$

Note from Eq. (7.1) that the coefficient of i_1 is the sum of the resistances in the first mesh, while the coefficient of i_2 is the negative of the resistance common to meshes 1 and 2. The same observation should be made in Eq. (7.2). This may serve as a shortcut way to write the mesh equations. We will exploit this idea in Section 7.6.

The third step is to solve for the mesh currents. Putting Eqs. (7.1) and (7.2) in matrix form yields

$$\begin{bmatrix} R_1 + R_3 & -R_3 \\ -R_3 & R_2 + R_3 \end{bmatrix} \begin{bmatrix} i_1 \\ i_2 \end{bmatrix} = \begin{bmatrix} V_1 \\ -V_2 \end{bmatrix} \tag{7.3}$$

which can be solved to obtain the mesh currents i_1 and i_2. We are at liberty to use any technique for solving simultaneous equations. According to Eq. (4.1), if a circuit has n nodes, b branches, and ℓ independent loops or meshes, then $\ell = b - n + 1$. Hence, ℓ independent simultaneous equations are required to solve the circuit using mesh analysis. For this circuit, $b = 5$, and $n = 4$; therefore $\ell = 2$.

Notice that the branch currents are not mesh currents unless there is no neighboring mesh. To distinguish between the two types of currents, we use i for a mesh current and I for a branch current. The current elements I_1, I_2, and I_3 are algebraic sums of the mesh currents. It is evident from Fig. 7.10 that

$$I_1 = i_1, \quad I_2 = i_2, \quad I_3 = i_1 - i_2 \tag{7.4}$$

The third part of Eq. (7.4) is obtained by applying KCL to node b (or d).

$$i_1 = i_2 + I_3 \quad \Rightarrow \quad I_3 = i_1 - i_2$$

Example 7.1

For the circuit in Fig. 7.11, find the branch currents I_1, I_2, and I_3 using mesh analysis.

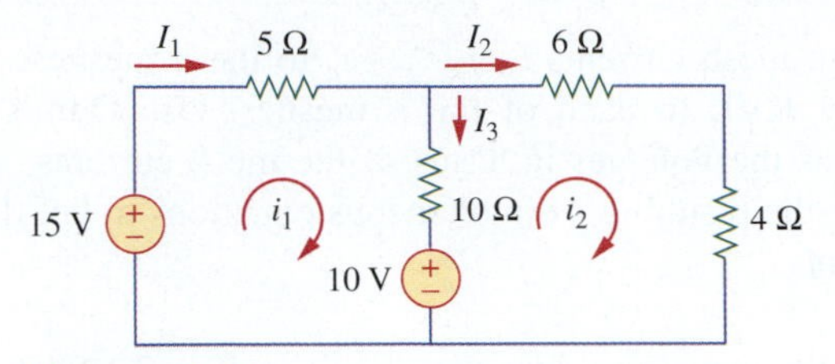

Figure 7.11
For Example 7.1.

Solution:

We first obtain the mesh currents using KVL. For mesh 1,

$$-15 + 5i_1 + 10(i_1 - i_2) + 10 = 0$$
$$15i_1 - 10i_2 = 5$$

Dividing by 5,

$$3i_1 - 2i_2 = 1 \tag{7.1.1}$$

For mesh 2,

$$6i_2 + 4i_2 + 10(i_2 - i_1) - 10 = 0$$
$$-10i_1 + 20i_2 = 10$$

Dividing by 10,

$$-i_1 + 2i_2 = 1 \tag{7.1.2}$$

■ **METHOD 1** Using the substitution method, we substitute Eq. (7.1.2) for i_1:

$$i_1 = 2i_2 - 1 \tag{7.1.2a}$$

Substitute this into Eq. (7.1.1),

$$6i_2 - 3 - 2i_2 = 1 \quad \Rightarrow \quad i_2 = 1 \text{ A}$$

From Eq. (7.1.2a),

$$i_1 = 2i_2 - 1 = 2 - 1 = 1 \text{ A}$$

Thus,

$$I_1 = i_1 = 1 \text{ A}, \quad I_2 = i_2 = 1 \text{ A}, \quad I_3 = i_1 - i_2 = 0$$

■ **METHOD 2** To use Cramer's rule, we cast Eqs. (7.1.1) and (7.1.2) in matrix form as

$$\begin{bmatrix} 3 & -2 \\ -1 & 2 \end{bmatrix} \begin{bmatrix} i_1 \\ i_2 \end{bmatrix} = \begin{bmatrix} 1 \\ 1 \end{bmatrix}$$

We obtain the determinants

$$\Delta = \begin{vmatrix} 3 & -2 \\ -1 & 2 \end{vmatrix} = 6 - 2 = 4$$

$$\Delta_1 = \begin{vmatrix} 1 & -2 \\ 1 & 2 \end{vmatrix} = 2 + 2 = 4, \qquad \Delta_2 = \begin{vmatrix} 3 & 1 \\ -1 & 1 \end{vmatrix} = 3 + 1 = 4$$

Thus,

$$i_1 = \frac{\Delta_1}{\Delta} = 1\text{ A}, \qquad i_2 = \frac{\Delta_2}{\Delta} = 1\text{ A}$$

as before.

Practice Problem 7.1

Calculate the mesh currents i_1 and i_2 in the circuit of Fig. 7.12.

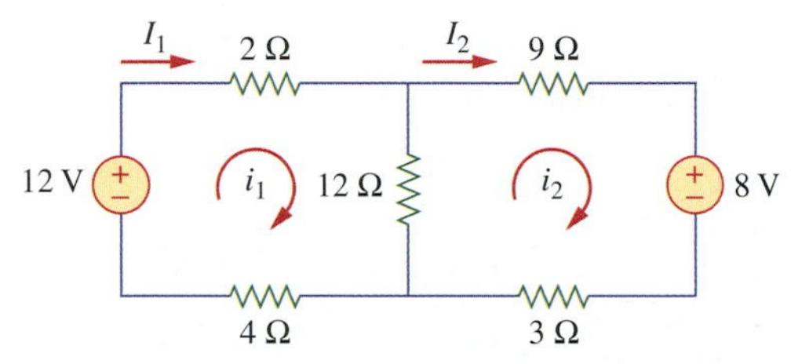

Figure 7.12
For Practice Problem 7.1.

Answer: $i_1 = 0.667$ A; $i_2 = 0$ A

Example 7.2

Use mesh analysis to find current I_o in the circuit in Fig. 7.13.

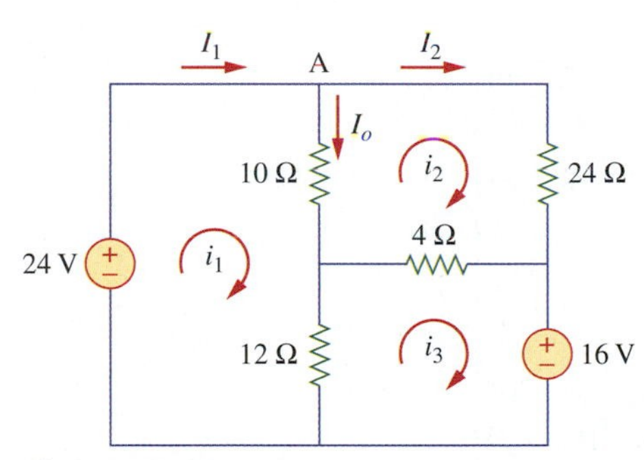

Figure 7.13
For Example 7.2.

Solution:
We apply KVL to the three meshes in turn. For mesh 1,

$$-24 + 10(i_1 - i_2) + 12(i_1 - i_3) = 0$$

or

$$11i_1 - 5i_2 - 6i_3 = 12 \tag{7.2.1}$$

For mesh 2,

$$24i_2 + 4(i_2 - i_3) + 10(i_2 - i_1) = 0$$

or

$$-5i_1 + 19i_2 - 2i_3 = 0 \tag{7.2.2}$$

For mesh 3,

$$+16 + 12(i_3 - i_1) + 4(i_3 - i_2) = 0$$

or

$$-3i_1 - i_2 + 4i_3 = -4 \tag{7.2.3}$$

In matrix form, Eqs. (7.2.1) through (7.2.3) become

$$\begin{bmatrix} 11 & -5 & -6 \\ -5 & 19 & -2 \\ -3 & -1 & 4 \end{bmatrix} \begin{bmatrix} i_1 \\ i_2 \\ i_3 \end{bmatrix} = \begin{bmatrix} 12 \\ 0 \\ -4 \end{bmatrix}$$

The determinants are obtained as

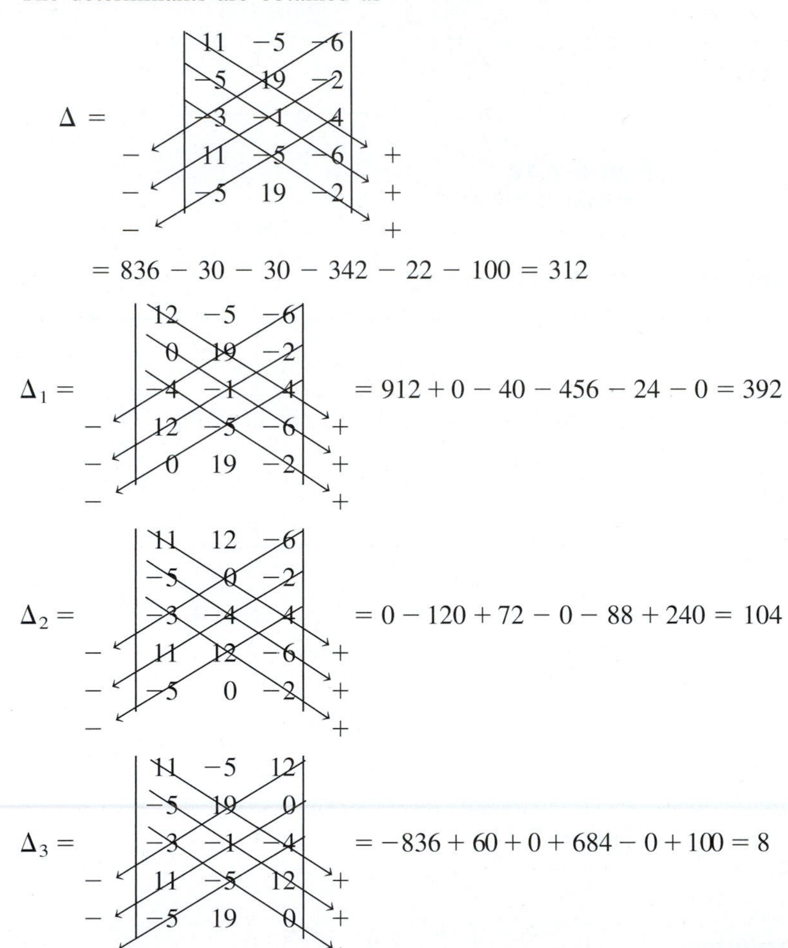

We calculate the mesh currents using Cramer's rule as

$$i_1 = \frac{\Delta_1}{\Delta} = \frac{392}{312} = 1.2564 \text{ A}$$

$$i_2 = \frac{\Delta_2}{\Delta} = \frac{104}{312} = 0.3333 \text{ A}$$

$$i_3 = \frac{\Delta_3}{\Delta} = \frac{8}{312} = 0.0256 \text{ A}$$

Thus,

$$I_o = i_1 - i_2 = 0.9231 \text{ A}$$

Practice Problem 7.2

Using mesh analysis, find I_o in the circuit in Fig. 7.14.

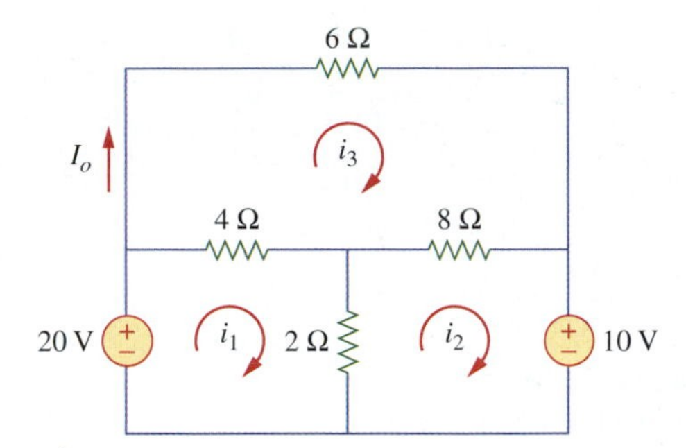

Figure 7.14
For Practice Problem 7.2.

Answer: 1.667 A

Example 7.3

Determine the mesh currents i_1 and i_2 in the circuit in Fig. 7.15.

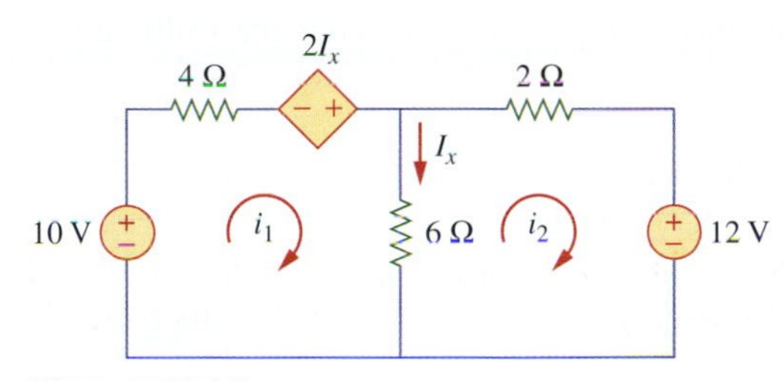

Figure 7.15
For Example 7.3.

Solution:
For mesh 1,

$$-10 - 2I_x + 10i_1 - 6i_2 = 0$$

But $I_x = i_1 - i_2$. Hence,

$$10 = -2i_1 + 2i_2 + 10i_1 - 6i_2 \quad \Rightarrow \quad 5 = 4i_1 - 2i_2 \qquad \textbf{(7.3.1)}$$

For mesh 2,

$$12 + 8i_2 - 6i_1 = 0 \quad \Rightarrow \quad 6 = 3i_1 - 4i_2 \qquad \textbf{(7.3.2)}$$

Solving Eqs. (7.3.1) and (7.3.2) leads to

$$i_1 = 0.8 \text{ A}, \qquad i_2 = -0.9 \text{ A}$$

The negative value for i_2 shows that the current is actually flowing counterclockwise.

Practice Problem 7.3

Apply mesh analysis to find the voltage V_o in the circuit in Fig. 7.16.

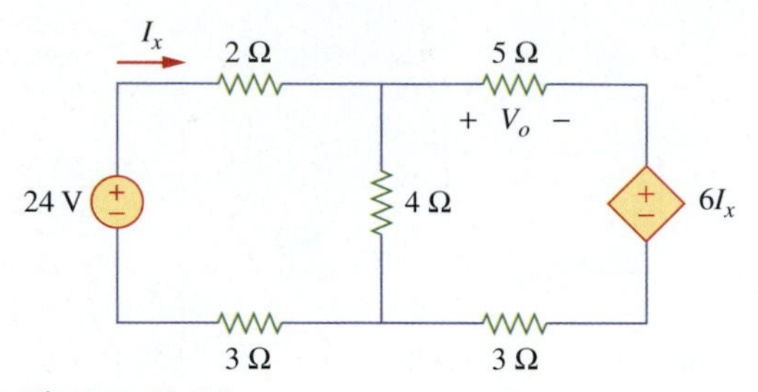

Figure 7.16
For Practice Problem 7.3.

Answer: 2.4 V

7.3 Mesh Analysis with Current Sources

Applying mesh analysis to circuits containing current sources may appear complicated. It is actually much easier than what we encountered in the previous section because the presence of the current sources reduces the number of equations. Consider the following two possible cases.

■ **CASE 1** A current source exists only in one mesh. Consider the circuit in Fig. 7.17, for example. We set $i_2 = -5$ A; it is negative because the direction of i_2 is opposite to the direction of the 5-A current source. We write a mesh equation for the other mesh in the usual way, that is,

$$-10 + 4i_1 + 6(i_1 - i_2) = 0$$

or

$$10i_1 = 6i_2 + 10 = -30 + 10 = -20 \quad \Rightarrow \quad i_1 = -2 \text{ A}$$

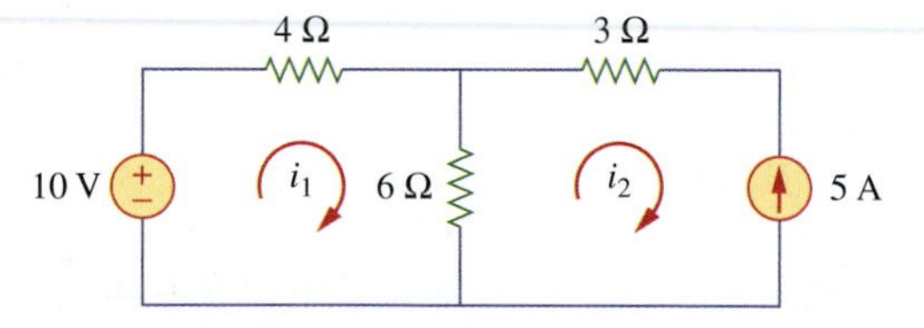

Figure 7.17
A circuit with a current source.

■ **CASE 2** A current source exists between two meshes. Consider the circuit in Fig. 7.18(a), for example. We create a *supermesh* by excluding the current source and any elements connected in series with it, as shown in Fig. 7.18(b). Thus,

> A **supermesh** results when two meshes have a current source in common.

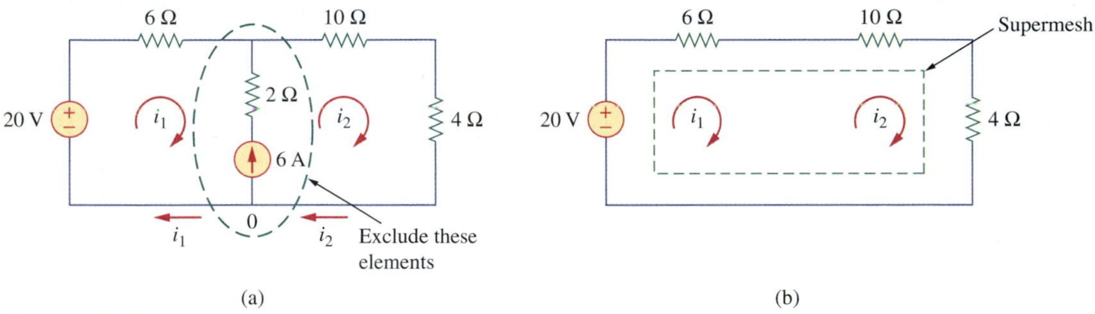

Figure 7.18
(a) Two meshes having a current source in common; (b) a supermesh, created by excluding the current sources.

A supermesh occurs when a current source (dependent or independent) is located between two meshes. As shown in Fig. 7.18(b), we create a supermesh as the periphery of the two meshes and treat it differently. (If a circuit has two or more supermeshes that intersect, they should be combined to form a larger supermesh.) Why treat the supermesh differently? Mesh analysis applies KVL, which requires that we know the voltage across each branch, and we do not know the voltage across a current source in advance. However, a supermesh must satisfy KVL like any other mesh.

To handle the supermesh, we treat it as if the current source is temporarily absent. This leads to one equation that incorporates two mesh currents. Therefore, applying KVL to the supermesh in Fig. 7.18(b) gives

$$-20 + 6i_1 + 10i_2 + 4i_2 = 0$$

or

$$6i_1 + 14i_2 = 20 \quad \textbf{(7.5)}$$

We apply KCL to a node in the branch where the two meshes intersect. Applying KCL to node 0 in Fig. 7.18(a) gives

$$i_2 = i_1 + 6 \quad \textbf{(7.6)}$$

Solving Eqs. (7.5) and (7.6), we get

$$i_1 = -3.2 \text{ A}, \qquad i_2 = 2.8 \text{ A} \quad \textbf{(7.7)}$$

Note the following properties of a supermesh:

1. The current source in the supermesh is not completely ignored; it provides the constraint equation necessary to solve for the mesh currents.
2. A supermesh has no current of its own.
3. A supermesh requires the application of both KVL and KCL.

Example 7.4

For the circuit in Fig. 7.19(a), find i_1 through i_3 using mesh analysis.

Solution:

We notice that meshes 2 and 3 form a supermesh because they have a current source in common. We also notice that $i_1 = 2$ A. We apply KVL to the supermesh shown in Fig. 7.19(b), and obtain

$$4(i_2 - i_1) + 10 + 5i_2 + 3i_3 = 0 \quad \Rightarrow \quad -4i_1 + 9i_2 + 3i_3 = -10$$

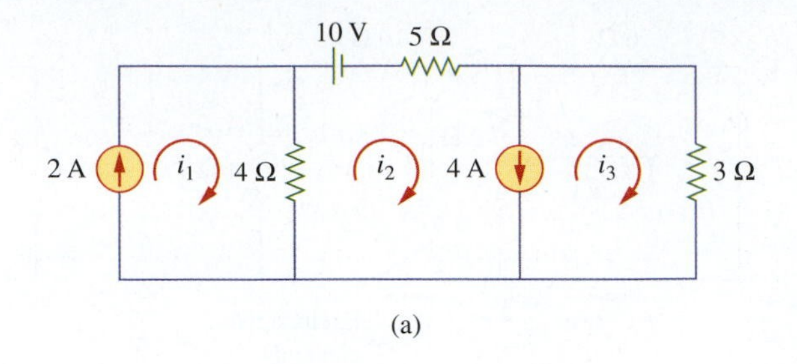

(a)

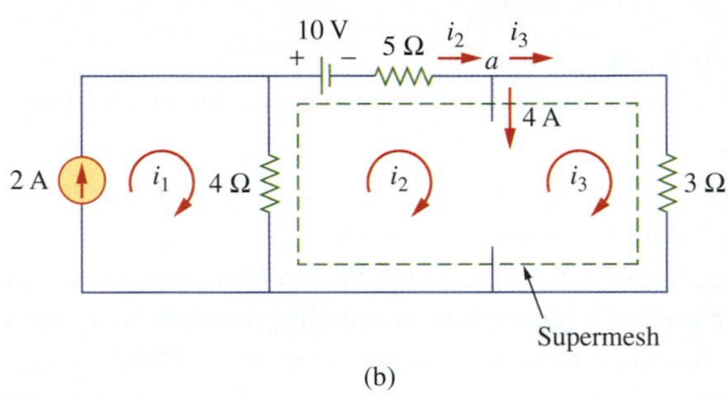

(b)

Figure 7.19
For Example 7.4.

But $i_1 = 2$ A, so

$$9i_2 + 3i_3 = -10 + 8 = -2 \qquad \textbf{(7.4.1)}$$

At node a in Fig. 7.19(b), KCL gives

$$i_2 = i_3 + 4 \qquad \textbf{(7.4.2)}$$

Substituting this into Eq. (7.4.1) gives

$$9(i_3 + 4) + 3i_3 = -2 \quad \Rightarrow \quad 12i_3 = -38$$

or

$$i_3 = -38/12 = -3.167 \text{ A}$$

From Eq. (7.4.2),

$$i_2 = i_3 + 4 = -3.167 + 4 = 0.833 \text{ A}$$

Thus,

$$i_1 = 2 \text{ A}, \qquad i_2 = 0.833 \text{ A}, \qquad i_3 = -3.167 \text{ A}$$

Practice Problem 7.4

Use mesh analysis to determine i_1 through i_3 in the circuit of Fig. 7.20.

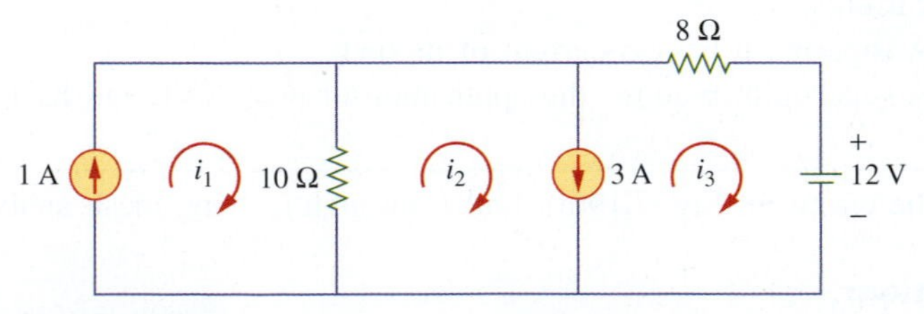

Figure 7.20
For Practice Problem 7.4.

Answer: $i_1 = 1$ A; $i_2 = 1.222$ A; $i_3 = -1.778$ A

7.4 Nodal Analysis

Mesh analysis applies KVL to find unknown currents, while nodal analysis applies KCL to determine unknown voltages. To simplify matters, we shall assume that circuits in this section do not contain voltage sources. Circuits that contain voltage sources will be analyzed in the next section.

In *nodal analysis*[3], we are interested in finding the node voltages. Given a circuit with n nodes without voltage sources, the nodal analysis of the circuit involves taking the following three steps:

1. Select a node as the reference node. Assign voltages $V_1, V_2, \dots, V_{n-1}$, to the remaining $n - 1$ nodes. The voltages are referenced with respect to the reference node.
2. Apply KCL to each of the $(n - 1)$ nonreference nodes. Use Ohm's law to express the branch currents in terms of node voltages. (Do not apply KCL to the reference node.)
3. Solve the resulting simultaneous equations to obtain the unknown node voltages.

We shall now explain how to apply these three steps.

The first step in nodal analysis is selecting a node as the *reference* or *datum node*. (We discussed ground in detail in Chapter 4, and we are just repeating it here for clarity.) The reference node is commonly called the *ground* because it is assumed to have zero potential. The reference node is usually selected for you. You will learn how to select it with experience. A reference node is indicated by any of the three symbols in Fig. 7.21. The type of ground shown in Fig. 7.21(c) is called a *chassis ground* and is used in devices where the case, enclosure, or chassis acts as a reference point for all circuits. When the potential of the earth is used a reference, we use the *earth ground* as shown in Fig. 7.21(a) or (b). We shall always use the symbol shown in Fig. 7.21(b).

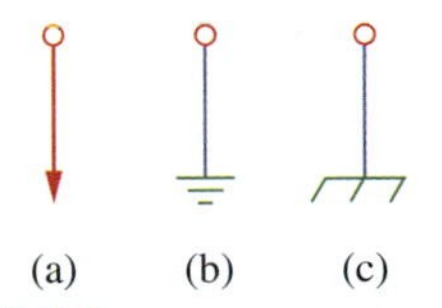

Figure 7.21
Common symbols for indicating the reference node. (a) earth ground; (b) earth ground; (c) chassis ground.

Once a reference node has been selected, we assign voltage designations to nonreference nodes. Consider, for example, the circuit in Fig. 7.22(a). Node 0 is the reference node ($V = 0$ V), while nodes 1 and 2 are assigned voltages V_1 and V_2, respectively. It should be kept in mind that the node voltages are defined with respect to the reference node. As illustrated in Fig. 7.22(a), each node voltage is the voltage rise from the reference node to the corresponding nonreference node or simply the voltage of that node with respect to the reference node.

As the second step, we apply Kirchhoff's current law (KCL) to each nonreference node in the circuit. To avoid putting too much information on the same circuit, the circuit in Fig. 7.22(a) is redrawn in Fig. 7.22(b), where we now add I_1, I_2, and I_3 as the currents through resistors R_1, R_2, and R_3, respectively.

How do we decide the direction of the currents? How do we know that the direction of I_2, for example, is from left to right? We assume the direction. If we get a positive result, then the assumption is right. If we get negative result, the assumed direction is opposite.

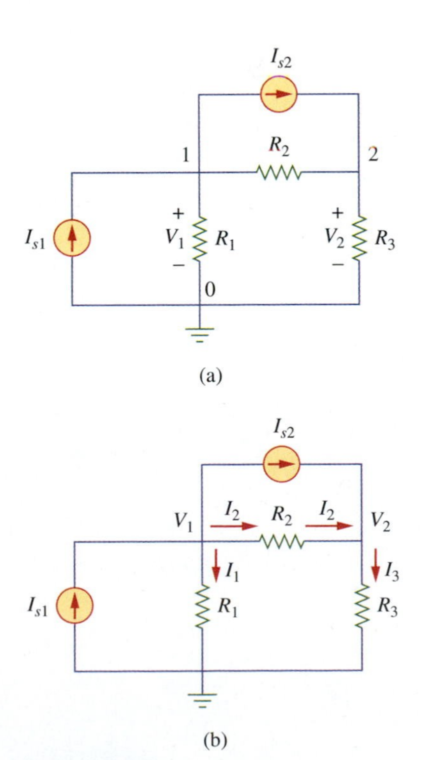

Figure 7.22
A typical circuit for nodal analysis.

[3] Note: The nodal analysis is also known as the *node-voltage method.*

At node 1, applying KCL gives

$$I_{s1} = I_{s2} + I_1 + I_2 \tag{7.8}$$

At node 2,

$$I_{s2} + I_2 = I_3 \tag{7.9}$$

Current flows from a higher potential to a lower potential in a resistor.

We now apply Ohm's law to express the unknown currents I_1, I_2, and I_3 in terms of node voltages. The key idea that we should bear in mind is that because resistance is a passive element (noted by the passive sign convention), current must always flow from a higher potential to a lower potential. (In other words, positive current flow is in the same direction as voltage drop.) We can express this principle as

$$\boxed{I = \frac{V_{\text{higher}} - V_{\text{lower}}}{R}} \tag{7.10}$$

Note that this principle is in agreement with the way we defined resistance in Chapter 2 (see Fig. 2.6).

With this in mind, we obtain from Fig. 7.22(b),

$$\begin{aligned} I_1 &= \frac{V_1 - 0}{R_1} \quad \text{or} \quad I_1 = G_1 V_1 \\ I_2 &= \frac{V_1 - V_2}{R_2} \quad \text{or} \quad I_2 = G_2(V_1 - V_2) \\ I_3 &= \frac{V_2 - 0}{R_3} \quad \text{or} \quad I_3 = G_3 V_2 \end{aligned} \tag{7.11}$$

Substituting Eq. (7.11) into Eqs. (7.8) and (7.9) results, respectively, in

$$I_{s1} = I_{s2} + \frac{V_1}{R_1} + \frac{V_1 - V_2}{R_2} \tag{7.12}$$

$$I_{s2} + \frac{V_1 - V_2}{R_2} = \frac{V_2}{R_3} \tag{7.13}$$

In terms of the conductances, Eqs. (7.12) and (7.13) become

$$I_{s1} = I_{s2} + G_1 V_1 + G_2(V_1 - V_2) \tag{7.14}$$

$$I_{s2} + G_2(V_1 - V_2) = G_3 V_2 \tag{7.15}$$

The third step in nodal analysis is to solve for the node voltages. If we apply KCL to $n - 1$ nonreference nodes, we obtain $n - 1$ simultaneous equations such as Eqs. (7.12) and (7.13) or (7.14) and (7.15). For the circuit of Fig. 7.22, we solve Eqs. (7.12) and (7.13) or (7.14) and (7.15) to obtain the node voltages V_1 and V_2 using any standard method such as the substitution method, elimination method, Cramer's rule,[4] or matrix inversion. To use either of the last two methods, the simultaneous equations need to be cast in standard matrix form. For example, Eqs. (7.14) and (7.15) can be rewritten as

$$(G_1 + G_2)V_1 - G_2 V_2 = I_{s1} - I_{s2}$$

$$-G_2 V_1 + (G_2 + G_3)V_2 = I_{s2}$$

[4] Note: How to use Cramer's rule is discussed in Appendix A.

These can be cast in matrix form as

$$\begin{bmatrix} G_1 + G_2 & -G_2 \\ -G_2 & G_2 + G_3 \end{bmatrix} \begin{bmatrix} V_1 \\ V_2 \end{bmatrix} = \begin{bmatrix} I_{s1} - I_{s2} \\ I_{s2} \end{bmatrix} \quad \textbf{(7.16)}$$

which can be solved to get V_1 and V_2.

The system of equations in Eq. (7.16) will be generalized in Section 7.6. The simultaneous equation may also be solved using calculators such as a TI-89 or an HP-48G11 or with software packages such as MATLAB, Mathcad, and Quattro Pro.

Example 7.5

Calculate the node voltages in the circuit shown in Fig. 7.23(a).

Solution:

Consider the circuit in Fig. 7.23(b), where the circuit in Fig. 7.23(a) has been prepared for nodal analysis. Notice how the currents are selected for the application of KCL. Except for the branches with current sources, the labeling of the currents is arbitrary but consistent. (By being consistent, we mean that if, for example, we assume that I_2 enters the 4-Ω resistor from the left-hand side, I_2 must leave the resistor from the right-hand side.) The reference node is selected, and the node voltages V_1 and V_2 are now to be determined.

At node 1, applying KCL and Ohm's law gives

$$I_1 = I_2 + I_3 \quad \Rightarrow \quad 5 = \frac{V_1 - V_2}{4} + \frac{V_1 - 0}{2}$$

Multiplying each term in the last equation by 4, we obtain

$$20 = V_1 - V_2 + 2V_1$$

or

$$3V_1 - V_2 = 20 \quad \textbf{(7.5.1)}$$

At node 2, we do the same thing and get

$$I_2 + I_4 = I_1 + I_5 \quad \Rightarrow \quad \frac{V_1 - V_2}{4} + 10 = 5 + \frac{V_2 - 0}{6}$$

Multiplying each term by 12 results in

$$3V_1 - 3V_2 + 120 = 60 + 2V_2$$

or

$$-3V_1 + 5V_2 = 60 \quad \textbf{(7.5.2)}$$

Now we have two simultaneous Eqs. (7.5.1) and (7.5.2). We can solve the equations using any method and obtain the values of V_1 and V_2.

■ **METHOD 1** Using the elimination technique, we add Eqs. (7.5.1) and (7.5.2).

$$4V_2 = 80 \quad \Rightarrow \quad V_2 = 20 \text{ V}$$

Substituting $V_2 = 20$ V in Eq. (7.5.1) gives

$$3V_1 - 20 = 20 \quad \Rightarrow \quad V_1 = \frac{40}{3} = 13.33 \text{ V}$$

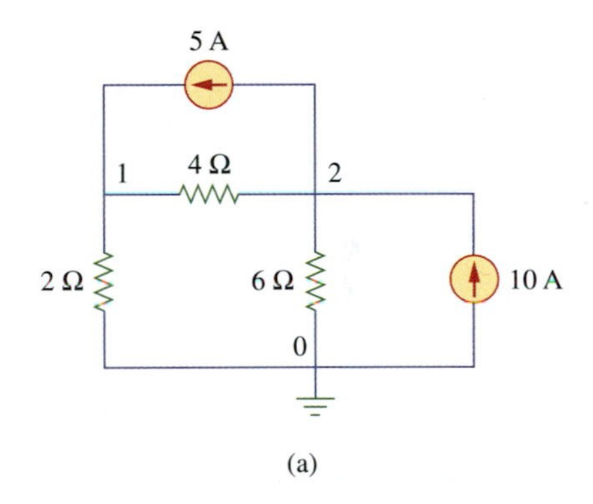

(a)

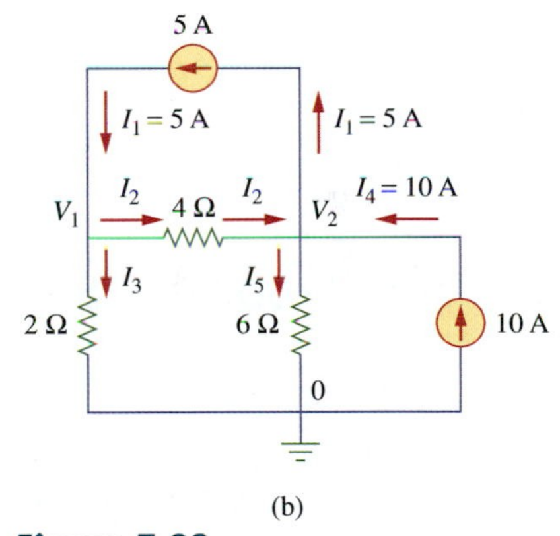

(b)

Figure 7.23
For Example 7.5: (a) original circuit; (b) circuit for analysis.

■ METHOD 2 To use Cramer's rule, we need to put Eqs. (7.5.1) and (7.5.2) in matrix form as

$$\begin{bmatrix} 3 & -1 \\ -3 & 5 \end{bmatrix} \begin{bmatrix} V_1 \\ V_2 \end{bmatrix} = \begin{bmatrix} 20 \\ 60 \end{bmatrix} \quad \textbf{(7.5.3)}$$

The determinant of the matrix is

$$\Delta = \begin{vmatrix} 3 & -1 \\ -3 & 5 \end{vmatrix} = 15 - 3 = 12$$

We now obtain V_1 and V_2 as

$$V_1 = \frac{\Delta_1}{\Delta} = \frac{\begin{vmatrix} 20 & -1 \\ 60 & 5 \end{vmatrix}}{\Delta} = \frac{100 + 60}{12} = 13.33 \text{ V}$$

$$V_2 = \frac{\Delta_2}{\Delta} = \frac{\begin{vmatrix} 3 & 20 \\ -3 & 60 \end{vmatrix}}{\Delta} = \frac{180 + 60}{12} = 20 \text{ V}$$

giving us the same result as the elimination method described earlier.

If we need the currents, we can easily calculate from the values of the nodal voltages.

$$I_1 = 5 \text{ A}, \qquad I_2 = \frac{V_1 - V_2}{4} = -1.6667 \text{ A},$$

$$I_3 = \frac{V_1}{2} = 6.667 \text{ A}, \qquad I_4 = 10 \text{ A}, \qquad I_5 = \frac{V_2}{6} = 3.333 \text{ A}$$

The fact that I_2 is negative shows that the current flows in the direction opposite to the one assumed.

Practice Problem 7.5

Obtain the node voltages in the circuit in Fig. 7.24.

Answer: $V_1 = -2$ V; $V_2 = -14$ V.

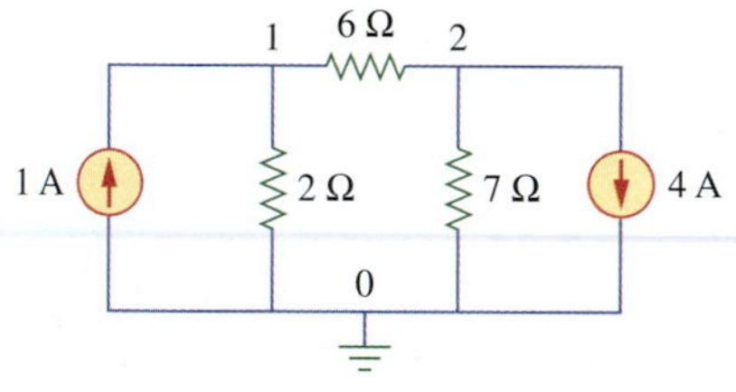

Figure 7.24
For Practice Problem 7.5.

Example 7.6

Determine the voltages at the nodes in Fig. 7.25(a).

Solution:
The circuit in this example has three nonreference nodes as opposed to the previous example, which has two nonreference nodes. We assign voltages to the three nodes as shown in Fig. 7.25(b) and label the currents.

At node 1,

$$3 = I_1 + I_2 \quad \Rightarrow \quad 3 = \frac{V_1 - V_3}{4} + \frac{V_1 - V_2}{2}$$

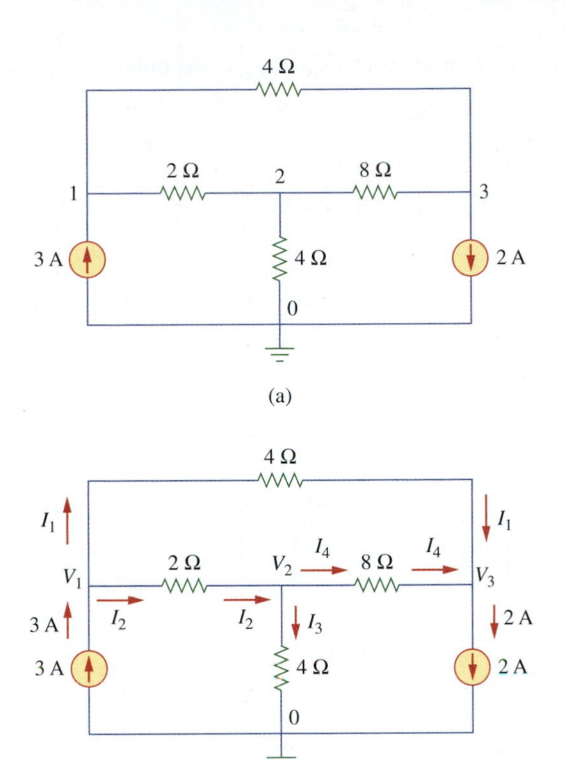

Figure 7.25
For Example 7.6: (a) original circuit; (b) circuit for analysis.

Multiplying by 4 and rearranging terms, we get

$$3V_1 - 2V_2 - V_3 = 12 \tag{7.6.1}$$

At node 2,

$$I_2 = I_3 + I_4 \quad \Rightarrow \quad \frac{V_1 - V_2}{2} = \frac{V_2 - 0}{4} + \frac{V_2 - V_3}{8}$$

Multiplying by 8 and rearranging terms, we get

$$-4V_1 + 7V_2 - V_3 = 0 \tag{7.6.2}$$

At node 3,

$$I_1 + I_4 = 2 \quad \Rightarrow \quad \frac{V_1 - V_3}{4} + \frac{V_2 - V_3}{8} = 2$$

Multiplying by 8 and rearranging terms, we get

$$2V_1 + V_2 - 3V_3 = 16 \tag{7.6.3}$$

We have three simultaneous equations to be solved to get the node voltages. We shall solve the equations in three ways.

METHOD 1 Using the elimination technique, we try to eliminate V_3. We subtract Eq. (7.6.2) from Eq. (7.6.1):

$$7V_1 - 9V_2 = 12 \tag{7.6.4}$$

Multiplying Eq. (7.6.1) by 3 and subtracting Eq. (7.6.3) from it gives

$$7V_1 - 7V_2 = 20 \tag{7.6.5}$$

Subtracting Eq. (7.6.4) from Eq. (7.6.5), we obtain

$$2V_2 = 8 \quad \Rightarrow \quad V_2 = 4 \text{ V}$$

From Eq. (7.6.4),

$$7V_1 = 12 + 9V_2 = 12 + 9 \times 4 = 48 \quad \Rightarrow \quad V_1 = \frac{48}{7} = 6.857 \text{ V}$$

From Eq. (7.6.1),

$$V_3 = 12 + 2V_2 - 3V_1 = 12 + 8 - 3 \times 6.857 = 0.571 \text{ V}$$

Thus,

$$V_1 = 6.857 \text{ V}, \qquad V_2 = 4 \text{ V}, \qquad V_3 = 0.571 \text{ V}$$

METHOD 2 To use Cramer's rule, we put Eqs. (7.6.1) through (7.6.3) in matrix form.

$$\begin{bmatrix} 3 & -2 & -1 \\ -4 & 7 & -1 \\ 2 & 1 & -3 \end{bmatrix} \begin{bmatrix} V_1 \\ V_2 \\ V_3 \end{bmatrix} = \begin{bmatrix} 12 \\ 0 \\ 16 \end{bmatrix} \tag{7.6.6}$$

From this, we obtain

$$V_1 = \frac{\Delta_1}{\Delta}, \qquad V_2 = \frac{\Delta_2}{\Delta}, \qquad V_3 = \frac{\Delta_3}{\Delta}$$

where Δ, Δ_1, Δ_2 and Δ_3 are the determinants calculated as follows. As explained in Appendix A, to calculate the determinant of a 3×3 matrix, we repeat the first two rows and cross-multiply.

$$\Delta = \begin{array}{c|ccc|c} & 3 & -2 & -1 & \\ & -4 & 7 & -1 & \\ & 2 & 1 & -3 & \\ - & 3 & -2 & -1 & + \\ - & -4 & 7 & -1 & + \\ - & & & & + \end{array}$$

$$= -63 + 4 + 4 + 14 + 3 + 24 = -14$$

$$\Delta_1 = \begin{array}{c|ccc|c} & 12 & -2 & -1 & \\ & 0 & 7 & -1 & \\ & 16 & 1 & -3 & \\ - & 12 & -2 & -1 & + \\ - & 0 & 7 & -1 & + \\ - & & & & + \end{array} = -252 - 0 + 32 + 112 + 12 + 0 = -96$$

$$\Delta_2 = \begin{array}{c|ccc|c} & 3 & 12 & -1 & \\ & -4 & 0 & -1 & \\ & 2 & 16 & -3 & \\ - & 3 & 12 & -1 & + \\ - & -4 & 0 & -1 & + \\ - & & & & + \end{array} = 0 + 64 - 24 - 0 + 48 - 144 = -56$$

$$\Delta_3 = \begin{array}{c|ccc|c} & 3 & -2 & 12 & \\ & -4 & 7 & 0 & \\ & 2 & 1 & 16 & \\ - & 3 & -2 & 12 & + \\ - & -4 & 7 & 0 & + \\ - & & & & + \end{array} = 336 - 48 + 0 - 168 - 0 - 128 = -8$$

Thus, we obtain

$$V_1 = \frac{\Delta_1}{\Delta} = \frac{-96}{-14} = 6.857 \text{ V}$$

$$V_2 = \frac{\Delta_2}{\Delta} = \frac{-56}{-14} = 4 \text{ V}$$

$$V_3 = \frac{\Delta_3}{\Delta} = \frac{-8}{-14} = 0.571 \text{ V}$$

as we obtained earlier.

METHOD 3 We can use MATLAB to solve the matrix equation. Equation (7.6.6) can be written as

$$AV = B \quad \Rightarrow \quad V = A^{-1}B$$

where **A** is the 3×3 square matrix, **B** is the column vector, and **V** is a column vector comprised of V_1, V_2, and V_3 that we want to determine. We use MATLAB to determine **V** as follows.

```
»A=[3  -2  -1;  -4  7  -1;  2  1  -3];
»B=[12  0  16];
»V=inv(A)*B'

V =
  6.8571
  4.0000
  0.5714
```

Thus,

$$V_1 = 6.857 \text{ V}, \qquad V_2 = 4 \text{ V}, \qquad V_3 = 0.571 \text{ V}$$

as obtained previously.

METHOD 4 We can use scientific calculators such as the TI-89 Titanium to solve simultaneous equations. We are trying to solve the simultaneous equation in Eq. (7.6.6), namely,

$$\begin{bmatrix} 3 & -2 & -1 \\ -4 & 7 & -1 \\ 2 & 1 & -3 \end{bmatrix} \begin{bmatrix} V_1 \\ V_2 \\ V_3 \end{bmatrix} = \begin{bmatrix} 12 \\ 0 \\ 16 \end{bmatrix}$$

To use the TI-89 Titanium calculator to solve the simultaneous equations, Press [2nd] [MATH]
Select 4: Matrix and Press [ENTER]
then press
Select 5: simult (and Press [ENTER]

On the entry line, type:

```
simult([3,-2,-1;-4,7,-1;2,1,-3],[12;0;16]),
```

(simult(will come up by itself and you type the remaining part.)
Press [♦] [ENTER]
The result:

$$V_1 = 6.857 \text{ V}, \quad V_2 = 4 \text{ V}, \quad V_3 = 0.5714 \text{ V}$$

as obtained earlier.

Practice Problem 7.6

Use any of the preceding methods to find the voltages at the three nonreference nodes in the circuit of Fig. 7.26.

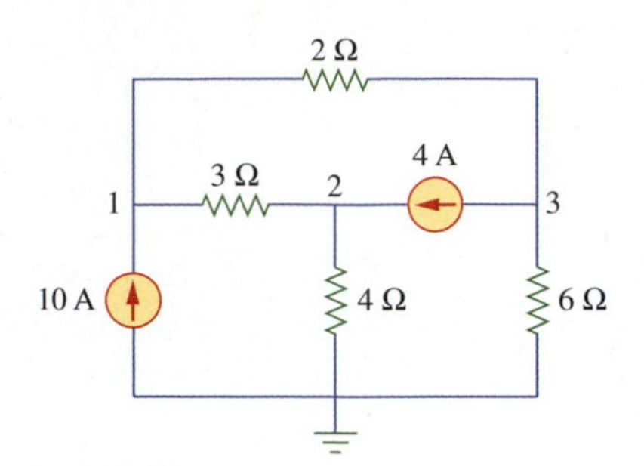

Figure 7.26
For Practice Problem 7.6.

Answer: $V_1 = 34.67$ V; $V_2 = 26.67$ V; $V_3 = 20$ V

7.5 Nodal Analysis with Voltage Sources

We now consider how voltage sources affect nodal analysis. We use the circuit in Fig. 7.27 for illustration. Consider the following two possibilities.

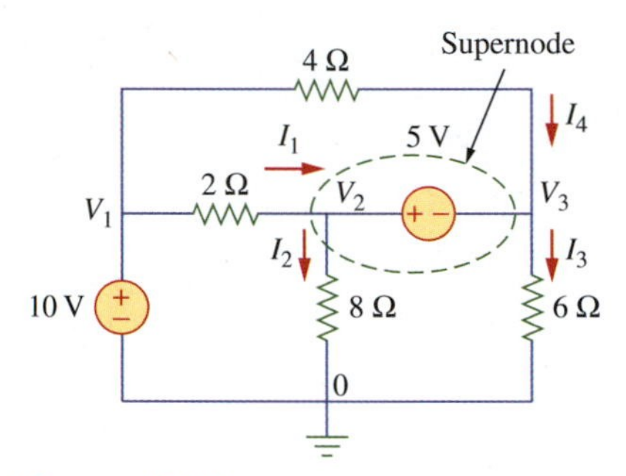

Figure 7.27
A circuit with a supernode.

■ **CASE 1** If a voltage source is connected between the reference node and a nonreference node, we simply set the voltage at the nonreference node equal to the voltage of the voltage source. In Fig. 7.27, for example,

$$V_1 = 10 \text{ V} \tag{7.17}$$

Thus, our analysis is somewhat simplified by this knowledge of the voltage at this node. Note that in this case, the voltage source must be connected directly to the reference node, without anything else in series with it.

■ **CASE 2** If the voltage source is connected between two nonreference nodes, the two nonreference nodes form a *generalized node* or *supernode;*[5] both KCL and KVL will be applied to determine the node voltages.

A **supernode** is formed by excluding a (dependent or independent) voltage source connected between two nonreference nodes and any elements connected in parallel with it.

A supernode is one in which voltage source is in between the two nonreference nodes. In Fig. 7.27, nodes 2 and 3 form a supernode. (We could have more than two nodes forming a single supernode. For example, if the 2-Ω resistor in Fig. 7.27 is replaced by a voltage source, nodes 1, 2, and 3 form a supernode.) We analyze a circuit with supernodes using the same three steps mentioned in the previous section except that the supernodes are treated differently. Why? An essential component of nodal analysis is applying KCL, which requires knowing the current through each element. There is no way of knowing the current through a voltage source in advance. However, Kirchhoff's current law must be satisfied at a supernode like at any other node. Hence, at the supernode in Fig. 7.27,

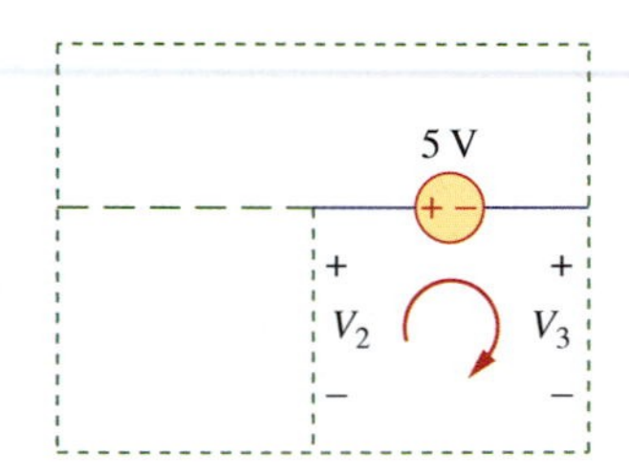

Figure 7.28
Applying KVL to a supernode.

$$I_1 + I_4 = I_2 + I_3 \tag{7.18}$$

or

$$\frac{V_1 - V_2}{2} + \frac{V_1 - V_3}{4} = \frac{V_2 - 0}{8} + \frac{V_3 - 0}{6} \tag{7.19}$$

To apply Kirchhoff's voltage law to the supernode in Fig. 7.27, we redraw the circuit as shown in Fig. 7.28. Going around the loop (that

[5] Note: A supernode may be regarded as a closed surface enclosing the voltage source.

contains the 5-V source) in the clockwise direction gives

$$-V_2 + 5 + V_3 = 0 \quad \Rightarrow \quad V_2 - V_3 = 5 \tag{7.20}$$

From Eqs. (7.17), (7.19), and (7.20), we obtain the node voltages. Note the following properties of a supernode:

1. The voltage source inside the supernode provides a constraint equation needed to solve for the node voltages.
2. A supernode has no voltage of its own.
3. A supernode requires the application of both KCL and KVL.

Example 7.7

Find V_o in the circuit of Fig. 7.29.

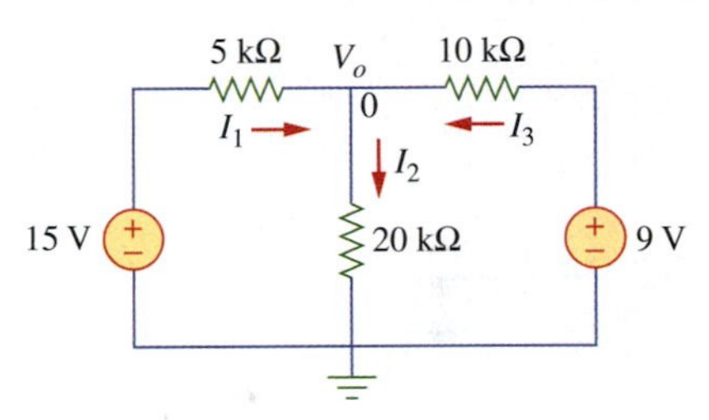

Figure 7.29
For Example 7.7.

Solution:
The circuit in Fig. 7.29 has two voltage sources that are connected to the reference node, but it has no supernode. At the node O, KCL gives

$$I_1 + I_3 = I_2$$

or

$$\frac{15 - V_o}{5k} + \frac{9 - V_o}{10k} = \frac{V_o - 0}{20k}$$

Multiplying through by $20k$,

$$60 - 4V_o + 18 - 2V_o = V_o$$

or

$$78 = 7V_o \quad \Rightarrow \quad V_o = \frac{78}{7} = 11.143 \text{ V}$$

Practice Problem 7.7

Determine V_x in the circuit of Fig. 7.30.

Answer: 20 V

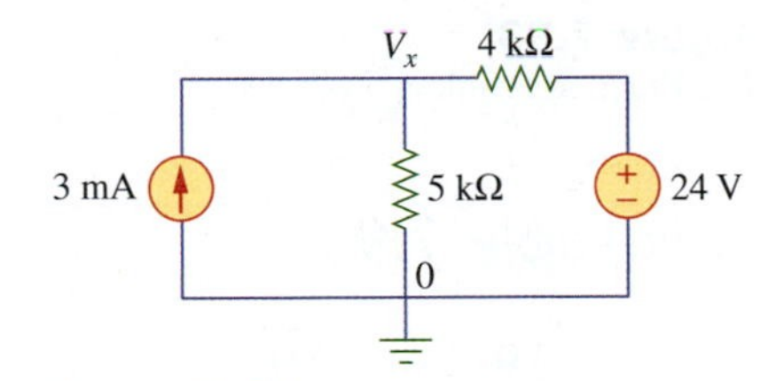

Figure 7.30
For Practice Problem 7.7.

Example 7.8

For the circuit shown in Fig. 7.31, find the node voltages.

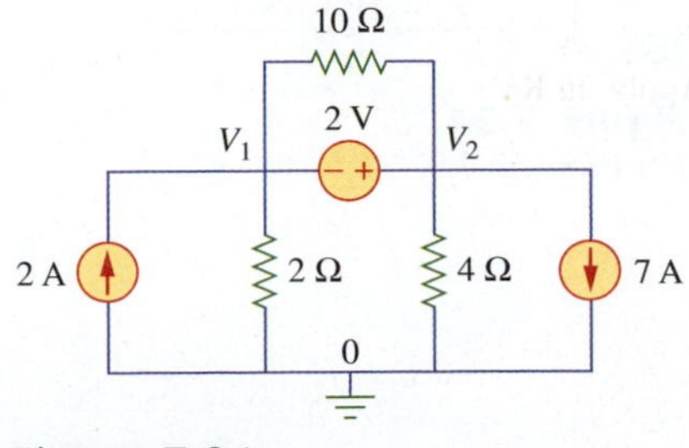

Figure 7.31
For Example 7.8.

Solution:
The supernode contains the 2-V source, nodes 1 and 2, and the 10-Ω resistor. Applying KCL to the supernode as shown in Fig. 7.32(a) gives

$$2 = I_1 + I_2 + 7$$

Expressing I_1 and I_2 in terms of the node voltages,

$$2 = \frac{V_1 - 0}{2} + \frac{V_2 - 0}{4} + 7 \quad \Rightarrow \quad 8 = 2V_1 + V_2 + 28$$

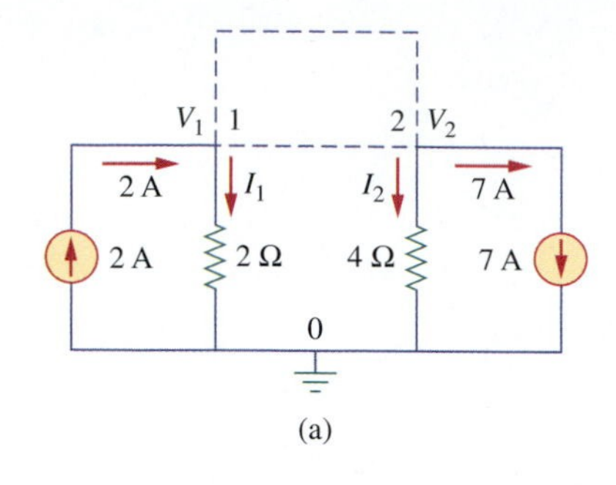

(a)

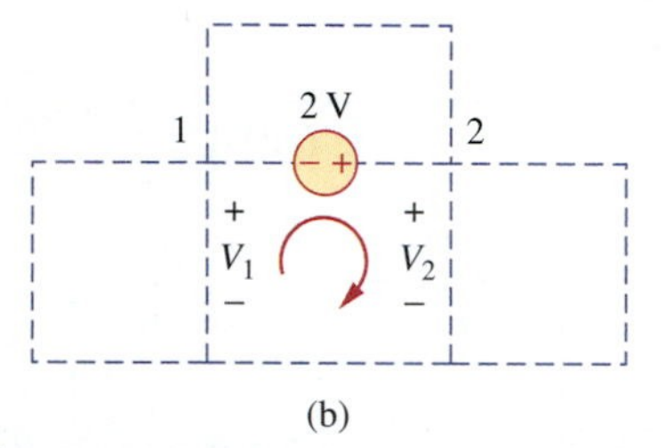

(b)

Figure 7.32
Applying (a) KCL to the supernode; (b) KVL to the loop.

or

$$V_2 = -20 - 2V_1 \tag{7.8.1}$$

To get the relationship between V_1 and V_2, we apply KVL to the circuit in Fig. 7.32(b). Going around the loop, we obtain

$$-V_1 - 2 + V_2 = 0 \quad \Rightarrow \quad V_2 = V_1 + 2 \tag{7.8.2}$$

From Eqs. (7.8.1) and (7.8.2), we get

$$V_2 = V_1 + 2 = -20 - 2V_1$$

or

$$3V_1 = -22 \quad \Rightarrow \quad V_1 = -7.333 \text{ V}$$

and

$$V_2 = V_1 + 2 = -5.333 \text{ V}$$

Note that the 10-Ω resistor does not affect branch variables elsewhere because it is connected to the supernode and subsumed by the supernode.

Practice Problem 7.8

Find V and I in the circuit in Fig. 7.33.

Answer: –0.2 V; 1.4 A

Figure 7.33
For Practice Problem 7.8.

Example 7.9

Using nodal analysis, find v_o in the circuit in Fig. 7.34.

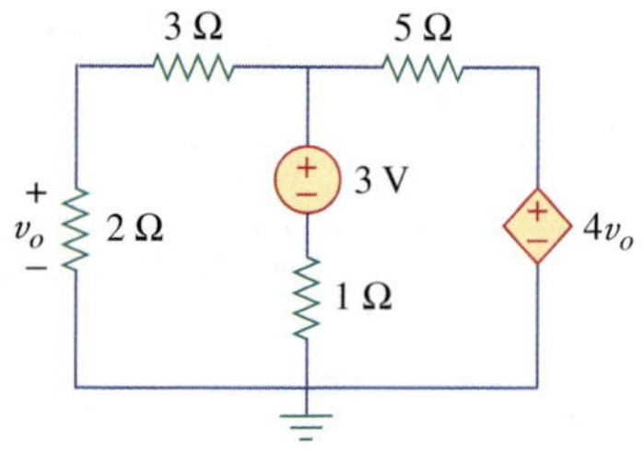

Figure 7.34
For Problem 7.9.

Solution:
Consider the circuit as shown in Fig. 7.35.

$$i_1 + i_2 + i_3 = 0 \quad \Rightarrow \quad \frac{v_1 - 0}{5} + \frac{v_1 - 3}{1} + \frac{v_1 - 4v_0}{5} = 0$$

Multiplying by 5 gives

$$v_1 + 5v_1 - 15 + v_1 - 4v_o = 0$$

But

$$v_0 = \frac{2}{5}v_1$$

(using voltage division) so that

$$7v_1 - 15 - \frac{8}{5}v_1 = 0$$

or

$$\frac{27}{5}v_1 = 15$$

$$v_1 = 15 \times 5/(27) = 2.778 \text{ V}$$

Therefore $v_o = 2v_1/5 = 1.111$ V

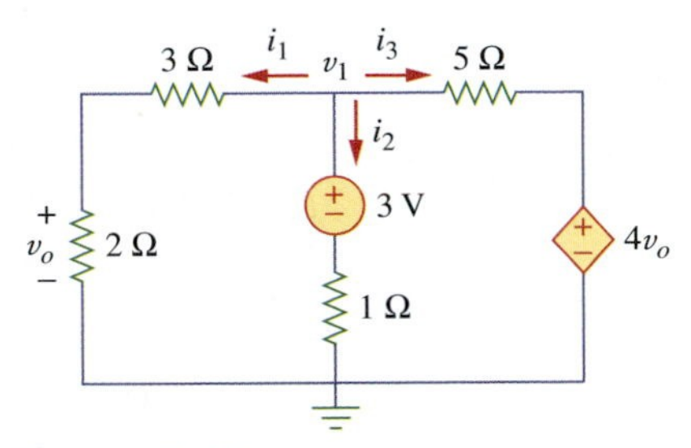

Figure 7.35
Analysis of the circuit in Figure 7.34.

Practice Problem 7.9

Determine I_b in the circuit in Fig. 7.36 using nodal analysis.

Answer: 79.34 mA

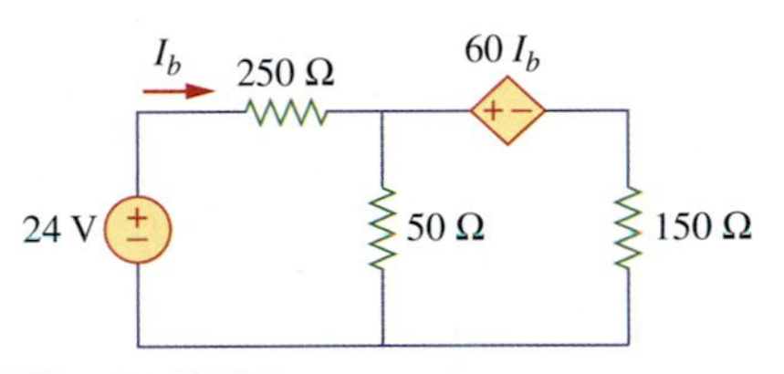

Figure 7.36
For Practice Problem 7.9.

7.6 †Mesh and Nodal Analysis by Inspection

This section presents a generalized procedure for nodal or mesh analysis. It is a shortcut approach based on a mere inspection of the circuit.

Mesh-current equations can be obtained by inspection when a linear resistive circuit has only independent voltage sources. For example, consider the circuit in Fig. 7.10, which is shown in Fig. 7.37(a) for convenience. The circuit has two nonreference nodes, and the node-equations were derived in Section 7.2 [see Eq. (7.3)] as

$$\begin{bmatrix} R_1 + R_3 & -R_3 \\ -R_3 & R_2 + R_3 \end{bmatrix} \begin{bmatrix} I_1 \\ I_2 \end{bmatrix} = \begin{bmatrix} V_1 \\ -V_2 \end{bmatrix} \tag{7.21}$$

We notice that each of the diagonal terms is the sum of the resistances in the related mesh, while each of the off-diagonal terms is the negative of the resistance common to meshes 1 and 2. Each term on the right-hand side of Eq. (7.21) is the algebraic sum taken clockwise of all independent voltage sources in the related mesh.

In general, if the circuit has N meshes, the mesh-current equations can be expressed in terms of the resistances as

$$\begin{bmatrix} R_{11} & R_{12} & \dots & R_{1N} \\ R_{21} & R_{22} & \dots & R_{2N} \\ \vdots & \vdots & \vdots & \vdots \\ R_{N1} & R_{N2} & \dots & R_{NN} \end{bmatrix} \begin{bmatrix} I_1 \\ I_2 \\ \vdots \\ I_N \end{bmatrix} = \begin{bmatrix} V_1 \\ V_2 \\ \vdots \\ V_N \end{bmatrix} \tag{7.22}$$

or simply

$$\mathbf{RI} = \mathbf{V} \tag{7.23}$$

where

R_{kk} = sum of the resistances in mesh k

$R_{kj} = R_{jk}$ = the negative of the sum of the resistances in common with meshes k and j, $k \neq j$

I_k = unknown mesh current for mesh k in the clockwise direction

V_k = algebraic sum taken clockwise of all independent voltage sources in mesh k, with voltage rise treated as positive

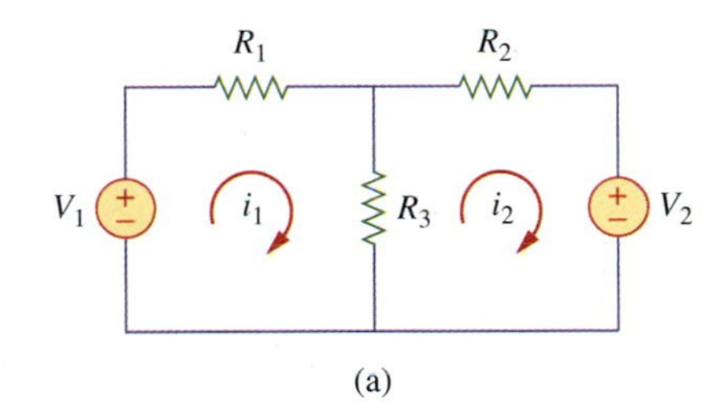

(a)

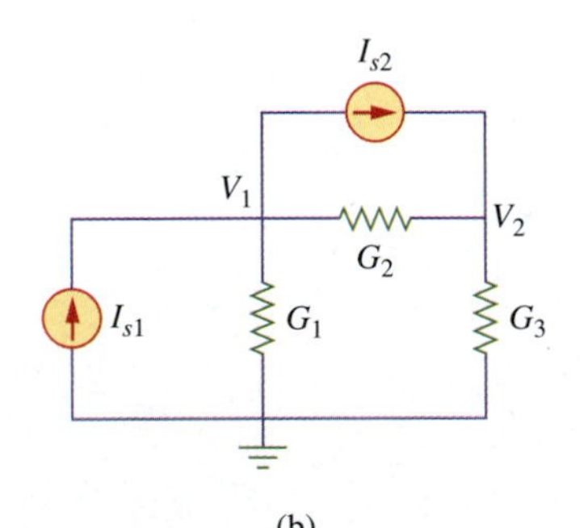

(b)

Figure 7.37
(a) The circuit in Figure 7.10; (b) the circuit in Figure 7.22.

$\mathbf{R}$ = the *resistance matrix*
$\mathbf{I}$ = output vector
$\mathbf{V}$ = input vector

We can solve Eq. (7.22) to obtain the unknown mesh currents. Note that we have assumed that all mesh currents flow clockwise. We should note Eq. (7.23) is valid for circuits with only independent voltage sources and linear resistors. A voltage source is independent if its value does not depend on branch variables elsewhere.

Similarly, when all sources in a circuit are independent current sources, we do not need to apply KCL to each node to obtain the node-voltage equations as we did in Section 7.4. The node-voltage equations can be obtained by mere inspection of the circuit. As an example, let us reexamine the circuit in Fig. 7.22, which is shown in Fig. 7.37(b) for convenience. The circuit has two nonreference nodes, and the node-equations were derived in Section 7.4 [see Eq. (7.16)] as

$$\begin{bmatrix} G_1 + G_2 & -G_2 \\ -G_2 & G_2 + G_3 \end{bmatrix} \begin{bmatrix} V_1 \\ V_2 \end{bmatrix} = \begin{bmatrix} I_{s1} - I_{s2} \\ I_{s2} \end{bmatrix} \tag{7.24}$$

We observe that each of the diagonal terms is the sum of the conductances connected directly to node 1 or 2, while the off-diagonal terms are the negatives of the conductances connected between the nodes. Also, each term on the right-hand side of Eq. (7.24) is the algebraic sum of the currents entering the node.

In general, if a circuit with independent current sources has N nonreference nodes, the node-voltage equations can be written in terms of the conductances as

$$\begin{bmatrix} G_{11} & G_{12} & \dots & G_{1N} \\ G_{21} & G_{22} & \dots & G_{2N} \\ \vdots & \vdots & \vdots & \vdots \\ G_{N1} & G_{N2} & \dots & G_{NN} \end{bmatrix} \begin{bmatrix} V_1 \\ V_2 \\ \vdots \\ V_N \end{bmatrix} = \begin{bmatrix} I_1 \\ I_2 \\ \vdots \\ I_N \end{bmatrix} \tag{7.25}$$

or simply

$$\mathbf{GV} = \mathbf{I} \tag{7.26}$$

where

G_{kk} = sum of the conductances connected to node k
$G_{kj} = G_{jk}$ = the negative of the sum of the conductances directly connecting nodes k and j, $k \neq j$
V_k = unknown voltage at node k
I_k = algebraic sum of all independent current sources directly connected to node k, with currents entering the node treated as positive
$\mathbf{G}$ = the *conductance matrix*
$\mathbf{V}$ = output vector
$\mathbf{I}$ = input vector

Equation (7.25) can be solved to obtain the unknown node voltages. We should keep in mind that this is valid for circuits with only independent current sources and linear resistors. A current source is independent if its value does not depend on branch variables elsewhere.

Example 7.10

By inspection, write the mesh-current equations for the circuit in Fig. 7.38.

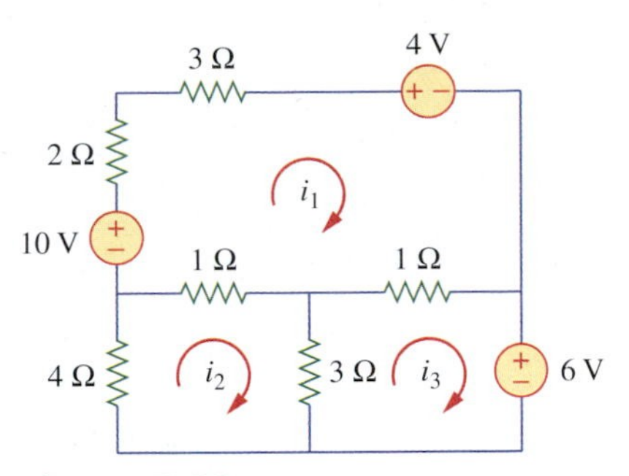

Figure 7.38
For Example 7.10.

Solution:
We have three meshes so that the resistance matrix is 3 × 3. The diagonal terms in ohms are:

$$R_{11} = 2 + 3 + 1 + 1 = 7$$
$$R_{22} = 4 + 1 + 3 = 8$$
$$R_{33} = 3 + 1 = 4$$

The off-diagonal terms are:

$$R_{12} = -1, \quad R_{13} = -1,$$
$$R_{21} = -1, \quad R_{23} = -3,$$
$$R_{31} = -1, \quad R_{32} = -3$$

The input voltage vector **v** has the following terms in volts:

$$V_1 = 10 - 4 = 6, \quad V_2 = 0, \quad V_3 = -6$$

Thus, the mesh-current equations are:

$$\begin{bmatrix} 7 & -1 & -1 \\ -1 & 8 & -3 \\ -1 & -3 & 4 \end{bmatrix} \begin{bmatrix} i_1 \\ i_2 \\ i_3 \end{bmatrix} = \begin{bmatrix} 6 \\ 0 \\ -6 \end{bmatrix}$$

From this, we can obtain mesh currents i_1, i_2, and i_3.

Practice Problem 7.10

By inspection, obtain the mesh-current equations for the circuit in Fig. 7.39.

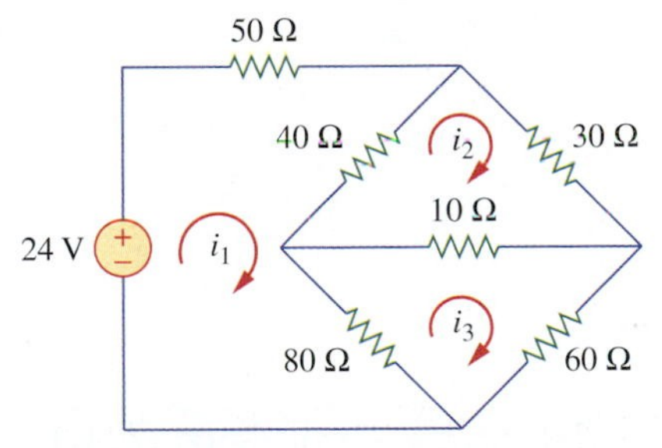

Figure 7.39
For Practice Problem 7.10.

Answer:

$$\begin{bmatrix} 170 & -40 & -80 \\ -40 & 80 & -10 \\ -80 & -10 & 150 \end{bmatrix} \begin{bmatrix} i_1 \\ i_2 \\ i_3 \end{bmatrix} = \begin{bmatrix} 24 \\ 0 \\ 0 \end{bmatrix}$$

Example 7.11

Write the node-voltage matrix equations for the circuit in Fig. 7.40 by inspection.

Solution:
The circuit in Fig. 7.40 has four nonreference nodes so that we need four node equations. This implies that the size of the conductance matrix **G** is 4 × 4. The diagonal terms of **G** in siemens are:

$$G_{11} = \frac{1}{5} + \frac{1}{10} = 0.3, \qquad G_{22} = \frac{1}{5} + \frac{1}{8} + \frac{1}{1} = 1.325,$$

$$G_{33} = \frac{1}{8} + \frac{1}{8} + \frac{1}{4} = 0.5, \qquad G_{44} = \frac{1}{8} + \frac{1}{2} + \frac{1}{1} = 1.625.$$

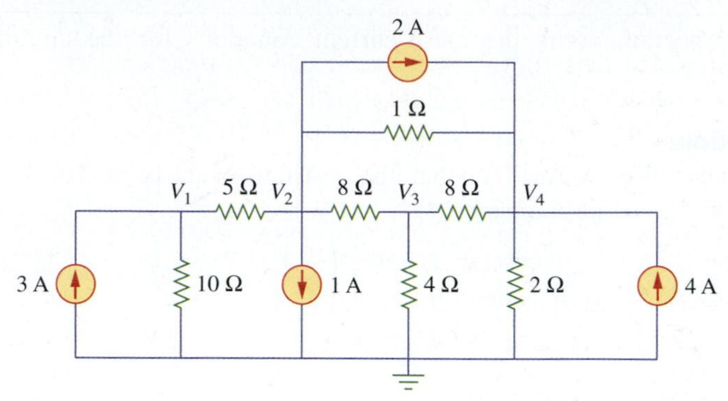

Figure 7.40
For Example 7.11.

The off-diagonal terms are:

$$G_{12} = -\frac{1}{5} = -0.2, \quad G_{13} = G_{14} = 0,$$

$$G_{21} = -0.2, \quad G_{23} = -\frac{1}{8} = -0.125, \quad G_{24} = -\frac{1}{1} = -1,$$

$$G_{31} = 0, \quad G_{32} = -0.125, \quad G_{34} = -\frac{1}{8} = -0.125,$$

$$G_{41} = 0, \quad G_{42} = -1, \quad G_{43} = -0.125.$$

The input current vector **I** has the following terms in amperes:

$$I_1 = 3, \quad I_2 = -1 - 2 = -3, \quad I_3 = 0, \quad I_4 = 2 + 4 = 6$$

Thus, the node-voltage equations are:

$$\begin{bmatrix} 0.3 & -0.2 & 0 & 0 \\ -0.2 & 1.325 & -0.125 & -1 \\ 0 & -0.125 & 0.5 & -0.125 \\ 0 & -1 & -0.125 & 1.625 \end{bmatrix} \begin{bmatrix} V_1 \\ V_2 \\ V_3 \\ V_4 \end{bmatrix} = \begin{bmatrix} 3 \\ -3 \\ 0 \\ 6 \end{bmatrix}$$

which can be solved to obtain the node voltages V_1, V_2, V_3, and V_4.

Practice Problem 7.11

By inspection, obtain the node-voltage equations for the circuit in Fig. 7.41.

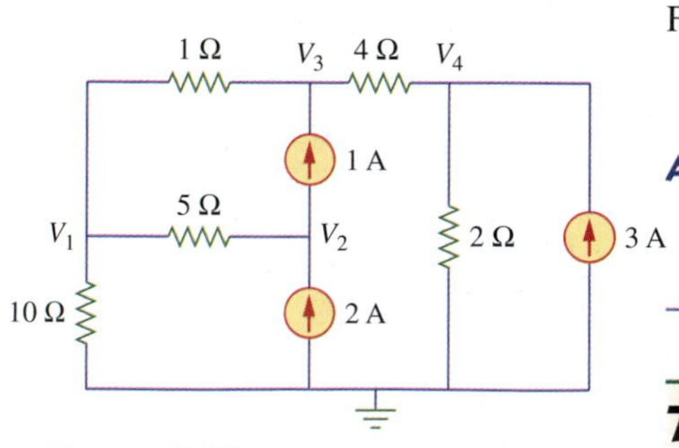

Figure 7.41
For Practice Problem 7.11.

Answer:

$$\begin{bmatrix} 1.3 & -0.2 & -1 & 0 \\ -0.2 & 0.2 & 0 & 0 \\ -1 & 0 & 1.25 & -0.25 \\ 0 & 0 & -0.25 & 0.75 \end{bmatrix} \begin{bmatrix} V_1 \\ V_2 \\ V_3 \\ V_4 \end{bmatrix} = \begin{bmatrix} 0 \\ 1 \\ 1 \\ 3 \end{bmatrix}$$

7.7 Mesh Versus Nodal Analysis

Both mesh and nodal analyses provide a systematic way of studying a complex network. Someone may ask: Given a network to be analyzed, how do we know which method is better or more efficient? The choice of the better method is dictated by two factors.

The first factor is the nature of the particular network. For example, networks that contain many series-connected elements, voltage sources, or supermeshes are more suitable for mesh analysis, whereas networks with parallel-connected elements, current sources, or supernodes are more suitable for nodal analysis. Also, a circuit with fewer nodes than meshes is better analyzed using nodal analysis, while a circuit with fewer meshes than nodes is better analyzed using mesh analysis. The key issue is to select the method that results in the smaller number of equations.

The second factor is the information required. If node voltages are required, it may be expedient to apply nodal analysis. If branch or mesh currents are required, it may be better to use mesh analysis.

It is helpful to be familiar with both methods of analysis for at least two reasons. First, one method can be used to check the results from the other method, if possible. Second, each method has its limitations. For this reason, only one method may be suitable for a particular problem. For example, mesh analysis is the only method to use in analyzing transistor circuits, as we shall see in Section 7.10. For nonplanar networks, nodal analysis is the only option we have because mesh analysis only applies to planar networks. Also, nodal analysis is more amenable to solution by computer because it is easy to program. This allows one to analyze complicated circuits that defy hand calculation. PSpice, a computer software that is based on nodal analysis, is covered in Section 7.9.1.

7.8 †Wye-Delta Transformations

Situations often arise in circuit analysis when the resistors are neither in parallel nor in series. For example, consider the bridge circuit in Fig. 7.42. How do we combine resistors R_1 through R_6 when the resistors are neither in series nor in parallel? Many circuits of the type shown in Fig. 7.42 can be simplified by using three-terminal equivalent networks. These are the wye (Y) or tee (T) network shown in Fig. 7.43 and the delta (Δ) or pi (Π) network shown in Fig. 7.44. The networks often occur by themselves, or they are part of a larger network. They occur in three-phase networks, electrical filters, and matching networks. Our main interest here is how to identify them when they occur as part of a network and how to apply wye-delta transformation in the analysis of that network.

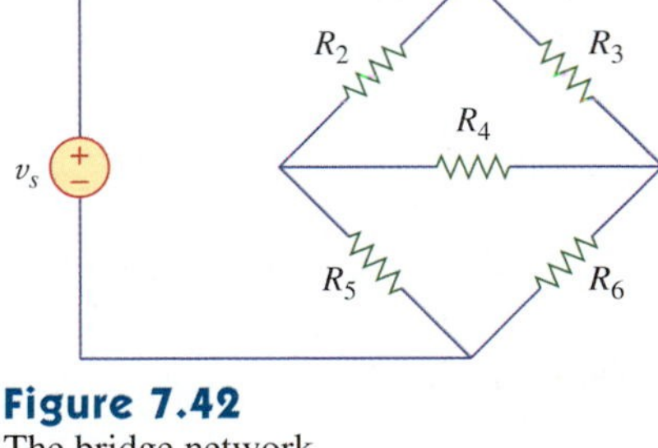

Figure 7.42
The bridge network.

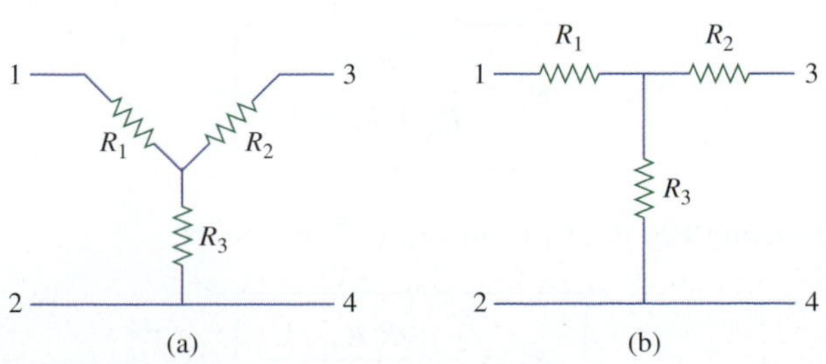

Figure 7.43
Two forms of the same network: (a) Y; (b) T.

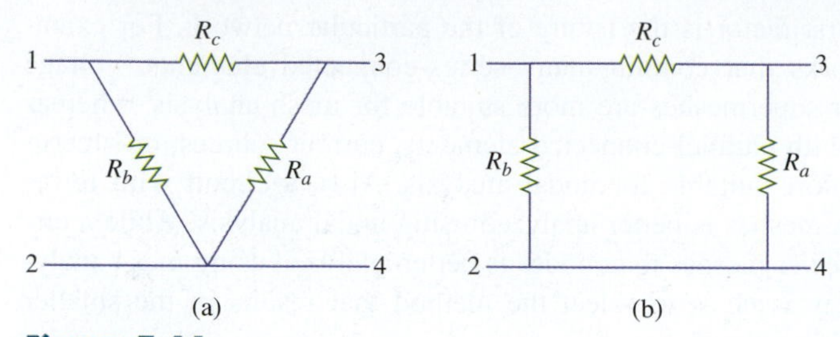

Figure 7.44
Two forms of the same network: (a) Δ; (b) Π.

7.8.1 Δ-to-Y Conversion

Suppose it is more convenient to work with a wye network in a place where the circuit contains a delta configuration. We superimpose a wye network on the existing delta network and find the equivalent resistances in the wye network. To obtain the equivalent resistances in the wye network, we compare the two networks and make sure that the resistance between each pair of nodes in the Δ (or Π) network is the same as the resistance between same pair of nodes in the Y (or T) network. For terminals 1 and 2 in Figures 7.43 and 7.44, for example,

$$R_{12}(Y) = R_1 + R_3 \tag{7.27}$$

$$R_{12}(\Delta) = R_b \parallel (R_a + R_c) \tag{7.28}$$

Setting $R_{12}(\Delta) = R_{12}(Y)$ gives

$$R_{12} = R_1 + R_3 = \frac{R_b(R_a + R_c)}{R_a + R_b + R_c} \tag{7.29a}$$

Similarly,

$$R_{13} = R_1 + R_2 = \frac{R_c(R_a + R_b)}{R_a + R_b + R_c} \tag{7.29b}$$

$$R_{34} = R_2 + R_3 = \frac{R_a(R_b + R_c)}{R_a + R_b + R_c} \tag{7.29c}$$

Subtracting Eq. (7.29c) from Eq. (7.29a), we get

$$R_1 - R_2 = \frac{R_c(R_b - R_a)}{R_a + R_b + R_c} \tag{7.30}$$

Adding Eqs. (7.29b) and (7.30) gives

$$\boxed{R_1 = \frac{R_b R_c}{R_a + R_b + R_c}} \tag{7.31}$$

and subtracting Eq. (7.30) from Eq. (7.29b) yields

$$\boxed{R_2 = \frac{R_c R_a}{R_a + R_b + R_c}} \tag{7.32}$$

Subtracting Eq.(7.31) from Eq. (7.29a), we obtain

$$R_3 = \frac{R_a R_b}{R_a + R_b + R_c} \tag{7.33}$$

We do not need to memorize Eqs. (7.31), (7.32), and (7.33). To transform a Δ network to Y, we create an extra node n as shown in Fig. 7.45 and follow this conversion rule:

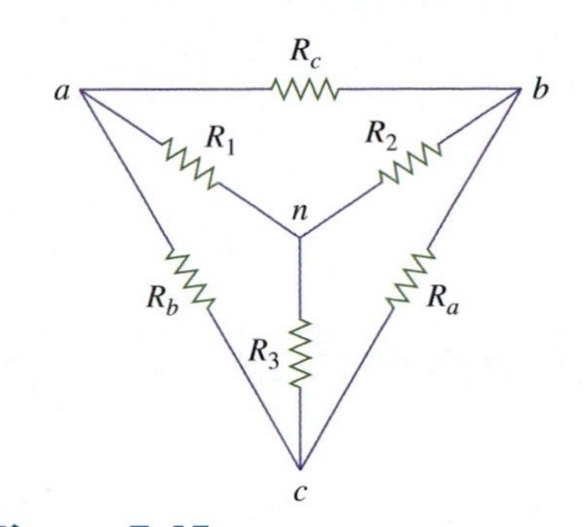

Figure 7.45
Superposition of Y and Δ networks as an aid in transforming one to another.

> Each resistor in the **Y network** is the product of the resistors in the two adjacent Δ branches, divided by the sum of the three Δ resistors.

7.8.2 Y-to-Δ Conversion

To obtain the conversion formulas for transforming a wye network to an equivalent delta network, we note from Eqs. (7.31), (7.32), and (7.33) that

$$\begin{aligned} R_1R_2 + R_2R_3 + R_3R_1 &= \frac{R_aR_bR_c(R_a + R_b + R_c)}{(R_a + R_b + R_c)^2} \\ &= \frac{R_aR_bR_c}{R_a + R_b + R_c} \end{aligned} \tag{7.34}$$

Dividing Eq. (7.34) by each of Eqs. (7.31), (7.32), and (7.33) leads to the following equations:

$$R_a = \frac{R_1R_2 + R_2R_3 + R_3R_1}{R_1} \tag{7.35}$$

$$R_b = \frac{R_1R_2 + R_2R_3 + R_3R_1}{R_2} \tag{7.36}$$

$$R_c = \frac{R_1R_2 + R_2R_3 + R_3R_1}{R_3} \tag{7.37}$$

From Eqs. (7.35), (7.36), and (7.37) and Fig. 7.45, the conversion rule for Y to Δ is as follows:

> Each resistor in the **Δ network** is the sum of all the three possible products of Y resistors taken two at a time, divided by the opposite Y resistor.

The Y and Δ networks are said to be balanced when

$$R_1 = R_2 = R_3 = R_Y, \quad R_a = R_b = R_c = R_\Delta \tag{7.38}$$

Under these conditions, conversion formulas become

$$R_Y = \frac{R_\Delta}{3} \quad \text{or} \quad R_\Delta = 3R_Y \tag{7.39}$$

Note that in making the transformation, we do not take anything out of the circuit and put in something new. We are merely substituting different but mathematically equivalent three-terminal network patterns to create a circuit in which resistors are either in series or parallel, allowing us to calculate R_{eq} if necessary.

Example 7.12

Convert the Δ network in Fig. 7.46(a) to an equivalent Y network.

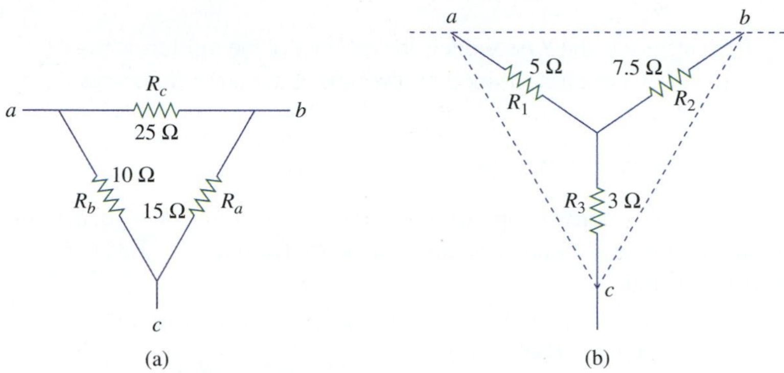

Figure 7.46
For Example 7.12: (a) original Δ network; (b) Y equivalent network.

Solution:
Using Eqs. (7.31), (7.32), and (7.33), we obtain

$$R_1 = \frac{R_b R_c}{R_a + R_b + R_c} = \frac{25 \times 10}{25 + 10 + 15} = \frac{250}{50} = 5\ \Omega$$

$$R_2 = \frac{R_c R_a}{R_a + R_b + R_c} = \frac{25 \times 15}{50} = 7.5\ \Omega$$

$$R_3 = \frac{R_a R_b}{R_a + R_b + R_c} = \frac{15 \times 10}{50} = 3\ \Omega$$

The equivalent Y network is shown in Fig. 7.46(b).

Practice Problem 7.12

Transform the Y network in Fig. 7.47 to a Δ network.

Answer: $R_a = 140\ \Omega$; $R_b = 70\ \Omega$; $R_c = 35\ \Omega$

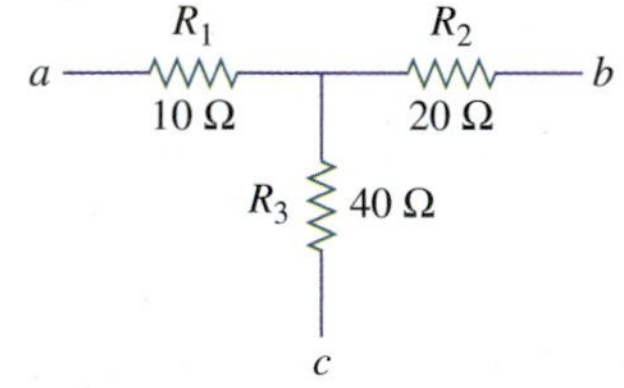

Figure 7.47
For Practice Problem 7.12.

Example 7.13

Obtain the equivalent resistance R_{ab} for the circuit in Fig. 7.48, and use it to find current i.

Solution:
In this circuit, there are two Y networks and one Δ network. Transforming just one of these will simplify the circuit. If we convert

the Y network comprised of the 5-Ω, 10-Ω, and 20-Ω resistors, we may select

$$R_1 = 10\ \Omega \qquad R_2 = 20\ \Omega \qquad R_3 = 5\ \Omega$$

Thus from Eqs. (7.35), (7.36), and (7.37),

$$R_a = \frac{R_1R_2 + R_2R_3 + R_3R_1}{R_1} = \frac{10 \times 20 + 20 \times 5 + 5 \times 10}{10}$$

$$= \frac{350}{10} = 35\ \Omega$$

$$R_b = \frac{R_1R_2 + R_2R_3 + R_3R_1}{R_2} = \frac{350}{20} = 17.5\ \Omega$$

$$R_c = \frac{R_1R_2 + R_2R_3 + R_3R_1}{R_3} = \frac{350}{5} = 70\ \Omega$$

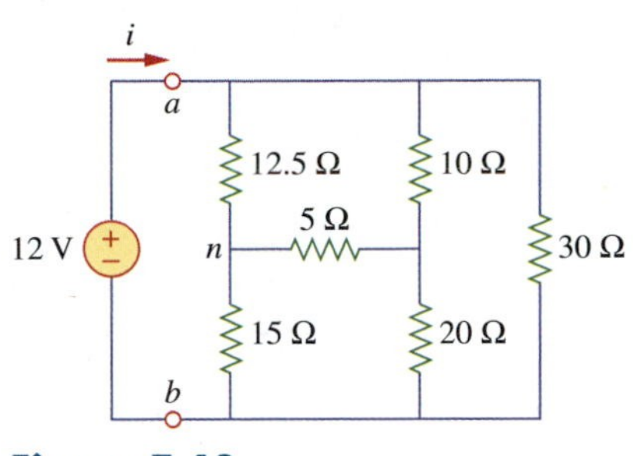

Figure 7.48
For Example 7.13.

With the Y converted to Δ, the equivalent circuit (with the voltage source removed for now) is shown in Fig. 7.49(a). Combining the three pairs of resistors in parallel, we obtain

$$70 \parallel 30 = \frac{70 \times 30}{70 + 30} = 21\ \Omega$$

$$12.5 \parallel 17.5 = \frac{12.5 \times 17.5}{12.5 + 17.5} = 7.2917\ \Omega$$

$$15 \parallel 35 = \frac{15 \times 35}{15 + 35} = 10.5\ \Omega$$

so that the equivalent circuit is shown in Fig. 7.44(b). Hence

$$R_{ab} = (7.292 + 10.5) \parallel 21 = \frac{17.792 \times 21}{17.792 + 21} = 9.632\ \Omega$$

Then

$$i = \frac{v_s}{R_{ab}} = \frac{12}{9.632} = 1.246\ \text{A}$$

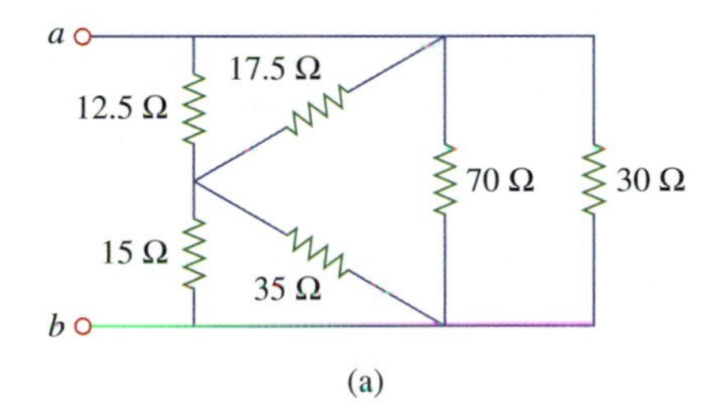

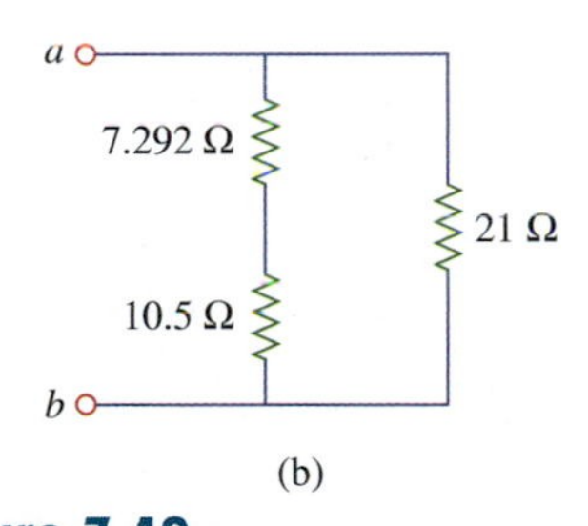

Figure 7.49
Equivalent circuits to Figure 7.48 with the voltage source removed.

Practice Problem 7.13

For the bridge network in Fig. 7.50, find R_{ab} and i.

Answer: 40 Ω; 2.5 A

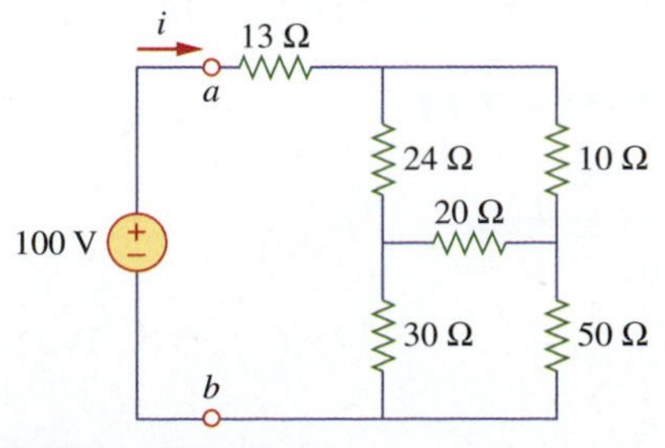

Figure 7.50
For Practice Problem 7.13.

7.9 Computer Analysis

Both PSpice and Multisim can be used to analyze the kinds of circuits we have studied in this chapter. This section demonstrates how.

7.9.1 PSpice

In fact, PSpice is based on nodal analysis as developed in this chapter. The reader is expected to review Sections C.1 through C.3 of Appendix C before proceeding with this section. It should be noted that PSpice is helpful in determining branch voltages and currents only when the numeric values of all the circuit components are known.

Example 7.14

Use PSpice to find the node voltages in the circuit of Fig. 7.51.

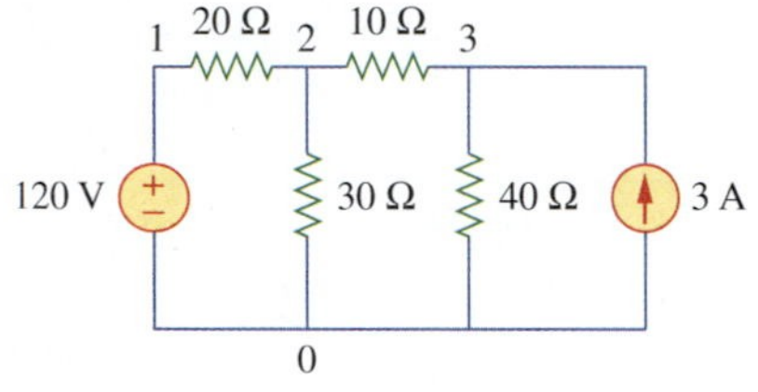

Figure 7.51
For Example 7.14.

Solution:
The first step is to draw the given circuit. Following the instructions given in Sections C.2 and C.3 in Appendix C, the schematic in Fig. 7.52 is produced. Because this is a dc analysis, we use voltage source VDC and current source IDC. Once the circuit is drawn and saved as exam714.dsn, we select **PSpice/New Simulation Profile**. This leads to the New Simulation dialog box. Type "exam714" as the name of the file and click Create. This leads to the Simulation Settings dialog box. Click OK. Then select **PSpice/Run**. The circuit is simulated and the results are displayed on the circuit as in Fig. 7.52. From Fig. 7.52, we notice that:

$$V_1 = 120 \text{ V}, \quad V_2 = 81.29 \text{ V}, \quad V_3 = 89.03 \text{ V}$$

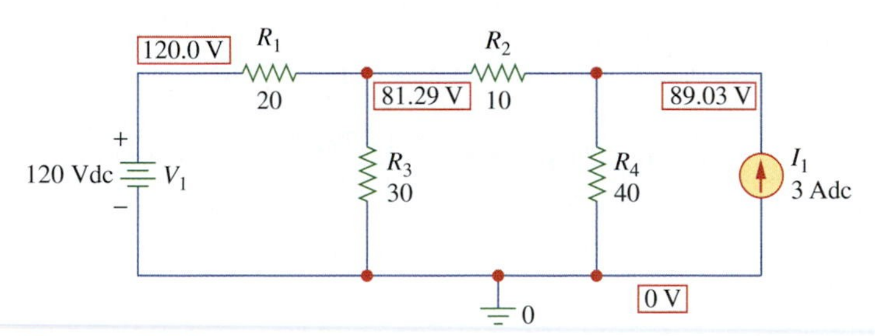

Figure 7.52
For Example 7.12; the schematic of the circuit in Fig. 7.51.

Practice Problem 7.14

For the circuit in Fig. 7.53, use PSpice to find the node voltages.

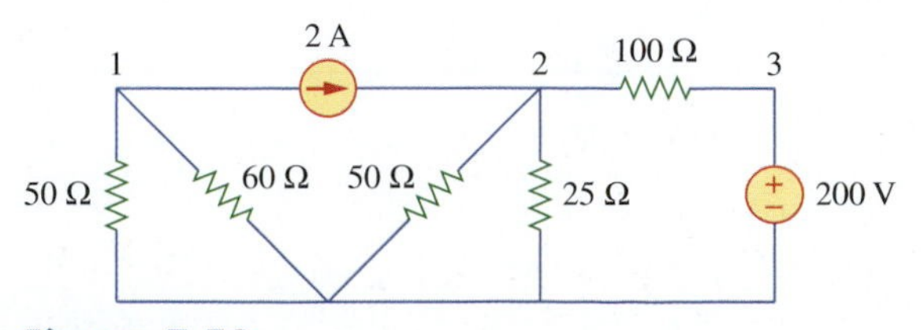

Figure 7.53
For Practice Problem 7.14.

Answer: $V_1 = -54.54$ V; $V_2 = 57.13$ V; $V_3 = 200$ V

7.9.2 Multisim

All the steps involved in creating a circuit are presented in Appendix D and will not be repeated here. The reader is encouraged to read Sections D.1 and D.2 before proceeding with this section.

Example 7.15

Using Multisim, determine V_o, I_1, and I_2 in the circuit of Fig. 7.54.

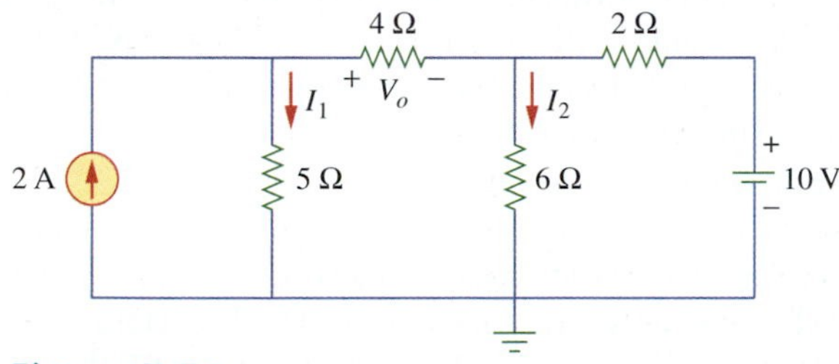

Figure 7.54
For Example 7.15.

Solution:
We first use Multisim to create the circuit as in Fig. 7.55. We connect a voltmeter in parallel with the 4-Ω resistor to measure V_o. We connect two ammeters to measure currents I_1 and I_2. We save the circuit and simulate it by turning on the power switch. After simulation, we obtain the results shown in Fig. 7.55, that is,

$$V_o = 0.952 \text{ V}, \qquad I_1 = 1.762 \text{ A}, \qquad I_2 = 1.310 \text{ A}$$

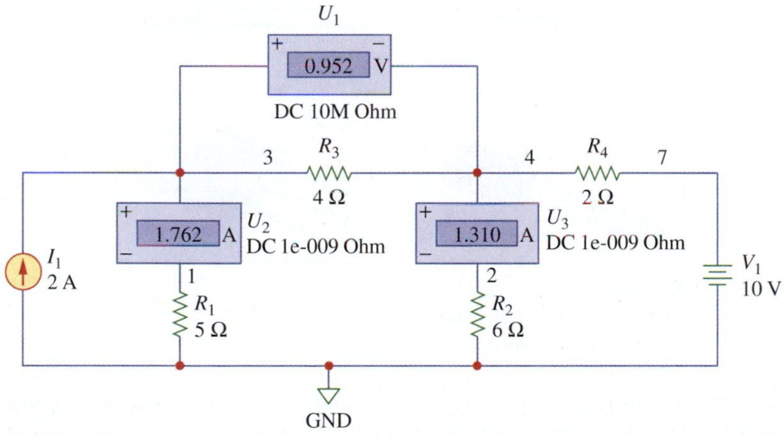

Figure 7.55
Simulation of the circuit in Figure 7.54.

Practice Problem 7.15

Use Multisim to find V_x and I_x in the circuit of Fig. 7.56.

Answer: 4.421 V; 2.211 A

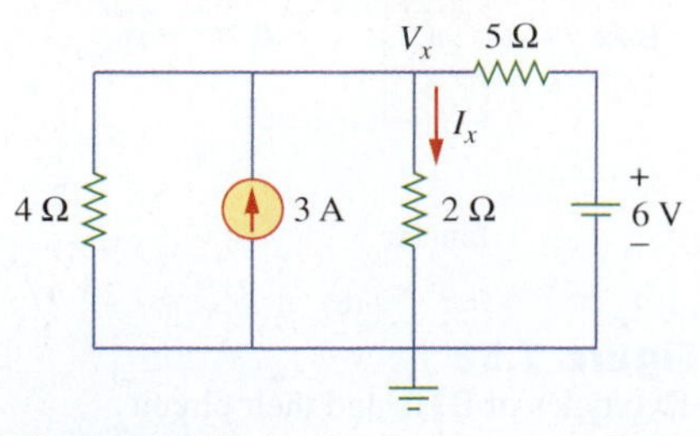

Figure 7.56
For Practice Problem 7.15.

7.10 †Applications: DC Transistor Circuits

Most of us deal with electronic products on a routine basis and have some experience with personal computers. A basic component for the integrated circuits found in these electronics and computers is the active, three-terminal device known as a *transistor*. A transistor is a semiconductor device that is used in a wide variety of applications including amplifiers, switches, voltage regulator, signal modulators, microprocessors, and oscillators. Invented in 1947 at Bell Labs, transistors may be regarded as the most important invention in the twentieth century.[6] Understanding the transistor is essential before we can start an electronic circuit design.

Figure 7.57 depicts various types of commercially available transistors. There are two basic types of transistors: *bipolar junction transistors* (BJTs) and *field-effect transistors* (FETs). Here, we consider only the BJTs, which were developed earlier than FETs and are still used today. Our objective is to present enough detail about BJTs to enable us to apply the techniques developed in this chapter to analyze dc transistor circuits.

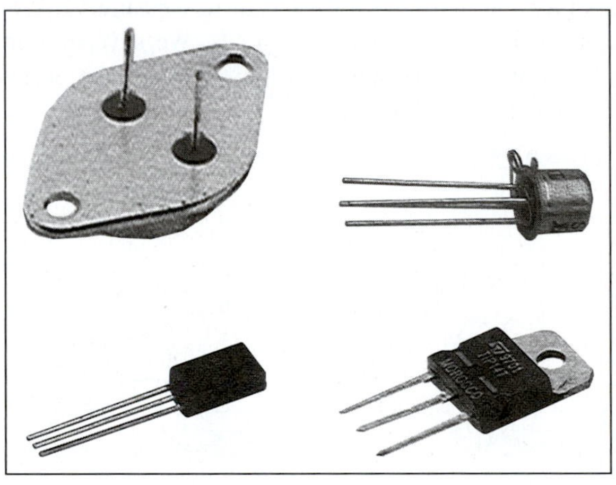

Figure 7.57
Various types of transistors.
Courtesy of Tech America

There are two types of BJTs: *npn* and *pnp*, with their circuit symbols as shown in Fig. 7.58. Each type has three terminals designated as emitter (E), base (B), and collector (C). If we consider the *npn* transistor, for example, the currents and voltages of the transistor are specified in Fig. 7.59. Applying KCL to Fig. 7.59(a) gives

$$I_E = I_B + I_C \tag{7.40}$$

where I_E, I_C, and I_B are the emitter, collector, and base currents, respectively. Similarly, applying KVL to Fig. 7.59(b) gives

$$V_{CE} + V_{EB} + V_{BC} = 0 \tag{7.41}$$

Collector
C
n
p
n
Base
B
Emitter
E
(a)

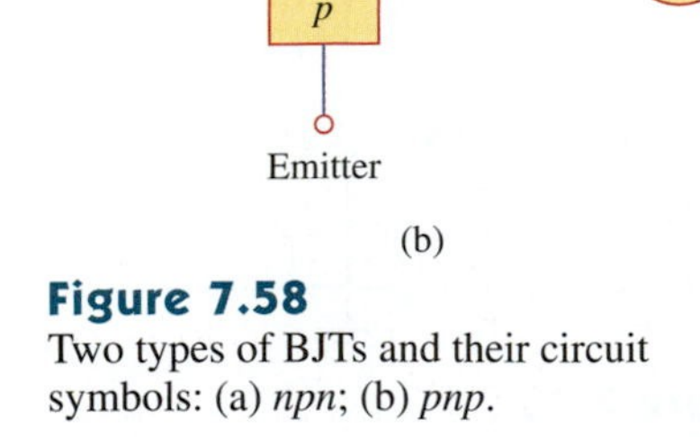

(b)

Figure 7.58
Two types of BJTs and their circuit symbols: (a) *npn*; (b) *pnp*.

[6] The scientists who were responsible for inventing the transistor in 1947 were John Bardeen, Walter Brattain, and William Shockley.

where V_{CE}, V_{EB}, and V_{BC} are the collector-emitter, emitter-base, and base-collector voltages, respectively. The BJT can operate in one of three modes: active, cutoff, or saturation. When silicon *npn* transistors operate in the active mode, typically $V_{BE} \approx 0.7$ V, and

$$I_C = \alpha I_E \tag{7.42}$$

where α is called the *common-base current gain*. It is evident from Eq. (7.42) that α denotes the fraction of electrons injected by the emitter that are collected by the collector. Also,

$$I_C = \beta I_B \tag{7.43}$$

where β is known as the *common-emitter current gain*. Here, α and β are characteristic properties of a given transistor and assume nearly constant values for that transistor. Typically, α takes on values in the range of 0.98 to 0.999, while β takes on values in the range 50 to 1,000. From Eqs. (7.40), (7.42), and (7.43), it is evident that

$$I_E = (1 + \beta)I_B \tag{7.44}$$

and

$$\beta = \frac{\alpha}{1 - \alpha} \tag{7.45}$$

or

$$\alpha = \frac{\beta}{\beta + 1} \tag{7.46}$$

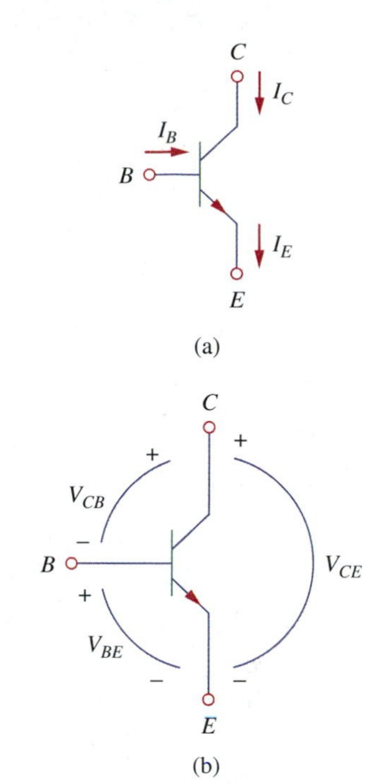

Figure 7.59
The terminal variables of an *npn* transistor: (a) currents; (b) voltages.

Example 7.16

Find I_B, I_C, and V_o in the transistor circuit of Fig. 7.60. Assume that the transistor operates in the active mode and that $\beta = 50$.

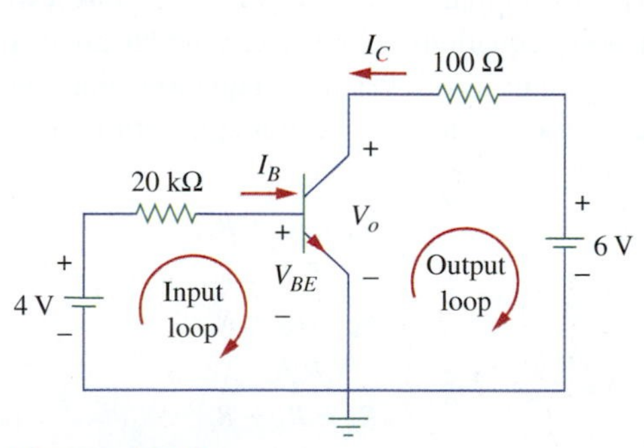

Figure 7.60
For Example 7.16.

Solution:
For the input loop, KVL gives

$$-4 + I_B(20 \times 10^3) + V_{BE} = 0$$

Because $V_{BE} = 0.7$ V in the active mode,

$$I_B = \frac{4 - 0.7}{20 \times 10^3} = 165\ \mu\text{A}$$

But

$$I_C = \beta I_B = 50 \times 165\ \mu\text{A} \ = \ 8.25\ \text{mA}$$

For the output loop, KVL gives

$$-V_o - 100I_C + 6 = 0$$

or

$$V_o = 6 - 100I_C = 6 - 0.825 = 5.175\ \text{V}$$

Note that $V_o = V_{CE}$ in this case.

Practice Problem 7.16

For the transistor circuit in Fig. 7.61, let $\beta = 100$ and $V_{BE} = 0.7$ V. Determine V_o and V_{CE}.

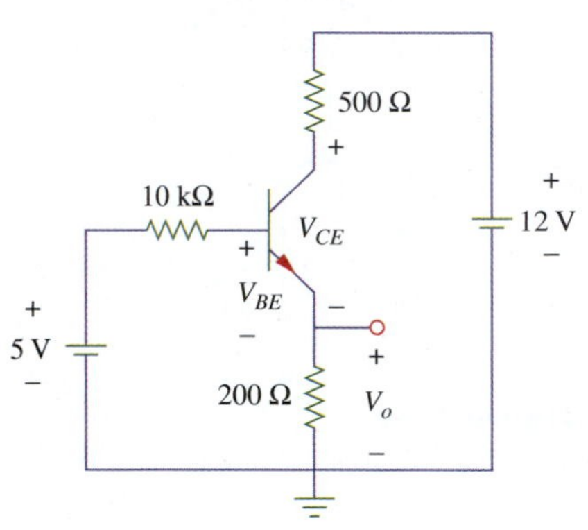

Figure 7.61
For Practice Problem 7.16.

Answer: 2.876 V; 2.004 V

7.11 Summary

1. Mesh analysis is the application of Kirchhoff's voltage law around meshes in a planar circuit. We express the result in terms of mesh currents. Solving the simultaneous equations yields the mesh currents.
2. A supermesh consists of two meshes that have a current source in common.
3. Nodal analysis is the application of Kirchhoff's current law at the nonreference nodes. (It is applicable to both planar and nonplanar circuits). We express the result in terms of the node voltages. Solving the simultaneous equations yields the node voltages.
4. A supernode consists of two nonreference nodes connected by a voltage source.
5. Mesh analysis is normally used when a circuit has fewer mesh equations than node equations. Conversely, nodal analysis is normally used when a circuit has fewer node equations than mesh equations.
6. The formulas for a delta-to-wye transformation are

$$R_1 = \frac{R_b R_c}{R_a + R_b + R_c}$$

$$R_2 = \frac{R_c R_a}{R_a + R_b + R_c}$$

$$R_3 = \frac{R_a R_b}{R_a + R_b + R_c}$$

7. The formulas for a wye-to-delta transformation are

$$R_a = \frac{R_1 R_2 + R_2 R_3 + R_3 R_1}{R_1}$$

$$R_b = \frac{R_1 R_2 + R_2 R_3 + R_3 R_1}{R_2}$$

$$R_c = \frac{R_1 R_2 + R_2 R_3 + R_3 R_1}{R_3}$$

8. Circuit analysis can be carried out using PSpice or Multisim.
9. DC transistor circuits can be analyzed using the techniques covered in this chapter.

Review Questions

7.1 The loop equation for the circuit in Fig. 7.62 is:

(a) $-10 + 4I + 6 + 2I = 0$

(b) $10 + 4I + 6 + 2I = 0$

(c) $10 + 4I - 6 + 2I = 0$

(d) $-10 + 4I - 6 + 2I = 0$

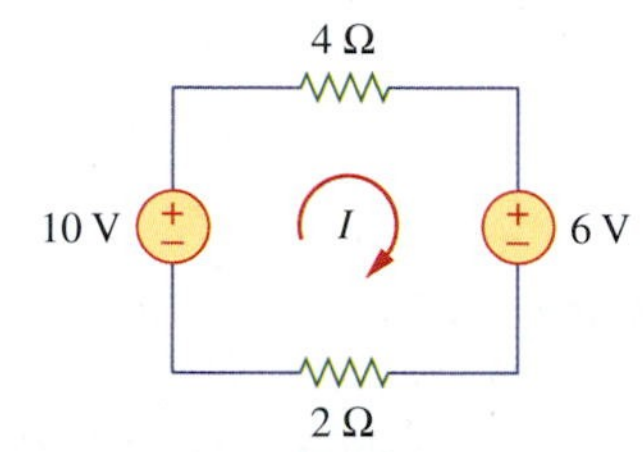

Figure 7.62
For Review Questions 7.1 and 7.2.

7.2 The current I in the circuit in Fig. 7.62 is:

(a) −2.667 A (b) −0.667 A

(c) 0.667 A (d) 2.667 A

7.3 In the circuit in Fig. 7.63, current I_1 is:

(a) 4 A (b) 3 A (c) 2 A (d) 1 A

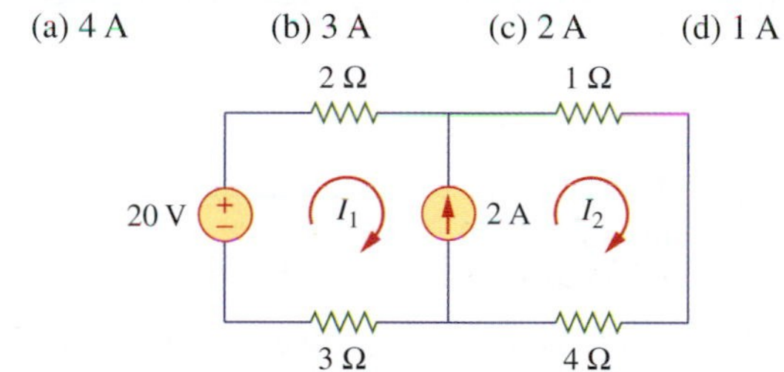

Figure 7.63
For Review Questions 7.3 and 7.4.

7.4 The voltage V across the current source in the circuit of Fig. 7.63 is:

(a) 20 V (b) 15 V (c) 10 V (d) 5 V

7.5 The mesh analysis mainly uses:

(a) Ohm's law and Kirchhoff's current law

(b) Kirchhoff's voltage law and Ohm's law

(c) Kirchhoff's voltage and current laws

(d) Ohm's law and Kirchhoff's voltage and current laws

7.6 At node 1 in the circuit in Fig. 7.64, applying KCL gives:

(a) $2 + \frac{12 - V_1}{3} = \frac{V_1}{6} + \frac{V_1 - V_2}{4}$

(b) $2 + \frac{V_1 - 12}{3} = \frac{V_1}{6} + \frac{V_2 - V_1}{4}$

(c) $2 + \frac{12 - V_1}{3} = \frac{0 - V_1}{6} + \frac{V_1 - V_2}{4}$

(d) $2 + \frac{V_1 - 12}{3} = \frac{0 - V_1}{6} + \frac{V_2 - V_1}{4}$

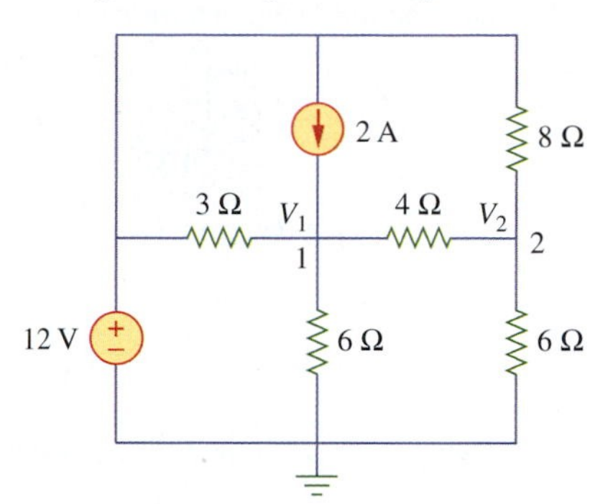

Figure 7.64
For Review Questions 7.6 and 7.7.

7.7 In the circuit in Fig. 7.64, applying KCL at node 2 gives:

(a) $\frac{V_2 - V_1}{4} + \frac{V_2}{8} = \frac{V_2}{6}$

(b) $\frac{V_1 - V_2}{4} + \frac{V_2}{8} = \frac{V_2}{6}$

(c) $\frac{V_1 - V_2}{4} + \frac{12 - V_2}{8} = \frac{V_2}{6}$

(d) $\frac{V_2 - V_1}{4} + \frac{V_2 - 12}{8} = \frac{V_2}{6}$

7.8 For the circuit in Fig. 7.65, V_1 and V_2 are related as:

(a) $V_1 = 6I + 8 + V_2$ (b) $V_1 = 6I - 8 + V_2$

(c) $V_1 = -6I + 8 + V_2$ (d) $V_1 = -6I - 8 + V_2$

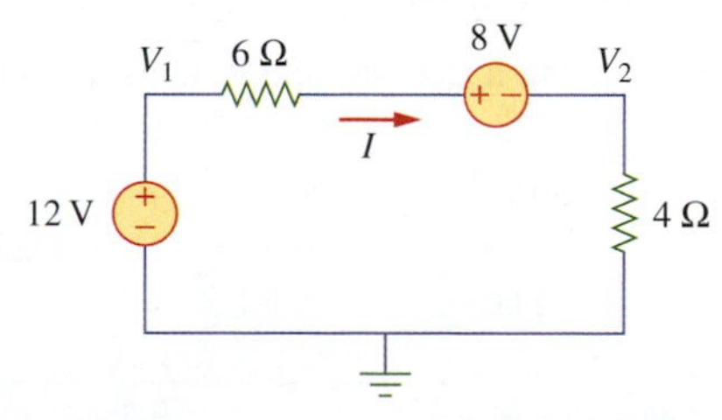

Figure 7.65
For Review Questions 7.8 and 7.9.

7.9 Referring to the circuit in Fig. 7.65, the voltage V_2 is:

(a) −8 V (b) −1.6 V (c) 1.6 V (d) 8 V

7.10 The PSpice part name for a current-controlled voltage source is:

(a) EX (b) FX (c) HX (d) GX

Answers: 7.1a, 7.2c, 7.3d, 7.4b, 7.5b, 7.6a, 7.7c, 7.8a, 7.9c, 7.10c

Problems

Sections 7.2 and 7.3 Mesh Analysis

7.1 Evaluate the following determinants:

(a) $\begin{vmatrix} 50 & -2 \\ 6 & 1 \end{vmatrix}$ (b) $\begin{vmatrix} 5 & 1 & 0 \\ 2 & -3 & 4 \\ 6 & 8 & 10 \end{vmatrix}$

7.2 Find the following determinants:

(a) $\begin{vmatrix} 4 & 8 \\ -3 & 5 \end{vmatrix}$ (b) $\begin{vmatrix} 5 & 3 & 7 \\ 1 & 1 & 4 \\ 2 & 2 & 8 \end{vmatrix}$

7.3 Determine I_1 and I_2 in the following set of equations.

$$2I_1 - I_2 = 4$$
$$8I_1 + 3I_2 = 5$$

7.4 Solve for V_1, V_2, and V_3 for the following set of equations.

$$3V_1 - V_2 + 2V_3 = 4$$
$$2V_1 + 3V_2 - V_3 = 14$$
$$7V_1 - 4V_2 + 3V_3 = -4$$

7.5 Use mesh analysis to find I in the circuit of Fig. 7.66.

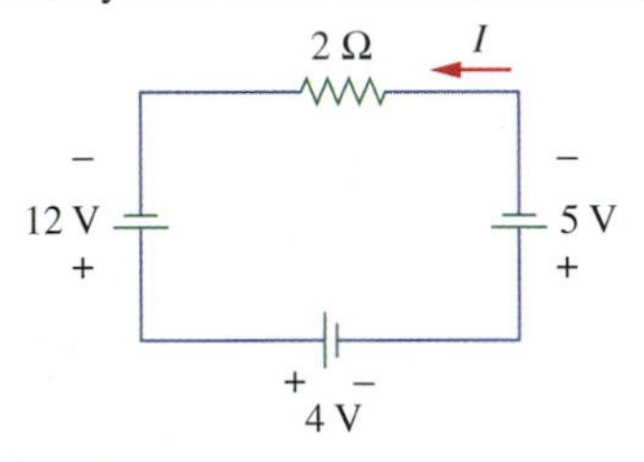

Figure 7.66
For Problem 7.5.

7.6 Using mesh analysis, find V_o in the circuit of Fig. 7.67.

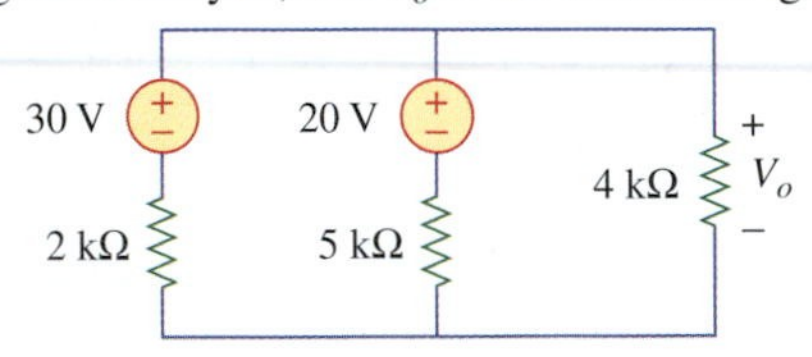

Figure 7.67
For Problems 7.6 and 7.27.

7.7 Use mesh analysis to find i_1 and i_2 in the circuit of Fig. 7.68.

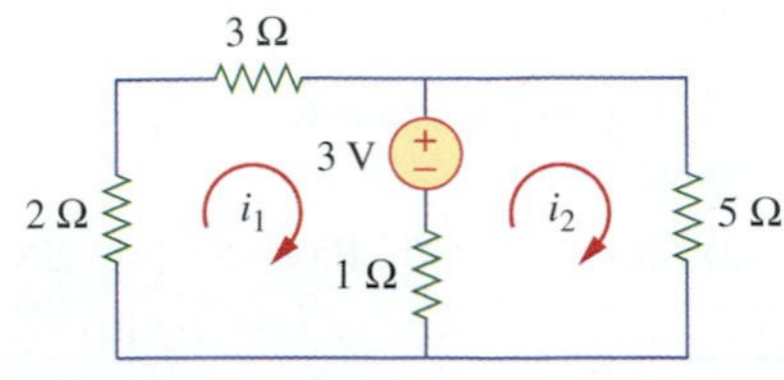

Figure 7.68
For Problems 7.7 and 7.60.

7.8 For the bridge circuit of Fig. 7.69, find the mesh currents.

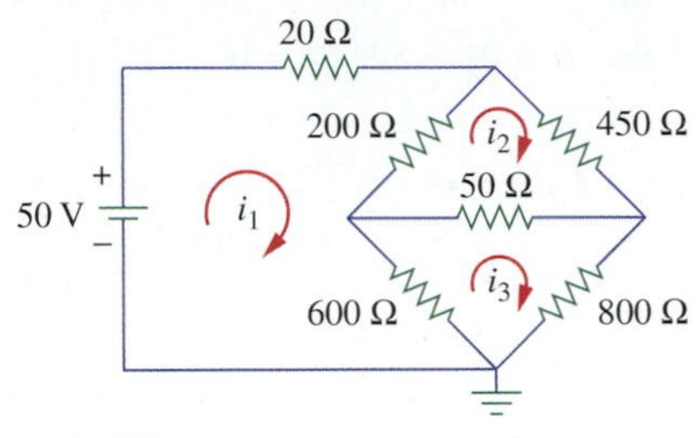

Figure 7.69
For Problems 7.8 and 7.61.

7.9 Apply mesh analysis to find I_x in Fig. 7.70.

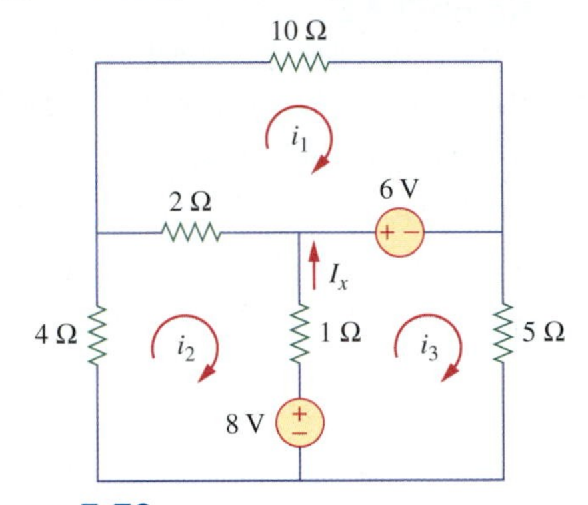

Figure 7.70
For Problem 7.9.

7.10 Apply mesh analysis to find V_o in Fig. 7.71.

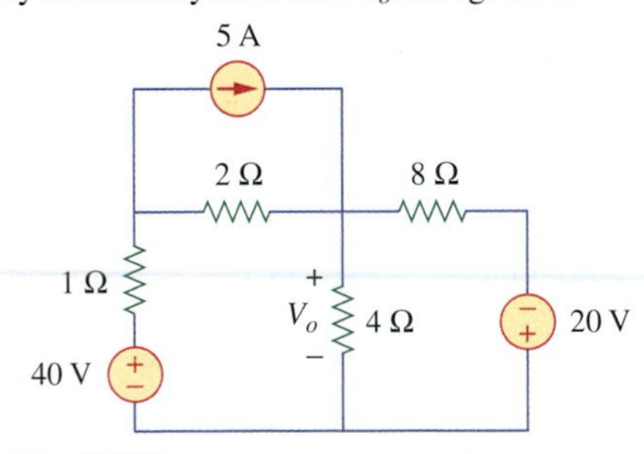

Figure 7.71
For Problems 7.10 and 7.65.

7.11 Find the mesh currents i_1, i_2, and i_3 in the circuit in Fig. 7.72.

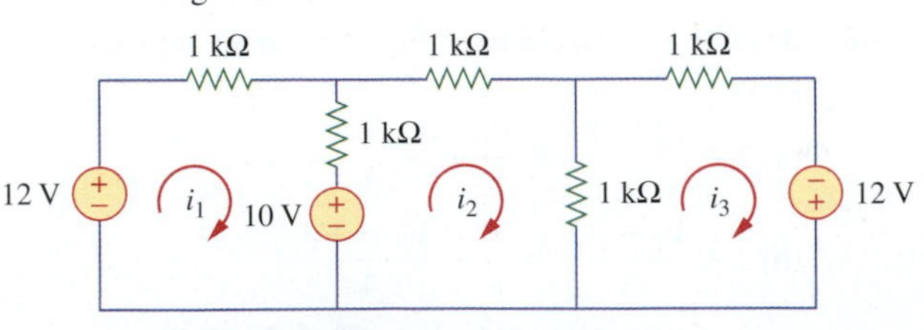

Figure 7.72
For Problem 7.11.

7.12 Use mesh analysis to find I_x in the circuit in Fig. 7.73.

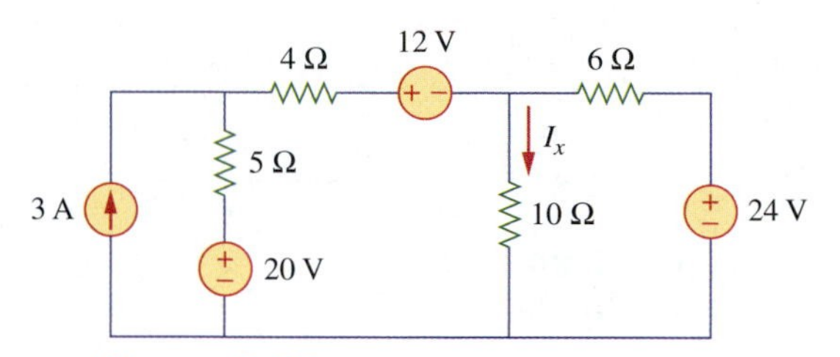

Figure 7.73
For Problem 7.12.

7.13 Use mesh analysis to determine I_a, I_b, and I_c in the circuit of Fig. 7.74. Assume all resistances are 20 Ω.

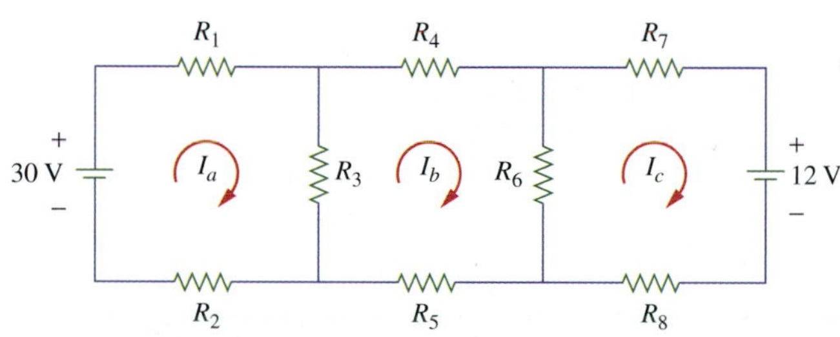

Figure 7.74
For Problem 7.13.

7.14 Use mesh analysis to solve for current I_o in Fig. 7.75.

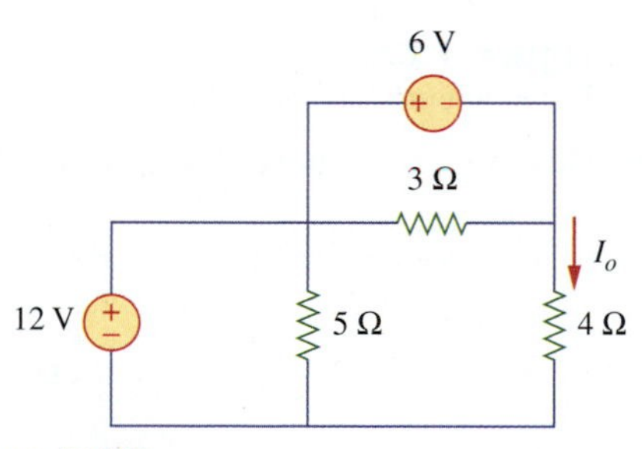

Figure 7.75
For Problem 7.14.

7.15 Obtain the mesh currents in the circuit shown in Fig. 7.76.

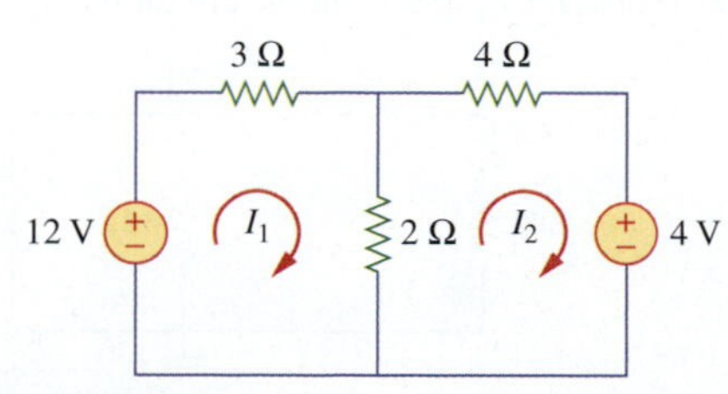

Figure 7.76
For Problem 7.15.

7.16 Write the mesh equations for the circuit in Fig. 7.77.

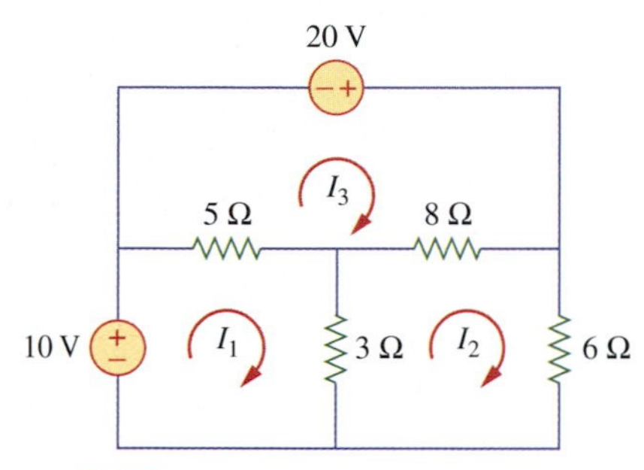

Figure 7.77
For Problem 7.16.

7.17 Write the mesh equations for the circuit in Fig. 7.78.

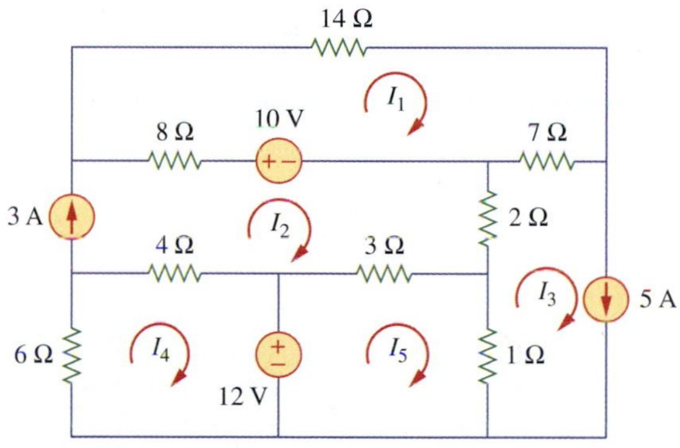

Figure 7.78
For Problem 7.17.

7.18 Using mesh analysis, find I and V in the circuit of Fig. 7.79.

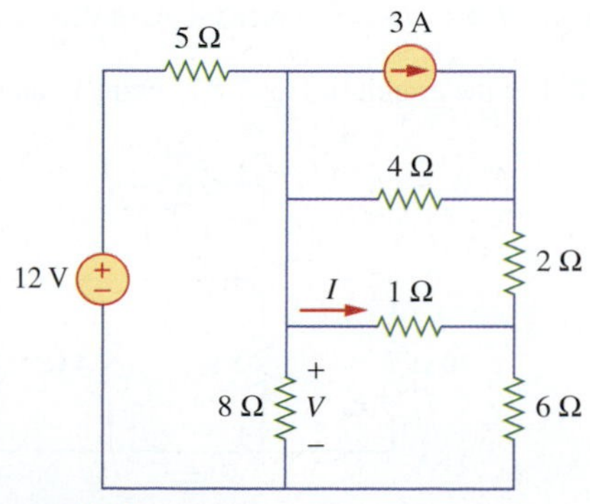

Figure 7.79
For Problem 7.18.

7.19 Find V_o and I_o in the circuit of Fig. 7.80.

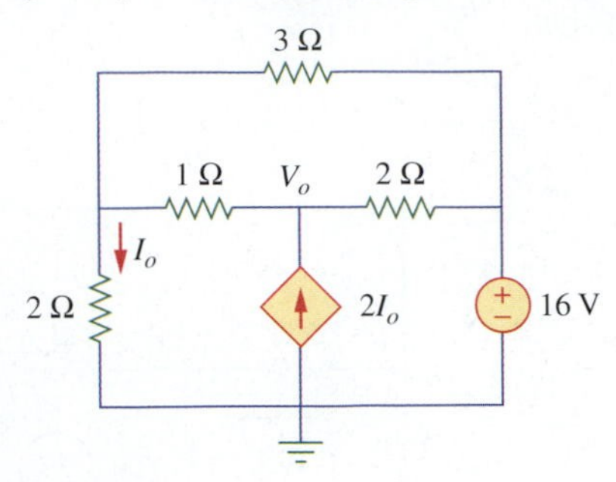

Figure 7.80
For Problem 7.19.

7.20 Use mesh analysis to find the current I_o in the circuit of Fig. 7.81.

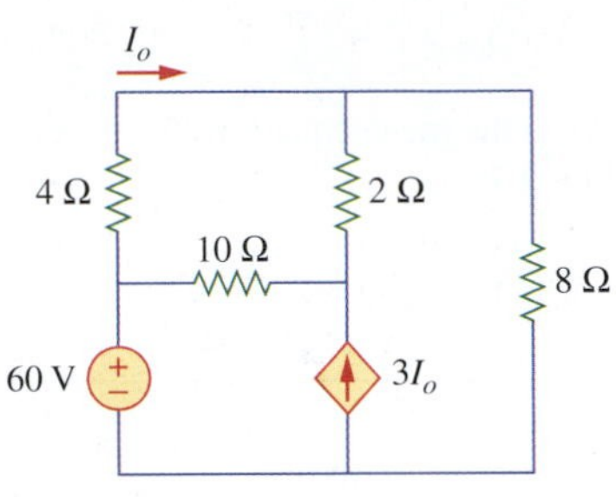

Figure 7.81
For Problem 7.20.

7.21 Use mesh analysis to find i_1, i_2, and i_3 in the circuit of Fig. 7.82.

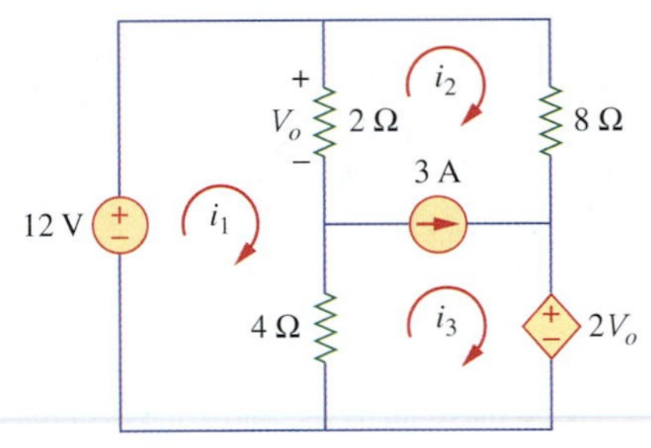

Figure 7.82
For Problem 7.21.

Sections 7.4 and 7.5 Nodal Analysis

7.22 For the circuit in Fig. 7.83 obtain V_1 and V_2.

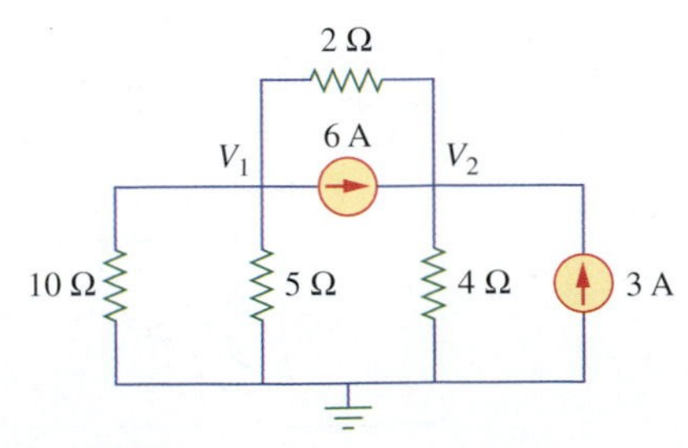

Figure 7.83
For Problem 7.22.

7.23 Determine the voltage V_o in the circuit of Fig. 7.84.

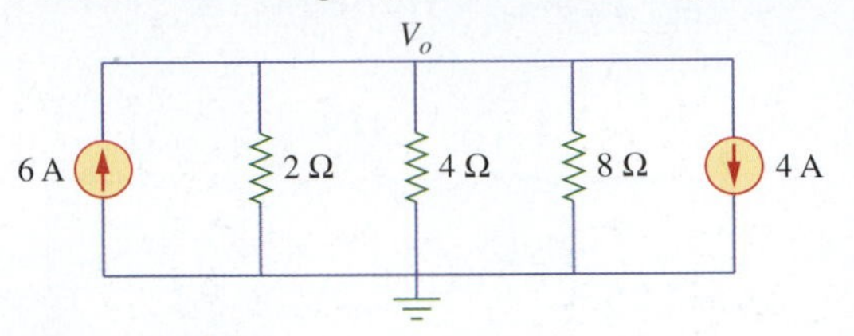

Figure 7.84
For Problem 7.23.

7.24 Using nodal analysis, find V_x in the circuit of Fig. 7.85.

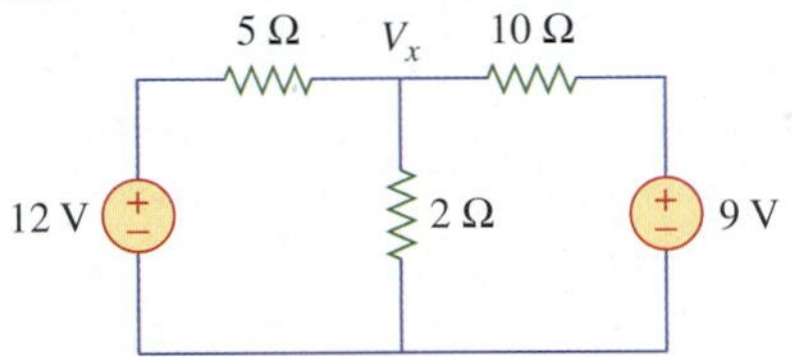

Figure 7.85
For Problem 7.24.

7.25 Find V_1, V_2, and V_3 in the circuit in Fig. 7.86.

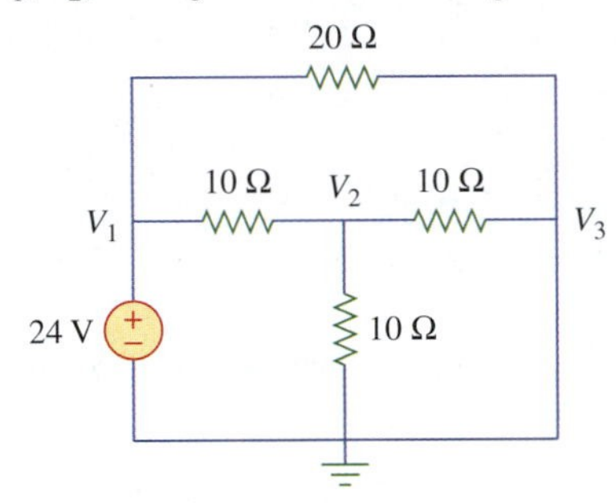

Figure 7.86
For Problems 7.25 and 7.62.

7.26 Given the circuit in Fig. 7.87, calculate V_1 and V_2.

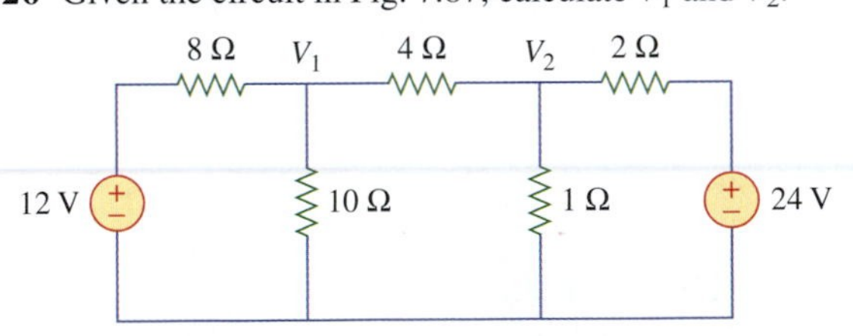

Figure 7.87
For Problem 7.26.

7.27 Obtain V_o in the circuit of Fig. 7.67.

7.28 Calculate V_1 and V_2 in the circuit of Fig. 7.88.

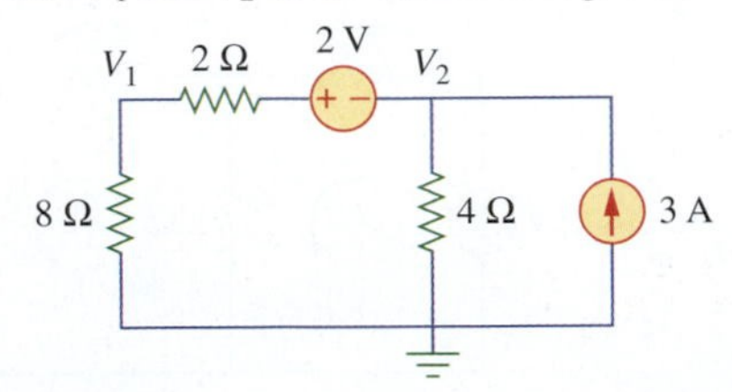

Figure 7.88
For Problem 7.28.

7.29 Use nodal analysis to find the current I_o in the circuit of Fig. 7.89.

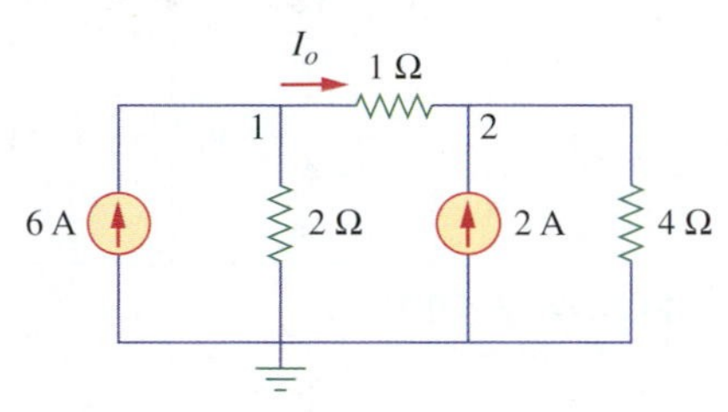

Figure 7.89
For Problem 7.29.

7.30 Determine the node voltages in the circuit of Fig. 7.90 using nodal analysis.

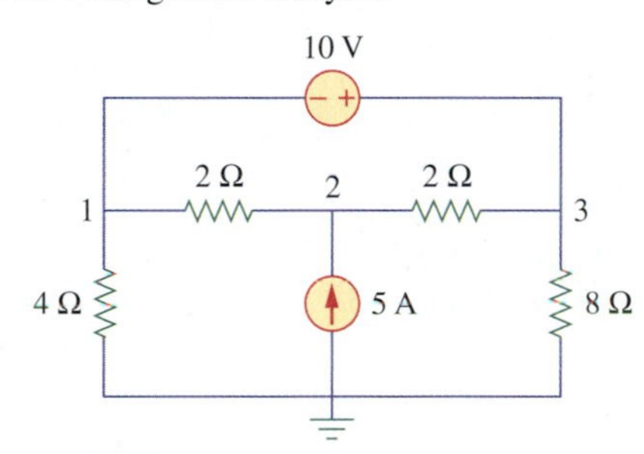

Figure 7.90
For Problem 7.30.

7.31 Obtain V_1 and V_2 in the circuit of Fig. 7.91.

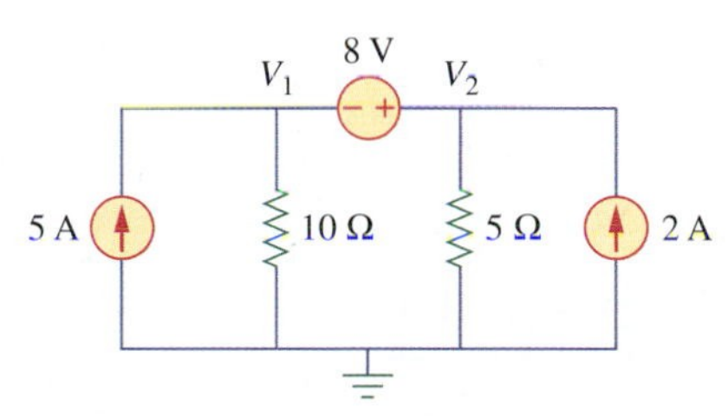

Figure 7.91
For Problems 7.31 and 7.64.

7.32 Find i_s in the circuit of Fig. 7.92.

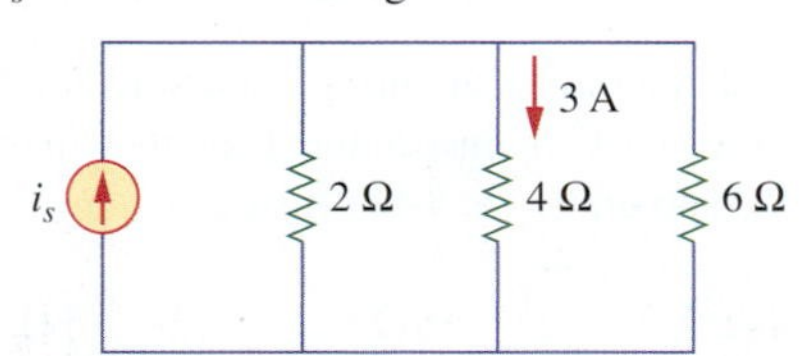

Figure 7.92
For Problem 7.32.

7.33 Calculate v_s in the circuit of Fig. 7.93.

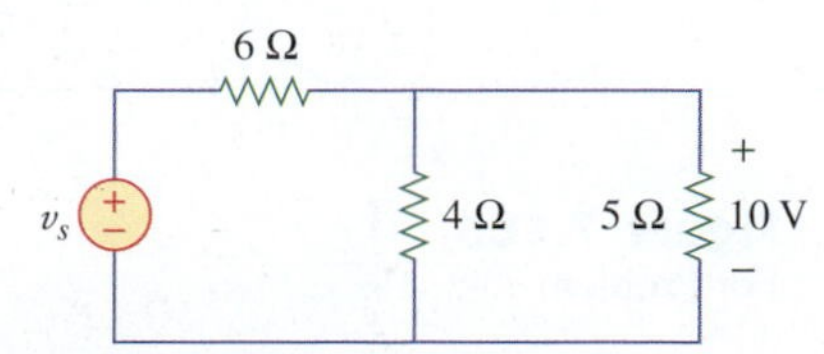

Figure 7.93
For Problem 7.33.

7.34 Use nodal analysis to find V_1 and V_2 in the circuit of Fig. 7.94.

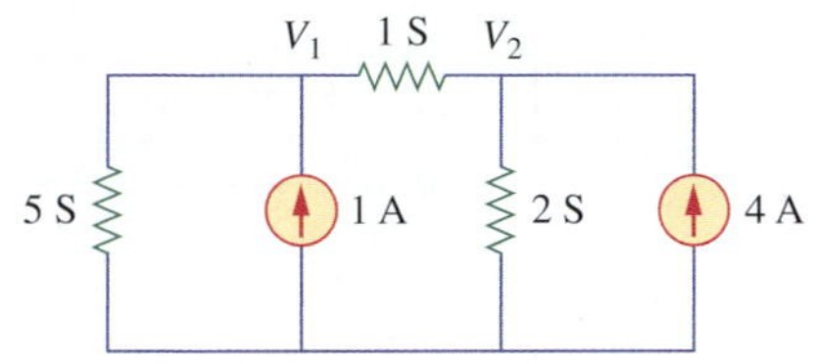

Figure 7.94
For Problem 7.34.

7.35 Use nodal analysis to find V_1 and V_2 in the circuit of Fig. 7.95.

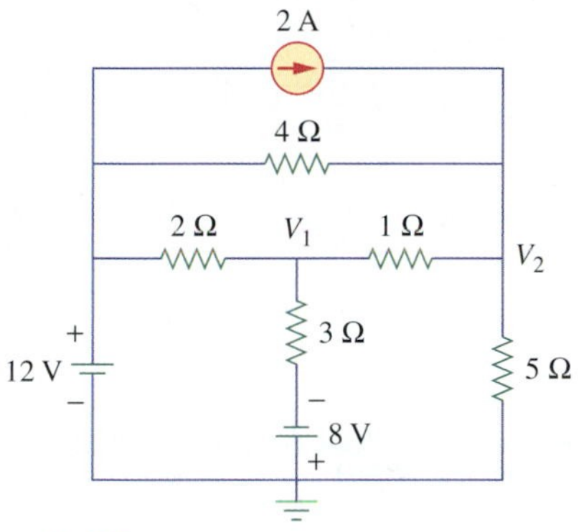

Figure 7.95
For Problem 7.35.

7.36 In Fig. 7.96, use nodal analysis to find V_o.

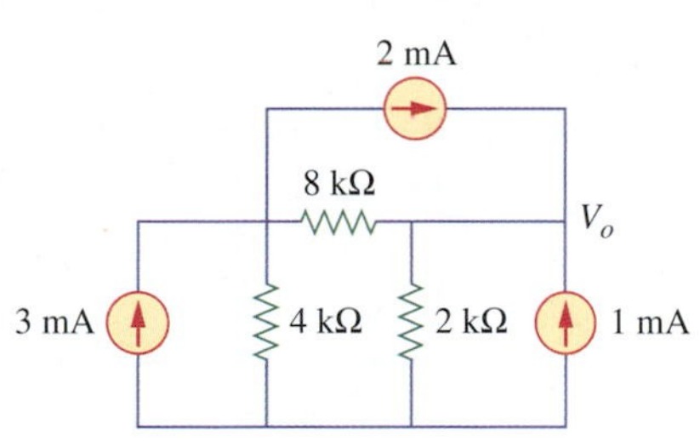

Figure 7.96
For Problem 7.36.

7.37 Determine the node voltages in the circuit of Fig. 7.97.

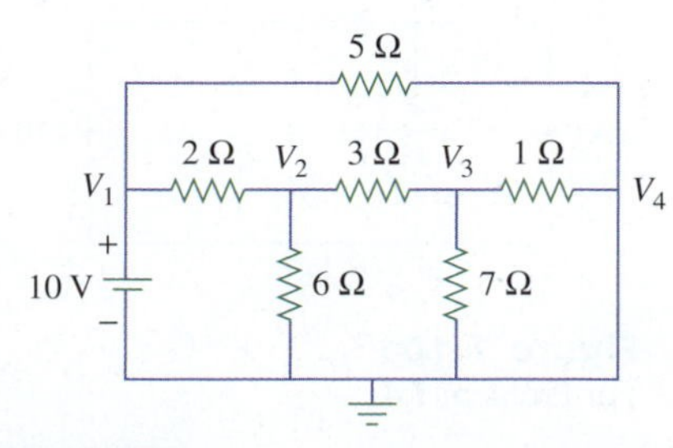

Figure 7.97
For Problem 7.37.

7.38 For the circuit in Fig. 7.98, use nodal analysis to find V and I.

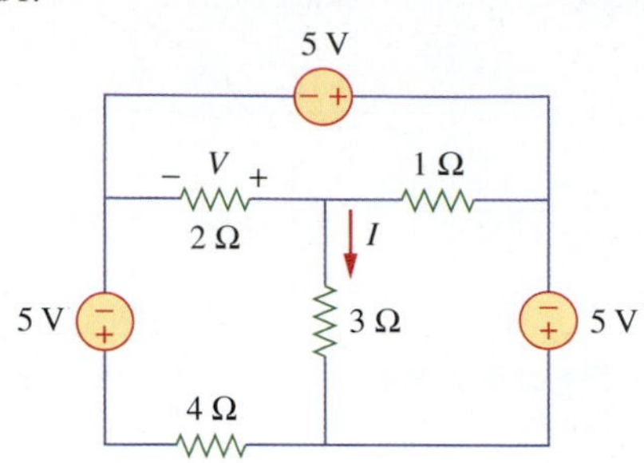

Figure 7.98
For Problem 7.38.

***7.39** Consider the circuit in Fig. 7.99. Use nodal analysis to find V and I.[7]

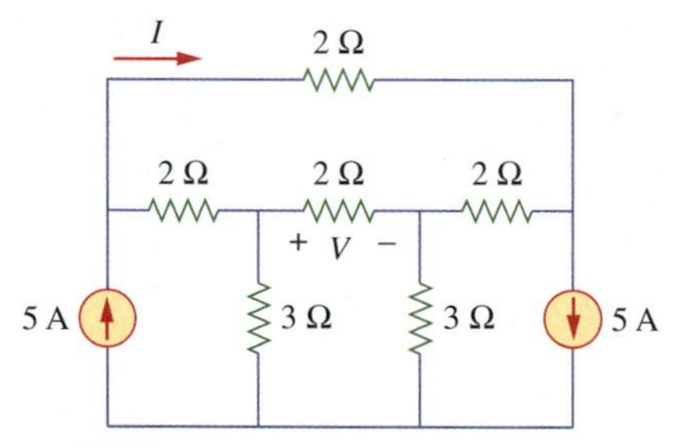

Figure 7.99
For Problem 7.39.

7.40 For the circuit in Fig. 7.100, use nodal analysis to find V and I.

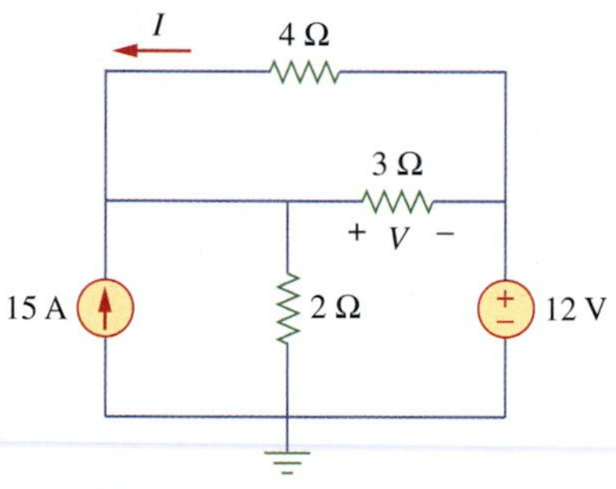

Figure 7.100
For Problem 7.40.

7.41 Determine V_1, V_2, and the power dissipated in all the resistors in the circuit of Fig. 7.101.

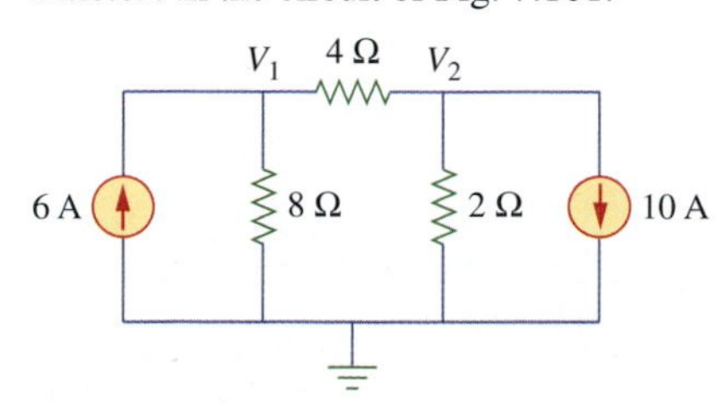

Figure 7.101
For Problem 7.41.

7.42 Apply nodal analysis to solve for V_x in the circuit in Fig. 7.102.

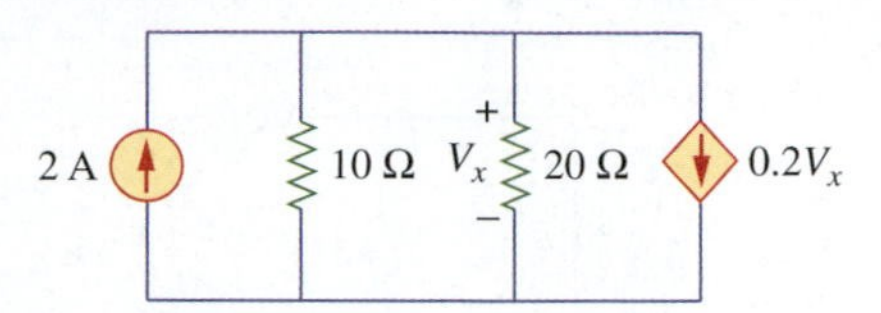

Figure 7.102
For Problem 7.42.

7.43 Determine I_b in the circuit of Fig. 7.103 using nodal analysis.

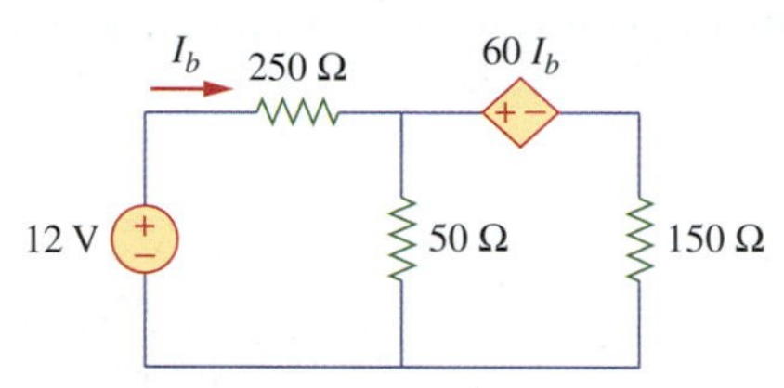

Figure 7.103
For Problem 7.43.

7.44 Find I_o in the circuit of Fig. 7.104.

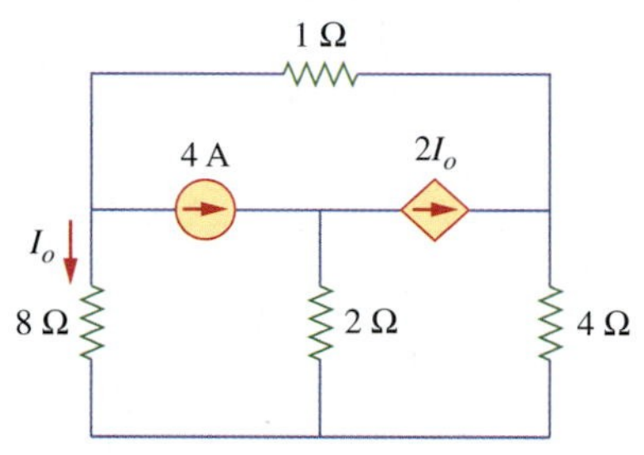

Figure 7.104
For Problem 7.44.

Section 7.6 Mesh and Nodal Analysis by Inspection

7.45 Obtain the mesh-current equations for the circuit in Fig. 7.105 by inspection. Calculate the power absorbed by the 8-Ω resistor.

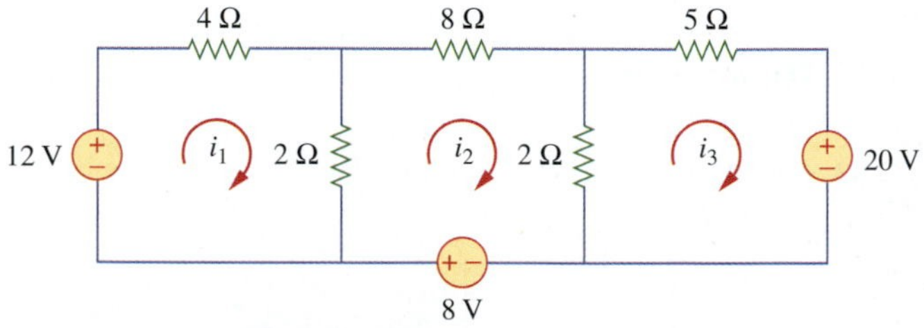

Figure 7.105
For Problem 7.45.

[7] An asterisk indicates a challenging problem.

7.46 By inspection, write the mesh-current equations for the circuit in Fig. 7.106.

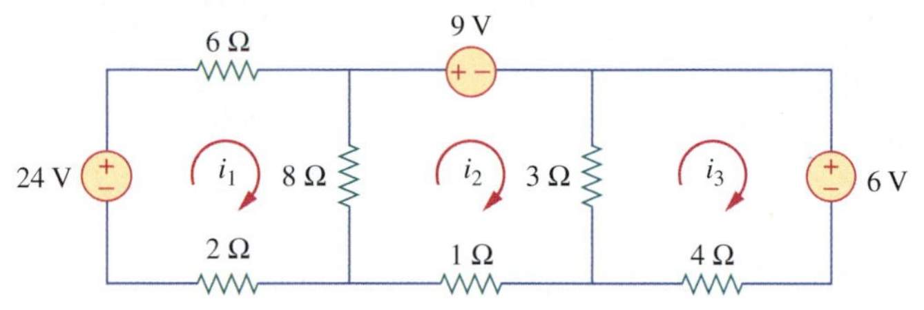

Figure 7.106
For Problem 7.46.

7.47 By inspection, obtain the mesh-current equations for the circuit in Fig. 7.107.

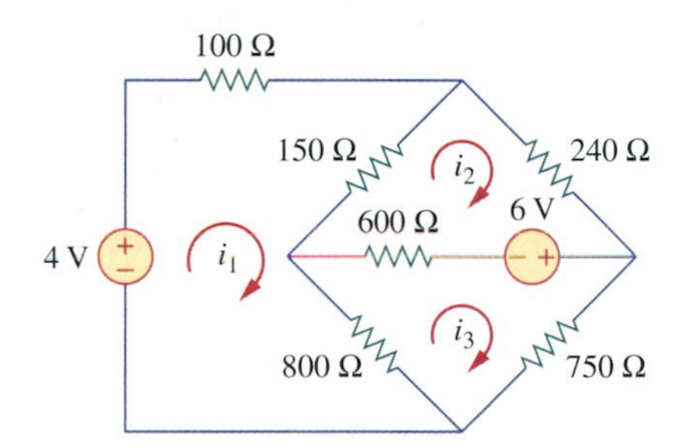

Figure 7.107
For Problem 7.47.

7.48 Obtain the node-voltage equations for the circuit in Fig. 7.108 by inspection. Determine the node voltages V_1 and V_2.

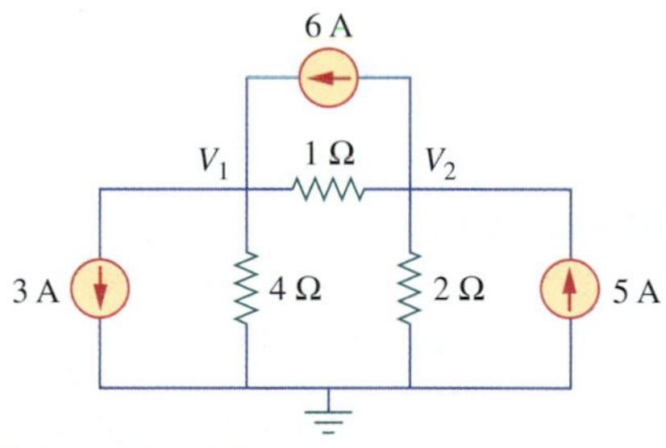

Figure 7.108
For Problem 7.48.

7.49 By inspection, write the node-voltage equations for the circuit in Fig. 7.109.

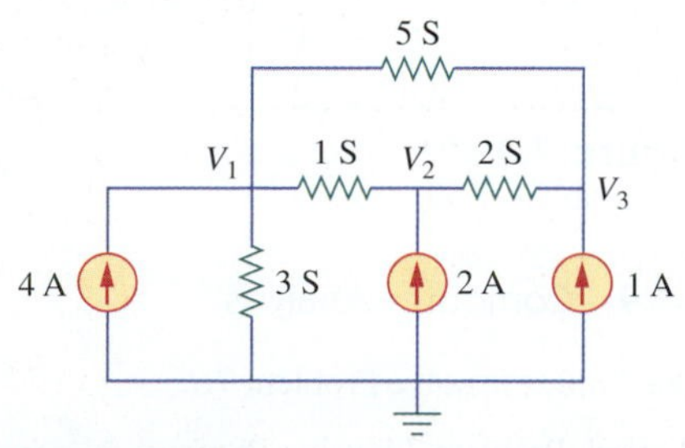

Figure 7.109
For Problem 7.49.

7.50 Write the node-voltage equations of the circuit in Fig. 7.110 by inspection.

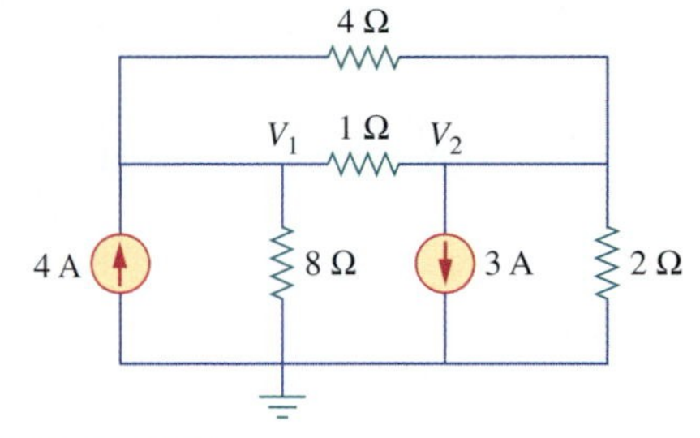

Figure 7.110
For Problem 7.50.

7.51 Obtain the node-voltage equations for the circuit in Fig. 7.111 by inspection.

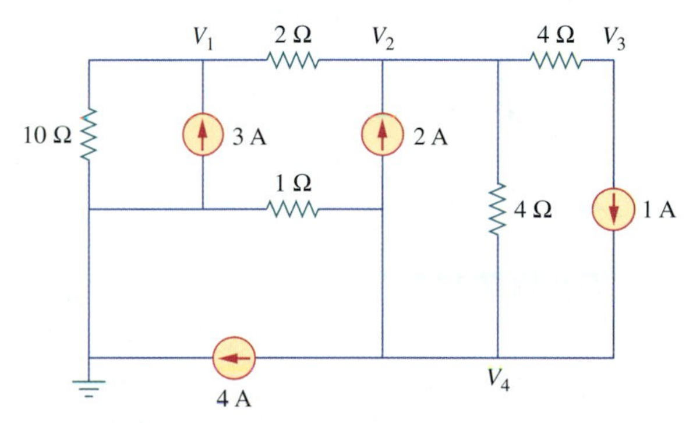

Figure 7.111
For Problems 7.51 and 7.63.

Section 7.8 Wye-Delta Transformations

7.52 Convert the circuits in Fig. 7.112 from Y to Δ.

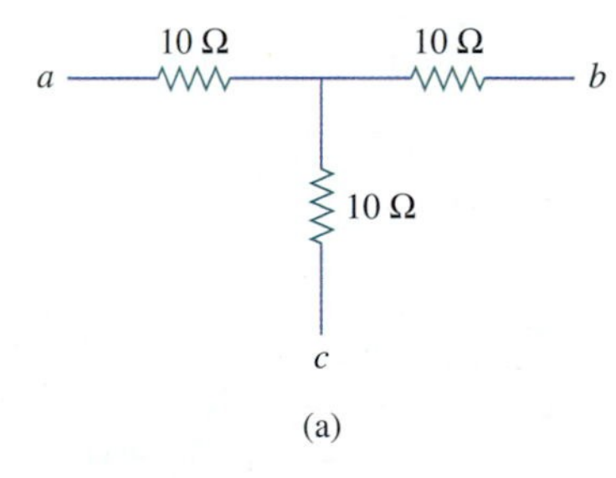

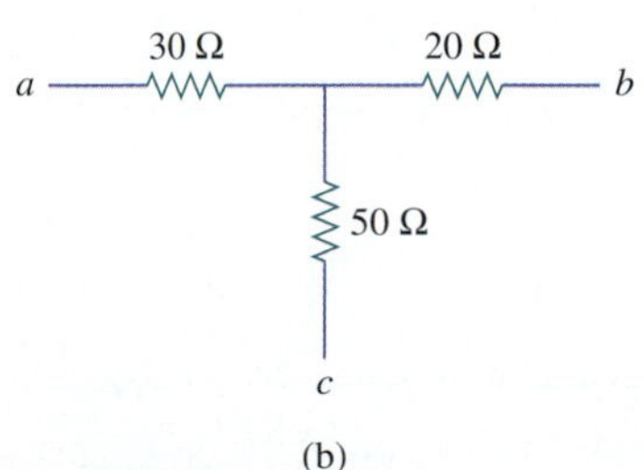

Figure 7.112
For Problem 7.52.

7.53 Transform the circuits in Fig. 7.113 from Δ to Y.

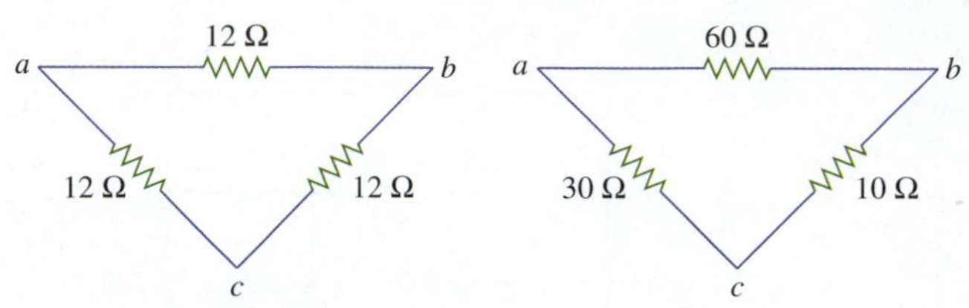

Figure 7.113
For Problem 7.53.

7.54 Obtain the equivalent resistance at the terminals a–b for the circuits in Fig. 7.114.

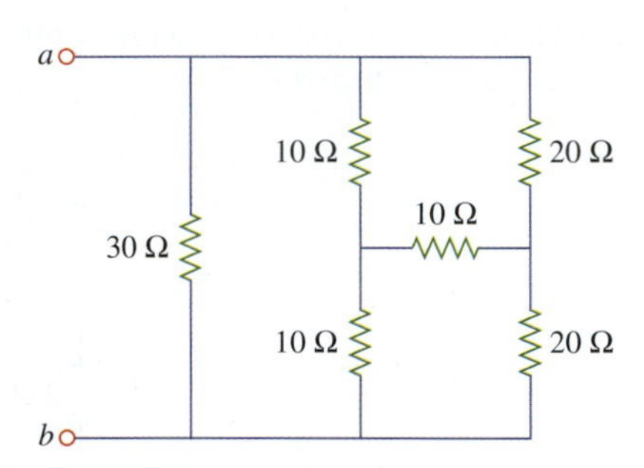

Figure 7.114
For Problem 7.54.

***7.55** Obtain the equivalent resistance R_{ab} in each of the circuits of Fig. 7.115. In (b), all resistors have a value of 30 Ω.

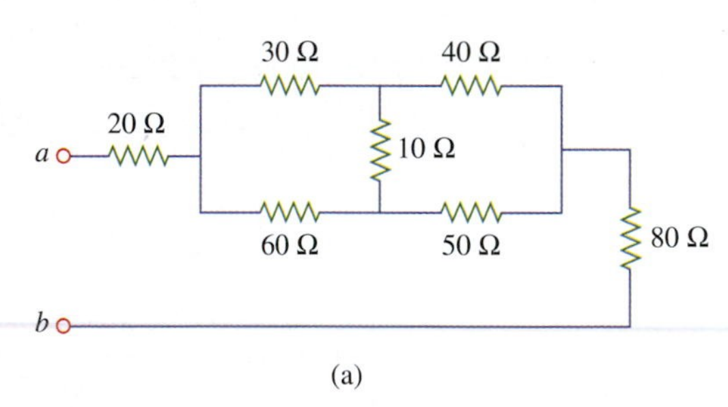

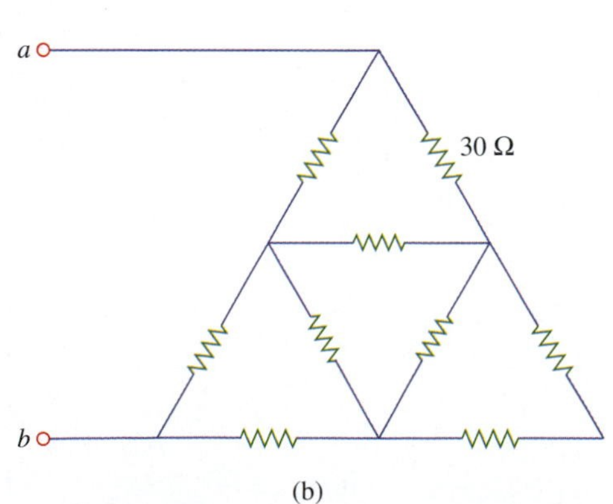

Figure 7.115
For Problem 7.55.

7.56 Consider the circuit in Fig. 7.116. Find the equivalent resistance at terminals (a) a–b and (b) c–d.

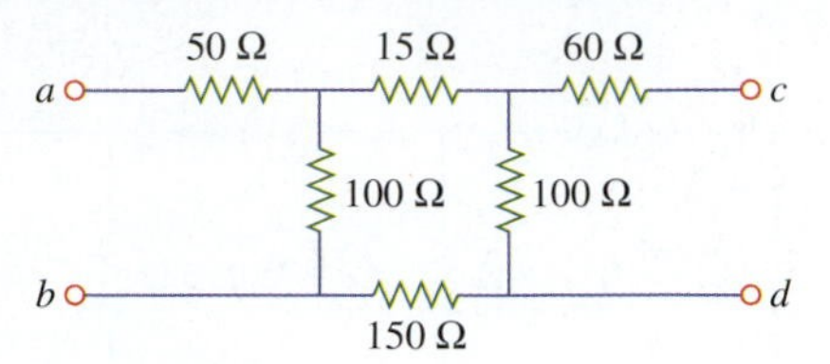

Figure 7.116
For Problem 7.56.

7.57 Calculate I_o in the circuit of Fig. 7.117.

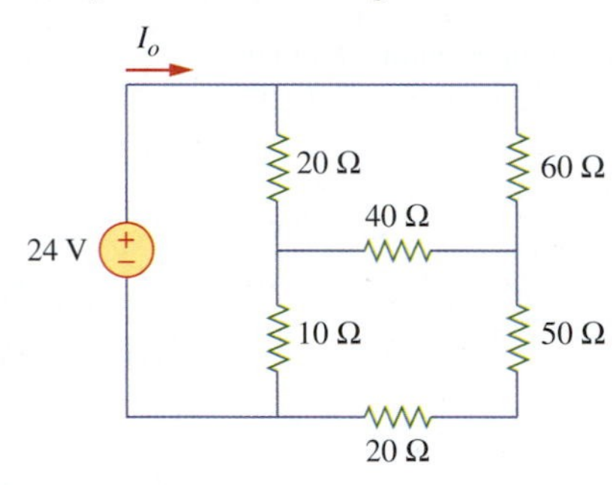

Figure 7.117
For Problem 7.57.

7.58 Determine V in the circuit of Fig. 7.118.

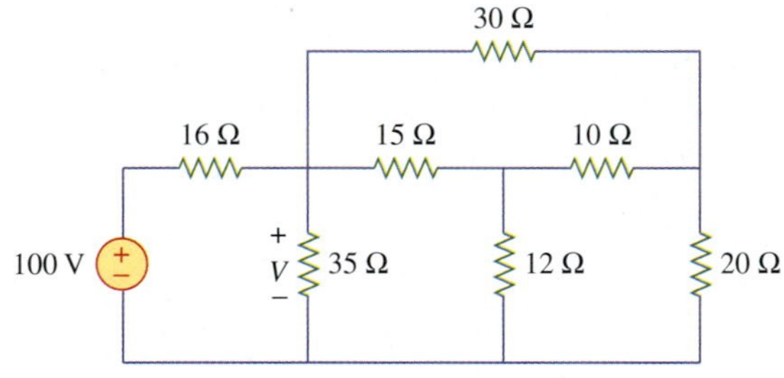

Figure 7.118
For Problem 7.58.

7.59 Calculate I_x in the circuit of Fig. 7.119.

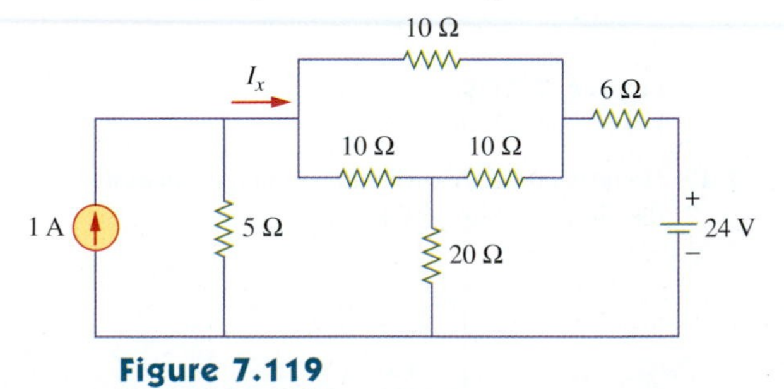

Figure 7.119
For Problem 7.59.

Section 7.9 Computer Analysis

7.60 Use PSpice to solve Problem 7.7.

7.61 Rework Problem 7.8 using PSpice.

7.62 Use PSpice to solve Problem 7.25.

7.63 Find the nodal voltages in the circuit of Fig. 7.111 using PSpice.

7.64 Rework Problem 7.31 using Multisim.

7.65 Use Multisim to solve Problem 7.10.

7.66 Solve Problem 7.59 using Multisim.

7.67 Using PSpice, find V_o in the circuit in Fig. 7.120.

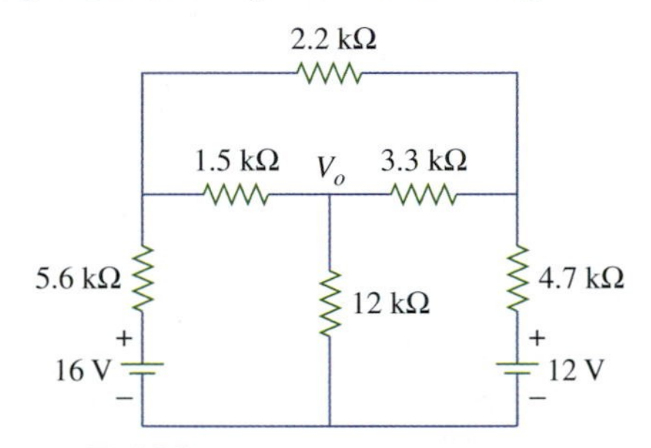

Figure 7.120
For Problem 7.67.

Section 7.10 Applications

7.68 Find I_C and V_{CE} for the circuit of Fig. 7.121. Assume $I_B \approx 0$ and $V_{BE} = 0.7$ V.

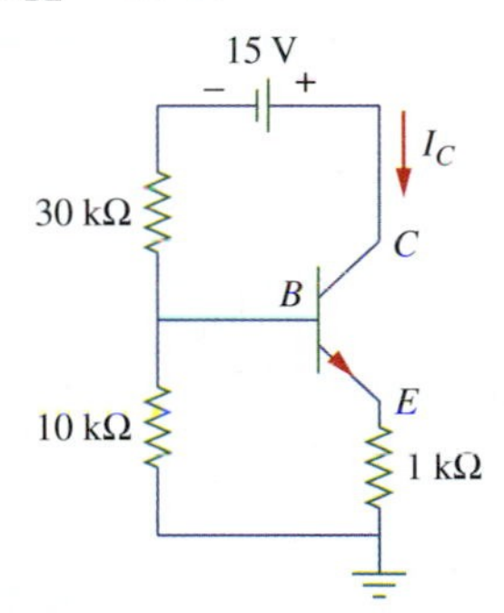

Figure 7.121
For Problem 7.68.

7.69 For the simple transistor circuit of Fig. 7.122, let $\beta = 75$ and $V_{BE} = 0.7$ V. What value of V_i is required to give a collector-emitter voltage of 2 V?

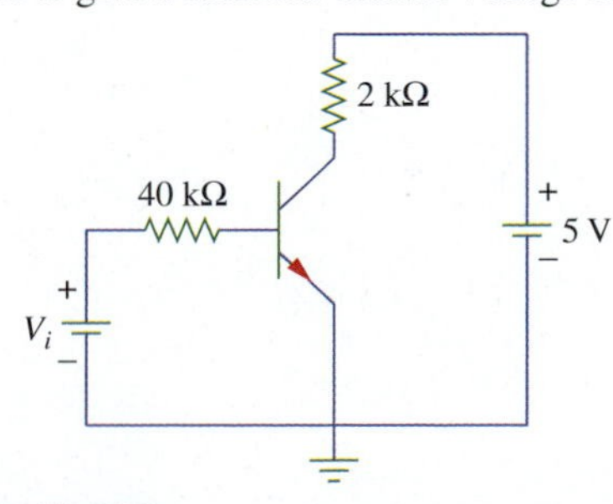

Figure 7.122
For Problem 7.69.

7.70 Calculate V_s for the transistor in Fig. 7.123 given that $V_o = 4$ V, $\beta = 150$, and $V_{BE} = 0.7$ V.

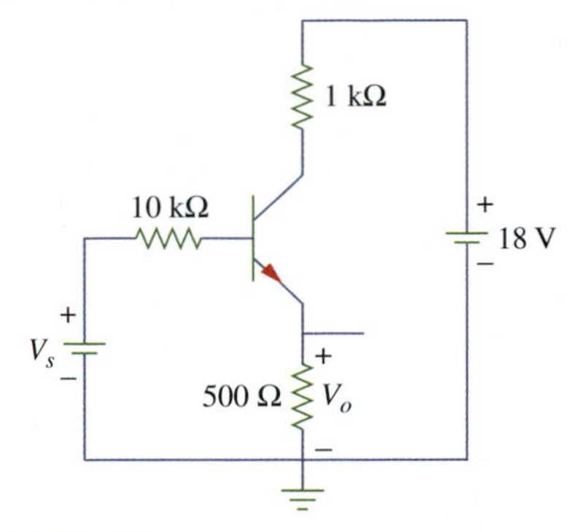

Figure 7.123
For Problem 7.70.

7.71 For the transistor circuit of Fig. 7.124, find I_B, V_{CE}, and V_o. Let $\beta = 200$ and $V_{BE} = 0.7$ V.

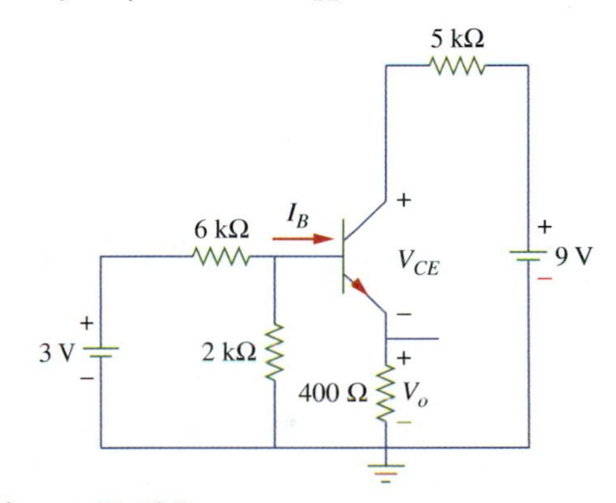

Figure 7.124
For Problem 7.71.

chapter

8

Circuit Theorems

Who does nothing, makes no mistakes, and who makes no mistakes never makes any progress.

—Paul Winkler

Enhancing Your Career

Careers in Computer Engineering Technology

Police officers use computers.
© The McGraw-Hill Companies, Inc./Kefover/Opatrany

Computer engineering technology education has gone through drastic changes in recent decades. Computers occupy a prominent place in modern society and education. They are helping to change the faces of research, development, production, business, and entertainment. The scientist, technologist, doctor, attorney, teacher, airline pilot, businessperson—almost everyone benefits from a computer's abilities to store large amounts of information and to process that information in very short periods of time. The Internet, a computer communication network, has become essential in industry, medicine, business, education, and library science, and computer usage is growing by leaps and bounds.

Three major disciplines study computer systems: computer science, computer engineering technology, and information management science. Computer engineering technology has grown so fast and wide that it is divorcing itself from electrical engineering technology. But in many schools of engineering technology, computer engineering technology is still an integral part of electrical engineering technology.

An education in computer engineering technology should provide breadth in software, hardware design, and basic modeling techniques. It should include courses in data structures, digital systems, computer architecture, microprocessors, interfacing, programming, software engineering, and operating systems. Electrical engineering technologists who specialize in computer engineering technology find jobs in computer industries and in most of the numerous fields where computers are being used. Companies that produce software are growing rapidly in number and size and are providing employment for those who are skilled in programming. An excellent way to advance your knowledge of computers is to join the IEEE Computer Society, which sponsors diverse magazines, journals, and conferences.

8.1 Introduction

A major advantage of analyzing circuits using Kirchhoff's laws as we did in the previous chapter is that we can analyze a circuit without tampering with its original configuration. A major disadvantage of this approach is that tedious computation is involved for a large, complex circuit.

The growth in the area of applications of electric circuits has led to an evolution from simple to complex circuits. To reduce the complexity of circuits, engineers over the years have developed theorems to simplify circuit analysis. Such theorems include Thevenin's, Norton's, Millman's, substitution, and reciprocity theorems.[1] These theorems may be regarded as special applications of mesh and nodal analyses discussed in the previous chapter. Because the theorems are applicable to linear circuits, we first discuss the concept of circuit linearity. In addition to circuit theorems, the concepts of superposition, source transformation, and maximum power transfer will be discussed in this chapter. The concepts developed in this chapter are then applied to source modeling.

8.2 Linearity Property

Linearity is the property of an element describing a constant ratio of incremental cause and effect. It is a combination of both the homogeneity property and the additivity property. Although this property applies to many circuit elements, we shall limit our discussion to its applicability to resistors in this chapter.

> **Linearity** is a condition in which the change in value of one quantity is directly proportional to that of another quantity.

The homogeneity property requires that if the input (also called the *excitation*) is multiplied by a constant, then the output (also called the *response*) is multiplied by the same constant. For a resistor, for example, Ohm's law relates the input I to the output V,

$$V = IR \tag{8.1}$$

If the current is increased by a constant k, then the voltage increases correspondingly by k, that is,

$$\boxed{kIR = kV} \tag{8.2}$$

The additivity property requires that the response to a sum of inputs is the sum of the responses to each input applied separately. Using the voltage-current relationship of a resistor, if

$$V_1 = I_1R \tag{8.3a}$$

and

$$V_2 = I_2R \tag{8.3b}$$

then applying $(I_1 + I_2)$ gives

$$\boxed{V = (I_1 + I_2)R = I_1R + I_2R = V_1 + V_2} \tag{8.4}$$

[1] One of the more common applications of Thevenin's theorem, for example, is in the analysis of an unbalanced Wheatstone bridge (see Fig. 6.41).

A **linear circuit** is one that contains linear elements.

We say that a resistor is a linear element because the voltage/current relationship satisfies both the homogeneity and the additivity properties. In general, a linear circuit consists of only linear elements and linear sources. A linear element is one in which the input/output relationship is linear. Examples of linear elements include resistors, capacitors, and inductors. Examples of nonlinear elements include diodes, transistors, and operational amplifiers.

Throughout this book, we will consider only linear circuits. Note that because $P = I^2R = V^2/R$ (making it a quadratic function rather than a linear one), the relationship between power and voltage (or current) is nonlinear. Therefore, the theorems covered in this chapter are not applicable to power.

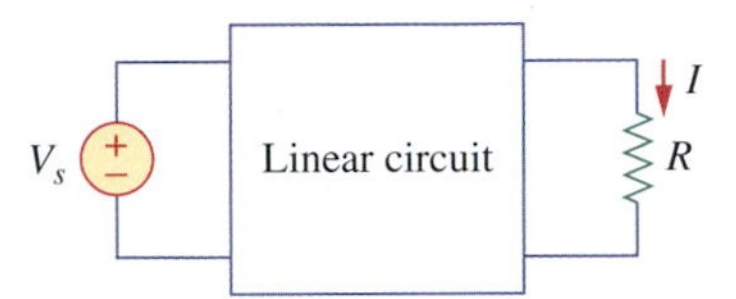

Figure 8.1
A linear circuit with input V_s and output I.

To understand the linearity principle, consider the linear circuit shown in Fig. 8.1. The linear circuit has no independent sources inside it. It is excited by a voltage source V_s, which serves as the input. The circuit is terminated by a load R. (The source may be a CD player, while the load may be a speaker.) We may take the current I through R as the output. Suppose $V_s = 10$ V gives $I = 2$ A. According to the linearity principle, $V_s = 1$ V will give $I = 0.2$ A. By the same token, $I = 1$ mA must be due to $V_s = 5$ mV.

Example 8.1

For the circuit in Fig. 8.2, find I_o when $V_s = 12$ V and $V_s = 24$ V.

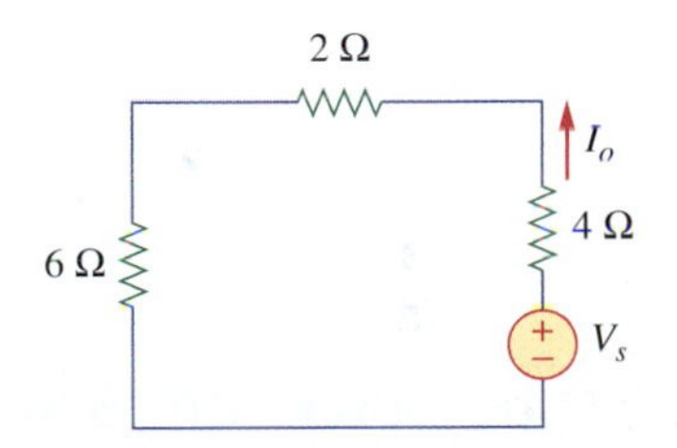

Figure 8.2
For Example 8.1.

Solution:
Applying KVL to the loop gives

$$(6 + 2 + 4)I_o - V_s = 0$$

or

$$I_o = \frac{V_s}{12}$$

When $V_s = 12$ V,

$$I_o = \frac{12}{12} = 1 \text{ A}$$

When $V_s = 24$ V,

$$I_o = \frac{24}{12} = 2 \text{ A}$$

showing that when the source doubles, I_o doubles.

Practice Problem 8.1

For the circuit in Fig. 8.3, find V_o when $I_s = 15$ A and $I_s = 30$ A.

Answer: 10 V; 20 V

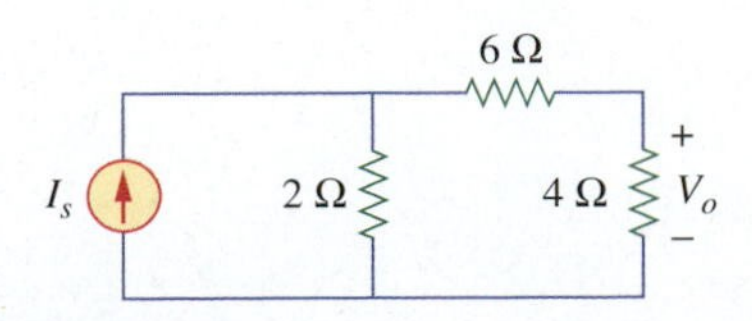

Figure 8.3
For Practice Problem 8.1.

Example 8.2

Assume $I_o = 1$ A and use linearity to find the actual value of I_o in the circuit in Fig. 8.4.

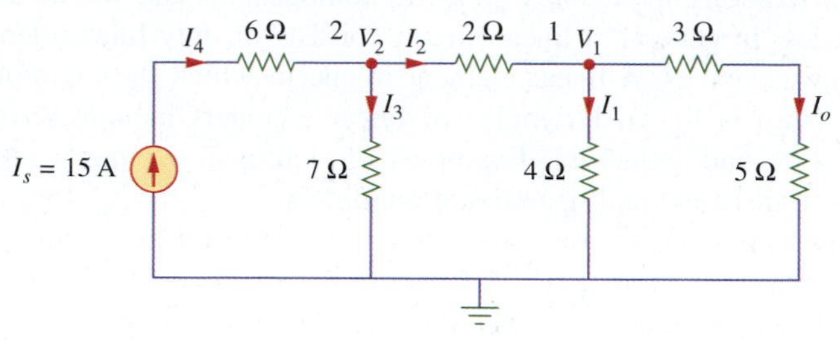

Figure 8.4
For Example 8.2.

Solution:
If $I_o = 1$ A, then $V_1 = (3 + 5)\, I_o = 8$ V and $I_1 = V_1/4 = 2$ A. Applying KCL at node 1 gives

$$I_2 = I_1 + I_o = 3 \text{ A}$$

Note that

$$V_2 = V_1 + 2I_2 = 8 + 2 \times 3 = 14 \text{ V}$$
$$I_3 = V_2/7 = 14/7 = 2 \text{ A}$$

Applying KCL at node 2 gives

$$I_4 = I_3 + I_2 = 5 \text{ A}$$

Therefore $I_s = 5$ A. This shows that assuming $I_o = 1$ A gives $I_s = 5$ A, the actual source current of 15 A will give $I_o = 3$ A as the actual value.

Practice Problem 8.2

Assume that $V_o = 1$ V and use linearity to calculate the actual value of V_o in the circuit of Fig. 8.5.

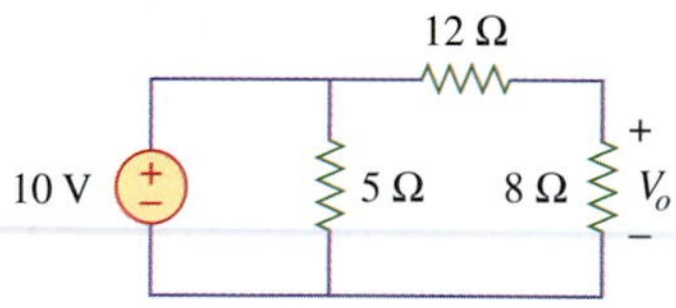

Answer: 4 V

Figure 8.5
For Practice Problem 8.2.

8.3 Superposition

The linearity property leads to the idea of *superposition.*[2] The principle of superposition helps us to analyze a linear circuit with more than one independent source by calculating the contribution of each independent source separately and later adding the contributions. However, to apply the superposition principle, we must keep one thing in mind. We consider one independent source at a time while all other sources are *turned off* or *reduced to zero.*[3] This implies that we replace every voltage source by 0 V, or a short circuit, and every current source by

[2] Note: Superposition is not limited to circuit analysis but is applicable in many fields where cause and effect bear a linear relationship to one another.

[3] Note: Other terms such as *killed, made inactive, deadened, suppressed,* or *set equal to zero* are often used instead of *turned off* to convey the same idea.

0 A, or an open circuit. (This is a theoretical technique and not a method to be used in the lab.) In this way, we obtain a simpler and more manageable circuit.

> The **superposition** principle states that the voltage across (or current through) an element in a linear circuit is the algebraic sum of the voltages across (or currents through) that element due to each independent source acting alone.

With this in mind, we apply the superposition principle by taking these steps:

Steps to Apply Superposition Principle

1. Turn off all independent sources except one source. Find the output (voltage or current) due to that active source using Kirchhoff's laws.
2. Repeat step 1 for each of the other independent sources.
3. Find the total contribution by adding algebraically the contributions of the independent sources.

Analyzing a circuit using superposition has one major disadvantage: it involves more work. For example, if the circuit has three independent sources, we have to analyze each of the three simpler circuits separately to find the contribution due to the respective individual source. However, superposition does help reduce a complex circuit to simpler circuits through replacement of voltage sources by short circuits and of current sources by open circuits.

Keep in mind that superposition is based on linearity. For this reason, as we mentioned earlier, it cannot be used to determine the effect on power due to each source, because the power absorbed by a resistor depends on the *square* of the voltage or current, making the equation nonlinear. If power is needed, the current through (or the voltage across) the element must be calculated first using superposition. Also, each time we switch a source on/off we are changing the circuit so that any previous calculations are no longer applicable and a new analysis is required.

Example 8.3

Use the superposition theorem to find V in the circuit in Fig. 8.6.

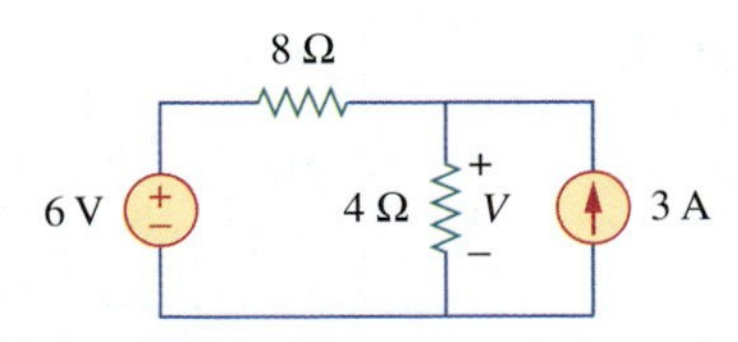

Figure 8.6
For Example 8.3.

Solution:
Because there are two sources, let

$$V = V_1 + V_2$$

where V_1 and V_2 are the contributions due to the 6-V voltage source and the 3-A current source, respectively. To obtain V_1, we set the current source to zero, as shown in Fig. 8.7(a). Applying KVL to the loop in Fig. 8.7(a) gives

$$12i_1 - 6 = 0 \quad \Rightarrow \quad i_1 = 0.5 \text{ A}$$

Thus,

$$V_1 = 4i_1 = 2 \text{ V}$$

We can also use voltage division to get V_1; that is,

$$V_1 = \frac{4}{4 + 8}(6) = 2 \text{ V}$$

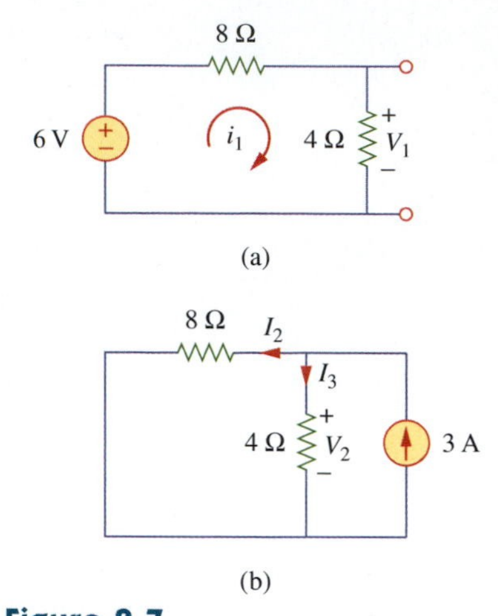

Figure 8.7
For Example 8.3: (a) calculating V_1; (b) calculating V_2.

To get V_2, we set the voltage source to zero, as shown in Fig. 8.7(b). Using current division,

$$I_3 = \frac{8}{4+8}(3) = 2 \text{ A}$$

Hence,

$$V_2 = 4I_3 = 8 \text{ V}$$

Thus

$$V = V_1 + V_2 = 2 + 8 = 10 \text{ V}$$

Practice Problem 8.3

Using the superposition theorem, find V_o in the circuit in Fig. 8.8.

Answer: 12 V.

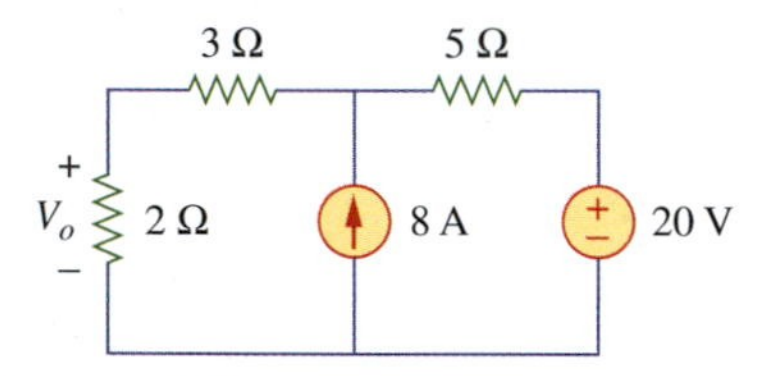

Figure 8.8
For Practice Problem 8.3.

Example 8.4

For the circuit in Fig. 8.9, use the superposition theorem to find I.

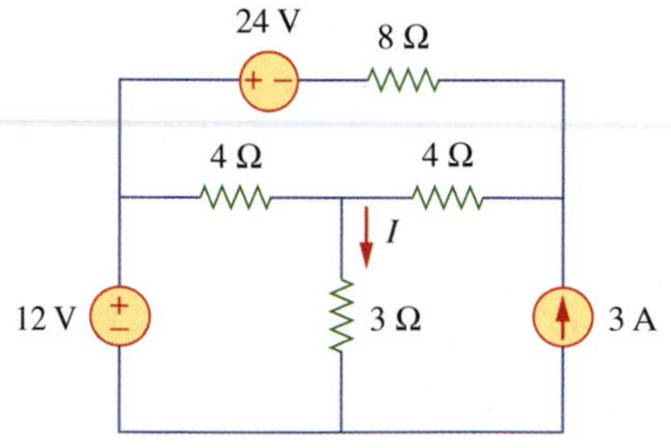

Figure 8.9
For Example 8.4.

Solution:
In this case, we have three sources. Let

$$I = I_1 + I_2 + I_3$$

where I_1, I_2 and I_3 are due to the 12-V, 24-V, and 3-A sources, respectively. To get I_1, consider the circuit in Fig. 8.10(a). Combining 4 Ω (on the right-hand side) in series with 8 Ω gives 12 Ω. The 12 Ω in parallel with 4 Ω gives 12 × 4/16 = 3 Ω. Thus,

$$I_1 = \frac{12}{6} = 2 \text{ A}$$

To get I_2, consider the circuit in Fig. 8.10(b). Applying mesh analysis,

$$16i_a - 4i_b + 24 = 0 \quad \Rightarrow \quad 4i_a - i_b = -6 \tag{8.4.1}$$

$$7i_b - 4i_a = 0 \quad \Rightarrow \quad i_a = \frac{7}{4}i_b \tag{8.4.2}$$

Substituting Eq. (8.4.2) into Eq. (8.4.1) gives

$$I_2 = i_b = -1 \text{ A}$$

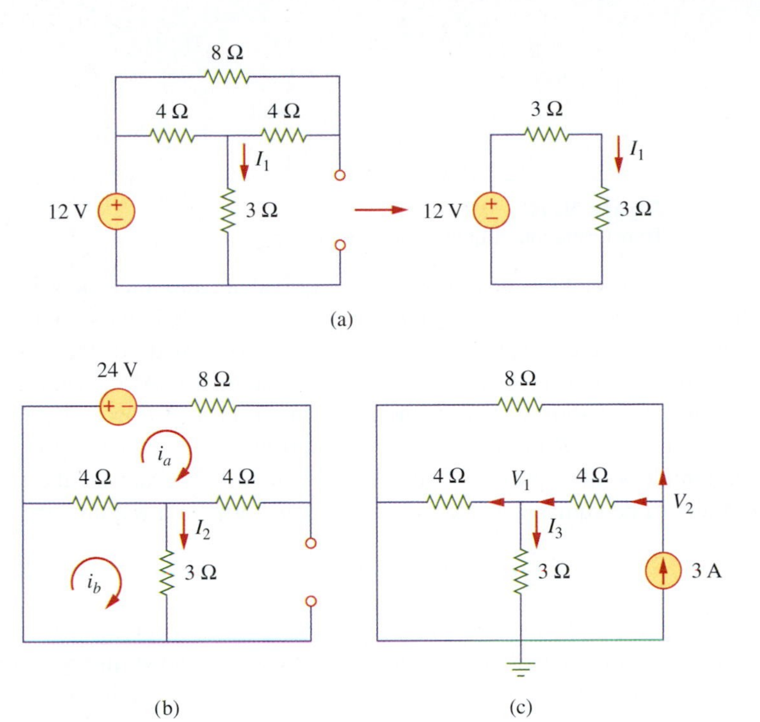

Figure 8.10
For Example 8.4.

To get I_3, consider the circuit in Fig. 8.10(c). Using nodal analysis,

$$3 = \frac{V_2}{8} + \frac{V_2 - V_1}{4} \quad \Rightarrow \quad 24 = 3V_2 - 2V_1 \tag{8.4.3}$$

$$\frac{V_2 - V_1}{4} = \frac{V_1}{4} + \frac{V_1}{3} \quad \Rightarrow \quad V_2 = \frac{10}{3}V_1 \tag{8.4.4}$$

Substituting Eq. (8.4.4) into Eq. (8.4.3) leads to $V_1 = 3$ V and

$$I_3 = \frac{V_1}{3} = 1 \text{ A}$$

Thus,

$$I = I_1 + I_2 + I_3 = 2 - 1 + 1 = 2 \text{ A}$$

Practice Problem 8.4

Find I in the circuit in Fig. 8.11 using the superposition principle.

Answer: 0.75 A

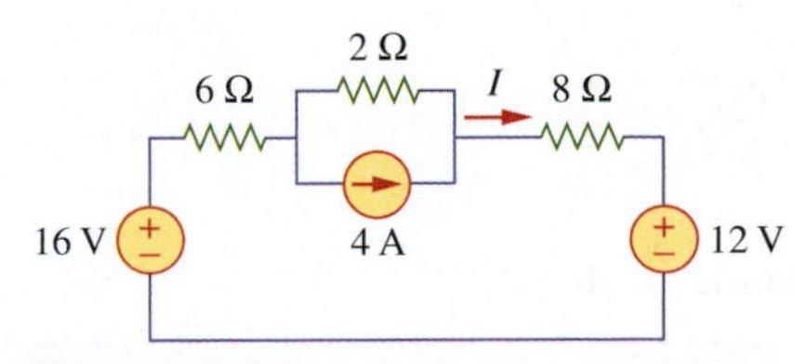

Figure 8.11
For Practice Problem 8.4.

8.4 Source Transformations

In Section 7.6, it was noted that node-voltage (or mesh-current) equations can be obtained by mere inspection of a circuit when the sources are all independent current (or all independent voltage) sources. It is therefore expedient in circuit analysis to be able to substitute a voltage source in series with a resistor for a current source in parallel with a resistor or vice versa, as shown in Fig. 8.12. Either substitution is known as a *source transformation.*

A **source transformation** is the process of replacing a voltage source V_s in series with a resistor R with a current source I_s in parallel with a resistor R or vice versa.

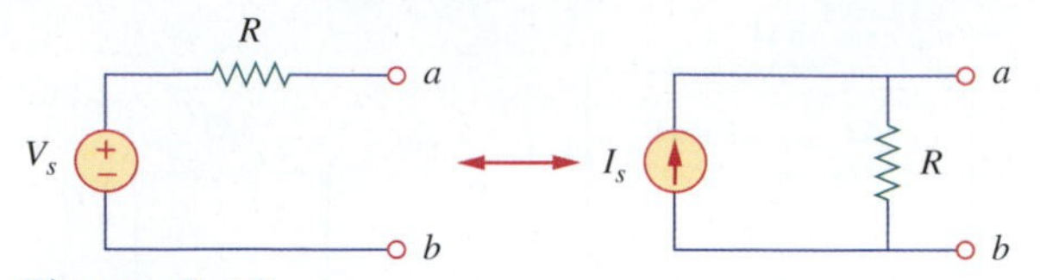

Figure 8.12
Transformation of independent sources.

The two circuits in Fig. 8.12 are equivalent, provided they have the same voltage/current relation at terminals *a-b*. It is easy to show that they are indeed equivalent. If the sources are turned off, the equivalent resistance at terminals *a-b* in both circuits is R. Also, when terminals *a-b* are short-circuited, the short-circuit current flowing from *a* to *b* is $I_{sc} = V_s/R$ in the circuit on the left-hand side and $I_{sc} = I_s$ for the circuit on the right-hand side. Thus $V_s/R = I_s$ in order for the two circuits to be equivalent. Hence, source transformation requires that

$$V_s = I_s R \qquad \text{or} \qquad I_s = \frac{V_s}{R} \tag{8.5}$$

Source transformation also applies to dependent sources, provided that we carefully handle the dependent variable. As shown in Fig. 8.13, a dependent voltage source in series with a resistor can be transformed to a dependent current source in parallel with the resistor or vice versa, where we make sure that Eq. (8.5) is satisfied.

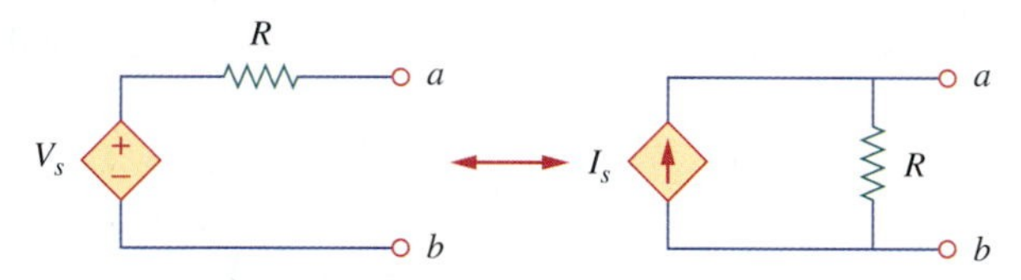

Figure 8.13
Transformation of dependent sources.

Like the wye-delta transformation we studied in Chapter 7, a source transformation does not affect the remaining part of the circuit. When applicable, source transformation is a powerful tool that allows circuit manipulations to ease circuit analysis. However, we should keep the following points in mind when dealing with source transformation.

1. Note from Fig. 8.12 that the arrow of the current source is directed toward the positive terminal of the voltage source.
2. Note from Eq. (8.5) that source transformation is not possible when $R = 0$, which is the case with an ideal voltage source. However, for a practical, nonideal voltage source, $R \neq 0$. Similarly, an ideal current source with $R = \infty$ cannot be replaced by a finite voltage source. More will be said on ideal and nonideal sources in Section 8.12.

Example 8.5

Find the current I in the circuit of Fig. 8.14(a) before and after source transformation.

Solution:

Before source transformation,

$$I = \frac{30}{6 + 4} = 3 \text{ A}$$

We now transform the 30-V voltage source in series with the 6-Ω resistor to become a 5-A (i.e., 30/6) current source in parallel with a 6-Ω resistor as shown in Fig. 8.14(b). Using the current divider rule,

$$I = \frac{6}{6+4}(5) = 3 \text{ A}$$

showing that the results are the same.

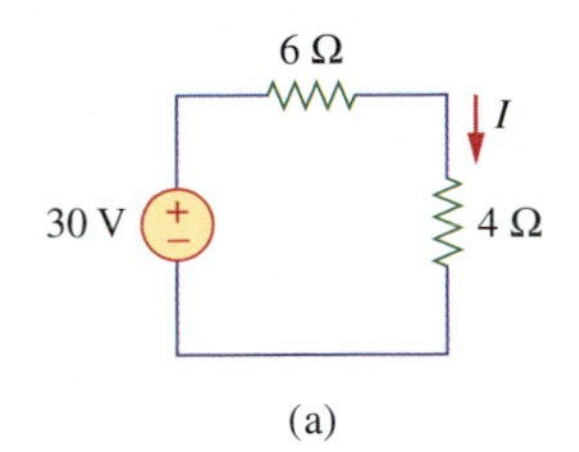

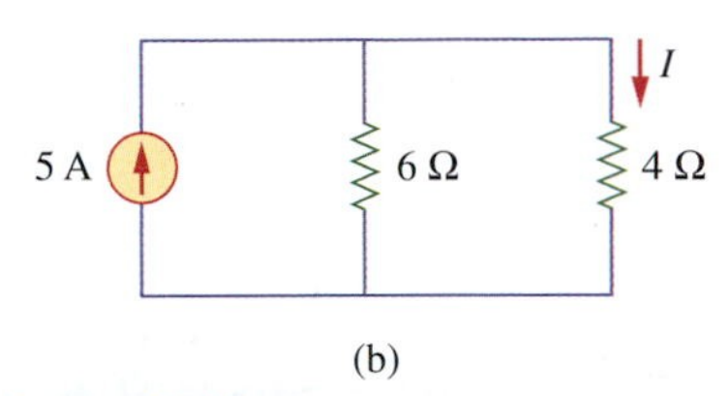

Figure 8.14
For Example 8.5.

Practice Problem 8.5

Consider the circuit in Fig. 8.15. Find the voltage V_o before and after source transformation.

Answer: 40 V; 40 V

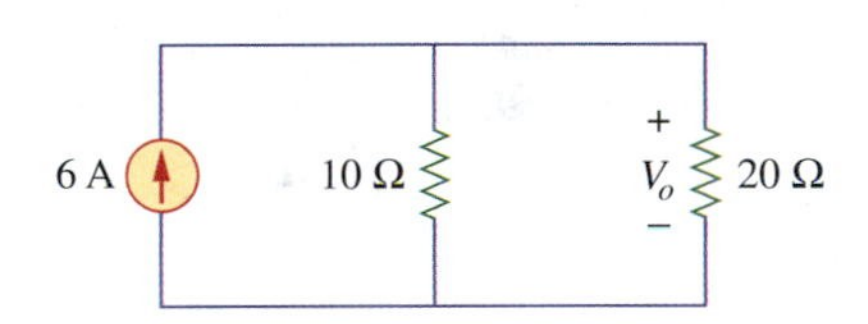

Figure 8.15
For Practice Problem 8.5.

Example 8.6

Use source transformation to find V_o in the circuit in Fig. 8.16.

Solution:
We first transform the current and voltage sources to obtain the circuit in Fig. 8.17(a). Combining the 4-Ω and 2-Ω resistors in series and transforming the 12-V voltage source gives us Fig. 8.17(b). We now combine the 3-Ω and 6-Ω resistors in parallel to get 2 Ω. We also combine the 2-A and 4-A current sources to get a 2-A current source. Thus, by repeatedly going back and forth in our transformations, we obtain the circuit seen in Fig. 8.17(c). (Notice that the 8 Ω is not transformed but kept intact because we are interested in the voltage across it.)

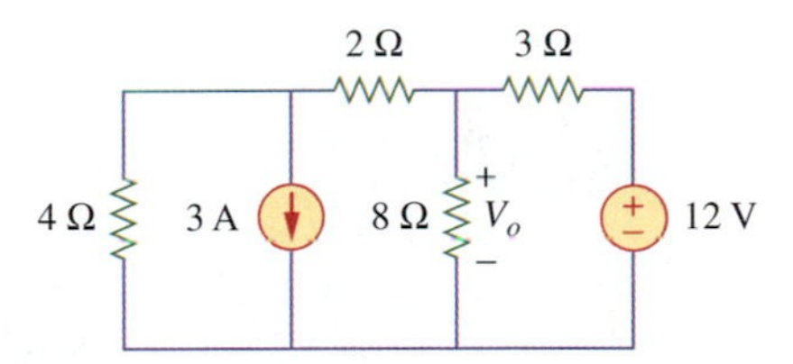

Figure 8.16
For Example 8.6.

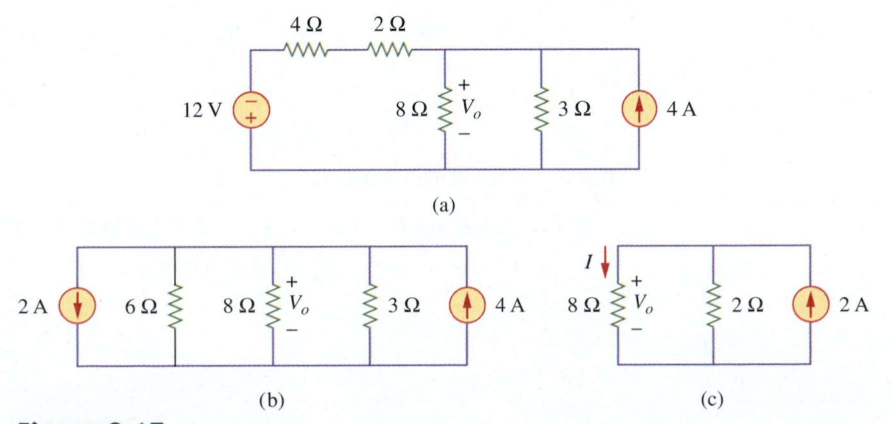

Figure 8.17
For Example 8.6.

We use current division in Fig. 8.17(c) to get

$$I = \frac{2}{2+8}(2) = 0.4 \text{ A}$$

$$V_o = 8I = 8(0.4) = 3.2 \text{ V}$$

Alternatively, because the 8-Ω and 2-Ω resistors in Fig. 8.17(c) are in parallel, they have the same voltage V_o across them. Hence

$$V_o = (8 \parallel 2)(2A) = \frac{8 \times 2}{10}(2) = 3.2 \text{ V}$$

Practice Problem 8.6

Find I_o in the circuit of Fig. 8.18 using source transformation.

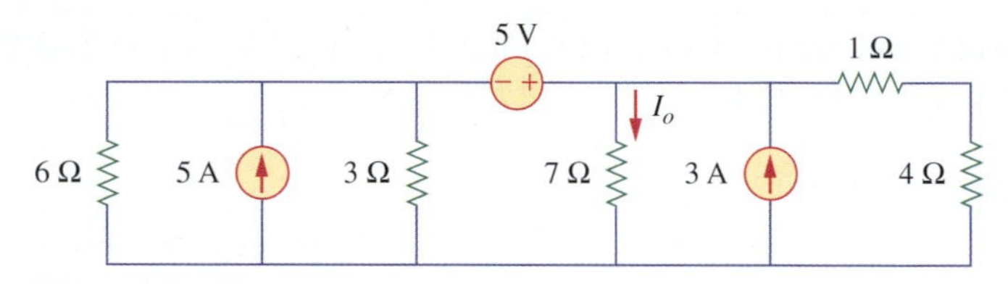

Figure 8.18
For Practice Problem 8.6.

Answer: 1.78 A

Example 8.7

Use source transformation to find the voltage V_x in the circuit of Fig. 8.19.

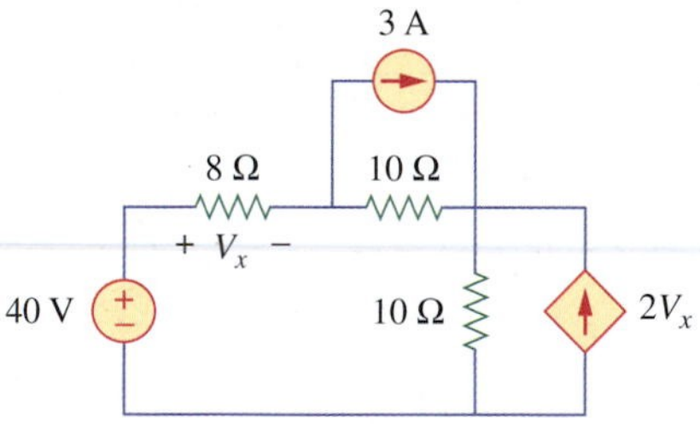

Figure 8.19
For Example 8.7.

Solution:
We transform the two current sources (both dependent and independent) in parallel with their resistors into their equivalent voltage sources as shown in Fig. 8.20. Applying KVL to the mesh in Fig. 8.20, we obtain

$$I(8 + 10 + 10) - 40 - 30 + 20V_x = 0$$

or

$$28I + 20V_x = 70$$

But $V_x = 8I$, which leads to

$$28I + 160I = 70 \qquad \text{or} \qquad I = 70/188 = 0.3723 \text{ A}$$

$$V_x = 8I = 2.978 \text{ V}$$

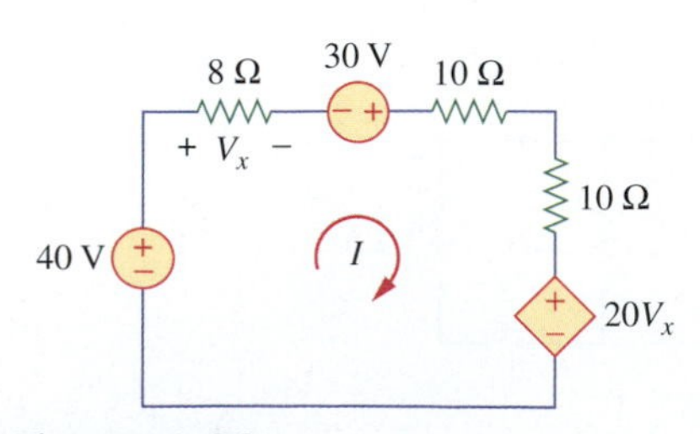

Figure 8.20
Analysis of the circuit in Fig. 8.19.

Practice Problem 8.7

Use source transformation to find I in Fig. 8.21.

Answer: 2 A

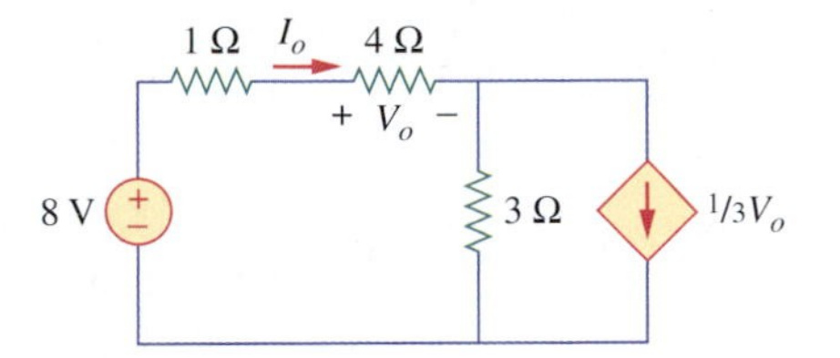

Figure 8.21
For Practice Problem 8.7.

8.5 Thevenin's Theorem

It often occurs in practice that a particular element in a circuit (sometimes called the *load*), is variable while other elements are fixed. As a typical example, a household outlet terminal may be connected to different appliances, so the load will vary depending on what is in use at any given time. Each time the variable element is changed, the entire circuit has to be analyzed all over again. To avoid this, Thevenin's theorem provides a technique by which the fixed part of the circuit is replaced by an equivalent circuit. Thevenin's theorem states that it is possible to simplify any linear circuit, no matter how complex, to an equivalent circuit with just a single voltage source and a series resistance.

> **Thevenin's theorem** states that a linear two-terminal circuit can be replaced by an equivalent circuit consisting of a voltage source V_{Th} in series with a resistor R_{Th}, where V_{Th} is the open-circuit voltage at the terminals and R_{Th} is the input or equivalent resistance at the terminals when the independent sources are turned off or reduced to zero.

According to Thevenin's theorem, the linear circuit in Fig. 8.22(a) can be replaced by that in Fig. 8.22(b). (The load in Fig. 8.22 may be a single resistor or another circuit.) The circuit to the left of the terminals a-b in Fig. 8.22(b) is known as the *Thevenin equivalent circuit.*[4]

Our major concern is how to find the Thevenin equivalent voltage V_{Th} and R_{Th}. To do so, suppose the two circuits in Fig. 8.22 are equivalent. Two circuits are said to be *equivalent* if they have the same voltage/current relation at their terminals. Let us find out what will make the two circuits in Fig. 8.22 equivalent. If the terminals a-b are open-circuited (i.e., by removing the load), no current flows; hence the open-circuit voltage across the terminals a-b in Fig. 8.22(a) must be equal to the voltage source V_{Th} in Fig. 8.22(b) because the two circuits are equivalent. Thus, V_{Th} is the open-circuit voltage at the terminals as shown in Fig. 8.23(a); that is,

$$V_{Th} = V_{oc} \tag{8.6}$$

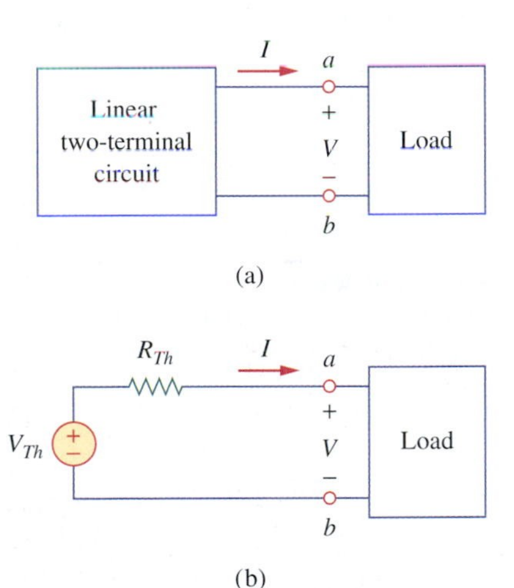

Figure 8.22
Replacing a linear two-terminal circuit with its Thevenin equivalent: (a) original circuit; (b) the Thevenin equivalent circuit.

Again, with the load disconnected and terminals a-b open-circuited, we turn off all independent sources. The input resistance (or equivalent resistance) of the dead circuit at the terminals a-b in Fig. 8.22(a) must be equal to R_{Th} in Fig. 8.22(b) because the two circuits are equivalent. Thus, R_{Th} is the input resistance at the terminals when the independent sources are turned off, as shown in Fig. 8.23(b); that is,

$$R_{Th} = R_{\text{in}} \tag{8.7}$$

Therefore, R_{Th} is the input resistance of the network looking between terminals a and b as shown in Fig. 8.23(b).

[4] The theorem was first discovered by German scientist Hermann von Helmholtz in 1853, but was then rediscovered in 1883 by Thevenin. For this reason, it is also known as Helmholtz's theorem.

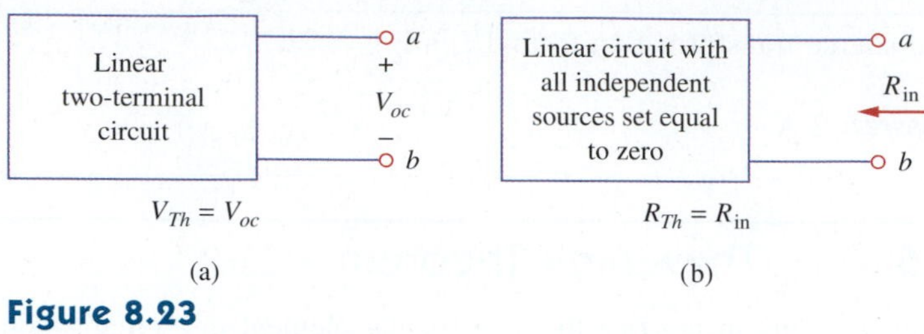

Figure 8.23
(a) Finding V_{Th}; (b) finding R_{Th}.

To apply this idea in finding the Thevenin equivalent R_{Th}, we need to consider two cases.

■ **CASE 1** If the network has no dependent sources, we turn off all independent sources. R_{Th} is the input resistance of the network looking between terminals a and b, as shown in Fig. 8.23(b).

■ **CASE 2** If the network has dependent sources, we turn off all independent sources. Dependent sources are not to be turned off because they are controlled by circuit variables. We apply a voltage source v_o at terminals a and b and determine the resulting current i_o. Then $R_{Th} = v_o/i_o$, as shown in Fig. 8.24(a). Alternatively, we may insert a current source i_o at terminals a-b as shown in Fig. 8.24(b) and find the terminal voltage v_o. Again, $R_{Th} = v_o/i_o$. Either of the two approaches will give the same result. In either approach we may assume any value of v_o and i_o. For example, we may use $v_o = 1$ V or $i_o = 1$ A, or even use unspecified values of v_o or i_o.

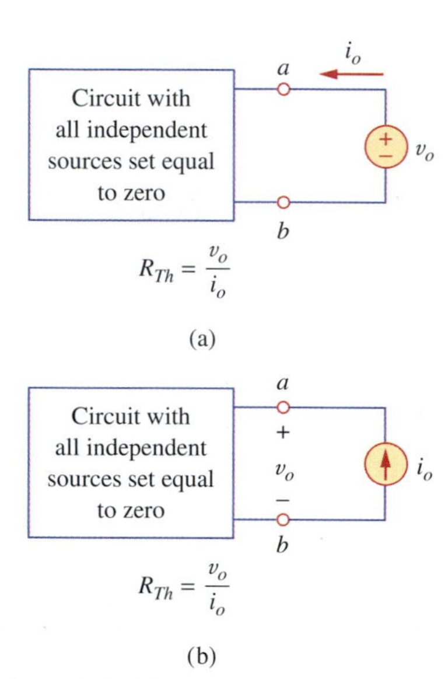

Figure 8.24
Finding R_{Th} when circuit has dependent sources.

Thevenin's theorem is very important in circuit analysis. First, it helps simplify a circuit. Second, it helps simplify circuit design. A large circuit may be replaced by a single independent voltage source and a single resistor. This replacement technique is a powerful tool in circuit design.

As mentioned earlier, a linear circuit with a variable load can be replaced by the Thevenin equivalent, exclusive of the load. The equivalent network behaves the same way externally as the original circuit. For example, consider a linear circuit terminated by a load R_L, as shown in Fig. 8.25(a). The current I_L through the load and the voltage V_L across the load are easily determined once the Thevenin equivalent of the circuit at the load's terminals is obtained, as shown in Fig. 8.25(b). From Fig. 8.25(b), we obtain

$$I_L = \frac{V_{Th}}{R_{Th} + R_L} \tag{8.8a}$$

$$V_L = R_L I_L = \frac{R_L}{R_{Th} + R_L} V_{Th} \tag{8.8b}$$

Note from Fig. 8.25(b) that the Thevenin equivalent is a simple voltage divider, and V_L can be obtained by mere inspection.

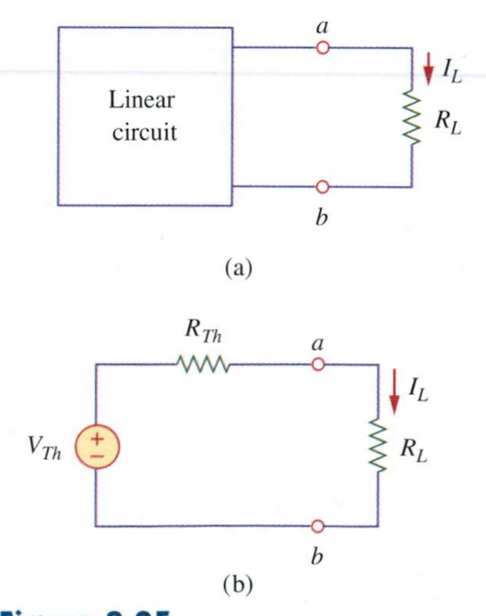

Figure 8.25
A circuit with a load: (a) original circuit; (b) Thevenin equivalent.

Applying Thevenin's theorem involves the following four steps:

1. Remove temporarily that portion of the circuit not to be replaced by a Thevenin equivalent circuit. Mark the terminals of the remaining portion.
2. Determine the Thevenin resistance R_{Th}, as the resistance appearing between the terminals with all sources set to zero (voltage sources replaced by short circuits and current sources replaced by open circuits).
3. Determine the Thevenin voltage V_{Th}, as the open-circuit (no-load) voltage between the terminals.
4. Form the Thevenin equivalent circuit by connecting V_{Th} and R_{Th} in series. Observe proper polarity for V_{Th}. Put back the portion of the circuit that was removed in step 1.

Example 8.8

Determine the Thevenin equivalent for the circuit to the left of terminals *a-b* in Fig. 8.26.

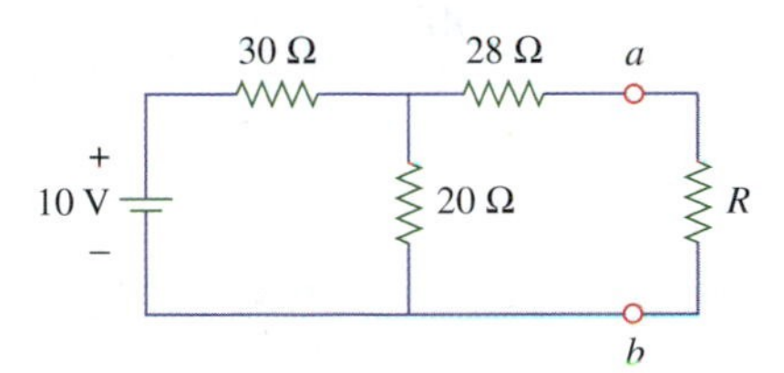

Figure 8.26
For Example 8.8.

Solution:
We need to find R_{Th} and V_{Th} at terminals *a-b* with resistor *R* removed. To find R_{Th}, we set the voltage source to zero by replacing it with a short circuit as shown in Fig. 8.27(a).

$$R_{Th} = 28 + 30 \parallel 20 = 28 + \frac{30 \times 20}{30 + 20} = 28 + 12 = 40\ \Omega$$

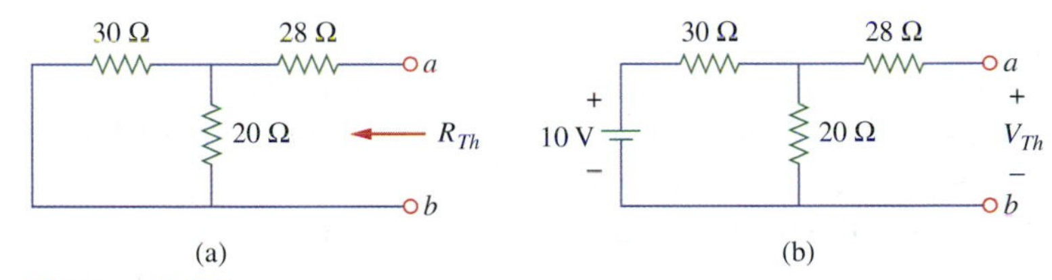

Figure 8.27
For Example 8.8: (a) finding R_{Th}; (b) finding V_{Th}.

To find V_{Th}, we consider the open-circuit voltage between terminals *a-b*, as shown in the circuit of Fig. 8.27(b). Because no current flows through the 28-Ω resistor, we apply voltage division rule to get V_{Th}.

$$V_{Th} = \frac{20}{20 + 30}(10) = 4\ \text{V}$$

The Thevenin equivalent with the resistor *R* in place is shown in Fig. 8.28.

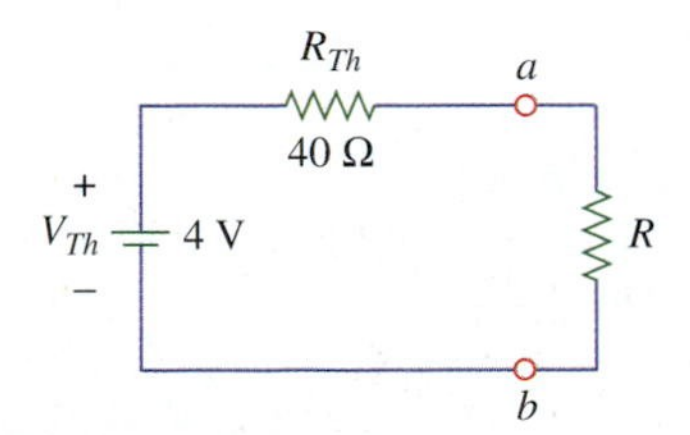

Figure 8.28
Thevenin equivalent circuit for Example 8.8.

Practice Problem 8.8

Obtain the Thevenin equivalent for the circuit to the left of terminals *a-b* in Fig. 8.29.

Answer: $R_{Th} = 3\ \Omega$; $V_{Th} = 6$ V

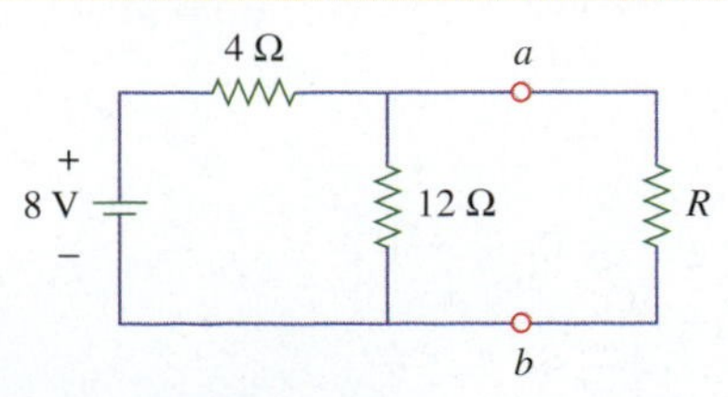

Figure 8.29
For Practice Problem 8.8.

Example 8.9

Find the Thevenin equivalent circuit of the circuit shown in Fig. 8.30, to the left of the terminals *a-b*. Then find the current through $R_L = 6$, 16, and 36 Ω.

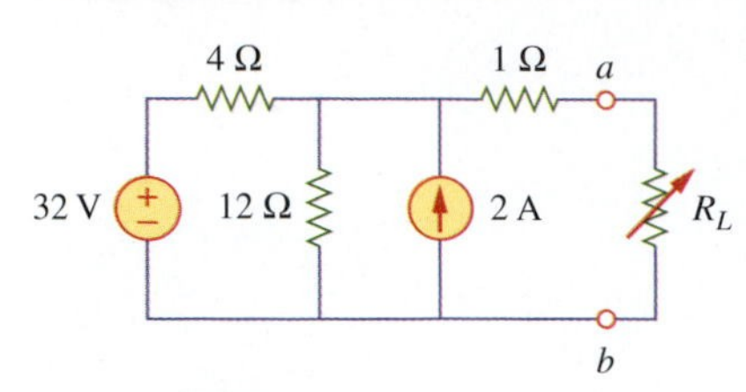

Figure 8.30
For Example 8.9.

Solution:
We find R_{Th} by turning off the 32-V voltage source by replacing it with a short circuit and the 2-A current source by replacing it with an open circuit. The circuit becomes what we have in Fig. 8.31(a). Thus,

$$R_{Th} = 4 \parallel 12 + 1 = \frac{4 \times 12}{16} + 1 = 4\ \Omega$$

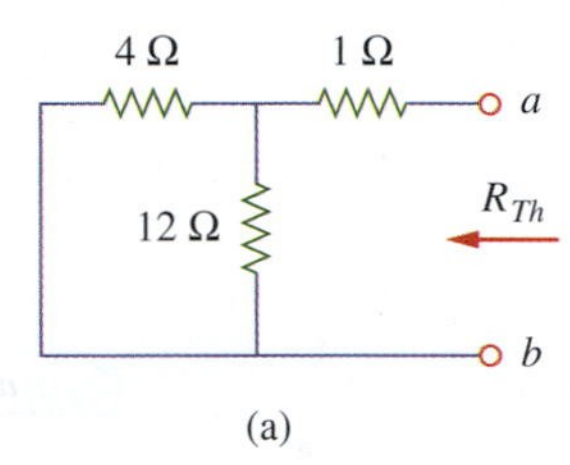

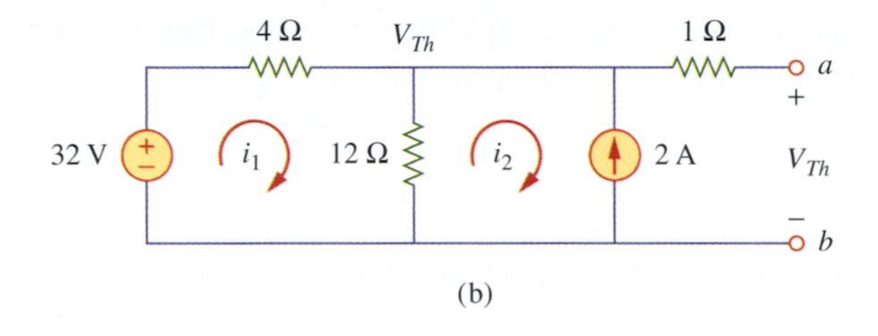

Figure 8.31
For Example 8.9: (a) finding R_{Th}; (b) finding V_{Th}.

To find V_{Th}, consider the circuit in Fig. 8.31(b). Applying mesh analysis to the two loops, we obtain

$$-32 + 4i_1 + 12(i_1 - i_2) = 0, \quad i_2 = -2\text{ A}$$

Solving for i_1, we get $i_1 = 0.5$ A. Thus,

$$V_{Th} = 12(i_1 - i_2) = 12(0.5 + 2.0) = 30\text{ V}$$

Alternatively, it is even easier to use nodal analysis. We ignore the 1-Ω resistor because no current flows through it. At the top node, KCL gives

$$\frac{32 - V_{Th}}{4} + 2 = \frac{V_{Th}}{12}$$

or

$$96 - 3V_{Th} + 24 = V_{Th} \quad \Rightarrow \quad V_{Th} = 30\text{ V}$$

as obtained before. We could also use source transformation to find V_{Th}. The Thevenin equivalent circuit is shown in Fig. 8.32. The current through R_L is

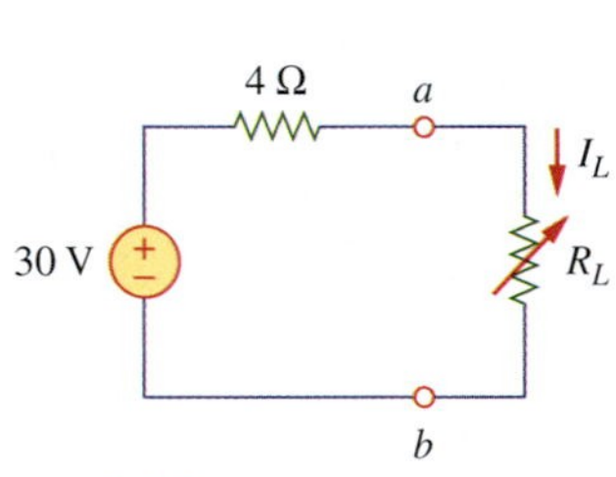

Figure 8.32
The Thevenin equivalent circuit for Example 8.9.

$$I_L = \frac{30}{4 + R_L}$$

When $R_L = 6\ \Omega$,

$$I_L = \frac{30}{4 + 6} = 3\text{ A}$$

When $R_L = 16\ \Omega$,

$$I_L = \frac{30}{4 + 16} = 1.5\text{ A}$$

When $R_L = 36\ \Omega$,

$$I_L = \frac{30}{4 + 36} = 0.75\text{ A}$$

Practice Problem 8.9

Using Thevenin's theorem, find the equivalent circuit to the left of terminal *a-b* in the circuit in Fig. 8.33. Then find I.

Answer: $V_{Th} = 6$ V; $R_{Th} = 3\ \Omega$; $I = 1.5$ A

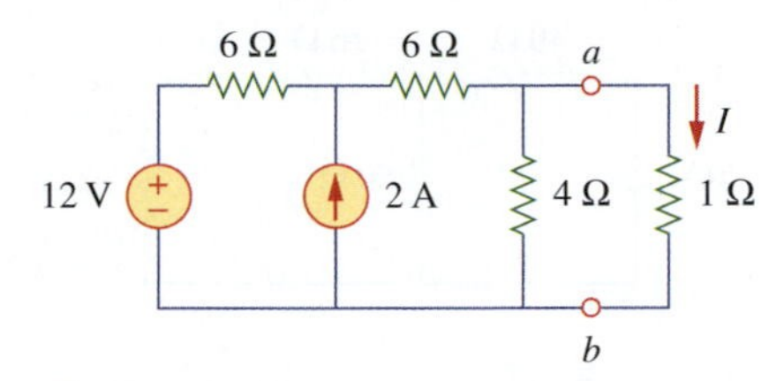

Figure 8.33
For Practice Problem 8.9.

Example 8.10

Find the Thevenin equivalent of the bridge circuit in Fig. 8.34 at terminals *a-b*. Use the equivalent to find the current through the 30-Ω resistor.

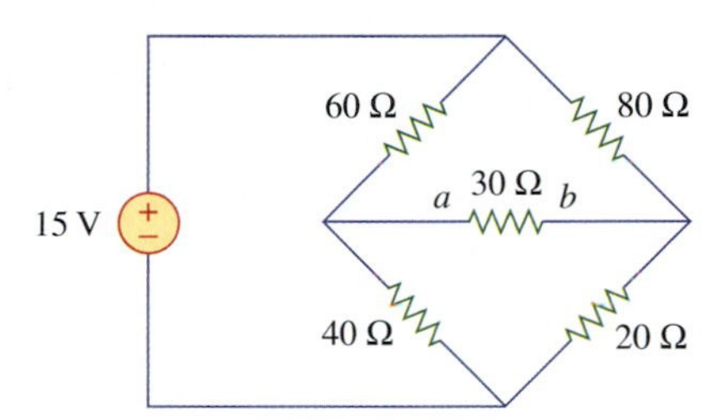

Figure 8.34
For Example 8.10.

Solution:

We remove the 30-Ω resistor because we are trying to find the equivalent circuit at terminals *a-b*. To find R_{Th}, we turn off the voltage source by replacing it with a short circuit, as shown in Fig. 8.35(a).

You will notice that the 40-Ω and 60-Ω resistors are in parallel, while the 20-Ω and 80-Ω resistors are also in parallel. The two combinations are in series. Hence,

$$R_{Th} = 40 \parallel 60 + 20 \parallel 80 = \frac{40 \times 60}{100} + \frac{20 \times 80}{100} = 24 + 16 = 40\ \Omega$$

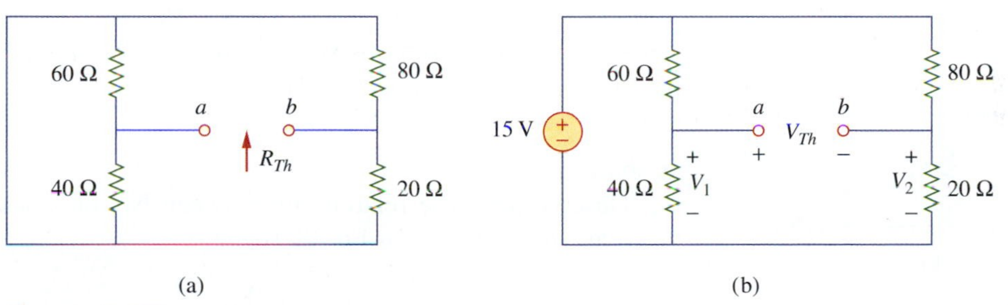

Figure 8.35
For Example 8.10.

To find V_{Th}, we use the circuit in Fig. 8.35(b). We can use voltage division principle to find V_1 and V_2.

$$V_1 = \frac{40}{40 + 60}(15) = 6 \text{ V}$$

$$V_2 = \frac{20}{20 + 80}(15) = 3 \text{ V}$$

Applying KVL around loop *aboa* gives

$$-V_1 + V_{Th} + V_2 = 0 \quad \Rightarrow \quad V_{Th} = V_1 - V_2 = 6 - 3 = 3 \text{ V}$$

Once we obtain R_{Th} and V_{Th}, we have the equivalent circuit as shown in Fig. 8.36. The current through the 30-Ω resistor is

$$\frac{V_{Th}}{R_{Th} + 30} = \frac{3}{40 + 30} = 42.86 \text{ mA}$$

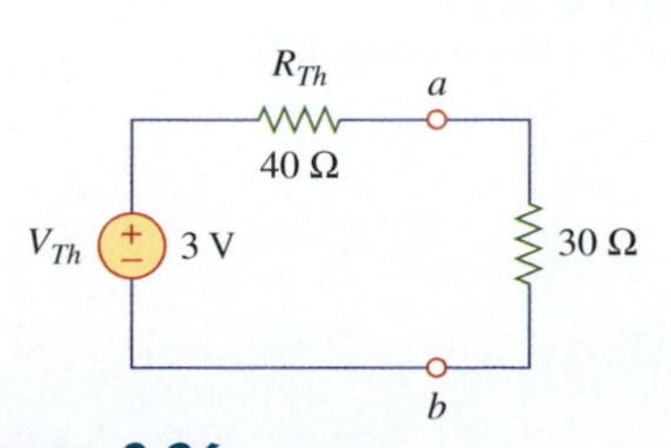

Figure 8.36
Equivalent circuit of that in Figure 8.34.

Practice Problem 8.10

Find the Thevenin equivalent circuit of the circuit in Fig. 8.37 to the left of terminal *a-b*. Use the equivalent circuit to find current I_x.

Answer: $V_{Th} = 12$ V; $R_{Th} = 40\ \Omega$; $I_x = 0.2$ A

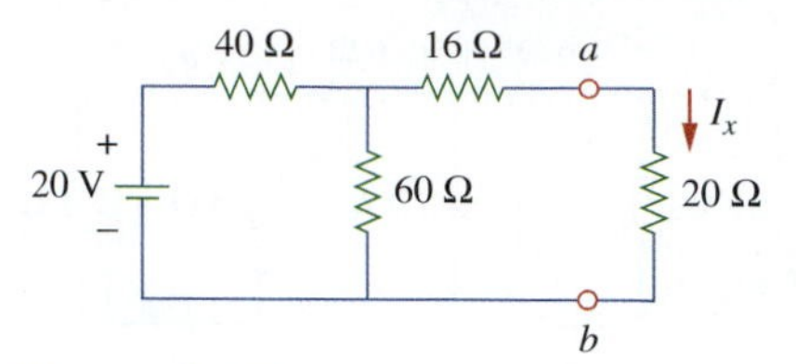

Figure 8.37
For Practice Problem 8.10 and Example 8.19.

8.6 Norton's Theorem

In 1926, about 43 years after Thevenin published his theorem, E. L. Norton, an American engineer at Bell Telephone Laboratories, proposed a theorem similar to Thevenin's.

> **Norton's theorem** states that a linear two-terminal circuit can be replaced by an equivalent circuit consisting of a current source I_N in parallel with a resistor R_N, where I_N is the short-circuit current at the terminals and R_N is the input or equivalent resistance at the terminals when the independent sources are turned off.

Thus, the circuit in Fig. 8.38(a) can be replaced by the one in Fig. 8.38(b).

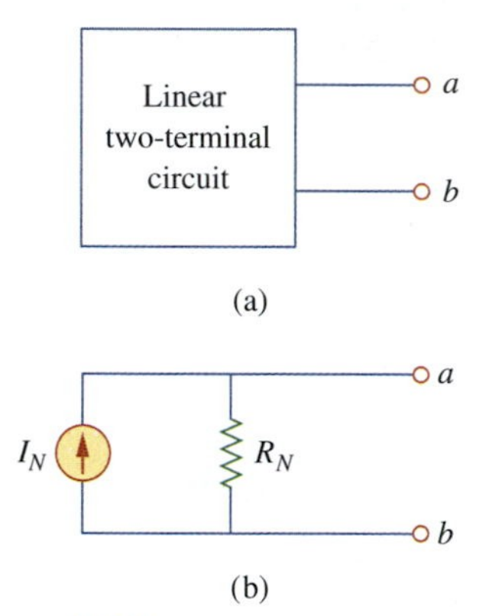

Figure 8.38
(a) Original circuit; (b) Norton equivalent circuit.

Our main concern is how to get R_N and I_N. We find R_N in the same way we find R_{Th}. In fact, from what we know about source transformation, the Thevenin and Norton resistances are equal; that is,

$$R_N = R_{Th} \tag{8.9}$$

To find the Norton current I_N, we determine the short-circuit current flowing from terminal *a* to *b* in both circuits in Fig. 8.38. It is evident that the short-circuit current in Fig. 8.38(b) is I_N. This must be the same short-circuit current from terminal *a* to *b* in Fig. 8.38(a) because the two circuits are equivalent. Thus,

$$I_N = I_{sc} \tag{8.10}$$

as shown in Fig. 8.39.

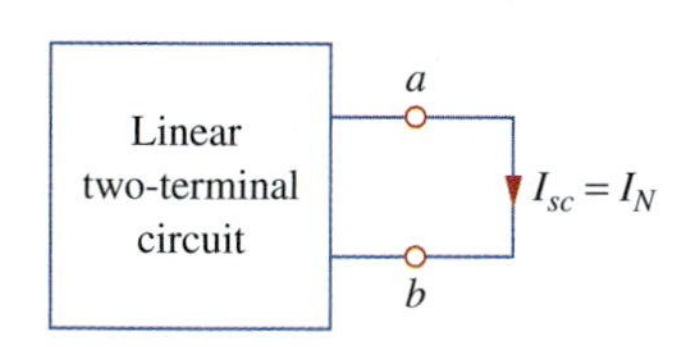

Figure 8.39
Finding Norton current I_N.

Observe the close relationship between Norton's and Thevenin's theorems: $R_N = R_{Th}$ as in Eq. (8.9) and

$$I_N = \frac{V_{Th}}{R_{Th}} \tag{8.11}$$

This is essentially source transformation. For this reason, source transformation is often called Thevenin-Norton transformation.

Applying Norton's theorem involves the following four steps:

1. Remove temporarily that portion of the circuit not to be replaced by a Norton equivalent circuit. Mark the terminals *a-b* of the remaining portion.
2. Determine the Norton resistance R_N as the resistance appearing between the terminals *a-b* with all sources set to zero (voltage sources replaced by short circuits and current sources replaced by open circuits).
3. Determine the Norton current I_N as the short-circuit current through the terminals *a-b*.
4. Form the Norton equivalent circuit by connecting I_N and R_N in parallel. Observe proper polarity for the Norton current source. Put back the portion of the circuit that was removed in step 1.

Because V_{Th}, I_N, and R_{Th} are related according to Eq. (8.11), in order to determine the Thevenin or Norton equivalent we need to find:

- The open-circuit voltage V_{oc} at terminals *a-b*.
- The short-circuit current I_{sc} at terminals *a-b*.
- The equivalent or input resistance R_{in} at terminals *a-b* when all independent sources are turned off.

We can calculate any two of the three using the method that takes the least effort and use the results to get the third using Ohm's law. This is illustrated in Example 8.11. Also, because

$$V_{Th} = V_{oc} \tag{8.12a}$$

$$I_N = I_{sc} \tag{8.12b}$$

$$R_{Th} = \frac{V_{oc}}{I_{sc}} = R_N \tag{8.12c}$$

the open-circuit and short-circuit tests are sufficient to find any Thevenin or Norton equivalent.

Example 8.11

Find the Norton equivalent of the circuit in Fig. 8.40.

Solution:

We find R_N in the same way we find R_{Th} in the Thevenin equivalent circuit. Set the independent sources equal to zero. This leads to the circuit in Fig. 8.41(a), from which we find R_N. Thus,

$$R_N = 5 \parallel (8 + 4 + 8) = 5 \parallel 20 = \frac{5 \times 20}{25} = 4\ \Omega$$

Figure 8.40
For Example 8.11.

To find I_N, we short-circuit terminals *a-b*, as shown in Fig. 8.41(b). We ignore the 5-Ω resistor because it has been short-circuited. Applying mesh analysis, we obtain

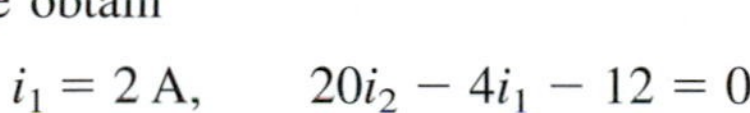

$$i_1 = 2\text{ A}, \qquad 20i_2 - 4i_1 - 12 = 0$$

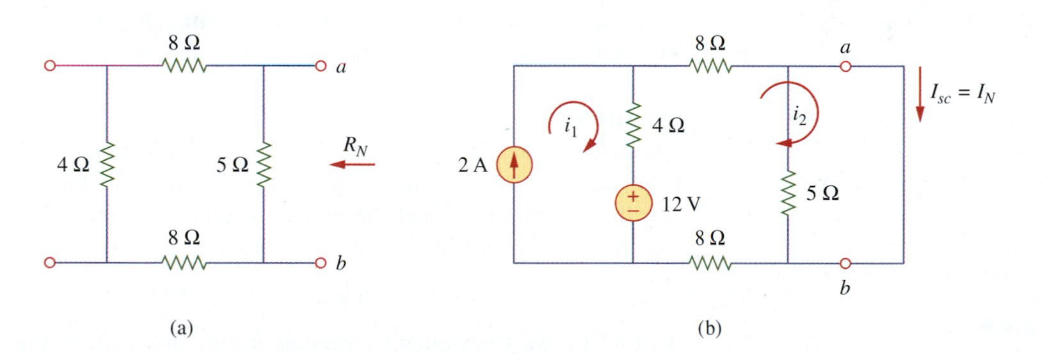

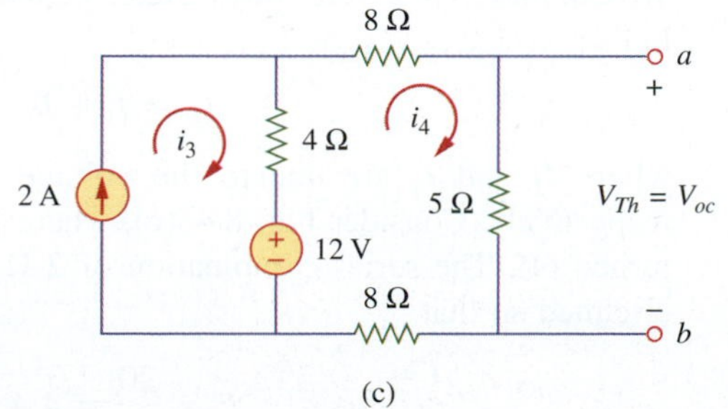

Figure 8.41
For Example 8.11: (a) R_N; (b) $I_N = I_{sc}$; (c) $V_{Th} = V_{oc}$.

From these equations, we obtain

$$i_2 = 1\text{ A} = I_{sc} = I_N$$

Alternatively, we can determine I_N from V_{Th}/R_T. We obtain V_{Th} as the open-circuit voltage across terminals *a-b* in Fig. 8.41(c). Using mesh analysis, we obtain

$$i_3 = 2\text{ A}$$

$$25i_4 - 4i_3 - 12 = 0 \quad \Rightarrow \quad i_4 = 0.8\text{ A}$$

and

$$V_{oc} = V_{Th} = 5i_4 = 4\text{ V}$$

Hence,

$$I_N = \frac{V_{Th}}{R_{Th}} = \frac{4}{4} = 1\text{ A}$$

as obtained previously. This also serves to confirm Eq. (8.12c) that

$$R_{Th} = \frac{V_{oc}}{I_{sc}} = \frac{4}{1} = 4\ \Omega$$

Thus, the Norton equivalent circuit is as shown in Fig. 8.42.

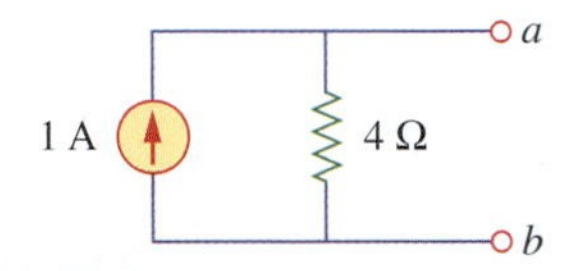

Figure 8.42
For Example 8.11: Norton equivalent of the circuit in Fig. 8.40.

Practice Problem 8.11

Find the Norton equivalent circuit for the circuit in Fig. 8.43.

Answer: $R_N = 3\ \Omega$; $I_N = 4.5$ A

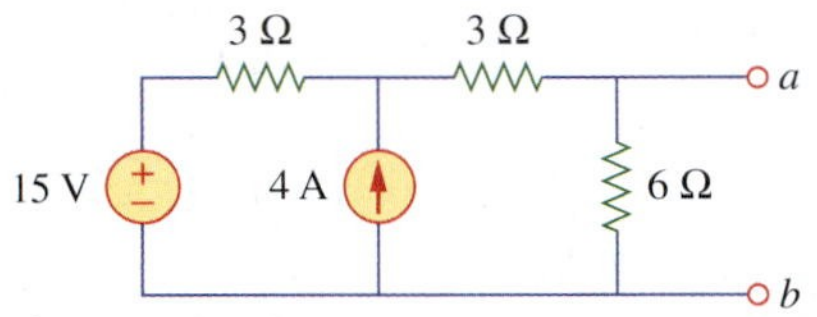

Figure 8.43
For Practice Problem 8.11.

Example 8.12

Find the Norton equivalent to the left of terminals *a-b* of the circuit in Fig. 8.44 and use it to find the current through the load $R_L = 10\ \Omega$.

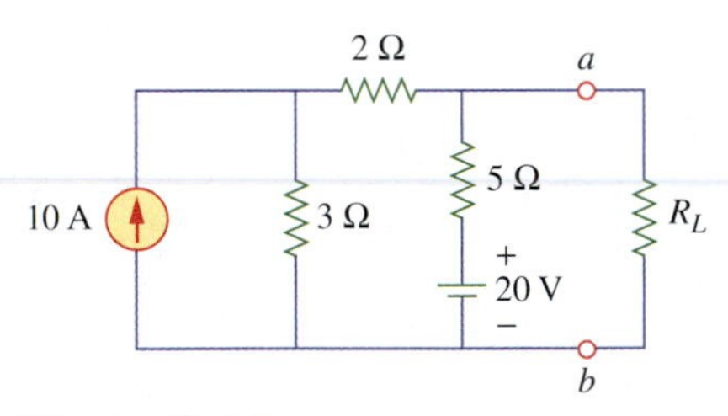

Figure 8.44
For Example 8.12.

Solution:

Because we are interested in the Norton equivalent to the left of terminals *a-b*, we can remove the load R_L for now. To find R_N, we turn off the current source and the voltage source and obtain the circuit in Fig. 8.45(a). From Fig. 8.45(a), we obtain

$$R_N = 5 \parallel (2 + 3) = 2.5\ \Omega$$

To find I_N, we short-circuit terminals *a-b* as indicated in Fig. 8.45(b). We can find I_N in several ways. Here, we use the superposition principle. Let

$$I_N = I_1 + I_2$$

where I_1 and I_2 are due to the voltage source and current source respectively. Consider Fig. 8.45(c), where the current source has been turned off. The series combination of 2-Ω and 3-Ω resistors is short-circuited so that

$$I_1 = \frac{20}{5} = 4\text{ A}$$

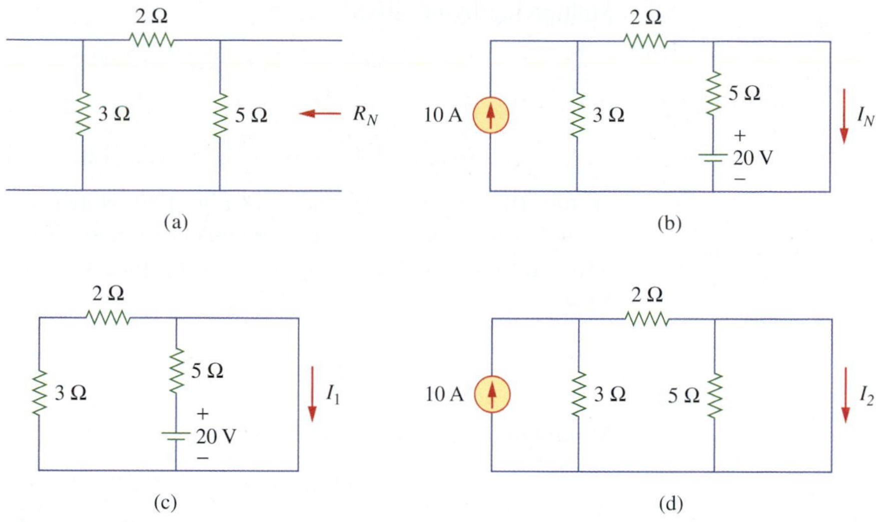

Figure 8.45
For Example 8.12.

To find I_2, consider the circuit in Fig. 8.45(d). The 5-Ω resistor can be ignored because it is short-circuited. Using current divider rule,

$$I_2 = \frac{3}{3+2}(10) = 6 \text{ A}$$

Hence,

$$I_N = I_1 + I_2 = 4 + 6 = 10 \text{ A}$$

The Norton equation circuit is shown in Fig. 8.46. To find the current through the load $R_L = 10\ \Omega$, we apply current divider rule,

$$I_L = \frac{2.5}{2.5+10}(10) = 2 \text{ A}$$

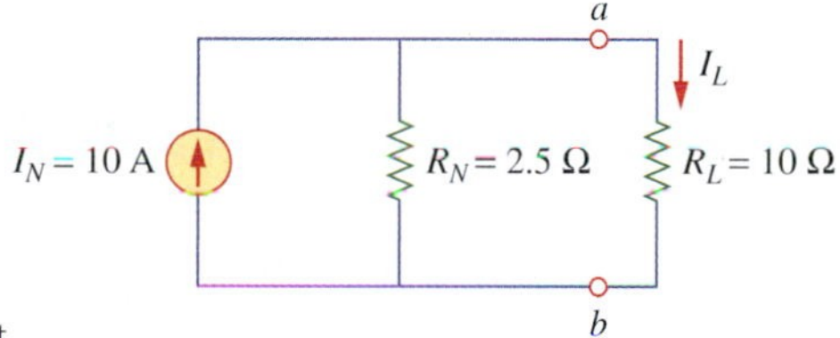

Figure 8.46
For Example 8.12.

Practice Problem 8.12

Find the Norton equivalent circuit of the circuit in Fig. 8.47 to the left of terminals *a-b*. Use the equivalent to find the current through the 2-Ω resistor.

Answer: $R_N = 0.8\ \Omega$; $I_N = 5$ A; 1.429 A

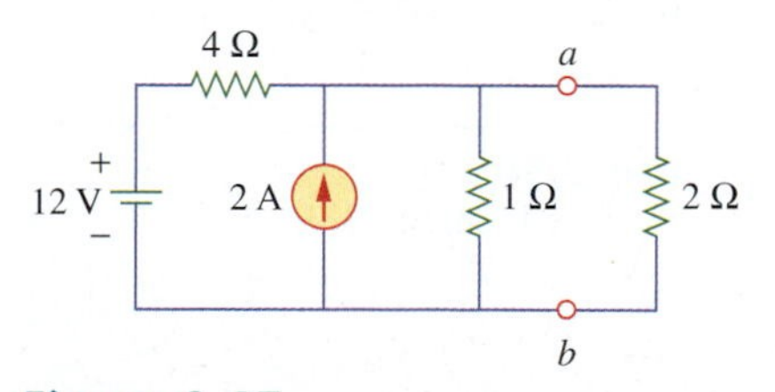

Figure 8.47
For Practice Problem 8.12.

Example 8.13

Obtain the Thevenin and Norton equivalent circuits of the circuit in Fig. 8.48 with respect to terminals *a* and *b*.

Solution:
Because $V_{Th} = V_{ab} = V_x$, we apply KCL at node *a* and obtain

$$\frac{30 - V_{Th}}{12} = \frac{V_{Th}}{60} + 2V_{Th}$$

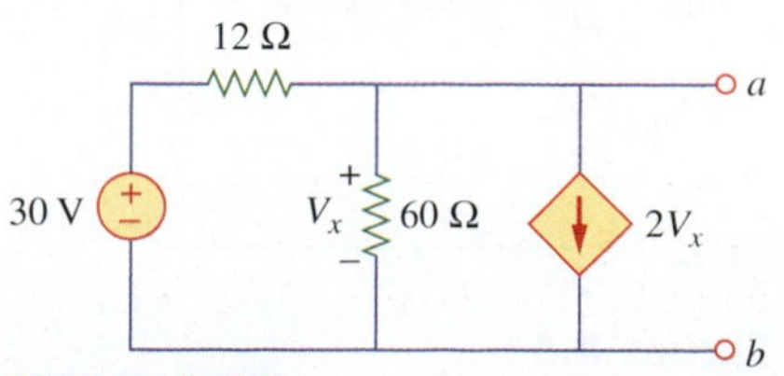

Figure 8.48
For Example 8.13.

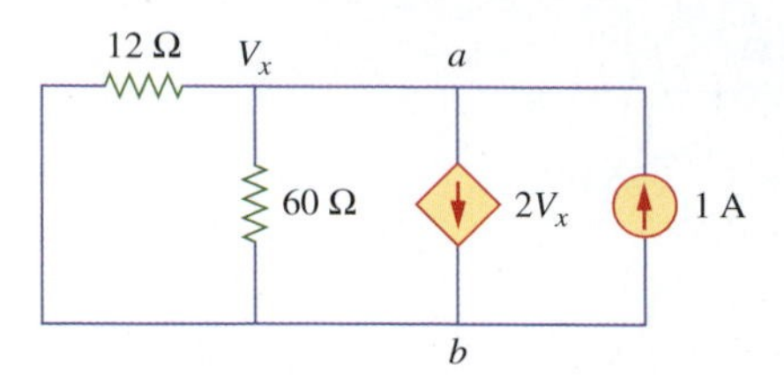

Figure 8.49
Finding R_{Th} in the circuit of Fig. 8.48.

Multiplying by 60 gives

$$150 - 5V_{Th} = V_{Th} + 120V_{Th}$$

or

$$126V_{Th} = 150 \quad \Rightarrow \quad V_{Th} = 150/126 = 1.19 \text{ V}$$

To find R_{Th}, consider the circuit in Fig. 8.49. Notice we have removed the independent source but the dependent source is left intact. Also, we have inserted a 1-A current source at terminals a and b. At node a, KCL gives

$$1 = 2V_x + \frac{V_x}{60} + \frac{V_x}{12}$$

Multiplying through by 60 yields

$$60 = 120V_x + V_x + 5V_x$$

or

$$126V_x = 60 \quad \Rightarrow \quad V_x = 60/126 = 0.4762 \text{ V}$$

$$R_{Th} = \frac{V_x}{1} = 0.4762\ \Omega, \qquad I_N = \frac{V_{Th}}{R_{Th}} = 1.19/0.4762 = 2.5 \text{ A}$$

Thus,

$$V_{Th} = 1.19 \text{ V}, \qquad R_{Th} = R_N = 0.4762\ \Omega, \qquad I_N = 2.5 \text{ A}$$

Practice Problem 8.13

Determine the Thevenin and Norton equivalent circuits at terminals a-b for the circuit in Fig. 8.50.

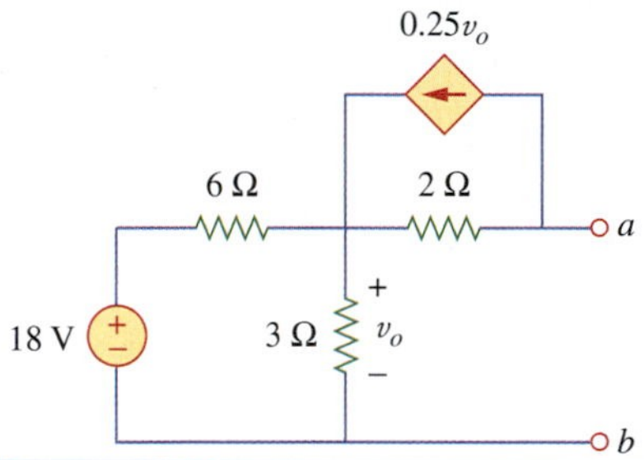

Figure 8.50
For Practice Problem 8.13.

Answer: $V_{Th} = 3$ V; $R_{Th} = R_N = 3\ \Omega$; $I_N = 1$ A

8.7 Maximum Power Transfer Theorem

The concept of matching a load to a source for maximum power transfer is very important in power systems, microwave ovens, stereo sound systems, electrical generating plants, solar cells, and hybrid electric cars. In many practical situations, a circuit is designed to provide power to a load (e.g., motor, heater, or audio amplifier). Such is the case with the electric utility companies, whose major concern is generating electric power and transmitting and distributing it to various users. Minimizing power loss in the process of transmission and distribution is critical for efficiency and economic reasons.

The Thevenin equivalent is useful in finding the maximum power a linear circuit can deliver to a load. We assume that we can adjust the load resistance R_L. If the entire circuit is replaced by its Thevenin equivalent except for the load, as shown in Fig. 8.51, the power delivered to the load is

$$P = i^2R = \left(\frac{V_{Th}}{R_{Th} + R_L}\right)^2 R_L \tag{8.13}$$

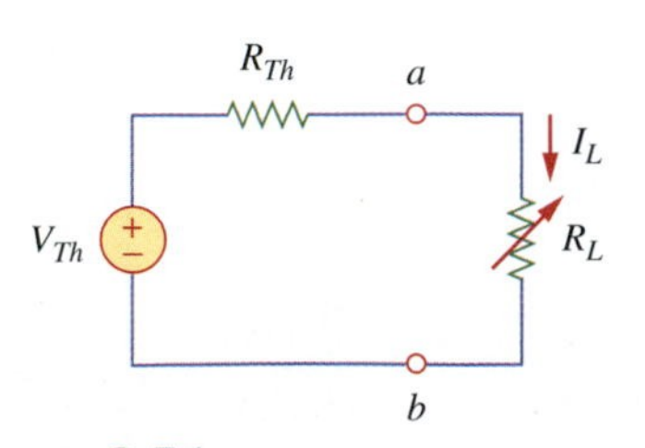

Figure 8.51
The circuit used for maximum power transfer.

For a given circuit, V_{Th} and R_{Th} are fixed. By varying the load resistance R_L, the power delivered to the load varies as sketched in

Fig. 8.52. We notice from Fig. 8.52 that the power is small for small or large values of R_L but maximum for some value of R_L between 0 and ∞. We can show that this maximum power occurs when R_L is equal to R_{Th},[5] as seen by the load; that is,

$$R_L = R_{Th} \tag{8.14}$$

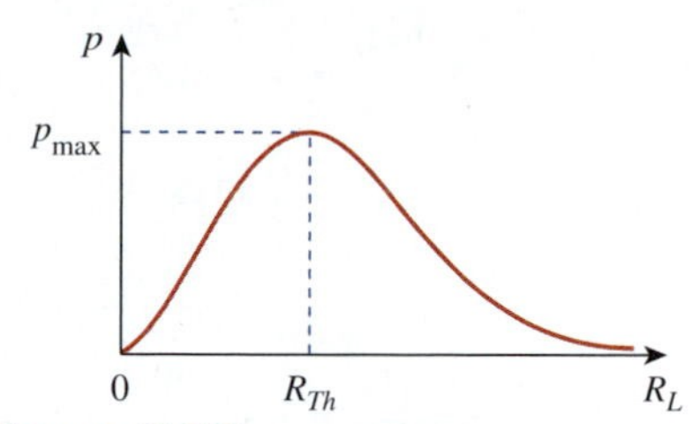

Figure 8.52
Power delivered to the load as a function of R_L.

This is known as the maximum power transfer theorem.

The maximum power transfer theorem is not really a means of analysis; it is an aid to system design. It was Moritz von Jacobi (1801–1874), a Russian engineer and physicist, who first discovered the maximum power transfer theorem known as Jacobi's law.

Maximum power is transferred to the load when the load resistance equals the Thevenin resistance as seen from the load ($R_L = R_{Th}$).

The maximum power transferred is obtained by substituting Eq. (8.14) into Eq. (8.13); that is,

$$P_{max} = \frac{V_{Th}^2}{4R_{Th}} \tag{8.15}$$

Equation (8.15) applies only when $R_L = R_{Th}$. (Any lower or higher value of R_L causes lesser power to be delivered to the load.) When $R_L \neq R_{Th}$, we compute the power delivered to the load using Eq. (8.13).

The power transfer efficiency η is given by

$$\eta = \frac{P_{out}}{P_{in}} = \frac{I^2R_L}{I^2R_L + I^2R_{Th}} = \frac{R_L}{R_L + R_{Th}} \tag{8.16}$$

Notice that efficiency is only 0.5 or 50% at maximum power transfer when $R_L = R_{Th}$. Efficiency increases toward 100% as the load resistance R_L increases toward infinity.

Example 8.14

Find the value of R_L for maximum power transfer in the circuit of Fig. 8.53. Find the maximum power.

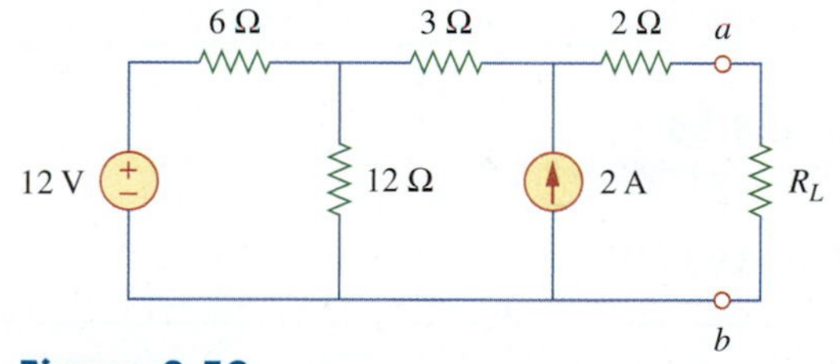

Figure 8.53
For Example 8.14.

Solution:

We need to find the Thevenin resistance R_{Th} and Thevenin voltage V_{Th} at the terminals *a-b*. To get R_{Th}, we use the circuit in Fig. 8.54(a) and obtain

$$R_{Th} = 2 + 3 + 6 \| 12 = 5 + \frac{6 \times 12}{18} = 9\ \Omega$$

To get V_{Th}, we consider the circuit in Fig. 8.54(b). Applying mesh analysis, we get

$$-12 + 18i_1 - 12i_2 = 0, \qquad i_2 = -2\text{ A}$$

[5] Note: The source and load are said to be *matched* when $R_L = R_{Th}$.

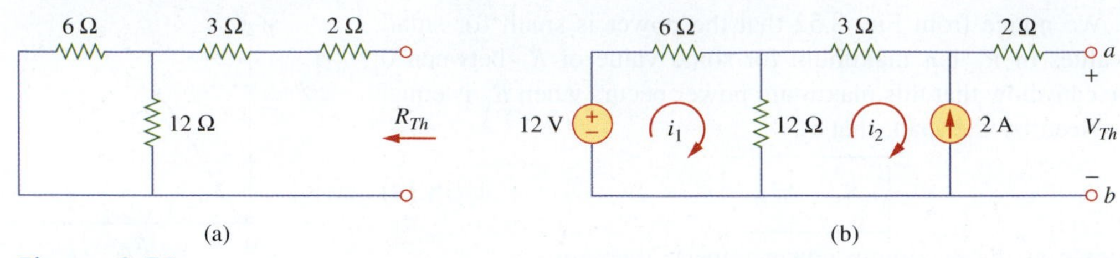

Figure 8.54
For Example 8.14: (a) finding R_{Th}; (b) finding V_{Th}.

Solving for i_1, we get $i_1 = -2/3$ A. Applying KVL around the outer loop to get V_{Th} at terminals *a-b*, we obtain

$$-12 + 6i_1 + 3i_2 + 2(0) + V_{Th} = 0 \quad \Rightarrow$$
$$V_{Th} = 12 + -6(-2/3) - 3(-2) = 22 \text{ V}$$

For maximum power transfer,

$$R_L = R_{Th} = 9\ \Omega$$

and the maximum power is

$$P_{\max} = \frac{V_{Th}^2}{4R_L} = \frac{22^2}{4 \times 9} = 13.44 \text{ W}$$

Practice Problem 8.14

Determine the value of R_L that will draw the maximum power from the rest of the circuit in Fig. 8.55. Calculate the maximum power.

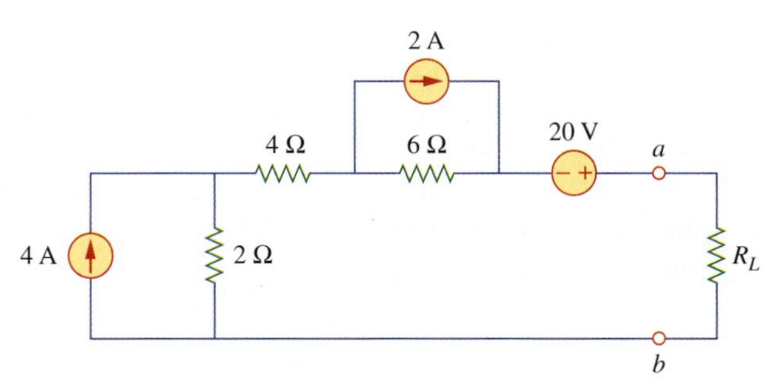

Figure 8.55
For Practice Problem 8.14.

Answer: 12 Ω; 33.33 W

8.8 †Millman's Theorem

Millman's theorem is named after Jacob Millman (1911–1991), a professor of electrical engineering at Columbia University. It is a combination of source transformation and Thevenin's and Norton's theorems. It is used to reduce a number of parallel voltage sources to an equivalent circuit containing only one source. It has the advantage of being easier to apply than nodal analysis, mesh analysis, or superposition.

> **Millman's theorem** states that any number of voltage sources (with series resistances) in parallel may be replaced by a single voltage source (with series resistance).

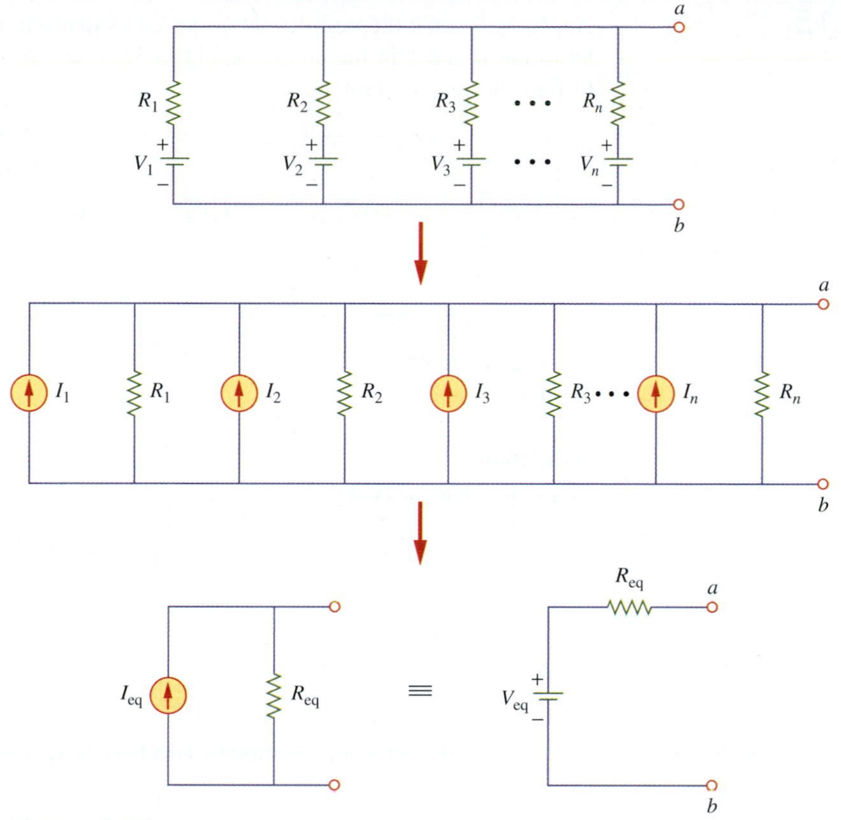

Figure 8.56
Illustrating Millman's theorem.

To apply Millman's theorem involves three steps:

1. Convert all voltage sources to current sources, as shown in Fig. 8.56.
2. Algebraically combine all parallel current sources and determine the equivalent resistance of the parallel resistances.
3. Convert the resulting current source to a voltage source. This gives the desired equivalent circuit.

In general, the equivalent current to be found in step 2 is calculated as

$$I_{eq} = I_1 + I_2 + I_3 + \cdots + I_n = \frac{V_1}{R_1} + \frac{V_2}{R_2} + \frac{V_3}{R_3} + \cdots + \frac{V_n}{R_n} \tag{8.17}$$

and the equivalent resistance as

$$R_{eq} = \frac{1}{\dfrac{1}{R_1} + \dfrac{1}{R_2} + \dfrac{1}{R_3} + \cdots + \dfrac{1}{R_n}} \tag{8.18}$$

Note that the summation in Eq. (8.17) is algebraic, based on the polarity of the voltage sources. Using Ohm's law,

$$V_{eq} = I_{eq}R_{eq} = \frac{\dfrac{V_1}{R_1} + \dfrac{V_2}{R_2} + \dfrac{V_3}{R_3} + \cdots + \dfrac{V_n}{R_n}}{\dfrac{1}{R_1} + \dfrac{1}{R_2} + \dfrac{1}{R_3} + \cdots + \dfrac{1}{R_n}} \tag{8.19}$$

Example 8.15

Apply Millman's theorem to obtain an equivalent circuit to the left of terminals a and b in the circuit of Fig. 8.57. Use the equivalent circuit to find the load current I_L.

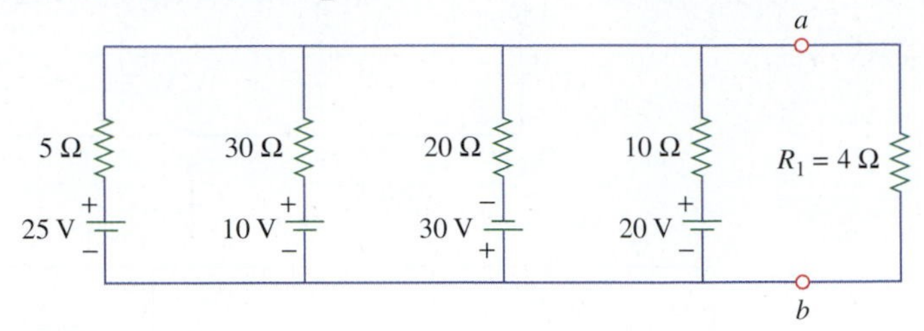

Figure 8.57
For Example 8.15.

Solution:
The equivalent resistance is

$$R_{eq} = \frac{1}{\frac{1}{5} + \frac{1}{30} + \frac{1}{20} + \frac{1}{10}} = 2.6087\ \Omega$$

The equivalent current is

$$I_{eq} = \frac{25}{5} + \frac{10}{30} - \frac{30}{20} + \frac{20}{10} = 5.833\ \text{A}$$

$$V_{eq} = I_{eq}R_{eq} = 2.6087 \times 5.833 = 15.216\ \text{V}$$

The equivalent circuit is shown in Fig. 8.58. Using the circuit, we have

$$I_L = \frac{V_{eq}}{R_{eq} + R_L} = \frac{15.216}{2.6087 + 4} = 2.303\ \text{A}$$

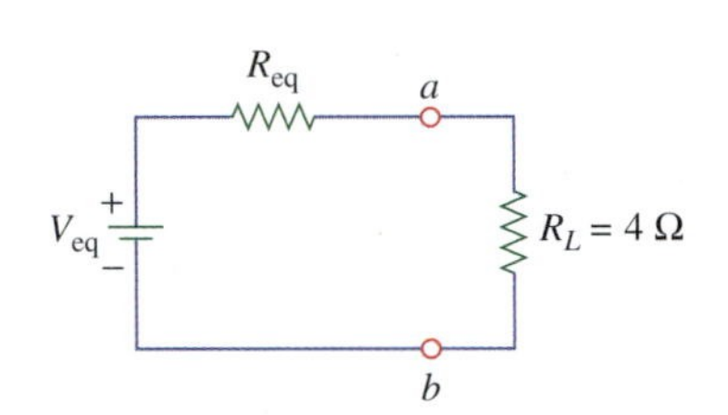

Figure 8.58
For Example 8.15.

Practice Problem 8.15

Use Millman's theorem to simplify the circuit in Fig. 8.59 and then obtain the load voltage.

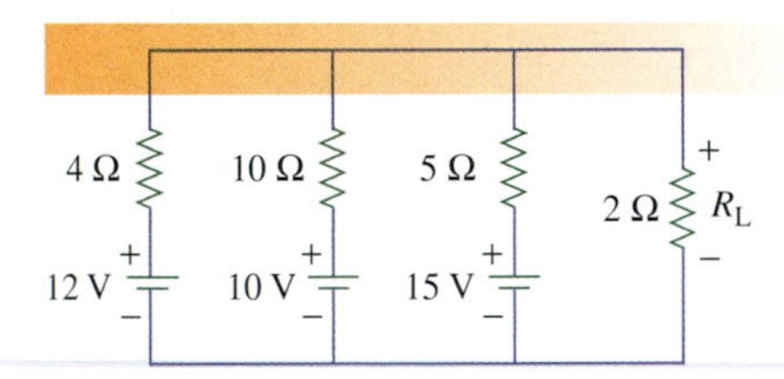

Figure 8.59
For Practice Problem 8.15.

Answer: 6.67 V

8.9 †Substitution Theorem

We are now familiar with the idea of replacing a network of resistors with its equivalent resistance and a circuit with its equivalent circuit. This substitution principle can be expanded into the substitution theorem. We illustrate the theorem by considering the branch a-b of the circuit in Fig. 8.60. The voltage V across and the current I through the branch are given by

$$V = \frac{2}{8+2}(20) = 4\ \text{V}, \qquad I = \frac{20}{8+2} = 2\ \text{A} \tag{8.20}$$

The **substitution theorem** states that in a linear network any branch may be replaced by any combination of circuit elements which produce the same voltage across and current through the branch.

According to the substitution theorem, the 2-Ω resistor may be replaced by any combination of circuit elements, provided that the

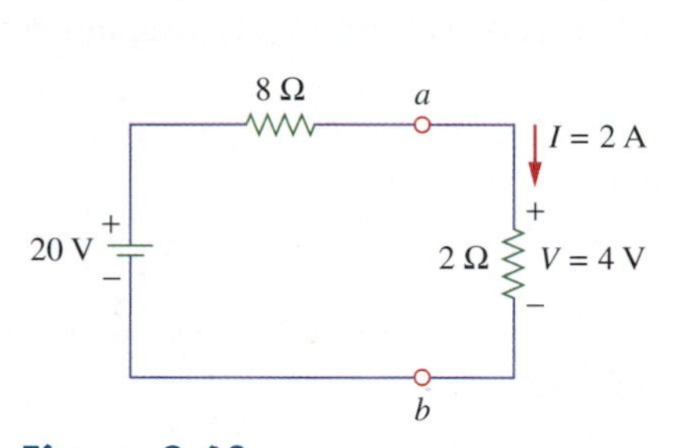

Figure 8.60
Illustration of the substitution theorem.

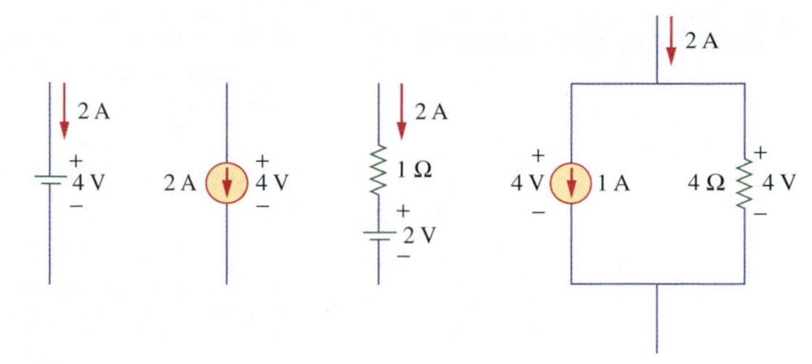

Figure 8.61
Equivalent branches for the branch a-b in Fig. 8.60.

substituted elements maintain the same voltage across and current through the branch. This is achieved by each branch in Fig. 8.61.

Because the substitution theorem requires that we know in advance the voltage across and the current through the selected branch, the theorem is not that useful for solving circuit problems. It is often used by circuit designers to optimize their design.

Example 8.16

Refer to the circuit in Fig. 8.62(a). If the branch a-b is to be replaced by a current source and a 20-Ω resistor as shown in Fig. 8.62(b), determine the magnitude and direction of the current source.

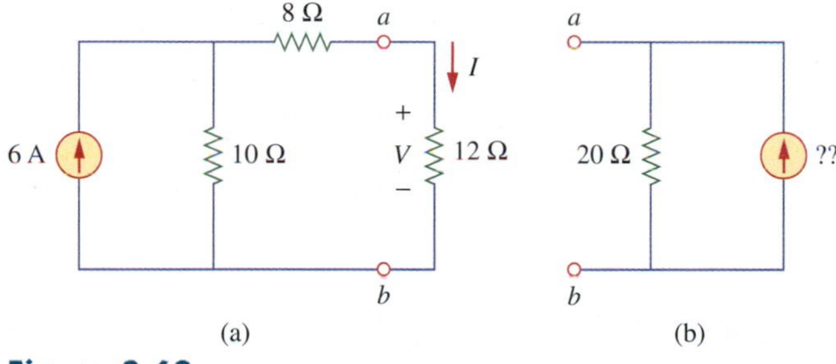

Figure 8.62
For Example 8.16.

Solution:
The current through branch a-b is obtained using current divider rule.

$$I = \frac{10}{10 + 8 + 12}(6) = 2 \text{ A}$$

The voltage across the branch is

$$V = 12I = 24 \text{ V}$$

The 20-Ω resistor in Fig. 8.54(b) must have the same voltage across it so that the current through this resistor is

$$I_{20} = \frac{24}{20} = 1.2 \text{ A}$$

To satisfy Kirchhoff's current law at node a in Fig. 8.63, the current source must have a magnitude of $2 - 1.2 = 0.8$ A, and its direction must be downward as shown in Fig. 8.63.

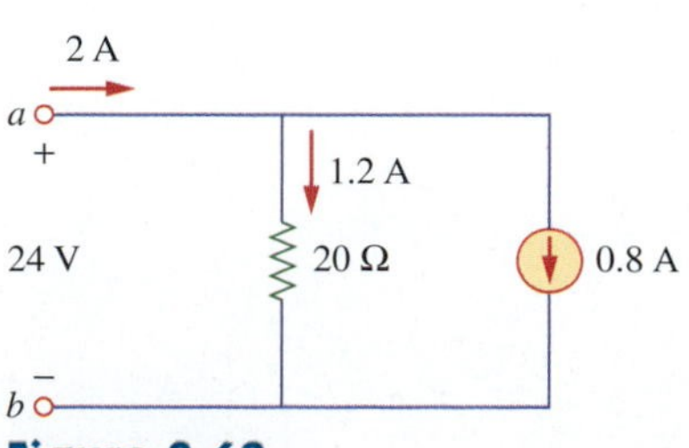

Figure 8.63
For Example 8.16.

Practice Problem 8.16

Refer to the circuit in Fig. 8.64. If the branch *a-b* is to be replaced by a voltage source and a series 10-Ω resistor, determine the magnitude and polarity of the voltage source.

Answer: 6 V with the positive polarity toward *a*

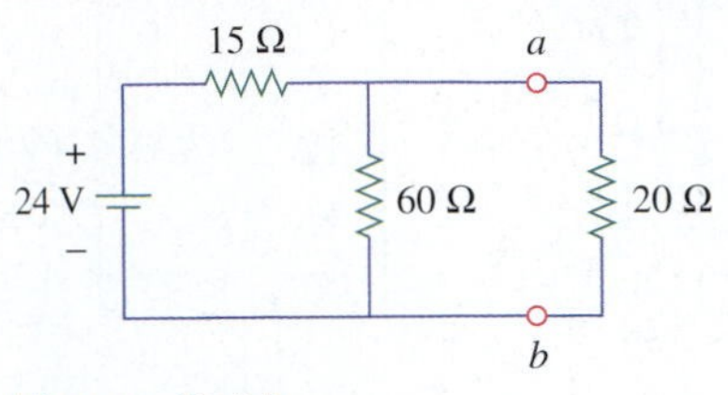

Figure 8.64
For Practice Problem 8.16.

8.10 †Reciprocity Theorem

This theorem applies only to single-source circuits. If a circuit has more than one source, then the reciprocity theorem does not apply. Because the source may be voltage or current, there are two special cases of the theorem.

CASE 1 Voltage Source

> The **reciprocity theorem** states that in a linear circuit with only one voltage source, if that source located in branch *A* causes a current *I* in branch *B*, then moving the voltage source to branch *B* will cause current *I* in branch *A*.

When the voltage source is moved to branch *B*, it must be connected in series with an element in branch *B* (if any) and replaced by a short circuit in its original location, that is, in branch *A*. Also, the polarity of voltage in branch *B* is such that the direction of the current in branch *B* remains the same. Although reciprocity theorem applies to a circuit with only one voltage or current source, its power can be demonstrated by considering a complex circuit such as the one such as in Fig. 8.65.

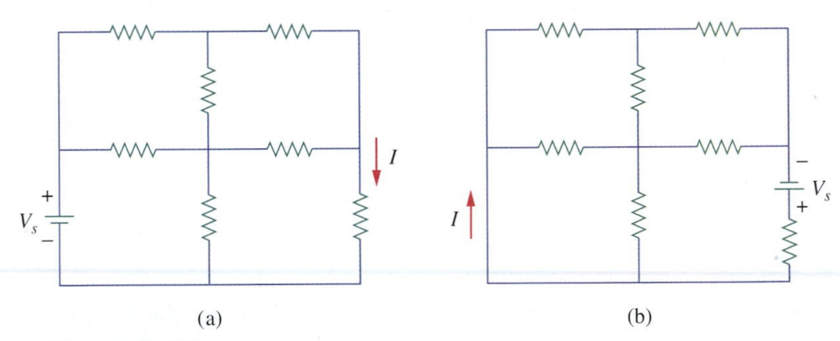

Figure 8.65
Illustration of the reciprocity theorem.

CASE 2 Current Source

> The **reciprocity theorem** states that in a linear circuit with only one current source, if that source located in branch *A* causes a voltage *V* in branch *B*, then moving the current source to branch *B* will cause voltage *V* in branch *A*.

This is the same as case 1. When the current source is moved to branch *B*, it must be replaced by an open circuit at its original location *A* and must be connected in parallel with any element in branch *B*. The direction of the current source in branch *B* is such that the polarity of the voltage in branch *B* remains the same.

Example 8.17

(a) Find I in the circuit of Fig. 8.66(a).
(b) Remove the voltage source and place it at the branch containing the 1-Ω resistor, as shown in Fig. 8.66(b), and determine the current I again.

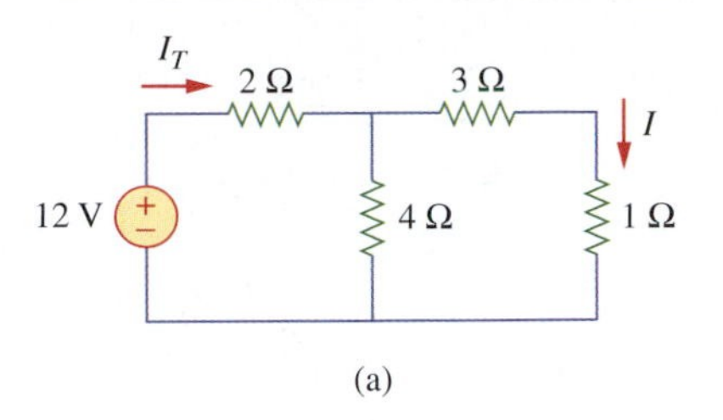

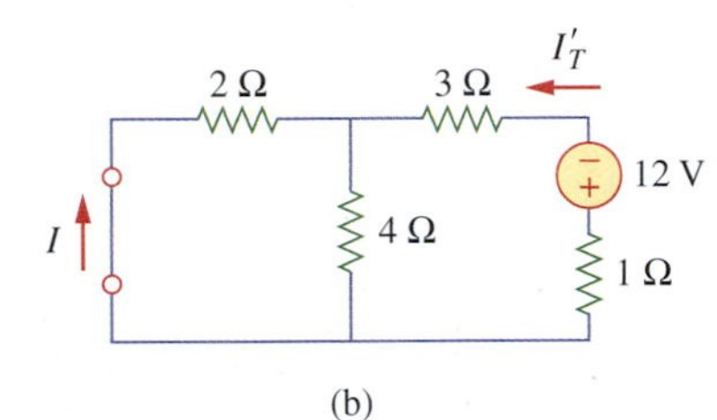

Figure 8.66
For Example 8.17.

Solution:
(a) The total resistance as seen by the voltage source is

$$R_T = 2 + 4 \,\|\, (3 + 1) = 2 + 2 = 4$$

The current I_T in Fig. 8.66(a) is given by $I_T = \dfrac{12}{4} = 3$ A. Hence, the desired current I is given by

$$I = \frac{1}{2} I_T = 1.5 \text{ A}$$

(b) As shown in Fig. 8.66(b), we replace the voltage source with a short circuit and place it in the branch that has the 1-Ω resistor. Notice that it is connected in series with the 1-Ω resistor and its polarity conforms with the direction of I in Fig. 8.66(a). Again, the total resistance as seen by the voltage source is

$$R'_T = 1 + 3 + 2 \,\|\, 4 = 16/3\ \Omega$$

The current I'_T is obtained as

$$I'_T = \frac{-12}{16/3} = -9/4 \text{ A}$$

Using current division principle,

$$I = -\frac{4}{4+2} I'_T = -\frac{4}{6}\left(-\frac{9}{4}\right) = 1.5 \text{ A}$$

as obtained previously. This validates the reciprocity theorem.

Practice Problem 8.17

(a) Calculate the current I in Fig. 8.67.
(b) Place the voltage source in the branch containing the 5-Ω resistor and determine I in the original location of the voltage source.

Answer: (a) 1.2 A; (b) 1.2 A

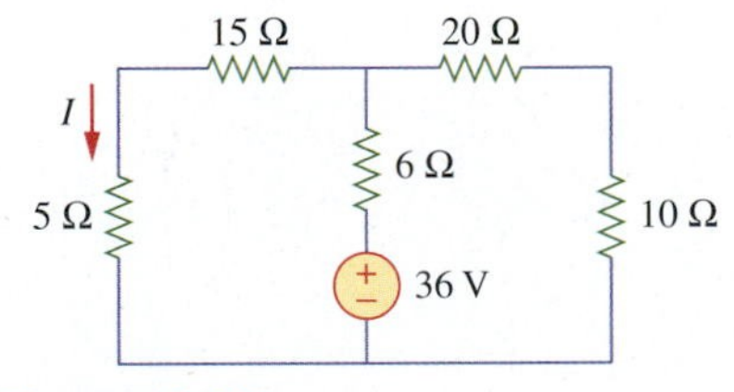

Figure 8.67
For Practice Problem 8.17.

Example 8.18

(a) Calculate V_o in the circuit of Fig. 8.68(a).
(b) Remove the current source and connect it in parallel with the 1-Ω resistor. Show that the voltage across the branch where the current was located is the same as V_o.

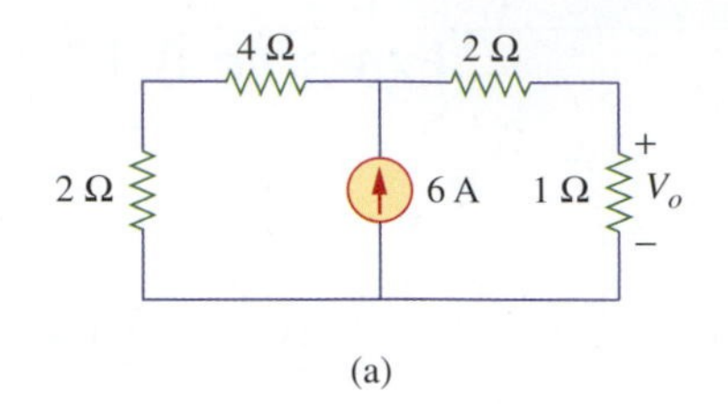

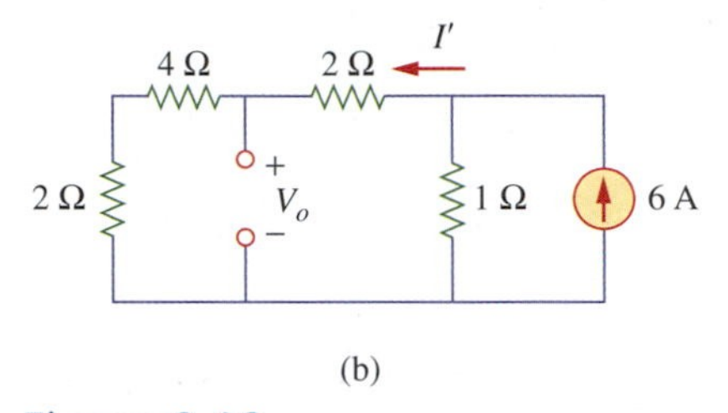

Figure 8.68
For Example 8.18.

Solution:

(a) Using current division principle, the current I through the 1-Ω resistor is

$$I = \frac{2+4}{2+4+1+2}(6) = 4 \text{ A}$$

Hence,

$$V_o = 1 \times I = 4 \text{ V}$$

(b) We now remove the current source and connect it across the 1-Ω resistor, as shown in Fig. 8.68(b). Using current divider rule,

$$I' = \frac{1}{1+2+4+2}(6) = \frac{2}{3} \text{ A}$$

Hence,

$$V_o = I'(4+2) = \frac{2}{3} \times 6 = 4 \text{ V}$$

Practice Problem 8.18

(a) Find V_x in the circuit of Fig. 8.69.

(b) Remove the current source and connect it in parallel with the 3-Ω resistor; calculate the voltage across the former location of the current source.

Answer: (a) 15 V; (b) 15 V

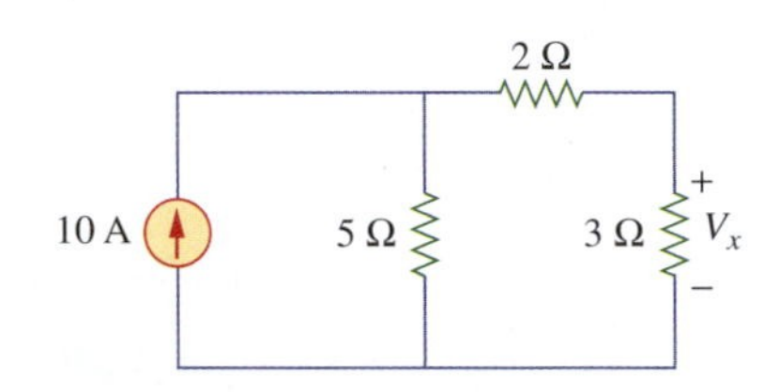

Figure 8.69
For Practice Problem 8.18.

8.11 Verifying Circuit Theorems with Computers

8.11.1 PSpice

In this section, we learn how to use PSpice to verify the theorems covered in this chapter. Specifically, we will consider using the DC Sweep analysis to find the Thevenin or Norton equivalent at any pair of nodes in a circuit and the maximum power transfer to a load. The reader is advised to read Section C.3 of Appendix C in preparation for this section.

To find the Thevenin equivalent of a circuit at a pair of open terminals using PSpice, we use the schematic editor to draw the circuit and insert an independent probing current source, say, *Ip*, at the terminals. The probing current source must have a part name ISRC. We then perform a DC Sweep on *Ip*, as discussed in Section C.3 of Appendix C. Typically, we may let the current through *Ip* vary from 0 to 1 A in 0.1-A increments. After simulating the circuit, we use PSpice A/D Demo to display a plot of the voltage across *Ip* versus the current through *Ip*. The zero intercept of the plot gives us the Thevenin equivalent voltage, while the slope of the plot is equal to the Thevenin resistance.

To find the Norton equivalent involves similar steps, except that we insert a probing independent voltage source (with a part name VSRC), say, *Vp*, at the terminals. We perform a DC Sweep on *Vp* and let *Vp* vary from 0 to 1 V in 0.1-V increments. A plot of the current through *Vp* versus the voltage across is obtained using the menu of

PSpice A/D Demo after simulation. The zero intercept is equal to the Norton current, while the slope of the plot is equal to the Norton conductance.

To find the maximum power transfer to a load using PSpice involves performing a dc parametric sweep on the component value of R_L in Fig. 8.51 and plotting the power delivered to the load a function of R_L. According to Fig. 8.52, the maximum power occurs when $R_L = R_{Th}$. This is best illustrated with an example, and we will do so in Example 8.19.

We use VSRC and ISRC part names for the independent voltage and current sources.

Example 8.19

Consider the circuit in Fig. 8.37 (see Practice Problem 8.10). Use PSpice to find the Thevenin and Norton equivalent circuits.

Solution:

(a) To find the Thevenin resistance R_{Th} and Thevenin voltage V_{Th} at the terminals *a-b* in the circuit in Fig. 8.37, we first use Orcad Capture to draw the circuit as shown in Fig. 8.70(a). Notice that a probing current source *I*1 is inserted at the terminals. Select **PSpice/New Simulation Profile**. In the New Simulation dialog box, enter the name of file (e.g., exam819) and click Create. In the Simulation Settings dialog box, select the following: DC Sweep under *Analysis Type*, Current source under *Sweep Variable,* Linear under *Sweep Type*, 0 as *Start Value*, 1 as *End Value*, 0.1 as *Increment*, and I1 under *Name* box. Click Apply, then click OK. Select **PSpice/Run.** The Probe window will pop up.

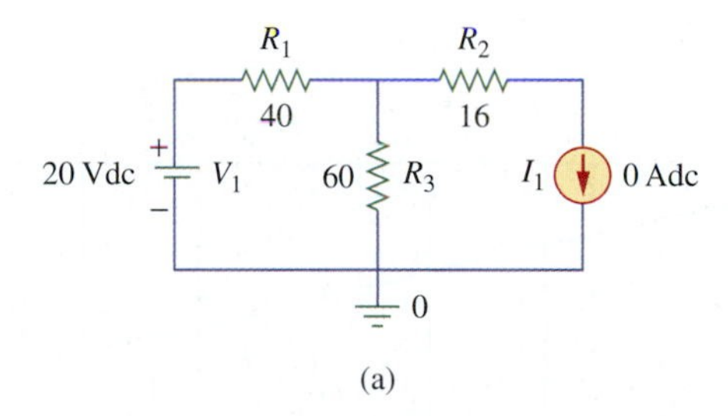

(a)

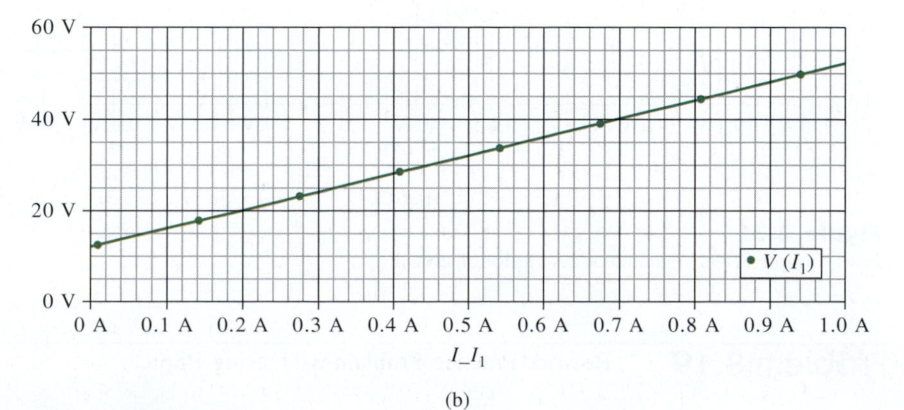

(b)

Figure 8.70
For Example 8.19.

Select **Trace/Add Trace** and choose V(I1:-). The voltage across *I*1 will be plotted as shown in Fig. 8.70(b). From the plot, we obtain

$$V_{Th} = \text{zero intercept} = 12\text{ V}, \quad R_{Th} = \text{slope} = \frac{52 - 12}{1} = 40\ \Omega$$

which agree with what we got analytically in Practice Problem 8.10.

(b) To find the Norton equivalent, we modify the schematic in Fig. 8.70(a) by replacing the probing current source with a probing voltage source *V*2. The result is the schematic in Fig. 8.71(a). Again, we select **PSpice/New Simulation**. In the New Simulation dialog box, type the name of the file and click Create. In the Simulation Settings dialog box, select Linear for the *Sweep Type* and Voltage Source for the *Sweep Var. Type*. We enter V2 under *Name* box, 0 as *Start Value*, 1 as *End Value*, and 0.1 as *Increment*. Click Apply, then OK. Select **PSpice/Run.** In the Probe window that appears, select **Trace/Add Trace** and choose trace I(V2). We obtain the plot in Fig. 8.71(b). From the plot, we obtain

$$I_N = \text{zero intercept} = 300\text{ mA}$$

$$G_N = \text{slope} = \frac{(300 - 275) \times 10^{-3}}{1} = 25\text{ mS}$$

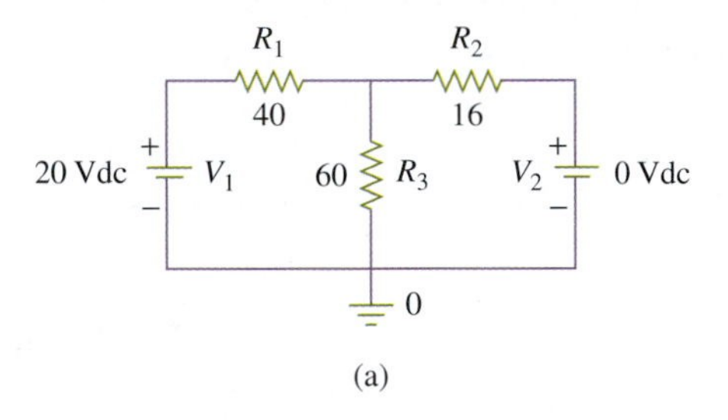

(a)

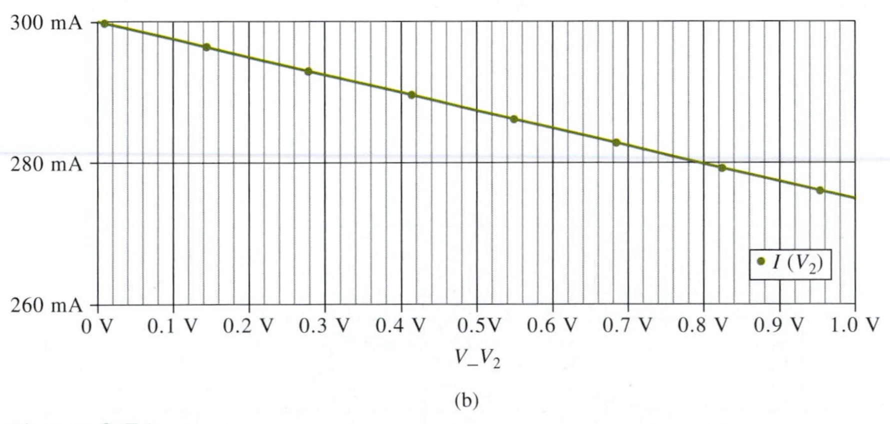

(b)

Figure 8.71
For Example 8.19: (a) schematic; (b) plot of I(V2).

Practice Problem 8.19

Rework Practice Problem 8.11 using PSpice.

Answer: $R_N = 3\ \Omega$; $I_N = 4.5$ A

8.11.2 Multisim

Multisim can be used to illustrate the theorems covered in this chapter. We will illustrate with an example of how to find the Thevenin and Norton equivalents of a circuit.

Example 8.20

Use Multisim to find the Thevenin and Norton equivalent circuits of the circuit in Fig. 8.40 (Example 8.11).

Solution:

To determine Thevenin resistance, we remove the current source and the voltage source. We connect terminals a and b to the multimeter. **DCLICKL** the multimeter and select the ohmmeter function. The Multisim circuit that results is shown in Fig. 8.72. After the circuit is saved and simulated by selecting **Simulate/Run**, we **DCLICKL** the multimeter to display the result. Thus,

$$R_{Th} = R_N = 4\ \Omega$$

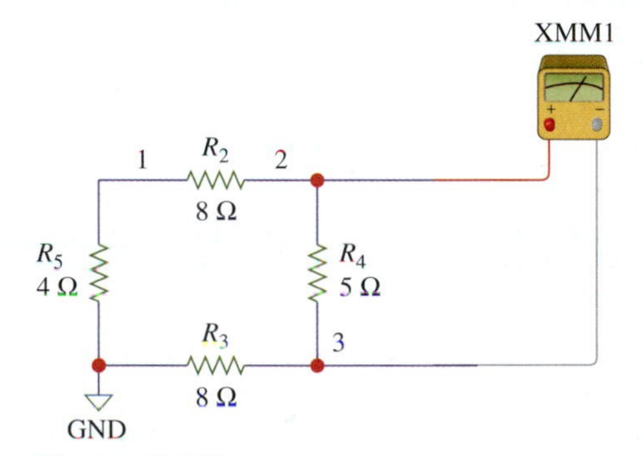

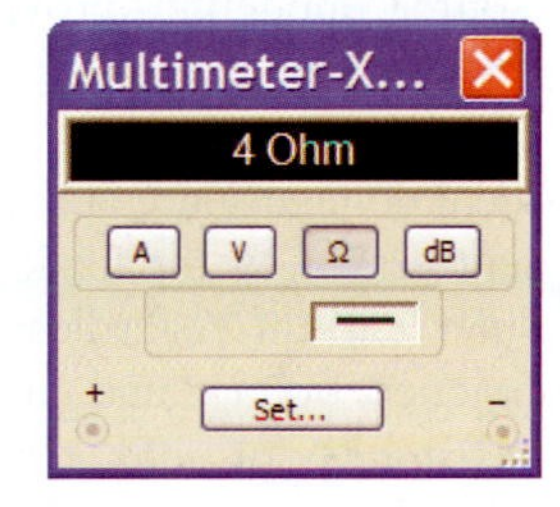

Figure 8.72
For Example 8.20: finding Thevenin resistance.

To find the Thevenin voltage, we first need to construct the circuit as shown in Fig. 8.73. We connect the multimeter to terminals a and b. We set it to measure voltage by selecting **DCLICKL** and then selecting the voltage function. After saving and simulating the circuit, we **DCLICKL** the multimeter to display the result. Thus,

$$V_{Th} = 4\ \text{V}$$

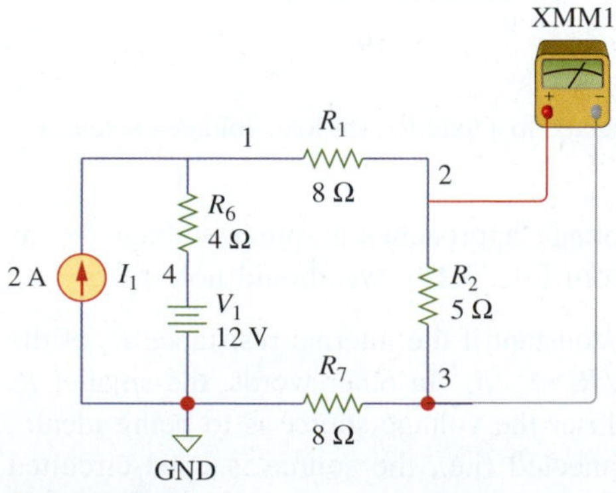

Figure 8.73
For Example 8.21: finding Thevenin voltage.

To obtain the Norton current, we connect the multimeter across terminals *a* and *b*. We set it to measure current by selecting **DCLICKL** and then selecting the current function. The circuit happens to be the same as that in Fig. 8.73 except that the multimeter is now set to measure current. Upon saving and simulating the circuit, we **DCLICKL** the multimeter to display the result.

$$I_N = 1\text{ A}$$

Notice that the values of the Norton resistance and Norton current agree with what we obtained in Example 8.11.

Practice Problem 8.20

For the circuit in Fig. 8.43 (Practice Problem 8.11), find the Thevenin and Norton equivalent circuits using Multisim.

Answer: $R_{Th} = R_N = 3\ \Omega$; $V_{Th} = 13.5$ V; $I_N = 4.5$ A

8.12 †Application: Source Modeling

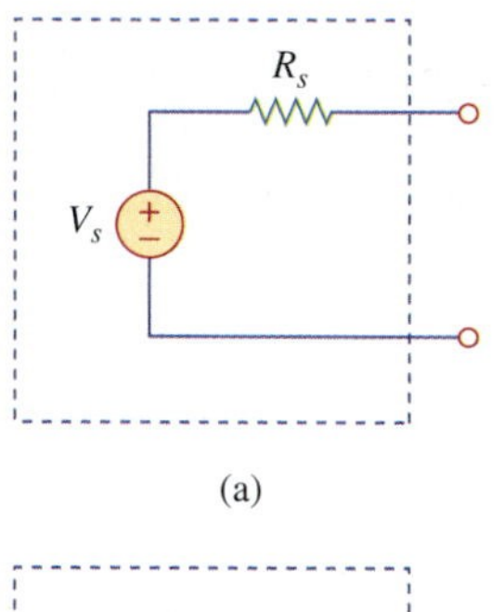

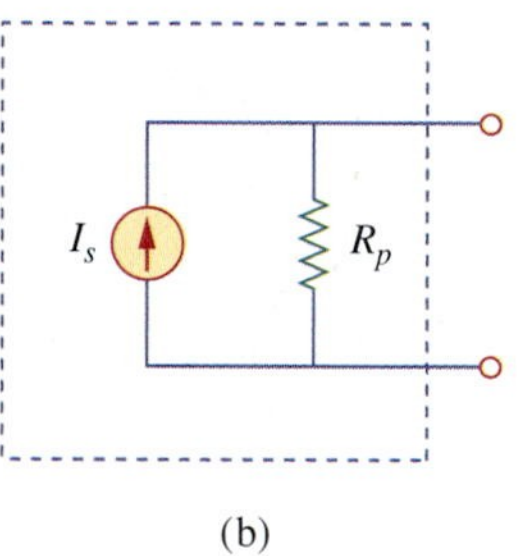

Figure 8.74
(a) Practical voltage source; (b) practical current source.

Source modeling provides an example of the usefulness of the Thevenin or Norton equivalent. An active source such as a battery is often characterized by its Thevenin or Norton equivalent circuit. An ideal voltage source provides a constant voltage irrespective of the current drawn by the load, while an ideal current source supplies a constant current regardless of the load voltage. As Fig. 8.74 shows, practical voltage and current sources are not ideal due to their *internal resistances* or *source resistances* R_s and R_p. They become ideal as $R_s \to 0$ and $R_p \to \infty$. To show that this is the case, consider the effect of the load on voltage sources, as shown in Fig. 8.75(a). By the voltage division principle, the load voltage is

$$V_L = \frac{R_L}{R_s + R_L} V_s \tag{8.21}$$

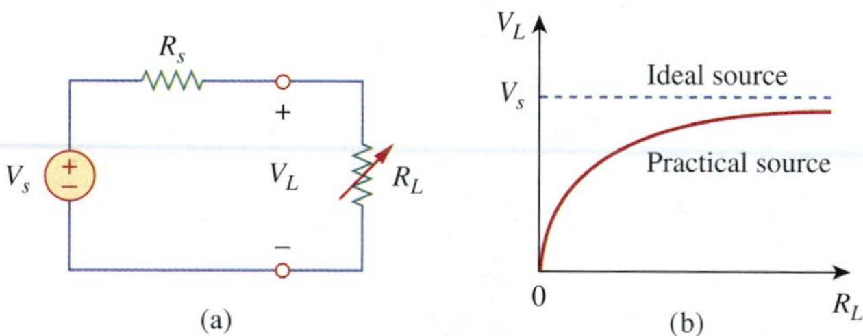

Figure 8.75
(a) Practical voltage source connected to a load R_L; (b) load voltage decreases as R_L decreases.

As R_L increases, the load voltage approaches a source voltage V_s, as illustrated in Fig. 8.75(b). From Eq. (8.21), we should note that:

1. The load voltage will be constant if the internal resistance R_s of the source is zero or at least $R_s \ll R_L$. In other words, the smaller R_s is compared to R_L, the closer the voltage source is to being ideal.
2. When the load is disconnected (i.e., the source is open-circuited so that $R_L \to \infty$), $V_{oc} = V_s$. Thus, V_s may be regarded as the unloaded source voltage. The connection of the load causes the

terminal voltage to drop in magnitude, and this is known as the *loading effect.*

The same argument can be made for a practical current source when connected to a load as shown in Fig. 8.76(a). By the current divider rule,

$$I_L = \frac{R_p}{R_p + R_L} I_s \tag{8.22}$$

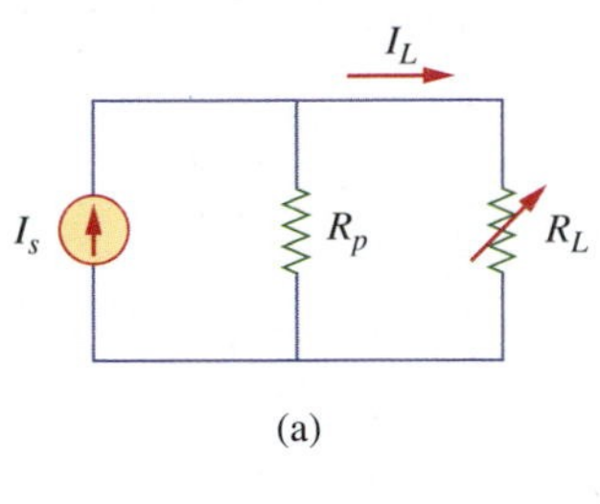

(a)

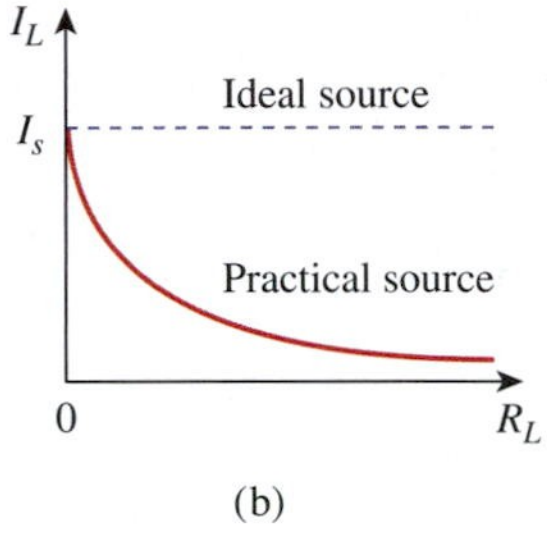

(b)

Figure 8.76
(a) Practice current source connected to a load R_L; (b) load current decreases R_L as increases.

Figure 8.76(b) shows the variation in the load current as the load resistance increases. Again, we notice a drop in current due to the load (loading effect), and the load current is constant (ideal current source) when the internal resistance is very large; that is $R_p \to \infty$ or at least $R_p \gg R_L$.

Sometimes we need to know the unloaded source voltage V_s and the internal resistance R_s of a voltage source. To find V_s and R_s, we use the circuit in Fig. 8.77. First, we measure the open-circuit voltage V_{oc} as in Fig. 8.77(a) and set

$$V_s = V_{oc} \tag{8.23}$$

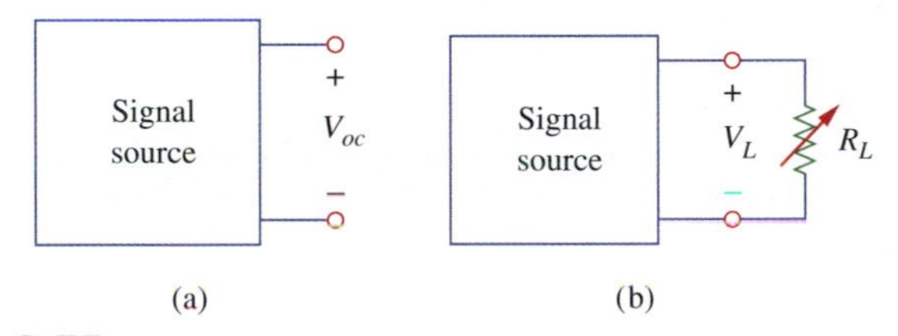

Figure 8.77
(a) Measuring V_{oc}; (b) measuring V_L.

Then, we connect a variable load R_L across the terminals as in Fig. 8.77(b). We adjust the resistance R_L until we measure a load voltage of exactly one-half of the open-circuit voltage; that is, $V_L = V_{oc}/2$, because now $R_L = R_{Th} = R_s$. At that point, we disconnect R_L and measure it. We set

$$R_s = R_L \tag{8.24}$$

For example, a car battery may have $V_s = 12$ V and $R_s = 0.05\ \Omega$.

Besides source modeling, another simple illustration of what is covered in this chapter is matching speakers (load) to the output resistance of an amplifier.

Example 8.21

The voltage of the terminals of a voltage source is 12 V when connected to a 2-W load. When the load is disconnected, the terminal voltage rises to 12.4 V. (a) Calculate the source voltage V_s and internal resistance R_s. (b) Determine the voltage when an 8-Ω load is connected to the source.

Solution:

(a) We replace the source by its Thevenin equivalent. The terminal voltage when the load is disconnected is the open-circuit voltage; that is,

$$V_s = V_{oc} = 12.4 \text{ V}$$

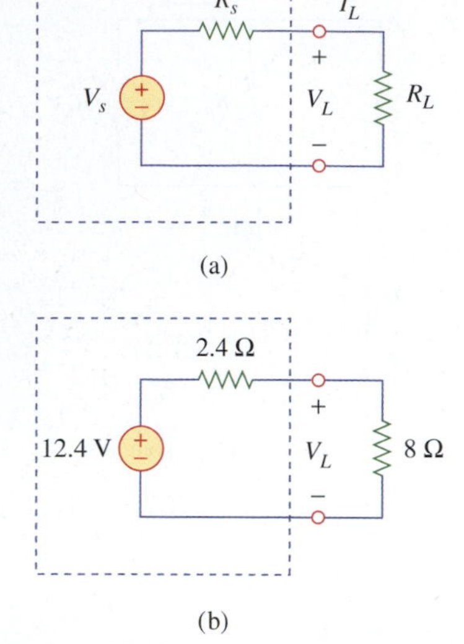

Figure 8.78
For Example 8.21.

When the load is connected, as shown in Fig. 8.78(a), $V_L = 12$ V and $P_L = 2$ W. Hence,

$$P_L = \frac{V_L^2}{R_L} \quad \Rightarrow \quad R_L = \frac{V_L^2}{P_L} = \frac{12^2}{2} = 72\ \Omega$$

The load current is

$$I_L = \frac{V_L}{R_L} = \frac{12}{72} = \frac{1}{6}\text{ A}$$

The voltage across R_s is the difference between the source voltage V_s and the load voltage V_L, or

$$12.4 - 12 = 0.4 = R_s I_L \quad \Rightarrow \quad R_s = \frac{0.4}{I_L} = 2.4\ \Omega$$

(b) Now that the Thevenin equivalent of the source is determined, we connect the 8-Ω load across the Thevenin equivalent as shown in Fig. 8.78(b). Using voltage division, we obtain

$$V_L = \frac{R_L}{R_s + R_L} V_{Th} = \frac{8}{8 + 2.4}(12.4) = 9.538\text{ V}$$

Practice Problem 8.21

The measured open-circuit voltage across a certain amplifier is 9 V. The voltage drops to 8 V when a 20-Ω loudspeaker is connected to the amplifier. Calculate the voltage when a 10-Ω loudspeaker is used instead.

Answer: 7.2 V

8.13 Summary

1. A linear network consists of linear elements, linear dependent sources, and linear independent sources.
2. Network theorems are used to reduce a complex circuit to a simpler one, thereby making circuit analysis much simpler.
3. The superposition principle states that for a circuit having multiple independent sources, the voltage across (or current through) an element is equal to the algebraic sum of all the individual voltages (or currents) due to each independent source acting one at a time.
4. Source transformation is a procedure for transforming a voltage source in series with a resistor to a current source in parallel with a resistor or vice versa.
5. Thevenin's and Norton's theorems allow us to isolate a portion of a network while the remaining portion of the network is replaced by an equivalent network. The Thevenin equivalent consists of a voltage source V_{Th} in series with a resistor R_{Th}, while the Norton equivalent consists of a current source I_N in parallel with a resistor R_N. The two theorems are related by source transformation.

$$R_N = R_{Th}, \qquad I_N = \frac{V_{Th}}{R_{Th}}$$

6. For a given Thevenin equivalent circuit, maximum power transfer occurs when $R_L = R_{Th}$; that is, the load resistance is equal to the Thevenin resistance.

7. Millman's theorem provides a method of combining several voltage sources in parallel.
8. The substitution theorem states that any branch of a linear circuit can be replaced by an equivalent branch that produces the same voltage across and current through the branch.
9. The reciprocity theorem states that in a linear circuit with only one voltage source, if that source located in branch A causes a current I in branch B, then moving the voltage source to branch B will cause current I in branch A.
10. PSpice and Multisim can be used to verify the circuit theorems covered in this chapter.
11. Source modeling provides applications for Thevenin's theorem.

Review Questions

8.1 The current through a branch in a linear network is 2 A when the input source voltage is 10 V. If the voltage is reduced to 1 V and the polarity is reversed, the current through the branch is:

(a) −2 A (b) −0.2 A (c) 0.2 A (d) 2 A (e) 20 A

8.2 For superposition, it is not required that only one independent source be considered at a time; any number of independent sources may be considered simultaneously.

(a) True (b) False

8.3 The superposition principle applies to power calculation.

(a) True (b) False

8.4 Refer to Fig. 8.79. The Thevenin resistance at terminals *a-b* is:

(a) 25 Ω (b) 20 Ω (c) 5 Ω (d) 4 Ω

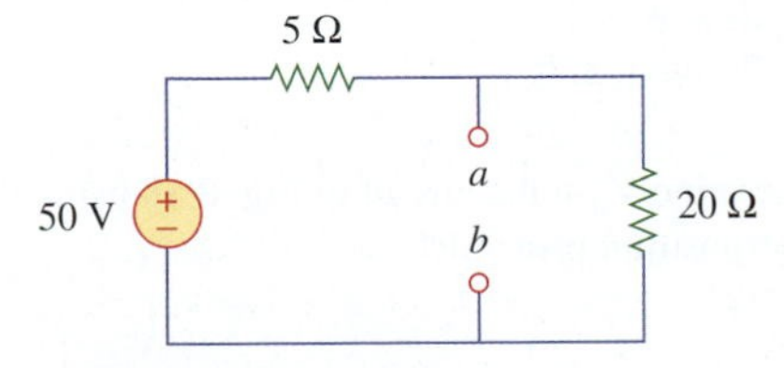

Figure 8.79
For Review Questions 8.4, 8.5, and 8.6.

8.5 The Thevenin voltage at terminals *a-b* of the circuit in Fig. 8.79 is

(a) 50 V (b) 40 V (c) 20 V (d) 10 V

8.6 The Norton current at terminals *a-b* of the circuit in Fig. 8.79 is

(a) 10 A (b) 2.5 A (c) 2 A (d) 0 A

8.7 The Norton resistance R_N is exactly equal to the Thevenin resistance R_{Th}.

(a) True (b) False

8.8 Which pair of circuits in Fig. 8.80 are equivalent?

(a) a and b (b) b and d (c) a and c (d) c and d

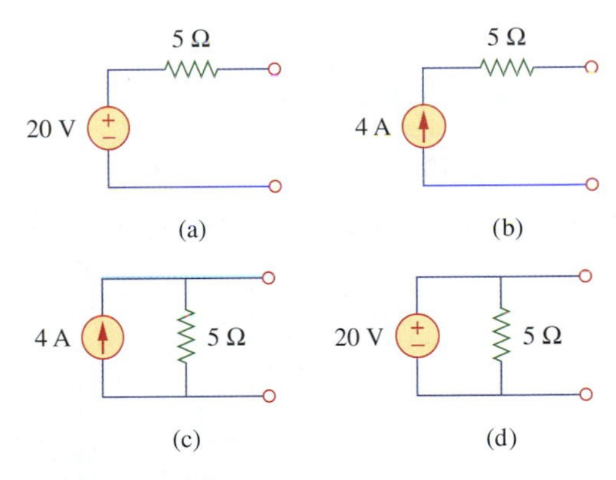

Figure 8.80
For Review Question 8.8.

8.9 A load is connected to a network. At the terminals to which the load is connected, $R_{Th} = 10\ \Omega$ and $V_{Th} = 40$ V. The maximum power supplied to the load is

(a) 160 W (b) 80 W (c) 40 W (d) 1 W

8.10 If a circuit has more than one source, then the reciprocity theorem does not apply.

(a) True (b) False

Answers: 8.1b, 8.2a, 8.3b, 8.4d, 8.5b, 8.6a, 8.7a, 8.8c, 8.9c, 8.10a

Problems

Section 8.2 Linearity Property

8.1 Calculate the current I_o in the circuit of Fig. 8.81. What does this current become when the input voltage is raised to 10 V?

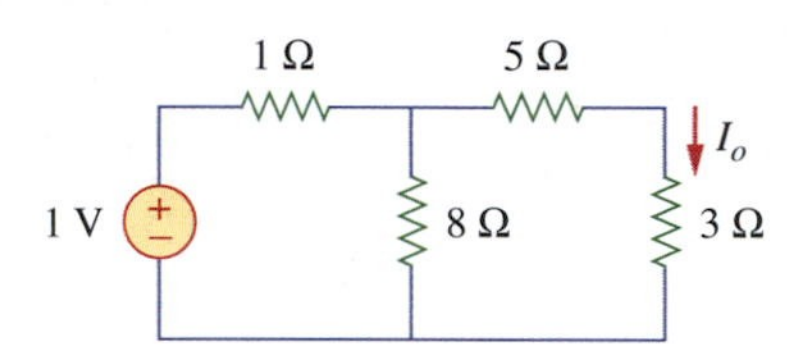

Figure 8.81
For Problem 8.1.

8.2 Find V_o in the circuit in Fig. 8.82. If the source current is reduced to 1 μA, what is V_o?

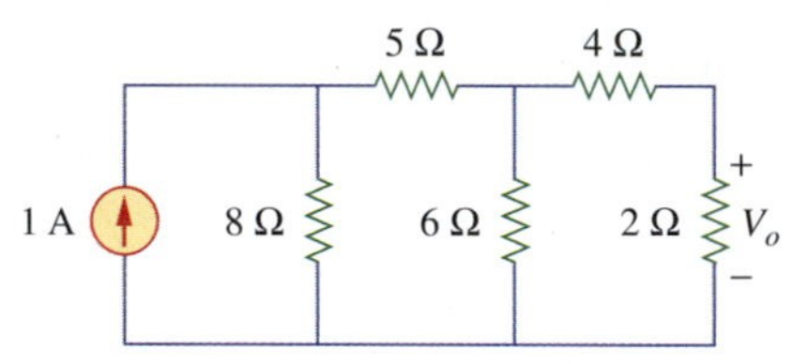

Figure 8.82
For Problem 8.2.

8.3 Use linearity to determine I_o in the circuit in Fig. 8.83.

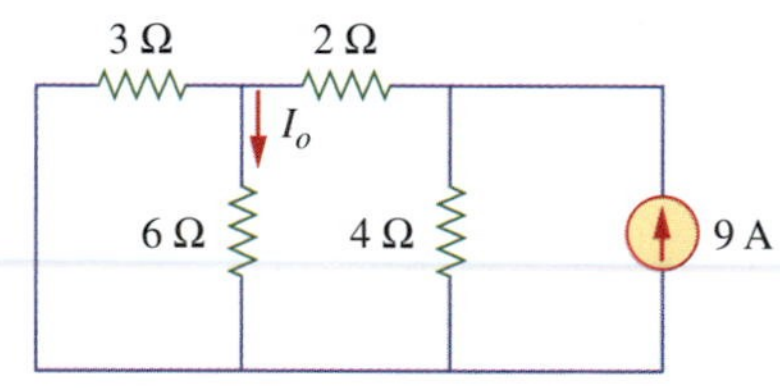

Figure 8.83
For Problem 8.3.

8.4 For the circuit in Fig. 8.84, assume $V_o = 1$ V and use linearity to find the actual value of V_o.

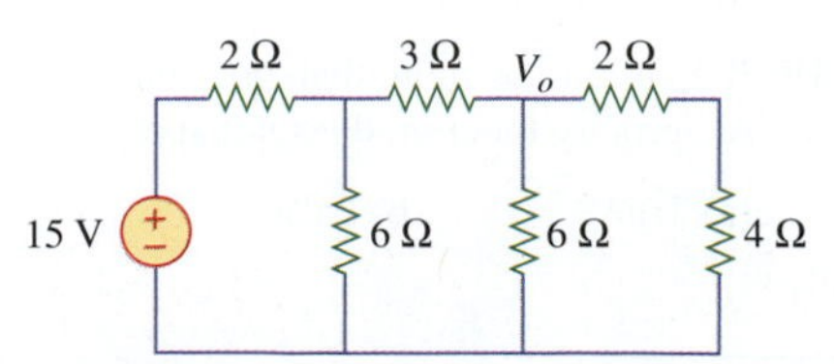

Figure 8.84
For Problem 8.4.

Section 8.3 Superposition

8.5 Apply superposition to find I in the circuit in Fig. 8.85.

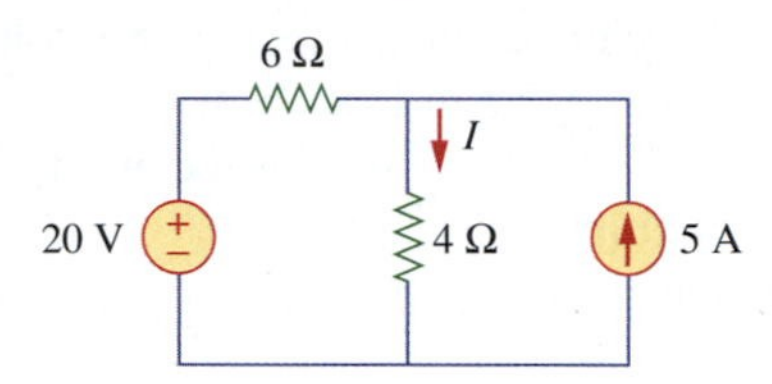

Figure 8.85
For Problem 8.5.

8.6 Given the circuit in Fig. 8.86, calculate I_x and the power dissipated by the 10-Ω resistor using superposition.

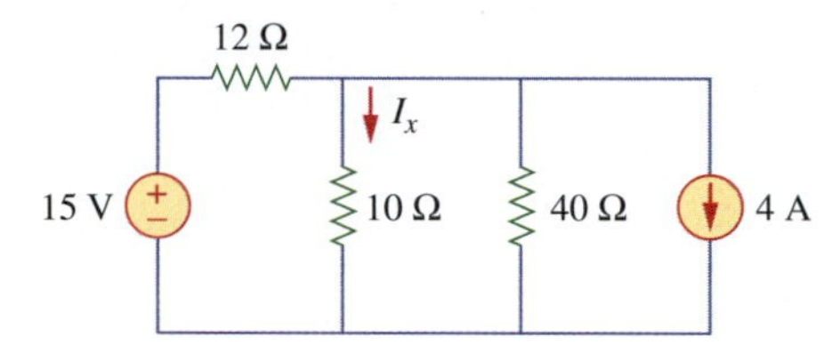

Figure 8.86
For Problem 8.6.

8.7 Use the superposition principle to find I in the circuit shown in Fig. 8.87.

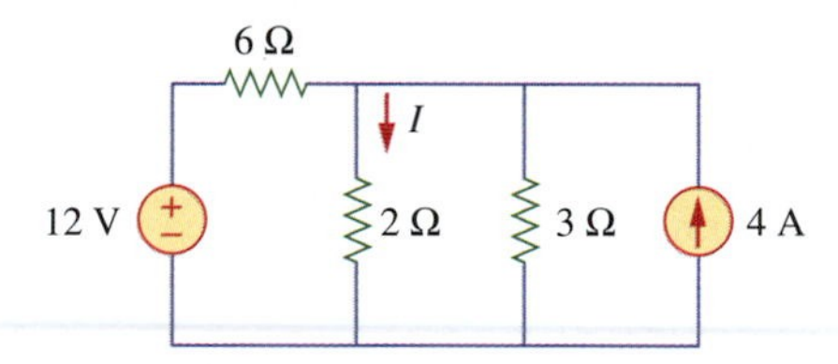

Figure 8.87
For Problem 8.7.

8.8 Determine V_o in the circuit in Fig. 8.88 using the superposition principle.

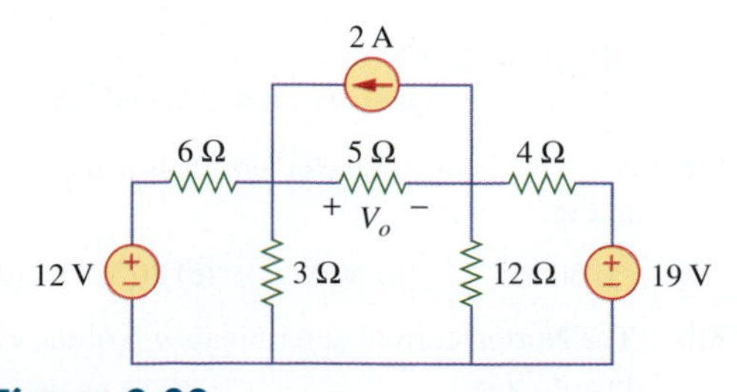

Figure 8.88
For Problem 8.8.

8.9 Using the superposition principle, find I_o in the circuit shown in Fig. 8.89.

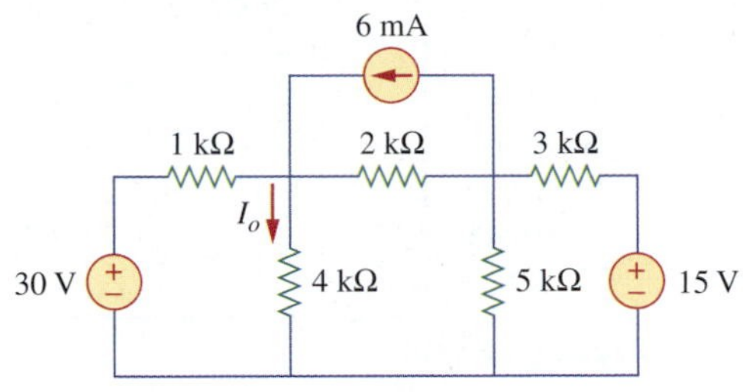

Figure 8.89
For Problem 8.9.

8.10 Apply the superposition principle to find V_o in the circuit in Fig. 8.90.

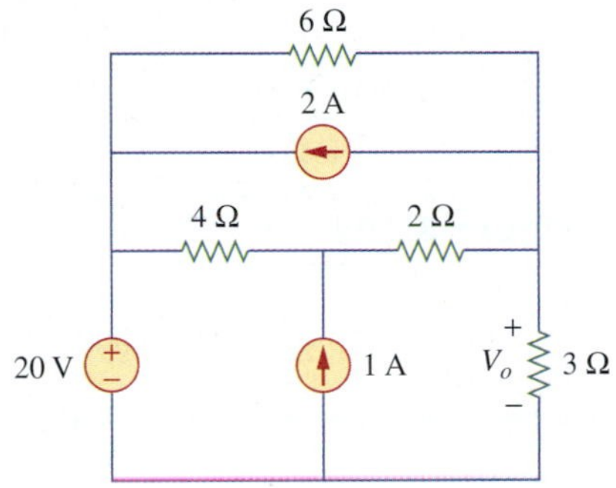

Figure 8.90
For Problem 8.10.

8.11 For the circuit in Fig. 8.91, use superposition to find I. Calculate the power delivered to the 3-Ω resistor.

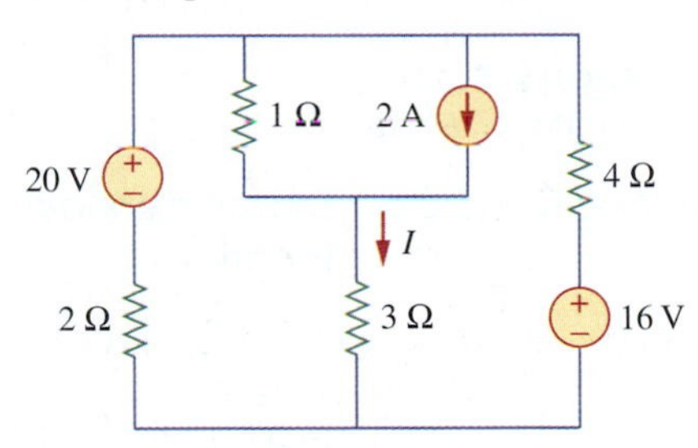

Figure 8.91
For Problem 8.11.

8.12 Given the circuit in Fig. 8.92, use superposition to get I_o.

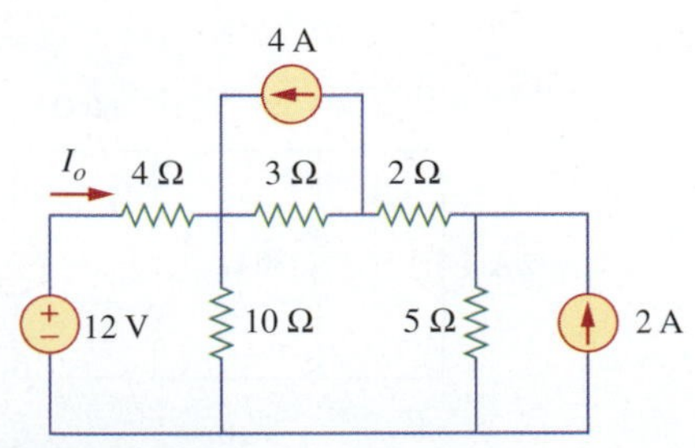

Figure 8.92
For Problem 8.12.

8.13 Apply the superposition principle to find V_o in the circuit of Fig. 8.93.

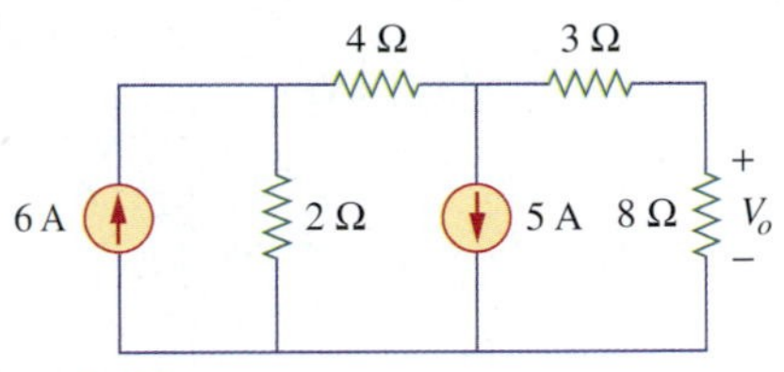

Figure 8.93
For Problem 8.13.

8.14 Use the superposition theorem to find V_o in the circuit of Fig. 8.94.

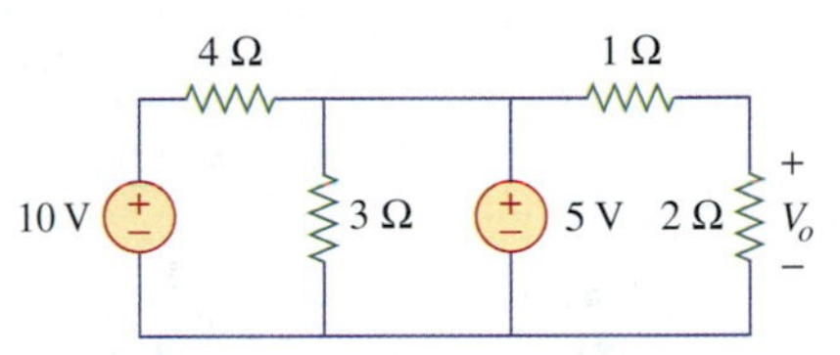

Figure 8.94
For Problem 8.14.

8.15 Refer to the circuit in Fig. 8.95. Use the superposition theorem to find V_o.

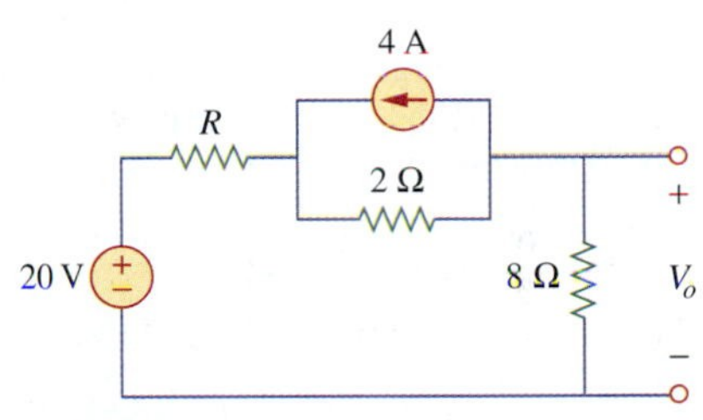

Figure 8.95
For Problem 8.15.

8.16 Obtain V_o in the circuit of Fig. 8.96 using the superposition principle.

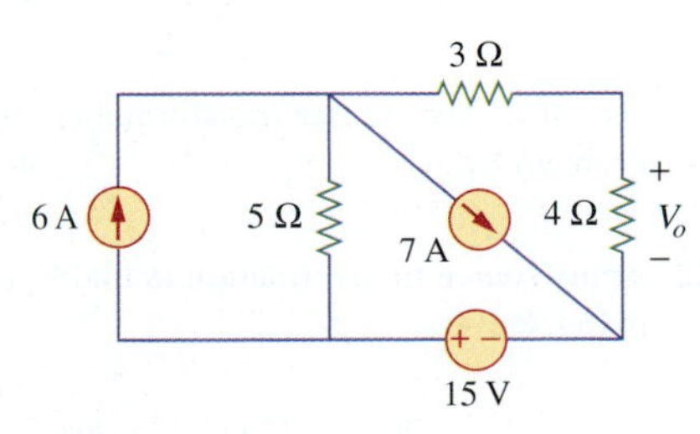

Figure 8.96
For Problem 8.16.

Section 8.4 Source Transformations

8.17 Find I in Problem 8.7 using source transformation.

8.18 Apply source transformation to determine V_o and I_o in the circuit in Fig. 8.97.

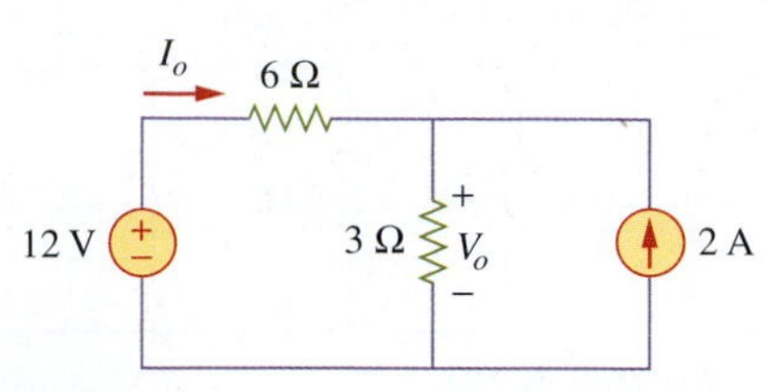

Figure 8.97
For Problem 8.18.

8.19 For the circuit in Fig. 8.98, use source transformation to find I.

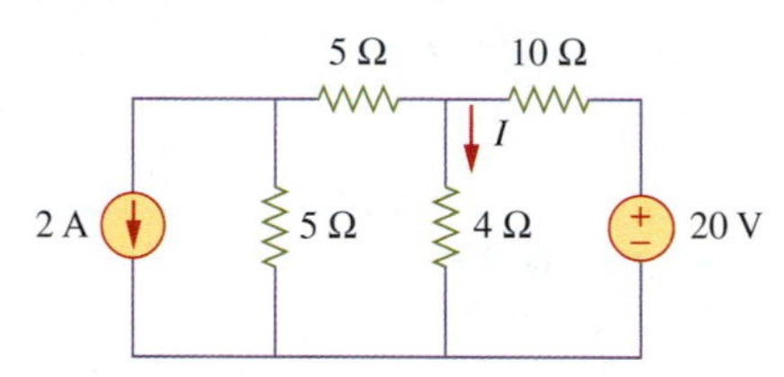

Figure 8.98
For Problem 8.19.

8.20 Referring to Fig. 8.99, use source transformation to determine the current and power in the 8-Ω resistor.

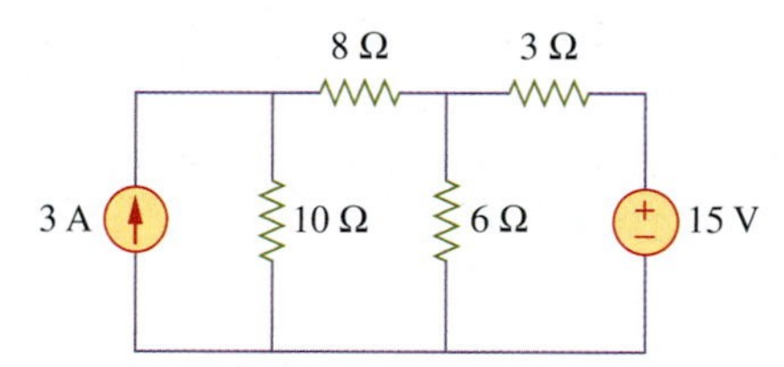

Figure 8.99
For Problem 8.20.

8.21 Use successive source transformations to find V_o in Problem 8.8.

8.22 Apply source transformation to find V_x in the circuit in Fig. 8.100.

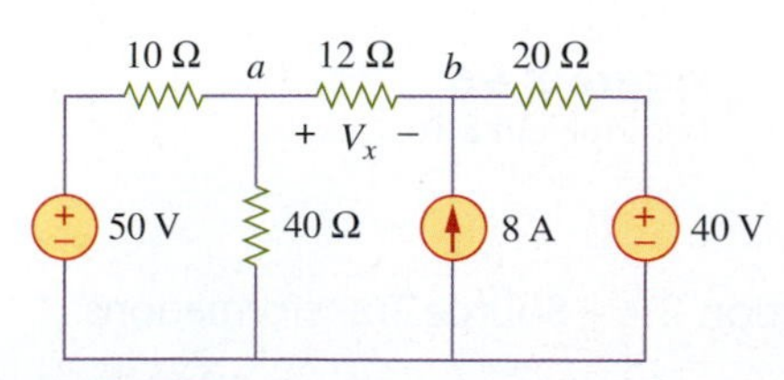

Figure 8.100
For Problems 8.22 and 8.31.

8.23 Use source transformation to find v_o in the circuit of Fig. 8.101.

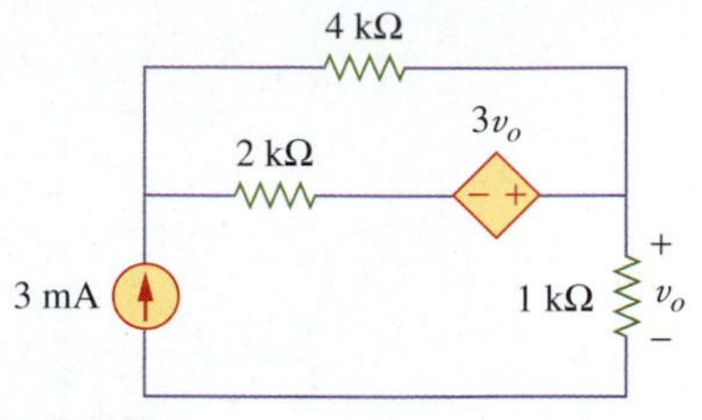

Figure 8.101
For Problem 8.23.

8.24 Use source transformation on the circuit shown in Fig. 8.102 to find i_x.

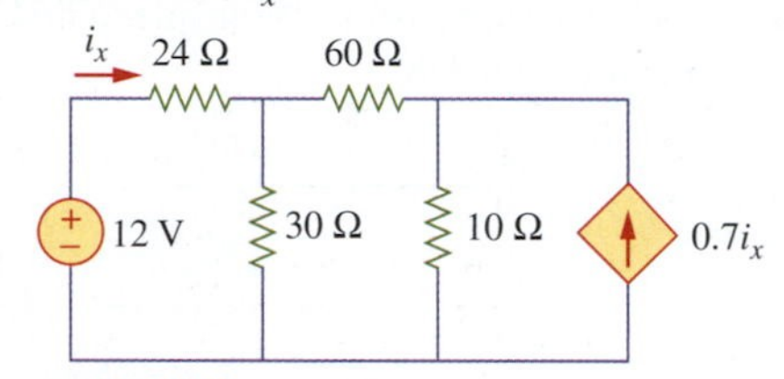

Figure 8.102
For Problem 8.24.

8.25 Use source transformation to find i_x in the circuit of Fig. 8.103.

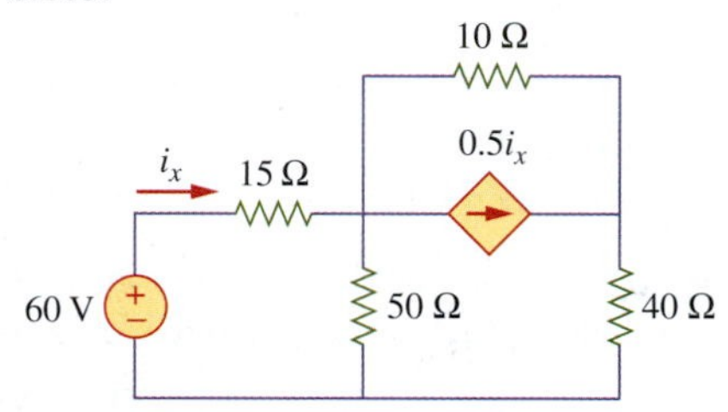

Figure 8.103
For Problem 8.25.

Sections 8.5 and 8.6 Thevenin's and Norton's Theorems

8.26 Determine R_{Th} and V_{Th} at terminals 1-2 of each of the circuits in Fig. 8.104.

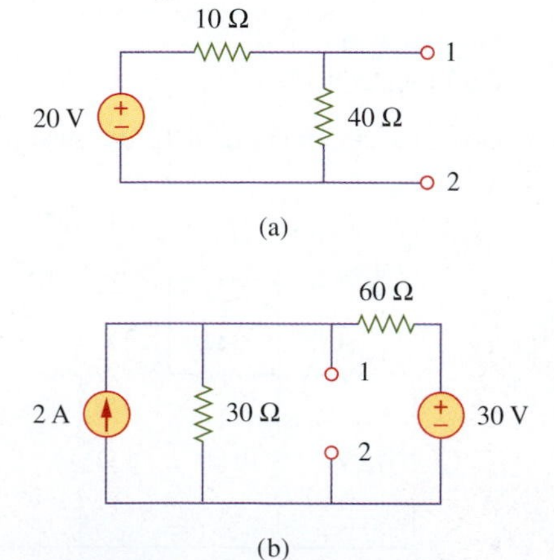

Figure 8.104
For Problem 8.26.

8.27 Find the Thevenin equivalent at terminals *a-b* of the circuit in Fig. 8.105.

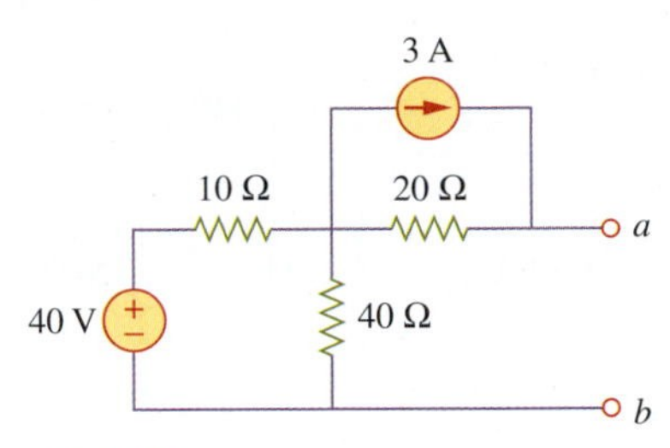

Figure 8.105
For Problems 8.27 and 8.36.

8.28 Use Thevenin's theorem to find V_o in Problem 8.8.

8.29 Solve for the current *I* in the circuit in Fig. 8.106 using Thevenin's theorem. (Hint: Find the Thevenin equivalent across the 12-Ω resistor.)

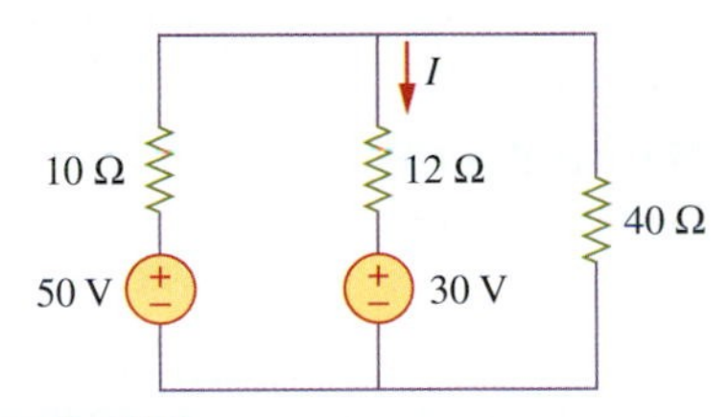

Figure 8.106
For Problem 8.29.

8.30 Apply Thevenin's theorem to find V_o in the circuit of Fig. 8.107.

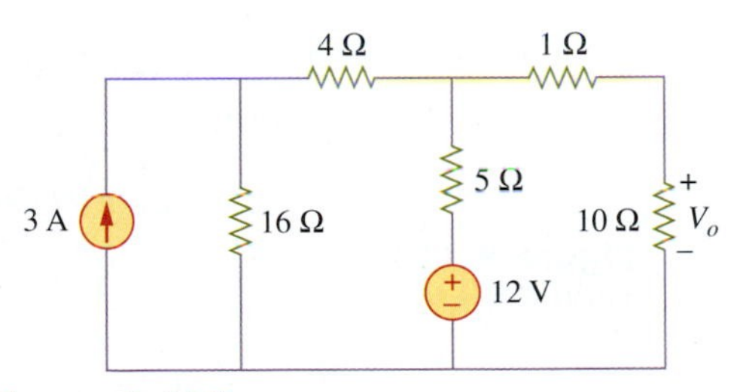

Figure 8.107
For Problem 8.30.

8.31 Given the circuit in Fig. 8.100, obtain the Thevenin equivalent at terminals *a-b* and use the result to find V_x.

8.32 Find the Thevenin and Norton equivalents at terminals *a-b* of the circuit in Fig. 8.108.

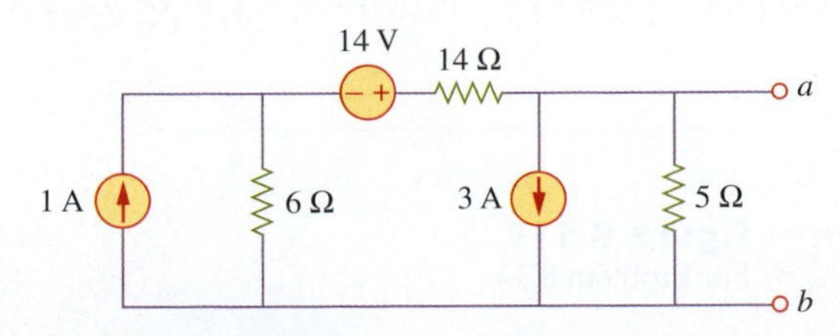

Figure 8.108
For Problem 8.32.

8.33 Find the Thevenin equivalent looking into terminals *a-b* of the circuit in Fig. 8.109, and solve for I_x.

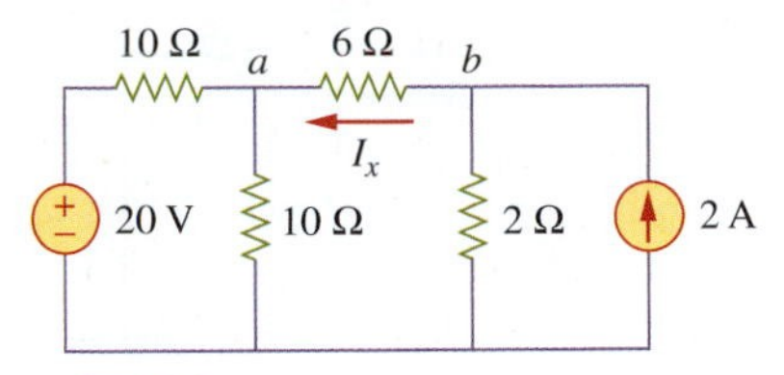

Figure 8.109
For Problem 8.33.

8.34 For the circuit in Fig. 8.110, obtain the Thevenin equivalent as seen from terminals: (a) *a-b* and (b) *b-c*.

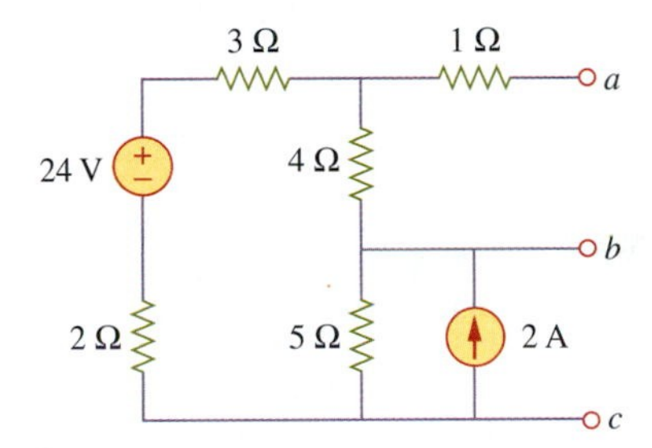

Figure 8.110
For Problem 8.34.

8.35 Find the Norton equivalent of the circuit in Fig. 8.111.

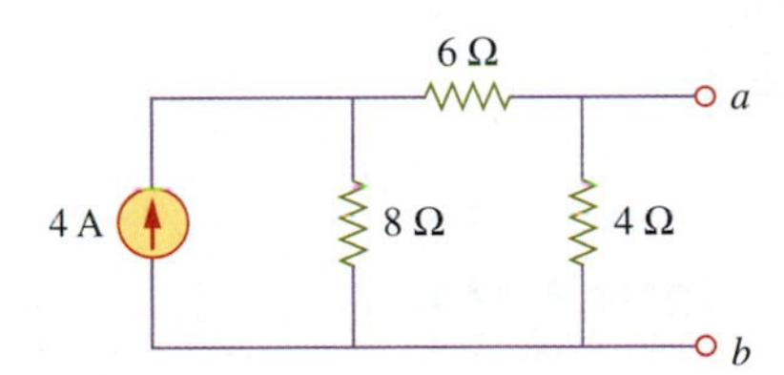

Figure 8.111
For Problems 8. 35 and 8.74.

8.36 Find the Norton equivalent looking into terminals *a-b* of the circuit in Fig. 8.105.

8.37 Obtain the Norton equivalent of the circuit in Fig. 8.112 to the left of terminals *a-b*. Use the result to find current *I*.

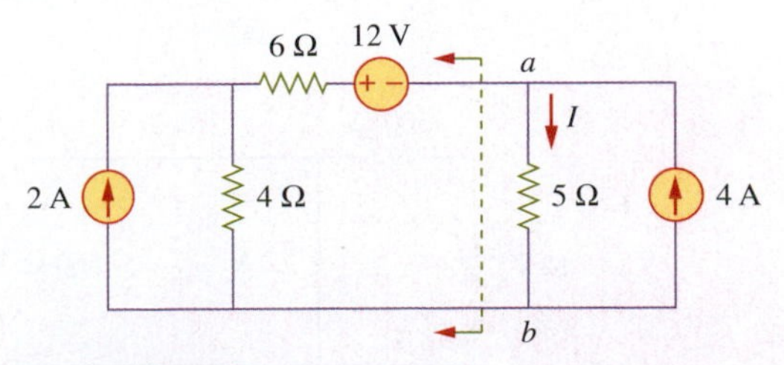

Figure 8.112
For Problem 8.37.

8.38 Given the circuit in Fig. 8.113, obtain the Norton equivalent as viewed from terminals (a) *a-b* and (b) *c-d*.

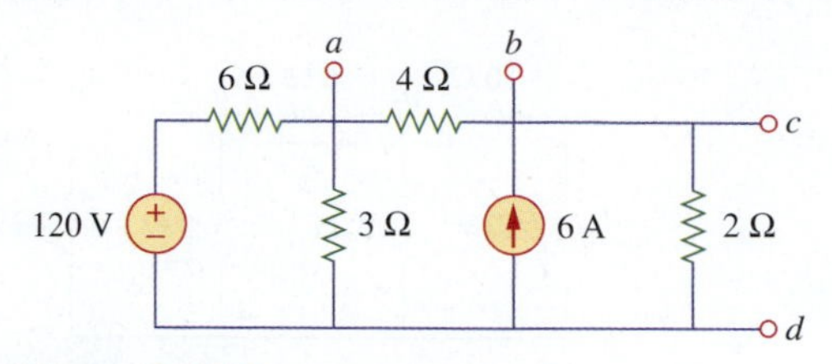

Figure 8.113
For Problem 8.38.

8.39 Determine the Thevenin and Norton equivalents at terminals *a-b* of the circuit in Fig. 8.114.

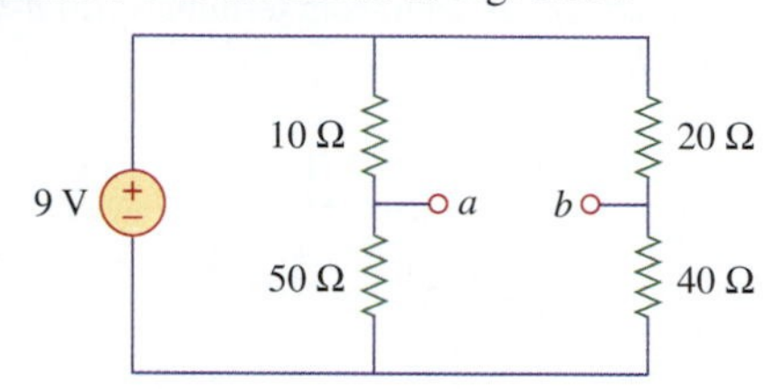

Figure 8.114
For Problem 8.39.

***8.40** Obtain the Thevenin and Norton equivalent circuits at terminals *a-b* of the circuit in Fig. 8.115.

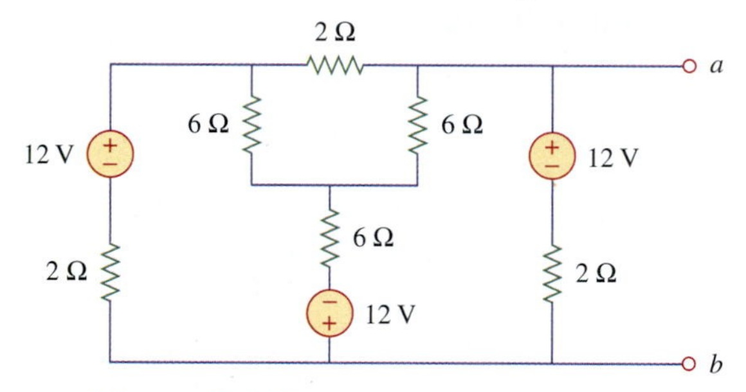

Figure 8.115
For Problem 8.40.

8.41 Find the Thevenin and Norton equivalent circuits at terminals *a-b* of the circuits in Fig. 8.116.

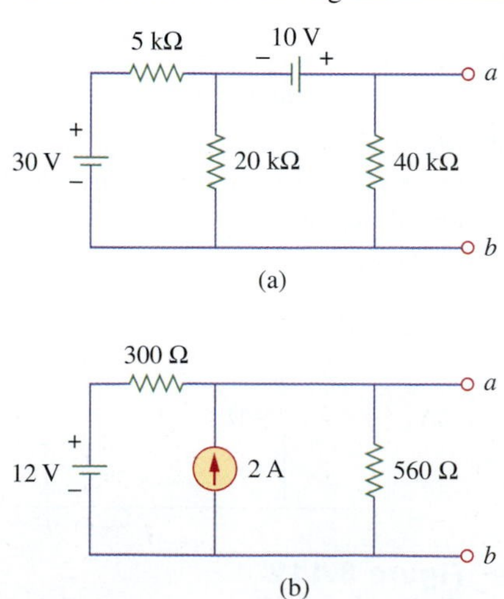

Figure 8.116
For Problem 8.41.

8.42 Obtain the Thevenin and Norton equivalent circuits at terminals *a-b* of the circuits in Fig. 8.117.

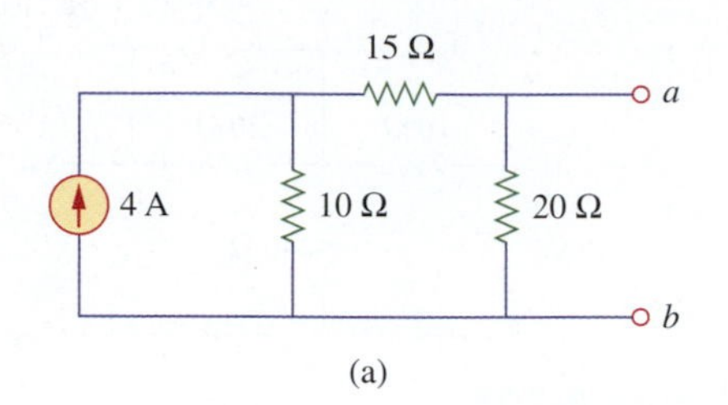

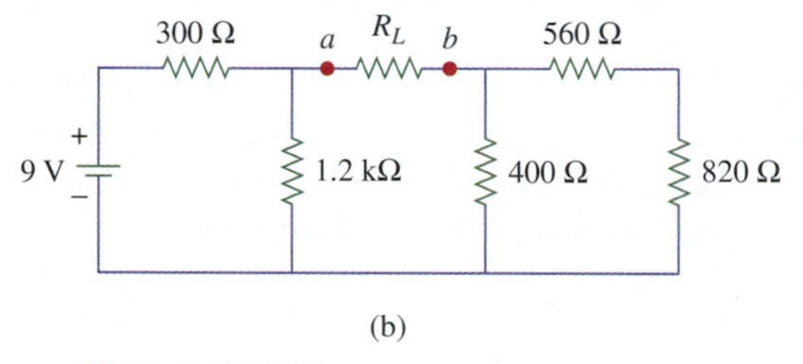

Figure 8.117
For Problem 8.42.

8.43 Using the Thevenin equivalent for the circuit shown in Fig. 8.118, calculate the range of V_L.

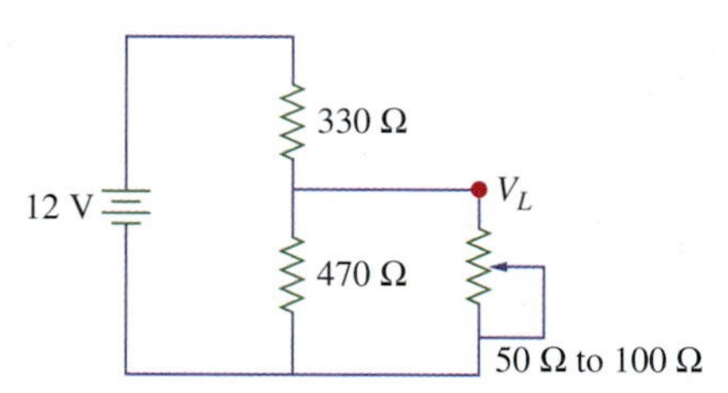

Figure 8.118
For Problem 8.43.

8.44 Find the Thevenin equivalent at terminals *a-b* of the circuit in Fig. 8.119. Use it to find V_o.

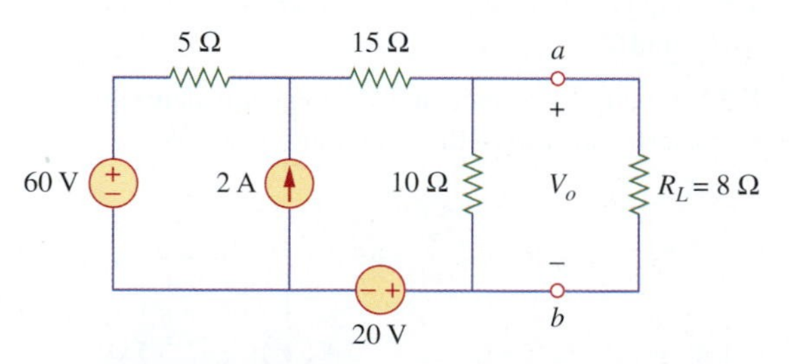

Figure 8.119
For Problem 8.44.

8.45 Determine the Thevenin and Norton equivalent circuits at terminals *a-b* of the circuit in Fig. 8.120.

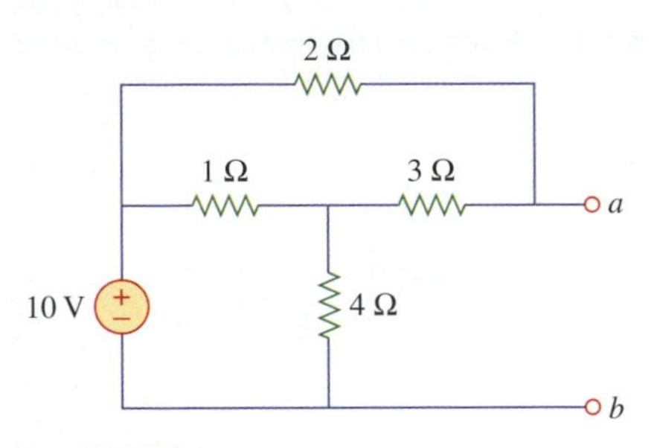

Figure 8.120
For Problem 8.45.

8.46 Determine the Norton equivalent at terminals *a-b* for the circuit in Fig. 8.121.

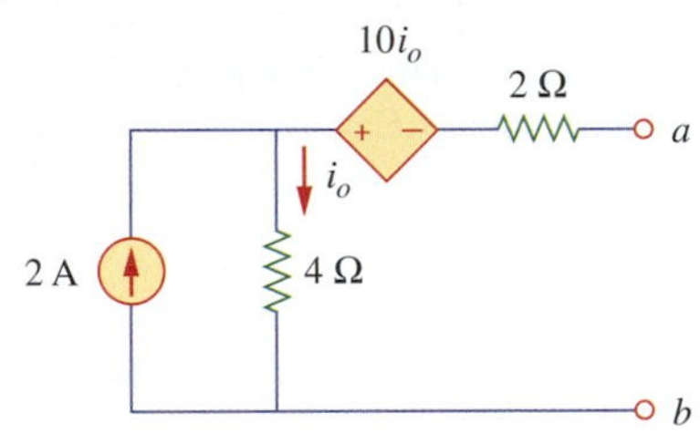

Figure 8.121
For Problem 8.46.

8.47 For the transistor model in Fig. 8.122, obtain the Thevenin equivalent at terminals *a-b*.

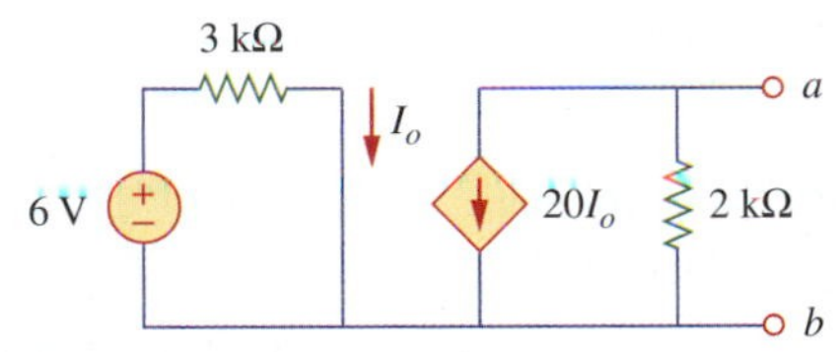

Figure 8.122
For Problem 8.47.

8.48 Find the Thevenin equivalent between terminals *a-b* of the circuit in Fig. 8.123.

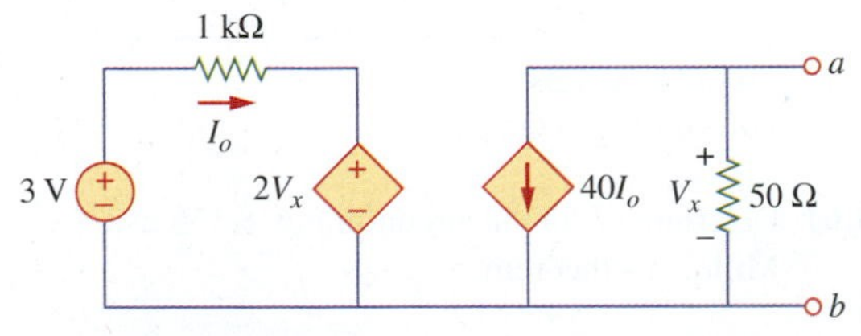

Figure 8.123
For Problem 8.48.

8.49 Obtain the Thevenin and Norton equivalent circuits at the terminals *a-b* for the circuit in Fig. 8.124.

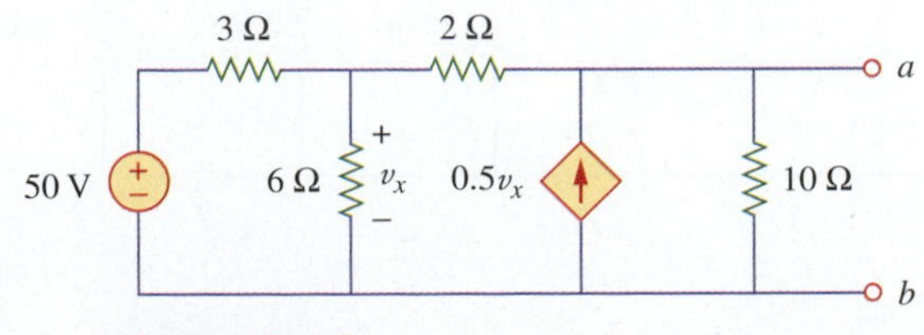

Figure 8.124
For Problem 8.49.

Section 8.7 Maximum Power Transfer Theorem

8.50 A network is reduced to $V_{eq} = 30$ V and $R_{eq} = 2$ kΩ. Calculate the maximum power the network can deliver.

8.51 Find the maximum power that can be delivered to the resistor *R* in the circuit in Fig. 8.125.

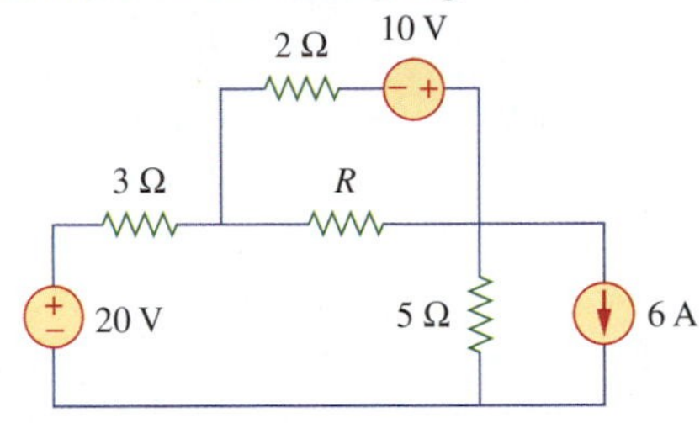

Figure 8.125
For Problem 8.51.

8.52 Refer to Fig. 8.126. For what value of *R* is the power dissipated in *R* maximum. Calculate that power.

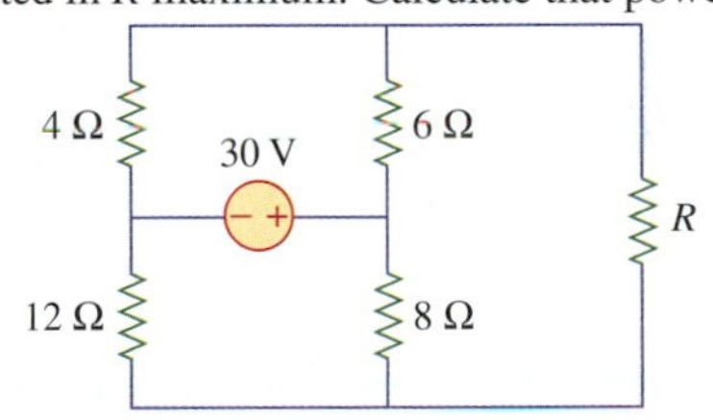

Figure 8.126
For Problem 8.52.

8.53 (a) For the circuit in Fig. 8.127, obtain the Thevenin equivalent at terminals *a-b*. (b) Calculate the current in $R_L = 8$ Ω. (c) Find R_L for maximum power deliverable to R_L. (d) Determine that maximum power.

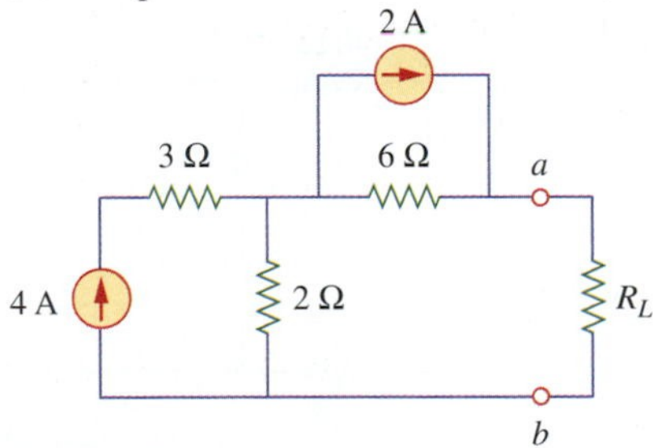

Figure 8.127
For Problem 8.53.

8.54 For the bridge circuit shown in Fig. 8.128, determine the maximum power that can be delivered to the variable resistor R_L.

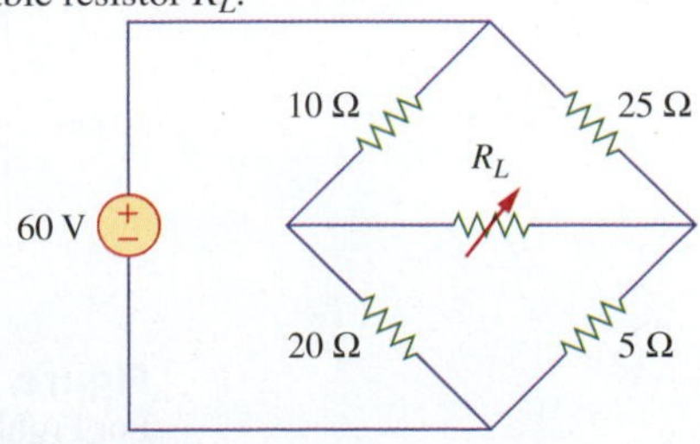

Figure 8.128
For Problem 8.54.

Section 8.8 Millman's Theorem

8.55 Find V_o in the circuit in Fig. 8.129 using Millman's theorem.

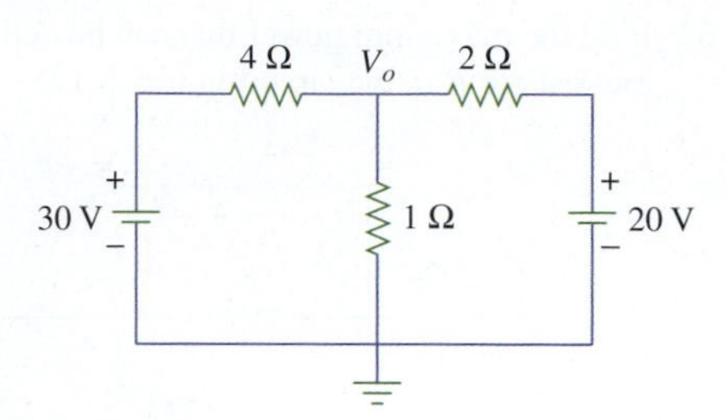

Figure 8.129
For Problem 8.55.

8.56 Apply Millman's theorem to find I_x in the circuit in Fig. 8.130.

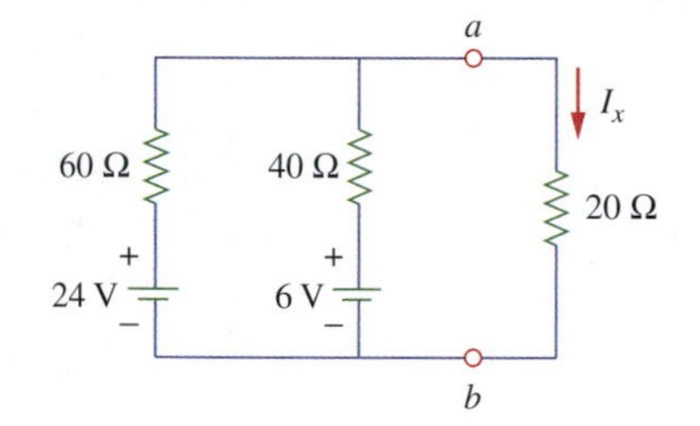

Figure 8.130
For Problem 8.56.

8.57 Use Millman's theorem to find V_o in the circuit in Fig. 8.131.

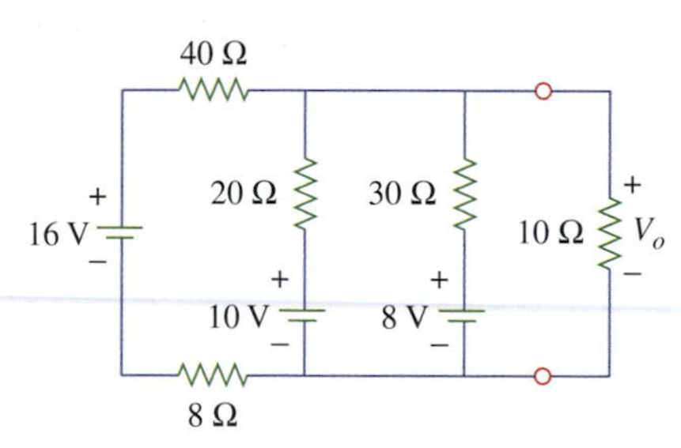

Figure 8.131
For Problem 8.57.

8.58 Use Millman's theorem to find the current I_o in the circuit in Fig. 8.132.

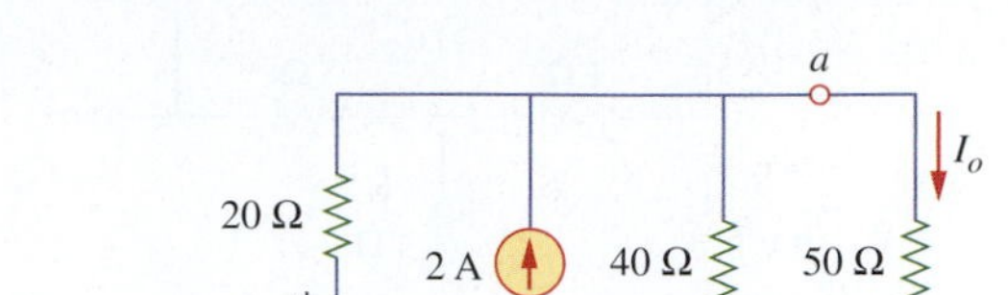

Figure 8.132
For Problem 8.58.

8.59 Apply Millman's theorem to reduce to the voltage and current sources in Fig. 8.133 to a single voltage source. Calculate the load current when $R_L = 100\ \Omega$.

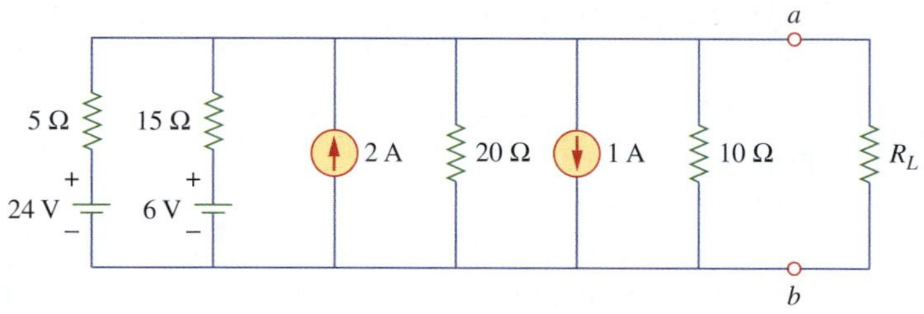

Figure 8.133
For Problem 8.59.

8.60 Using Millman's theorem, find V and I in the circuit of Fig. 8.134.

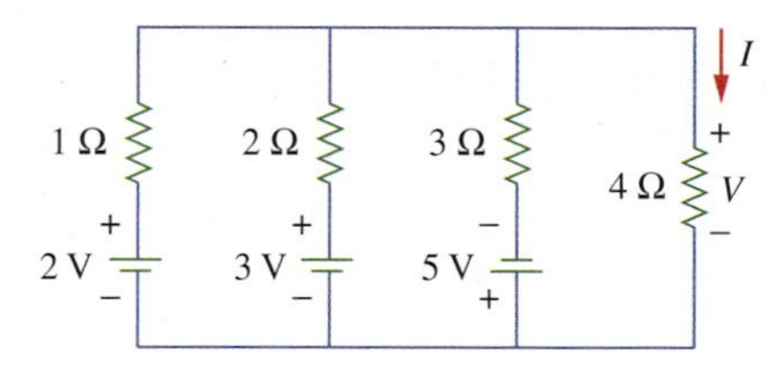

Figure 8.134
For Problem 8.60.

8.61 Determine I_x in the circuit in Fig. 8.135 using Millman's theorem.

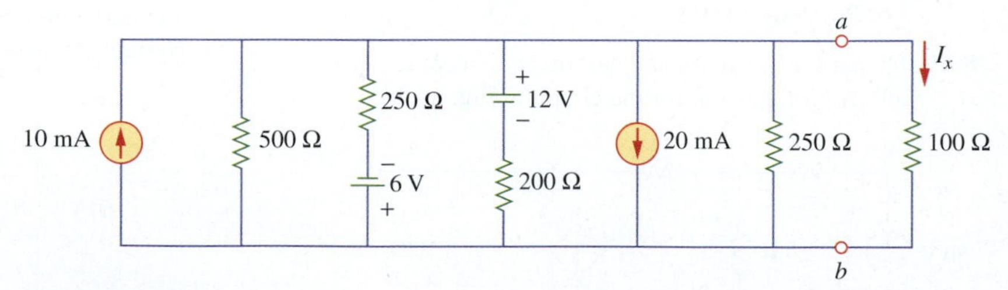

Figure 8.135
For Problem 8.61.

Section 8.9 Substitution Theorem

8.62 Refer to the circuit in Fig. 8.136. If the 10-Ω resistor is to be replaced with a voltage source and a 20-Ω series resistor, determine the magnitude of the voltage source.

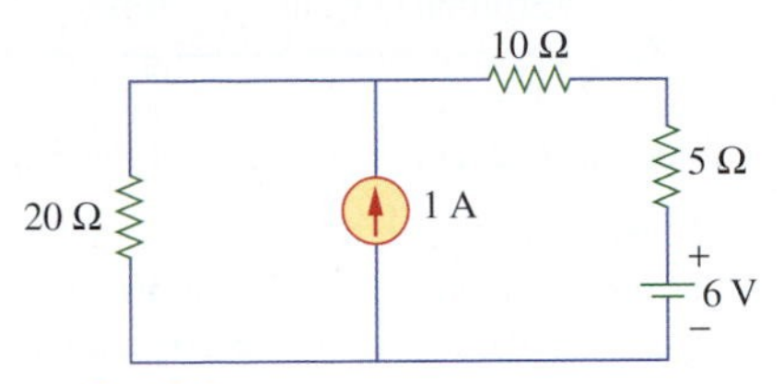

Figure 8.136
For Problems 8.62 and 8.63.

8.63 If the 10-Ω resistor in Fig. 8.136 is to be replaced with a current source and a 20-Ω shunt resistor, determine the magnitude of the current source.

8.64 Use the substitution theorem to draw two equivalent branches for branch *a-b* of the circuit in Fig. 8.137.

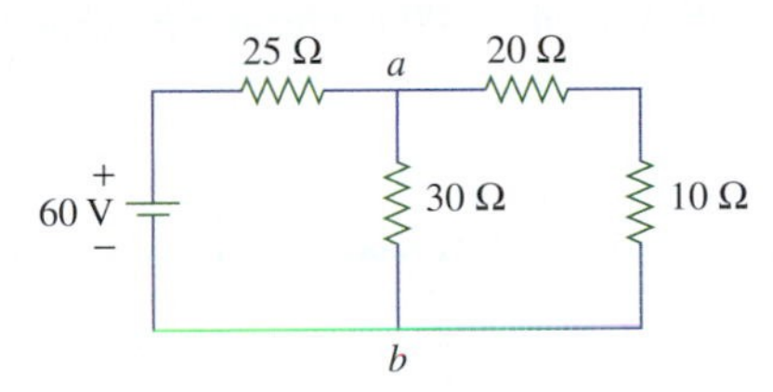

Figure 8.137
For Problem 8.64.

Section 8.10 Reciprocity Theorem

8.65 Determine the current I_x in the circuit in Fig. 8.138. Show that the reciprocity theorem is satisfied.

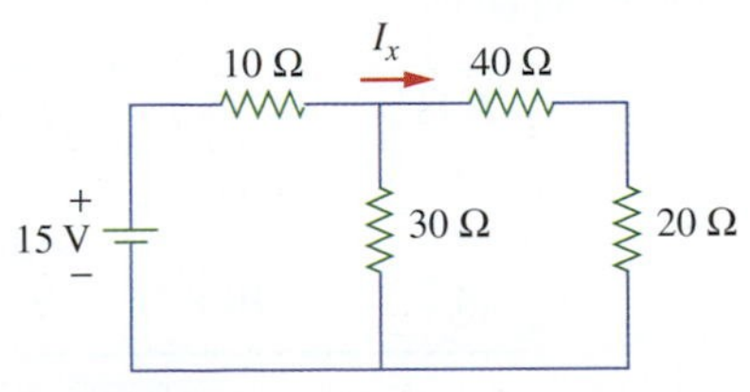

Figure 8.138
For Problem 8.65.

8.66 Find voltage V_o in the circuit in Fig. 8.139. Show that reciprocity applies to the circuit.

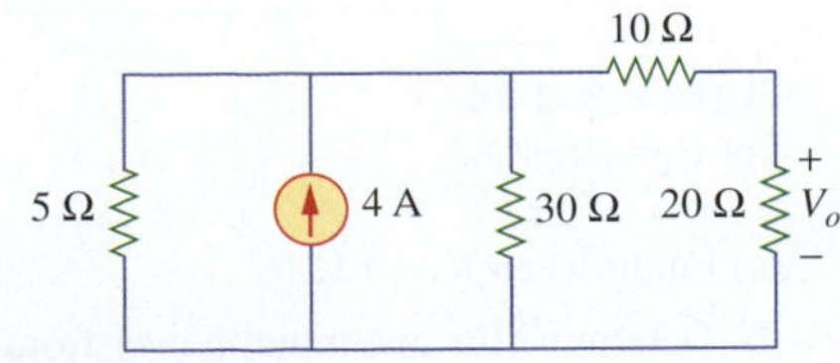

Figure 8.139
For Problem 8.66.

8.67 (a) For the circuit in Fig. 8.140(a), determine the current *I*.

(b) Repeat part (a) for the circuit in Fig. 8.140(b).

(c) Is the reciprocity theorem satisfied?

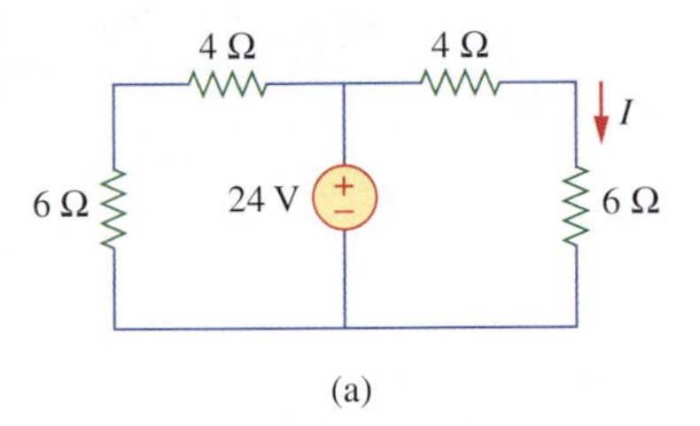

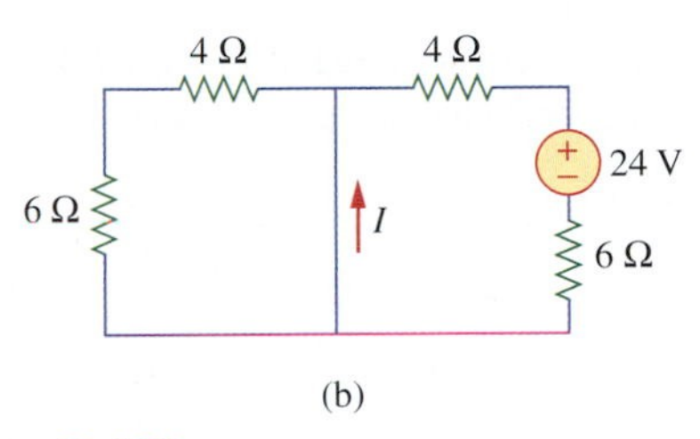

Figure 8.140
For Problem 8.67.

8.68 (a) Find *V* in the circuit in Fig. 8.141(a).

(b) Determine *V* in the circuit in Fig. 8.141(b).

(c) Is the reciprocity theorem satisfied?

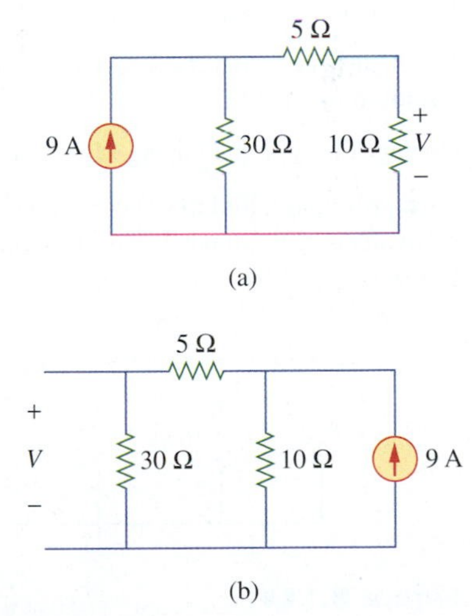

Figure 8.141
For Problems 8.68 and 8.71.

Section 8.11 Verifying Circuit Theorems with Computers

8.69 Solve Problem 8.39 using PSpice.

8.70 Use PSpice to solve 8.61

8.71 Use PSpice to solve Problem 8.68.

8.72 Obtain the Thevenin equivalent of the circuit in Fig. 8.142 using PSpice.

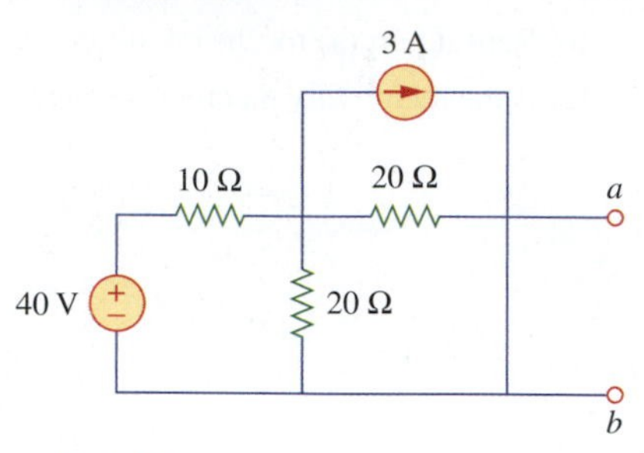

Figure 8.142
For Problems 8.72 and 8.75.

8.73 For the circuit in Fig. 8.143, use PSpice to find the Thevenin equivalent at terminals *a*-*b*.

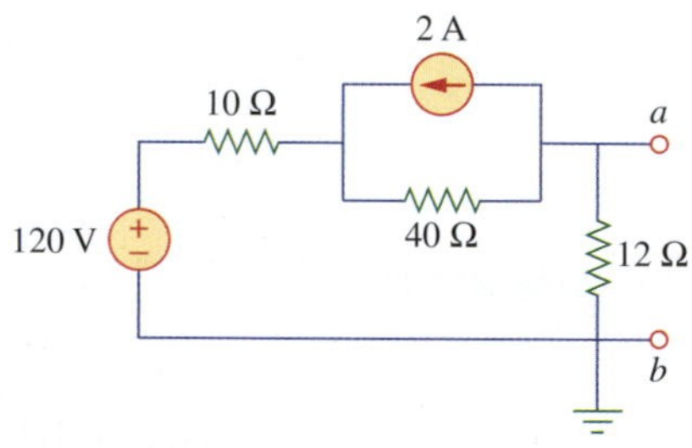

Figure 8.143
For Problems 8.73 and 8.76.

8.74 Find the Norton equivalent of the circuit in Fig. 8.111 (Problem 8.35) using Multisim.

8.75 Use Multisim to find the Thevenin equivalent of the circuit in Fig. 8.142.

8.76 Rework Problem 8.73 using Multisim.

8.77 Using Multisim, find the Thevenin and Norton equivalents at terminals *a*-*b* of the circuit in Fig. 8.144.

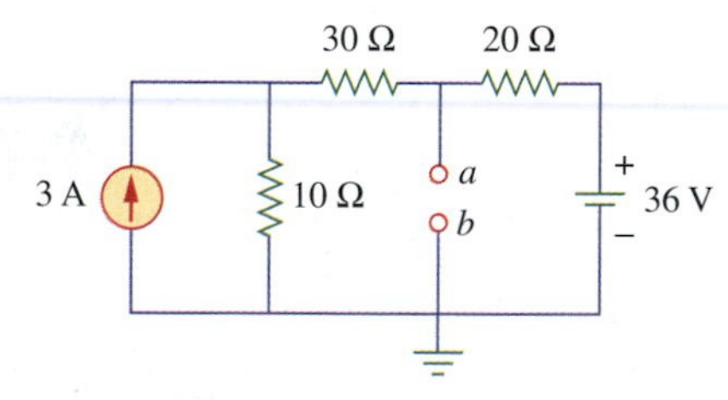

Figure 8.144
For Problem 8.77.

Section 8.12 Applications: Source Modeling

8.78 A battery has a short-circuit current of 20 A and an open-circuit voltage of 12 V. If the battery is connected to an electric bulb with resistance of 2 Ω, calculate the power dissipated by the bulb.

8.79 The following results were obtained from measurements taken between the two terminals of a resistive network.

Terminal voltage	12 V	0 V
Terminal current	0 A	1.5 A

Find the Thevenin equivalent of the network.

8.80 When connected to a 4-Ω resistor, a battery has a terminal voltage of 10.8 V but produces 12 V on an open circuit. Determine the Thevenin equivalent circuit for the battery.

8.81 The Thevenin equivalent at terminals *a*-*b* of the linear network shown in Fig. 8.145 is to be determined by measurement. When a 10-kΩ resistor is connected to terminals *a*-*b*, the voltage V_{ab} is measured as 6 V. When a 30-kΩ resistor is connected to the terminals, V_{ab} is measured as 12 V. Determine: (a) the Thevenin equivalent at terminals *a*-*b* and (b) V_{ab} when a 20-kΩ resistor is connected to terminals *a*-*b*.

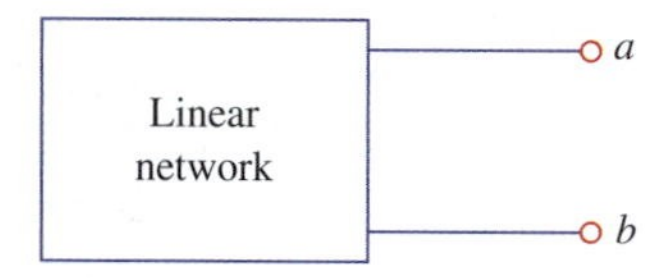

Figure 8.145
For Problem 8.81.

8.82 A black box with a circuit in it is connected to a variable resistor. An ideal ammeter (with zero resistance) and an ideal voltmeter (with infinite resistance) are used to measure current and voltage as shown in Fig. 8.146. The results are shown in the following table.

R **(Ω)**	***V*** **(V)**	***I*****(A)**
2	3	1.5
8	8	1.0
14	10.5	0.75

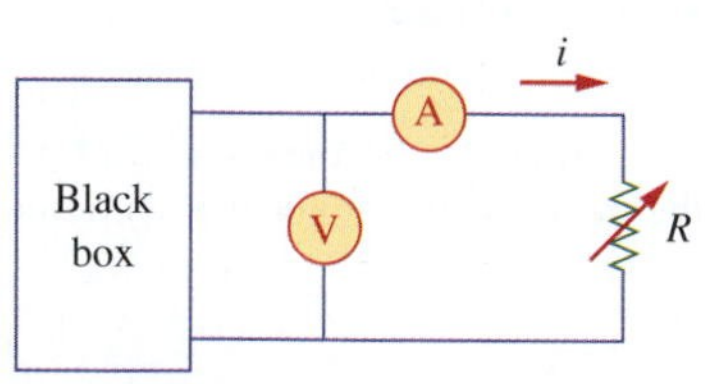

Figure 8.146
For Problem 8.82.

(a) Find I when $R = 4$ Ω.

(b) Determine the maximum power from the box.

chapter

9

Capacitance

We look forward to a world founded upon four essential human freedoms. The first is freedom of speech and expression—everywhere in the world. The second is freedom of every person to worship God in his own way—everywhere in the world. The third is freedom from want—everywhere in the world. The fourth is freedom from fear—anywhere in the world.

—Franklin D. Roosevelt

Historical Profiles

Michael Faraday (1791–1867), an English chemist and physicist, was probably the greatest experimentalist who ever lived.

Born near London, Faraday realized his boyhood dream by working with the great chemist Sir Humphry Davy at the Royal Institution, where he worked for 54 years. He made several contributions in all areas of physical science and coined such words as *electrolysis, anode,* and *cathode*. His discovery of electromagnetic induction in 1831 was a major breakthrough in engineering because it provided a way of generating electricity. The electric motor and generator operate on this principle. The unit of capacitance, the farad, was chosen in his honor.

Michael Faraday
© The Huntington Library, Burndy Library, San Marino, California

Benjamin Franklin (1706–1790), an American inventor, scientist, philosopher, economist, and statesman, was one of the most extraordinary human beings the world has ever known.

Although he was born in Boston, the city of Philadelphia is remembered as the home of Franklin. In Philadelphia, you can find the Benjamin Franklin National Memorial. A list of Benjamin Franklin's inventions reveals a man of many talents and interests. It was the scientist in him that brought out the inventor. Among other things, he invented the lightning rod, which protected buildings and ships from lightning damage.

Franklin had a simple formula for success. He believed that successful people worked just a little harder than other people. He helped found a new nation and defined the American character. Franklin was one of the founding fathers of the United States of America.

Benjamin Franklin
Library of Congress Prints and Photographs Division (LC-USZ62-25564)

9.1 Introduction

So far we have limited our study to resistive circuits. In this chapter and the next, we introduce two new and important passive linear circuit elements: the capacitor and the inductor. Unlike resistors, which dissipate energy, *capacitors* and *inductors* do not dissipate energy but store energy that can be retrieved at a later time. For this reason, capacitors and inductors are called *storage elements*.[1]

The application of resistive circuits is quite limited, although resistive circuits are prevalent in digital circuits. For example, they are used in the tuning circuit of a radio receiver and as dynamic memory elements in computer systems. With the introduction of capacitors in this chapter and inductors in the next, we will be able to analyze more important and practical circuits. It is reassuring to note that some of the circuit analysis techniques covered in Chapters 7 and 8 are equally applicable to circuits with capacitors and inductors.

Capacitors may be the unsung heroes in the electronic components world. A capacitor is one of the most widely used electronic components due to its characteristics of opposing a change in voltage, of blocking dc current, and of storing electrical charge or energy. For example, the proliferation of cell phones over recent years has brought about not only convenient communication and obnoxious ring tones, but also capacitor innovation.

We begin the chapter by introducing capacitors and describing how they store energy in their electric fields. We consider different types of capacitors that are available commercially. We examine how capacitors combine in series or parallel. We analyze *RC* circuits (circuits containing resistance *R* and capacitance *C*) and discuss how computers can be used to model them. We finally discuss two applications of capacitors in real life.

9.2 Capacitors

Besides resistors, capacitors are the most common electrical components. A capacitor is a passive element designed to store energy in its electric field. Capacitors find extensive usage in electronics, communications, computers, and power systems. For example, they are used in the tuning circuit of a radio receiver and as dynamic memory elements in computer systems. A capacitor is typically constructed as depicted in Fig. 9.1.

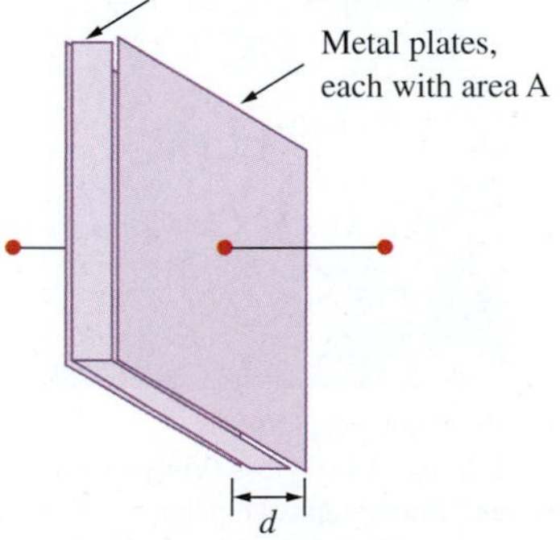

> A **capacitor** consists of two conducting plates separated by an insulator (or dielectric).

In many practical applications, the plates may be aluminum foil while the dielectric material may be glass, mica, ceramic, paper, or plastics such as polyethylene or polycarbonate. Even air can be used as the dielectric. Capacitor types are named after the dielectric. Thus, we have ceramic, mica, polyester, paper, and air capacitors. The dielectric prevents the plates from touching each other. Usually a capacitor has more than two plates.

Dielectric with permittivity ϵ

Metal plates, each with area A

d

Figure 9.1
A typical capacitor.

[1] Note: As opposed to a resistor, which absorbs or dissipates energy irreversibly, an inductor or capacitor stores or releases energy; that is, it has a memory.

When a voltage source V is connected to the capacitor, as shown in Fig. 9.2, the source deposits a positive charge $+Q$ on one plate and a negative charge $-Q$ on the other so that the capacitor is neutral. The capacitor is said to store the electric charge. The amount of charge stored, represented by Q, is directly proportional to the applied voltage V so that

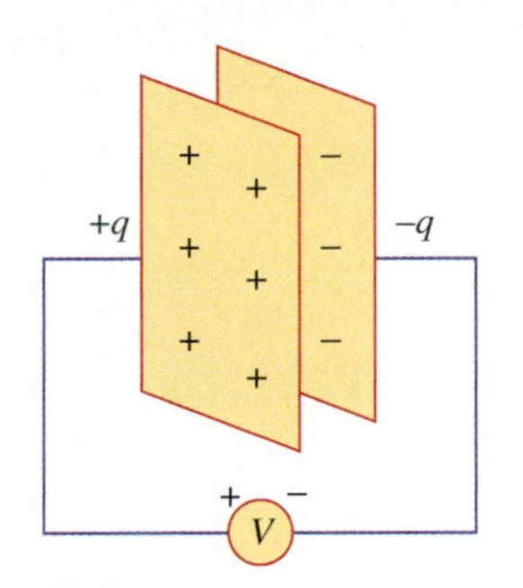

Figure 9.2
A capacitor with applied voltage V.

$$\boxed{Q = CV} \tag{9.1}$$

where C, the constant of proportionality, is known as the *capacitance* of the capacitor. (C here should not be confused with Coulomb, the unit of charge.) The capacitance is the ability of the capacitor to store energy in its field. The unit of capacitance is the farad (F), in honor of the English physicist Michael Faraday (1791–1867). However 1 F is very large and most capacitors are only a fraction of a farad. Thus, prefixes (multipliers) are used to show the smaller values. For example, 1 μF is one-millionth of 1 F. From Eq. (9.1),

$$1\ C = 1\text{F} - V \tag{9.2}$$

By rearranging the terms of Eq. (9.1), we can rewrite the formula as

$$C = \frac{Q}{V} \tag{9.3a}$$

$$V = \frac{Q}{C} \tag{9.3b}$$

From Eq. (9.3a), we may derive the following definition:[2]

> **Capacitance** is the ratio of the charge on one plate of a capacitor to the voltage difference between the two plates, measured in farads (F).

Although the capacitance C of a capacitor is the ratio of the charge Q per plate to the applied voltage V, it does not depend on Q or V. It depends on the physical dimensions of the capacitor. This will be made clear in the next section.

We can find the energy stored in the electrostatic field of a capacitor. We recall that

$$C = \frac{Q}{V}, \qquad Q = It, \qquad C = \frac{It}{V} \tag{9.4}$$

Hence,

$$I = \frac{CV}{t} \tag{9.5}$$

The electrical energy is

$$\begin{aligned} W &= (\text{Average voltage}) \times \text{Current} \times \text{Time} \\ &= \frac{(V - 0)}{2} It = \frac{1}{2} V \left(\frac{CV}{t}\right) t \end{aligned}$$

---

[2] Note: Alternatively, capacitance is the amount of charge stored per plate for a unit voltage difference in a capacitor.

or

$$W = \frac{1}{2}CV^2 \tag{9.6}$$

where V is in volts and C is in farads, so that W is in joules. We have applied the fact that the capacitor voltage rises from zero to its final value of V so that the average voltage is $V/2$. If we introduce $V = Q/C$, Eq. (9.6) may be written as

$$W = \frac{Q^2}{2C} \tag{9.7}$$

This is another way of expressing Eq. (9.6).

Example 9.1

(a) Calculate the charge stored on a 3-pF capacitor with 20 V across it.
(b) Find the energy stored in the capacitor.

Solution:
(a) Because $Q = CV$,

$$Q = 3 \times 10^{-12} \times 20 = 60 \text{ pC}$$

(b) The energy stored is

$$W = \frac{1}{2}CV^2 = \frac{1}{2} \times 3 \times 10^{-12} \times 400 = 600 \text{ pJ}$$

Practice Problem 9.1

What is the voltage across a 3.3-μF capacitor if the charge on one plate is 0.12 mC? How much energy is stored?

Answer: 36.36 V; 2.182 mJ

9.3 Electric Fields

A capacitor stores energy in its electric field, which is established by the opposite charges on its plates. The electric fields are the force fields that exist wherever there are charged bodies. For a parallel-plate capacitor, the electric field is represented by the lines of force as shown in Fig. 9.3. The *electric field strength, electric flux*, and *electric flux density* can all be calculated.

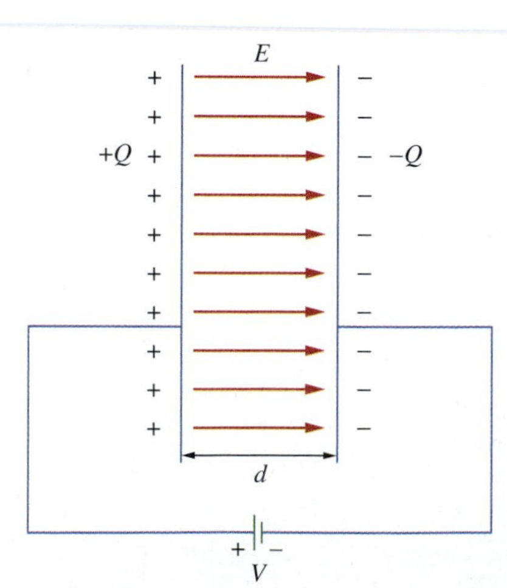

Figure 9.3
Electric field inside a capacitor.

As we saw in Chapter 1, electric charge is measured in coulombs. The electric flux is a measure of the number of electric field lines passing through an area. It is also measured in coulombs. Therefore, if a capacitor has Q coulombs of charge, the total electric flux Ψ between its plates is Q; that is,

$$\Psi = Q \tag{9.8}$$

The electric flux density D is the electric flux per unit area, i.e.

$$D = \frac{\Psi}{A} = \frac{Q}{A} \tag{9.9}$$

The electric field strength E (or electric field intensity) is the ratio between the applied voltage V and the distance d between the plates; that is,

$$E = \frac{V}{d} \tag{9.10}$$

Both the electric field strength E (in V/m) and the electric flux density D (in C/m^2) increase as the charge on the capacitor plates increases and they are related.

$$D = \varepsilon E \tag{9.11}$$

where ε is called the permittivity of the dielectric material. The permittivity specifies the ease with which electric flux can pass through the material.

For the parallel-plate capacitor (such as shown in Figs. 9.1 through 9.3), because $\Psi = DA$, $V = Ed$ and $D/E = \varepsilon$, it follows that

$$C = \frac{Q}{V} = \frac{\Psi}{V} = \frac{DA}{Ed} = \frac{\varepsilon A}{d} \tag{9.12}$$

or

$$\boxed{C = \frac{\varepsilon A}{d}} \tag{9.13}$$

where A is the surface area of each plate, d is the distance between the plates, and ε is the permittivity of the dielectric material between the plates. Notice from Eqs. (9.12) and (9.13) that the capacitance does not depend on Q and V but on their ratio and the physical dimensions of the capacitor. Although Eq. (9.13) applies to only parallel-plate capacitors, we may infer from it that, in general, three factors determine the value of the capacitance:

1. The surface area of the plates—the larger the area the greater the capacitance.
2. The spacing between the plates—the smaller the spacing the greater the capacitance.
3. The permittivity of the material—the higher the permittivity the greater the capacitance.

The permittivity of a dielectric material may be written as

$$\varepsilon = \varepsilon_o \varepsilon_r \tag{9.14}$$

where $\varepsilon_o = 8.85 \times 10^{-12}$ farads per meter (F/m) is the permittivity of a vacuum and $\varepsilon_r = \dfrac{\varepsilon}{\varepsilon_o}$ is the relative permittivity or dielectric constant of the material.

The dielectric constants of common materials are presented in Table 9.1. The dielectric constants listed in Table 9.1 are approximate; the dielectric constant varies widely for a given material. We should keep in mind that the dielectric constant of a material is dimensionless because it is a relative measure.

If the voltage V in Fig. 9.3 is increased beyond a certain value, the dielectric material separating the plates may break down. The electric

TABLE 9.1

Dielectric constants of common materials.

Material	Dielectric constant (ε_r)
Vacuum	1.0
Air	1.0006
Teflon	2.0
Paper (dry)	2.5
Polystyrene	2.5
Rubber	3.0
Oil (transformer)	4.0
Mica	5.0
Porcelain	6.0
Glass	7.5
Tantalum oxide	30
Water (distilled)	80
Ceramic	7,500

TABLE 9.2

Dielectric strengths of common materials.

Material	Dielectric strength (kV/cm)
Air	30
Ceramic	30
Porcelain	70
Paper	500
Teflon (plastic)	600
Glass	1,200
Mica	2,000

field intensity at breakdown is known as the *dielectric strength* of the material. The dielectric strength of a given material is the voltage per unit thickness at which the material may break down. The dielectric strengths of common materials are shown in Table 9.2. Again, the dielectric strengths listed in Table 9.2 are approximate; the dielectric strength varies widely for a given material. Typically, the breakdown voltage can be increased by increasing the separation distance between the plates (the thickness of the dielectric), but this trades off capacitance.

Example 9.2

(a) Calculate the capacitance of a parallel-plate capacitor with a plate area of 4 cm^2 and a plate separation of 0.3 cm. Assume that the insulator is air.
(b) Repeat your calculation in part (a) if the dielectric is ceramic.

Solution:

(a) For air, $\varepsilon_r \approx 1$ so that

$$C = \varepsilon_o \frac{A}{d} = 8.85 \times 10^{-12} \frac{4 \times 10^{-4}}{0.3 \times 10^{-2}} = 1.18 \text{ pF}$$

(b) For ceramic, $\varepsilon_r = 7{,}500$ so that

$$C = \varepsilon_o \varepsilon_r \frac{A}{d} = 8.85 \times 10^{-12} \times 7{,}500 \frac{4 \times 10^{-4}}{0.3 \times 10^{-2}}$$

$$= 8{,}850 \text{ pF} = 8.85 \text{ nF}$$

showing that the nature of dielectric can make a significant difference in the value of the capacitance.

Practice Problem 9.2

Determine the capacitance of a parallel-plate capacitor with a plate area of 0.02 m^2 and plate separation of 5 mm. Assume that the dielectric separating the plates is Teflon.

Answer: 70.8 pF

9.4 Types of Capacitors

Capacitors are commercially available in different values and types. Typically capacitors have values in the picofarad (pF) to microfarad (μF) range. They are described by the dielectric material they are made of and by whether they are a fixed or variable type. Figure 9.4 shows the circuit symbols for fixed and variable capacitors. Note that according to the passive sign convention, current is considered to flow into the positive terminal of the capacitor when the capacitor is being charged and out of the positive terminal when the capacitor is discharging.

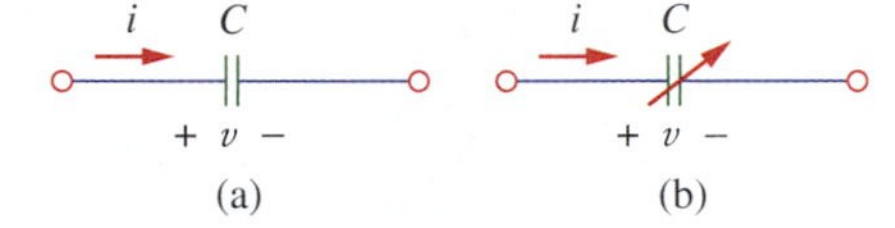

Figure 9.4
Circuit symbols for capacitors: (a) fixed capacitor; (b) variable capacitor.

Common types of fixed-value capacitors are shown in Fig. 9.5. Polyester capacitors are light in weight, stable, and their change with temperature is predictable. Instead of polyester, other dielectric materials such as mica, ceramic, and polystyrene may be used. As noticed from Table 9.1, ceramic provides a very high dielectric constant and high dielectric strength. As a result, a relatively high capacitance can be achieved in a small space. Film capacitors are rolled and housed in metal or plastic films. Electrolytic capacitors produce very high capacitance but have relative low breakdown voltages.

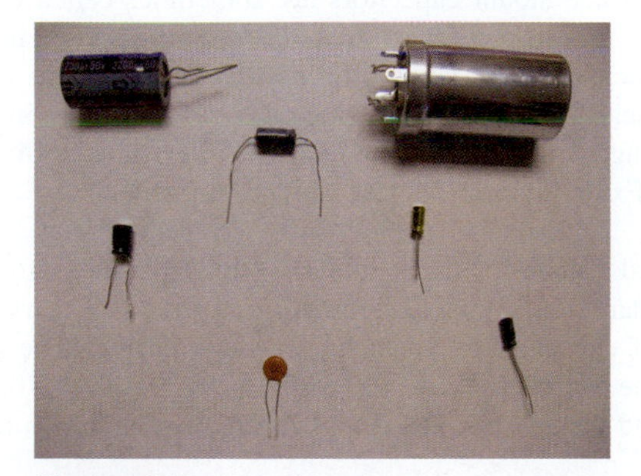

Figure 9.5
Different types of fixed capacitors.
© Sarhan M. Musa

Variable or adjustable capacitors are used in circuits that require adjustable capacitance. Examples of such circuits are found in radio receivers, TV tuners, or impedance matching in antenna tuners. Figure 9.6 shows the most common types of variable capacitors. The capacitance of a trimmer (or padder) capacitor or a glass piston capacitor is varied by turning the screw. The trimmer capacitor is often placed in parallel with another capacitor so that the equivalent capacitance can be varied slightly. The capacitance of the variable

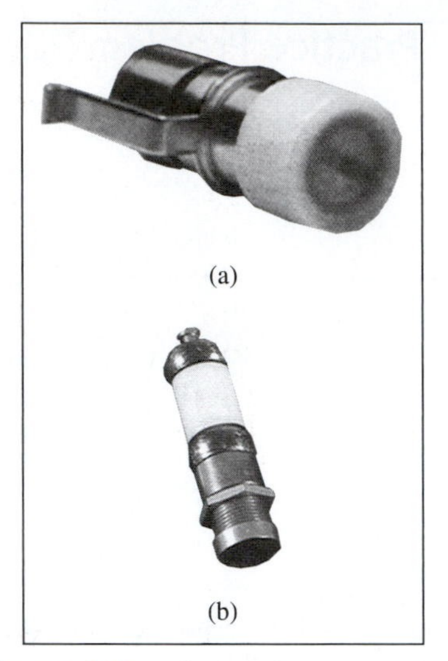

Figure 9.6
Variable capacitors: (a) trimmer capacitor, (b) filmtrim capacitor.
© Johanson Manufacturing Corporation

air capacitor (meshed plates) is varied by turning the shaft. Variable capacitors are used in radio receivers allowing one to tune to various stations.

Capacitance value and voltage rating are usually indicated on the body of the capacitor. Capacitance value tells how large an electrical charge the capacitor can store [see Eq. (9.1)]. Voltage rating indicates how much voltage the capacitor can withstand. Like some resistors, some capacitors are color-coded due to their small size. Some of the color codes are no longer in common use. They can be found in older textbooks and reference handbooks. Although it is not important that we learn the color-code markings at this point, you should be aware of their existence. Other capacitors, such as electrolytic capacitors, are large enough that the information about their capacitance, voltage ratings, and tolerance ratings are printed on them. Most capacitors actually have the numeric values stamped on them; however, some are color-coded and some have alphanumeric codes.

Electrolytic capacitors are those in which one or both of the "plates" is a nonmetallic substance, an electrolyte. Electrolytes have lower conductivity than metals, so they are only used in capacitors when a metallic plate is not practical, such as when the dielectric surface is fragile or rough in shape. Electrolytic capacitors are shown in Fig. 9.7. They usually possess the largest values of capacitance with values ranging from 0.1 to 200,000 μF. Electrolytic capacitors are usually polarized with one plate being positive and the other negative. For this reason, failure to observe proper polarity can have destructive (explosive) results. Nonelectrolytic capacitors can be connected either way in a circuit without worrying about polarity.

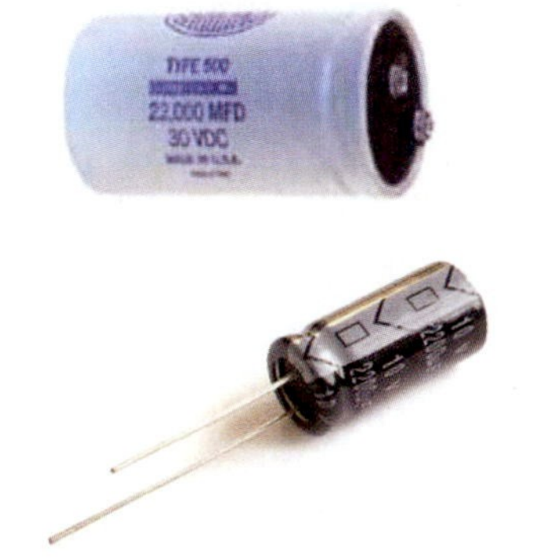

Figure 9.7
Electrolytic capacitors.
Courtesy of Surplus Sales of Nebraska

Capacitors are available as surface-mounted components just like resistors. Surface-mount capacitors are sometimes called chip capacitors. They are designed for applications requiring stable temperature and frequency characteristics similar to polyester film capacitors. They are ideal for applications such as electromagnetic interference, noise filtering, power supply input/output filters, and audio or signal coupling. Examples of surface-mount capacitors are shown in Fig. 9.8.

Standard values for capacitors are similar to those for resistors. These standard capacitors have been designed as primary reference standards of capacitance to which working values can be compared. They include values such as 10, 100, 150, 220, 330, 470, 560, and 1,000 pF and then 1, 1.5, 2.2, 3.3, 4.7, 5.6, 10 μF, and so on.

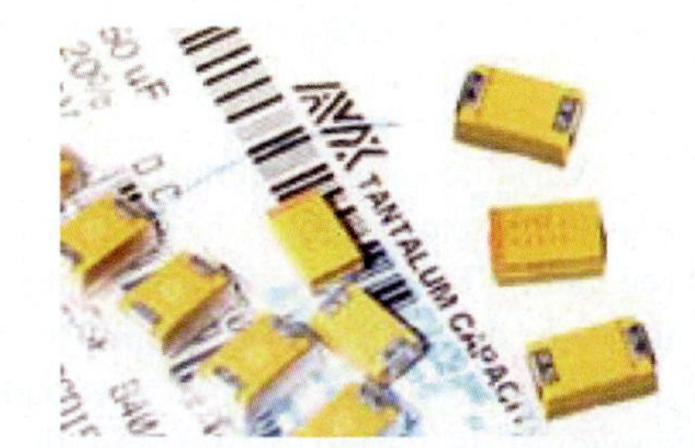

Figure 9.8
Surface-mount capacitors.
Courtesy of Surplus Sales of Nebraska

Capacitors are used for several purposes:

- To smooth the output of power supplies.
- To block the flow of direct current while allowing alternating current to pass.
- To store energy such as in a camera flash circuit.
- For timing, such as in a 555 timer IC controlling the charging and discharging.
- For coupling, such as in between stages of an audio system and to connect a loudspeaker.
- For filtering, such as in the tone control of an audio system.
- For tuning, such as in a radio system.

9.5 Series and Parallel Capacitors

Series-parallel connections of capacitors are sometimes encountered. We desire to replace these capacitors by a single equivalent capacitor C_{eq}.

In order to obtain the equivalent capacitor C_{eq} of N capacitors in parallel, consider the circuit in Fig. 9.9(a). The equivalent circuit is shown in Fig. 9.9(b). Note that the capacitors have the same voltage V across them but the total charge is the sum of the individual charges,

$$Q_T = Q_1 + Q_2 + Q_3 + \cdots + Q_N \tag{9.15}$$

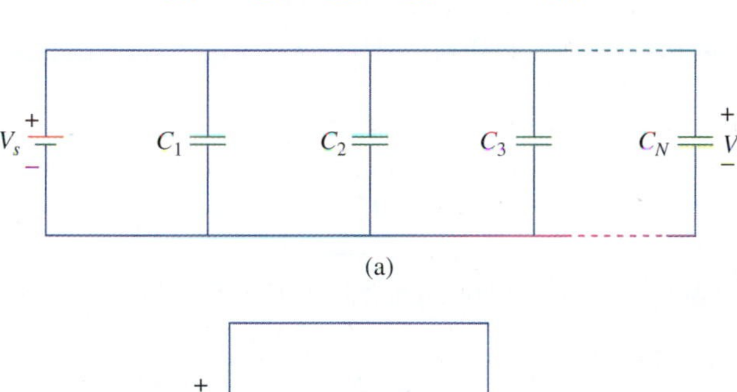

(a)

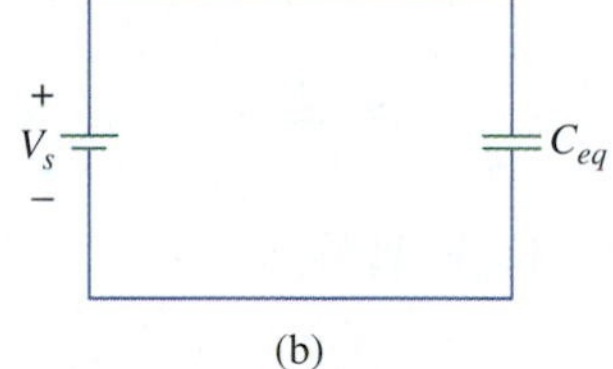

(b)

Figure 9.9
(a) Parallel-connected N capacitors; (b) equivalent circuit for the parallel capacitors.

But $Q = CV$. Hence,

$$C_{eq}V = C_1V + C_2V + C_3V + \cdots + C_NV$$

or

$$\boxed{C_{eq} = C_1 + C_2 + C_3 + \cdots + C_N} \tag{9.16}$$

We observe that capacitors in parallel combine in the same manner as resistors in series.

> The **equivalent capacitance** of N parallel-connected capacitors is the sum of the individual capacitances.

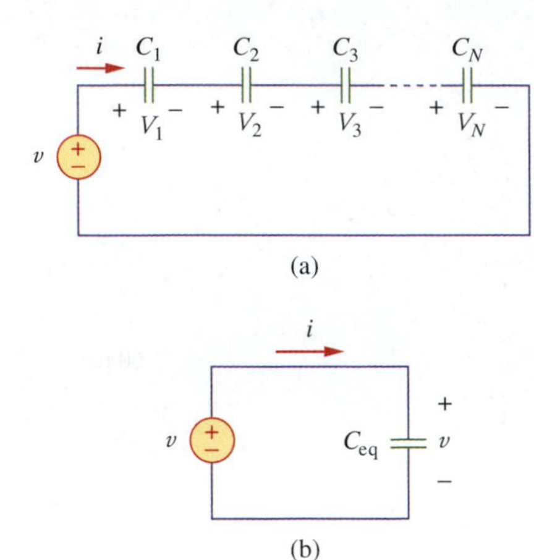

Figure 9.10
(a) Series-connected N capacitors; (b) equivalent circuit for the series capacitor.

We now obtain C_{eq} for N capacitors connected in series by comparing the circuit in Fig. 9.10(a) with the equivalent circuit in Fig. 9.10(b). Note that the same current i (and consequently the same charge builds up on the plates of each capacitor) flows through the capacitors. Applying KVL to the loop in Fig. 9.10(a),

$$V = V_1 + V_2 + V_3 + \cdots + V_N \tag{9.17}$$

But $V = Q/C$. Therefore,

$$\frac{Q}{C_{eq}} = \frac{Q}{C_1} + \frac{Q}{C_2} + \frac{Q}{C_3} + \cdots + \frac{Q}{C_N}$$

or

$$\boxed{\frac{1}{C_{eq}} = \frac{1}{C_1} + \frac{1}{C_2} + \frac{1}{C_3} + \cdots + \frac{1}{C_N}} \tag{9.18}$$

The **equivalent capacitance** of series-connected capacitors is the reciprocal of the sum of the reciprocals of the individual capacitances.

Note that capacitors in series combine in the same manner as resistors in parallel. For $N = 2$ (i.e., two capacitors in series), Eq. (9.18) becomes

$$\frac{1}{C_{eq}} = \frac{1}{C_1} + \frac{1}{C_2}$$

or

$$\boxed{C_{eq} = \frac{C_1 C_2}{C_1 + C_2}} \tag{9.19}$$

The voltage across each capacitor in Fig. 9.10(a) can be found as follows.

$$V_1 = \frac{Q}{C_1}, \quad V_2 = \frac{Q}{C_2}, \ldots, \quad V_N = \frac{Q}{C_N}$$

But $V = Q/C_{eq}$ or $Q = C_{eq}V$. Hence,

$$V_1 = \frac{C_{eq}}{C_1}V, \quad V_2 = \frac{C_{eq}}{C_2}V, \ldots, \quad V_N = \frac{C_{eq}}{C_N}V \tag{9.20}$$

Notice that the capacitors act as voltage dividers.

Example 9.3

Three capacitors $C_1 = 4\ \mu$F, $C_2 = 5\ \mu$F, and $C_3 = 10\ \mu$F are connected in parallel across a 110-V supply.
(a) Determine the total capacitance. (b) Find the total energy stored.

Solution:
(a) Because the capacitors are connected in parallel, the equivalent or total capacitance is

$$C_{eq} = C_1 + C_2 + C_3 = 4 + 5 + 10 = 19\ \mu\text{F}$$

(b) The total energy stored is

$$W = \frac{1}{2}C_{eq}V^2 = \frac{1}{2} \times 19 \times 10^{-6} \times 110^2 = 0.115\ \text{J}$$

Practice Problem 9.3

Rework Example 9.3 if the capacitors are now connected in series.

Answer: 1.818 μF; 11.01 mJ

Example 9.4

Find the equivalent capacitance seen at terminals *a-b* of the circuit in Fig. 9.11.

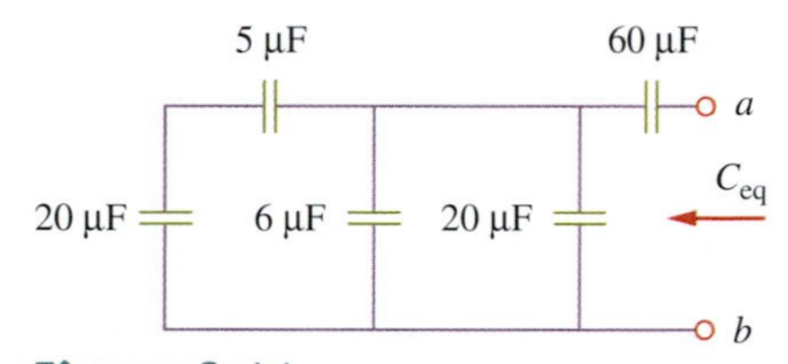

Figure 9.11
For Example 9.4.

Solution:
The 20-μF and 5-μF capacitors are in series; their equivalent capacitance is

$$\frac{20 \times 5}{20 + 5} = 4\ \mu\text{F}$$

This 4-μF capacitor is in parallel with the 6-μF and 20-μF capacitors; their combined capacitance is

$$4 + 6 + 20 = 30\ \mu\text{F}$$

This 30-μF capacitor is in series with the 60-μF capacitor. Hence, the equivalent capacitance for the entire circuit is

$$C_{\text{eq}} = \frac{30 \times 60}{30 + 60} = 20\ \mu\text{F}$$

Practice Problem 9.4

Find the equivalent capacitance seen at terminals *a-b* of the circuit in Fig. 9.12.

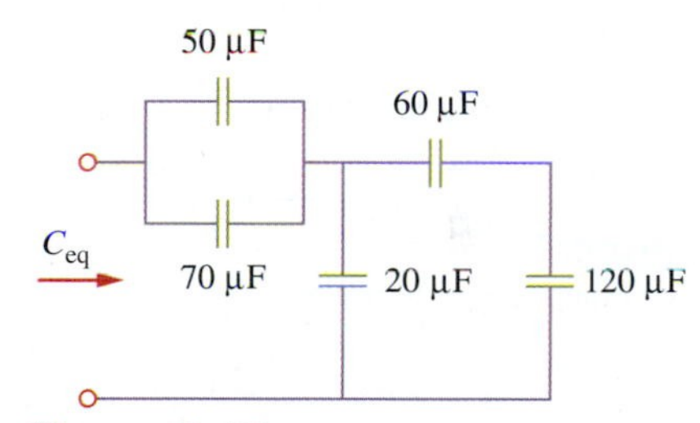

Figure 9.12
For Practice Problem 9.4.

Answer: 40 μF

Example 9.5

For the circuit in Fig. 9.13, find the voltage across each capacitor.

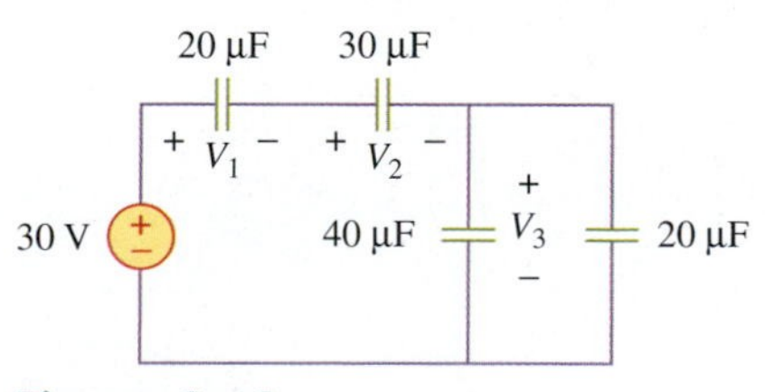

Figure 9.13
For Example 9.5.

Solution:
We first find the equivalent capacitance C_{eq}, shown in Fig. 9.14. The two parallel capacitors in Fig. 9.13 can be combined to get $40 + 20 = 60\ \mu$F. This 60-μF capacitor is in series with the 20-μF and 30-μF capacitors. Hence,

$$C_{\text{eq}} = \frac{1}{\dfrac{1}{60} + \dfrac{1}{30} + \dfrac{1}{20}}\ \mu\text{F} = 10\ \mu\text{F}$$

The total charge is

$$Q = C_{\text{eq}}V = 10 \times 10^{-6} \times 30 = 0.3\ \text{mC}$$

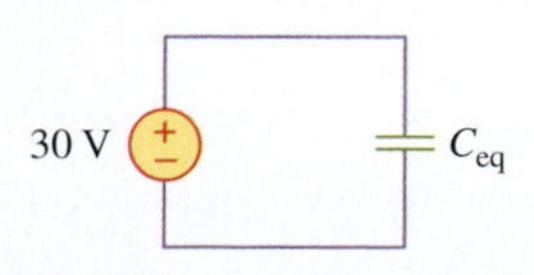

Figure 9.14
Equivalent circuit for that in Figure 9.12.

This is the charge on the 20-μF and 30-μF capacitors because they are in series with the 30-V source. (A crude way to see this is to imagine that charge acts like current because $I = Q/t$.) Therefore,

$$V_1 = \frac{Q}{C_1} = \frac{0.3 \times 10^{-3}}{20 \times 10^{-6}} = 15 \text{ V}$$

$$V_2 = \frac{Q}{C_2} = \frac{0.3 \times 10^{-3}}{30 \times 10^{-6}} = 10 \text{ V}$$

Having determined V_1 and V_2, we now use KVL to determine V_3; that is,

$$V_3 = 30 - V_1 - V_2 = 5 \text{ V}$$

Alternatively, because the 40-μF and 20-μF capacitors are in parallel, they have the same voltage V_3 and their combined capacitance is $40 + 20 = 60$ μF. This combined capacitance is in series with the 20-μF and 30-μF capacitors and consequently has the same charge on it. Hence,

$$V_3 = \frac{Q}{60\ \mu\text{F}} = \frac{0.3 \times 10^{-3}}{60 \times 10^{-6}} = 5 \text{ V}$$

Practice Problem 9.5

Find the voltage across each of the capacitors in Fig. 9.15.

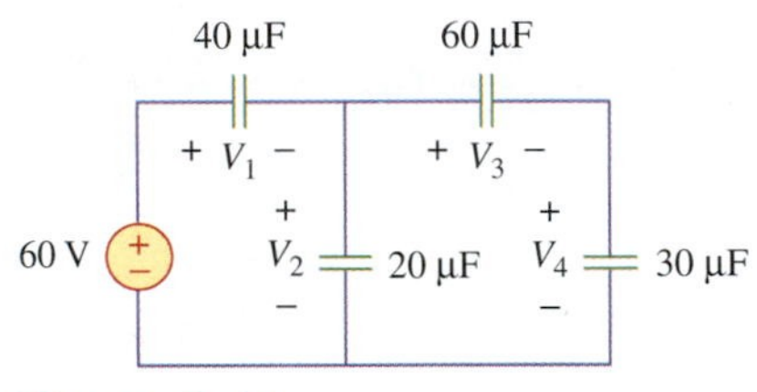

Figure 9.15
For Practice Problem 9.5.

Answer: $V_1 = 30$ V; $V_2 = 30$ V; $V_3 = 10$ V; $V_4 = 20$ V

9.6 Current–Voltage Relationship

The relationship between charge and voltage for a capacitor is given by Eq. (9.1); that is,

$$Q = Cv \tag{9.1}$$

To obtain the current–voltage relationship for the capacitor, we notice that Eq. (9.1) implies that

$$\text{Rate of change of charge } Q = C \times \text{Rate of change of voltage } v$$

because the capacitance C is constant. But

$$\text{Rate of change of } Q = \text{Current} = i$$

that is,

$$i = C \times \text{Rate of change of } v$$

Hence, using calculus notation,

$$\boxed{i = C\frac{dv}{dt}} \tag{9.21}$$

where dv/dt is the rate of change in v or the derivative of v. Equation (9.21) is the current–voltage relationship for a capacitor, assuming the passive sign convention (see Fig. 9.4). Equation (9.21) implies that the faster the voltage of a capacitor changes, the larger the current, and vice versa. If the voltage increases with time, the derivative dv/dt is positive. It is negative if the voltage decreases with time. If the voltage remains constant (no change), the current is zero.

Note that we use lowercase v and i to designate instantaneous voltage and current, while uppercase V and I for dc voltage and current.

We should note the following important properties of a capacitor:

1. Note from Eq. (9.21) that when the voltage across a capacitor is not changing with time (i.e., dc voltage), the current through the capacitor is zero. Thus,

   A capacitor is an open circuit to dc.

   However, if a battery (dc voltage) is connected across a capacitor, the capacitor charges.
2. The voltage on the capacitor must be continuous; that is,

   The voltage on a capacitor cannot change abruptly.

   The capacitor resists an abrupt change in the voltage across it. According to Eq. (9.21), an instantaneous change in voltage requires an infinite current, which is physically impossible. For example, the voltage across a capacitor may take the form shown in Fig. 9.16(a), whereas it is not physically possible for the capacitor voltage to take the form shown in Fig. 9.16(b) because of the abrupt changes. Conversely, the current through a capacitor can change instantaneously.
3. The ideal capacitor does not dissipate energy. It retrieves power from the circuit when storing energy in its field and returns previously stored energy when delivering power to the circuit.
4. A real, nonideal capacitor has a parallel-model leakage resistance, as shown in Fig. 9.17. The leakage resistance may be as high as 100 MΩ and can be neglected for most practical applications. For this reason, we will assume ideal capacitors in this book.

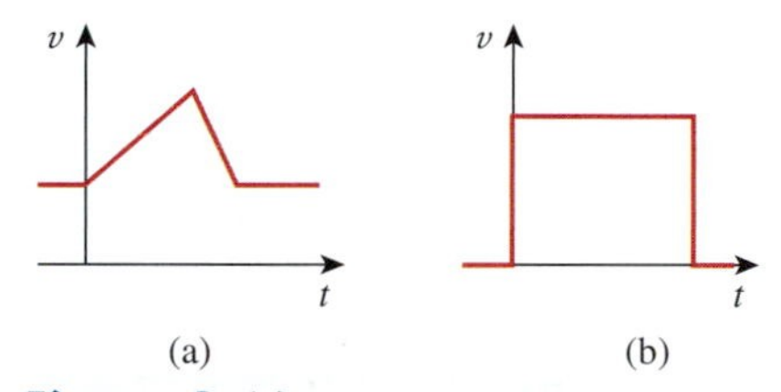

Figure 9.16
Voltage across a capacitor: (a) allowed; (b) not allowed. Note: An abrupt change is not possible.

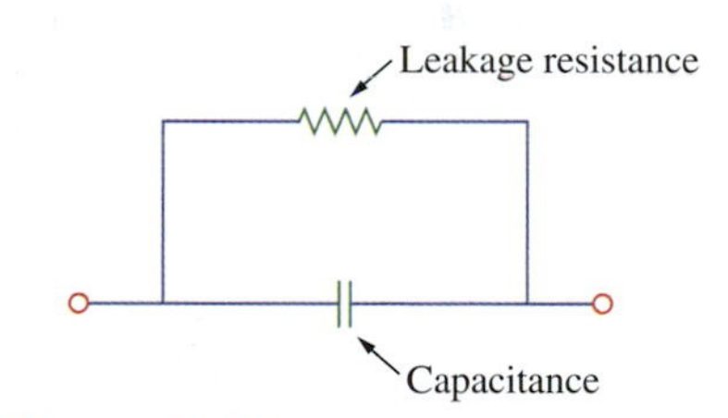

Figure 9.17
Circuit model of a nonideal capacitor.

Example 9.6

Determine the current through a 200-μF capacitor whose voltage is shown in Fig. 9.18.

Solution:
Because $i(t) = C\,\Delta v/\Delta t$ and $C = 200\ \mu$F, we take the derivative or slope of v to obtain $i(t)$.

From $t = 0$ s to $t = 1$ s, $\Delta v = 50$ V, while $\Delta t = 1$ s so that slope $= \Delta v/\Delta t = 50/1 = 50$ V/s.

From $t = 1$ s to $t = 3$ s, $\Delta v = -100$ V, while $\Delta t = 2$ s so that slope $= \Delta v/\Delta t = -100/2 = -50$ V/s.

From $t = 3$ s to $t = 4$ s, $\Delta v = 50$ V, while $\Delta t = 1$ s so that slope $= \Delta v/\Delta t = 50/1 = 50$ V/s.

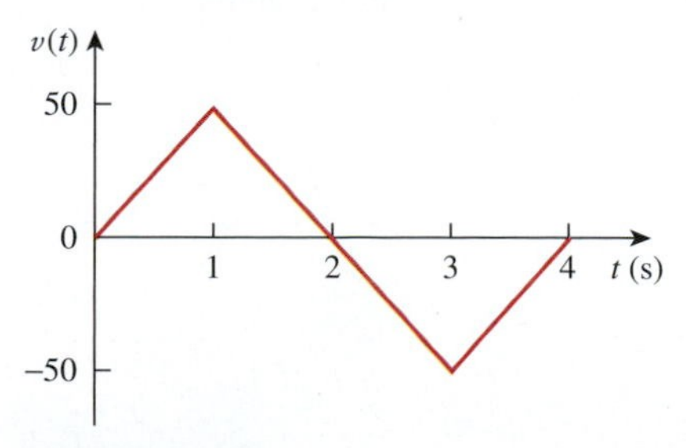

Figure 9.18
For Example 9.6.

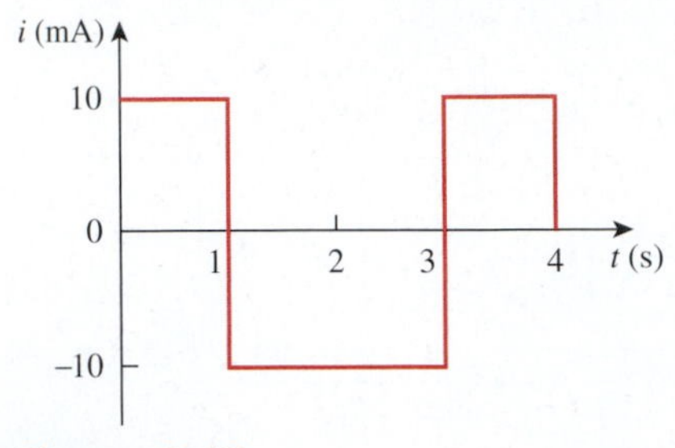

Figure 9.19
For Example 9.6.

Thus, $i(t) = C\,\Delta v/\Delta t$ becomes

$$i(t) = 200 \times 10^{-6} \times \begin{cases} 50, & 0 < t < 1 \\ -50, & 1 < t < 3 \\ 50, & 3 < t < 4 \\ 0, & \text{otherwise} \end{cases}$$

$$= \begin{cases} 10 \text{ mA}, & 0 < t < 1 \\ -10 \text{ mA}, & 1 < t < 3 \\ 10 \text{ mA}, & 3 < t < 4 \\ 0, & \text{otherwise} \end{cases}$$

The current waveform is as shown in Fig. 9.19.

Practice Problem 9.6

An initially uncharged 1-mF capacitor has the current shown in Fig. 9.20 through it. Calculate the voltage across it at $t = 2$ ms and at $t = 5$ ms.

Answer: 100 mV; 400 mV

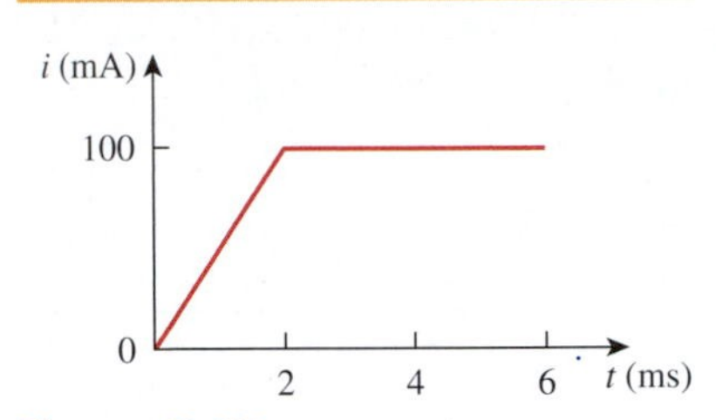

Figure 9.20
For Practice Problem 9.6.

Example 9.7

Obtain the energy stored in each capacitor in Fig. 9.21(a) under dc conditions.

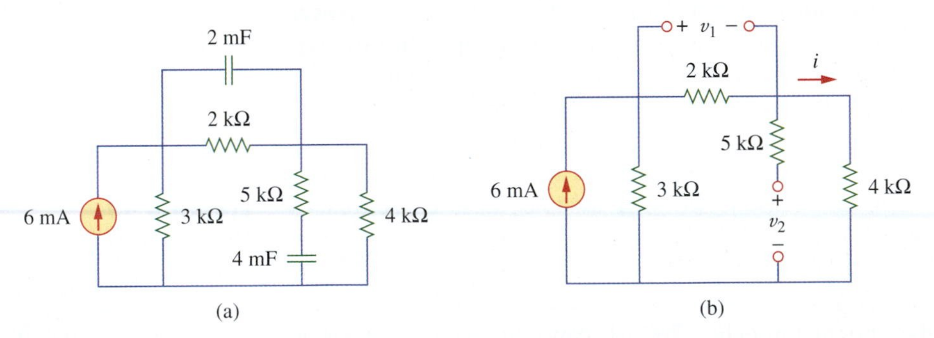

Figure 9.21
For Example 9.7.

Solution:
Under dc conditions, we replace each capacitor with an open circuit, as shown in Fig. 9.21(b). The current through the series combination of the 2-kΩ and 4-kΩ resistors is obtained by current division as

$$i = \frac{3}{3 + 2 + 4}(6 \text{ mA}) = 2 \text{ mA}$$

Hence, the voltages v_1 and v_2 across the capacitors are

$$v_1 = 2{,}000i = 4 \text{ V}, \qquad v_2 = 4{,}000i = 8 \text{ V}$$

and the energies stored in them are

$$w_1 = \frac{1}{2}C_1v_1^2 = \frac{1}{2}(2 \times 10^{-3})(4)^2 = 16 \text{ mJ}$$

$$w_2 = \frac{1}{2}C_2v_2^2 = \frac{1}{2}(4 \times 10^{-3})(8)^2 = 128 \text{ mJ}$$

Practice Problem 9.7

Under dc conditions, find the energy stored in the capacitors in Fig. 9.22.

Answer: 405 μJ; 90 μJ

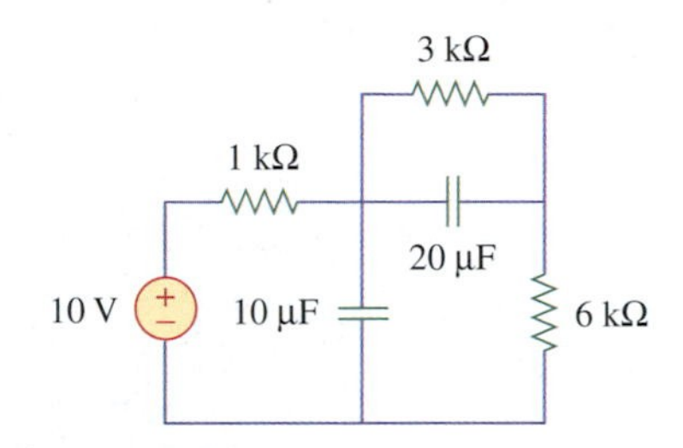

Figure 9.22
For Practice Problem 9.7.

9.7 Charging and Discharging a Capacitor

We now consider how a capacitor charges or discharges. In either case, the capacitor charges or discharges through a resistor.

9.7.1 Charging Cycle

Consider a battery connected to a series combination of a resistor and a capacitor, as shown in Fig. 9.23. (In general, the resistor and capacitor in Fig. 9.23 may be the equivalent resistance and capacitance of combinations of resistors and capacitors.) Assume that the capacitor in Fig. 9.23 is not charged and that the switch is turned on at $t = 0$. We start charging the capacitor by turning the switch on in Fig. 9.23. Applying KVL to the circuit in Fig. 9.23,

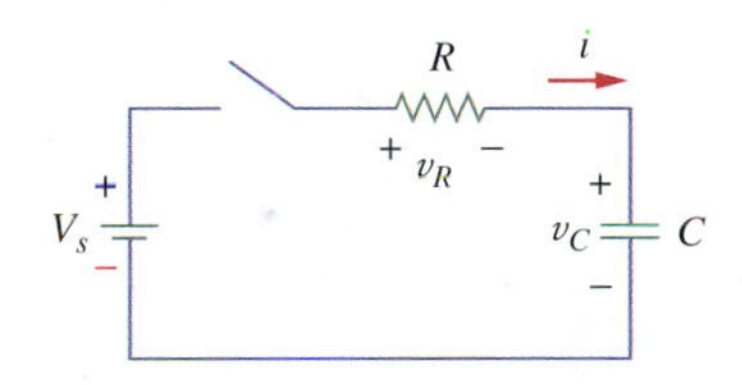

Figure 9.23
Charging circuit.

$$V_s = v_R + v_C \tag{9.22}$$

The capacitor voltage at any time t can be derived by differential calculus as

$$v_C = V_s - (V_s - V_o)e^{-t/RC} \tag{9.23}$$

where V_s is the supply voltage and V_o is the initial voltage. In our case, the capacitor is initially uncharged so that $V_o = 0$ and

$$v_C = V_s(1 - e^{-t/RC}) \tag{9.24}$$

where charging the capacitor stores energy in the electric field between the capacitor plates.

The capacitor (C) in the circuit diagram is being charged from a supply voltage (V_s) with the current passing through a resistor (R). The voltage across the capacitor (v_C) is assumed to be initially zero, but it increases as the capacitor charges, as shown in Fig. 9.24. The capacitor is fully charged when $v_C = V_s$.

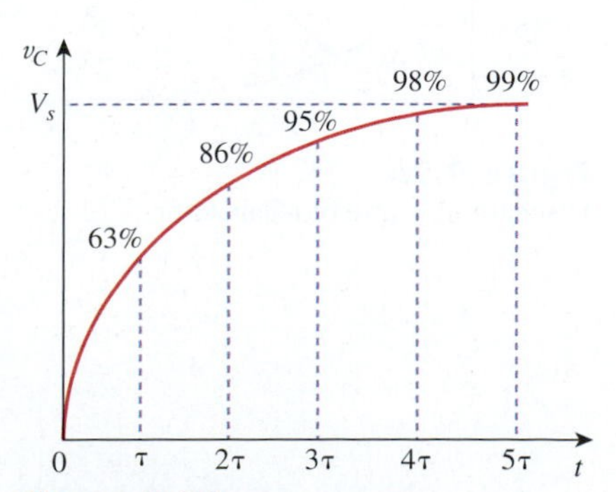

Figure 9.24
Charging curve for a capacitor.

The charging current i is determined by the voltage across the resistor from Eqs. (9.22) and (9.24):

$$v_R = V_s - v_C = V_s - V_s(1 - e^{-t/RC}) = V_se^{-t/RC}$$

The charging current is

$$i = \frac{v_R}{R} = \frac{V_s}{R}e^{-t/RC} \tag{9.25}$$

As t increases, the current decreases toward zero. The rapidity with which the current decreases is expressed in terms of the *time constant,* denoted by τ, the Greek letter tau.

> The **time constant** (in seconds) τ of a circuit is the time required for the response (current) to decay by a factor of $1/e$ or 36.8 percent of its initial value.

This implies that at $t = \tau$, Eq. (9.25) becomes

$$\frac{V_s}{R}e^{-\tau/RC} = \frac{V_s}{R}e^{-1} = 0.368\frac{V_s}{R}$$

or

$$\boxed{\tau = RC} \tag{9.26}$$

The time constant may also be regarded as the amount of time the circuit would have taken to reach its final state if it had kept going at its initial rate of change. In terms of the time constant, Eq. (9.24) can be written as

$$\boxed{v_C(t) = V_s(1 - e^{-t/\tau}), \qquad \tau = RC} \tag{9.27}$$

9.7.2 Discharging Cycle

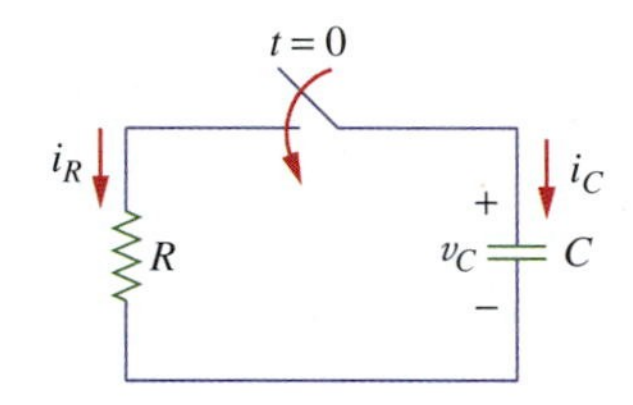

Figure 9.25
Discharging circuit.

For discharging, we consider the source-free RC circuit in Fig. 9.25 and assume that the switch is closed at $t = 0$. Because the capacitor is initially charged, we can assume that at time $t = 0$, the initial voltage is

$$v_C(0) = V_o$$

We now apply Eq. (9.23) with $V_s = 0$ and we get

$$\boxed{v_C(t) = V_o e^{-t/\tau}, \qquad \tau = RC} \tag{9.28}$$

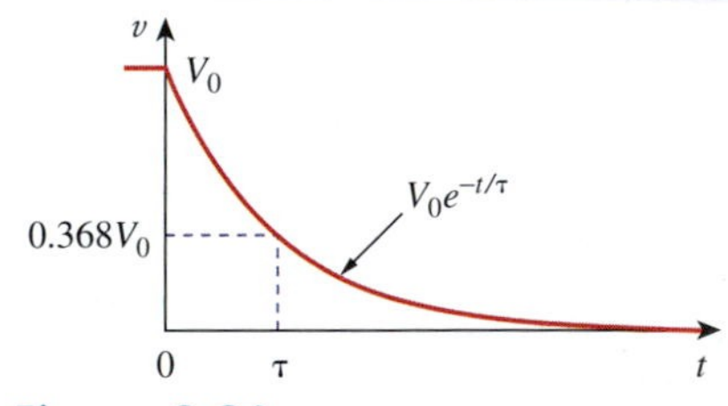

Figure 9.26
Discharging curve of a capacitor.

This shows that the charging process is an exponential decay of the initial voltage, as illustrated in Fig. 9.26. This implies that at $t = \tau$, Eq. (9.28) becomes

$$V_o e^{-\tau/RC} = V_o e^{-1} = 0.368V_o$$

With a calculator, it is easy to show that the value of $v(t)/V_o$ is as shown in Table 9.3. It is evident from Table 9.3 that the voltage $v(t)$ is less than 1 percent of V_o after 5τ (five time constants). Five time constants is regarded as the *transient time.* Thus, it is customary to assume (for all practical purposes) that the capacitor is fully discharged (or charged) after five time constants. In other words, it takes approximately 5τ for the circuit to reach its final or steady state when no changes take place with time.

As illustrated in Fig. 9.27, a circuit with a small time constant ($\tau = RC$, either R or C or both may change to decrease or increase the

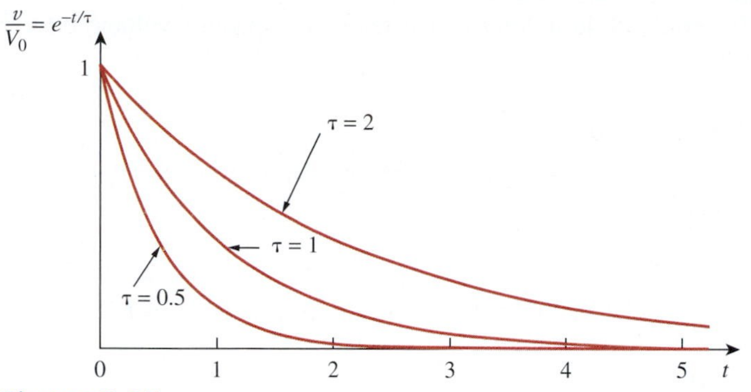

Figure 9.27
Plot of $v/V_o = e^{-t/\tau}$ for various values of the time constant.

TABLE 9.3

Values of $v(t)/V_o = e^{-t/\tau}$

t	$v(t)/V_o$	Percentage
0	1	100
τ	0.36788	36.79
2τ	0.13534	13.53
3τ	0.04979	4.98
4τ	0.01832	1.83
5τ	0.00674	0.67

time constant) gives a fast response in that it reaches the steady state (or final state) quickly due to quick dissipation of energy stored, whereas a circuit with a large time constant gives a slow response because it takes a relatively long time to reach steady state. At any rate, whether the time constant is small or large, the circuit reaches steady state in five time constants.

The idea developed in the section can be extended to the general case of a source-free *RC* transient analysis. With the initial value and the time constant specified, we can obtain the response of the capacitor voltage as:

$$v_C(t) = v(0)e^{-t/\tau}$$

Once the capacitor voltage is first obtained, other variables (capacitor current i_C, resistor voltage v_R, and resistor current i_R) can be determined. In finding the time constant $\tau = RC$, R is often the Thevenin equivalent resistance at the terminals of the capacitor; that is, we take out the capacitor C and find $R = R_{Th}$ at its terminals.[3]

Example 9.8

For the circuit in Fig. 9.23, let $R = 500\ \Omega$, $C = 10\ \mu F$, and $V_s = 15$ V.

(a) Calculate the transient period;
(b) find v_C, v_R, and i.

Solution:

(a) The time constant is $RC = \tau = 500 \times 10 \times 10^{-6} = 5$ ms. The transient time is $5\tau = 25$ ms.
(b) From Eq. (9.27),

$$v_C(t) = V_s(1 - e^{-t/\tau}) = 15(1 - e^{-t/\tau})\ V$$

$$v_R(t) = V_S - v_C = V_s e^{-t/\tau} = 15e^{-t/\tau}$$

$$i(t) = \frac{v_R}{R} = \frac{V_s}{R}e^{-t/\tau} = \frac{15}{500}e^{-t/\tau}$$

[3] Note: When a circuit contains a single capacitor and several resistors, the Thevenin equivalent can be found at the terminals of the capacitor to form a simple *RC* circuit. Also, the Thevenin theorem can be used when several capacitors can be combined to form a single equivalent capacitor.

Practice Problem 9.8

In Example 9.8, how long will it take the capacitor voltage v to reach 10 V?

Answer: 5.493 ms

Example 9.9

In Fig. 9.28, let $v_C(0) = 15$ V. Find v_C, v_x, and i_x for $t > 0$.

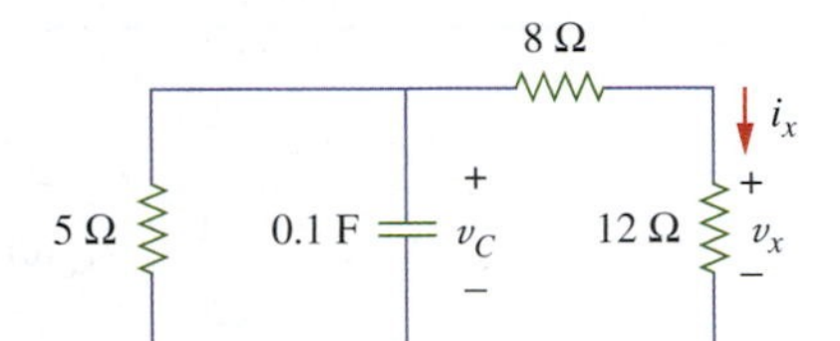

Figure 9.28
For Example 9.8.

Solution:
We first need to make the circuit in Fig. 9.28 conform with the standard *RC* circuit in Fig. 9.25. We find the equivalent resistance or the Thevenin resistance at the capacitor terminals. Our objective is always to first obtain capacitor voltage v_C. From this, v_x and i_x can be determined.

The 8-Ω and 12-Ω resistors in series can be combined to give a 20-Ω resistance. This 20-Ω resistance in parallel with 5-Ω resistor can be combined so that the equivalent resistance is

$$R_{eq} = \frac{20 \times 5}{20 + 5} = 4\ \Omega$$

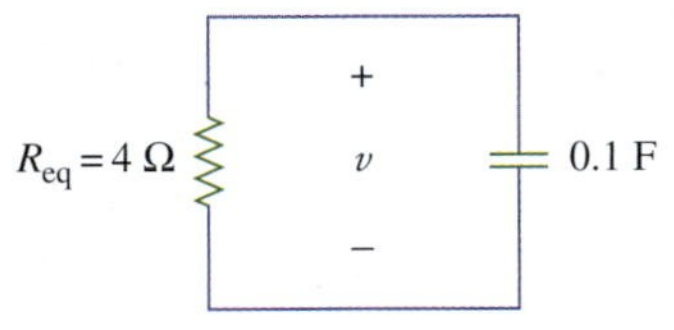

Figure 9.29
Equivalent circuit for the circuit in Fig. 9.28.

Hence, the equivalent circuit is shown in Fig. 9.29, which is similar to Fig. 9.25. The time constant is

$$\tau = R_{eq}C = 4(0.1) = 0.4 \text{ s}$$

Thus,

$$v(t) = v(0)e^{-t/\tau} = 15e^{-t/0.4} \text{ V}, \qquad v_C(t) = v(t) = 15e^{-2.5t} \text{ V}$$

From Fig. 9.28, we can use voltage division to get v_x; that is,

$$v_x = \frac{12}{12 + 8} v_C = 0.6(15e^{-2.5t}) = 9e^{-2.5t} \text{ V}$$

Finally,

$$i_x = \frac{v_x}{12} = 0.75e^{-2.5t} \text{ A}$$

Practice Problem 9.9

Refer to the circuit in Fig. 9.30. Let $v_C(0) = 30$ V. Determine v_C, v_x, and i_x for $t \geq 0$.

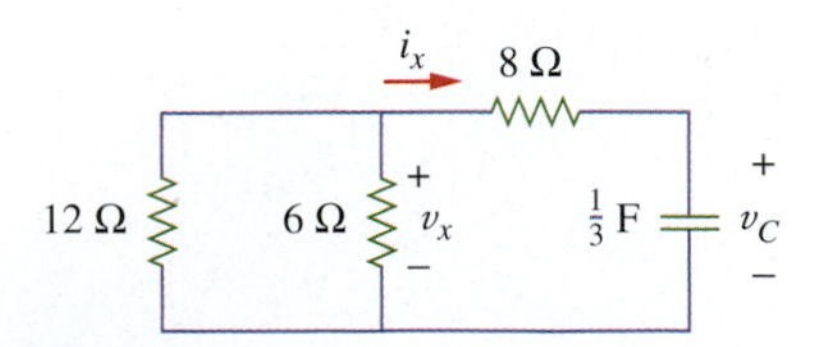

Figure 9.30
For Practice Problem 9.9 and Problem 9.68.

Answer: $30e^{-0.25t}$ V; $10e^{-0.25t}$ V; $-2.5e^{-0.25t}$ A

Example 9.10

The switch in the circuit in Fig. 9.31 has been closed for a long time, and it is opened at $t = 0$. Find $v(t)$ for $t \geq 0$. Calculate the initial energy stored in the capacitor.

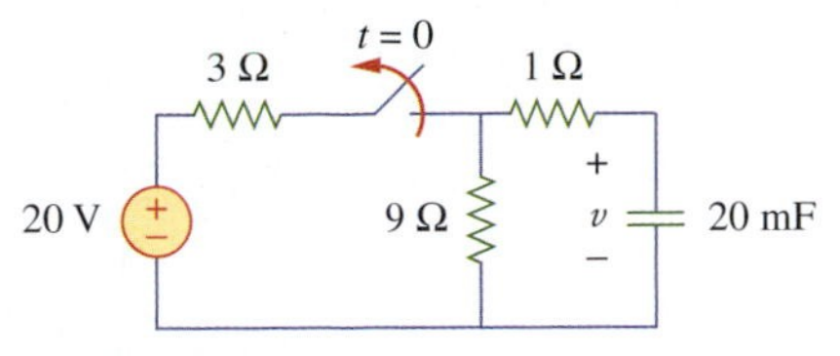

Figure 9.31
For Example 9.10.

Solution:
For $t < 0$, the switch is closed; the capacitor is an open circuit to dc, as represented in Fig. 9.32 (a). Using voltage division

$$v_C(t) = \frac{9}{9+3}(20) = 15 \text{ V}, \qquad t < 0$$

Because the voltage across a capacitor cannot change instantaneously, the voltage across the capacitor at $t = 0^-$ (just before $t = 0$) is the same as at $t = 0$; that is,

$$v_C(0) = V_o = 15 \text{ V}$$

For $t > 0$, the switch is opened, and we have the *RC* circuit shown in Fig. 9.32(b). [Notice that the *RC* circuit in Fig. 9.32(b) is source free; the independent source in Fig. 9.31 is needed only to provide V_o or the initial energy in the capacitor.] The 1-Ω and 9-Ω resistors in series give

$$R_{\text{eq}} = 1 + 9 = 10\ \Omega$$

The time constant is

$$\tau = R_{\text{eq}}C = 10 \times 20 \times 10^{-3} = 0.2 \text{ s}$$

Thus, the voltage across the capacitor for $t \geq 0$ is

$$v(t) = v_C(0)e^{-t/\tau} = 15e^{-t/0.2} \text{ V}$$

or

$$v(t) = 15e^{-5t} \text{ V}$$

The initial energy stored in the capacitor is

$$w_C(0) = \frac{1}{2}Cv_C^2(0) = \frac{1}{2} \times 20 \times 10^{-3} \times 15^2 = 2.25 \text{ J}$$

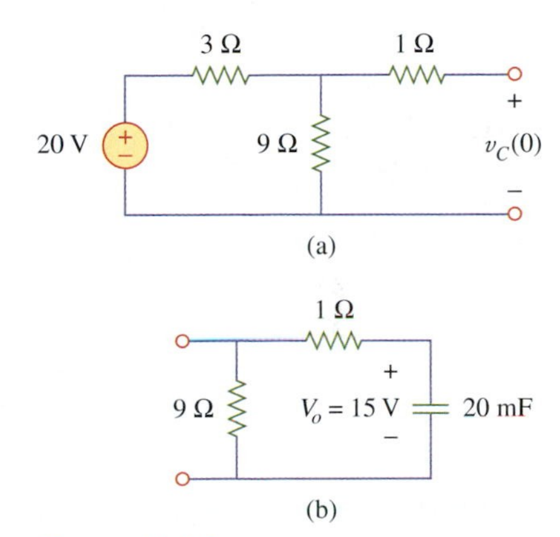

Figure 9.32
For Example 9.10: (a) $t < 0$; (b) $t > 0$.

Practice Problem 9.10

If the switch in Fig. 9.33 opens at $t = 0$, find $v(t)$ for $t \geq 0$ and $w_C(0)$.

Answer: $8e^{-2t}$ V; 5.33 J

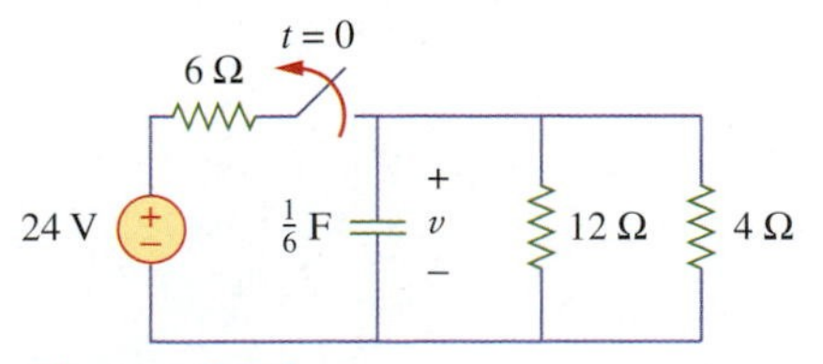

Figure 9.33
For Practice Problem 9.10.

9.8 Computer Analysis

9.8.1 PSpice

PSpice can be used to obtain the response of a circuit with storage elements. Section C.4 in Appendix C provides a review of transient analysis using PSpice for Windows. It is recommended that Section C.4 be reviewed before continuing with this section.

If necessary, dc PSpice analysis is first carried out to determine the initial conditions. Then, the initial conditions are used in the transient PSpice analysis to obtain the transient responses. It is recommended but not necessary that during this dc analysis, all capacitors should be open-circuited while all inductors should be short-circuited.

Example 9.11

In the circuit in Fig. 9.34, determine the response $v(t)$.

Solution:

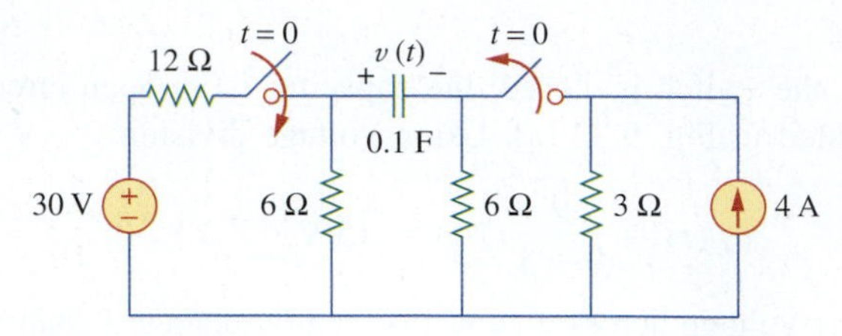

Figure 9.34
For Example 9.11.

There are two ways of solving this using PSpice.

■ **METHOD 1** One way is to first do the dc PSpice analysis to determine the initial capacitor voltage. The schematic of the relevant portion of the circuit is in Fig. 9.35(a). Because this is a dc analysis, we use current source IDC. Once the circuit is drawn and saved as exam911a.dsn, we select **PSpice/New Simulation Profile**. This leads to the New Simulation dialog box. Type "exam911a" as the name of the file and click Create. This leads to the Simulation Settings dialog box. Click OK, then **PSpice/Run**. When the circuit is simulated, we obtained the displayed values in Fig. 9.35(a) as $V_1 = 0$ V and $V_2 = 8$ V. Thus, the initial capacitor voltage is $v(0) = V_1 - V_2 = -8$ V. This value along with the schematic in Fig. 9.35(b) is used in the PSpice transient analysis.

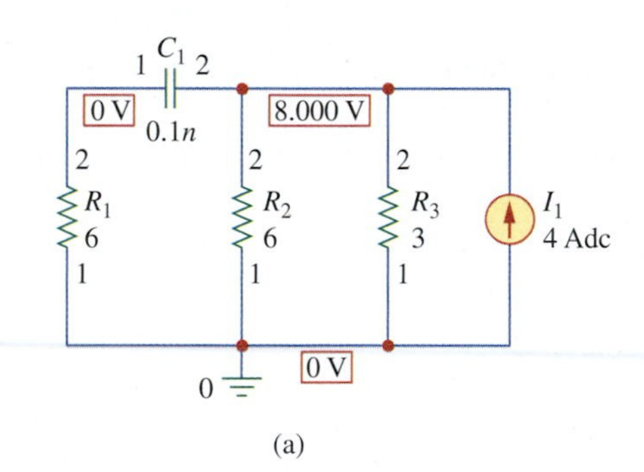

(a)

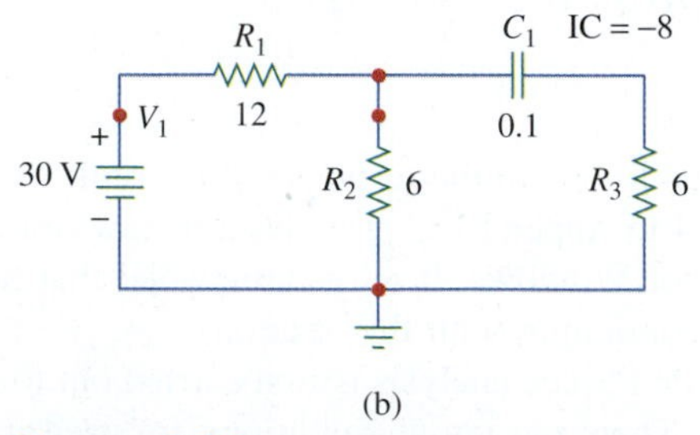

(b)

Figure 9.35
(a) Schematic for dc analysis to get $v(0)$;
(b) schematic for transient analysis used to get the response $v(t)$.

Once the circuit in Fig. 9.35(b) is drawn, we insert the capacitor initial voltage as $IC = -8$. We do this by double clicking on the capacitor symbol and typing -8 under IC. We select **PSpice/New Simulation Profile.** In the New Simulation dialog box, type "exam911b" under Name and click Create. In Simulation Settings, select Time Domain (Transient) under *Analysis Type* and $4\tau = 4$s as *Run to time*. Click Apply and then click OK. After saving the circuit, we select **PSpice/Run** to simulate the circuit. In the window that pops up, we select **Trace/Add** and display V(R2:2) − V(R3:2) or V(C1:1) − V(C1:2) as the capacitor voltage $v(t)$. The plot of $v(t)$ is shown in Fig. 9.36. This agrees with the result obtained by hand calculation, $v(t) = 10 - 18e^{-t}$ V.

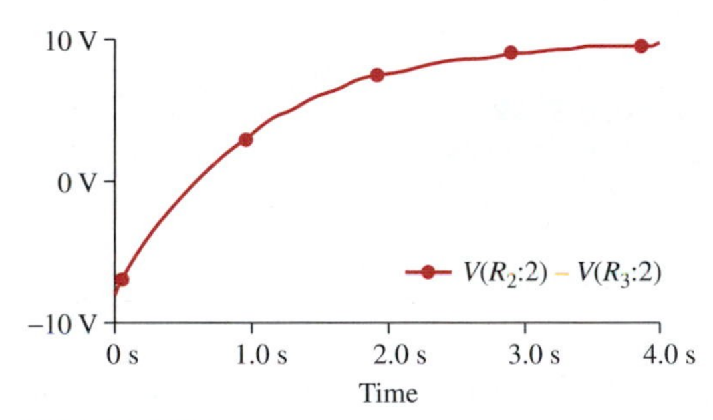

Figure 9.36
Response $v(t)$ for the circuit in Fig. 9.34.

METHOD 2 We can simulate the circuit in Fig. 9.34 directly because PSpice can handle the open and closed switches and determine the initial conditions automatically. (The part names for open and close switches are Sw_topen and Sw_tclose, respectively.) Using this approach, the schematic is drawn as shown in Fig. 9.37. After drawing the circuit, we select **PSpice/New Simulation Profile**. This leads to a New Simulation dialog box. Type "exam911a" as the name of the file and click Create. This leads to the Simulation Settings dialog box. In Simulation Settings, select Time Domain (Transient) under *Analysis Type* and $4\tau = 4$s as *Run to time*. Click Apply and then click OK. After saving the circuit, we select **PSpice/Run** to simulate the circuit. In the window that pops up, we select **Trace/Add Trace** and display V(R2:2) − V(R3:2) or V(C1:1) − V(C2:2) as the capacitor voltage $v(t)$. The plot of $v(t)$ is the same as that shown in Fig. 9.36.

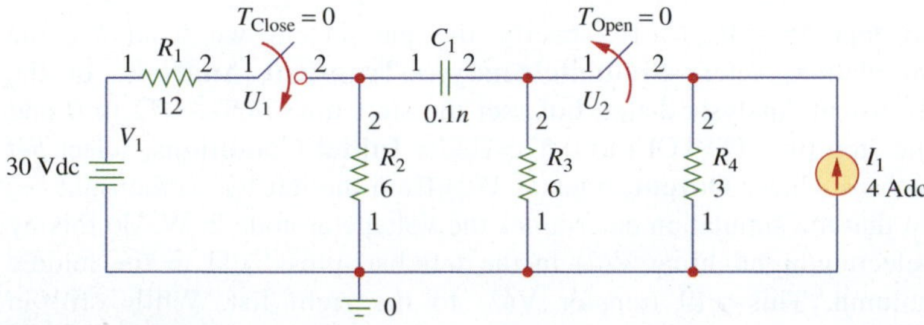

Figure 9.37
For Example 9.11.

Practice Problem 9.11

The switch in Fig. 9.38 has been open for a long time but is closed at $t = 0$. Use PSpice to find $v(t)$ for $t > 0$.

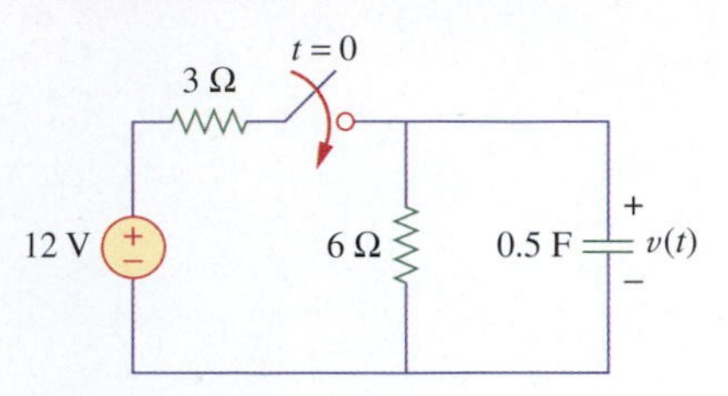

Figure 9.38
For Practice Problem 9.11.

Answer: See Fig. 9.39

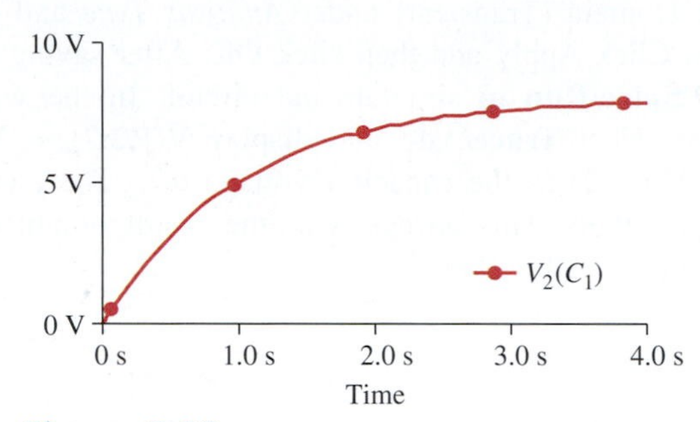

Figure 9.39
For Practice Problem 9.11.

9.8.2 Multisim

We can use Multisim to analyze the *RC* circuits considered in this chapter. It is recommended that you read Section D.3 in Appendix D on transient analysis before proceeding with this section.

Example 9.12

Use Multisim to determine the response v_o of the circuit in Fig. 9.40.

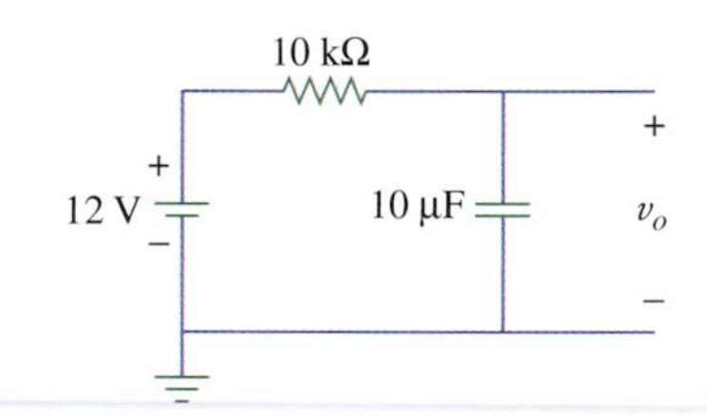

Figure 9.40
For Example 9.12.

Solution:
We first use Multisim to draw the circuit as shown in Fig. 9.41. Multisim automatically puts a number on the nodes. In case the node numbers are not shown, we select **Options/Sheet Properties.** Under Net Names, select "Show all." This will label the nodes as shown in Fig. 9.41.

We need to determine how long the simulation must run. A reasonable value is 5τ, where τ is the time constant of the circuit. In our case,

$$\tau = RC = 10 \times 10^3 \times 10 \times 10^{-6} = 0.1 \text{ s}$$

so that $5\tau = 0.5$ s. To specify the parameters we need for the simulation, select **Simulate/Analyses/Transient Analysis**. In the Transient Analysis dialog box, set the start time (TSTART) to 0 and the stop time (TSTOP) to 0.5 s. Under **Initial Conditions,** select *Set to zero*. Under **Output,** transfer *V*(2) from the left list to the right list so that the simulation determines the voltage at node 2. We do this by selecting/highlighting V(2) in the left list, press Add in the middle column. This will transfer V(2) to the right list. While still in the Transient Analysis dialog box, we finally select **Simulate** and the output will automatically be displayed, as shown in Fig. 9.42.

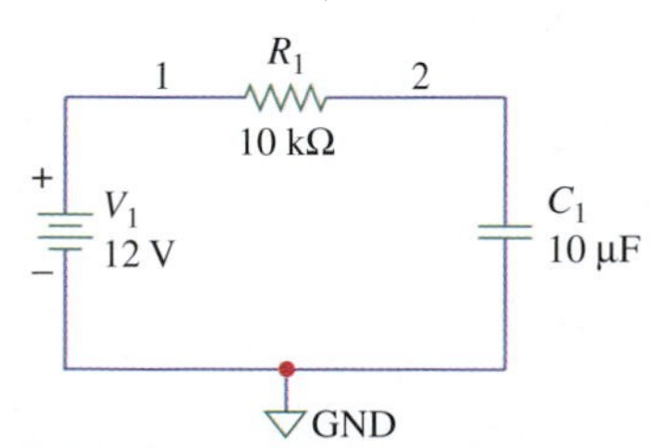

Figure 9.41
Simulation of the circuit in Fig. 9.40.

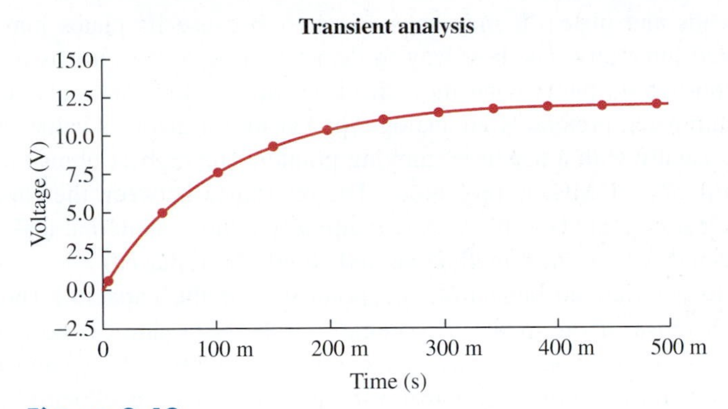

Figure 9.42
Response v_o of the circuit in Fig. 9.40.

Practice Problem 9.12

Find V_x in the circuit of Fig. 9.43 using Multisim.

Answer: See Fig. 9.44

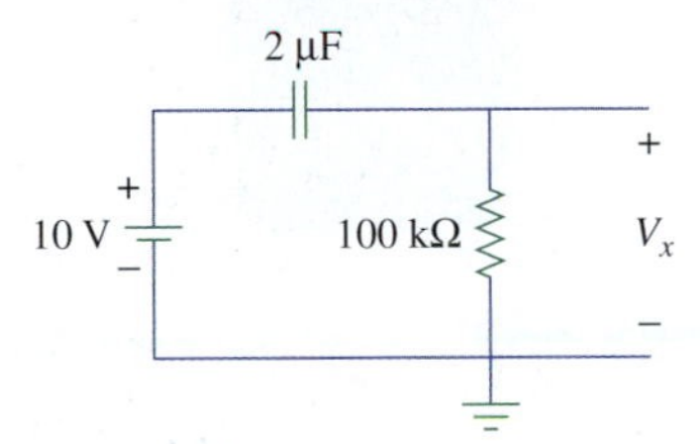

Figure 9.43
For Practice Problem 9.12.

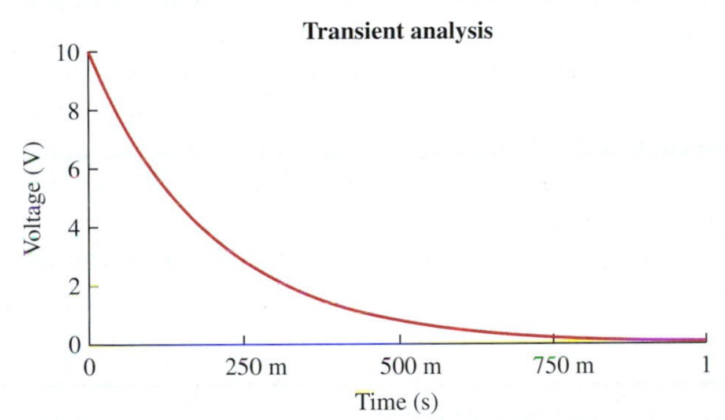

Figure 9.44
For Practice Problem 9.12.

9.9 Troubleshooting

The skill of locating areas in a circuit is very valuable. Because learning to troubleshoot is really both an art and a science, learning it can only be done by doing. Although one may begin with the schematic, one should keep in mind that the schematic is not the actual circuit. You must always begin at a known point and proceed in a cause-and-effect manner to locate the fault.

From experience, capacitors develop problems more frequently than resistors. Second only to power cords, capacitors are the most failure-prone components in old radios. Some kinds of capacitors—paper, molded paper, and electrolytics—are failure-prone and need to be replaced after some time. Other kinds, such as mica and ceramic, almost never need replacement. A capacitor may fail due to environmental temperature, age, a short, or an open. Capacitors are definitely affected by the temperature of their surroundings. They deteriorate with age. A capacitor may act as open due to a broken connection between its leads and plates. It may act as a short because its plates may be shorted internally. The best way to detect fault in a capacitor is to use an appropriate meter. One may check capacitors for continuity using an ohmmeter, preferably an analog type because it displays values on a dial, usually with a needle or moving pointer. The highest ohms range, such as $R \times 1\ \text{M}\Omega$, is preferable. The resistance between the capacitor's leads should be infinite. Any continuity indicates internal leakage; the capacitor becomes ineffective and should be replaced.

In a polarized capacitor, the polarities of the capacitor should match those of the meter. The negative (cathode) side of the polarized capacitor should always be connected to ground. (Explosions can result when polarized capacitors are connected "backward" in a circuit.) To protect the meter, ensure that the capacitor is fully discharged before testing. The capacitor is discharged by shorting its leads. Be aware that very large capacitors or capacitors charged to a high voltage contain a large amount of energy. Simply shorting the leads may cause an unpleasant result such as sparking or lead melting. Also, capacitors used in high-voltage applications can retain their charge for very long times. Before working on such a circuit, disconnect the power source and give the circuit sufficient time to discharge.

Figure 9.45
Digital capacitance meter.
© Mastech

Although capacitance meters as stand-alone units are available, many digital multimeters (DMMs) are capable of measuring a wide range of capacitance values. An example of a digital capacitance meter is shown in Fig. 9.45. A DMM or volt-ohm-milliammeter (VOM) can be used to test a capacitor. More sophisticated testers are available for measuring capacitance value, leakage, dielectric absorption, and the like.

9.10 †Applications

Circuit elements such as resistors and capacitors are commercially available in either discrete form or integrated-circuit (IC) form. Capacitors are used in many applications such as timing circuits, tuned circuits, power supplies, filters, and computer memories. For example, capacitors may be used to block the flow of direct current (dc) while allowing alternating current (ac) to pass. Capacitors, along with resistors and inductors, are found in electronic and microelectronic devices. Analog circuits commonly contain resistors and capacitors as well.

Capacitors (and inductors) possess the following three special properties that make them very useful in electric circuits:

1. The capacity to store energy makes them useful as temporary voltage or current sources. Thus, they can be used for generating a large amount of current or voltage for a short period of time.
2. Capacitors oppose any change in voltage.
3. Capacitors are frequency sensitive. This property makes them useful for frequency discrimination.

The first two properties are put to use in dc circuits, while the third one is taken advantage of in ac circuits. We will see how useful these properties are used in later chapters. For now, we consider two simple applications involving capacitors.

9.10.1 Delay Circuits

RC circuits find applications in several devices. These include filtering in dc power supplies, smoothing circuits in digital communications, differentiators, integrators, delay circuits, and relay circuits. Some of these applications take advantage of the short or long time constants of the *RC* circuits.

An *RC* circuit can be used to provide various time delays. Such a circuit is shown in Fig. 9.46. It consists of an *RC* circuit with the capacitor connected in parallel with a neon lamp. The voltage source can provide enough voltage to fire the lamp. When the switch is closed, the capacitor voltage increases gradually toward 120 V at a rate determined by the circuit's time-constant $(R_1 + R_2)C$. The lamp will act as an open circuit and not emit light until the voltage across it exceeds a particular level, say, 70 V. When the voltage level is reached, the lamp fires (lights up), and the capacitor discharges through it. Due to the low resistance of the lamp when it is on, the capacitor voltage drops quickly and the lamp turns off. The lamp acts again as an open circuit and the capacitor recharges.

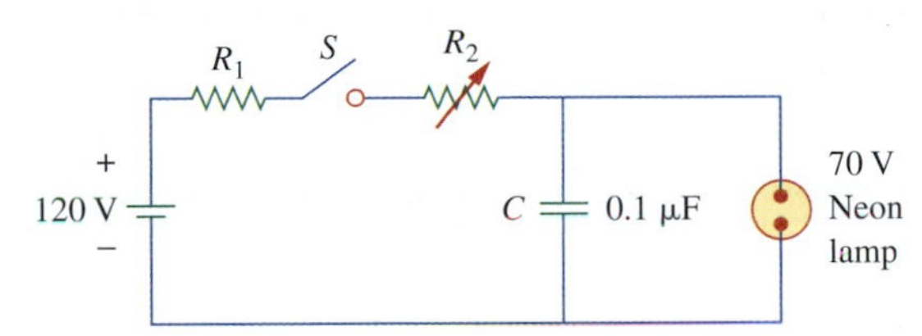

Figure 9.46
An *RC* delay circuit.

By adjusting R_2, we can introduce either short or long time delays into the circuit and make the lamp fire, recharge, and fire repeatedly every time constant $\tau = (R_1 + R_2)C$ because it takes a time period τ to get the capacitor voltage high enough to fire or low enough to turn off. Such an *RC* delay circuit finds applications in warning blinkers commonly found on road construction sites.

Example 9.13

Consider the circuit in Fig. 9.46 and assume that $R_1 = 1.5\ \text{M}\Omega$ and $0 < R_2 < 2.5\ \text{M}\Omega$. Calculate the extreme limits of the time constant of the circuit.

Solution:
The smallest value for R_2 is $0\ \Omega$, and the corresponding time constant for the circuit is

$$\tau = (R_1 + R_2)C = (1.5 \times 10^6 + 0) \times 0.1 \times 10^{-6} = 0.15\ \text{s}$$

The largest value for R_2 is $2.5\ \text{M}\Omega$, and the corresponding time constant for the circuit is

$$\tau = (R_1 + R_2)C = (1.5 + 2.5) \times 10^6 \times 0.1 \times 10^{-6} = 0.4\ \text{s}$$

Thus, by proper circuit design, the time constant can be adjusted to introduce a proper time delay in the circuit.

Practice Problem 9.13

The RC circuit in Fig. 9.47 is designed to sound an alarm that operates when the current through it exceeds 120 μA. If $0 \leq R \leq 6$ kΩ, find the range of the time delay that the circuit can cause.

Answer: Between 47.32 and 79.4 ms

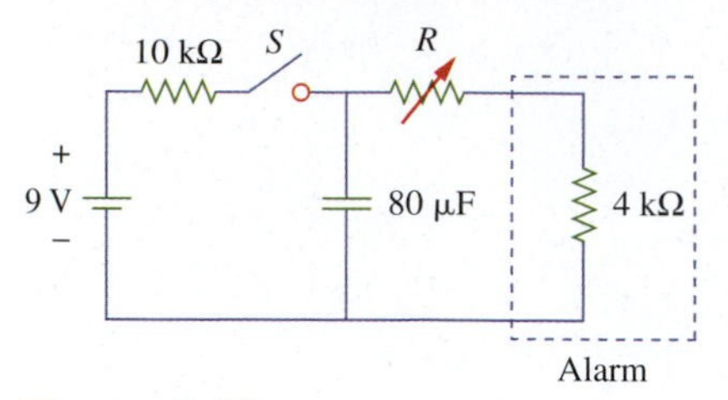

Figure 9.47
For Practice Problem 9.13.

9.10.2 Photoflash Unit

A common application of an RC circuit is found in an electronic flash unit. This application exploits the ability of the capacitor to oppose any abrupt change in voltage. A simplified circuit is shown in Fig. 9.48. It consists of a high-voltage dc supply, a current-limiting large value resistor R_1, and a capacitor C in parallel with the flash lamp of low resistance R_2. When the switch is in position 1, the capacitor charges slowly due to the large time constant ($\tau_1 = R_1C$). As shown in Fig. 9.49(a), the capacitor voltage rises gradually from zero to V_s while its current decreases gradually from $I_1 = V_s/R_1$ to zero. The charging time is approximately five times the time constant; that is,

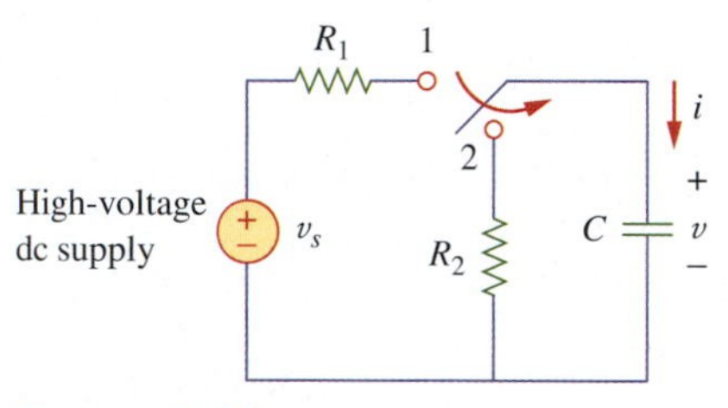

Figure 9.48
Circuit for a flash unit providing slow charge in position 1 and fast discharge in position 2.

$$t_{\text{charge}} = 5R_1C \tag{9.29}$$

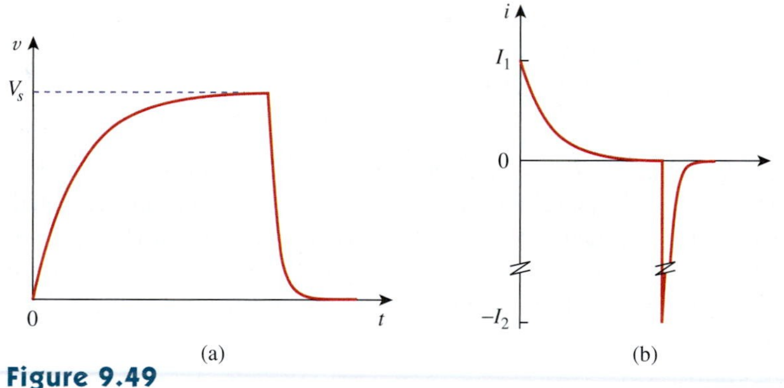

Figure 9.49
(a) Capacitor voltage showing slow charge and fast discharge; (b) capacitor current showing low peak charging current $I_1 = V_s/R_1$ and high peak discharge current $I_2 = V_s/R_2$.

With the switch in position 2, the capacitor voltage is discharged. The low resistance R_2 of the flash lamp permits a high discharge current with peak current $I_2 = V_s/R_2$ in a short duration, as depicted in Fig. 9.49(b). Discharging takes place at approximately five times the time constant; that is,

$$t_{\text{discharge}} = 5R_2C \tag{9.30}$$

Thus, the simple RC circuit of Fig. 9.48 provides a short-duration, high-current pulse. Such a circuit also finds applications in electric spot welding and with radar transmitter tubes.

Example 9.14

An electronic photoflash has a current-limiting 6-kΩ resistor and 2,000-μF electrolytic capacitor charged to 240 V. If the lamp resistance is 12-Ω, find:

(a) the peak charging current,
(b) the time required for the capacitor to fully charge,
(c) the peak discharging current,
(d) the total energy stored in the capacitor, and
(e) the average power dissipated by the lamp.

Solution:

We use the circuit of Fig. 9.48.

(a) The peak charging current is

$$I_1 = \frac{V_s}{R_1} = \frac{240}{6 \times 10^3} = 40 \text{ mA}$$

(b) From Eq. (9.29),

$$t_{\text{charge}} = 5R_1C = 5 \times 6 \times 10^3 \times 2{,}000 \times 10^{-6} = 60 \text{ s} = 1 \text{ min}$$

(c) The peak discharging current is

$$I_2 = \frac{V_s}{R_2} = \frac{240}{12} = 20 \text{ A}$$

(d) The energy stored is

$$W = \frac{1}{2}CV_s^2 = \frac{1}{2} \times 2{,}000 \times 10^{-6} \times 240^2 = 57.6 \text{ J}$$

(e) The energy stored in the capacitor is dissipated across the lamp during the discharging period. From Eq. (9.30),

$$t_{\text{discharge}} = 5R_2C = 5 \times 12 \times 2{,}000 \times 10^{-6} = 0.12 \text{ s}$$

Thus, the average power dissipated is

$$P = \frac{W}{t_{\text{discharge}}} = \frac{57.6}{0.12} = 480 \text{ W}$$

Practice Problem 9.14

The flash unit of a camera has a 2-mF capacitor charged to 80 V.

(a) How much charge is on the capacitor?
(b) What is the energy stored in the capacitor?
(c) If the flash fires in 0.8 ms, what is the average current through the flashtube?
(d) How much power is delivered to the flashtube?
(e) After a picture has been taken, the capacitor needs to be recharged by a power unit that supplies a maximum of 5 mA. How much time does it take to charge the capacitor?

Answer: (a) 0.16 C; (b) 6.4 J; (c) 200 A; (d) 8 kW; (e) 32 s

9.11 Summary

1. A capacitor consists of two (or more) parallel plates separated by a dielectric material.
2. The capacitance of a capacitor is the ratio of the charge on one plate to the voltage across the plates.
$$C = \frac{Q}{V}$$
3. A capacitor stores energy in the electric field between its plates. At any given time t, the energy stored in a capacitor is
$$W = \frac{1}{2}CV^2$$
4. Capacitors in series and parallel are combined in the same way as conductances.
5. The current through a capacitor is directly proportional to the time rate of change of the voltage across it; that is,
$$i = C\frac{dv}{dt}$$
6. The current through a capacitor is zero unless the voltage is changing. Thus, a capacitor acts like an open circuit to a dc source.
7. The voltage across a capacitor cannot change instantly.
8. A capacitor charges and discharges according to exponential curves. The natural response (the discharging equation in this case) has the form
$$v_C(t) = V_o e^{-t/\tau}, \qquad \tau = RC$$
where V_o is the initial voltage and τ is the time constant, which is the time required for the response to decay to $1/e$ of its initial value.
9. Troubleshooting a circuit with capacitors is a skill that one develops by doing.
10. *RC* circuits can be analyzed using PSpice and Multisim.
11. Two application circuits considered in this chapter are those of delay circuits and the photoflash unit.

Review Questions

9.1 What charge is on a 5-mF capacitor when it is connected across a 120-V source?

(a) 600 mC (b) 300 mC
(c) 24 mC (d) 12 mC

9.2 Capacitance is measured in:

(a) coulombs (b) joules
(c) henrys (d) farads

9.3 When the total charge in a capacitor is doubled, the energy stored:

(a) remains the same
(b) is halved
(c) is doubled
(d) is quadrupled

9.4 Polarized ceramic capacitors do not exist.

(a) True (b) False

9.5 Can the voltage waveform in Fig. 9.50 be associated with a capacitor?

(a) Yes (b) No

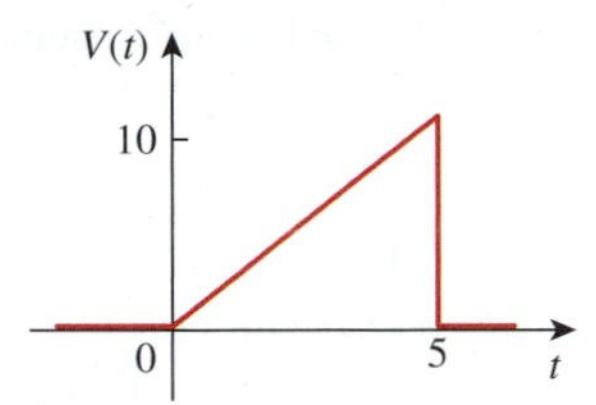

Figure 9.50
For Review Question 9.5.

9.6 The total capacitance of two 40-mF series-connected capacitors in parallel with a 4-mF capacitor is:

(a) 3.8 mF (b) 5 mF
(c) 24 mF (d) 44 mF (e) 84 mF

9.7 The capacitance of a parallel-plate capacitor is increased by:

(a) decreasing the dielectric constant
(b) decreasing the plate area
(c) decreasing the separation distance
(d) decreasing the charge per plate

9.8 An *RC* circuit has $R = 2\ \Omega$ and $C = 4$ F. The time constant is:

(a) 0.5 s (b) 2 s
(c) 4 s (d) 8 s (e) 15 s

9.9 A capacitor in an *RC* circuit with $R = 2\ \Omega$ and $C = 4$ F is being charged. The time required for the capacitor voltage to reach 63.2 percent of its steady-state value is:

(a) 2 s (b) 4 s
(c) 8 s (d) 16 s (e) none of the above

9.10 In the circuit of Fig. 9.51, the capacitor voltage just before $t = 0$ is:

(a) 10 V (b) 7 V
(c) 6 V (d) 4 V (e) 0 V

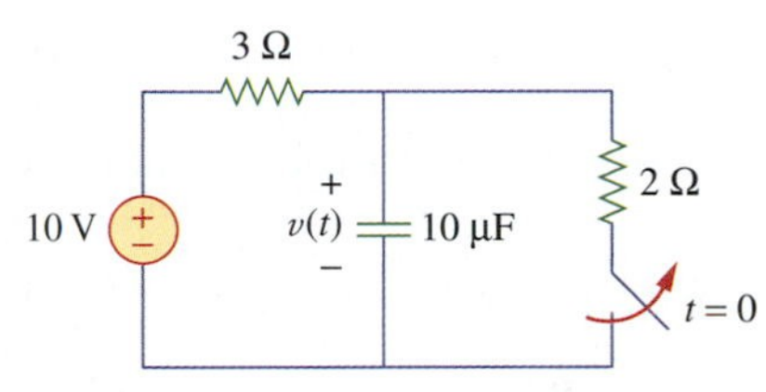

Figure 9.51
For Review Question 9.10.

Answers: 9.1a, 9.2d, 9.3d, 9.4a, 9.5b, 9.6c, 9.7c, 9.8d, 9.9c, 9.10d

Problems

Section 9.1 Capacitors

9.1 (a) Convert 268 pF to μF.

(b) Convert 0.045 μF to pF.

(c) Convert 0.0024 nF to pF.

9.2 (a) Find the charge given that $C = 2\ \mu$F and $V = 100$ V.

(b) Determine the voltage when $C = 40\ \mu$F and $Q = 30$ mC.

(c) Calculate the capacitance when $Q = 50$ mC and $V = 2$ kV.

9.3 A 20-μF capacitor has a 450-μC charge on the plates. Determine the voltage across the capacitor.

9.4 Determine the charge and energy stored in each of the following capacitors:

(a) a 5-μF capacitor with a voltage of 20 V
(b) a 6-nF capacitor with a voltage of 9 V

9.5 Applying 30 V across the plates of a parallel-plate capacitor results in a charge of 450 μC on its plates. Find the capacitance of the capacitor.

9.6 The terminals of a 20-mF capacitor are maintained at 12 V. Calculate the energy stored in the capacitor.

9.7 Two capacitors are identical except that the second is charged to five times the voltage of the first. Compare the energy stored in the capacitors.

9.8 A 2-μC charge raises the potential difference of a capacitor to 100 V. (a) Find the capacitance of the capacitor. (b) Determine how much charge must be removed to drop the voltage to 20 V. (c) What is the potential difference across the capacitor when the charge is increased to 4-μC?

Section 9.3 Electric Fields

9.9 A ceramic parallel-plate capacitor has plates each with an area of 0.2 m^2 and separation distance of 0.5 mm. Calculate the capacitance.

9.10 Calculate the capacitance of a parallel-plate capacitor when the area of each plate is 40 cm^2, the plates are 0.25 mm apart, and the dielectric is: (a) air, (b) mica, and (c) ceramic.

9.11 A parallel-plate capacitor has its plates 1 cm apart with an area of 0.02 m^2 and the plates are separated by mica. If 120 V is applied across the plates, determine:

(a) the capacitance

(b) the electric field intensity between the plates

(c) the charge on each plate

9.12 A capacitor has 56 μC of charge on each plate. If the electric flux density is 2 mC/m^2, find the plate area.

9.13 Find the electric field strength in a capacitor when 75 V is applied to its plates. Take the dielectric thickness as 0.2 mm.

9.14 A parallel-plate capacitor has dimensions 1.2 cm × 1.6 cm and separation of 0.15 mm. If the plates are separated by mica, calculate the capacitance.

9.15 When a capacitor uses air as a dielectric, its capacitance is 4 μF. When a dielectric material is inserted between the plates, its capacitance becomes 12 μF. Find the dielectric constant of the material.

9.16 The capacitance of an air capacitor is 10 nF. If the distance between its plates is doubled and mica ($\varepsilon_r = 5.0$) is inserted between the plates, determine the new capacitance of the capacitor.

9.17 Compare the capacitances of two parallel-plate capacitors C_1 and C_2 that are identical—except that C_1 is air-filled while C_2 has porcelain as the dielectric.

9.18 A 5-μF capacitor is to be constructed from a rolled sheet of aluminum foil separated by a layer of dry paper 0.2 mm thick. Calculate the area for each sheet of foil.

Section 9.5 Series and Parallel Capacitors

9.19 What is the total capacitance of four 30-mF capacitors connected: (a) in parallel and (b) in series?

9.20 Two capacitors (20 μF and 30 μF) are connected to a 100-V source. Find the energy stored in each capacitor if they are connected: (a) in parallel and (b) in series.

9.21 The equivalent capacitance at terminals *a-b* in the circuit of Fig. 9.52 is 30 μF. Calculate the value of C.

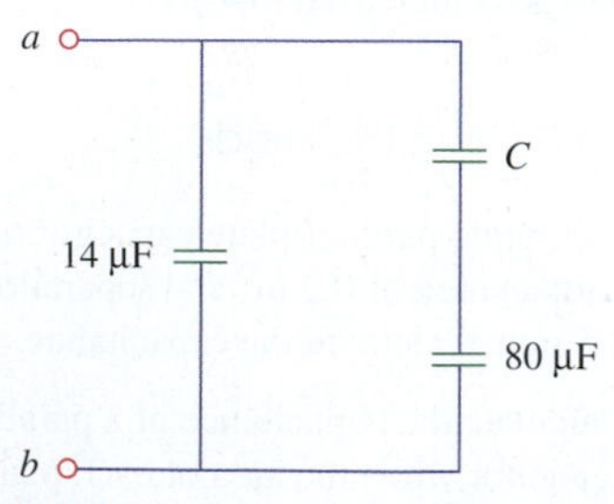

Figure 9.52
For Problem 9.21.

9.22 Determine the equivalent capacitance for each of the circuits in Fig. 9.53.

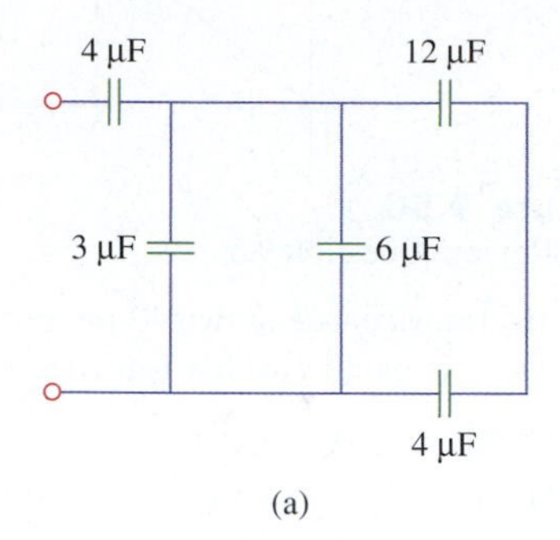

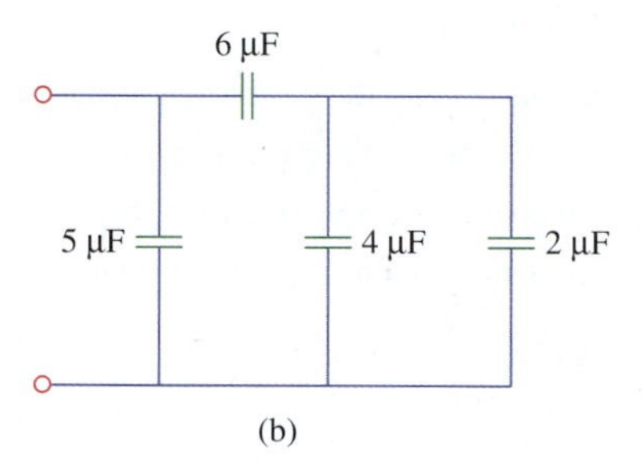

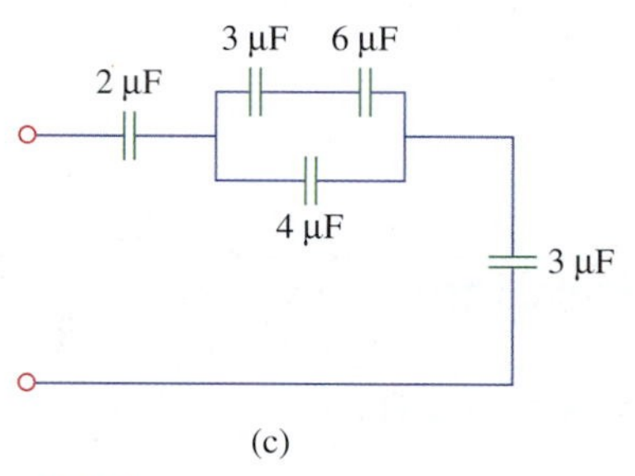

Figure 9.53
For Problem 9.22.

9.23 Find C_{eq} for the circuit in Fig. 9.54.

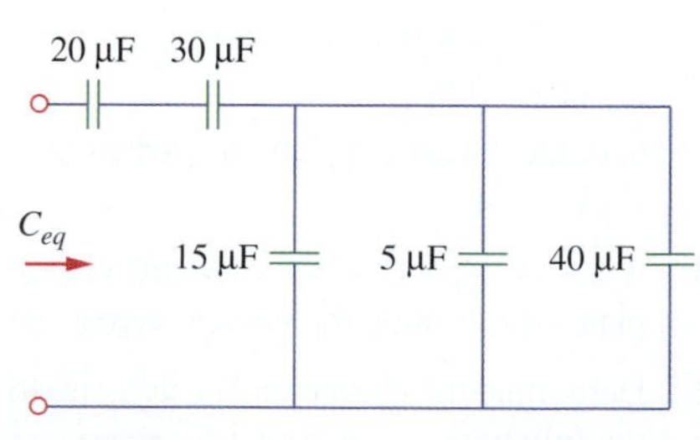

Figure 9.54
For Problem 9.23.

9.24 Find the equivalent capacitance between terminals *a* and *b* in the circuit of Fig. 9.55. All capacitances are in μF.

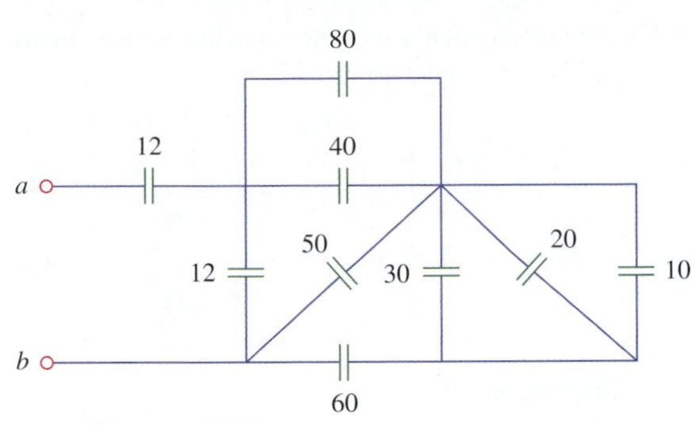

Figure 9.55
For Problem 9.24.

9.25 Calculate the equivalent capacitance for the circuit in Fig. 9.56. All capacitances are in mF.

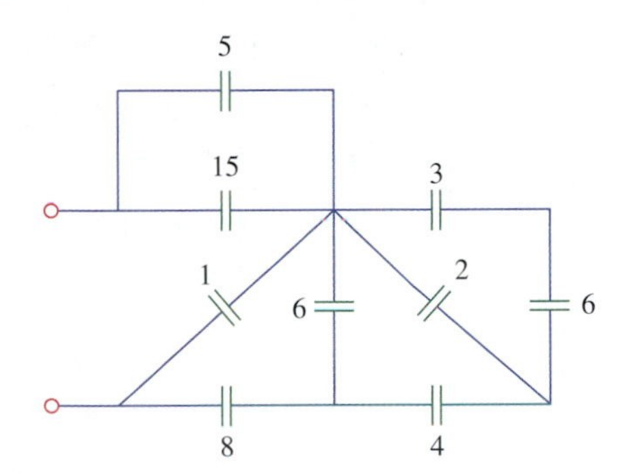

Figure 9.56
For Problem 9.25.

9.26 For the circuit in Fig. 9.57, determine: (a) the voltage across each capacitor and (b) the energy stored in each capacitor.

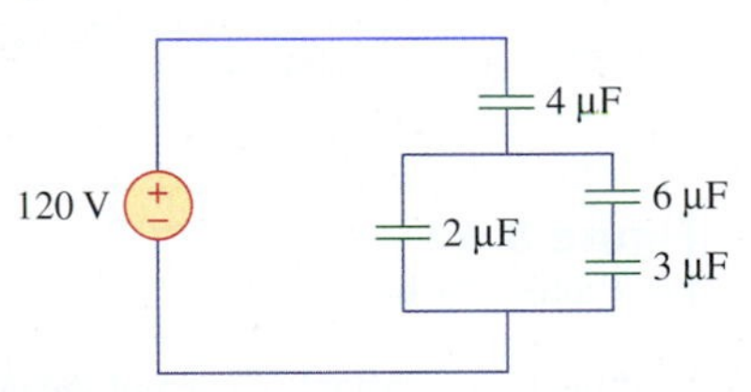

Figure 9.57
For Problem 9.26.

9.27 Three capacitors $C_1 = 5\ \mu F$, $C_2 = 10\ \mu F$, and $C_3 = 20\ \mu F$ are connected in parallel across a 150-V source. Determine: (a) the total capacitance, (b) the charge on each capacitor, and (c) the total energy stored in the parallel combination.

9.28 The three capacitors in Problem 9.27 are placed in series with a 200-V source. Compute: (a) the total capacitance, (b) the charge on each capacitor, and (c) the total energy stored in the series combination.

9.29 A 20-μF capacitor and a 50-μF capacitor are connected in parallel to a 200-V source. (a) Find the total capacitance. (b) Determine the magnitude of the charge stored by each capacitor. (c) Calculate the voltage across each capacitor.

9.30 The voltage across the 10-μF capacitor in Fig. 9.58 is 20 V. (a) What is the source voltage V_s? (b) Determine the total charge on the capacitors.

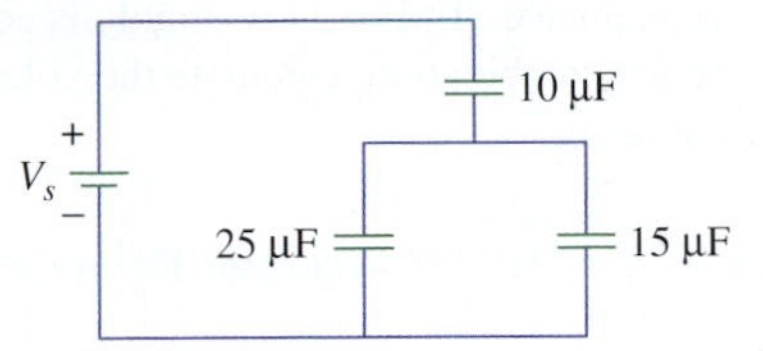

Figure 9.58
For Problem 9.30.

9.31 Three capacitors are connected in series such that the equivalent capacitance is 2.4 nF. If $C_1 = 2C_2$ and $C_3 = 10C_1$, determine the values of C_1, C_2, and C_3.

9.32 A series circuit contains 10-, 40-, and 60-μF capacitors across a dc source of 120 V. Calculate the voltage across the 40-μF capacitor.

9.33 Find the total capacitance of the series-parallel connection of capacitors in Fig. 9.59.

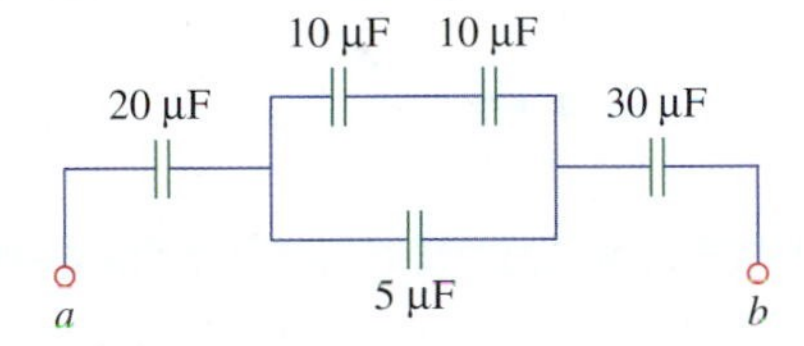

Figure 9.59
For Problem 9.33.

9.34 Points A and B in Fig. 9.60 are connected to a 120-V dc source. Find the amount of the total energy stored.

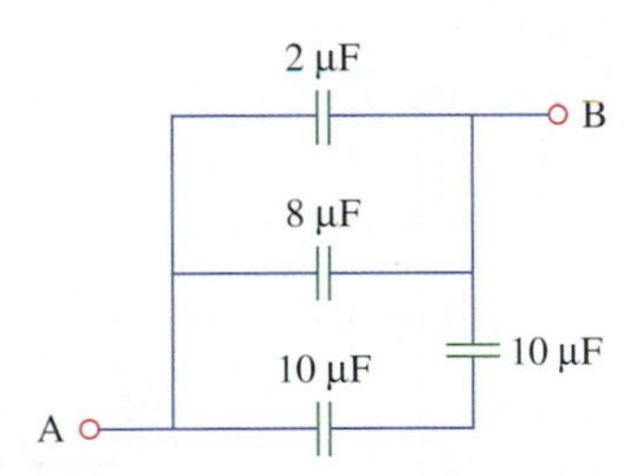

Figure 9.60
For Problem 9.34.

9.35 Calculate the equivalent capacitance for the circuit in Fig. 9.61.

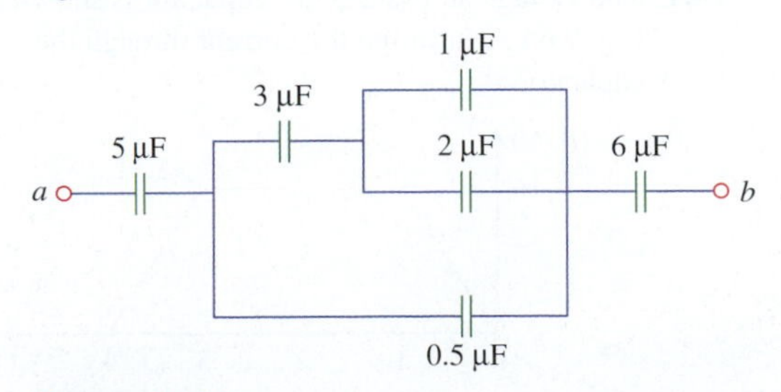

Figure 9.61
For Problem 9.35.

9.36 An 80-μF capacitor is connected in parallel with a 40-μF capacitor. The two are connected in series with a 30-μF capacitor. (a) Determine the total capacitance. (b) If a 24-V supply is connected to the series combination, calculate the voltage across each capacitor.

Section 9.6 Current–Voltage Relationship

9.37 In 5 s, the voltage across a 40-mF capacitor changes from 160 to 220 V. Calculate the average current through the capacitor.

9.38 If the voltage waveform in Fig. 9.62 is applied to a 20-μF capacitor, find the current $i(t)$ through the capacitor.

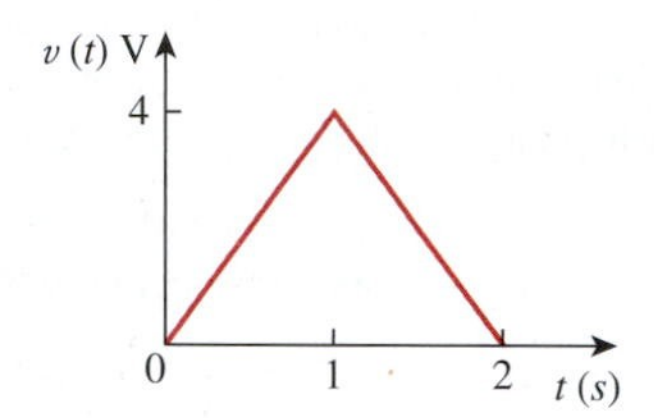

Figure 9.62
For Problem 9.38.

9.39 A 20-μF capacitor is connected in parallel with a 40-μF capacitor and across a time-varying voltage source. If the total current supplied by the source is 5 A at a certain instant, what are the instantaneous currents through the individual capacitors?

9.40 The voltage waveform in Fig. 9.63 is applied across a 30-μF capacitor. Draw the current waveform through it.

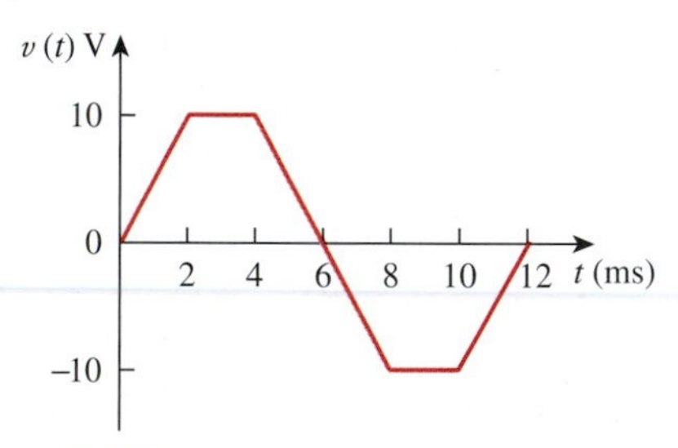

Figure 9.63
For Problem 9.40.

9.41 The voltage across a 2-mF capacitor is shown in Fig. 9.64. Determine the current through the capacitor.

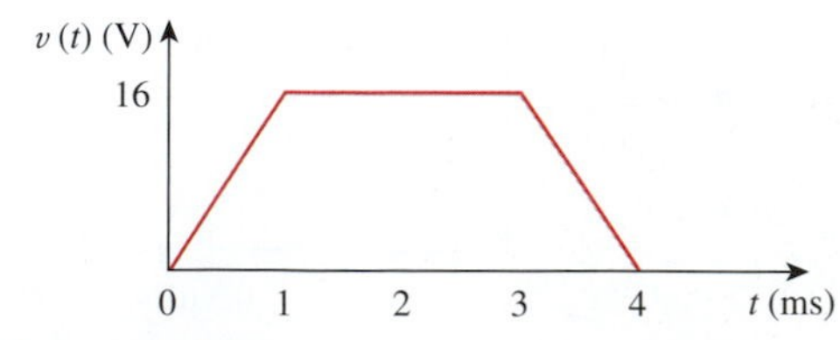

Figure 9.64
For Problem 9.41.

9.42 Find the voltage across the capacitors in the circuit of Fig. 9.65 under dc conditions.

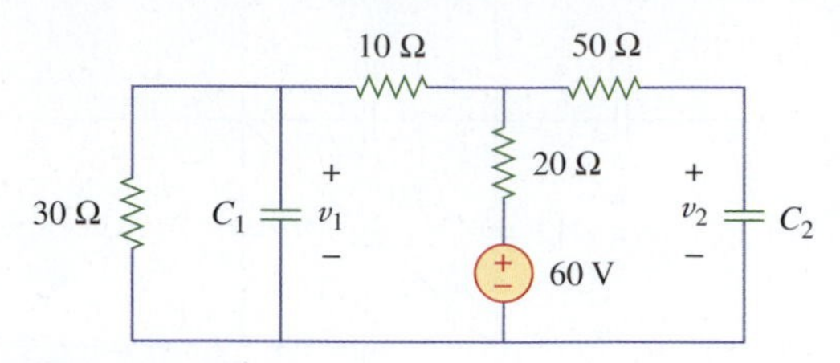

Figure 9.65
For Problem 9.42.

9.43 The voltage across a 100-μF capacitor is shown in Fig. 9.66. Determine the current $i(t)$ through it.

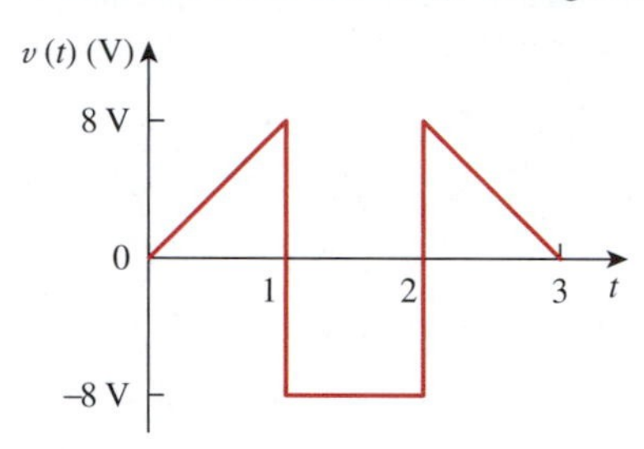

Figure 9.66
For Problem 9.43.

Section 9.7 Charging and Discharging a Capacitor

9.44 Find the time constant for the RC circuit in Fig. 9.67.

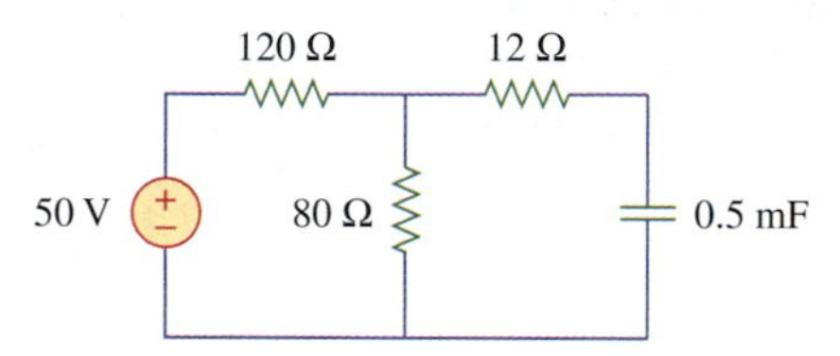

Figure 9.67
For Problem 9.44.

9.45 Find the time constant of each of the circuits in Fig. 9.68.

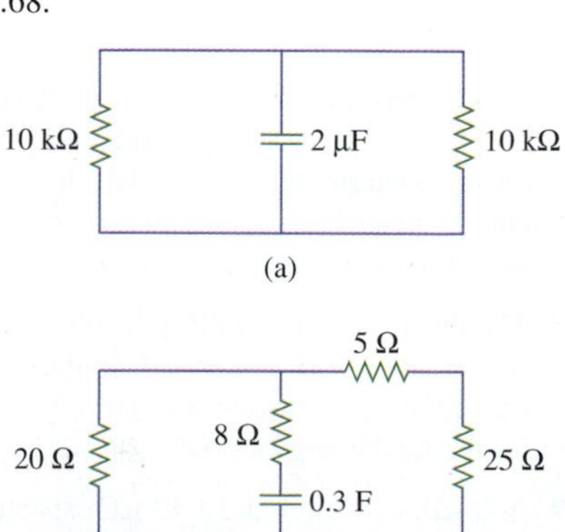

Figure 9.68
For Problem 9.45.

9.46 Calculate the time constant for each of the following *RC* circuits:

(a) $R = 56\ \Omega$, $C = 2\ \mu$F

(b) $R = 6.4\ \text{M}\Omega$, $C = 50$ pF

9.47 The switch in Fig. 9.69 has been closed for a long time, and it opens at $t = 0$. Find $v(t)$ for $t \geq 0$.

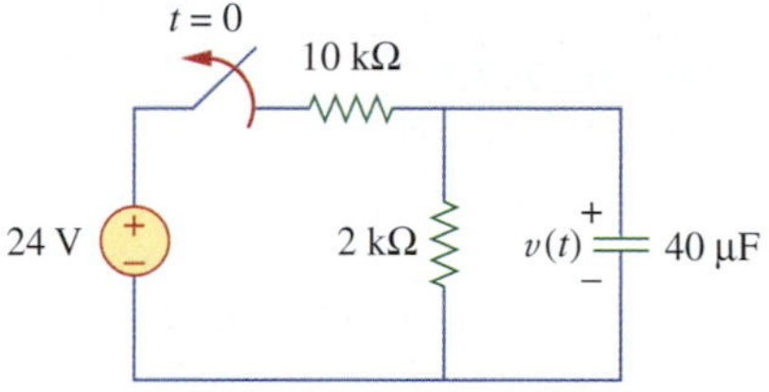

Figure 9.69
For Problems 9.47 and 9.67.

9.48 For the circuit in Fig. 9.70,

$$v(t) = 10e^{-4t}\ \text{V} \quad \text{and} \quad i(t) = 0.2e^{-4t}\ A, \quad t > 0$$

(a) find *R* and *C*,

(b) determine the time constant.

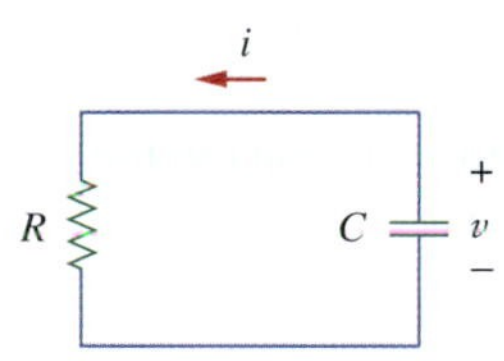

Figure 9.70
For Problem 9.48.

9.49 In the circuit of Fig. 9.71, $v(0) = 20$ V. Find $v(t)$ for $t > 0$.

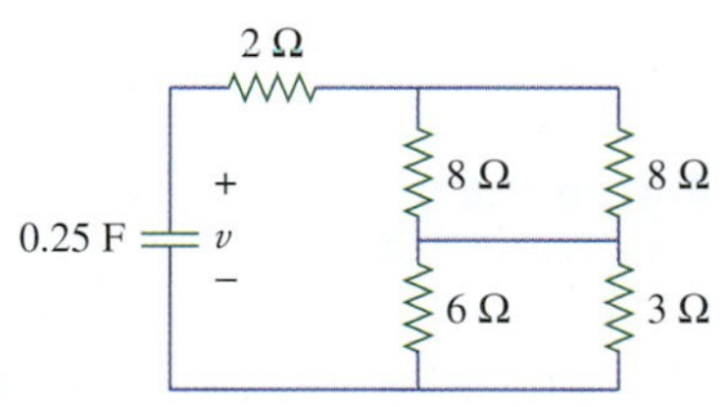

Figure 9.71
For Problem 9.49.

9.50 Given that $i(0) = 3$ A, find $i(t)$ for $t > 0$ in the circuit in Fig. 9.72.

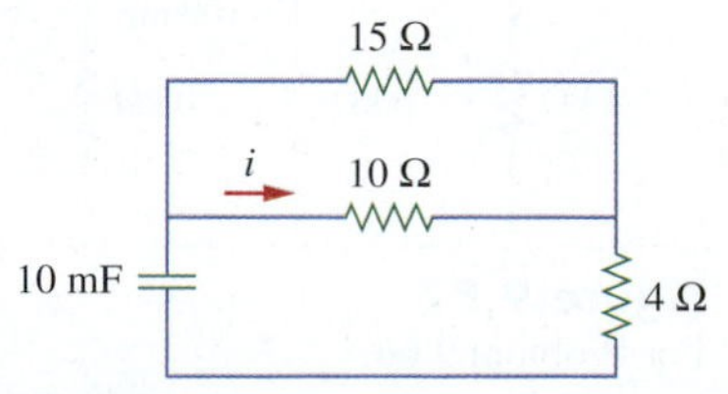

Figure 9.72
For Problem 9.50.

9.51 The switch in Fig. 9.73 is turned on at $t = 0$. Find v_C and i_C for $t > 0$.

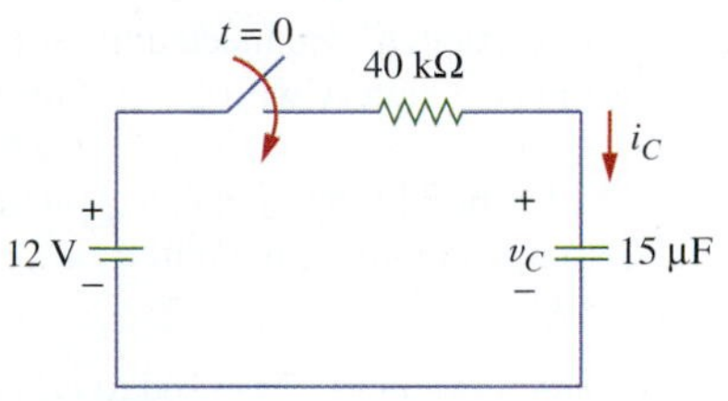

Figure 9.73
For Problem 9.51.

9.52 The switch in Fig. 9.74 has been in position 1 for a long time. If the switch is thrown into position 2 at $t = 0$, determine v_C and i_C.

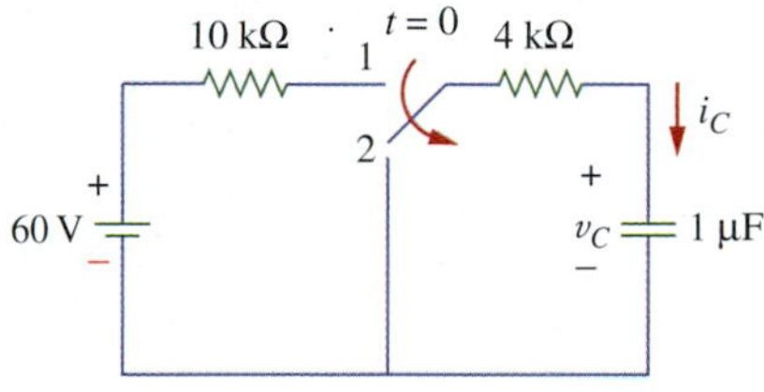

Figure 9.74
For Problem 9.52.

9.53 After being open for a long time, the switch in Fig. 9.75 is closed at $t = 0$. Write the equation for v_C.

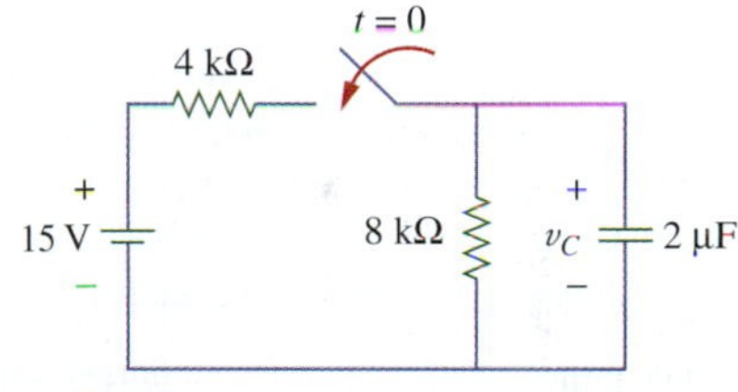

Figure 9.75
For Problem 9.53.

9.54 How long would it take to discharge a 500-pF capacitor from 120 to 80 V through a 200-kΩ resistor?

9.55 Determine how long it would take to discharge a 40-μF capacitor from 40 to 10 V through a 100-kΩ resistor.

9.56 A 1-nF capacitor and a 200-kΩ resistor are connected in series to an 80-V dc source. How long would it take the capacitor to charge from 0 to 40 V?

9.57 Determine the time it will take the capacitor in Fig. 9.76 to discharge to 20 V after the switch is closed.

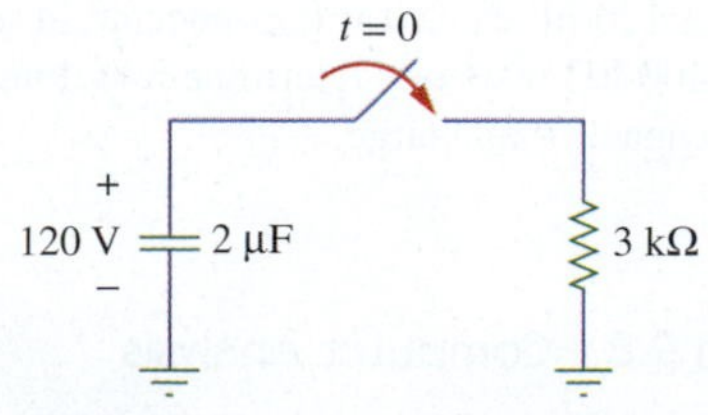

Figure 9.76
For Problem 9.57.

9.58 A 2-μF capacitor is connected in series with a 6-MΩ resistor and a battery with 24 V. If the capacitor is uncharged at time $t = 0$, find: (a) the time constant, (b) the fraction of the final charge at time $t = 34$ s, and (c) the fraction of the initial current that remains at $t = 34$ s.

9.59 The circuit in Fig. 9.77 is used to charge two capacitors. Find: (a) the charge on each capacitor after the switch has been closed for a long time, (b) the potential across each capacitor after the switch has been closed for a long time, and (c) the time it takes for the charges and potentials to reach half their final values.

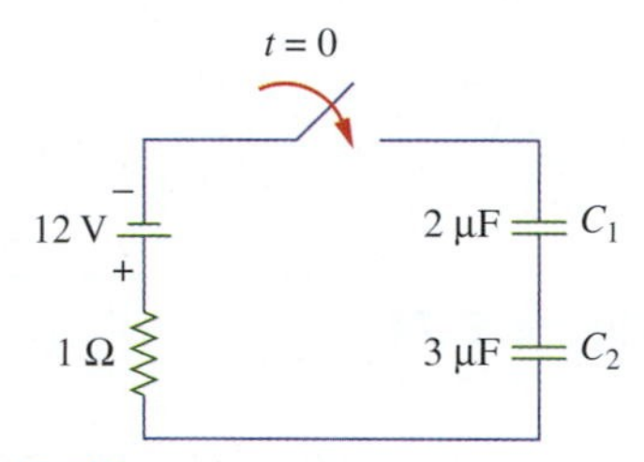

Figure 9.77
For Problem 9.59.

9.60 A 0.2-μF capacitor has an initial voltage of $v_C(0) = 10$ V. The voltage across the capacitor is $v_C(t) = ke^{-\beta t}$ V, while the current through the capacitor is $i_C(t) = 2e^{-\beta t}$ mA for $t > 0$. Determine the values of the constants k and β.

9.61 The capacitor in Fig. 9.78 is charged to 12 V. How long after the switch is opened would the capacitor voltage reach zero?

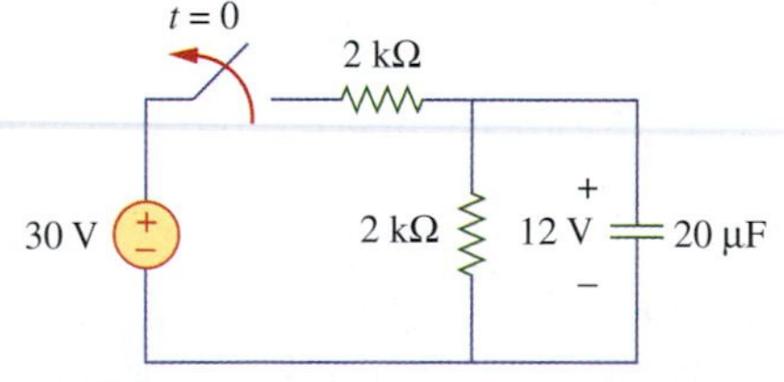

Figure 9.78
For Problem 9.61.

9.62 A 120-nF capacitor is connected in series with a 400-kΩ resistor. Determine how long it takes the capacitor to charge.

Section 9.8 Computer Analysis

9.63 Use PSpice to determine $v(t)$ for $t > 0$ in the circuit of Fig. 9.79.

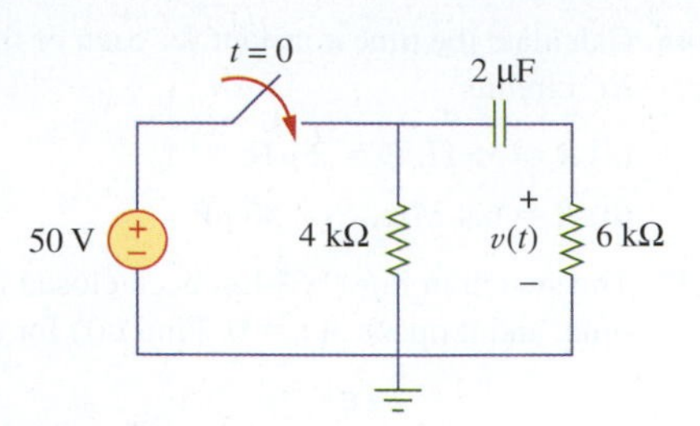

Figure 9.79
For Problem 9.63.

9.64 The switch in Fig. 9.80 moves from point A to B at $t = 0$. Use PSpice to find $v(t)$ for $t > 0$.

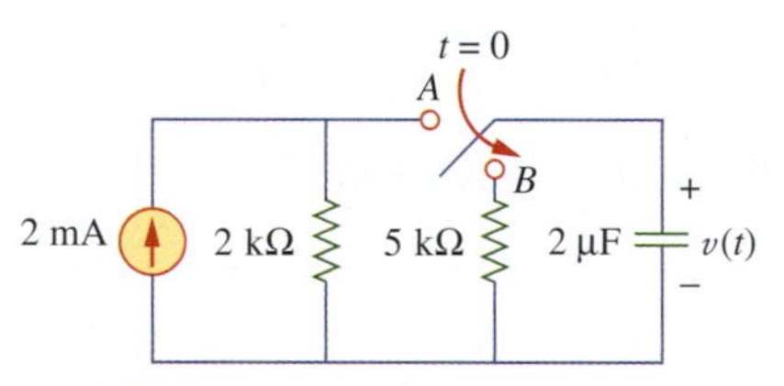

Figure 9.80
For Problem 9.64.

9.65 Use PSpice to find $v(t)$ for $t > 0$ for the circuit in Fig. 9.81.

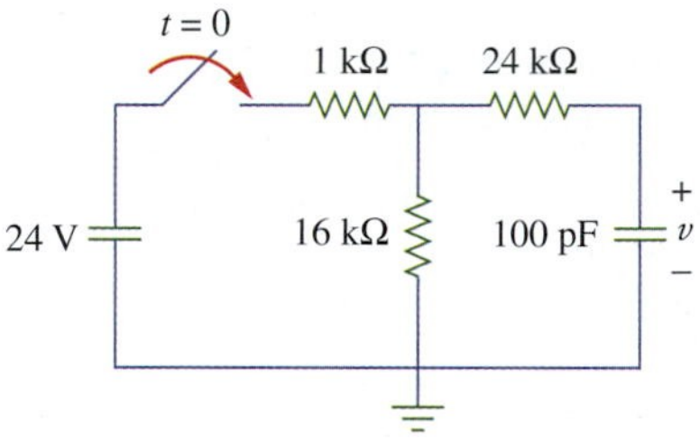

Figure 9.81
For Problem 9.65.

9.66 The switch in Fig. 9.82 closes at $t = 0$. Use PSpice to determine $v(t)$ for $t > 0$.

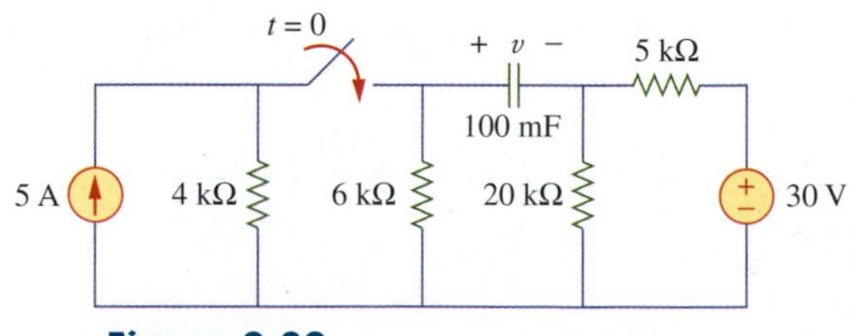

Figure 9.82
For Problem 9.66.

9.67 Use PSpice to solve Problem 9.47.

9.68 Use Multisim to find v_c in Fig. 9.30 (for Practice 9.9).

9.69 Use Multisim to determine $v(t)$ in the circuit of Fig. 9.83.

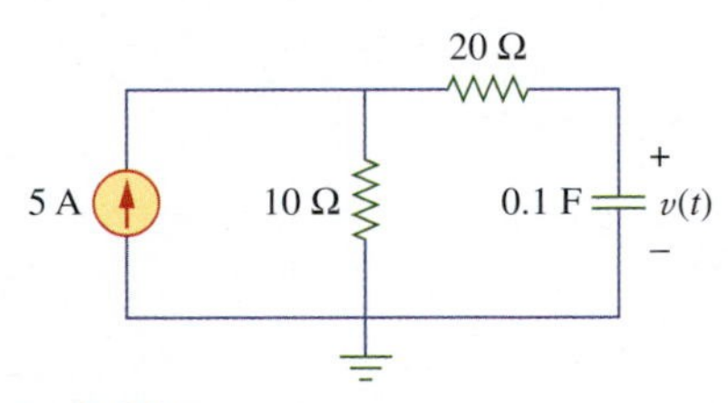

Figure 9.83
For Problem 9.69.

9.70 The voltage source in Fig. 9.84 produces a square wave that cycles between 0 and 10 V with a frequency of 2,000 Hz. Use Multisim to find $v_o(t)$.

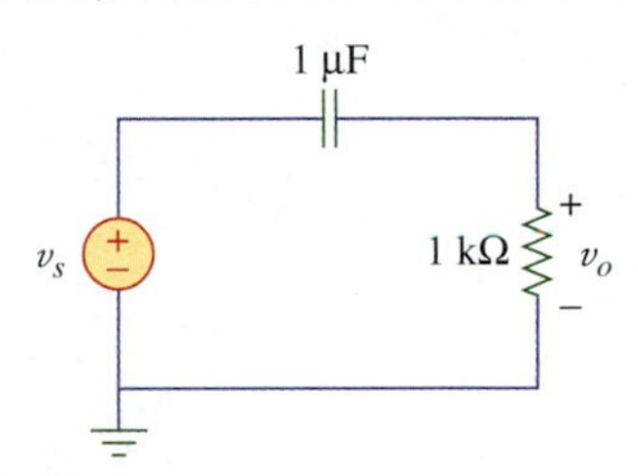

Figure 9.84
For Problem 9.70.

Section 9.9 Troubleshooting

9.71 You are testing a capacitor with an ohmmeter. After connecting the leads across the capacitor, you find that the reading goes to zero and stays there. What could be wrong with the capacitor?

Section 9.10 Applications

9.72 In designing a signal switching circuit, it was found that a 100-μF capacitor was needed for a time constant of 3 ms. What value resistor is necessary for the circuit?

9.73 Your laboratory has a large number of 10-μF capacitors available that are rated at 300 V. To design a capacitor bank of 40-μF rated at 600 V, how many 10-μF capacitors are needed, and how would you connect them?

9.74 When a capacitor is connected to a dc source, its voltage rises from 20 to 35 V in 4 μs with an average charging current of 0.6 A. Determine the value of the capacitance.

9.75 In an electric power plant substation, a capacitor bank is made of 10 capacitor strings connected in parallel. Each string consists of eight 1,000-μF capacitors connected in series with each capacitor charged to 100 V. (a) Calculate the total capacitance of the bank. (b) Determine the total energy stored in the bank.

chapter

10

Inductance

Marriage is like a cage; one sees the birds outside desperate to get in, and those inside equally desperate to get out.

—Michel de Montaigne

Historical Profiles

Joseph Henry (1797–1878), an American physicist, discovered inductance and constructed an electric motor.

Born in Albany, New York, Henry graduated from Albany Academy and taught philosophy at Princeton University from 1832 to 1846. He was the first secretary of the Smithsonian Institution. He conducted several experiments on electromagnetism and developed powerful electromagnets that could lift objects weighing thousands of pounds. It is interesting to note that Joseph Henry discovered electromagnetic induction before Faraday but failed to publish his findings. The unit of inductance, the henry, was named after him.

Joseph Henry
National Oceanic and Atmospheric Administration/Department of Commerce

Heinrich Lenz (1804–1865), a Russian physicist, formulated Lenz's law, a fundamental law of electromagnetism.

Born and educated in Dorpat (now Tartu, Estonia), Lenz studied chemistry and physics at the University of Dorpat. He traveled with Otto von Kotzebue on his third expedition around the world from 1823 to 1826. After the voyage, Lenz began working and studying electromagnetism at the University of St. Petersburg. Through several experiments, Lenz discovered the principle of electromagnetism, which defines the polarity of an induced voltage in a coil. Lenz also studied the relationship between heat and current. Independently of English physicist James Joule, in 1842 Lenz discovered the law that we now call Joule's law.

Heinrich Lenz
AIP Emilio Segre Visual Archives, E. Scott Barr Collection

10.1 Introduction

So far we have considered two types of passive elements: resistors and capacitors. In this chapter, we consider the third passive element—the inductor. An inductor is an electrical component that stores energy in its magnetic field. An inductor is a simple electronic component—in its simplest form, an inductor is simply a coil of wire. It turns out, however, that a coil of wire can do some very interesting things because of the magnetic properties of a coil. Inductors are used extensively in analog circuits. Although applications using inductors are less common than those using capacitors, inductors are very common in high frequency circuits. Inductors are not often used in integrated circuit (IC) chips because it is difficult to fabricate them onto the chips.

We begin the chapter by introducing electromagnetic induction, the basis of Faraday's and Lenz's laws. We describe inductors as storage elements. We examine different types of inductors and how to combine them in series or parallel. Later, we consider *RL* circuits (circuits containing resistor and inductor) and how to model them using PSpice and Multisim. As typical applications of inductors, we discuss relay and automobile ignition circuits.

10.2 Electromagnetic Induction

A magnet is surrounded by a magnetic field. A magnetic field can be thought of consisting of lines of force, or flux lines. The forces of magnetic attraction and repulsion move along the lines of force. An electromotive force (emf) is generated when a magnetic flux passes through a coil or conductor. This is known as *electromagnetic induction*. Michael Faraday and Joseph Henry independently observed that when the magnetic flux linking a conductor changes, voltage is induced across the terminals of the conductor. This leads to Faraday's law:

> **Faraday's law** states that the voltage induced in a circuit is proportional to the rate of change of the magnetic flux linking the circuit.

Mathematically,

$$v = N\frac{d\phi}{dt} = N \times \text{Rate of change in flux} \tag{10.1}$$

where v = induced voltage in volts (V), N is the number of turns of the coil, and $d\phi/dt$ is rate of change in the magnetic flux in webers per second (Wb/s). As illustrated in Fig. 10.1, an emf is induced in the coil

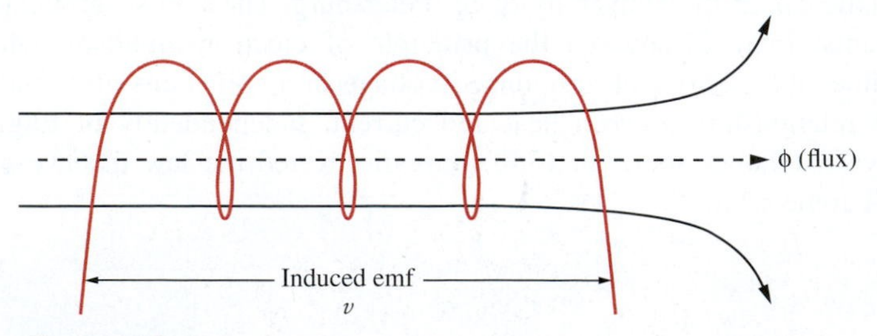

Figure 10.1
Externally produced flux linking a coil of *N* turns.

only when the magnetic flux linking the coil changes; that is, $d\phi/dt \neq 0$. When the flux does not change with time, $d\phi/dt = 0$ and $v = 0$.

The polarity of the induced emf and the direction of current flow can be determined by applying Lenz's law:

> **Lenz's law** states that the induced current always develops a flux that opposes the change producing the current.

We apply Faraday's law to determine the magnitude of the induced voltage and use Lenz's law to determine its polarity.

Example 10.1

A coil with 200 turns is located in a magnetic field that changes at the rate of 30 mWb/s. Find the induced voltage.

Solution:

$$v = N\frac{d\phi}{dt} = N \times \text{Rate of change in flux} = 200 \times 30 \times 10^{-3} = 6 \text{ V}$$

Practice Problem 10.1

The flux linking a coil is increased from 0 to 40 mWb in 2 seconds. Find the number of turns if the induced voltage is 2.5 V.

Answer: 125

10.3 Inductors

An inductor is a passive element designed to store energy in its magnetic field. Inductors find numerous applications in electronic and power systems. For example, they are used in power supplies, transformers, radios, TVs, radar, and electric motors.

Any conductor of electric current has inductive properties and may be regarded as an inductor. But in order to enhance the inductive effect, a practical inductor is usually formed into a cylindrical coil with many turns of conducting wire, as shown in Fig. 10.2. Thus,

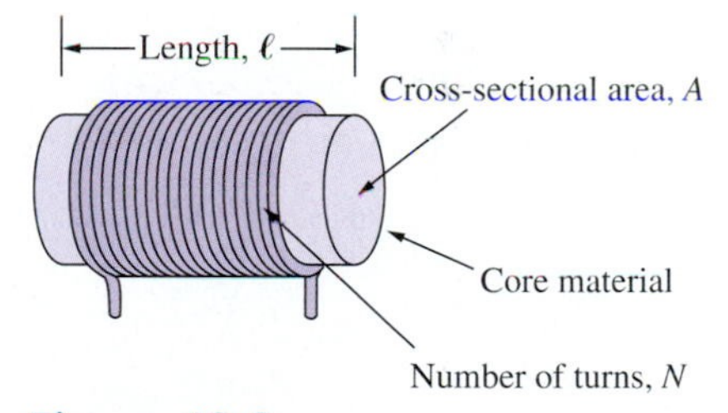

Figure 10.2
Typical form of an inductor.

> An **inductor** consists of a coil of conducting wire wound around some core which may be air or some magnetic material.

(This definition is true for most inductors but not true for some inductors.) The circuit symbol for an inductor is shown in Fig. 10.3, it looks like a coil of wire.

Figure 10.3
Circuit symbol for an ideal inductor.

If current is allowed to pass through an inductor, it is found that the voltage across the inductor is directly proportional to the time rate of change of the current. Using the passive sign convention,[1] the voltage across an inductor is given as

$$v = L\frac{di}{dt} = L \times \text{Rate of change of current} \tag{10.2}$$

[1] Note: In view of Eq. (10.2), for an inductor to have voltage across its terminals, its current must vary with time. Hence, $v = 0$ for constant current through the inductor.

where L is the constant of proportionality called the *inductance* of the inductor[2] and di/dt is the instantaneous time rate of change of current through the inductor. The unit of inductance is the henry (H) in honor of the American inventor Joseph Henry (1797–1878).

Inductance is a measure of an inductor's ability to exhibit opposition to the change of current flowing through it, measured in henrys (H).

The inductance of an inductor depends on its physical dimension and construction. Formulas for calculating the inductance of inductors of different shapes are derived from electromagnetic theory and can be found in standard electrical engineering handbooks. For example, for the inductor (solenoid[3]) shown in Fig. 10.2,

$$L = \frac{\mu_o \mu_r N^2 A}{\ell} \tag{10.3}$$

where

N = the number of turns of wire
ℓ = the length of the coil in meters
A = the cross sectional area in square meters
μ_o = the permeability of the air or free space = $4\pi \times 10^{-7}$ H/m
μ_r = the relative permeability of the core

Permeability ($\mu = \mu_o \mu_r$) is the ability of a material to support magnetic flux. In other words, it is a material property that describes the ease with which a magnetic flux is established in the material. The relative permeability (μ_r) is the ratio of the material's permeability (μ) to the permeability of free space (μ_o). For air and nonmagnetic metals such as copper, gold, and silver, $\mu_r = 1$. Table 10.1 presents values of relative permeability of some common materials. We can see from Eq. (10.3) that inductance can be increased by increasing the number of turns of coil, using material with higher permeability as core, increasing the cross-section area, or reducing the length of the coil.

TABLE 10.1

Relative permeability of some common materials*.

Material	Relative permeability (μ_r)
Air	1
Cobalt	250
Nickel	600
Soft iron	5,000
Silicon-iron	7,000
Nonmagnetic metals	1

*The values given are typical; they vary from one published source to another owing to different varieties of most materials.

[2] Both voltage v and rate of current change di/dt are *instantaneous:* which means in relation to a specific point in time; thus the lowercase letters v and i.

[3] A solenoid is a loop of wire, wrapped around a metallic core (see Fig. 10.2), which produces a magnetic field when an electrical current is passed through it. Solenoids have an enormous number of practical applications.

The circuit symbol shown in Fig. 10.3 assumes that the inductor is ideal. Like capacitors, inductors are not ideal. Practically, the coil is made of a conducting wire and therefore has a very small but finite resistance (R_w). Hence, the practical equivalent model of an inductor is shown in Fig. 10.4. Also, the conductive turns of the coil are separated and therefore a small stray capacitance (C_s) is accounted in the model in Fig. 10.4. Unless otherwise stated, both R_w and C_s can be ignored. C_s only becomes significant in very high frequency applications. We will assume ideal inductors in this book.

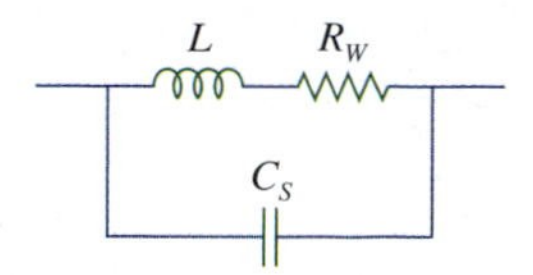

Figure 10.4
Circuit symbol for a practical inductor.

Example 10.2

The current through a 100-mH inductor changes from 300 to 500 mA in 2 ms. Find the voltage across the inductor.

Solution:
Because $v = L\, di/dt$ and $L = 100$ mH $= 0.1$ H,

$$v = 0.1 \times \frac{(500 - 300) \times 10^{-3}}{2 \times 10^{-3}} = 10 \text{ V}$$

Practice Problem 10.2

If the current in a 2-mH coil grows from 0 to 8 A in 0.4 s, find the induced voltage.

Answer: 40 mV

Example 10.3

An air-cored solenoid has diameter 1.2 cm and length 18 cm. If it has 500 turns, find its inductance. Take the permeability of air as $4\pi \times 10^{-7}$ H/m.

Solution:
The area is obtained

$$A = \pi(d/2)^2 = \pi(0.6 \times 10^{-2})^2 = 1.131 \times 10^{-4} \text{ m}^2$$

Because the solenoid is air-cored, its permeability is that of a vacuum. Hence,

$$L = \frac{N^2 \mu A}{\ell} = \frac{500^2 \times 4\pi \times 10^{-7} \times 1.131 \times 10^{-4}}{18 \times 10^{-2}} = 0.1974 \text{ mH}$$

Practice Problem 10.3

A 15-cm long inductor has 40 turns, a cross-sectional area of 0.02 m^2, and permeability of 0.3×10^{-4} H/m. Find the inductance.

Answer: 6.4 mH

10.4 Energy Storage and Steady-State DC

The inductor is designed to store energy in its magnetic field. To find the energy stored ($W = Pt = vit$), we assume a linearly increasing current and select a final value of current I_m flowing through the inductor.

The average value of the current as it rises from zero to I_m is $0.5I_m$. Therefore,

$$W = v \times 0.5I_m \times t \tag{10.4}$$

But from Eq. (10.2),

$$v = L\frac{di}{dt} = L \times \text{Rate of change of current} \tag{10.2}$$

Because the rate of change of current is constant,

$$v = L \times \frac{I_m}{t}$$

Substituting this in Eq. (10.4) gives the energy stored as

$$W = L \times \frac{I_m}{t} \times 0.5I_m \times t$$

or

$$\boxed{W = \frac{1}{2}LI_m^2} \tag{10.5}$$

We should note the following important properties of an inductor.

1. Note from Eq. (10.2) that the voltage across an inductor is zero when the current is constant. Thus, at steady state,

   An inductor acts like a short circuit to dc.

2. An important property of the inductor is its opposition to the change in current flowing through it; that is,

   The current through an inductor cannot change abruptly.

   According to Eq. (10.2), a discontinuous change in the current through an inductor requires an infinite voltage, which is not physically possible. Thus, an inductor opposes an abrupt change in the current through it. For example, the current through an inductor may take the form shown in Fig. 10.5(a), whereas the inductor current cannot take the form shown in Fig. 10.5(b) in real-life situations due to the discontinuities. However, the voltage across an inductor can change abruptly.
3. Like the ideal capacitor, the ideal inductor does not dissipate energy. The energy stored in it can be retrieved at a later time. The inductor retrieves power from the circuit when storing energy and delivers power to the circuit when returning previously stored energy.
4. A practical, nonideal inductor has a fairly significant resistive and capacitive components, as mentioned before and shown in Fig. 10.4.

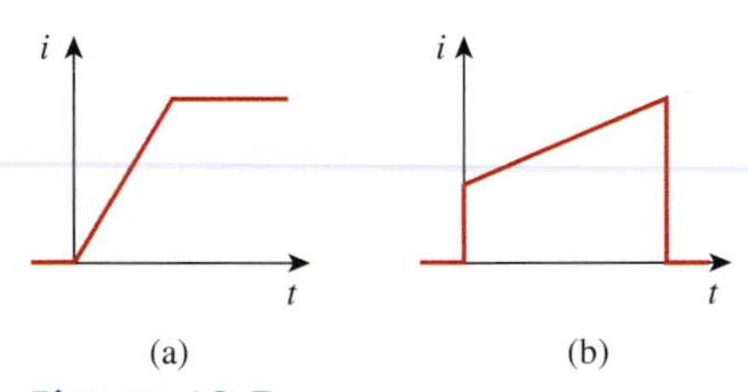

Figure 10.5
Current through an inductor: (a) allowed; (b) not allowable. Note: An abrupt change is not possible.

Example 10.4

Consider the circuit in Fig. 10.6(a). Under dc conditions, find:
(a) i, v_C, and i_L, and
(b) the energy stored in the capacitor and in the inductor.

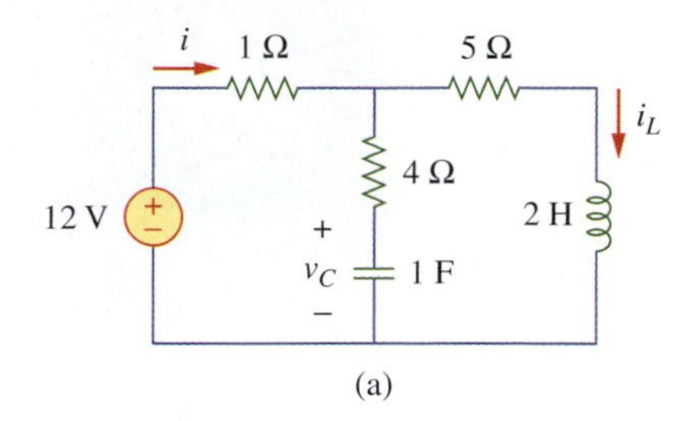

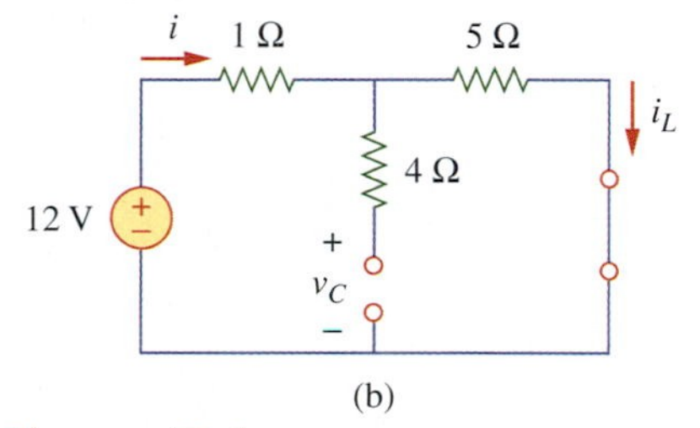

Figure 10.6
For Example 10.4.

Solution:
(a) Under dc conditions, we replace the capacitor with an open circuit and the inductor with a short circuit, as in Fig. 10.6(b). It is evident from Fig. 10.6(b) that

$$i = i_L = \frac{12}{1+5} = 2 \text{ A}$$

The voltage v_C is the same as the voltage across the 5-Ω resistor. Hence,

$$v_C = 5i = 10 \text{ V}$$

(b) The energy in the capacitor is

$$W_C = \frac{1}{2}Cv_C^2 = \frac{1}{2}(1)(10^2) = 50 \text{ J}$$

and that in the inductor is

$$W_L = \frac{1}{2}Li_L^2 = \frac{1}{2}(2)(2^2) = 4 \text{ J}$$

Practice Problem 10.4

Determine v_C, i_L, and the energy stored in the capacitor and inductor in the circuit of Fig. 10.7 under dc conditions.

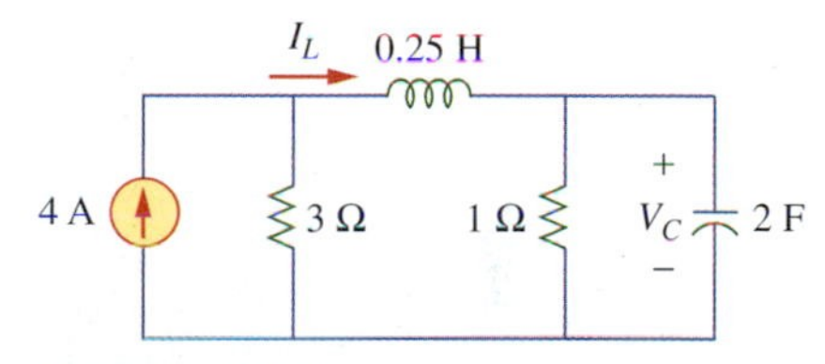

Figure 10.7
For Practice Problem 10.4.

Answer: 3 V; 3 A; 9 J; 1.125 J

10.5 Types of Inductors

Like capacitors, commercially available inductors come in different values and types. Typical practical inductors have inductance values ranging from a few microhenrys (μH) as in communication systems to tens of henrys (H) as in power systems. Inductors may be fixed or variable (adjustable). The circuit symbols for fixed and variable inductors are shown in Fig. 10.8. Both fixed and variable inductors may be classified according to the type of core material used. Common cores may be made of air, iron, or ferrite. The circuit symbols for these cores are shown in Fig. 10.9. In order to minimize losses, the core can be laminated steel plate insulated from each other. The type of core used depends on the intended application and frequency range of the inductor. Iron-core inductors have large inductance and are used in audio or power supply applications; they typically have inductances in the henrys ranges. Air- or ferrite-core inductors are generally used for radio-frequency applications. Air-core inductors are typically in the

Figure 10.8
Symbols for (a) fixed and (b) variable inductors.

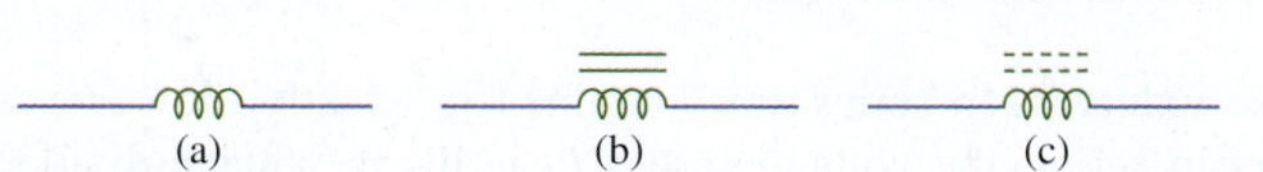

Figure 10.9
Symbols for various inductors: (a) air core; (b) iron core; (c) ferrite core.

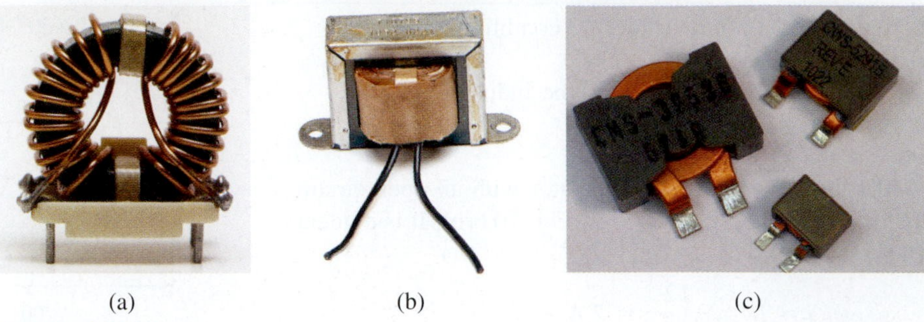

(a) (b) (c)

Figure 10.10
Various types of inductors: (a) toroidal inductor; (b) iron-core inductor; (c) surface-mount inductor.
(a) © GIPhotoStock/Photo Researchers (b) © The McGraw-Hill Companies, Inc./Cindy Schroeder, photographer (c) © Photo courtesy of Coil Winding Specialists, Inc.

microhenrys ranges. The terms *coil* and *choke* are also used for inductors. Common inductors are shown in Fig. 10.10. Laboratory-type inductors can be made in form of a *decade box*, which is an assembly of precision resistors, inductors, or capacitors whose individual values vary in multiples of 10.

Small inductors are often color coded like resistors. Some appear in standard values such as 0.1, 0.12, 0.15, 0.18, 0.22, and 0.27 μHs and 1, 1.2, 1.5, 2.2, 2.7, and so on.

10.6 Series and Parallel Inductors

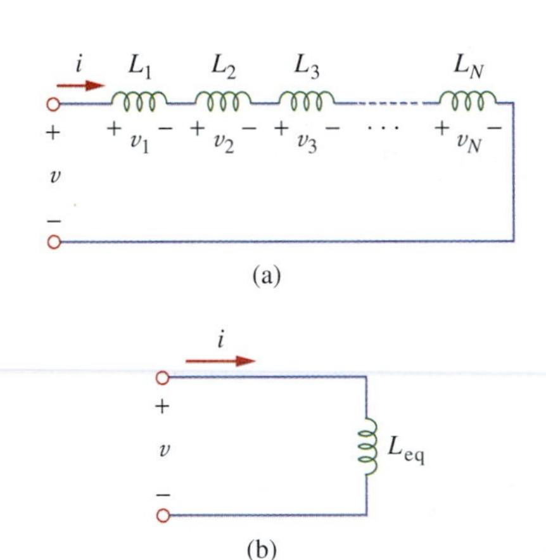

Figure 10.11
(a) A series connection of N inductors; (b) equivalent circuit for the series inductors.

In practical circuits, we may have inductors in series or in parallel. It is therefore important that we know how to find the equivalent inductance of a series-connected or parallel-connected set of inductors.

Consider a series connection of N inductors, as shown in Fig. 10.11(a), with the equivalent circuit shown in Fig. 10.11(b). The inductors have the same current through them. Applying KVL to the loop,

$$v = v_1 + v_2 + v_3 + \cdots + v_N \tag{10.6}$$

where $v_k = L_k\, di/dt$. Inductors in series are combined in exactly the same way as resistors in series. Hence,

$$\boxed{L_{eq} = L_1 + L_2 + L_3 + \cdots + L_N} \tag{10.7}$$

Thus,

> The **equivalent inductance** of series-connected inductors is the sum of the individual inductances.

Because inductors in series are combined in exactly the same way as resistors in series, the voltage across L_k is like a voltage divider

$$v_k = \frac{L_k}{L_{eq}} v \tag{10.8}$$

We now consider a parallel connection of N inductors, as shown in Fig. 10.12(a), with the equivalent circuit shown in Fig. 10.12(b). The inductors have the same voltage across them. Using KCL,

$$i = i_1 + i_2 + i_3 + \cdots + i_N \tag{10.9}$$

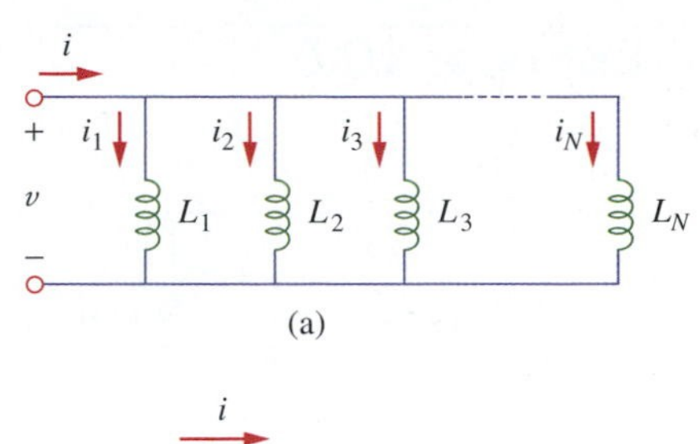

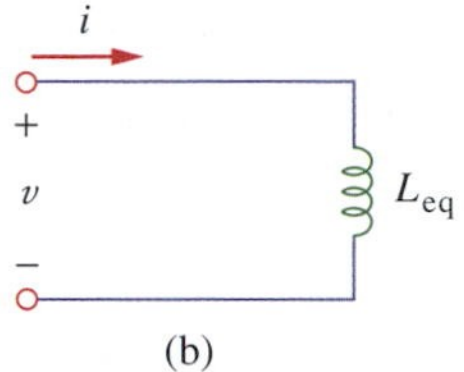

Figure 10.12
(a) A parallel connection of N inductors; (b) equivalent circuit for the parallel inductors.

Because the inductors are in parallel, they are combined in exactly the same way as resistors in parallel,

$$\frac{1}{L_{eq}} = \frac{1}{L_1} + \frac{1}{L_2} + \frac{1}{L_3} + \cdots + \frac{1}{L_N} \tag{10.10}$$

For two inductors in parallel ($N = 2$), Eq. (10.10) becomes

$$\frac{1}{L_{eq}} = \frac{1}{L_1} + \frac{1}{L_2}$$

or

$$L_{eq} = \frac{L_1 L_2}{L_1 + L_2} \tag{10.11}$$

Note that Eq. (10.11) is only valid for two inductors.

The **equivalent inductance** of parallel inductors is the reciprocal of the sum of the reciprocals of the individual inductances.

Example 10.5

Find the equivalent inductance of the circuit shown in Fig. 10.13.

Solution:

The 10-, 12-, and 20-H inductors are in series so that combining them gives a 42-H inductance. This 42-H inductor is in parallel with the 7-H inductor so that they are combined to give

$$\frac{7 \times 42}{7 + 42} = 6 \text{ H}$$

This 6-H inductor is in series with the 4- and 8-H inductors. Hence,

$$L_{eq} = 6 + 4 + 8 = 18 \text{ H}$$

Figure 10.13
For Example 10.5.

Practice Problem 10.5

Calculate the equivalent inductance for the inductive ladder network in Fig. 10.14.

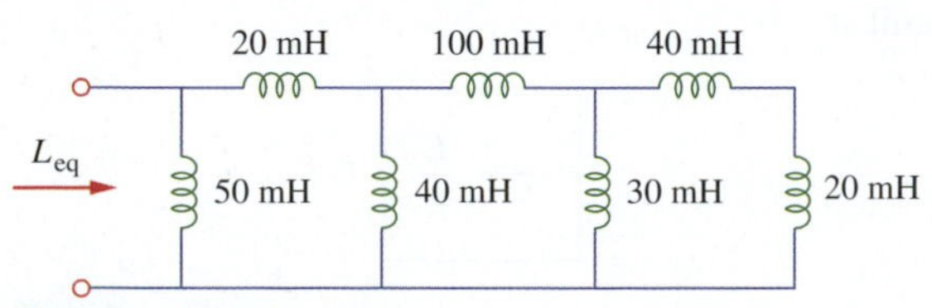

Figure 10.14
For Practice Problem 10.5.

Answer: 25 mH

Example 10.6

For the circuit in Fig. 10.15, $i(t) = 4(2 - e^{-10t})$ mA. If $i_2(0) = -1$ mA, find:

(a) $i_1(0)$; (b) L_{eq}

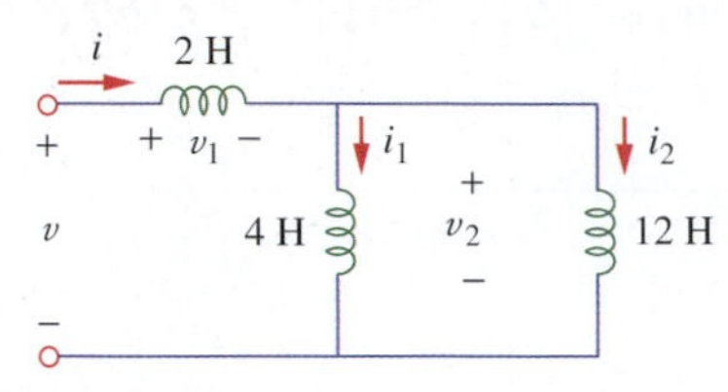

Figure 10.15
For Example 10.6.

Solution:

(a) From $i(t) = 4(2 - e^{-10})$ mA, $i(0) = 4(2 - 1) = 4$ mA. Because $i = i_1 + i_2$,

$$i_1(0) = i(0) - i_2(0) = 4 - (-1) = 5 \text{ mA}$$

(b) The equivalent inductance is

$$L_{eq} = 2 + 4 \parallel 12 = 2 + 3 = 5 \text{ H}$$

Practice Problem 10.6

In the circuit of Fig. 10.16, $i_1(t) = 0.6\, e^{-2t}$ A. If $i(0) = 1.4$ A, find $i_2(0)$.

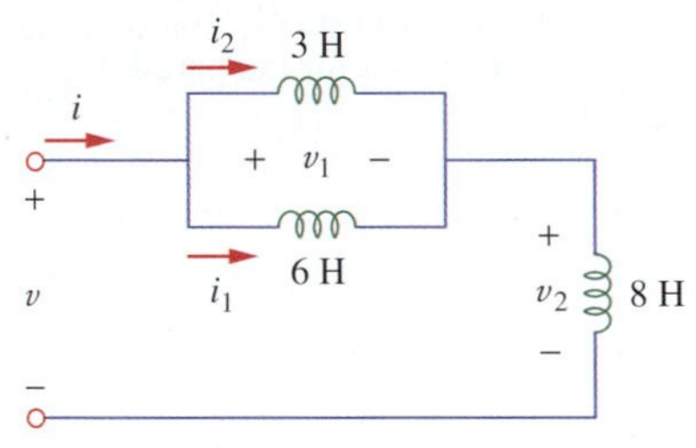

Figure 10.16
For Practice Problem 10.6.

Answer: 0.8 A

10.7 Transient *RL* Circuits

An *RL* circuit is one that contains only resistors and inductors. A typical *RL* circuit consists of the series connection of a resistor and an inductor, as shown in Fig. 10.17. Our goal is to determine the circuit response, which will be assumed to be the current $i(t)$ through the inductor. We select the inductor current as the response to be able to enforce the idea that the inductor current cannot change instantaneously. At $t = 0$, we assume that the inductor has an initial current I_o; that is,

$$i(0) = I_o \tag{10.12}$$

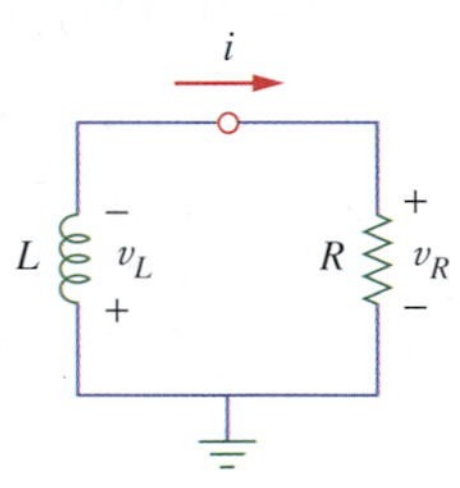

Figure 10.17
A source-free *RL* circuit.

Applying KVL around the loop in Fig. 10.17,

$$v_L + v_R = 0 \tag{10.13}$$

We can apply differential calculus to show that the current through the *RL* circuit is

$$i(t) = I_o e^{-Rt/L} \tag{10.14}$$

This shows that the natural response of the *RL* circuit is an exponential decay of the initial current. The current response is shown in Fig. 10.18. It is evident from Eq. (10.14) that the time constant[4] for the *RL* circuit is

$$\boxed{\tau = \frac{L}{R}} \tag{10.15}$$

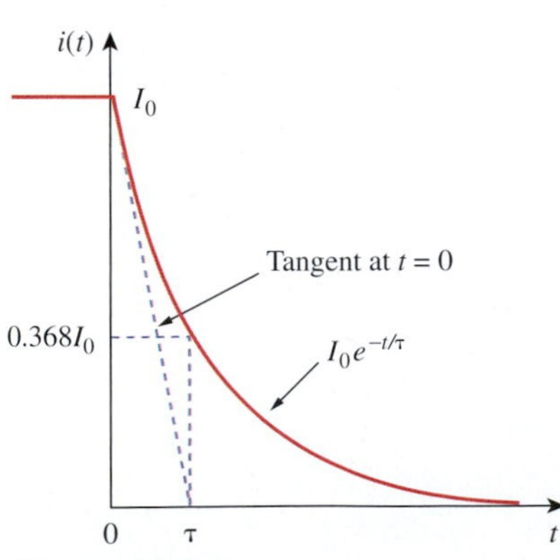

Figure 10.18
The current response of the *RL* circuit.

[4] Reminder: The smaller the time constant τ of a circuit, the faster the rate of decay of the response. However, the larger the time constant, the slower the rate of decay of the response. At any rate, the response decays to less than 1 percent of its initial value (i.e., reaches steady state) after 5τ.

with τ again having the unit of seconds, while R is in ohms and L is in henrys. Thus, Eq. (10.14) may be written as

$$i(t) = I_o e^{-t/\tau} \qquad \textbf{(10.16)}$$

With the current in Eq. (10.16), we can find the voltage across the resistor as

$$v_R(t) = iR = I_o R e^{-t/\tau} \qquad \textbf{(10.17)}$$

In summary,

Key to Working with a Source-Free *RL* Circuit:

1. Find the initial current $i(0) = I_o$ through the inductor.
2. Find the time constant τ of the circuit.

With these two items, we obtain the response as the inductor current $i_L(t) = i(t) = i(0)\ e^{-t/\tau}$. Once we determine the inductor current i_L, other variables (inductor voltage v_L, resistor voltage v_R, and resistor current i_R) can be obtained. Note that in general, R in Eq. (10.17) is the Thevenin resistance as seen by the terminals of the inductor.[5]

Example 10.7

The switch in the circuit of Fig. 10.19 has been closed for a long time. At $t = 0$, the switch is opened. Calculate $i(t)$ for $t > 0$.

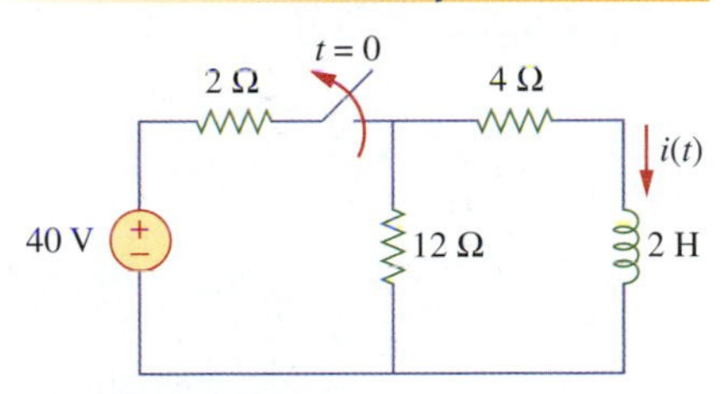

Figure 10.19
For Example 10.7.

Solution:
When $t < 0$, the switch is closed and the inductor acts as a short-circuit to dc because the circuit has reached steady state. The resulting circuit is shown in Fig. 10.20(a). To get i_1 in Fig. 10.20(a), we combine the 4- and 12-Ω resistors in parallel to get

$$\frac{4 \times 12}{4 + 12} = 3\ \Omega$$

Hence,

$$i_1 = \frac{40}{2+3} = 8\ \text{A}$$

We obtain $i(t)$ from i_1 in Fig. 10.20(a) using the current divider rule.

$$i(t) = \frac{12}{12+4} i_1 = 6\ \text{A}, \quad t < 0$$

Because the current through an inductor cannot change instantaneously,

$$i(0) = 6\ \text{A}$$

(a)

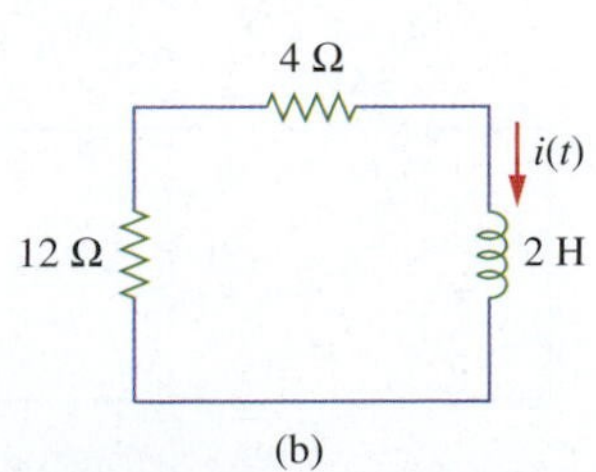

(b)

Figure 10.20
Solving the circuit of Fig. 10.19: (a) $t < 0$, (b) $t > 0$.

[5] Note: When a circuit has a single inductor and several resistors and dependent sources, the Thevenin equivalent can be found at the terminals of the inductor to form a simple *RL* circuit. Also, the Thevenin theorem can be used when several inductors can be combined to form a single equivalent inductor.

When $t > 0$, the switch is open and the voltage source is disconnected. We now have the source-free RL circuit in Fig. 10.20(b). Combining the resistors, we have

$$R_{eq} = 12 + 4 = 16\ \Omega$$

The time constant is

$$\tau = \frac{L}{R_{eq}} = \frac{2}{16} = \frac{1}{8}\text{ s}$$

Thus,

$$i(t) = i(0)e^{-t/\tau} = 6e^{-8t}\text{ A}$$

Practice Problem 10.7

For the circuit in Fig. 10.21, find $i_o(t)$ for $t > 0$.

Answer: $1.4118e^{-3t}$ A, $t > 0$

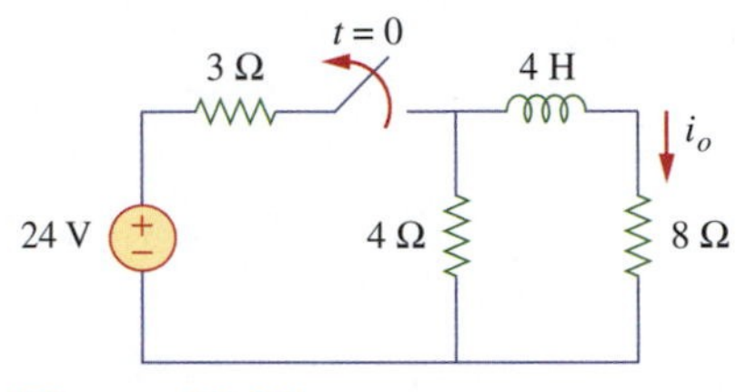

Figure 10.21
For Practice Problem 10.7.

Example 10.8

In the circuit shown in Fig. 10.22, find v_o and i for all time, assuming that the switch was open for a long time.

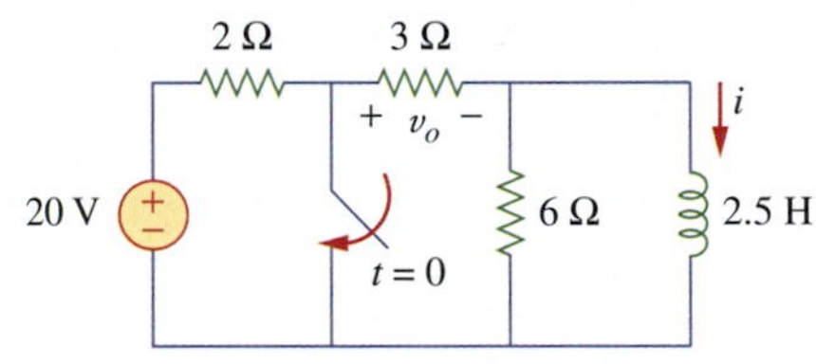

Figure 10.22
For Example 10.8.

Solution:

It is better to first find the inductor current i and then obtain other quantities from it.

For $t < 0$, the switch is open. Because the inductor acts like a short-circuit to dc, the 6-Ω resistor is short-circuited so that we have the circuit shown in Fig. 10.23(a). Hence,

$$i(t) = \frac{20}{2+3} = 4\text{ A}, \qquad t < 0$$

$$v_o(t) = 3i(t) = 12\text{ V}, \qquad t < 0$$

Thus $i(0) = 4$ A.

For $t > 0$, the switch is closed so that the voltage source is short-circuited. We now have a source-free RL circuit as shown in Fig. 10.23(b). At the inductor terminals,

$$R_{Th} = 3 \parallel 6 = 2\ \Omega$$

so that the time constant

$$\tau = \frac{L}{R_{Th}} = \frac{2.5}{2} = 1.25\text{ s}$$

Hence,

$$i(t) = i(0)e^{-t/\tau} = 4e^{-t/1.25} = 4e^{-0.8t}\text{ A}, \qquad t > 0$$

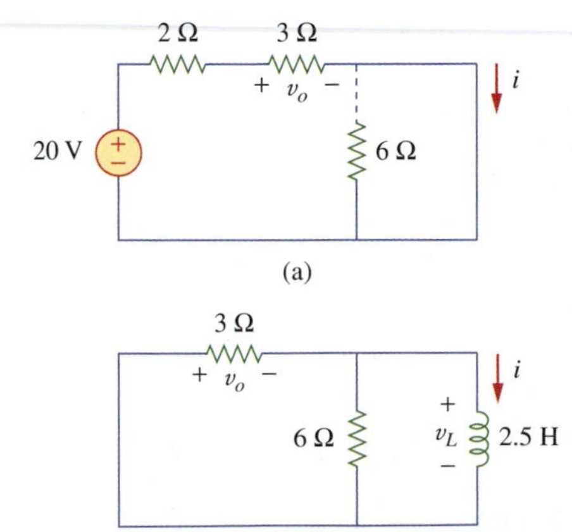

Figure 10.23
The circuit in Fig. 10.22: (a) $t < 0$, (b) $t > 0$.

Thus, for all time,

$$i(t) = \begin{cases} 4 \text{ A}, & t < 0 \\ 4e^{-0.8t} \text{ A}, & t > 0 \end{cases}$$

We notice that the inductor current is continuous at $t = 0$; $i(t)$ is sketched in Fig. 10.24.

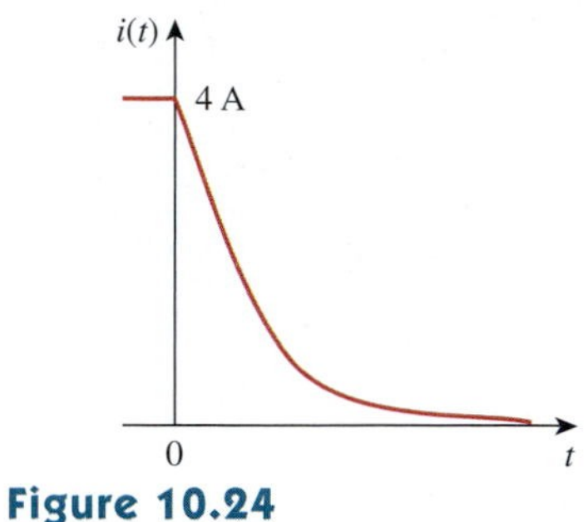

Figure 10.24
A plot of $i(t)$.

Practice Problem 10.8

Determine $i(t)$ for all t in the circuit shown in Fig. 10.25. Assume that the switch was closed for a long time.

Answer: $i(t) = \begin{cases} 4 \text{ A}, & t < 0 \\ 4e^{-2t} \text{ A}, & t > 0 \end{cases}$

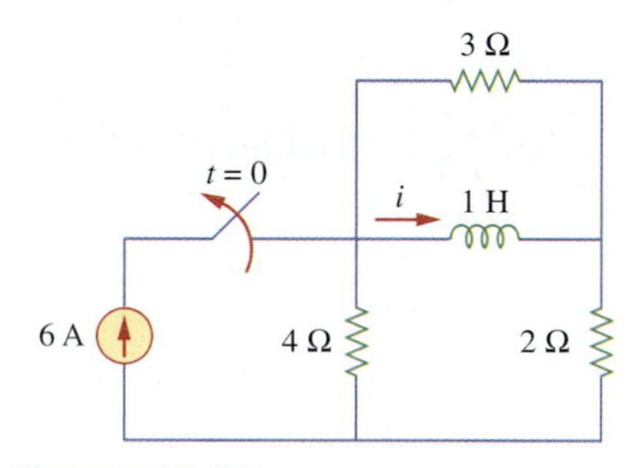

Figure 10.25
For Practice Problem 10.8.

10.8 Computer Analysis

10.8.1 PSpice

Just as we used PSpice to model an *RC* circuit in Section 9.8, we can use PSpice to model an *RL* circuit. PSpice can be used to obtain the transient response of an *RL* circuit. Section C.4 in Appendix C provides a review of transient analysis using PSpice for Windows. It is recommended that Section C.4 be reviewed before proceeding with this section.

If necessary, dc PSpice analysis is first carried out to determine the initial conditions. Then the initial conditions are used in the transient PSpice analysis to obtain the transient responses. It is recommended but not necessary that during this dc analysis, all inductors should be short-circuited.

Example 10.9

Use PSpice to find the response $i(t)$ for $t > 0$ in the circuit of Fig. 10.26.

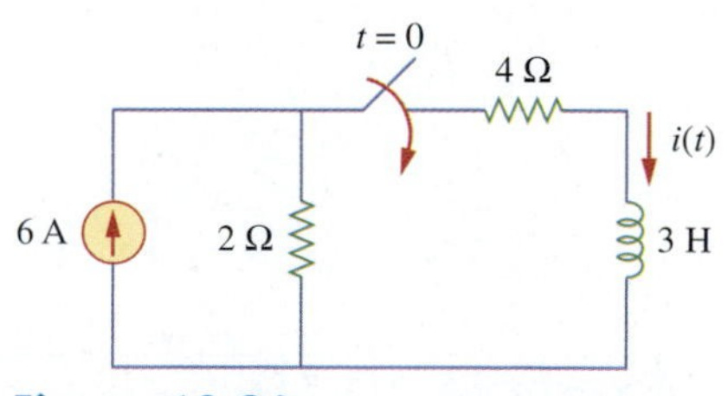

Figure 10.26
For Example 10.9.

Solution:
To use PSpice, we first draw the schematic as shown in Fig. 10.27. We recall that the part name for a close switch is Sw_tclose. We do not need to specify the initial condition of the inductor because PSpice will determine that from the circuit. After drawing the circuit, we select **PSpice/New Simulation Profile**. This leads to the New Simulation dialog box. Type "exam109" as the name of the file and click Create. This leads to the Simulation Settings dialog box. In Simulation Settings, select Time Domain (Transient) under *Analysis Type* and $5\tau = 2.5$ s as *Run to time*. Click Apply and then click OK. After saving the circuit, we select **PSpice/Run** to simulate the circuit. In the window that pops up, we select **Trace/Add Trace** and display I(L1) as the current through the inductor. The plot of $i(t)$ is shown in Fig. 10.28.

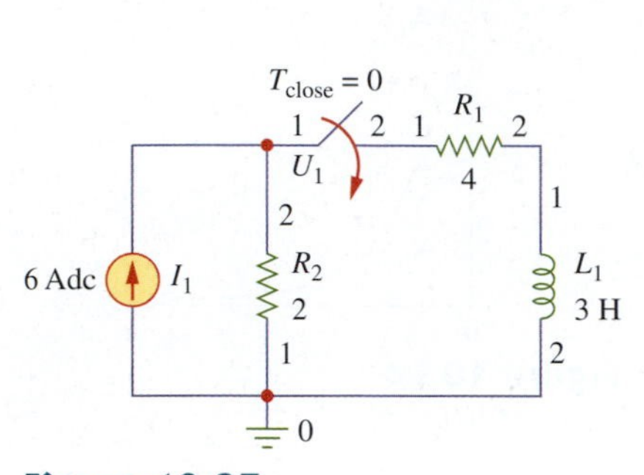

Figure 10.27
The schematic of the circuit in Fig. 10.26.

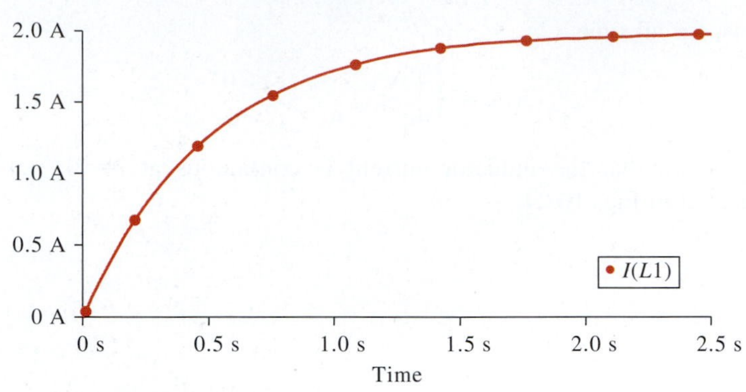

Figure 10. 28
For Example 10.9; the response of the circuit in Fig. 10.26.

Practice Problem 10.9

The switch in Fig. 10.29 has been open for a long time but closed at $t = 0$. Find $i(t)$ using PSpice.

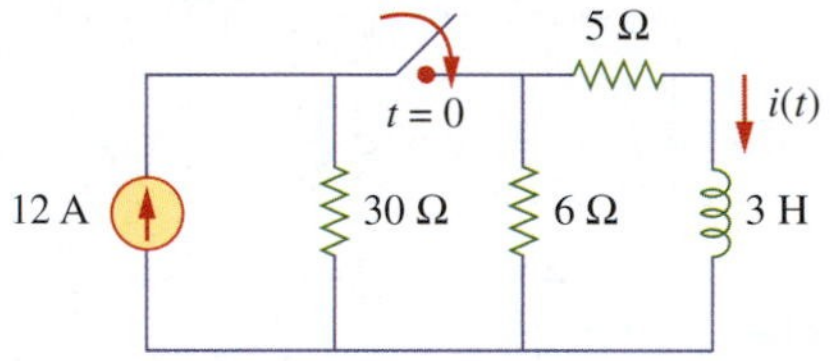

Figure 10.29
For Practice Problem 10.9.

Answer: The plot of $i(t)$ is shown in Fig. 10.30.

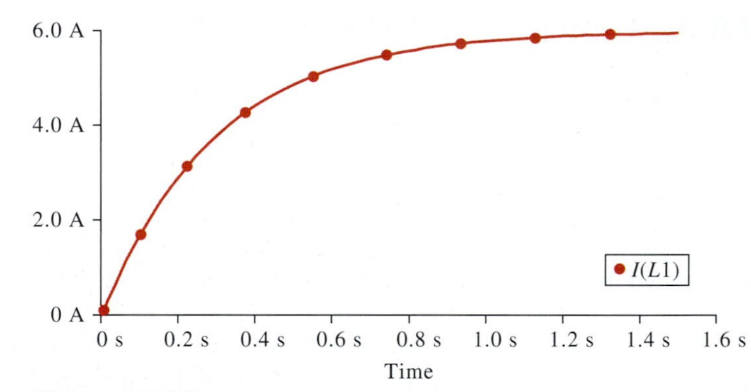

Figure 10.30
For Practice Problem 10.9.

10.8.2 Multisim

Multisim can be used to model an *RL* circuit just as we did for *RC* circuits in Section 9.7. (You are advised to read Section D.3 in Appendix D before proceeding with this section.) The only problem is that Multisim has no direct way of plotting current. In order to determine current, you need to apply Ohm's law to the corresponding voltage waveform.

Example 10.10

Use Multisim to find v_o in the circuit in Fig. 10.31.

Solution:
We first create the circuit as shown in Fig. 10.32. The time constant in this case is

$$\tau = \frac{L}{R} = \frac{10^{-3}}{2 \times 10^{3}} = 0.5\ \mu\text{s}$$

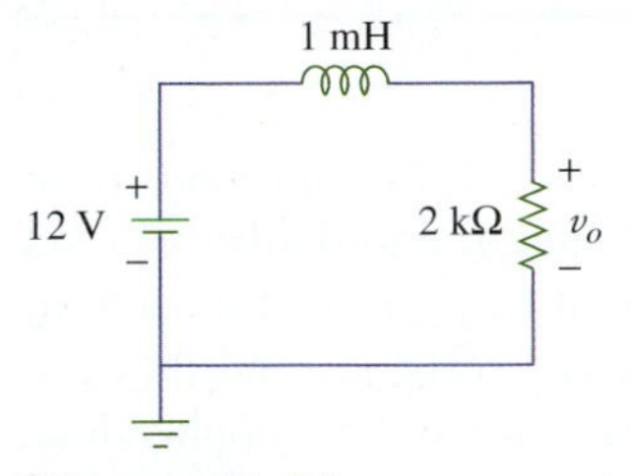

Figure 10.31
For Example 10.10.

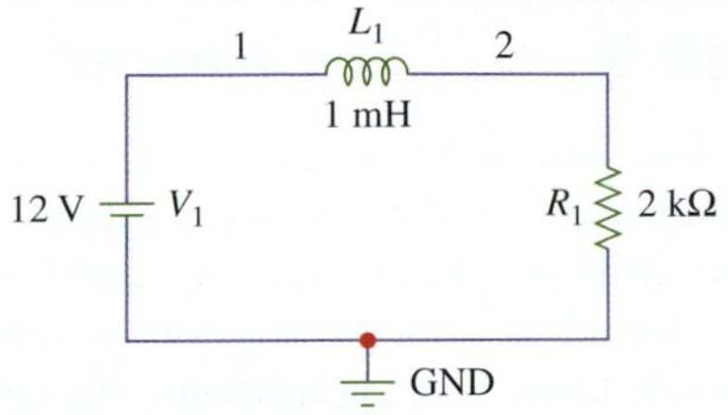

Figure 10.32
Simulation of the circuit in Fig. 10.31.

so that $5\tau = 2.5\ \mu s$. To input the simulation parameters, select **Simulation/Analyses/Transient Analysis.** In the Transient Analysis dialog box that comes up, select *Set to zero* under Initial Condition. Set TSTART to 0 or 0.000001 and TSTOP to 2.5 μs or 0.0000025 s. For the Output, move *V*(2) from the left list to the right list so that the voltage at node 2 is displayed after simulation. Finally, select **Simulate**. The output is shown in Fig. 10.33.

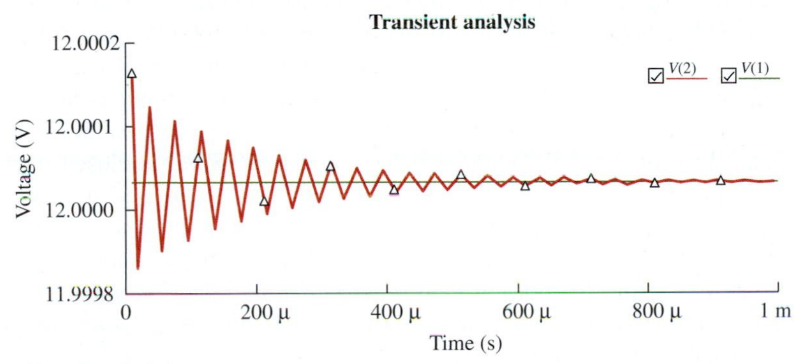

Figure 10.33
Response of the circuit in Fig. 10.31.

Practice Problem 10.10

Refer to the *RL* circuit of Fig. 10.34. Use Multisim to determine $v(t)$.

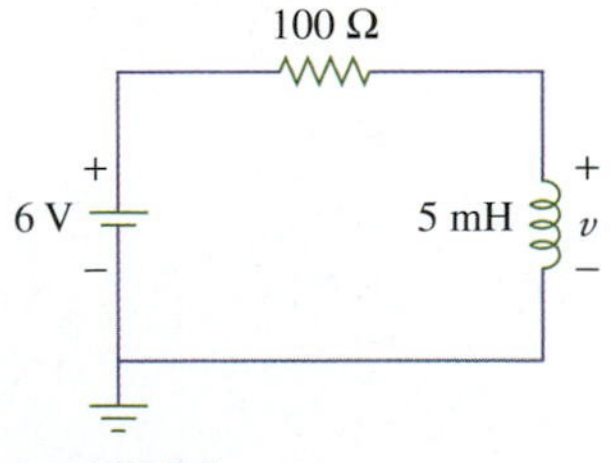

Answer: See Fig. 10.35.

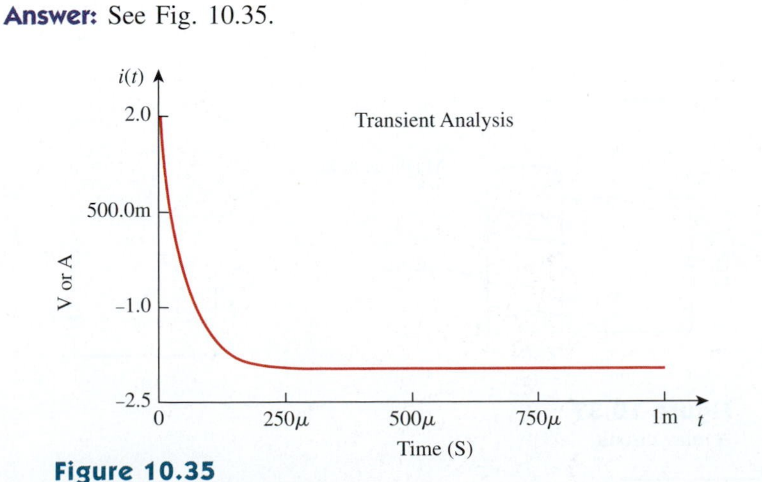

Figure 10.35
For Practice Problem 10.10.

Figure 10.34
For Practice Problem 10.10.

Figure 10.36
Typical relays.
© Sarhan M. Musa

10.9 †Applications

Most inductors (coils) usually come in discrete form and tend to be more bulky and expensive. (This is typically not true for inductors used in switching power supplies and EMI/RFI[6] filtering.) For this reason, inductors are not as versatile as capacitors and resistors, and they are more limited in applications. However, there are several applications in which inductors have no practical substitute. They are routinely used in relays, delays, sensing devices, pick-up heads, telephone circuits, tuned circuits, radio and TV receivers, power supplies, filters, electric motors, microphones, and loudspeakers, to mention a few. Here, we will we consider two simple applications involving inductors.

10.9.1 Relay Circuits

A magnetically controlled switch is called a *relay*. Typical relays are shown in Fig. 10.36. A relay is an electromagnetic device used to open or close a switch that controls another circuit. A typical relay circuit is shown in Fig. 10.37(a). Notice that a relay uses an electromagnet. This is a device that consists of a coil of wire wrapped around an iron core. When the coil is energized, by passing current through it, it becomes magnetized—hence the term *electromagnet*. The coil circuit is an *RL* circuit as shown in Fig. 10.37(b), where R and L are the resistance and inductance of the coil. When switch S_1 in Fig. 10.37(a) is closed, the coil circuit is energized. The coil current gradually increases and produces a magnetic field. Eventually, the magnetic field is sufficiently strong to pull the movable contact in the other circuit and close switch S_2. At this point, the relay is said to be *pulled in*. The time interval t_d between the closure of switches S_1 and S_2 is called the *relay delay time*. A formula for finding t_d is

$$t_d = \tau \ln \frac{i(0) - i(\infty)}{i(t_d) - i(\infty)} \tag{10.18}$$

where ln stands for natural logarithm, $i(0)$ is the initial inductor current, $i(\infty)$ is the final value of the inductor current, and $i(t_d)$ is the inductor current at $t = t_d$.

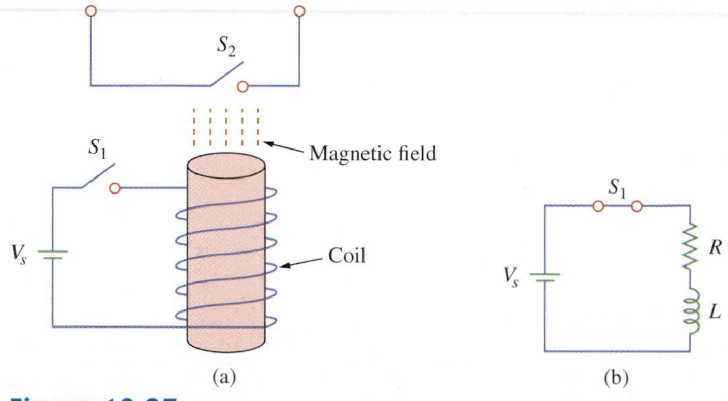

Figure 10.37
A relay circuit.

[6] EMI stands for electromagnetic interference, while RFI denotes radio-frequency interference.

Relays are very useful when we have a need to control a large amount of current and/or voltage with a small electrical signal. Such a high current/voltage is generated by the collapsing relay field. Relays were used in the earliest digital circuits and are still used for switching high-power circuits.

Example 10.11

The coil of a certain relay is operated by a 12-V battery. If the coil has a resistance of 150 Ω and an inductance of 30 mH and the current needed to pull in is 50 mA, calculate the relay delay time.

Solution:

To use Eq. (10.18), we first need to calculate the following items.

$$i(0) = 0, \qquad i(\infty) = \frac{12}{150} = 80 \text{ mA}$$

$$\tau = \frac{L}{R} = \frac{30 \times 10^{-3}}{150} = 0.2 \text{ ms}$$

If $i(t_d) = 50$ mA, then

$$t_d = 0.2 \ln \frac{(0 - 80) \text{ mA}}{(50 - 80) \text{ mA}} \text{ ms} = 0.1962 \text{ ms} = 196.2 \ \mu\text{s}$$

Practice Problem 10.11

A relay has a resistance of 200 Ω and an inductance of 500 mH. The relay contacts close when the current through the coil reaches 350 mA. What time elapses between the application of 110 V to the coil and contact closure?

Answer: 2.529 ms

10.9.2 Automobile Ignition Circuits

The ability of inductors to oppose rapid changes in current makes them useful for arc or spark generation. This feature is applied in an automobile ignition system.

A gasoline engine of an automobile requires that the fuel–air mixture in each cylinder be ignited at proper times. This is achieved by means of a spark plug, shown in Fig. 10.38, which consists of a pair of electrodes separated by an air gap. By creating a large voltage (thousands of volts) between the electrodes, a spark is formed across the air gap, thereby igniting the fuel. But how can such a large voltage be obtained from the car battery, which supplies only 12 V? This is achieved by means of an inductor (the spark coil) L. Because the voltage across the inductor is $v = L\,\Delta i/\Delta t$, we can make $\Delta i/\Delta t$ large by creating a large change in current in a very short time. When the ignition switch in Fig. 10.38 is closed, the current through the inductor increases gradually and reaches the final value of $i = V_s/R$, where $V_s =$ 12 V. Again, the time taken for the inductor to charge is five times the time constant of the circuit ($\tau = L/R$); that is,

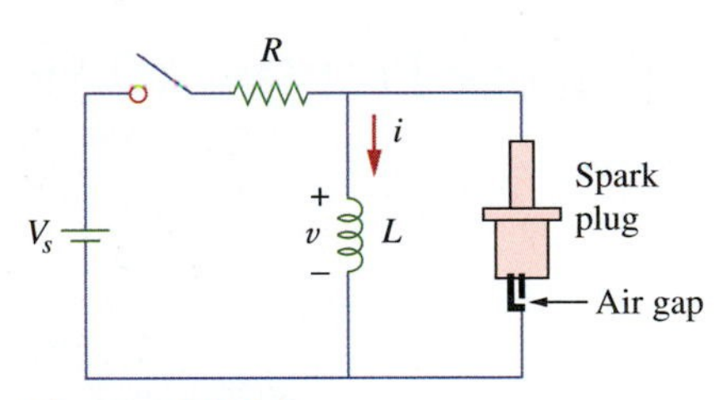

Figure 10.38
Circuit for an automobile ignition system.

$$t_{\text{charge}} = 5\frac{L}{R} = 5\tau \qquad \textbf{(10.19)}$$

Because at steady state, i is constant, $\Delta i/\Delta t = 0$ and the inductor voltage $v = 0$. When the switch suddenly opens, a large voltage is developed across the inductor because, by definition, at the instant the switch is opened i has to go to 0 (causing a rapidly collapsing field) causing a spark or arc in the air gap. The spark continues until the energy stored in the inductor is dissipated in the spark discharge. This same effect can cause a very nasty shock in laboratories when one is working with inductive circuits, so exercise caution.

Example 10.12

A solenoid with resistance of 4 Ω and an inductance of 6 mH is used in an automobile ignition circuit similar to that in Fig. 10.38. If the battery supplies 12 V, determine:
(a) the final current through the solenoid when the switch is closed,
(b) the energy stored in the coil, and
(c) the voltage across the air gap assuming that the switch takes 1 μs to open.

Solution:
(a) The final current through the coil is

$$I = \frac{V_s}{R} = \frac{12}{4} = 3 \text{ A}$$

(b) The energy stored in the coil is

$$w = \frac{1}{2}LI^2 = \frac{1}{2} \times 6 \times 10^{-3} \times 3^2 = 27 \text{ mJ}$$

(c) The voltage across the gap is

$$v = L\frac{\Delta I}{\Delta t} = 6 \times 10^{-3} \times \frac{3}{1 \times 10^{-6}} = 18 \text{ kV}$$

Practice Problem 10.12

The spark coil of an automobile ignition system has 20 mH of inductance and 5-Ω resistance. With a supply voltage of 12 V, calculate:
(a) the time needed for the coil to fully charge,
(b) the energy stored in the coil, and
(c) the voltage developed at the spark gap if the switch opens in 2 μs.

Answer: (a) 20 ms; (b) 57.6 mJ; (c) 24 kV

10.10 Summary

1. Faraday's law establishes the relationship between induced voltage in a circuit and the rate of change of the magnetic flux linking with the circuit.

$$v = N\frac{d\phi}{dt}$$

Lenz's law establishes the polarity of the induced voltage.
2. The inductance L is the property of the electric circuit element that exhibits opposition to the change of current flowing in it. Inductance is measured in henrys (H).

3. The voltage across an inductor is directly proportional to the time rate of change of the current through it; that is,

$$v = L\frac{di}{dt}$$

4. The voltage across an inductor is zero unless the current through it is changing. Thus, an inductor acts like a short circuit to a dc source.
5. The current through an inductor cannot change instantly.
6. An inductor stores energy in its magnetic field when the current is rising and returns energy when the current is falling. The energy stored is

$$W = \frac{1}{2}Li^2$$

7. Inductors come in a variety of forms and sizes.
8. Inductors in series and parallel are combined in the same way resistors in series and parallel are combined.
9. In an *RL* circuit, the time constant is $\tau = L/R$. If the initial current is I_o, the current through the inductor is

$$i(t) = I_o e^{-t/\tau}$$

10. We learned how to model *RL* circuits using PSpice and Multisim.
11. Two application circuits of the *RL* circuits considered in this chapter are those of the relay and automobile ignition circuits.

Review Questions

10.1 The voltage induced in a coil is related to:

(a) the resistance of the coil
(b) the initial energy in the coil
(c) the inductance of the coil
(d) the rate of change of current

10.2 The inductance of an air-core inductor increases when:

(a) wire of larger diameter is used
(b) the length is increased
(c) the number of turns is increased
(d) the core is replaced by material with higher permeability

10.3 A 5-H inductor changes its current by 3 A in 0.2 s. The voltage produced at the terminals of the inductor is:

(a) 75 V (b) 8.888 V
(c) 3 V (d) 1.2 V

10.4 If the current through a 10-mH inductor increases from 0 to 2 A, how much energy is stored in the inductor?

(a) 40 mJ (b) 20 mJ
(c) 10 mJ (d) 5 mJ

10.5 Inductors in parallel can be combined just like resistors in parallel.

(a) True (b) False

10.6 The total inductance of inductors connected in series is calculated the same way as:

(a) resistances in series
(b) resistances in parallel
(c) capacitors in series
(d) none of the above

10.7 If 40- and 60-mH inductors are connected in series, their combined equivalent inductance is

(a) 2400 mH (b) 100 mH
(c) 80 mH (d) 24 mH

10.8 If 40- and 60-mH inductors are connected in parallel, their combined equivalent inductance is

(a) 24 mH (b) 80 mH
(c) 100 mH (d) 2400 mH

10.9 The final (steady-state) value of the current *i* in Fig. 10.39 is:

(a) 2.8 A (b) 2 A
(c) 1.75 A (d) 0 A

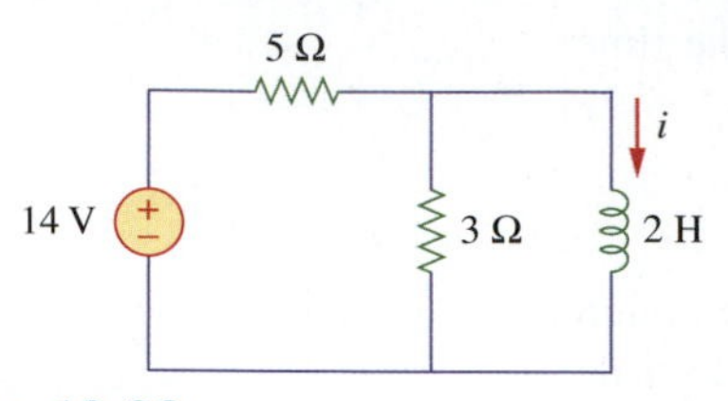

Figure 10.39
For Review Question 10.9.

10.10 If an *RL* circuit has $R = 2\ \Omega$ and $L = 8$ mH, the time constant of the circuit is:

(a) 16 ms (b) 8 ms
(c) 4 ms (d) 0.25 ms

Answers: 10.1c,d, 10.2a,c,d, 10.3a, 10.4b, 10.5a, 10.6a, 10.7b, 10.8a, 10.9a, 10.10c

Problems

Section 10.2 Electromagnetic Induction

10.1 Determine the voltage induced in a 300-turn coil when the magnetic flux changes from 0.6 μWb to 0.8 μWb in 2 ms.

10.2 A coil has 250 turns and the rate of flux change is 2 Wb/s. Calculate the induced voltage.

10.3 5 V is induced in a coil when the flux changes from 0.3 to 0.8 mWb in 5 ms. Calculate the number of turns.

10.4 The flux through a 400-turn coil changes from 2 to 7 Wb in 1 s. What is the voltage induced?

10.5 The flux linking a coil changes at a rate of 40 mWb/s. If the coil has 60 turns, determine the induced voltage across the coil.

10.6 A coil with 500 turns is exposed to a magnetic flux of 200 Wb in 5 μs. Calculate the voltage induced in the coil.

Section 10.3 Inductors

10.7 The current through a 0.25-H inductor is shown in Fig. 10.40. Sketch the inductor voltage.

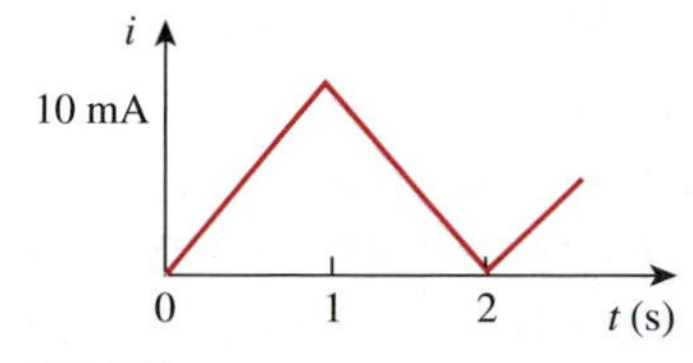

Figure 10.40
For Problem 10.7.

10.8 The plot of i versus t for a 4-H inductor is shown in Fig. 10.41. Construct the v versus t plot.

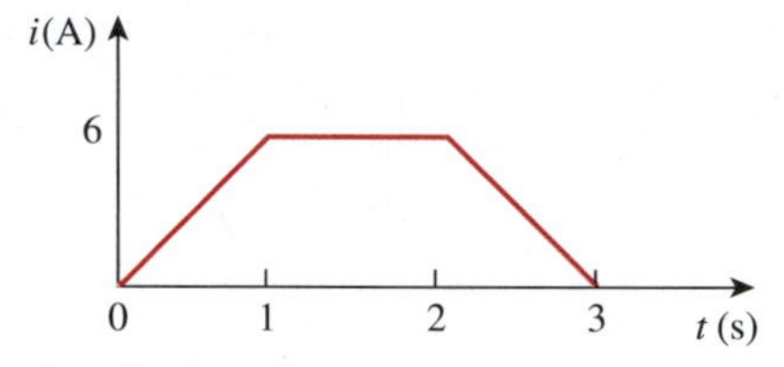

Figure 10.41
For Problem 10.8.

10.9 A 40-mH inductor has its number of turns doubled, while keeping its length, cross section, and core material the same. What is the new inductance value?

10.10 The current flowing through an inductor changes at the rate of 50 mA/s, while the induced voltage is 10 mV. Calculate the inductance of the inductor.

10.11 Calculate the inductance of a coil whose induced voltage is 6 V when there is a 0.2-A/s change in current.

10.12 The current in an inductor rises from 0 to 4 A in 1s. If the induced voltage is 6.5 V, determine: (a) the rate of change of current and (b) the inductance of the inductor.

10.13 Find the induced voltage across a 200-mH coil when the change in current is 240 mA/s.

10.14 The current shown in Fig. 10.42 flows through a 20-mH inductor. Plot the inductor voltage.

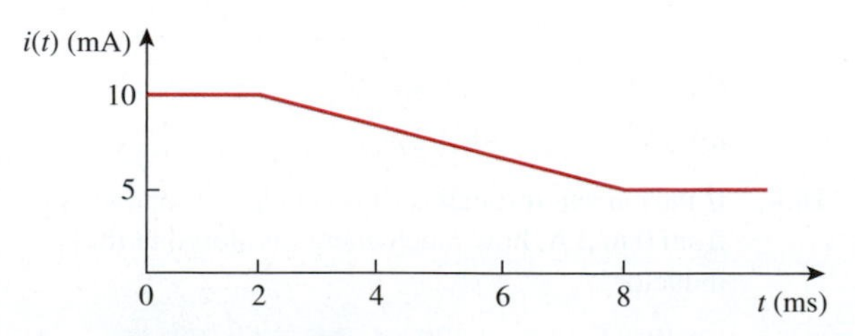

Figure 10.42
For Problem 10.14.

10.15 A voltmeter connected across a coil reads 26 mV. If the current is increasing by 12 mA every 1.5 ms, determine the inductance of the coil.

10.16 A solenoid with air as core is to be constructed with a diameter of 1.2 cm and length of 8 cm. If the inductance of the coil is to be 800 μH, determine the number of turns required.

10.17 An air-core coil is 4 cm long and 2 cm in diameter. How many turns must we wind to obtain an inductance of 50 μH?

10.18 A 4-mH inductance has 500 turns. How many turns must we add to raise the inductance to 6 mH?

10.19 The toroid in Fig. 10.43 has a circular cross section. Determine the number of turns of wire required to produce an inductance of 400 mH. Assume that the cross-sectional area is 3.5 cm^2, the average path length is 54.2 cm, and the permeability of cast-steel core is 7.6×10^{-4} H/m.

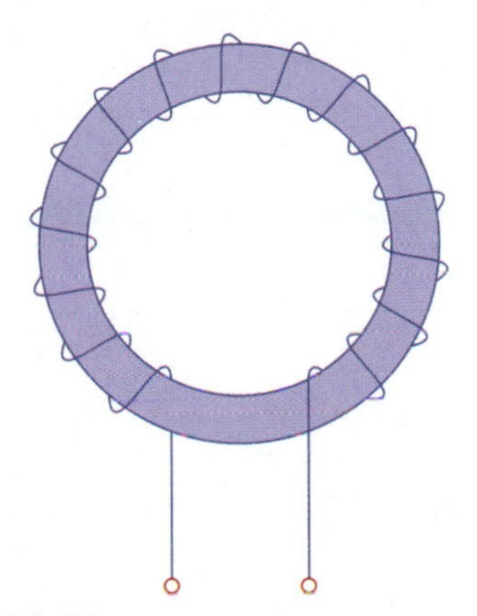

Figure 10.43
For Problem 10.19.

10.20 Two air-core coils have the same cross-sectional area, but the second coil has three times the number of turns and half the length of the first coil. Determine the relationship between their inductances.

10.21 A voltmeter connected across a 4-H coil reads 2.5 V. Determine how fast the current through the coil is changing with time.

Section 10.4 Energy Storage and Steady-State DC

10.22 If a 10-H inductor is conducting 2 A of current, how much energy is the inductor storing?

10.23 If a 400-mH inductor is to store 0.25 J, how much current is required?

10.24 Determine how much energy is stored in a 60-mH inductor with a current of 2 A.

10.25 For the circuit of Fig. 10.44, the voltages and currents have reached the final value. Find v, i_1, and i_2.

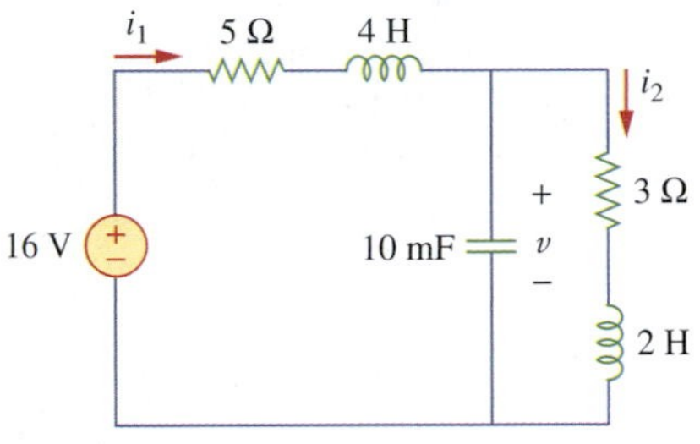

Figure 10.44
For Problem 10.25.

10.26 Find v_C, i_L, and the energy stored in the capacitor and in the inductor in the circuit of Fig. 10.45 under steady-state dc conditions.

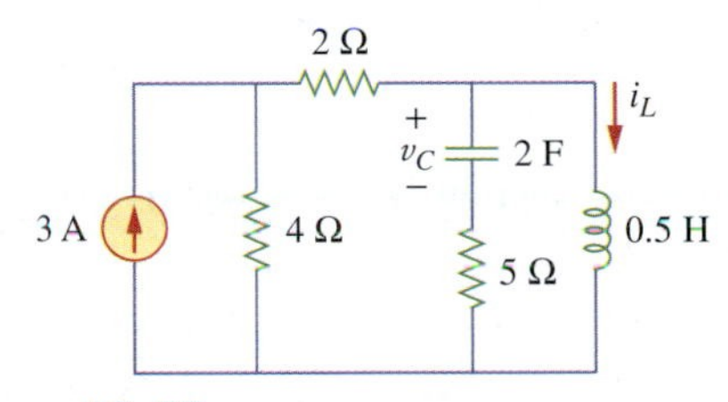

Figure 10.45
For Problem 10.26.

10.27 Under steady-state dc conditions, find the voltage across the capacitors and the current through the inductors in the circuit of Fig. 10.46.

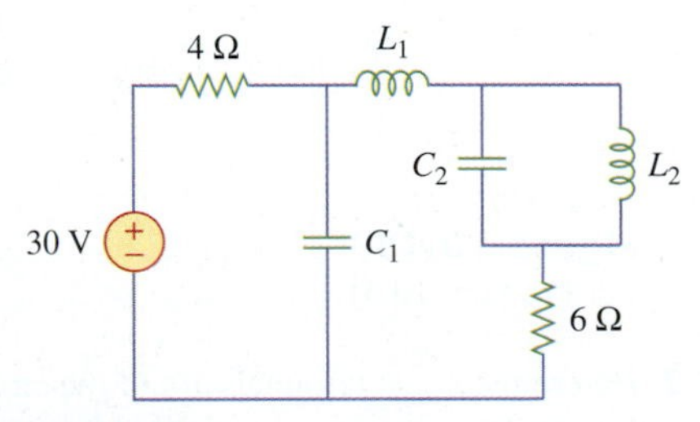

Figure 10.46
For Problem 10.27.

Section 10.6 Series and Parallel Inductors

10.28 Five inductors, each with an inductance of 80 mH, are connected in series. What is the total inductance?

10.29 A 200-μH and a 800-μH inductor are connected in parallel. What is the equivalent total inductance?

10.30 Find the equivalent inductance for each circuit in Fig. 10.47.

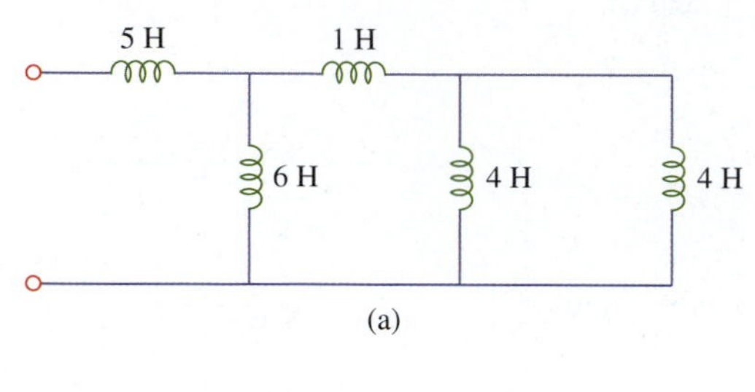

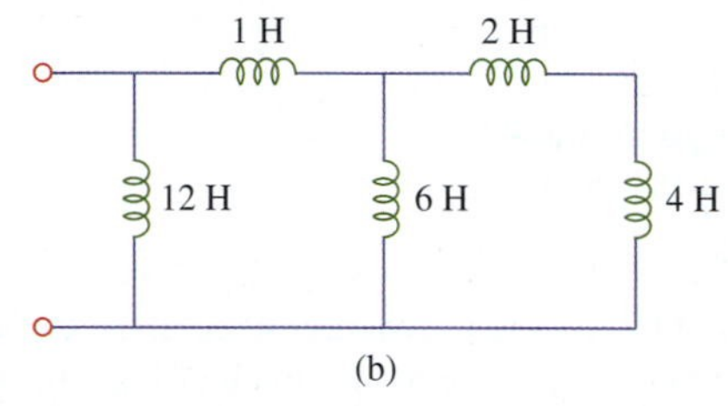

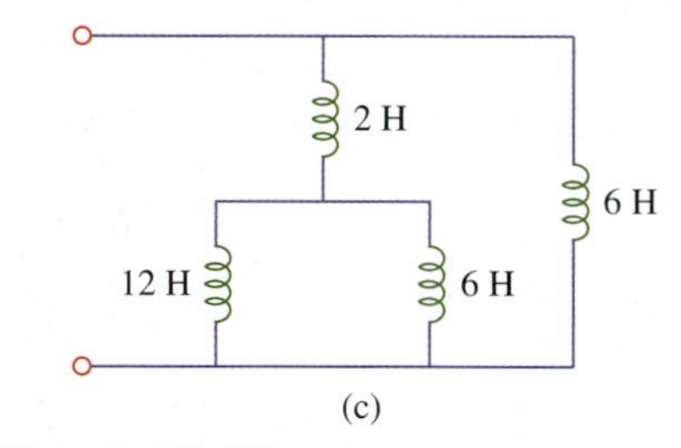

Figure 10.47
For Problem 10.30.

10.31 Obtain L_{eq} for the inductive circuit of Fig. 10.48.

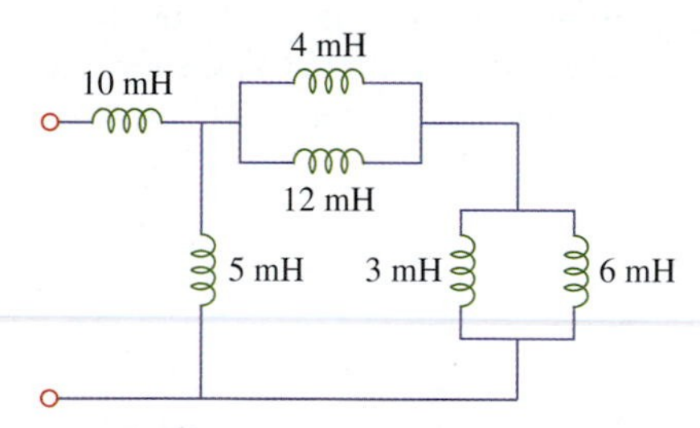

Figure 10.48
For Problem 10.31.

10.32 Determine L_{eq} at terminals a-b of the circuit in Fig. 10.49.

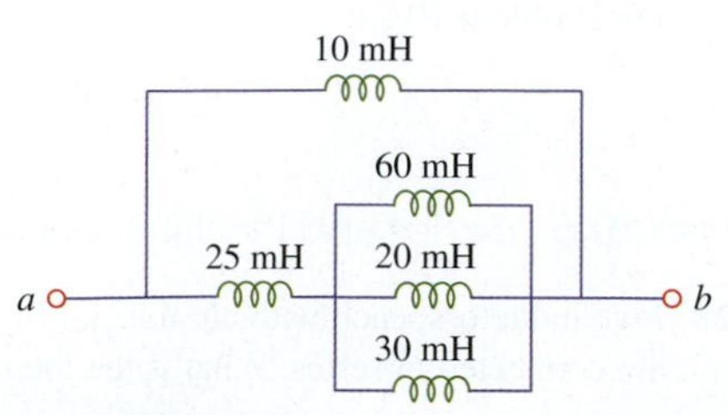

Figure 10.49
For Problem 10.32.

10.33 Find L_{eq} at the terminals a-b of the circuit in Fig. 10.50.

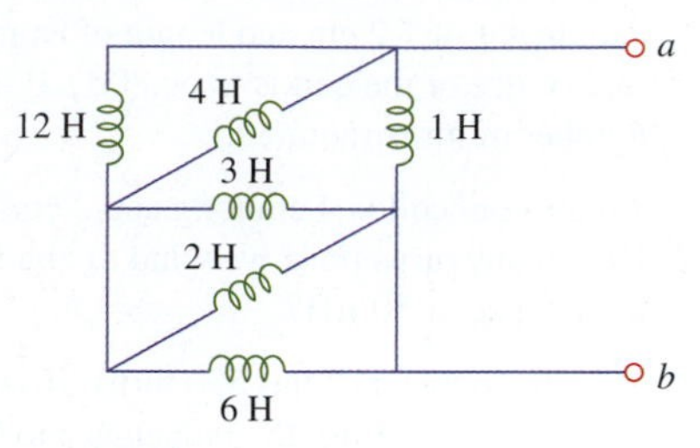

Figure 10.50
For Problem 10.33.

10.34 Find the equivalent inductance looking into terminals a-b of the circuit in Fig. 10.51.

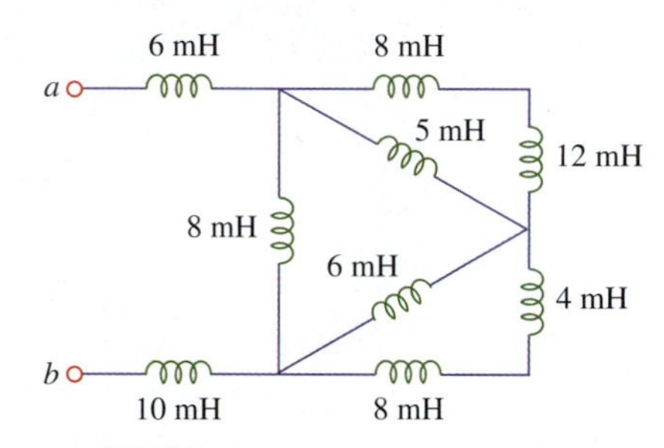

Figure 10.51
For Problem 10.34.

10.35 Find L_{eq} in the circuit in Fig. 10.52.

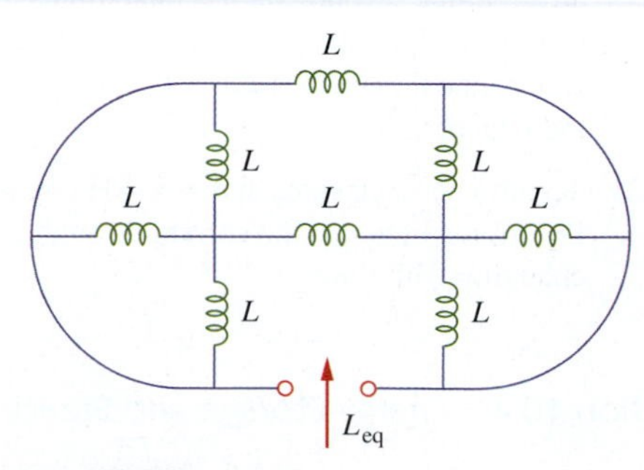

Figure 10.52
For Problem 10.35.

10.36 Determine the equivalent inductance L_{eq} of the circuits shown in Fig. 10.53.

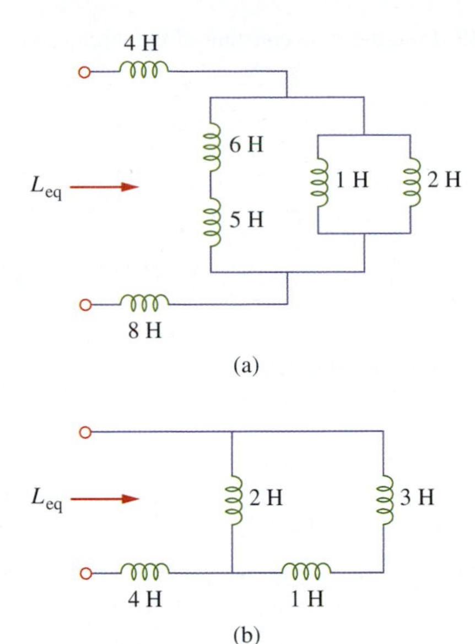

Figure 10.53
For Problem 10.36.

Section 10.7 Transient *RL* Circuits

10.37 Calculate the value of $e^{-tR/L}$ for the following values of R, L, and t.

(a) $R = 1\ \Omega$, $L = 5$ H, and $t = 1$ s

(b) $R = 2\ \Omega$, $L = 10$ H, and $t = 1$ s

(c) $R = 5\ \Omega$, $L = 5$ H, and $t = 2$ s

10.38 A 3-A current flows through a 4-H coil the instant before the switch is opened. If a 2-Ω resistor is placed in series with the coil when the switch is opened, determine the current 5 s later.

10.39 The switch in the circuit in Fig. 10.54 has been closed for a long time. At $t = 0$, the switch is opened. Calculate $i(t)$ for $t > 0$.

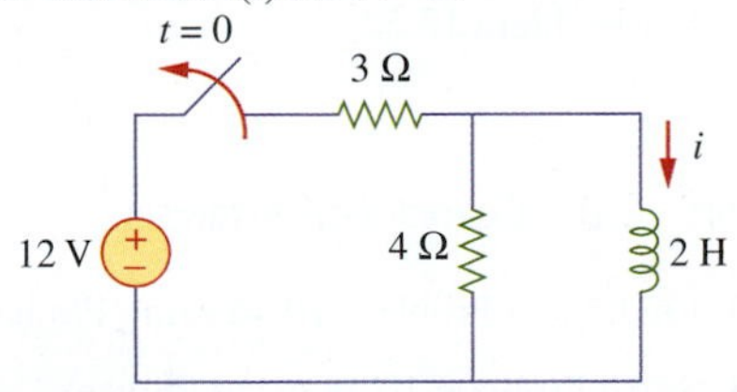

Figure 10.54
For Problem 10.39.

10.40 For the circuit shown in Fig. 10.55, calculate the time constant.

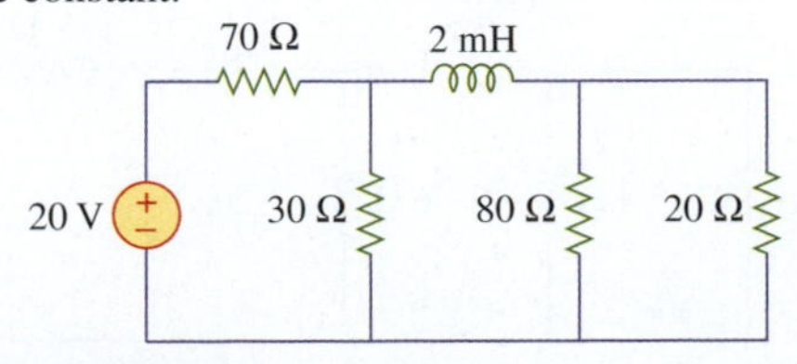

Figure 10.55
For Problem 10.40.

10.41 Obtain the time constant of the circuit in Fig. 10.56.

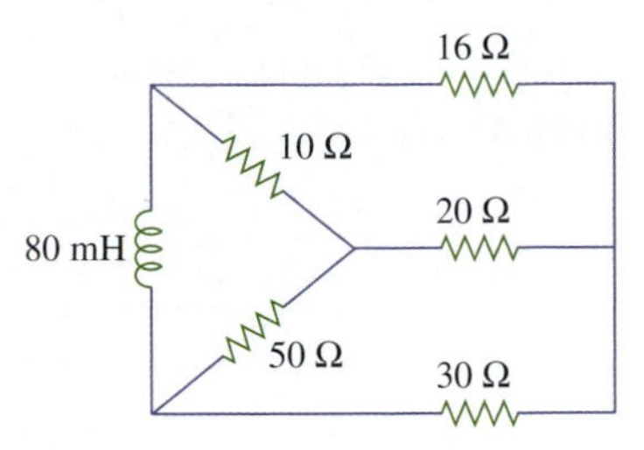

Figure 10.56
For Problem 10.41.

10.42 Find the time constant for each of the circuits in Fig. 10.57.

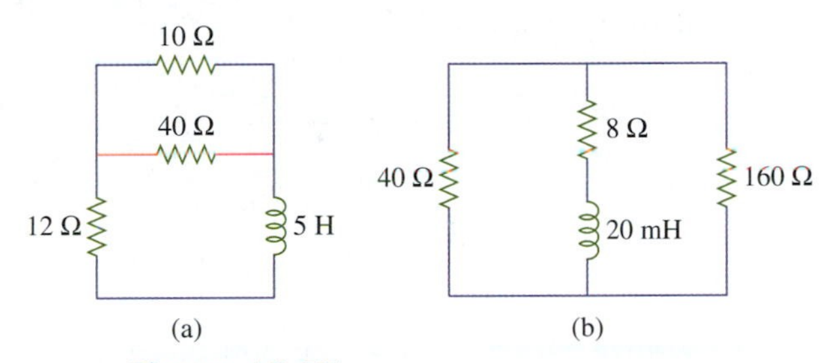

Figure 10.57
For Problem 10.42.

10.43 Consider the circuit of Fig. 10.58. Find $v_o(t)$ if $i(0) = 2$ A and $v(t) = 0$.

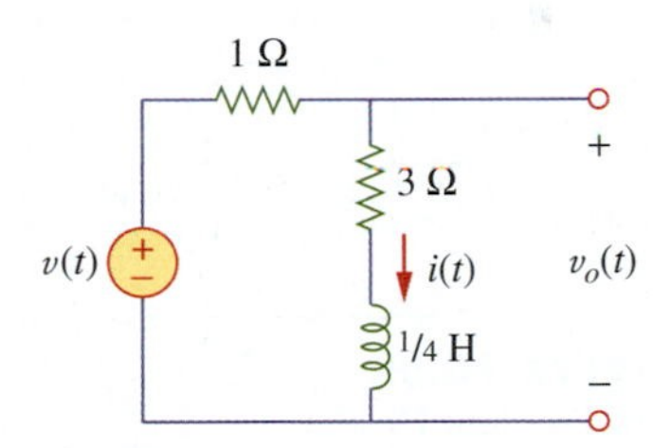

Figure 10.58
For Problem 10.43.

10.44 For the circuit in Fig. 10.59, determine $v_o(t)$ when $i(0) = 1$ A and $v(t) = 0$.

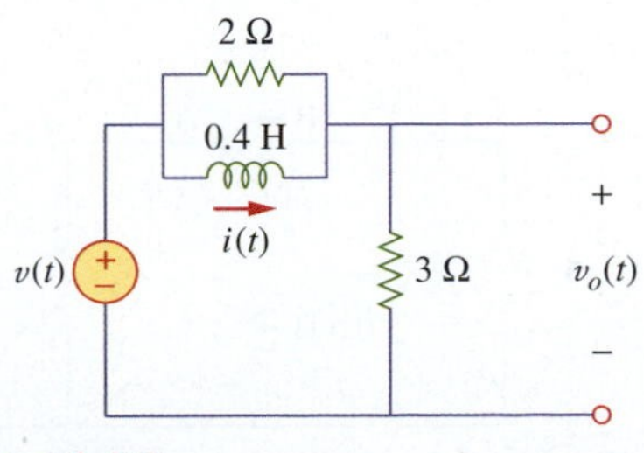

Figure 10.59
For Problem 10.44.

10.45 For the circuit in Fig. 10.60, $v(t) = 120e^{-50t}$ V and $i = 30e^{-50t}$ A, $t > 0$.

(a) Determine the time constant.

(b) Find L and R.

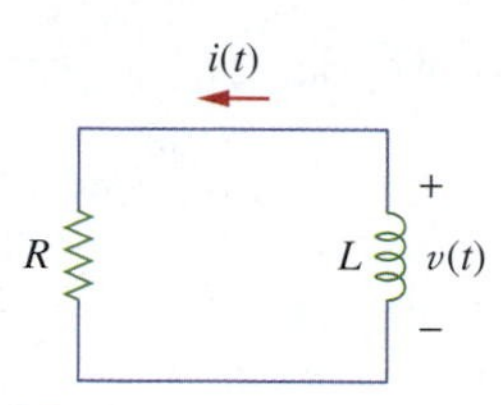

Figure 10.60
For Problem 10.45.

10.46 Find $i(t)$ and $v(t)$ for $t > 0$ in the circuit of Fig. 10.61 if $i(0) = 10$ A.

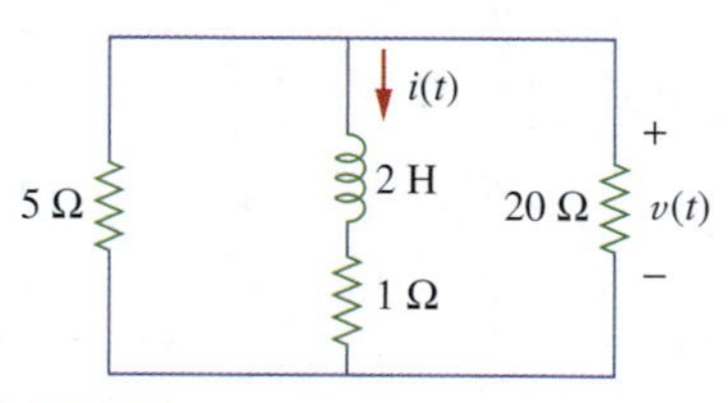

Figure 10.61
For Problems 10.46 and 10.53.

10.47 Consider the circuit in Fig. 10.62. Given that $v_o(0) = 2$ V, find v_o and v_x for $t > 0$.

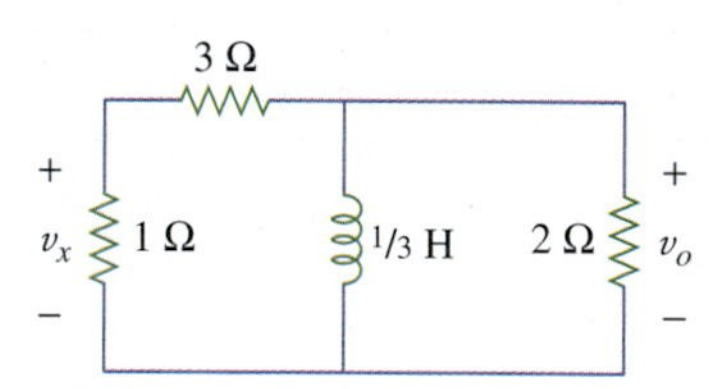

Figure 10.62
For Problem 10.47.

10.48 The switch in Fig. 10.63 has been in position A for a long time. At $t = 0$, the switch moves from position A to B. The switch is a make-before-break type, so that there is no interruption in the inductor current. Find $i(t)$ for $t > 0$.

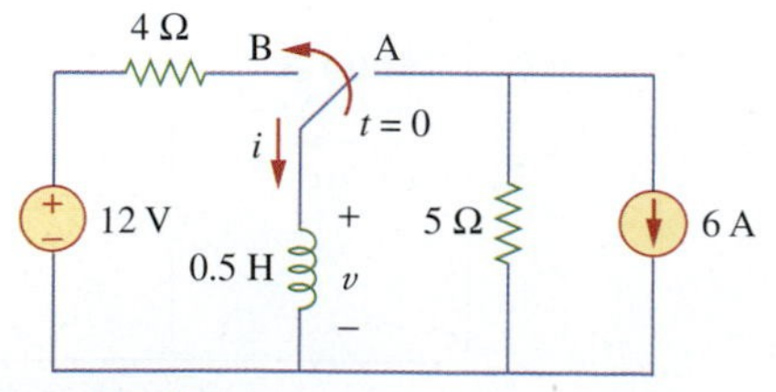

Figure 10.63
For Problems 10.48 and 10.54.

10.49 Find the time constant of the circuit in Fig. 10.64.

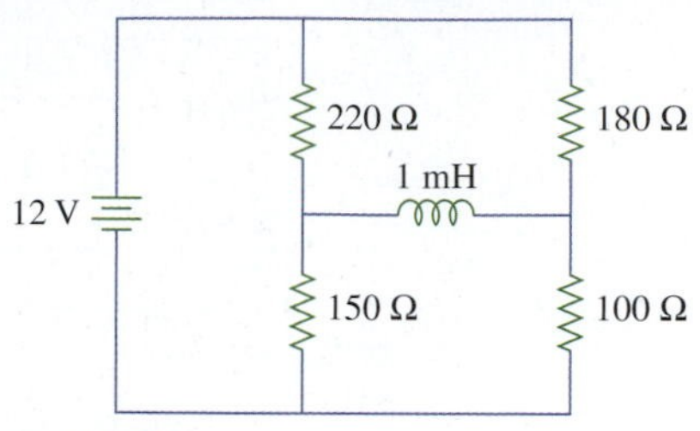

Figure 10.64
For Problem 10.49.

10.50 A 50-mH inductor has a wire resistance of 2 Ω. Calculate how long it takes the current through it to reach its full value.

10.51 An RL circuit is connected to a 6-V source and has $R = 120$ Ω. If it takes 40 μs for current to build up from 0 to 5 mA, calculate: (a) the final current, (b) the inductance L, and (c) the time constant.

10.52 In the RL circuit in Fig. 10.65, switch S_2 is opened, while switch S_1 is closed and left until a constant current is reached. Then S_2 is closed, while S_1 is opened.

(a) Determine the initial current in the resistor just after S_2 is closed and S_1 is opened.

(b) Find the time it takes the current to decrease to half its initial value.

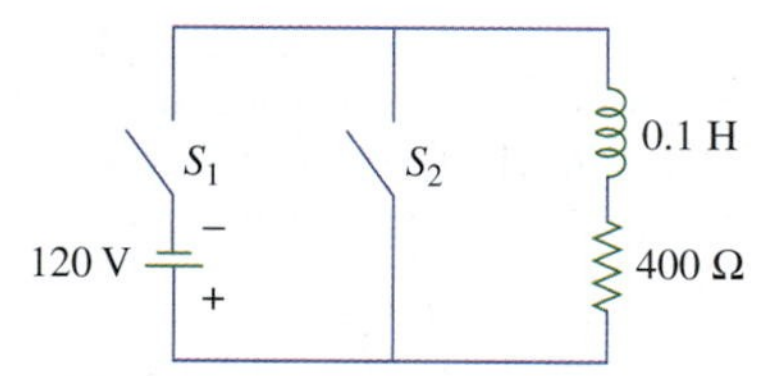

Figure 10.65
For Problem 10.52.

Section 10.8 Computer Analysis

10.53 Find $i(t)$ in Problem 10.46 using PSpice.

10.54 Solve Problem 10.48 using PSpice.

10.55 The switch in Fig. 10.66 opens at $t = 0$. Use PSpice to determine $v(t)$ for $t > 0$.

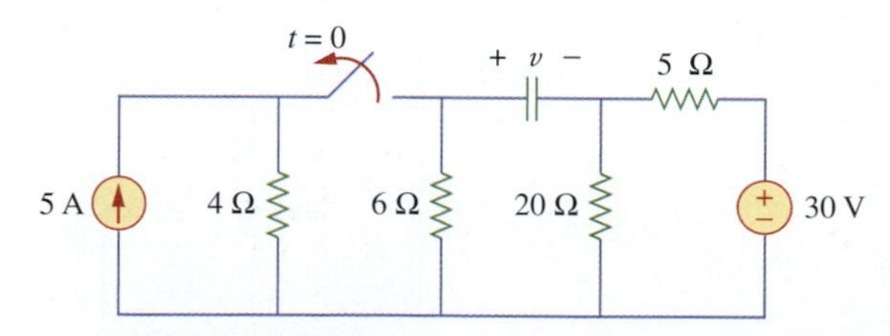

Figure 10.66
For Problem 10.55.

10.56 The switch in Fig. 10.67 moves from position a to b at $t = 0$. Use PSpice to find $i(t)$ for $t > 0$.

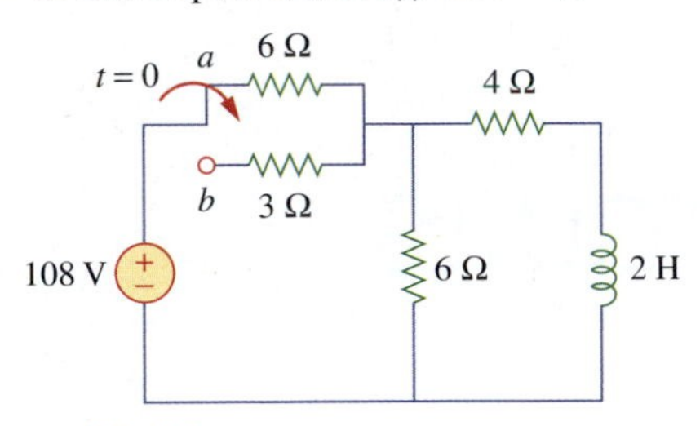

Figure 10.67
For Problem 10.56.

10.57 In the circuit of Fig. 10.68, the switch has been in position a for a long time but moves instantaneously to position b at $t = 0$. Determine $i_o(t)$ using Multisim.

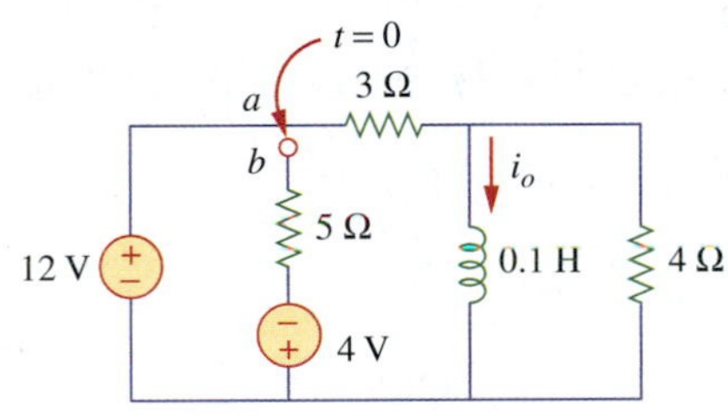

Figure 10.68
For Problem 10.57.

10.58 Determine $v_o(t)$ in the circuit of Fig. 10.69 using Multisim.

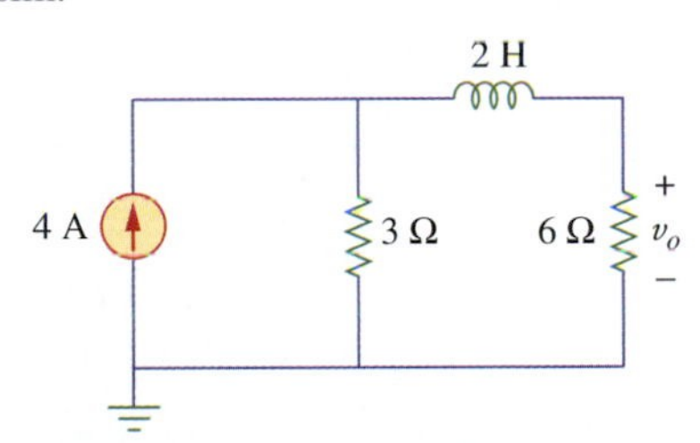

Figure 10.69
For Problem 10.58.

10.59 The voltage source v_s in the circuit in Fig. 10.70 produces a square wave that cycles between 0 and 12 V with a frequency of 2 kHz. Use Multisim to find $v_o(t)$.

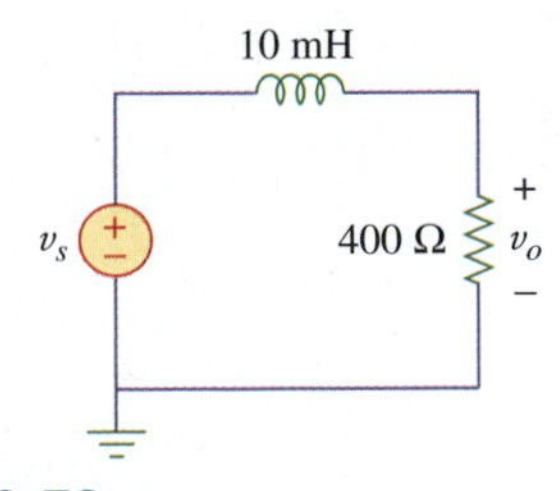

Figure 10.70
For Problem 10.59.

Section 10.9 Applications

10.60 The resistance of a 160-mH coil is 8 Ω. Find the time required for the current to build up to 60 percent of the final value when voltage is applied to the coil.

10.61 A 120-V dc generator energizes a motor whose coil has an inductance of 50 H and a resistance of 100 Ω. A field discharge resistor of 400 Ω is connected in parallel with the motor to avoid damage to the motor, as shown in Fig. 10.71. The system is at steady state. Find the current through the discharge resistor 100 ms after the breaker is tripped.

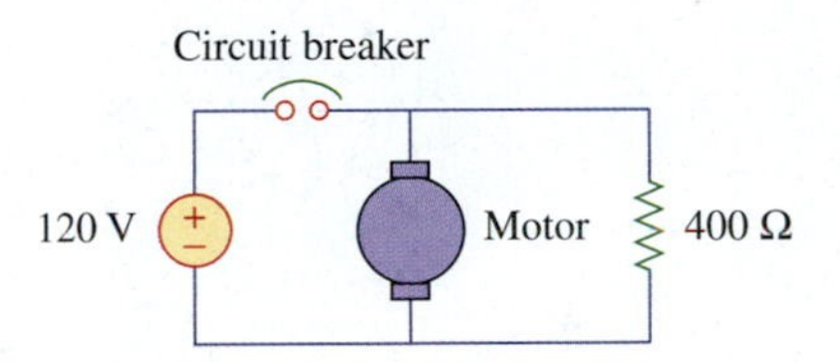

Figure 10.71
For Problem 10.61.

10.62 An *RL* circuit may be used as a differentiator if the output is taken across the inductor and $\tau \ll T$ (say, $\tau < 0.1T$), where T is the width of the input pulse. If R is fixed at 200 kΩ, determine the maximum value of L required to differentiate a pulse with $T = 10\ \mu$s.

10.63 The circuit in Fig. 10.72 is used by a biology student to study "frog kick." She noticed that the frog kicked a little when the switch was closed but kicked violently for 5 s when the switch was opened. Model the frog as a resistor, and calculate its resistance. Assume that it takes 10 mA for the frog to kick violently.

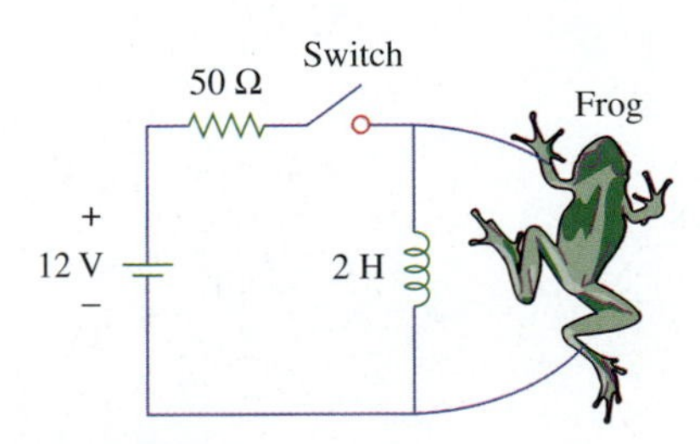

Figure 10.72
For Problem 10.63.

PART TWO

AC Circuits

chapter

11

AC Voltage and Current

Nature, time, and patience are the three great physicians.

—Proverb

Historical Profiles

Heinrich Rudorf Hertz (1857–1894), a German experimental physicist, demonstrated that electromagnetic waves obey the same fundamental laws as light. His work confirmed James Clerk Maxwell's celebrated 1864 theory and prediction that such waves existed.

Born into a prosperous family in Hamburg, Germany, Hertz attended the University of Berlin and did his doctorate under the prominent physicist Hermann von Helmholtz. He became a professor at Karlsruhe, where he began his quest for electromagnetic waves. He successfully generated and detected electromagnetic waves. He was the first to show that light is electromagnetic energy. In 1887, Hertz noted for the first time the photoelectric effect of electrons in a molecular structure. Although Hertz died early at the age of 37, his discovery of electromagnetic waves paved the way for the practical use of such waves in radio, television, and other communication systems. The unit of frequency, the hertz, bears his name.

Heinrich Rudorf Hertz
© The Huntington Library, Burndy Library, San Marino, California

Charles Proteus Steinmetz (1865–1923), a German Austrian mathematician and engineer, introduced the phasor method (covered in the next chapter) in ac circuit analysis. He is also noted for his work on the theory of hysteresis.

Born in Breslau, Germany, Steinmetz lost his mother at the age of 1. He was forced to leave Germany due to his political activities when he was just about to complete his doctoral dissertation in mathematics at the University of Breslau. He migrated to Switzerland and later to the United States. He was employed by General Electric in 1893. The same year, he published a paper in which complex numbers were used to analyze ac circuits for the first time. This led to one of his many textbooks, such as *Theory and Calculation of AC Phenomena* published by McGraw-Hill in 1897. In 1901, he became the president of the American Institute of Electrical Engineers which later became the IEEE.

Charles Proteus Steinmetz
© Bettmann/Corbis

11.1 Introduction

Thus far, our analysis has been limited to dc circuits, that is, circuits excited by constant or time-invariant sources. We have restricted our study to dc sources for the sake of simplicity, for pedagogic reasons, and also for historic reasons. Historically, dc sources were the main means of providing electricity up to the late 1800s. At the end of the nineteenth century, the battle of direct current (dc) versus alternating current (ac) began. Both had their advocates among the electrical engineers of that time. Because ac is more efficient and economical to generate and transmit over long distances than dc, ac systems turned out to be the winner. Thus, it is in keeping with the historical sequence of events that we considered dc sources first.

We now begin the analysis of circuits in which the source voltage or current is time-varying. In this chapter, we are particularly interested in sinusoidally time-varying excitation or simply excitation by a *sinusoid*. A sinusoidal current is usually referred to as *alternating current* (*ac*). Such a current reverses at regular time intervals and has alternately positive and negative values. Circuits driven by sinusoidal current or voltage sources are called *ac circuits*.

A **sinusoid** is a signal in the form of the sine or cosine function.

We are interested in sinusoids for a number of reasons. First, nature itself is characteristically sinusoidal. We experience sinusoidal variation in the motion of a pendulum, the vibration of a string, the ripples on the ocean surface, and the natural response of second-order systems, to mention but a few. Second, a sinusoidal signal is easy to generate and transmit. It is the form of voltage generated throughout the world and supplied to homes, factories, laboratories, and so on. Third, it is the dominant form of signal in the communications and electric power industries. Fourth, sinusoids play an important role in the analysis of periodic signals. Lastly, a sinusoid is easy to handle mathematically. For these and other reasons, the sinusoid is an extremely important function in circuit analysis.

We begin with the generation of ac voltage. We then discuss sinusoids and phase relations. We introduce the concepts of average and rms values. We finally talk about time-domain measurements using the oscilloscopes.

11.2 AC Voltage Generator

We recall that dc voltage is generated by a battery, which is able to maintain a constant terminal voltage for a considerable amount of time. The battery does not provide a suitable means of generating large amounts of energy required for both home and industry. In the last chapter, we discussed Faraday's law of electromagnetic induction as a means of generating ac voltage. This involves rotating a coil of wire in a static magnetic field, as typically shown in Fig. 11.1. Because the voltage generated by the rotating coil is alternately positive and negative, it is referred to as an alternating voltage. It can be expressed mathematically as

$$v = V_m \sin\theta \qquad \textbf{(11.1)}$$

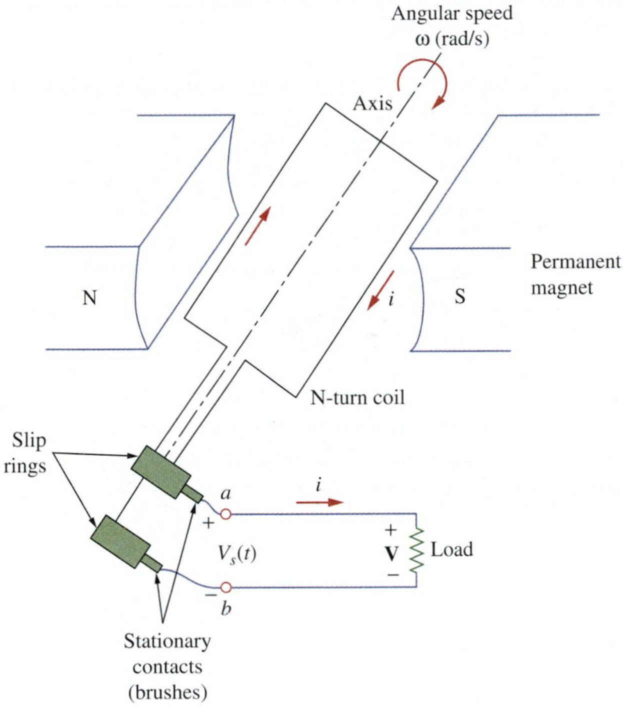

Figure 11.1
AC generator.

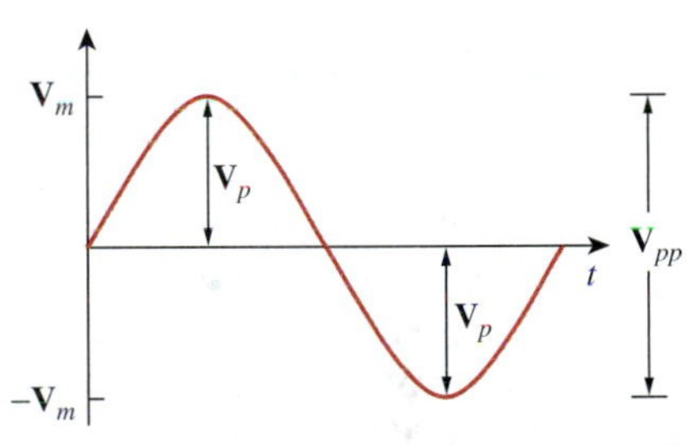

Figure 11.2
Alternating voltage.

where V_m is the maximum value and θ is the angle (in radians) of rotation of the coil. The value of voltage v at any point in time is called its *instantaneous value*. The instantaneous value is usually represented by the lowercase designation $v(t)$. As shown in Fig. 11.2, the peak value V_p is the maximum value; that is, $V_p = V_m$.[1] The peak-to-peak value V_{pp} is the value of the voltage from the positive peak to the negative peak; that is, $V_{pp} = 2V_p$.

Example 11.1

An ac voltage is represented by 40 $\sin(10^3 t)$V.
(a) Find the instantaneous value at 0.2 ms.
(b) Determine the peak-to-peak value.

Solution:
(a) At $t = 0.2$ ms,

$$v = 40 \sin 10^3 \times 0.2 \times 10^{-3} = 40 \sin 0.2 = 7.947 \text{ V}$$

(b) The peak-to-peak is

$$V_{pp} = 2 \times 40 = 80 \text{ V}$$

Practice Problem 11.1

A voltage generator produces 25 $\sin(10^6 t)$V.
(a) Determine the peak value.
(b) Calculate the instantaneous value at 3 μs.

Answer: (a) 25 V; (b) 3.528 V

[1] If there is dc offset, then $V_p \neq V_m$.

11.3 Sinusoids

If we replace θ with ωt in Eq. (11.1), the sinusoidal voltage becomes

$$v(t) = V_m \sin \omega t \tag{11.2}$$

where

V_m = the *amplitude* of the sinusoid
ω = the *angular frequency* in radians per second
t = *time* in seconds
ωt = the *argument* of the sinusoid in radians

The sinusoid is shown in Fig. 11.3(a) as a function of its argument and in Fig. 11.3(b) as a function of time. It is evident that the wave repeats itself every T seconds; hence, T is called the *period* of the wave. From the two plots in Fig. 11.3, we observe that $\omega T = 2\pi$,

$$\boxed{T = \frac{2\pi}{\omega}} \tag{11.3}$$

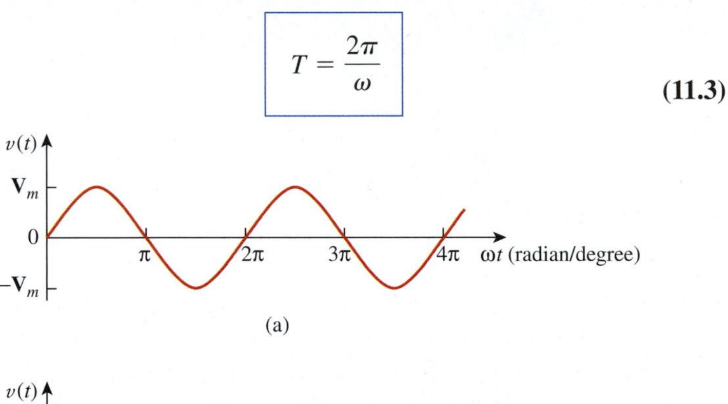

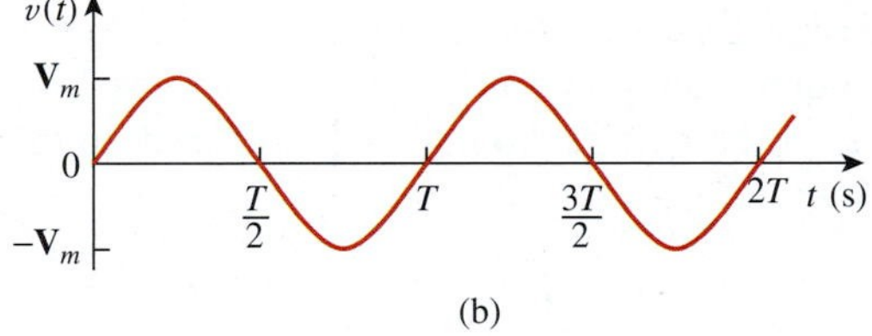

Figure 11.3
A sketch of $V_m \sin(\omega t)$: (a) as a function of ωt; (b) as a function of t.

The fact that $v(t)$ repeats itself every T seconds is shown by replacing t by $t + T$ in Eq. (11.2).

$$v(t+T) = V_m \sin \omega (t+T) = V_m \sin\left(\omega t + \omega \frac{2\pi}{\omega}\right)$$
$$= V_m \sin(\omega t + 2\pi) = V_m \sin \omega t = v(t)$$

Thus, $v(t+T) = v(t)$; that is, v has the same value at $t + T$ as it does at t, and $v(t)$ is said to be *periodic*.

As mentioned earlier, the period T of the periodic function is the time of one complete cycle or the number of seconds per cycle. The reciprocal of this quantity is the number of cycles per second and is known as the *frequency* f of the wave. Thus,

$$\boxed{f = \frac{1}{T}} \tag{11.4}$$

From Eqs. (11.3) and (11.4), it is clear that

$$\omega = 2\pi f \tag{11.5}$$

where ω is in radians per second (rad/s) and f is in hertz[2] (Hz).

From Fig. 11.3, we notice that the horizontal axis can be in time, degrees, or radians. It is appropriate at this point to establish the relationship between degrees and radians. A 360-degree revolution corresponds to 2π radians; that is,

$$2\pi \text{ rad} \equiv 360^\circ \tag{11.6a}$$

or

$$1 \text{ rad} \equiv \frac{360^\circ}{2\pi} = 57.3^\circ \tag{11.6b}$$

The conversion formulas between degrees and radians are:

$$\text{Radians} = \left(\frac{\pi}{180^\circ}\right) \times \text{Degrees} \tag{11.7}$$

$$\text{Degrees} = \left(\frac{180^\circ}{\pi}\right) \times \text{Radians} \tag{11.8}$$

TABLE 11.1

Some degrees and their corresponding radians

Degrees (°)	Radians (rad)
0	0
30	$\pi/6$
45	$\pi/4$
60	$\pi/3$
90	$\pi/2$
135	$3\pi/4$
180	π
225	$5\pi/4$
270	$3\pi/2$
315	$7\pi/4$
360	2π

Table 11.1 provides different values of degrees and the corresponding radians.

If the sine wave is not zero at $t = 0$ as in Fig. 11.3, it has a *phase shift* (also called *phase* or *phase angle*). As shown in Fig. 11.4, the sine wave may be shifted to the right or to the left. If the wave is shifted to the left, as in Fig. 11.4(a),

$$v(t) = V_m \sin(\omega t + \theta) \tag{11.9a}$$

where θ is the phase shift in radians or degrees. If the wave is shifted to the right, as in Fig. 11.4(b),

$$v(t) = V_m \sin(\omega t - \theta) \tag{11.9b}$$

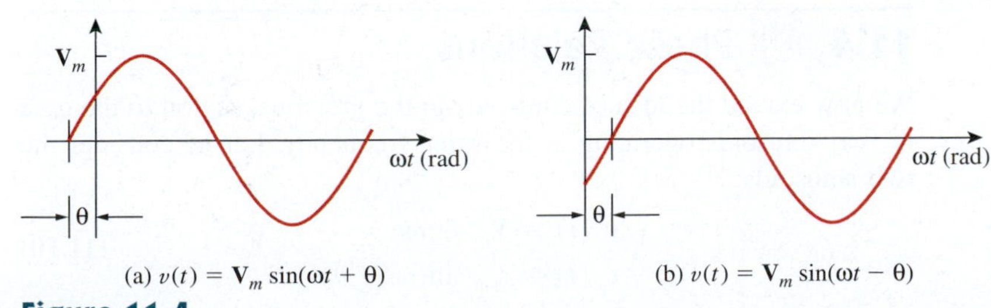

Figure 11.4
Phase-shifted sine waves.

[2] Historical note: Named after the German physicist Heinrich R. Hertz (1857–1894).

Example 11.2

Find the amplitude, phase, angular frequency, period, and frequency of the following sinusoid:

$$v(t) = 12\cos(50t + 10°)$$

Solution:
The amplitude is $V_m = 12$ V.
The phase is $\theta = 10°$.
The angular frequency is $\omega = 50$ rad/s.

The period is $T = \dfrac{2\pi}{\omega} = \dfrac{2\pi}{50} = 0.1257$ s.

The frequency is $f = \dfrac{1}{T} = 7.958$ Hz.

Practice Problem 11.2

Given the sinusoid $5\sin(4\pi t - 60°)$, calculate its amplitude, phase, angular frequency, period, and frequency.

Answer: 5; −60 degrees; 12.566 rad/s; 0.5 s; 2 Hz

Example 11.3

(a) Convert 150 degrees to radians.
(b) Convert $4\pi/3$ radians to degrees.

Solution:
(a) Using Eq. (11.7),

$$\text{Radians} = \left(\frac{\pi}{180°}\right) \times 150° = 2.618 \text{ rad}$$

(b) Using eq. (11.8),

$$\text{Degrees} = \left(\frac{180°}{\pi}\right) \times \frac{4\pi}{3} = 240°$$

Practice Problem 11.3

(a) Convert 210 degrees to radians.
(b) Convert $5\pi/6$ radians to degrees.

Answer: (a) 3.6652 rad; (b) 150 degrees

11.4 Phase Relations

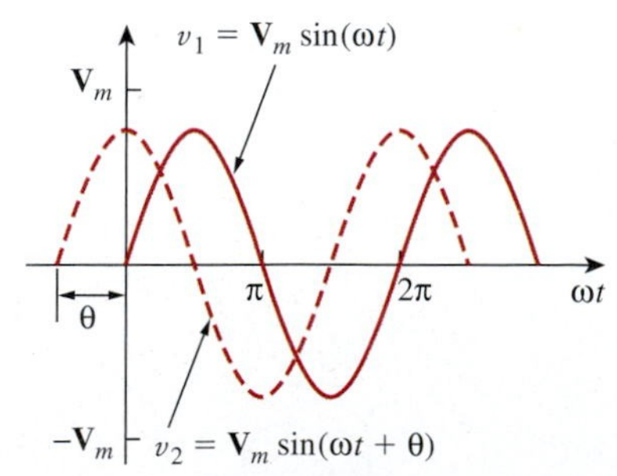

Figure 11.5
Two sinusoids with different phases.

We now extend the idea of sinusoids in the previous section to the case of two sinusoids operating at the same frequency. Let us consider the two sinusoids

$$v_1(t) = V_m \sin\omega t$$
$$v_2(t) = V_m \sin(\omega t + \theta) \tag{11.10}$$

shown in Fig. 11.5. The starting point of v_2 in Fig. 11.5 occurs first in time. Therefore, we say that v_2 *leads* v_1 by θ or that v_1 *lags* v_2 by θ. If $\theta \neq 0$, we also say that v_1 and v_2 are *out of phase*. If $\theta = 0$, then v_1 and v_2 are said to be *in phase*; they reach their minima and maxima at exactly the same time. We can compare v_1 and v_2 in this manner because they operate at the same frequency; they do not need to have the same amplitude. The difference angle must be a magnitude less than 180 degrees to determine leading or lagging. If the phase difference is 180 degrees, then neither can be said to be leading or lagging.

So far, we have expressed sinusoids as functions of sine. A sinusoid can also be expressed in cosine form. When comparing two sinusoids, it is expedient to express both as either sine or cosine with positive amplitudes. This is achieved by using the following trigonometric identities:

$$\begin{aligned}
\sin(-\omega t) &= -\sin(\omega t) \\
\cos(-\omega t) &= \cos(\omega t) \\
\sin(\omega t \pm 180^\circ) &= -\sin \omega t \\
\cos(\omega t \pm 180^\circ) &= -\cos \omega t \\
\sin(\omega t \pm 90^\circ) &= \pm\cos \omega t \\
\cos(\omega t \pm 90^\circ) &= \mp\sin \omega t
\end{aligned} \tag{11.11}$$

Using these relationships, we can easily transform a sinusoid from sine form to cosine form or vice versa.

Example 11.4

Calculate the phase angle between the following expressions:

$$v_1(t) = -10\cos(\omega t + 50^\circ)$$
$$v_2(t) = 12\sin(\omega t - 10^\circ)$$

State which sinusoid is leading.

Solution:

The phase will be calculated in two ways.

METHOD 1 In order to compare v_1 and v_2, we must express them in the same form. If we express them in cosine form with positive amplitudes,

$$v_1 = -10\cos(\omega t + 50^\circ) = 10\cos(\omega t + 50^\circ - 180^\circ)$$

or

$$v_1 = 10\cos(\omega t - 130^\circ) = 10\cos(\omega t + 230^\circ) \tag{11.4.1}$$

and

$$v_2 = 12\sin(\omega t - 10^\circ) = 12\cos(\omega t - 10^\circ - 90^\circ)$$

or

$$v_2 = 12\cos(\omega t - 100^\circ) \tag{11.4.2}$$

It can be deduced from Eqs. (11.4.1) and (11.4.2) that the phase difference between v_1 and v_2 is 30 degrees. We can write v_2 as

$$v_2 = 12\cos(\omega t - 130^\circ + 30^\circ)$$

or

$$v_2 = 12\cos(\omega t + 260^\circ) = 12\cos(\omega t + 230^\circ + 30^\circ) \tag{11.4.3}$$

Comparing Eqs. (11.4.1) and (11.4.3) shows clearly that v_2 leads v_1 by 30 degrees.

METHOD 2 Alternatively, we may express v_1 in sine form:

$$\begin{aligned} v_1 &= -10\cos(\omega t + 50^\circ) = 10\sin(\omega t + 50^\circ - 90^\circ) \\ &= 10\sin(\omega t - 40^\circ) = 10\sin(\omega t - 10^\circ - 30^\circ) \end{aligned}$$

But

$$v_2 = 12\sin(\omega t - 10^\circ)$$

Comparing the two shows that v_1 lags v_2 by 30 degrees. This is the same as saying that v_2 leads v_1 by 30 degrees.

Practice Problem 11.4

Find the phase angle between the following expressions:

$$\begin{aligned} i_1 &= -4\sin(377t + 25^\circ) \\ i_2 &= 5\cos(377t - 40^\circ) \end{aligned}$$

Does i_1 lead or lag i_2?

Answer: 155 degrees, i_1 leads i_2

11.5 Average and RMS Values

For any periodic function $f(t)$ (which may be current or voltage) with period T, the average value is defined as

$$F_{\text{ave}} = \frac{1}{T} \times [\text{Area under } f(t) \text{ curve over one period}] \tag{11.12}$$

Keep in mind that areas above the axis are positive, while areas below the axis are negative.

For a sinusoidal waveform, the net area over one period is always zero due to the symmetrical nature of the waveform, as illustrated in Fig. 11.6(a). As evident in Fig. 11.6(a), for every positive value of the waveform during the one half-cycle, there is a similar negative value during the next half-cycle, which cancels the positive value. Thus, for an ac voltage $v(t) = V_m \sin \omega t$,

$$V_{\text{ave}} = 0 \tag{11.13}$$

Similarly, for an ac current $i(t) = I_m \sin \omega t$,

$$I_{\text{ave}} = 0 \tag{11.14}$$

For the full-wave-rectified sine waveform shown in Fig. 11.6(b), by applying Eq. (11.12), we obtain

$$V_{\text{ave}} = 0.637V_m \tag{11.15}$$

$$I_{\text{ave}} = 0.637I_m \tag{11.16}$$

The **effective (or rms) value** of a periodic current (or voltage) is the dc current (or voltage) that delivers the same average power to a resistor as the periodic current (or voltage).

Another important property of a periodic waveform is the *effective* or *rms (root mean square)* value. The idea of effective value arises from the need to measure the effectiveness of a voltage or current source in delivering power to a resistive load.

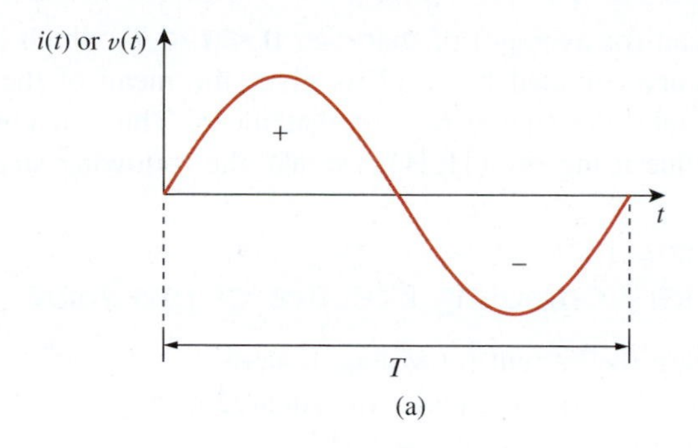

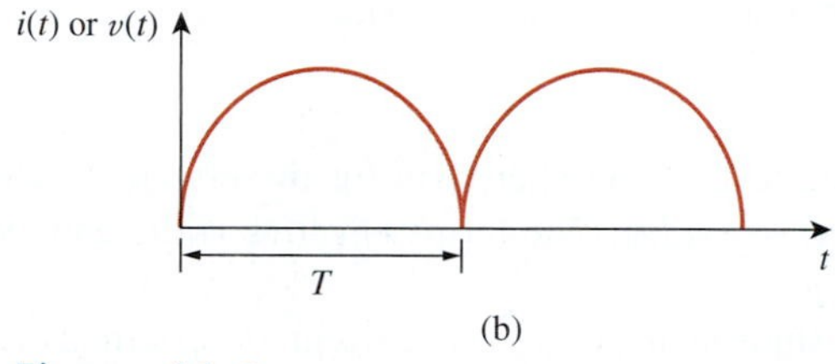

Figure 11.6
(a) Sine waveform; (b) full-wave-rectified sine waveform.

The **effective** (or equivalent dc) **value** of a sinusoidal current or voltage is 0.707 of its amplitude.

The effective value of a sinusoidal quantity may be regarded as the equivalent dc value. As shown in Fig. 11.7, the circuit in (a) is ac while that of (b) is dc. Our objective is to find I_{eff} that will transfer the same power to resistor R as the sinusoid $i(t)$. The average power absorbed by the resistor in the ac circuit is

$$I_{\text{eff}}^2 R = P_{\text{ave}} = \frac{1}{2} I_m^2 R \tag{11.17}$$

Hence, for a sinusoidal current $i(t)$,

$$I_{\text{eff}} = I_{\text{rms}} = \frac{I_m}{\sqrt{2}} = 0.707 I_m \tag{11.18}$$

Similarly, for a sinusoidal voltage $v(t)$,

$$V_{\text{eff}} = V_{\text{rms}} = \frac{V_m}{\sqrt{2}} = 0.707 V_m \tag{11.19}$$

The average power in a given load R is given by

$$P_{\text{ave}} = V_{\text{eff}} I_{\text{eff}} = \frac{V_{\text{eff}}^2}{R} = I_{\text{eff}}^2 R \tag{11.20}$$

that is, the same as in the dc case.

The factor of 0.707 is valid only for sinusoids. For other periodic functions, we need a more general formula. For any periodic waveform $f(t)$,

$$F_{\text{eff}} = \sqrt{\frac{1}{T} \times \pi[\text{Area under the square of } f(t)]} \tag{11.21}$$

where $f(t)$ may be current or voltage. Equation (11.21) states that to find the effective value of $f(t)$, we first find its square $f^2(t)$ and then

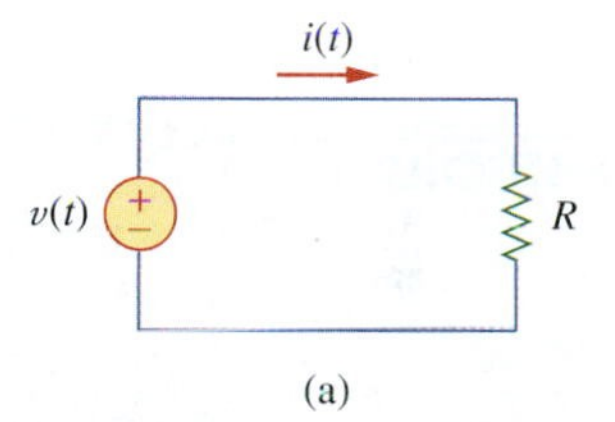

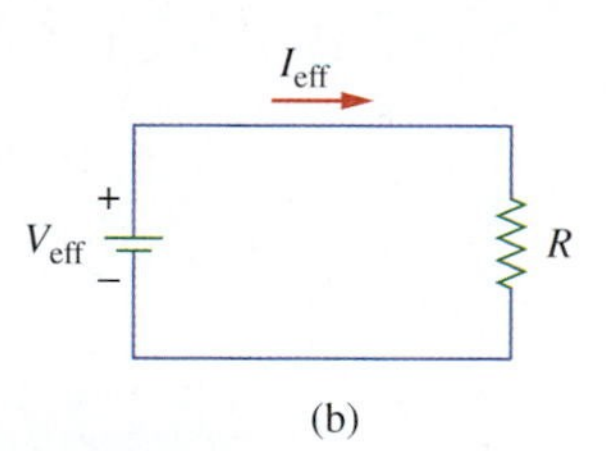

Figure 11.7
Finding the effective current: (a) ac circuit; (b) dc circuit.

find the mean (or average) of that over $0 < t < T$, which is the area under the curve divided by T. (This gives the mean of the squared). We finally take the square root of that mean. Thus, to compute the effective value using Eq. (11.21), we take the following steps:

Steps for Computing Effective or rms Value

1. Square the current (or voltage) curve.
2. Determine the area under the squared curve for one period.
3. Divide the area by the period T.
4. Find the square root of the result.

This procedure leads to another term for the effective value, *the root-mean-square (rms) value.* The terms *effective value* and *rms value* are synonymous.

When a sinusoidal voltage or current is specified, it is often in terms of its maximum (or peak or amplitude) value or its rms value because its average value is zero. In power industries, voltages and currents are specified in terms of their rms values rather than peak values. For instance, the 120-V available at every household is the rms value of the voltage from the power company. It is convenient in power analysis to express voltage and current in their rms values. Also, analog voltmeters and ammeters are designed to read directly the rms value of voltage and current, respectively.

Example 11.5

The current through a resistor is

$$i(t) = 4\sin(377t + 30°)\text{ A}$$

while the voltage through it is

$$v(t) = 60\sin(377t + 30°)\text{ A}$$

Calculate the average power dissipated in the resistor.

Solution:
The rms values of the current and voltage are:

$$I_{\text{rms}} = \frac{I_m}{\sqrt{2}} = \frac{4}{\sqrt{2}} = 2.828\text{ A}$$

$$V_{\text{rms}} = \frac{V_m}{\sqrt{2}} = \frac{60}{\sqrt{2}} = 42.43\text{ V}$$

Power is

$$P_{\text{ave}} = V_{\text{rms}}I_{\text{rms}} = 42.43 \times 2.828 = 120\text{ W}$$

Practice Problem 11.5

The voltage across an 8-Ω resistor is $v(t) = 50\sin(628t + 25°)$ V. Find the rms value of the voltage and the average power dissipated in the resistor.

Answer: 35.35 V; 156.25 W

Example 11.6

Find the average and the rms value of the signal shown in Fig. 11.8.

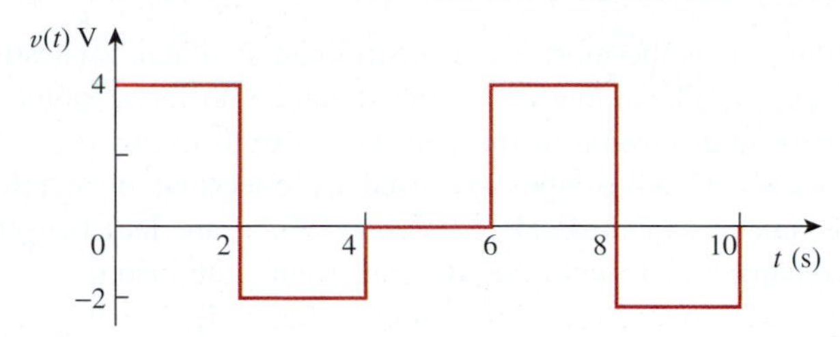

Figure 11.8
For Example 11.6.

Solution:

We first notice that the signal repeats itself after 6s, that is, the period $T = 6$s. The average value is found using Eq. (11.12).

$$V_{ave} = \frac{1}{T} \times [\text{Area under } v(t) \text{ curve over one period}]$$

$$= \frac{1}{6} \times [4 \times 2 + (-2) \times 2 + 0 \times 2] = \frac{4}{6} = 0.6667 \text{ V}$$

We find the rms using the four steps given earlier. We first square $v(t)$ as shown in Fig. 11.9.

Second, we find the area under $v^2(t)$ over one period, i.e.

$$\text{Area} = 16 \times 2 + 4 \times 2 + 0 \times 2 = 40$$

Third, divide the area by T, that is, $40/6$. Lastly, take the square root. Hence,

$$V_{rms} = \sqrt{\frac{40}{6}} = 2.582 \text{ V}$$

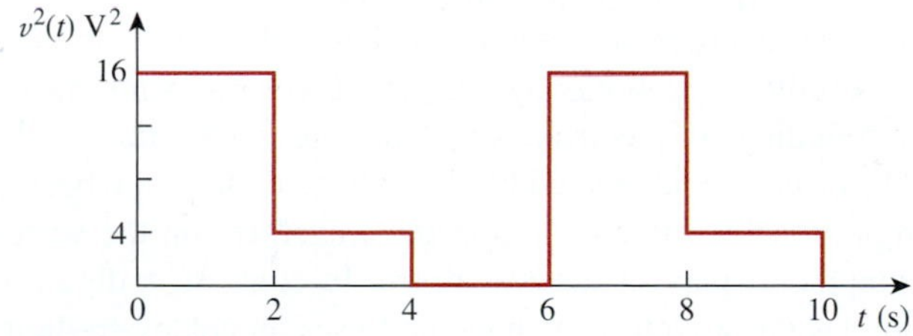

Figure 11.9
Square of $v(t)$ in Fig. 11.8.

Practice Problem 11.6

Refer to Fig. 11.10.
(a) Find the period of $i(t)$. (b) Calculate its average value.
(c) Determine its rms value.

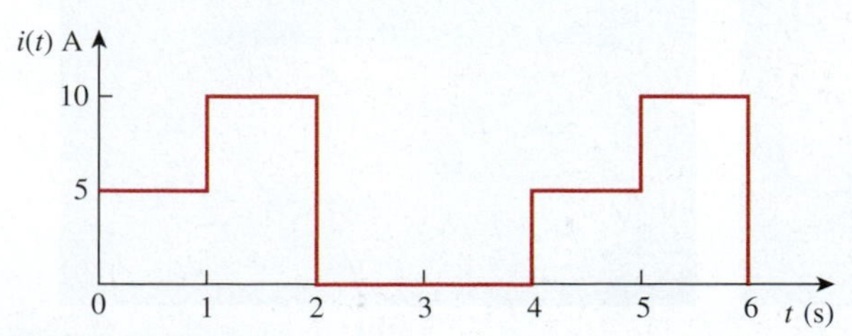

Figure 11.10
For Practice Problem 11.6.

Answer: (a) 4s; (b) 3.75 A; (c) 5.59 A

11.6 Oscilloscopes

An oscilloscope is the most useful instrument available for testing circuits because it allows one to *see* the signals at different points in the circuit. It is an electronic instrument widely used in making electrical measurements. Oscilloscopes are used by everyone from television repair technicians to medical researchers. They are indispensable for anyone repairing or troubleshooting electronic equipment.

An oscilloscope directly measures voltage. To use it to measure current, one would need a current-to-voltage converter (aka a resistor).

The oscilloscope is basically a graph-displaying device—it displays the graph of an electrical signal. The graph can tell us many things about a signal. For example:

- We can determine the time and voltage values of a signal.
- We can calculate the frequency of a signal.
- We can measure the dc level or average value of a signal.
- We can tell if a malfunctioning component is causing distortion of the signal.
- We can find out how much of a signal is direct current (dc) or alternating current (ac).
- We can tell how much of the signal is noise and whether the noise is changing with time.

An oscilloscope (or scope) looks like a small television set, except that it has a grid drawn on its screen and more controls than a television. Most scopes have dual-trace capabilities, meaning that they can simultaneously display two waveforms.

Like most measuring devices, oscilloscopes come in analog and digital types. Both types are shown in Fig. 11.11. They differ in the way the input signal is processed before it is being displayed on the screen. An analog oscilloscope works by applying a voltage being measured to plates surrounding an electron beam moving across the oscilloscope screen. The electric field created by the voltage deflects the beam up and down proportionally, thereby tracing the waveform on the screen. This gives a graphic display of the waveform. In contrast, a digital oscilloscope samples the waveform voltage and uses an analog-to-digital converter (or ADC) to convert the voltage being measured into digital

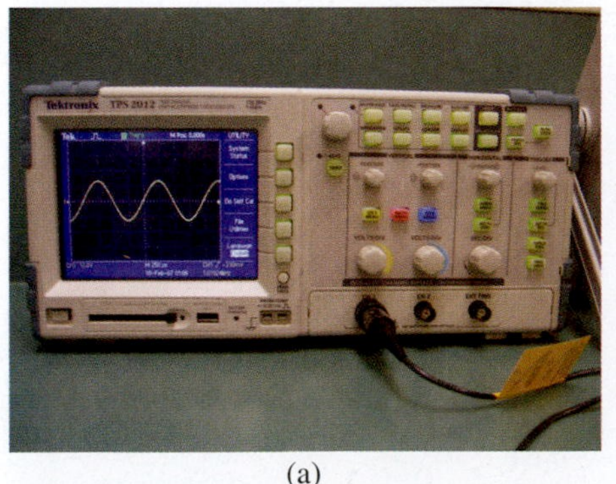

(a)

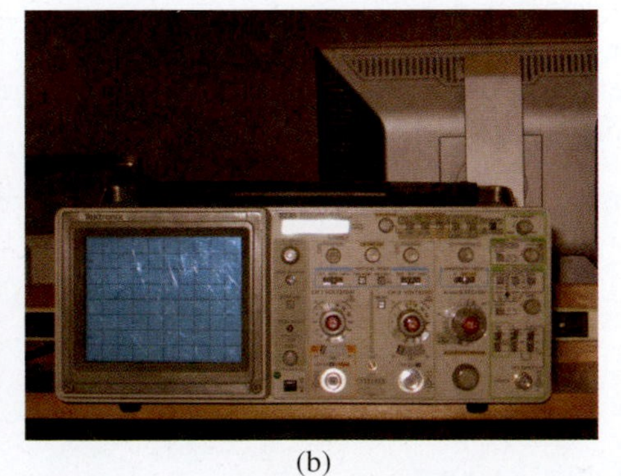

(b)

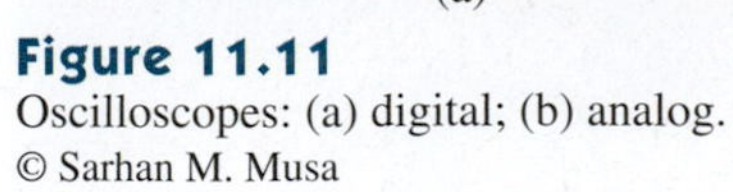

Figure 11.11
Oscilloscopes: (a) digital; (b) analog.
© Sarhan M. Musa

information. It then uses this digital information to reconstruct the waveform on the screen. For many applications, either an analog or digital oscilloscope will do. However, the digital scope is becoming the instrument of choice because of its expanded capabilities, such as waveform storage and waveform measurements and calculations.

An oscilloscope can be used to measure voltage and frequency of a signal. It is calibrated for volts per division on the vertical scale and seconds per division on the horizontal scale. Voltage is read directly from the vertical scale. To measure the frequency, we first obtain the period from the horizontal scale. The period is calculated as follows.

$$T = (\text{Division/Cycle}) \times (\text{Time/Division}) \tag{11.22}$$

The frequency of the signal is computed as

$$f = \frac{1}{T} \tag{11.23}$$

Become familiar with the operation of the oscilloscope. Practice working with it and read more about it. Soon its operation will be second nature to you.

11.7 True RMS Meters

As mentioned earlier, power is the product of current and voltage. In direct current or dc circuits, power is simple to measure, because both voltage and current are constant. But in an ac circuit, voltage and current are continually varying, from zero up to their maximum value, then back to zero again and then to their maximum value. Therefore, the effective voltage and current will be less than their maximum values. The effective values (root-mean-square, or rms, values) are 0.707 times the maximum values.

True RMS meters use an integrated circuit, which computes the true rms value of complex signals such as square waves or ac signals which have been half-wave rectified or chopped. They look exactly the same as a normal DMM, but the price is higher. Many DMMs are designed to read the rms value of only sinusoidal waveforms. The true RMS meter can read the rms value of any waveform, and it is not limited to sinusoidal waveform. A true RMS meter will have a much wider range, and the manual will probably tell you what that range is so you know when to stop trusting it. A typical true rms multimeter is shown in Fig. 11.12.

Figure 11.12
True RMS multimeter.
© Sarhan M. Musa

11.8 Summary

1. AC voltage is generated based on Faraday's law of electromagnetic induction.
2. A sinusoid is a signal in the form of the sine or cosine function. It has the general form

$$v(t) = V_m \sin(\omega t + \phi)$$

where V_m is the magnitude, $\omega = 2\pi f$ is the angular frequency, $(\omega t + \phi)$ is the argument, and ϕ is the phase.

3. The average value of a periodic signal is equal to the total algebraic area enclosed by the signal and the horizontal axis over one period divided by the period.
4. The effective (or rms) value of a periodic signal is its equivalent dc value that would produce the same power in a given resistance as the original signal would produce in that resistance.
5. The oscilloscope is a time-domain voltmeter used to study, measure, and display the parameters of a time-varying waveform.
6. The true rms meter measures the rms value of any waveform.

Review Questions

11.1 Which of these statements is true for only dc?

(a) Current flows in one direction only.
(b) Current periodically alternates.
(c) Voltage can be stepped up or down.
(d) DC is an efficient means to transfer power over long distances.
(e) DC is useful for all purposes.

11.2 A function that repeats itself after fixed intervals is said to be:

(a) a phasor (b) harmonic
(c) periodic (d) reactive

11.3 Which of these frequencies has the shorter period?

(a) 1 krad/s (b) 1 kHz

11.4 A sine wave with a period of 5ms has its frequency as:

(a) 5 Hz (b) 100 Hz
(c) 200 Hz (d) 2 kHz

11.5 A sine wave has a frequency of 60 Hz. In 2s, it goes through

(a) $\frac{1}{30}$ cycle (b) 60 cycles
(c) 120 cycles (d) 600 cycles

11.6 If $v_1 = 30 \sin(\omega t + 10°)$ V and $v_2 = 20 \sin(\omega t + 50°)$ V, which of these statements are true?

(a) v_1 leads v_2 (b) v_2 leads v_1
(c) v_2 lags v_1 (d) v_1 lags v_2
(e) v_1 and v_2 are in phase

11.7 The average value of $10 \sin(\omega t - 30°)$ is:

(a) 0 (b) 5 (c) 7.07
(d) 10 (e) 14.14

11.8 The effective value of $10 \sin(\omega t - 30°)$ is:

(a) 0 (b) 5 (c) 7.07
(d) 10 (e) 14.14

11.9 An oscilloscope directly measures current.

(a) True (b) False

11.10 The oscilloscope cannot measure this quantity.

(a) voltage (b) frequency
(c) phase shift (d) power

Answers: 11.1a, 11.2c, 11.3b, 11.4c, 11.5c, 11.6b,d, 11.7a, 11.8c, 11.9b, 11.10d

Problems

Section 11.2 AC Voltage Generator

11.1 A sinusoid has a peak value of 1.2 V. Calculate the peak-to-peak value.

11.2 An alternating voltage is described by $v = 120 \sin(2000t)$ V. Find: (a) the peak value, (b) the peak-to-peak value, and (c) the instantaneous voltage at $t = 1$ ms.

11.3 Given that $i(t) = 24 \cos(377t)$ mA, find $i(t)$ at $t = 0$, 10, and 40 ms.

11.4 A certain alternating current is described by $i(t) = I_m \sin(754t)$ A. If the current is 10 A when $t = 2$ ms, what is the peak current?

11.5 A sinusoidal voltage is given by $v = 10 \sin(337t)$ V. Calculate its instantaneous values at $t = 2$ ms, 14.5 ms, and 25.2 ms.

Section 11.3 Sinusoids

11.6 Convert the following degrees to radians.

(a) 22.5 (b) 65 (c) 122 (d) 270

11.7 Convert the following radians to degrees.

(a) $\pi/5$ (b) $6\pi/7$ (c) 2.368 (d) 4.5

11.8 Determine the period of each of the following values of frequency.

(a) 50 Hz (b) 600 Hz (c) 2 kHz (d) 1 MHz

11.9 Determine the frequency of each of the following values of period.

(a) 0.4 s (b) 2 ms (c) 30 μs

11.10 Calculate the amplitude and frequency of the following waveforms.

(a) $5\sin(2\pi t)$ (b) $10\sin(377t)$
(c) $30\sin(10^6 t)$ (d) $0.04\sin(42.56t)$

11.11 For the waveform shown in Fig. 11.13:

(a) Determine the period.

(b) Find the frequency.

(c) How many cycles are shown?

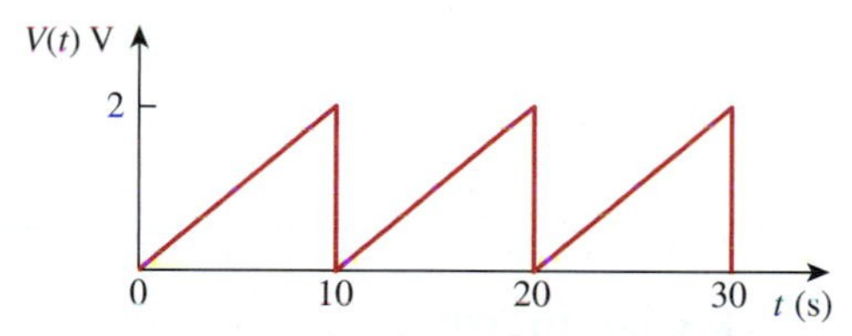

Figure 11.13
For Problem 11.11.

11.12 A periodic waveform has a frequency of 60 Hz. How long will it take to complete 10 cycles?

11.13 A periodic waveform completes 100 cycles in 2 s. Determine the frequency of the waveform.

11.14 What is the frequency of an ac generator that produces 1 cycle in 2 ms?

11.15 A square wave is at a voltage level of +20 V for a time interval of 40 ms and at a voltage level of −20 V for 40 ms and then repeats. Determine the frequency of the square wave.

11.16 The frequency of an ac signal doubles. Determine what happens to the period of the signal.

11.17 A waveform has a period of 4 ms. Determine the angular frequency of the waveform.

11.18 A triangular current waveform is shown in Fig. 11.14. Find: (a) the period, (b) the frequency, (c) the radian frequency, (d) the peak current, and (e) the peak-to-peak current.

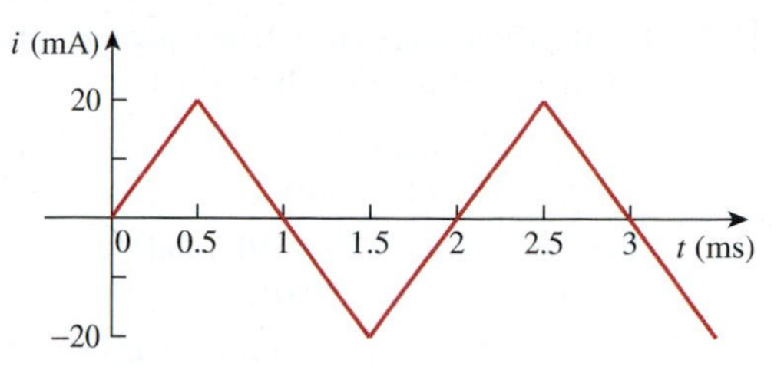

Figure 11.14
For Problem 11.18.

11.19 In a circuit, the voltage source is $v_s(t) = 12\sin(10^3 t + 24°)$ V.

(a) What is the angular frequency of the voltage?

(b) What is the frequency of the source?

(c) Find the period of the voltage.

(d) Determine v_s at $t = 2.5$ ms.

11.20 A current source in a circuit has $i_s(t) = 8\cos(500\pi t - 25°)$ A.

(a) What is the amplitude of the current?

(b) What is the angular frequency?

(c) Find the frequency of the current.

(d) Calculate i_s at $t = 2$ms.

11.21 Express the following functions in cosine form:

(a) $4\sin(\omega t - 30°)$

(b) $-2\sin(6t)$

(c) $-10\sin(\omega t + 20°)$

11.22 Express the following functions in sine form:

(a) $2\cos(\omega t)$

(b) $10\cos(\omega t + 20°)$

(c) $-70\cos(\omega t + 30°)$

11.23 (a) Express $v(t) = 8\cos(7t + 15°)$ in sine form.

(b) Convert $i(t) = -10\sin(3t - 85°)$ to cosine form.

11.24 A voltage sine wave $v = A\sin\theta$ has an instantaneous value of 8V at $\theta = 30°$. Find the value of the wave when $\theta = 120°$.

11.25 A voltage waveform is given by $v = 200\sin(521t + 25°)$. Find: (a) the amplitude of the waveform, (b) the frequency, (c) the period, and (d) the phase angle.

Section 11.4 Phase Relations

11.26 Given $v_1(t) = 20\sin(\omega t + 60°)$ and $v_2(t) = 60\cos(\omega t - 10°)$, determine the phase angle between the two sinusoids and which one lags the other.

11.27 For the following pairs of sinusoids, determine which one leads and by how much.

(a) $v(t) = 10\cos(4t - 60°)$ and $i(t) = 4\sin(4t + 50°)$

(b) $v_1(t) = 4\cos(377t + 10°)$ and $v_2(t) = -20\cos(377t)$

(c) $x(t) = 13\cos(2t) + 5\sin(2t)$ and $y(t) = 15\cos(2t - 11.8°)$

11.28 For the following pairs of sinusoids, determine which one leads or lags and by how much.

(a) $v = 20\sin(\omega t - 30°)$ and $i = 2\sin(\omega t - 90°)$

(b) $v = 20\sin(\omega t + 15°)$ and $i = 2\sin(\omega t + 60°)$

(c) $v = 20\sin(\omega t + 45°)$ and $i = 2\sin(\omega t - 45°)$

Section 11.5 Average and RMS Values

11.29 What is the rms value of a sine wave that has a peak-to-peak value of 10 V?

11.30 A current waveform is shown over its period in Fig. 11.15. Determine the average value.

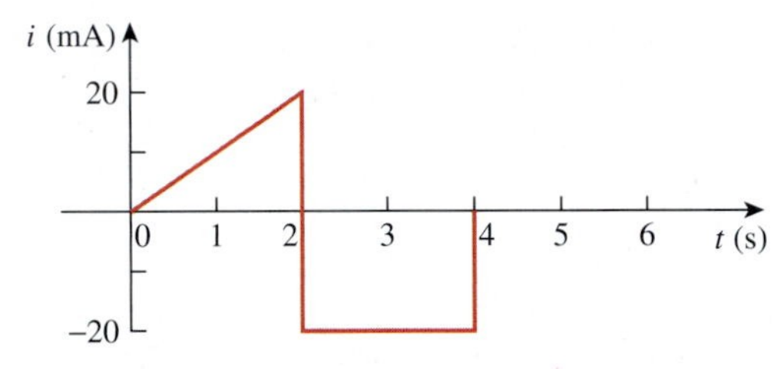

Figure 11.15
For Problem 11.30.

11.31 Determine the rms values of the following sinusoids.

(a) $v = 10\sin(377t)$ V

(b) $i = 2\sin(200t + 30°)$ mA

11.32 If $V_{rms} = 6$ V for a sine wave, what is its amplitude?

11.33 Find the average and rms values of the periodic signal in Fig. 11.16.

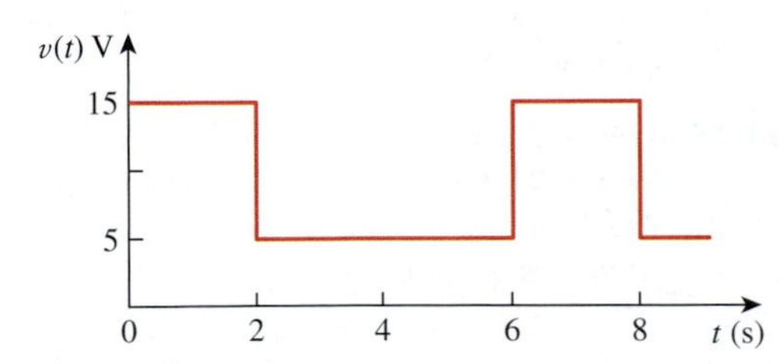

Figure 11.16
For Problem 11.33.

11.34 Determine the rms value of the waveform in Fig. 11.17.

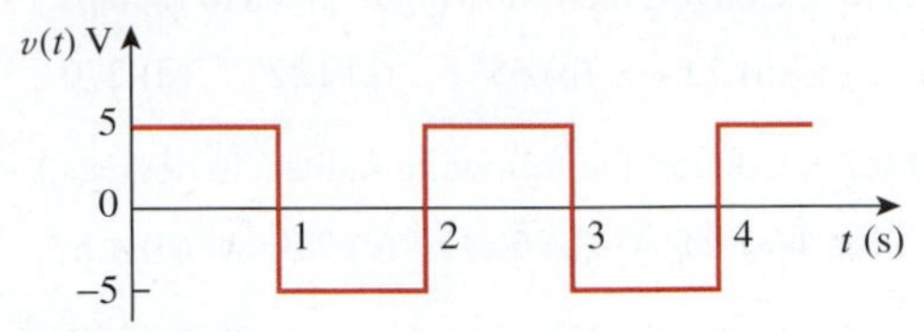

Figure 11.17
For Problem 11.34.

11.35 Find the effective value of the voltage waveform in Fig. 11.18.

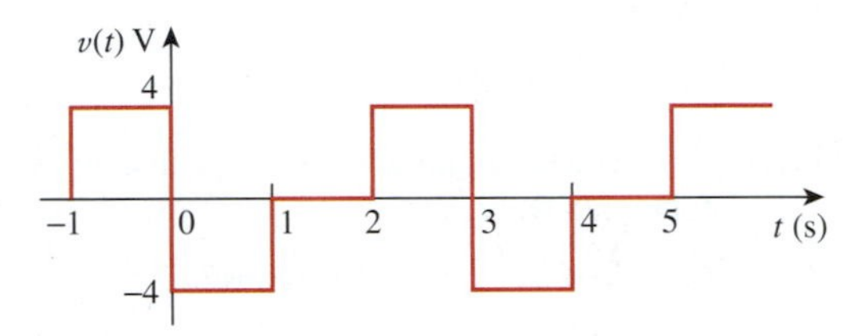

Figure 11.18
For Problem 11.35.

11.36 Calculate the average and rms values of the current waveform of Fig. 11.19.

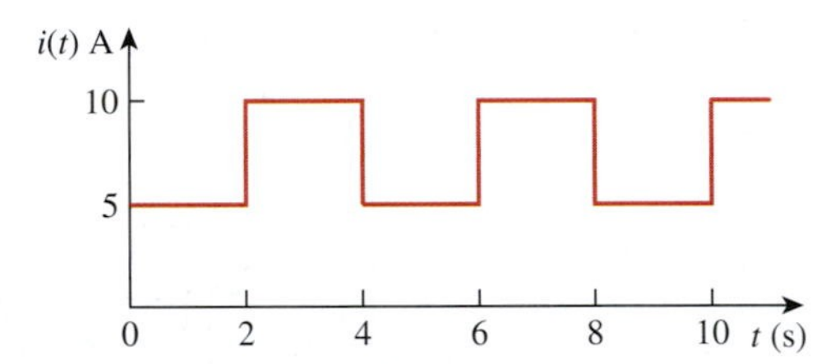

Figure 11.19
For Problem 11.36.

11.37 A fruit juicer is connected to a 120-V rms source and is rated at 1.6 kW. Determine: (a) the resistance of the mixer and (b) the rms value of the current.

11.38 A PC draws 3.2 A from a 120-V source. Find: (a) the average current, (b) the rms current, (c) the average value of the current squared, and (d) the current amplitude.

Section 11.6 Oscilloscopes

11.39 Explain how an oscilloscope is used to measure voltage and frequency.

11.40 The horizontal sweep line of an oscilloscope represents 0.2 ms per division. Calculate the frequency of a signal if one displayed cycle uses four horizontal divisions.

chapter

12

Phasors and Impedance

Leadership is the art of accomplishing more than the science of management says is possible.

—Colin Powell

Enhancing Your Career

Codes of Ethics

Engineering is a profession that makes significant contributions to the economic and social well-being of people all over the world. As members of this important profession, engineers are expected to exhibit the highest standards of honesty and integrity. Unfortunately, the engineering curriculum is so crowded that there is no room for a course on ethics in most schools. The Institute of Electrical and Electronics Engineers (or IEEE) Code of Ethics is presented here to acquaint students with ethical behavior in engineering professions.

Engineering technologists at work.
© Getty RF

> We, the members of the IEEE, in recognition of the importance of our technologies in affecting the quality of life throughout the world, and in accepting a personal obligation to our profession, its members and the communities we serve, do hereby commit ourselves to the highest ethical and professional conduct and agree:

1. to accept responsibility in making engineering decisions consistent with the safety, health, and welfare of the public, and to disclose promptly factors that might endanger the public or the environment;
2. to avoid real or perceived conflicts of interest whenever possible, and to disclose them to affected parties when they do exist;
3. to be honest and realistic in stating claims or estimates based on available data;
4. to reject bribery in all its forms;
5. to improve the understanding of technology, its appropriate application, and potential consequences;
6. to maintain and improve our technical competence and to undertake technological tasks for others only if qualified by training or experience, or after full disclosure of pertinent limitations;
7. to seek, accept, and offer honest criticism of technical work, to acknowledge and correct errors, and to credit properly the contributions of others;
8. to treat fairly all persons regardless of such factors as race, religion, gender, disability, age, or national origin;
9. to avoid injuring others, their property, reputation, or employment by false or malicious action;
10. to assist colleagues and co-workers in their professional development and to support them in following this code of ethics.

—Courtesy of IEEE. © 2011 IEEE.
Reprinted with permission of the IEEE.

12.1 Introduction

In the previous chapter, we learned about sinusoids. Sinusoids are easily expressed in terms of *phasors,* which are more convenient to work with than sine and cosine functions.

> A **phasor** is a complex number that represents the amplitude and phase of a sinusoid.

Sinusoids are defined by three properties (magnitude, phase, and frequency). Often a circuit involves voltages and currents of the same frequency, and thus we would benefit from defining voltages and currents using a single number containing two measurements (magnitude and phase on the sinusoid) such as a phasor.

Phasors provide a simple means of analyzing linear circuits excited by sinusoidal sources; solutions of such circuits would be time-consuming otherwise. The notion of solving ac circuits using phasors was first introduced by Charles Steinmetz[1] in 1893. We will notice that the techniques presented in this chapter (Ohm's law, KCL, KVL, current divider, voltage divider, and the like) are an extension of the techniques used for dc circuits. The only difference is that ac quantities are complex, while dc quantities are real. While this complicates matters for ac circuits, it does not alter the basic underlying principles.

12.2 Phasors and Complex Numbers

Before we completely define phasors and apply them to circuit analysis, we need to be thoroughly familiar with complex numbers.

A complex number[2] z can be written in rectangular form as

$$z = x + jy \tag{12.1}$$

where $j = \sqrt{-1}$; x is the real part of z; y is the imaginary part[3] of z.

The complex number z can also be written in polar form as

$$z = r\underline{/\phi} \tag{12.2}$$

where r is the magnitude of z and ϕ is the phase of z. We notice that z can be represented in two ways:

$$\begin{aligned} z &= x + jy \qquad \text{(Rectangular form)} \\ z &= r\underline{/\phi} \qquad \text{(Polar form)} \end{aligned} \tag{12.3}$$

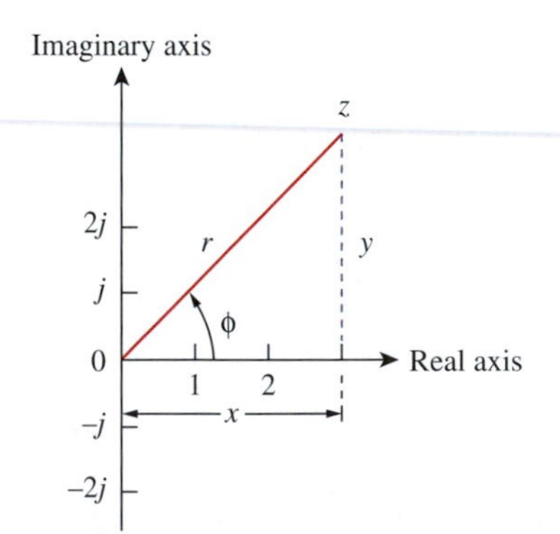

Figure 12.1
Representation of a complex number $z = x + jy = r\underline{/\phi}$.

The relationship between rectangular form and the polar form is shown in Fig. 12.1, where the x-axis represents the real part and the y-axis represents the imaginary part of a complex number. Given x and y, we can get r and ϕ as

$$r = \sqrt{x^2 + y^2}, \qquad \phi = \tan^{-1}\frac{y}{x} \tag{12.4}$$

---

[1] Historical Note: Charles Proteus Steinmetz (1865–1923) was a German Austrian mathematician and electrical engineer.

[2] Note: A short tutorial on complex numbers is presented in Appendix B.

[3] In mathematics, the symbol i is used to indicate an imaginary number. However, in engineering, the symbol j is substituted because i is used for time-varying current.

[We must exercise care in determining the correct value of ϕ in Eq. (12.4). See Appendix B for details.] On the other hand, if we know r and ϕ, x and y can be obtained as

$$x = r\cos\phi, \qquad y = r\sin\phi$$

Thus, z may be written as

$$z = x + jy = r\underline{/\phi} = r(\cos\phi + j\sin\phi) \tag{12.5}$$

Addition and subtraction of complex numbers are better performed in rectangular form; multiplication and division are better done in polar form. Given the complex numbers:

$$\begin{aligned} z &= x + jy = r\underline{/\phi} \\ z_1 &= x_1 + jy_1 = r_1\underline{/\phi_1} \\ z_2 &= x_2 + jy_2 = r_2\underline{/\phi_2} \end{aligned}$$

the following operations are important.

Addition:

$$z_1 + z_2 = (x_1 + x_2) + j(y_1 + y_2) \tag{12.6a}$$

Subtraction:

$$z_1 - z_2 = (x_1 - x_2) + j(y_1 - y_2) \tag{12.6b}$$

Multiplication:

$$z_1 z_2 = r_1 r_2\underline{/(\phi_1 + \phi_2)} \tag{12.6c}$$

Division:

$$\frac{z_1}{z_2} = \frac{r_1}{r_2}\underline{/(\phi_1 - \phi_2)} \tag{12.6d}$$

Reciprocal:

$$\frac{1}{z} = \frac{1}{r}\underline{/-\phi} \tag{12.6e}$$

Square root:

$$\sqrt{z} = \sqrt{r}\underline{/\phi/2} \tag{12.6f}$$

Complex conjugate:

$$z^* = x - jy = r\underline{/-\phi} \tag{12.6g}$$

Note that from Eq. (12.6e),

$$\frac{1}{j} = -j \tag{12.7a}$$

and

$$j^2 = -1 \tag{12.7b}$$

[We may also derive Eq. (12.7b) from the fact that $j = \sqrt{-1}$.] These are the basic properties of complex numbers we need. Other properties of complex numbers can be found in Appendix B.

The idea of phasor representation is based on Euler's identity. Euler's identity is a representation of z in terms of its polar coordinates; that is,

$$e^{\pm j\phi} = \cos\phi \pm j\sin\phi \tag{12.8}$$

which shows that we may regard $\cos\phi$ and $\sin\phi$ as the real and imaginary parts of $e^{j\phi}$; that is,

$$\cos\phi = \text{Re}(e^{j\phi}) \tag{12.9a}$$

$$\sin\phi = \text{Im}(e^{j\phi}) \tag{12.9b}$$

where Re and Im stand for the *real part of* and the *imaginary part of*, respectively. Given a sinusoid

$$v(t) = V_m \sin(\omega t + \phi),$$

it can be rewritten as

$$v(t) = V_m \sin(\omega t + \phi) = \text{Im}(V_m e^{j(\omega t+\phi)}) = \text{Im}(V_m e^{j\phi} e^{j\omega t})$$

Thus,

$$v(t) = \text{Im}(\mathbf{V}e^{j\omega t}) \tag{12.10}$$

where

$$\mathbf{V} = V_m e^{j\phi} = V_m\underline{/\phi} \tag{12.11}$$

and $\mathbf{V}$ is defined the *phasor representation* of the sinusoid $v(t)$. In other words, a phasor is a complex representation of the magnitude and phase of a sinusoid. A phasor may also be regarded as a mathematical equivalent of a sinusoid with the time dependence dropped.

One way of looking at Eq. (12.10) is to consider the plot of the sinor $\mathbf{V}e^{j\omega t} = V_m e^{j(\omega t+\theta)}$ on the complex plane. As time increases, the sinor rotates on a circle of radius V_m at an angular velocity ω in the counterclockwise direction, as shown in Fig. 12.2(a). We may regard $v(t)$ as the projection of the sinor $\mathbf{V}e^{j\omega t}$ on the imaginary axis, as shown in Fig. 12.2(b). The value of the sinor at time $t = 0$ is the phase $\mathbf{V}$ of the sinusoid $v(t)$. The sinor may be regarded as a rotating phasor. Thus, whenever a sinusoid is expressed as a phasor, the term $e^{j\omega t}$ is implicitly present.

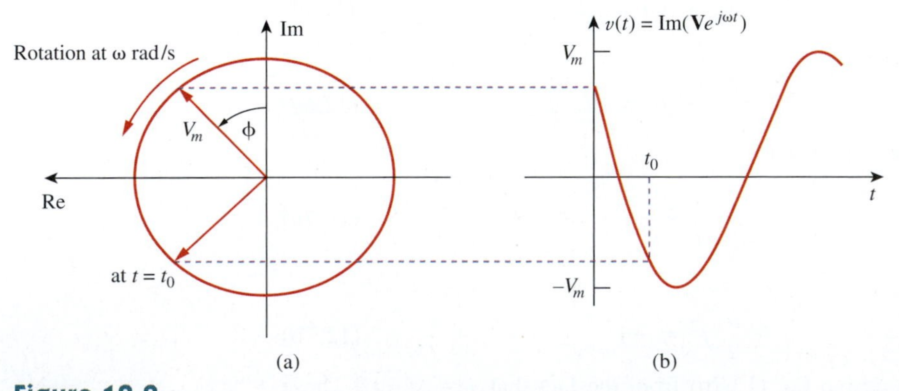

Figure 12.2
Representation of $Ve^{j\omega t}$: (a) sinor rotating counterclockwise; (b) its projection on the real axis, as a function of time.

Therefore, it is important when dealing with phasors, to keep in mind the frequency ω of the phasor; otherwise, we can make serious mistakes.

Equation (12.10) states that to obtain the sinusoid corresponding to a given phasor $\mathbf{V}$, multiply the phasor by the time factor $e^{j\omega t}$ and take the imaginary part. As a complex quantity, a phasor may be expressed in rectangular form, polar form, or exponential form. Because a phasor has magnitude and phase ("direction"), it behaves as a vector and is printed in boldface. For example, phasors $\mathbf{V} = V_m\underline{/\phi}$ and $\mathbf{I} = I_m\underline{/-\phi}$ are graphically represented in Fig. 12.3. Such a graphical representation of phasors is known as a *phasor diagram*.

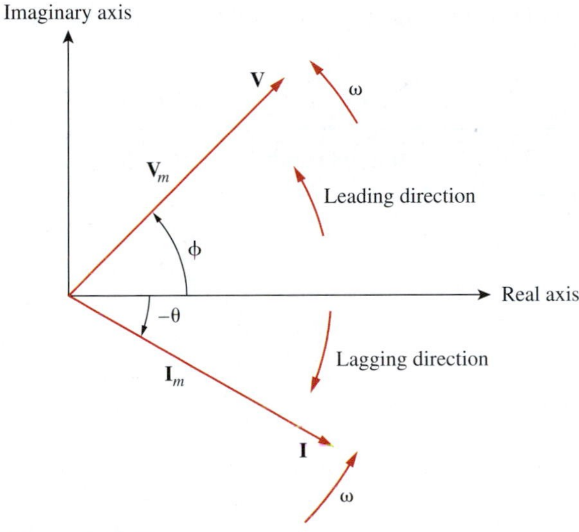

Figure 12.3
A phasor diagram showing $\mathbf{V} = \mathbf{V}_m\underline{/\phi}$ and $\mathbf{I} = \mathbf{I}_m\underline{/-\theta}$.

Equations (12.10) and (12.11) reveal that to get the phasor corresponding to a sinusoid, we first express the sinusoid in the sine form so that the sinusoid can be written as the imaginary part of a complex number. Then we take out the time factor $e^{j\omega t}$, and whatever is left is the phasor corresponding to the sinusoid. This transformation is summarized as follows:

$$\underset{\text{(Time-domain representation)}}{v(t) = V_m \sin(\omega t + \phi)} \quad \Leftrightarrow \quad \underset{\text{(Phasor-domain representation)}}{\mathbf{V} = V_m\underline{/\phi}} \tag{12.12}$$

Note that in Eq. (12.12), the frequency (or time) factor $e^{j\omega t}$ is suppressed and the frequency is not explicitly shown in the phasor-domain representation because ω is constant. However, the response depends on ω.

The differences between $v(t)$ and $\mathbf{V}$ should be emphasized:

1. $v(t)$ is the *instantaneous or time-domain* representation, while $\mathbf{V}$ is the *frequency or phasor-domain* representation.
2. $v(t)$ is time-dependent, while $\mathbf{V}$ is not. (This fact is often forgotten by students.)
3. $v(t)$ is always real with no complex term, while $\mathbf{V}$ is generally complex.

We should bear in mind that phasor analysis applies only when frequency is constant; it applies in manipulating two or more sinusoidal signals only if they are of the same frequency.

Finally, the complex algebra covered in this section can be handled easily with a TI-89 Titanium calculator. A complex number $x + jy$ is entered into the calculator as (x, y) in rectangular form or as $(r\angle\phi)$. In polar form, we press MODE to select the mode setting. This determines whether the angle ϕ is in *radian* or *degree* and whether the final result of the calculation is in rectangular or polar. Once the mode is selected, we must press ENTER to accept it. For example, if we select RECTANGULAR for the complex format, complex numbers are displayed in rectangular form, regardless of the form in which we enter them.

Example 12.1

If $z_1 = 3 + j5$ and $z_2 = 6 - j8$, find:
(a) $z_1 + z_2$, (b) $z_1 z_2$.

Solution:

(a) $z_1 + z_2 = (3 + j5) + (6 - j8) = 9 - j3$
(b) $z_1 z_2 = (3 + j5)(6 - j8) = 18 - j24 + j30 - j^2\,40 = 18 + j6 + 40 = 58 + j6$ where $j^2 = -1$

Practice Problem 12.1

If $z_1 = 2 - j4$ and $z_2 = 4 + j7$, find:
(a) $z_1 - z_2$, (b) z_1/z_2.

Answer:
(a) $-2 - j11$; (b) $-0.3077 - j0.4615$

Example 12.2

Evaluate these complex numbers:
(a) $(40\angle 50^\circ + 20\angle -30^\circ)^{1/2}$
(b) $\dfrac{10\angle -30^\circ + (3 - j4)}{(2 + j4)(3 - j5)}$

Solution:

(a) We can evaluate the complex numbers in two ways.

METHOD 1 With hand calculation, and using polar to rectangular transformation,

$$40\angle 50^\circ = 40(\cos 50^\circ + j\sin 50^\circ) = 25.71 + j30.64$$

$$20\angle -30^\circ = 20[\cos(-30^\circ) + j\sin(-30^\circ)] = 17.32 - j10$$

Adding them up gives

$$40\angle 50^\circ + 20\angle -30^\circ = 43.03 + j20.64 = 47.72\angle 25.63^\circ$$

Taking the square root of this,

$$(40\angle 50^\circ + 20\angle -30^\circ)^{1/2} = \sqrt{47.72}\angle 25.63^\circ/2 = 6.91\angle 12.82^\circ$$

■ **METHOD 2** Alternatively, we can use the TI-89 Titanium calculator. Because we want the final result to be in polar form, we press MODE. As the complex format, select POLAR and then press ENTER twice.

Then we type in the numbers in the entry line as:

$$\sqrt{((40\angle 50^\circ) + (20\angle -30^\circ))}$$

After pressing ♦ ENTER, the result is displayed as:

$$6.908\angle 12.81$$

(b) Using polar-rectangular transformations,

$$\frac{10\angle -30^\circ + (3 - j4)}{(2 + j4)(3 - j5)} = \frac{8.66 - j5 + (3 - j4)}{(2 + j4)(3 + j5)} = \frac{11.66 - j9}{-14 + j22}$$

$$= \frac{14.73\angle -37.66^\circ}{26.08\angle 122.47^\circ} = 0.565\angle -160.13^\circ$$

Alternatively, we can use the TI-89 Titanium calculator. We type in the numbers as:

$$((10\angle -30^\circ) + (3 - 4*i))/((2 + 4*i)*(3 + 5*i))$$

We press ♦ ENTER

The result is displayed as:

$$.5648\angle -160.134$$

Practice Problem 12.2

Evaluate the following complex numbers:

(a) $[(5 + j2)(-1 + j4) - 5\angle 60^\circ]^*$

(b) $\dfrac{10 + j5 + 3\angle 40^\circ}{-3 + j4} + 10\angle 30^\circ$

Answer: (a) $-15.5 - j13.67$; (b) $8.292 + j2.2$

Example 12.3

Transform these sinusoids to phasors:

(a) $v(t) = 4\sin(30t + 20^\circ)$

(b) $i(t) = 6\cos(50t - 40^\circ)$

Solution:

(a) The phasor form of $v(t)$ is

$$\mathbf{V} = 4\angle 20^\circ$$

(b) We first express cosine in form of sine using Eq. (11.11)

$$i(t) = 6\cos(50t - 40^\circ) = 6\sin(50t - 40^\circ + 90^\circ)$$
$$= 6\sin(50t + 50^\circ)$$

Hence, the phasor form of $i(t)$ is

$$\mathbf{I} = 6\angle 50^\circ$$

Notice that a phasor is a complex number, which is expressed here in polar form.

Practice Problem 12.3

Express these sinusoids as phasors:
(a) $i(t) = 4 \sin(10t + 10°)$ A
(b) $v(t) = -7 \cos(2t + 40°)$ V

Answer: (a) $4\underline{/-10°}$ A (b) $7\underline{/-50°}$ V

Example 12.4

Find the sinusoids represented by these phasors:
(a) $\mathbf{I} = -3 + j4$ A
(b) $\mathbf{V} = j8e^{-j20°}$ V

Solution:

(a) $\mathbf{I} = -3 + j4 = 5\underline{/126.87°}$. Transforming this to the time domain gives

$$i(t) = 5 \sin(\omega t + 126.87°) \text{ A}$$

(b) Because $j = 1\underline{/90°}$,

$$\mathbf{V} = j8\underline{/-20°} = (1\underline{/90°})(8\underline{/-20°}) = 8\underline{/(90° - 20°)} = 8\underline{/70°} \text{ V}$$

Converting this to the time domain gives us

$$v(t) = 8 \sin(\omega t + 70°) \text{ V}$$

Practice Problem 12.4

Find the sinusoids corresponding to these phasors:
(a) $\mathbf{V} = -10 \underline{/30°}$
(b) $\mathbf{I} = j(5 - j12)$

Answer: (a) $v(t) = 10 \sin(\omega t + 210°)$; (b) $i(t) = 13 \sin(\omega t + 22.62°)$

Example 12.5

Given $i_1(t) = 4 \cos(\omega t + 30°)$ and $i_2(t) = 5 \sin(\omega t - 20°)$, find their sum.

Solution:

Here is an important use of phasors—summing sinusoids of the same frequency. Current $i_1(t)$ is not in the standard form, which is the sine form. The rule for converting cosine to sine is to add 90°. But

$$i_1(t) = 4 \cos(\omega t + 30°) = 4 \sin(\omega t + 30° + 90°)$$

Hence, its phasor is

$$\mathbf{I}_1 = 4\underline{/120°}$$

and $i_2(t)$ is already in the standard form. Hence,

$$\mathbf{I}_2 = 5\underline{/-20°}$$

If we let $i = i_1 + i_2$, then

$$\begin{aligned}\mathbf{I} = \mathbf{I}_1 + \mathbf{I}_2 &= 4\underline{/120°} + 5\underline{/-20°}\\ &= -2 + j3.464 + 4.698 - j1.71 = 2.698 + j1.754\\ &= 3.218\underline{/33.03°} \text{ A}\end{aligned}$$

Transforming this to the time domain, we get

$$i(t) = 3.218 \cos(\omega t + 33.03°) \text{ A}$$

Practice Problem 12.5

If $v_1 = -10 \sin(\omega t + 30°)$ and $v_2 = 20 \cos(\omega t - 45°)$, find $v_1 + v_2$.

Answer: $v(t) = 10.66 \sin(\omega t + 59.05°)$ V

12.3 Phasor Relationships for Circuit Elements

Now that we know how to represent a voltage or current in the phasor or frequency domain, one may legitimately ask how we do apply this to circuits involving the passive elements R, L, and C. What we need to do is to transform the voltage–current relationship from the time domain to the phasor domain for each element. Again, we will assume the passive sign convention.

We begin with the resistor. If the current through a resistor R is i, the voltage across it is given by Ohm's law as

$$v = iR = RI_m \sin(\omega t + \phi) \tag{12.13}$$

The phasor form of this voltage is

$$\mathbf{V} = RI_m \underline{/\phi} \tag{12.14}$$

But the phasor representation of the current is $\mathbf{I} = I_m \underline{/\phi}$. Hence,

$$\mathbf{V} = R\mathbf{I} \tag{12.15}$$

showing that the voltage–current relation for the resistor in the phasor domain continues to obey Ohm's law, as it does in the time domain. This is illustrated in Fig. 12.4. We should note from Eq. (12.15) that voltage and current are in phase, as illustrated in the phasor diagram in Fig. 12.5.

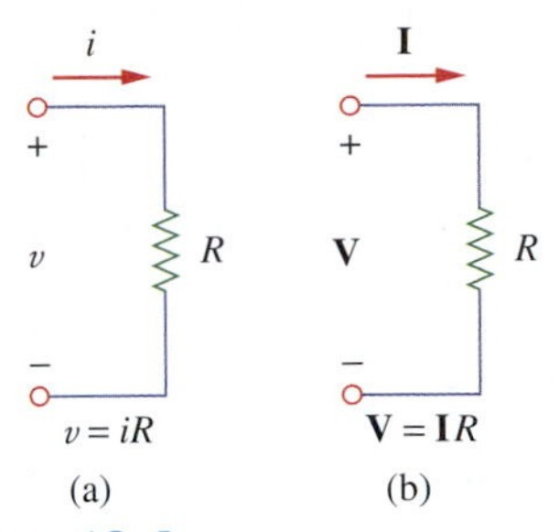

Figure 12.4
Voltage–current relations for a resistor: (a) time domain; (b) frequency domain.

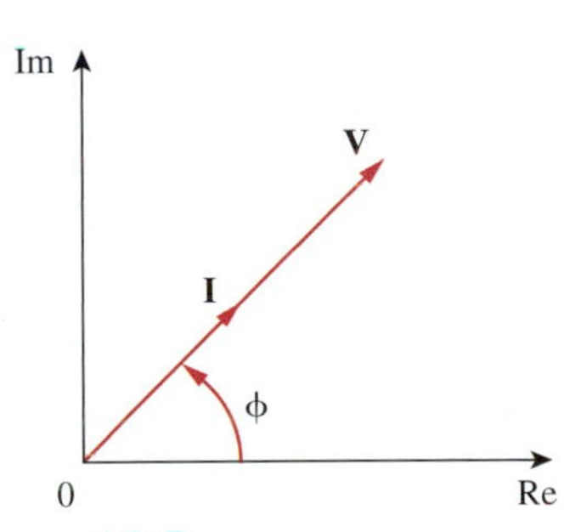

Figure 12.5
Phasor diagram for the resistor.

For the inductor L, assume the current through it is $i(t) = I_m \sin(\omega t + \phi)$. The voltage across the inductor is

$$v = L\frac{di}{dt} = \omega L I_m \cos(\omega t + \phi) \tag{12.16}$$

Recall from trigonometry that $\cos A = \sin(A + 90°)$. We can write the voltage as

$$v = \omega L I_m \sin(\omega t + \phi + 90°) \tag{12.17}$$

which transforms to the phasor

$$\mathbf{V} = \omega L I_m e^{j(\phi + 90°)} = \omega L I_m e^{j\phi} e^{j90°} = \omega L I_m \underline{/\phi} e^{j90°} \tag{12.18}$$

But $I_m \underline{/\phi} = \mathbf{I}$ and from Eq. (12.8), $e^{j90°} = j$. Hence,

$$\boxed{\mathbf{V} = j\omega L\mathbf{I}} \tag{12.19}$$

showing that the voltage has a magnitude of $\omega L I_m$ and a phase of $\phi + 90°$. Thus the voltage and current are 90° out of phase. Specifically, the current lags the voltage by 90°. The voltage−current relations for the inductor are shown in Fig. 12.6. The phasor diagram is in Fig. 12.7.

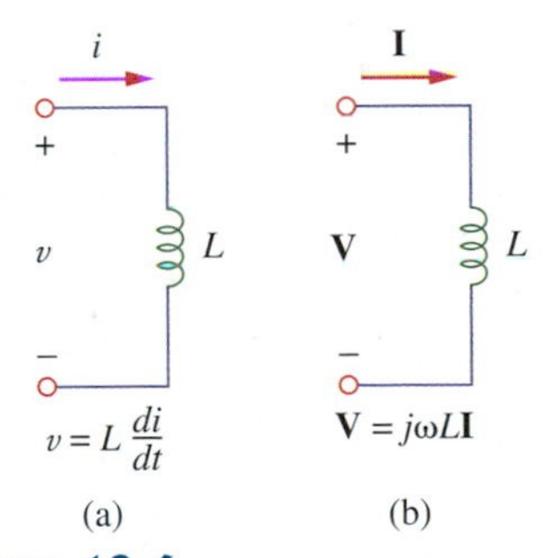

Figure 12.6
Voltage–current relations for an inductor: (a) time domain; (b) frequency domain.

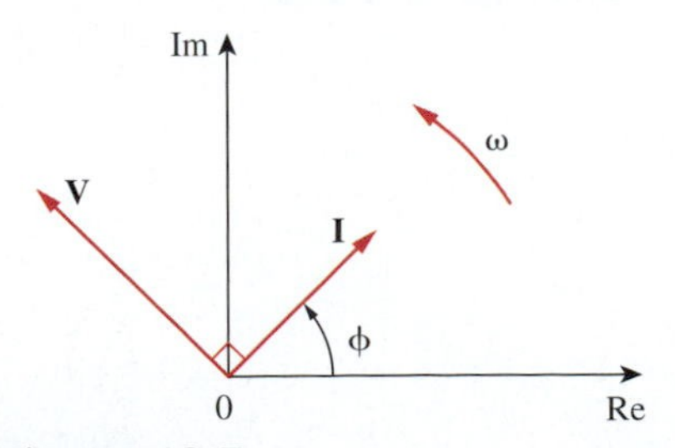

Figure 12.7
Phasor diagram for the inductor; **I** lags **V** by 90°.

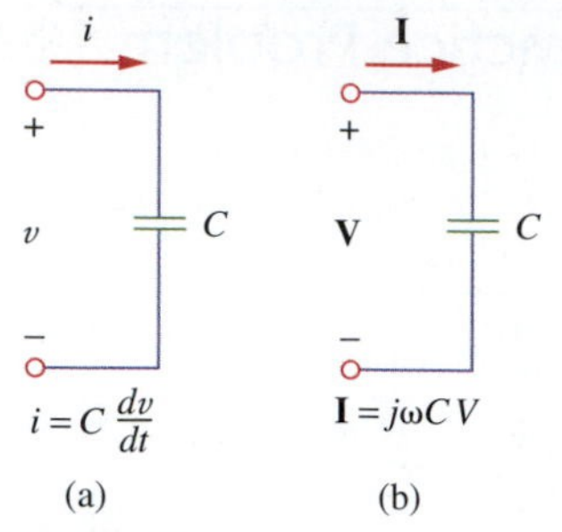

Figure 12.8
Voltage–current relations for an inductor: (a) time domain; (b) frequency domain.

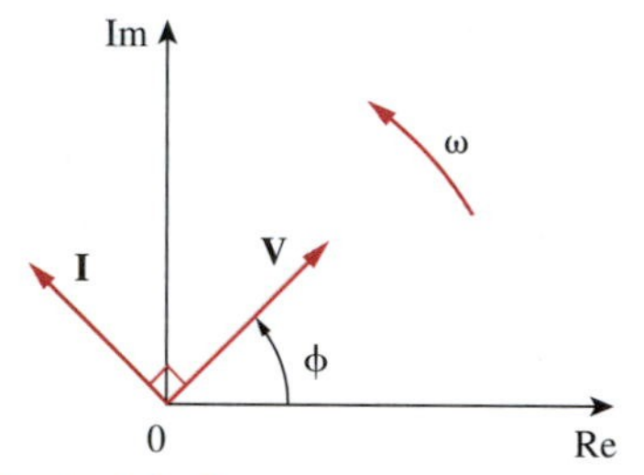

Figure 12.9
Phasor diagram for the capacitor; **I** leads **V** by 90°.

For the capacitor C, assume the voltage across it is $v(t) = V_m \sin(\omega t + \phi)$. The current through the capacitor is

$$i = C\frac{dv}{dt} \tag{12.20}$$

By following through the same steps as taken for the inductor, we obtain

$$\mathbf{I} = j\omega C\mathbf{V}$$

or

$$\boxed{\mathbf{V} = \mathbf{I}/j\omega C} \tag{12.21}$$

showing that the current and voltage are 90° out of phase. To be specific, the current leads the voltage by 90°. The voltage−current relationships for the capacitor are shown in Fig. 12.8; while the phasor diagram is shown in Fig. 12.9. The time-domain and phasor-domain representations of the circuit elements are summarized in Table 12.1. Note that unless a problem states otherwise, the answer should be given in form of the question: phasor to phasor, sinusoid to sinusoid, rectangular to rectangular, and polar to polar.

TABLE 12.1

Summary of voltage–current relationships.

Element	Time domain	Frequency domain
R	$v = Ri$	$\mathbf{V} = R\mathbf{I}$
L	$v = L\frac{di}{dt}$	$\mathbf{V} = j\omega L\mathbf{I}$
C	$i = C\frac{dv}{dt}$	$\mathbf{V} = \mathbf{I}/j\omega C$

For remembering the relationship between current and voltage in the inductor and capacitor, just think "ELI the ICE man."

E = Voltage (emf); I = Current; L = Inductor; C = Capacitor

- *ELI* (Inductive circuit) − Voltage leads Current
- *ICE* (Capacitive circuit) − Current leads Voltage

Example 12.6

The voltage $v(t) = 12 \sin(60t + 45°)$ is applied to a 0.1-H inductor. Find the steady-state current through the inductor.

Solution:

For the inductor, $\mathbf{V} = j\omega L\mathbf{I}$, where $\omega = 60$ rad/s and $\mathbf{V} = 12\underline{/45°}$ V. Hence,

$$\mathbf{I} = \frac{\mathbf{V}}{j\omega L} = \frac{12\underline{/45°}}{j60 \times 0.1} = \frac{12\underline{/45°}}{6\underline{/90°}} = 2\underline{/-45°}\ \text{A}$$

Converting this to the time domain,

$$i(t) = 2\sin(60t - 45°)\ \text{A}$$

Practice Problem 12.6

If voltage $v(t) = 6\cos(100t - 30°)$ is applied to a 50-μF capacitor, calculate the current through the capacitor.

Answer: $30\cos(100t + 60°)$ mA

12.4 Impedance and Admittance

In the preceding section, we obtained the voltage–current relations for the three passive elements as

$$\mathbf{V} = R\mathbf{I}, \qquad \mathbf{V} = j\omega L\mathbf{I}, \qquad \mathbf{V} = \mathbf{I}/j\omega C \tag{12.22}$$

These equations may be written in terms of the ratio of the phasor voltage to the phasor current; that is,

$$\frac{\mathbf{V}}{\mathbf{I}} = R, \qquad \frac{\mathbf{V}}{\mathbf{I}} = j\omega L, \qquad \frac{\mathbf{V}}{\mathbf{I}} = \frac{1}{j\omega C} \tag{12.23}$$

From these three expressions, we obtain Ohm's law in phasor form for any type of element as

$$\boxed{\mathbf{Z} = \mathbf{V}/\mathbf{I} \qquad \text{or} \qquad \mathbf{V} = \mathbf{Z}\mathbf{I}} \tag{12.24}$$

where $\mathbf{Z}$ is a frequency-dependent quantity known as *impedance*, measured in ohms.

> The **impedance Z** of a circuit is the ratio of the phasor voltage **V** to the phasor current **I**, measured in ohms (Ω).

The impedance represents the opposition that the circuit exhibits to the flow of sinusoidal current. Although the impedance is the ratio of two phasors, it is not a phasor because it does not correspond to a sinusoidally varying quantity.

The impedances of resistors, inductors, and capacitors can be readily obtained from Eq. (12.23). They are summarized in Table 12.2. From Table 12.2, we notice that $\mathbf{Z}_L = j\omega L$ for an inductor and $\mathbf{Z}_C = -j/\omega C$ for a capacitor. Consider two extreme cases of ω. When $\omega = 0$ (i.e., for dc sources), $\mathbf{Z}_L = 0$ and $\mathbf{Z}_C \rightarrow \infty$ confirming what we already know—that the inductor acts like a short circuit while the capacitor acts like an open circuit. When $\omega \rightarrow \infty$ (i.e., for high frequencies), $\mathbf{Z}_L \rightarrow \infty$ and $\mathbf{Z}_C = 0$, indicating that the inductor is an open circuit to high frequencies while the capacitor is a short circuit. This is illustrated in Fig. 12.10.

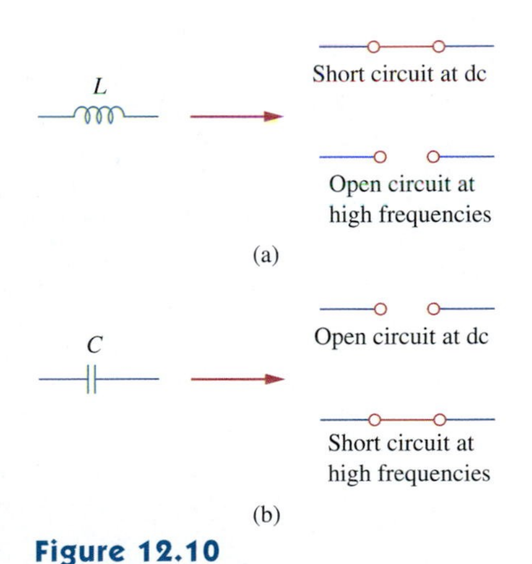

Figure 12.10
Equivalent circuits at dc and high frequencies: (a) inductor; (b) capacitor.

As a complex quantity, the impedance may be expressed in rectangular form as

$$\mathbf{Z} = R + jX \tag{12.25}$$

where $R = \text{Re}(\mathbf{Z})$ is the *resistance* and $X = \text{Im}(\mathbf{Z})$ is the *reactance*. The reactance X may be positive or negative. We say that the impedance is *inductive* when X is positive and *capacitive* when X is negative. Thus, impedance $\mathbf{Z} = R + jX$ is said to be inductive or lagging (because current lags voltage in inductors), while impedance $\mathbf{Z} = R - jX$ is capacitive or leading (because current leads voltage in capacitors). The impedance, resistance, and reactance are all measured in ohms. The impedance may be also be expressed in polar form as

$$\mathbf{Z} = |\mathbf{Z}|\underline{/\theta} \tag{12.26}$$

Comparing Eqs. (12.25) and (12.26), we infer that

$$\boxed{\mathbf{Z} = R + jX = |\mathbf{Z}|\angle\theta} \tag{12.27}$$

where

$$|Z| = \sqrt{R^2 + X^2}, \qquad \theta = \tan^{-1}\frac{X}{R} \tag{12.28a}$$

and

$$R = |\mathbf{Z}|\cos\theta, \qquad X = |\mathbf{Z}|\sin\theta \tag{12.28b}$$

It is sometimes convenient to work with the reciprocal of impedance, known as *admittance*. The admittance **Y** of a circuit is the ratio of the phasor current through it to the phasor voltage across it; that is,

$$\boxed{\mathbf{Y} = \frac{1}{\mathbf{Z}} = \frac{\mathbf{I}}{\mathbf{V}}} \tag{12.29}$$

The **admittance Y** is the reciprocal of impedance, measured in siemens (S).

The admittances of resistors, inductors, and capacitors can be obtained from Eq. (12.23). They are also summarized in Table 12.2.

As a complex quantity, we may write **Y** as

$$\boxed{\mathbf{Y} = G + jB} \tag{12.30}$$

TABLE 12.2

Impedances and admittances of passive elements.

Element	Impedance	Admittance
R	$\mathbf{Z} = R$	$\mathbf{Y} = 1/R$
L	$\mathbf{Z} = j\omega L$	$\mathbf{Y} = 1/j\omega L$
C	$\mathbf{Z} = 1/j\omega C$	$\mathbf{Y} = j\omega C$

where $G = \text{Re}(\mathbf{Y})$ is the *conductance* and $B = \text{Im}(\mathbf{Y})$ is the *susceptance*. Admittance, conductance, and susceptance are all expressed in the units of siemens. From Eqs. (12.25) and (12.30),

$$G + jB = \frac{1}{R + jX} \tag{12.31}$$

By rationalization,

$$G + jB = \frac{1}{R + jX} \cdot \frac{R - jX}{R - jX} = \frac{R - jX}{R^2 + X^2} \tag{12.32}$$

Equating the real and imaginary parts gives

$$G = \frac{R}{R^2 + X^2}, \qquad B = -\frac{X}{R^2 + X^2} \tag{12.33}$$

showing that $G \neq 1/R$ as it is in resistive circuits. Of course, if $X = 0$, then $G = 1/R$. Note that in Eq. (12.33) with a negative sign tells that an impedance Z with a positive Im part will result in an admittance Y with a negative Im part and vice versa.

Example 12.7

Find $v(t)$ and $i(t)$ in the circuit shown in Fig. 12.11.

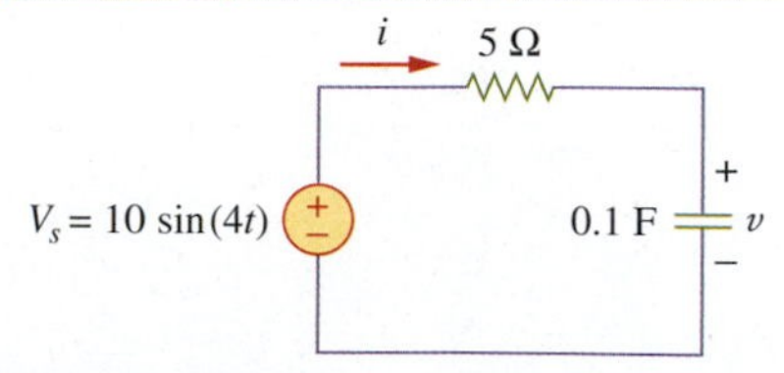

Figure 12.11
For Example 12.7.

Solution:

From the voltage source $10\sin(4t)$, $\omega = 4$,

$$\mathbf{V}_s = 10\angle 0^\circ \text{ V}$$

The impedance is

$$\mathbf{Z} = R + \frac{1}{j\omega C} = 5 + \frac{1}{j4 \times 0.1} = 5 - j2.5\ \Omega$$

Hence, the current is

$$\mathbf{I} = \frac{\mathbf{V}_s}{\mathbf{Z}} = \frac{10\angle 0^\circ}{5 - j2.5} = \frac{10(5 + j2.5)}{(5 - j2.5)(5 + j2.5)}$$

$$= \frac{10(5 + j2.5)}{5^2 + 2.5^2}$$

$$= 1.6 + j0.8 = 1.789\angle 26.57^\circ \text{ A} \qquad \textbf{(12.13.1)}$$

and the voltage across the capacitor is

$$\mathbf{V} = \mathbf{I}\mathbf{Z}_C = \frac{\mathbf{I}}{j\omega C} = \frac{1.789\angle 26.57^\circ}{j4 \times 0.1}$$

$$= \frac{1.789\angle 26.57^\circ}{0.4\angle 90^\circ}$$

$$= 4.47\angle -63.43^\circ \text{ V} \qquad \textbf{(12.13.2)}$$

Converting **I** and **V** in Eqs. (12.13.1) and (12.13.2) to the time domain, we get

$$i(t) = 1.789 \sin(4t + 26.57^\circ) \text{ A}$$
$$v(t) = 4.47 \sin(4t - 63.43^\circ) \text{ V}$$

Notice that $i(t)$ leads $v(t)$ by 90° as expected. Also, using a calculator could have made life easier.

Practice Problem 12.7

Refer to Fig. 12.12. Determine $v(t)$ and $i(t)$.

Answer: $2.236 \sin(10t + 63.44^\circ)$ V; $1.118 \sin(10t - 26.56^\circ)$ A

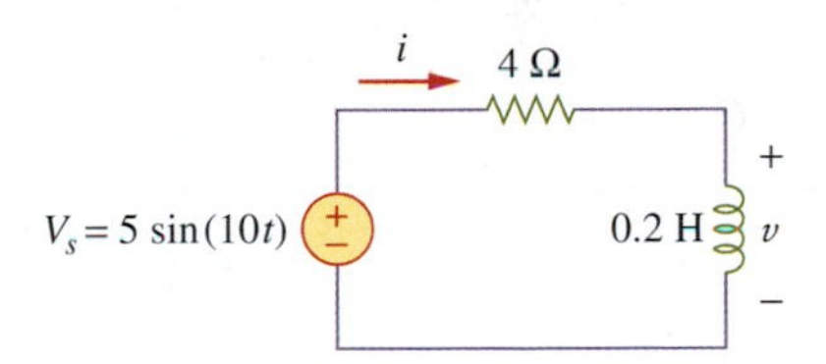

Figure 12.12
For Practice Problem 12.7.

12.5 Impedance Combinations

Consider the N series-connected impedances shown in Fig. 12.13. The same current **I** flows through the impedances. Applying KVL around the loop gives

$$\mathbf{V} = \mathbf{V}_1 + \mathbf{V}_2 + \cdots + \mathbf{V}_N = \mathbf{I}(\mathbf{Z}_1 + \mathbf{Z}_2 + \cdots + \mathbf{Z}_N) \qquad \textbf{(12.34)}$$

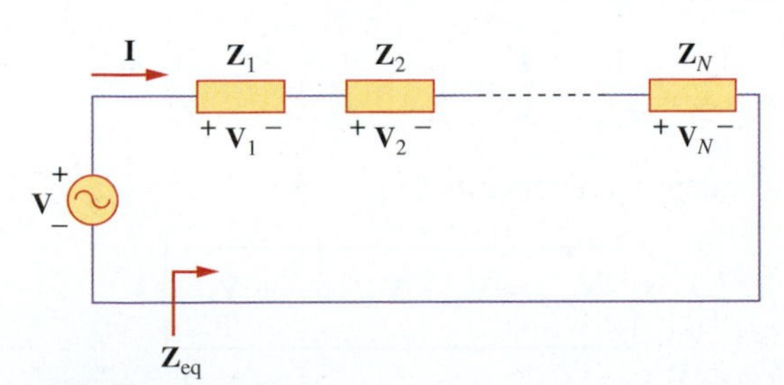

Figure 12.13
N impedances in series.

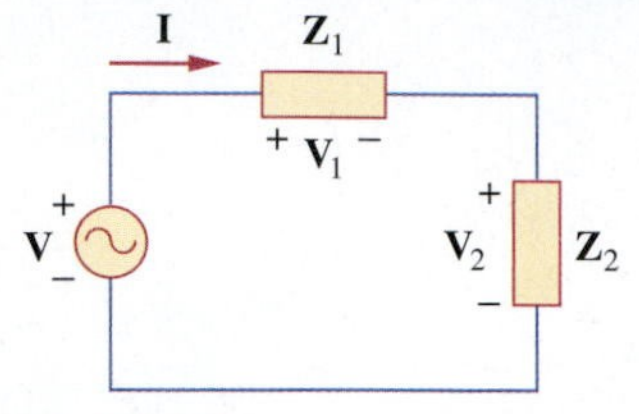

Figure 12.14
Voltage divider rule.

The equivalent impedance at the input terminals is

$$\mathbf{Z}_{eq} = \mathbf{V}/\mathbf{I} = \mathbf{Z}_1 + \mathbf{Z}_2 + \cdots + \mathbf{Z}_N$$

or

$$\boxed{\mathbf{Z}_{eq} = \mathbf{Z}_1 + \mathbf{Z}_2 + \cdots + \mathbf{Z}_N} \tag{12.35}$$

showing that the total or equivalent impedance of series-connected impedances is the sum of the individual impedances. This is similar to the series connection of resistances.

If $N = 2$, as shown in Fig. 12.14, the current through the impedances is

$$\mathbf{I} = \frac{\mathbf{V}}{\mathbf{Z}_1 + \mathbf{Z}_2} \tag{12.36}$$

Because $\mathbf{V}_1 = \mathbf{Z}_1\mathbf{I}$ and $\mathbf{V}_2 = \mathbf{Z}_2\mathbf{I}$, then

$$\boxed{\mathbf{V}_1 = \frac{\mathbf{Z}_1}{\mathbf{Z}_1 + \mathbf{Z}_2}\mathbf{V}, \qquad \mathbf{V}_2 = \frac{\mathbf{Z}_2}{\mathbf{Z}_1 + \mathbf{Z}_2}\mathbf{V}} \tag{12.37}$$

which is the *voltage divider rule.*

In the same manner, we can obtain the equivalent impedance or admittance of the N parallel-connected impedances shown in Fig. 12.15. The voltage across each impedance is the same. Applying KCL at the top node,

$$\mathbf{I} = \mathbf{I}_1 + \mathbf{I}_2 + \cdots + \mathbf{I}_N = \mathbf{V}\left(\frac{1}{\mathbf{Z}_1} + \frac{1}{\mathbf{Z}_2} + \cdots + \frac{1}{\mathbf{Z}_N}\right) \tag{12.38}$$

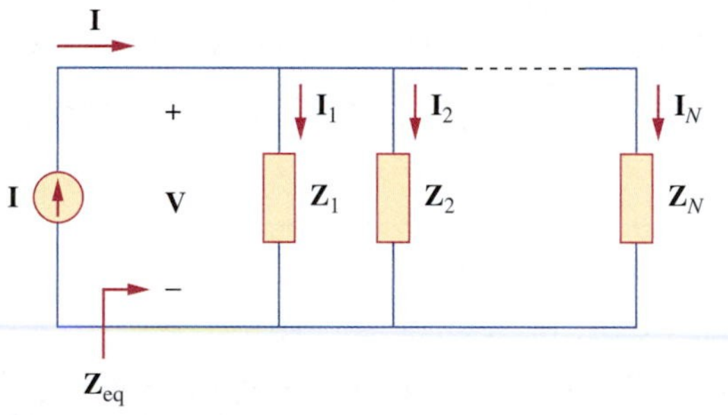

Figure 12.15
N impedances in parallel.

The equivalent impedance is

$$\frac{1}{\mathbf{Z}_{eq}} = \frac{\mathbf{I}}{\mathbf{V}} = \frac{1}{\mathbf{Z}_1} + \frac{1}{\mathbf{Z}_2} + \cdots + \frac{1}{\mathbf{Z}_N} \tag{12.39}$$

and the equivalent admittance is

$$\boxed{\mathbf{Y}_{eq} = \mathbf{Y}_1 + \mathbf{Y}_2 + \cdots + \mathbf{Y}_N} \tag{12.40}$$

This indicates that the equivalent admittance of a parallel connection of admittances is the sum of the individual admittances.

When $N = 2$, as shown in Fig. 12.16, the equivalent impedance becomes

$$\mathbf{Z}_{eq} = 1/\mathbf{Y}_{eq} = \frac{1}{\mathbf{Y}_1 + \mathbf{Y}_2} = \frac{1}{1/\mathbf{Z}_1 + 1/\mathbf{Z}_2} = \frac{\mathbf{Z}_1\mathbf{Z}_2}{\mathbf{Z}_1 + \mathbf{Z}_2} \tag{12.41}$$

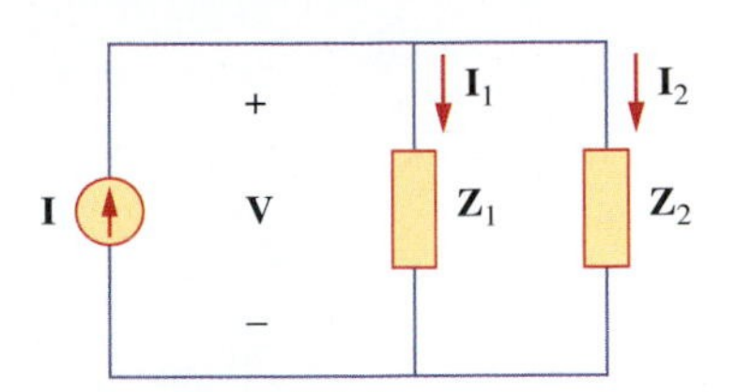

Figure 12.16
Current divider rule.

Also, because

$$\mathbf{V} = \mathbf{I}\mathbf{Z}_{eq} = \mathbf{I}_1\mathbf{Z}_1 = \mathbf{I}_2\mathbf{Z}_2$$

the currents in the impedances are

$$\mathbf{I}_1 = \frac{\mathbf{Z}_2}{\mathbf{Z}_1 + \mathbf{Z}_2}\mathbf{I}, \qquad \mathbf{I}_2 = \frac{\mathbf{Z}_1}{\mathbf{Z}_1 + \mathbf{Z}_2}\mathbf{I} \tag{12.42}$$

which is the *current divider rule*.

The delta-to-wye and wye-to-delta transformations applied to resistive circuits are also valid for impedances. With reference to Fig. 12.17, the conversion formulas are as follows.

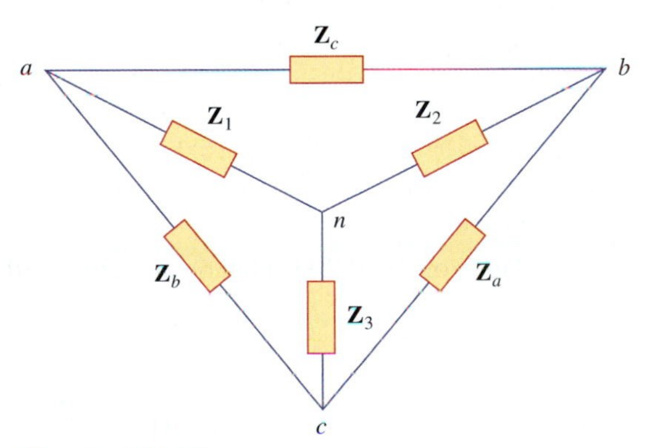

Figure 12.17
Superimposed Y and Δ networks.

Y-Δ Conversion

$$\mathbf{Z}_a = \frac{\mathbf{Z}_1\mathbf{Z}_2 + \mathbf{Z}_2\mathbf{Z}_3 + \mathbf{Z}_3\mathbf{Z}_1}{\mathbf{Z}_1}$$
$$\mathbf{Z}_b = \frac{\mathbf{Z}_1\mathbf{Z}_2 + \mathbf{Z}_2\mathbf{Z}_3 + \mathbf{Z}_3\mathbf{Z}_1}{\mathbf{Z}_2} \tag{12.43}$$
$$\mathbf{Z}_c = \frac{\mathbf{Z}_1\mathbf{Z}_2 + \mathbf{Z}_2\mathbf{Z}_3 + \mathbf{Z}_3\mathbf{Z}_1}{\mathbf{Z}_3}$$

Δ-*Y* Conversion

$$\mathbf{Z}_1 = \frac{\mathbf{Z}_b\mathbf{Z}_c}{\mathbf{Z}_a + \mathbf{Z}_b + \mathbf{Z}_c}$$
$$\mathbf{Z}_2 = \frac{\mathbf{Z}_c\mathbf{Z}_a}{\mathbf{Z}_a + \mathbf{Z}_b + \mathbf{Z}_c} \tag{12.44}$$
$$\mathbf{Z}_3 = \frac{\mathbf{Z}_a\mathbf{Z}_b}{\mathbf{Z}_a + \mathbf{Z}_b + \mathbf{Z}_c}$$

A delta or wye circuit is said to be **balanced** if it has equal impedances in all three branches.

When a delta-wye circuit is balanced, Eqs. (12.43) and (12.44) become

$$\mathbf{Z}_\Delta = 3\mathbf{Z}_Y \quad \text{or} \quad \mathbf{Z}_Y = \frac{1}{3}\mathbf{Z}_\Delta \tag{12.45}$$

where $\mathbf{Z}_Y = \mathbf{Z}_1 = \mathbf{Z}_2 = \mathbf{Z}_3$ and $\mathbf{Z}_\Delta = \mathbf{Z}_c = \mathbf{Z}_b = \mathbf{Z}_c$

As we have noticed in this section, the principles of voltage division, current division, circuit reduction, impedance equivalence, and Y-Δ transformation all apply to ac circuits. As we shall see in the next chapter, other circuit techniques such as superposition, nodal analysis, mesh analysis, source transformation, Thevenin theorem, and Norton theorem are all applied to ac circuits in a manner similar to their application to dc circuits.

Example 12.8

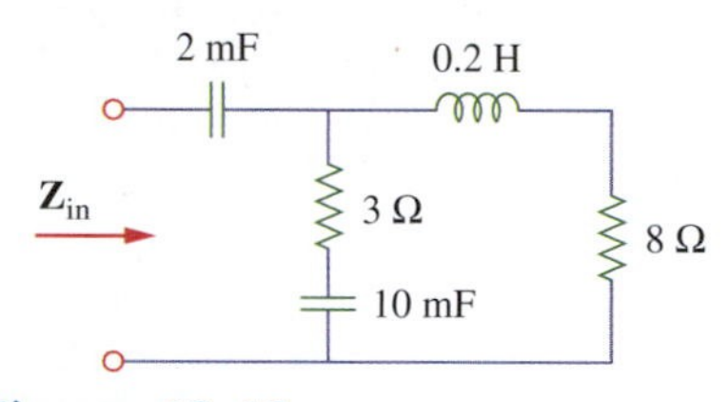

Figure 12.18
For Example 12.8.

Find the input impedance of the circuit in Fig. 12.18. Assume that the circuit operates at $\omega = 50$ rad/s.

Solution:
Let

$\mathbf{Z}_1$ = the impedance of the 2-mF capacitor
$\mathbf{Z}_2$ = the impedance of the 3-Ω resistor in series with the 10-mF capacitor
$\mathbf{Z}_3$ = the impedance of the 0.2-H inductor in series with the 8-Ω resistor

Then

$$\mathbf{Z}_1 = \frac{1}{j\omega C} = \frac{1}{j50 \times 2 \times 10^{-3}} = -j10\ \Omega$$

$$\mathbf{Z}_2 = 3 + \frac{1}{j\omega C} = 3 + \frac{1}{j50 \times 10 \times 10^{-3}} = (3 - j2)\ \Omega$$

$$\mathbf{Z}_3 = 8 + j\omega L = 8 + j50 \times 0.2 = (8 + j10)\ \Omega$$

The input impedance is

$$\begin{aligned}\mathbf{Z}_{\text{in}} &= \mathbf{Z}_1 + \mathbf{Z}_2 \parallel \mathbf{Z}_3 = -j10 + \frac{(3 - j2)(8 + j10)}{11 + j8} \\ &= -j10 + \frac{(44 + j14)(11 - j8)}{11^2 + 8^2} \\ &= -j10 + 3.22 - j1.07\ \Omega\end{aligned}$$

Thus,

$$\mathbf{Z}_{\text{in}} = 3.22 - j12.07\ \Omega.$$

Practice Problem 12.8

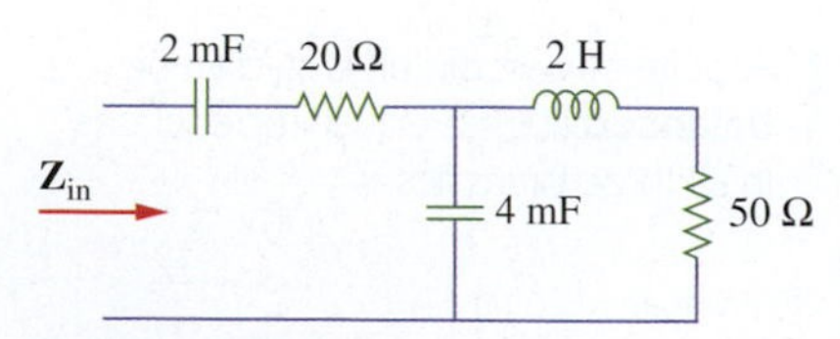

Figure 12.19
For Practice Problem 12.8.

Determine the input impedance of the circuit in Fig. 12.19 at $\omega =$ 10 rad/s.

Answer: $32.376 - j73.76\ \Omega$

Example 12.9

Determine $v_o(t)$ in the circuit in Fig. 12.20.

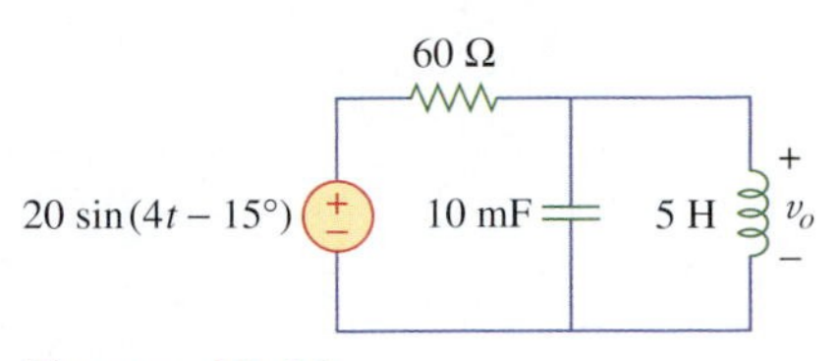

Figure 12.20
For Example 12.9.

Solution:
To do the analysis in the frequency domain, we must first transform the time-domain circuit in Fig. 12.20 to the phasor-domain equivalent in Fig. 12.21. Thus, the transformation produces:

$$v_s = 20\sin(4t - 15^\circ) \quad \Rightarrow \quad \mathbf{V}_s = 20\underline{/-15^\circ}\text{ V}, \qquad \omega = 4\text{ rad/s}$$

$$10\text{ mF} \quad \Rightarrow \quad \frac{1}{j\omega C} = \frac{1}{j4 \times 10 \times 10^{-3}} = -j25\ \Omega$$

$$5\text{ H} \quad \Rightarrow \quad j\omega L = j4 \times 5 = j20\ \Omega$$

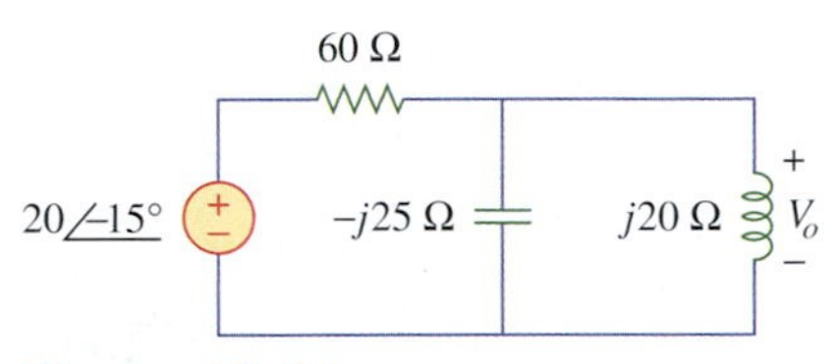

Figure 12.21
The frequency-domain equivalent of the circuit in Fig. 12.19.

Let

$\mathbf{Z}_1$ = the impedance of the 60-Ω resistor

$\mathbf{Z}_2$ = the impedance of the parallel combination of the 10-mF capacitor and the 5-H inductor

Then

$$\mathbf{Z}_1 = 60\ \Omega \quad \text{and}$$

$$\mathbf{Z}_2 = -j25 \parallel j20 = \frac{-j25 \times j20}{-j25 + j20} = j100\ \Omega$$

By the voltage divider rule,

$$\mathbf{V}_o = \frac{\mathbf{Z}_2}{\mathbf{Z}_1 + \mathbf{Z}_2}\mathbf{V}_s = \frac{j100}{60 + j100}(20\underline{/-15^\circ})$$
$$= (0.8575\underline{/30.96^\circ})(20\underline{/-15^\circ}) = 17.15\underline{/15.96^\circ}\text{ V}$$

We convert this to the time domain and obtain

$$v_o(t) = 17.15\sin(4t + 15.96^\circ)\text{ V}$$

Practice Problem 12.9

Calculate v_o in the circuit in Fig. 12.22.

Answer: $v(t) = 7.071\cos(10t - 60^\circ)$ V.

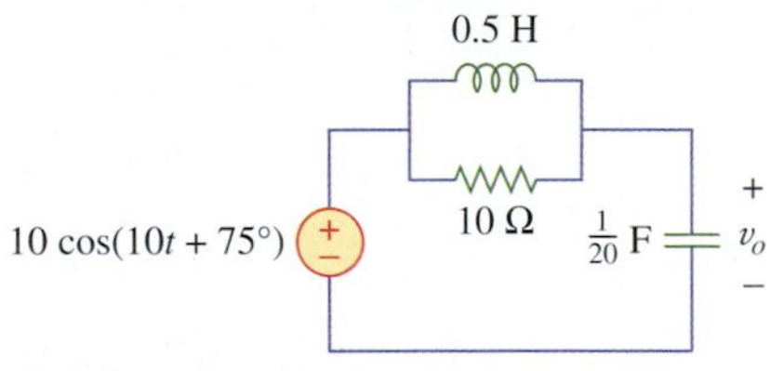

Figure 12.22
For Practice Problem 12.9.

Example 12.10

Find current **I** in the circuit in Fig. 12.23.

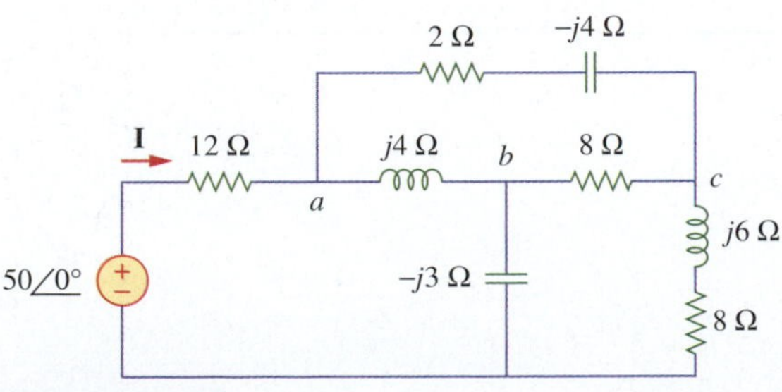

Figure 12.23
For Example 12.10.

Solution:
The delta network connected to nodes a, b, and c can be converted to the wye network. We obtain the network in Fig. 12.24. The wye impedances are obtained as follows using Eq. (12.44).

$$\mathbf{Z}_{an} = \frac{j4(2 - j4)}{j4 + 2 - j4 + 8} = \frac{4(4 + j2)}{10} = (1.6 + j0.8)\ \Omega$$

$$\mathbf{Z}_{bn} = \frac{j4(8)}{10} = j3.2\ \Omega$$

$$\mathbf{Z}_{cn} = \frac{8(2 - j4)}{10} = (1.6 - j3.2)\ \Omega$$

The total impedance at the source terminals is

$$\begin{aligned} \mathbf{Z} &= 12 + \mathbf{Z}_{an} + (\mathbf{Z}_{bn} - j3) \parallel (\mathbf{Z}_{cn} + j6 + 8) \\ &= 12 + 1.6 + j0.8 + (j0.2) \parallel (9.6 + j2.8) \\ &= 13.6 + j0.8 + \frac{j0.2(9.6 + j2.8)}{9.6 + j2.8 + j0.2} \\ &= 13.6 + j1 = 13.64\underline{/4.2^\circ}\ \Omega \end{aligned}$$

The desired current is

$$\mathbf{I} = \frac{\mathbf{V}}{\mathbf{Z}} = \frac{50\underline{/0^\circ}}{13.64\underline{/4.2^\circ}} = 3.666\underline{/-4.2^\circ}\ \text{A}$$

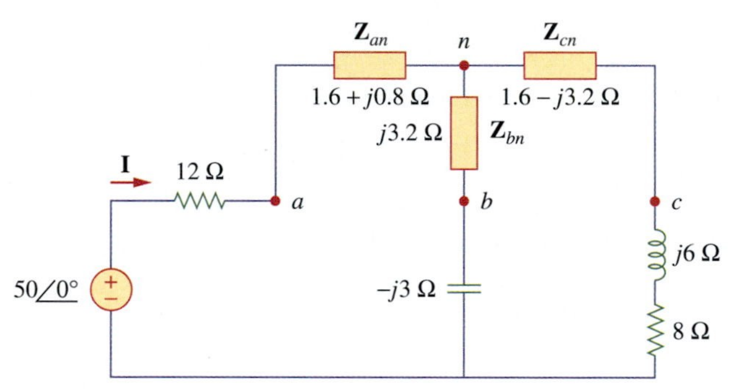

Figure 12.24
The circuit in Fig. 12.22 after delta-to-wye transformation.

Practice Problem 12.10

Find **I** in the circuit in Fig. 12.25.

Answer: 6.364 $\underline{/4.22^\circ}$ A.

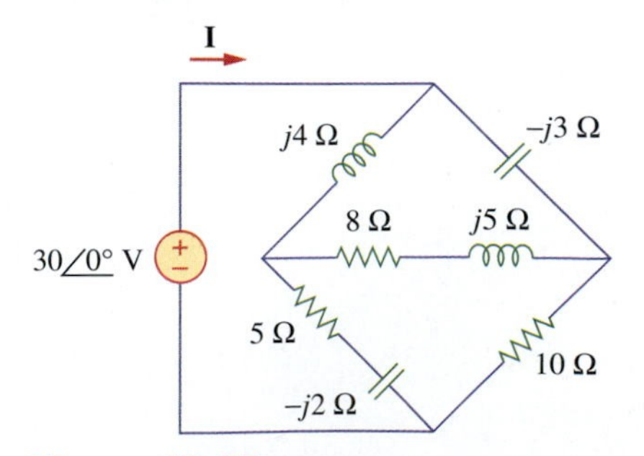

Figure 12.25
For Practice Problem 12.10.

12.6 Computer Analysis

We will consider MATLAB and PSpice as two computer tools that can be used to handle the material covered in this chapter.

12.6.1 MATLAB

MATLAB can be used to plot a sinusoid. It can also be used to handle the complex algebra encountered in this chapter. For example, to plot $v(t) = 10 \sin(2\pi t - 20°)$ V for $0 < t < 2$, we use the following commands:

```
»t=0:0.01:2;        % start t from 0, increment by
                    0.01, and stop at 2
»ang=20*pi/180      % converts angle to radian;
»v=10*sin(2*pi*t - ang);
»plot(t,v)
```

and the sinusoid $v(t)$ is plotted in Fig. 12.26.

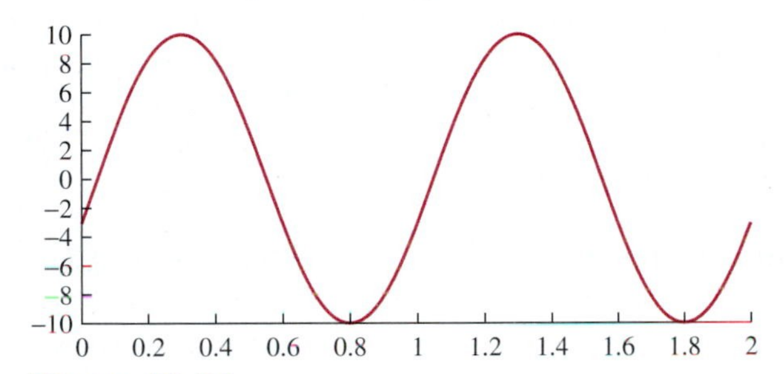

Figure 12.26
Plot of $v(t) = 10 \sin(2\pi t - 20°)$.

MATLAB can handle complex numbers. For example, to evaluate the complex number in Example 12.1, we proceed as follows.

```
»z1= 3  + j*5;
»z2= 6  - j*8;
»z3= z1 + z2
z3 =
    9.0000  - 3.0000i
»z4=z1*z2
z4 =
    58.0000 + 6.0000i
```

confirming what we had before. MATLAB allows using either i or j to represent $\sqrt{-1}$. Although using MATLAB as a calculator may not be allowed in exams, you can use it to check your homework assignments.

12.6.2 PSpice

PSpice can be used to perform phasor analysis. The reader is advised to read Section C.5 in Appendix C for a review of PSpice concepts for ac analysis. Although ac analysis with PSpice involves using ac sweep, our analysis here requires a single frequency. AC circuit analysis is done in the phasor or frequency domain, and all sources must have the same frequency, $f = \omega/2\pi$. The following example illustrates using PSpice for phasor analysis.

Example 12.11

Obtain v_o and i_o in the circuit of Fig. 12.27 using PSpice.

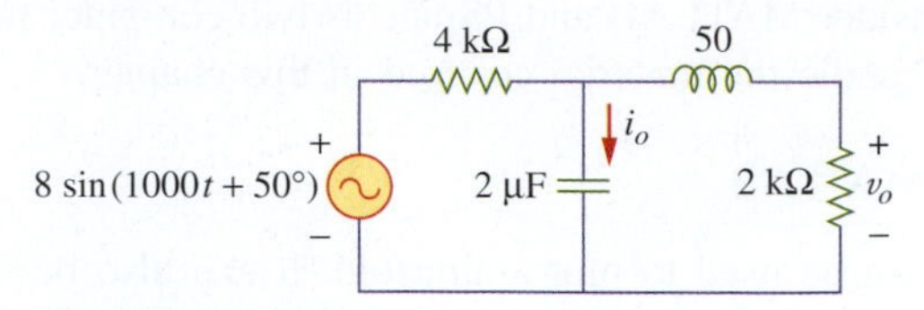

Figure 12.27
For Example 12.11.

Solution:
The frequency f is obtained from ω as

$$f = \frac{\omega}{2\pi} = \frac{1000}{2\pi} = 159.155 \text{ Hz}$$

The schematic for the circuit is shown in Fig. 12.28. We use part name VAC for the voltage source. We double click on VAC and enter ACMAG = 8 V and ACPHASE = 50. Because we only want the magnitude and phase of v_o and i_o, we set the attributes of IPRINT AND VPRINT1 each to *AC* = *yes*, *MAG* = *yes*, *PHASE* = *yes*. Once the circuit is drawn and saved as exam1211.dsn, we select **PSpice/New Simulation Profile**. This leads to the New Simulation dialog box. Type "exam1211" as the name of the file and click Create. This leads to the Simulation Settings dialog box. We select AC Sweep/Noise for *Analysis type*. As a single frequency analysis, we type 159.155 as the *Start Freq*, 159.155 as the *Final Freq*, and 1 as *Total Point*. After saving the schematic, we simulate it by selecting **PSpice/Run**. This leads to Probe, the graphical processor. We go back to the schematic window and select **PSpice/View Output File**. The output file includes the source frequency in addition to the attributes checked for the pseudocomponents IPRINT and VPRINT1; that is,

```
FREQ          VM(3)          VP(3)
1.592E+02     9.412E-01      -2.077E+01

FREQ          IM(V_PRINT1)   IP(V_PRINT1)
1.592E+02     1.883E-03      7.067E+01
```

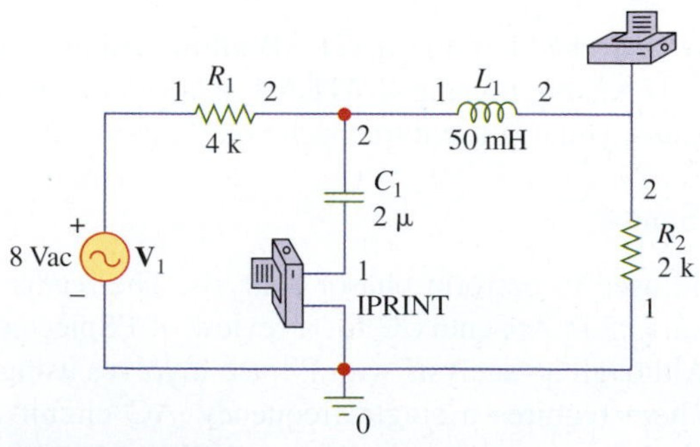

Figure 12.28
Schematic for Example 12.11.

From this output file, we obtain

$$\mathbf{V}_o = 0.941 \underline{/-20.71^\circ} \text{ V}, \qquad \mathbf{I}_o = 1.883 \underline{/70.67^\circ} \text{ mA}$$

which are the phasors for

$$v_o(t) = 0.941 \sin(1000t - 20.71^\circ) \text{ V}$$

and

$$i_o(t) = 1.883 \sin(1000t + 70.67^\circ) \text{ mA}$$

Practice Problem 12.11

Use PSpice to obtain v_o and i_o in the circuit in Fig. 12.29.

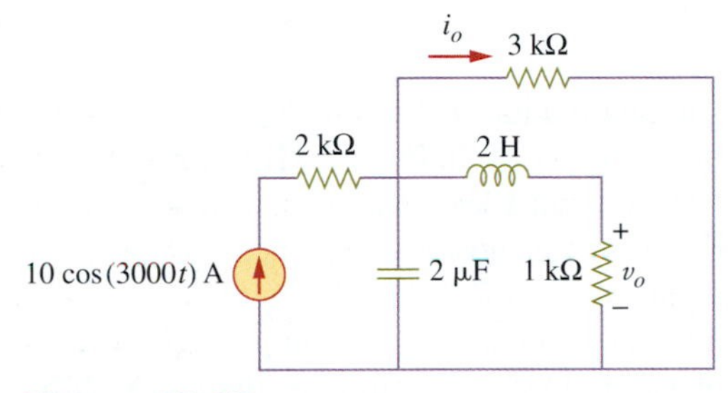

Figure 12.29
For Practice Problem 12.11.

Answer: $0.2772 \cos(3000t - 153.7^\circ)$ V;
$0.562 \cos(3000t - 77.13^\circ)$ mA

12.7 †Applications

In Chapters 9 and 10, *RC* and *RL* circuits were shown to be useful in certain dc applications. These circuits also have ac applications such as coupling circuits, phase-shifting circuits, filters, resonant circuits, ac bridge circuits, and transformers (to mention just a few of them). Here, we will consider two simple applications: *RC* phase-shifting circuits and ac bridge circuits.

12.7.1 Phase-Shifters

A phase-shifting circuit is often employed to correct an undesirable phase shift already present in a circuit or to produce special desired effects. An *RC* circuit is suitable for this purpose because its capacitor causes the circuit current to lead the applied voltage. Two commonly used *RC* circuits are shown in Fig. 12.30. (*RL* circuits or any reactive circuits could also serve the same purpose.)

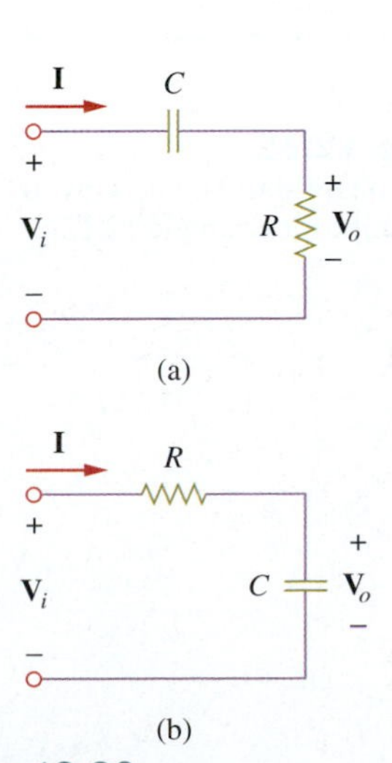

Figure 12.30
Series *RC* shift circuits: (a) leading output; (b) lagging output.

In Fig. 12.30(a), the circuit current **I** leads the applied voltage $\mathbf{V}_i$ by some phase angle θ, where $0 < \theta < 90^\circ$ depending on the values of R and C. If $X_C = 1/\omega C$, then the total impedance is $\mathbf{Z} = R + jX_C$ and the phase shift is given by

$$\theta = \tan^{-1}\frac{X_C}{R} \tag{12.46}$$

This shows that the amount of phase shift depends on the values of R, C, and the operating frequency. Because the output voltage $\mathbf{V}_o$ across

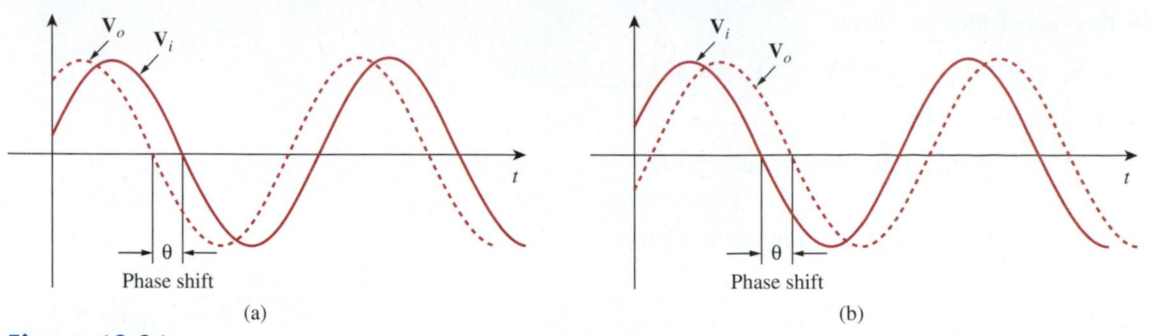

Figure 12.31
Phase shift in *RC* circuits: (a) leading output; (b) lagging output.

the resistor is in phase with the current, $\mathbf{V}_o$ leads (positive phase shift) $\mathbf{V}_i$ as shown in Fig. 12.31(a). In Fig. 12.31(b), the output is taken across the capacitor. The current $\mathbf{I}$ leads the input voltage $\mathbf{V}_i$ by θ, but the output voltage $\mathbf{V}_o$ across the capacitor lags (negative phase shift) the input voltage $\mathbf{V}_i$ as illustrated in Fig. 12.31(b).

We should keep in mind that the simple *RC* circuits in Fig. 12.30 also act as voltage dividers. Therefore, as the phase shift θ approaches 90°, the output voltage $\mathbf{V}_o$ approaches zero. For this reason, these simple *RC* circuits are used only when small amounts of phase shift are required. If phase shifts greater than 60° are called for, simple *RC* networks are cascaded, thereby providing a total phase shift equal to the sum of the individual phase shifts.

Example 12.12

Design an *RC* circuit to provide a phase of 90° leading.

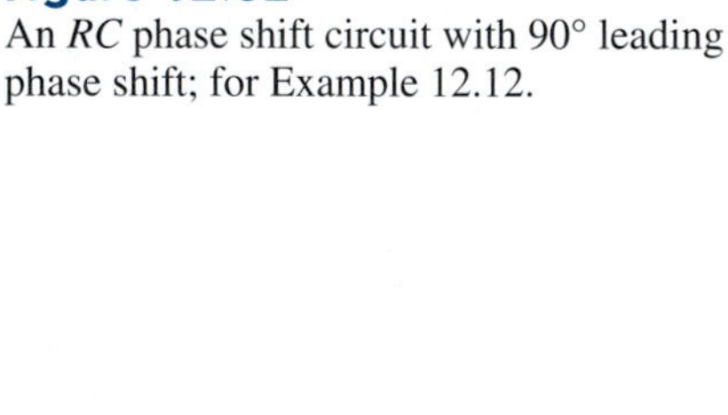

Figure 12.32
An *RC* phase shift circuit with 90° leading phase shift; for Example 12.12.

Solution:

If we select circuit components of equal ohmic value, say, $R = |X_C| = 20\ \Omega$, at a particular frequency, according to Eq. (12.46), the phase shift is exactly 45°. By cascading two similar *RC* circuits in Fig. 12.30(a), we obtain the circuit in Fig. 12.32, providing a positive or leading phase shift of 90°, as we shall show. Using the series-parallel combination techniques, $\mathbf{Z}$ in Fig. 12.32 is obtained as

$$\mathbf{Z} = 20 \parallel (20 - j20) = \frac{20(20 - j20)}{40 - j20} = 12 - j4\ \Omega \qquad \textbf{(12.12.1)}$$

Using the voltage divider rule,

$$\mathbf{V}_1 = \frac{\mathbf{Z}}{\mathbf{Z} - j20}\mathbf{V}_i = \frac{12 - j4}{12 - j24}\mathbf{V}_i = \frac{\sqrt{2}}{3}\underline{/45^\circ}\ \mathbf{V}_i \qquad \textbf{(12.12.2)}$$

and

$$\mathbf{V}_o = \frac{20}{20 - j20}\mathbf{V}_1 = \frac{\sqrt{2}}{2}\underline{/45^\circ}\ \mathbf{V}_1 \qquad \textbf{(12.12.3)}$$

Substituting Eq. (12.12.2) into Eq. (12.12.3) yields

$$\mathbf{V}_o = \left(\frac{\sqrt{2}}{2}\underline{/45^\circ}\right)\left(\frac{\sqrt{2}}{3}\underline{/45^\circ}\ \mathbf{V}_i\right) = \frac{1}{3}\underline{/90^\circ}\ \mathbf{V}_i$$

Thus, the output leads the input by 90°, but its magnitude is only about 33 percent of the input.

Practice Problem 12.12

Design an RC circuit to provide a 90° lagging phase shift. If a voltage of 10 V is applied, what is the output voltage?

Answer: Figure 12.33 shows a typical design; 3.33 V

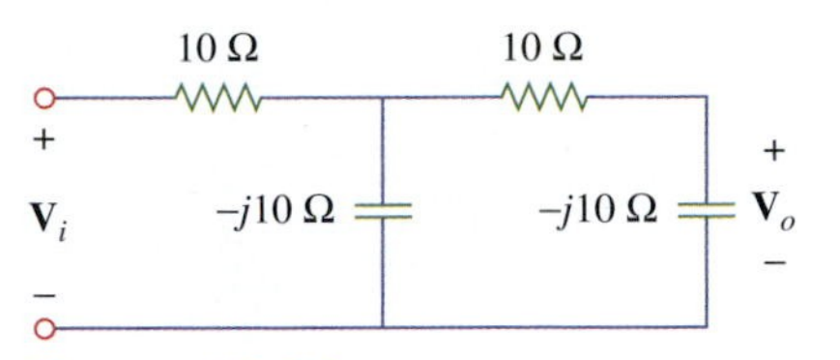

Figure 12.33
For Practice Problem 12.12.

12.7.2 AC Bridges

An ac bridge circuit is used in measuring the inductance L of an inductor or the capacitance C of a capacitor. It is similar in form to the Wheatstone bridge, also used to measure an unknown resistance (discussed in Section 6.5) and follows the same principle. To measure L and C, however, an ac source is needed, and an ac meter replaces the galvanometer used there. The ac meter may be a sensitive ac ammeter or voltmeter.

Consider the general ac bridge circuit displayed in Fig. 12.34. The bridge is *balanced* when no current flows through the meter. This means that $\mathbf{V}_1 = \mathbf{V}_2$. Applying the voltage divider rule,

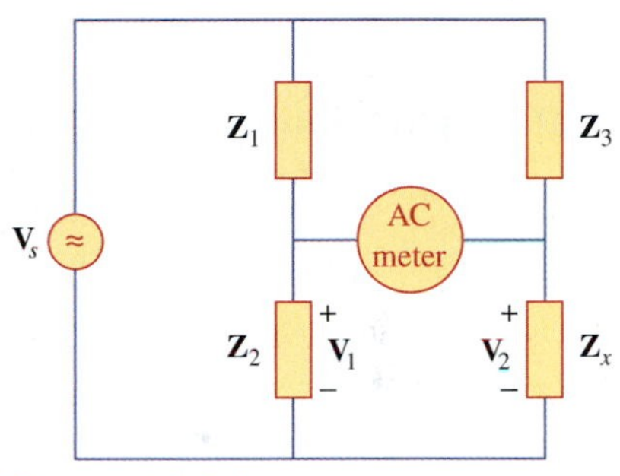

Figure 12.34
A general ac bridge.

$$\mathbf{V}_1 = \frac{\mathbf{Z}_2}{\mathbf{Z}_1 + \mathbf{Z}_2}\mathbf{V}_s = \mathbf{V}_2 = \frac{\mathbf{Z}_x}{\mathbf{Z}_3 + \mathbf{Z}_x}\mathbf{V}_s \tag{12.47}$$

Thus,

$$\frac{\mathbf{Z}_2}{\mathbf{Z}_1 + \mathbf{Z}_2} = \frac{\mathbf{Z}_x}{\mathbf{Z}_3 + \mathbf{Z}_x} \quad \Rightarrow \quad \mathbf{Z}_2\mathbf{Z}_3 = \mathbf{Z}_1\mathbf{Z}_x \tag{12.48}$$

or

$$\boxed{\mathbf{Z}_x = \frac{\mathbf{Z}_3}{\mathbf{Z}_1}\mathbf{Z}_2} \tag{12.49}$$

This is the balanced equation for the ac bridge and is similar to Eq. (6.3) for the resistance bridge except that R_2 are replaced by **Z**s.

Specific ac bridges for measuring L and C are shown in Fig. 12.35, where L_x and C_x are the unknown inductance and capacitance to be measured, while L_s and C_s are a standard inductance and capacitance (the values of which are known to great precision). In each case, two

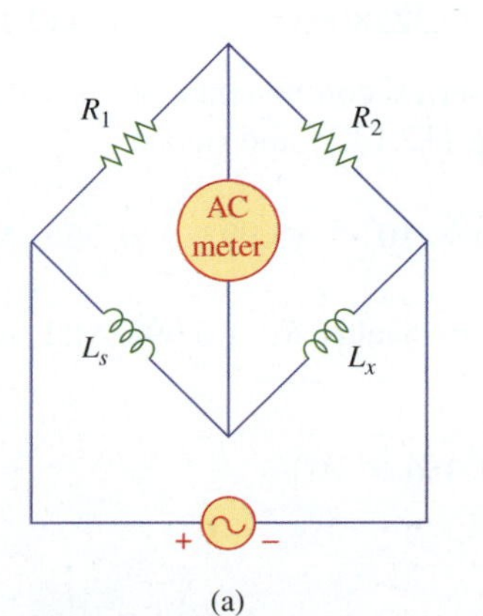

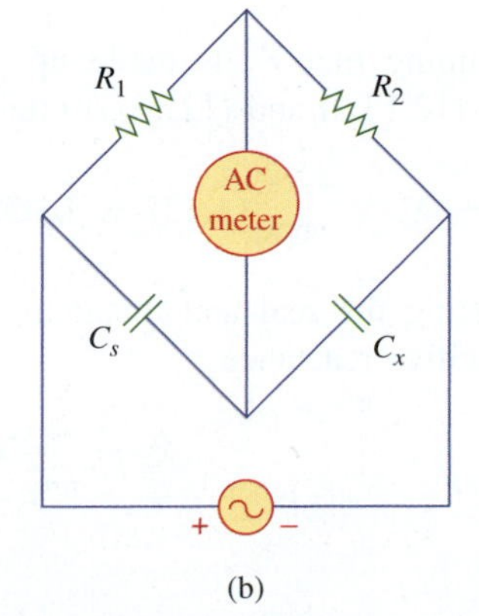

Figure 12.35
Specific ac bridges: (a) for measuring L; (b) for measuring C.

resistors R_1 and R_2 are varied until the ac meter reads zero. Then the bridge is balanced. From Eq. (12.49), we obtain

$$L_x = \frac{R_2}{R_1} L_s \tag{12.50}$$

and

$$C_x = \frac{R_1}{R_2} C_s \tag{12.51}$$

Notice that the balancing of the ac bridges in Fig. 12.35 does not depend on the frequency f of the ac source because f does not appear in the relationships in Eqs. (12.50) and (12.51).

Example 12.13

The ac bridge circuit of Fig. 12.34 balances when $\mathbf{Z}_1$ is a 1-kΩ resistor, $\mathbf{Z}_2$ is a 4.2-kΩ resistor, $\mathbf{Z}_3$ is a parallel combination of a 1.5-MΩ resistor and a 12-pF capacitor, and $f = 2$ kHz. Find the series components that make up $\mathbf{Z}_x$.

Solution:
From Eq. (12.49),

$$\mathbf{Z}_x = \frac{\mathbf{Z}_3}{\mathbf{Z}_1} \mathbf{Z}_2 \tag{12.13.1}$$

where $\mathbf{Z}_x = R_x + jX_x$

$$\mathbf{Z}_1 = 1000\ \Omega, \quad \mathbf{Z}_2 = 4200\ \Omega \tag{12.13.2}$$

and

$$\mathbf{Z}_3 = R_3 \parallel \frac{1}{j\omega C} = \frac{\dfrac{R_3}{j\omega C_3}}{R_3 + 1/j\omega C_3} = \frac{R_3}{1 + j\omega R_3 C_3}$$

Because $R_3 = 1.5$ MΩ and $C_3 = 12$ pF,

$$\mathbf{Z}_3 = \frac{1.5 \times 10^6}{1 + j2\pi \times 2 \times 10^3 \times 1.5 \times 10^6 \times 12 \times 10^{-12}} = \frac{1.5 \times 10^6}{1 + j0.2262}$$

or

$$\mathbf{Z}_3 = 1.427 - j0.3228\ \text{M}\Omega \tag{12.13.3}$$

Assuming that $\mathbf{Z}_x$ is made up of series components, we substitute Eqs. (12.13.2) and (12.13.3) into Eq. (12.13.1) and obtain

$$R_x + jX_x = \frac{4200}{1000}(1.427 - j0.3228) \times 10^6 = (5.993 - j1.356)\ \text{M}\Omega$$

Equating the real and imaginary parts yields $R_x = 5.993$ MΩ and a capacitive reactance

$$X_x = \frac{1}{\omega C} = 1.356 \times 10^6$$

or

$$C = \frac{1}{\omega X_x} = \frac{1}{2\pi \times 2 \times 10^3 \times 1.356 \times 10^6} = 58.69\ \text{pF}$$

Practice Problem 12.13

In the ac bridge circuit of Fig. 12.34, suppose that balance is achieved when $\mathbf{Z}_1$ is a 4.8-kΩ resistor, $\mathbf{Z}_2$ is a 10-Ω resistor in series with a 0.25-μH inductor, $\mathbf{Z}_3$ is a 12-kΩ resistor, and $f = 6$ MHz. Determine the series components that make up $\mathbf{Z}_x$.

Answer: 25-Ω resistor in series with a 0.625-μH inductor

12.8 Summary

1. A phasor is a complex quantity that represents both the magnitude and the phase of a sinusoid. Given the sinusoid $v(t) = V_m \sin(\omega t + \phi)$ its phasor $\mathbf{V}$ is $\mathbf{V} = V_m/\underline{\phi}$.
2. In ac circuits, voltage and current phasors always have a fixed relation to one another at any moment of time. If $v(t) = V_m \sin(\omega t + \phi_v)$ represents the voltage across an element and $i(t) = I_m \sin(\omega t + \phi_i)$ represents the current through the element, then $\phi_i = \phi_v$ if the element is a resistor, ϕ_i leads ϕ_v by 90° if the element is a capacitor, and ϕ_i lags ϕ_v by 90° if the element is an inductor.
3. Basic circuit laws (Ohm's and Kirchhoff's) apply to ac circuits in the same manner as they do for dc circuits; i.e.,

$$\mathbf{V} = \mathbf{Z}\mathbf{I}$$

$$\sum \mathbf{I}_k = 0 \quad \text{(KCL)}$$

$$\sum \mathbf{V}_k = 0 \quad \text{(KVL)}$$

4. The impedance $\mathbf{Z}$ of a circuit is the ratio of the phasor voltage across it to the phasor current through it:

$$\mathbf{Z} = \frac{\mathbf{V}}{\mathbf{I}} = R(\omega) + jX(\omega)$$

   where $R = \text{Re}(\mathbf{Z})$ is the *resistance* and $X = \text{Im}(\mathbf{Z})$ is the *reactance*. We say that the impedance is *inductive* when X is positive and *capacitive* when X is negative.
5. The admittance $\mathbf{Y}$ is the reciprocal of impedance:

$$\mathbf{Y} = \frac{1}{\mathbf{Z}} = G(\omega) + jB(\omega)$$

   Impedances are combined in series or in parallel just like resistances; in this case, impedances in series add while admittances in parallel add.
6. For a resistor $\mathbf{Z} = R$, for an inductor $\mathbf{Z} = jX = j\omega L$, and for a capacitor $\mathbf{Z} = -jX = 1/j\omega C$.
7. The techniques of voltage/current division, series/parallel combination of impedance/admittance, circuit reduction, and Y-Δ transformation all apply to ac circuit analysis.
8. MATLAB and PSpice are common computer tools for analyzing ac circuits.
9. AC circuits are applied in phase-shifters and bridges.

Review Questions

12.1 The negative angle $-50°$ is equivalent to:

(a) 40° (b) 50° (c) 130° (d) 310°

12.2 The complex number $6 + j6$ is equivalent to:

(a) $6\angle 45°$ (b) $36\angle 0°$
(c) $8.485\angle 45°$ (d) $8.485\angle 135°$

12.3 $(12 + j5) + (3 - j6)$ is equal to:

(a) $15 - j$ (b) $17 - j3$
(c) $18 - j6$ (d) $17 - j3$

12.4 $(10\angle 30)(2\angle -15°)$ can be expressed as:

(a) $12\angle -45°$ (b) $20\angle 15°$ (c) $5\angle -55°$

12.5 The voltage across an inductor leads the current through it by 90°.

(a) True (b) False

12.6 The imaginary part of impedance is called:

(a) resistance (b) admittance
(c) susceptance (d) conductance
(e) reactance

12.7 The impedance of a capacitor increases with increasing frequency.

(a) True (b) False

12.8 The phasor does not provide the frequency of the sinusoid it represents.

(a) True (b) False

12.9 A series *RC* circuit has $V_R = 12$ V and $V_C = 5$ V. The supply voltage is:

(a) -7 V (b) 7 V
(c) 13 V (d) 17 V

12.10 A series *RCL* circuit has $R = 30\ \Omega$, $X_C = 50\ \Omega$, $X_L = 90\ \Omega$. The impedance of the circuit is:

(a) $30 + j140\ \Omega$ (b) $30 + j40\ \Omega$
(c) $30 - j40\ \Omega$ (d) $-30 - j40\ \Omega$
(e) $-30 + j40\ \Omega$

Answers: 12.1d, 12.2c, 12.3a, 12.4b, 12.5a, 12.6e, 12.7b, 12.8a, 12.9c, 12.10b

Problems

Section 12.2 Phasors and Complex Numbers

12.1 Simplify and express the result of each the following operations in rectangular form.

(a) $(5 + j6) - (2 - j3)$

(b) $(25 + j7)^* (1 + j2)$

(c) $20\angle 30° - 10\angle 45°$

(d) $\dfrac{26\angle 40° + 5\angle -10°}{6 + j8}$

12.2 Simplify each the following complex expressions and express your result in rectangular form.

(a) $(2 + j) + (4 - j7)$ (b) $(j3)^*(3 + j5)$

(c) $\dfrac{4 + j3}{1 - j2}$ (d) $\dfrac{2 - j5}{2 + j4}$

12.3 Calculate these complex numbers and express your results in rectangular form:

(a) $\dfrac{15\angle 45°}{3 - j4} + j2$

(b) $\dfrac{8\angle -20°}{(2 + j)(3 - j4)} + \dfrac{10}{-5 + j12}$

(c) $10 + (8\angle 50°)(5 - j12)$

12.4 Given that $z_1 = 6 - j8$, $z_2 = 10\angle -30°$, and $z_3 = 8\angle -120°$, find:

(a) $z_1 + z_2 + z_3$, (b) $\dfrac{Z_1 Z_2}{Z_3}$

12.5 Given the complex numbers $z_1 = -3 + j4$ and $z_2 = 12 + j5$, find:

(a) $z_1 z_2$, (b) $\dfrac{Z_1}{Z_2^*}$, (c) $\dfrac{Z_1 + Z_2}{Z_1 - Z_2}$

12.6 Let $\mathbf{X} = 8\angle 40°$ and $\mathbf{Y} = 10\angle -30°$. Evaluate the following quantities and express your results in polar form.

(a) $(\mathbf{X} + \mathbf{Y})\mathbf{X}^*$, (b) $(\mathbf{X} - \mathbf{Y})^*$, (c) $(\mathbf{X} + \mathbf{Y})/\mathbf{X}$

12.7 Evaluate the following complex numbers:

(a) $\dfrac{2 + j3}{1 - j6} + \dfrac{7 - j8}{-5 + j11}$

(b) $\dfrac{(5\angle 10°)(10\angle -40°)}{(4\angle -80°)(-6\angle 50°)}$

(c) $\begin{vmatrix} 2 + j3 & -j2 \\ -j2 & 8 - j5 \end{vmatrix}$

12.8 Obtain the sinusoids corresponding to each of the following phasors:

(a) $\mathbf{V}_1 = 60\ \underline{/15^\circ}$ V, $\omega = 1$

(b) $\mathbf{V}_2 = 6 + j8$ V, $\omega = 40$

(c) $\mathbf{I}_1 = 2.8\underline{/-\pi/3}$ A, $\omega = 377$

(d) $\mathbf{I}_2 = -0.5 - j1.2$ A, $\omega = 10^3$

12.9 Find a single sinusoid corresponding to each of these phasors:

(a) $\mathbf{V} = 40\underline{/-60^\circ}$ V

(b) $\mathbf{V} = -30\ \underline{/10^\circ} + 50\underline{/60^\circ}$ V

(c) $\mathbf{I} = j6\underline{/-10^\circ}$ A

(d) $\mathbf{I} = \dfrac{2}{j} + 10\underline{/-45^\circ}$ A

12.10 Express the following sinusoids in phasor form.

(a) $v = 10\sin(\omega t + 20^\circ)$ V

(b) $v = 25\sin(\omega t - 30^\circ)$ V

(c) $i = 40\sin(\omega t + 270^\circ)$ A

(d) $i = 50\cos(\omega t - 33^\circ)$ A

12.11 Express the following sinusoids in phasor form.

(a) $v = 20\sin(\omega t - 60^\circ)$ V

(b) $i = -5\cos(\omega t + 70^\circ)$ A

12.12 Using phasors, find:

(a) $3\cos(20t + 10^\circ) - 5\cos(20t - 30^\circ)$,

(b) $40\sin(50t) + 30\cos(50t - 45^\circ)$,

(c) $20\sin(400t) + 10\cos(400t + 60^\circ) - 5\sin(400t - 20^\circ)$

Section 12.3 Phasor Relationships for Circuit Elements

12.13 Determine the current that flows through an 8-Ω resistor connected to a voltage source $v_s(t) = 110\cos(377t)$ V.

12.14 Calculate the reactance of a 2-H inductor at:

(a) 60 Hz (b) 1 MHz (c) 600 rad/s

12.15 A 40-mH inductor is supplied from a 110-V source with a frequency of 60 Hz. Calculate the reactance and the current.

12.16 A 220-V, 60-Hz supply is applied to a 50-μF capacitor. Determine the reactance and the current that flows through the capacitor.

12.17 What is the instantaneous voltage across a 2-μF capacitor when the current through it is $i(t) = 4\sin(10^6 t + 25^\circ)$ A?

12.18 The voltage across a 4-mH inductor is $v(t) = 60\cos(500t - 65^\circ)$ V. Find the instantaneous current through it.

12.19 A current source of $i(t) = 10\sin(377t + 30^\circ)$ A is applied to a single-element load. The resulting voltage across the element is $v(t) = -65\cos(377t + 120^\circ)$ V. What type of element is this? Calculate its value.

12.20 Two elements are connected in series as shown in Fig. 12.36. If $i = 12\cos(2t - 30^\circ)$ A, find the element values.

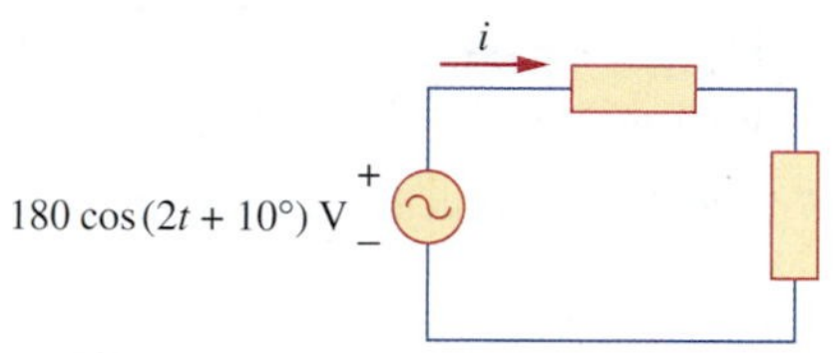

Figure 12.36
For Problem 12.20.

12.21 A series *RL* circuit is connected to a 110-V ac source. If the voltage across the resistor is 85 V, find the voltage across the inductor.

12.22 What value of ω will cause the response v_o in Fig. 12.37 to be zero?

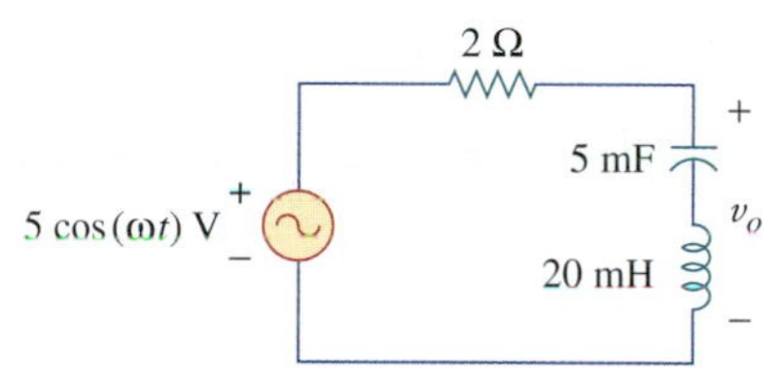

Figure 12.37
For Problem 12.22.

12.23 An applied voltage of $120\sin(1000t + 40^\circ)$ V is connected to a series combination of a 50-Ω resistor and a 40-μF capacitor. Find the sinusoidal expression for the current.

Section 12.4 Impedance and Admittance

12.24 If $v_s(t) = 5\cos(2t)$ V in the circuit of Fig. 12.38, find v_o.

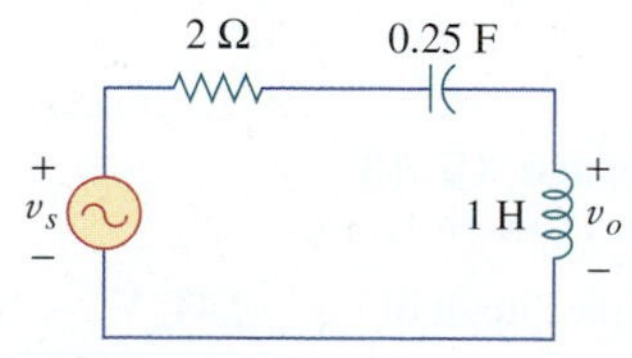

Figure 12.38
For Problems 12.24 and 12.53.

12.25 Find $i(t)$ and $v_o(t)$ in the circuit of Fig. 12.39.

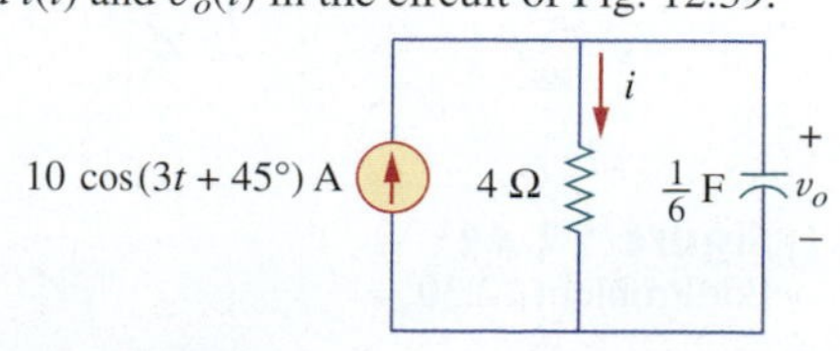

Figure 12.39
For Problem 12.25.

12.26 Calculate $i_1(t)$ and $i_2(t)$ in the circuit of Fig. 12.40 if the source frequency is 60 Hz.

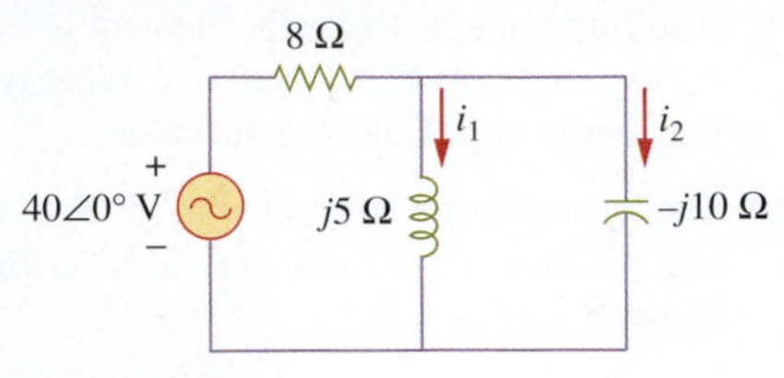

Figure 12.40
For Problem 12.26.

12.27 Determine $i_o(t)$ in the *RLC* circuit of Fig. 12.41.

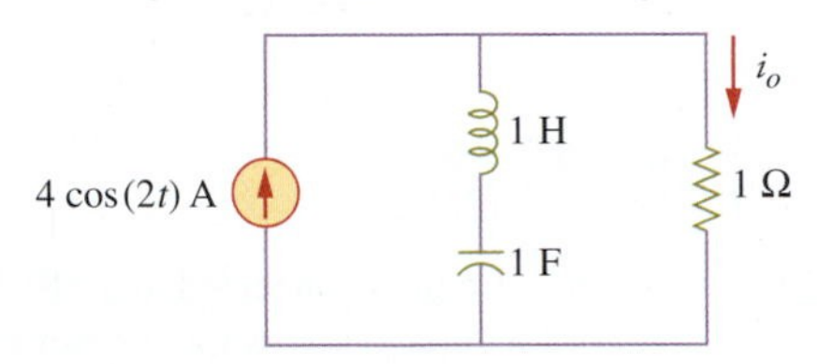

Figure 12.41
For Problem 12.27.

12.28 Determine $v_x(t)$ in the circuit of Fig. 12.42. Let $i_s(t) = 5\cos(100t + 40°)$ A.

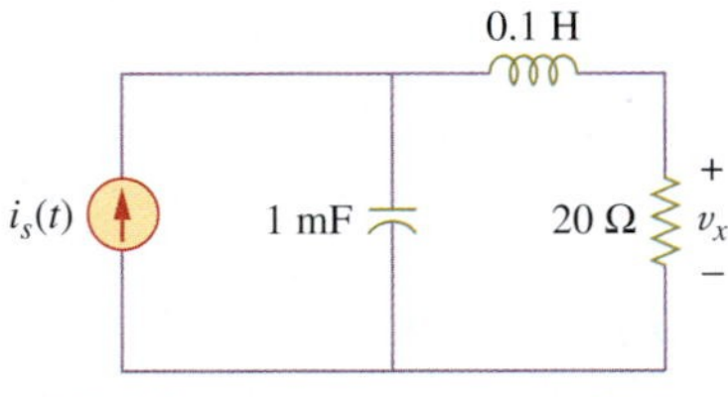

Figure 12.42
For Problem 12.28.

12.29 If the voltage $v_o(t)$ across the 2-Ω resistor in the circuit of Fig. 12.43 is $10\cos(2t)$ V, obtain i_s.

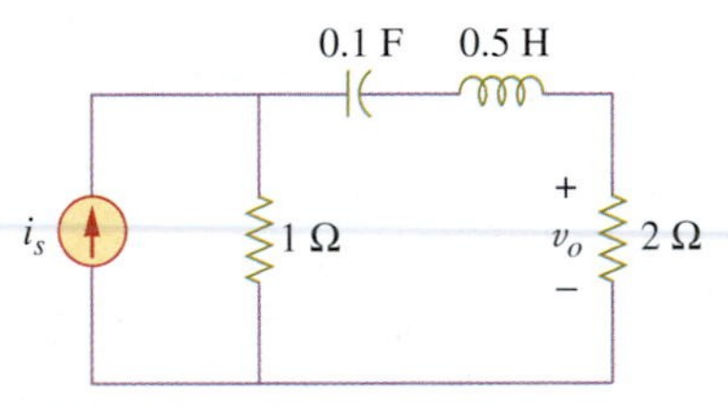

Figure 12.43
For Problem 12.29.

12.30 In the circuit of Fig. 12.44, $\mathbf{V}_s = 10\underline{/0°}$ V. Find *I*.

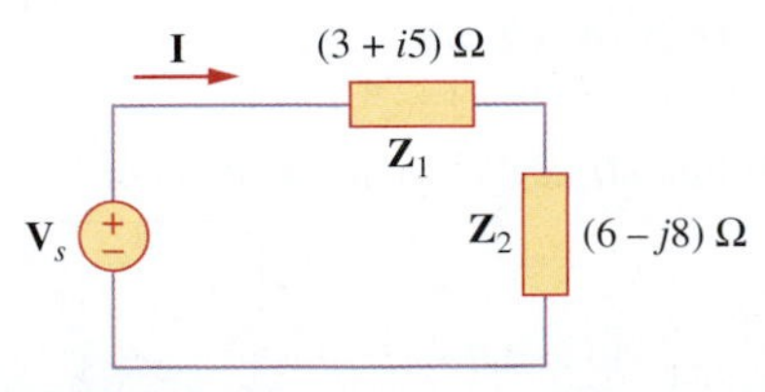

Figure 12.44
For Problem 12.30.

12.31 What is the peak current through a 10-μF capacitor at 10 MHz if the peak voltage across it is 0.42 mV?

12.32 Determine the peak current through a 80-mH inductor when the peak voltage is 12 mV. Take the operating frequency as 2 kHz.

Section 12.5 Impedance Combinations

12.33 Determine $\mathbf{Z}_{eq}$ for the circuit in Fig. 12.45.

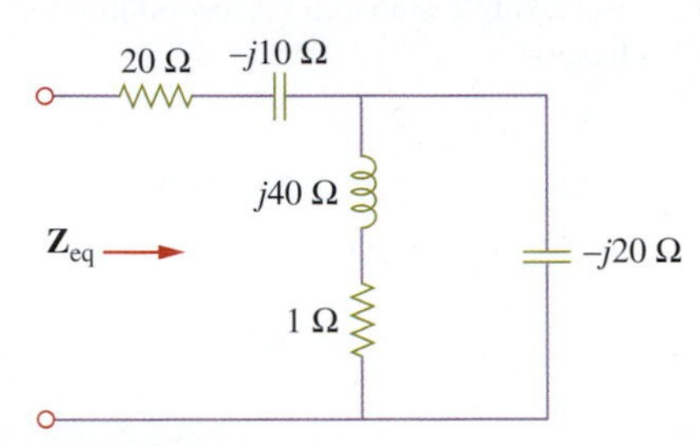

Figure 12.45
For Problem 12.33.

12.34 At $\omega = 50$ rad/s, determine $\mathbf{Z}_{in}$ for the circuit in Fig. 12.46.

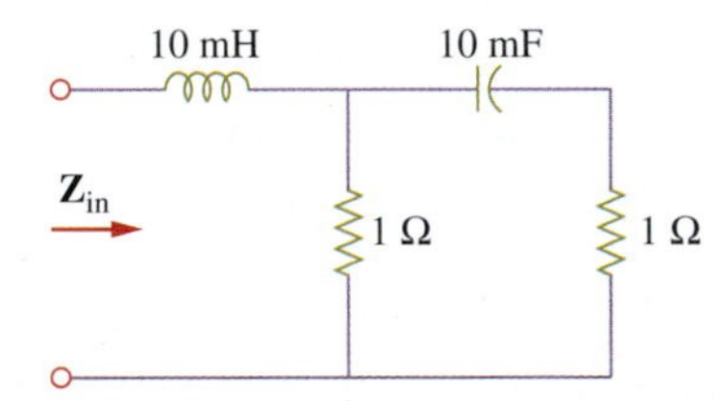

Figure 12.46
For Problem 12.34.

12.35 At $\omega = 1$ rad/s, obtain the input admittance in the circuit of Fig. 12.47.

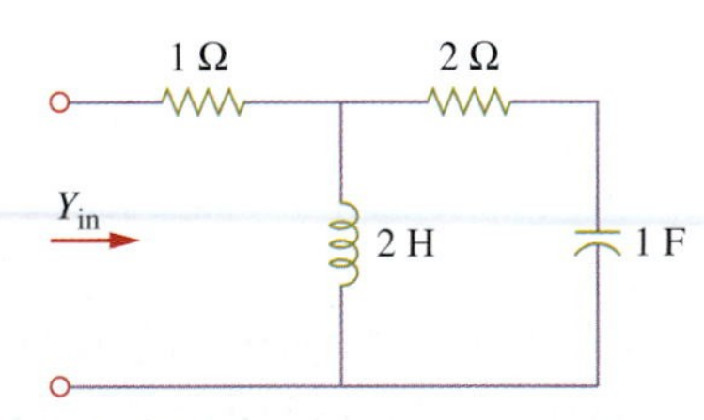

Figure 12.47
For Problem 12.35.

12.36 Calculate $\mathbf{Z}_{eq}$ for the circuit in Fig. 12.48.

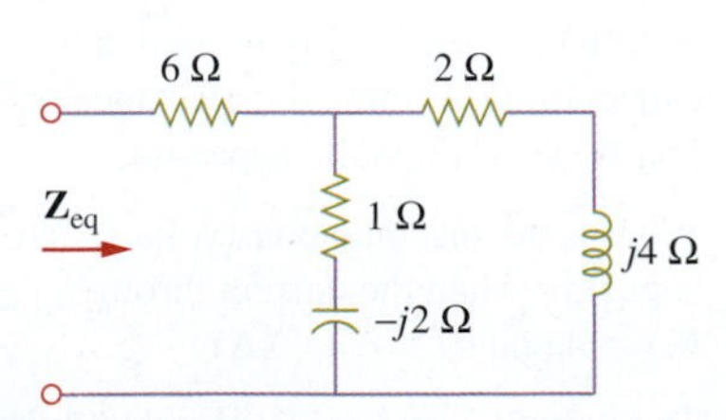

Figure 12.48
For Problem 12.36.

12.37 Find $\mathbf{Z}_{eq}$ in the circuit of Fig. 12.49.

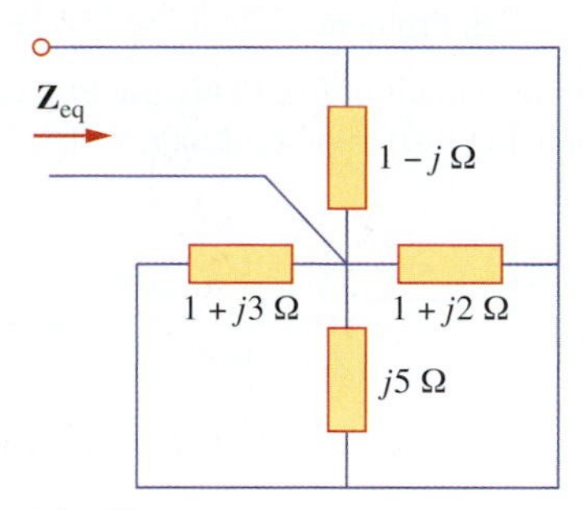

Figure 12.49
For Problem 12.37.

12.38 For the circuit in Fig. 12.50, find the value of $\mathbf{Z}_T$.

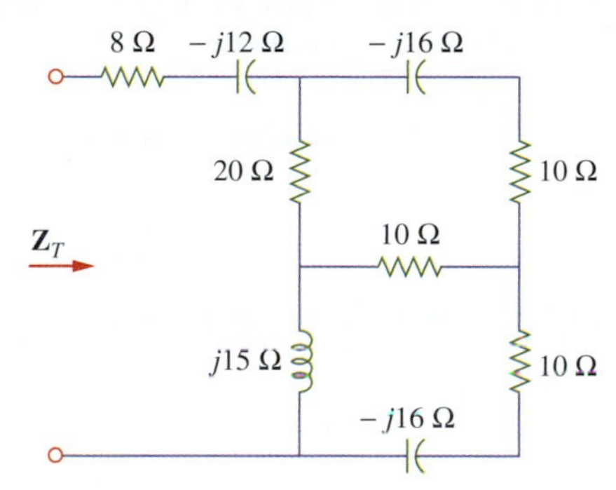

Figure 12.50
For Problem 12.38.

12.39 Find $\mathbf{Z}_T$ and $\mathbf{I}$ in the circuit of Fig. 12.51.

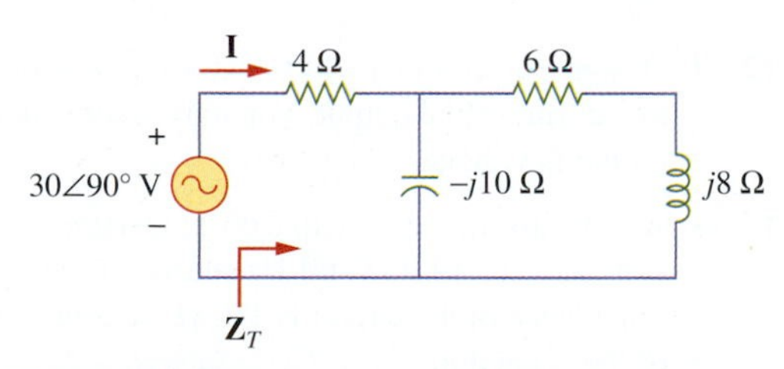

Figure 12.51
For Problem 12.39.

12.40 At $\omega = 10^3$ rad/s, find the input admittance of the circuit in Fig. 12.52.

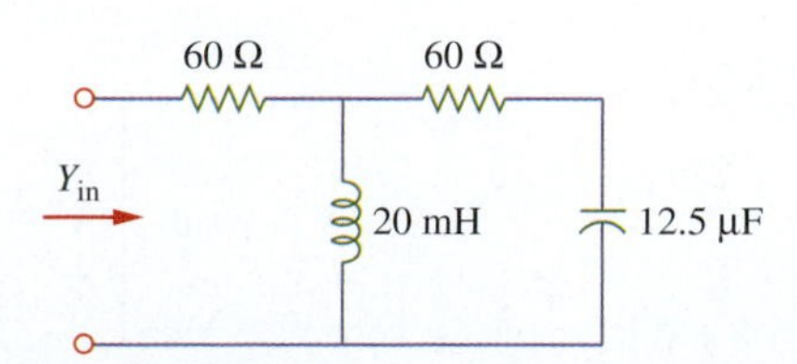

Figure 12.52
For Problem 12.40.

12.41 Determine $\mathbf{Y}_{eq}$ for the circuit in Fig. 12.53.

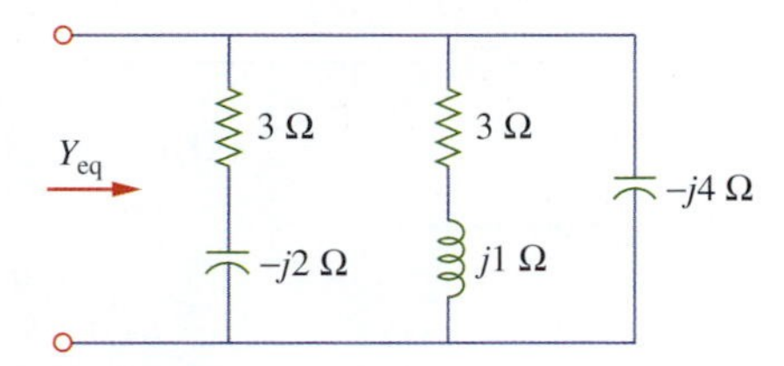

Figure 12.53
For Problem 12.41.

12.42 Find the equivalent impedance of the circuit in Fig. 12.54.

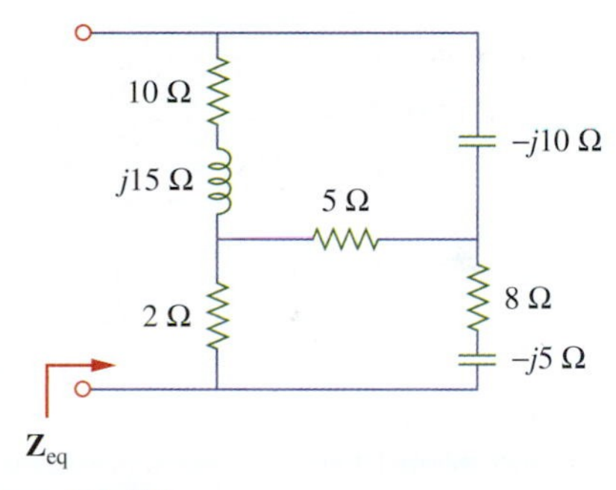

Figure 12.54
For Problem 12.42.

12.43 Calculate the value of $\mathbf{Z}_{ab}$ in the network of Fig. 12.55.

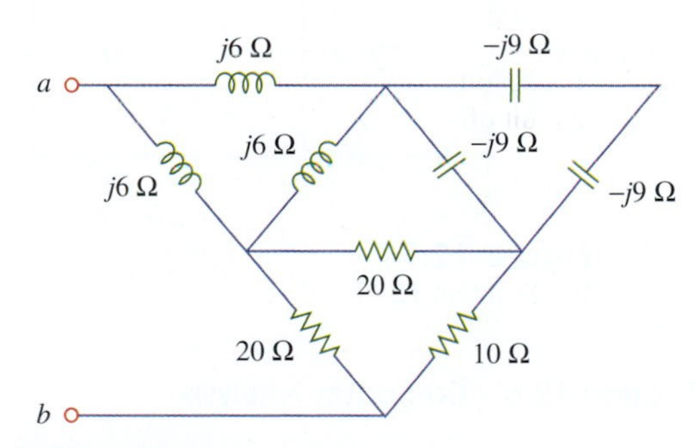

Figure 12.55
For Problem 12.43.

12.44 Use the voltage division rule to find $\mathbf{V}_{AB}$ in Fig. 12.56.

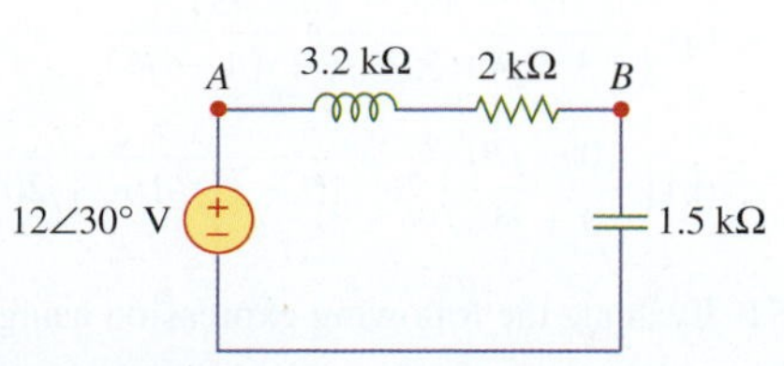

Figure 12.56
For Problem 12.44.

12.45 Determine $\mathbf{Z}_{ab}$ in the circuit in Fig. 12.57.

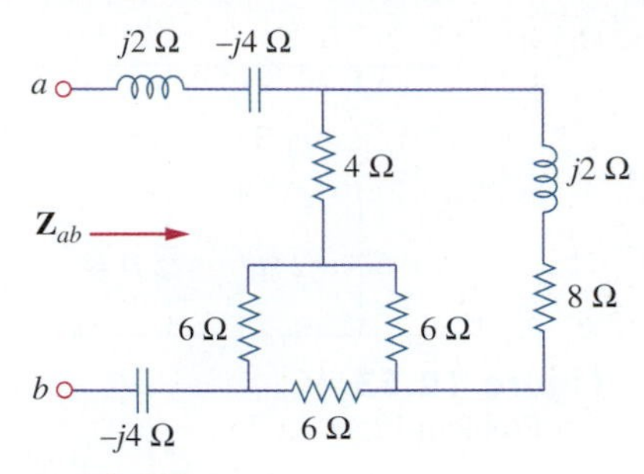

Figure 12.57
For Problem 12.45.

12.46 Find $\mathbf{Z}_{in}$ in the circuit in Fig. 12.58.

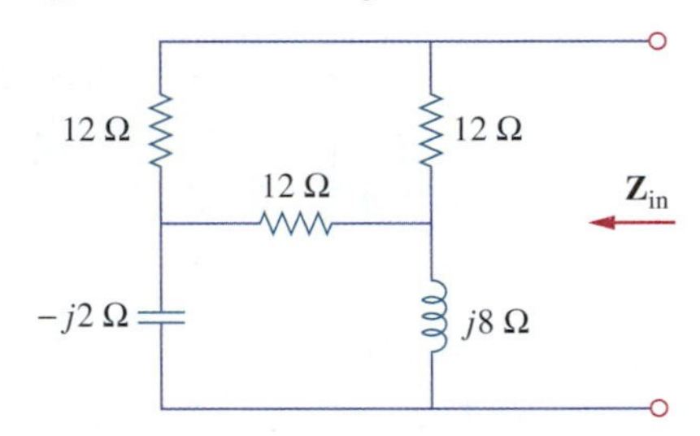

Figure 12.58
For Problem 12.46.

12.47 Refer to the circuit in Fig. 12.59. Find the input impedance Z_i and the input current I_i.

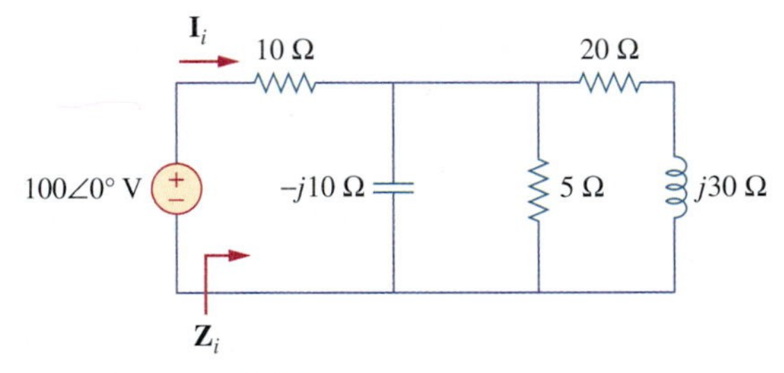

Figure 12.59
For Problem 12.47.

Section 12.6 Computer Analysis

12.48 Use MATLAB to plot $v(t) = 60 \sin(377t + 60°)$ V.

12.49 Plot $i(t) = 10 \cos(4\pi t - 25°)$ A using MATLAB.

12.50 Use MATLAB to evaluate the following expressions:

(a) $$\frac{(5 - j6) - (2 + j8)}{(-3 + j4)(5 - j) + (4 - j6)}$$

(b) $$\left(\frac{10 + j20}{3 + j4}\right)^2 \sqrt{(10 + j5)(16 - j20)}$$

12.51 Evaluate the following expression using MATLAB.

$$\frac{[(10 + j12) + (30 - j40)] \times (15 + j18)}{(50 - j60) - (35 + j80)}$$

12.52 Rework Problem 12.28 using PSpice.

12.53 Rework Problem 12.24 using PSpice.

12.54 In the circuit of Fig.12.60, use PSpice to determine $i(t)$. Let $v_s(t) = 60 \cos(200t - 10°)$ V.

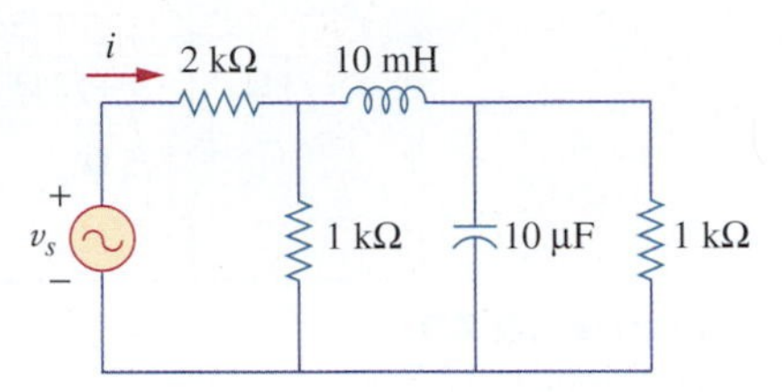

Figure 12.60
For Problem 12.54.

12.55 Find $i_x(t)$ using PSpice when $i_s(t) = 2 \sin(5t)$ A is supplied to the circuit of Fig. 12.61.

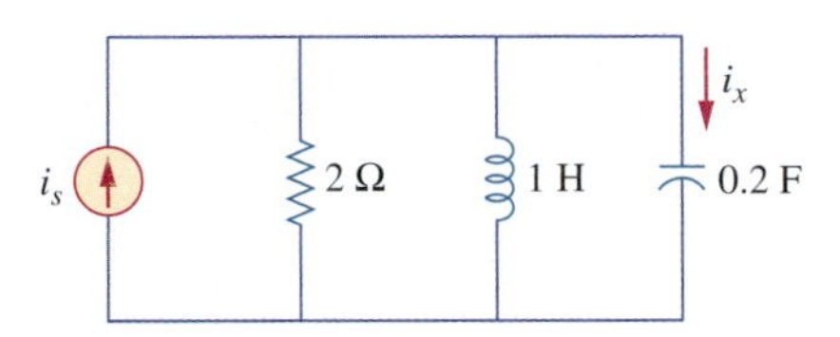

Figure 12.61
For Problem 12.55.

Section 12.7 Applications

12.56 Design an *RL* circuit to provide a 90° leading phase shift.

12.57 Design a circuit that will transform a sinusoidal input to a cosinusoidal output. You may assume that the output is voltage.

12.58 A capacitor in series with a 66-Ω resistor is connected to a 120-V, 60-Hz source. If the impedance of the circuit is 116 Ω, determine the size of the capacitor.

12.59 Refer to the *RC* circuit in Fig. 12.62.

(a) Calculate the phase shift at 2 MHz.

(b) Find the frequency at which the phase shift is 45°.

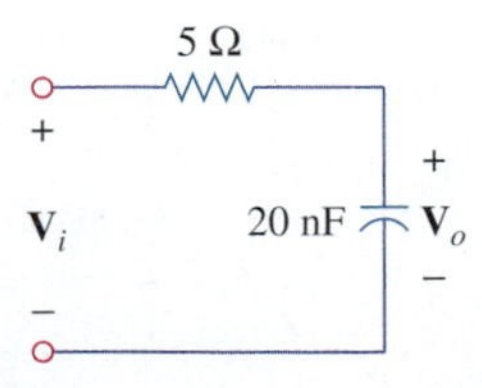

Figure 12.62
For Problem 12.59.

12.60 Consider the phase-shifting circuit in Fig. 12.63. Let $\mathbf{V}_i = 120$ V operating at 60 Hz. Find: (a) $\mathbf{V}_o$ when R is maximum, (b) $\mathbf{V}_o$ when R is minimum, and (c) the value of R that will produce a phase shift of 45°.

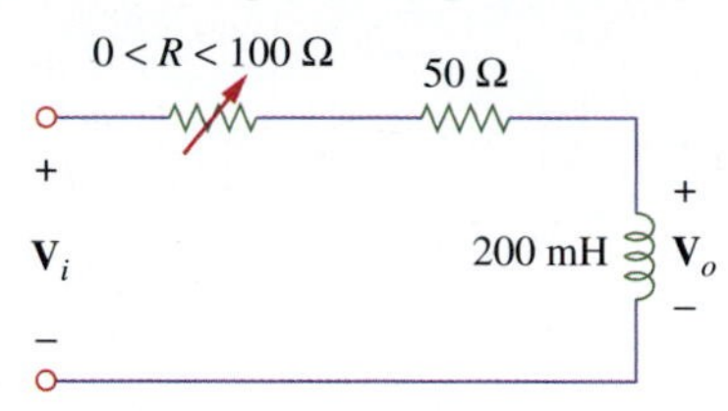

Figure 12.63
For Problem 12.60.

12.61 The ac bridge in Fig. 12.34 is balanced when $R_1 = 400\ \Omega$, $R_2 = 600\ \Omega$, $R_3 = 1.2\ \text{k}\Omega$, and $C_2 = 0.3\ \mu$F. Find R_x and C_x. Assume R_2 and C_2 are in series and that R_x and C_x are also in series.

12.62 A capacitance bridge balances when $R_1 = 100\ \Omega$, $R_2 = 2\ \text{k}\Omega$, and $C_s = 40\ \mu$F. What is C_x, the capacitance of the capacitor under test?

12.63 An inductive bridge balances when $R_1 = 1.2\ \text{k}\Omega$, $R_2 = 500\ \Omega$, and $L_s = 250$ mH. What is the value of L_x, the inductance of the inductor under test?

12.64 The circuit shown in Fig. 12.64 is used in a television receiver. What is the total impedance of this circuit?

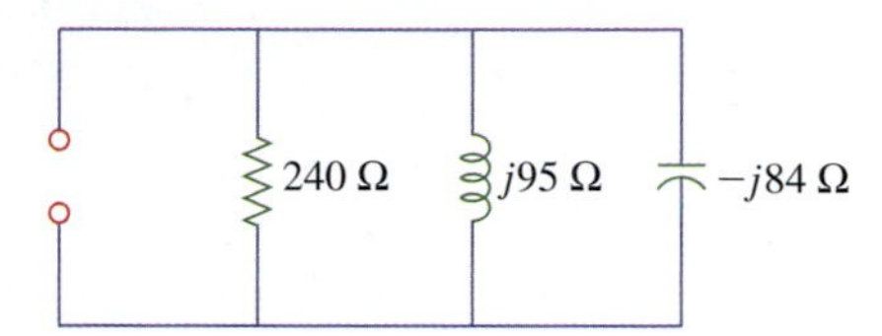

Figure 12.64
For Problem 12.64.

12.65 A transmission line has a series impedance of $\mathbf{Z} = 100\underline{/75^\circ}\ \Omega$ and a shunt admittance of $\mathbf{Y} = 450\underline{/48^\circ}\ \mu$S. Find: (a) the characteristic impedance $\mathbf{Z}_o = \sqrt{\mathbf{Z}/\mathbf{Y}}$ and (b) the propagation constant $\gamma = \sqrt{\mathbf{ZY}}$.

chapter

13

Sinusoidal Steady-State Analysis

A leader is a man who has the ability to get other people to do what they don't want to do, and like it.

—Henry S. Truman

Enhancing Your Career

Career in Software Engineering Technology

Software engineering technology is that aspect of learning that deals with the practical application of scientific knowledge in the design, construction, and validation of computer programs and the associated documentation required to develop, operate, and maintain them. It is a branch of electrical engineering technology that is becoming increasingly important as more and more disciplines require one form of software package or another to perform routine tasks and as programmable microelectronic systems are used in more and more applications.

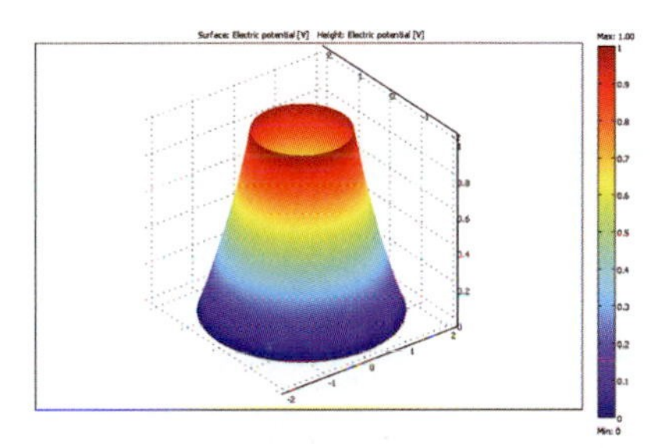

Output of a modeling software by COMSOL Inc. © Sarhan M. Musa

The role of a software engineering technologist should not be confused with that of a computer scientist: the software technologist is a practitioner, not a theoretician. A software engineering technologist should have good computer programming skills and be familiar with programming languages, in particular C^{++}, which is becoming increasingly popular. Because hardware and software are closely interlinked, it is essential that a software engineering technologist have a thorough understanding of hardware design and be able to troubleshoot circuits. Most important, the software engineering technologist should have some specialized knowledge of the area in which the software development skill is to be applied.

All in all, the field of software engineering technology offers a great career to those who enjoy programming and developing software packages. The higher rewards will go to those having the best preparation, with the most interesting and challenging opportunities going to those with a graduate education.

13.1 Introduction

From the preceding chapter, we have learned that the response of circuits to sinusoidal inputs can be obtained by using phasors. We also know that Ohm's and Kirchhoff's laws are applicable to ac circuits. In this chapter, we want to see how mesh analysis, nodal analysis, Thevenin's theorem, Norton's theorem, superposition, and source transformations are applied in analyzing ac circuits. Because these techniques were already introduced for dc circuits, our major effort here will be to illustrate with examples.

Analyzing ac circuits usually requires three steps:

Steps to Analyze ac Circuits

1. Transform the circuit to the phasor or frequency domain.
2. Solve the problem using circuit techniques such as the current divider rule, the voltage divider rule, nodal analysis, mesh analysis, and superposition.
3. Transform the resulting phasor to the time domain.

Step 1 is not necessary if the problem is specified in the frequency domain. In step 2, the analysis is performed in the same manner as dc circuit analysis except that complex numbers are involved. Having read chapter 12, we are adept at handling step 3.[1] Toward the end of the chapter, we learn how to apply PSpice to solve ac circuit problems.

13.2 Mesh Analysis

Kirchhoff's voltage law (KVL) forms the basis of mesh analysis. The validity of KVL for ac circuits was demonstrated in Chapter 12. As mentioned in Section 7.2, in the mesh analysis of an ac circuit with n meshes, we take the following three steps:

1. Assign mesh currents $\mathbf{I}_1, \mathbf{I}_2, \dots, \mathbf{I}_n$, in to the n meshes.
2. Apply KVL to each of the n meshes. Use Ohm's law to express the voltages in terms of the mesh currents.
3. Solve the resulting n simultaneous equations to get the mesh currents.

These steps will be illustrated in the following examples.

[1] Note: Frequency-domain analysis of an ac circuit via phasors is much easier than analysis of the circuit in the time domain.

Example 13.1

Determine the current $\mathbf{I}_o$ in the circuit of Fig. 13.1 using mesh analysis.

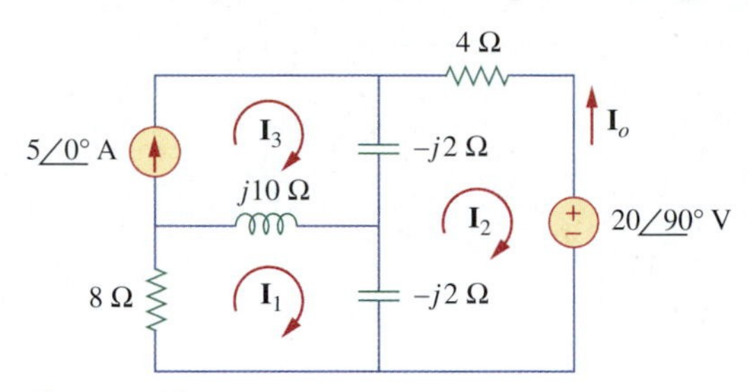

Figure 13.1
For Examples 13.1 and 13.6.

Solution:
Applying KVL to mesh 1, we obtain

$$(8 + j10 - j2)\mathbf{I}_1 - (-j2)\mathbf{I}_2 - j10\mathbf{I}_3 = 0 \qquad \textbf{(13.1.1)}$$

For mesh 2,

$$(4 - j2 - j2)\mathbf{I}_2 - (-j2)\mathbf{I}_1 - (-j2)\mathbf{I}_3 + 20\underline{/90^\circ} = 0 \qquad \textbf{(13.1.2)}$$

For mesh 3, $\mathbf{I}_3 = 5$. Substituting this in Eqs. (13.1.1) and (13.1.2), we get

$$(8 + j8)\mathbf{I}_1 + j2\mathbf{I}_2 = j50 \qquad \textbf{(13.1.3)}$$

$$j2\mathbf{I}_1 + (4 - j4)\mathbf{I}_2 = -j20 - j10 \qquad \textbf{(13.1.4)}$$

Equations (13.1.3) and (13.1.4) can be put in matrix form as

$$\begin{bmatrix} 8+j8 & j2 \\ j2 & 4-j4 \end{bmatrix}\begin{bmatrix} \mathbf{I}_1 \\ \mathbf{I}_2 \end{bmatrix} = \begin{bmatrix} j50 \\ -j30 \end{bmatrix}$$

from which we obtain the determinants

$$\Delta = \begin{vmatrix} 8+j8 & j2 \\ j2 & 4-j4 \end{vmatrix} = 32(1+j)(1-j) + 4 = 68$$

$$\Delta_2 = \begin{vmatrix} 8+j8 & j50 \\ j2 & -j30 \end{vmatrix} = 340 - j240 = 416.17\underline{/-35.22^\circ}$$

$$\mathbf{I}_2 = \frac{\Delta_2}{\Delta_1} = \frac{416.17\underline{/-35.22^\circ}}{68} = 6.12\underline{/-35.22^\circ} \text{ A}$$

The desired current is

$$\mathbf{I}_o = -\mathbf{I}_2 = 6.12\underline{/(-35.22^\circ + 180^\circ)} = 6.12\underline{/144.78^\circ} \text{ A}$$

Practice Problem 13.1

Find $\mathbf{I}_o$ in Fig. 13.2 using mesh analysis.

Answer: 3.582 $\underline{/65.45^\circ}$ A

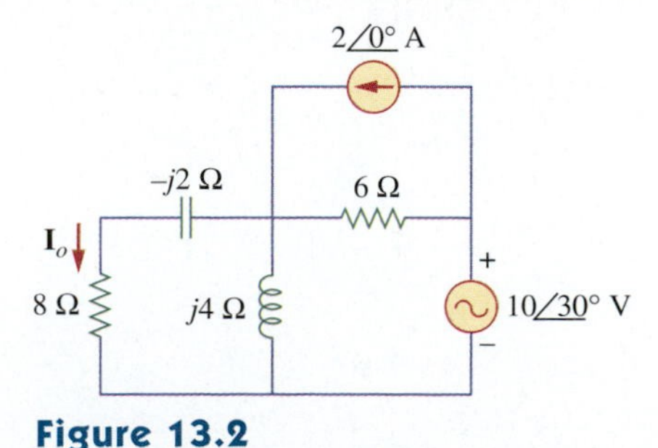

Figure 13.2
For Practice Problem 13.1.

Example 13.2

Solve for $\mathbf{V}_o$ in the circuit in Fig. 13.3 using mesh analysis.

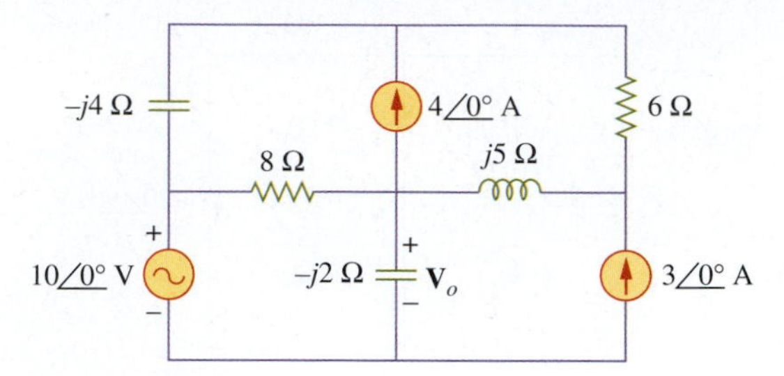

Figure 13.3
For Example 13.2.

Solution:
As shown in Fig. 13.4, meshes 3 and 4 form a supermesh due to the current source between the meshes. For mesh 1, KVL gives

$$-10 + (8 - j2)\mathbf{I}_1 - (-j2)\mathbf{I}_2 - 8\mathbf{I}_3 = 0$$

or

$$(8 - j2)\mathbf{I}_1 + j2\mathbf{I}_2 - 8\mathbf{I}_3 = 10 \quad \textbf{(13.2.1)}$$

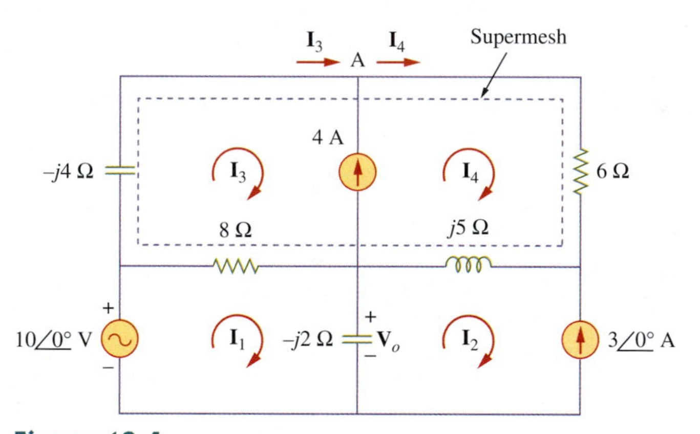

Figure 13.4
Analysis of the circuit in Fig. 13.3.

For mesh 2,

$$\mathbf{I}_2 = -3 \quad \textbf{(13.2.2)}$$

For the supermesh,

$$(8 - j4)\mathbf{I}_3 - 8\mathbf{I}_1 + (6 + j5)\mathbf{I}_4 - j5\mathbf{I}_2 = 0 \quad \textbf{(13.2.3)}$$

Consider the current source between meshes 3 and 4. At node A,

$$\mathbf{I}_4 = \mathbf{I}_3 + 4 \quad \textbf{(13.2.4)}$$

We can solve the preceding equations in two ways.

■ **METHOD 1** Rather than solving the preceding four equations, we reduce them to two by elimination. Combining Eqs. (13.2.1) and (13.2.2),

$$(8 - j2)\mathbf{I}_1 - 8\mathbf{I}_3 = 10 + j6 \quad \textbf{(13.2.5)}$$

Combining Eqs. (13.2.2) through (13.2.4),

$$-8\mathbf{I}_1 + (14 + j)\mathbf{I}_3 = -24 - j35 \qquad \textbf{(13.2.6)}$$

From Eqs. (13.2.5) and (13.2.6), we obtain the matrix equation

$$\begin{bmatrix} 8 - j2 & -8 \\ -8 & 14 + j \end{bmatrix} \begin{bmatrix} \mathbf{I}_1 \\ \mathbf{I}_3 \end{bmatrix} = \begin{bmatrix} 10 + j6 \\ -24 - j35 \end{bmatrix}$$

We obtain the following determinants

$$\Delta = \begin{vmatrix} 8 - j2 & -8 \\ -8 & 14 + j \end{vmatrix} = 112 + j8 - j28 + 2 - 64 = 50 - j20$$

$$\Delta_1 = \begin{vmatrix} 10 + j6 & -8 \\ -24 - j35 & 14 + j \end{vmatrix} = 140 + j10 + j84 - 6 - 192 - j280$$

$$= -58 - j186$$

Current $\mathbf{I}_1$ is obtained as

$$\mathbf{I}_1 = \frac{\Delta_1}{\Delta} = \frac{-58 - j186}{50 - j20} = 3.618\underline{/274.5^\circ}\text{ A}$$

The required voltage $\mathbf{V}_o$ is

$$\mathbf{V}_o = -j2(\mathbf{I}_1 - \mathbf{I}_2) = -j2(3.618\underline{/274.5^\circ} + 3)$$
$$= -7.2134 - j6.568$$
$$= 9.756\underline{/222.32^\circ}\text{ V}$$

METHOD 2 We can use MATLAB to solve Eqs. (13.2.1) to (13.2.4). We first cast the equations as

$$\begin{bmatrix} 8 - j2 & j2 & -8 & 0 \\ 0 & 1 & 0 & 0 \\ -8 & -j5 & 8 - j4 & 6 + j5 \\ 0 & 0 & -1 & 1 \end{bmatrix} \begin{bmatrix} \mathbf{I}_1 \\ \mathbf{I}_2 \\ \mathbf{I}_3 \\ \mathbf{I}_4 \end{bmatrix} = \begin{bmatrix} 10 \\ -3 \\ 0 \\ 4 \end{bmatrix} \qquad \textbf{(13.2.7)}$$

or

$$\mathbf{AI} = \mathbf{B}$$

By inverting $\mathbf{A}$, we can obtain $\mathbf{I}$ as

$$\mathbf{I} = \mathbf{A}^{-1}\mathbf{B}$$

We now apply MATLAB as follows:

```
>> A = [(8-j*2) j*2   -8       0;
        0       1     0        0;
        -8      -j*5  (8-j*4)  (6+j*5);
        0       0     -1       1];
>> B = [10 -3 0 4]';
>> I = inv(A)*B

I =
  0.2828 - 3.6069i
  -3.0000
  -1.8690 - 4.4276i
   2.1310 - 4.4276i
>> Vo = -2*j*(I(1) - I(2))

Vo =
  -7.2138 - 6.5655i
```

as obtained previously.

■ **METHOD 3** Using the TI-89 Titanium calculator, we can solve Eq. (13.2.7.) as follows. Because we want the final result to be in rectangular form, we first need to set the mode.

We press MODE

And change the complex format to RECTANGULAR and press ENTER

We press 2nd MATH

Select 4: Matrix and press ENTER

Select 5: simult(ENTER

Type the following:

```
simult([8-2*i,2*i,-8,0;0,1,0,0;-8,-5*i,8-4*i,
6+5*i;0,0,-1,1],[10;-3;0;4])
```

and press ◆ ENTER

The result:

$$\mathbf{I}_1 = 0.2828 - j3.607, \qquad \mathbf{I}_2 = -3 + j0,$$
$$\mathbf{I}_3 = -1.869 - 4.427, \qquad 2.131 - j4.428$$

which is essentially what we had before under method 2.

Practice Problem 13.2

Calculate current $\mathbf{I}_o$ in the circuit of Fig. 13.5.

Answer: $0.197\underline{/-5.84^\circ}$ A

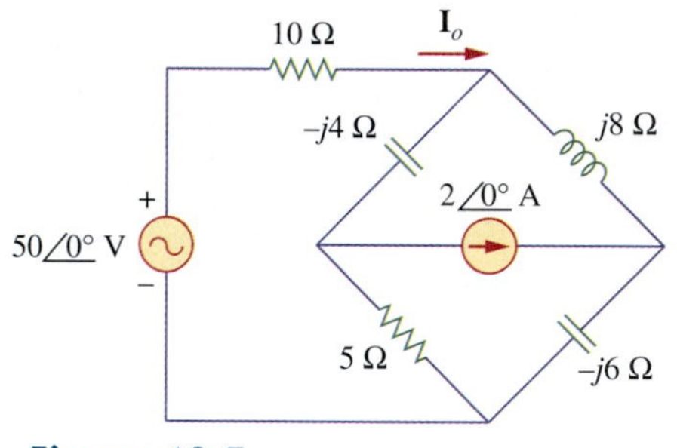

Figure 13.5
For Practice Problem 13.2.

13.3 Nodal Analysis

The basis of nodal analysis is Kirchhoff's current law (KCL). Because KCL is valid for phasors, as was demonstrated in Chapter 12, we can analyze ac circuits by nodal analysis. As mentioned in Section 7.4, the nodal analysis of an ac circuit involves taking the following three steps:

1. Select a node as the reference node (or ground). Assign voltages $\mathbf{V}_1, \mathbf{V}_2, \ldots, \mathbf{V}_{n-1}$, to the remaining $n - 1$ nodes. The voltages are referenced with respect to the reference node.
2. Apply KCL to each of the $(n - 1)$ nonreference nodes. Use Ohm's law to express the branch currents in terms of node voltages.
3. Solve the resulting simultaneous equations to obtain the unknown node voltages.

We will illustrate these steps with the following examples.

Example 13.3

Find i_x in the circuit of Fig. 13.6 using nodal analysis.

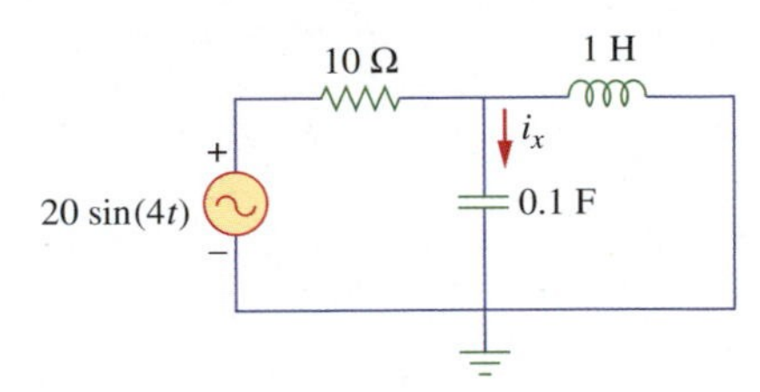

Figure 13.6
For Example 13.3.

Solution:
We first convert the circuit to the frequency domain.

$$20\sin(4t) \quad \Rightarrow \quad 20\angle 0^\circ, \qquad \omega = 4 \text{ rad/s}$$

$$1\text{ H} \quad \Rightarrow \quad j\omega L = j4\ \Omega$$

$$0.1\text{ F} \quad \Rightarrow \quad \frac{1}{j\omega C} = -j2.5\ \Omega$$

Thus, the frequency-domain equivalent circuit is as shown in Fig. 13.7.

Applying KCL at the top node,

$$\frac{20 - \mathbf{V}}{10} = \frac{\mathbf{V}}{-j2.5} + \frac{\mathbf{V}}{j4}$$

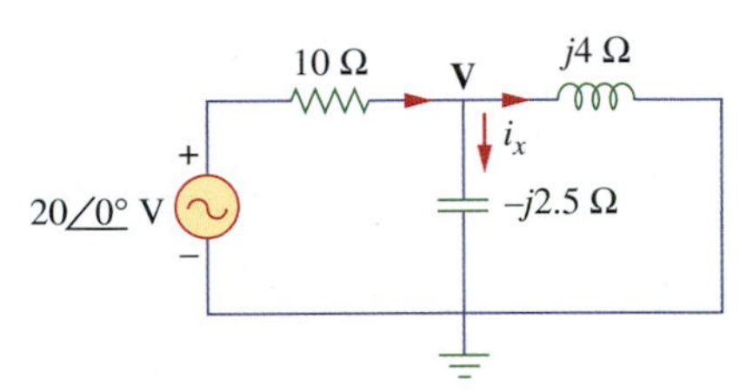

Figure 13.7
Frequency-domain equivalent of the circuit of Figure 13.6.

Multiplying through by 10,

$$20 - \mathbf{V} = j4\,\mathbf{V} - j2.5\,\mathbf{V} \qquad \text{or} \qquad 20 = \mathbf{V}(1 + j1.5)$$

Thus,

$$\mathbf{V} = \frac{20}{1 + j1.5}$$

But

$$\mathbf{I}_x = \frac{\mathbf{V}}{-j2.5} = \frac{20}{-j2.5(1 + j1.5)} = 3.6923 + j2.4615$$
$$= 4.438\angle 33.69^\circ \text{ A}$$

Transforming this to the time domain

$$i_x(t) = 4.438\sin(4t + 33.69^\circ) \text{ A}$$

Practice Problem 13.3

Using nodal analysis, find v_x in the circuit of Fig. 13.8.

Answer: $12\sin(2t + 53.13^\circ)$ V

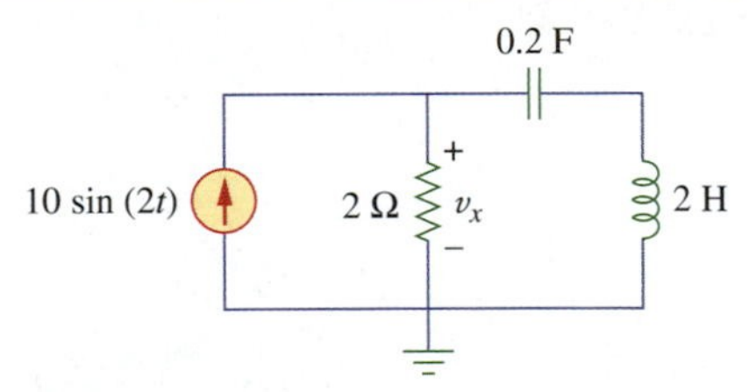

Figure 13.8
For Practice Problem 13.3.

Example 13.4

Compute $\mathbf{V}_1$ and $\mathbf{V}_2$ in the circuit of Fig. 13.9.

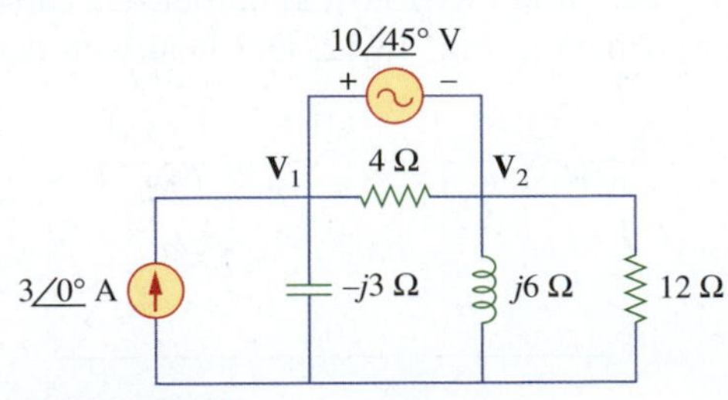

Figure 13.9
For Example 13.4.

Solution:

Nodes 1 and 2 form a supernode as shown in Fig. 13.10. Applying KCL at the supernode gives

$$3 = \frac{\mathbf{V}_1}{-j3} + \frac{\mathbf{V}_2}{j6} + \frac{\mathbf{V}_2}{12}$$

or

$$36 = j4\mathbf{V}_1 + (1 - j2)\mathbf{V}_2 \tag{13.4.1}$$

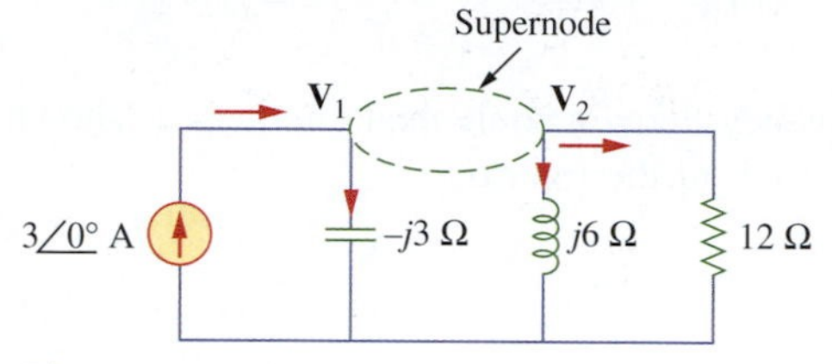

Figure 13.10
A supernode in the circuit of Fig. 13.9.

But a voltage source is connected between nodes 1 and 2 so that

$$\mathbf{V}_1 = \mathbf{V}_2 + 10\angle 45^\circ \tag{13.4.2}$$

Substituting Eq. (13.4.2) in Eq. (13.4.1) results in

$$36 - 40\angle 135^\circ = (1 + j2)\mathbf{V}_2 \quad \Rightarrow \quad \mathbf{V}_2 = 31.41\angle -87.18^\circ \text{ V}$$

From Eq. (13.4.2),

$$\mathbf{V}_1 = \mathbf{V}_2 + 10\angle 45^\circ = 25.78\angle -70.48^\circ \text{ V}$$

Practice Problem 13.4

Calculate $\mathbf{V}_1$ and $\mathbf{V}_2$ in the circuit shown in Fig. 13.11.

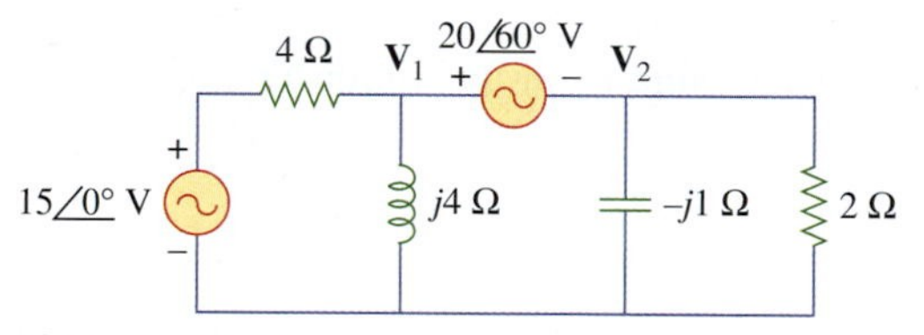

Figure 13.11
For Practice Problem 13.4.

Answer: $\mathbf{V}_1 = 17.81\angle 67.8^\circ$ V; $\mathbf{V}_2 = 3.376\angle 165.72^\circ$ V

Example 13.5

Let us modify the circuit in Fig. 13.6 and introduce a current-controlled current source as shown in Fig. 13.12. Our goal is to determine i_x.

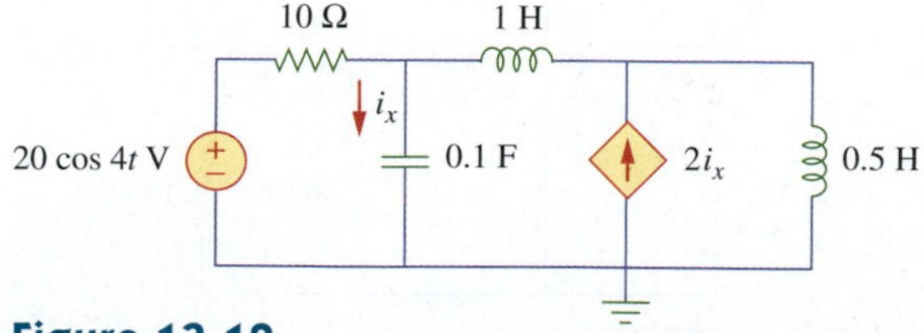

Figure 13.12
For Example 13.5.

Solution:

As usual, we first convert the circuit to the frequency domain:

$$20\cos(4t) \quad \Rightarrow \quad 20\angle 0^\circ, \quad \omega = 4 \text{ rad/s}$$

$$1\text{ H} \quad \Rightarrow \quad j\omega L = j4\ \Omega$$

$$0.5\text{ H} \quad \Rightarrow \quad j\omega L = j2\ \Omega$$

$$0.1\text{ F} \quad \Rightarrow \quad \frac{1}{j\omega C} = -j2.5\ \Omega$$

The frequency-domain equivalent circuit is as shown in Fig. 13.13.

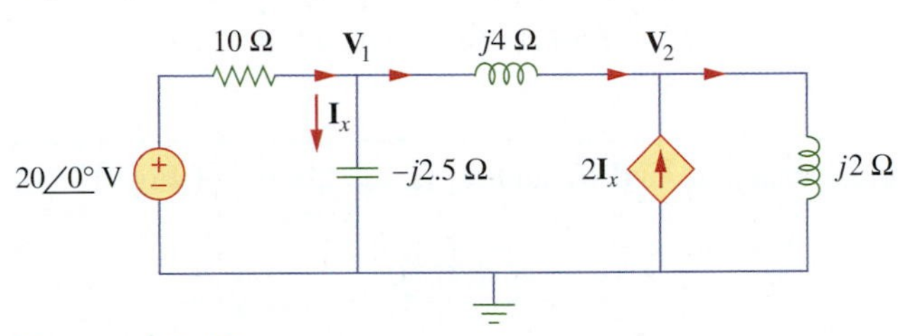

Figure 13.13
Frequency-domain equivalent of the circuit in Fig. 13.12.

Applying KCL at node 1,

$$\frac{20 - \mathbf{V}_1}{10} = \frac{\mathbf{V}_1}{-j2.5} + \frac{\mathbf{V}_1 - \mathbf{V}_2}{j4}$$

or

$$(1 + j1.5)\mathbf{V}_1 + j2.5\mathbf{V}_2 = 20 \qquad \textbf{(13.5.1)}$$

At node 2,

$$2\mathbf{I}_x + \frac{\mathbf{V}_1 - \mathbf{V}_2}{j4} = \frac{\mathbf{V}_2}{j2}$$

But

$$\mathbf{I}_x = \frac{\mathbf{V}_1}{-j2.5}$$

Substituting this gives

$$\frac{2\mathbf{V}_1}{-j2.5} + \frac{\mathbf{V}_1 - \mathbf{V}_2}{j4} = \frac{\mathbf{V}_2}{j2}$$

By simplifying, we get

$$11\mathbf{V}_1 + 15\mathbf{V}_2 = 0 \qquad \textbf{(13.5.2)}$$

Equations (13.5.1) and (13.5.2) can be put in matrix form as

$$\begin{bmatrix} 1 + j1.5 & j2.5 \\ 11 & 15 \end{bmatrix} \begin{bmatrix} \mathbf{V}_1 \\ \mathbf{V}_2 \end{bmatrix} = \begin{bmatrix} 20 \\ 0 \end{bmatrix}$$

We obtain the determinants as

$$\Delta = \begin{vmatrix} 1 + j1.5 & j2.5 \\ 11 & 15 \end{vmatrix} = 15 - j5$$

$$\Delta_1 = \begin{vmatrix} 20 & j2.5 \\ 0 & 15 \end{vmatrix} = 300, \quad \Delta_2 = \begin{vmatrix} 1 + j1.5 & 20 \\ 11 & 0 \end{vmatrix} = -220$$

$$\mathbf{V}_1 = \frac{\Delta_1}{\Delta} = \frac{300}{15 - j5} = 18.97\underline{/18.43^\circ} \text{ V}$$

$$\mathbf{V}_2 = \frac{\Delta_2}{\Delta} = \frac{-220}{15 - j5} = 13.91\underline{/198.3^\circ} \text{ V}$$

The current $\mathbf{I}_x$ is given by

$$\mathbf{I}_x = \frac{\mathbf{V}_1}{-j2.5} = \frac{18.97\underline{/18.43^\circ}}{2.5\underline{/-90^\circ}} = 7.59\underline{/108.4^\circ} \text{ A}$$

Transforming this to the time domain,

$$i_x = 7.50\cos(4t + 108.4^\circ) \text{ A}$$

Practice Problem 13.5

Using nodal analysis, find v_1 and v_2 in the circuit of Fig. 13.14.

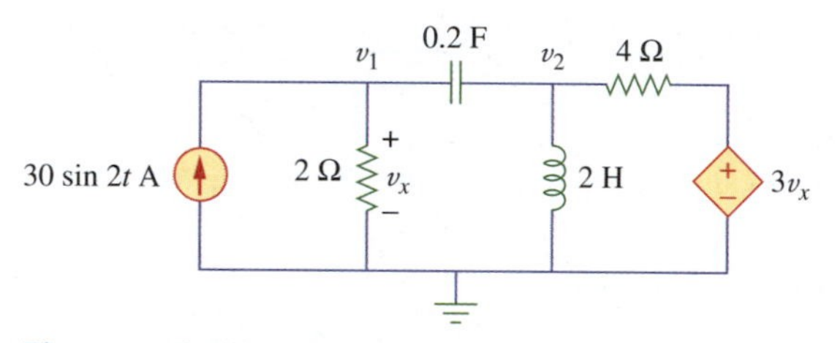

Figure 13.14
For Practice Problem 13.5.

Answer: $v_1(t) = 33.96\sin(2t + 60.01^\circ)$ V;
$v_2(t) = 99.6\sin(2t + 57.12^\circ)$ V

13.4 Superposition Theorem

Because ac circuits are linear, the superposition theorem applies to ac circuits the same way it applies to dc circuits. The theorem becomes important if the circuit has sources operating at *different* frequencies. In this case, because the impedances depend on frequency, we must have a different frequency-domain circuit for each frequency. The total response must be obtained by adding the individual responses in the *time* domain. It is incorrect to try to add the responses in the phasor or frequency domain. Why? Because the exponential time factor $e^{j\omega t}$ is implicit in sinusoidal analysis and that factor would change for each ω. Therefore, it would not make sense to add responses at different frequencies in the phasor domain. Thus, when a circuit has sources operating at different frequencies, the responses due to the individual frequencies must be added in the time domain.

Example 13.6

Use the superposition theorem to find $\mathbf{I}_o$ in the circuit in Fig. 13.1.

Solution:
Let

$$\mathbf{I}_o = \mathbf{I}'_o + \mathbf{I}''_o \tag{13.6.1}$$

where $\mathbf{I}_o'$ and $\mathbf{I}_o''$ are due to the voltage and current sources, respectively. To find $\mathbf{I}_o'$, consider the circuit in Fig. 13.15(a). If we let $\mathbf{Z}$ be the parallel combination of $-j2$ and $8 + j10$, then

$$\mathbf{Z} = \frac{-j2(8+j10)}{-2j+8+j10} = 0.25 - j2.25$$

and current $\mathbf{I}_o'$ is

$$\mathbf{I}_o' = \frac{j20}{4 - j2 + \mathbf{Z}} = \frac{j20}{4.25 - j4.25}$$

or

$$\mathbf{I}_o' = -2.353 + j2.353 \tag{13.6.2}$$

To get $\mathbf{I}_o''$, consider the circuit in Fig. 13.15(b). For mesh 1,

$$(8 + j8)\mathbf{I}_1 - j10\mathbf{I}_3 + j2\mathbf{I}_2 = 0 \tag{13.6.3}$$

For mesh 2,

$$(4 - j4)\mathbf{I}_2 + j2\mathbf{I}_1 + j2\mathbf{I}_3 = 0 \tag{13.6.4}$$

For mesh 3,

$$\mathbf{I}_3 = 5 \tag{13.6.5}$$

From Eqs. (13.6.4) and (13.6.5),

$$(4 - j4)\mathbf{I}_2 + j2\mathbf{I}_1 + j10 = 0$$

Expressing $\mathbf{I}_1$ in terms of $\mathbf{I}_2$ gives

$$\mathbf{I}_1 = (2 + j2)\mathbf{I}_2 - 5 \tag{13.6.6}$$

Substituting Eqs. (13.6.5) and (13.6.6) into Eq. (13.6.3), we get

$$(8 + j8)[(2 + j2)\mathbf{I}_2 - 5] - j50 + j2\mathbf{I}_2 = 0$$

or

$$\mathbf{I}_2 = \frac{90 - j40}{34} = 2.647 - j1.176$$

Current $\mathbf{I}_o''$ is obtained as

$$\mathbf{I}_o'' = -\mathbf{I}_2 = -2.647 + j1.176 \tag{13.6.7}$$

From Eq. (13.6.2) and (13.6.7),

$$\mathbf{I}_o = \mathbf{I}_o' + \mathbf{I}_o'' = -5 + j3.529 = 6.12\underline{/144.78^\circ} \text{ A}$$

It should be noted that applying the superposition theorem is not the best way to solve this problem. It seems that we have made the problem twice as hard as the original one by using superposition. However, in Example 13.7, superposition is clearly the easiest approach.

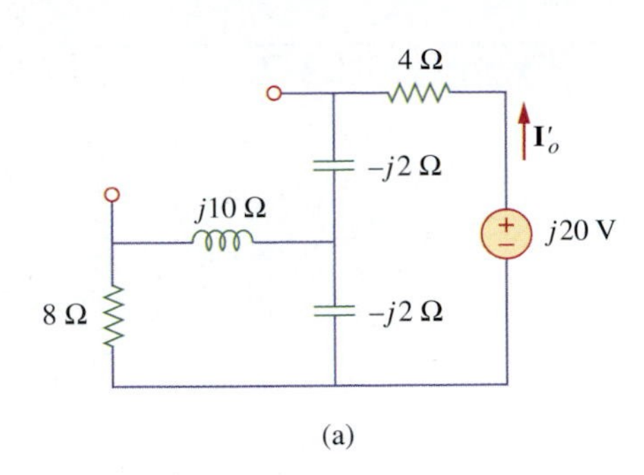

(a)

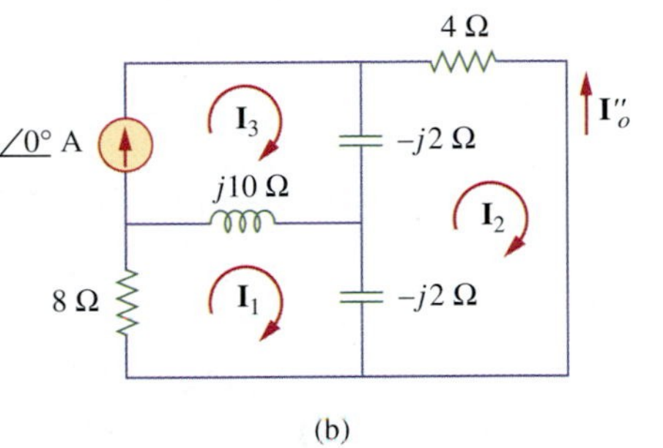

(b)

Figure 13.15
Solution of Example 13.6.

Practice Problem 13.6

Find current $\mathbf{I}_o$ in the circuit of Fig. 13.2 using the superposition theorem.

Answer: $3.582\underline{/65.45^\circ}$ A

Example 13.7

Find v_o in the circuit in Fig. 13.16 using the superposition theorem.

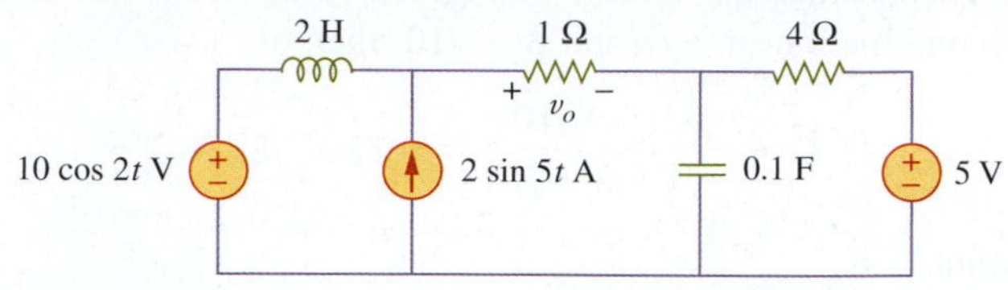

Figure 13.16
For Example 13.7.

Solution:

Because the circuit operates at three different frequencies ($\omega = 0$ for the dc voltage source), one way to obtain a solution is to use superposition, which breaks the problem into single-frequency problems. So we let

$$v_o = v_1 + v_2 + v_3 \tag{13.7.1}$$

where v_1 is due to the 5-V dc voltage source, v_2 is due to the 10 cos(2t) V voltage source, and v_3 is due to the 2 sin(5t) A current source.

To find v_1, we set to zero all sources except the 5-V dc source. We recall that at steady state, a capacitor is an open circuit to dc, while an inductor is a short circuit to dc. There is an alternative way of looking at this. Because $\omega = 0$, $j\omega L = 0$, $1/j\omega C = \infty$. Either way, the equivalent circuit is as shown in Fig. 13.17(a). By voltage division,

$$-v_1 = \frac{1}{1+4}(5) = 1 \text{ V} \tag{13.7.2}$$

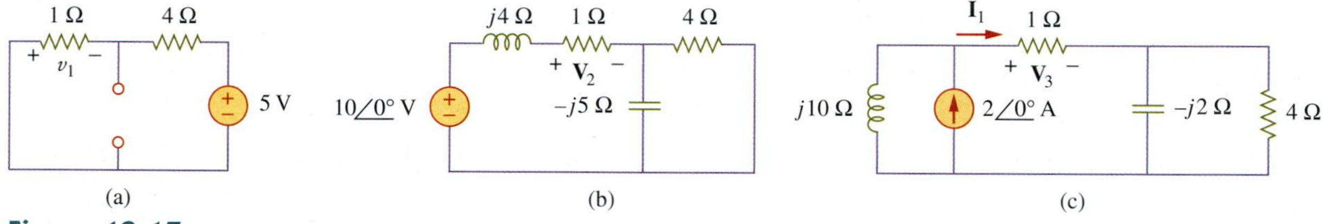

Figure 13.17
Solution of Example 13.7: (a) setting all sources to zero except the 5-V dc source; (b) setting all sources to zero except the ac voltage source; (c) setting all sources to zero except the ac current source.

To find v_2, we set to zero both the 5-V dc source and the 2 sin(5t) current source and transform the circuit to the frequency domain.

$$10\cos(2t) \quad \Rightarrow \quad 10\underline{/90^\circ}\text{ A}, \qquad \omega = 2 \text{ rad/s}$$

$$2 \text{ H} \quad \Rightarrow \quad j\omega L = j4\ \Omega$$

$$0.1 \text{ F} \quad \Rightarrow \quad \frac{1}{j\omega C} = -j5\ \Omega$$

The equivalent circuit is as shown in Fig. 13.17(b). Let

$$\mathbf{Z} = -j5 \parallel 4 = \frac{-j5 \times 4}{4 - j5} = 2.439 - j1.951$$

By voltage division

$$\mathbf{V}_2 = \frac{1}{1 + j4 + \mathbf{Z}}(10\underline{/90^\circ}) = \frac{10\underline{/90^\circ}}{3.439 + j2.049} = 2.498\underline{/59.21^\circ}$$

In the time domain,

$$v_2(t) = 2.498 \sin(2t + 59.21°) = 2.498 \cos(2t - 30.79°) \quad \textbf{(13.7.3)}$$

To obtain v_3, we set the voltage sources to zero and transform what is left to the frequency domain.

$$2 \sin(5t) \quad \Rightarrow \quad 2\underline{/0°} \text{ A}, \qquad \omega = 5 \text{ rad/s}$$

$$2 \text{ H} \quad \Rightarrow \quad j\omega L = j10\ \Omega$$

$$0.1 \text{ F} \quad \Rightarrow \quad \frac{1}{j\omega C} = -j2\ \Omega$$

The equivalent circuit is in Fig. 13.17(c). Let

$$\mathbf{Z}_1 = -j2 \parallel 4 = \frac{-j2 \times 4}{4 - j2} = 0.8 - j1.6\ \Omega$$

By current division,

$$\mathbf{I}_1 = \frac{j10}{j10 + 1 + \mathbf{Z}_1}(2\underline{/0°}) \text{ A}$$

$$\mathbf{V}_3 = \mathbf{I}_1 \times 1 = \frac{j10}{1.8 + j8.4}(2) = 2.328\underline{/12.09°} \text{ V}$$

In the time domain,

$$v_3(t) = 2.33 \sin(5t + 12.09°) \text{ V} \quad \textbf{(13.7.4)}$$

Substituting Eqs. (13.7.2) through (13.7.4) into Eq. (13.7.1), we have

$$v_o(t) = -1 + 2.498 \cos(2t - 30.79°) + 2.33 \sin(5t + 12.09°) \text{ V}$$

Practice Problem 13.7

Calculate v_o in the circuit of Fig 13.18 using the superposition theorem.

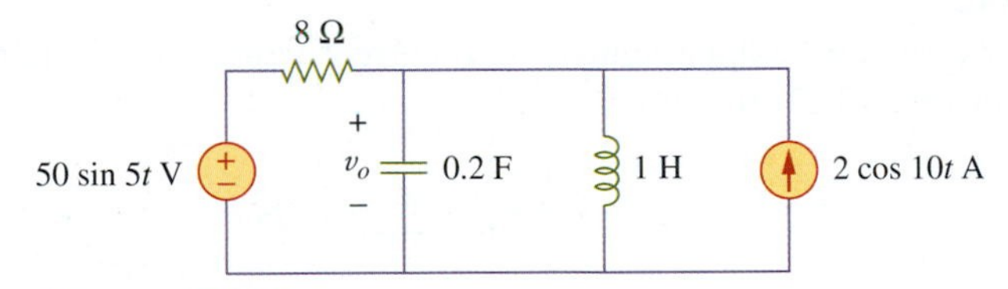

Figure 13.18
For Practice Problem 13.7.

Answer: $4.631 \sin(5t - 81.12°) + 0.42 \cos(10t - 86.24°)$ V

13.5 Source Transformation

As shown in Fig. 13.19, source transformation in the frequency domain involves transforming a voltage source in series with an impedance to a current source in parallel with an impedance or vice versa. As we go from one source type to another, we must keep the following relationship in mind:

$$\boxed{\mathbf{V}_s = \mathbf{Z}_s \mathbf{I}_s \quad \Leftrightarrow \quad \mathbf{I}_s = \mathbf{V}_s/\mathbf{Z}_s} \quad \textbf{(13.1)}$$

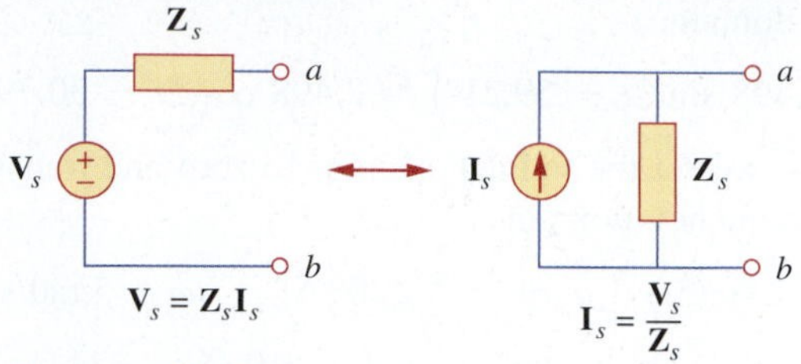

Figure 13.19
Source transformation.

Remember that the positive terminal of the voltage source must correspond with the arrowhead of the current source, as shown in Fig. 13.19. Source transformation applies to dependent sources as well (see Fig. 8.13).

Example 13.8

Calculate $\mathbf{V}_x$ in the circuit of Fig. 13.20 using the method of source transformation.

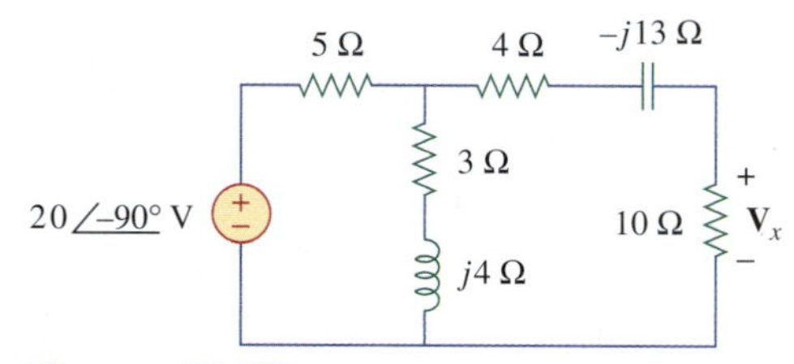

Figure 13.20
For Example 13.8, Fig. 13.19 source transformation.

Solution:
We transform the voltage source to a current source and obtain the circuit in Fig. 13.21(a), where

$$\mathbf{I}_s = \frac{20\angle -90^\circ}{5} = 4\angle -90^\circ = -j4 \text{ A}$$

The parallel combination of the 5-Ω resistance and $(3 + j4)$ Ω impedance gives

$$\mathbf{Z}_1 = \frac{5(3 + j4)}{8 + j4} = 2.5 + j1.25 \ \Omega$$

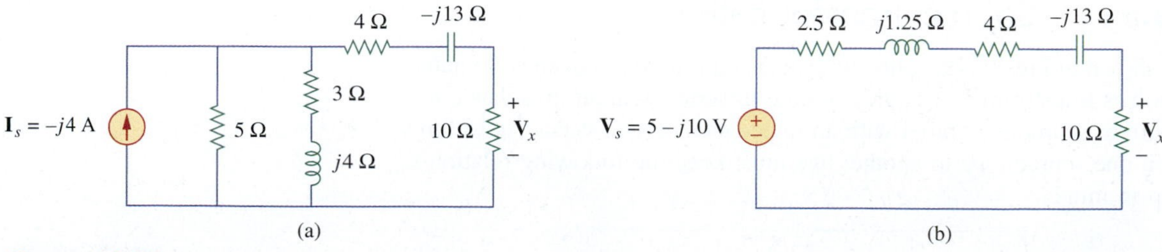

Figure 13.21
Solution of the circuit in Fig. 13.19.

Converting the current source to a voltage source yields the circuit in Fig. 13.21(b), where

$$\mathbf{V}_s = \mathbf{I}_s\mathbf{Z}_1 = -j4(2.5 + j1.25) = 5 - j10 \text{ V}$$

By voltage division,

$$\mathbf{V}_x = \frac{10}{10 + 2.5 + j1.25 + 4 - j13}(5 - j10) = 5.519\underline{/-28^\circ} \text{ V}$$

Practice Problem 13.8

Find $\mathbf{I}_o$ in the circuit of Fig. 13.22 using the concept of source transformation.

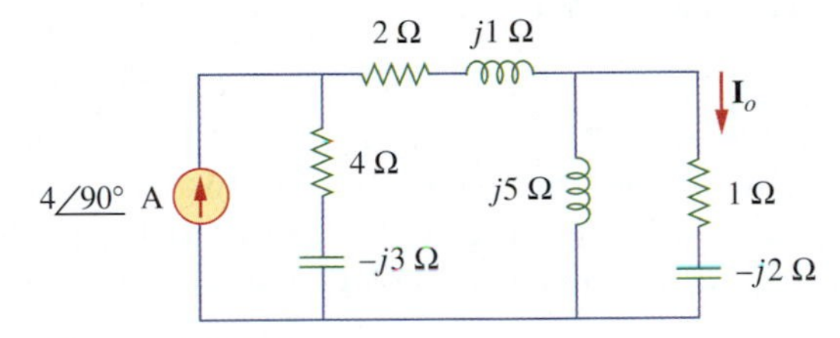

Figure 13.22
For Practice Problem 13.8.

Answer: $2.21\underline{/19.8^\circ}$ A

13.6 Thevenin and Norton Equivalent Circuits

Thevenin's and Norton's theorems are applied to ac circuits in the same way as they are to dc circuits. The only additional effort arises from the need to manipulate complex numbers. The frequency domain version of a Thevenin equivalent circuit is depicted in Fig. 13.23, where a linear circuit is replaced by a voltage source in series with an impedance. The Norton equivalent circuit is illustrated in Fig. 13.24, where a linear circuit is replaced by a current source in parallel with an impedance. Keep in mind that the two equivalent circuits are related as

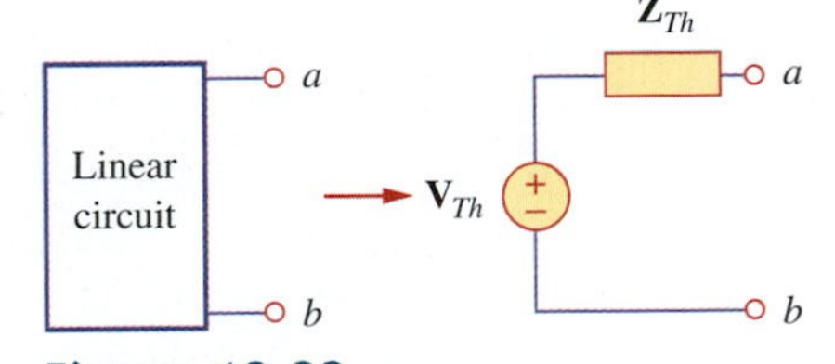

Figure 13.23
Thevenin equivalent.

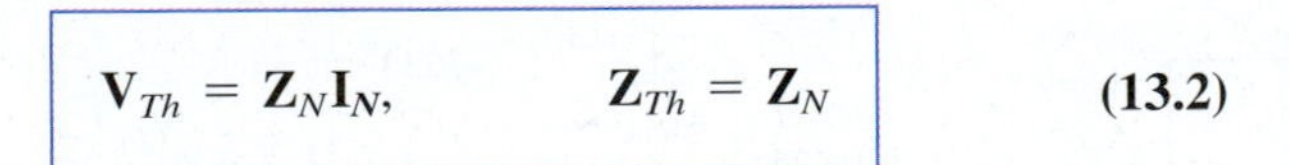

$$\mathbf{V}_{Th} = \mathbf{Z}_N\mathbf{I}_N, \qquad \mathbf{Z}_{Th} = \mathbf{Z}_N \tag{13.2}$$

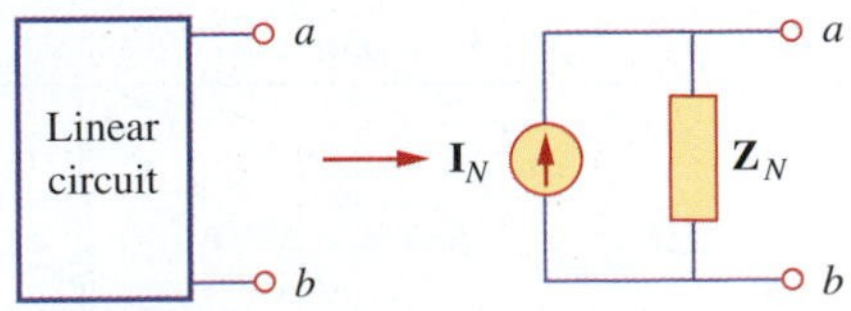

Figure 13.24
Norton equivalent.

just as in source transformation; $\mathbf{V}_{Th}$ is the open-circuit voltage, while $\mathbf{I}_N$ is the short-circuit current.

If the circuit has sources operating at different frequencies (e.g., see Example 13.7), Thevenin's or Norton's equivalent circuit must be determined at each frequency. This leads to entirely different equivalent circuits, one for each frequency, not one equivalent circuit with equivalent sources and equivalent impedances.

Example 13.9

Obtain the Thevenin equivalent at terminals *a-b* of the circuit in Fig. 13.25.

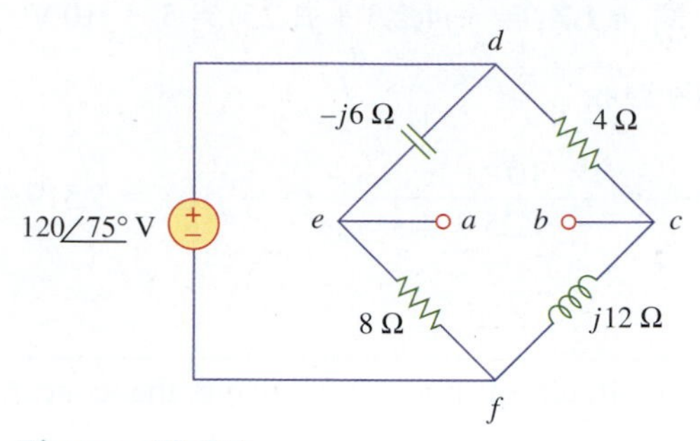

Figure 13.25
For Example 13.9.

Solution:

We find $\mathbf{Z}_{Th}$ by setting the voltage source to zero. As shown in Fig. 13.26(a), the 8-Ω resistance is now in parallel with the $-j6$-Ω reactance so that their combination gives

$$\mathbf{Z}_1 = -j6 \parallel 8 = \frac{-j6 \times 8}{8 - j6} = 2.88 - j3.84\ \Omega$$

Similarly, the 4-Ω resistance is in parallel with the $j12$ Ω reactance and their combination gives

$$\mathbf{Z}_2 = 4 \parallel j12 = \frac{j12 \times 4}{4 + j12} = 3.6 + j1.2\ \Omega$$

The Thevenin impedance is the series combination of $\mathbf{Z}_1$ and $\mathbf{Z}_2$; that is,

$$\mathbf{Z}_{Th} = \mathbf{Z}_1 + \mathbf{Z}_2 = 6.48 - j2.64\ \Omega$$

To find $\mathbf{V}_{Th}$, consider the circuit in Fig. 13.26(b). Currents $\mathbf{I}_1$ and $\mathbf{I}_2$ are obtained as

$$\mathbf{I}_1 = \frac{120\underline{/75^\circ}}{8 - j6}\ \text{A}, \qquad \mathbf{I}_2 = \frac{120\underline{/75^\circ}}{4 + j12}\ \text{A}$$

Applying KVL around loop *bcdeab* in Fig. 13.26(b) gives

$$\mathbf{V}_{Th} - 4\mathbf{I}_2 + (-j6)\mathbf{I}_1 = 0$$

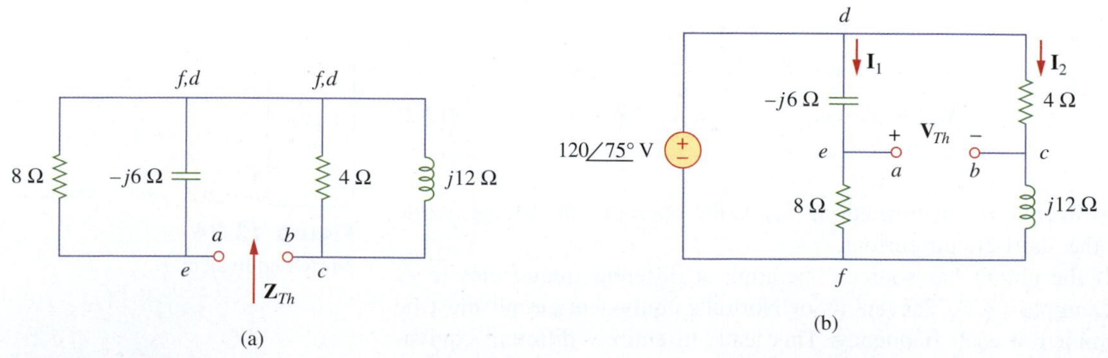

Figure 13.26
Solution of the circuit in Fig. 13.25: (a) finding $\mathbf{Z}_{Th}$; (b) finding $\mathbf{V}_{Th}$.

or

$$\mathbf{V}_{Th} = 4\mathbf{I}_2 + j6\mathbf{I}_1 = \frac{480\underline{/75^\circ}}{4 + j12} + \frac{720\underline{/75^\circ + 90^\circ}}{8 - j6}$$
$$= 37.95\underline{/3.43^\circ} + 72\underline{/201.87^\circ} = -28.936 - j24.55$$
$$= 37.9\underline{/220.31^\circ}\text{ V}$$

Practice Problem 13.9

Find the Thevenin equivalent at terminals *a-b* of the circuit in Fig. 13.27.

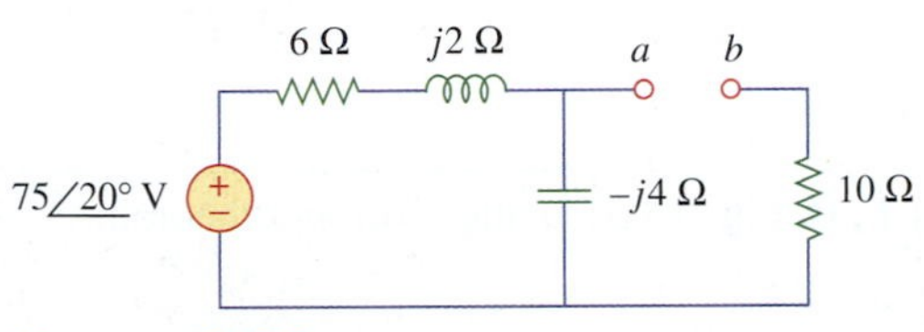

Figure 13.27
For Practice Problem 13.9.

Answer: $\mathbf{Z}_{Th} = 12.4 - j3.2\ \Omega$, $\mathbf{V}_{Th} = 47.42\underline{/-51.57^\circ}\text{ V}$

Example 13.10

Find the Thevenin equivalent of the circuit in Fig. 13.28 as seen from terminals *a-b*.

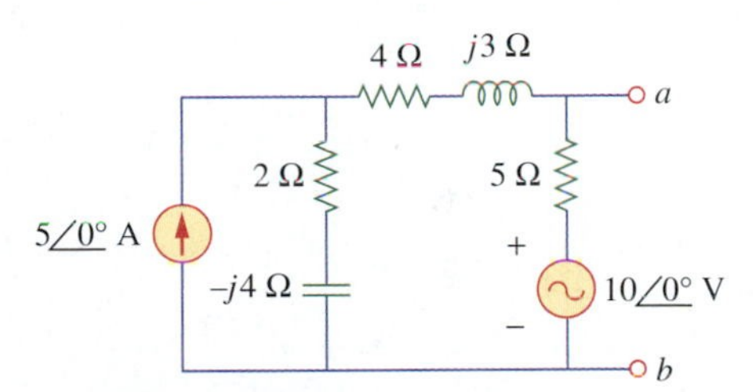

Figure 13.28
For Example 13.10.

Solution:
To find $\mathbf{Z}_{Th}$, we turn off the sources so that we obtain the circuit in Fig. 13.29(a).

$$\mathbf{Z}_{Th} = 5 \parallel (4 + j3 + 2 - j4) = \frac{5(6 - j)}{5 + 6 - j} = 2.746 - j0.205\ \Omega$$

To obtain $\mathbf{V}_{Th}$, we convert the current source to its equivalent voltage source as shown in Fig. 13.29(b). Applying KVL to the loop gives

$$-(10 - j20) + \mathbf{I}(2 - j4 + 4 + j3 + 5) + 10 = 0 \quad \Rightarrow \quad \mathbf{I} = \frac{-j20}{11 - j}$$

But

$$\mathbf{V}_{Th} = 5\mathbf{I} + 10 = \frac{-j100}{11 - j} + 10 = 10.82 - j9.016 = 14.08\underline{/-39.81^\circ}$$

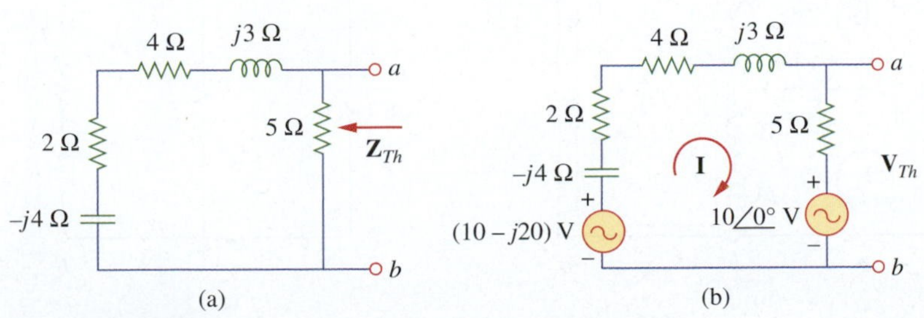

Figure 13.29
Solution of the problem in Fig. 13.28: (a) finding $\mathbf{Z}_{Th}$; (b) finding $\mathbf{V}_{Th}$.

Practice Problem 13.10

Determine the Thevenin equivalent of the circuit in Fig. 13.30 as seen from the terminals *a-b*.

Answer: $\mathbf{Z}_{Th} = 4.024 - j0.2195\ \Omega$; $\mathbf{V}_{Th} = 10.748\underline{/15.8^\circ}$ V

Figure 13.30
For Practice Problem 13.10.

Example 13.11

Obtain current $\mathbf{I}_o$ in Fig. 13.31 using Norton's theorem.

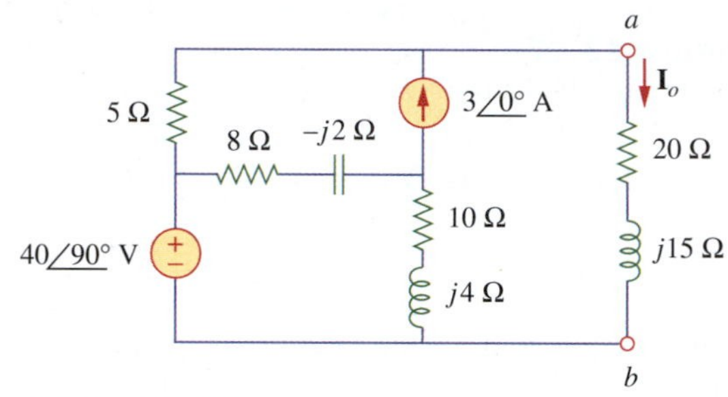

Figure 13.31
For Example 13.11.

Solution:
Our first objective is to find the Norton equivalent at terminals *a-b*. $\mathbf{Z}_N$ is found in the same way as $\mathbf{Z}_{Th}$. We set the sources to zero as shown in Fig. 13.32(a). As evident from the figure, the $(8 - j2)$ and $(10 + j4)$ impedances are short-circuited so that

$$\mathbf{Z}_N = 5\ \Omega$$

To get $\mathbf{I}_N$, we short-circuit terminals *a-b* as in Fig. 13.32(b) and apply mesh analysis. Notice that meshes 2 and 3 form a supermesh because of the current source linking them. For mesh 1,

$$-j40 + (18 + j2)\mathbf{I}_1 - (8 - j2)\mathbf{I}_2 - (10 + j4)\mathbf{I}_3 = 0 \quad \textbf{(13.13.1)}$$

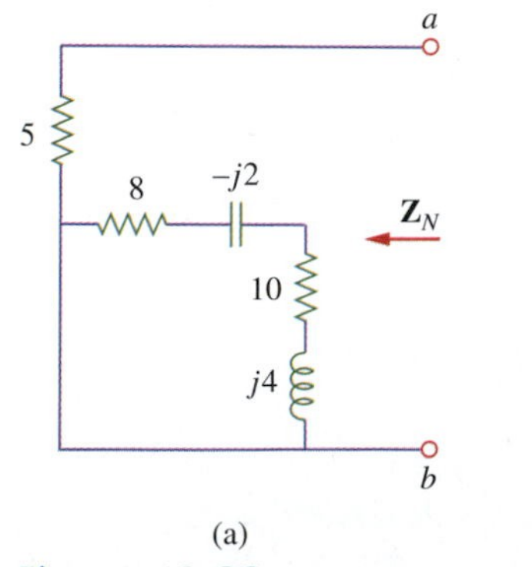

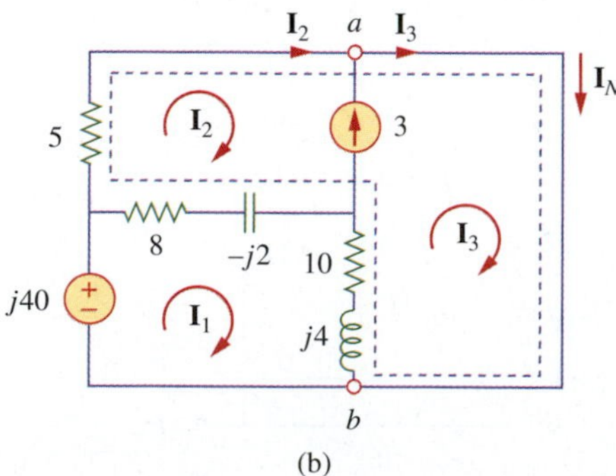

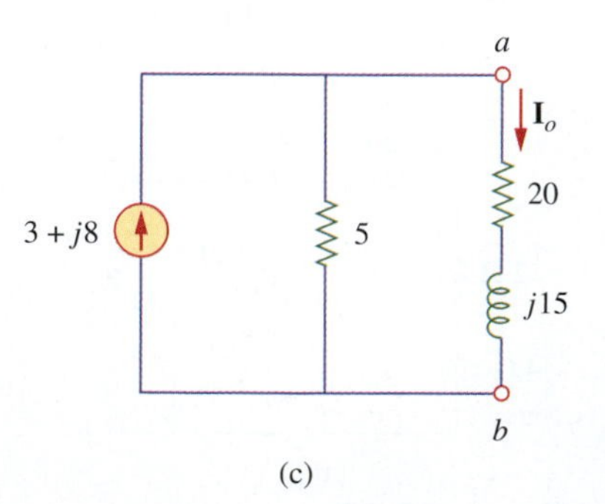

Figure 13.32
Solution of the circuit in Fig. 13.30: (a) finding $\mathbf{Z}_N$; (b) finding $\mathbf{I}_N$; (c) calculating $\mathbf{I}_o$.

For the supermesh,

$$(13 - j2)\mathbf{I}_2 + (10 + j4)\mathbf{I}_3 - (18 + j2)\mathbf{I}_1 = 0 \quad \textbf{(13.13.2)}$$

At node a, due to the current source between meshes 2 and 3,

$$\mathbf{I}_3 = \mathbf{I}_2 + 3 \quad \textbf{(13.13.3)}$$

Adding Eqs. (13.13.1) and (13.13.2) gives

$$-j40 + 5\mathbf{I}_2 = 0 \quad \Rightarrow \quad \mathbf{I}_2 = j8$$

From Eq. (13.13.3),

$$\mathbf{I}_3 = \mathbf{I}_2 + 3 = 3 + j8$$

The Norton current is

$$\mathbf{I}_N = \mathbf{I}_3 = (3 + j8)\text{ A}$$

The Norton equivalent circuit along with the impedance at terminals a-b is shown in Fig. 13.32(c). By the current divider rule,

$$\mathbf{I}_o = \frac{5}{5 + 20 + j15}\mathbf{I}_N = \frac{3 + j8}{5 + j3} = 1.465\underline{/38.48^\circ}\text{ A}$$

Practice Problem 13.11

Determine the Norton equivalent of the circuit in Fig. 13.33 as seen from terminals a-b. Use the equivalent circuit to find $\mathbf{I}_o$.

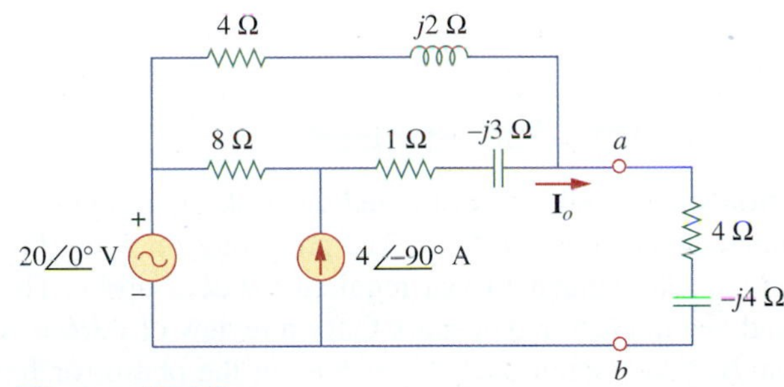

Figure 13.33
For Practice Problem 13.11.

Answer: $\mathbf{Z}_N = 3.176 + j0.706\ \Omega$; $\mathbf{I}_N = 8.395\underline{/-7.62^\circ}$ A; $\mathbf{I}_o = 1.971\underline{/22.95^\circ}$ A

Example 13.12

Obtain the Thevenin and Norton equivalents at terminals a-b of the circuit in Fig. 13.34.

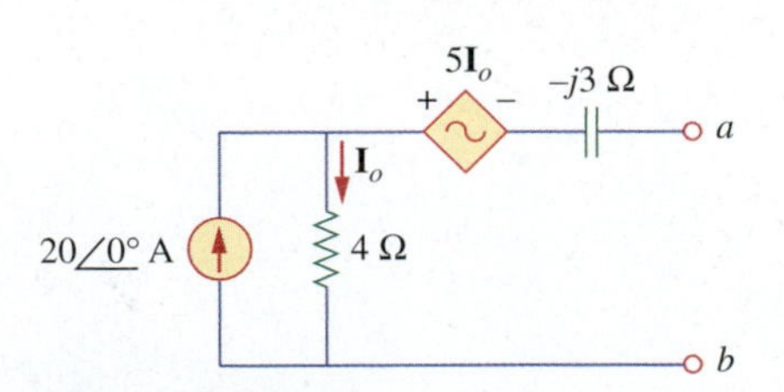

Figure 13.34
For Example 13.12.

Solution:
To find $\mathbf{V}_{Th}$, consider the circuit in Fig. 13.35(a). Because the terminals a-b are open,

$$\mathbf{I}_o = 2\text{ A}$$

We apply KVL to the right-hand loop in Fig. 13.35(a).

$$-\mathbf{V}_{Th} + 0(-j3) - 5\mathbf{I}_o + 4\mathbf{I}_o = 0 \quad \Rightarrow \quad V_{Th} = -\mathbf{I}_o = -2\text{ V}$$

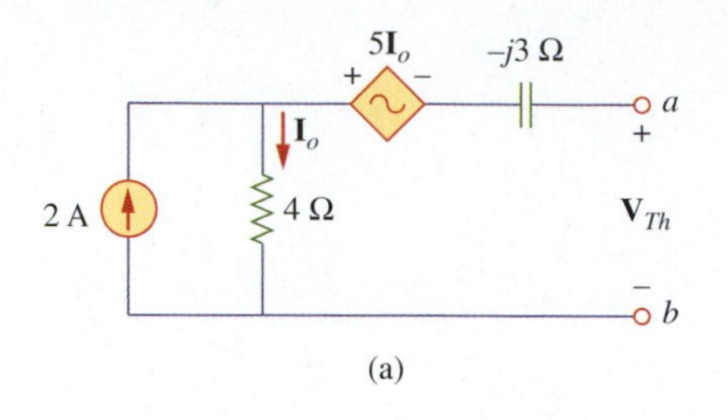

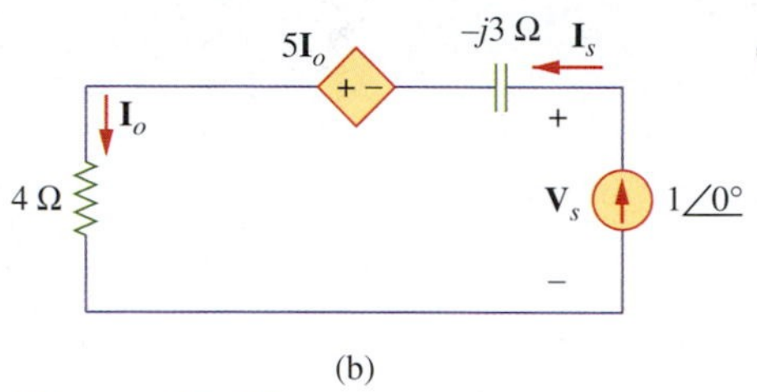

Figure 13.35
For Example 13.12: (a) finding $\mathbf{V}_{Th}$; (b) finding $\mathbf{Z}_{Th}$.

To find $\mathbf{Z}_{Th}$, we remove the independent current source and leave the dependent source intact. Due to the dependent source, we connect a 1-A source at terminals *a-b*, as shown in Fig. 13.35(b). It is evident that

$$\mathbf{I}_o = \mathbf{I}_s = 1 \text{ A}$$

Applying KVL around the loop gives

$$-\mathbf{V}_s - 5\mathbf{I}_o + \mathbf{I}_o(4 - j3) = 0 \quad \Rightarrow \quad \mathbf{V}_s = -1 - j3$$

Thus,

$$\mathbf{Z}_{Th} = \mathbf{Z}_N = \frac{\mathbf{V}_s}{\mathbf{I}_s} = \mathbf{V}_s = -1 - j3 = 3.162\underline{/251.57^\circ} \text{ V}$$

$$\mathbf{I}_N = \frac{\mathbf{V}_{Th}}{\mathbf{Z}_{Th}} = \frac{-2}{3.162\underline{/251.57^\circ}} = 0.632\underline{/(180^\circ - 251.57^\circ)}$$

$$= 0.632\underline{/-71.57^\circ} \text{ A}$$

Practice Problem 13.12

Find the Thevenin and Norton equivalents at terminals *a-b* of the circuit in Fig. 13.36.

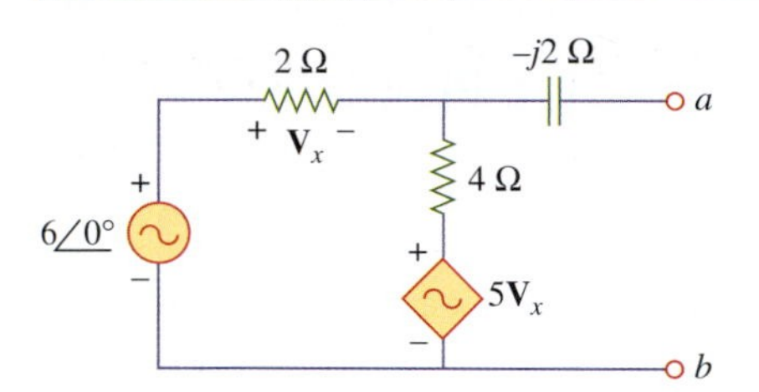

Figure 13.36
For Practice Problem 13.12

Answer: $\mathbf{V}_{Th} = 5.25$ V; $\mathbf{Z}_{Th} = \mathbf{Z}_N = 2.828\underline{/225^\circ}\ \Omega$; $\mathbf{I}_N = 1.856\underline{/-225^\circ}$ A

13.7 Computer Analysis

PSpice affords a big relief from the tedious task of manipulating complex numbers in ac circuit analysis. The procedure for using PSpice for ac analysis is quite similar to that required for dc analysis. The reader should read Section C.5 in Appendix C for a review of PSpice concepts for ac analysis. AC circuit analysis is done in the phasor or frequency domain, and all sources must have the same frequency. Although ac analysis with PSpice involves using AC Sweep, our analysis in this chapter requires a single frequency $f = \omega/2\pi$. The output file of PSpice contains voltage and current phasors. If necessary, the impedances can be calculated using the voltages and currents in the output file.

Example 13.13

Find $\mathbf{V}_1$ and $\mathbf{V}_2$ in the circuit of Fig. 13.37.

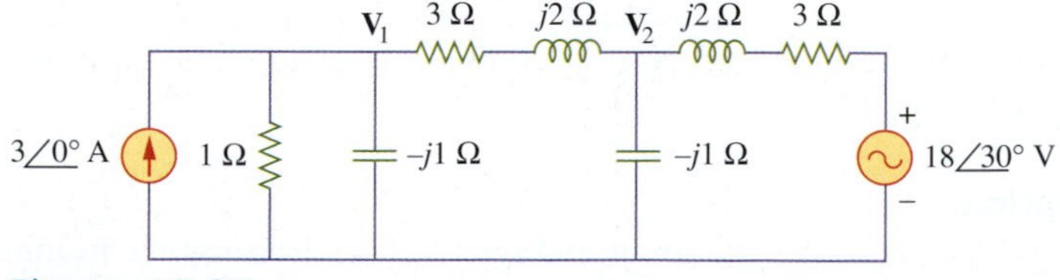

Figure 13.37
For Example 13.13.

Solution:
The way we use PSpice here is similar to how we used it in Chapter 12 (see Example 12.11). The circuit in Fig. 12.27 is in the time domain, whereas the one in Fig. 13.37 is in the frequency domain. The only

difference here is that the circuit is more complicated. Because we are not given a particular frequency and PSpice requires one, we select any frequency consistent with the given impedances. For example, if we select $\omega = 1$ rad/s, the corresponding frequency is $f = \omega/2\pi = 0.159155$ Hz. We obtain the values of the capacitance ($C = 1/\omega X_C$) and inductances ($L = X_L/\omega$). Making these changes results in the schematic in Fig. 13.38.

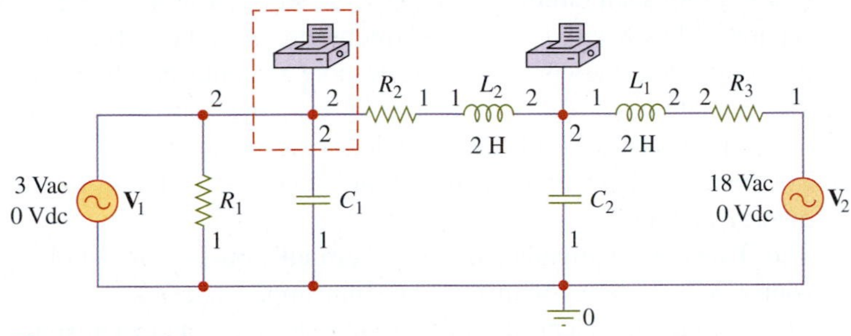

Figure 13.38
Schematic for the circuit in Fig. 13.37.

Once the circuit is drawn and saved as exam1313.dsn, we select **PSpice/New Simulation Profile**. This leads to the New Simulation dialog box. Type "exam1313" as the name of the file and click Create. This leads to the Simulation Settings dialog box. We select AC Sweep/Noise for *Analysis type*. As a single frequency analysis, we type 0.159155 as the *Start Freq*, 0.159155 as the *Final Freq*, and 1 as *Total Point*. After saving the schematic, we simulate it by selecting **PSpice/Run**. This leads to Probe, the graphical processor. We go back to the schematic window and select **PSpice/View Output File**. The output file includes the source frequency in addition to the attributes checked for the pseudocomponents VPRINT1; that is,

```
FREQ          VM(1)         VP(1)
1.592E-01     2.230E+00     1.724E+02

FREQ          VM(2)         VP(2)
1.592E-01     5.430E+00     -5.521E+01
```

from which we obtain

$$\mathbf{V}_1 = 2.23\underline{/172.4^\circ} \text{ V}, \qquad \mathbf{V}_2 = 5.43\underline{/-55.21^\circ} \text{ V}$$

Practice Problem 13.13

Obtain $\mathbf{V}_x$ and $\mathbf{I}_x$ in the circuit depicted in Fig. 13.39.

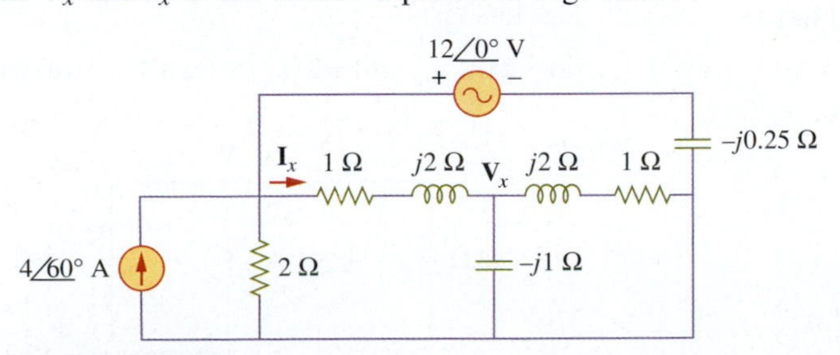

Figure 13.39
For Practice Problem 13.13.

Answer: $13.02\underline{/103.9^\circ}$ V; $8.234\underline{/175.5^\circ}$ A

13.8 Summary

1. We apply nodal and mesh analysis to ac circuits by applying KCL and KVL to the phasor form of the circuits.
2. In solving for the steady-state response of a circuit having independent sources with different frequencies, each independent source *must* be considered separately. The most natural approach to analyzing such circuits is to apply the superposition theorem. A separate phasor circut for each frequency *must* be solved independently, and the corresponding response should be obtained in the time domain. The overall response is the sum of the time-domain responses of all the individual phasor circuits.
3. The concept of source transformation is also applicable in the frequency domain.
4. The Thevenin equivalent of an ac circuit consists of a voltage source $\mathbf{V}_{TH}$ in series with the Thevenin impedance $\mathbf{Z}_{TH}$.
5. The Norton equivalent of an ac circuit consists of a current source $\mathbf{I}_N$ in parallel with the Norton impedance $\mathbf{Z}_N$ $(=\mathbf{Z}_{TH})$.
6. PSpice is a simple and powerful tool for solving ac circuit problems. It relieves us of the tedious task of working with complex numbers involved in steady-state analysis.

Review Questions

13.1 In the circuit of Fig. 13.40, current $i(t)$ is:

(a) $10\cos(t)$ A (b) $10\sin(t)$ A
(c) $5\cos(t)$ A (d) $5\sin(t)$ A
(e) $4.472\cos(t - 63.43°)$ A

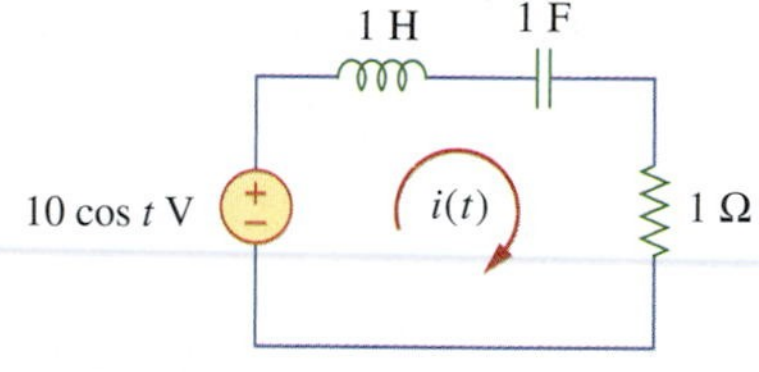

Figure 13.40
For Review Question 13.1.

13.2 The voltage $\mathbf{V}_o$ across the capacitor in Fig. 13.41 is:

(a) $5\underline{/0°}$ V (b) $7.071\underline{/45°}$ V
(c) $7.071\underline{/-45°}$ V (d) $5\underline{/-45°}$ V

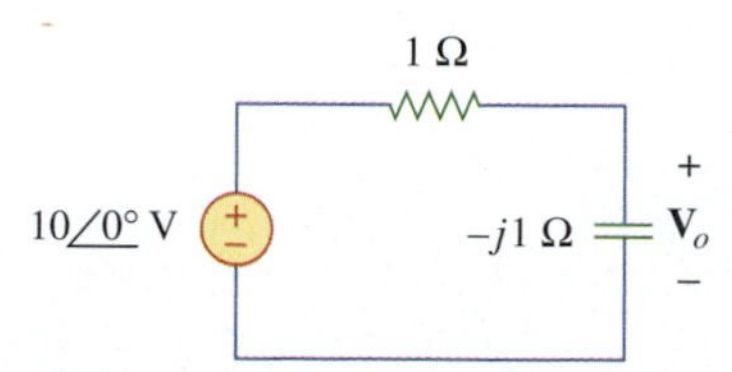

Figure 13.41
For Review Question 13.2.

13.3 The value of the current $\mathbf{I}_o$ in the circuit in Fig. 13.42 is:

(a) $4\underline{/0°}$ A (b) $2.4\underline{/-90°}$ A
(c) $0.6\underline{/0°}$ A (d) -1 A

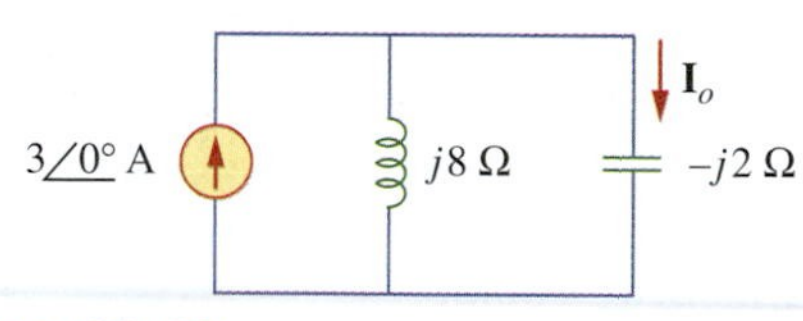

Figure 13.42
For Review Question 13.3.

13.4 Using nodal analysis, the value of $\mathbf{V}_o$ in the circuit of Fig. 13.43 is:

(a) -24 V (b) -8 V (c) 8 V (d) 24 V

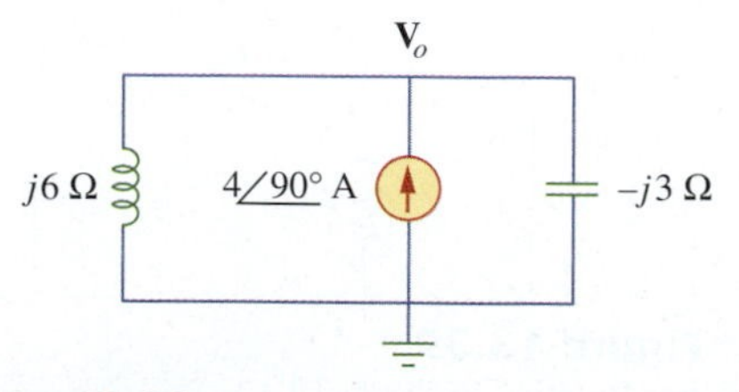

Figure 13.43
For Review Question 13.4.

13.5 Refer to the circuit in Fig. 13.44 and observe that the two sources do not have the same frequency. The current $i_x(t)$ can be obtained by:

(a) source transformation

(b) the superposition theorem

(c) PSpice

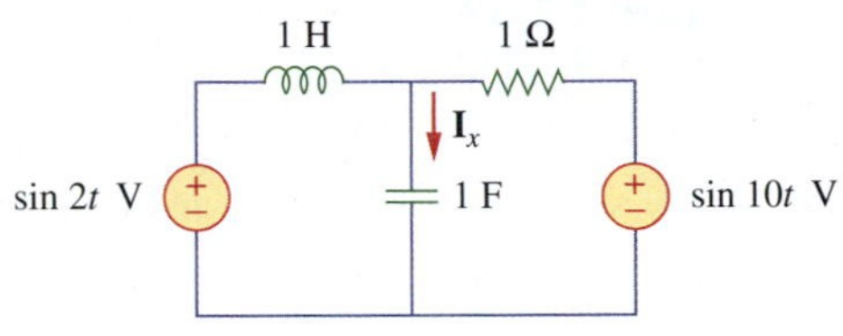

Figure 13.44
For Review Question 13.5.

13.6 For the circuit in Fig. 13.45, the Thevenin impedance at terminals *a-b* is:

(a) 1 Ω (b) $0.5 - j0.5\ \Omega$ (c) $0.5 + j0.5\ \Omega$

(d) $1 + j2\ \Omega$ (e) $1 - j2\ \Omega$

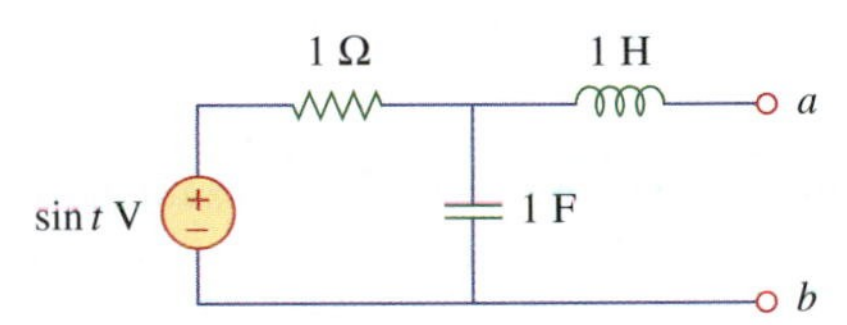

Figure 13.45
For Review Questions 13.6 and 13.7.

13.7 In the circuit of Fig. 13.45 the Thevenin voltage at terminals *a-b* is:

(a) $0.7071\angle{-45^\circ}$ V (b) $7.071\angle{45^\circ}$ V

(c) $0.3535\angle{-45^\circ}$ V (d) $0.3535\angle{45^\circ}$ V

13.8 Refer to the circuit in Fig. 13.46. The Norton equivalent impedance at terminals *a-b* is:

(a) $-j4\ \Omega$ (b) $-j2\ \Omega$ (c) $j2\ \Omega$ (d) $j4\ \Omega$

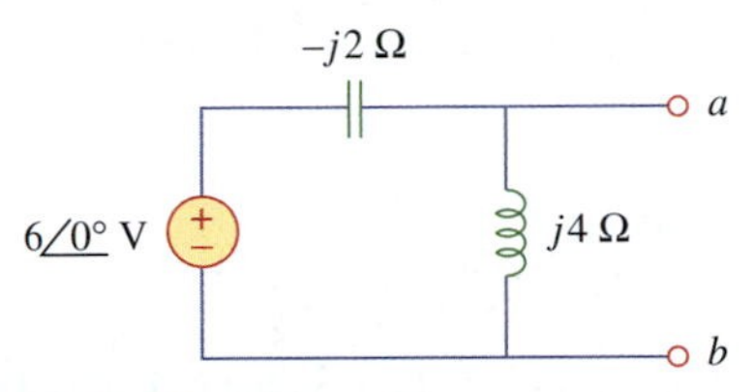

Figure 13.46
For Review Questions 13.8 and 13.9.

13.9 The Norton current at terminals *a-b* in the circuit of Fig. 13.46 is:

(a) $1\angle{0^\circ}$ A (b) $1.5\angle{-90^\circ}$ A

(c) $1.5\angle{90^\circ}$ A (d) $3\angle{90^\circ}$ A

13.10 PSpice can handle a circuit with two independent sources of different frequencies.

(a) True (b) False

Answers: 13.1a, 13.2c, 13.3a, 13.4d, 13.5b, 13.6c, 13.7a, 13.8a, 13.9d, 13.10b

Problems

Section 13.2 Mesh Analysis

13.1 Solve for i_o in Fig. 13.47 using mesh analysis.

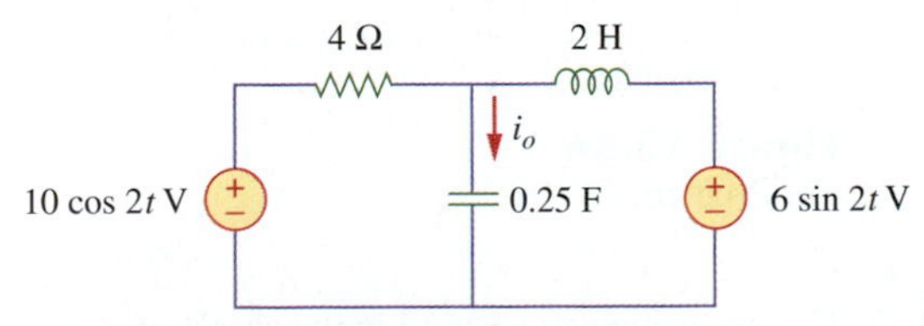

Figure 13.47
For Problem 13.1.

13.2 Using mesh analysis, find $\mathbf{I}_1$ and $\mathbf{I}_2$ in the circuit of Fig. 13.48.

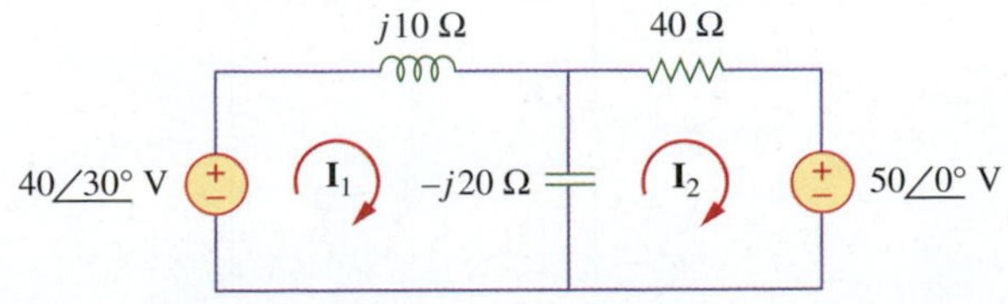

Figure 13.48
For Problems 13.2 and 13.36.

13.3 By using mesh analysis, find $\mathbf{I}_1$ and $\mathbf{I}_2$ in the circuit depicted in Fig. 13.49.

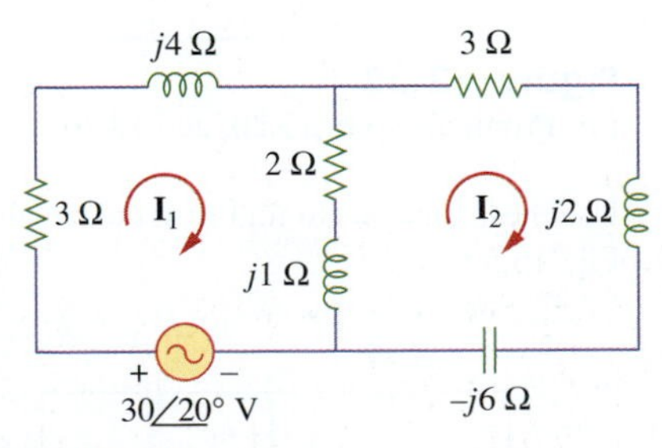

Figure 13.49
For Problem 13.3.

13.4 Using mesh analysis, find i_o in the circuit of Fig. 13.50.

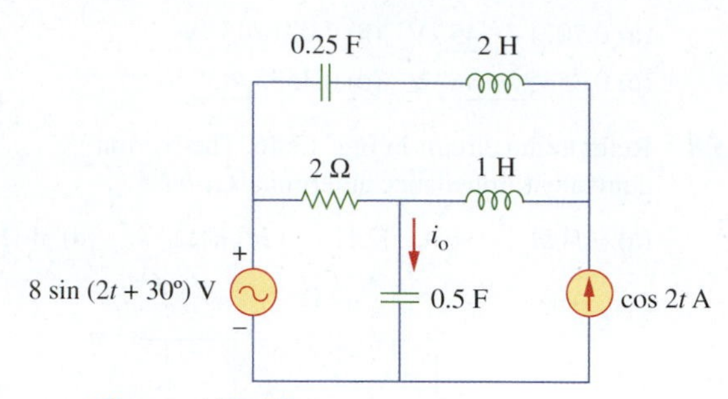

Figure 13.50
For Problems 13.4 and 13.14.

13.5 Find $\mathbf{I}_1$, $\mathbf{I}_2$, $\mathbf{I}_3$, and $\mathbf{I}_x$ in the circuit of Fig. 13.51.

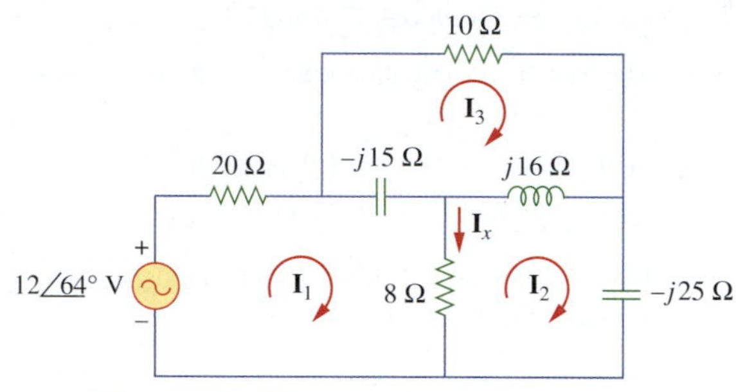

Figure 13.51
For Problem 13.5.

13.6 Use mesh analysis to find $\mathbf{V}_o$ in the circuit of Fig. 13.52.

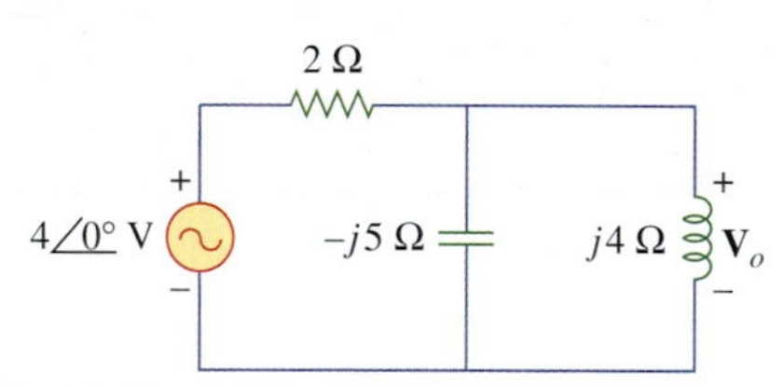

Figure 13.52
For Problems 13.6, 13.20, and 13.36.

13.7 Use mesh analysis to find i_o in the circuit in Fig. 13.53.

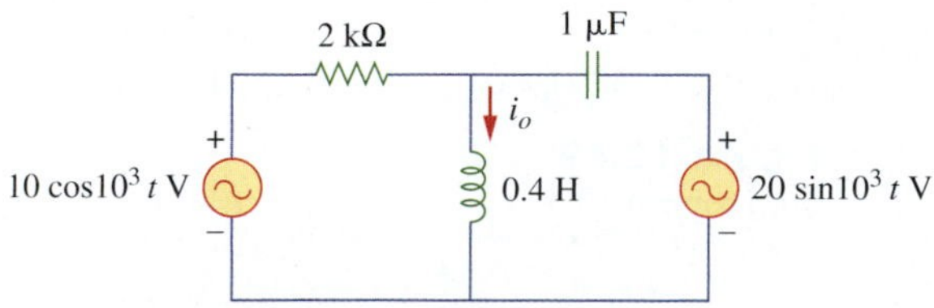

Figure 13.53
For Problem 13.7.

13.8 Use mesh analysis to find $\mathbf{V}_o$ in the circuit of Fig. 13. 54. Let $v_{s1} = 120\cos(100t + 90°)$ V, $v_{s2} = 80\cos 100t$ V.

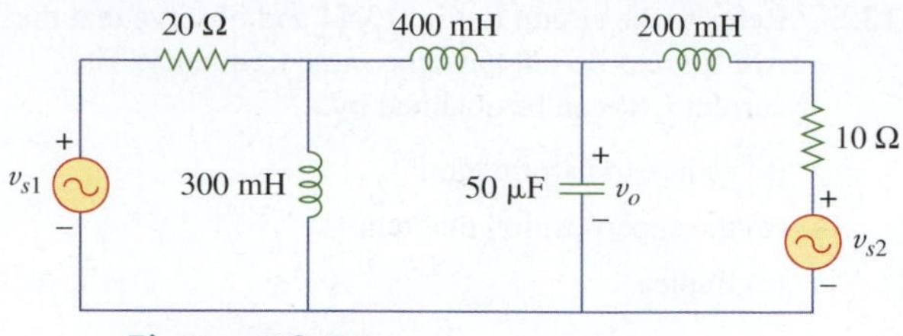

Figure 13.54
For Problem 13.8.

13.9 Apply mesh analysis to find $\mathbf{I}_1$ and $\mathbf{I}_2$ in the circuit of Fig. 13.55.

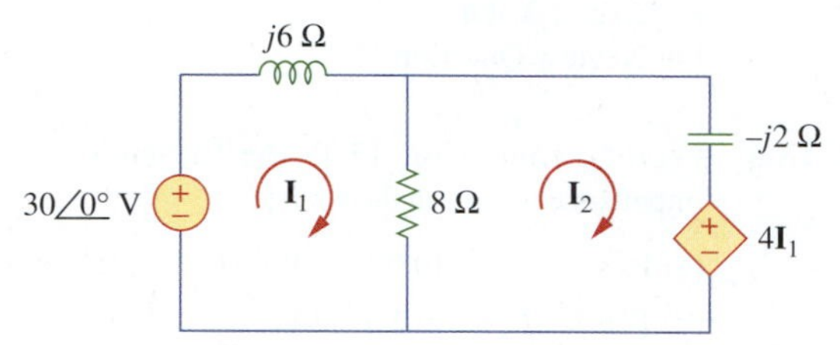

Figure 13.55
For Problem 13.9.

13.10 Write the four mesh equations for the circuit in Fig. 13.56. You do not need to solve them.

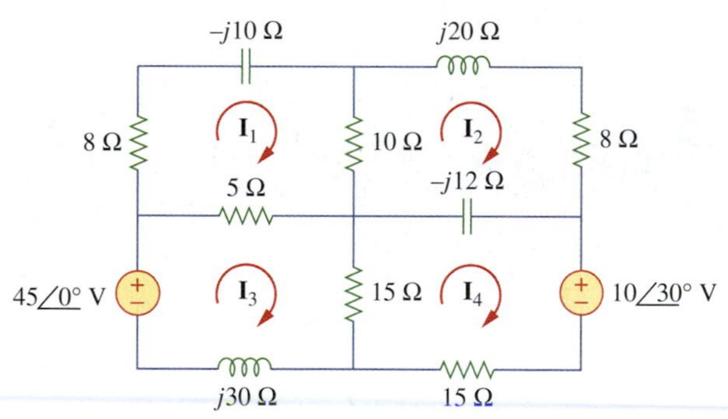

Figure 13.56
For Problem 13.10.

13.11 Use mesh analysis to find **I** in the circuit of Fig. 13.57.

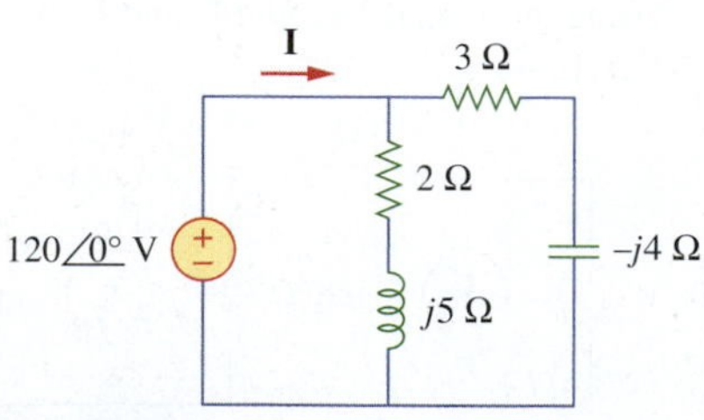

Figure 13.57
For Problem 13.11.

Section 13.3 Nodal Analysis

13.12 Find v_x in the circuit in Fig. 13.58.

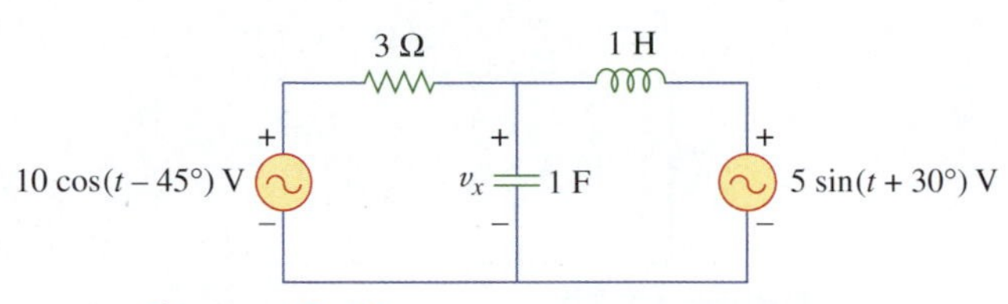

Figure 13.58
For Problem 13.12.

13.13 Use nodal analysis to find $\mathbf{V}_o$ in the circuit of Fig. 13.59.

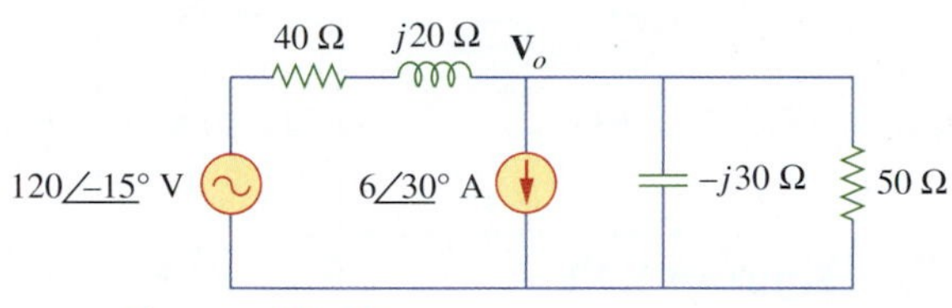

Figure 13.59
For Problem 13.13.

13.14 Using nodal analysis, find i_o in the circuit of Fig. 13.50.

13.15 Determine $\mathbf{V}_x$ in the circuit of Fig. 13.60 using any method of your choice.

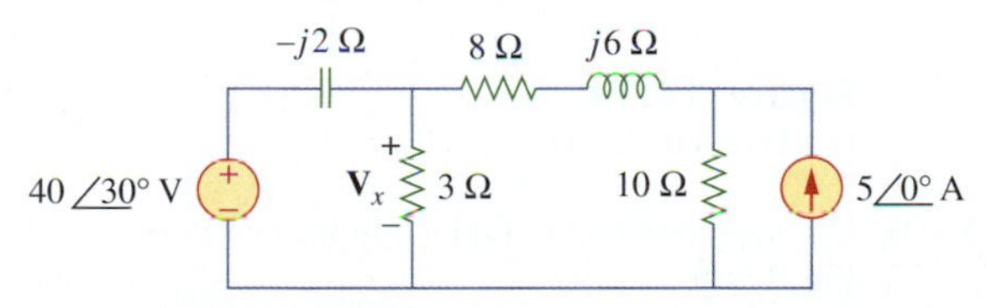

Figure 13.60
For Problem 13.15.

13.16 Calculate the voltages at nodes 1 and 2 in the circuit of Fig. 13.61 using nodal analysis.

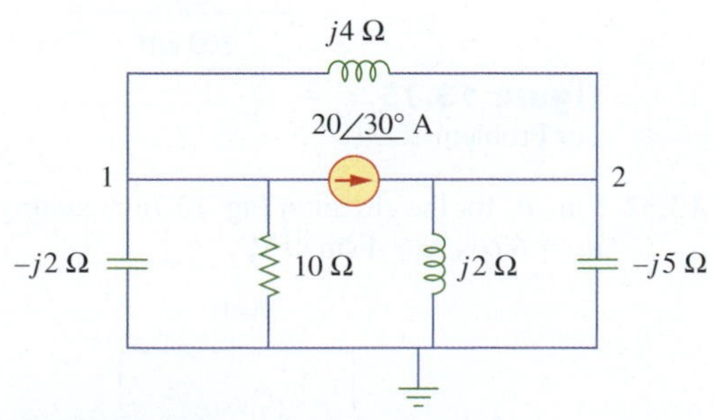

Figure 13.61
For Problems 13.16 and 13.53.

13.17 Using nodal analysis, find $\mathbf{V}_1$ and $\mathbf{V}_2$ in the circuit of Fig. 13.62.

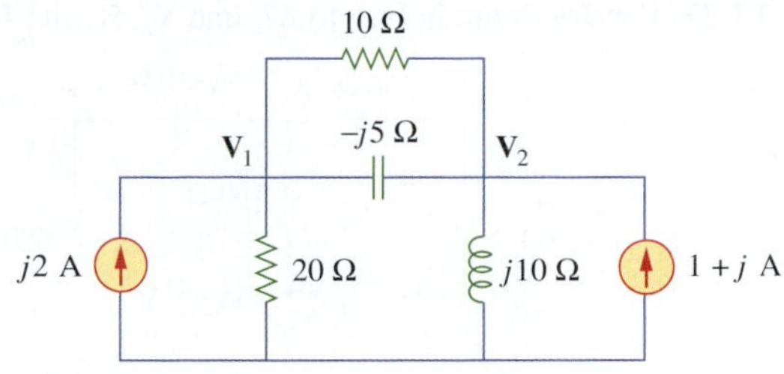

Figure 13.62
For Problem 13.17.

13.18 By nodal analysis, obtain current $\mathbf{I}_o$ in the circuit in Fig 13.63.

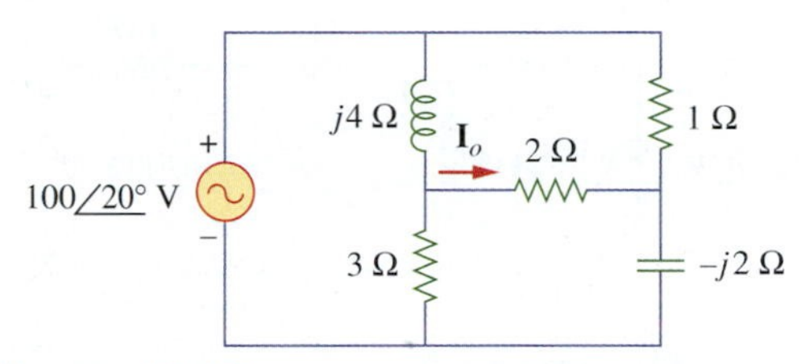

Figure 13.63
For Problem 13.18.

13.19 Use nodal analysis to find $\mathbf{V}_x$ in the circuit shown in Fig. 13.64.

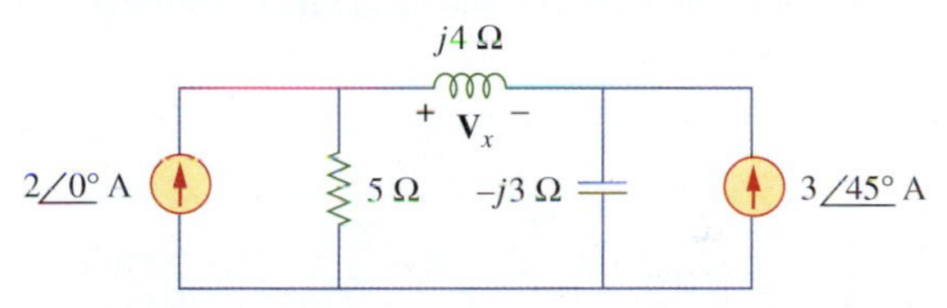

Figure 13.64
For Problems 13.19 and 13.54.

13.20 Find $\mathbf{V}_o$ in the circuit shown in Fig. 13.52 using nodal analysis.

13.21 Compute the value of $\mathbf{I}_x$ in Fig. 13.65.

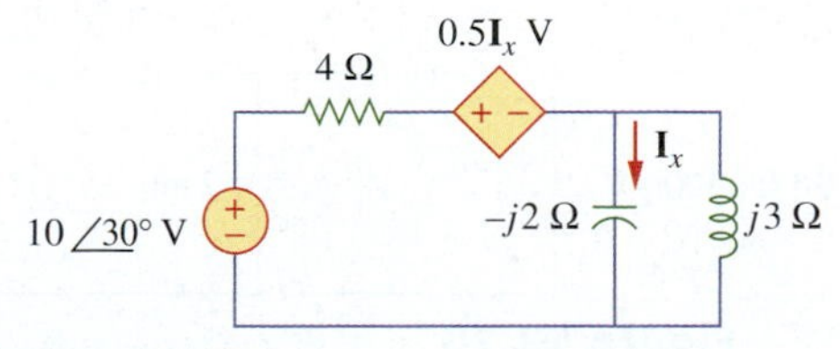

Figure 13.65
For Problem 13.21.

13.22 Determine $\mathbf{V}_x$ in Fig. 13.66.

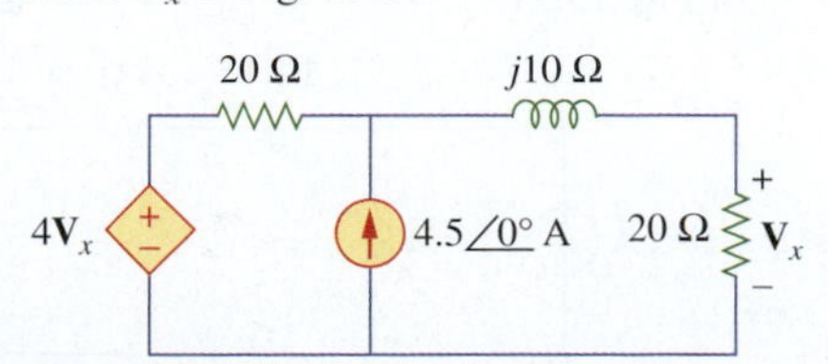

Figure 13.66
For Problem 13.22.

13.23 For the circuit in Fig. 13.67, find $\mathbf{V}_o$, $\mathbf{I}_1$, and $\mathbf{I}_2$.

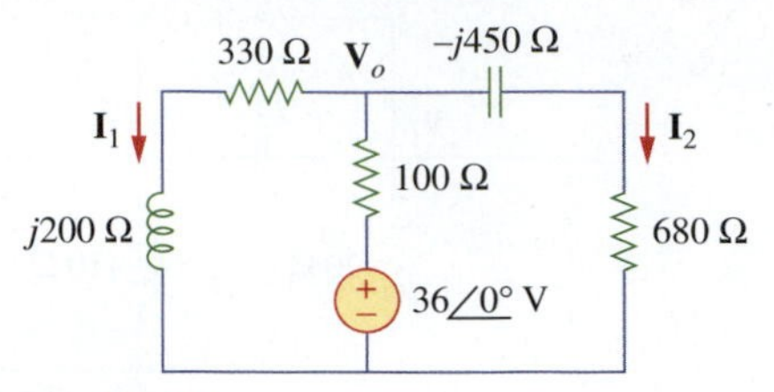

Figure 13.67
For Problem 13.23.

13.24 Use nodal analysis to find **I** in the circuit of Fig. 13.68.

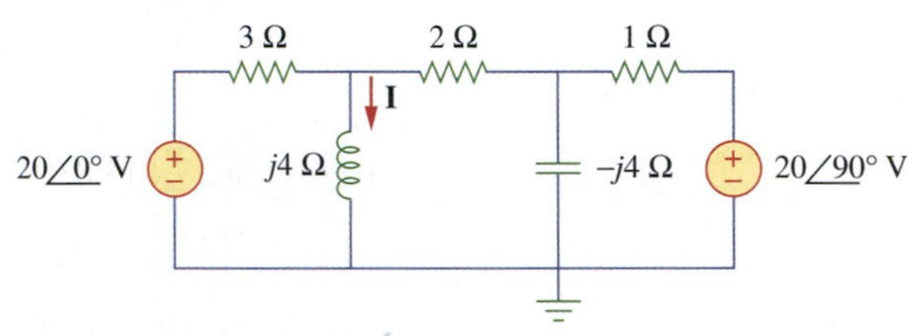

Figure 13.68
For Problem 13.24.

Section 13.4 Superposition Theorem

13.25 Find i_o in the circuit shown in Fig. 13.69 using superposition.

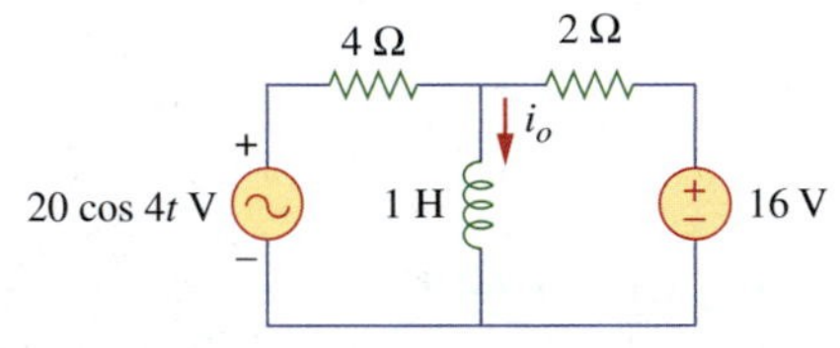

Figure 13.69
For Problem 13.25.

13.26 Use the superposition principle to calculate v_x in the circuit of Fig. 13.70.

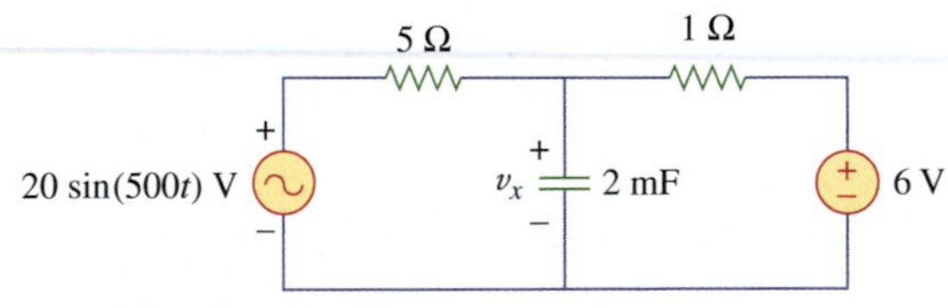

Figure 13.70
For Problem 13.26.

13.27 Using the superposition theorem, find i_x in the circuit of Fig. 13.71.

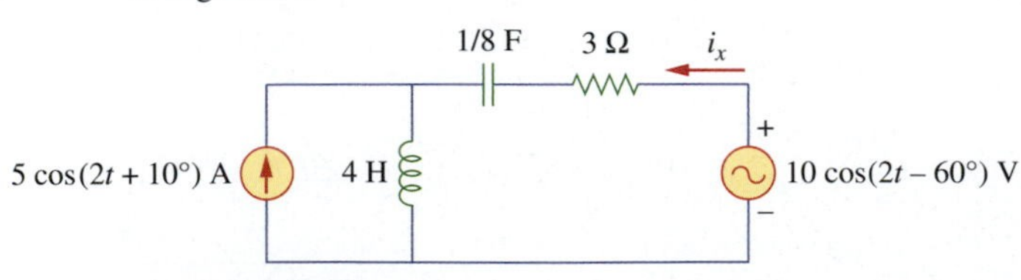

Figure 13.71
For Problem 13.27.

13.28 Use the superposition principle to obtain v_x in the circuit of Fig. 13.72. Let $v_s = 50 \sin(2t)$ V and $i_s = 12 \cos(6t + 10)$ A.

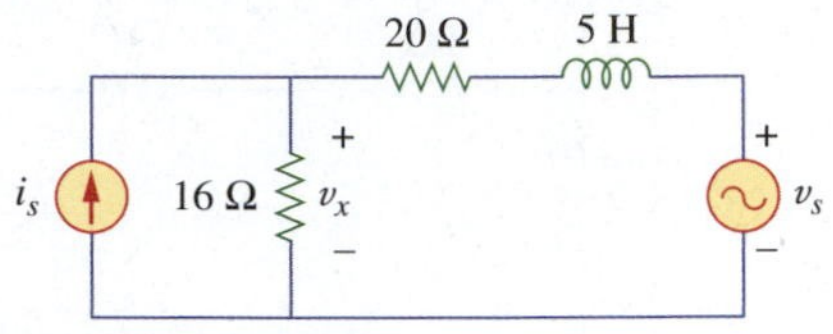

Figure 13.72
For Problem 13.28.

13.29 Solve for $v_o(t)$ in the circuit of Fig. 13.73 using the superposition principle.

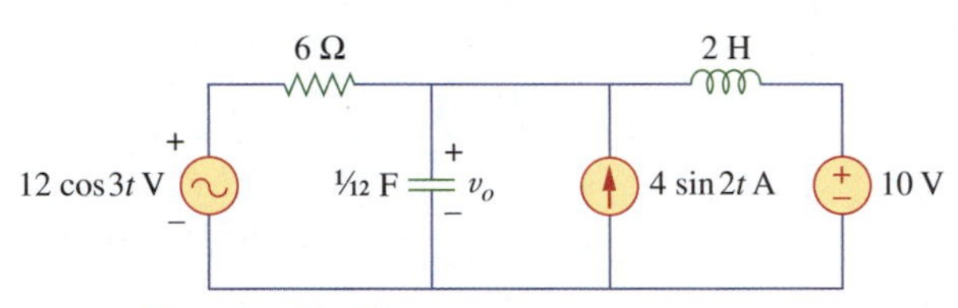

Figure 13.73
For Problem 13.29.

13.30 Determine i_o in the circuit of Fig. 13.74.

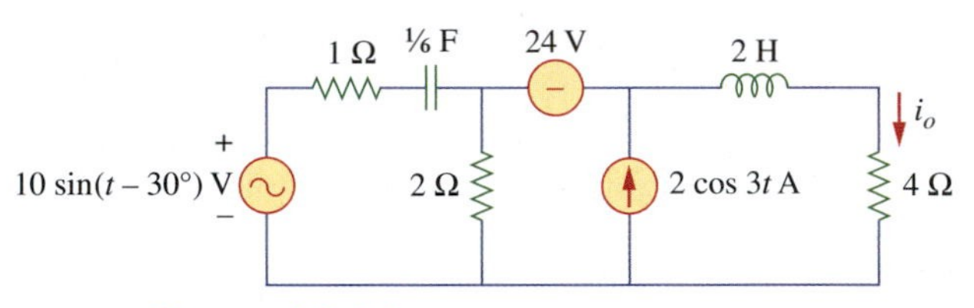

Figure 13.74
For Problem 13.30.

13.31 Use superposition to find $i(t)$ in the circuit of Fig. 13.75.

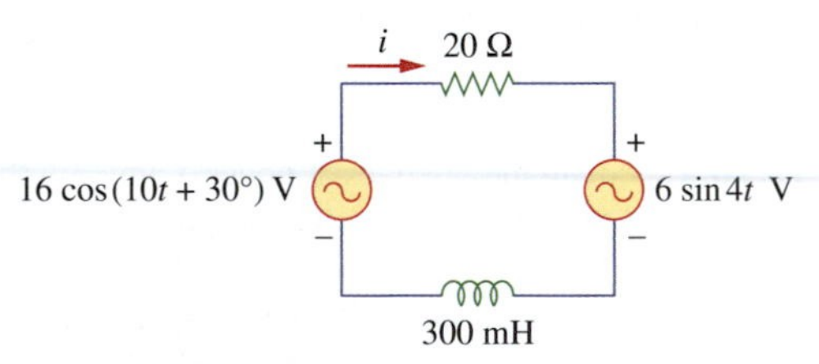

Figure 13.75
For Problem 13.31.

13.32 Find v_o for the circuit in Fig. 13.76 assuming that $v_s = 6 \cos 2t + 4 \sin 4t$ V.

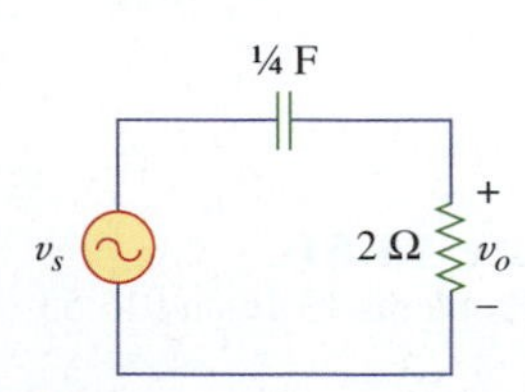

Figure 13.76
For Problem 13.32.

13.33 Using Thevenin theorem, find $\mathbf{V}_o$ in the circuit of Fig. 13.77.

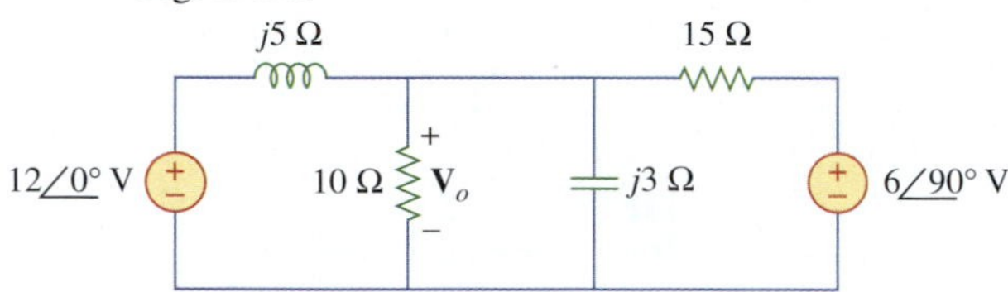

Figure 13.77
For Problem 13.33.

Section 13.5 Source Transformation

13.34 Using source transformation, find i in the circuit of Fig. 13.78.

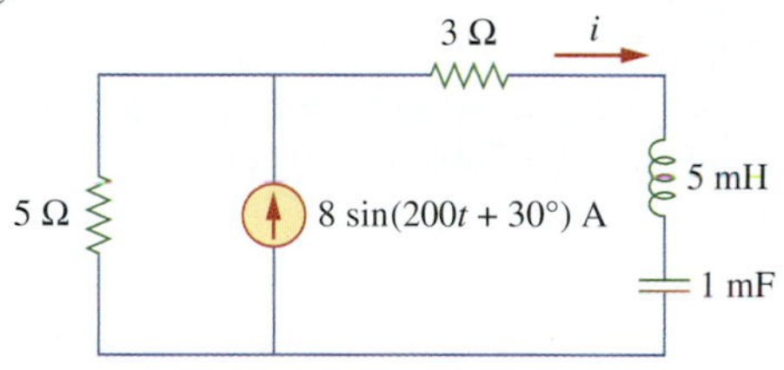

Figure 13.78
For Problem 13.34.

13.35 Use source transformation to find v_o in the circuit of Fig. 13.79.

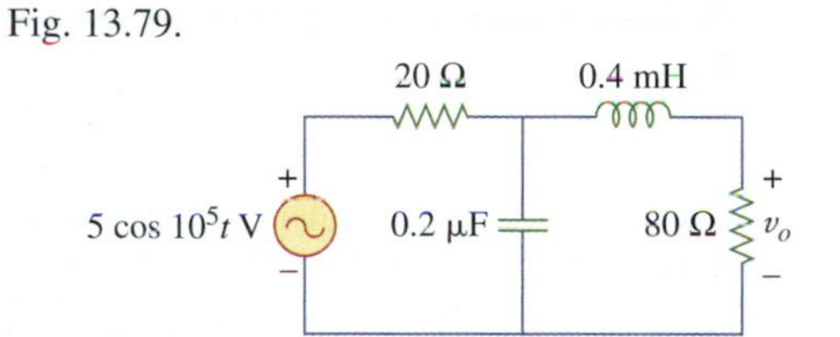

Figure 13.79
For Problem 13.35.

13.36 Solve Problem 13.2 using source transformation.

13.37 Use the concept of source transformation to find $\mathbf{V}_o$ in the circuit of Fig. 13.80.

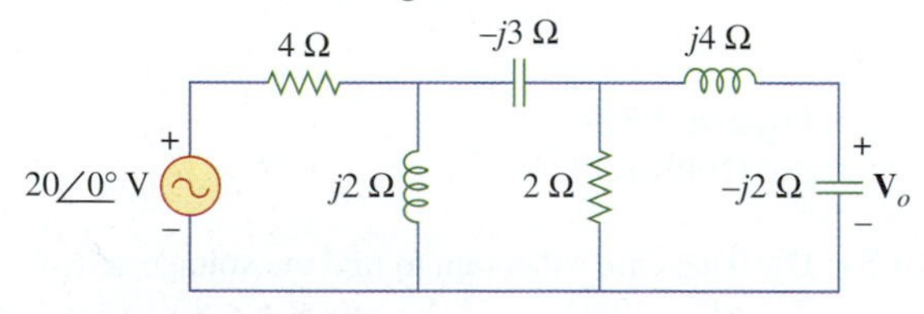

Figure 13.80
For Problem 13.37.

13.38 Use source transformation to find $\mathbf{I}_o$ in the circuit of Fig. 13.81.

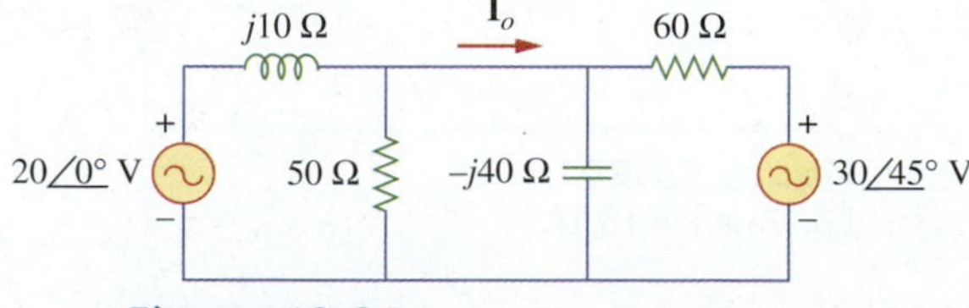

Figure 13.81
For Problem 13.38.

13.39 Apply source transformation to find $\mathbf{V}_o$ in the circuit of Fig. 13.82.

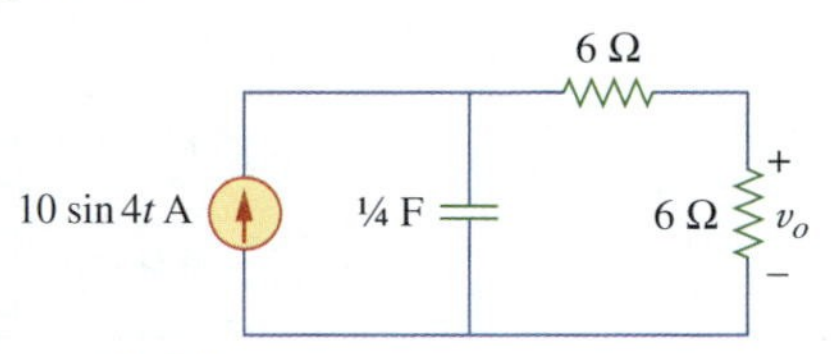

Figure 13.82
For Problem 13.39.

Section 13.6 Thevenin and Norton Equivalent Circuits

13.40 Find the Thevenin and Norton equivalent circuits at terminals a-b for each of the circuits in Fig. 13.83.

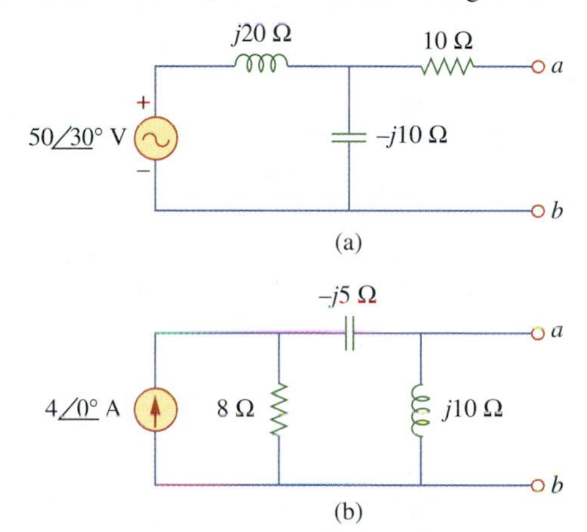

Figure 13.83
For Problem 13.40.

13.41 For the circuit shown in Fig. 13.84, obtain Thevenin and Norton equivalent circuits at terminals a-b.

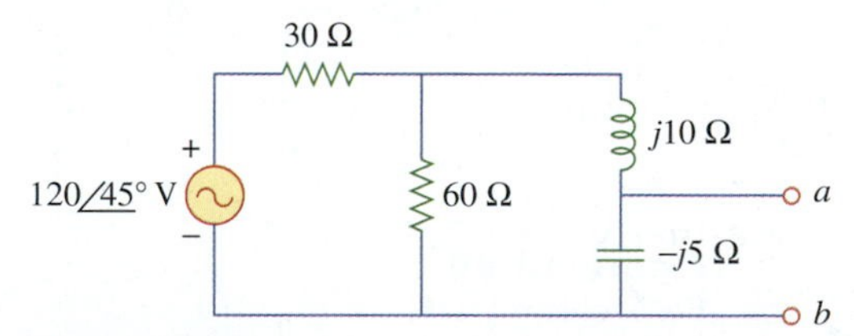

Figure 13.84
For Problem 13.41.

13.42 Find the Thevenin and Norton equivalent circuits for the circuit shown in Fig. 13.85.

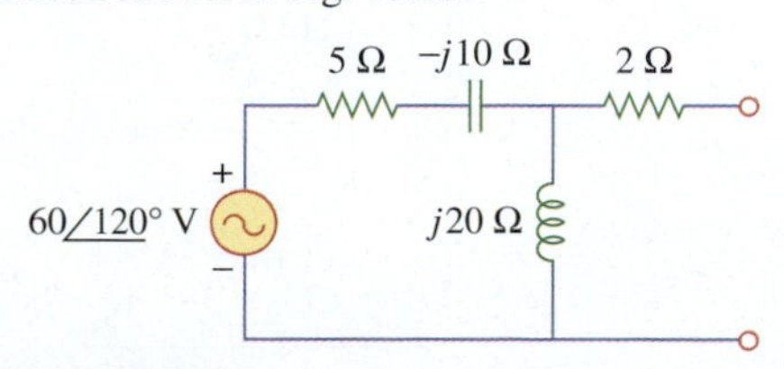

Figure 13.85
For Problem 13.42.

13.43 For the circuit shown in Fig. 13.86, find the Thevenin equivalent circuit at terminals *a-b*.

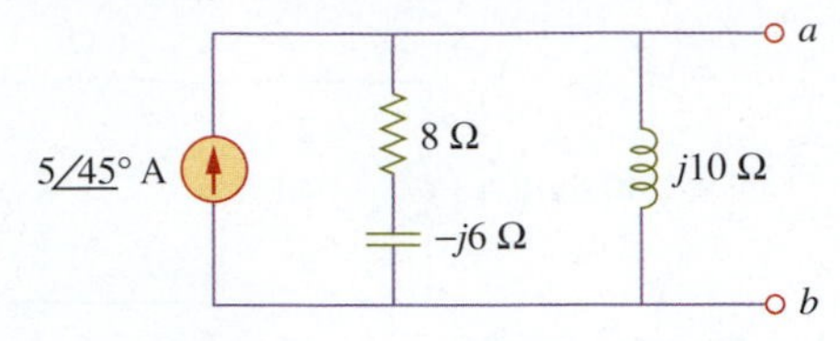

Figure 13.86
For Problem 13.43.

13.44 Find the Thevenin equivalent of the circuit in Fig. 13.87 as seen from: (a) terminals *a-b* and (b) terminals *c-d*.

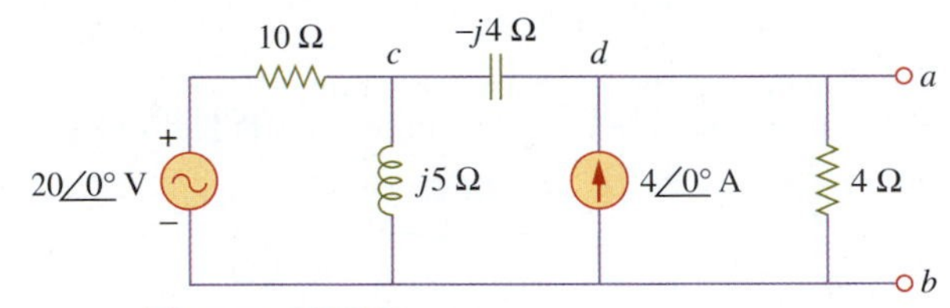

Figure 13.87
For Problem 13.44.

13.45 For the circuit shown in Fig. 13.88, find the Norton equivalent circuit at terminals *a-b*.

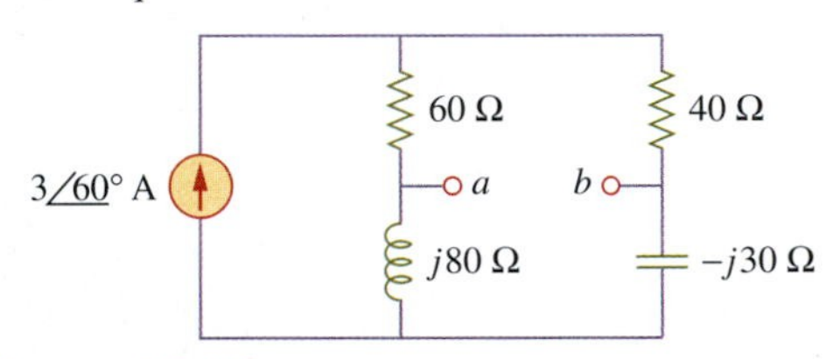

Figure 13.88
For Problem 13.45.

13.46 Compute i_o in Fig. 13.89 using Norton's theorem.

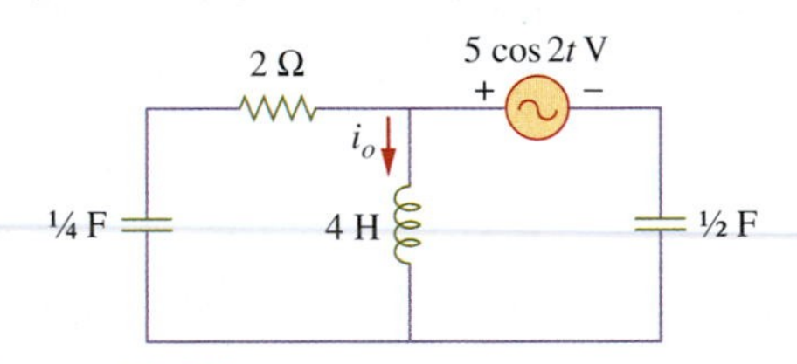

Figure 13.89
For Problem 13.46.

13.47 Find the Thevenin and Norton equivalent circuits at terminals *a-b* in the circuit of Fig. 13.90.

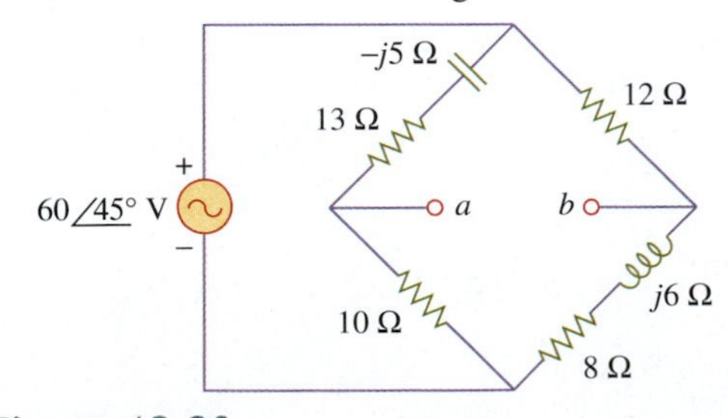

Figure 13.90
For Problem 13.47.

13.48 Using Thevenin's theorem, find v_o in the circuit in Fig. 13.91.

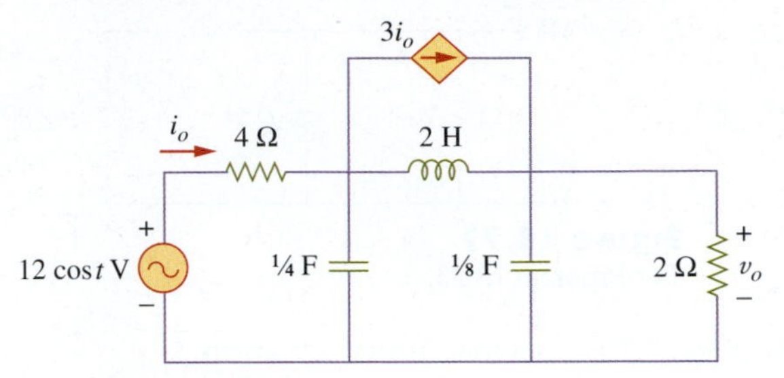

Figure 13.91
For Problem 13.48.

13.49 At terminals *a-b*, obtain Thevenin and Norton equivalent circuits for the circuit depicted in Fig. 13.92. Take $\omega = 10$ rad/s.

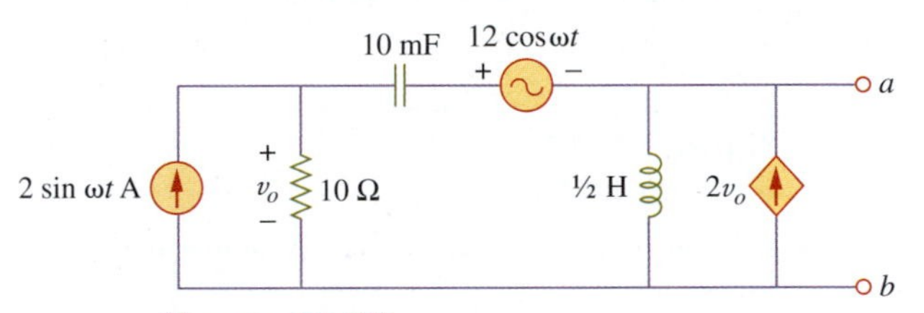

Figure 13.92
For Problem 13.49.

13.50 Find the Thevenin equivalent circuit at terminals *a-b* in the circuit shown in Fig. 13.93.

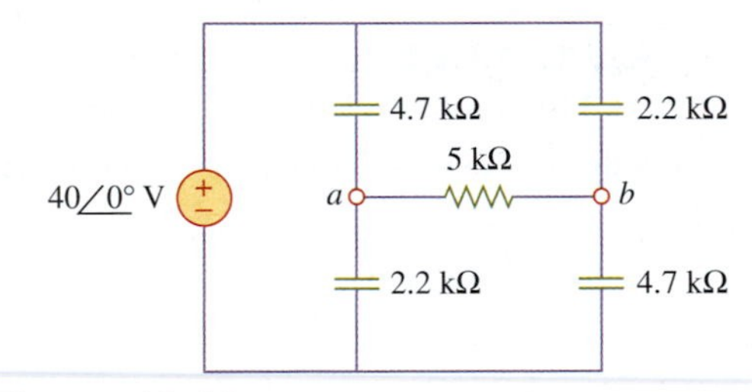

Figure 13.93
For Problem 13.50.

13.51 Use Thevenin's theorem to find the voltage across $\mathbf{Z}_L$ in Fig. 13.94.

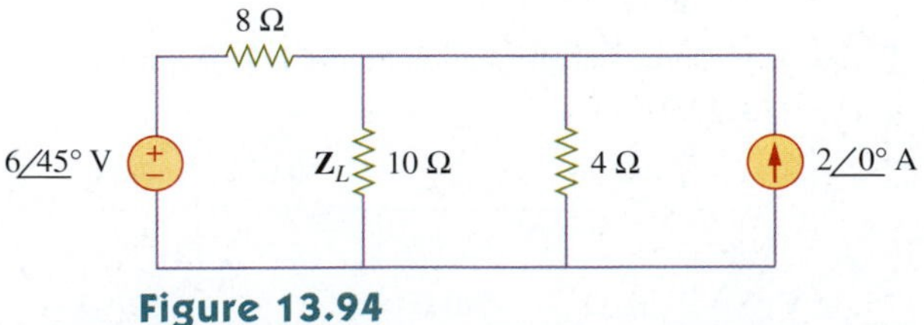

Figure 13.94
For Problem 13.51.

13.52 Apply Thevenin's theorem to find $\mathbf{V}_o$ in the circuit of Fig. 13.95.

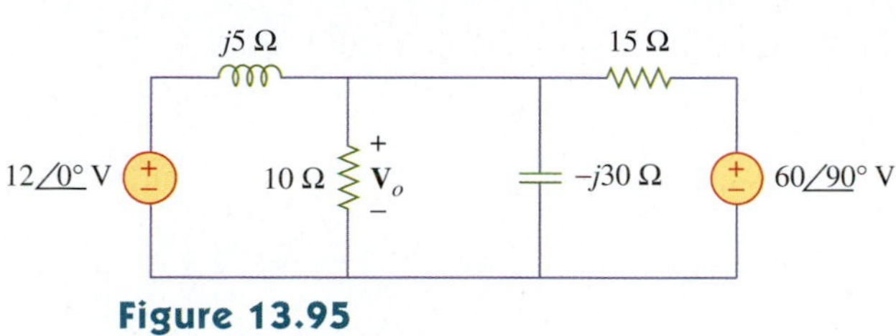

Figure 13.95
For Problem 13.52.

Section 13.7 Computer Analysis

13.53 Use PSpice to solve Problem 13.16.

13.54 Rework Problem 13.19 using PSpice.

13.55 Use PSpice to find v_o in the circuit of Fig. 13.96. Let $i_s = 2\cos(10^3 t)$ A.

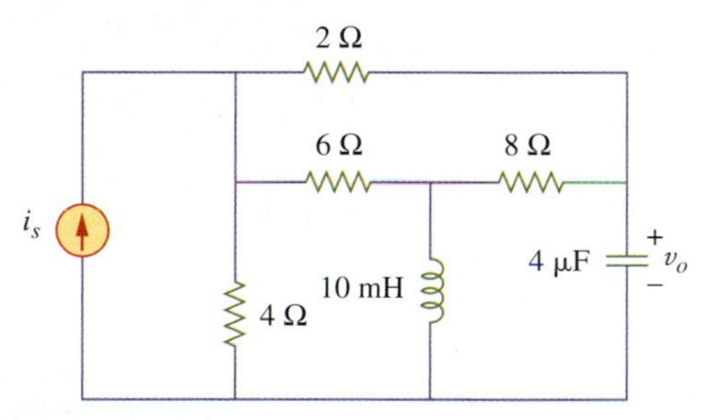

Figure 13.96
For Problem 13.55.

13.56 Use PSpice to find $\mathbf{V}_1$, $\mathbf{V}_2$, and $\mathbf{V}_3$ in the network of Fig. 13.97.

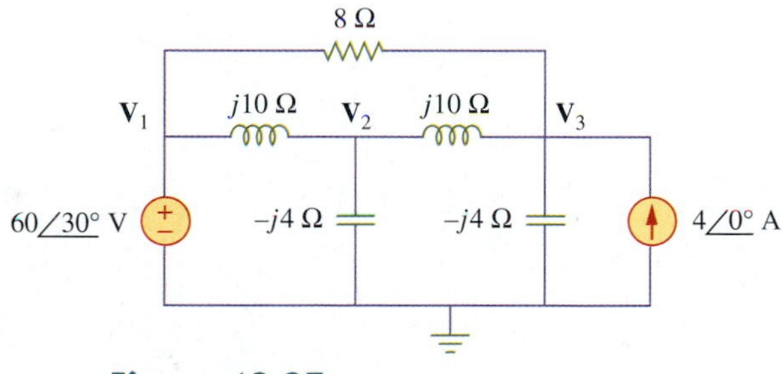

Figure 13.97
For Problem 13.56.

13.57 Determine $\mathbf{V}_1$, $\mathbf{V}_2$, and $\mathbf{V}_3$ in the circuit of Fig. 13.98 using PSpice.

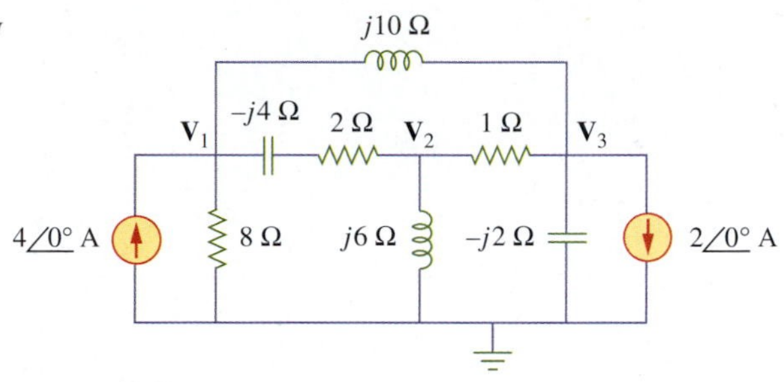

Figure 13.98
For Problem 13.57.

13.58 Use PSpice to determine $\mathbf{V}_o$ in the circuit of Fig. 13.99. Assume $\omega = 1$ rad/s.

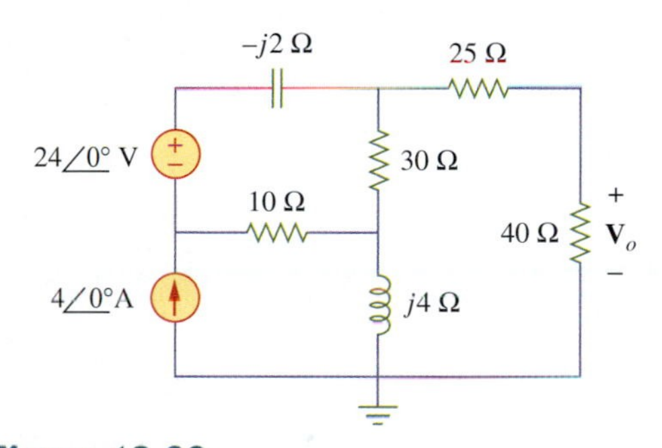

Figure 13.99
For Problem 13.58.

chapter 14

AC Power Analysis

The best career advice to give to the young is, "Find out what you like doing best and get someone to pay you for doing it."

—Katharine Whitehorn

Enhancing Your Career

Careers in Power Systems

Electric circuits analysis is applied in many areas in electrical engineering technology. One such area is power systems. The discovery of the principle of the ac generator by Michael Faraday in 1831 was a major breakthrough in engineering; it led to the development of a convenient way of generating the electric power that is now needed in every electronic, electrical, or electromechanical device.

Electric power lines.
© Sarhan M. Musa

Electric power is obtained by converting energy from different sources such as fossil fuels (gas, oil, and coal), nuclear fuel (uranium), hydro energy (water falling through a height differential), geothermal energy (hot water, steam), wind energy, tidal energy, and biomass energy (wastes). These various ways of generating electric power are studied in detail in power engineering, which has become an indispensable subdiscipline in electrical engineering technology. An electrical engineering technologist should be familiar with the analysis, generation, transmission, distribution, and cost of electric power.

The electric power industry is a very large employer of electrical engineering technologists. The industry includes thousands of electric utility systems ranging from large, interconnected systems serving large regional areas to small power companies serving individual communities or factories. Due to the complexity of the power industry, there are numerous electrical engineering technology jobs in different areas of the industry: power plants (generation), transmission and distribution, maintenance, research, data acquisition and flow control, and management. Because electric power is used everywhere, electric utility companies are also found everywhere, offering exciting training and steady employment for men and women in several thousand communities throughout the world.

14.1 Introduction

Our effort in ac circuit analysis has been focused mainly on calculating voltage and current. Our major concern in this chapter is power analysis of ac circuits.

Power analysis is of paramount importance. Power is the most important quantity in electric utilities, electronic, and communication systems because these systems all involve transmission of power from one point to another. Also, every industrial and household electrical device (e.g., fans, motors, lamps, irons, TVs, and personal computers) has a power rating that indicates how much power the piece of equipment requires; exceeding the power rating can do permanent damage to an appliance.

Electric power transmission refers to the bulk transfer of electrical power from one place to another. This is typically between the power plant and a substation near a populated area. This is distinct from electricity distribution, which deals with the delivery from the substation to the consumers. Because large amounts of power are involved, transmission normally takes place at high voltage (110 kV or above). Power is usually transmitted over long distances through overhead power transmission lines as shown in Fig. 14.1. Table 14.1 summarizes information on the electrical systems in use in some countries of the world.

We will begin by defining and deriving *instantaneous power* and *average power*. We will then introduce other power concepts. As practical applications of these concepts, we will discuss how power is measured and reconsider how electric utility companies charge their customers.

TABLE 14.1

Electricity around the world.

Country	Voltage (rms)	Frequency
Australia	240 V	50 Hz
Bangladesh	240 V	50 Hz
Brazil	110/220 V	60 Hz
Canada	120 V	60 Hz
China	220 V	50 Hz
France	230 V	50 Hz
Egypt	220 V	50 Hz
Germany	230 V	50 Hz
India	230 V	50 Hz
Israel	220 V	50 Hz
Japan	100 V	50/60 Hz
Nigeria	240 V	50 Hz
Russia	220 V	50 Hz
South Africa	220/230 V	50 Hz
Spain	220 V	50 Hz
United Kingdom	230 V	50 Hz
United States	120 V	60 Hz

Figure 14.1
A typical power distribution system.
© Sarhan M. Musa

14.2 Instantaneous and Average Power

The *instantaneous power* $p(t)$ absorbed by an element is the product of the instantaneous voltage $v(t)$ across the element and the instantaneous current $i(t)$ through it.

$$p(t) = v(t)i(t) \tag{14.1}$$

It is the rate at which an element absorbs energy.[1]

The **instantaneous power** is power absorbed by an element at any given instant of time.

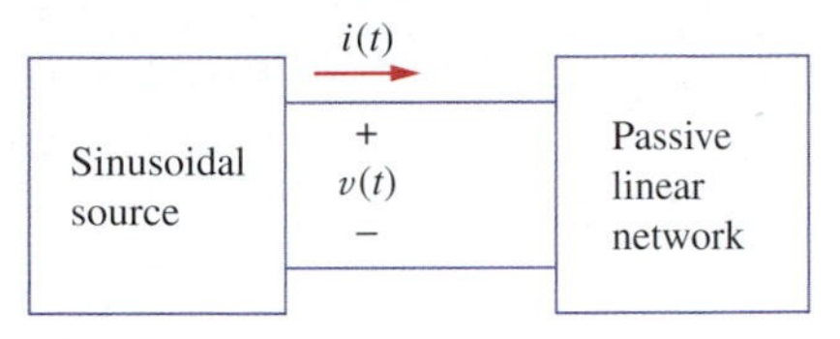

Figure 14.2
Sinusoidal source and passive linear circuit.

Consider now the general case of instantaneous power absorbed by an arbitrary combination of circuit elements under sinusoidal excitation as shown in Fig. 14.2. Let the voltage and current at the terminals of the circuit be

$$v(t) = V_m \cos(\omega t + \theta_v) \tag{14.2a}$$

$$i(t) = I_m \cos(\omega t + \theta_i) \tag{14.2b}$$

where V_m and I_m are the amplitudes (or peak values), and θ_v and θ_i are the phase angles of the voltage and current, respectively. The instantaneous power absorbed by the circuit is

$$p(t) = v(t)i(t) = V_m I_m \cos(\omega t + \theta_v)\cos(\omega t + \theta_i) \tag{14.3}$$

Using a trigonometric identity,

$$\cos A \cos B = \frac{1}{2}[\cos(A - B) + \cos(A + B)],$$

$$p(t) = \frac{1}{2}V_m I_m \cos(\theta_v - \theta_i) + \frac{1}{2}V_m I_m \cos(2\omega t + \theta_v + \theta_i) \tag{14.4}$$

The instantaneous power changes with time and is therefore difficult to measure. The *average* power is more convenient to measure. In fact, the wattmeter, the instrument for measuring power, responds to average power.

The **average power** is the average of the instantaneous power over one period.

A closer look at the instantaneous power as in Eq. (14.4) shows that $p(t)$ has two terms. The first term is constant, while the second term varies with time. Because the average of a cosine wave over a period is zero, the average of the second term is zero. Thus, the average power is given by

$$P = \frac{1}{2}V_m I_m \cos(\theta_v - \theta_i) \tag{14.5}$$

Because $\cos(\theta_v - \theta_i) = \cos(\theta_i - \theta_v)$, what is important is the difference in the phase of the voltage and current.

Note that $p(t)$ is time-varying, while P does not depend on time. To find the instantaneous power, we must necessarily have $v(t)$ and $i(t)$ in the time domain. But the average power can be obtained when

[1] Note: The instantaneous power may also be considered as the power absorbed by the element at a specific instant of time. As mentioned in Chapter 1, instantaneous or time-varying quantities are denoted by lowercase letters.

voltage and current are expressed in the time domain as in Eq. (14.2) or when they are expressed in the frequency domain. The phasor forms of $v(t)$ and $i(t)$ in Eq. (14.2) are $\mathbf{V} = V_m \underline{/\theta_v}$ and $\mathbf{I} = I_m \underline{/\theta_i}$, respectively. P is calculated using Eq. (14.5) or using phasors $\mathbf{V}$ and $\mathbf{I}$. To use phasors, we notice that

$$\begin{aligned}\frac{1}{2}\mathbf{VI}^* &= \frac{1}{2}V_m I_m \underline{/(\theta_v - \theta_i)} \\ &= \frac{1}{2}V_m I_m[\cos(\theta_v - \theta_i) + j\sin(\theta_v - \theta_i)]\end{aligned} \tag{14.6}$$

[The asterisk (*) refers to the complex conjugation, as in Eq. (12.6g).]

We recognize the real part of this expression as the average power P according to Eq. (14.5). Thus,

$$P = \frac{1}{2}\text{Re}[\mathbf{VI}^*] = \frac{1}{2}V_m I_m \cos(\theta_v - \theta_i) \tag{14.7}$$

Consider two special cases of Eq. (14.7).

CASE 1 When $\theta_v = \theta_i$, the voltage and current are in phase. This implies a purely resistive circuit or resistive load R and

$$P = \frac{1}{2}V_m I_m = \frac{1}{2}I_m^2 R = \frac{1}{2}|\mathbf{I}|^2 R \tag{14.8}$$

where $|\mathbf{I}|^2 = \mathbf{I} \times \mathbf{I}^*$. Equation (14.8) shows that a purely resistive circuit absorbs power at all times.

CASE 2 When $\theta_v - \theta_i = \pm 90°$, we have a purely reactive circuit and

$$P = \frac{1}{2}V_m I_m \cos 90° = 0 \tag{14.9}$$

showing that a purely reactive circuit absorbs no average power. In summary,

> A resistive load (R) absorbs power at all times, while a reactive load (L or C) absorbs zero average power.

Example 14.1

Given that

$$v(t) = 120 \cos(377t + 45°) \text{ V and } i(t) = 10 \cos(377t - 10°) \text{ A}$$

find the instantaneous power and the average power absorbed by the passive linear network of Fig. 14.2.

Solution:

The instantaneous power is given by

$$p(t) = v(t)i(t) = 1200 \cos(377t + 45°)\cos(377t - 10°)$$

Applying the trigonometric identity

$$\cos A \cos B = \frac{1}{2}[\cos(A + B) + \cos(A - B)]$$

gives

$$p = 600[\cos(754t + 35°) + \cos 55°]$$

or

$$p(t) = 344.15 + 600\cos(754t + 35^\circ)\ \text{W}$$

The average power is

$$P = \frac{1}{2}V_m I_m \cos(\theta_v - \theta_i) = \frac{1}{2}120(10)\cos[45^\circ - (-10^\circ)]$$
$$= 600\cos 55^\circ = 344.15\ \text{W}$$

which is the constant part of $p(t)$ given earlier.

Practice Problem 14.1

Calculate the instantaneous power and average power absorbed by the passive linear network of Fig. 14.2 if

$$v(t) = 80\cos(10t + 20^\circ)\ \text{V} \quad \text{and} \quad i(t) = 15\sin(10t + 60^\circ)\ \text{A}$$

Answer: $385.67 + 600\cos(20t - 10^\circ)$ W; 385.67 W

Example 14.2

Calculate the average power absorbed by an impedance $\mathbf{Z} = 30 - j70\ \Omega$ when a voltage $\mathbf{V} = 120\underline{/0^\circ}$ V is applied across it.

Solution:
The current through the impedance is

$$\mathbf{I} = \mathbf{V}/\mathbf{Z} = \frac{120\underline{/0^\circ}}{30 - j70} = \frac{120\underline{/0^\circ}}{76.16\underline{/-66.8^\circ}} = 1.576\underline{/66.8^\circ}\ \text{A}$$

The average power is

$$P = \frac{1}{2}V_m I_m \cos(\theta_v - \theta_i) = \frac{1}{2}(120)(1.576)\cos(0^\circ - 66.8^\circ) = 37.25\ \text{W}$$

Practice Problem 14.2

A current $\mathbf{I} = 10\underline{/30^\circ}$ A flows through an impedance $\mathbf{Z} = 20\underline{/22^\circ}\ \Omega$. Find the average power delivered to the impedance.

Answer: 927.18 W

Example 14.3

For the circuit shown in Fig. 14.3, find the average power supplied by the source and the average power absorbed by the resistor.

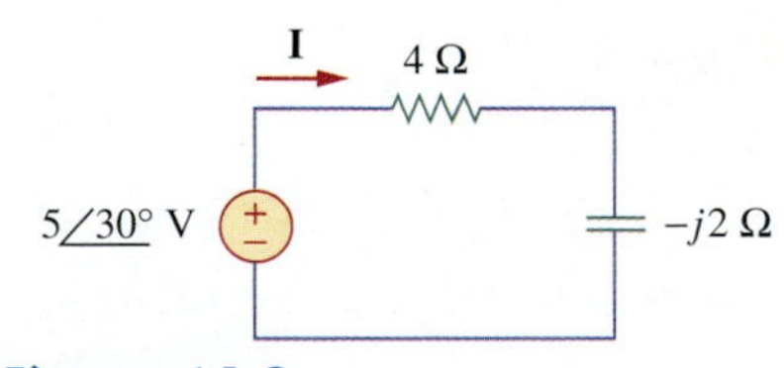

Figure 14.3
For Example 14.3.

Solution:
The current **I** is given by

$$\mathbf{I} = \frac{5\underline{/30^\circ}}{4 - j2} = \frac{5\underline{/30^\circ}}{4.472\underline{/-26.57^\circ}} = 1.118\underline{/56.57^\circ}\ \text{A}$$

The average power supplied by the voltage source is

$$P = \frac{1}{2}(5)(1.118)\cos(30^\circ - 56.57^\circ) = 2.5\ \text{W}$$

The current through the resistor is

$$\mathbf{I}_R = \mathbf{I} = 1.118\underline{/56.57^\circ}\ \text{A}$$

and the voltage across it is

$$\mathbf{V}_R = 4\mathbf{I}_R = 4.471\underline{/56.57^\circ}\ \text{V}$$

The average power absorbed by the resistor is

$$P = \frac{1}{2}(4.472)(1.118) = 2.5 \text{ W}$$

which is the same as the average power supplied. Zero average power is absorbed by the capacitor.

Practice Problem 14.3

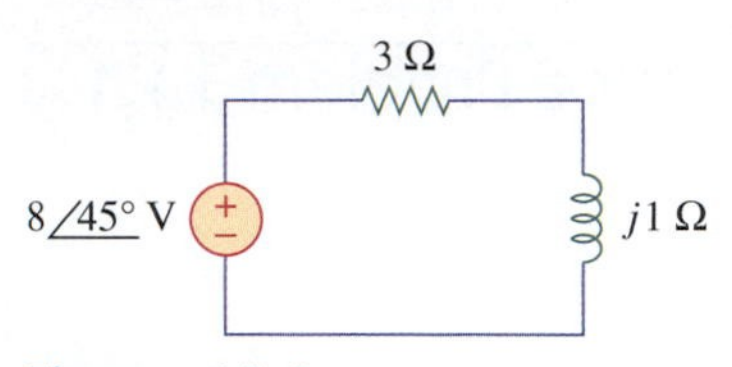

Figure 14.4
For Practice Problem 14.3.

In the circuit of Fig. 14.4, calculate the average power absorbed by the resistor and inductor. Find the average power supplied by the voltage source.

Answer: 9.6 W; 0 W, 9.6 W

14.3 Maximum Average Power Transfer

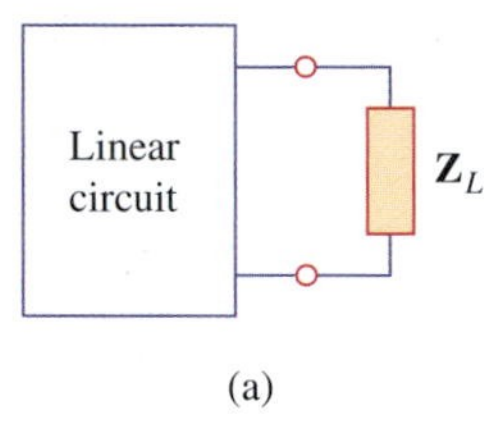

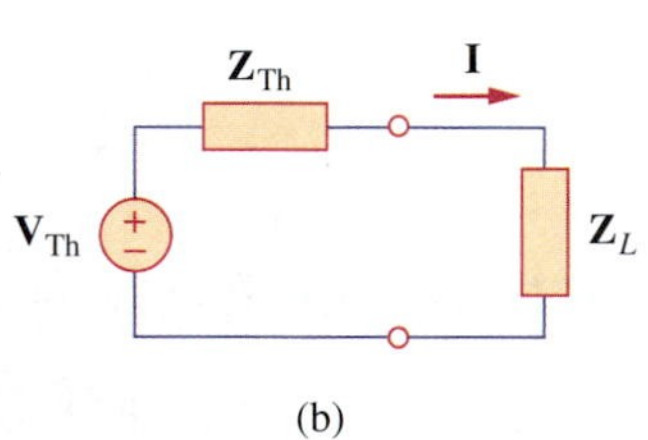

Figure 14.5
Finding the maximum average power transfer: (a) circuit with a load; (b) the Thevenin equivalent.

The problem of maximizing the power delivered by a power-supplying resistive network to a load R_L was solved in Section 8.7. By representing the circuit by its Thevenin equivalent, we proved that the maximum power transfer takes place when the load resistance is equal to the Thevenin resistance; that is, when $R_L = R_{Th}$. We now extend that result to ac circuits.

Consider the circuit in Fig. 14.5(a), where an ac circuit is represented by its Thevenin equivalent and is connected to a load $\mathbf{Z}_L$ as shown in Fig. 14.5(b). The load is usually represented by an impedance, which may model an electric motor, an antenna, a TV, or the like. In rectangular form, the Thevenin impedance $\mathbf{Z}_{Th}$ and load impedance $\mathbf{Z}_L$ are expressed as

$$\mathbf{Z}_{Th} = R_{Th} + jX_{Th} \tag{14.10}$$

$$\mathbf{Z}_L = R_L + jX_L \tag{14.11}$$

The current through the load is

$$\mathbf{I} = \frac{\mathbf{V}_{Th}}{\mathbf{Z}_{Th} + \mathbf{Z}_L} = \frac{\mathbf{V}_{Th}}{(R_{Th} + jX_{Th}) + (R_L + jX_L)} \tag{14.12}$$

From Eq. (14.8), the average power delivered to the load is

$$P = \frac{1}{2}|\mathbf{I}|^2 R = \frac{|\mathbf{V}_{Th}|^2 R_L/2}{(R_{Th} + R_L)^2 + (X_{Th} + X_L)^2} \tag{14.13}$$

Our objective is to adjust the load parameters R_L and X_L so that P is maximum. We can be sure that the circuit current is maximum when the total impedance is minimum. This is achieved by setting

$$X_L = -X_{Th} \tag{14.14}$$

and

$$R_L = \sqrt{R_{Th}^2 + (X_{Th} + X_L)^2} \tag{14.15}$$

Combining Eqs. (14.14) and (14.15) leads to the conclusion that for maximum average power transfer, $\mathbf{Z}_L$ must be selected so that $X_L = -X_{Th}$ and $R_L = R_{Th}$; that is,

$$\mathbf{Z}_L = \mathbf{Z}_{Th}^* \quad \text{or} \quad R_L + jX_L = R_{Th} - jX_{Th} \tag{14.16}$$

Thus,[2]

For **maximum average power transfer**, the load impedance $\mathbf{Z}_L$ must be equal to the complex conjugate of the Thevenin impedance $\mathbf{Z}_{Th}$.

This result is known as the *maximum average power transfer theorem* for the sinusoidal steady state. Setting $R_L = R_{Th}$ and $X_L = -X_{Th}$ in Eq. (14.13) gives us the maximum average power as

$$P_{max} = \frac{|\mathbf{V}_{Th}|^2}{8R_{Th}} \tag{14.17}$$

In a situation in which the load is purely real (or resistive), the condition for maximum power transfer is obtained from Eq. (14.15) by setting $X_L = 0$; that is,

$$R_L = \sqrt{R_{Th}^2 + X_{Th}^2} = |\mathbf{Z}_{Th}| \tag{14.18}$$

This means that for maximum average power transfer to a purely resistive load, the load impedance (or resistance) is equal to the magnitude of the Thevenin impedance.

Example 14.4

Determine the load impedance $\mathbf{Z}_L$ that maximizes the average power drawn from the circuit of Fig. 14.6. What is the maximum average power?

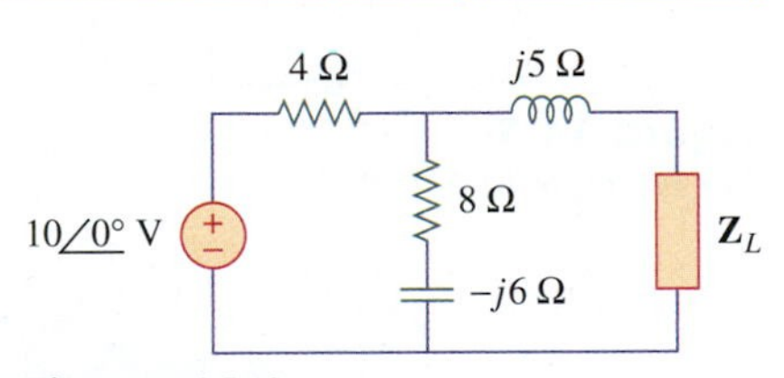

Figure 14.6
For Example 14.4.

Solution:
We first obtain the Thevenin equivalent at the load terminals. To get $\mathbf{Z}_{Th}$, consider the circuit shown in Fig. 14.7(a).

$$\mathbf{Z}_{Th} = j5 + 4 \parallel (8 - j6) = j5 + \frac{4(8 - j6)}{4 + 8 - j6} = 2.933 + j4.4667\ \Omega$$

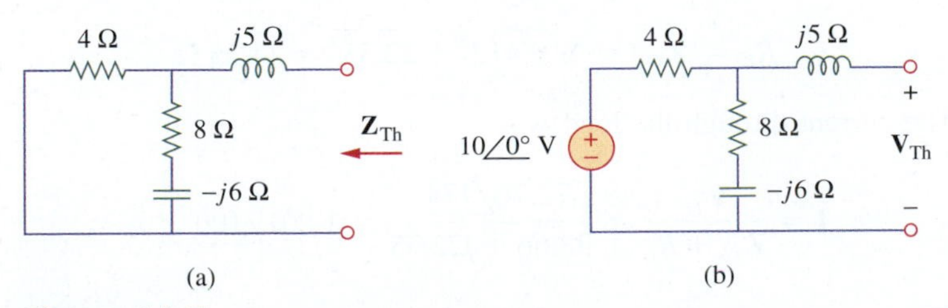

Figure 14.7
Finding the Thevenin equivalent of the circuit in Fig. 14.6.

[2] Note: When $\mathbf{Z}_L = \mathbf{Z}_{Th}^*$, we say that the load is matched to the source.

To find $\mathbf{V}_{Th}$, consider the circuit in Fig. 14.7(b). By the voltage divider rule,

$$\mathbf{V}_{Th} = \frac{8 - j6}{4 + 8 - j6}(10) = 7.454\underline{/-10.3^\circ}$$

The load impedance draws the maximum power from the circuit when

$$\mathbf{Z}_L = \mathbf{Z}_{Th}^* = 2.933 - j4.4667\ \Omega$$

According to Eq. (14.17), the maximum average power is

$$P_{\max} = \frac{|\mathbf{V}_{Th}|^2}{8R_{Th}} = \frac{(7.454)^2}{8(2.933)} = 2.368\text{ W}$$

Practice Problem 14.4

For the circuit shown in Fig. 14.8, find the load impedance $\mathbf{Z}_L$ that absorbs the maximum average power. Calculate that maximum average power.

Answer: $3.413 - j0.731\ \Omega$; 1.431 W

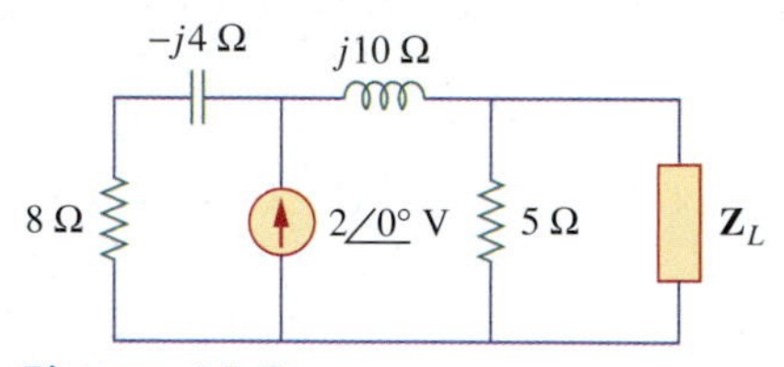

Figure 14.8
For Practice Problem 14.4.

Example 14.5

In the circuit of Fig. 14.9, find the value of R_L that will absorb the maximum average power. Calculate that power.

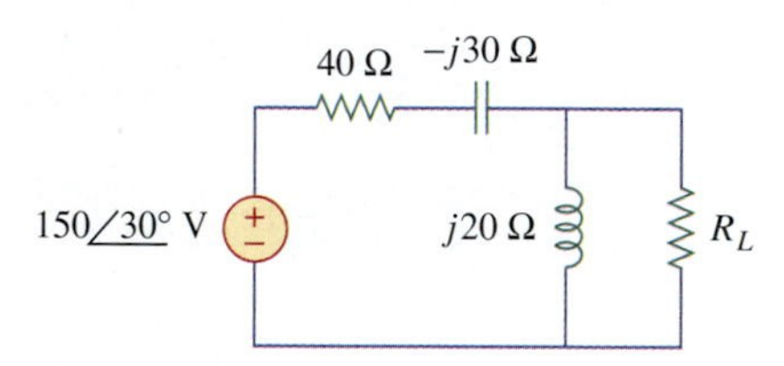

Figure 14.9
For Example 14.5.

Solution:
We first need to find the Thevenin equivalent.

$$\mathbf{Z}_{Th} = (40 - j30) \parallel j20 = \frac{j20(40 - j30)}{j20 + 40 - j30} = 9.412 + j22.35\ \Omega$$

By the voltage divider rule,

$$\mathbf{V}_{Th} = \frac{j20}{j20 + 40 - j30}(150\underline{/30^\circ}) = 72.76\underline{/134^\circ}\text{ V}$$

The value of R_L that will absorb the maximum average power is

$$R_L = |\mathbf{Z}_{Th}| = \sqrt{9.412^2 + 22.35^2} = 24.25\ \Omega$$

The current through the load is

$$\mathbf{I} = \frac{\mathbf{V}_{Th}}{\mathbf{Z}_{Th} + R_L} = \frac{72.76\underline{/134^\circ}}{33.66 + j22.35} = 1.801\underline{/100.2^\circ}\text{ A}$$

The maximum average power absorbed by R_L is

$$P_{\max} = \frac{1}{2}|\mathbf{I}|^2 R = \frac{1}{2}(1.801)^2(24.25) = 39.33\text{ W}$$

Practice Problem 14.5

In Fig. 14.10, the resistor R_L is adjusted until it absorbs the maximum average power. Calculate R_L and the maximum average power absorbed by it.

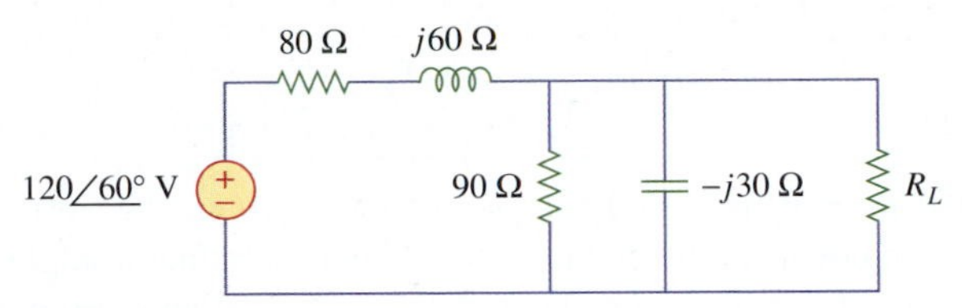

Figure 14.10
For Practice Problem 14.5.

Answer: 29.98 Ω; 5.472 W

14.4 Apparent Power and Power Factor

In Section 14.2 we saw that if the voltage and current at the terminals of a circuit are

$$v(t) = V_m \cos(\omega t + \theta_v) \quad \text{and} \quad i(t) = I_m \cos(\omega t + \theta_i) \tag{14.19}$$

or, in phasor form,

$$\mathbf{V} = V_m \underline{/\theta_v} \quad \text{and} \quad \mathbf{I} = I_m \underline{/\theta_i}$$

the average power is

$$P = \frac{1}{2} V_m I_m \cos(\theta_v - \theta_i) \tag{14.20}$$

The root mean square voltage (V_{rms}) is the effective value of a time-varying voltage. It is the equivalent steady dc value which gives the same effect. The rms voltage or current is 0.7 of the peak voltage (V_m) or peak current (I_m). The average power in Eq. (14.20) can be written in terms of the rms values as

$$\begin{aligned} P &= \frac{1}{2} V_m I_m \cos(\theta_v - \theta_i) = \frac{V_m}{\sqrt{2}} \frac{I_m}{\sqrt{2}} \cos(\theta_v - \theta_i) \\ &= V_{\text{rms}} I_{\text{rms}} \cos(\theta_v - \theta_i) \end{aligned} \tag{14.21}$$

or

$$P = S \cos(\theta_v - \theta_i) \tag{14.22}$$

where

$$\boxed{S = V_{\text{rms}} I_{\text{rms}}} \tag{14.23}$$

From Eq. (14.22), we notice that the average power is a product of two terms. The product $V_{\text{rms}} I_{\text{rms}}$ is known as the *apparent power S* and the factor $\cos(\theta_v - \theta_i)$ is called the *power factor (pf)*.

> The **apparent power** (in voltamperes, VA) is the product of the rms values of voltage and current.

The apparent power is so called because it seems apparent that the power should be the voltage–current product, by analogy with dc resistive circuits. It is measured in voltamperes (or VA) to distinguish it from the average or real power, which is measured in watts. The power factor is dimensionless because it is the ratio of the average power to the apparent power; that is,

$$pf = \frac{P}{S} = \cos(\theta_v - \theta_i) \tag{14.24}$$

The angle $\theta_v - \theta_i$ is called the *power factor angle* because it is the angle whose cosine is the power factor. The power factor angle is equal to the angle of the load impedance if **V** is the voltage across the load and **I** is the current through it. This is evident from the fact that

$$\mathbf{Z} = \mathbf{V}/\mathbf{I} = \frac{V_m/\theta_v}{I_m/\theta_i} = \frac{V_m}{I_m}/(\theta_v - \theta_i) \tag{14.25}$$

Alternatively, because

$$\mathbf{V}_{\text{rms}} = \mathbf{V}/\sqrt{2} = V_{\text{rms}}/\theta_v \tag{14.26a}$$

and

$$\mathbf{I}_{\text{rms}} = \mathbf{I}/\sqrt{2} = I_{\text{rms}}/\theta_i \tag{14.26b}$$

the impedance is

$$\mathbf{Z} = \mathbf{V}/\mathbf{I} = \mathbf{V}_{\text{rms}}/\mathbf{I}_{\text{rms}} = \frac{V_{\text{rms}}}{I_{\text{rms}}}/(\theta_v - \theta_i) \tag{14.27}$$

The **power factor** is the cosine of the phase difference between voltage and current. It is also the cosine of the angle of the load impedance.[3]

From Eq. (14.24), the power factor may also be defined as that factor by which the apparent power must be multiplied to obtain the real or average power. As shown in Table 14.2, the value of *pf* ranges between zero and unity. For a purely resistive load, the voltage and current are in phase so that $\theta_v - \theta_i = 0$ and $pf = 1$. This implies that the apparent power is equal to the average power. For a purely reactive load, $\theta_v - \theta_i = \pm 90°$ and $pf = 0$. In this case, the average power is zero. In between these two extreme cases, *pf* is said to be *leading* or *lagging*. Leading power factor means that current leads voltage, implying a capacitive load. Lagging power factor means that current lags voltage, implying an inductive load. Power factor affects the electric bills paid by industrial consumers. This will be discussed in Section 14.8.2.

TABLE 14.2

Power factor for typical power factor angles.

Power angle $(\theta_v - \theta_i)$	Power factor $[\cos(\theta_v - \theta_i)]$
+90°	0
+60°	0.5
+45°	0.7071
+30°	0.8660
0°	1.0
−30°	0.8660
−45°	0.07071
−60°	0.5
−90°	0

Example 14.6

A series-connected load draws a current $i(t) = 4\cos(100\pi t + 10°)$ A when the applied voltage is $v(t) = 120\cos(100\pi t - 20°)$ V. Find the apparent power and the power factor of the load. Determine the element values that form the series-connected load.

Solution:

The apparent power is

$$S = V_{\text{rms}} I_{\text{rms}} = \frac{120}{\sqrt{2}}\frac{4}{\sqrt{2}} = 240 \text{ VA}$$

[3] Note: From Eq. (14.24), the power factor may also be regarded as the ratio of the real power (P) dissipated in the load to the apparent power (S) of the load. Although P and S have different units, they are still powers distinguished only by the units and their ratio P/S is dimensionless. In fact, the unit for all forms of power is the watt (W). In practice, however, this is generally reserved for the real power.

The power factor is

$$pf = \cos(\theta_v - \theta_i) = \cos(-20^\circ - 10^\circ) = 0.866$$

The pf is leading because the current leads the voltage. The pf may also be obtained from the load impedance.

$$\mathbf{Z} = \mathbf{V}/\mathbf{I} = \frac{120\angle -20^\circ}{4\angle 10^\circ} = 30\angle -30^\circ = 25.98 - j15\ \Omega$$

$$pf = \cos(-30^\circ) = 0.866 \qquad \text{(leading)}$$

The load impedance $\mathbf{Z}$ can be modeled by a 25.98-Ω resistor in series with a capacitor, with

$$X_C = 15 = \frac{1}{\omega C}$$

or

$$C = \frac{1}{15\omega} = \frac{1}{15 \times 100\pi} = 212.2\ \mu\text{F}$$

Practice Problem 14.6

Obtain the power factor and the apparent power of a load whose impedance is $\mathbf{Z} = 60 + j40\ \Omega$ when the applied voltage is $v(t) = 150 \cos(377t + 10^\circ)$ V.

Answer: 0.832 lagging; 156 VA

Example 14.7

Determine the power factor of the entire circuit of Fig. 14.11 as seen by the source. Calculate the average power delivered by the source.

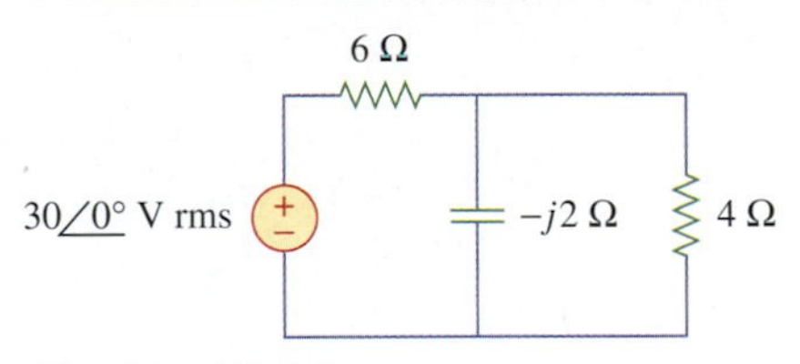

Figure 14.11
For Example 14.7.

Solution:
The total impedance is

$$\mathbf{Z} = 6 + 4 \parallel (-j2) = 6 + \frac{-j2 \times 4}{4 - j2} = 6.8 - j1.6 = 7\angle -13.24^\circ\ \Omega$$

The power factor is

$$pf = \cos(-13.24^\circ) = 0.9734 \qquad \text{(leading)}$$

It is leading because the impedance is capacitive. The rms value of the current is

$$\mathbf{I}_{\text{rms}} = \mathbf{V}_{\text{rms}}/\mathbf{Z} = \frac{30\angle 0^\circ}{7\angle -13.24^\circ} = 4.294\angle 13.24^\circ$$

The average power supplied by the source is

$$P = V_{\text{rms}} I_{\text{rms}}\, pf = (30)(4.294)0.9734 = 125\text{ W}$$

or

$$P = I_{\text{rms}}^2 R = (4.294)^2(6.8) = 125\text{ W}$$

where R is the resistive part of $\mathbf{Z}$.

Practice Problem 14.7

Calculate the power factor of the entire circuit in Fig. 14.12 as seen by the source. What is the average power supplied by the source?

Answer: 0.936 lagging; 118 W

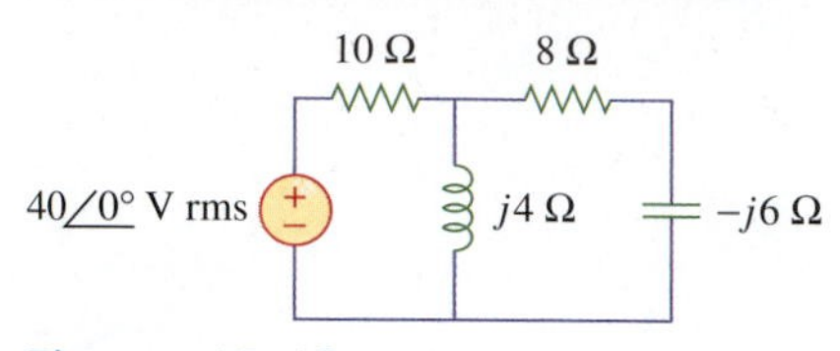

Figure 14.12
For Practice Problem 14.7.

14.5 Complex Power

Considerable effort has been expended over the years to express power relations as simply as possible. Power engineers have coined the term *complex power*, which is used in finding the total effect of parallel loads. Complex power is important in power analysis because it contains *all* the information pertaining to the power absorbed by a given load.[4]

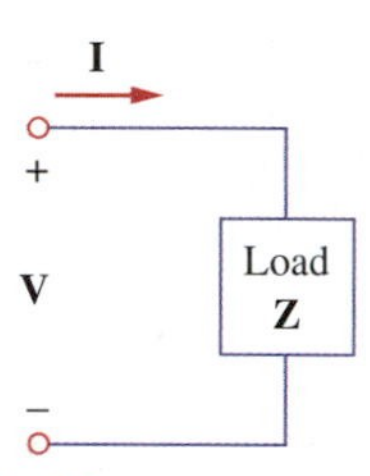

Figure 14.13
The voltage and current phasors associated with a load.

Consider the ac load in Fig. 14.13. Given the phasor form $\mathbf{V} = V_m \underline{/\theta_v}$ and $\mathbf{I} = I_m = \underline{/\theta_i}$ of voltage $v(t)$ and current i(i), the *complex power* **S** absorbed by the ac load is the product of the voltage and the complex conjugate of the current; that is,

$$\mathbf{S} = \frac{1}{2}\mathbf{V}\mathbf{I}^* \tag{14.28}$$

assuming the passive sign convention (see Fig. 14.13). In terms of the rms values

$$\mathbf{S} = \mathbf{V}_{\text{rms}}\mathbf{I}^*_{\text{rms}} \tag{14.29}$$

where

$$\mathbf{V}_{\text{rms}} = \mathbf{V}/\sqrt{2} = V_{\text{rms}} \underline{/\theta_v} \tag{14.30}$$

and

$$\mathbf{I}_{\text{rms}} = \mathbf{I}/\sqrt{2} = I_{\text{rms}} \underline{/\theta_I} \tag{14.31}$$

Thus we may write Eq. (14.29) as[5]

$$\begin{aligned}\mathbf{S} &= V_{\text{rms}}I_{\text{rms}} \underline{/(\theta_v - \theta_i)} \\ &= V_{\text{rms}}I_{\text{rms}}\cos(\theta_v - \theta_i) + jV_{\text{rms}}I_{\text{rms}}\sin(\theta_v - \theta_i)\end{aligned} \tag{14.32}$$

This equation can also be obtained from Eq. (14.6). We notice from Eq. (14.32) that the magnitude of the complex power is the apparent power; hence, the complex power is measured in voltamperes (VA). Also, we notice that the angle of the complex power is the power factor angle.

The complex power may be expressed in terms of the load impedance **Z**. From Eq. (14.27), the load impedance **Z** may be written as

$$\mathbf{Z} = \mathbf{V}/\mathbf{I} = \mathbf{V}_{\text{rms}}/\mathbf{I}_{\text{rms}} = \frac{V_{\text{rms}}}{I_{\text{rms}}} \underline{/\theta_v - \theta_i} \tag{14.33}$$

[4] Complex power has no physical significance; it is purely a mathematical concept that helps in understanding power analysis.

[5] Note: When working with the rms values of currents or voltages, we may drop the subscript rms if no confusion will be caused by doing so.

Thus, $\mathbf{V}_{\text{rms}} = \mathbf{Z}\mathbf{I}_{\text{rms}}$. Substituting this into Eq. (14.28) gives

$$\mathbf{S} = I_{\text{rms}}^2\mathbf{Z} = \frac{V_{\text{rms}}^2}{\mathbf{Z}^*} = \mathbf{V}_{\text{rms}}\mathbf{I}_{\text{rms}}^* \tag{14.34}$$

Because $\mathbf{Z} = R + jX$, Eq. (14.34) becomes

$$\mathbf{S} = I_{\text{rms}}^2 (R + jX) = P + jQ \tag{14.35}$$

where P and Q are the real and imaginary parts of the complex power, respectively; that is,

$$P = \text{Re}(\mathbf{S}) = I_{\text{rms}}^2 R \tag{14.36}$$

$$Q = \text{Im}(\mathbf{S}) = I_{\text{rms}}^2 X \tag{14.37}$$

Here, P is the average or real power, and it depends on the load's resistance R; Q depends on the load's reactance X and is called the *reactive* (or *quadrature*) *power*.

Comparing Eq. (14.32) with Eq. (14.35), we notice that

$$P = V_{\text{rms}} I_{\text{rms}} \cos(\theta_v - \theta_i), \quad Q = V_{\text{rms}} I_{\text{rms}} \sin(\theta_v - \theta_i) \tag{14.38}$$

The unit of Q is the *voltampere reactive* (VAR) to distinguish it from the real power whose unit is the watt. The real power P is the average power delivered to a load; it is the only useful power. It is the actual power dissipated by the load. The reactive power is being transferred back and forth between the load and the source. It serves as a measure of the energy storage capability of the reactive component of the load. It represents a lossless interchange between the load and the source. Notice that:

1. $Q = 0$ for resistive loads (unity *pf*).
2. $Q < 0$ for capacitive loads (leading *pf*).
3. $Q > 0$ for inductive loads (lagging *pf*).

Thus,

> **Complex power** (in VA) is the product of rms voltage phasor and the complex conjugate of rms current phasor. As a complex quantity, its real part is real power P, and its imaginary part is reactive power Q.

Introducing the complex power enables us to obtain the real and reactive powers directly from voltage and current phasors. In summary,

$$\begin{aligned}
\text{Complex power} &= \mathbf{S} = P + jQ = \frac{1}{2}\mathbf{V}\mathbf{I}^* \\
&= V_{\text{rms}} I_{\text{rms}} \underline{/(\theta_v - \theta_i)} \\
\text{Apparent power} &= S = |\mathbf{S}| = V_{\text{rms}} I_{\text{rms}} = \sqrt{P^2 + Q^2} \\
\text{Real power} &= P = \text{Re}(\mathbf{S}) = S\cos(\theta_v - \theta_i) \\
\text{Reactive power} &= Q = \text{Im}(\mathbf{S}) = S\sin(\theta_v - \theta_i) \\
\text{Power factor} &= \frac{P}{S} = \cos(\theta_v - \theta_i)
\end{aligned} \tag{14.39}$$

which shows how the complex power contains *all* the relevant power information in a given load.[6]

It is a standard practice to represent S, P, and Q in the form of a triangle, known as the *power triangle*, as shown in Fig. 14.14(a). This is similar to the impedance triangle showing the relationship among $\mathbf{Z}$, R, and X, as illustrated in Fig. 14.14(b). The power triangle has four items (the apparent/complex power, real power, reactive power, and the power angle). Given two of these items, the other two can easily be obtained from the triangle. As shown in Fig. 14.15, when $\mathbf{S}$ lies in the first quadrant, we have an inductive load and a lagging *pf*. When $\mathbf{S}$ lies in the fourth quadrant, the load is capacitive and the *pf* is leading. It is also possible for the complex power to lie in the second or third quadrant. This requires that the load impedance have a negative resistance, which is possible with active circuits. (An active circuit is one that has active elements such as transistors and operational amplifiers.)

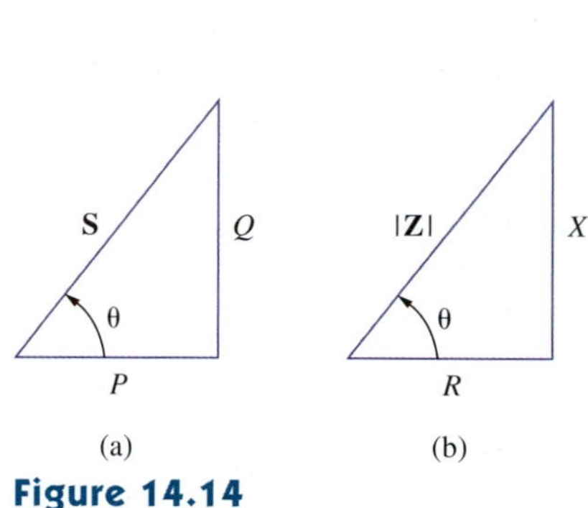

Figure 14.14
(a) Power triangle;
(b) impedance triangle.

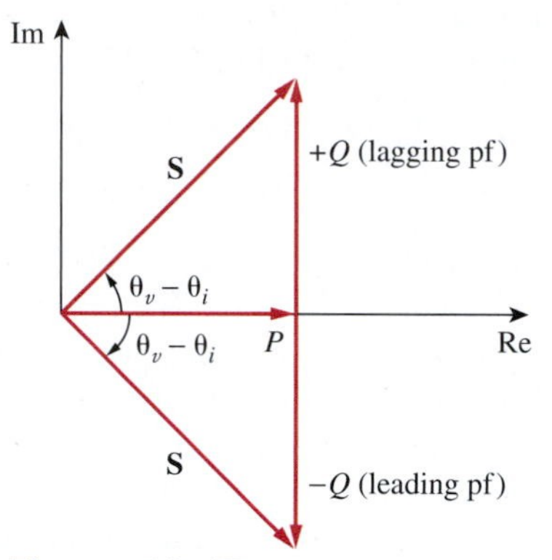

Figure 14.15
Power triangle.

Example 14.8

The voltage across a load is $v(t) = 60\cos(\omega t - 10^\circ)$ V and the current through the load in the direction of voltage drop is $i(t) = 1.5\cos(\omega t + 50^\circ)$ A. Find:
(a) the complex and apparent powers,
(b) the real and reactive powers, and
(c) the power factor and the load impedance.

Solution:

(a) The rms values of the voltage and current are given by

$$\mathbf{V}_{\text{rms}} = \frac{60}{\sqrt{2}}\underline{/-10^\circ} \qquad \mathbf{I}_{\text{rms}} = \frac{1.5}{\sqrt{2}}\underline{/+50^\circ}$$

The complex power is

$$\mathbf{S} = \mathbf{V}_{\text{rms}}\mathbf{I}^*_{\text{rms}} = \left(\frac{60}{\sqrt{2}}\underline{/-10^\circ}\right)\left(\frac{1.5}{\sqrt{2}}\underline{/-50^\circ}\right) = 45\underline{/-60^\circ} \text{ VA}$$

[6] Note: Note that $\mathbf{S}$ contains *all* power information of a load. The real part of $\mathbf{S}$ is the real power P, its imaginary part is the reactive power Q, its magnitude is the apparent power S, and the cosine of its phase angle is the power factor *pf*.

The apparent power is

$$S = |\mathbf{S}| = 45 \text{ VA}$$

(b) We can express the complex power in rectangular form as

$$\mathbf{S} = 45\angle{-60^\circ} = 45\,[\cos(-60^\circ) + j\sin(-60^\circ)] = 22.5 - j38.97$$

Because $\mathbf{S} = P + jQ$, the real power is

$$P = 22.5 \text{ W}$$

while the reactive power is

$$Q = -38.97 \text{ VAR}$$

(c) The power factor is

$$pf = \cos(-60^\circ) = 0.5 \text{ (leading)}$$

It is leading because the reactive power is negative. The load impedance is

$$\mathbf{Z} = \mathbf{V}/\mathbf{I} = \frac{60\angle{-10^\circ}}{1.5\angle{+50^\circ}} = 40\angle{-60^\circ}\ \Omega$$

which is a capacitive impedance.

Practice Problem 14.8

For a load, $\mathbf{V}_{\text{rms}} = 110\angle{85^\circ}$ V, $\mathbf{I}_{\text{rms}} = 0.4\angle{15^\circ}$ A. Determine:
(a) the complex and apparent powers,
(b) the real and reactive powers, and
(c) the power factor and the load impedance.

Answer: (a) $44\angle{70^\circ}$ VA, 44 VA; (b) 15.05 W, 41.35 VAR; (c) 0.342 lagging, $94.06 + j258.4\ \Omega$

Example 14.9

If a load $\mathbf{Z}$ draws 12 kVA at a power factor of 0.856 lagging from a 120-V rms sinusoidal source, calculate:
(a) the average and reactive powers delivered to the load,
(b) the peak current, and
(c) the load impedance.

Solution:

(a) Given that $pf = \cos\theta = 0.856$, we obtain the power angle as

$$\theta = \cos^{-1} 0.856 = 31.13^\circ$$

If the apparent power is $S = 12{,}000$ VA, then average or real power is

$$P = S\cos\theta = 12{,}000 \times 0.856 = 10.272 \text{ kW}$$

while the reactive power

$$Q = S\sin\theta = 12{,}000 \times 0.517 = 6.204 \text{ kVA}$$

(b) Because the pf is lagging, the complex power is

$$\mathbf{S} = P + jQ = 10.272 + j6.204 \text{ kVA}$$

From $\mathbf{S} = \mathbf{V}_{\text{rms}}\mathbf{I}^*_{\text{rms}}$, we obtain

$$\mathbf{I}^*_{\text{rms}} = \mathbf{S}/\mathbf{V}_{\text{rms}} = \frac{10{,}272 + j6204}{120\underline{/0^\circ}} = 85.6 + j51.7 \text{ A} = 100\underline{/31.13^\circ}\text{A}$$

Thus $\mathbf{I}_{\text{rms}} = 100\underline{/-31.13^\circ}$ and the peak current is

$$I_p = \text{I}_m = \sqrt{2}I_{\text{rms}} = \sqrt{2}(100) = 141.4 \text{ A}$$

(c) The load impedance

$$\mathbf{Z} = \mathbf{V}_{\text{rms}}/\,\mathbf{I}_{\text{rms}} = \frac{120\underline{/0^\circ}}{100\underline{/-31.13^\circ}} = 1.2\underline{/31.13^\circ}\ \Omega$$

which is an inductive impedance.

Practice Problem 14.9

A sinusoidal source supplies 10 kVA to a load $\mathbf{Z} = 250\underline{/-75^\circ}\ \Omega$. Determine:

(a) the power factor,
(b) the apparent power delivered to the load, and
(c) the peak voltage.

Answer: (a) 0.2588 (leading); (b) 10.353 kVA; (c) 2.275 kV

14.6 †Conservation of AC Power

The principle of conservation of power applies to ac circuits as it does to dc circuits (see Section 1.7).[7] Whether the loads are connected in series or in parallel (or other configurations), the total power *supplied* by the source equals the total power *delivered* to the load. Thus, in general, for a source connected to N loads,

$$\mathbf{S} = \mathbf{S}_1 + \mathbf{S}_2 + \mathbf{S}_3 + \cdots + \mathbf{S}_N \tag{14.40}$$

(Keep in mind that the loads can be connected in any configuration.) This means that the total complex power in a network is the sum of the complex powers of the individual components. (This is also true of real power and reactive power, but not true of apparent power.) This expresses the principle of conservation of ac power:

From this we imply that the real (or reactive) power from sources in a network equals the real (or reactive) power into the other elements in the network.

> The complex, real, and reactive powers of the source equal the respective sums of the complex, real, and reactive powers of the individual loads.

Example 14.10

Figure 14.16 shows a load being fed by a voltage source through a transmission line. The impedance of the line is represented by the $(4 + j2)\ \Omega$ impedance and a return path. Find the real power and reactive power absorbed by:

(a) the source,
(b) the line, and
(c) the load.

[7] Note: In fact, we already saw in Example 14.3 that average power is conserved in ac circuits.

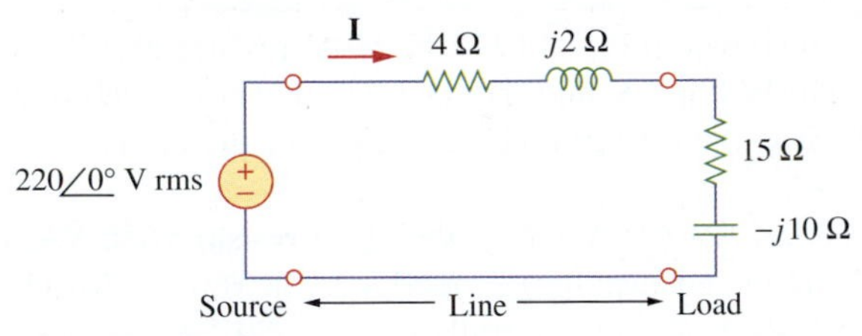

Figure 14.16
For Example 14.10.

Solution:

The total impedance is

$$\mathbf{Z} = (4 + j2) + (15 - j10) = 19 - j8 = 20.62\underline{/-22.83^\circ}\ \Omega$$

The current through the circuit is

$$\mathbf{I} = \mathbf{V}_s/\mathbf{Z} = \frac{220\underline{/0^\circ}}{20.62\underline{/-22.83^\circ}} = 10.67\underline{/22.83^\circ}\ \text{A rms}$$

(a) For the source, the complex power is

$$\mathbf{S}_s = \mathbf{V}_s\mathbf{I}^* = (220\underline{/0^\circ})(10.67\underline{/-22.83^\circ})$$
$$= 2347.4\underline{/-22.83^\circ} = (2163.5 - j910.8)\ \text{VA}$$

From this, we obtain the real power as 2163.5 W and the reactive power as 910.8 VAR (leading).

(b) For the line, the voltage is

$$\mathbf{V}_{\text{line}} = (4 + j2)\mathbf{I} = (4.472\underline{/26.57^\circ})(10.67\underline{/22.83^\circ})$$
$$= 47.72\underline{/49.4^\circ}\ \text{V rms}$$

The complex power absorbed by the line is

$$\mathbf{S}_{\text{line}} = \mathbf{V}_{\text{line}}\mathbf{I}^* = (47.72\underline{/49.4^\circ})(10.67\underline{/-22.83^\circ})$$
$$= 509.2\underline{/26.57^\circ} = 455.4 + j227.76\ \text{VA}$$

or

$$\mathbf{S}_{\text{line}} = |\mathbf{I}|^2\mathbf{Z}_{\text{line}} = (10.67)^2\,(4 + j2) = 455.4 + j227.7\ \text{VA}$$

that is, the real power is 455.4 W and the reactive power is 227.76 VAR (lagging).

(c) For the load, the voltage is

$$\mathbf{V}_L = (15 - j10)\mathbf{I} = (18.03\underline{/-33.7^\circ})(10.67\underline{/22.83^\circ})$$
$$= 192.38\underline{/-10.87^\circ}\ \text{V rms}$$

The complex power absorbed by the load is

$$\mathbf{S}_L = \mathbf{V}_L\mathbf{I}^* = (192.38\underline{/-10.87^\circ})(10.67\underline{/-22.83^\circ})$$
$$= 2053\underline{/-33.7^\circ} = (1708 - j1139)\ \text{VA}$$

The real power is 1708 W and the reactive power is 1139 VAR (leading). Note that

$$\mathbf{S}_s = \mathbf{S}_{\text{line}} + \mathbf{S}_L$$

as expected. We have used the rms values of voltages and currents throughout this problem.

Practice Problem 14.10

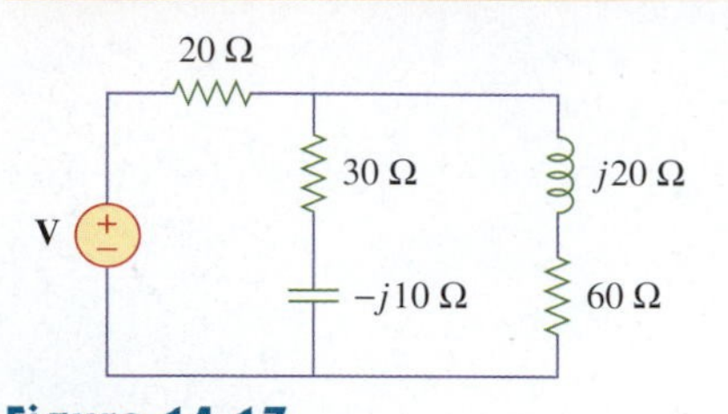

Figure 14.17
For Practice Problem 14.10.

In the circuit shown in Fig. 14.17, the 60-Ω resistor absorbs an average power of 240 W. Find **V** and the complex power of each branch of the circuit. What is the overall complex power of the circuit?

Answer: $240.67\underline{/21.45^\circ}$ V (rms); the 20-Ω resistor: 656 VA; the $(30 - j10)$ Ω impedance: $480 - j160$ VA; the $(60 + j20)$ Ω impedance: $240 + j80$ VA; overall: $1376 - j80$ VA

Example 14.11

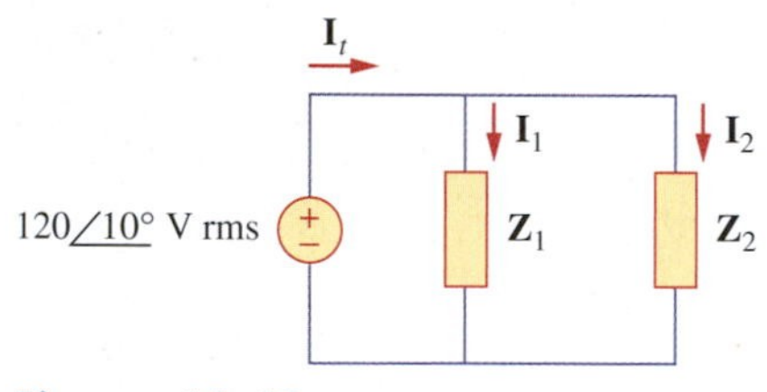

Figure 14.18
For Example 14.11.

In the circuit of Fig. 14.18, $\mathbf{Z}_1 = 60\underline{/-30^\circ}$ Ω and $\mathbf{Z}_2 = 40\underline{/45^\circ}$ Ω. Calculate the total:

(a) apparent power,
(b) real power,
(c) reactive power, and
(d) *pf*.

Solution:

The current through $\mathbf{Z}_1$ is

$$\mathbf{I}_1 = \mathbf{V}/\mathbf{Z}_1 = \frac{120\underline{/10^\circ}}{60\underline{/-30^\circ}} = 2\underline{/40^\circ}\text{ A rms}$$

while the current through $\mathbf{Z}_2$ is

$$\mathbf{I}_2 = \mathbf{V}/\mathbf{Z}_2 = \frac{120\underline{/10^\circ}}{40\underline{/45^\circ}} = 3\underline{/-35^\circ}\text{ A rms}$$

The complex powers absorbed by the impedances are

$$\mathbf{S}_1 = \frac{V_{\text{rms}}^2}{\mathbf{Z}_1^*} = \frac{120^2}{60\underline{/30^\circ}} = 240\underline{/-30^\circ} = 207.85 - j120\text{ VA}$$

$$\mathbf{S}_2 = \frac{V_{\text{rms}}^2}{\mathbf{Z}_2^*} = \frac{120^2}{40\underline{/-45^\circ}} = 360\underline{/45^\circ} = 254.6 + j254.6\text{ VA}$$

The total complex power is

$$\mathbf{S}_t = \mathbf{S}_1 + \mathbf{S}_2 = 462.45 + j134.6\text{ VA}$$

(a) The total apparent power is

$$|\mathbf{S}_t| = \sqrt{462.45^2 + 134.6^2} = 481.64\text{ VA}$$

(b) The total real power is

$$P_t = \text{Re}(\mathbf{S}_t) = 462.45\text{ W} \quad \text{or} \quad P_t = P_1 + P_2$$

(c) The total reactive power is

$$Q_t = \text{Im}(\mathbf{S}_t) = 134.6\text{ VAR} \quad \text{or} \quad Q_t = Q_1 + Q_2$$

(d) The $pf = P_t/|\mathbf{S}_t| = 462.45/481.64 = 0.96$ (lagging).

We may cross check the result by finding the complex power $\mathbf{S}_s$ supplied by the source.

$$\begin{aligned}\mathbf{I}_t &= \mathbf{I}_1 + \mathbf{I}_2 = (1.532 + j1.286) + (2.457 - j1.721)\\ &= 4 - j0.435 = 4.024\underline{/-6.21^\circ}\text{ A rms}\end{aligned}$$

$$\begin{aligned}\mathbf{S}_s &= \mathbf{V}\mathbf{I}_t^* = (120\underline{/10^\circ})(4.024\underline{/6.21^\circ})\\ &= 482.88\underline{/16.21^\circ} = 463 + j135\text{ VA}\end{aligned}$$

which is the same as given earlier.

Practice Problem 14.11

Two loads connected in parallel are respectively 2 kW at a *pf* of 0.75 leading and 4 kW at *pf* of 0.95 lagging. Calculate the *pf* of the two loads. Find the complex power supplied by the source.

Answer: 0.9972 (leading); $6 - j0.4495$ kVA

14.7 Power Factor Correction

Most residential loads (such as washing machines, air conditioners, and refrigerators) and industrial loads (such as induction motors) are inductive and therefore operate at a lagging power factor. Although the inductive nature of the load cannot be changed, we can increase its power factor.

> The process of increasing the power factor without altering the real power of the original load is known as **power factor correction**.[8]

Because most loads are inductive,[9] as shown in Fig. 14.19(a), a load's power factor is improved or corrected by deliberately installing a capacitor in parallel with the load, as shown in Fig. 14.19(b). The effect of adding the capacitor can be illustrated using either the power triangle or the phasor diagram of the currents involved. The latter is shown in Fig. 14.20, where it is assumed that the circuit in Fig. 14.19(a) has a power factor of $\cos\theta_1$, while the one in Fig. 14.19(b) has a power factor of $\cos\theta_2$. It is evident from Fig. 14.20 that adding the capacitor has caused the phase angle between the supplied voltage and current to reduce from θ_1 to θ_2, thereby increasing the power factor.

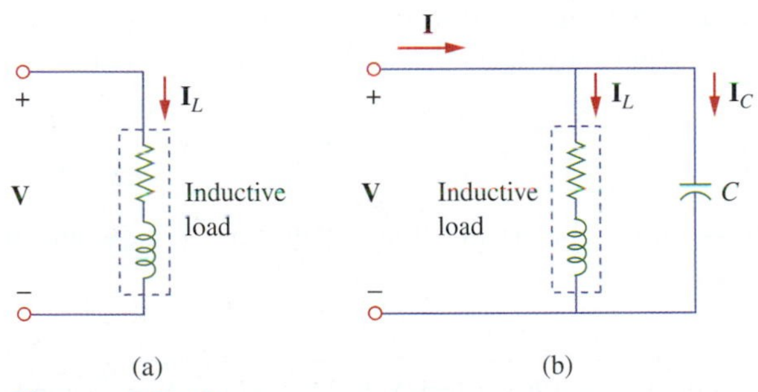

Figure 14.19
Power factor correction: (a) original inductive load; (b) inductive load with improved power factor.

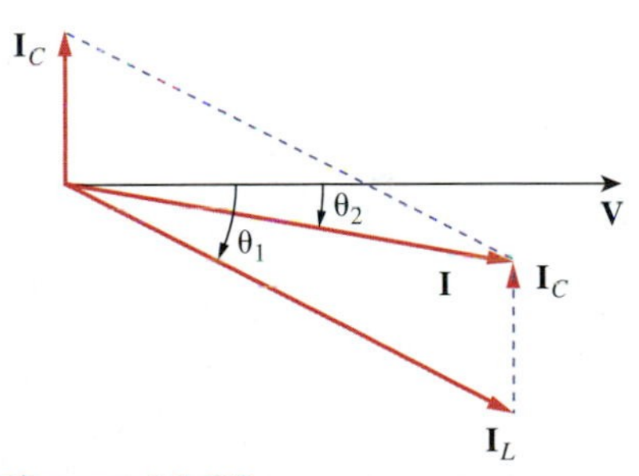

Figure 14.20
Phasor diagram showing the effect of adding a capacitor in parallel with the inductive load.

We also notice from the magnitudes of the vectors in Fig. 14.20 that with the same supplied voltage, the circuit in Fig. 14.19(a) draws a larger current I_L than the current I drawn by the circuit in Fig. 14.19(b). Power companies charge more for larger currents because they result in increased power losses (by a squared factor because $P = I_L^2 R$). Therefore, it is mutually beneficial to both the power company and the consumer that every effort be made to minimize current level or keep the power factor as close to unity as possible. By choosing a capacitor of a

[8] Note: Alternatively, power factor correction may be viewed as the addition of a reactive element (usually a capacitor) in parallel with the load in order to make the power factor closer to unity.

[9] Reminder: An inductive load is modeled as a series combination of an inductor and a resistor.

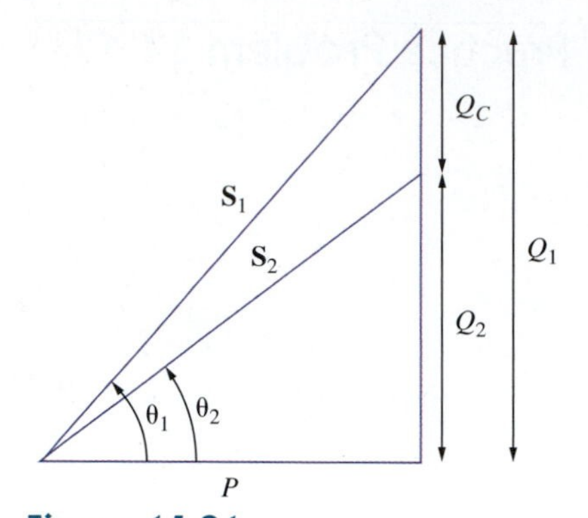

Figure 14.21
Power triangle illustrating power factor correction.

suitable size, the current can be made to be completely in phase with the voltage, implying unity power factor.

We can look at the power factor correction from another perspective. Consider the power triangle in Fig. 14.21. If the original inductive load has apparent power S_1, then

$$P = S_1 \cos\theta_1, \qquad Q_1 = S_1 \sin\theta_1 \tag{14.41}$$

If we desire to increase the power factor from $\cos\theta_1$ to $\cos\theta_2$ without altering the real power—that is, $P = S_2 \cos\theta_2$—then the new reactive power is

$$Q_2 = P\tan\theta_2 \tag{14.42}$$

The reduction in the reactive power is caused by the shunt capacitor; that is,

$$Q_C = Q_1 - Q_2 = P(\tan\theta_1 - \tan\theta_2) \tag{14.43}$$

But from Eq. (14.34), $Q_C = V_{\text{rms}}^2/X_C = \omega C V_{\text{rms}}^2$. The value of the required shunt capacitance C is determined as

$$C = \frac{Q_C}{\omega V_{\text{rms}}^2} = \frac{P(\tan\theta_1 - \tan\theta_2)}{\omega V_{\text{rms}}^2} \tag{14.44}$$

Note that the real power P dissipated by the load is not affected by the power factor correction.

Although the most common situation in practice is that of an inductive load, it is also possible that the load is capacitive, that is, that the load is operating at a leading power factor. In this case, an inductor should be connected across the load for power factor correction. The required shunt inductance L can be calculated from

$$Q_L = \frac{V_{\text{rms}}^2}{X_L} = \frac{V_{\text{rms}}^2}{\omega L} \quad \Rightarrow \quad L = \frac{V_{\text{rms}}^2}{\omega Q_L} \tag{14.45}$$

where $Q_L = Q_1 - Q_2$, the difference between the new and old reactive powers.

Example 14.12

When connected to a 120-V (rms), 60-Hz power line, a load absorbs 4 kW at a lagging power factor of 0.8. Find the value of capacitance necessary to raise the *pf* to 0.95.

Solution:

If $pf = 0.8$, then

$$\cos\theta_1 = 0.8 \quad \Rightarrow \quad \theta_1 = 36.87^\circ$$

where θ_1 is the phase difference between voltage and current. We obtain the apparent power from the real power and the *pf*; that is,

$$S_1 = \frac{P}{\cos\theta_1} = \frac{4000}{0.8} = 5000 \text{ VA}$$

The reactive power is

$$Q_1 = S_1 \sin\theta_1 = 5000 \sin 36.87 = 3000 \text{ VAR}$$

When the *pf* is raised to 0.95,

$$\cos\theta_2 = 0.95 \quad \Rightarrow \quad \theta_2 = 18.19^\circ$$

The real power P has not changed because the average power due to the capacitance is zero. But the apparent power has changed; its new value is

$$S_2 = \frac{P}{\cos\theta_2} = \frac{4000}{0.95} = 4210.5 \text{ VA}$$

The new reactive power is

$$Q_2 = S_2 \sin\theta_2 = 1314.4 \text{ VAR}$$

The difference between the new and old reactive powers is due to the parallel addition of the capacitor to the load. The reactive power due to the capacitor is

$$Q_C = Q_1 - Q_2 = 3000 - 1314.4 = 1685.6 \text{ VAR}$$

and

$$C = \frac{Q_C}{\omega V_{\text{rms}}^2} = \frac{1685.6}{2\pi \times 60 \times 120^2} = 310.5\ \mu\text{F}$$

Practice Problem 14.12

Find the value of parallel capacitance needed to correct a load of 140 kVAR at 0.85 lagging *pf* to unity *pf*. Assume that the load is supplied by a 110-V (rms), 60-Hz line.

Answer: 30.69 mF

14.8 †Applications

In this section, we consider three important application areas: how power is measured, how electric utility companies determine the cost of electricity consumption, and power in CPUs.

14.8.1 Power Measurement

The average power absorbed by a load is measured by an instrument called the *wattmeter*.[10]

> The **wattmeter** is the instrument for measuring the average power.

As shown in Fig. 14.22, a wattmeter essentially consists of two coils:[11] the current coil and the voltage coil. A current coil with very low impedance (ideally zero) is connected in series with the load as shown in Fig. 14.23 and responds to the load current. The voltage coil with very high impedance (ideally infinite) is connected in parallel with the load as shown in Fig. 14.23 and responds to the load voltage. The current coil acts like a short circuit because of its low impedance, while the voltage coil behaves like an open circuit because of its high impedance. As a result, ideally, the presence of the wattmeter does not disturb the circuit or have an effect on the power measurement.

When the two coils are energized, the mechanical inertia of the moving system produces a deflection angle that is proportional to the average value of the product $v(t)i(t)$. If the voltage and current of the load

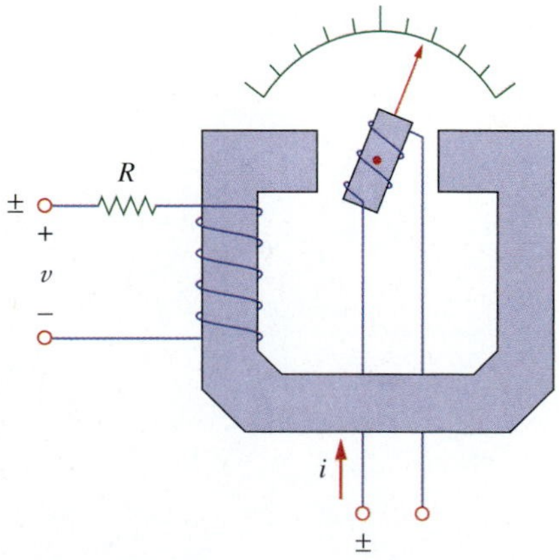

Figure 14.22
A wattmeter.

[10] Note: Reactive power is measured by an instrument called the *varmeter*. The varmeter is often connected to the load in the same way as the wattmeter.

[11] Note: Some wattmeters do not have coils; the wattmeter considered here is the electromagnetic type.

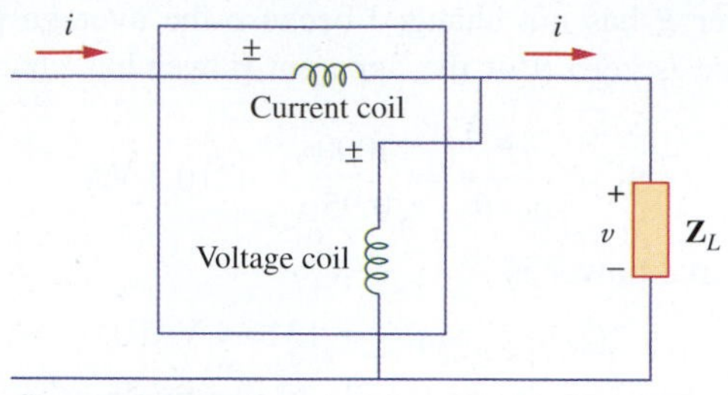

Figure 14.23
The wattmeter connected to the load.

are $v(t) = V_m \cos(\omega t + \theta_v)$ and $i(t) = I_m \cos(\omega t + \theta_i)$, their corresponding rms phasors are

$$\mathbf{V}_{\text{rms}} = \frac{V_m}{\sqrt{2}}\underline{/\theta_v} \quad \text{and} \quad \mathbf{I}_{\text{rms}} = \frac{I_m}{\sqrt{2}}\underline{/\theta_i} \tag{14.46}$$

and the wattmeter measures the average power given by

$$P = |\mathbf{V}_{\text{rms}}|\,|\mathbf{I}_{\text{rms}}|\cos(\theta_v - \theta_I) = \frac{1}{2}V_m I_m \cos(\theta_v - \theta_I) \tag{14.47}$$

As shown in Fig. 14.22, each wattmeter coil has two terminal pairs with one of each marked ± (plus/minus). To ensure upscale deflection, the plus/minus terminal of the current coil is placed closer to the source, while the plus/minus terminal of the voltage coil is connected to the same line as the other end of the current coil, as shown in Fig. 14.23. Reversing both coil connections still results in upscale deflection. However, reversing one coil and not the other results in downscale deflection and no wattmeter reading.

Example 14.13

Find the wattmeter reading of the circuit in Fig. 14.24.

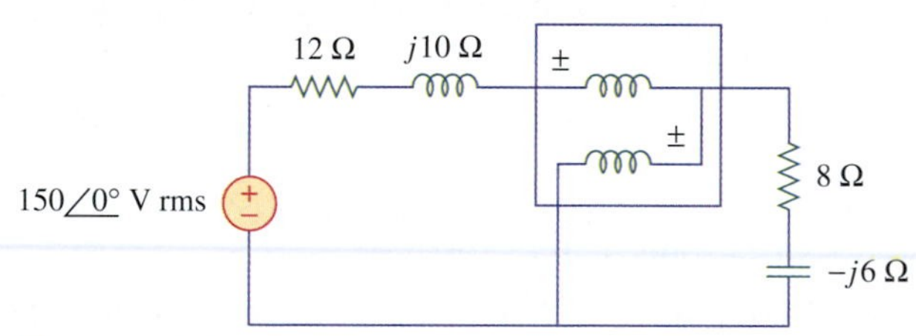

Figure 14.24
For Example 14.14.

Solution:
In Fig. 14.24, the wattmeter reads the average power absorbed by the $(8 - j6)\ \Omega$ impedance because the current coil is in series with the impedance while the voltage coil is in parallel with it. The current through the circuit is

$$\mathbf{I}_{\text{rms}} = \frac{150\underline{/0^\circ}}{(12 + j10) + (8 - j6)} = \frac{150}{20 + j4}\ \text{A}$$

The voltage across the $(8 - j6)\ \Omega$ impedance is

$$\mathbf{V}_{\text{rms}} = \mathbf{I}_{\text{rms}}(8 - j6) = \frac{150(8 - j6)}{20 + j4}\ \text{V}$$

The complex power is

$$\mathbf{S} = \mathbf{V}_{\text{rms}}\mathbf{I}^*_{\text{rms}} = \frac{150(8 - j6)}{20 + j4} \cdot \frac{150}{20 - j4} = \frac{150^2(8 - j6)}{20^2 + 4^2}$$

$$= 423.7 - j324.5 \text{ VA}$$

The wattmeter reads

$$P = \text{Re}(\mathbf{S}) = 423.7 \text{ W}$$

Practice Problem 14.13

For the circuit in Fig. 14.25, find the wattmeter reading.

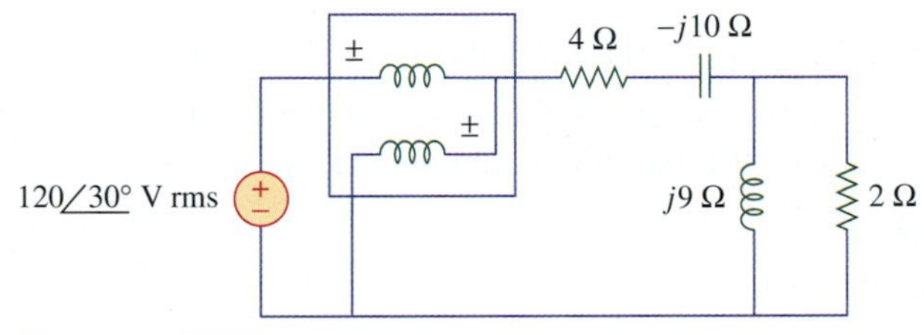

Figure 14.25
For Practice Problem 14.14.

Answer: 1437.3 W

14.8.2 Electricity Consumption

In Section 1.7, we considered a simplified model of the way the cost of electricity consumption is determined. But the concept of power factor was not included in the calculations. Now we consider the importance of power factor in electricity consumption cost.

Loads with low power factors are costly to serve because they require large currents, as explained in Section 14.7. The ideal situation would be to draw minimum current from a supply so that $S = P$, $Q = 0$, and $pf = 1$. A load with nonzero Q means that energy flows forth and back between the load and the source, giving rise to additional power losses. In view of this, power companies often encourage their commercial and industrial customers to have power factors as close to unity as possible and penalize some customers who do not improve their load power factors.

Utility companies divide their customers into categories as residential (domestic), commercial, and industrial or as small, medium, and large power. They have different rate structures for each category. The amount of energy consumed in units of kWh (kilowatt-hours) is measured using a kilowatt-hour meter (discussed in Chapter 3, Section 3.8; see Figure 3.15), installed at the customer's premises.

Although utility companies use different methods for charging customers, the tariff or charge to a consumer often has two parts. The first part is fixed and corresponds to the cost of generation, transmission, and distribution of electricity to meet the load requirements of the industrial consumers. This part of the tariff is generally expressed as a certain price per kW of maximum demand. Or it may instead be based on kVA of maximum demand to account for the power factor (*pf*) of the consumer. A *pf* penalty charge may be imposed on the consumer, such that a certain percentage of kW or kVA maximum demand is charged for every 0.01 fall in *pf* below a prescribed value, say, 0.85 or 0.9. On the other hand, a *pf* credit may be given for every 0.01 that the *pf* exceeds the prescribed value. The second part is proportional to the energy consumed in kWh; it may be in graded form (e.g. first 100 kWH at 16 cents/kWh, next 200 kWh at 10 cents/kWh, etc.). Thus, the bill is determined based on the following equation:

$$\text{Total cost} = \text{Fixed cost} + \text{Cost of energy} \tag{14.48}$$

Example 14.14

A manufacturing industry consumes 200 MWh in one month. If the maximum demand is 1600 kW, calculate the electricity bill based on the following two-part tariff:

Demand charge: $5.00 a month/kW of billing demand

Energy charge: 8 cents/kWh for the first 50,000 kWh, 5 cents/kWh for the remaining energy

Solution:

The demand charge is:

$$\$5.00 \times 1600 = \$8,000 \tag{14.14.1}$$

The energy charge for the first 50,000 kWh is:

$$\$0.08 \times 50,000 = \$4,000 \tag{14.14.2}$$

The remaining energy is

$$200,000 \text{ kWh} - 50,000 \text{ kWh} = 150,000 \text{ kWh}$$

and the corresponding energy charge is:

$$\$0.05 \times 150,000 = \$7,500 \tag{14.14.3}$$

Adding Eqs.(14.15.1), (14.15.2), and (14.15.3) gives the total bill for the month

$$\$8,000 + \$4,000 + \$7,500 = \$19,500$$

It may appear that the cost of electricity is too high. But this is often a small fraction of the overall cost of production of the goods manufactured or the selling price of the finished product.

Practice Problem 14.14

The monthly reading of a paper mill's meter is as follows:

Maximum demand: 32,000 kW

Energy consumed: 500 MWh

Using the two-part tariff in Example 14.14, calculate the monthly bill for the paper mill.

Answer: $186,500

Example 14.15

A 300-kW load supplied at 13 kV (rms) operates 520 hours a month at an 80 percent power factor. Calculate the average cost per month based on this simplified tariff:

Energy charge: 6 cents/kWh

Power-factor penalty: 0.1 percent of energy charge for every 0.01 that *pf* falls below 0.85

Power-factor credit: 0.1 percent of energy charge for every 0.01 that *pf* exceeds 0.85

Solution:
The energy consumed is

$$W = 300 \text{ kW} \times 520 \text{ h} = 156{,}000 \text{ kWh}$$

The operating power factor $pf = 80\% = 0.8$ is 5×0.01 below the prescribed power factor of 0.85. Because there is a 0.1 percent energy charge for every 0.01, there is a power-factor penalty charge of 0.5 percent. This amounts to energy charge of

$$\Delta W = 156{,}000 \times \frac{5 \times 0.1}{100} = 780 \text{ kWh}$$

The total energy is

$$W_t = W + \Delta W = 156{,}000 + 780 = 156{,}780 \text{ kWh}$$

The cost per month is given by

$$\text{Cost} = 6 \text{ cents} \times W_t = \$0.06 \times 156{,}780 = \$9{,}406.80$$

Practice Problem 14.15

A 800-kW induction furnace at a 0.88 power factor operates 20 hours per day for 26 days in a month. Determine the electricity bill per month based on the billing schedule presented in Example 14.15.

Answer: \$24,885.12

14.8.3 Power in CPUs

Central processing units (CPUs) found in computers consume some amount of electric power. This power is dissipated both by the switching devices contained in the CPU (such as transistors) as well as energy lost in the form of heat due to the resistivity of the electrical circuits. On the one hand, low-power CPUs, such as found in mobile phones, use very little power. On the other hand, CPUs in general-purpose microcomputers dissipate significantly more power because of their complexity and speed. Early CPUs implemented with vacuum tubes consumed power in the order of many kilowatts.

14.9 Summary

1. The instantaneous power absorbed by an element is the product of the element's terminal voltage and the current through the element: $p = vi$.
2. Average or real power P (in watts) is the average of instantaneous power $p(t)$. If $v(t) = V_m \cos(\omega t + \theta_v)$ and $i(t) = I_m \cos(\omega t + \theta_i)$, then $V_{\text{rms}} = V_m/\sqrt{2}$, $I_{\text{rms}} = I_m/\sqrt{2}$, and

$$P = \frac{1}{2} V_m I_m \cos(\theta_v - \theta_i)$$

3. Inductors and capacitors absorb no average power, while the average power absorbed by a resistor is

$$\frac{1}{2} I_m^2 R = I_{\text{rms}}^2 R$$

4. Maximum average power is transferred to a load when the load impedance is the complex conjugate of the Thevenin impedance as seen from the load terminals; that is,

$$\mathbf{Z}_L = \mathbf{Z}_{Th}^*$$

5. The power factor is the cosine of the phase difference between voltage and current: $pf = \cos(\theta_v - \theta_i)$. It is also the cosine of the angle of the load impedance or the ratio of real power to apparent power. The *pf* is lagging if the current lags voltage (inductive load) and is leading when the current leads voltage (capacitive load).
6. Apparent power S (in VA) is the product of the rms values of voltage and current:

$$S = V_{\text{rms}} I_{\text{rms}}$$

It is also given by $S = |\mathbf{S}| = \sqrt{P^2 + Q^2}$, where Q is the reactive power.
7. Reactive power (in VAR) is:

$$Q = \frac{1}{2} V_m I_m \sin(\theta_v - \theta_i) = V_{\text{rms}} I_{\text{rms}} \sin(\theta_v - \theta_i)$$

8. Complex power $\mathbf{S}$ (in VA) is the product of the rms voltage phasor and the complex conjugate of the rms current phasor. It is also the complex sum of real power P and reactive power Q.

$$\mathbf{S} = \mathbf{V}_{\text{rms}} \mathbf{I}_{\text{rms}}^* = V_{\text{rms}} I_{\text{rms}} \underline{/(\theta_v - \theta_i)} = P + jQ$$

Also,

$$\mathbf{S} = I_{\text{rms}}^2 \mathbf{Z} = V_{\text{rms}}^2 / \mathbf{Z}^*$$

9. The total complex power in a network is the sum of the complex powers of the individual components. Total real power and reactive power are also the sums of the individual real powers and reactive powers, respectively, but the total apparent power is not calculated by this summing process.
10. Power-factor correction is necessary for economic reasons; it is the process of improving the power factor of a load by reducing the overall reactive power.
11. The wattmeter is the instrument for measuring the average power. Energy consumed is measured with a kilowatt-hour meter.

Review Questions

14.1 The average power absorbed by an inductor is zero.

(a) True (b) False

14.2 The Thevenin impedance of a network seen from the load terminals is $80 + j55\ \Omega$. For maximum power transfer, the load impedance must be:

(a) $-80 + j55\ \Omega$ (b) $-80 - j55\ \Omega$
(c) $80 - j55\ \Omega$ (d) $80 + j55\ \Omega$.

14.3 The amplitude of the voltage available in the 60-Hz, 120-V power outlet in your home is:

(a) 110 V (b) 120 V (c) 170 V (d) 210 V

14.4 If the load impedance is $20 - j20$, the power factor is

(a) $\underline{/-45^\circ}$ (b) 0 (c) 1
(d) 0.7071 (e) none of the above

14.5 A quantity that contains all the power information in a given load is

(a) power factor
(b) apparent power
(c) average power
(d) reactive power
(e) complex power

14.6 Reactive power is measured in:

(a) watts (b) VA
(c) VAR (d) none of the above

14.7 In the power triangle shown in Fig. 14.26(a), the reactive power is:

(a) 1000 VAR leading
(b) 1000 VAR lagging
(c) 866 VAR leading
(d) 866 VAR lagging

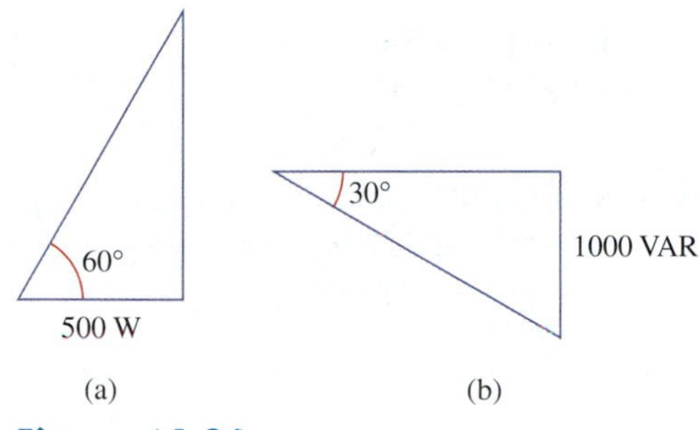

Figure 14.26
For Review Questions 14.7 and 14.8.

14.8 For the power triangle in Fig. 14.26(b), the apparent power is:

(a) 2000 VAR (b) 1000 VAR
(c) 866 VAR (d) 500 VAR

14.9 A source is connected to three loads $\mathbf{Z}_1$, $\mathbf{Z}_2$, and $\mathbf{Z}_3$ in parallel. Which of these is not true?

(a) $P = P_1 + P_2 + P_3$
(b) $Q = Q_1 + Q_2 + Q_3$
(c) $S = S_1 + S_2 + S_3$
(d) $\mathbf{S} = \mathbf{S}_1 + \mathbf{S}_2 + \mathbf{S}_3$

14.10 The instrument for measuring average power is:

(a) voltmeter
(b) ammeter
(c) wattmeter
(d) varmeter
(e) kilowatt-hour meter.

Answers: 14.1a, 14.2c, 14.3c, 14.4d, 14.5e, 14.6c, 14.7d, 14.8a, 14.9c, 14.10c

Problems

Section 14.2 Instantaneous and Average Power

14.1 If $v(t) = 160 \cos(50t)$ V and $i(t) = -20 \sin(50t - 30°)$ A, calculate the instantaneous power and the average power.

14.2 At $t = 2$s, find the instantaneous power on each of the elements in the circuit of Fig. 14.27.

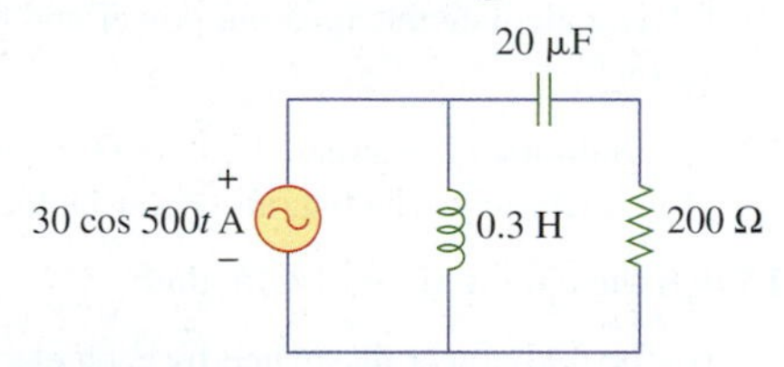

Figure 14.27
For Problem 14.2.

14.3 Refer to the circuit depicted in Fig. 14.28. Find the average power absorbed by each element.

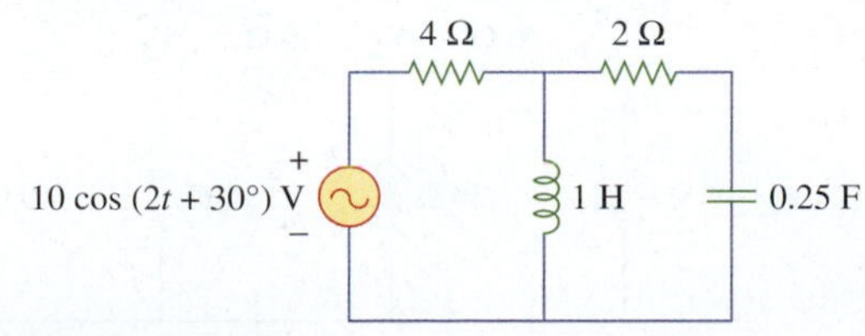

Figure 14.28
For Problem 14.3.

14.4 Given the circuit in Fig. 14.29, find the average power absorbed by each of the elements.

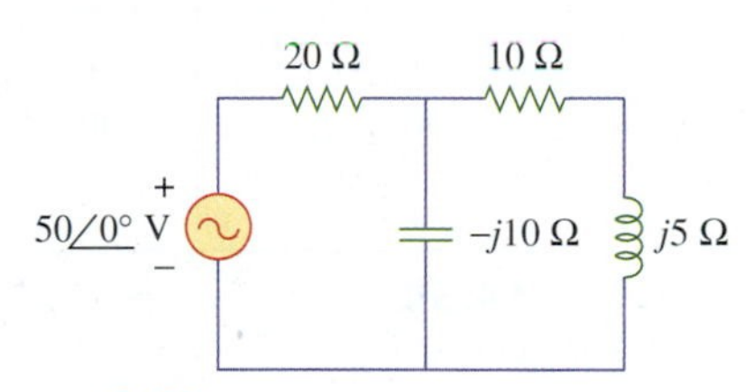

Figure 14.29
For Problem 14.4.

14.5 Assuming that $v_s(t) = 8 \cos(2t - 40°)$ V in the circuit in Fig. 14.30, find the average power delivered to each of the passive elements.

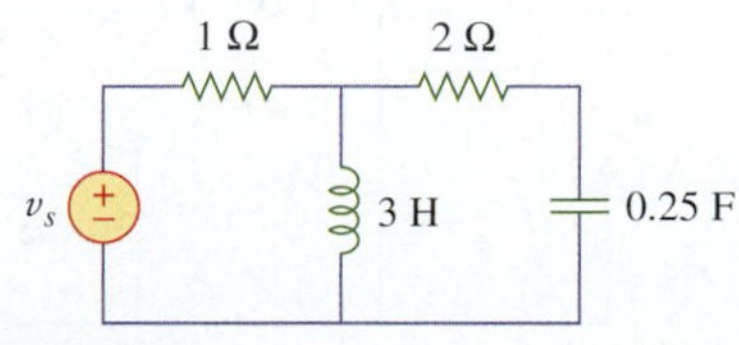

Figure 14.30
For Problem 14.5.

Section 14.3 Maximum Average Power Transfer

14.6 For each of the circuits in Fig. 14.31, determine the value of the load **Z** for maximum power transfer and the maximum average power transferred.

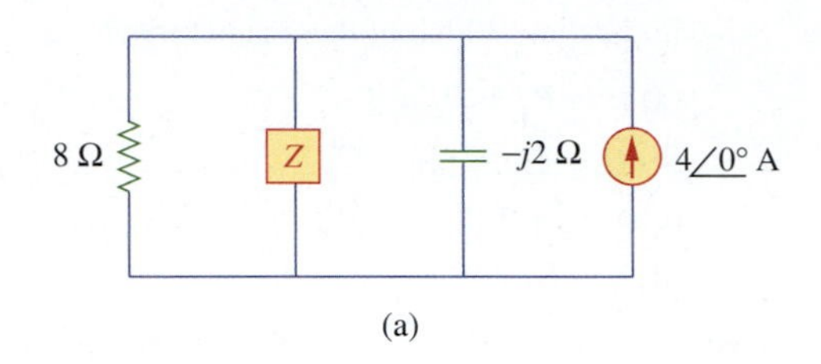

(a)

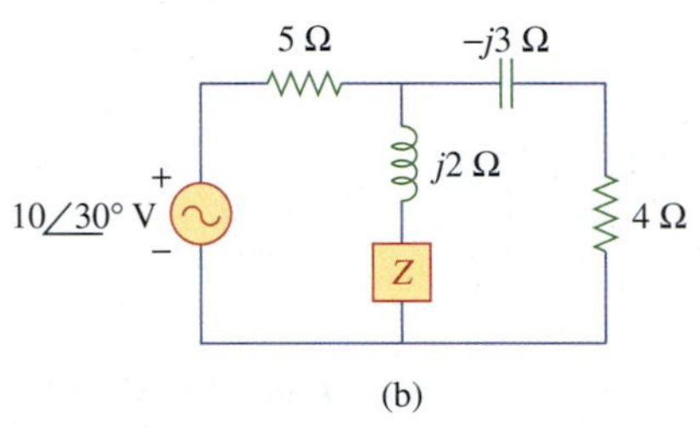

(b)

Figure 14.31
For Problem 14.6.

14.7 For the circuit in Fig. 14.32, find: (a) the value of the load impedance that absorbs the maximum average power and (b) the value of the maximum average power absorbed.

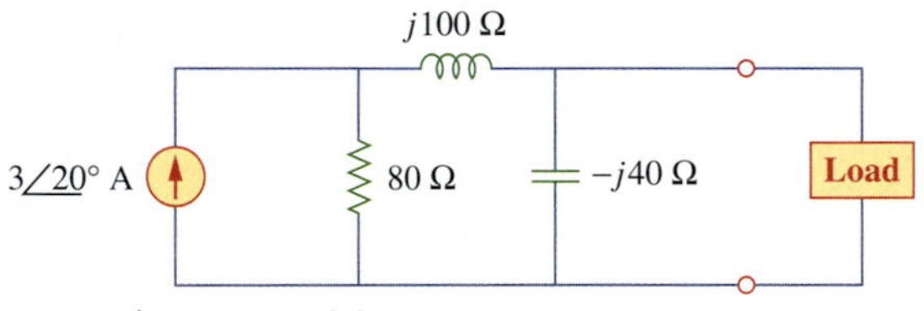

Figure 14.32
For Problem 14.7.

14.8 In the circuit of Fig. 14.33, find the value of **Z** that will result in the maximum power being delivered to **Z**. Calculate the maximum power delivered to **Z**.

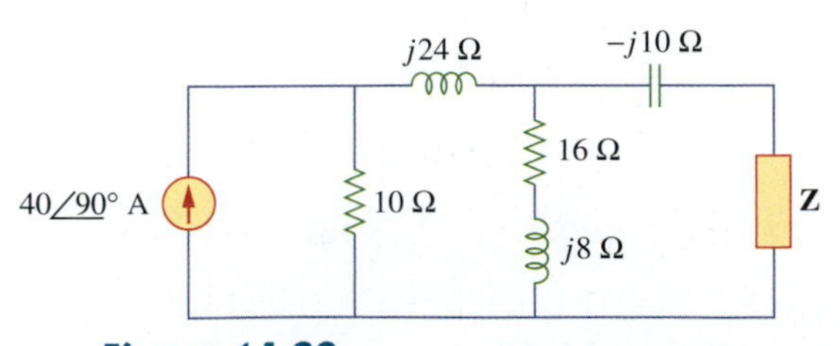

Figure 14.33
For Problem 14.8.

14.9 Calculate the value of $\mathbf{Z}_L$ in the circuit of Fig. 14.34 in order for $\mathbf{Z}_L$ to receive the maximum average power. What is the maximum average power received by $\mathbf{Z}_L$?

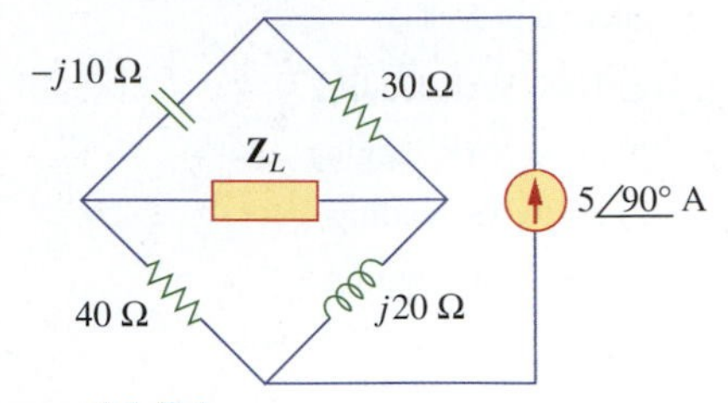

Figure 14.34
For Problem 14.9.

14.10 The variable resistor R in the circuit of Fig. 14.35 is adjusted until it absorbs the maximum average power. Find R and the maximum average power absorbed.

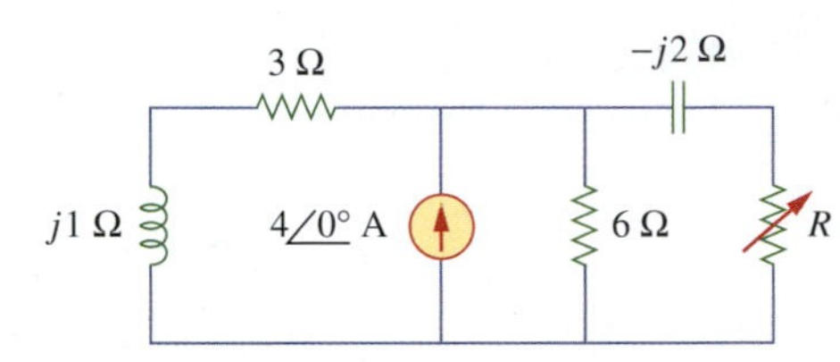

Figure 14.35
For Problem 14.10.

Section 14.4 Apparent Power and Power Factor

14.11 A relay coil is connected to a 210-V, 50-Hz supply. If it has a resistance of 30 Ω and an inductance of 0.5 H, calculate the apparent power and the power factor.

14.12 A certain load comprises $12 - j8\ \Omega$ in parallel with $j4\ \Omega$. Determine the overall power factor.

14.13 For the circuit in Fig. 14.36, find:

(a) the real power dissipated by each element

(b) the total apparent power supplied by the circuit

(c) the power factor of the circuit

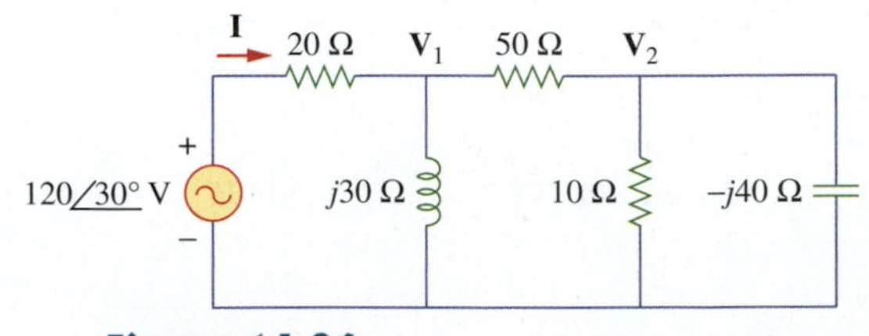

Figure 14.36
For Problem 14.13.

14.14 Obtain the power factor for each of the circuits in Fig. 14.37. Specify each power factor as leading or lagging.

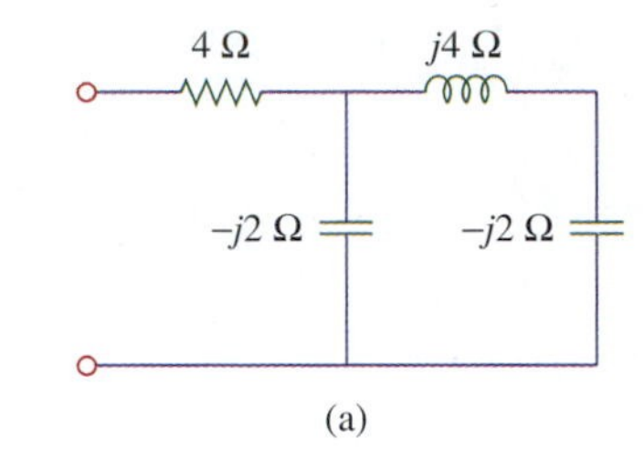

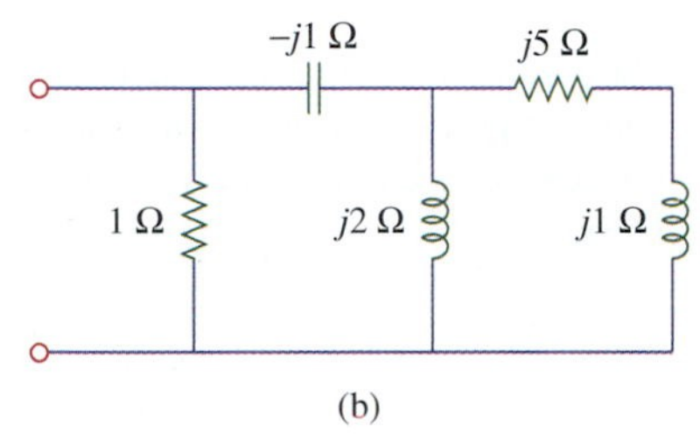

Figure 14.37
For Problem 14.14.

Section 14.5 Complex Power

14.15 In a series *RL* circuit, $V_R = 220$ V (rms), $V_L = 150$ V (rms), and the current is 6 A (rms). Calculate the real power, reactive power, and apparent power.

14.16 In a circuit, the real power is 4.2 W, while the reactive power is 6.2 VAR. Calculate the apparent power.

14.17 For the circuit in Fig. 14.38, find the complex power delivered by the source. Let $v = 20\cos(10t)$ V.

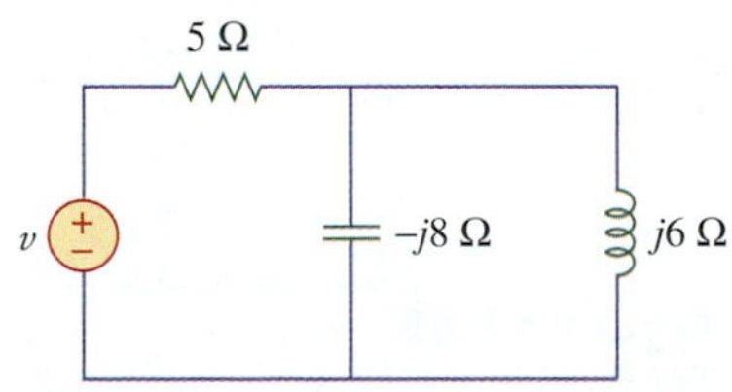

Figure 14.38
For Problem 14.17.

14.18 A 240-V (rms) source is connected to an inductive ac motor. If the motor draws 100 A and is rated 10 kW, determine: (a) the power factor *pf*, (b) the reactive power *Q*, and (c) the motor's Z_{eq}.

14.19 A 240-V (rms) motor has an output real power of 4 hp. If the motor draws a current of 15 A, find: (a) the apparent power, (b) the power factor, and (c) the reactive power.

14.20 A load draws 5 kVAR at a power factor of 0.86 (leading) from a 220-V rms source. Calculate the peak current and the apparent power supplied to the load.

14.21 A source delivers 50 kVA to a load with a power factor of 65 percent lagging. Find the load's average and reactive powers.

14.22 For the following voltage and current phasors, calculate the complex power, apparent power, real power, and reactive power. Specify whether the *pf* is leading or lagging.

(a) $\mathbf{V} = 220\angle 30^\circ$ V (rms), $\mathbf{I} = 0.5\angle 60^\circ$ A (rms)

(b) $\mathbf{V} = 250\angle -10^\circ$ V (rms), $\mathbf{I} = 6.2\angle -25^\circ$ A (rms)

(c) $\mathbf{V} = 120\angle 0^\circ$ V (rms), $\mathbf{I} = 2.4\angle -15^\circ$ A (rms)

(d) $\mathbf{V} = 160\angle 45^\circ$ V (rms), $\mathbf{I} = 8.5\angle 90^\circ$ A (rms)

14.23 For each of the following cases, find the complex power, the average power, and the reactive power:

(a) $v(t) = 112\cos(\omega t + 10^\circ)$ V, $i(t) = 4\cos(\omega t - 50^\circ)$ A

(b) $v(t) = 160\cos 377t$ V, $i(t) = 4\cos(377t + 45^\circ)$ A

(c) $\mathbf{V} = 80\angle 60^\circ$ V (rms), $\mathbf{Z} = 50\angle 30^\circ$ Ω

(d) $\mathbf{I} = 10\angle 60^\circ$ V (rms), $\mathbf{Z} = 100\angle 45^\circ$ Ω

14.24 Determine the complex power for the following cases:

(a) $P = 269$ W, $Q = 150$ VAR (capacitive)

(b) $Q = 2000$ VAR, $pf = 0.9$ (leading)

(c) $S = 600$ VA, $Q = 450$ VAR (inductive)

(d) $V_{rms} = 220$ V, $P = 1$ kW, $|\mathbf{Z}| = 40$ Ω (inductive)

14.25 Obtain the overall impedance for the following cases:

(a) $P = 1000$ W, $pf = 0.8$ (leading), $V_{rms} = 220$ V

(b) $P = 1500$ W, $Q = 2000$ VAR (inductive), $I_{rms} = 12$ A

(c) $\mathbf{S} = 4500\angle 60^\circ$ VA, $\mathbf{V} = 120\angle 45^\circ$ V

14.26 For the entire circuit in Fig. 14.39, calculate:

(a) the power factor

(b) the average power delivered by the source

(c) the reactive power

(d) the apparent power

(e) the complex power

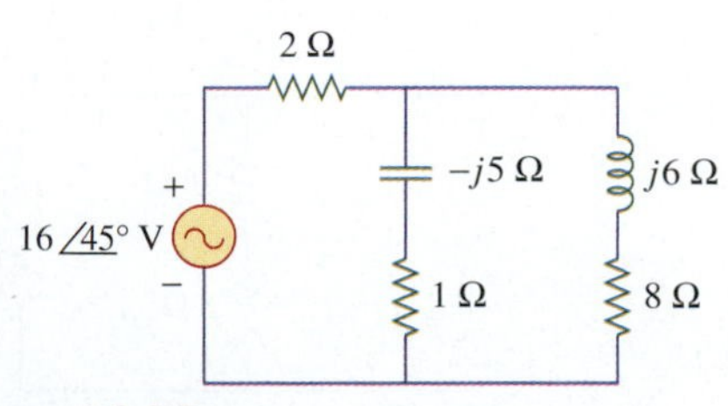

Figure 14.39
For Problem 14.26.

14.27 In the circuit of Fig. 14.40, device *A* receives 2 kW at 0.8 *pf* lagging, device *B* receives 3 kVA at 0.4 *pf* leading, while device *C* is inductive and consumes 1 kW and receives 500 VAR. (a) Determine the power factor of the entire system. (b) Find **I** given that $\mathbf{V}_s = 120\underline{/45^\circ}$ V rms.

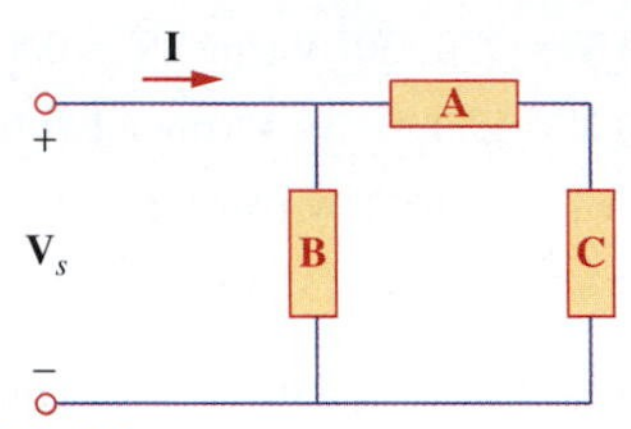

Figure 14.40
For Problem 14.27.

14.28 An *RL* series circuit has a real power of 100 W and an apparent power of 240 VA. What is the reactive power?

14.29 A 60-Hz voltage source gives a current of $4\underline{/30^\circ}$ A (rms) in a load $Z = 100\underline{/45^\circ}\ \Omega$. (a) Find the apparent power *S*. (b) Calculate the real power *P*. (c) Determine the reactive power *Q*. (d) Draw the power triangle.

14.30 Refer to Fig. 14.41. (a) Find the total apparent power delivered by the source. (b) Sketch the power triangle.

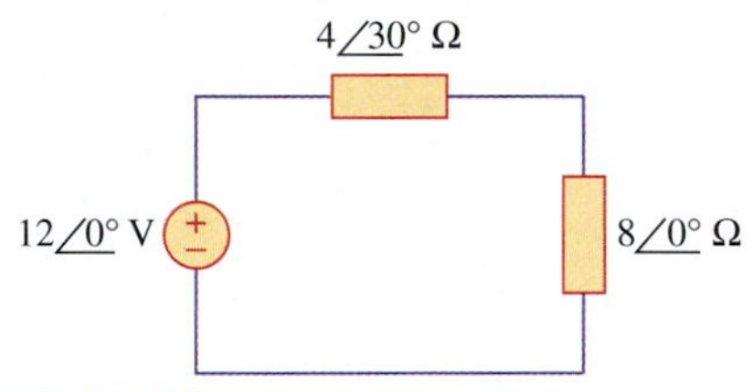

Figure 14.41
For Problem 14.30.

14.31 Determine the unknown impedance *Z* in Fig. 14.42 if the circuit has an overall apparent power of 600 VA and overall lagging power factor of 0.84. Let $V = 50\underline{/0^\circ}$ V(rms).

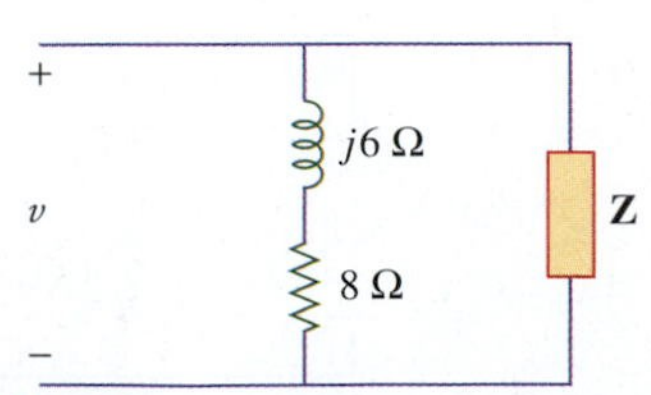

Figure 14.42
For Problem 14.31.

14.32 For the circuit in Fig. 14.43, find: (a) *I* (rms), (b) the real power *P*, (c) the reactive power *Q*.

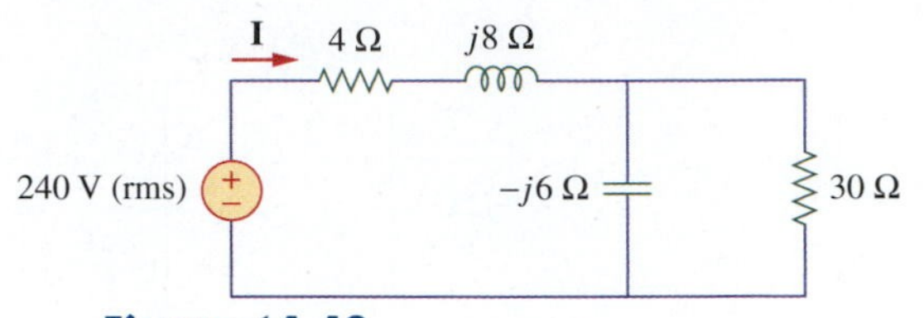

Figure 14.43
For Problem 14.32.

14.33 For the circuit in Fig. 14.44, determine: (a) the complex power delivered to *RL* and (b) the impedance Z_L that will draw the maximum average power.

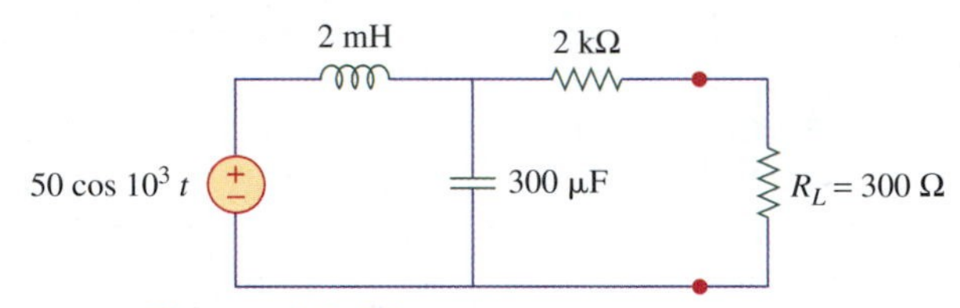

Figure 14.44
For Problem 14.33.

Section 14.6 Conservation of AC Power

14.34 For the network in Fig. 14.45, find the complex power absorbed by each element.

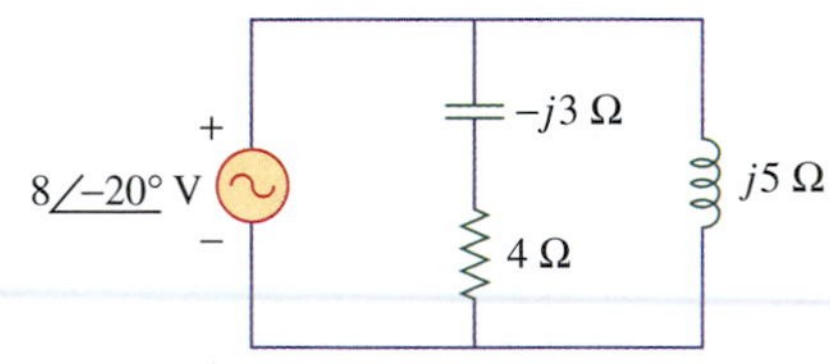

Figure 14.45
For Problem 14.34.

14.35 Find the complex power absorbed by each of the five elements in the circuit in Fig. 14.46.

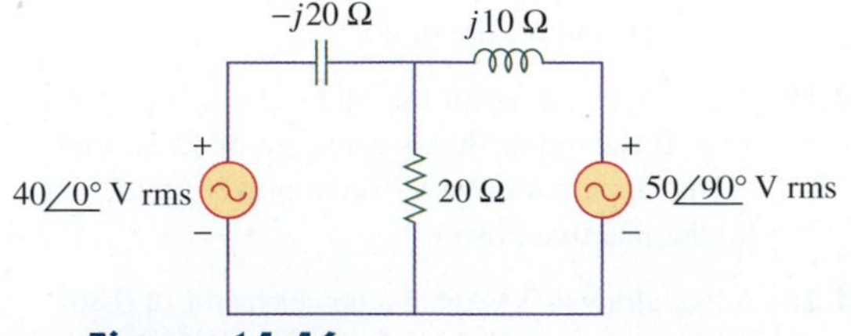

Figure 14.46
For Problem 14.35.

14.36 Obtain the complex power delivered by the source in the circuit in Fig. 14.47.

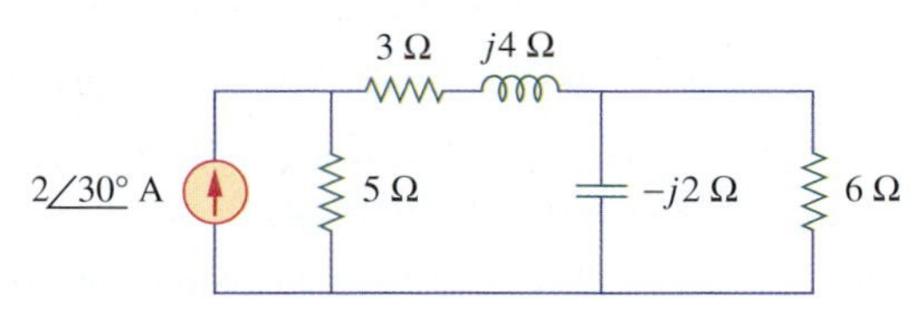

Figure 14.47
For Problem 14.36.

14.37 For the circuit in Fig. 14.48, find $\mathbf{V}_o$ and the input power factor.

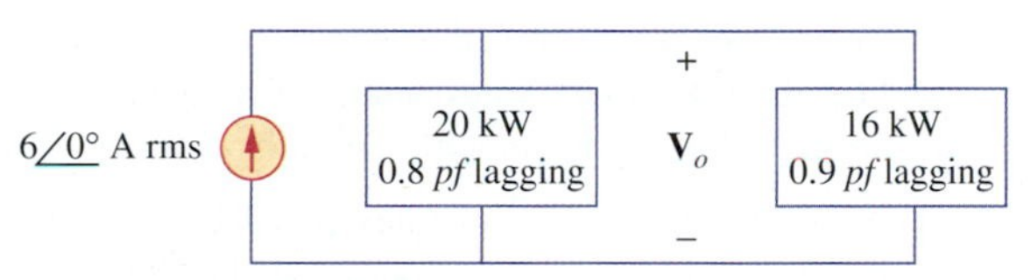

Figure 14.48
For Problem 14.37.

14.38 Given the circuit in Fig. 14.49, find $\mathbf{I}_o$ and the overall complex power.

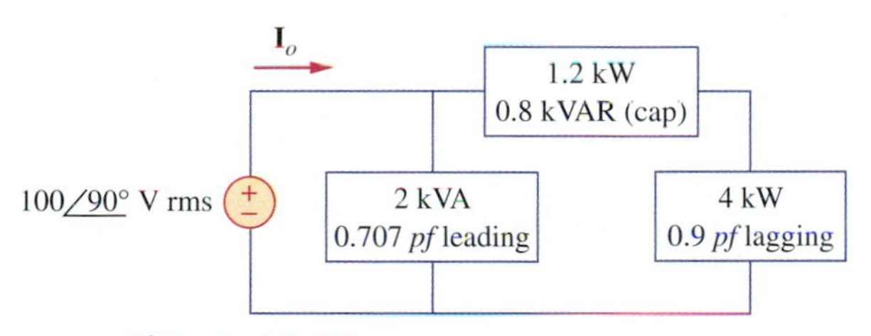

Figure 14.49
For Problem 14.38.

Section 14.7 Power Factor Correction

14.39 Refer to the circuit shown in Fig. 14.50.
(a) What is the power factor?
(b) What is the average power dissipated?
(c) What is the value of the capacitance that will give a unity power factor when connected to the load?

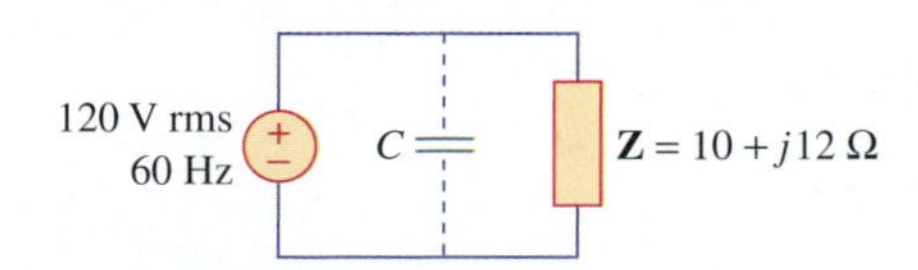

Figure 14.50
For Problem 14.39.

14.40 An 880-VA, 220-V, 50-Hz load has a power factor of 0.8 lagging. What value of parallel capacitance will correct the load power factor to unity?

14.41 Two loads are placed in parallel across a 120-V rms 60-Hz line. The first load draws 150 VA at a lagging power factor of 0.707, while the second load draws 50 VAR at a leading power factor of 0.8. A third load is purely capacitive and is placed in parallel across the 120-V line in order to make the *pf* of the system equal to unity. Calculate the value of the capacitance.

14.42 A 40-kW induction motor, with a lagging power factor of 0.76, is supplied by a 120-V rms 60-Hz sinusoidal voltage source. Find the capacitance needed in parallel with the motor to raise the power factor to: (a) 0.9 lagging and (b) 1.0.

14.43 A 240-V rms 60-Hz supply serves a load that is 10 kW (resistive), 15 kVAR (capacitive), and 22 kVAR (inductive). Find: (a) the apparent power, (b) the current drawn from the supply, (c) the kVAR rating and capacitance required to improve the power factor to 0.96 lagging, and (d) the current drawn from the supply under the new power-factor conditions.

14.44 A 50-V (rms) 400-Hz supply is connected to a 6-kW load with a 75 percent lagging power factor. Calculate the capacitance needed for power factor correction when the power factor is corrected to 95 percent lagging.

Section 14.8 Applications

14.45 Determine the wattmeter reading for the circuit in Fig. 14.51.

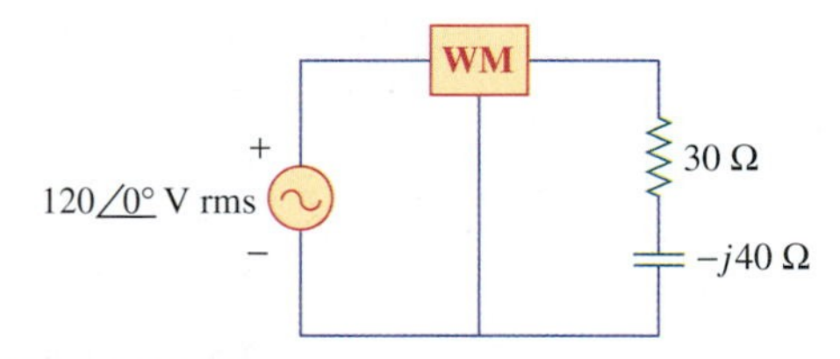

Figure 14.51
For Problem 14.45.

14.46 What is the reading of the wattmeter in the network of Fig. 14.52?

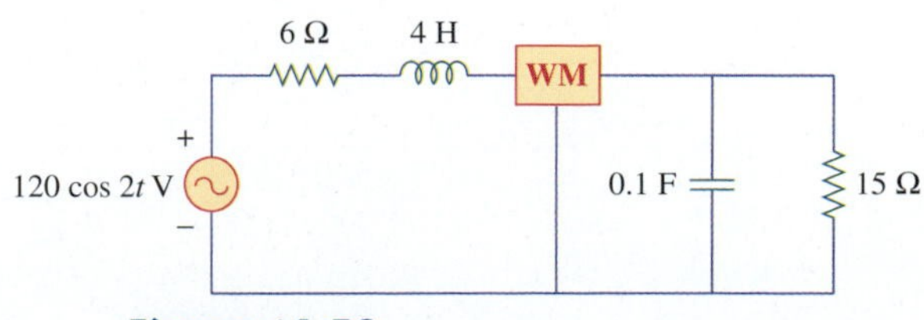

Figure 14.52
For Problem 14.46.

14.47 The kilowatt-hourmeter of a home is read once a month. For a particular month, the previous and present readings are as follows:

Previous reading: 3246 kWh

Present reading: 4017 kWh

Calculate the electricity bill for that month based on the following residential rate schedule:

Minimum monthly charge—$12.00

First 100 kWh per month at 16 cents/kWh

Next 200 kWh per month at 10 cents/kWh

More than 200 kWh per month at 6 cents/kWh

14.48 A 240-V rms 60-Hz source supplies a parallel combination of a 5-kW heater and a 30-kVA induction motor whose power factor is 0.82. Determine: (a) the system apparent power, (b) the system reactive power, (c) the kVA rating of a capacitor required to adjust the system power factor to 0.9 lagging, and (d) the value of the capacitor required.

14.49 A consumer has an annual consumption of 1,200 MWh with a maximum demand of 2.4 MVA. The maximum demand charge is $30 per kVA per annum and the energy charge per kWh is 4 cents.
(a) Determine the annual cost of energy.
(b) Calculate the charge per kWh with a flat-rate tariff if the revenue to the utility company is to remain the same as for the two-part tariff.

14.50 A transmitter delivers maximum power to an antenna when the antenna is adjusted to represent a load of 75-Ω resistance in series with an inductance of 4 μH. If the transmitter operates at 4.12 MHz, find its internal impedance.

14.51 An industrial heater has a nameplate which reads:

210 V   60 Hz   12 kVA   0.78 *pf* (lagging)

Determine: (a) the apparent and the complex power, (b) the impedance of the heater.

14.52 The nameplate of an electric motor has the following information:

Line voltage: 220 V rms

Line current: 15 A rms

Line frequency: 60 Hz

Power: 2700 W

Determine the power factor (lagging) of the motor. Find the value of the capacitance C that must be connected across the motor to raise the *pf* to unity.

14.53 A power transmission system is modeled as shown in Fig. 14.53. If $\mathbf{V}_s = 240\angle 0^\circ$ V rms, find the average power absorbed by the load.

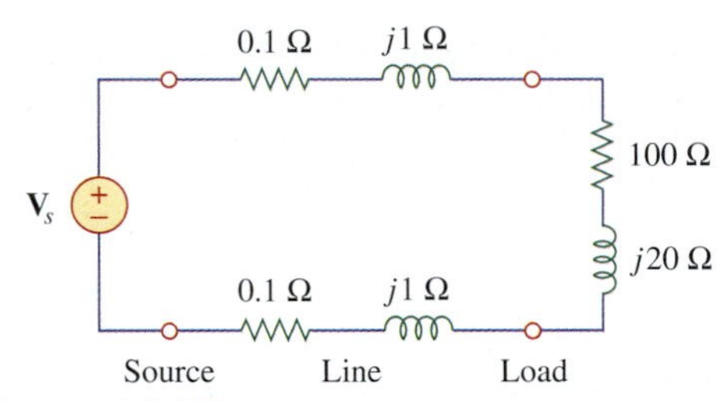

Figure 14.53
For Problem 14.53.

chapter

15

Resonance

Pay no attention to what the critics say. No statue has ever been put up to a critic.

—Jean Sibelius

Enhancing Your Career

Enhancing Your Communication Skills

Taking a class in circuit analysis is a means of preparing yourself for a career in electrical engineering technology. Enhancing your communication skills while in school should also be part of that preparation because, as a technologist or technician, a large part of your time will be spent communicating.

People in industry have complained again and again that graduating technologists are ill-prepared in written and oral communication. They want technologists who communicate effectively because they become assets.

Ability to communicate effectively is regarded by many as the most important step to an executive promotion.
© IT Stock/Punchstock RF

You probably can speak or write easily and quickly. But how effectively do you communicate? The art of effective communication is of the utmost importance to your success as a technologist. For technologists in industry, communication is key to your promotability. Consider the result of a survey of U.S. corporations that asked what factors influence managerial promotion. The survey includes a listing of 22 personal qualities and their importance in advancement. You will be surprised to note that "technical skill based on experience" placed fourth from the bottom. Attributes such as self-confidence, ambition, flexibility, maturity, ability to make sound decisions, getting things done with and through people, and capacity for hard work all ranked higher. At the top of the list was "ability to communicate." As your professional career progresses, you will need to communicate more frequently. Therefore, you should regard effective communication as an important tool in your engineering tool chest.

Learning to communicate effectively is a lifelong task you should work toward. The best place to begin in developing your communication skills is while in school. Look for continuing opportunities to develop and strengthen your reading, writing, listening, and speaking skills. You can do this through classroom presentations, team projects, active participation in student organizations, and enrollment in communication courses. Learn to write good laboratory reports in terms of grammar, spelling, accuracy, and coherence. The risks are less here than in the workplace.

15.1 Introduction

In our sinusoidal circuit analysis, we have learned how to find voltages and currents in a circuit due to a constant frequency source. We now turn our attention to ac circuits in which the source frequency varies. If we let the amplitude of the sinusoidal source remain constant and vary the frequency, we obtain the circuit's *frequency response*. The frequency response may be regarded as a complete description of the sinusoidal steady-state behavior of a circuit as a function of frequency.

The most prominent feature of the frequency response of a resonant circuit may be the peak (or *resonant peak*) exhibited in its amplitude characteristic. The concept of resonance applies in several areas of science and engineering. Without resonance we wouldn't have radio, television, or music. Of course, resonance also has its dark side; it occasionally causes a bridge to collapse or a helicopter to break apart.

Resonance is the cause of oscillations of stored energy from one form to another. It is the phenomenon that allows frequency selection in communications networks. Electronic resonance occurs in any circuit that has at least one inductor and one capacitor. Resonant circuits (series or parallel) are useful for constructing filters because their frequency response can be highly frequency selective. They are also used in selecting the desired stations in radio and TV receivers.

We begin with a series *RLC* resonant circuit. We discuss the importance of the quality factor. We later consider a parallel *RLC* resonant circuit. We use PSpice to obtain the frequency response of *RLC* circuits. We finally consider the radio receiver as a practical application of resonant circuits.

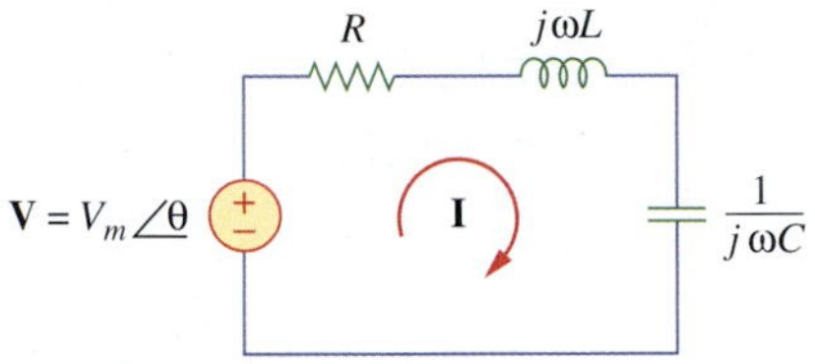

Figure 15.1
The series resonant circuit.

15.2 Series Resonance

The basic configuration of a series *RLC* circuit is shown in Fig. 15.1. The resonant circuit consists of a variable-frequency ac source in series with a resistor (optional), a capacitor, and an inductor. Such a circuit will cause resonance as the frequency varies.

> **Resonance** is a condition in an *RLC* circuit in which the capacitive and inductive reactances are equal in magnitude but, because they are opposite in sign, result in a purely resistive impedance.

Consider the series *RLC* circuit shown in Fig. 15.1 in the frequency domain. The input impedance is

$$\mathbf{Z} = \frac{\mathbf{V}}{\mathbf{I}} = R + j\omega L + \frac{1}{j\omega C} \tag{15.1}$$

or

$$\mathbf{Z} = R + j\left(\omega L - \frac{1}{\omega C}\right) \tag{15.2}$$

The graph in Fig. 15.2 shows the variation of the imaginary part of the input impedance; that is,

$$X(\omega) = \omega L - \frac{1}{\omega C} \tag{15.3}$$

Resonance results when X, the imaginary part of the input impedance, is zero; that is,

$$\text{Im}(\mathbf{Z}) = X(\omega) = \omega L - \frac{1}{\omega C} = 0 \tag{15.4}$$

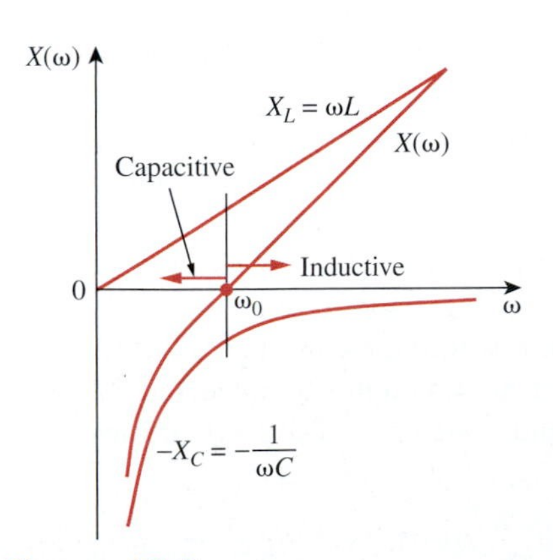

Figure 15.2
Plot of X, X_L, and X_C.

The value of ω that satisfies this condition is called the *resonant frequency*. Thus, the resonant condition is

$$\omega_0 L = \frac{1}{\omega_0 C} \tag{15.5}$$

or

$$\boxed{\omega_0 = \frac{1}{\sqrt{LC}} \text{ (rad/s)}} \tag{15.6}$$

Because $\omega_0 = 2\pi f_0$

$$f_0 = \frac{1}{2\pi\sqrt{LC}} \text{ (Hz)} \tag{15.7}$$

Note that at resonance:

1. The impedance is purely resistive; that is, $\mathbf{Z} = R$. In other words, the LC series combination acts like a short circuit and the entire source voltage is across R.
2. The voltage $\mathbf{V}_s$ and the current $\mathbf{I}$ are in phase so that the power factor is unity.
3. The magnitude of the impedance $\mathbf{Z}(\omega)$ is minimum.
4. The inductor voltage and capacitor voltage can be much more than the source voltage

$$\left(V_L = IX_L = \frac{V_m}{R}\omega_o L, \qquad V_C = -IX_C = -\frac{V_m}{R}\frac{1}{\omega_o C}\right)$$

The frequency response of the circuit's current magnitude

$$I = |\mathbf{I}| = \frac{V_m}{\sqrt{R^2 + (\omega L - 1/\omega C)^2}} \tag{15.8}$$

is shown in Fig. 15.3; the plot only shows the symmetry illustrated in the graph when the frequency axis scale is logarithmic. The average power dissipated by the RLC circuit is

$$P(\omega) = \frac{1}{2}I^2 R \tag{15.9}$$

The highest power dissipated occurs at resonance, when $I = V_m/R$, so that

$$P(\omega_0) = \frac{1}{2}\frac{V_m^2}{R} \tag{15.10}$$

At certain frequencies $\omega = \omega_1, \omega_2$, the dissipated power is half the maximum value; that is,

$$P(\omega_1) = P(\omega_2) = \frac{(V_m/\sqrt{2})^2}{2R} = \frac{V_m^2}{4R} \tag{15.11}$$

Hence, ω_1 and ω_2 are called the *half-power frequencies*.

The half-power frequencies are obtained by setting $\mathbf{Z}$ equal to $\sqrt{2}R$:

$$\sqrt{R^2 + \left(\omega L - \frac{1}{\omega C}\right)^2} = \sqrt{2}R \tag{15.12}$$

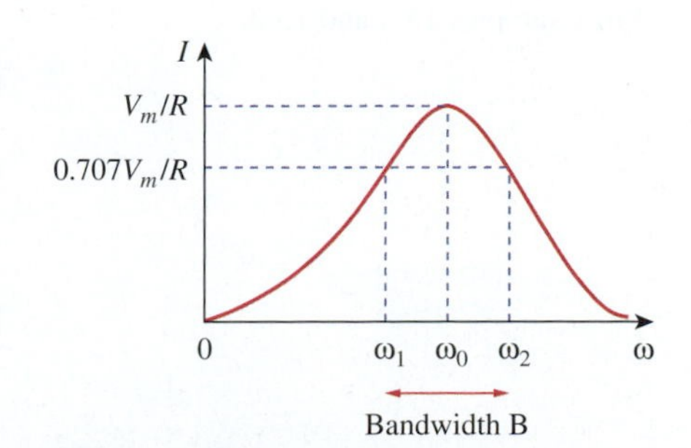

Figure 15.3
The current amplitude versus frequency for the series resonant circuit of Fig. 15.1.

Solving for ω, we obtain

$$\omega_1 = -\frac{R}{2L} + \sqrt{\left(\frac{R}{2L}\right)^2 + \frac{1}{LC}} \qquad \textbf{(15.13a)}$$

$$\omega_2 = \frac{R}{2L} + \sqrt{\left(\frac{R}{2L}\right)^2 + \frac{1}{LC}} \qquad \textbf{(15.13b)}$$

We can relate the half-power frequencies to the resonant frequency. From Eqs. (15.6) and (15.13),

$$\omega_0 = \sqrt{\omega_1 \omega_2} \qquad \textbf{(15.14)}$$

showing that the resonant frequency is the geometric mean of the half-power frequencies. Notice that ω_1 and ω_2 are, in general, not equally distant from the resonant frequency ω_0 because the frequency response is not generally symmetrical. However, as will be explained shortly, arithmetic symmetry of the half-power frequencies around the resonant frequency is often a reasonable approximation.

Although the height of the curve in Fig. 15.3 is determined by R, the width of the curve depends on other factors such as the half-power frequencies. The *bandwidth BW* is defined as the difference between the two half-power frequencies; that is,

$$BW = \omega_2 - \omega_1 \qquad \textbf{(15.15)}$$

Bandwidth (*BW*) is the difference between the two half-power frequencies.

This definition of bandwidth is just one of several definitions that are commonly used.[1] Strictly speaking, BW in Eq. (15.15) is a half-power bandwidth because it is the width of the frequency band between the half-power frequencies.

Example 15.1

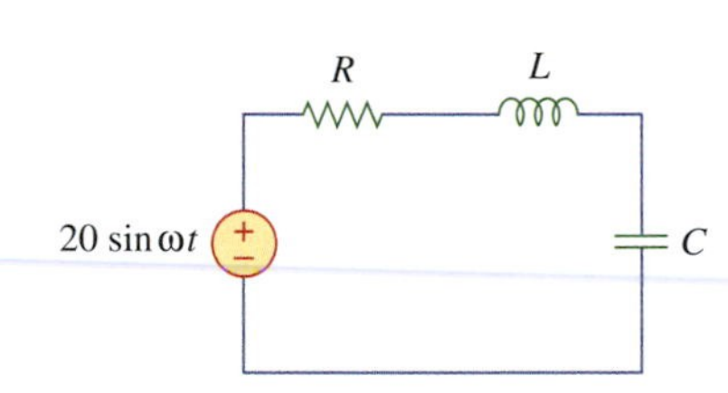

Figure 15.4
For Examples 15.1 and 15.2.

In the circuit in Fig. 15.4, $R = 2\ \Omega$, $L = 1$ mH, and $C = 0.4\ \mu$F.
(a) Find the resonant frequency and the half-power frequencies.
(b) Find the bandwidth.
(c) Determine the amplitude of the current at ω_0, ω_1, and ω_2.
(d) Calculate the power dissipated at resonance and at the half-power frequencies.

Solution:

(a) The resonant frequency is

$$\omega_0 = \frac{1}{\sqrt{LC}} = \frac{1}{\sqrt{10^{-3} \times 0.4 \times 10^{-6}}} = 50 \text{ krad/s}$$

The lower half-power frequency is

$$\begin{aligned}\omega_1 &= -\frac{R}{2L} + \sqrt{\left(\frac{R}{2L}\right)^2 + \frac{1}{LC}} \\ &= -\frac{2}{2 \times 10^{-3}} + \sqrt{(10^3)^2 + (50 \times 10^3)^2} \\ &= -1 + \sqrt{1 + 2500} \text{ krad/s} \\ &= 49 \text{ krad/s}\end{aligned}$$

[1] In computer communication networks, bandwidth is the data transfer rate minus the amount of data that can be carried from one point to another in 1 s.

Similarly, the upper half-power frequency is

$$\omega_2 = 1 + \sqrt{1 + 2500} \text{ krad/s}$$
$$= 51 \text{ krad/s}$$

(b) The bandwidth is

$$BW = \omega_2 - \omega_1 = 51 - 49 = 2 \text{ krad/s}$$

(c) At $\omega = \omega_0$,

$$I = \frac{V_m}{R} = \frac{20}{2} = 10 \text{ A}$$

At $\omega = \omega_1, \omega_2$

$$I = \frac{V_m}{\sqrt{2}R} = \frac{10}{\sqrt{2}} = 7.071 \text{ A}$$

(d) The power dissipated at half-power frequencies is

$$P = \frac{V_m^2}{4R} = \frac{20^2}{4 \times 2} = 50 \text{ W}$$

Practice Problem 15.1

A series connected circuit has $R = 4\ \Omega$, $C = 0.625\ \mu\text{F}$, and $L = 25$ mH.
(a) Calculate the resonant frequency.
(b) Find ω_1, ω_2, and BW.
(c) Determine the average power dissipated at $\omega = \omega_0, \omega_1, \omega_2$. Take $V_m = 100$ V.

Answer: (a) 8 krad/s; (b) 7920 rad/s, 8080 rad/s, 160 rad/s; (c) 1.25 kW, 0.625 kW, 0.625 kW

15.3 Quality Factor

The "sharpness" of the resonance in a resonant circuit is measured quantitatively by the *quality factor* Q.[2] At resonance, the reactive energy in the circuit oscillates between the inductor and the capacitor. For any resonant circuit, the quality factor is the ratio of the reactive power to the average power at the resonant frequency; that is,

$$Q = \frac{\text{Reactive power}}{\text{Average power}} \tag{15.16}$$

In the series RLC circuit,

$$Q = \frac{I^2 X_L}{I^2 R} = \frac{\omega_o L}{R} = \frac{I^2 X_c}{I^2 R} = \frac{1}{\omega_o C R} \tag{15.17}$$

or

$$\boxed{Q = \frac{\omega_0 L}{R} = \frac{1}{\omega_0 C R}} \tag{15.18}$$

[2] NOTE: Although the same symbol Q is used for the reactive power, the two are not equal and should not be confused. Their units will help distinguish between the two.

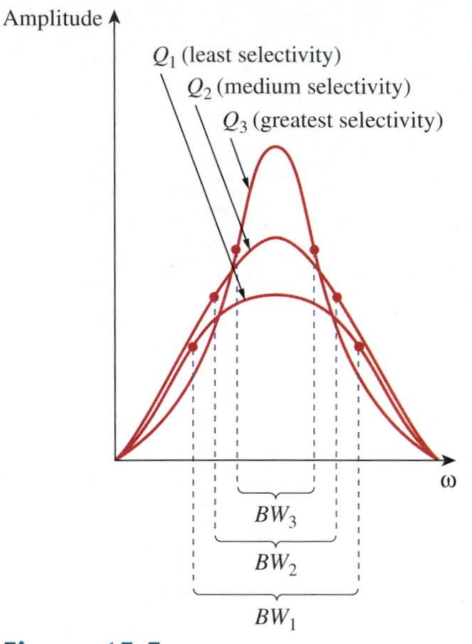

Figure 15.5
The higher the circuit Q, the smaller the bandwidth.

Notice that quality factor is dimensionless. The relationship between the bandwidth BW and the quality factor Q is obtained by substituting Eq. (15.13) into Eq. (15.15). The result is

$$BW = \frac{R}{L}$$

Utilizing Eq. (15.18), we get

$$BW = \frac{R}{L} = \frac{\omega_0}{Q} \quad \Rightarrow \quad Q = \frac{\omega_0}{BW} \tag{15.19}$$

or

$$BW = \omega_0^2 CR.$$

Thus

> The **quality factor** of a resonant circuit is the ratio of its resonant frequency to its bandwidth.

The higher the Q, the greater the current at resonance, and the sharper the curve. As illustrated in Fig. 15.5, the higher the value of Q, the more selective the circuit is but the smaller the bandwidth. The *selectivity* of an *RLC* circuit is the ability of the circuit to respond to a certain frequency and discriminate against all other frequencies.[3] If the band of frequencies to be selected or rejected is narrow, the quality factor of the resonant circuit must be high. If the band of frequencies of interest is wide, the quality factor must be low.

A resonant circuit is designed to operate at or near its resonant frequency. It is said to be a *high-Q circuit* when its quality factor is equal to or greater than 10. For high-Q circuits ($Q \geq 10$), the half-power frequencies are, for all practical purposes, arithmetically symmetrical around the resonant frequency and can be approximated as

$$\omega_1 \cong \omega_0 - \frac{BW}{2}, \qquad \omega_2 \cong \omega_0 + \frac{BW}{2} \tag{15.20}$$

High-Q circuits are used often in communications networks.

We note that a resonant circuit is characterized by five related parameters: the two half-power frequencies ω_1 and ω_2, the resonant frequency ω_0, the bandwidth BW, and the quality factor Q. Note also that Eqs. (15.18) and (15.19) only apply to series resonant circuits.

Example 15.2

In the circuit of Fig. 15.4, $R = 2\ \Omega$, $L = 1$ mH, and $C = 0.4\ \mu$F. Calculate the quality factor and obtain the half-power frequencies.

Solution:
This can be done in two ways.

METHOD 1 From Example 15.1, $BW = 2$ krad/s or

$$BW = \frac{R}{L} = \frac{2}{10^{-3}} = 2 \text{ krad/s}$$

[3] NOTE: The quality factor is a measure of the selectivity (or "sharpness" of resonance) of the circuit.

The quality factor is

$$Q = \frac{\omega_0}{BW} = \frac{50 \times 10^3}{2 \times 10^3} = 25$$

The lower half-power frequency is

$$\begin{aligned}\omega_1 &= -\frac{R}{2L} + \sqrt{\left(\frac{R}{2L}\right)^2 + \frac{1}{LC}} \\ &= -\frac{2}{2 \times 10^{-3}} + \sqrt{\left(\frac{2}{2 \times 10^3}\right)^2 + \frac{1}{10^{-3} \times 0.4 \times 10^{-6}}} \\ &= -10^3 + \sqrt{(10^3)^2 + (50 \times 10^3)^2} \\ &= -1 + \sqrt{1 + 2500} \text{ krad/s} = 49 \text{ krad/s}\end{aligned}$$

Similarly, the lower half-power frequency is

$$\omega_2 = 1 + \sqrt{1 + 2500} \text{ krad/s} = 51 \text{ krad/s}$$

■ **METHOD 2** Alternatively, we can find

$$Q = \frac{\omega_0 L}{R} = \frac{50 \times 10^3 \times 10^{-3}}{2} = 25$$

From Q, we find

$$BW = \frac{\omega_0}{Q} = \frac{50 \times 10^3}{25} = 2 \text{ krad/s}$$

Because $Q > 10$, this is a high-Q circuit and we can obtain the half-power frequencies as

$$\omega_1 = \omega_0 - \frac{BW}{2} = 50 - 1 = 49 \text{ krad/s}$$

$$\omega_1 = \omega_0 + \frac{BW}{2} = 50 + 1 = 51 \text{ krad/s}$$

Practice Problem 15.2

A series-connected circuit has $R = 4\ \Omega$ and $L = 25$ mH.
(a) Calculate the value of C that will produce a quality factor of 50.
(b) Find ω_0, ω_1, ω_2, and BW.

Answer: (a) 0.625 μF; (b) 7920 rad/s, 8080 rad/s, 160 rad/s

15.4 Parallel Resonance

The parallel RLC circuit in Fig. 15.6 is the dual of the series RLC circuit. So we will avoid needless repetition. The admittance is

$$\mathbf{Y} = \mathbf{I}/\mathbf{V} = \frac{1}{R} + j\omega C + \frac{1}{j\omega L} \tag{15.21}$$

or

$$\mathbf{Y} = \frac{1}{R} + j\left(\omega C - \frac{1}{\omega L}\right) \tag{15.22}$$

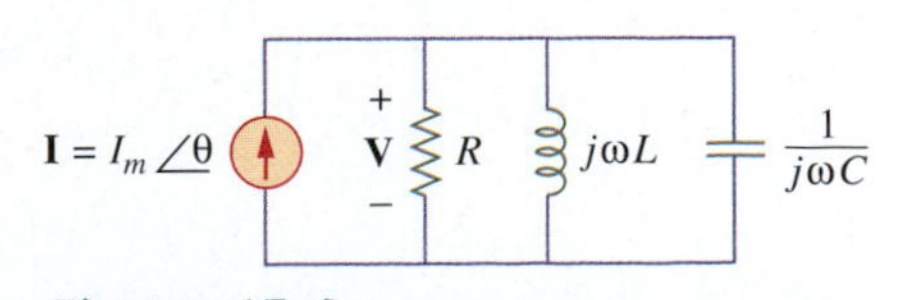

Figure 15.6
The parallel resonant circuit.

Resonance occurs when the imaginary part of **Y** is zero; that is,

$$\omega C - \frac{1}{\omega L} = 0 \tag{15.23}$$

or

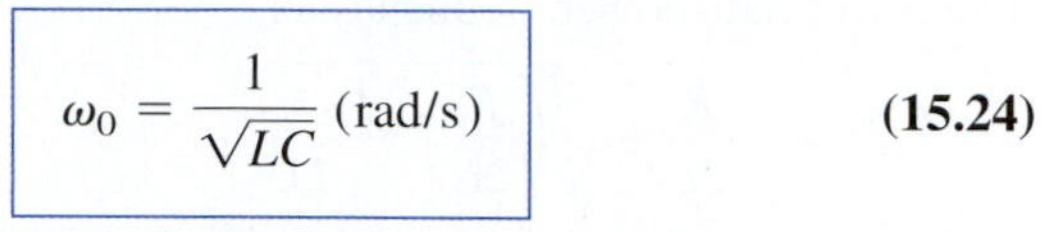

$$\omega_0 = \frac{1}{\sqrt{LC}} \text{ (rad/s)} \tag{15.24}$$

which is the same as Eq. (15.6) for the series resonant circuit. The voltage $|\mathbf{V}|$ is sketched in Fig. 15.7 as a function of frequency. Notice that at resonance, the parallel LC combination acts like an open circuit so that the entire current flows through R.

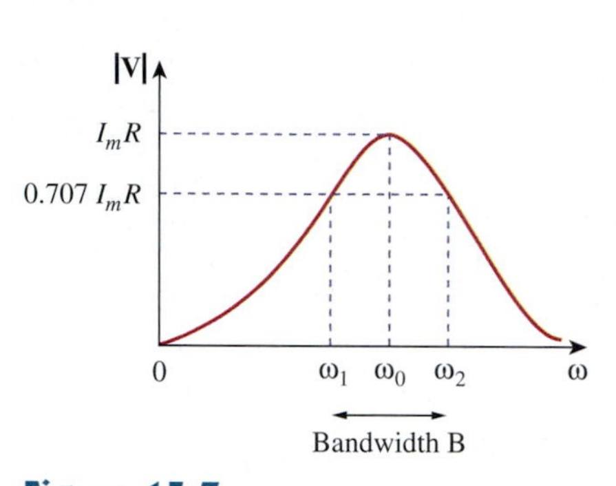

Figure 15.7
The voltage amplitude versus frequency for parallel resonant circuit Fig. 15.6.

We exploit the duality between Figs. 15.1 and 15.6 by comparing Eq. (15.21) with Eq. (15.2). By replacing R, L, and C in the expressions for the series circuit with $1/R$, C, and L, respectively, we obtain for the parallel circuit

$$\omega_1 = -\frac{1}{2RC} + \sqrt{\left(\frac{1}{2RC}\right)^2 + \frac{1}{LC}} \tag{15.25a}$$

$$\omega_2 = \frac{1}{2RC} + \sqrt{\left(\frac{1}{2RC}\right)^2 + \frac{1}{LC}} \tag{15.25b}$$

$$BW = \omega_2 - \omega_1 = \frac{1}{RC} \tag{15.26}$$

$$Q = \frac{\omega_0}{BW} = \omega_0 RC = \frac{R}{\omega_0 L} \tag{15.27}$$

Again, for high-Q circuits ($Q \geq 10$),

$$\omega_1 \cong \omega_0 - \frac{BW}{2}, \qquad \omega_2 \cong \omega_0 + \frac{BW}{2}$$

A summary of the characteristics of the series and parallel resonant circuits is presented in Table 15.1. Besides the series and parallel RLC considered here and in the previous section, other resonant circuits exist. A typical example is treated in Example 15.4.

TABLE 15.1

Summary of the characteristics of resonant RLC circuits.*

Characteristic	Series circuit	Parallel circuit
Resonant frequency, ω_0	$\frac{1}{\sqrt{LC}}$	$\frac{1}{\sqrt{LC}}$
Quality factor, Q	$\frac{\omega_0 L}{R}$ or $\frac{1}{\omega_0 RC}$	$\frac{R}{\omega_0 L}$ or $\omega_0 RC$
Bandwidth, BW	$\frac{\omega_0}{Q}$	$\frac{\omega_0}{Q}$
Half-power frequencies ω_1, ω_2	$\omega_0 \pm \frac{BW}{2}$	$\omega_0 \pm \frac{BW}{2}$

*The half-power frequencies are only valid for high Q.

Our discussion about resonance would be incomplete without mentioning ringing. *Ringing* is an unwanted oscillation or near-oscillation of a voltage or current in an electric circuit. It is caused when an electrical pulse causes the parasitic capacitances and inductances in the circuit (i.e., those that are just by-products of the materials used to construct the circuit but not part of the design) to resonate. Ringing is undesirable because it causes extra current to flow, thereby wasting energy and causing extra heating of the components.

Example 15.3

In the parallel *RLC* circuit of Fig. 15.8, let $R = 8$ kΩ, $L = 0.2$ mH, and $C = 8\ \mu$F.
(a) Calculate ω_0, Q, and BW. (b) Find ω_1 and ω_2.
(c) Determine the power dissipated at ω_0, ω_1, and ω_2.

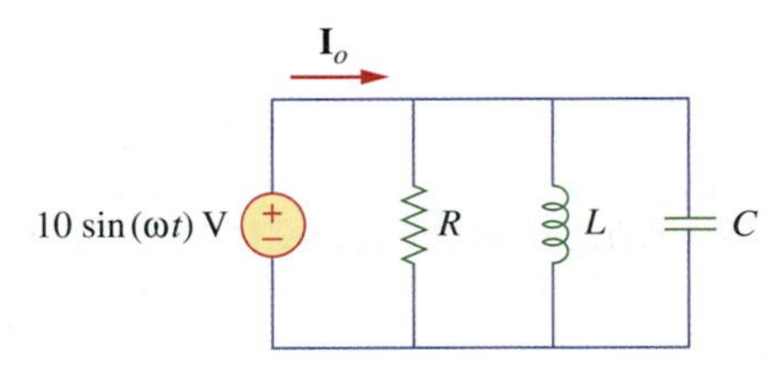

Figure 15.8
For Example 15.3.

Solution:

(a)

$$\omega_0 = \frac{1}{\sqrt{LC}} = \frac{1}{\sqrt{0.2 \times 10^{-3} \times 8 \times 10^{-6}}} = \frac{10^5}{4} = 25 \text{ krad/s}$$

$$Q = \frac{R}{\omega_0 L} = \frac{8 \times 10^3}{25 \times 10^3 \times 0.2 \times 10^{-3}} = 1{,}600$$

$$BW = \frac{\omega_0}{Q} = 15.625 \text{ rad/s}$$

(b) Due to the value of Q, we can regard this as a high-Q circuit. Hence,

$$\omega_1 = \omega_0 - \frac{BW}{2} = 25{,}000 - 7.812 = 24{,}992 \text{ rad/s}$$

$$\omega_2 = \omega_0 + \frac{BW}{2} = 25{,}000 + 7.812 = 25{,}008 \text{ rad/s}$$

(c) At $\omega = \omega_0$, $\mathbf{Y} = 1/R$ or $\mathbf{Z} = R = 8$ kΩ. Hence,

$$\mathbf{I}_o = \mathbf{V}/\mathbf{Z} = \frac{10\underline{/0^\circ}}{8{,}000} = 1.25\underline{/0^\circ} \text{ mA}$$

Because the entire current flows through R at resonance, the average power dissipated at $\omega = \omega_0$ is

$$P = \frac{1}{2}|\mathbf{I}_o|^2 R = \frac{1}{2}(1.25 \times 10^{-3})^2(8 \times 10^3) = 6.25 \text{ mW}$$

or

$$P = \frac{V_m^2}{2R} = \frac{10^2}{2 \times 8 \times 10^3} = 6.25 \text{ mW}$$

At $\omega = \omega_1, \omega_2$,

$$P = \frac{V_m^2}{4R} = 3.125 \text{ mW}$$

Practice Problem 15.3

A parallel resonant circuit has $R = 100$ kΩ, $L = 20$ mH, and $C = 5$ nF. Calculate ω_0, ω_1, ω_2, Q, and BW.

Answer: 100 krad/s; 99 krad/s; 101 krad/s; 50; 2 krad/s

Example 15.4

Determine the resonant frequency of the circuit in Fig. 15.9.

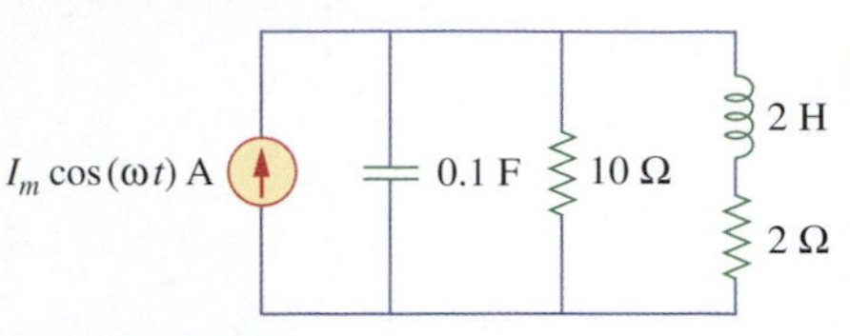

Figure 15.9
For Example 15.4.

Solution:
The input admittance is

$$\mathbf{Y} = j\omega 0.1 + \frac{1}{10} + \frac{1}{2 + j\omega 2} = 0.1 + j\omega 0.1 + \frac{2 - j\omega 2}{4 + 4\omega^2}$$

$$= \left(0.1 + \frac{1}{2 + 2\omega^2}\right) + j\omega\left(0.1 - \frac{1}{2 + 2\omega^2}\right)$$

At resonance, $\text{Im}(\mathbf{Y}) = 0$; that is,

$$\omega_0 0.1 - \frac{\omega_0}{2 + 2\omega_0^2} = 0 \qquad \Rightarrow \qquad \omega_0 = 2 \text{ rad/s}$$

Practice Problem 15.4

Calculate the resonant frequency of the circuit in Fig. 15.10.

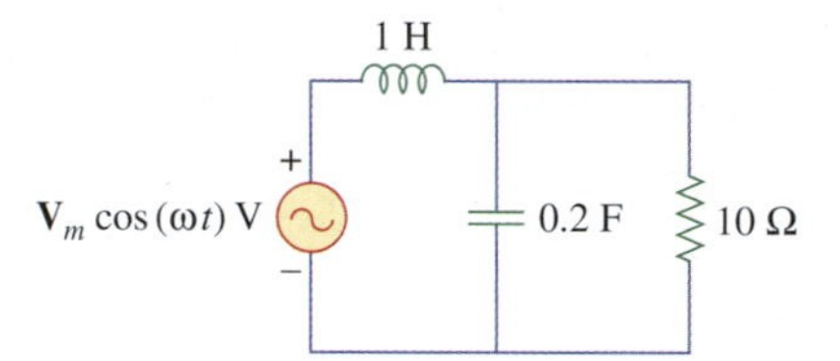

Figure 15.10
For Example 15.4.

Answer: 2.179 rad/s

15.5 Computer Analysis

15.5.1 PSpice

PSpice is a useful tool in the hands of the modern circuit designer for obtaining the frequency response of circuits. The frequency response is obtained using the AC Sweep as discussed in Section C.5 (Appendix C). This requires that we specify in the AC Sweep dialog box *Total Pts, Start Freq, End Freq*, and the sweep type. *Total Pts* is the number of points in the frequency sweep, and *Start Freq* and *End Freq* are the starting and final frequencies, respectively, in hertz. In order to know what frequencies to select for *Start Freq* and *End Freq*, one must have an idea of the frequency range of interest by making a rough sketch of the frequency response. In a complex circuit where this may not be possible, one may use a trial-and-error approach.

There are three types of sweeps:

- **Linear:** The frequency is varied linearly from *Start Freq* to *End Freq* with *Total* equally spaced points (or responses).
- **Octave:** The frequency is swept logarithmically by octaves from *Start Freq* to *End Freq* with *Total* points per octave. Two frequencies are an octave apart if the ratio of the higher frequency to the lower frequency is 2:1, for example, from 100 to 200 Hz, from 200 Hz to 400 Hz, from 400 to 800 Hz, and so on.
- **Decade:** The frequency is varied logarithmically by decades from *Start Freq* to *End Freq* with *Total* points per decade. A decade is a factor of 10, for example, from 1 to 10 Hz, from 10 to 100 Hz, from 100 to 1 kHz, and so forth.

With the preceding specifications, PSpice performs a steady-state sinusoidal analysis of the circuit as the frequency of all the independent sources is varied (or swept) from *Start Freq* to *End Freq*.

The Probe window is used to produce a graphical output. The output data type may be specified in the *Trace Command Box* by adding one of the following suffixes to V or I:

M Amplitude of the sinusoid

P Phase of the sinusoid

dB Amplitude of the sinusoid in decibels, for example, $20 \log_{10}$ (amplitude)

Example 15.5

Use PSpice to determine the frequency response of the parallel resonant circuit of Fig. 15.11.

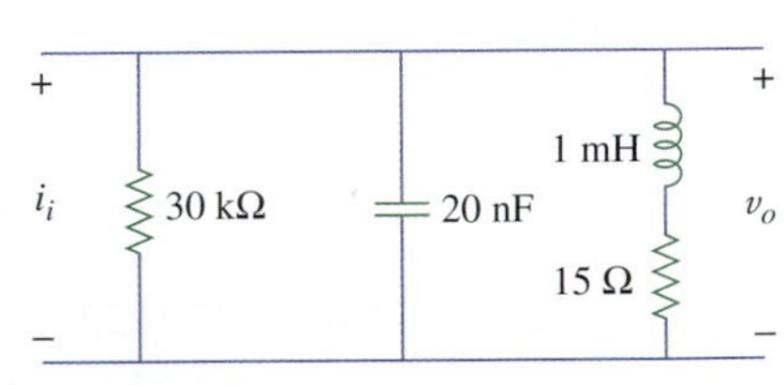

Figure 15.11
For Example 15.5.

Solution:

The schematic is shown in Fig. 15.12. We use the ac current source IAC as the input with magnitude 1 mA and phase 0°. The frequency response is taken as the output voltage v_o in Fig. 15.11. Once the circuit is drawn and saved as exam155.dsn, we select **PSpice/New Simulation Profile**. This leads to the New Simulation dialog box. Type "exam155" as the name of the file and click Create. This leads to the Simulation Settings dialog box. In the Simulation Settings dialog box, select AC Sweep/Noise under *Analysis Type*, select Linear under *AC Sweep Type*. We enter 10 k as the *Start Freq*, 80 k as the *Final Freq*, and 10 k as *Total Points*.

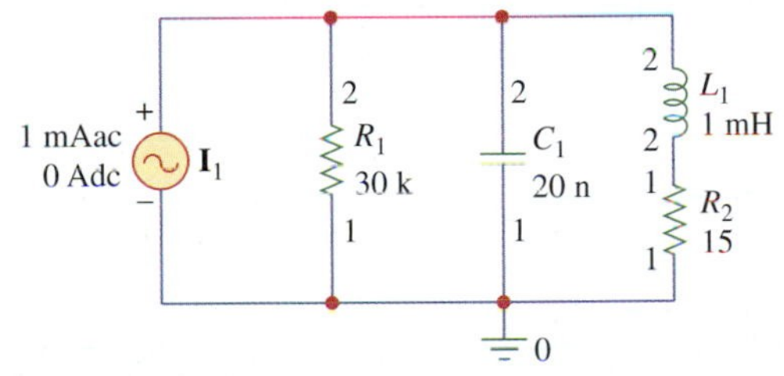

Figure 15.12
PSpice schematic for Example 15.5.

We are now ready to perform AC sweep for 10 kHz $< f <$ 80 kHz. We select **PSpice/Run** to simulate the circuit. The Probe window will automatically come up. We select **Trace/Add Trace** and select V(C1:2). (We could achieve the same thing by placing a voltage marker at node 2 in the schematic.) The result is shown in Fig. 15.13. You will notice that the resonant frequency is roughly 36 kHz.

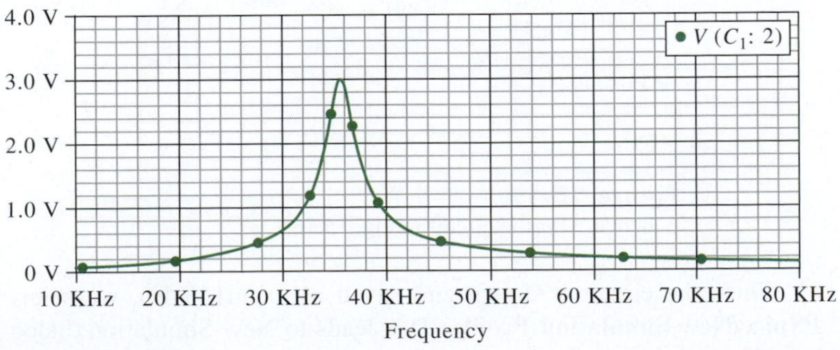

Figure 15.13
Frequency response of the circuit in Fig. 15.11.

Practice Problem 15.5

Obtain the frequency response (i_o) of the circuit in Fig. 15.14 using PSpice.

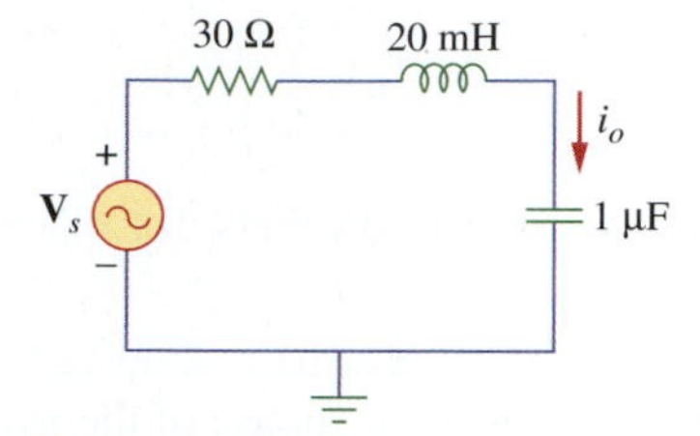

Figure 15.14
For Practice Problem 15.5 and Example 15.7.

Answer: See the plot in Fig. 15.15.

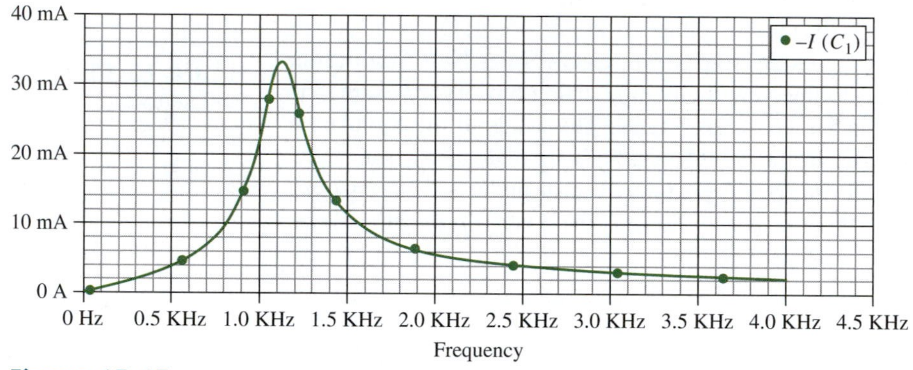

Figure 15.15
Frequency response of the circuit in Fig. 15.14.

Example 15.6

Determine the frequency response of the circuit shown in Fig. 15.16.

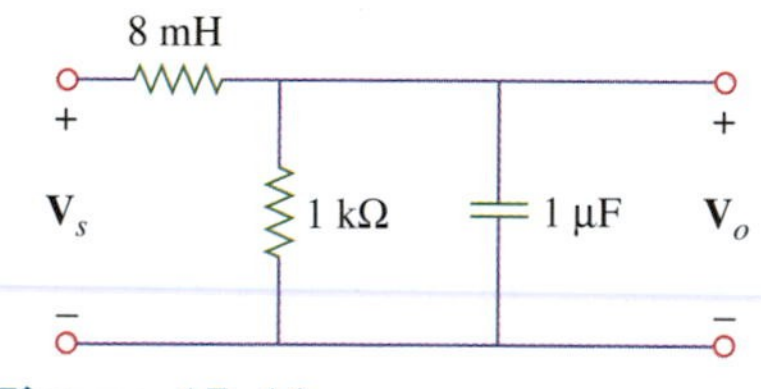

Figure 15.16
For Example 15.6.

Solution:

While the circuits in the previous example and practice problem are parallel and series, respectively, the one in Fig. 15.16 is neither parallel nor series. We let the input voltage v_s be a sinusoid of amplitude 1 V and phase 0°. The schematic for the circuit is in Fig. 15.17. The voltage marker is inserted to the output voltage across the capacitor.

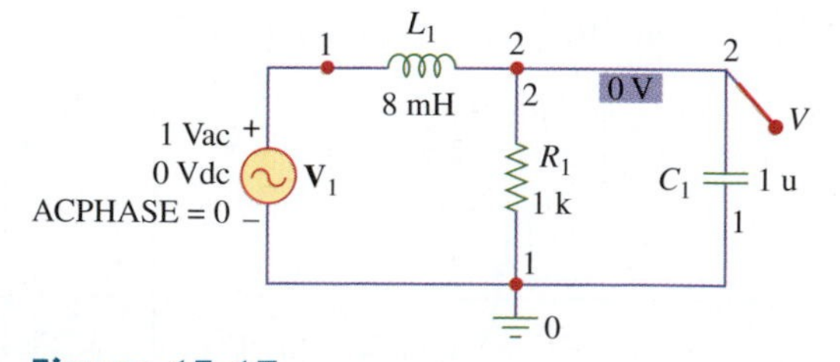

Figure 15.17
Schematic for the circuit in Fig. 15.16.

Once the circuit is drawn and saved as exam156.dsn, we select **PSpice/New Simulation Profile**. This leads to New Simulation dialog box. Type "exam156" as the name of the file and click Create. This leads to the Simulation Settings dialog box. In the Simulation Settings dialog box, select AC Sweep/Noise under *Analysis Type*, select Linear under *AC Sweep Type*. We enter 500 as the *Start Freq*, 4k as the *Final Freq*, and 10 as *Points/Decade*.

We are now ready to perform AC sweep for 500 kHz $< f <$ 4 kHz. We select **PSpice/Run** to simulate the circuit. The Probe window will automatically come up. We select **Trace/Add Trace** and select V(C1:2) or VM(C1:2). The result is shown in Fig. 15.18(a). You will notice that the resonant frequency is roughly 1.8 kHz. To obtain the phase, select **Trace/Add Trace** and type VP(C1:1) in the **Trace Command** box. The result is shown in Fig. 15.18(b).

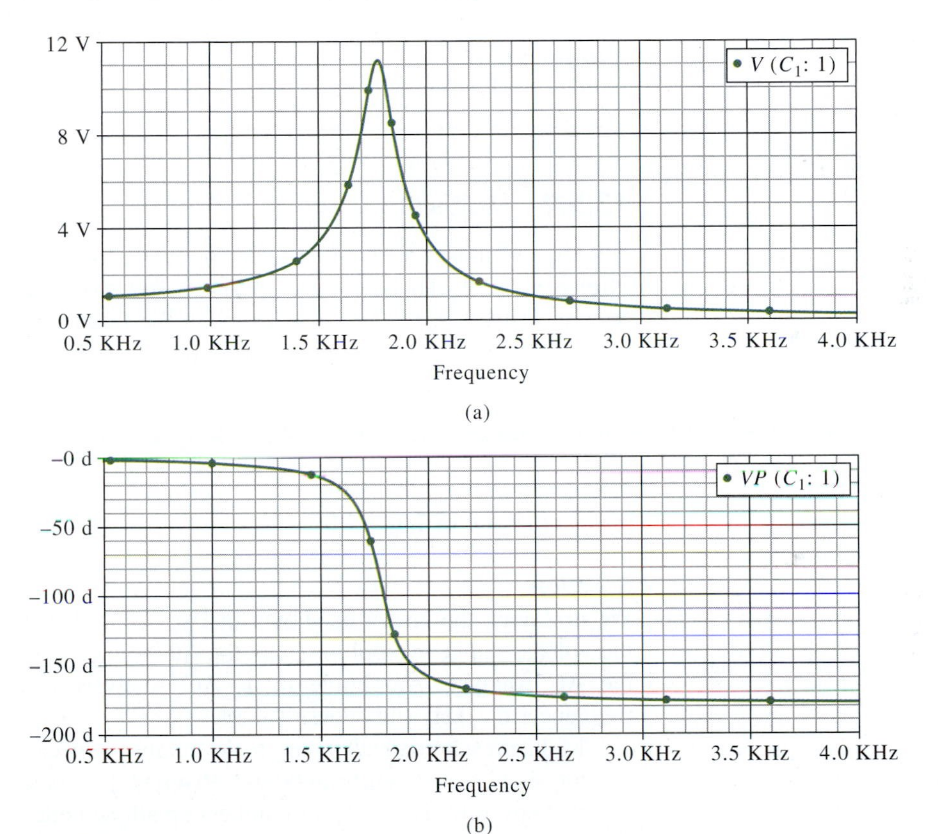

Figure 15.18
For Example 15.6: (a) magnitude plot; (b) phase plot of the frequency response.

Practice Problem 15.6

Obtain the frequency response of the circuit in Fig. 15.19 using PSpice. Use a linear frequency sweep and consider $1 < f < 10{,}000$ Hz with 1,001 points.

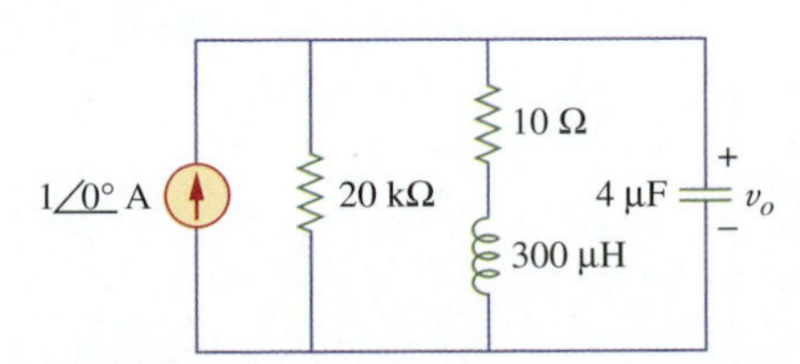

Figure 15.19
For Practice Problem 15.6.

Answer: See Fig. 15.20.

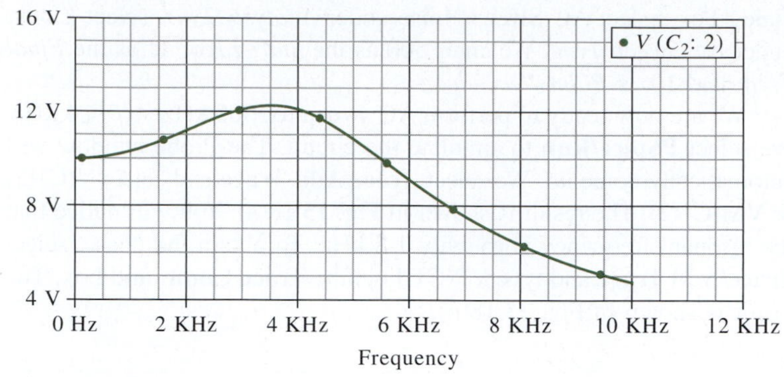

Figure 15.20
For Practice Problem 15.6.

15.5.2 Multisim

Although Multisim does not provide for single-frequency analysis like PSpice does, it can be used to perform AC sweep as discussed in Section D.4 (of Appendix D) and obtain the frequency response of a circuit. The reader is encouraged to read Section D.4 before proceeding with this section.

Example 15.7

Use Multisim to determine the frequency response of the series resonant circuit of Fig. 15.14.

Solution:
We first draw the circuit as shown in Fig. 15.21. We double-click on the voltage source symbol to obtain the AC Voltage dialog box. We set the AC Analysis Amplitude to 1, AC Analysis Phase to 0, and frequency to 1 kHz (this doesn't matter).

Multisim automatically puts node numbers. In case the numbers do not show, select **Options/Sheet Properties**. Under *Net Names*, select "Show all." This will put numbers on all the nodes. Our response is taken from node 3. To specify the frequency range for the simulation, we select **Simulate/Analyses/AC Analysis**. In the AC Analysis dialog box, we select *FSTART* as 1 Hz, *FSTOP* as 8 kHz, *Sweep type* as Decade, *Number of points per decade* as 100, and *Vertical scale* as

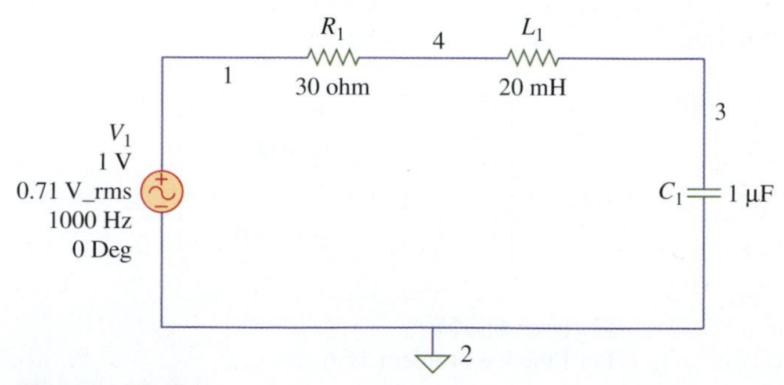

Figure 15.21
For Example 15.7.

Linear. (We know the frequency range from Fig. 15.15. If we do not know it in advance, we select it by trial and error.) Under output variables, we move V(3) from the left list to the right list so that the voltage at node 3 is displayed. We finally select **Simulate**. The frequency response (both magnitude and phase) is shown in Fig. 15.22.

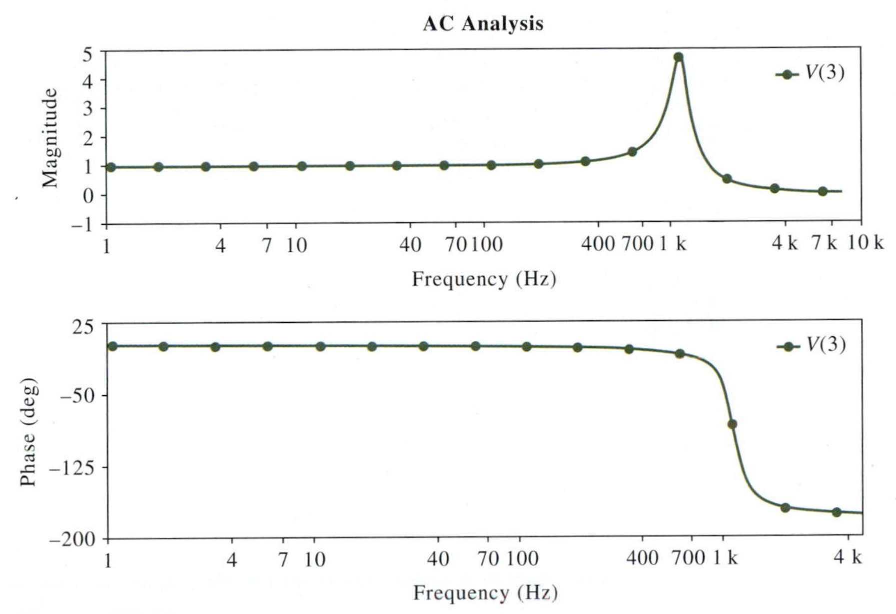

Figure 15.22
Frequency response of the circuit in Fig. 15.21.

Practice Problem 15.7

Determine the frequency response of the circuit in Fig. 15.11 using Multisim.

Answer: See Fig. 15.13.

15.6 †Applications

Resonant circuits are widely used, particularly in electronics, power systems, and communications systems. In this section, we consider one practical application of resonant circuits. The focus of the application is not to understand the details of how the device works, but to see how the circuits considered in this chapter are applied in the practical devices.

Series and parallel resonant circuits are commonly used in radio and TV receivers to tune in stations and to separate the audio signal from the radio-frequency carrier wave. As an example, consider the block diagram of an AM radio receiver shown in Fig. 15.23. Incoming amplitude-modulated radio waves (thousands of them at different frequencies from different broadcasting stations) are received by the antenna. A resonant circuit (or a bandpass filter) is needed to select just one of the incoming waves. (This is part of the RF amplifier in Fig. 15.23.) The selected signal is very weak and is amplified in stages in order to generate an audible audio-frequency wave. Thus, we have

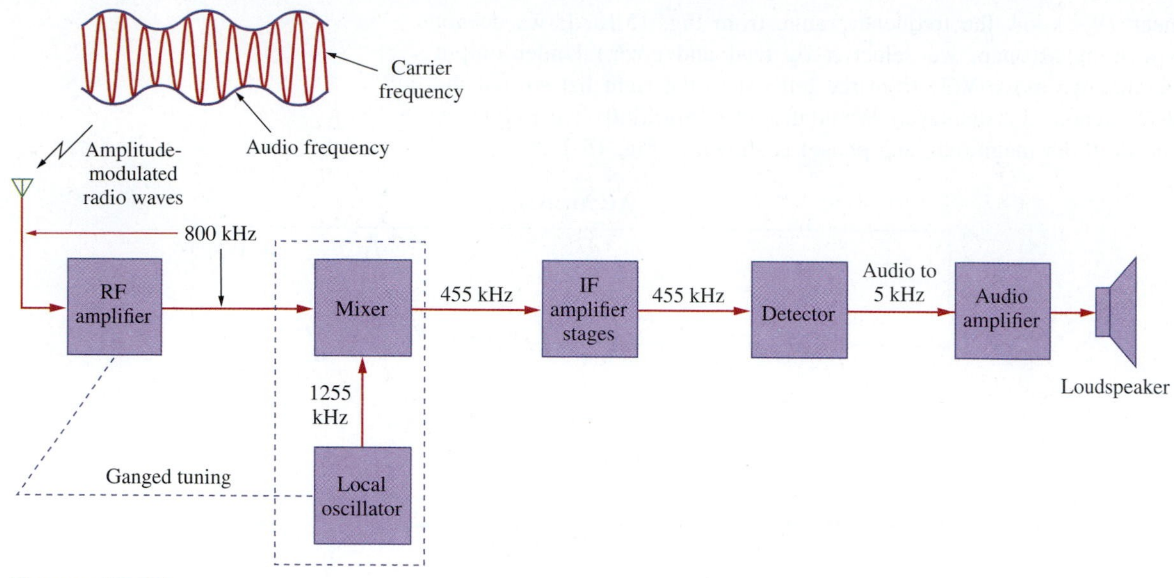

Figure 15.23
A simplified block diagram of a superheterodyne AM radio receiver.

the radio-frequency (RF) amplifier to amplify the selected broadcast signal, the intermediate-frequency (IF) amplifier to amplify an internally generated signal based on the RF signal, and the audio amplifier to amplify the audio signal just before it reaches the loudspeaker. It is much easier to amplify the signal in three stages than to build an amplifier to provide the same amplification for the entire band.

The type of AM receiver shown in Fig. 15.23 is known as the *superheterodyne receiver*. In the early development of radio, each amplification stage had to be tuned to the frequency of the incoming signal. This way, each stage must have several tuned circuits to cover the entire AM band (530 to 1,600 kHz). To avoid the problem of having several resonant circuits, modern receivers use a *frequency mixer* or *heterodyne* circuit, which always produces the same IF signal (455 kHz) but retains the audio frequencies carried on the incoming signal. To produce the constant IF frequency, the rotors of two separate variable capacitors are mechanically coupled with one another so that they can be rotated simultaneously with a single control; this is called *ganged tuning*. A *local oscillator* ganged with the RF amplifier produces an RF signal that is combined with the incoming wave by the frequency mixer to produce an output signal that contains the sum and the difference of the two signals. For example, if the resonant circuit is tuned to receive an 800-kHz incoming signal, the local oscillator must produce a 1,255-kHz signal so that the sum (1,255 + 800 = 2,055 kHz) and the difference frequency (1,255 − 800 = 455 kHz) of frequencies are available at the output of the mixer. However, only the difference (455 kHz) is used in practice. This is the only frequency to which all the IF amplifier stages are tuned regardless of the station dialed. The original audio signal (containing the "intelligence") is extracted in the detector stage. The detector basically removes the IF signal, leaving the audio signal. The audio signal is amplified to drive the loudspeaker, which acts as a transducer converting the electrical signal to sound.

Our example here is the tuning circuit for an AM radio receiver. The operation of AM radio receivers discussed here applies to FM radio receivers but in a much different band of frequencies.

Example 15.8

The resonant or tuner circuit of an AM radio is portrayed in Fig. 15.24. Given that $L = 1\ \mu$H, what must be the range of C to have the resonant frequency adjustable from one end of the AM band to another?

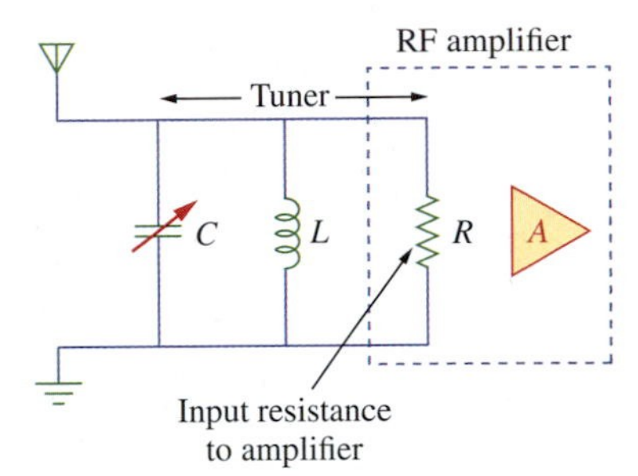

Figure 15.24
For Example 15.8.

Solution:
The frequency range for AM broadcasting is 530 to 1,600 kHz. We consider the low and high ends of the band. Because the resonant circuit in Fig. 15.24 is a parallel type, we apply the ideas in Section 15.4. From Eq. (15.24),

$$\omega_0 = 2\pi f_0 = \frac{1}{\sqrt{LC}}$$

or

$$C = \frac{1}{4\pi^2 f_0^2 L}$$

For the high end of the AM band, $f_0 = 1{,}600$ kHz and the corresponding C is

$$C_1 = \frac{1}{4\pi^2 \times 1{,}600^2 \times 10^6 \times 10^{-6}} = 9.9 \text{ nF}$$

For the low end of the AM band, $f_0 = 530$ kHz and the corresponding C is

$$C_2 = \frac{1}{4\pi^2 \times 530^2 \times 10^6 \times 10^{-6}} = 90.18 \text{ nF}$$

Thus, C must be an adjustable (gang) capacitor varying from 9.9 to 90.2 nF.

Practice Problem 15.8

For an FM radio receiver, the incoming wave is in the frequency range from 88 to 108 MHz. The tuner circuit is a parallel RLC circuit with a 4-μH coil. If the local oscillator frequency must always be 10.7 MHz above the carrier frequency, calculate the range of the variable capacitor necessary to cover the entire band.

Answer: From 0.543 to 0.8177 pF

15.7 Summary

1. The resonant frequency is that frequency at which the imaginary part of an impedance or an admittance vanishes. For series and parallel RLC circuits,

$$\omega_0 = \frac{1}{\sqrt{LC}} = 2\pi f_o$$

2. The half-power frequencies (ω_1, ω_2) are those frequencies at which the power dissipated is one-half of that dissipated at the resonant

frequency. The geometric mean of the half-power frequencies is the resonant frequency; that is,

$$\omega_0 = \sqrt{\omega_1 \omega_2}$$

3. The bandwidth is the frequency band between half-power frequencies.

$$BW = \omega_2 - \omega_1$$

4. The quality factor is a measure of the sharpness of the resonance peak. It is the ratio of the resonant (angular) frequency to the bandwidth,

$$Q = \frac{\omega_0}{BW}$$

5. PSpice and Multisim can be used to obtain the frequency response of a circuit if a frequency range for the response and the desired number of points within the range are specified in the AC Sweep.
6. The radio receiver is treated as one practical application of resonant circuits. It employs a resonant circuit to tune in one frequency among all the broadcast signals picked up by the antenna.

Review Questions

15.1 The resonant frequency for a 40-pF capacitor in series with a 90-μH inductor is:

(a) 36 MHz (b) 24.5 MHz
(c) 16.67 MHz (d) 2.65 MHz

15.2 How much inductance is needed to resonate at 5 kHz with a capacitance of 12 nF?

(a) 2652 H (b) 11.844 H
(c) 3.333 H (d) 84.43 mH

15.3 The frequency at which the voltage across the inductor in a series *RLC* circuit has fallen to 0.707 of its value at resonance is called:

(a) resonant frequency
(b) cutoff frequency
(c) half-power frequency
(d) bandwidth

15.4 The difference between the half-power frequencies is called the:

(a) quality factor
(b) resonant frequency
(c) bandwidth
(d) cutoff frequency

15.5 In a series *RLC* circuit, the quality factor is directly proportional to *R*.

(a) True (b) False

15.6 In a series *RLC* circuit, which of these quality factors has the steepest curve at resonance?

(a) $Q = 20$ (b) $Q = 12$
(c) $Q = 8$ (d) $Q = 4$

15.7 What is the quality factor for a series resonant circuit that has 40 Ω of resistance and an X_L of 2,800 Ω?

(a) 70 (b) 1.75 (c) 0.5714

15.8 In a parallel *RLC* circuit, the bandwidth *BW* is directly proportional to *R*.

(a) True (b) False

15.9 Two parameters that are equal when a parallel resonant circuit is operating at its resonant frequency are:

(a) *L* and *C* (b) *R* and *C*
(d) *R* and *L* (d) X_C and X_L

15.10 A resonant circuit is said to be high-*Q* when its quality factor is greater than:

(a) 1 (b) 10
(c) 100 (d) 1000

Answers: 15.1d, 15.2d, 15.3c, 15.4c, 15.5b, 15.6a, 15.7a, 15.8b, 15.9d, 15.10b

Problems

Sections 15.2 and 15.3 Series Resonance and Quality Factor

15.1 A series *RLC* network has $R = 2\ \text{k}\Omega$, $L = 40$ mH, and $C = 1\ \mu\text{F}$. Calculate the impedance at resonance and at one-fourth, one-half, twice, and four times the resonant frequency.

15.2 A series *RLC* circuit has $R = 0.1\ \text{k}\Omega$, $L = 10$ mH, and $C = 5$ nF. Find: (a) the resonant frequency, (b) the bandwidth, and (c) the quality factor.

15.3 Design a series *RLC* circuit that will have an impedance of 10 Ω at the resonant frequency of $\omega_0 = 50$ rad/s and a quality factor of 80. Find the bandwidth.

15.4 Design a series *RLC* circuit with $BW = 20$ rad/s and $\omega_0 = 1000$ rad/s. Find the circuit's Q.

15.5 For the circuit in Fig. 15.25, find the frequency ω_0 for which $v(t)$ and $i(t)$ are in phase.

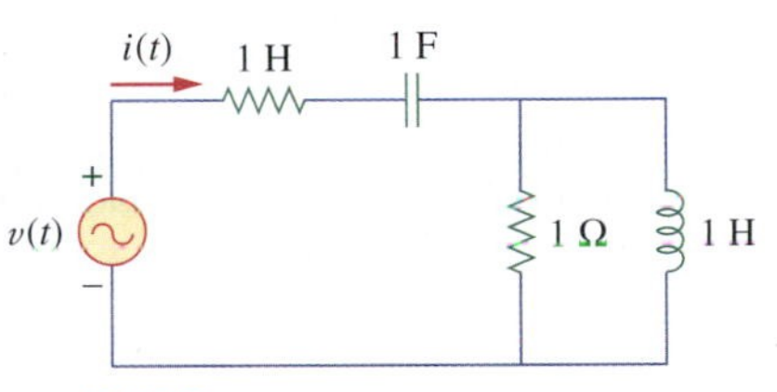

Figure 15.25
For Problem 15.5.

15.6 For the series *RLC* circuit in Fig. 15.26, find: (a) the value of X_C at resonance, (b) the magnitude of the current $i(t)$ at resonance, (c) the quality factor, and (d) the bandwidth.

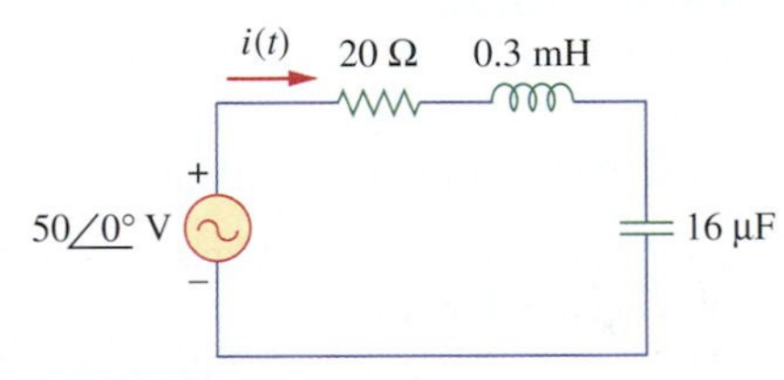

Figure 15.26
For Problem 15.6.

15.7 A series resonant circuit has a resonant frequency of 8 kHz and its quality factor is 20. (a) Find the bandwidth. (b) Calculate the half-power frequencies. (c) Determine X_C if $R = 10\ \Omega$.

15.8 A circuit consisting of a coil with an inductance of 10 mH and a resistance of 20 Ω is connected in series with a capacitor and a generator with an rms voltage of 120 V. Find: (a) the value of the capacitance that will cause the circuit to be in resonance at 15 kHz, (b) the current through the coil at resonance, and (c) the Q of the circuit.

15.9 A series *RLC* circuit has $R = 4\ \text{k}\Omega$, $X_L = 40\ \text{k}\Omega$, and $X_C = 30\ \text{k}\Omega$ at $f = 4$ MHz. Determine the bandwidth and the quality factor.

15.10 A 60-Hz supply voltage with $V = 12\underline{/0^\circ}$ V is applied to a series *RLC* circuit. Let $R = 10\ \Omega$, $X_L = 160\ \Omega$. (a) Determine the value of C to produce a series resonance. (b) Find the maximum current at resonance. (c) Calculate the voltage across the inductor at resonance.

15.11 What value of capacitance is needed to produce a series resonance with a 2.4-mH coil at 4.5 kHz?

15.12 A series *RLC* circuit has a bandwidth of 6 Mrad/s and an impedance of 20 Ω at the resonant frequency of 40 Mrad/s. Calculate: (a) the inductance L, (b) the capacitance C, (c) the quality factor Q, and (d) the upper and lower cutoff frequencies.

15.13 A 10-mH inductor that has an internal resistance of 5 Ω is connected in series with a capacitor and a voltage source with Thevenin equivalent resistance of 15 Ω. Find: (a) the capacitance value that will produce resonance at 2 krad/s, (b) quality factor of the circuit, (c) the bandwidth of the circuit.

Section 15.4 Parallel Resonance

15.14 Design a parallel resonant *RLC* circuit with $\omega_0 = 10$ rad/s and $Q = 20$. Calculate the bandwidth of the circuit. Select $R = 10\ \Omega$.

15.15 At 10 MHz, a parallel circuit has $R = 5.6\ \text{k}\Omega$, $X_L = 40\ \text{k}\Omega$, $X_C = 40\ \text{k}\Omega$. Calculate the bandwidth.

15.16 A parallel resonant circuit with a quality factor of 120 has a resonant frequency of 6×10^6 rad/s. Calculate the bandwidth and the half-power frequencies.

15.17 A parallel *RLC* circuit is resonant at 5.6 MHz, has a Q of 80, and has a resistive branch of 40 kΩ. Determine the values of L and C in the other two branches.

15.18 A parallel *RLC* circuit has $R = 5\ \text{k}\Omega$, $L = 8$ mH, and $C = 60\ \mu\text{F}$. Determine: (a) the resonant frequency, (b) the bandwidth, and (c) the quality factor.

15.19 It is expected that a parallel *RLC* resonant circuit has a midband admittance of 25 mS, a quality factor of 80, and a resonant frequency of 200 krad/s. Calculate the values of R, L, and C. Find the bandwidth and the half-power frequencies.

15.20 Rework Problem 15.1 if the elements are connected in parallel.

15.21 For the "tank" circuit in Fig. 15.27, find the resonant frequency.

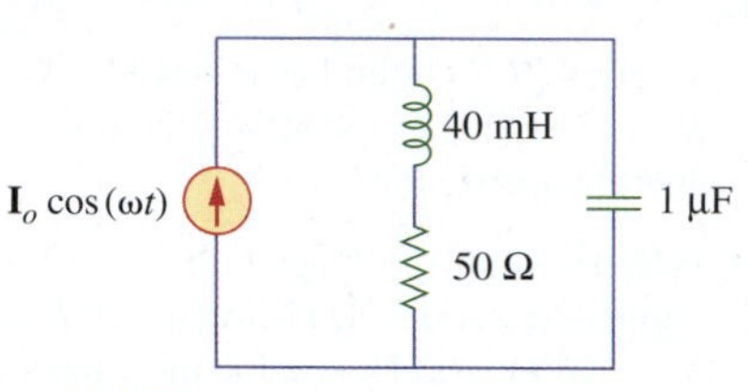

Figure 15.27
For Problems 15.21 and 15.31.

15.22 A parallel resonant circuit has a resistance of 2 kΩ and half-power frequencies of 86 kHz and 90 kHz. Determine: (a) the capacitance, (b) the inductance, (c) the resonant frequency, (d) the bandwidth, and (e) the quality factor.

15.23 For the circuits in Fig. 15.28, find the resonant frequency ω_0, the quality factor Q, and the bandwidth BW.

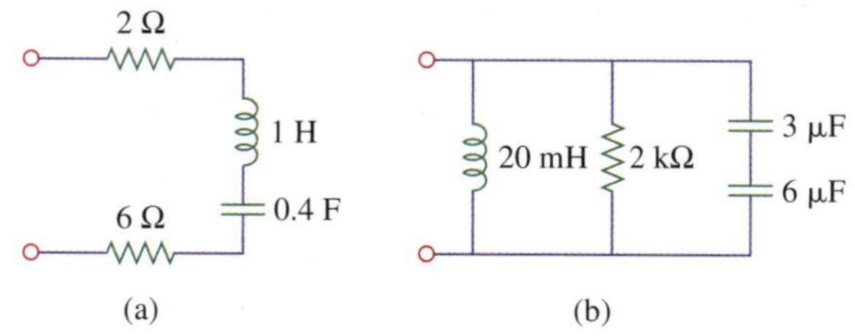

Figure 15.28
For Problem 15.23.

15.24 For the parallel *RLC* circuit in Fig. 15.29, determine: (a) ω_o, (b) the total impedance at ω_o, (c) Q, and (d) bandwidth.

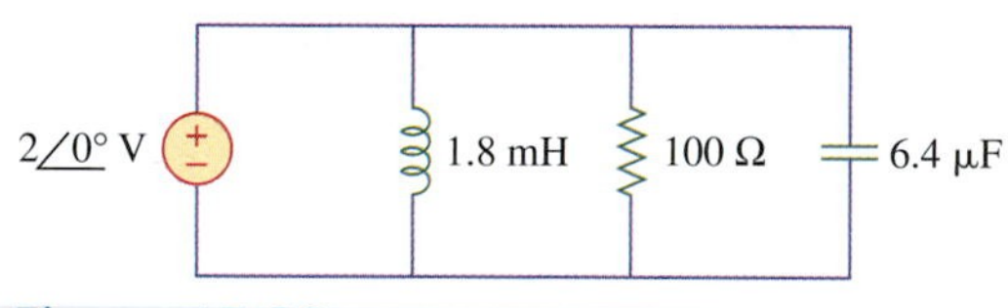

Figure 15.29
For Problem 15.24.

15.25 A parallel *RLC* circuit is resonant at 2 MHz and has a $Q = 80$ and $R = 300$ kΩ. Calculate the values of L and C.

15.26 A parallel *RLC* circuit has a resistance of 2 kΩ and is resonant at 300 kHz. Find Q, L, and C required to provide a bandwidth of 5 kHz.

15.27 What value of inductance is required to produce a parallel resonance with a 5-pF capacitor at 200 Hz?

15.28 Refer to the circuit in Fig. 15.30. Let $v = 10 \sin(\omega t)$. (a) Find the resonant frequency. (b) Calculate the power delivered by the voltage source at resonance. (c) Obtain the quality factor and the bandwidth.

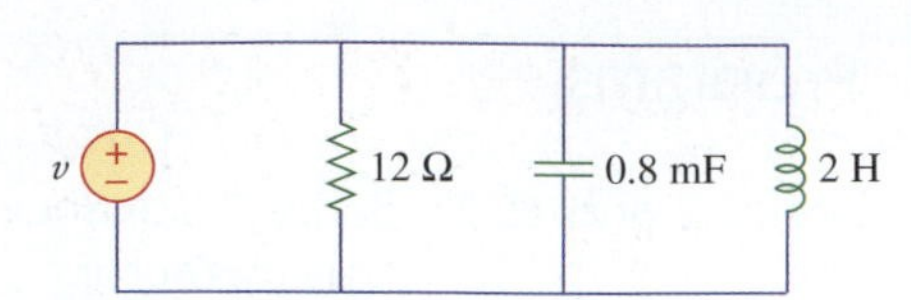

Figure 15.30
For Problem 15.28.

Section 15.5 Computer Analysis

15.29 Obtain the frequency response of the circuit in Fig. 15.31 using PSpice. Determine the range of frequency by trial and error.

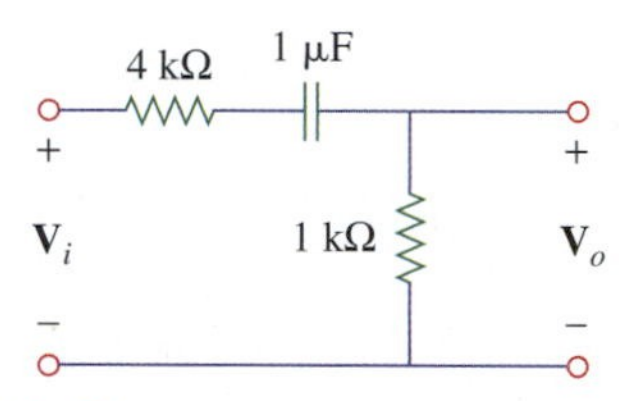

Figure 15.31
For Problems 15.29 and 15.33.

15.30 Use PSpice to provide the frequency response (magnitude and phase of $\mathbf{V}_o$) of the circuit in Fig. 15.32. Use a linear frequency sweep from 1 to 1000 Hz.

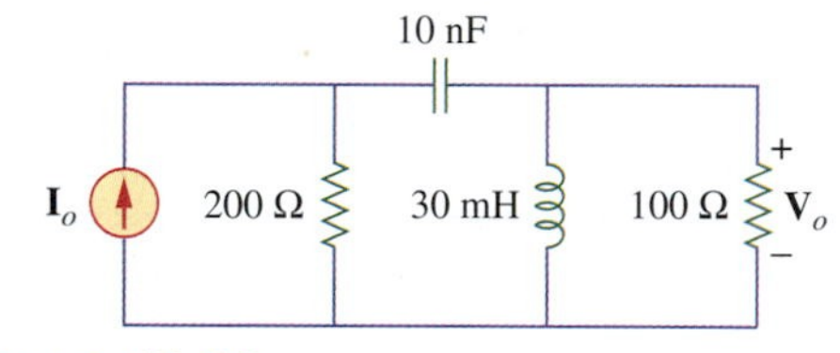

Figure 15.32
For Problem 15.30.

15.31 Solve Problem 15.21 using PSpice.

15.32 Use PSpice to determine the resonant frequency of the circuit in Fig. 15.33.

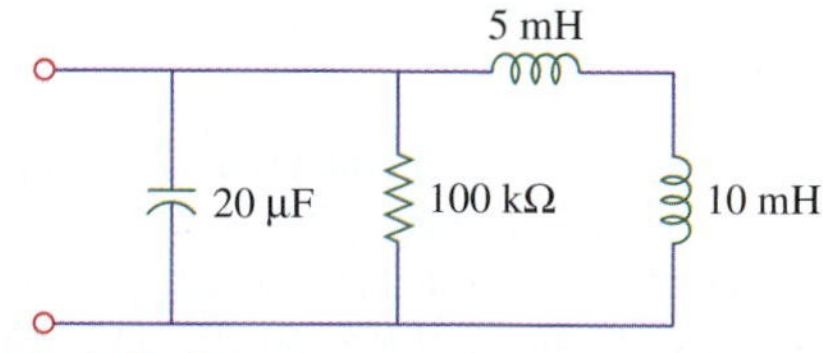

Figure 15.33
For Problem 15.32.

15.33 Repeat Problem 15.29 using Multisim.

15.34 Use Multisim to obtain the frequency response of the circuit shown in Fig. 15.34. Let 10 Hz $< f <$ 10 kHz.

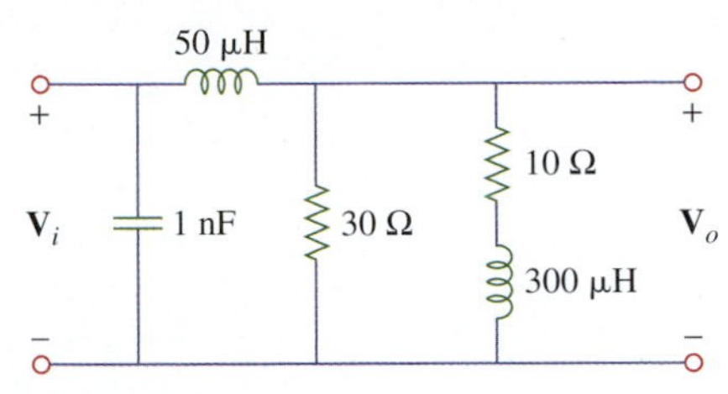

Figure 15.34
For Problem 15.34.

15.35 Determine the frequency response of the circuit in Fig. 15.35 using Multisim.

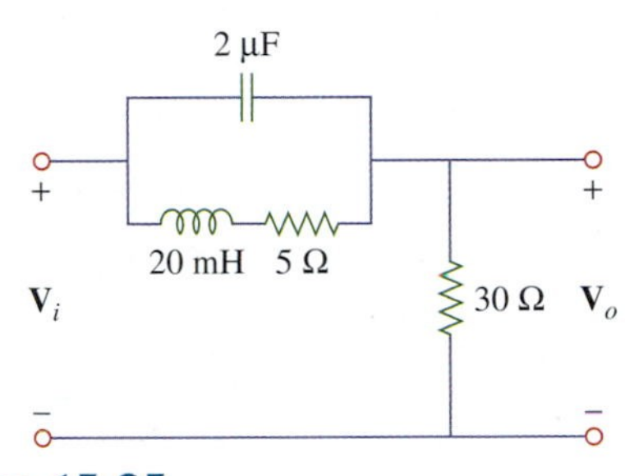

Figure 15.35
For Problem 15.35.

Section 15.6 Applications

15.36 The resonant circuit for a radio broadcast consists of a 120-pF capacitor in parallel with a 240-μH inductor. If the inductor has an internal resistance of 400 Ω, what is the resonant frequency of the circuit? What would be the resonant frequency if the inductor resistance were reduced to 40 Ω?

15.37 A series-tuned antenna circuit consists of a variable capacitor (40 to 360 pF) and a 240-μH antenna coil that has a dc resistance of 12 Ω. (a) Find the frequency range of radio signals to which the radio is tunable. (b) Determine the value of Q at each end of the frequency range.

15.38 A certain electronic test circuit produced a resonant curve with half-power points at 432 and 454 Hz. If $Q = 20$, what is the resonant frequency of the circuit?

15.39 In an electronic device, a series circuit is employed that has a resistance of 100 Ω, a capacitive reactance of 5 kΩ, and an inductive reactance of 300 Ω when used at 2 MHz. Find the resonant frequency and bandwidth of the circuit.

chapter

16

Filters and Bode Plots

Accomplishments have no color.

—Leontyne Price

Historical Profiles

Alexander Graham Bell (1847–1922), inventor of the telephone, was a Scottish American scientist.

Bell was born in Edinburgh, Scotland, the son of Alexander Melville Bell, a well-known speech teacher. Alexander the younger also became a speech teacher after graduating from the University of Edinburgh and the University of London. In 1866, he became interested in transmitting speech electrically. After his older brother died of tuberculosis, his father decided to move to Canada. Alexander was asked to come to Boston to work at the School of the Deaf. There he met Thomas A. Watson, who became his assistant in his electromagnetic transmitter experiment. On March 10, 1876, Alexander sent the famous first telephone message, "Watson, come here, I want you." The bel, the logarithmic unit introduced in this chapter, is named in his honor.

Alexander Graham Bell
© Ingram Publishing

Hendrik W. Bode (1905–1982) was an American engineer who invented Bode plots, discussed in this chapter.

Born in Madison, Wisconsin, he received his B.A. and M.A. degrees from Ohio State University. He worked at Bell Telephone Laboratories, where he began his career with electric filter and equalizer design. While employed at Bell Laboratories, he received his Ph.D. degree from Columbia University. Bode's work in electric filters and equalizers resulted in the publication of his book, *Network Analysis and Feedback Amplifier Design*, which is considered a classic in its field. Bode retired from Bell Telephone Laboratories at the age of 61 and was immediately elected Gordon McKay Professor of Systems Engineering at Harvard University. In his new career, Bode taught and directed graduate student research. Bode was a member or fellow of a number of scientific and engineering societies.

Hendrik W. Bode
AIP Emilio Segre Visual Archives, Physics Today Collection

16.1 Introduction

The sinusoidal steady-state frequency responses of circuits are important in many applications, especially in communications and control systems. One specific application is in electronic filters, which block out or eliminate signals with unwanted frequencies and pass the signals of the desired frequencies. Filters are found in radio, TV, and cellular telephone systems to separate one broadcast frequency from another.

We begin this chapter by learning how to express power gain and voltage gain in decibels. We then consider transfer function and Bode plots, which are the industry-standard way of presenting frequency response. We discuss different kinds of filters—lowpass, highpass, bandpass, and bandstop filters. In the last section, we will consider two practical applications of filters.

16.2 The Decibel Scale

A systematic way of obtaining the frequency response is to use Bode plots. Before we begin to construct Bode plots, we should take care of two important issues: the use of logarithms and decibels in expressing gain. The logarithmic scale is necessary because it makes it convenient to display the wide range of frequencies involved in plotting frequency responses.

Because Bode plots are based on logarithms, it is important that we keep the following properties of logarithms in mind:

1. $\log P_1 P_2 = \log P_1 + \log P_2$ **(16.1)**
2. $\log \dfrac{P_1}{P_2} = \log P_1 - \log P_2$ **(16.2)**
3. $\log P^n = n \log P$ **(16.3)**
4. $\log 1 = 0$ **(16.4)**

Equations (16.1) though (16.4) apply for any logarithm to any base. The ratios expressing voltage or power gain may be either too small or too large to display or discuss easily. For this reason, gain in communications systems is measured in *bels*.[1] Historically, the bel is used in measuring the ratio of two levels of power or power gain G_p; that is,

$$G_p = \text{Number of bels} = \log_{10} \frac{P_2}{P_1} \qquad \textbf{(16.5)}$$

The decibel (dB) is one-tenth of a bel and is therefore given by

$$G_p(\text{dB}) = \text{Number of decibels} = 10 \log_{10} \frac{P_2}{P_1} \qquad \textbf{(16.6)}$$

[1] Historical Note: Named after Alexander Graham Bell, the inventor of the telephone.

When $P_1 = P_2$, there is no change in power and the gain is 0 dB. If $P_2 = 2P_1$, the power gain is

$$G_p(\text{dB}) = 10 \log_{10} 2 = 3.01 \text{ dB} \tag{16.7}$$

and when $P_2 = 0.5P_1$, the gain is

$$G_p(\text{dB}) = 10 \log_{10} 0.5 = -3.01 \text{ dB} \tag{16.8}$$

Equations (16.7) and (16.8) show another reason logarithm is often used; that is, the logarithm of the reciprocal of a quantity is simply the negative of the logarithm of that quantity. A positive dB value indicates gain, while a negative dB value indicates loss or attenuation.

Alternatively, the gain G can be expressed in terms of voltage or current ratio. To do so, consider the network shown in Fig. 16.1. If P_1 is the input power, P_2 is the output (load) power, R_1 is the input resistance, and R_2 is the load resistance, then

$$P_1 = \frac{V_1^2}{R_1} \tag{16.9}$$

and

$$P_2 = \frac{V_2^2}{R_2} \tag{16.10}$$

and Eq. (16.6) becomes

$$\begin{aligned} G_p(dB) &= 10 \log_{10} \frac{P_2}{P_1} = 10 \log_{10} \frac{V_2^2/R_2}{V_1^2/R_1} \\ &= 20 \log_{10} \frac{V_2}{V_1} - 10 \log_{10} \frac{R_2}{R_1} \end{aligned} \tag{16.11}$$

For the case when $R_2 = R_1$ (a condition which is often assumed, though often not true), when comparing voltage levels, we obtain the voltage gain in dB from Eq. (16.11) as

$$G_v(\text{dB}) = 20 \log_{10} \frac{V_2}{V_1} \tag{16.12}$$

Table 16.1 specifies dB for various voltage gain.

Alternatively, if $P_1 = I_1^2 R_1$ and $P_2 = I_2^2 R_2$, and when $R_2 = R_1$, we obtain the current gain in dB as

$$G_I(\text{dB}) = 20 \log_{10} \frac{I_2}{I_1} \tag{16.13}$$

Two things are important to note from Eqs. (16.6), (16.12), and (16.13):

1. 10 log is used for power gain, while 20 log is used for voltage or current gain, because of the square relationship between them ($P = V^2/R = I^2 R$).
2. The dB value is a logarithmic measurement of the *ratio* of one variable to another *of the same type*.

The decibel measure can also be used to indicate absolute levels of power with respect to some reference level. The dBm is the decibel power level with respect to 1 mW; that is,

$$\text{dBm} = 10 \log_{10} \left(\frac{\text{Actual power level in watts}}{1 \text{ mW}} \right) \tag{16.14}$$

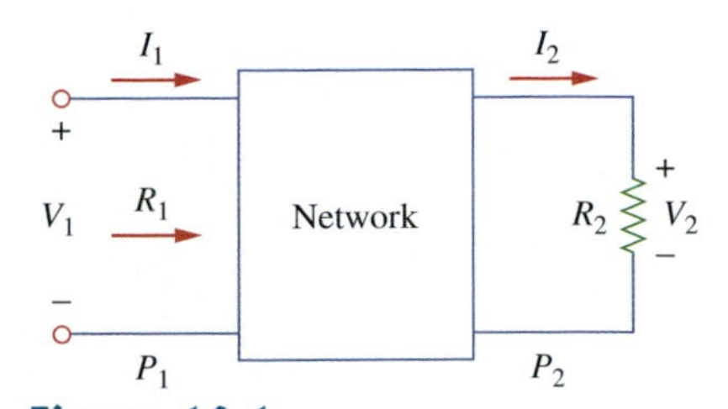

Figure 16.1
Voltage–current relationships for a four-terminal network.

TABLE 16.1
dB for various voltage gain.

Gain (V_o/V_1)	dB = 20 log (V_o/V_1)
1	0
2	6
10	20
20	26
100	40
1,000	60
10,000	80
100,000	100

TABLE 16.2

Some typical useful cases of dBm.

dBm level	Power	Application
80 dBm	100,000 W	Transmission power of FM radio station
60 dBm	1,000 W	RF power inside microwave oven
27 dBm	500 mW	Cellular phone transmission power
26 dBm	400 mW	Maximum output power of an 1,800-MHz mobile phone
20 dBm	100 mW	Bluetooth class 1 radio, 100-m range
−70 dBm	0.0000001 mW	Average typical of wireless signal over a network
−127.5 dBm	0.00000000000018 mW	Typical received signal power from GPS satellite

If the actual power is P, then

$$\text{dBm} = 10 \log_{10}\left(\frac{P}{1\text{ mW}}\right) \quad \textbf{(16.15a)}$$

or

$$P = (1\text{ mW}) \times 10^{\text{dBm}/10} \quad \textbf{(16.15b)}$$

The suffix m after dB means that 1 milliwatt is the reference value for 0 dB; dBm (or dBmW) is used in radio, microwave and fiber-optic networks as a convenient measure of absolute power. Table 16.2 summarizes useful cases of dBm.

Example 16.1

Determine the following logarithmic expressions.

(a) $\log_{10}(0.6)(400)$

(b) $\log_{10}\dfrac{8 \times 10^3}{10^{-2}}$

(c) $\log_{10} 200^3$

Solution:

(a) $\log_{10}(0.6)(400) = \log_{10} 0.6 + \log_{10}(400) = -0.2218 + 2.602 = 2.3802$

Alternatively, using the TI-89 Titanium calculator, we first select the function log. Press [CATALOG] and scroll down until you see log(and press [ENTER]

Type in the number.

```
Log (0.6*400)
```

Then press [♦] [ENTER]

The result is 2.3802

(b) $\log_{10}\dfrac{8 \times 10^3}{10^{-2}} = \log_{10} 8 + \log_{10} 10^3 - \log_{10} 10^{-2}$

$= 0.903 + 3 - (-2) = 5.903$

Or using the TI-89 Titanium calculator,

```
log (8*10^3/10^-2)=5.903
```

(c) $\log_{10} 200^3 = 3 \log_{10} 200 = 3(2.301) = 6.903$
Using the TI-89 Titanium calculator,

```
log(200^3)=6.903
```

Practice Problem 16.1

Evaluate the following logarithmic expressions:
(a) $\log_{10} (0.001)(56)$
(b) $\log_{10} (0.62)^4$
(c) $\log_{10} \dfrac{58{,}000}{48}$

Answer: (a) -1.2518 (b) -0.8304 (c) 3.082

Example 16.2

Find the gain in dB of a system with the following conditions:
(a) $P_{in} = 2$ mW, $P_{out} = 40$ mW
(b) $P_{in} = 6\ \mu$W, $P_{out} = 1\ \mu$W
(c) $V_{in} = 0.4$ mV, $V_{out} = 2.1$ mV
(d) $V_{in} = 1$ V, $V_{out} = 0.8$ V

Solution:

(a) $G_p(\text{dB}) = 10 \log_{10}\dfrac{P_{out}}{P_{in}} = 10 \log_{10}\dfrac{40 \text{ mW}}{2 \text{ mW}} = 13.01 \text{ dB}$

(b) $G_p(\text{dB}) = 10 \log_{10}\dfrac{P_{out}}{P_{in}} = 10 \log_{10}\dfrac{1\ \mu\text{W}}{6\ \mu\text{W}} = -7.782 \text{ dB}$

(c) $G_v(\text{dB}) = 20 \log_{10}\dfrac{V_{out}}{V_{in}} = 20 \log_{10}\dfrac{2.1 \text{ mV}}{0.4 \text{ mV}} = 14.403 \text{ dB}$

(d) $G_v(\text{dB}) = 20 \log_{10}\dfrac{V_{out}}{V_{in}} = 20 \log_{10}\dfrac{0.8 \text{ V}}{1 \text{ V}} = -1.9382 \text{ dB}$

Practice Problem 16.2

(a) Determine the power gain in dB of a system where the input power is 1 mW and the output power is 60 mW.
(b) Find the voltage gain in dB if the applied voltage is 10 mV while the output voltage is 2.4 V.

Answer: (a) 17.781 dB; (b) 47.604 dB

Example 16.3

Convert the following powers in dB into ratio.
(a) 42.3 dB
(b) -26.5 dB

Solution:
(a) If

$$42.3 \text{ dB} = 10 \log_{10}\frac{P_2}{P_1},$$

then

$$42.3/10 = 4.23 = \log_{10}\frac{P_2}{P_1} \quad \Rightarrow \quad \frac{P_2}{P_1} = 10^{4.23} = 16{,}982.44$$

(c) Similarly,

$$-26.5/10 = -2.65 = \log_{10}\frac{P_2}{P_1} \quad \Rightarrow \quad \frac{P_2}{P_1} = 10^{-2.65} = 0.00224$$

Practice Problem 16.3

Determine the ratios corresponding to the following powers in dB:
(a) 16.5 dB (b) −47.6 dB

Answer: (a) 44.67; (b) 1.737×10^{-5}

Example 16.4

A network has a voltage gain of 25 dB. Find the output voltage when an input of 2 mV is applied.

Solution:
We first convert the gain in dB to a ratio of two voltages.

$$25 \text{ dB} = 20 \log_{10}\frac{V_{out}}{V_{in}} \quad \Rightarrow \quad \log 10 \frac{V_{out}}{V_{in}} = 25/20 = 1.25$$

Hence,

$$\frac{V_{out}}{V_{in}} = 10^{1.25} = 17.78$$

$$V_{out} = 17.78\, V_{in} = 17.78\,(2 \text{ mV}) = 35.56 \text{ mV}$$

Practice Problem 16.4

If a system has a voltage gain of 32 dB, determine the applied input voltage when the output voltage is 4.6 V.

Answer: 0.1155 V

Example 16.5

(a) Convert the following power levels to decibel levels referenced to 1 mW: 1 nW, 0.5 W, and 12 W.
(b) Find the power levels in units of mW of the following dBm values: −13, 5, and 30 dBm.

Solution:
(a) This can be solved in two ways:

METHOD 1 Using Eq. (16.15a)

For 1 nW,

$$1 \text{ nW} = 10^{-9} \text{ W} = 10^{-6} \text{ mW}$$

$$\text{dBm} = 10 \log_{10}\left(\frac{10^{-6}\text{mW}}{1 \text{ mW}}\right) = 10(-6) = -60 \text{ dBm}$$

For 0.5 W,

$$0.5\text{ W} = 500\text{ mW}$$

$$\text{dBm} = 10\log_{10}\left(\frac{500\text{ mW}}{1\text{ mW}}\right) = 10(2.7) = 27\text{ dBm}$$

For 12 W,

$$12\text{ W} = 12{,}000\text{ mW}$$

$$\text{dBm} = 10\log_{10}\left(\frac{12000\text{ mW}}{1\text{ mW}}\right) = 10(4.079) = 40.79\text{ dBm}$$

METHOD 2 Equation (16.15a) can be written as

$$\text{dBm} = 10\log_{10}\left(\frac{P(\text{in W})}{10^{-3}}\right) = 30 + 10\log_{10}(P(\text{in W}))$$

For 1 nW,

$$\text{dBm} = 30 + 10\log_{10}(10^{-9}) = 30 - 90 = -60$$

For 0.5 W,

$$\text{dBm} = 30 + 10\log_{10}(0.5) = 30 - 3 = 27$$

For 12 W,

$$\text{dBm} = 30 + 10\log_{10}(12) = 30 + 10.79 = 40.79$$

(b) Using Eq. (16.15b)

$$P = (1\text{ mW}) \times 10^{\text{dBm}/10}$$

For -13 dBm,

$$P = 10^{-13/10} = 10^{-1.3} = 50.12 \times 10^{-3}\text{ mW}$$

For 5 dBm,

$$P = 10^{5/10} = 10^{0.5} = 3.16\text{ mW}$$

For 30 dBm,

$$P = 10^{30/10} = 10^{3} = 1000\text{ mW}$$

Practice Problem 16.5

(a) Express the optical output power 300 μW in dBm units.
(b) What is the power level in mW of 25 dBm?

Answer: (a) -5.23 dBm; (b) 316.23 mW

16.3 Transfer Function

Having a good grasp of frequency response is important in many areas. The transfer function $\mathbf{H}(\omega)$ (also called the *network function*) is a useful analytical tool for finding the frequency response of a circuit. In fact, the frequency response of a circuit is the plot of the circuit's transfer function $\mathbf{H}(\omega)$ versus ω with ω varying from $\omega = 0$ to $\omega = \infty$.

A transfer function is the frequency-dependent ratio of the output to the input. The idea of a transfer function was implicit when we used

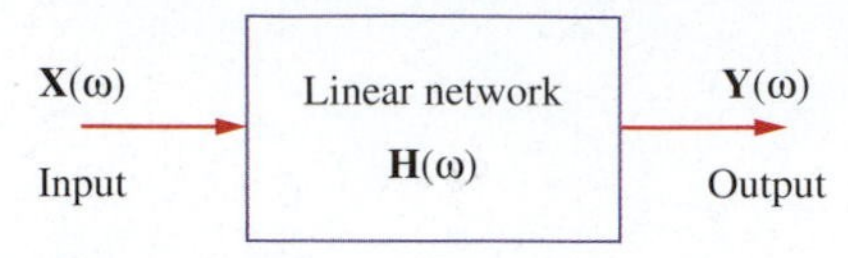

Figure 16.2
A block diagram representation of a linear network.

The **transfer function** $\mathbf{H}(\omega)$ of a circuit is the frequency-dependent ratio of a phasor output $\mathbf{Y}(\omega)$ (an element voltage or current) to a phasor input $\mathbf{X}(\omega)$ (source voltage or current).

the concepts of impedance and admittance to relate voltage and current. In general, a linear network can be represented by the block diagram shown in Fig. 16.2.

Thus,[2]

$$\mathbf{H}(\omega) = \mathbf{Y}(\omega)/\mathbf{X}(\omega) \tag{16.16}$$

In the general definition of a transfer function, zero initial conditions is assumed. But in ac circuit analysis, we are dealing with steady state and initial conditions are irrelevant.

Because the input and output can be either voltage or current at any place in the circuit, there are four possible transfer functions:

$$\mathbf{H}(\omega) = \text{Voltage gain} = \frac{\mathbf{V}_o(\omega)}{\mathbf{V}_i(\omega)} \tag{16.17a}$$

$$\mathbf{H}(\omega) = \text{Current gain} = \frac{\mathbf{I}_o(\omega)}{\mathbf{I}_i(\omega)} \tag{16.17b}$$

$$\mathbf{H}(\omega) = \text{Transfer impedance} = \frac{\mathbf{V}_o(\omega)}{\mathbf{I}_i(\omega)} \tag{16.17c}$$

$$\mathbf{H}(\omega) = \text{Transfer admittance} = \frac{\mathbf{I}_o(\omega)}{\mathbf{V}_i(\omega)} \tag{16.17d}$$

where subscripts i and o denote input and output values, respectively. Being a complex quantity, $\mathbf{H}(\omega)$ has a magnitude $H(\omega)$ and a phase ϕ; that is, $\mathbf{H}(\omega) = H(\omega)\underline{/\phi}$. Note that the transfer functions in Eqs. (16.17c) and (16.17d) cannot be expressed in dB because they are ratios of different parameters.

To obtain the transfer function using Eq. (16.17), we first obtain the frequency-domain equivalent of the circuit by replacing resistors, inductors, and capacitors with their impedances R, $j\omega L$, and $1/j\omega C$. We then use any circuit technique(s) to obtain the appropriate quantity in Eq. (16.17). We can obtain the frequency response of the circuit by plotting the magnitude and phase of the transfer function as the frequency varies. A computer is a real time-saver for plotting the transfer function as we will see in Section 16.6, where PSpice and Multisim will be used.

To avoid complex algebra, it is expedient to replace $j\omega$ temporarily with s when working with $\mathbf{H}(\omega)$ and replace s with $j\omega$ at the end.

Example 16.6

For the RC circuit shown in Fig. 16.3(a), obtain the transfer function $\mathbf{V}_o(\omega)/\mathbf{V}_s(\omega)$ and its frequency response. Let $v_s(t) = V_m \cos(\omega t)$.

Solution:

The frequency-domain equivalent of the circuit is shown in Fig. 16.3(b). By voltage division, the transfer function is given by

$$\mathbf{H}(\omega) = \mathbf{V}_o(\omega)/\mathbf{V}_s(\omega) = \frac{1/j\omega C}{R + 1/j\omega C} = \frac{1}{1 + j\omega RC}$$

[2]Note: In this context, $\mathbf{X}(\omega)$ and $\mathbf{Y}(\omega)$ denote the input and output phasors of a network, respectively; they should not be confused with the same symbolism used for reactance and admittance. The multiple usage of symbols is conventionally permissible due to lack of enough letters in the English language to express all circuit variables distinctly.

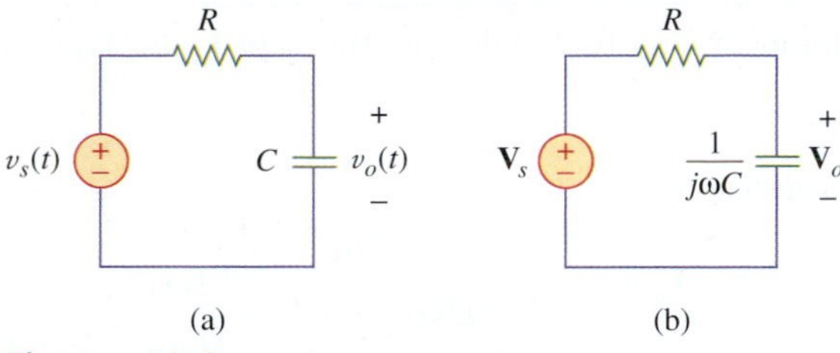

Figure 16.3
For Example 16.6: (a) time-domain *RC* circuit; (b) frequency-domain *RC* circuit.

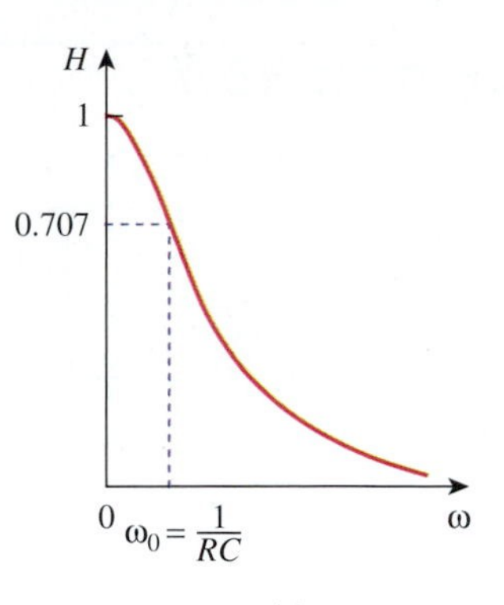

(a)

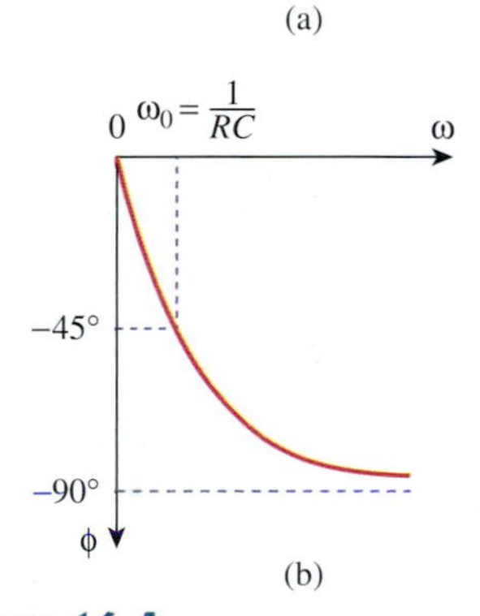

(b)

Figure 16.4
Frequency response of the *RC* circuit of Fig. 16.3: (a) amplitude response; (b) phase response.

The magnitude and phase of $\mathbf{H}(\omega)$ are

$$H = \frac{1}{\sqrt{1 + (\omega/\omega_0)^2}}, \qquad \phi = -\tan^{-1}\frac{\omega}{\omega_0}$$

where $\omega_0 = 1/RC$. To plot H and ϕ for $0 < \omega < \infty$, we obtain their values at some critical points and then sketch.

At $\omega = 0$, $H = 1$ and $\phi = 0°$. At $\omega = \infty$, $H = 0$ and $\phi = -90°$. Also, at $\omega = \omega_0$, $H = 1/\sqrt{2}$ and $\phi = -45°$. With these and a few more points as shown in Table 16.3, the frequency response is as shown in Fig. 16.4. Additional features of the frequency response in Fig. 16.4 will be explained in Section 16.5.1 on lowpass filters.

TABLE 16.3

For Example 16.6.

ω/ω_0	H	ϕ	ω/ω_0	H	ϕ
0	1	0	10	0.1	−84°
1	0.71	−45°	20	0.05	−87°
2	0.45	−63°	100	0.01	−89°
3	0.32	−72°	∞	0	−90°

Practice Problem 16.6

Obtain the transfer function $\mathbf{V}_o(\omega)/\mathbf{V}_s(\omega)$ of the *RL* circuit in Fig. 16.5, assuming $v_s = V_m \cos(\omega t)$. Sketch its frequency response.

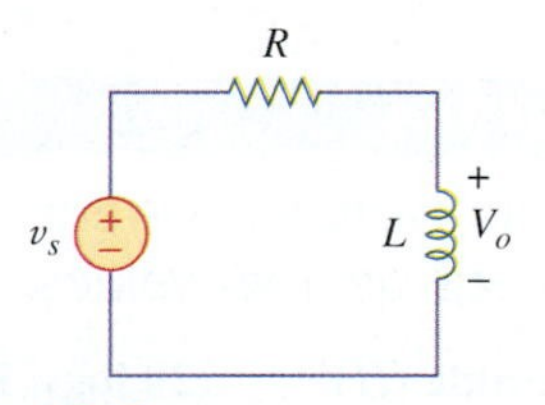

Figure 16.5
RL circuit for Practice Problem 16.6.

Answer: $j\omega L/(R + j\omega L)$; see Fig. 16.6 for the response.

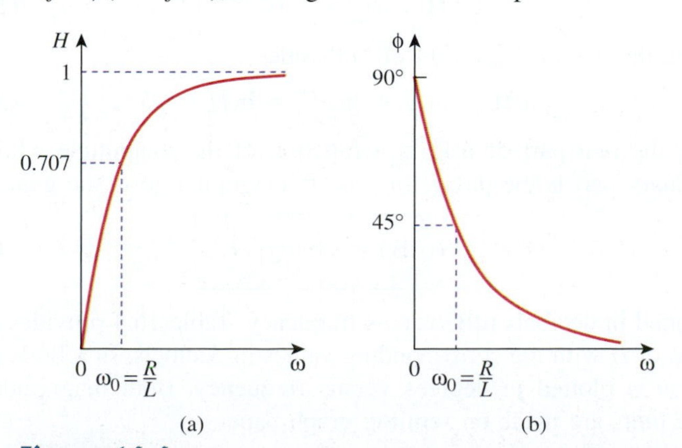

Figure 16.6
Frequency response of the *RL* circuit in Fig. 16.5.

Example 16.7

For the circuit in Fig. 16.7, calculate the gain $\mathbf{I}_o(\omega)/\mathbf{I}_i(\omega)$.

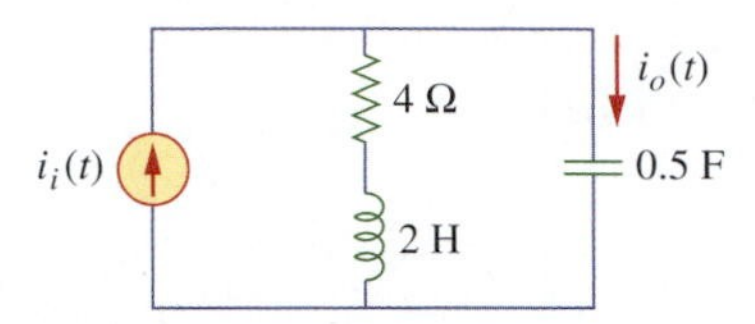

Figure 16.7
For Example 16.7.

Solution:
By current division,

$$\mathbf{I}_o(\omega) = \frac{4 + j2\omega}{4 + j2\omega + 1/j0.5\omega}\mathbf{I}_i(\omega)$$

or

$$\mathbf{I}_o(\omega)/\mathbf{I}_i(\omega) = \frac{j0.5\omega(4 + j2\omega)}{1 + j2\omega + (j\omega)^2} = \frac{s(s + 2)}{s^2 + 2s + 1}, \qquad s = j\omega$$

Practice Problem 16.7

Find the transfer function $\mathbf{V}_o(\omega)/\mathbf{I}_i(\omega)$ for the circuit of Fig. 16.8.

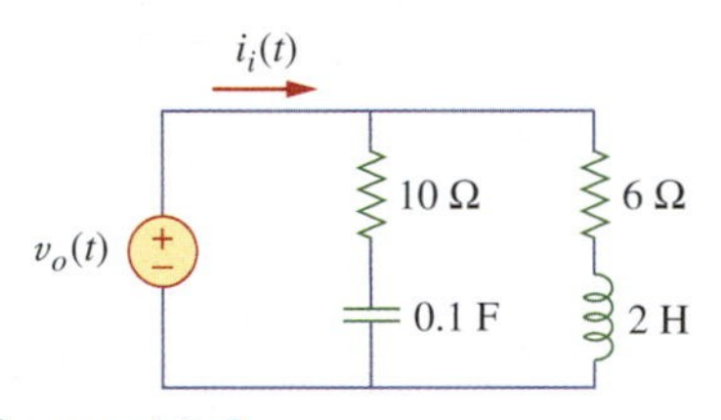

Figure 16.8
For Practice Problem 16.7.

Answer: $\dfrac{10(s + 1)(s + 3)}{s^2 + 8s + 5}, \qquad s = j\omega$

16.4 Bode Plots

Obtaining the frequency response from the transfer function as we did in the previous section is an uphill task. The frequency range required in the frequency response is often so wide that it is inconvenient to use a linear scale for the frequency axis. Also, there is a more systematic way of locating the important features of the magnitude and phase plots of the transfer function. For this reason, it has become standard practice to use a logarithmic scale for the frequency axis and a linear scale in each of the separate plots of magnitude (in dB) and phase. Such semilogarithmic plots of the transfer function are known as *Bode plots,*[3] which have become the industry standard. Bode plots contain the same information as the nonlogarithmic plots discussed in the previous section, but they are much easier to construct, as we shall see shortly.

> **Bode plots** are semilog plots of the magnitude (in decibels) and phase (in degrees) of a transfer function versus frequency.

The transfer function can be written as

$$\mathbf{H} = H\underline{/\phi} = He^{j\phi} \tag{16.18}$$

Taking the natural logarithm of both sides

$$\ln \mathbf{H} = \ln H + \ln e^{j\phi} = \ln H + j\phi \tag{16.19}$$

Thus, the real part of $\ln \mathbf{H}$ is a function of the magnitude while the imaginary part is the phase. In a Bode magnitude plot, the gain

$$H(\text{dB}) = 20 \log_{10} H \tag{16.20}$$

is plotted in decibels (dB) versus frequency. Table 16.4 provides a few values of H with the corresponding values in decibels. In a Bode phase plot, ϕ is plotted in degrees versus frequency. Both magnitude and phase plots are made on semilog graph paper.

TABLE 16.4

Specific (voltage/current) gains and their decibel values.

Magnitude (H)	$20 \log_{10} H(DB)$
0.001	−60
0.01	−40
0.1	−20
0.5	−6
$1/\sqrt{2}$	−3
1	0
$\sqrt{2}$	3
2	6
10	20
20	26
100	40
1,000	60

[3] Historical Note: Named after Hendrik W. Bode (1905–1982), an engineer with the Bell Telephone Laboratories, for his pioneering work in the 1930s and 1940s.

A transfer function in the form of Eq. (16.16) may be written in terms of factors that have real and imaginary parts. One such representation might be

$$\mathbf{H}(\omega) = \frac{K(j\omega)^{\pm 1}(1 + j\omega/z_1)[1 + j2\zeta_1\omega/\omega_k + (j\omega/\omega_k)^2]}{(1 + j\omega/p_1)[1 + j2\zeta_2\omega/\omega_n + (j\omega/\omega_n)^2]} \quad \textbf{(16.21)}$$

The representation of $\mathbf{H}(\omega)$ as in Eq. (16.21) is called the *standard form*. Equation (16.21) may look complex right now, but when we break it up into seven individual factors, it is less intimidating. In this particular case, $\mathbf{H}(\omega)$ has seven different factors that can appear in various combinations in a transfer function. These are:

1. A gain K
2. A factor $(j\omega)^{-1}$ or $(j\omega)$ at the origin[4]
3. A linear factor $1/(1 + j\omega/p_1)$ or $(1 + j\omega_0/z_1)$ on the frequency axis
4. A quadratic factor $1/[1 + j2\,\zeta_2\omega/\omega_n + (j\omega/\omega_n)^2]$ or $[1 + j2\,\zeta_1\omega/\omega_k + (j\omega/\omega_k)^2]$

In constructing a Bode plot, we plot each factor separately and then combine them graphically. The factors can be considered one at a time and then combined algebraically because of the logarithms involved. It is the mathematical convenience of the logarithm that makes the Bode plots a powerful engineering tool.

We will now make straight-line plots of the seven factors. We shall find that these straight-line plots, known as Bode plots, approximate the actual plots to a surprising degree of accuracy.

1. **Constant term:** For the gain K, the magnitude is $20 \log_{10} K$ and the phase is 0°; both are constant with frequency. Thus, the magnitude and phase plots of the gain are shown in Fig. 16.9. If K is negative, the magnitude remains $20 \log_{10}|K|$ but the phase is $\pm 180°$.

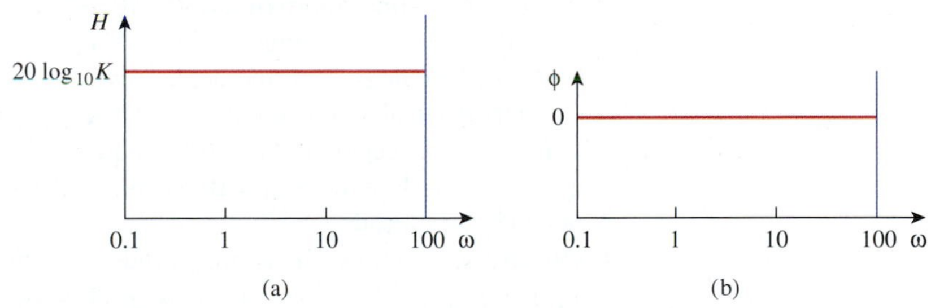

Figure 16.9
Bode plots for gain K: (a) magnitude plot; (b) phase plot.

2. **Factors that pass through the origin:** For the factor $(j\omega)$ at the origin, the magnitude is $20 \log_{10} \omega$ and the phase is 90°. These are plotted in Fig. 16.10, where we notice that the slope of the magnitude plot is 20 dB/decade[5] while the phase is constant with frequency.

   The Bode plots for the factor $(j\omega)^{-1}$ are similar except that the slope of the magnitude plot is -20 dB/decade while the phase is $-90°$. In general, for $(j\omega)^N$, where N is an integer, the magnitude plot will have a slope of $20N$ dB/decade, while the phase is $90N°$, where N can be positive or negative.

[4] Note: The origin is where $\omega = 1$ and the log of the gain is zero.

[5] Note: A decade is an interval between two frequencies with a ratio of 10; e.g., between ω_0 and $10\omega_0$ or between 10 and 100 Hz.

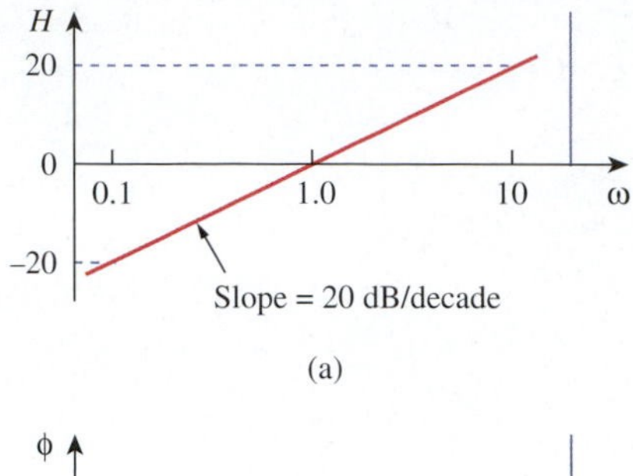

(a)

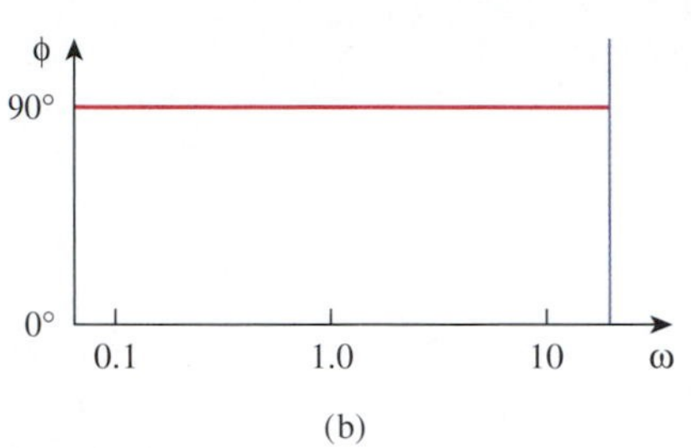

(b)

Figure 16.10
Bode plot for $(j\omega)$: (a) magnitude plot; (b) phase plot.

3. **Linear factors on frequency axis:** For the factor $(1 + j\omega/z_1)$, the magnitude is $20 \log_{10} |1 + j\omega/z_1|$ and the phase is $\tan^{-1} \omega/z_1$. We notice that

$$H(\text{dB}) = 20 \log_{10} \left|1 + \frac{j\omega}{z_1}\right| \quad \Rightarrow \quad 20 \log_{10} 1 = 0 \quad \text{as} \quad \omega \to 0 \tag{16.22}$$

$$H(\text{dB}) = 20 \log_{10} \left|1 + \frac{j\omega}{z_1}\right| \quad \Rightarrow \quad 20 \log_{10} \frac{\omega}{z_1} \quad \text{as} \quad \omega \to \infty \tag{16.23}$$

showing that we can approximate the magnitude as zero (a straight line with zero slope) for small values of ω and by a straight line with slope 20 dB/decade for large values of ω. The frequency $\omega = z_1$ where the two asymptotic lines meet is called the *corner frequency* or *break frequency*. Thus, the approximate magnitude plot is shown in Fig. 16.11(a), where the actual plot is also shown. Notice that the approximate plot is close to the actual plot except near the break frequency, where $\omega = z_1$ and the deviation is

$$20 \log_{10} |(1 + j_1)| = 20 \log_{10} \sqrt{2} = 3 \text{ dB}$$

The phase $\tan^{-1} (\omega/z_1)$ can be expressed as

$$\phi = \tan^{-1}\left(\frac{\omega}{z_1}\right) = \begin{cases} 0, & \omega = 0 \\ 45°, & \omega = z_1 \\ 90°, & \omega \to \infty \end{cases} \tag{16.24}$$

As a straight-line approximation, we let $\phi \approx 0$ for $\omega \le z_1/10$; $\phi \cong 45°$ for $\omega = z_1$ and $\phi \cong 90°$ for $\omega \ge 10z_1$ as shown in Fig. 16.11(b) along with the actual plot.

The Bode plots for the factor $1/(1 + j\omega p_1)$ are similar to those in Fig. 16.11 except that the corner frequency is at $\omega = p_1$, the magnitude has a slope of -20 dB/decade, and the phase has a slope of $-45°$ per decade.

4. **Quadratic factors:** The magnitude of the quadratic factor $1/[1 + j2\zeta_2\omega/\omega_n + (j\omega/\omega_n)^2]$ is $-20 \log_{10}|[1 + j2\zeta_2\omega/\omega_n + (j\omega/\omega_n)^2|$ and the phase is $-\tan^{-1}(2\zeta_2\omega/\omega_n)/(1 - \omega^2/\omega_n^2)$.

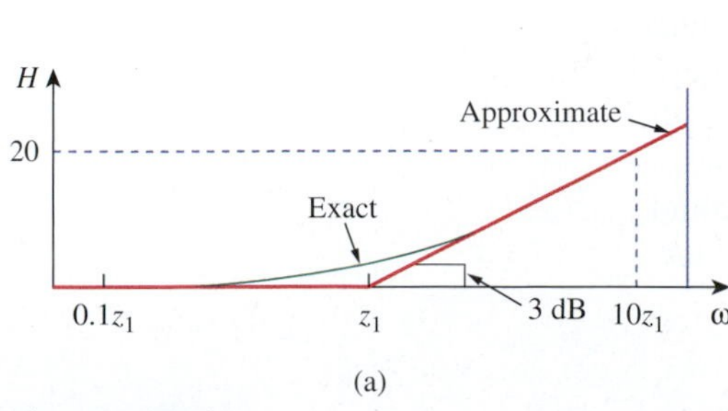

(a)

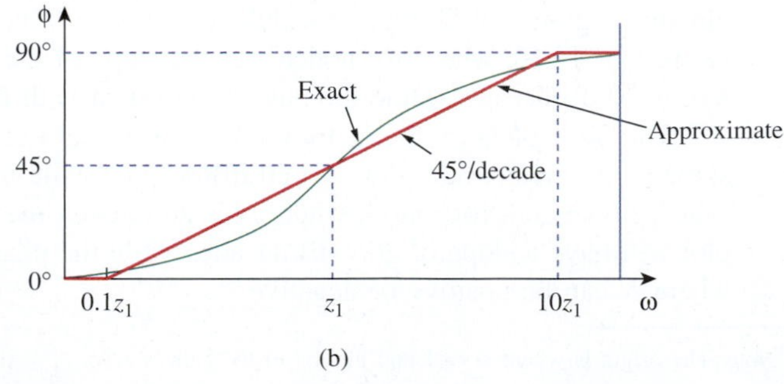

(b)

Figure 16.11
Bode plots of factor $(1 + j\omega/z_1)$: (a) magnitude plot; (b) phase plot.

But

$$H_{\text{dB}} = -20 \log_{10} \left| 1 + \frac{j2\zeta_2\omega}{\omega_n} + \left(\frac{j\omega}{\omega_n}\right)^2 \right| \quad \Rightarrow \quad 0 \quad \text{as} \quad \omega \to 0 \tag{16.25}$$

and

$$H_{\text{dB}} = -20 \log_{10} \left| 1 + \frac{j2\zeta_2\omega}{\omega_n} + \left(\frac{j\omega}{\omega_n}\right)^2 \right| \quad \Rightarrow \quad -40 \log_{10} \frac{\omega}{\omega_n} \quad \text{as} \quad \omega \to \infty \tag{16.26}$$

Thus, the amplitude plot consists of two straight asymptotic lines: one with zero slope for $\omega < \omega_n$, and the other with slope -40 dB/decade for $\omega > \omega_n$, with ω_n as the corner frequency. The approximate and actual amplitude plots are shown in Fig. 16.12(a). Note that the actual plot depends on the damping factor ζ_2 as well as the corner frequency ω_n. The significant peaking in the neighborhood of the corner frequency should be added to the straight-line approximation if a high level of accuracy is desired. However, we will use the straight-line approximation for the sake of simplicity. The phase can be expressed as

$$\phi = -\tan^{-1} \frac{2\zeta_2\omega/\omega_n}{1 - \omega^2/\omega_n^2} = \begin{cases} 0, & \omega = 0 \\ -90^\circ, & \omega = \omega_n \\ -180^\circ, & \omega \to \infty \end{cases} \tag{16.27}$$

The phase plot is a straight line with a slope of $-90°$ per decade starting at $\omega_n/10$ and ending at $10\omega_n$, as shown in Fig. 16.12(b). We again notice that the difference between the actual plot and the straight-line plot is due to the damping factor. The straight-line approximation is reasonable in the neighborhood of the corner frequency. Notice that the straight-line approximations for both magnitude and phase plots for the quadratic factor are the same as those for a linear factor; that is, we should expect this because the double factor $(1 + j\omega/\omega_n)^2$ equals the quadratic factor $1/[1 + j2\zeta_2\omega/\omega_n + (j\omega/\omega_n)^2]$ when $\zeta_2 = 1$. Thus, the quadratic factor can be treated as the square of a linear factor as far as straight-line approximation is concerned.

For the quadratic factor $[1 + j2\zeta_1\omega/\omega_k + (j\omega/\omega_k)^2]$, the plots in Fig. 16.12 are inverted because the magnitude plot has a slope of 40 dB/decade while the phase plot has a slope of $+90°$ per decade.

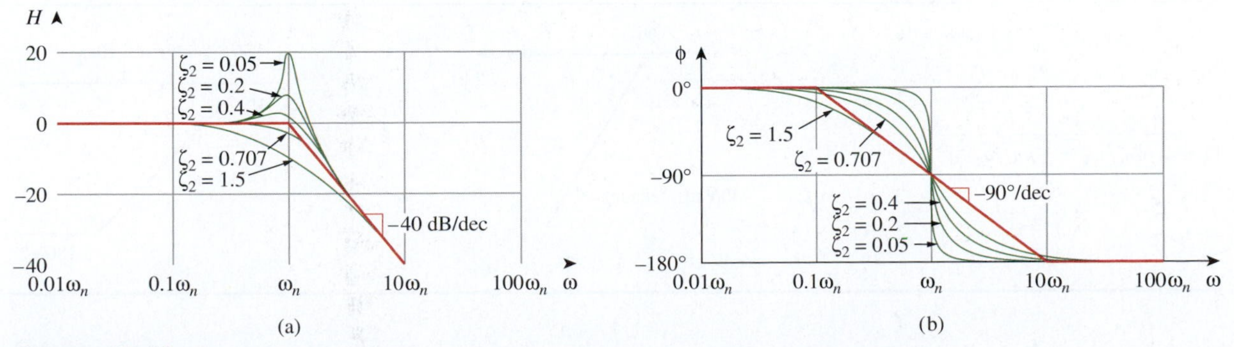

Figure 16.12
Bode plots of quadratic factor $[1 + j2\zeta\omega/\omega_n - (\omega/\omega_n)^2]^{-1}$: (a) magnitude plot; (b) phase plot.

TABLE 16.5

Summary of Bode straight-line magnitude and phase plots.

Factor	Magnitude	Phase
K	$20\log_{10}K$; ω	0°; ω
$(j\omega)^N$	20N dB/decade; 1; ω	90N°; ω
$\dfrac{1}{(j\omega)^N}$	1; ω; −20N dB/decade	ω; −90N°
$\left(1+\dfrac{j\omega}{z}\right)^N$	20N dB/decade; z; ω	90N°; 0°; $\frac{z}{10}$; z; 10z; ω
$\dfrac{1}{(1+j\omega/p)^N}$	p; ω; −20N dB/decade	$\frac{p}{10}$; p; 10p; 0°; ω; −90N°
$\left[1+\dfrac{2j\omega\zeta}{\omega_n}+\left(\dfrac{j\omega}{\omega_n}\right)^2\right]^N$	40N dB/decade; ω_n; ω	180N°; 0°; $\frac{\omega_n}{10}$; ω_n; 10ω_n; ω
$\dfrac{1}{[1+2j\omega\zeta/\omega_k+(j\omega/\omega_k)^2]^N}$	ω_k; ω; −40N dB/decade	$\frac{\omega_k}{10}$; ω_k; 10ω_k; 0°; ω; −180N°

A summary of Bode plots for the seven factors is presented in Table 16.5. To sketch the Bode plots for a function $\mathbf{H}(\omega)$ in the form of Eq. (16.21), for example, we first record the corner frequencies on the semilog graph paper, sketch the factors one at a time as discussed earlier, and then add the graphs of the factors. The combined graph is often drawn from left to right, changing slopes appropriately each time a corner frequency is encountered. This procedure will be illustrated with the following examples.

Example 16.8

Construct the Bode plots for the transfer function

$$\mathbf{H}(\omega) = \frac{200\,j\omega}{(j\omega + 2)(j\omega + 10)}$$

Solution:

We first put $\mathbf{H}(\omega)$ in the standard form as follows.

$$\mathbf{H}(\omega) = \frac{10\,j\omega}{(1 + j\omega/2)(1 + j\omega/10)}$$

$$= \frac{10|j\omega|}{|1 + j\omega/2||1 + j\omega/10|} \angle(90^\circ - \tan^{-1}\omega/2 - \tan^{-1}\omega/10)$$

Hence, the magnitude and phase are

$$H_{\text{dB}} = 20\log_{10}10 + 20\log_{10}|j\omega| - 20\log_{10}\left|1 + j\frac{\omega}{2}\right| - 20\log_{10}\left|1 + j\frac{\omega}{10}\right|$$

$$\phi = 90^\circ - \tan^{-1}\frac{\omega}{2} - \tan^{-1}\frac{\omega}{10}$$

We notice that there are two corner frequencies at $\omega = 2$, 10 rad/s. For both the magnitude and phase plots, we sketch each term as shown by the dotted lines in Fig. 16.13. Starting from the y-axis, we add them up graphically to obtain the overall plots shown by the solid curves.

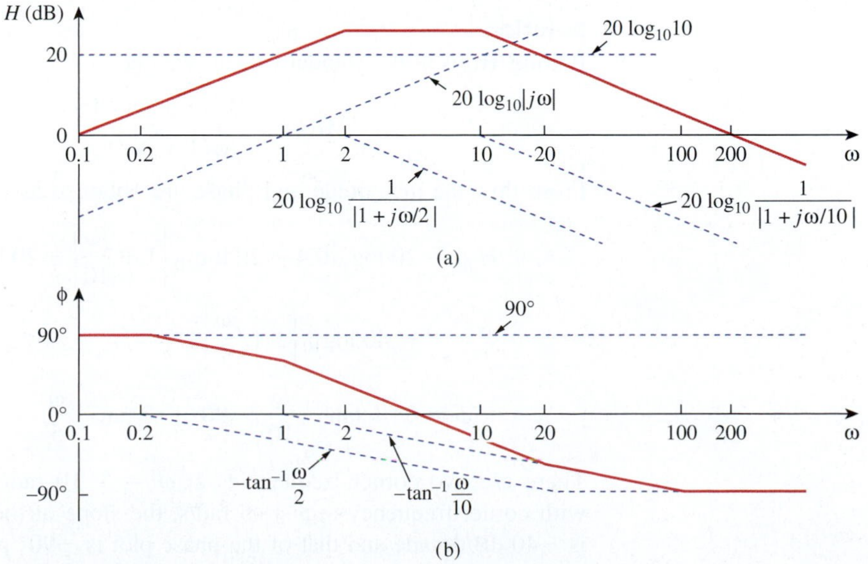

Figure 16.13
For Example 16.8: (a) magnitude plot; (b) phase plot.

Practice Problem 16.8

Draw the Bode plots for the transfer function

$$\mathbf{H}(\omega) = \frac{5(j\omega + 2)}{j\omega(j\omega + 10)}$$

Answer: See Fig. 16.14.

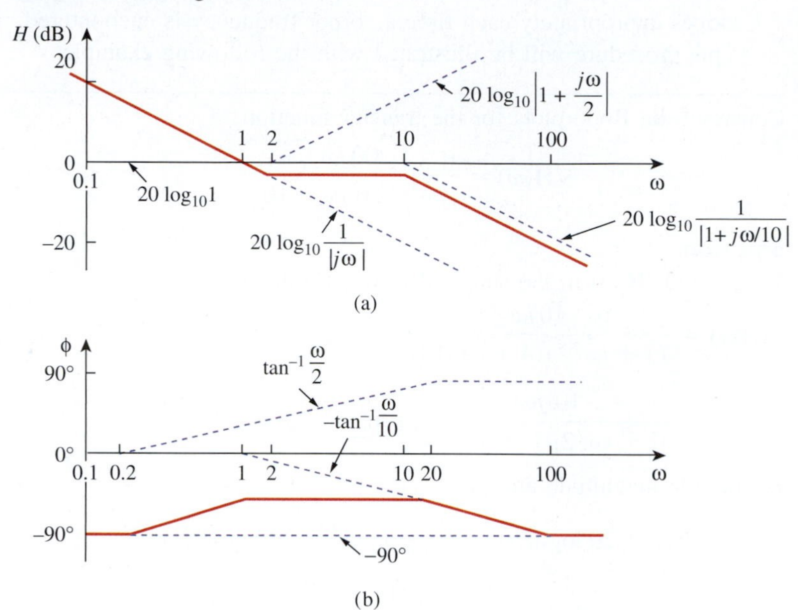

Figure 16.14
For Practice Problem 16.8: (a) magnitude plot; (b) phase plot.

Example 16.9

Obtain the Bode plots for

$$\mathbf{H}(\omega) = \frac{j\omega + 10}{j\omega(j\omega + 5)^2}$$

Solution:
Putting $\mathbf{H}(\omega)$ in the standard form, we get

$$\mathbf{H}(\omega) = \frac{0.4(1 + j\omega/10)}{j\omega(1 + j\omega/5)^2}$$

From this, the magnitude and phase are obtained as

$$H_{\text{dB}} = 20 \log_{10} 0.4 + 20 \log_{10} \left|1 + \frac{j\omega}{10}\right| - 20 \log_{10} |j\omega| - 40 \log_{10} \left|1 + \frac{j\omega}{5}\right|$$

$$\phi = 0° + \tan^{-1} \frac{\omega}{10} - 90° - 2 \tan^{-1} \frac{\omega}{5}$$

There are two corner frequencies at $\omega = 5, 10$ rad/s. For the factor with corner frequency at $\omega = 5$ rad/s, the slope of the magnitude plot is -40 dB/decade and that of the phase plot is $-90°$ per decade due to the power of 2. The magnitude and the phase plots for the individual terms (in dotted lines) and the entire $\mathbf{H}(\omega)$ (in solid lines) are in Fig. 16.15.

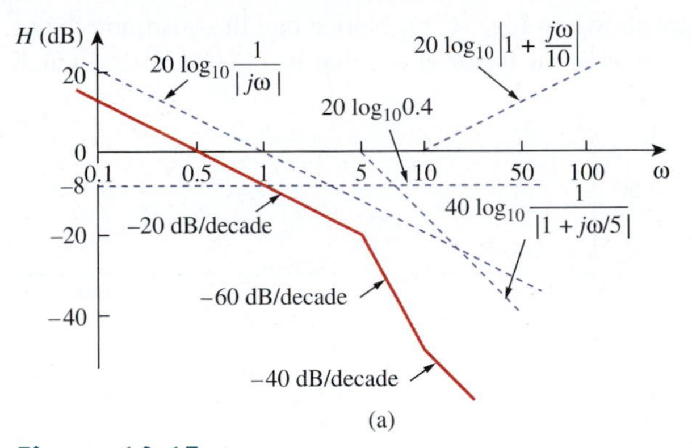

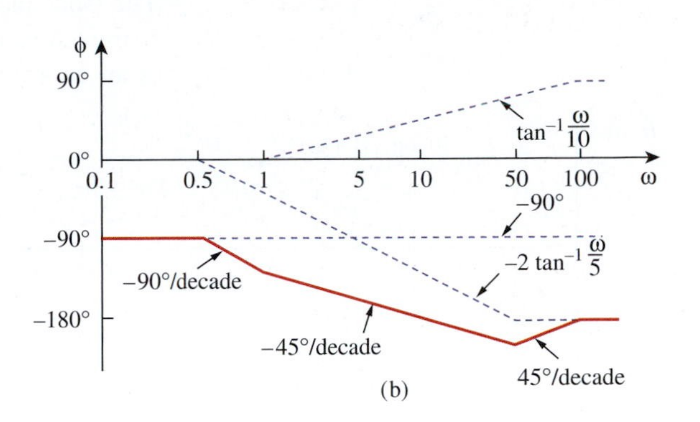

Figure 16.15
Bode plots for Example 16.9: (a) magnitude plot; (b) phase plot.

Practice Problem 16.9

Sketch the Bode plots for

$$\mathbf{H}(\omega) = \frac{50\,j\omega}{(j\omega + 4)(j\omega + 10)^2}$$

Answer: See Fig. 16.16.

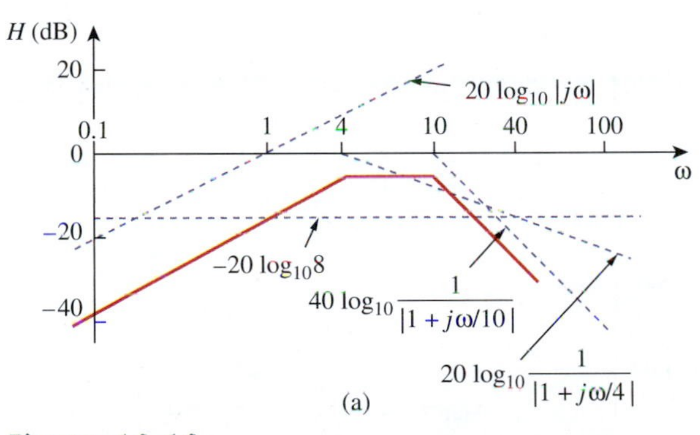

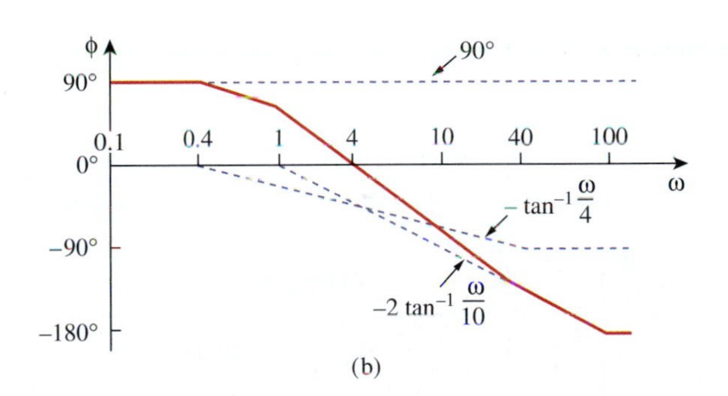

Figure 16.16
For Practice Problem 16.9: (a) magnitude plot; (b) phase plot.

Example 16.10

Draw the Bode plots for

$$\mathbf{H}(s) = \frac{s + 1}{s^2 + 60s + 100}, \qquad s = j\omega$$

Solution:
We express $\mathbf{H}(s)$ as

$$\mathbf{H}(\omega) = \frac{1/100(1 + j\omega)}{1 + j\omega 6/10 + (j\omega/10)^2}$$

For the quadratic factor, $\omega_n = 10$ rad/s, which serves as the corner frequency. The magnitude and phase are

$$H_{\text{dB}} = -20 \log_{10} 100 + 20 \log_{10}|1 + j\omega| - 20 \log_{10}\left|1 + \frac{j6\omega}{10} - \frac{\omega^2}{100}\right|$$

$$\phi = 0° + \tan^{-1}\omega - \tan^{-1}\left[\frac{\omega 6/10}{1 - \omega^2/100}\right]$$

The Bode plots are shown in Fig. 16.17. Notice that the quadratic factor is treated as a repeated linear factor at ω_k; that is, $(1 + j\omega/\omega_k)^2$, which is an approximation.

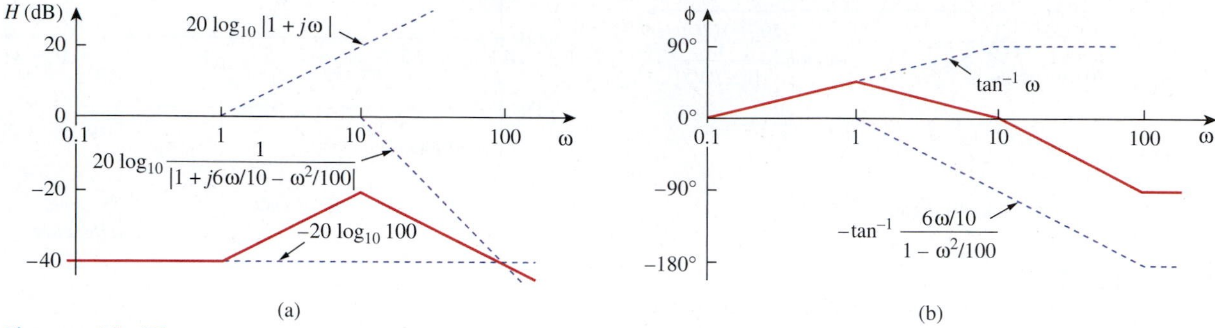

Figure 16.17
For Example 16.10: (a) magnitude plot; (b) phase plot.

Practice Problem 16.10

Construct the Bode plots for

$$H(s) = \frac{10}{s(s^2 + 80s + 400)}$$

Answer: See Fig. 16.18.

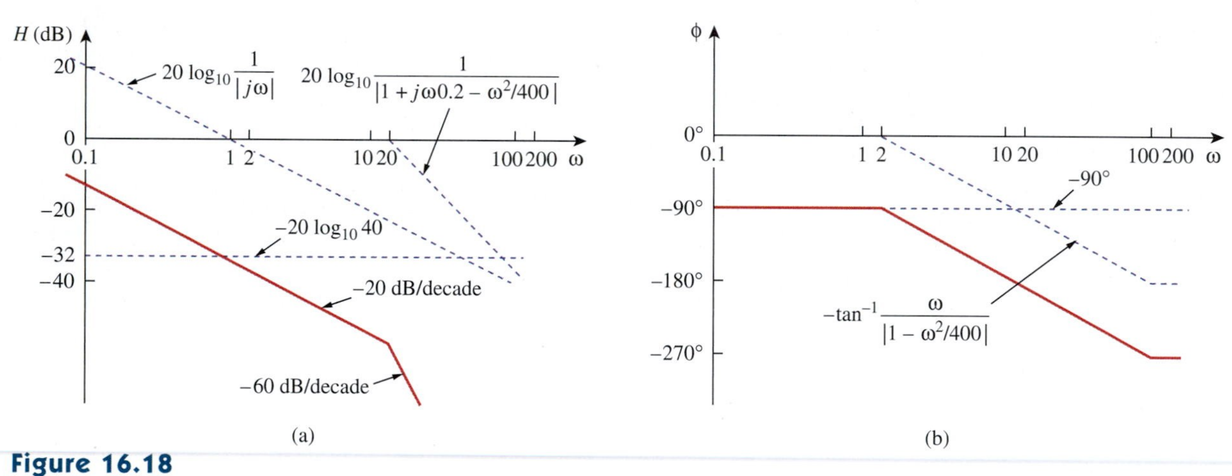

Figure 16.18
For Practice Problem 16.10: (a) magnitude plot; (b) phase plot.

16.5 Filters

The concept of filters has been an integral part of the evolution of electrical engineering technology from the beginning. Several technological achievements would not have been possible without electrical filters. Because of this prominent role of filters, much effort has been expended on the theory, design, and construction of filters, and many articles and books have been written on them. Our discussion in this chapter should be considered introductory.

> A **filter** is a circuit that is designed to pass signals with desired frequencies and reject (attenuate) others or to reject (attenuate) signals with undesired frequencies and allow others to pass.

As a frequency-selective device, a filter can be used to limit the frequency spectrum of a signal to some specified band of frequencies. Filters are the circuits used in radio and TV receivers to allow us to select one desired signal out of a multitude of broadcast signals in the environment. Besides the filters we will study in these sections, there are other kinds of filters such as digital filters, electromechanical filters, and microwave filters, which are beyond the level of the text.

As shown in Fig. 16.19, there are four types of filters:

1. A *lowpass filter* passes low frequencies and stops high frequencies, as shown ideally in Fig. 16.19(a).
2. A *highpass filter* passes high frequencies and rejects low frequencies, as shown ideally in Fig. 16.19(b).
3. A *bandpass filter* passes frequencies within a frequency band and blocks or attenuates frequencies outside the band, as shown ideally in Fig. 16.19(c).
4. A *bandstop filter* passes frequencies outside a frequency band and blocks or attenuates frequencies within the band, as shown ideally in Fig. 16.19(d).

A summary of the characteristics of these filters is presented in Table 16.6. It should be emphasized that the characteristics in Table 16.6 are only valid for simple filters and one should not have the impression that only these kinds of filters exist. We will now consider typical circuits for realizing the filters shown in Table 16.6.

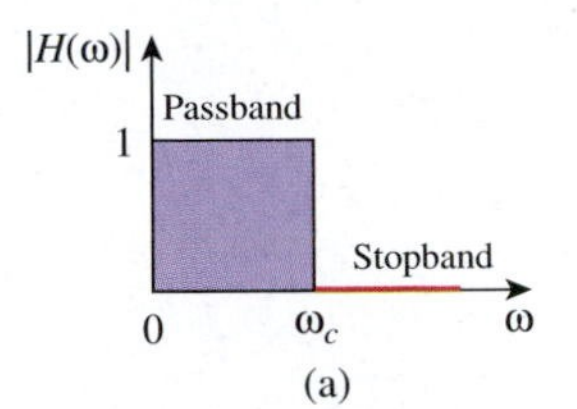

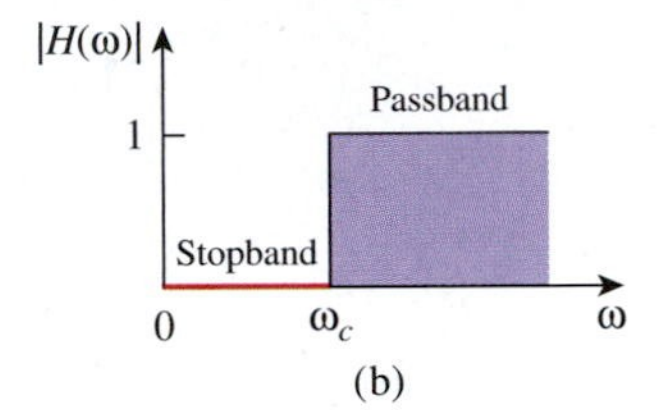

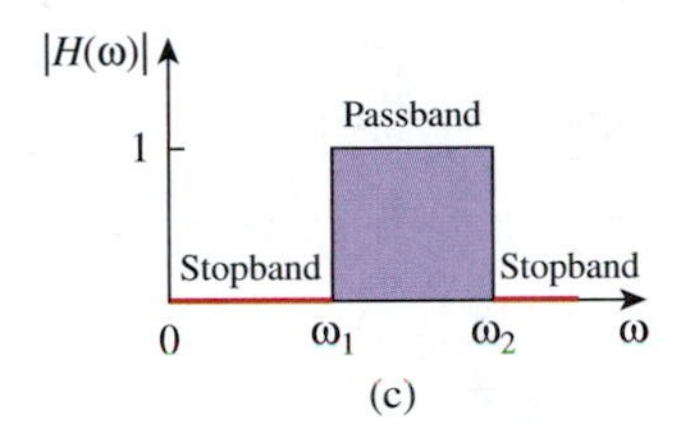

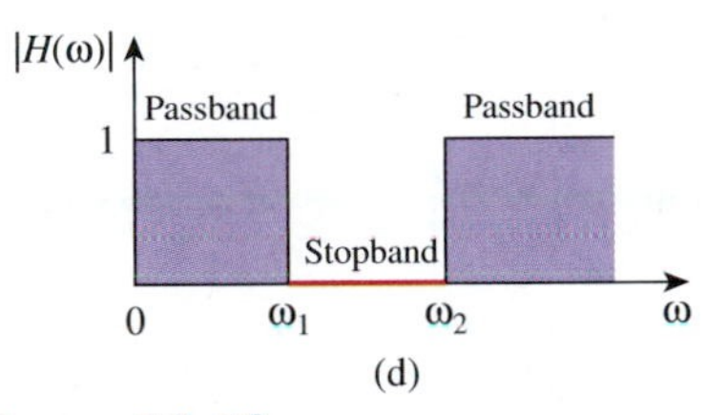

Figure 16.19
Ideal frequency response of four types of filters: (a) lowpass filter; (b) highpass filter; (c) bandpass filter; (d) bandstop filter.

TABLE 16.6

Summary of the characteristics of filters.

Type of filter	$H(0)$	$H(\infty)$	$H(\omega_c)$ or $H(\omega_0)$
Lowpass	1	0	$1/\sqrt{2}$
Highpass	0	1	$1/\sqrt{2}$
Bandpass	0	0	1
Bandstop	1	1	0

Note: ω_c is the cutoff frequency for lowpass and highpass filters; ω_0 is the center frequency for bandpass and bandstop frequency.

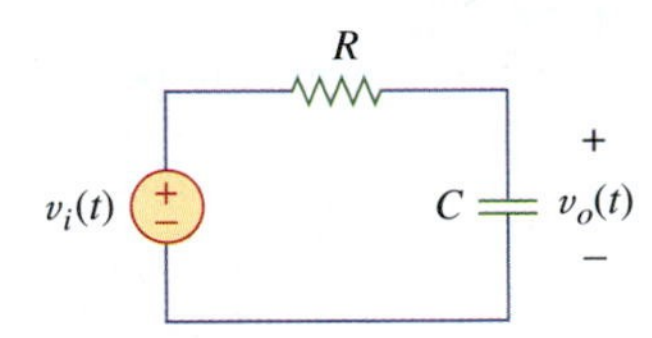

Figure 16.20
A lowpass filter.

16.5.1 Lowpass Filter

A typical lowpass filter is formed when the output of a series *RC* circuit is taken from the capacitor as shown in Fig. 16.20. The transfer function (see also Example 16.6) is

$$\mathbf{H}(\omega) = \mathbf{V}_o/\mathbf{V}_i = \frac{1/j\omega C}{R + 1/j\omega C}$$

$$\mathbf{H}(\omega) = \frac{1}{1 + j\omega RC} \tag{16.28}$$

Note that $\mathbf{H}(0) = 1$, $\mathbf{H}(\infty) = 0$. The plot of $|H(\omega)|$ is shown in Fig. 16.21, where the ideal characteristic is also shown. The half-power frequency, which is equivalent to the corner frequency on the Bode plots but in the context of filters, is usually known as the *cutoff*

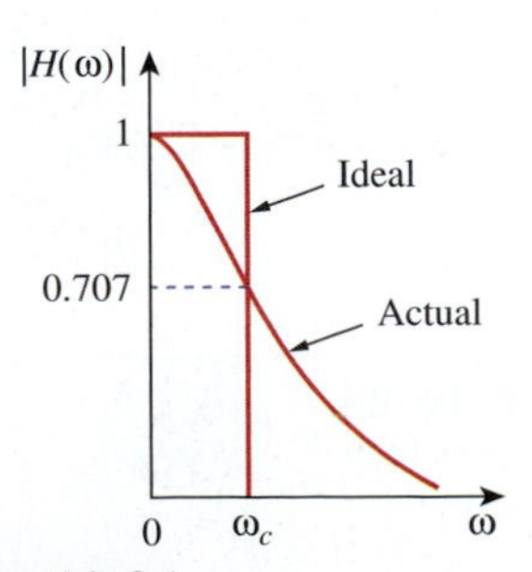

Figure 16.21
Ideal and actual frequency response of a lowpass filter.

frequency ω_c, and is obtained by setting the magnitude of $\mathbf{H}(\omega)$ equal to $1/\sqrt{2}$; that is,

$$H(\omega_c) = \frac{1}{\sqrt{1 + \omega_c^2 R^2 C^2}} = \frac{1}{\sqrt{2}} \tag{16.29}$$

where

$$\omega_c = \frac{1}{RC} \tag{16.30}$$

The cutoff frequency[6] is also called the *rolloff frequency*. Thus,

> A **lowpass filter** is designed to pass only frequencies from dc up to the cutoff frequency ω_c.

A lowpass filter can also be formed when the output of a series *RL* circuit is taken off the resistor. Of course, there are many other circuits for lowpass filters.

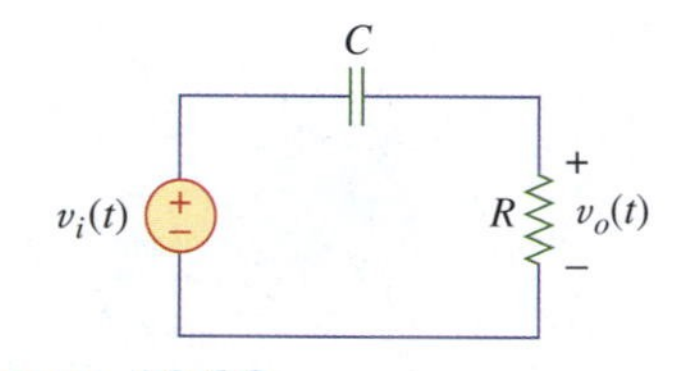

Figure 16.22
A highpass fliter.

16.5.2 Highpass Filter

A highpass filter is formed when the output of a series *RC* circuit is taken from the resistor as shown in Fig. 16.22. The transfer function is

$$\mathbf{H}(\omega) = \mathbf{V}_o/\mathbf{V}_i = \frac{R}{R + 1/j\omega C}$$

$$\mathbf{H}(\omega) = \frac{j\omega RC}{1 + j\omega RC} \tag{16.31}$$

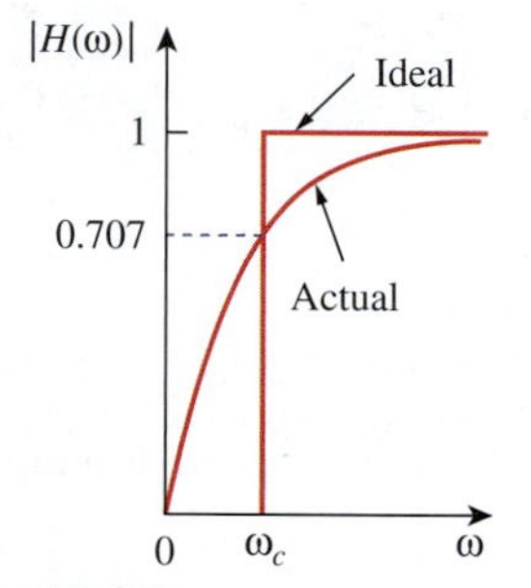

Figure 16.23
Ideal and actual frequency response of a highpass filter.

Note that $\mathbf{H}(0) = 0$, $\mathbf{H}(\infty) = 1$. The plot of $|\mathbf{H}(\omega)|$ is shown in Fig. 16.23. Again, the corner or cutoff frequency is

$$\omega_c = \frac{1}{RC} \tag{16.32}$$

Thus,

> A **highpass filter** is designed to pass all frequencies above its cutoff frequency ω_c.

A highpass filter can also be formed when the output of a series *RL* circuit is taken from the inductor.

16.5.3 Bandpass Filter

The *RLC* series resonant circuit provides a bandpass filter when the output is taken from the resistor as shown in Fig. 16.24. The transfer function is

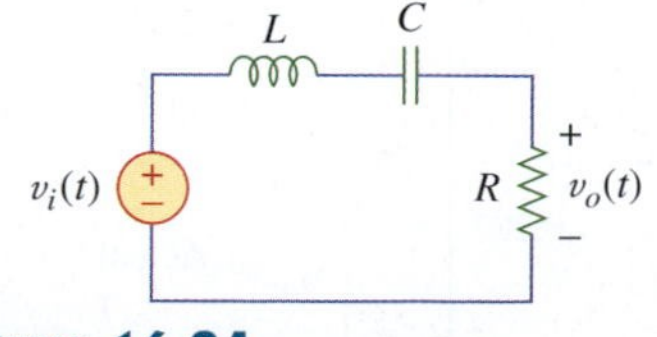

Figure 16.24
A bandpass filter.

$$\mathbf{H}(\omega) = \mathbf{V}_o/\mathbf{V}_i = \frac{R}{R + j(\omega L - 1/\omega C)} \tag{16.33}$$

[6] Note: The cutoff frequency is the frequency at which the transfer function **H** drops in magnitude to 70.71 percent of its maximum value. It is also the frequency at which the power dissipated in a circuit is half of its maximum value.

We observe that $\mathbf{H}(0) = 0$ and $\mathbf{H}(\infty) = 0$. The plot of $|\mathbf{H}(\omega)|$ is shown in Fig. 16.25. The bandpass filter passes a band of frequencies ($\omega_1 < \omega < \omega_2$) centered on ω_0, the *center frequency*, which is given by

$$\omega_0 = \frac{1}{\sqrt{LC}} \tag{16.34}$$

Thus,

> A **bandpass filter** is designed to pass all frequencies within a band of frequencies, $\omega_1 < \omega < \omega_2$.

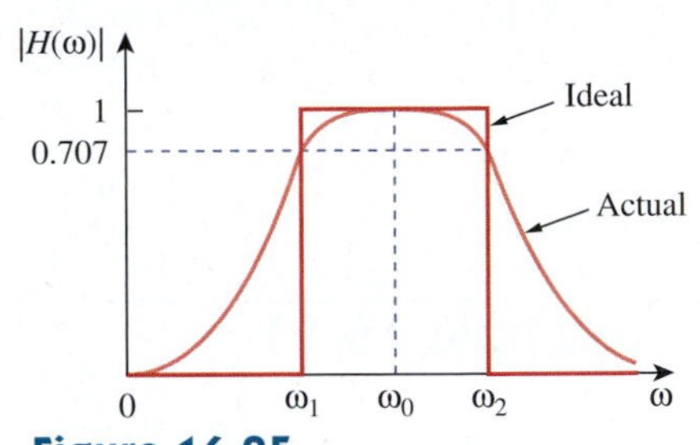

Figure 16.25
Ideal and actual frequency response of a bandpass filter.

Because the bandpass filter in Fig. 16.24 is a series resonant circuit, the half-power frequencies, the bandwidth, and the quality factor are determined as in Section 15.2.

A bandpass filter can also be formed by cascading a lowpass filter with $\omega_2 = \omega_c$ in Fig. 16.20 with a highpass filter with $\omega_1 = \omega_c$ in Fig. 16.22.

16.5.4 Bandstop Filter

A filter that prevents a band of frequencies between two designated values (ω_1 and ω_2) from passing is variably known as a *bandstop*, *bandreject*, or *notch* filter. A bandstop filter is formed when the output of an *RLC* series resonant circuit is taken from the *LC* series combination as shown in Fig. 16.26. The transfer function is

$$\mathbf{H}(\omega) = \mathbf{V}_o/\mathbf{V}_i = \frac{j(\omega L - 1/\omega C)}{R + j(\omega L - 1/\omega C)} \tag{16.35}$$

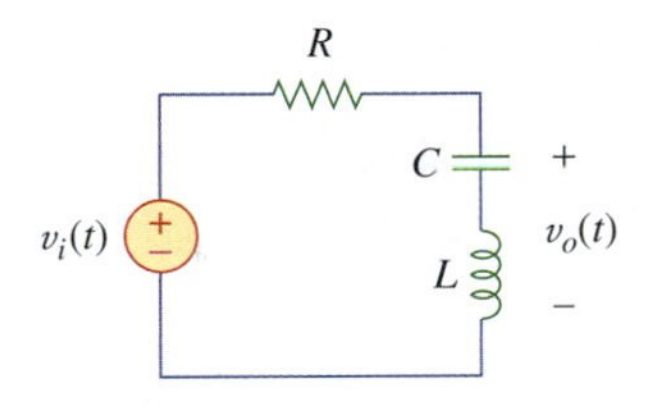

Figure 16.26
A bandstop filter; for Example 16.12.

Notice that $\mathbf{H}(0) = 1$ and $\mathbf{H}(\infty) = 1$. The plot of $|\mathbf{H}(\omega)|$ is shown in Fig. 16.27. Again, the center frequency is given by

$$\omega_0 = \frac{1}{\sqrt{LC}} \tag{16.36}$$

while the half-power frequencies, the bandwidth, and the quality factor are calculated using the formulas in Section 15.2 for a series resonant circuit. Here, ω_0 is called the *frequency of rejection*, while the corresponding bandwidth ($BW = \omega_2 - \omega_1$) is known as the *bandwidth of rejection*. Thus,

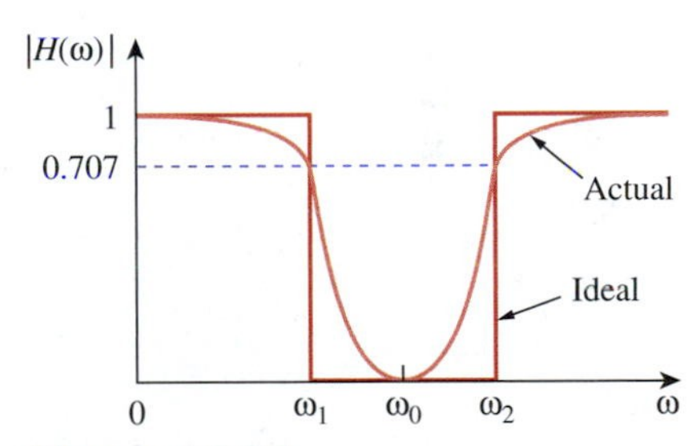

Figure 16.27
Ideal and actual frequency response of a bandstop filter.

> A **bandstop filter** is designed to stop or eliminate all frequencies within a band of frequencies, $\omega_1 < \omega < \omega_2$.

Notice that adding the transfer functions of the bandpass and the bandstop gives unity at any frequency for the same values of *R*, *L*, and *C*. Of course, this is not true in general but true for the dual circuits treated here. This is due to the fact that the characteristic of one is the inverse of the other.

In concluding this section, we should note the following:

1. From Eqs. (16.29), (16.31), (16.33), and (16.35), the maximum gain of a passive filter is unity (or 0 dB). To generate a gain greater than unity, an active filter should be used. Active filters are beyond the scope of this textbook.

2. There are other ways to get the types of filters treated in this section.
3. The filters treated here are the simple types. There are many other filters with sharper and more complex frequency responses.

Example 16.11

Determine what type of filter is shown in Fig. 16.28. Calculate the corner or cutoff frequency. Take $R = 2\ \text{k}\Omega$, $L = 2$ H, and $C = 2\ \mu$F.

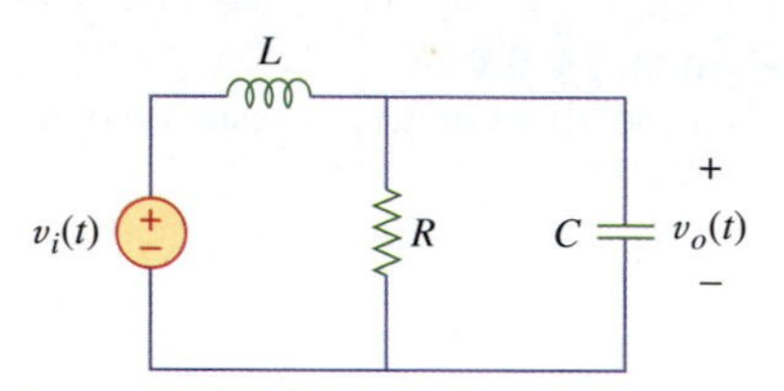

Figure 16.28
For Example 16.11.

Solution:
The transfer function is

$$\mathbf{H}(s) = \mathbf{V}_o/\mathbf{V}_i = \frac{R \parallel 1/sC}{sL + R \parallel 1/sC}, \qquad s = j\omega \qquad \textbf{(16.11.1)}$$

But

$$R \left\| \frac{1}{sC} = \frac{R/sC}{R + 1/sC} = \frac{R}{1 + sRC} \right.$$

Substituting this in Eq. (16.10.1) gives

$$\mathbf{H}(s) = \frac{R/(1 + sRC)}{sL + R/(1 + sRC)} = \frac{R}{s^2RLC + sL + R}, \qquad s = j\omega$$

or

$$\mathbf{H}(\omega) = \frac{R}{-\omega^2RLC + j\omega L + R} \qquad \textbf{(16.11.2)}$$

Because $\mathbf{H}(0) = 1$, $\mathbf{H}(\infty) = 0$, we conclude from Table 16.6 that the circuit in Fig. 16.28 is a second-order lowpass filter. The magnitude of $\mathbf{H}$ is

$$H = \frac{R}{\sqrt{(R - \omega^2RLC)^2 + \omega^2L^2}} \qquad \textbf{(16.11.3)}$$

The corner frequency is the same as the half-power frequency, that is, where $\mathbf{H}$ is reduced by a factor of $1/\sqrt{2}$. Because the dc value of $H(\omega)$ is 1, at the corner frequency, Eq. (16.11.3) becomes after squaring

$$H^2 = \frac{1}{2} = \frac{R^2}{(R - \omega_c^2RLC)^2 + \omega_c^2L^2}$$

or

$$2 = (1 - \omega_c^2LC)^2 + \left(\frac{\omega_c LC}{R}\right)^2$$

Substituting the values of R, L, and C, we obtain

$$2 = (1 - \omega_c^2\, 4 \times 10^{-6})^2 + (\omega_c\, 10^{-3})^2$$

Assuming that ω_c is in krad/s,

$$2 = (1 - 4\omega_c)^2 + \omega_c^2 \qquad \Rightarrow \qquad 16\omega_c^4 - 7\omega_c^2 - 1 = 0$$

Solving the quadratic equation in ω_c^2, we get $\omega_c^2 = 0.5509$ or $\omega_c = 0.742$ krad/s $= 742$ rad/s. Because $\omega_c = 2\pi f_c$, this is the same as $f_c = 118.6$ Hz.

Practice Problem 16.11

For the circuit in Fig. 16.29, obtain the transfer function $V_o(\omega)/V_i(\omega)$. Identify the type of filter the circuit represents and determine the corner frequency. Take $R_1 = 100\ \Omega = R_2$, $L = 2$ mH.

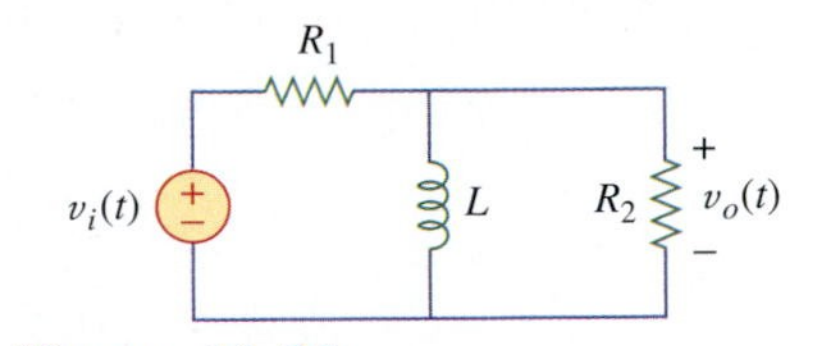

Figure 16.29
For Practice Problem 16.11.

Answer: $\dfrac{R_2}{R_1 + R_2}\left(\dfrac{j\omega}{j\omega + \omega_c}\right)$; highpass filter;

$$\omega_c = \frac{R_1 R_2}{(R_1 + R_2)L} = 25 \text{ krad/s}.$$

Example 16.12

If the bandstop filter in Fig. 16.26 is to reject a 200-Hz sinusoid while passing other frequencies, calculate the values of L and C. Take $R = 150\ \Omega$ and the bandwidth as 100 Hz.

Solution:
We use the formulas for series resonant circuit in Section 15.2.

$$BW = 2\pi\,(100) = 200\pi \text{ rad/s}$$

But

$$BW = \frac{R}{L} \quad \Rightarrow \quad L = \frac{R}{BW} = \frac{150}{200\pi} = 0.2387 \text{ H}$$

Rejection of the 200-Hz sinusoid means that f_0 is 200 Hz so that ω_0 in Fig. 16.27 is

$$\omega_0 = 2\pi f_0 = 2\pi\,(200) = 400\pi$$

Because $\omega_0 = 1/\sqrt{LC}$,

$$C = \frac{1}{\omega_0^2 L} = \frac{1}{(400\pi)^2(0.2387)} = 2.66\ \mu\text{F}$$

Practice Problem 16.12

Design a bandpass filter of the form in Fig. 16.24 with a lower cutoff frequency of 20.1 kHz and an upper cutoff frequency of 20.3 kHz. Take $R = 20$ kΩ. Calculate L, C, and Q.

Answer: 7.96 H; 3.9 pF; 101

16.6 Computer Analysis

16.6.1 PSpice

PSpice can be used in a way similar to the way it was used in the previous chapter. The only difference here is introducing the use of PSpice to obtain the Bode magnitude and phase plots.

As mentioned in Section 15.5, there are three types of sweeps:

- **Linear:** The frequency is varied linearly from *Start Freq* to *End Freq* with *Total* equally spaced points (or responses).
- **Octave:** The frequency is swept logarithmically by octaves from *Start Freq* to *End Freq* with *Total* points per octave. A factor of 2

in frequency is called an octave. For example, the frequency range from 20 to 40 kHz is an octave.

- **Decade:** The frequency is varied logarithmically by decades from *Start Freq* to *End Freq* with *Total* points per decade. An interval between two frequencies with a ratio of 1:10 is called a decade. For example, the frequency range from 20 to 200 kHz is a decade.

With the preceding specifications, PSpice performs a steady-state sinusoidal analysis of the circuit as the frequency of all the independent sources is varied (or swept) from *Start Freq* to *End Freq*.

The PSpice A/D window is used to produce a graphical output. The output data type may be specified in the *Trace Command Box* by adding one of the following suffixes to V or I:

M Amplitude of the sinusoid (peak value)

P Phase of the sinusoid

dB Amplitude of the sinusoid in decibels; that is, 20 $\log_{10}$ (amplitude)

Example 16.13

Use PSpice to generate the gain and phase Bode plots of V_o in the circuit of Fig. 16.30.

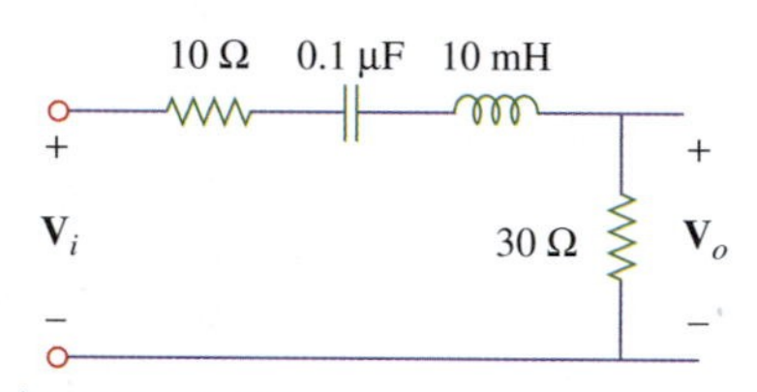

Figure 16.30
For Example 16.13 and Practice Problem 16.15.

Solution:

Because we are interested in Bode plots, we use the decade frequency sweep for 300 Hz $< f <$ 30 kHz with 101 points per decade. We select this range because we know that the resonant frequency of the circuit is within the range. Recall that

$$\omega_0 = \frac{1}{\sqrt{LC}} = 31.62 \text{ krad/s} \qquad \text{or} \qquad f_0 = \frac{\omega_0}{2\pi} = 5.03 \text{ kHz}$$

After drawing the circuit as in Fig. 16.31 and saved as exam1613.dsn, we select **PSpice/New Simulation Profile**. This leads to the New Simulation dialog box. Type "exam1613" as the name of the file and click Create. This leads to the Simulation Settings dialog box. In the Simulation Settings dialog box, select AC Sweep/Noise under *Analysis Type*, select Logarithmic/Decade under *AC Sweep Type*. We type 300 as the *Start Freq*, 30k as the *Final Freq*, and 101 as *Points/Decade*.

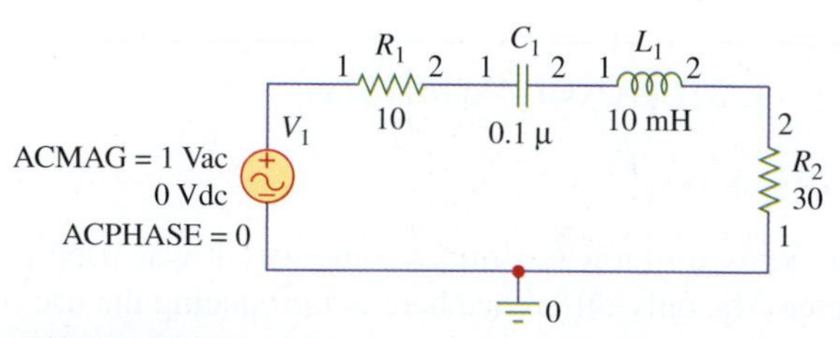

Figure 16.31
The schematic for the circuit in Fig. 16.30.

We simulate it by selecting **PSpice/Run**. This will automatically bring up the Probe window if there are no errors. Because we are interested in Bode plot, we select **Trace/Add** in the PSpice A/D

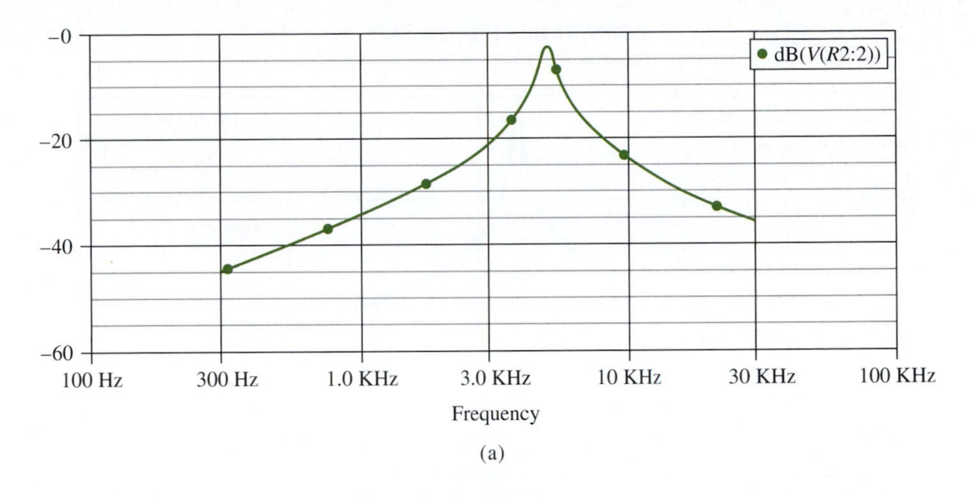

(a)

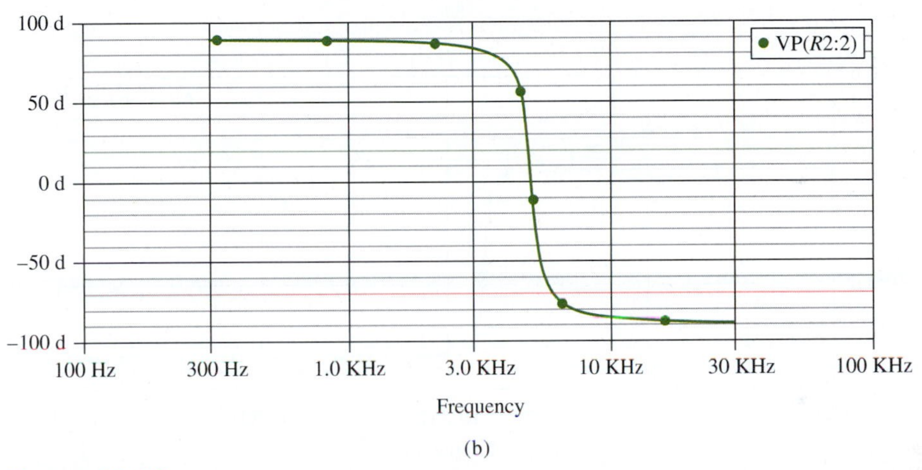

(b)

Figure 16.32
For Example 16.13: (a) Bode plot; (b) phase plot of the response.

window and type dB(V(R2:2)) in the **Trace Command** box. The result is the Bode magnitude plot in Fig. 16.32(a). For the phase plot, we select **Trace/Add Trace** in the Probe window and type VP(R2:2) in the **Trace Command** box. The result is the Bode phase plot of Fig. 16.32(b). Notice from Fig. 16.32 that the circuit is a bandpass filter.

Practice Problem 16.13

Consider the network in Fig. 16.33. Use PSpice to obtain the Bode plots for $\mathbf{V}_o$ over a frequency from 1 to 100 kHz using 20 points per decade.

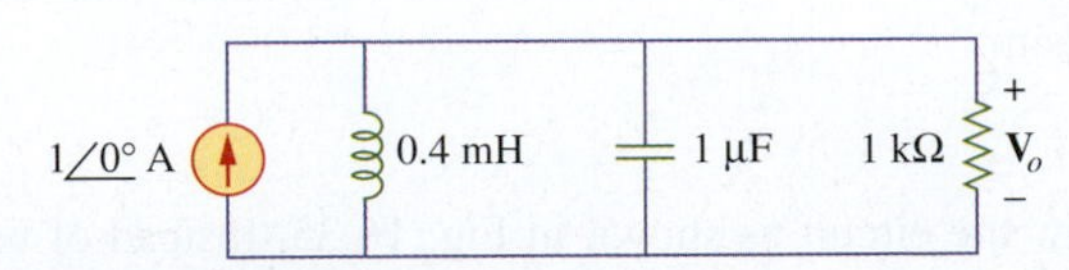

Figure 16.33
For Practice Problems 16.13 and 16.14.

Answer: See Fig. 16.34.

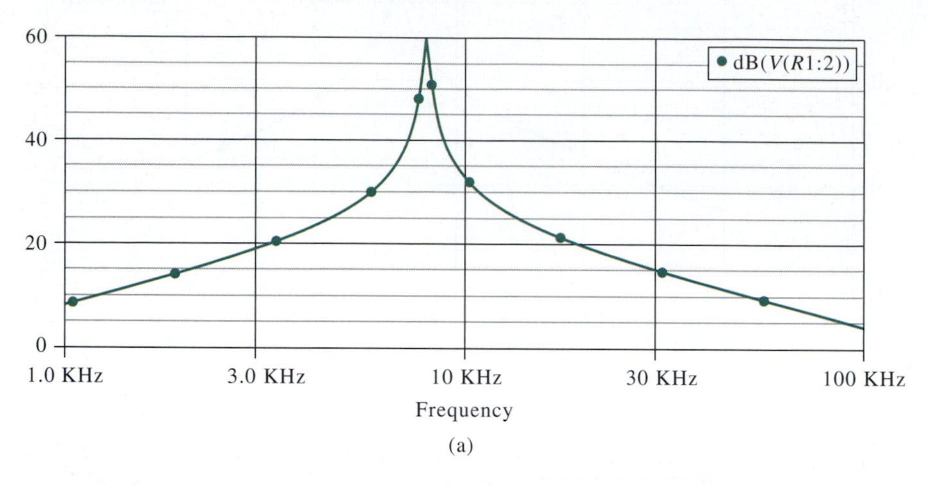

(a)

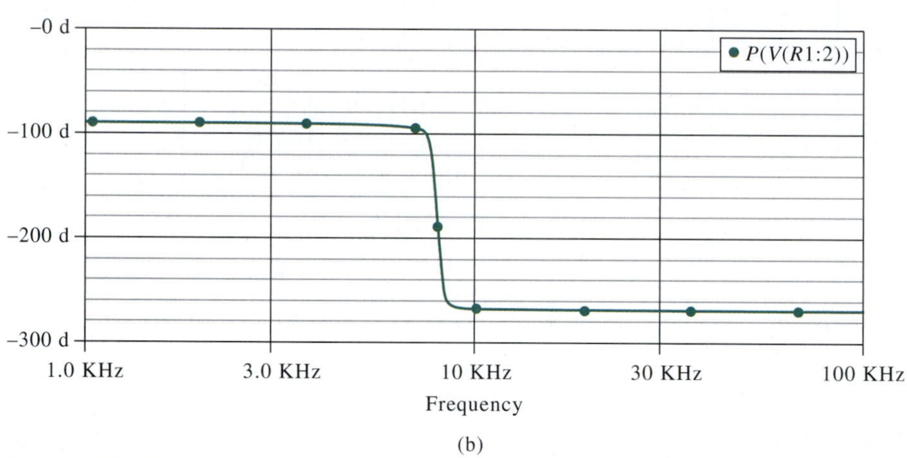

(b)

Figure 16.34
For Practice Problem 16.13: (a) Bode magnitude plot; (b) Bode phase plot.

16.6.2 Multisim

Multisim has a special instrument for Bode plots. The Bode plotter (an instrument) is often used to provide a display of either the voltage gain response or the phase response. When the magnitude mode is selected, the Bode plotter measures the ratio of the magnitudes (in decibels) between two points. When the phase mode is selected, it measures the phase shift (in degrees) between two points. The plotter plots gain or phase shift against frequency (in hertz). The following example illustrates how to use the Bode plotter.

Example 16.14

Construct the Bode plots for the circuit in Fig. 15.14 (see Chapter 15) using Multisim.

Solution:
We first draw the circuit as shown in Fig. 16.35. Instead of using an ac source voltage as we did in Fig. 15.21, we can use a function generator. By double-clicking the symbol of function generator (XFG1), we

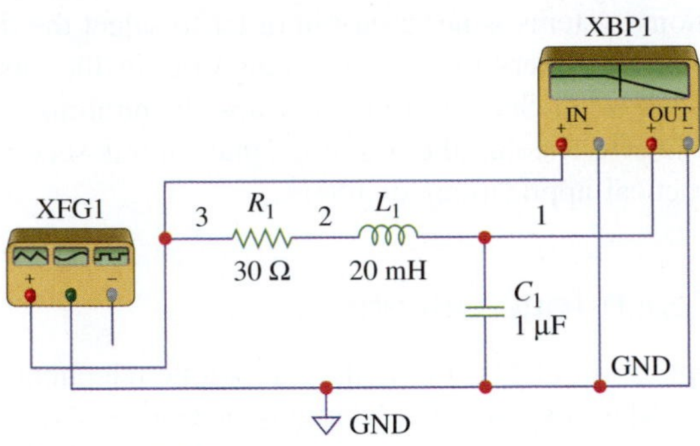

Figure 16.35
For Example 16.14.

change *Frequency* to 1 kHz and *Amplitude* to 1 V. Similarly, we double-click the symbol for the Bode plotter (XBP1), we choose *Magnitude* as the mode, *Log* as vertical scale from −40 to 40 dB, *Log* as horizontal scale from 1 to 10 kHz. We save the file and select **Simulate/Run**. While the circuit is being simulated, we double-click the symbol for Bode plotter, and the magnitude plot will be displayed. We change the mode to *Phase* and have the phase plot displayed. The magnitude and phase plots are shown in Fig. 16.36.

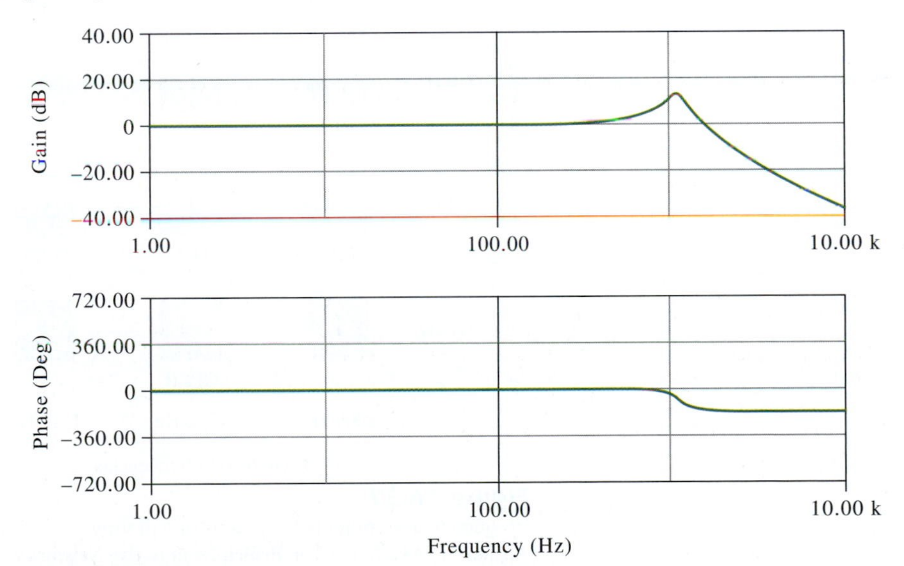

Figure 16.36
For Example 16.14; Bode magnitude and phase plots.

Practice Problem 16.14

Repeat Practice Problem 16.13 using Multisim.

Answer: See Fig. 16.34.

16.7 Applications

Filters are widely used, particularly in electronics, power systems, and communications systems. For example, a highpass filter with cutoff frequency above 60 Hz may be used to eliminate the 60-Hz power line noise in various communications electronics. Filtering of signals in communications systems is necessary in order to select the desired signal from a host of others in the same range (as in the case of radio receivers discussed in Section 15.6) and also to minimize the effects of noise and interference on the desired signal. In this section, we consider two practical applications of filters.

16.7.1 Touch-Tone Telephone

A typical application of filtering is the touch-tone telephone set shown in Fig. 16.37. The keypad has 12 buttons arranged in four rows and three columns. The arrangement provides 12 distinct signals by using seven tones divided into two groups: the low frequency group (697 to 941 Hz) and the high-frequency group (1,209 to 1,477 Hz). Pressing a button generates a sum of two sinusoids corresponding to its unique pair of frequencies. For example, pressing the number 6 button generates sinusoidal tones with the frequencies 770 and 1,477 Hz.

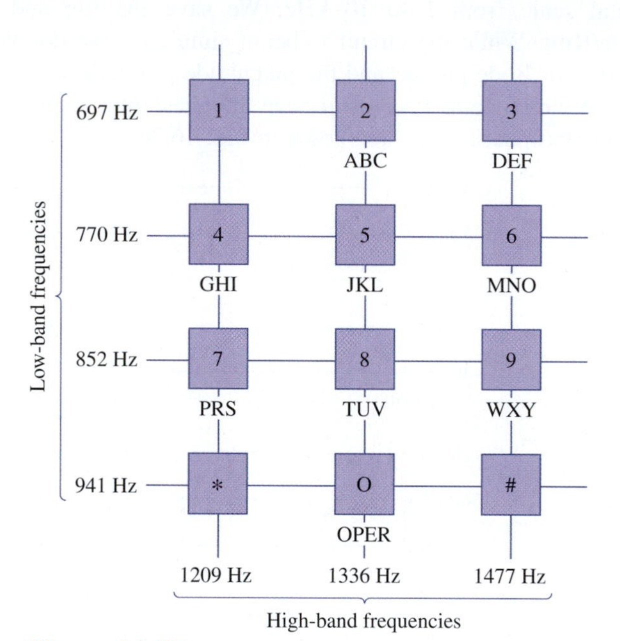

Figure 16.37
Frequency assignments for touch-tone dialing.
Adapted from G. Daryanani, **Principles of Active Network Synthesis and Design** (New York: John Wiley & Sons), 1976, p. 79.

When a caller enters a telephone number, the set of signals is transmitted to the telephone office, where the touch-tone signals are decoded by detecting the frequencies they contain. The block diagram for the detection scheme is shown in Fig. 16.38. The signals are first amplified and separated into their respective groups by the lowpass (LP) and highpass (HP) filters. The limiters (L) are used to convert the separated tones into square waves. The individual tones are identified using seven bandpass (BP) filters, with each filter passing one tone and rejecting the other tones. Each filter is followed by a detector (D), which is energized when its input voltage exceeds a certain level. The outputs of the detectors provide the required dc signals needed by the switching system to connect the caller to the party being called.

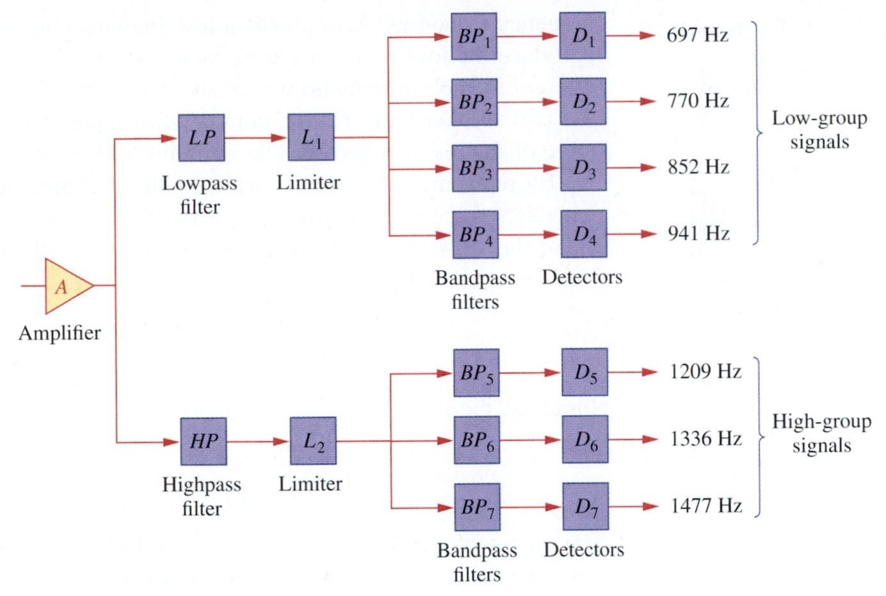

Figure 16.38
Block diagram of detection scheme.
G. Daryanani, **Principles of Active Network Synthesis and Design** (New York: John Wiley & Sons), 1976, p. 79.

Example 16.15

Using the standard 600-Ω resistor used in telephone circuits and a series *RLC* circuit, design the bandpass filter BP_2 in Fig. 16.38.

Solution:
The bandpass filter is the series *RLC* circuit in Fig. 16.24. Because BP_2 passes frequencies 697 Hz to 852 Hz and is centered at $f_0 = 770$ Hz, its bandwidth is

$$BW = 2\pi(f_2 - f_1) = 2\pi(852 - 697) = 973.89 \text{ rad/s}$$

From Eq. (15.19),

$$L = \frac{R}{BW} = \frac{600}{973.89} = 0.616 \text{ H}$$

From Eq. (15.6),

$$C = \frac{1}{\omega_0^2 L} = \frac{1}{4\pi^2 f_0^2 L} = \frac{1}{4\pi^2 \times 770^2 \times 0.616} = 69.36 \text{ nF}$$

Practice Problem 16.15

Repeat Example 16.13 for bandpass filter BP_6.

Answer: 0.356 H; 39.86 nF

16.7.2 Crossover Network

Another typical application of filters is the *crossover network* that couples an audio amplifier output to woofer and tweeter speakers, as shown in Fig. 16.39(a). The network basically consists of one highpass *RC* filter and one lowpass *RL* filter. It routes frequencies higher than a prescribed crossover frequency f_c to the tweeter (high-frequency speaker) and frequencies below f_c into the woofer (low-frequency speaker). These speakers have been designed to accommodate certain

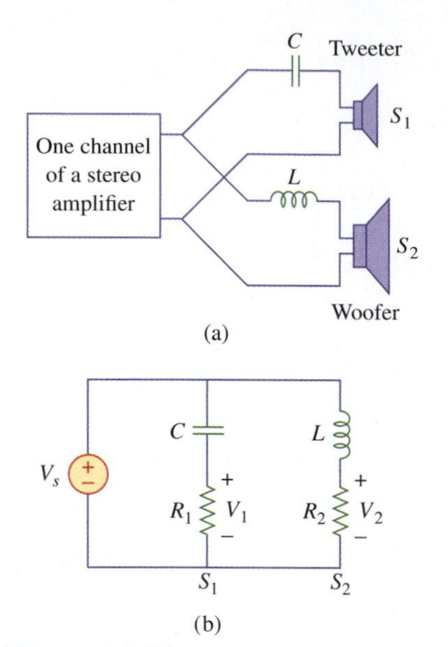

Figure 16.39
(a) A crossover network for two speakers; (b) equivalent model.

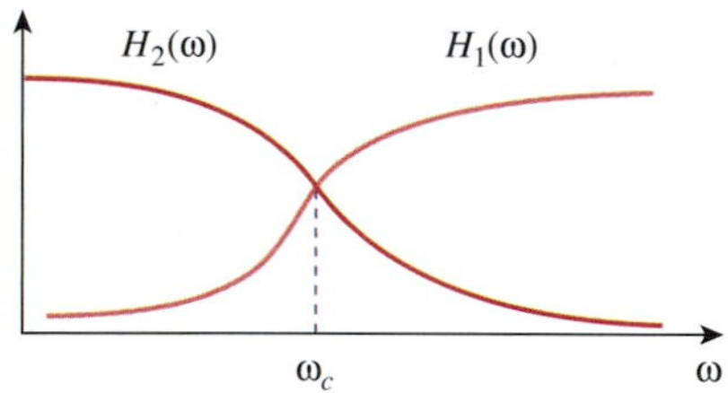

Figure 16.40
Frequency responses of the crossover network in Fig. 16.39.

frequency responses. A woofer is a low-frequency speaker designed to reproduce the lower part of the frequency range, up to about 3 kHz. A tweeter can reproduce audio frequencies from about 3 to about 20 kHz. The two speaker types can be combined to reproduce the entire audio range of interest and provide the optimum in frequency response.

By replacing the amplifier with a voltage source, the approximate equivalent circuit of the crossover network is shown in Fig. 16.39(b), where the speakers are modeled by resistors. As a highpass filter, the transfer function V_1/V_s is given by

$$H_1(\omega) = \frac{V_1}{V_s} = \frac{j\omega R_1 C}{1 + j\omega R_1 C} \tag{16.37}$$

Similarly, the transfer function of the lowpass filter is given by

$$H_2(\omega) = \frac{V_2}{V_s} = \frac{R_2}{R_2 + j\omega L} \tag{16.38}$$

The values of R_1, R_2, L, and C may be selected such that the two filters have the same cutoff frequency, known as the *crossover frequency*, as shown in Fig. 16.40.

The principle behind the crossover network is also used in the resonant circuit for a TV receiver, where it is necessary to separate the video and audio bands of RF carrier frequencies. The lower-frequency band (picture information in the range from about 30 Hz to about 4 MHz) is channeled into the receiver's video amplifier, while the high-frequency band (sound information around 4.5 MHz) is channeled to the receiver's sound amplifier.

Example 16.16

In the crossover network of Fig. 16.39, suppose each speaker acts as a 6-Ω resistance. Find C and L if the crossover frequency is 2.5 kHz.

Solution:

For the highpass filter,

$$\omega_c = 2\pi f_c = \frac{1}{R_1 C}$$

or

$$C = \frac{1}{2\pi f_c R_1} = \frac{1}{2\pi \times 2.5 \times 10^3 \times 6} = 10.61\ \mu\text{F}$$

For the lowpass filter,

$$\omega_c = 2\pi f_c = \frac{R_2}{L}$$

or

$$L = \frac{R_2}{2\pi f_c} = \frac{6}{2\pi \times 2.5 \times 10^3} = 382\ \mu\text{H}$$

Practice Problem 16.16

If each speaker in Fig. 16.39 has 8-Ω resistance and $C = 10\ \mu$F, find L and the crossover frequency.

Answer: 0.64 mH; 1.989 kHz

16.8 Summary

1. The transfer function $\mathbf{H}(\omega)$ is the ratio of the output response $\mathbf{Y}(\omega)$ to the input excitation $\mathbf{X}(\omega)$, that is, $\mathbf{H}(\omega) = \mathbf{Y}(\omega)/\mathbf{X}(\omega)$.
2. The frequency response is the variation of the transfer function with frequency.
3. The decibel is the unit of logarithmic gain. For power gain G, its decibel equivalent is

$$G_{\text{dB}} = 10 \log_{10} G$$

For a voltage (or current) gain G, its decibel equivalent is

$$G_{\text{dB}} = 20 \log_{10} G$$

Note that dBm is an abbreviation for the power ratio in decibel (dB) of the measured power referenced to one milliwatt (1mW).
4. Bode plots are straight-line semilog plots of the magnitude and phase of the transfer function as it varies with frequency. The straight-line approximations of H (in dB) and ϕ (in degrees) are constructed using the corner frequencies defined by the factors of $\mathbf{H}(\omega)$.
5. A filter is a circuit that is designed to pass (or reject) a band of frequencies and reject (or allow) others. Passive filters are constructed with resistors, capacitors, and inductors.
6. Four common types of filters are lowpass, highpass, bandpass, and bandstop.
7. A lowpass filter passes only signals whose frequencies are below the cutoff frequency ω_c.
8. A highpass filter passes only signals whose frequencies are above the cutoff frequency ω_c.
9. A bandpass filter passes only signals whose frequencies are within a prescribed range ($\omega_1 < \omega < \omega_2$).
10. A bandstop filter passes only signals whose frequencies are outside a prescribed range ($\omega_1 < \omega < \omega_2$).
11. PSpice can be used to obtain the frequency response of a circuit if a frequency range for the response and the desired number of points within the range are specified in the AC Sweep.
12. The touch-tone telephone and the crossover network are presented as two typical applications of filters. The touch-tone telephone system employs filters to separate tones of different frequencies to activate electronic switches. Crossover networks separate signals in different frequency ranges so that they can be delivered to different devices such as tweeters and woofers in loudspeaker systems.

Review Questions

16.1 If the ratio of the output power to the input power is 1,000:1. This ratio can be expressed in dB as:

(a) 3 dB (b) 30 dB
(c) 300 dB (d) 1000 dB

16.2 In a certain network, the input voltage is 2mV, while the output voltage is 4 V. The voltage gain expressed in dB is

(a) -33 dB (b) 33 dB
(c) 66 dB (d) 152 dB

16.3 On the Bode magnitude plot, the slope of the factor $\dfrac{1}{(5 + j\omega)^2}$ is

(a) 20 dB/decade (b) 40 dB/decade
(c) -40 dB/decade (d) -20 dB/decade

16.4 On the Bode phase plot, the slope of $[1 + j10\omega - \omega^2/25]^2$ is

(a) 45°/decade (b) 90°/decade
(c) 135°/decade (d) 180°/decade

16.5 If the maximum output voltage from a highpass filter is 1 V, the output voltage at the cutoff frequency is:

(a) 0 V (b) 0.707 V
(c) 1 V (d) 1.414 V

16.6 At the cutoff frequency, the output of a filter is down from its maximum value by

(a) −10 dB (b) −3 dB
(c) 0 dB (d) 3 dB

16.7 A filter that passes only signals whose frequencies are above a particular frequency is called:

(a) lowpass (b) highpass
(c) bandpass (d) bandstop

16.8 What kind of filter can be used to select a signal of one particular radio station?

(a) lowpass (b) highpass
(c) bandpass (d) bandstop

16.9 A voltage source supplies a signal of constant amplitude, from 0 to 40 kHz, to an *RC* lowpass filter. The load resistor experiences the maximum voltage at:

(a) dc (b) 10 kHz
(c) 20 kHz (d) 40 kHz

16.10 A bandstop filter attenuates signals of any frequency except the frequency to which it is tuned.

(a) True (b) False

Answers: 16.1b, 16.2c, 16.3c, 16.4d, 16.5b, 16.6b, 16.7b, 16.8c, 16.9a, 16.10a

Problems

Section 16.2 The Decibel Scale

16.1 Find $\log_{10} X$ given that X is:

(a) 10^{-4} (b) 46,000
(c) 10^8 (d) 0.2114

16.2 If $Y = \log_{10} X$, determine X given that Y is:

(a) 4 (b) 0.003
(c) 6.5 (d) −2.3

16.3 Determine the number of decibels of gain for the following conditions:

(a) $P_{in} = 6$ mW, $P_{out} = 100$ mW
(b) $V_{in} = 3$ mV, $V_{out} = 40$ V
(c) $P_{in} = 10\ \mu$W, $P_{out} = 60$ mW
(d) $V_{in} = 300\ \mu$V, $V_{out} = 8$ V

16.4 The output voltage of an amplifier is 3.8 V when the input voltage is 20 mV. What is the amplifier gain both as a ratio and in dB?

16.5 Calculate $|\mathbf{H}(\omega)|$ if H_{dB} equals

(a) 0.05 dB (b) −6.2 dB (b) 104.7 dB

16.6 Determine the magnitude (in dB) and the phase (in degrees) of $\mathbf{H}(\omega)$ at $\omega = 1$ rad/s if $\mathbf{H}(\omega)$ equals

(a) 0.05 (b) 125

(c) $\dfrac{10j\omega}{2 + j\omega}$ (d) $\dfrac{3}{1 + j\omega} + \dfrac{6}{2 + j\omega}$

16.7 A power meter is connected to the output circuit of a transmitter. The meter reads 24 W. What is the power in dBm?

16.8 Express the following powers in dBm: 10 μW, 13 mW, and 50 W.

16.9 Obtain the power levels of the following dBm values: −5, 6, and 40 dBm.

16.10 Consider the system in Fig. 16.41. Determine the overall gain of the system in dB.

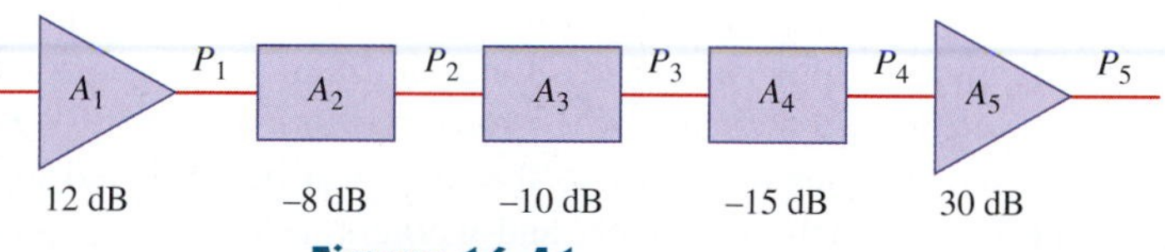

Figure 16.41
For Problem 16.10.

16.11 The system shown in Fig. 16.42 consists of the three stages. Determine the output power and the overall power gain of the system in dB.

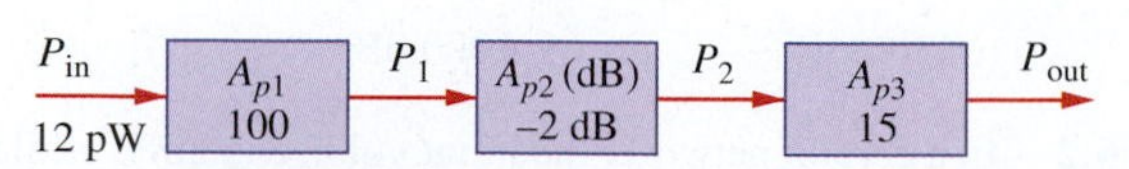

Figure 16.42
For Problem 16.11.

16.12 A certain amplifier has an input power of 60 mW and an output power of 10 mW. Calculate the attenuation in decibels.

16.13 At a particular frequency, the ratio V_{out}/V_{in} is 0.2. Express the ratio in dB.

16.14 An amplifier has a power gain of 4. Express the power gain in dB.

Section 16.3 Transfer Function

16.15 Find the transfer function $\mathbf{V}_o/\mathbf{V}_i$ of the *RC* circuit in Fig. 16.43. Express it using $\omega_o = 1/RC$.

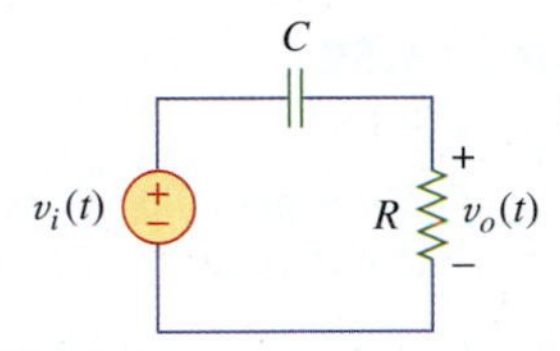

Figure 16.43
For Problem 16.15.

16.16 Obtain the transfer function $\mathbf{V}_o/\mathbf{V}_i$ of the *RL* circuit of Fig. 16.44. Express it using $\omega_o = R/L$.

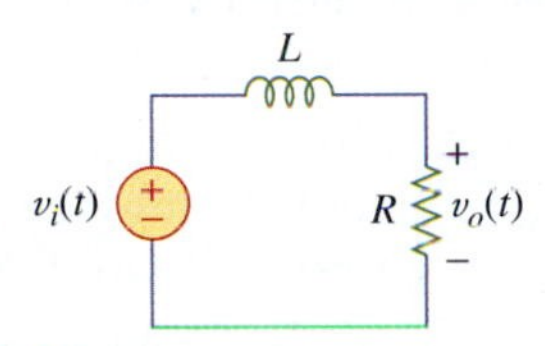

Figure 16.44
For Problems 16.16 and 16.34.

16.17 Given the circuit in Fig. 16.45, determine the transfer function $H(s) = V_o(s)/V_i(s)$.

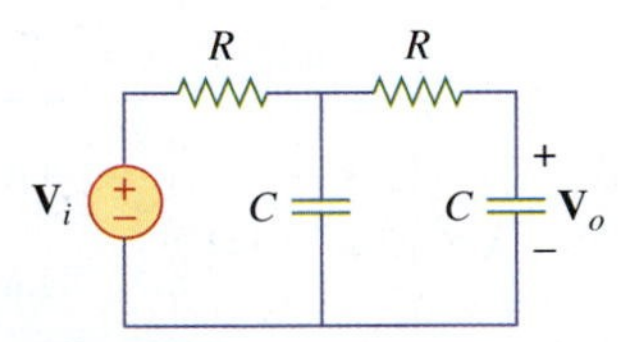

Figure 16.45
For Problem 16.17.

16.18 Find the transfer function $\mathbf{H}(\omega)$ of the circuit shown in Fig. 16.46.

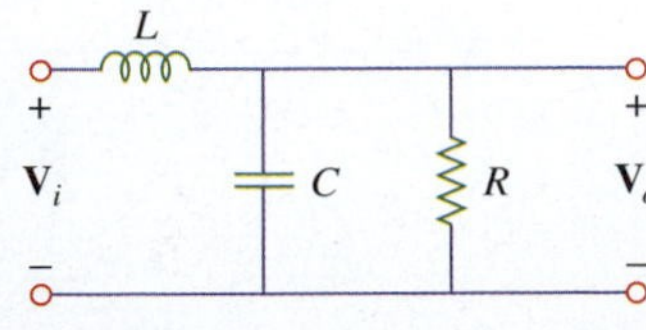

Figure 16.46
For Problem 16.18.

16.19 Repeat the previous problem for the circuit in Fig. 16.47.

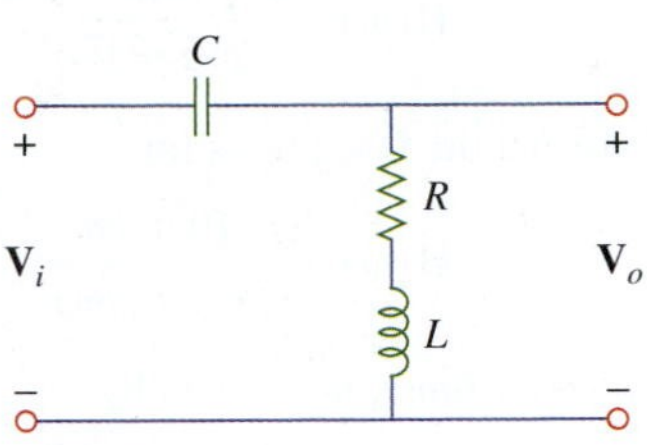

Figure 16.47
For Problem 16.19.

16.20 Determine the transfer function $\mathbf{V}_o/\mathbf{V}_i$ of the *RLC* circuit in Fig. 16.48.

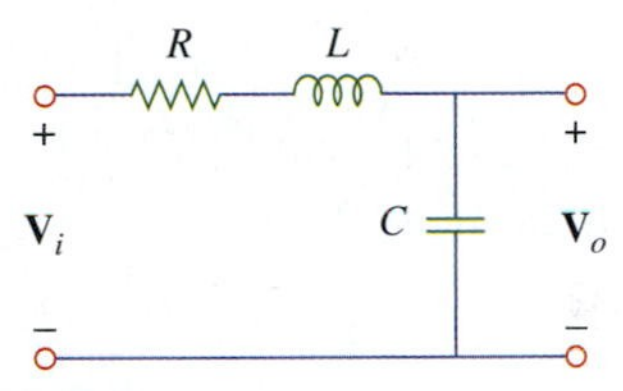

Figure 16.48
For Problem 16.20.

16.21 Repeat Problem 16.13 for the circuit in Fig. 16.49.

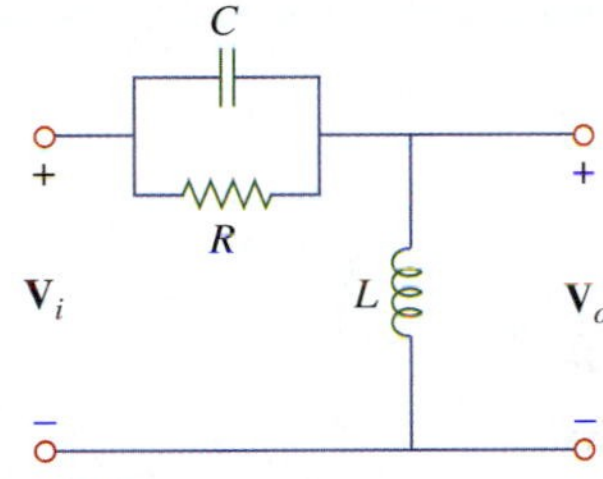

Figure 16.49
For Problem 16.21.

16.22 Compute the transfer function $H(\omega)$ for the circuit in Fig. 16.50.

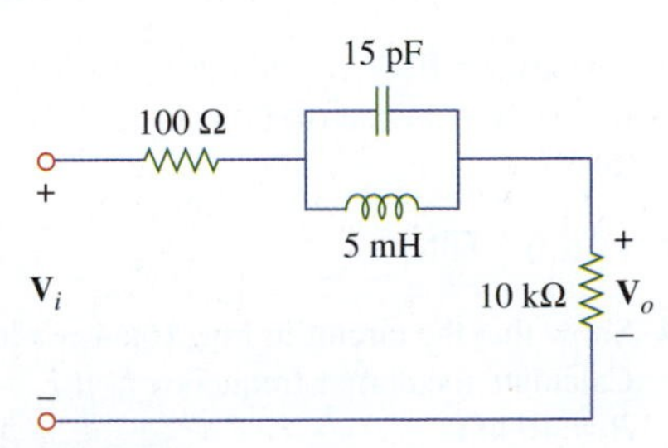

Figure 16.50
For Problem 16.22.

Section 16.4 Bode Plots

16.23 A ladder network has a voltage gain of

$$\mathbf{H}(\omega) = \frac{10}{(1 + j\omega)(10 + j\omega)}$$

Sketch the Bode plots for the gain.

16.24 Sketch the Bode plots for

$$\mathbf{H}(\omega) = \frac{50}{j\omega(5 + j\omega)}$$

16.25 Construct the Bode plots for

$$\mathbf{H}(\omega) = \frac{10 + j\omega}{j\omega(2 + j\omega)}$$

16.26 A transfer function is given by

$$T(s) = \frac{s + 1}{s(s + 10)}, \qquad s = j\omega$$

Sketch the magnitude and phase Bode plots.

16.27 Construct the Bode plots for

$$G(s) = \frac{s + 1}{s^2(s + 10)}$$

16.28 Draw the Bode plots for

$$\mathbf{H}(\omega) = \frac{50(j\omega + 1)}{j\omega(-\omega^2 + 10j\omega + 25)}$$

16.29 Construct the Bode magnitude and phase plots for

$$H(s) = \frac{40(s + 1)}{(s + 2)(s + 10)}, \qquad s = j\omega$$

16.30 Sketch the Bode plots for

$$G(s) = \frac{s}{(s + 2)^2(s + 1)}, \qquad s = j\omega$$

16.31 Draw the Bode plots for

$$G(s) = \frac{(s + 2)^2}{s(s + 5)^2(s + 10)}, \qquad s = j\omega$$

16.32 Construct the Bode plots for

$$\mathbf{T}(\omega) = \frac{10j\omega(1 + j\omega)}{(10 + j\omega)(100 + 10j\omega - \omega^2)}$$

16.33 Construct a Bode magnitude plot for $H(s)$ equal to: (a) $10/(s + 1)$ and (b) $(s + 1)/(s + 10)$.

Section 16.5 Filters

16.34 Show that the circuit in Fig. 16.44 is a lowpass filter. Calculate the corner frequency f_C if $L = 2$ mH and $R = 10$ kΩ.

16.35 Find the transfer function $\mathbf{V}_o/\mathbf{V}_i$ of the circuit in Fig. 16.51. Show that the circuit is a lowpass filter.

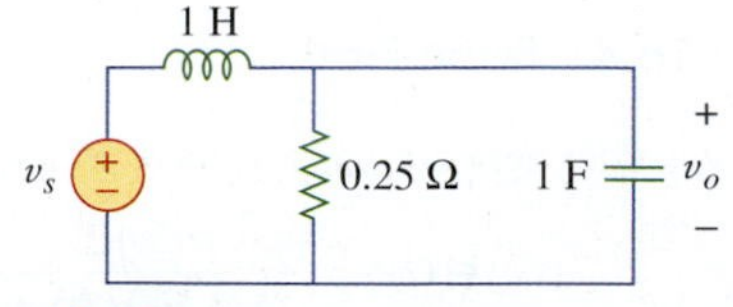

Figure 16.51
For Problem 16.35.

16.36 Determine the cutoff frequency of the lowpass filter described by

$$\mathbf{H}(\omega) = \frac{4}{2 + j\omega 10}$$

Find the gain in dB and phase of $\mathbf{H}(\omega)$ at $\omega = 2$ rad/s.

16.37 Determine what type of filter is shown in Fig. 16.52. Calculate the corner frequency f_c.

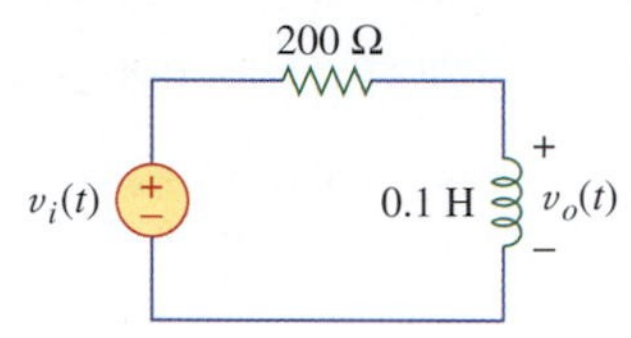

Figure 16.52
For Problem 16.37.

16.38 Design an *RL* lowpass filter that uses a 40-mH coil and has a cutoff frequency of 5 kHz.

16.39 In a highpass *RL* filter with a cutoff frequency of 100 kHz, $L = 40$ mH, find R.

16.40 Design a series *RLC* type bandpass filter with cutoff frequencies of 10 kHz and 11 kHz. Assuming $C = 80$ pF, find R, L, and Q.

16.41 Design an *RC* highpass filter that has a cutoff frequency of 2 kHz and uses a 300-pF capacitor.

16.42 Determine the range of frequencies that will be passed by a series *RLC* bandpass filter with $R = 10$ Ω, $L = 25$ mH, and $C = 0.4$ μF. Find the quality factor.

16.43 The circuit parameters for a series *RLC* bandstop filter are $R = 2$ kΩ, $L = 0.1$ H, $C = 40$ pF. Calculate: (a) the center frequency, (b) the half-power frequencies, and (c) the quality factor.

16.44 Find the bandwidth and center frequency of the bandstop filter of Fig. 16.53.

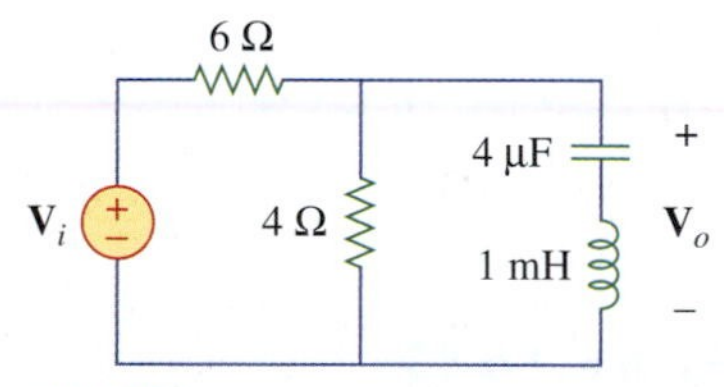

Figure 16.53
For Problem 16.44.

16.45 Find the resonant frequency of the filter shown in Fig. 16.54. Is it a bandpass or bandstop filter?

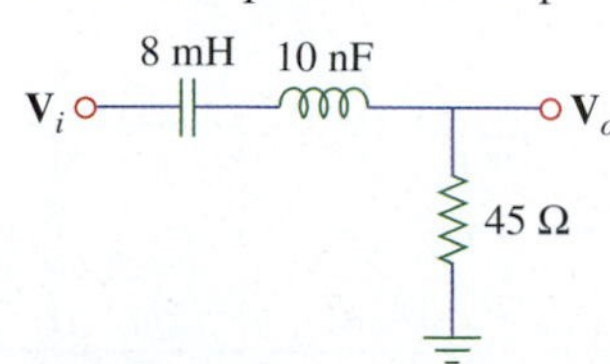

Figure 16.54
For Problem 16.45.

16.46 A lowpass filter is constructed by a 1.8-kΩ resistor placed in series with a capacitor. What is the capacitance reactance if $V_o/V_i = 0.2$?

16.47 Calculate the bandwidth of the bandpass filter shown in Fig. 16.55.

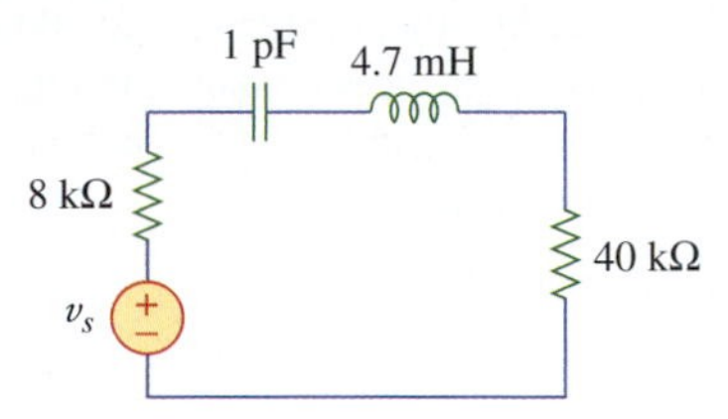

Figure 16.55
For Problem 16.47.

16.48 Construct an *RC* lowpass filter that will have a cutoff frequency at 500 Hz. Let $C = 0.45\ \mu$F. Confirm your design using computer simulation.

16.49 A lowpass filter has an output voltage of 800 μV with an input of 20 mV. Determine the gain of the filter in dB.

Section 16.6 Computer Analysis

16.50 Obtain the frequency response of the circuit in Fig. 16.56 using PSpice.

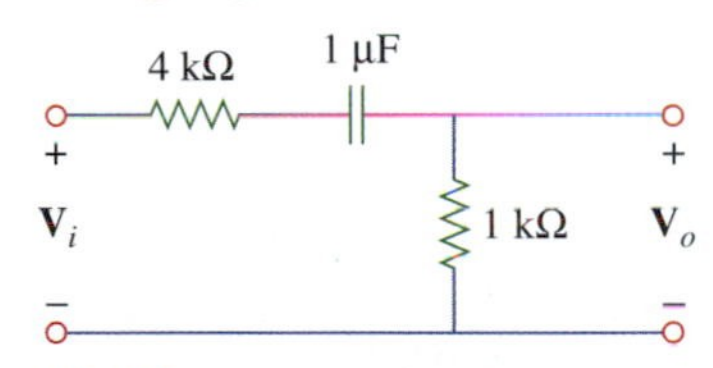

Figure 16.56
For Problems 16.50 and 16.56.

16.51 Use PSpice to obtain the magnitude and phase plots of $\mathbf{V}_o/\mathbf{I}_i$ of the circuit in Fig. 16.57.

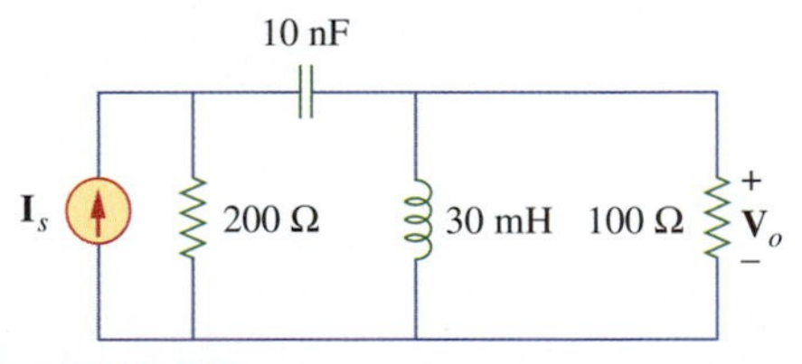

Figure 16.57
For Problems 16.51 and 16.57.

16.52 In the interval $0.1 < f < 100$ Hz, plot the response of the network in Fig. 16.58. Classify this filter and obtain ω_o.

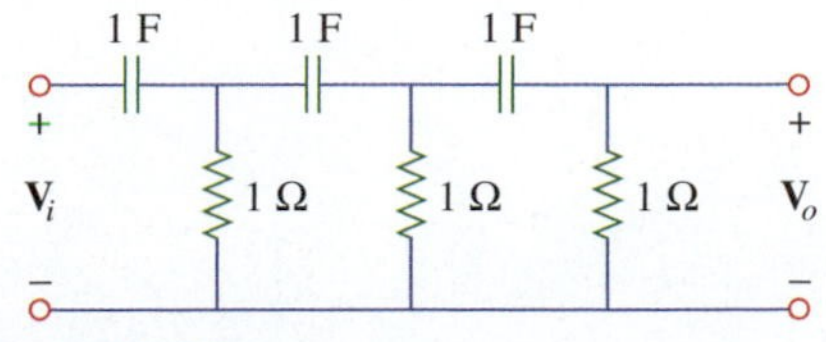

Figure 16.58
For Problem 16.52.

16.53 Use PSpice to generate the magnitude and phase Bode plots of $\mathbf{V}_o$ in the circuit of Fig. 16.59.

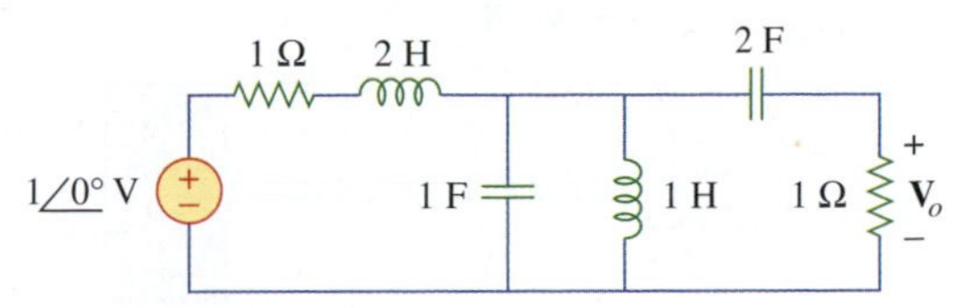

Figure 16.59
For Problem 16.53.

16.54 Using PSpice, plot the magnitude of the frequency response of the circuit in Fig. 16.60.

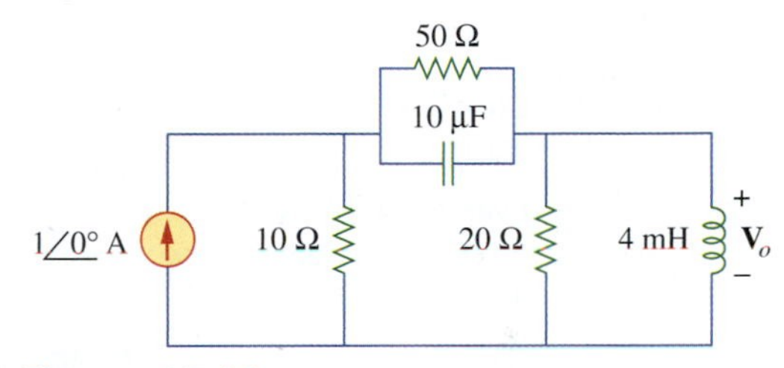

Figure 16.60
For Problem 16.54.

16.55 Using PSpice, plot the magnitude of the frequency response of the circuit in Fig. 16.61.

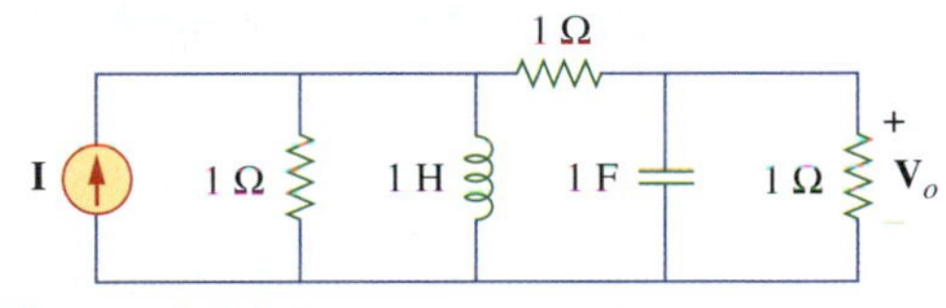

Figure 16.61
For Problem 16.55.

16.56 Use Multisim to construct the magnitude Bode plot of the circuit in Fig. 16.56.

16.57 Rework Problem 16.51 using Multisim.

16.58 Obtain the Bode plots for the circuit in Fig. 16.62 using Multisim.

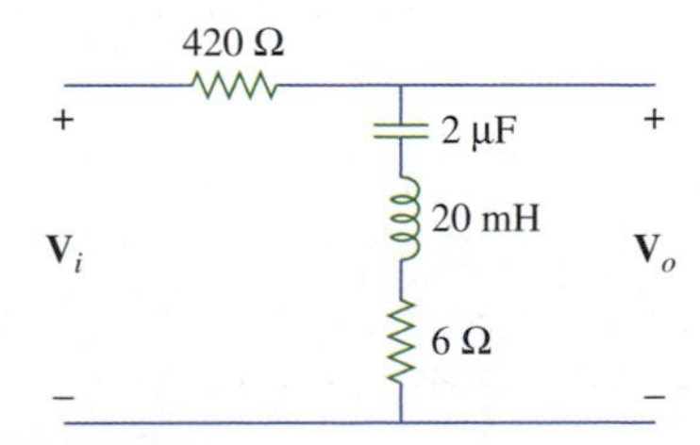

Figure 16.62
For Problem 16.58.

Section 16.7 Applications

16.59 For an emergency situation, an engineer needs to make an *RC* highpass filter. He has one 10-pF capacitor, one 30-pF capacitor, one 1.8-kΩ resistor, and one 3.3-kΩ resistor available. Find the greatest cutoff frequency possible using these elements.

16.60* The crossover circuit in Fig. 16.63 is a lowpass filter that is connected to a woofer. Find the transfer function $\mathbf{H}(\omega) = \mathbf{V}_o/\mathbf{V}_i$.

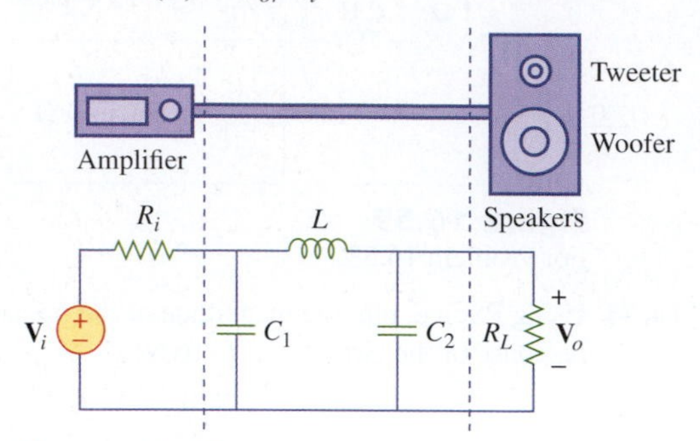

Figure 16.63
For Problem 16.60.

16.61* The crossover circuit in Fig. 16.64 is a highpass filter that is connected to a tweeter. Determine the transfer function $\mathbf{H}(\omega) = \mathbf{V}_o/\mathbf{V}_i$.

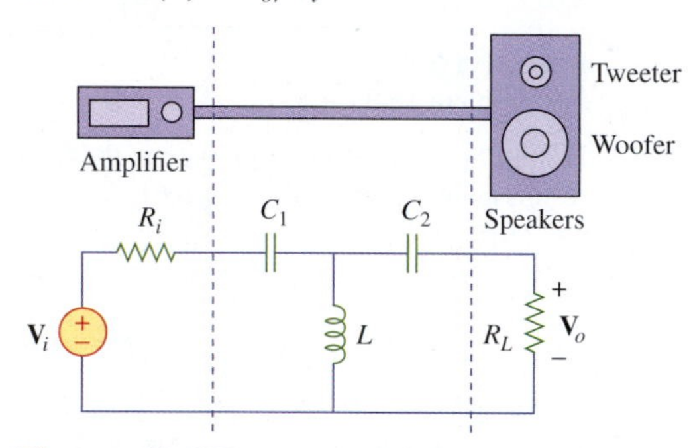

Figure 16.64
For Problem 16.61.

16.62 In a certain application, a simple *RC* lowpass filter is designed to reduce high-frequency noise. If the desired corner frequency is 20 kHz and $C = 0.5\ \mu$F, find the value of *R*.

16.63 In an amplifier circuit, a simple *RC* highpass filter is needed to block the dc component while passing the time-varying component. If the desired rolloff frequency is 15 Hz and $C = 10\ \mu$F, find the value of *R*.

16.64 Practical *RC* filter design should allow for source and load resistances as shown in Fig. 16.65. Let $R = 4$ kΩ and $C = 40$ nF. Obtain the cutoff frequency when:

(a) $R_s = 0$, $R_L = \infty$

(b) $R_s = 1$ kΩ, $R_L = 5$ kΩ

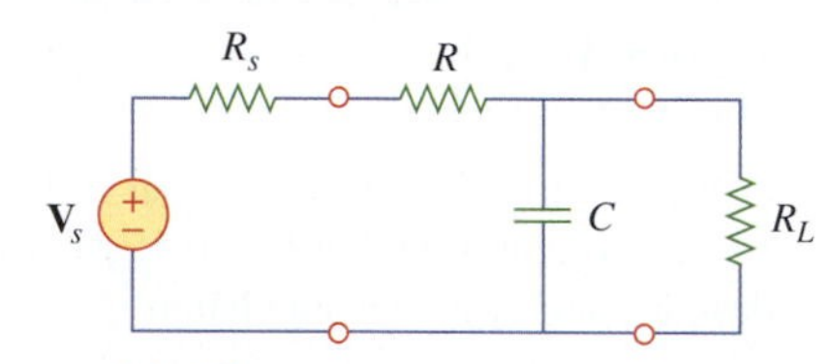

Figure 16.65
For Problem 16.64.

16.65 A low-quality factor, double-tuned bandpass filter is shown in Fig. 16.66. Use PSpice to generate the magnitude plot of $\mathbf{V}_o(\omega)$.

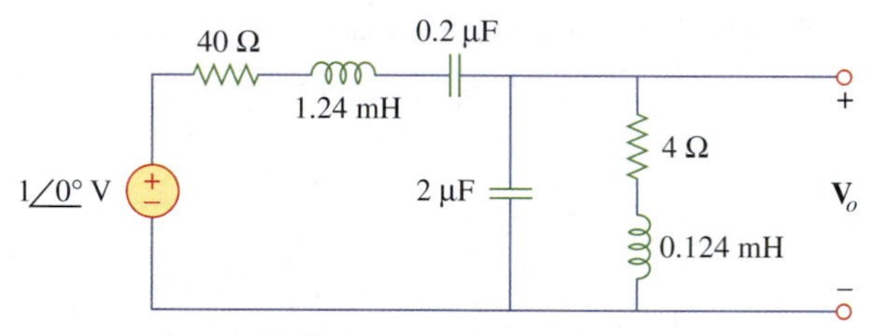

Figure 16.66
For Problem 16.65.

16.66 A satellite transmitter has an output power of 2 kW and an input power of 1 W. Determine the power gain in dB.

chapter

Three-Phase Circuits

17

Society is never prepared to receive any invention. Every new thing is resisted, and it takes years for the inventor to get people to listen to him and years more before it can be introduced.

— Thomas Alva Edison

Historical Profiles

Thomas Alva Edison (1847–1931) was perhaps the greatest American inventor. He patented 1,093 inventions, including such history-making inventions as the incandescent electric bulb, the phonograph, and the first commercial motion picture.

Born in Milan, Ohio, the youngest of seven children, Edison received only three months of formal education because he hated school. He was home-schooled by his mother and soon began to read on his own. In 1868, Edison read one of Faraday's books and found his calling. He moved to Menlo Park, New Jersey, in 1876, where he managed a well-staffed research laboratory. Most of his inventions came out of this laboratory. His laboratory served as a model for modern research organizations.

Edison established manufacturing companies for making the devices he invented and designed the first electric power station to supply electric light. Formal electrical engineering education began in the mid-1880s with Edison as a role model and leader.

Thomas Alva Edison
Library of Congress

Nikola Tesla (1856–1943) was a Croatian American engineer whose inventions, such as the induction motor and the first polyphase ac power system, greatly influenced the ac versus dc debate in favor of ac. He was also responsible for adoption of 60 Hz as the standard for ac power systems in the United States.

Born in Austria-Hungary (now Croatia) to a clergyman, Tesla had an incredible memory and a keen affinity for mathematics. He moved to the United States in 1884 and first worked for Thomas Edison. At that time, the country was in the "battle of the currents" with George Westinghouse (1846–1914) promoting ac and Thomas Edison rigidly leading the dc forces. Tesla left Edison and joined Westinghouse because of his interest in ac. Through Westinghouse, Tesla gained a reputation and acceptance of his polyphase ac generation, transmission, and distribution system. He held 700 patents in his lifetime. His inventions include ac induction motors, the tesla coil (a high-voltage transformer), and a wireless transmission system. The unit of magnetic flux density, the tesla, was named after him.

Nikola Tesla
Courtesy Smithsonian Institution

17.1 Introduction

So far in this book, we have dealt with single-phase circuits. A single-phase ac power system consists of a generator connected through a pair of wires (transmission line) to a load. A single-phase two-wire system is depicted in Fig. 17.1(a), where V_p (peak, not rms) is the magnitude of the source voltage and ϕ is the phase. What is more common in practice is a single-phase three-wire system, shown in Fig. 17.1(b). It contains two identical sources (equal magnitude and the same phase) that are connected to two loads by two outer wires and a neutral line. For example, the normal household system is a single-phase three-wire system in which the terminal voltages have the same magnitude and the same phase.[1]

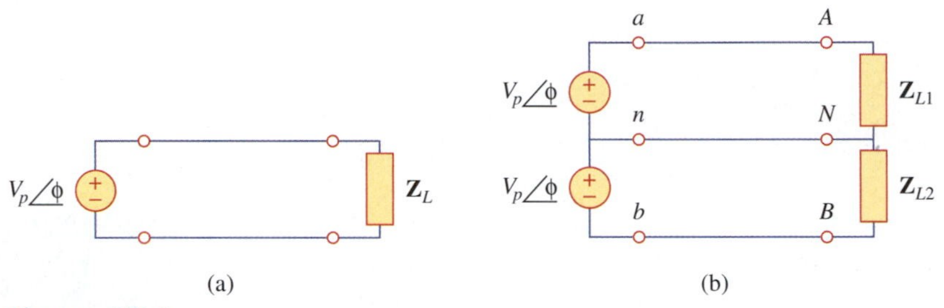

Figure 17.1
Single-phase systems: (a) two-wire type; (b) three-wire type.

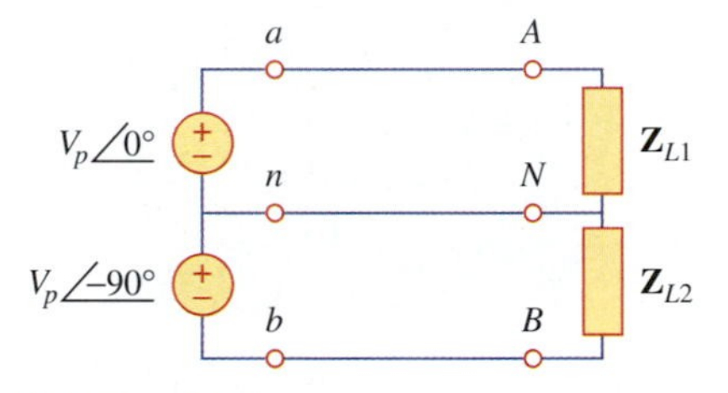

Figure 17.2
Two-phase three-wire system.

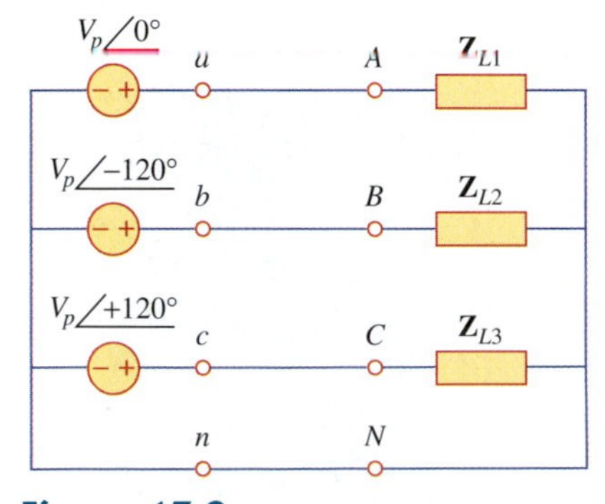

Figure 17.3
Three-phase four-wire system.

Circuits or systems in which the ac sources operate at the same frequency but different phases are known as *polyphase*. A two-phase three-wire system is shown in Fig. 17.2, while a three-phase four-wire system is shown in Fig. 17.3. As distinct from a single-phase system, a two-phase system is produced by a generator consisting of two coils placed perpendicular to each other so that the voltage generated by one lags the other by 90°. By the same token, a three-phase system is produced by a generator consisting of three sources having the same amplitude and frequency but out of phase with each other by 120°. Because the three-phase system is by far the most prevalent and most economical polyphase system, discussion in this chapter centers mainly on three-phase systems.

Three-phase systems are important for at least three reasons. First, nearly all electric power is generated and distributed in a three-phase system at the operating frequency of 60 Hz (or $\omega = 377$ rad/s) in the United States or 50 Hz (or $\omega = 314$ rad/s) in some other parts of the world. (See Table 14.1.) When one-phase or two-phase inputs are required, they are taken from the three-phase system rather than generated independently. Even when more than three phases are needed, such as in the aluminum industry where 48 phases are required for melting purposes, they can be provided by manipulating the three phases supplied. Second, the instantaneous power in a three-phase system can be constant (not pulsating), as will be shown in Section 17.8. This results in uniform power transmission and less vibration of

[1] Historical Note: Thomas Edison invented the *three-wire system,* using three wires instead of four.

three-phase systems. Third, for the same amount of power the three-phase system is more economical than the single-phase. The amount of wire required for a three-phase system is substantially less than that required for an equivalent single-phase system.

We begin with a discussion of balanced three-phase voltages. Then we analyze each of the four possible configurations of balanced three-phase systems. The analysis of unbalanced three-phase systems is also discussed. We learn how to use PSpice to analyze a balanced or unbalanced three-phase system. Finally, we apply the concepts developed in this chapter to three-phase power measurement and residential electrical wiring.

17.2 Three-Phase Generator

Three-phase voltages are often produced with a three-phase ac generator (often called an *alternator*) whose cross-sectional view is shown in Fig. 17.4. The generator in essence consists of a rotating magnet (called the *rotor*) surrounded by a stationary winding (called the *stator*). Three separate windings or coils with terminals a-a', b-b', and c-c' are physically placed 120° apart around the stator. Terminals a and a', for example, stand for one of the ends of coils going into and the other end coming out of the page. As the rotor rotates, its magnetic field "cuts" the three coils and induces voltages in the coils. Because the coils are placed 120° apart, the induced voltages in the coils are equal in magnitude but out of phase by 120° as illustrated in Fig. 17.5. Because each coil can be regarded as a single-phase generator by itself, the three-phase generator can supply power to both single-phase and three-phase loads. See Fig. 17.6 for a typical three-phase electric power transmission configuration.

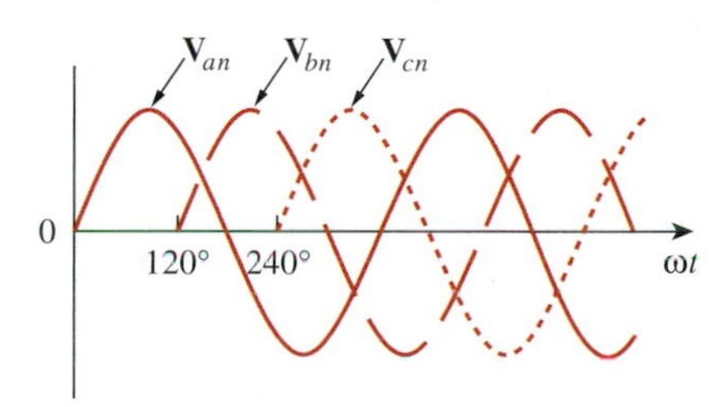

Figure 17.5
The generated voltages are 120° apart from each other.

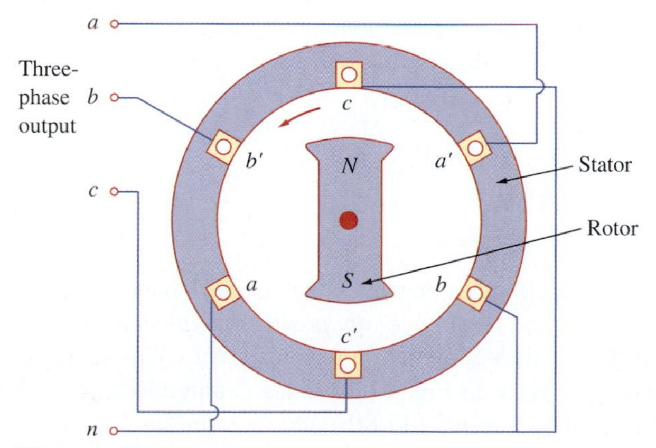

Figure 17.4
A three-phase generator.

Figure 17.6
Three-phase electric power transmission.
© Sarhan M. Musa

A typical three-phase system consists of three voltage sources connected to loads by three or four wires (or transmission lines). (Three-phase current sources are very scarce.) A three-phase system is equivalent to three single-phase circuits. The voltage sources can be either wye-connected as shown in Fig. 17.7(a) or delta-connected as in Fig. 17.7(b).

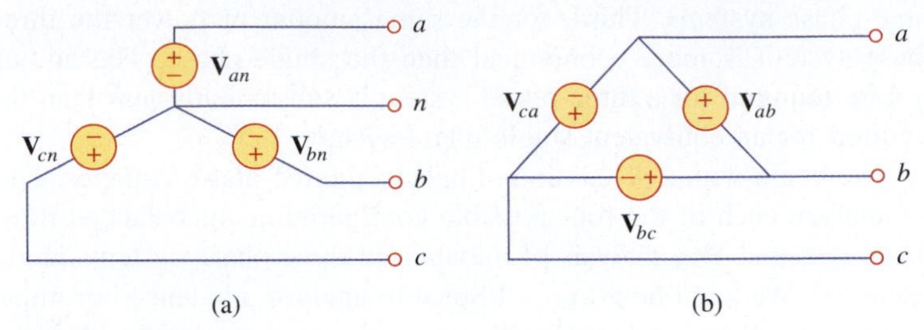

Figure 17.7
Three-phase voltage sources: (a) *Y*-connected source; (b) Δ-connected source.

17.3 Balanced Three-Phase Voltages

For now, let us consider the wye-connected voltages in Fig. 17.7(a). The voltages $\mathbf{V}_{an}$, $\mathbf{V}_{bn}$, and $\mathbf{V}_{cn}$ are between lines *a*, *b*, and *c*, respectively, and the neutral line *n*. These voltages are called *phase voltages*. If the voltage sources have the same amplitude and frequency ω and are out of phase with each other by 120°, the voltages are said to be *balanced*. This implies that:

$$\mathbf{V}_{an} + \mathbf{V}_{bn} + \mathbf{V}_{cn} = 0 \tag{17.1}$$

$$|\mathbf{V}_{an}| = |\mathbf{V}_{bn}| = |\mathbf{V}_{cn}| \tag{17.2}$$

Thus,

> **Balanced phase voltages** are equal in magnitude and are out of phase with each other by 120°.

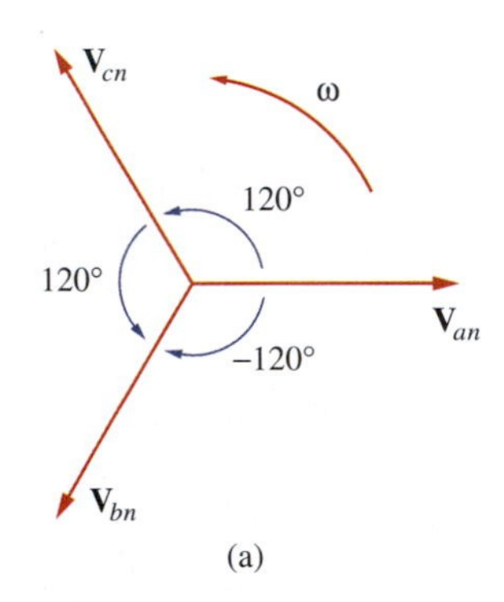

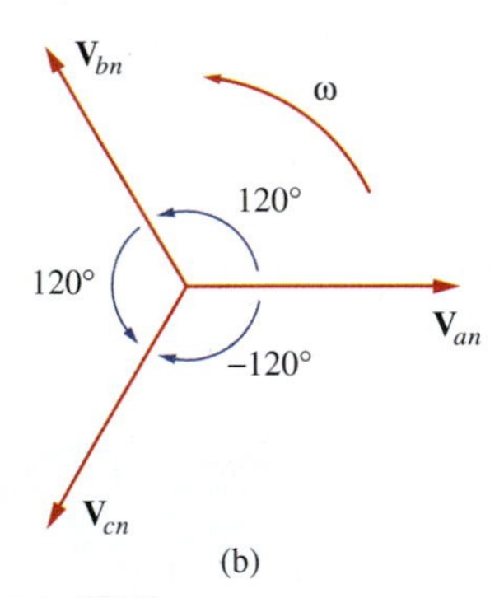

Figure 17.8
Phase sequences: (a) *abc* or positive sequence; (b) *acb* or negative sequence.

Because the three-phase voltages are 120° out of phase with each other, there are two possible combinations. One possibility is shown in Fig. 17.8(a) and expressed mathematically as

$$\begin{aligned}\mathbf{V}_{an} &= V_p\underline{/0^\circ}\\ \mathbf{V}_{bn} &= V_p\underline{/-120^\circ}\\ \mathbf{V}_{cn} &= V_p\underline{/-240^\circ} = V_p\underline{/+120^\circ}\end{aligned} \tag{17.3}$$

where V_p is the effective or rms value[2] of the phase voltages. This is known as the *abc sequence* or *positive sequence*. In this phase sequence, $\mathbf{V}_{an}$ leads $\mathbf{V}_{bn}$, which in turn leads $\mathbf{V}_{cn}$. This sequence is produced when the rotor in Fig. 17.4 rotates counterclockwise. The other possibility is shown in Fig. 17.8(b) and is given by

$$\begin{aligned}\mathbf{V}_{an} &= V_p\underline{/0^\circ}\\ \mathbf{V}_{cn} &= V_p\underline{/-120^\circ}\\ \mathbf{V}_{bn} &= V_p\underline{/-240^\circ} = V_p\underline{/+120^\circ}\end{aligned} \tag{17.4}$$

[2] Note: As a common tradition in power systems, voltage and current in this chapter are in rms values unless otherwise stated.

This is called the *acb sequence* or *negative sequence*. For this phase sequence, $\mathbf{V}_{an}$ leads $\mathbf{V}_{cn}$, which in turn leads $\mathbf{V}_{bn}$. The *acb* sequence is produced when the rotor in Fig. 17.4 rotates in the clockwise direction. It is easy to show that the voltages in Eqs. (17.3) or (17.4) satisfy Eqs. (17.1) and (17.2). For example, from Eq. (17.3),

$$\begin{aligned}\mathbf{V}_{an} + \mathbf{V}_{bn} + \mathbf{V}_{cn} &= V_p\underline{/0^\circ} + V_p\underline{/-120^\circ} + V_p\underline{/+120^\circ} \\ &= V_p\,(1.0 - 0.5 - j0.866 - 0.5 + j0.866) \\ &= 0\end{aligned} \tag{17.5}$$

The **phase sequence** is the time order in which the voltages pass through their respective maximum values.

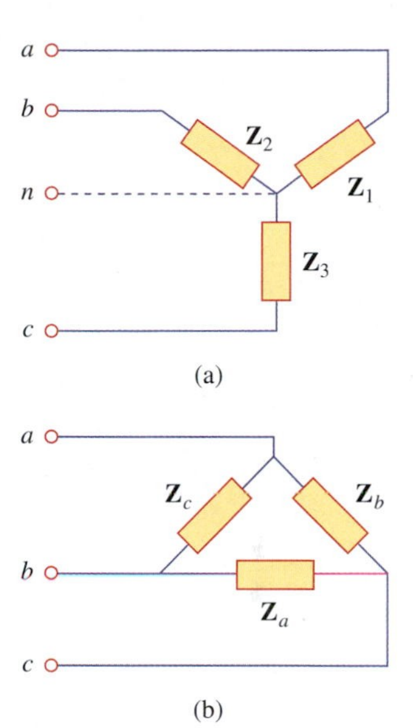

Figure 17.9
Two possible three-phase load configurations: (a) a *Y*-connected load; (b) a Δ-connected load.

The phase sequence is determined by the order in which the phasors pass through a fixed point in the phase diagram.[3]

In Fig. 17.8(a), as the phasors rotate in the counterclockwise direction with frequency ω, they pass through the horizontal axis in a sequence *abcabca* ... Thus, the sequence is *abc* or *bca* or *cab*. Similarly, for the phasors in Fig. 17.8(b), as they rotate in the counterclockwise direction, they pass the horizontal axis in a sequence *acbacba* ... This describes the *acb* sequence. The phase sequence is important in three-phase power distribution because it determines the direction of the rotation of a motor connected to the power source, for example.[4]

Like the generator connections, a three-phase load can be either wye-connected or delta-connected, depending on the end application. A wye-connected load is shown in Fig. 17.9(a), while a delta-connected load is shown in Fig. 17.9(b). The neutral line in Fig. 17.9(a) may or may not be there depending on whether the system is four- or three-wire. (And, of course, a neutral connection is topologically impossible for a delta connection.) A wye- or delta-connected load is said to be *unbalanced* if the phase impedances are not equal in magnitude or phase.

A **balanced load** is one in which the phase impedances are equal in magnitude and in phase.

For a balanced *wye*-connected load,[5]

$$\mathbf{Z}_1 = \mathbf{Z}_2 = \mathbf{Z}_3 = \mathbf{Z}_Y \tag{17.6}$$

where $\mathbf{Z}_Y$ is the load impedance per phase. For a balanced delta-connected load,

$$\mathbf{Z}_a = \mathbf{Z}_b = \mathbf{Z}_c = \mathbf{Z}_\Delta \tag{17.7}$$

where $\mathbf{Z}_\Delta$ is the load impedance per phase in this case. We recall from Eq. (11.70) that

$$\mathbf{Z}_\Delta = 3\mathbf{Z}_Y \qquad \text{or} \qquad \mathbf{Z}_Y = \frac{1}{3}\mathbf{Z}_\Delta \tag{17.8}$$

so we see that a wye-connected load can be transformed into a delta-connected load or vice versa using Eq. (17.8).

---

[3] Note: The phase sequence may also be regarded as the order in which the phase voltages reach their peak (or maximum) values with respect to time.

[4] Reminder: As time increases, each phasor (or sinor) rotates at an angular velocity ω.

[5] Reminder: A wye-connected load consists of three impedances connected to a neutral node, while a delta-connected load consists of three impedances connected around a loop. In both cases, the load is balanced when three impedances are equal.

Because either the three-phase source or the three-phase load can be either wye- or delta-connected, we have four possible source/load connections:

- Wye-wye connection—that is, wye-connected source with a wye-connected load.
- Wye-delta connection.
- Delta-delta connection.
- Delta-wye connection.

In subsequent sections, we will consider each of these configurations.

Example 17.1

Determine the phase sequence of the set of voltages

$$v_{an}(t) = 200\cos(\omega t + 10^\circ)$$
$$v_{bn}(t) = 200\cos(\omega t - 230^\circ)$$
$$v_{cn}(t) = 200\cos(\omega t - 110^\circ)$$

Solution:

The voltages can be expressed in phasor form as

$$\mathbf{V}_{an} = 200\underline{/10^\circ}, \quad \mathbf{V}_{bn} = 200\underline{/-230^\circ}, \quad \mathbf{V}_{cn} = 200\underline{/-110^\circ}$$

We notice that $\mathbf{V}_{an}$ leads $\mathbf{V}_{cn}$ by 120° and $\mathbf{V}_{cn}$ in turn leads $\mathbf{V}_{bn}$ by 120°. Hence, we have an *acb* sequence.

Practice Problem 17.1

Given that $\mathbf{V}_{bn} = 110\underline{/30^\circ}$, find $\mathbf{V}_{an}$ and $\mathbf{V}_{cn}$ assuming positive (*abc*) sequence.

Answer: $110\underline{/150^\circ}$; $110\underline{/-90^\circ}$

17.4 Balanced Wye-Wye Connection

We begin with the *Y-Y* system because any balanced three-phase system can be reduced to an equivalent *Y-Y* system, which is the most easily analyzed. For this reason, analysis of this system can be regarded as the key to solving all balanced three-phase systems.

> A **balanced *Y-Y* system** is a three-phase system with a balanced *Y*-connected source and a balanced *Y*-connected load.

Wye-wye systems are mostly used by utility companies. There are two reasons for this. First, the wye-wye connection provides a convenient point of grounding the neutral at source, regardless of which direction the power is flowing. Second, the wye-wye connection costs a little less than a delta-wye or delta-delta connection.

Consider the balanced four-wire *Y-Y* system shown in Fig. 17.10, where a *Y*-connected load is connected to a *Y*-connected source. We assume a balanced load so that load impedances are equal. Although the impedance $\mathbf{Z}_Y$ is the load impedance per phase, it may also be regarded as the sum of the source impedance $\mathbf{Z}_s$, line impedance $\mathbf{Z}_\ell$, and the load impedance $\mathbf{Z}_L$ for each phase because these impedances are in series. As illustrated in Fig. 17.10, $\mathbf{Z}_s$ denotes the internal impedance of the phase winding of the generator; $\mathbf{Z}_\ell$ is the impedance of the

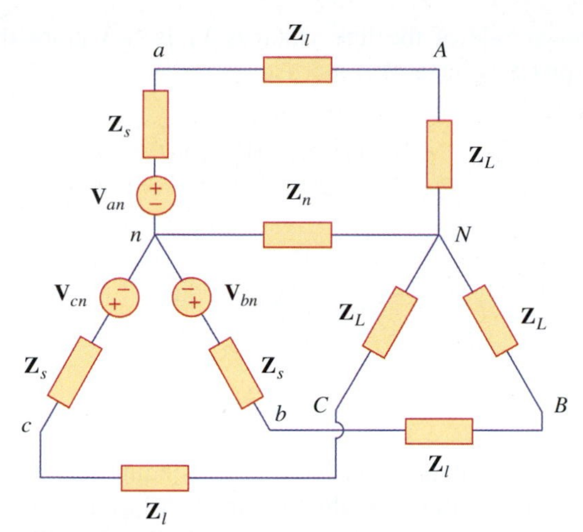

Figure 17.10
A balanced, four-wire *Y-Y* system, showing the source, line, and load impedances.

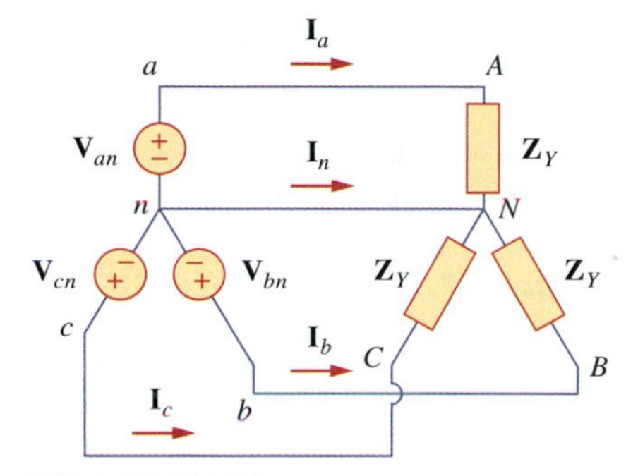

Figure 17.11
A balanced *Y-Y* connection.

line joining a phase of the source with a phase of the load; $\mathbf{Z}_L$ represents the impedance of each phase of the load; and $\mathbf{Z}_n$ is the impedance of the neutral line. Thus, in general

$$\mathbf{Z}_Y = \mathbf{Z}_s + \mathbf{Z}_\ell + \mathbf{Z}_L \tag{17.9}$$

Note that $\mathbf{Z}_s$, $\mathbf{Z}_\ell$, and $\mathbf{Z}_n$ are often very small compared with $\mathbf{Z}_L$ so that we can assume that $\mathbf{Z}_Y = \mathbf{Z}_L$ when no source or line impedance is given. In any event, by lumping the impedances together, the *Y-Y* system in Fig. 17.10 can be simplified to that shown in Fig. 17.11.

Assuming the positive sequence, the *phase* voltages (or line-to-neutral voltages)[6] are

$$\begin{aligned} \mathbf{V}_{an} &= V_p\underline{/0^\circ} \\ \mathbf{V}_{bn} &= V_p\underline{/-120^\circ} \\ \mathbf{V}_{cn} &= V_p\underline{/+120^\circ} \end{aligned} \tag{17.10}$$

The *line-to-line* voltages (called simply *line* voltages) $\mathbf{V}_{ab}$, $\mathbf{V}_{bc}$, and $\mathbf{V}_{ca}$ are related to the phase voltages.[7] For example,

$$\begin{aligned} \mathbf{V}_{ab} &= \mathbf{V}_{an} + \mathbf{V}_{nb} = \mathbf{V}_{an} - \mathbf{V}_{bn} = V_p\underline{/0^\circ} - V_p\underline{/-120^\circ} \\ &= V_p\left(1 + \frac{1}{2} + j\frac{\sqrt{3}}{2}\right) = \sqrt{3}V_p\underline{/30^\circ} \end{aligned} \tag{17.11a}$$

Similarly, we can obtain

$$\mathbf{V}_{bc} = \mathbf{V}_{bn} - \mathbf{V}_{cn} = \sqrt{3}V_p\underline{/-90^\circ} \tag{17.11b}$$

$$\mathbf{V}_{ca} = \mathbf{V}_{cn} - \mathbf{V}_{an} = \sqrt{3}V_p\underline{/-210^\circ} \quad \text{or} \quad 150^\circ \tag{17.11c}$$

---

[6] We use the term *phase* to refer to each branch, whether a source or a load, for both the wye and delta connections.

[7] Line voltages are voltages between lines, while phase voltages are voltages across the phase impedances. The phase voltages are referenced to the neutral point, while a line voltage is referenced to another line.

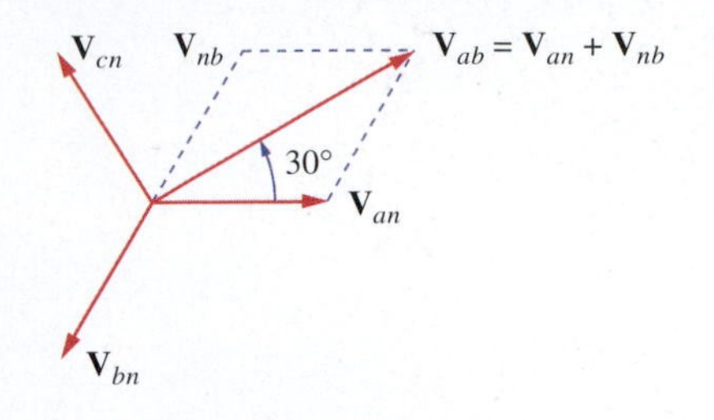

(a)

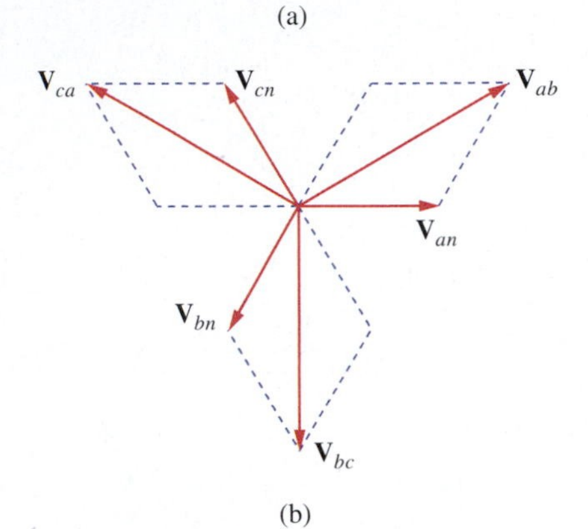

(b)

Figure 17.12
Phasor diagrams illustrating the relationship between line voltages and phase voltages.

Thus, the magnitude of the line voltages V_L is $\sqrt{3}$ times the magnitude of the phase voltages V_p; that is,

$$V_L = \sqrt{3}V_p \tag{17.12}$$

where

$$V_p = |\mathbf{V}_{an}| = |\mathbf{V}_{bn}| = |\mathbf{V}_{cn}| \tag{17.13}$$

and

$$V_L = |\mathbf{V}_{ab}| = |\mathbf{V}_{bc}| = |\mathbf{V}_{ca}| \tag{17.14}$$

Also the line voltages lead their corresponding phase voltages by 30°. Figure 17.12(a) illustrates this. Figure 17.12(a) also shows how to determine $\mathbf{V}_{ab}$ from the phase voltages, while Fig. 17.12(b) shows the same for all three line voltages. Notice that $\mathbf{V}_{ab}$ leads $\mathbf{V}_{bc}$ by 120 and $\mathbf{V}_{bc}$ leads $\mathbf{V}_{ca}$ by 120° so that line voltages sum up to zero, as do the phase voltages.

Applying KVL to each phase in Fig. 17.11, we obtain the line currents as

$$\mathbf{I}_a = \mathbf{V}_{an}/\mathbf{Z}_Y = \frac{V_p \angle 0^\circ}{Z_Y}$$

$$\mathbf{I}_b = \mathbf{V}_{bn}/\mathbf{Z}_Y = \frac{V_{an} \angle -120^\circ}{Z_Y} = \mathbf{I}_a \angle -120^\circ \tag{17.15}$$

$$\mathbf{I}_c = \mathbf{V}_{cn}/\mathbf{Z}_Y = \frac{V_{an} \angle -240^\circ}{Z_Y} = \mathbf{I}_a \angle -240^\circ \quad \text{or} \quad \mathbf{I}_a < +120^\circ$$

We can readily infer that the line currents add up to zero; that is,

$$\mathbf{I}_a + \mathbf{I}_b + \mathbf{I}_c = 0 \tag{17.16}$$

so that

$$\mathbf{I}_n = -(\mathbf{I}_a + \mathbf{I}_b + \mathbf{I}_c) = 0 \tag{17.17a}$$

or

$$\mathbf{V}_{nN} = \mathbf{Z}_n \mathbf{I}_n = 0 \tag{17.17b}$$

that is, the voltage across the neutral wire is zero. The neutral line can thus be removed from the configuration without affecting the system. In fact, in long-distance power transmission, conductors in multiples of three are used with the earth itself acting as the neutral conductor. Power systems designed in this way are well grounded at all critical points to ensure safety.

While the *line* current is the current in each line, the *phase* current is the current in each phase of the source or load. In the *Y*-*Y* system, the line current is the same as the phase current. We will use single subscripts for line currents because it is natural and conventional to assume that line currents flow from the source to the load.

An alternative way of analyzing a balanced *Y*-*Y* system is to do so on a "per-phase" basis. We look at one phase—say, phase *a*—and analyze the single-phase equivalent circuit in Fig. 17.13. The single-phase analysis yields the line current $\mathbf{I}_a$ as

$$\mathbf{I}_a = \mathbf{V}_{an}/\mathbf{Z}_Y \quad (17.18)$$

From $\mathbf{I}_a$, we use the phase sequence to obtain other line currents. Thus, as long as the system is balanced, we need only analyze one phase. We may do this even if the neutral line is absent, that is, for the three-wire system.

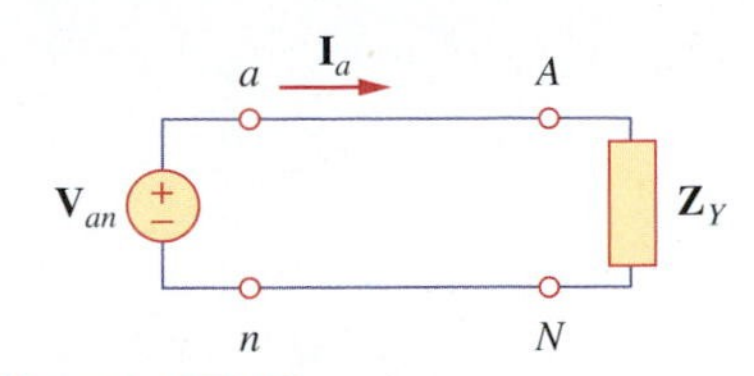

Figure 17.13
A single-phase equivalent circuit.

Example 17.2

Calculate the line currents in the three-wire Y-Y system of Fig. 17.14.

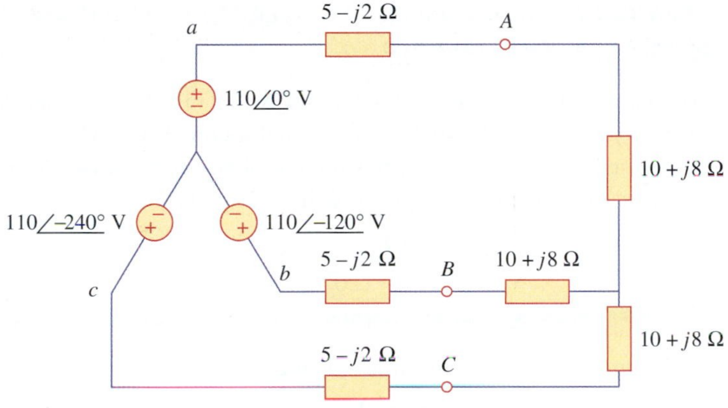

Figure 17.14
Three-wire Y-Y system; for Examples 17.2 and 17.6.

Solution:
The three-phase circuit in Fig. 17.14 is balanced; therefore we may replace it with its single-phase equivalent circuit such as in Fig. 17.13. We obtain $\mathbf{I}_a$ from the single-phase analysis as

$$\mathbf{I}_a = \mathbf{V}_{an}/\mathbf{Z}_Y$$

where $\mathbf{Z}_Y = (5 - j2) + (10 + j8) = 15 + j6 = 16.155\underline{/21.8^\circ}\ \Omega$. Hence,

$$\mathbf{I}_a = \frac{110\underline{/0^\circ}}{16.155\underline{/21.8^\circ}} = 6.81\underline{/-21.8^\circ}\ \text{A}$$

Because the source voltages in Fig. 17.14 are in positive sequence, the line currents are also in positive sequence,

$$\mathbf{I}_b = \mathbf{I}_a\underline{/-120^\circ} = 6.81\underline{/-141.8^\circ}\ \text{A}$$
$$\mathbf{I}_c = \mathbf{I}_a\underline{/-240^\circ} = 6.81\underline{/-261.8^\circ}\ \text{A} = 6.81\underline{/98.2^\circ}\ \text{A}$$

Practice Problem 17.2

A Y-connected balanced three-phase generator with an impedance of $0.4 + j0.3\ \Omega$ per phase is connected to a Y-connected balanced load with an impedance of $24 + j19\ \Omega$ per phase. The line joining the generator and the load has an impedance of $0.6 + j0.7\ \Omega$ per phase.

Assuming positive sequence for the source voltages and assuming $\mathbf{V}_{an} = 120\underline{/30^\circ}$ V, find: (a) the line voltages and (b) the line currents.

Answer: (a) $207.85\underline{/60^\circ}$ V, $207.85\underline{/-60^\circ}$ V, $207.85\underline{/-180^\circ}$ V; (b) $3.75\underline{/-8.66^\circ}$ A, $3.75\underline{/-128.66^\circ}$ A, $3.75\underline{/-248.66^\circ}$ A

17.5 Balanced Wye-Delta Connection

A typical application of the wye-delta connection by the utility companies is the construction of wye-delta transformer bank. (Transformers will be covered in the next chapter.) This connection has the ability to simultaneously serve single-phase and three-phase loads.

> A **balanced *Y*-Δ system** consists of a balanced *Y*-connected source feeding a balanced Δ-connected load.

The balanced wye-delta system[8] is shown in Fig. 17.15, where the source is wye-connected and the load is delta-connected. There is, of course, no neutral connection from source to load for this case. Assuming the positive sequence, the phase voltages again are

$$\begin{aligned} \mathbf{V}_{an} &= V_p\underline{/0^\circ} \\ \mathbf{V}_{bn} &= V_p\underline{/-120^\circ} \\ \mathbf{V}_{cn} &= V_p\underline{/+120^\circ} \end{aligned} \qquad \textbf{(17.19)}$$

As shown in Section 17.4, the line voltages are

$$\begin{aligned} \mathbf{V}_{ab} &= \sqrt{3}V_p\underline{/30^\circ} = \mathbf{V}_{AB} \\ \mathbf{V}_{bc} &= \sqrt{3}V_p\underline{/-90^\circ} = \mathbf{V}_{BC} \\ \mathbf{V}_{ca} &= \sqrt{3}V_p\underline{/-210^\circ} \quad \text{or} \quad \sqrt{3}V_p\underline{/+150^\circ} = \mathbf{V}_{CA} \end{aligned} \qquad \textbf{(17.20)}$$

showing that the line voltages are equal to the voltages across the load impedances for this system configuration. From these voltages, we can obtain the phase currents as

$$\mathbf{I}_{AB} = \mathbf{V}_{AB}/\mathbf{Z}_\Delta, \qquad \mathbf{I}_{BC} = \mathbf{V}_{BC}/\mathbf{Z}_\Delta, \qquad \mathbf{I}_{CA} = \mathbf{V}_{CA}/\mathbf{Z}_\Delta \qquad \textbf{(17.21)}$$

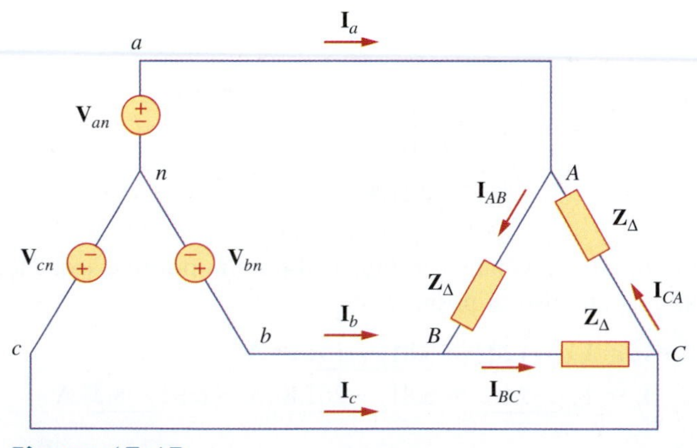

Figure 17.15
Balanced *Y*-Δ connection.

[8] Note: This is perhaps the most practical three-phase system because the three-phase sources are usually wye-connected, while the three-phase loads are usually delta-connected.

These currents have the same magnitude but are out of phase with each other by 120°.

Another way to get these phase currents is to apply KVL. For example, applying KVL around loop *aABbna* gives

$$-\mathbf{V}_{an} + \mathbf{Z}_\Delta \mathbf{I}_{AB} + \mathbf{V}_{bn} = 0$$

or

$$\mathbf{I}_{AB} = \frac{\mathbf{V}_{an} - \mathbf{V}_{bn}}{\mathbf{Z}_\Delta} = \frac{\mathbf{V}_{ab}}{\mathbf{Z}_\Delta} = \frac{\mathbf{V}_{AB}}{\mathbf{Z}_\Delta} \tag{17.22}$$

which is the same as that obtained earlier. This is the more general way of finding the phase currents.

The line currents are obtained from the phase currents by applying KCL at nodes A, B, and C. Thus,

$$\begin{aligned} \mathbf{I}_a &= \mathbf{I}_{AB} - \mathbf{I}_{CA} \\ \mathbf{I}_b &= \mathbf{I}_{BC} - \mathbf{I}_{AB} \\ \mathbf{I}_c &= \mathbf{I}_{CA} - \mathbf{I}_{BC} \end{aligned} \tag{17.23}$$

Because

$$\begin{aligned} \mathbf{I}_{CA} &= \mathbf{I}_{AB} \underline{/-240^\circ}, \\ \mathbf{I}_a &= \mathbf{I}_{AB} - \mathbf{I}_{CA} = \mathbf{I}_{AB}(1 - 1\underline{/-240^\circ}) \\ &= \mathbf{I}_{AB}(1 + 0.5 - j0.866) \\ &= \mathbf{I}_{AB}\sqrt{3}\underline{/-30^\circ} \text{ A} \end{aligned} \tag{17.24}$$

we can show that the magnitude I_L of the line current is $\sqrt{3}$ times the magnitude I_p of the phase current; that is,

$$\boxed{I_L = \sqrt{3} I_p} \tag{17.25}$$

where

$$I_L = |\mathbf{I}_a| = |\mathbf{I}_b| = |\mathbf{I}_c| \tag{17.26}$$

and

$$I_p = |\mathbf{I}_{AB}| = |\mathbf{I}_{BC}| = |\mathbf{I}_{CA}| \tag{17.27}$$

Also the line currents lag the corresponding phase currents by 30°, assuming the positive sequence. Figure 17.16 is a phasor diagram illustrating the relationship between the phase and line currents.

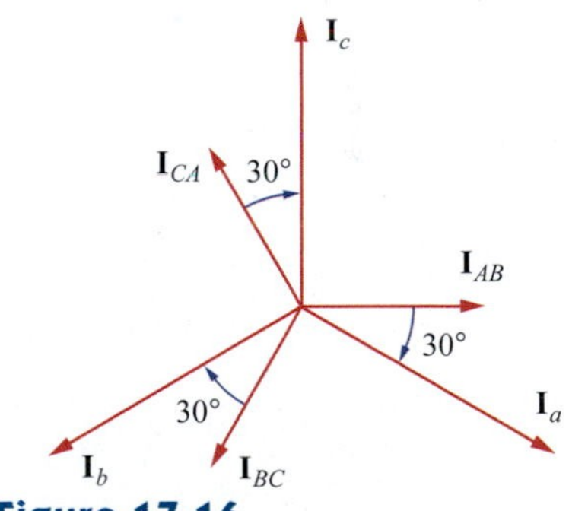

Figure 17.16
Phasor diagram illustrating the relationship between phase and line currents.

An alternative way of analyzing the wye-delta circuit is to transform the delta-connected load to an equivalent wye-connected load. Using the delta-wye transformation formula in Eq. (11.70), we get

$$\boxed{\mathbf{Z}_Y = \frac{1}{3}\mathbf{Z}_\Delta} \tag{17.28}$$

After this transformation, we now have a wye-wye system as in Fig. 17.11. The three-phase wye-delta system in Fig. 17.15 can be replaced by the single-phase equivalent circuit in Fig. 17.17. This allows us to calculate only the line currents. The phase currents are obtained using Eq. (17.25) and, utilizing that fact, the phase currents lead the corresponding line currents by 30°.

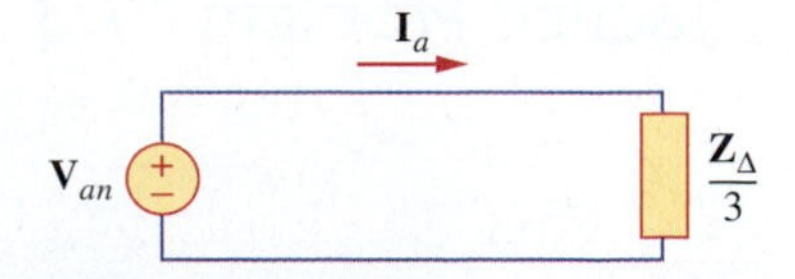

Figure 17.17
A single-phase equivalent circuit of a balanced Y-Δ circuit.

Example 17.3

A balanced *abc*-sequence wye-connected source with $\mathbf{V}_{an} = 100\underline{/10^\circ}$ V is connected to a delta-connected balanced load of $(8 + j4)\ \Omega$ per phase. Calculate the phase and line currents.

Solution:
This can be solved in two ways.

■ **METHOD 1** The load impedance is

$$\mathbf{Z}_\Delta = 8 + j4 = 8.944\underline{/26.57^\circ}\ \Omega$$

If the phase voltage $\mathbf{V}_{an} = 100\underline{/10^\circ}$, then the line voltage is

$$\mathbf{V}_{ab} = \mathbf{V}_{an}\sqrt{3}\underline{/30^\circ} = 100\sqrt{3}\underline{/(10^\circ + 30^\circ)} = \mathbf{V}_{AB}$$

or

$$\mathbf{V}_{AB} = 173.2\underline{/40^\circ}\text{ V}$$

The phase currents are

$$\mathbf{I}_{AB} = \mathbf{V}_{AB}/\mathbf{Z}_\Delta = \frac{173.2\underline{/40^\circ}}{8.944\underline{/26.57^\circ}} = 19.36\underline{/13.43^\circ}\text{ A}$$

$$\mathbf{I}_{BC} = \mathbf{I}_{AB}\underline{/-120^\circ} = 19.36\ \underline{/-106.57^\circ}\text{ A}$$

$$\mathbf{I}_{CA} = \mathbf{I}_{AB}\underline{/+120^\circ} = 19.36\ \underline{/\ 133.43^\circ}\text{ A}$$

The line currents are

$$\mathbf{I}_a = \mathbf{I}_{AB}\sqrt{3}\underline{/-30^\circ} = \sqrt{3}(19.36)\underline{/(13.43^\circ - 30^\circ)}$$
$$= 33.53\underline{/-17.57^\circ}\text{ A}$$

$$\mathbf{I}_b = \mathbf{I}_a\underline{/-120^\circ} = 33.53\underline{/-136.57^\circ}\text{ A}$$

$$\mathbf{I}_c = \mathbf{I}_a\underline{/+120^\circ} = 33.53\underline{/103.43^\circ}\text{ A}$$

■ **METHOD 2** Alternatively, using single-phase analysis

$$\mathbf{I}_a = \frac{\mathbf{V}_{an}}{\mathbf{Z}_\Delta/3} = \frac{100\underline{/10^\circ}}{2.981\underline{/26.57^\circ}} = 33.54\ \underline{/-16.57^\circ}\text{ A}$$

as demonstrated earlier. Other line currents are obtained using the *abc* phase sequence.

Practice Problem 17.3

One line voltage of a balanced wye-connected source is $\mathbf{V}_{ab} = 180\underline{/-20^\circ}$ V. If the source is connected to a delta-connected load of $20\underline{/40^\circ}\ \Omega$, find the phase and line currents. Assume the *abc* sequence.

Answer: $9\underline{/-60^\circ}$; $9\underline{/-180^\circ}$; $9\underline{/60^\circ}$; $15.59\underline{/-90^\circ}$; $15.59\underline{/-210^\circ}$; $15.59\underline{/-30^\circ}$

17.6 Balanced Delta-Delta Connection

The source as well as the load may be delta-connected, as shown in Fig. 17.18. Our goal is to obtain the phase and line currents, as usual. Assuming a positive sequence, the phase voltages for a delta-connected source are

$$\mathbf{V}_{ab} = V_p\angle 0^\circ$$
$$\mathbf{V}_{bc} = V_p\angle -120^\circ \qquad (17.29)$$
$$\mathbf{V}_{ca} = V_p\angle +120^\circ$$

A balanced Δ-Δ system is one in which both the balanced source and balanced load are Δ-connected.

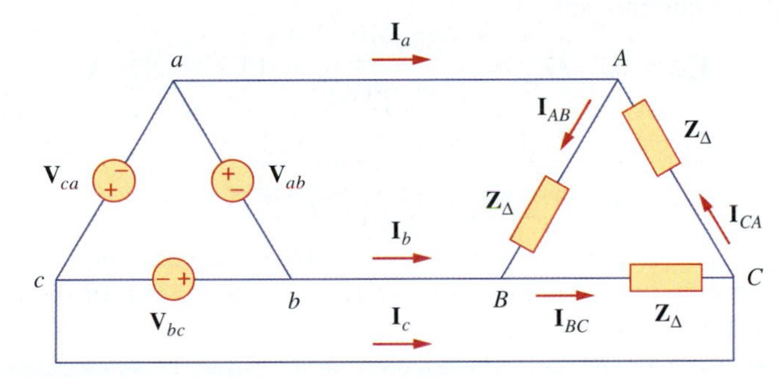

Figure 17.18
A balanced Δ-Δ connection.

The line voltages are the same as the phase voltages. From Fig. 17.18, assuming there is no line impedances, the phase voltages of the delta-connected source are equal to the voltages across the impedances; i.e.,

$$\mathbf{V}_{ab} = \mathbf{V}_{AB}, \qquad \mathbf{V}_{bc} = \mathbf{V}_{BC}, \qquad \mathbf{V}_{ca} = \mathbf{V}_{CA} \qquad (17.30)$$

Hence, the phase currents are

$$\mathbf{I}_{AB} = \mathbf{V}_{AB}/\mathbf{Z}_\Delta = \mathbf{V}_{ab}/\mathbf{Z}_\Delta$$
$$\mathbf{I}_{BC} = \mathbf{V}_{BC}/\mathbf{Z}_\Delta = \mathbf{V}_{bc}/\mathbf{Z}_\Delta \qquad (17.31)$$
$$\mathbf{I}_{CA} = \mathbf{V}_{CA}/\mathbf{Z}_\Delta = \mathbf{V}_{ca}/\mathbf{Z}_\Delta$$

Because the load is delta-connected, as in the previous section, some of the formulas derived there apply here. The line currents are obtained from the phase currents by applying KCL at nodes *A*, *B*, and *C*, as we did in the previous section:

$$\mathbf{I}_a = \mathbf{I}_{AB} - \mathbf{I}_{CA}$$
$$\mathbf{I}_b = \mathbf{I}_{BC} - \mathbf{I}_{AB} \qquad (17.32)$$
$$\mathbf{I}_c = \mathbf{I}_{CA} - \mathbf{I}_{BC}$$

Also, as shown in the last section, each line current lags the corresponding phase current by 30°; the magnitude I_L of the line current is $\sqrt{3}$ times the magnitude I_p of the phase current; that is,

$$I_L = \sqrt{3}I_p \qquad (17.33)$$

An alternative way of analyzing the Δ-Δ circuit is to convert both the source and the load to their *Y* equivalents. For the delta-connected load, we already know that $\mathbf{Z}_Y = \frac{1}{3}\mathbf{Z}_\Delta$. To convert a delta-connected source to a wye-connected source, see the next section.

The delta-delta connection is suitable for three-wire 240/120-V service. This connection is often served by three single-phase units. Generally, it should not be used unless the three-phase load is much larger than single-phase load.

Example 17.4

A balanced delta-connected load having an impedance of $20 - j15\ \Omega$ is connected to a delta-connected, positive-sequence generator having $\mathbf{V}_{ab} = 330\underline{/0^\circ}$ V. Calculate the phase currents of the load and the line currents.

Solution:

The load impedance per phase is

$$\mathbf{Z}_\Delta = 20 - j15 = 25\underline{/-36.87^\circ}\ \Omega$$

The phase currents are

$$\mathbf{I}_{AB} = \mathbf{V}_{AB}/\mathbf{Z}_\Delta = \frac{330\underline{/0^\circ}}{25\underline{/-36.87^\circ}} = 13.2\underline{/36.87^\circ}\ \text{A}$$

$$\mathbf{I}_{BC} = \mathbf{I}_{AB}\underline{/-120^\circ} = 13.2\underline{/-83.13^\circ}\ \text{A}$$

$$\mathbf{I}_{CA} = \mathbf{I}_{AB}\underline{/+120^\circ} = 13.2\underline{/156.87^\circ}\ \text{A}$$

For a delta load, the line current always lags the corresponding phase current by 30° and has a magnitude $\sqrt{3}$ times that of the phase current. Hence, the line currents are:

$$\mathbf{I}_a = \mathbf{I}_{AB}\sqrt{3}\underline{/-30^\circ} = (13.2\underline{/36.87^\circ})(\sqrt{3}\underline{/-30^\circ}) = 22.86\underline{/6.87^\circ}\ \text{A}$$

$$\mathbf{I}_b = \mathbf{I}_a\underline{/-120^\circ} = 22.86\underline{/-113.13^\circ}\ \text{A}$$

$$\mathbf{I}_c = \mathbf{I}_a\underline{/+120^\circ} = 22.86\underline{/126.87^\circ}\ \text{A}$$

Practice Problem 17.4

A positive-sequence, balanced wye-connected source supplies a balanced delta-connected load. If the impedance per phase of the load is $(18 + j12)\ \Omega$ and $\mathbf{I}_a = 22.5\underline{/35^\circ}$ A, find $\mathbf{I}_{AB}$ and $\mathbf{V}_{AB}$.

Answer: $13\underline{/65^\circ}$ A; $281.2\underline{/98.69^\circ}$ V

17.7 Balanced Delta-Wye Connection

The delta-wye connection is popular with three-phase transformer connection in power distribution. This is because you can use the secondary circuit to provide a neutral point for supplying line-to-neutral power to serve single-phase loads and ground the neutral point for safety reasons.

> A **balanced Δ-Y system** consists of a balanced Δ-connected source feeding a balanced Y-connected load.

Consider the Δ-Y circuit in Fig. 17.19. Again, assuming *abc* sequence, the phase voltages of a delta-connected source are

$$\begin{aligned} \mathbf{V}_{ab} &= V_p\underline{/0^\circ} \\ \mathbf{V}_{bc} &= V_p\underline{/-120^\circ} \\ \mathbf{V}_{ca} &= V_p\underline{/+120^\circ} \end{aligned} \tag{17.34}$$

These are the line voltages as well as the phase voltages.

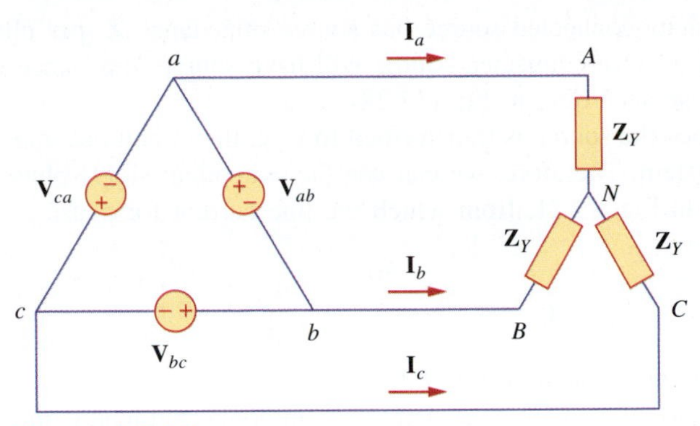

Figure 17.19
A balanced Δ-*Y* connection.

We can obtain the line currents in many ways. One way is to apply KVL to loop *aANBba* in Fig. 17.19.

$$-\mathbf{V}_{ab} + \mathbf{Z}_Y\mathbf{I}_a - \mathbf{Z}_Y\mathbf{I}_b = 0$$

or

$$\mathbf{Z}_Y(\mathbf{I}_a - \mathbf{I}_b) = \mathbf{V}_{ab} = V_p\underline{/0^\circ}$$

Thus,

$$\mathbf{I}_a - \mathbf{I}_b = \frac{V_p\underline{/0^\circ}}{\mathbf{Z}_Y} \tag{17.35}$$

But $\mathbf{I}_b$ lags $\mathbf{I}_a$ by 120° because we assumed *abc* sequence; that is, $\mathbf{I}_b = \mathbf{I}_a\underline{/-120^\circ}$. Hence,

$$\begin{aligned}\mathbf{I}_a - \mathbf{I}_b &= \mathbf{I}_a(1 - 1\underline{/-120^\circ}) = \mathbf{I}_a\left(1 + \frac{1}{2} + j\frac{\sqrt{3}}{2}\right) \\ &= \mathbf{I}_a\sqrt{3}\underline{/30^\circ}\end{aligned} \tag{17.36}$$

Substituting Eq. (17.36) into Eq. (17.35) gives

$$\mathbf{I}_a = \frac{\dfrac{V_p}{\sqrt{3}}\underline{/-30^\circ}}{Z_Y} = \frac{V_p}{\sqrt{3}Z_Y}\underline{/-30^\circ} \tag{17.37}$$

From this, we obtain the other line currents $\mathbf{I}_b$ and $\mathbf{I}_c$ using the positive phase sequence; that is, $\mathbf{I}_b = \mathbf{I}_a\underline{/-120^\circ}$, $\mathbf{I}_c = \mathbf{I}_a\underline{/+120^\circ}$. The phase currents are equal to the line currents.

Another way to obtain the line currents is to replace the delta-connected source with its equivalent wye-connected source, as shown in Fig. 17.20. In Section 17.4, we found that the line-to-line voltages of a wye-connected source lead their corresponding phase voltages by 30°. Therefore, we obtain each phase voltage of the equivalent wye-connected source by dividing the corresponding line voltage of the delta-connected source by $\sqrt{3}$ and shifting its phase by −30°. Thus, the equivalent wye-connected source has phase voltages

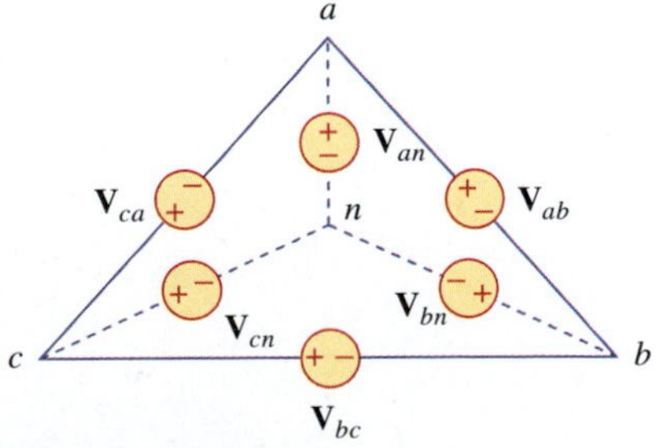

Figure 17.20
Transforming a Δ-connected source to an equivalent *Y*-connected source.

$$\begin{aligned}\mathbf{V}_{an} &= \frac{V_p}{\sqrt{3}}\underline{/-30^\circ} \\ \mathbf{V}_{bn} &= \frac{V_p}{\sqrt{3}}\underline{/-150^\circ} \\ \mathbf{V}_{cn} &= \frac{V_p}{\sqrt{3}}\underline{/+90^\circ}\end{aligned} \tag{17.38}$$

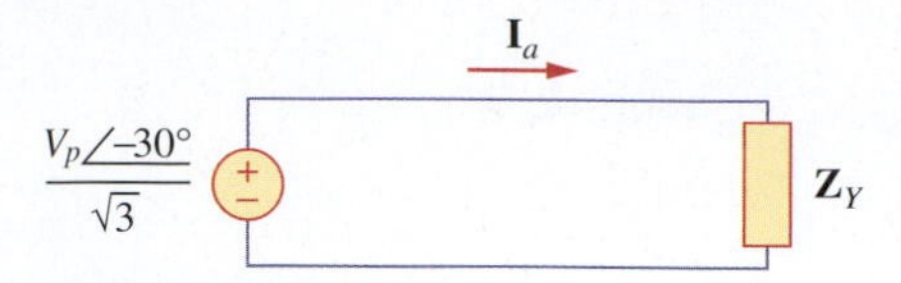

Figure 17.21
The single-phase equivalent circuit.

If the delta-connected source has source impedance $\mathbf{Z}_s$ per phase, the equivalent wye-connected source will have source impedance of $\mathbf{Z}_s/3$ per phase according to Eq. (17.28).

Once the source is transformed to wye, the circuit becomes a wye-wye system. Therefore, we can use the equivalent single-phase circuit shown in Fig. 17.21, from which the line current for phase a is

$$\mathbf{I}_a = \frac{\frac{V_p}{\sqrt{3}}\underline{/-30^\circ}}{Z_Y} = \frac{V_p}{\sqrt{3}Z_Y}\underline{/-30^\circ} \qquad \textbf{(17.39)}$$

which is the same as Eq. (17.37).

It should be mentioned that the delta-connected load is more desirable than the wye-connected load. It is easier to alter the loads in any one phase of the delta-connected loads because the individual loads are connected directly across the lines. However, the delta-connected source is rarely used in practice because any slight imbalance in the phase voltages will result in unwanted circulating currents.

A summary of the formulas for phase currents and voltages and line currents and voltages for the four connections is presented in Table 17.1.

TABLE 17.1

Summary of phase and line voltages/currents for balanced three-phase systems.

Connection	Phase voltages/ currents	Line voltages/ currents
Y-Y	$\mathbf{V}_{an} = V_p\underline{/0^\circ}$	$\mathbf{V}_{ab} = \sqrt{3}V_p\underline{/30^\circ}$
	$\mathbf{V}_{bn} = V_p\underline{/-120^\circ}$	$\mathbf{V}_{bc} = \mathbf{V}_{ab}\underline{/-120^\circ}$
	$\mathbf{V}_{cn} = V_p\underline{/+120^\circ}$	$\mathbf{V}_{ca} = \mathbf{V}_{ab}\underline{/+120^\circ}$
	Same as line currents	$\mathbf{I}_a = \mathbf{V}_{an}/\mathbf{Z}_Y$
		$\mathbf{I}_b = \mathbf{I}_a\underline{/-120^\circ}$
		$\mathbf{I}_c = \mathbf{I}_a\underline{/+120^\circ}$
Y-Δ	$\mathbf{V}_{an} = V_p\underline{/0^\circ}$	$\mathbf{V}_{ab} = \mathbf{V}_{AB} = \sqrt{3}V_p\underline{/30^\circ}$
	$\mathbf{V}_{bn} = V_p\underline{/-120^\circ}$	$\mathbf{V}_{bc} = \mathbf{V}_{BC} = \mathbf{V}_{ab}\underline{/-120^\circ}$
	$\mathbf{V}_{cn} = V_p\underline{/+120^\circ}$	$\mathbf{V}_{ca} = \mathbf{V}_{CA} = \mathbf{V}_{ab}\underline{/+120^\circ}$
	$\mathbf{I}_{AB} = \mathbf{V}_{AB}/\mathbf{Z}_\Delta$	$\mathbf{I}_a = \mathbf{I}_{AB}\sqrt{3}\underline{/-30^\circ}$
	$\mathbf{I}_{BC} = \mathbf{V}_{BC}/\mathbf{Z}_\Delta$	$\mathbf{I}_b = \mathbf{I}_a\underline{/-120^\circ}$
	$\mathbf{I}_{CA} = \mathbf{V}_{CA}/\mathbf{Z}_\Delta$	$\mathbf{I}_c = \mathbf{I}_a\underline{/+120^\circ}$
Δ-Δ	$\mathbf{V}_{ab} = V_p\underline{/0^\circ}$	Same as phase voltages
	$\mathbf{V}_{bc} = V_p\underline{/-120^\circ}$	
	$\mathbf{V}_{ca} = V_p\underline{/+120^\circ}$	
	$\mathbf{I}_{AB} = \mathbf{V}_{ab}/\mathbf{Z}_\Delta$	$\mathbf{I}_a = \mathbf{I}_{AB}\sqrt{3}\underline{/-30^\circ}$
	$\mathbf{I}_{BC} = \mathbf{V}_{bc}/\mathbf{Z}_\Delta$	$\mathbf{I}_b = \mathbf{I}_a\underline{/-120^\circ}$
	$\mathbf{I}_{CA} = \mathbf{V}_{ca}/\mathbf{Z}_\Delta$	$\mathbf{I}_c = \mathbf{I}_a\underline{/+120^\circ}$
Δ-Y	$\mathbf{V}_{ab} = V_p\underline{/0^\circ}$	Same as phase voltages
	$\mathbf{V}_{bc} = V_p\underline{/-120^\circ}$	
	$\mathbf{V}_{ca} = V_p\underline{/+120^\circ}$	
	Same as line currents	$\mathbf{I}_a = \frac{V_p\underline{/-30^\circ}}{\sqrt{3}\,Z_Y}$
		$\mathbf{I}_b = \mathbf{I}_a\underline{/-120^\circ}$
		$\mathbf{I}_c = \mathbf{I}_a\underline{/+120^\circ}$

Students are advised not to memorize the formulas but to understand how they are derived. The formulas can always be obtained by directly applying KCL and KVL to the appropriate three-phase circuits.

Example 17.5

A balanced Y-connected load with a phase resistance of 40 Ω and a reactance of 25 Ω is supplied by a balanced positive sequence Δ-connected source with a line voltage of 210 V. Calculate the phase currents. Use $\mathbf{V}_{ab}$ as a reference.

Solution:
The load impedance is

$$\mathbf{Z}_Y = 40 + j25 = 47.17\underline{/32^\circ}\ \Omega$$

and the source voltage is

$$\mathbf{V}_{ab} = 210\underline{/0^\circ}\ \text{V}$$

When the Δ-connected source is transformed to a Y-connected source,

$$\mathbf{V}_{an} = \frac{\mathbf{V}_{ab}}{\sqrt{3}}\underline{/-30^\circ} = 121.3\underline{/-30^\circ}\ \text{V}$$

The line currents are

$$\mathbf{I}_a = \mathbf{V}_{an}/\mathbf{Z}_Y = \frac{121.2\underline{/-30^\circ}}{47.17\underline{/32^\circ}} = 2.57\underline{/-62^\circ}\ \text{A}$$

$$\mathbf{I}_b = \mathbf{I}_a\underline{/-120^\circ} = 2.57\underline{/-182^\circ}\ \text{A}$$

$$\mathbf{I}_c = \mathbf{I}_a\underline{/120^\circ} = 2.57\underline{/58^\circ}\ \text{A}$$

which are the same as the phase currents.

Practice Problem 17.5

In a balanced Δ-Y circuit, $\mathbf{V}_{ab} = 240\underline{/15^\circ}$ and $\mathbf{Z}_Y = (12 + 15)\ \Omega$. Calculate the line currents.

Answer: $7.21\underline{/-66.34^\circ}$; $7.21\underline{/-186.34^\circ}$; $7.21\underline{/53.66^\circ}$ A

17.8 Power in a Balanced System

We now consider the power in a balanced three-phase system. We begin by examining the instantaneous power absorbed by the load. This requires that the analysis be done in the time domain. For a Y-connected load, the phase voltages are

$$\begin{aligned} v_{AN} &= \sqrt{2}V_p \cos(\omega t) \\ v_{BN} &= \sqrt{2}V_p \cos(\omega t - 120^\circ) \\ v_{CN} &= \sqrt{2}V_p \cos(\omega t + 120^\circ) \end{aligned} \tag{17.40}$$

where the factor $\sqrt{2}$ is necessary because V_P has been defined as the rms value of the phase voltage. If $\mathbf{Z}_Y = Z\underline{/\theta}$, the phase currents lag behind their corresponding phase voltages by θ. The total instantaneous power in the load is the sum of the instantaneous powers in the three phases; that is,

$$\begin{aligned} p = p_a + p_b + p_c &= v_{AN}i_a + v_{BN}i_b + v_{CN}i_c \\ &= 2V_pI_p\{\cos \omega t \cos(\omega t - \theta) \\ &\quad + \cos(\omega t - 120°)\cos(\omega t - \theta - 120°) \\ &\quad + \cos(\omega t + 120°)\cos(\omega t - \theta + 120°)\} \end{aligned} \tag{17.41}$$

where I_p is the rms value of the phase current. By applying some trigonometric identities, we can show that

$$p = 3V_pI_p \cos\theta \tag{17.42}$$

Thus, the total instantaneous power in a balanced three-phase system is constant rather than changing with time as the instantaneous power of each phase does. This result is true whether the load is Y- or Δ-connected. This is one important reason for using a three-phase system to generate and distribute power. We will look into another reason a little later.

Because the total instantaneous power is independent of time, the average power per phase P_P for the Δ-connected load or Y-connected load is p/3; that is,

$$P_P = V_PI_P \cos\theta \tag{17.43}$$

and reactive power per phase is

$$Q_P = V_PI_P \sin\theta \tag{17.44}$$

The apparent power per phase is

$$S_p = V_p\, I_p \tag{17.45}$$

The complex power per phase is

$$\mathbf{S}_p = P_P + jQ_p = \mathbf{V}_p\, \mathbf{I}_P^* \tag{17.46}$$

where $\mathbf{V}_p$ and $\mathbf{I}_p$ are the phase voltage and phase current with magnitudes V_p and I_p, respectively. The total average power is the sum of the average powers in the phases; that is,

$$P = P_a + P_b + P_c = 3P_P = 3V_p\, I_p \cos\theta = \sqrt{3}V_LI_L \cos\theta \tag{17.47}$$

For a Y-connected load, $I_L = I_p$ but $V_L = \sqrt{3}V_p$, whereas for a Δ-connected load, $I_L = \sqrt{3}I_p$ but $V_L = V_p$. Thus Eq. (17.47) applies for both Y-connected load and Δ-connected load. Similarly, the total reactive power is

$$Q = 3V_p\, I_p \sin\theta = 3Q_p = \sqrt{3}V_LI_L \sin\theta \tag{17.48}$$

and the total complex power is

$$\boxed{\mathbf{S} = 3\mathbf{S}_p = 3\mathbf{V}_p\mathbf{I}_P^* = 3\mathbf{I}_p^2\mathbf{Z}_p = 3V_p^2/\mathbf{Z}_p^*} \tag{17.49}$$

where $\mathbf{Z}_p = Z_p \underline{/\theta}$ is the load impedance per phase ($\mathbf{Z}_p$ could be $\mathbf{Z}_Y$ or $\mathbf{Z}_\Delta$). Alternatively, we may write Eq. (17.49) as

$$\mathbf{S} = P + jQ = \sqrt{3}V_LI_L \underline{/\theta} \tag{17.50}$$

Remember that V_p, I_p, V_L, and I_L are all rms values and that θ is the angle of the load impedance or the angle between the phase voltage and the phase current.

A second major reason three-phase systems are used for power distribution is that the three-phase system uses a lesser amount of wire (or copper) than the single-phase system for the same line voltage V_L and the same absorbed power P_L. It can be shown that the single-phase system uses 33 percent more material than the three-phase system or that the three-phase system uses only 75 percent of the material used in the equivalent single-phase system. (Instead of copper, one can make a similar case for aluminum.) In other words, considerably less material is needed to deliver the same power with a three-phase system than is required for a single-phase system.

Example 17.6

Refer to the circuit in Fig. 17.14 (in Example 17.2). Determine the total average power, reactive power, and the complex power at the source and at the load.

Solution:

It is sufficient to consider one phase because the system is balanced. For phase a,

$$\mathbf{V}_p = 110\underline{/0^\circ}\text{ V}$$
$$\mathbf{I}_p = 6.81\underline{/-21.8^\circ}\text{ A}$$

Thus, at the source, the complex power supplied is

$$\mathbf{S}_s = -3\mathbf{V}_p\mathbf{I}_p^* = 3(110\underline{/0^\circ})(6.81\underline{/21.8^\circ})$$
$$= -2247.4\underline{/21.8^\circ} = -(2{,}086.6 + j834.6)\text{ VA}$$

The real or average power supplied is $-2{,}086.6$ W and the reactive power is -834.6 VAR.

At the load, the complex power absorbed is

$$\mathbf{S}_L = 3|\mathbf{I}_p|^2\mathbf{Z}_p$$

where

$$\mathbf{Z}_p = 10 + j8 = 12.81\underline{/38.66^\circ}$$
$$\mathbf{I}_p = \mathbf{I}_a = 6.81\underline{/-21.8^\circ}$$

Hence,

$$\mathbf{S}_L = 3(6.81)^2\ 12.81\underline{/38.66^\circ} = 1782.2\underline{/38.66^\circ}$$
$$= (1391.7 + j1113.3)\text{ VA}$$

The real power absorbed is 1,391.7 W and the reactive power absorbed is 1,113.3 VAR. The difference between the two complex powers is absorbed by the line impedance $(5 - j2)\ \Omega$. To show that this is the case, we find the complex power absorbed by the line as

$$\mathbf{S}_\ell = 3|\mathbf{I}_p|^2\ \mathbf{Z}_\ell = 3(6.81)^2(5 - j2) = 695.64 - j278.3\text{ VA}$$

which is the difference between $\mathbf{S}_s$ and $\mathbf{S}_L$; that is, $\mathbf{S}_s + \mathbf{S}_\ell + \mathbf{S}_L = 0$, as expected.

Practice Problem 17.6

For the *Y-Y* circuit in Practice Problem 17.2, calculate the complex power at the source and at the load.

Answer: $(1054.2 + j843.3)$ VA; $(1017.45 + j801.6)$ VA

Example 17.7

A three-phase motor can be regarded as a balanced Y-load. A three-phase motor draws 5.6 kW when the line voltage is 220 V and the line current is 18.2 A. Determine the power factor of the motor.

Solution:
The apparent power is

$$S = \sqrt{3}V_L I_L = \sqrt{3}(220)(18.2) = 6935.13 \text{ VA}$$

Since the real power is

$$P = S\cos\theta = 5{,}600 \text{ W}$$

the power factor is

$$pf = \cos\theta = \frac{P}{S} = \frac{5600}{6935.13} = 0.8075$$

Practice Problem 17.7

Calculate the line current required for a 30-kW three-phase motor having a power factor of 0.85 lagging if it is connected to a balanced source with line voltage of 440 V.

Answer: 46.31 A

Example 17.8

Two balanced loads are connected to a 240-kV rms 60-Hz line, as shown in Fig. 17.22(a). Load 1 draws 30 kW at a power factor of 0.6 lagging, while load 2 draws 45 kVAR at a power factor of 0.8 lagging. Assuming the *abc* sequence, determine:
(a) the complex, real, and reactive powers absorbed by the combined load;
(b) the line currents;
(c) the kVAR rating of the three capacitors delta-connected in parallel with the load that will raise the power factor to 0.9 lagging. Also calculate the capacitance of each capacitor.

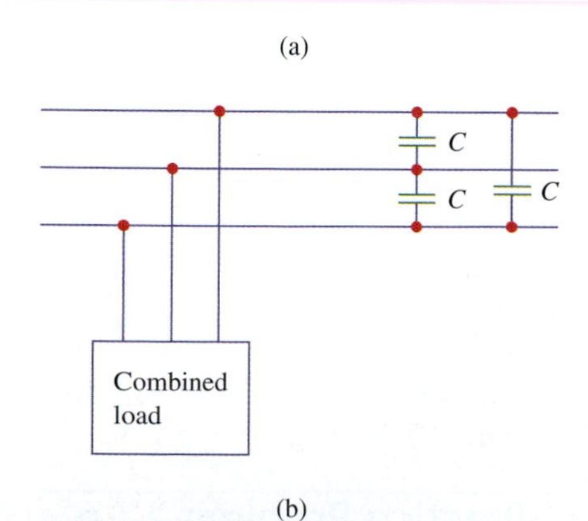

Figure 17.22
For Example 17.8: (a) the original balanced loads; (b) the combined load with improved power factor.

Solution:
(a) For load 1, given that $P_1 = 30$ kW, and $\cos\theta_1 = 0.6$, then $\sin\theta_1 = 0.8$. Hence,

$$S_1 = \frac{P_1}{\cos\theta_1} = \frac{30 \text{ kW}}{0.6} = 50 \text{ kVA}$$

and $Q_1 = S_1\sin\theta_1 = 50(0.8) = 40$ kVAR. Thus, the complex power due to load 1 is

$$\mathbf{S}_1 = P_1 + jQ_1 = 30 + j40 \text{ kVA} \qquad \textbf{(17.8.1)}$$

For load 2, if $Q_2 = 45$ kVA and $\cos\theta_2 = 0.8$, then $\sin\theta_2 = 0.6$. We find

$$S_2 = \frac{Q_2}{\sin\theta_2} = \frac{45 \text{ kVAR}}{0.6} = 75 \text{ kVA}$$

and $P_2 = S_2\cos\theta_2 = 75(0.8) = 60$ kW. Hence, the complex power due to load 2 is

$$\mathbf{S}_2 = P_2 = jQ_2 = 60 + j45 \text{ kVA} \qquad \textbf{(17.8.2)}$$

From Eqs. (17.8.1) and (17.8.2), the total complex power absorbed by the load is

$$\mathbf{S} = \mathbf{S}_1 + \mathbf{S}_2 = 90 + j85 \text{ kVA} = 123.8\underline{/43.36^\circ} \text{ kVA} \quad \textbf{(17.8.3)}$$

which has a power factor of $\cos 43.36^\circ = 0.727$ lagging. The real power is 90 kW, while the reactive power is 85 kVAR.

(b) Because $S = \sqrt{3}V_L I_L$, the line current is

$$I_L = \frac{S}{\sqrt{3}V_L} \quad \textbf{(17.8.4)}$$

We apply this to each load keeping in mind that for both loads, $V_L = 240$ kV. For load 1,

$$I_{L1} = \frac{50{,}000}{\sqrt{3} \times 240{,}000} = 120.28 \text{ mA}$$

Because the power factor is lagging, the line current lags the line voltage by $\theta_1 = \cos^{-1} 0.6 = 53.13^\circ$. Thus,

$$\mathbf{I}_{a1} = 120.28\underline{/-53.13^\circ} \text{ mA}$$

For load 2,

$$I_{L1} = \frac{75{,}000}{\sqrt{3}\,240{,}000} = 180.42 \text{ mA}$$

and the line current lags the line voltage by $\theta_2 = \cos^{-1} 0.8 = 36.87^\circ$. Hence,

$$\mathbf{I}_{a2} = 180.42\underline{/-36.87^\circ} \text{ mA}$$

The total line current is

$$\begin{aligned}\mathbf{I}_a &= \mathbf{I}_{a1} + \mathbf{I}_{a2} = 120.28\underline{/-53.13^\circ} + 180.42\underline{/-36.87^\circ} \\ &= (72.168 - j96.224) + (144.336 - j108.252) \\ &= 217.5 - j204.47 = 297.8\underline{/-43.36^\circ} \text{ mA}\end{aligned}$$

Alternatively, we could obtain the current from the total complex power using Eq. (17.8.4)

$$I_L = \frac{123{,}800}{\sqrt{3}240{,}000} = 297.82 \text{ mA}$$

and

$$\mathbf{I}_a = 297.82\underline{/-43.36^\circ} \text{ mA}$$

which is the same as that derived earlier.

Other line currents $\mathbf{I}_b$ and $\mathbf{I}_c$ can be obtained according to *abc* sequence; that is,

$$\mathbf{I}_b = 297.82\underline{/-163.36^\circ} \text{ mA} \quad \text{and} \quad \mathbf{I}_c = 297.82\underline{/76.64^\circ} \text{ mA}$$

(c) The reactive power needed to bring the power factor to 0.9 lagging is found using Eq. (14.43); that is,

$$Q_c = P(\tan\theta_{\text{old}} - \tan\theta_{\text{new}})$$

where $P = 90$ kW, $\theta_{old} = 43.36°$ and $\theta_{new} = \cos^{-1} 0.9 = 25.84°$. Hence,

$$Q_C = 90{,}000\,(\tan 43.36° - \tan 25.84°) = 41.4 \text{ kVAR}$$

This reactive power is for the three capacitors. For each capacitor, the rating $Q'_C = 13.8$ kVAR. From Eq. (14.44), the required capacitance is

$$C = \frac{Q'_C}{\omega V_{rms}^2}$$

Because the capacitors are delta-connected as shown in Fig. 17.22(b), V_{rms} in the preceding formula is the line-to-line or line voltage, which is 240 kV. Hence,

$$C = \frac{13{,}800}{(2\pi 60)(240{,}000)^2} = 635.5 \text{ pF}$$

Practice Problem 17.8

Assume that the two balanced loads in Fig. 17.22(a) are supplied by an 840-V rms 60-Hz line. Load 1 is wye-connected with $30 + j40\ \Omega$ per phase, while load 2 is a balanced three-phase motor drawing 48 kW at a power factor of 0.8 lagging. Assuming the *abc* sequence, calculate:
(a) the complex power absorbed by the combined load,
(b) the kVAR rating of the three capacitors delta-connected in parallel with the load to raise the power factor to unity,
(c) the current drawn from the supply at unity power factor condition.

Answer: (a) $56.47 + j47.29$ kVA; (b) 15.7 kVAR; (c) 38.813 A

17.9 †Unbalanced Three-Phase Systems

This chapter would be incomplete without mentioning unbalanced three-phase systems. An unbalanced system is caused by two possible situations: (1) the source voltages are not equal in magnitude and/or differ in phase by angles that are not equal or (2) load impedances are not equal. Thus,

> An **unbalanced system** is due to unbalanced voltage sources or an unbalanced load.

However, to simplify analysis, we will assume balanced source voltages, but an unbalanced load.

Unbalanced three-phase systems are solved by direct application of mesh and nodal analysis, discussed in Chapter 13 as applied to ac systems. Figure 17.23 shows an example of an unbalanced three-phase system, which consists of balanced source voltages (not shown in Fig. 17.23) and an unbalanced *Y*-connected load (shown in Fig. 17.23). Because the load is unbalanced, $\mathbf{Z}_A$, $\mathbf{Z}_B$, and $\mathbf{Z}_C$ are not equal. The line currents are determined by Ohm's law as

$$\mathbf{I}_a = \mathbf{V}_{AN}/\mathbf{Z}_A, \quad \mathbf{I}_b = \mathbf{V}_{BN}/\mathbf{Z}_B, \quad \mathbf{I}_c = \mathbf{V}_{CN}/\mathbf{Z}_C \tag{17.51}$$

This set of unbalanced line currents produces current in the neutral line, which is not zero as in a balanced system. Applying KCL at node N gives the neutral line current as

$$\mathbf{I}_n = -(\mathbf{I}_a + \mathbf{I}_b + \mathbf{I}_c) \tag{17.52}$$

In the case whereby the neutral line is absent so that we have a three-wire system, we can still find the line currents $\mathbf{I}_a$, $\mathbf{I}_b$, and $\mathbf{I}_c$ using mesh analysis. At node N, KCL must be satisfied so that $\mathbf{I}_a + \mathbf{I}_b + \mathbf{I}_c = 0$ in this case. The same could be done for a Δ-Y, Y-Δ, or Δ-Δ three-wire system. As mentioned earlier, long-distance power transmission conductors in multiples of three (multiple three-wire system) are used with the Earth itself acting as the neutral conductor.

To calculate power in an unbalanced three-phase system requires that we find the power in each phase using Eqs. (17.43) through (17.46). The total power is not simply three times the power in one phase, but the sum of the powers in the three phases.

Example 17.9

The unbalanced Y-load of Fig. 17.23 has balanced voltages of 100 V and the acb sequence. Calculate the line currents and the neutral current. Take $\mathbf{Z}_A = 15\ \Omega$, $\mathbf{Z}_B = 10 + j5\ \Omega$, and $\mathbf{Z}_C = 6 - j8\ \Omega$.

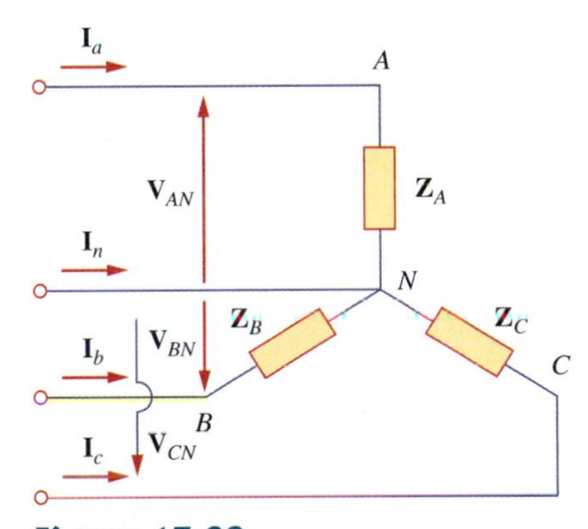

Figure 17.23
Unbalanced three-phase Y-connected load; for Examples 17.9 and 17.13.

Solution:
Using Eq. (17.51), the line currents are

$$\mathbf{I}_a = \frac{100\underline{/0^\circ}}{15} = 6.67\underline{/0^\circ}\text{ A}$$

$$\mathbf{I}_b = \frac{100\underline{/120^\circ}}{10 + j5} = \frac{100\underline{/120^\circ}}{11.18\underline{/26.56^\circ}} = 8.94\underline{/93.44^\circ}\text{ A}$$

$$\mathbf{I}_c = \frac{100\underline{/-120^\circ}}{6 - j8} = \frac{100\underline{/-120^\circ}}{10\underline{/-53.13^\circ}} = 10\underline{/-66.87^\circ}\text{ A}$$

Using Eq. (17.52), the current in the neutral line is

$$\begin{aligned}\mathbf{I}_n = -(\mathbf{I}_a + \mathbf{I}_b + \mathbf{I}_c) &= -(6.67 - 0.54 + j8.92 + 3.93 - j9.2)\\ &= -10.06 + j0.28\\ &= 10.06\underline{/178.4^\circ}\text{ A}\end{aligned}$$

Practice Problem 17.9

The unbalanced Δ-load of Fig. 17.24 is supplied by balanced voltages of 200 V in the positive sequence. Find the line currents. Take $\mathbf{V}_{ab}$ as a reference.

Answer: $15.08\underline{/-15^\circ}$; $29.15\underline{/220.2^\circ}$; $24\underline{/71.33^\circ}$ A

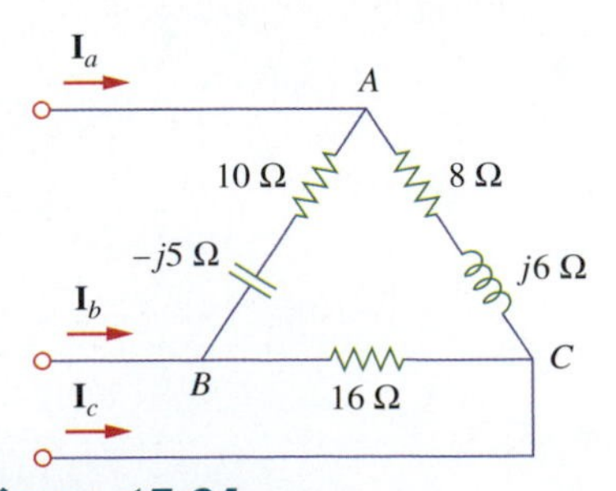

Figure 17.24
Unbalanced Δ-load; for Practice Problems 17.9 and 17.13.

Example 17.10

For the unbalanced circuit in Fig. 17.25, find:
(a) the line currents,
(b) the total complex power absorbed by the load,
(c) the total complex power supplied by the source.

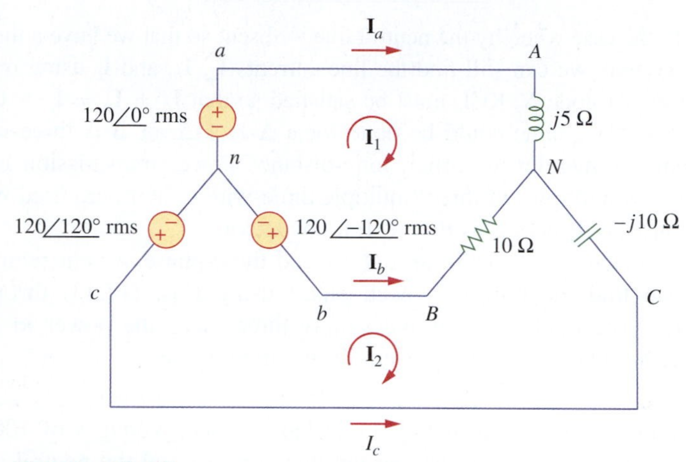

Figure 17.25
For Example 17.10.

Solution:

(a) We use mesh analysis to find the required currents. For mesh 1,

$$120\angle -120^\circ - 120\angle 0^\circ + (10 + j5)\mathbf{I}_1 - 10\mathbf{I}_2 = 0$$

or

$$(10 + j5)\mathbf{I}_1 - 10\mathbf{I}_2 = 120\sqrt{3}\angle 30^\circ \qquad \textbf{(17.9.1)}$$

For mesh 2,

$$120\angle 120^\circ - 120\angle -120^\circ + (10 - j10)\mathbf{I}_2 - 10\mathbf{I}_1 = 0$$

or

$$-10\mathbf{I}_1 + (10 - j10)\mathbf{I}_2 = 120\sqrt{3}\angle -90^\circ \qquad \textbf{(17.9.2)}$$

Equations (17.9.1) and (17.9.2) form a matrix equation:

$$\begin{bmatrix} 10 + j5 & -10 \\ -10 & 10 - j10 \end{bmatrix} \begin{bmatrix} I_1 \\ I_2 \end{bmatrix} = \begin{bmatrix} 120\sqrt{3}\angle 30^\circ \\ 120\sqrt{3}\angle -90^\circ \end{bmatrix}$$

The determinants are

$$\Delta = \begin{vmatrix} 10 + j5 & -10 \\ -10 & 10 - j10 \end{vmatrix} = 50 - j50 = 70.71\angle -45^\circ$$

$$\Delta_1 = \begin{vmatrix} 120\sqrt{3}\angle 30^\circ & -10 \\ 120\sqrt{3}\angle -90^\circ & 10 - j10 \end{vmatrix} = 207.85(13.66 - j13.66)$$

$$= 4015\angle -45^\circ$$

$$\Delta_2 = \begin{vmatrix} 10 + j5 & 120\sqrt{3}\underline{/30^\circ} \\ -10 & 120\sqrt{3}\underline{/-90^\circ} \end{vmatrix} = 207.85(13.66 - j5)$$

$$= 3023\underline{/-20.1^\circ}$$

The mesh currents are

$$\mathbf{I}_1 = \frac{\Delta_1}{\Delta} = \frac{4015.23\underline{/-45^\circ}}{70.71\underline{/-45^\circ}} = 56.78 \text{ A}$$

$$\mathbf{I}_2 = \frac{\Delta_2}{\Delta} = \frac{3023.4\underline{/-20.1^\circ}}{70.71\underline{/-45^\circ}} = 42.75\underline{/24.9} \text{ A}$$

The line currents are

$$\mathbf{I}_a = \mathbf{I}_1 = 56.78 \text{ A}$$
$$\mathbf{I}_c = -\mathbf{I}_2 = 42.75\underline{/-155.1^\circ} \text{ A}$$
$$\mathbf{I}_b = \mathbf{I}_2 - \mathbf{I}_1 = 38.78 + j18 - 56.78 = 25.46\underline{/135^\circ} \text{ A}$$

(b) We can now calculate the complex power absorbed by the load. For phase A,

$$\mathbf{S}_A = |\mathbf{I}_a|^2\mathbf{Z}_A = (56.78)^2(j5) = j16{,}120 \text{ VA}$$

For phase B,

$$\mathbf{S}_B = |\mathbf{I}_b|^2\mathbf{Z}_B = (25.46)^2(10) = 6{,}480 \text{ VA}$$

For phase C,

$$\mathbf{S}_C = |\mathbf{I}_c|^2\mathbf{Z}_C = (42.75)^2\ (-j10) = -j18{,}276 \text{ VA}$$

The total complex power absorbed by the load is

$$\mathbf{S}_L = \mathbf{S}_A + \mathbf{S}_B + \mathbf{S}_C = 6{,}480 - j2{,}159 \text{ VA}$$

(c) We check the preceding result by finding the power supplied by the source. For the voltage source in phase a,

$$\mathbf{S}_a = -\mathbf{V}_{an}\,\mathbf{I}_a^* = -(120\underline{/0^\circ})(56.78) = -6{,}813.6 \text{ VA}$$

For the source in phase b,

$$\mathbf{S}_b = -\mathbf{V}_{bn}\,\mathbf{I}_b^* = -(120\underline{/-120^\circ})(25.46\underline{/-135^\circ})$$
$$= -3{,}055.2\underline{/105^\circ}$$
$$= 790 - j2951.1 \text{ VA}$$

For the source in phase c,

$$\mathbf{S}_c = -\mathbf{V}_{cn}\,\mathbf{I}_c^* = -(120\underline{/0^\circ})(42.75\underline{/155.1^\circ}) = -5{,}130\underline{/275.1^\circ}$$
$$= -456.03 + j5{,}109.7 \text{ VA}$$

The total complex power supplied by the three-phase source is

$$\mathbf{S}_s = \mathbf{S}_a + \mathbf{S}_b + \mathbf{S}_c = -6{,}480 + j2{,}156 \text{ VA}$$

showing that $\mathbf{S}_s + \mathbf{S}_L = 0$ and confirming the principle of conservation of ac power.

Practice Problem 17.10

Find the line currents in the unbalanced three-phase circuit of Fig. 17.26 and the real power absorbed by the load.

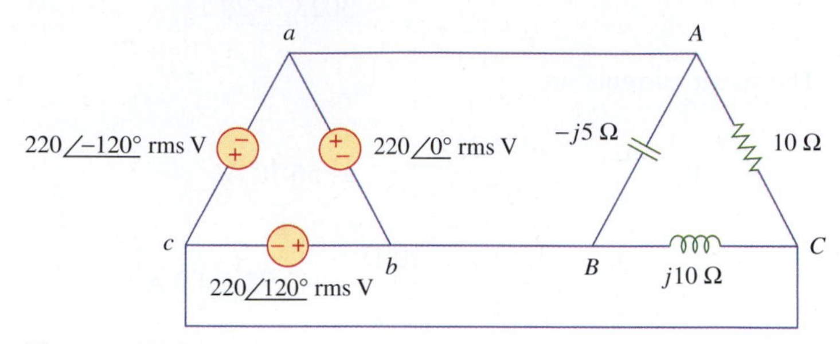

Figure 17.26
For Practice Problem 17.10.

Answer: $64\angle 80.1^\circ$; $38.1\angle -60^\circ$; $42.5\angle 225^\circ$ A; 4.84 kW

17.10 Computer Analysis

PSpice can be used to analyze three-phase balanced or unbalanced circuits in the same way it is used to analyze single phase ac circuits. However, a delta-connected source presents two major problems to PSpice. First, a delta-connected source is a loop of voltage sources that PSpice does not like. To avoid this problem, we insert a resistor of negligible resistance (say, 1 $\mu\Omega$ per phase) into each phase of the delta-connected source. Second, the delta-connected source does not provide a convenient node for the 0 ground node, which is necessary to run PSpice. This problem can be eliminated by inserting balanced wye-connected large resistors (say, 1 MΩ per phase) in the delta-connected source so that the neutral node of the wye-connected resistors serves as the ground node 0. This will be illustrated in Example 17.12.

Example 17.11

For the balanced *Y*-Δ circuit in Fig. 17.27, use PSpice to find the line current $\mathbf{I}_{aA}$, the phase voltage $\mathbf{V}_{AB}$, and the phase current $\mathbf{I}_{AC}$. Assume that the source frequency is 60 Hz.

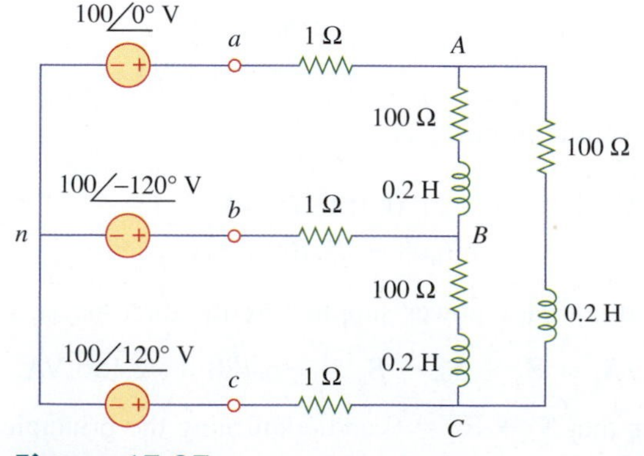

Figure 17.27
For Example 17.11.

Solution:

The schematic is shown in Fig. 17.28. The pseudocomponents IPRINT are inserted in the appropriate lines to obtain $\mathbf{I}_{aA}$ and $\mathbf{I}_{AC}$, while VPRINT2 is inserted between nodes A and B to print differential voltage $\mathbf{V}_{AB}$. We set the attributes of IPRINT and VPRINT2 each to AC = *yes*, MAG = *yes*, $PHASE$ = *yes* to print only the magnitude and phase of the currents and voltages.

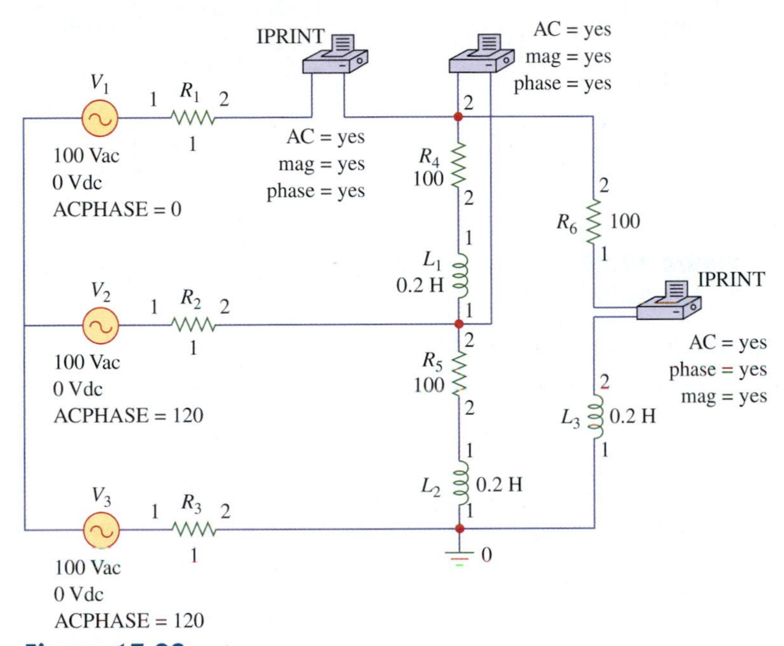

Figure 17.28
Schematic for the circuit in Fig. 17.27.

After drawing the circuit as in Fig. 17.31 and saved as exam1711.dsn, we select **PSpice/New Simulation Profile**. This leads to the New Simulation dialog box. Type "exam1711" as the name of the file and click Create. This leads to the Simulation Settings dialog box. In the Simulation Settings dialog box, select AC Sweep/Noise under *Analysis Type*, select Linear under *AC Sweep Type*. We type 60 as the *Start Freq*, 60 as the *Final Freq*, and 1 as *Total Points*.

We simulate the circuit by selecting **PSpice/Run**. We obtain the output file by selecting **PSpice/View Output file**. The output file includes the following:

```
FREQ          V(A,B)         VP(A,B)
6.000E+01     1.699E+02      3.081E+01

FREQ          IM(V_PRINT2)   IP(V_PRINT2)
6.000E+01     2.350E+00      -3.620E+01

FREQ          IM(V_PRINT3)   IP(V_PRINT3)
6.000E+01     1.357E+00      -6.620E+01
```

From this, we obtain

$$\mathbf{I}_{aA} = 2.35\underline{/-36.2^\circ} \text{ A}$$
$$\mathbf{V}_{AB} = 169.9\underline{/30.81^\circ} \text{ V}$$
$$\mathbf{I}_{AC} = 1.357\underline{/-66.2^\circ} \text{ A}$$

Practice Problem 17.11

Refer to the balanced *Y-Y* circuit of Fig. 17.29. Use PSpice to find the line current $\mathbf{I}_{aA}$ and the phase voltage $\mathbf{V}_{AN}$. Take $f = 100$ Hz.

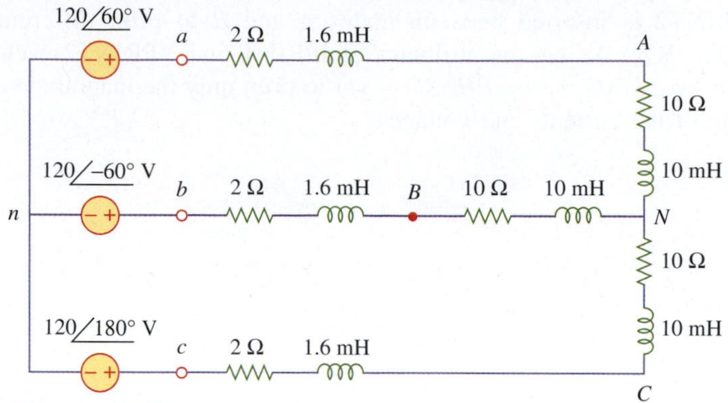

Figure 17.29
For Practice Problem 17.11.

Answer: $8.547\underline{/-91.27^\circ}$ A; $100.9\underline{/60.87^\circ}$ V

Example 17.12

Consider the unbalanced Δ-Δ circuit in Fig. 17.30. Use PSpice to find the generator current $\mathbf{I}_{ab}$, the line current $\mathbf{I}_{bB}$, and the phase current $\mathbf{I}_{BC}$.

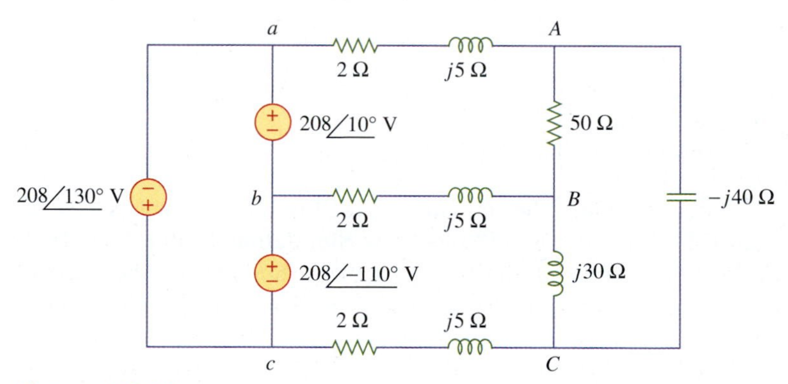

Figure 17.30
For Example 17.12.

Solution:
As mentioned earlier, we avoid the loop of voltage sources by inserting a 1-$\mu\Omega$ resistor in each of the delta-connected sources. To provide a ground node 0, we insert a balanced wye-connected resistor (1 MΩ per phase) in each of the delta-connected sources as shown in the schematic in Fig. 17.31. Three IPRINT pseudocomponents with their attributes are inserted to be able to get the required currents $\mathbf{I}_{ab}$, $\mathbf{I}_{bB}$, and $\mathbf{I}_{BC}$. Because the operating frequency is not given and the inductances and capacitances should be specified instead of impedances, we assume $\omega = 1$ rad/s so that $f = 1/2\pi = 0.159155$ Hz. Thus,

$$L = \frac{X_L}{\omega} \quad \text{and} \quad C = \frac{1}{\omega X_C}$$

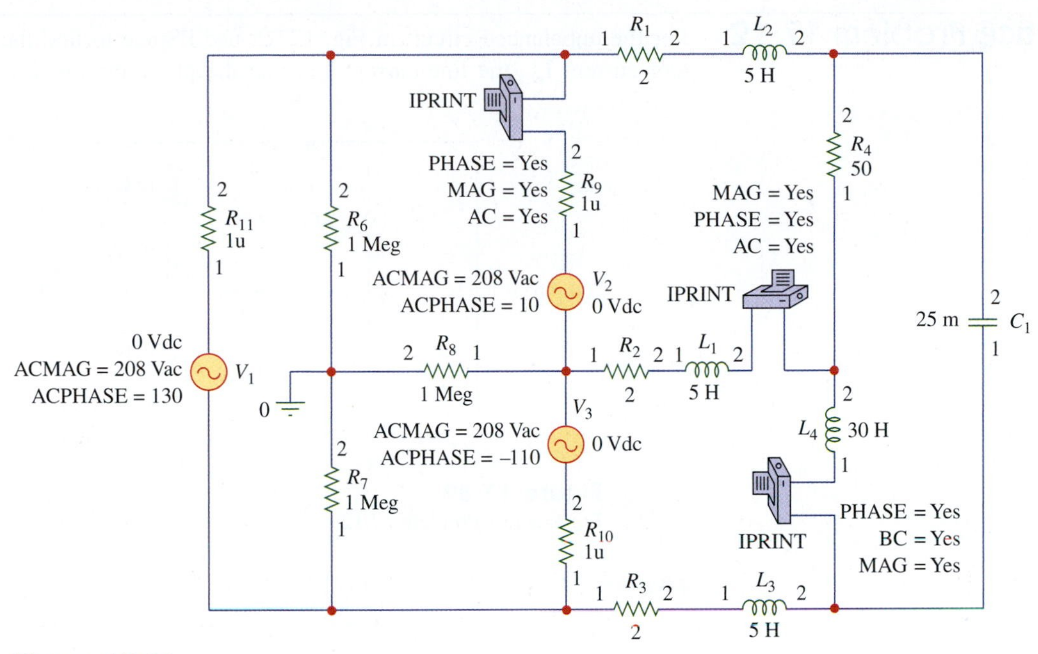

Figure 17.31
Schematic for the circuit in Fig. 17.30.

After drawing the circuit as in Fig. 17.31 and saved as exam1712.dsn, we select **PSpice/New Simulation Profile**. This leads to the New Simulation dialog box. Type "exam1712" as the name of the file and click Create. This leads to the Simulation Settings dialog box. In the Simulation Settings dialog box, select AC Sweep/Noise under *Analysis Type*, select Linear under *AC Sweep Type*. We type 0.159155 as the *Start Freq*, 0.159155 as the *Final Freq*, and 1 as *Total Points*.

We simulate the circuit by selecting **PSpice/Run**. We obtain the output file by selecting **PSpice/View Output file**. The output file includes the following:

```
FREQ          IM(V_PRINT1)    IP(V_PRINT1)
1.592E-01     9.106E+00       1.685E+02

FREQ          IM(V_PRINT2)    IP(V_PRINT2)
1.592E-01     5.959E+00       2.821E+00

FREQ          IM(V_PRINT3)    IP(V_PRINT3)
1.592E-01     5.500E+00       -7.532E+00
```

from which we get

$$\mathbf{I}_{ab} = 5.96\,\underline{/2.82^\circ}\text{ A}$$

$$\mathbf{I}_{bB} = 9.106\,\underline{/168.5^\circ}\text{ A}$$

$$\mathbf{I}_{BC} = 5.5\,\underline{/-7.53^\circ}\text{ A}$$

Practice Problem 17.12

For the unbalanced circuit in Fig. 17.32, use PSpice to find the generator current $\mathbf{I}_{ca}$, the line current $\mathbf{I}_{cC}$, and the phase current $\mathbf{I}_{AB}$.

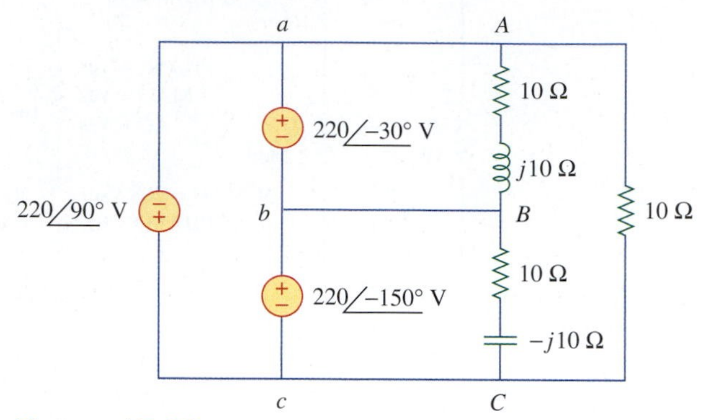

Figure 17.32
For Practice Problem 17.12.

Answer: $24.68\angle{-90^\circ}$ A; $15.56\angle{105^\circ}$ A; $37.24\angle{83.79^\circ}$ A

17.11 †Applications

Both wye and delta source connections find important practical applications. The wye source connection is used for long-distance transmission of electric power, where resistive losses (I^2R) need to be minimal. This is due to the fact that the wye connection gives a line voltage which is $\sqrt{3}$ greater than the delta connection; hence, for the same power, the line current is $\sqrt{3}$ smaller. In addition, delta connected are also undesirable to the potential of having disastrous circulating currents. Sometimes, using transformers, we create the equivalent of delta connect source. This conversion from three-phase to single-phase is required in residential wiring, because household lighting and appliances use single-phase power. Three-phase power is used in industrial wiring where a large power is required. In some applications, it is immaterial whether the load is wye- or delta-connected. For example, either connection can be used with induction motors. In fact, some manufacturers connect a motor in delta for 220 V and in wye for 311 V (in ratio 1:$\sqrt{3}$) so that one line of motors can be readily adapted to two different voltages.

Here, we consider two practical applications of what has been covered in this chapter: power measurement in three-phase circuits and residential wiring.

17.11.1 Three-Phase Power Measurement

Section 3.7 presented the wattmeter as the instrument for measuring the average (or real) power in single-phase circuits. A single wattmeter can also measure the average power in a three-phase system that is balanced, so that $P_1 = P_2 = P_3$; the total power is three times the reading of that one wattmeter. However, two or three single-phase wattmeters are necessary to measure power if the system is unbalanced. The *three-wattmeter method* of power measurement, shown in Fig. 17.33, will work whether the load is balanced or unbalanced, wye- or delta-connected. The three-wattmeter method is well suited for

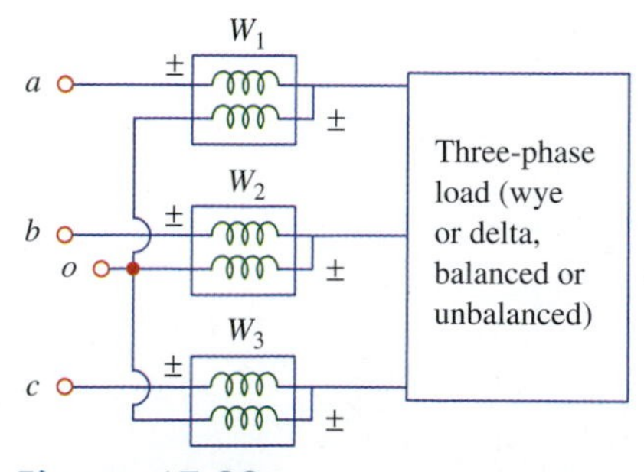

Figure 17.33
Three-wattmeter method for measuring three-phase power.

power measurement in a three-phase system where the power factor is constantly changing. The total average power is the algebraic sum of the three wattmeter readings; that is,

$$P_T = P_1 + P_2 + P_3 \tag{17.53}$$

where P_1, P_2, and P_3 correspond to the readings of wattmeters W_1, W_2, and W_3, respectively. Notice that the common or reference point o in Fig. 17.33 is selected arbitrarily. If the load is wye-connected, point o can be connected to the neutral point n. For a delta-connected load, point o can be connected to any point. If point o is connected to point b, for example, the voltage coil in wattmeter W_2 reads zero and $P_2 = 0$, indicating that wattmeter W_2 is not necessary. Thus, two wattmeters are sufficient to measure the total power.

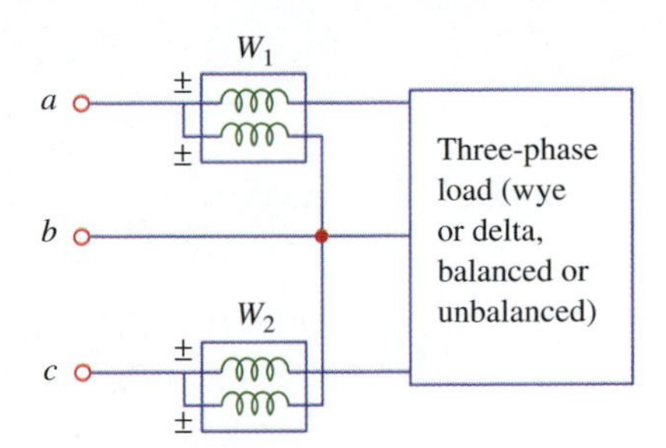

Figure 17.34
Two-wattmeter method for measuring three-phase power.

The *two-wattmeter method* is the most commonly used method for three-phase power measurement. The two wattmeters must be properly connected to any two phases, typically as shown in Fig. 17.34. Notice that the current coil of each wattmeter measures the line current, while the respective voltage coil is connected between the line and the third line and measures the line voltage. Also notice that the ± (plus/minus) terminal of the voltage coil is connected to the line to which the corresponding current coil is connected. Although the individual wattmeters no longer read the power taken by any particular phase, the algebraic sum of the two wattmeter readings equals the total average power absorbed by the load, regardless of whether it is wye- or delta-connected, balanced or unbalanced. The total real power is equal to the algebraic sum of the two wattmeter readings; that is,

$$\boxed{P_T = P_1 + P_2} \tag{17.54}$$

The difference in the wattmeter readings is proportional to the total reactive power, or

$$\boxed{Q_T = \sqrt{3}(P_2 - P_1)} \tag{17.55}$$

We can show that the method works for a balanced three-phase system. Dividing Eq. (17.55) by Eq. (17.54) gives the tangent of the power angle as

$$\tan\theta = \frac{Q_T}{P_T} \tag{17.56}$$

We should note that:

1. If $P_2 = P_1$, the load is resistive.
2. If $P_2 > P_1$, the load is inductive.
3. If $P_2 < P_1$, the load is capacitive.

Example 17.13

Three wattmeters W_1, W_2, and W_3 are connected respectively to phases a, b, and c to measure the total power absorbed by the unbalanced wye-connected load in Example 17.9 (see Fig. 17.23).
(a) Predict the wattmeter readings.
(b) Find the total power absorbed.

Solution:
Part of the problem is already solved in Example 17.9. Assume that the wattmeters are properly connected as shown in Fig. 17.35.

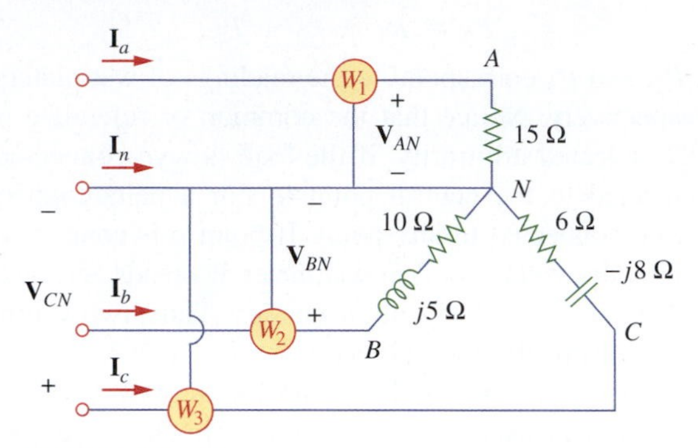

Figure 17.35
For Example 17.13.

(a) From Example 17.9,

$$\mathbf{V}_{AN} = 100\underline{/0^\circ}$$
$$\mathbf{V}_{BN} = 100\underline{/120^\circ}$$
$$\mathbf{V}_{CN} = 100\underline{/-120^\circ}\text{ V}$$

while

$$\mathbf{I}_a = 6.67\underline{/0^\circ}$$
$$\mathbf{I}_b = 8.94\underline{/93.44^\circ}$$
$$\mathbf{I}_c = 10\underline{/-66.87^\circ}\text{ A}$$

The wattmeter readings are calculated as follows:

$$\begin{aligned} P_1 &= \text{Re}(\mathbf{V}_{AN}\mathbf{I}_a^*) = V_{AN}I_a\cos(\theta_{\mathbf{V}_{AN}} - \theta_{\mathbf{I}_a}) \\ &= 100 \times 6.67 \times \cos(0^\circ - 0^\circ) = 667\text{ W} \\ P_2 &= \text{Re}(\mathbf{V}_{BN}\mathbf{I}_b^*) = V_{BN}I_b\cos(\theta_{\mathbf{V}_{BN}} - \theta_{\mathbf{I}_b}) \\ &= 100 \times 8.94 \times \cos(120^\circ - 93.44^\circ) = 800\text{ W} \\ P_3 &= \text{Re}(\mathbf{V}_{CN}\mathbf{I}_c^*) = V_{CN}I_c\cos(\theta_{\mathbf{V}_{CN}} - \theta_{\mathbf{I}_c}) \\ &= 100 \times 10 \times \cos(-120^\circ + 66.87^\circ) = 600\text{ W} \end{aligned}$$

(b) The total power absorbed is

$$P_T = P_1 + P_2 + P_3 = 667 + 800 + 600 = 2{,}067\text{ W}$$

We may find the power absorbed by the resistors in Fig. 17.35 and use that to check or confirm this result.

$$\begin{aligned} P_T &= |\mathbf{I}_a|^2(15) + |\mathbf{I}_b|^2(10) + |\mathbf{I}_c|^2(6) \\ &= 6.67^2(15) + 8.94^2(10) + 10^2(6) \\ &= 667 + 800 + 600 = 2{,}067\text{ W} \end{aligned}$$

which is exactly the same thing.

Practice Problem 17.13

Repeat Example 17.13 for the network in Fig. 17.24 (see Practice Problem 17.9). Hint: Connect the reference point o in Fig. 17.32 to point B.

Answer: (a) 2913.23 W, 0 W, 4706.46 W; (b) 7619.29 W

Example 17.14

The two-wattmeter method produces wattmeter readings $P_1 = 1560$ W and $P_2 = 2100$ W when connected to a delta-connected load. If the line voltage is 220 V, calculate:
(a) the per-phase average power,
(b) the per-phase reactive power,
(c) the power factor,
(d) the phase impedance.

Solution:
We can apply the preceding results to the delta-connected load.
(a) The total real or average power is

$$P_T = P_1 + P_2 = 1{,}560 + 2{,}100 = 3{,}660 \text{ W}$$

The per-phase average power is

$$P_p = \frac{1}{3}P_T = 1{,}220 \text{ W}$$

(b) The total reactive power is

$$Q_T = \sqrt{3}(P_2 - P_1) = \sqrt{3}(2{,}100 - 1{,}560) = 935.3 \text{ VAR}$$

so that the per-phase reactive power is

$$Q_p = \frac{1}{3}Q_T = 311.77 \text{ VAR}$$

(c) The power angle is

$$\theta = \tan^{-1}\frac{Q_T}{P_T} = \tan^{-1}\frac{935.3}{3660} = 14.33^\circ$$

Hence, the power factor is

$$\cos\theta = 0.9689 \text{ (lagging)}$$

It is a lagging *pf* because Q_T is positive or $P_2 > P_1$.
(d) The phase impedance is $\mathbf{Z}_p = Z_p\underline{/\theta}$. We know that θ is the same as the *pf* angle; that is, $\theta = 14.33^\circ$.

$$Z_p = \frac{V_p}{I_p}$$

We recall that for a delta-connected load, $V_p = V_L = 220$ V. From Eq. (17.43),

$$P_p = V_p I_p \cos\theta \qquad \Rightarrow \qquad I_p = \frac{1220}{220 \times 0.9689} = 5.723 \text{ A}$$

Hence,

$$Z_p = \frac{V_p}{I_p} = \frac{220}{5.723} = 38.44\ \Omega$$

and

$$\mathbf{Z}_p = 38.39\underline{/14.33^\circ}\ \Omega$$

Practice Problem 17.14

Let the line voltage $V_L = 208$ V and the wattmeter readings of the balanced system in Fig. 17.34 be $P_1 = -560$ W (negative sign indicates that the current coils are reversed) and $P_2 = 800$ W. Determine:
(a) the total average power,
(b) the total reactive power,

(c) the power factor,
(d) the phase impedance. Is the impedance inductive or capacitive?

Answer: (a) 240 W; (b) 2355.6 VAR; (c) 0.9948; (d) $179.1\underline{/5.8174^\circ}$; inductive

Example 17.15

The three-phase balanced load in Fig. 17.36 has impedance per phase $\mathbf{Z}_Y = 8 + j6\ \Omega$. If the load is connected to 208-V lines, predict the readings of wattmeters W_1 and W_2. Find P_T and Q_T.

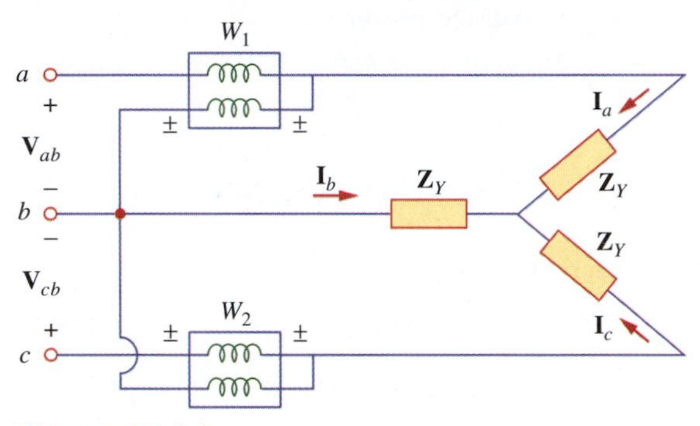

Figure 17.36
Two-wattmeter method applied to a balanced wye load.

Solution:
The impedance per phase is

$$\mathbf{Z}_Y = 8 + j6 = 10\underline{/36.87^\circ}\ \Omega$$

so that the *pf* angle is 36.87°.
Since the line voltage is $V_L = 208$ V, the line current is

$$I_L = \frac{V_p}{|\mathbf{Z}_Y|} = \frac{208/\sqrt{3}}{10} = 12 \text{ A}$$

Then

$$\begin{aligned} P_1 &= V_L I_L \cos(\theta + 30^\circ) = 208 \times 12 \times \cos(36.87^\circ + 30^\circ) \\ &= 980.48 \text{ W} \\ P_2 &= V_L I_L \cos(\theta - 30^\circ) = 208 \times 12 \times \cos(36.87^\circ - 30^\circ) \\ &= 2478.1 \text{ W} \end{aligned}$$

Thus, wattmeter 1 reads 980.48 W while wattmeter 2 reads 2,478.1 W. Because $P_2 > P_1$, the load is inductive. This is evident from the load $\mathbf{Z}_Y$ itself.

$$P_T = P_1 + P_2 = 3{,}458.1 \text{ kW}$$

and

$$Q_T = \sqrt{3}(P_2 - P_1) = \sqrt{3}(1{,}497.62) \text{ VAR} = 2.594 \text{ kVAR}$$

Practice Problem 17.15

If the load in Fig. 17.36 is delta-connected with impedance per phase of $\mathbf{Z}_Y = 30 - j40\ \Omega$ and $V_L = 440$ V, predict the readings of the wattmeters W_1 and W_2. Calculate P_T and Q_T.

Answer: 6.166 kW; 0.8021 kW; 6.968 kW; −9.291 kVAR

17.11.2 Residential Electrical Wiring

In the United States, most household lighting and appliances operate on 120-V, 60-Hz single-phase alternating current. (The electricity may also be supplied at 110, 115, or 117 V, depending on the location.) The local power company supplies the house with a three-wire ac system. As shown in Fig. 17.37, the line voltage of, say, 12,000 V is typically stepped down to 120/240 V with a transformer (more details on transformers in the next chapter). The three wires coming from the transformer are typically colored red (hot), black (hot), and white (neutral). As shown in Fig. 17.38, the two 120-V voltages are opposite in phase and, hence, add up to zero. That is,

$$\mathbf{V}_W = 0\underline{/0^\circ}, \quad \mathbf{V}_B = 120\underline{/0^\circ}, \quad \mathbf{V}_R = 120\underline{/180^\circ} = -\mathbf{V}_B$$
$$\mathbf{V}_{BR} = \mathbf{V}_B - \mathbf{V}_R = \mathbf{V}_B - (-\mathbf{V}_B) = 2\mathbf{V}_B = 240\underline{/0^\circ} \quad \textbf{(17.57)}$$

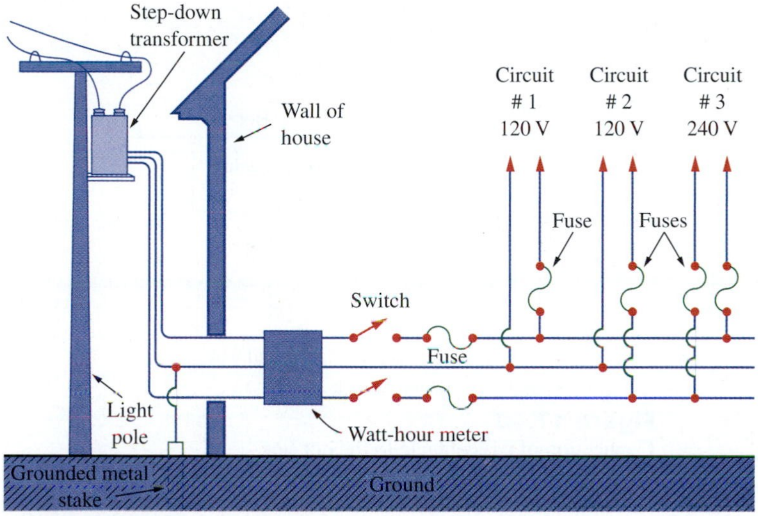

Figure 17.37
A 120/240 household power system.
A. Marcus and C. M. Thomson, *Electricity for Technicians,* 2nd edition, © 1975, p. 324. Reprinted by permission of Pearson Education, Inc., Upper Saddle River, NJ.

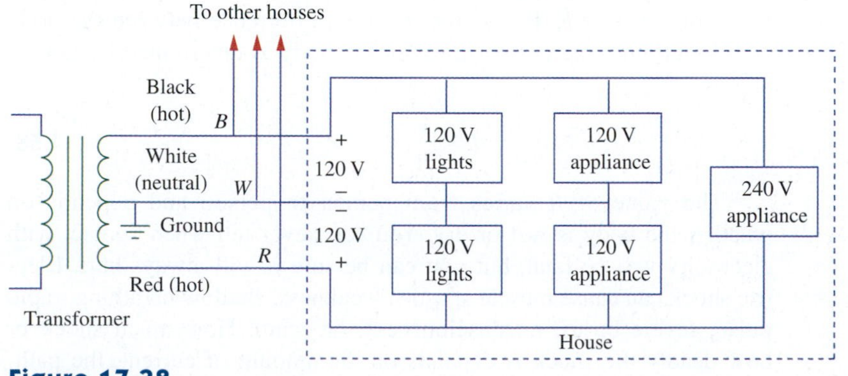

Figure 17.38
Single-phase three-wire residential writing.

Because most appliances are designed to operate with 120 V, the lighting and appliances in a room are connected to the 120-V lines, as illustrated in Fig. 17.39. Notice in Fig. 17.38 that all appliances are connected in parallel. Heavy appliances (those that consume large amount of power)—such as air conditioners, dishwashers, ovens, laundry machines, and dryers—are connected to the 240-V power line.

Because of the potential dangers of electricity, house wiring is carefully regulated by a code drawn by local ordinances and by the National Electrical Code (NEC) in the United States. [In Canada, electrical work is governed by the Canadian Electrical Code (CEC)]. To avoid trouble, insulation, grounding, fuses, and circuit breakers are used. Modern wiring codes require a third wire for a separate ground. The ground wire does not carry current like the neutral wire but enables appliances to have a separate ground connection. For example, the connection of the receptacle to a 120-V rms line and to the ground is shown in Fig. 17.40. As shown in the figure, the neutral line is connected to the ground (the earth) at some critical locations. Although the ground line seems redundant, grounding is important for many reasons.

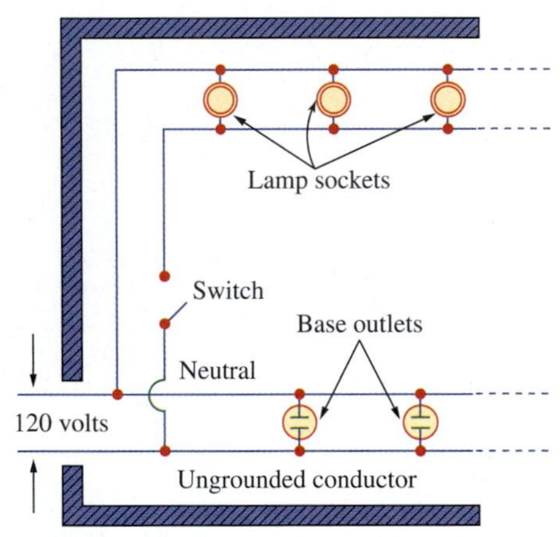

Figure 17.39
A typical wiring diagram of a room.
A. Marcus and C. M. Thomson, *Electricity for Technicians,* 2nd edition, © 1975, p. 325. Reprinted by permission of Pearson Education, Inc., Upper Saddle River, NJ.

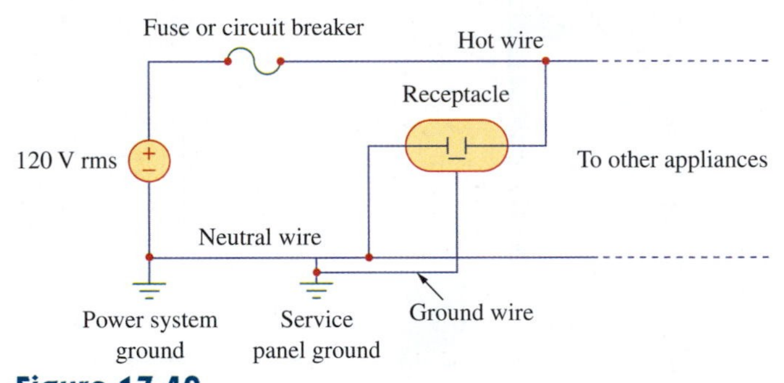

Figure 17.40
Connection of a receptacle to the hot line and to the ground.

First, it is required by NEC. Second, grounding provides a convenient path to ground for lightning that strikes the power line. Third, ground wires minimize the risk of electric shock. What causes shock is current passing from one part of the body to another. The human body is like a big resistor R. If V is the potential difference between the body and the ground, the current through the body is determined by Ohm's law as

$$I = \frac{V}{R} \tag{17.58}$$

The value of R varies from person to person and depends on whether the body is wet or dry. You can never tell when contact with electricity will be fatal, but you can be sure it will always hurt. Electric shock can cause muscle spasms, weakness, shallow breathing, rapid pulse, severe burns, unconsciousness, or death. How much shock or how deadly the shock is depends on the amount of current, the pathway of the current through the body, and how long the body is exposed to the current. Currents less than 1 mA may not be harmful to the body, but currents greater than 10 mA can cause severe shock. An approximate general framework for shock effects is shown in Table 17.2.

A modern safety device is the *ground-fault circuit interrupter* (GFCI), used in outdoor circuits and in bathrooms, where the risk of

TABLE 17.2

Electric shock

Electric Current	Physiological effect
Less than 1mA	No sensation or feeling
1 mA	Tingling sensation
5–20 mA	Involuntary muscle contraction
20–100mA	Loss of breathing, fatal if continued

electric shock is greatest. It is essentially a circuit breaker that opens when the sum of the currents *iR*, *iW*, and *iB* through the red, white, and the black lines is not equal to zero (which results if there is another path for the current, e.g., through a person); that is,

$$i_R + i_W + i_B \neq 0$$

The best way to avoid electric shock is to follow safety guidelines. Here are some of them:

- Never assume that an electrical circuit is dead. Always check to be sure.
- When necessary wear suitable clothing (insulated shoes, gloves, etc.)
- Never use two hands when testing high-voltage circuits because the current through one hand to the other hand has a direct path through your chest and heart.
- Do not touch an electrical appliance when you are wet. Remember that water conducts electricity.
- Be extremely careful when working with electronic appliances such as radios and TVs because these appliances have large capacitors in them, potentially charged to very high voltages and containing large amounts of energy. The capacitors take time to discharge after the power is disconnected.
- Always have another person present in case of an accident.
- In case you find someone experiencing electric shock, do not touch the victim still in contact with an electrical source to avoid endangering yourself. Use a wooden pole or board to separate the victim from the power source. Turn off the source of power if possible and call 911 for emergency immediately.

17.12 Summary

1. The phase sequence is the order in which the phase voltages of a three-phase generator occur with respect to time. In an *abc* sequence of balanced source voltages, $\mathbf{V}_{an}$ leads $\mathbf{V}_{bn}$ by 120°, which in turn leads $\mathbf{V}_{cn}$ by 120°. In an *acb* sequence balanced voltages, $\mathbf{V}_{an}$ leads $\mathbf{V}_{cn}$ by 120°, which in turn leads $\mathbf{V}_{bn}$ by 120°.
2. A balanced wye- or delta-connected load is one in which the three-phase impedances are equal.

3. The easiest way to analyze a balanced three-phase circuit is to transform both the source and the load to a *Y-Y* system and then analyze the single-phase equivalent circuit. Table 17.1 presents a summary of the formulas for phase currents and voltages and line currents and voltages for the four configurations for balanced systems.
4. The line current I_L is the current flowing from the generator to the load in each transmission line in a three-phase system. The line voltage V_L is the voltage between each pair of lines, excluding the neutral line if it exists. The phase current I_p is the current flowing through each phase in a three-phase load. The phase voltage V_p is the voltage of each phase.

   For a wye-connected load,

$$V_L = \sqrt{3}V_P \qquad \text{and} \qquad I_L = I_p$$

   For a delta-connected load,

$$V_L = V_P \qquad \text{and} \qquad I_L = \sqrt{3}I_p$$

5. The total instantaneous power in a balanced three-phase system is constant and equal to the average power.
6. The total complex power absorbed by a balanced three-phase *Y*-connected or Δ-connected load is

$$\mathbf{S} = P + jQ = \sqrt{3}V_L I_L \underline{/\theta}$$

   where θ is the angle of the load impedances.
7. An unbalanced three-phase system can be analyzed using nodal or mesh analysis.
8. PSpice is used to analyze three-phase circuits in the same way as it is used for analyzing single phase circuits.
9. The total real power is measured in three-phase systems using either the three-wattmeter method or the two-wattmeter method.
10. Residential wiring in the United States and Canada uses a 120/240-V, single-phase, three-wire system.

Review Questions

17.1 What is the phase sequence of a three-phase motor for which $\mathbf{V}_{AN} = 220\underline{/-100^\circ}$ V and $\mathbf{V}_{BN} = 220\underline{/140^\circ}$ V?

(a) *abc* (b) *acb*

17.2 If in an *acb* phase sequence, $\mathbf{V}_{an} = 100\underline{/-20^\circ}$, then $\mathbf{V}_{cn}$ is:

(a) $100\underline{/-140^\circ}$ (b) $100\underline{/100^\circ}$

(c) $100\underline{/-50^\circ}$ (d) $100\underline{/10^\circ}$

17.3 Which of these is not a required condition for a balanced system?

(a) $|\mathbf{V}_{an}| = |\mathbf{V}_{bn}| = |\mathbf{V}_{cn}|$

(b) $\mathbf{I}_a + \mathbf{I}_b + \mathbf{I}_c = 0$

(c) $\mathbf{V}_{an} + \mathbf{V}_{bn} + \mathbf{V}_{ac} = 0$

(d) Source voltages are 120° out of phase with each other.

(e) Load impedances for the three phases are equal.

17.4 In a Y-connected load, the line current and phase current are equal.

(a) True (b) False

17.5 In a Δ-connected load, the line current and phase current are equal.

(a) True (b) False

17.6 In a *Y-Y* system, a line voltage of 220 V produces a phase voltage of:

(a) 381 V (b) 311 V (c) 220 V

(d) 156 V (e) 127 V

17.7 In a Δ-Δ system, a phase voltage of 100 V produces a line voltage of:

(a) 58 V (b) 71 V
(c) 100 V (d) 173 V
(e) 141 V

17.8 When a *Y*-connected load is supplied by voltages in *abc* phase sequence, the line voltages lag the corresponding phase voltages by 30°.

(a) True (b) False

17.9 In a balanced three-phase circuit, the total instantaneous power is equal to the average power.

(a) True (b) False

17.10 The total power supplied to a balanced delta-load is found in the same way as for a balanced *Y*-load.

(a) True (b) False

Answers: 17.1a, 17.2a, 17.3c, 17.4a, 17.5b, 17.6e, 17.7c, 17.8b, 17.9a, 17.10a

Problems

Section 17.3 Balanced Three-Phase Voltages

17.1 If $\mathbf{V}_{ab} = 400$ V in a balanced *Y*-connected three-phase generator, find the phase voltages assuming the phase sequence is: (a) *abc* and (b) *acb*.

17.2 What is the phase sequence of a balanced three-phase circuit for which $\mathbf{V}_{an} = 160\underline{/30^\circ}$ V and $\mathbf{V}_{cn} = 160\underline{/-90^\circ}$ V? Find $\mathbf{V}_{bn}$.

17.3 Determine the phase sequence of a balanced three-phase circuit in which $\mathbf{V}_{bn} = 208\underline{/130^\circ}$ V and $\mathbf{V}_{cn} = 208\underline{/10^\circ}$ V? Obtain $\mathbf{V}_{an}$.

17.4 Assuming the *abc* sequence, if $\mathbf{V}_{ca} = 208\underline{/20^\circ}$ V in a balanced three-phase circuit, find $\mathbf{V}_{ab}$, $\mathbf{V}_{bc}$, $\mathbf{V}_{an}$, and $\mathbf{V}_{bn}$.

17.5 If $\mathbf{V}_{An} = 440\underline{/30^\circ}$ V, determine $\mathbf{V}_{nA}$.

17.6 For a certain network, we know that $\mathbf{V}_{12} = 120\underline{/30^\circ}$ V, $\mathbf{V}_{42} = 60\underline{/-60^\circ}$ V, and $\mathbf{V}_{45} = 40\underline{/90^\circ}$ V. Find $\mathbf{V}_{14}$ and $\mathbf{V}_{25}$.

Section 17.4 Balanced Wye-Wye Connection

17.7 A three-phase wye-connected generator has a line voltage of 440 V. Determine the phase voltage.

17.8 A three-phase wye-connected load has a phase voltage of 127 V. Find the line voltage.

17.9 A three-phase wye-connected generator has an *abc* phase sequence. If one phase voltage is $\mathbf{V}_{AN} = 230\underline{/15^\circ}$ V, find: (a) the remaining phase voltages and (b) the line voltages $\mathbf{V}_{AB}$, $\mathbf{V}_{BC}$, and $\mathbf{V}_{CA}$.

17.10 For the *Y-Y* circuit of Fig. 17.41, find the line currents, the line voltages, and the load voltages.

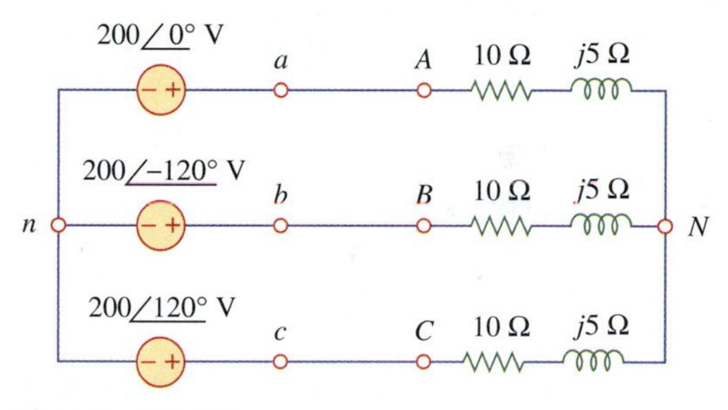

Figure 17.41
For Problem 17.10.

17.11 Calculate the line currents in the three-phase circuit of Fig. 17.42.

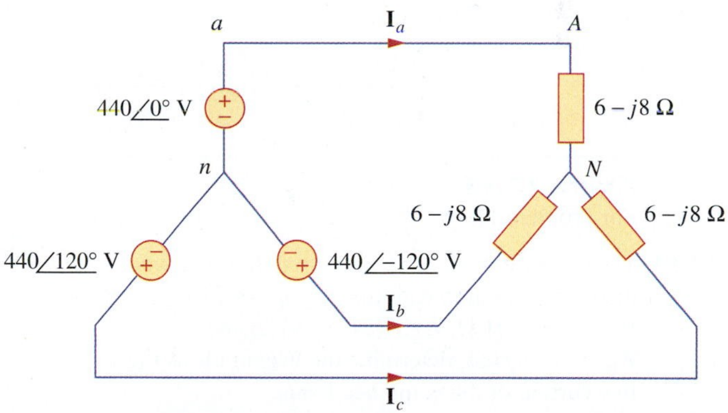

Figure 17.42
For Problems 17.11. and 17.44.

17.12 A balanced *Y*-connected load with a phase impedance of $16 + j9\ \Omega$ is connected to a balanced three-phase source with a line voltage of 220 V. Calculate the line current I_L.

17.13 A balanced *Y-Y* four-wire system has phase voltages $\mathbf{V}_{an} = 120\underline{/0^\circ}$, $\mathbf{V}_{bn} = 120\underline{/-120^\circ}$, $\mathbf{V}_{cn} = 120\underline{/120^\circ}$ V. The load impedance per phase is $19 + j13\ \Omega$ and the line impedance per phase is $1 + j2\ \Omega$. Solve for the line currents and neutral current.

17.14 In a four-wire wye-wye system, the line currents are $8\underline{/-30^\circ}$ A, $12\underline{/60^\circ}$ A, and $-j16$ A. Find the current in the neutral line.

Section 17.5 Balanced Wye-Delta Connection

17.15 For the three-phase circuit of Fig. 17.43, $\mathbf{I}_{bB} = 30\underline{/60^\circ}$ A and $\mathbf{V}_{BC} = 220\,\underline{/10^\circ}$ V. Find $\mathbf{V}_{an}$, $\mathbf{V}_{AB}$, $\mathbf{I}_{AC}$, and $\mathbf{Z}$.

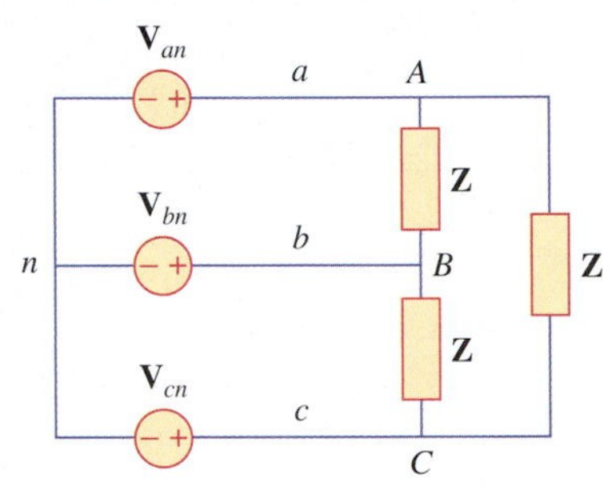

Figure 17.43
For Problem 17.15.

17.16 Solve for the line currents in the Y-Δ circuit of Fig. 17.44. Take $\mathbf{Z}_\Delta = 60\underline{/45^\circ}\ \Omega$.

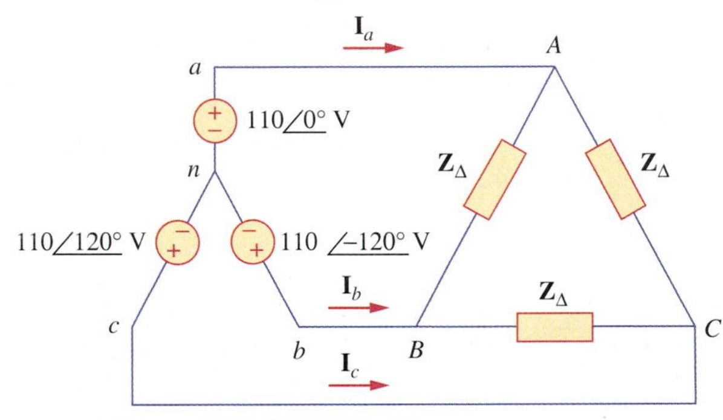

Figure 17.44
For Problem 17.16.

17.17 The circuit in Fig. 17.45 is excited by a balanced three-phase source with line voltage of 210 V. If $\mathbf{Z}_\ell = 1 + j1\ \Omega$, $\mathbf{Z}_\Delta = 24 - j30\ \Omega$, and $\mathbf{Z}_Y = 12 + j5\ \Omega$, determine the magnitude of the line current of the combined loads.

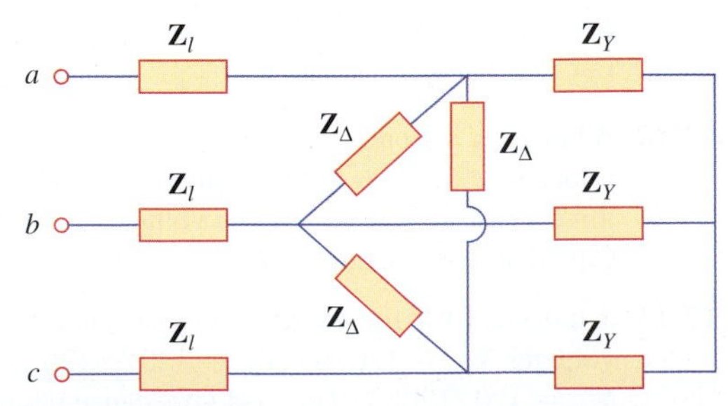

Figure 17.45
For Problem 17.17.

17.18 A balanced delta-connected load has a phase current $\mathbf{I}_{AC} = 10\underline{/-30^\circ}$ A.

(a) Determine the three line currents assuming that the circuit operates in the positive phase sequence.

(b) Calculate the load impedance if the line voltage is $\mathbf{V}_{AB} = 110\underline{/0^\circ}$ V.

17.19 A 240-V three-phase power system supplied 25 kW to a delta-connected balanced load with a power factor of 0.8, lagging. Find: (a) the line currents, (b) the phase currents, and (c) the phase impedance.

Section 17.6 Balanced Delta-Delta Connection

17.20 For the Δ-Δ circuit of Fig. 17.46, calculate the phase and line currents.

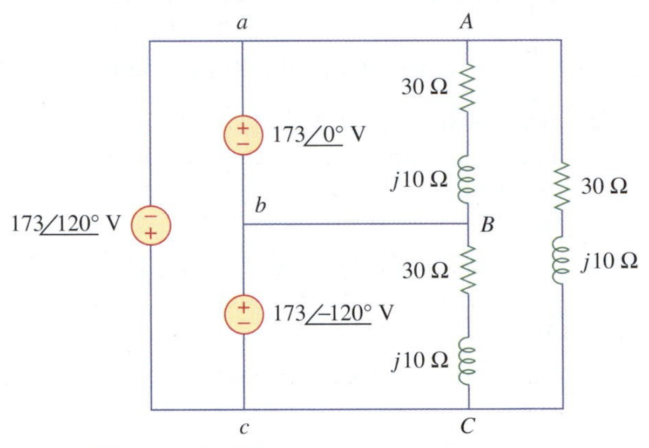

Figure 17.46
For Problem 17.20.

17.21 Refer to the Δ-Δ circuit shown in Fig. 17.47. Find the line and phase currents. Assume that the load impedance is $12 + j9\ \Omega$ per phase.

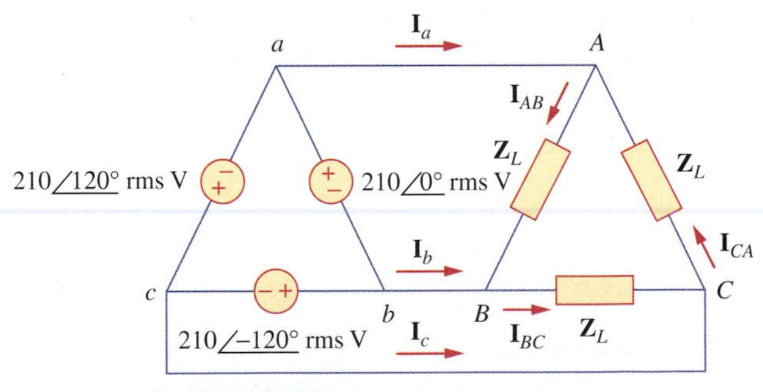

Figure 17.47
For Problem 17.21.

17.22 A three-phase balanced system with a line voltage of 208 V rms feeds a delta-connected load with $\mathbf{Z}_p = 25\underline{/60^\circ}\ \Omega$. (a) Find the line currents.
(b) Determine the total power supplied to the load using two wattmeters connected to the A and C lines.

17.23 A balanced delta-connected source has phase voltage $\mathbf{V}_{ab} = 416\underline{/30^\circ}$ V and a positive phase sequence. If this is connected to a balanced delta-connected load, find the line and phase currents. Take the load impedance per phase as $60\underline{/30^\circ}\ \Omega$ and the line impedance per phase as $1 = j1\ \Omega$.

17.24 Refer to the three-phase load shown in Fig. 17.48. Let $\mathbf{V}_{AB} = 120\underline{/0^\circ}$ V, $\mathbf{V}_{BC} = 120\underline{/-120^\circ}$ V, and $\mathbf{V}_{CA} = 120\underline{/120^\circ}$ V.

(a) Calculate $\mathbf{V}_{an}$, $\mathbf{V}_{bn}$, and $\mathbf{V}_{cn}$.

(b) Determine $\mathbf{I}_{an}$, $\mathbf{I}_{bn}$, and $\mathbf{I}_{cn}$.

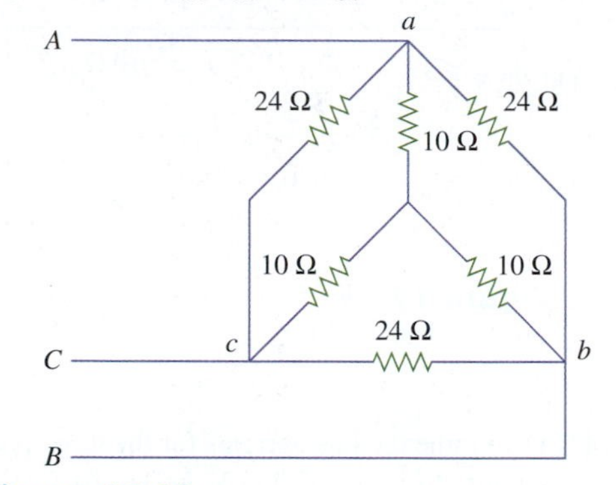

Figure 17.48
For Problems 17.24 and 17.30.

Section 17.7 Balanced Delta-Wye Connection

17.25 In the circuit of Fig. 17.49, if $\mathbf{V}_{ab} = 440\underline{/10^\circ}$, $\mathbf{V}_{bc} = 440\underline{/250^\circ}$, $\mathbf{V}_{ca} = 440\underline{/130^\circ}$ V, find the line currents.

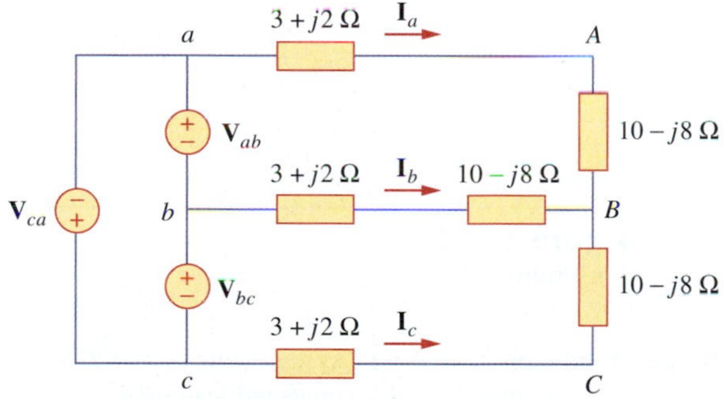

Figure 17.49
For Problem 17.25.

17.26 For the balanced circuit in Fig. 17.50, $\mathbf{V}_{ab} = 125\underline{/0^\circ}$ V. Find the line currents $\mathbf{I}_{aA}$, $\mathbf{I}_{bB}$, and $\mathbf{I}_{cC}$.

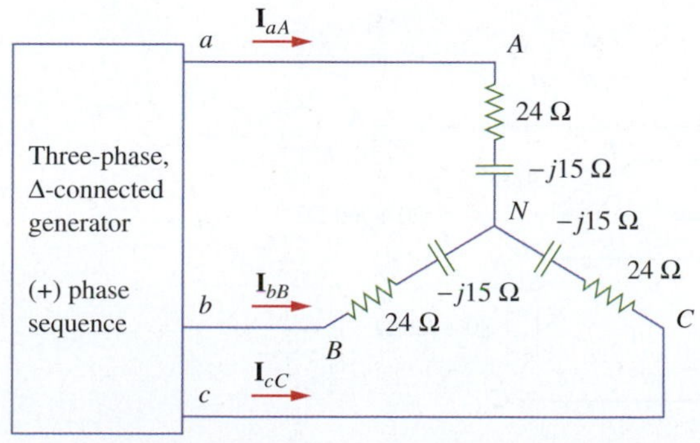

Figure 17.50
For Problem 17.26.

17.27 In a balanced three-phase Δ-Y circuit, the source is connected in the positive sequence with $\mathbf{V}_{ab} = 220\underline{/20^\circ}$ V and $\mathbf{Z}_Y = 20 + j15$ Ω. Find the line currents.

17.28 A delta-connected generator supplies a balanced wye-connected load with an impedance of $30\underline{/-60^\circ}$ Ω. If the line voltages of the generator have a magnitude of 400 V and are in the positive phase sequence, find the line current I_L and phase voltage V_p at the load.

17.29 A balanced delta-wye system is shown in Fig. 17.51. Find the magnitude of the line current.

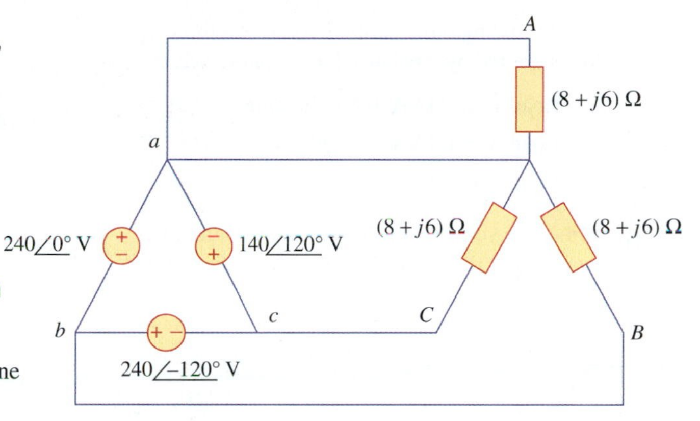

Figure 17.51
For Problem 17.29.

Section 17.8 Power in a Balanced System

17.30 For the system in Fig. 17.48 (Problem 17.24), find:
(a) the power delivered to thc 10-Ω wye load,
(b) the power delivered to the 24-Ω delta load.

17.31 Determine the maximum total instantaneous power supplied by a three-phase generator if each phase winding delivers an average of 4 kW. Assume a balanced, three-phase load.

17.32 A balanced wye-connected load absorbs a total power of 5 kW at a leading power factor of 0.6 when connected to a line voltage of 240 V. Find the impedance of each phase and the total complex power of the load.

17.33 A balanced wye-connected load absorbs 50 kVA at 0.6 lagging power factor when the line voltage is 440 V. Find the line current and the phase impedance.

17.34 A three-phase source delivers 4800 VA to a wye-connected load with a phase voltage of 208 V and a power factor of 0.9 lagging. Calculate the source line current and the source line voltage.

17.35 A balanced wye-connected load with a phase impedance of $10 - j16$ Ω is connected to a balanced three-phase generator with a line voltage of 220 V. Determine the line current and the complex power absorbed by the load.

17.36 Find the real power absorbed by the load in Fig. 17.52.

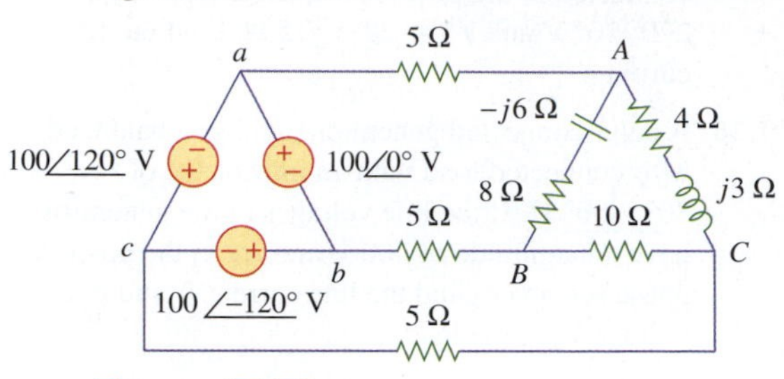

Figure 17.52
For Problem 17.36.

17.37 The following three parallel-connected three-phase loads are fed by a balanced three-phase source.

Load 1: 250 kVA, 0.8 *pf* lagging

Load 2: 300 kVA, 0.95 *pf* leading

Load 3: 450 kVA, unity *pf*

If the line voltage is 13.8 kV, calculate the line current and the power factor of the source. Assume that the line impedance is zero.

Section 17.9 Unbalanced Three-Phase Systems

17.38 For the circuit in Fig. 17.53, $\mathbf{Z}_a = 6 - j8\ \Omega$, $\mathbf{Z}_b = 12 + j9\ \Omega$, and $\mathbf{Z}_c = 15\ \Omega$. Find the line currents $\mathbf{I}_a$, $\mathbf{I}_b$, and $\mathbf{I}_c$.

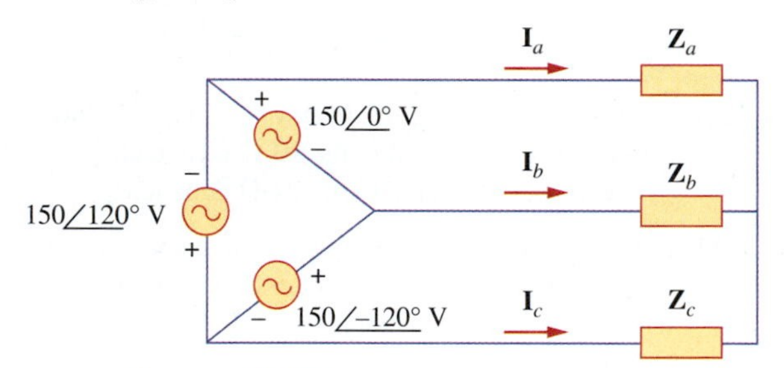

Figure 17.53
For Problem 17.38.

17.39 A delta-connected load whose phase impedances are $\mathbf{Z}_{AB} = 50\ \Omega$, $\mathbf{Z}_{BC} = -j50\ \Omega$, and $\mathbf{Z}_{CA} = j50\ \Omega$ is fed by a balanced wye-connected three-phase source with $V_p = 100$ V. Find the phase currents.

17.40 Refer to the unbalanced circuit in Fig. 17.54. Calculate: (a) the line currents, (b) the real power absorbed by the load, (c) the total complex power supplied by the source.

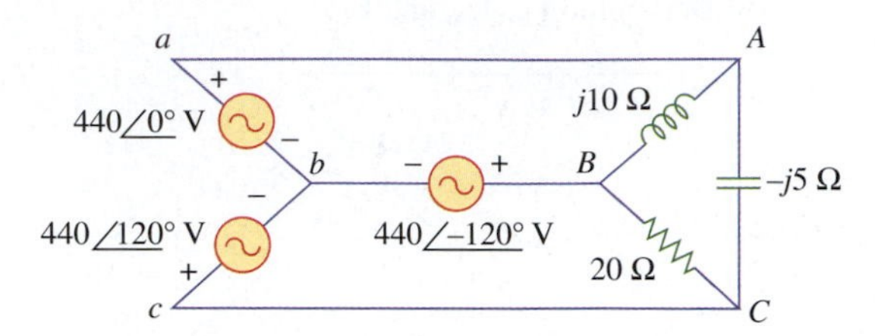

Figure 17.54
For Problem 17.40.

17.41 Determine the line currents for the three-phase circuit of Fig. 17.55. Let $\mathbf{V}_a = 110\angle 0^\circ$, $\mathbf{V}_b = 110\angle -120^\circ$, $\mathbf{V}_c = 110\angle 120^\circ$ V.

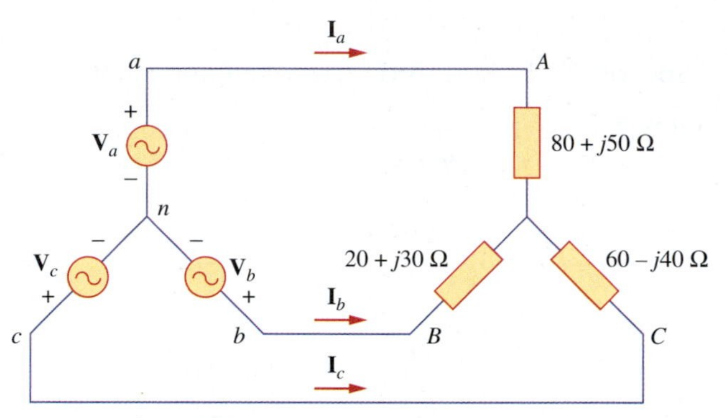

Figure 17.55
For Problem 17.41.

17.42 A three-phase four wire system with $V_L = 120$ V feeds an unbalanced Y-connected load with $Z_A = 4\angle 90^\circ\ \Omega$, $Z_B = 10\angle 60^\circ\ \Omega$, and $Z_C = 8\angle 0^\circ\ \Omega$. Calculate the four line currents.

17.43 Find the complex power delivered to the three-phase load in the wye-wye system shown in Fig. 17.56.

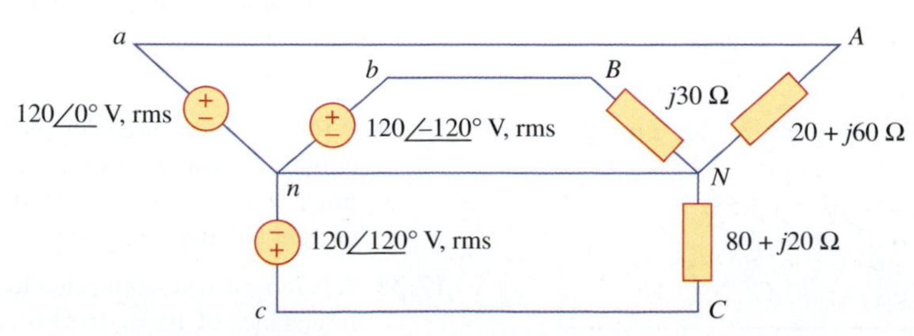

Figure 17.56
For Problem 17.43.

Section 17.10 Computer Analysis

17.44 Solve Problem 17.11 using PSpice.

17.45 The source in Fig. 17.57 is balanced and exhibits a positive phase sequence. If $f = 60$ Hz, use PSpice to find $\mathbf{V}_{AN}$, $\mathbf{V}_{BN}$, and $\mathbf{V}_{CN}$.

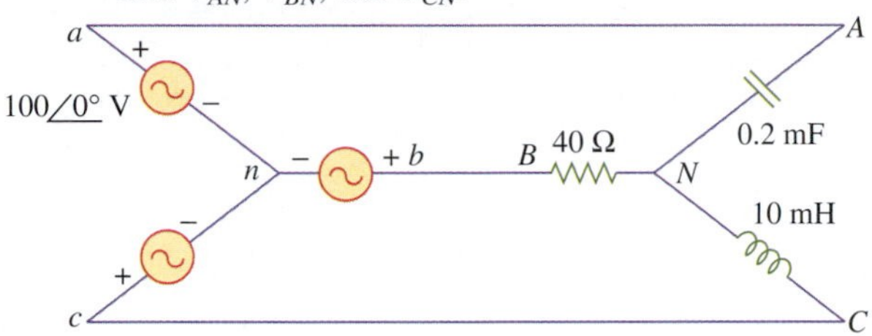

Figure 17.57
For Problem 17.45.

17.46 Use PSpice to determine $\mathbf{I}_o$ in the single-phase, three-wire circuit of Fig. 17.58. Let $\mathbf{Z}_1 = 15 - j10\ \Omega$, $\mathbf{Z}_2 = 30 + j20\ \Omega$ and $\mathbf{Z}_3 = 12 + j5\ \Omega$.

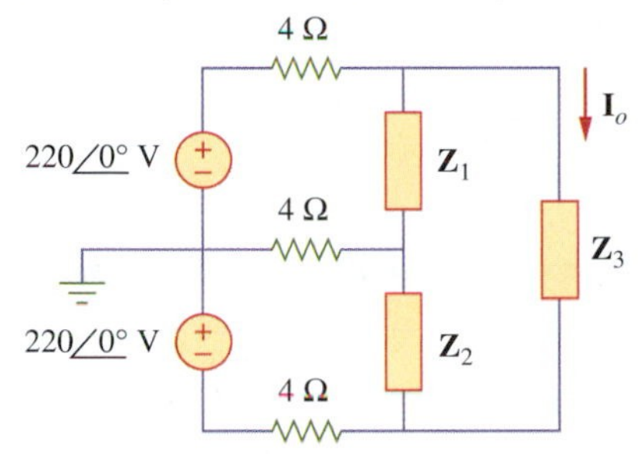

Figure 17.58
For Problem 17.46.

17.47 The circuit in Fig. 17.59 operates at 60 Hz. Use PSpice to find the source current $\mathbf{I}_{ab}$ and the line current $\mathbf{I}_{bB}$.

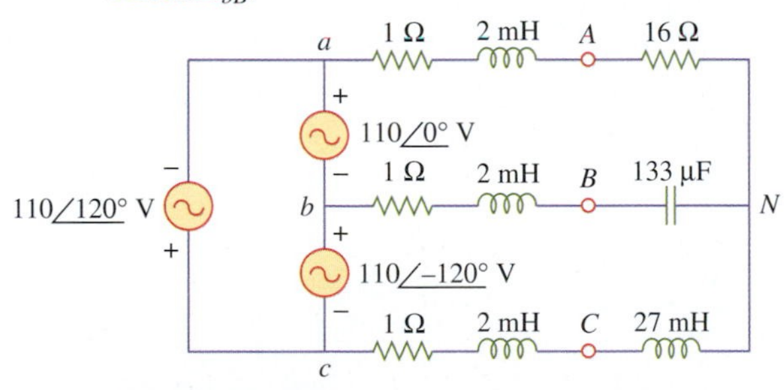

Figure 17.59
For Problem 17.47.

Section 17.11 Applications

17.48 A professional center is supplied by a balanced three-phase source. The center has four balanced three-phase loads as follows:

Load 1: 150 kVA at 0.8 *pf* leading

Load 2: 100 kW at unity *pf*

Load 3: 200 kVA at 0.6 *pf*

Load 4: 80 kW and 95 kVAR (inductive)

If the line impedance is $0.02 + j0.05\ \Omega$ per phase and the line voltage at the loads is 480 V, find the magnitude of the line voltage at the source.

17.49 The two-wattmeter method gives $P_1 = 1200$ W and $P_2 = -400$ W for a three-phase motor running on a 240-V line. Assume that the motor load is wye-connected and that it draws a line current of 6 A. Calculate the *pf* of the motor and its phase impedance.

17.50 For the circuit displayed in Fig. 17.60, find the wattmeter readings.

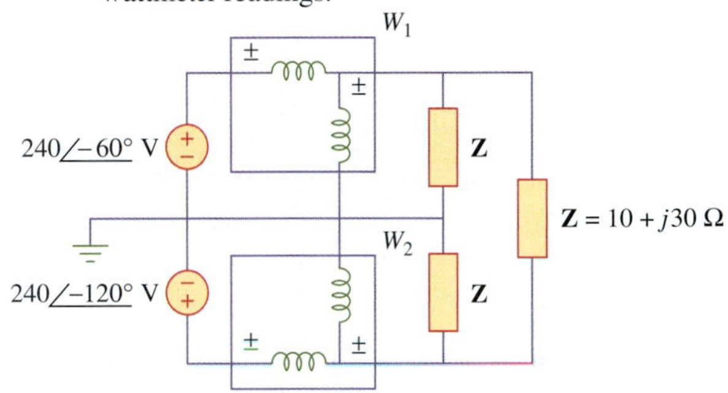

Figure 17.60
For Problem 17.50.

17.51 Predict the wattmeter readings for the circuit in Fig. 17.61.

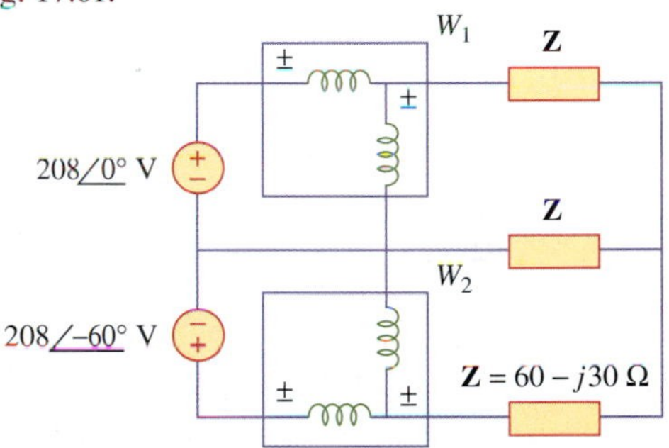

Figure 17.61
For Problem 17.51.

17.52 For the single-phase three-wire system in Fig.17.62, find currents $\mathbf{I}_{aA}$, $\mathbf{I}_{bB}$, and $\mathbf{I}_{nN}$.

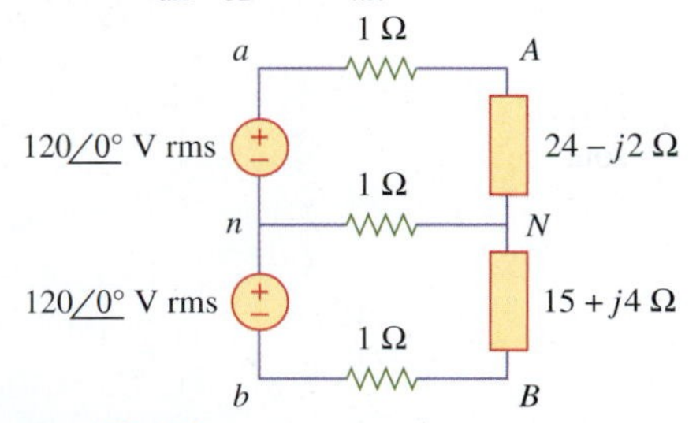

Figure 17.62
For Problem 17.52.

17.53 A 480-V source is used in powering a three-phase industrial plant that consumes 160 kVA at a power factor of 0.8 lagging. Find: (a) the current in each line, (b) the total power delivered, and (c) the total reactive power in the load.

chapter

18

Transformers and Coupled Circuits

He who adds not to his learning diminishes it.

—Talmud

Enhancing Your Career

Career in Control Systems

Transformers mounted on a pole.
© Sarhan M. Musa

Circuit analysis is also important in the area of control systems. A control system is designed to regulate the behavior of one or more variables in some desired manner. A lot of control systems are sensor based. Control systems play major roles in our everyday life. For example, household appliances such as heating and air-conditioning systems, switch-controlled thermostats, washers, and dryers, cruise controllers in automobiles, elevators, traffic lights, manufacturing plants, and navigation systems all utilize control systems. Control systems are widely used in various fields. In aerospace, for example, precision guidance of space probes, the wide range of operational modes of the space shuttle, and the ability to maneuver space vehicles remotely from Earth all require knowledge of control systems. In the manufacturing sector, repetitive production line operations are increasingly being taken over by robots, which are programmable control systems designed to mimic humans and operate for hours without fatigue. From modern agribusiness to sewage treatment, from subways to skyways, no field of modern endeavor operates without some dependence on control systems.

Control engineering technology integrates circuit theory and communication theory. It is not limited to any specific engineering discipline but may involve environmental, chemical, aeronautical, mechanical, civil, and electrical engineering. For example, a typical task for a control system technologist might be to design a speed regulator for a compact disk. Another one may be to develop a terahertz (THz) sensor system capable of remotely identifying suspicious objects when concealed on people or within packages and capable of spectrally identifying explosives and drugs. (That system would be based on continuous-wave THz technology.)

A thorough understanding of control systems techniques is essential to the electrical technologies and is of great value in designing control systems to perform a desired task.

18.1 Introduction

The circuits we have considered so far may be regarded as *conductively coupled* because one loop affects the neighboring loop through current conduction through a wire. When two loops with or without contacts between them affect each other through the magnetic field generated by one of them, they are said to be *magnetically coupled.*

The transformer is an electrical device that takes advantage of the principles of magnetic coupling. It uses magnetically coupled coils to transfer energy from one circuit to another without a wired electrical connection. Transformers are important circuit elements. They are used in power systems for stepping up or stepping down ac voltages or currents. They are used in electronic circuits such as in radio and television receivers for such purposes as impedance matching and isolating one part of a circuit from another, as well as for stepping up or down ac voltages and currents.

We will begin with the concept of mutual inductance and introduce the dot convention used for determining the voltage polarities of inductively coupled components. We will consider the linear transformer, the ideal transformer, and the ideal autotransformer. Finally, as important applications of transformers, we look at transformers as isolating and matching devices and at their use in power distribution.

18.2 Mutual Inductance

When two coils are close to each other, a change in the current of one coil affects the current and voltage in the second coil. This is quantified in the property called *mutual inductance.* The mutual inductance of the two circuits is dependent on the geometrical arrangement of both circuits.

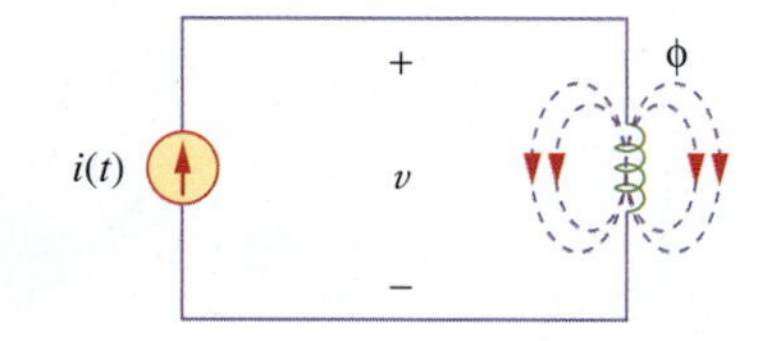

Figure 18.1
Magnetic flux produced by a single coil with N turns.

Let us first consider a single inductor, a coil with N turns. When current i flows through the coil, a magnetic flux ϕ is produced around it, as illustrated in Fig. 18.1. According to Faraday's law, the voltage v induced in the coil is proportional to the number of turns N and the time rate of change of the magnetic flux ϕ; that is,

$$v = N\frac{d\phi}{dt} \tag{18.1}$$

But the flux ϕ is produced by current i so that any change in flux is caused by a change in the current. Hence Eq. (18.1) can be written as

$$v = N\frac{d\phi}{di}\frac{di}{dt} = L\frac{di}{dt} \tag{18.2}$$

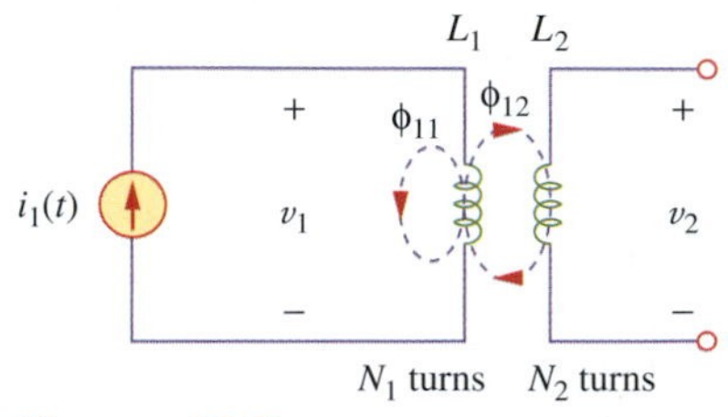

Figure 18.2
Mutual inductance M_{21} of coil 2 with respect to coil 1.

where $L = N(d\phi/dt)$ is the inductance of the inductor. Equation (18.2) is the voltage–current relationship for the inductor. The inductance L is commonly called *self-inductance* because it relates the voltage induced in a coil by a time-varying current in the same coil.

> **Self-inductance** is the ability of a wire to induce voltage in itself, measured in henrys (H).

Now we consider two coils with self-inductances L_1 and L_2, which are in close proximity with each other, as shown in Fig. 18.2. Coil 1 has N_1 turns, while coil 2 has N_2 turns. For the sake of simplicity, we will assume that the second inductor carries no current. The magnetic flux ϕ_1 emanating from coil 1 has two components: one component ϕ_{11} links only coil 1, and another component ϕ_{12} links both coils. Hence,

$$\phi_1 = \phi_{11} + \phi_{12} \tag{18.3}$$

Although the two coils are physically separated, they are said to be *magnetically coupled*. Because the entire flux ϕ_1 links coil 1, the voltage induced in coil 1 is

$$v_1 = N_1 \frac{d\phi_1}{dt} = N_1 \frac{d\phi_1}{di_1}\frac{di_1}{dt} = L_1 \frac{di_1}{dt} \tag{18.4}$$

where $L_1 = N_1 d\phi_1/di_1$ is the self-inductance of coil 1. Only flux ϕ_{12} links coil 2 so that the voltage induced in coil 2 is

$$v_2 = N_2 \frac{d\phi_{12}}{dt} = N_2 \frac{d\phi_{12}}{di_1}\frac{di_1}{dt} = M_{21} \frac{di_1}{dt} \tag{18.5}$$

where $M_{21} = N_2(d\phi_{12}/di_1)$ is known as the *mutual inductance* of coil 2 with respect to coil 1. Subscript 21 indicates that the inductance M_{21} relates the voltage induced in coil 2 to the current in coil 1. Thus, the open-circuit *mutual voltage* (or induced voltage) across coil 2 is

$$\boxed{v_2 = M_{21} \frac{di_1}{dt}} \tag{18.6}$$

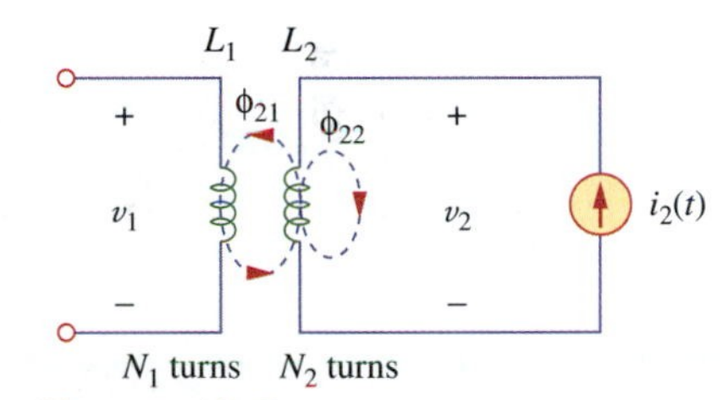

Figure 18.3
Mutual inductance M_{12} of coil 1 with respect to coil 2.

Suppose we now let current i_2 flow in coil 2, while coil 1 carries no current, as shown in Fig. 18.3. The magnetic flux ϕ_2 emanating from coil 2 comprises flux ϕ_{22} that links only coil 2 and flux ϕ_{21} that links both coils. Hence,

$$\phi_2 = \phi_{21} + \phi_{22} \tag{18.7}$$

The entire flux ϕ_2 links coil 2 so that the voltage induced in coil 2 is

$$v_2 = N_2 \frac{d\phi_2}{dt} = N_2 \frac{d\phi_2}{di_2}\frac{di_2}{dt} = L_2 \frac{di_2}{dt} \tag{18.8}$$

where $L_2 = N_2(d\phi_2/di_2)$ is the self-inductance of coil 2. Because only flux ϕ_{21} links coil 1, the voltage induced in coil 1 is

$$v_1 = N_1 \frac{d\phi_{21}}{dt} = N_1 \frac{d\phi_{21}}{di_2}\frac{di_2}{dt} = M_{12} \frac{di_2}{dt} \tag{18.9}$$

where $M_{12} = N_1(d\phi_{21}/di_2)$ is the *mutual inductance* of coil 1 with respect to coil 2. Thus, the open-circuit *mutual voltage* across coil 1 is

$$\boxed{v_1 = M_{12} \frac{di_2}{dt}} \tag{18.10}$$

It can be shown that M_{12} and M_{21} are equal; that is,

$$M_{12} = M_{21} = M \tag{18.11}$$

and we refer to M as the mutual inductance between the two coils. Like self-inductance L, mutual inductance M is measured in henrys (H). Keep in mind that mutual coupling only exists when the inductors or coils are in relatively close proximity and the circuits are driven by time-varying sources. We recall that inductors act like short-circuits to dc.

From the two cases in Figs. 18.2 and 18.3, we conclude that mutual voltage results if a voltage is induced by a time-varying current in another circuit. Mutual inductance is the property of an inductor to produce a voltage in reaction to a time-varying current in another inductor near it. Thus,

> **Mutual inductance** is the ability of one inductor to induce a voltage across a neighboring inductor, measured in henrys (H).

Although mutual inductance M is always a positive quantity, the mutual voltage $M(di/dt)$ may be negative or positive just like self-induced voltage $L(di/dt)$. However, unlike self-induced voltage $L(di/dt)$, whose polarity is determined by the reference direction of the current and the reference polarity of the voltage (according to the passive sign convention), the polarity of mutual voltage $M(di/dt)$ is not easy to determine because four terminals are involved. The choice of the correct polarity for $M(di/dt)$ is made by examining the orientation or particular way in which both coils are physically wound and applying Lenz's law in conjunction with the right-hand rule. Because it is not convenient to show the construction details of coils on a circuit schematic, we apply the *dot convention* in circuit analysis. By this convention, a dot is placed at one end of each of the two magnetically coupled coils to indicate the direction of the magnetic flux if current enters that dotted terminal of the coil. This is illustrated in Fig. 18.4. The dots appear on actual physical transforms. Given a circuit, the dots are already placed beside the coils so that we need not bother about how to place them. The dots are used along with the dot convention to determine the polarity of the mutual voltage. The dot convention is stated as follows:

> If a current **enters** the dotted terminal of one coil, the reference polarity of the mutual voltage in the second coil is **positive** at the dotted terminal of the second coil.

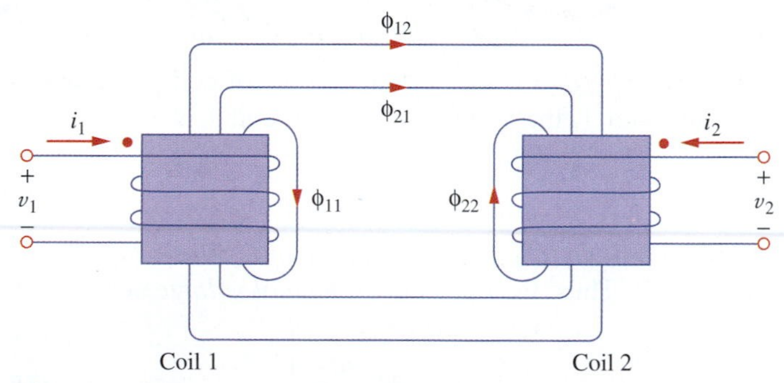

Figure 18.4
Illustration of the dot convention.

Alternatively,

> If a current **leaves** the dotted terminal of one coil, the reference polarity of the mutual voltage in the second coil is **negative** at the dotted terminal of the second coil.

Thus, the reference polarity of the mutual voltage depends on the reference direction of the inducing current and the dots on the coupled coils. Application of the dot convention is illustrated in the four pairs of mutually coupled coils in Fig. 18.5. For the coupled coils in

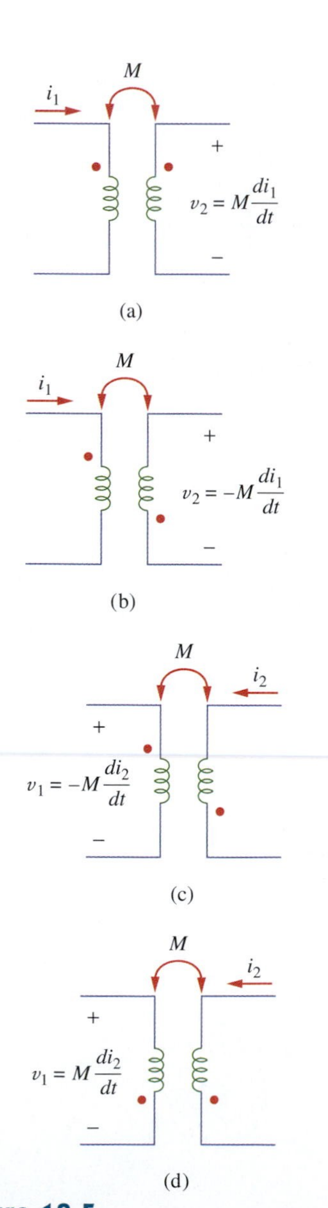

Figure 18.5
Examples illustrating how to apply the dot convention.

Fig. 18.5(a), the sign of the mutual voltage v_2 is determined by the reference polarity for v_2 and the direction of i_1. Because i_1 enters the dotted terminal of coil 1 and v_2 is positive at the dotted terminal of coil 2, the mutual voltage is $+M(di_1/dt)$. For the coils in Fig. 18.5(b), the current i_1 enters the dotted terminal of coil 1 and v_2 is negative at the dotted terminal of coil 2. Hence, the mutual voltage is $-M(di_1/dt)$. The same reasoning applies to the coils in Figs. 18.5(c) and 18.5(d). Figure 18.6 also shows the dot convention for coupled coils in series. For the coils in Fig. 18.6(a), the total inductance is

$$L = L_1 + L_2 + 2M \qquad \text{(Series-aiding connection)} \tag{18.12}$$

For the coil in Fig. 18.6(b),

$$L = L_1 + L_2 - 2M \qquad \text{(Series-opposing connection)} \tag{18.13}$$

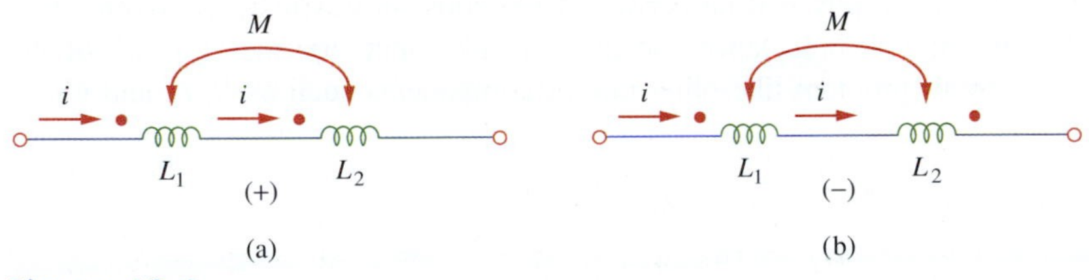

Figure 18.6
Dot convention for coils in series; the sign indicates the polarity of the mutual voltage: (a) series-aiding connection; (b) series-opposing connection.

For series-aiding connection, the magnetic field from one coil adds to that of the other coil so that the coupling is additive. For series-opposing connection, the magnetic field from each coil opposes that of the other coil so that the coupling is subtractive.

Now that we know how to determine the polarity of the mutual voltage, we are prepared to analyze circuits involving mutual inductance. As the first example, consider the circuit in Fig. 18.7. Applying KVL to coil 1 gives

$$v_1 = i_1R_1 + L_1\frac{di_1}{dt} + M\frac{di_2}{dt} \tag{18.14a}$$

For coil 2, KVL gives

$$v_2 = i_2R_2 + L_2\frac{di_2}{dt} + M\frac{di_1}{dt} \tag{18.14b}$$

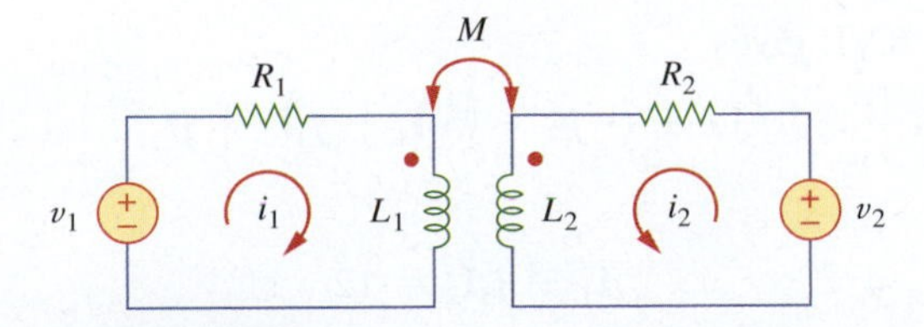

Figure 18.7
Time-domain analysis of a circuit containing coupled coils.

Equation (18.14) can be written in frequency domain as

$$\mathbf{V}_1 = (R_1 + j\omega L_1)\mathbf{I}_1 + j\omega M\mathbf{I}_2 \tag{18.15a}$$

$$\mathbf{V}_2 = j\omega M\mathbf{I}_1 + (R_2 + j\omega L_2)\mathbf{I}_2 \tag{18.15b}$$

As a second example, consider the circuit in Fig. 18.8. We analyze this in frequency domain. Applying KVL to coil 1, we get

$$\mathbf{V} = (\mathbf{Z}_1 + j\omega L_1)\mathbf{I}_1 - j\omega M\mathbf{I}_2 \tag{18.16a}$$

For coil 2, KVL yields

$$0 = -j\omega M\mathbf{I}_1 + (Z_L + j\omega L_2)\mathbf{I}_2 \tag{18.16b}$$

Equations (18.15) and (18.16) are solved in the usual manner to determine the currents.

To conclude this section, it should be said that at this introductory level, we are not concerned with the determination of the mutual inductances of the coils and their dot placements. Like R, L, and C, calculation of M would involve applying the theory of electromagnetism to the actual physical properties of the coils. In this text, we assume that the mutual inductance and the dot placements are the "givens" of the circuit problem like other circuit components such as R, L, and C.

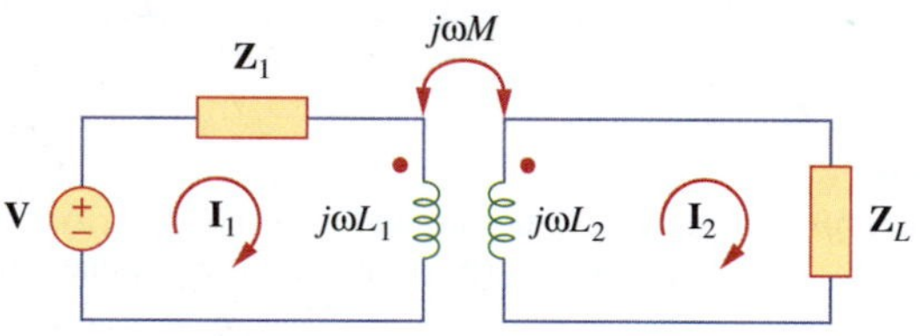

Figure 18.8
Frequency-domain analysis of a circuit containing coupled coils.

Example 18.1

Calculate the phasor currents $\mathbf{I}_1$ and $\mathbf{I}_2$ in the circuit of Fig. 18.9.

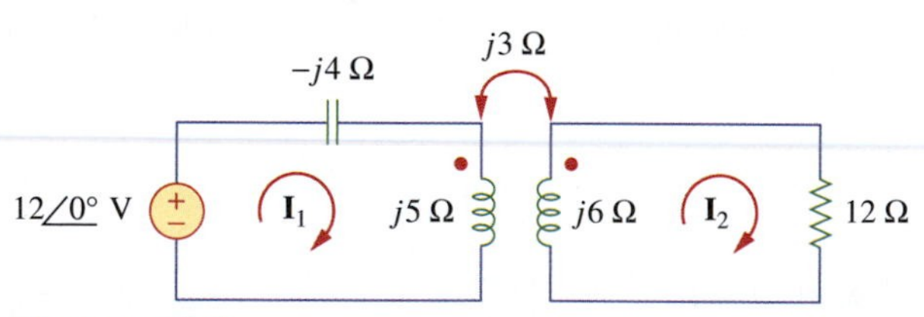

Figure 18.9
For Example 18.1, Practice Problem 18.6, and Problem 18.20.

Solution:

For coil 1, KVL gives

$$-12 + (-j4 + j5)\mathbf{I}_1 - j3\mathbf{I}_2 = 0$$

or

$$j\mathbf{I}_1 - j3\mathbf{I}_2 = 12 \tag{18.1.1}$$

For coil 2, KVL gives

$$-j3\mathbf{I}_1 + (12 + j6)\mathbf{I}_2 = 0$$

or

$$\mathbf{I}_1 = (12 + j6)\mathbf{I}_2/j3 = (2 - j4)\mathbf{I}_2 \qquad \textbf{(18.1.2)}$$

Substituting this in Eq. (18.1.1), we get

$$(j2 + 4 - j3)\mathbf{I}_2 = (4 - j)\mathbf{I}_2 = 12$$

or

$$\mathbf{I}_2 = \frac{12}{4 - j} = 2.91\underline{/14.04^\circ}\text{ A} \qquad \textbf{(18.1.3)}$$

From Eqs. (18.1.2) and (18.1.3),

$$\mathbf{I}_1 = (2 - j4)\mathbf{I}_2 = (4.472\underline{/-63.43^\circ})(2.91\underline{/14.04^\circ}) = 13\underline{/-49.4^\circ}\text{ A}$$

Practice Problem 18.1

Determine the voltage $\mathbf{V}_o$ in the circuit of Fig. 18.10.

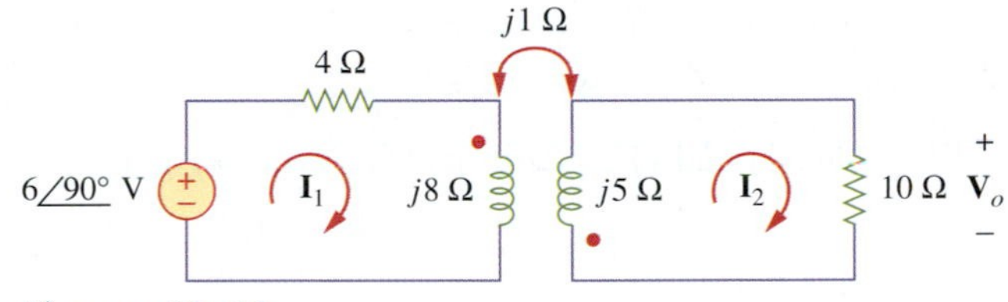

Figure 18.10
For Practice Problem 18.1.

Answer: $0.6\underline{/-90^\circ}$ V

Example 18.2

Calculate the mesh currents in the circuit shown in Fig. 18.11.

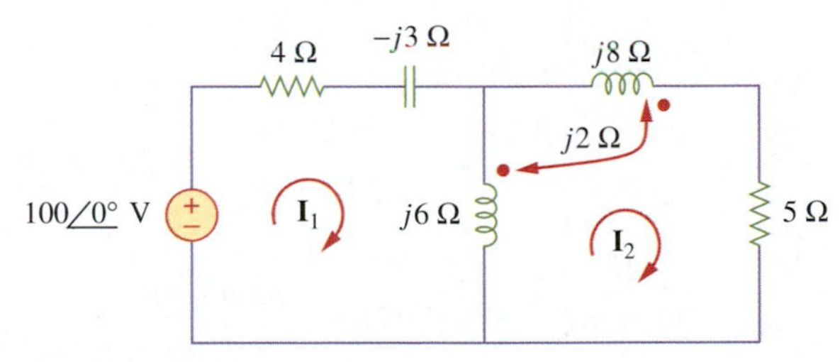

Figure 18.11
For Example 18.2.

Solution:
The key to analyzing a magnetically coupled circuit is knowing the polarity of the mutual voltage. We need to apply the dot rule stated earlier. In Fig. 18.11, suppose coil 1 is the one whose reactance is 6 Ω and coil 2 is the one whose reactance is 8 Ω. To figure out the polarity of the mutual voltage in coil 1 due to current $\mathbf{I}_2$, we observe that $\mathbf{I}_2$ leaves the dotted terminal of coil 2. Because we are applying KVL in the clockwise direction, it implies that the reference polarity of the mutual voltage in coil 1 is positive at the dotted terminal of that coil. Hence, the mutual voltage is negative according to the dot convention—that is, $-j2\mathbf{I}_2$.

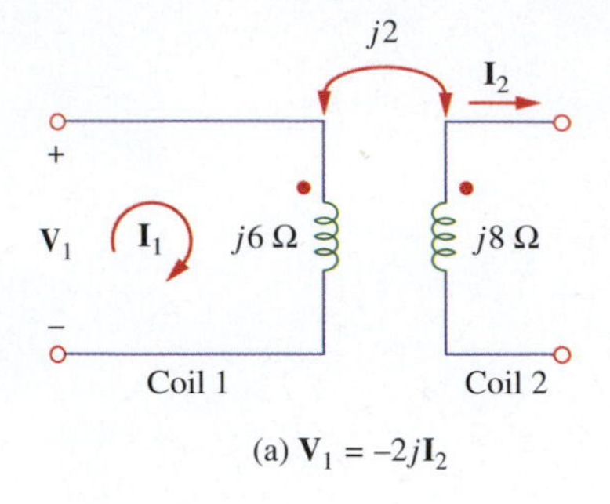

(a) $V_1 = -2jI_2$

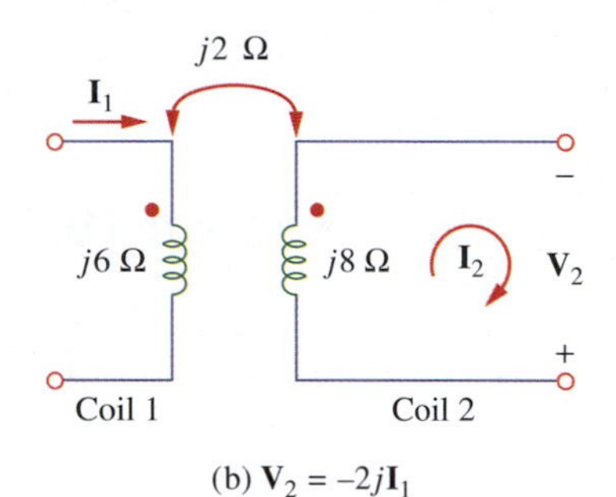

(b) $V_2 = -2jI_1$

Figure 18.12
For Example 18.2; redrawing the relevant portion of the circuit in Fig. 18.11 to find mutual voltages by the dot connection.

Alternatively, it is perhaps best to figure out the mutual voltage by redrawing the relevant portion of the circuit, as shown in Fig. 18.12(a), where it becomes clear that the mutual voltage is $\mathbf{V}_1 = -2j\mathbf{I}_2$, as obtained earlier.

Hence, for mesh 1, KVL gives

$$-100 + \mathbf{I}_1(4 - j3 + j6) - j6\mathbf{I}_2 - j2\mathbf{I}_2 = 0$$

or

$$100 = (4 + j3)\mathbf{I}_1 - j8\mathbf{I}_2 \tag{18.2.1}$$

Similarly, to figure out the mutual voltage in coil 2 due to current $\mathbf{I}_1$, consider the relevant portion of the circuit, as shown in Fig. 18.12(b). Applying the dot convention gives the mutual voltage as $\mathbf{V}_2 = -2j\mathbf{I}_1$. Also, current $\mathbf{I}_2$ sees the two coupled coils in series in Fig. 18.11; because it leaves the dotted terminals in both coils, Eq. (18.18) applies. Therefore, for mesh 2, KVL gives

$$0 = -2j\mathbf{I}_1 - j6\mathbf{I}_1 + (j6 + j8 + j2 \times 2 + 5)\mathbf{I}_2$$

or

$$0 = -j8\mathbf{I}_1 + (5 + j18)\mathbf{I}_2 \tag{18.2.2}$$

Putting Eqs. (18.2.1) and (18.2.2) in matrix form, we get

$$\begin{bmatrix} 100 \\ 0 \end{bmatrix} = \begin{bmatrix} 4 + j3 & -j8 \\ -j8 & 5 + j18 \end{bmatrix} \begin{bmatrix} I_1 \\ I_2 \end{bmatrix}$$

The determinants are:

$$\Delta = \begin{vmatrix} 4 + j3 & -j8 \\ -j8 & 5 + j18 \end{vmatrix} = 30 + j87$$

$$\Delta_1 = \begin{vmatrix} 100 & -j8 \\ 0 & 5 + j18 \end{vmatrix} = 100(5 + j18)$$

$$\Delta_2 = \begin{vmatrix} 4 + j3 & 100 \\ -j8 & 0 \end{vmatrix} = j800$$

Thus, the mesh currents are obtained as

$$\mathbf{I}_1 = \frac{\Delta_1}{\Delta} = \frac{100(5 + j18)}{30 + j87} = \frac{1{,}868.2\underline{/74.5^\circ}}{92.03\underline{/71^\circ}} = 20.3\underline{/3.5^\circ}\text{ A}$$

$$\mathbf{I}_2 = \frac{\Delta_2}{\Delta} = \frac{j800}{30 + j87} = \frac{800\underline{/90^\circ}}{92.03\underline{/71^\circ}} = 8.693\underline{/19^\circ}\text{ A}$$

Practice Problem 18.2

Determine the phasor currents $\mathbf{I}_1$ and $\mathbf{I}_2$ in the circuit of Fig. 18.13.

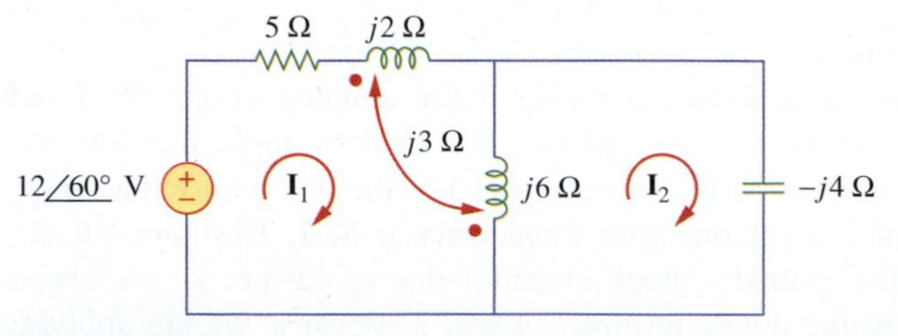

Figure 18.13
For Practice Problem 18.2.

Answer: $2.15\underline{/86.56^\circ}$; $3.23\underline{/86.56^\circ}$ A

18.3 Energy in a Coupled Circuit

In Chapter 10, it was shown that the energy stored in an inductor is given by

$$w = \frac{1}{2}Li^2 \tag{18.17}$$

where L is the self-inductance and i is the instantaneous current. To determine the energy stored in magnetically coupled coils, consider the circuit in Fig. 18.14. The instantaneous energy stored in the circuit has the general expression

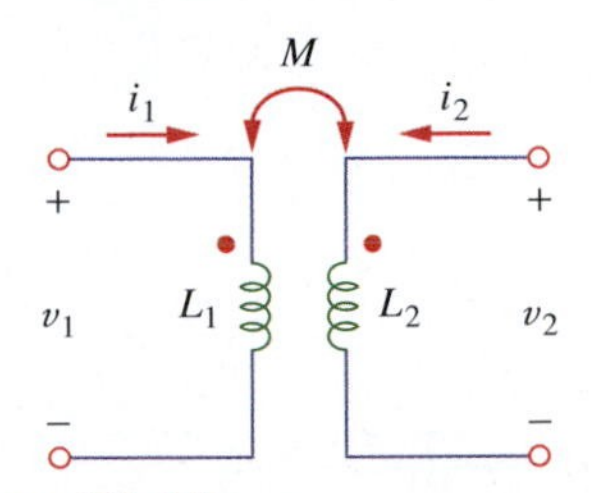

Figure 18.14
The circuit for deriving energy stored in a coupled circuit.

$$w = \frac{1}{2}L_1 i_1^2 + \frac{1}{2}L_2 i_2^2 \pm M i_1 i_2 \tag{18.18}$$

The positive sign is selected for the mutual inductance if both currents enter or leave the dotted terminals of the coils; the negative sign is selected otherwise.

The mutual inductance cannot be greater than the geometric mean of the self-inductances of the coils; that is,

$$M \leq \sqrt{L_1 L_2} \tag{18.19}$$

The extent to which the mutual inductance M approaches the upper limit is specified by the *coefficient of coupling* k given by

$$k = \frac{M}{\sqrt{L_1 L_2}} \tag{18.20}$$

or

$$\boxed{M = k\sqrt{L_1 L_2}} \tag{18.21}$$

where $0 \leq k \leq 1$ or equivalently $0 \leq M \leq \sqrt{L_1 L_2}$. The coupling coefficient is the fraction of the total flux emanating from one coil that links the other coil. For example, in Fig. 18.2,

$$k = \frac{\phi_{12}}{\phi_1} = \frac{\phi_{12}}{\phi_{11} + \phi_{12}} \tag{18.22}$$

and in Fig. 18.3,

$$k = \frac{\phi_{21}}{\phi_2} = \frac{\phi_{21}}{\phi_{21} + \phi_{22}} \tag{18.23}$$

If the entire flux produced by one coil links another coil, then $k = 1$ and we have 100 percent coupling, or the coils are said to be *perfectly coupled*. Thus,

> The **coupling coefficient** k is a measure of the magnetic coupling between two coils; $0 < k < 1$.

For $k < 0.5$, coils are said to be *loosely coupled*; and for $k > 0.5$, they are said to be *tightly coupled*.

We expect k to depend on the closeness of the two coils, their core, their orientation relative to each other, and their windings. Figure 18.15 shows loosely coupled windings and tightly coupled windings. Also,

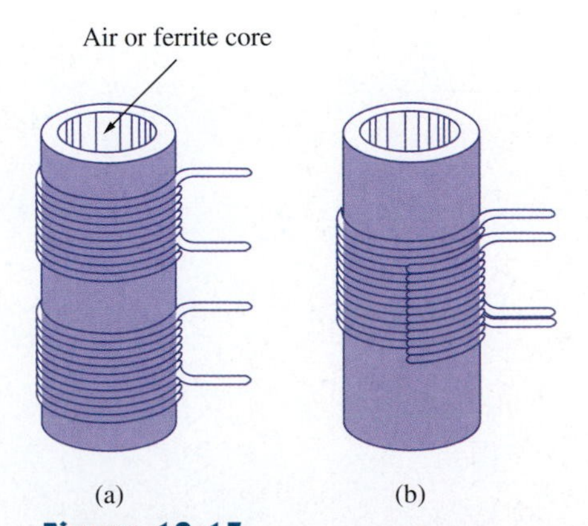

Figure 18.15
Windings: (a) loosely coupled; (b) tightly coupled.

air-core transformers used in radio frequency circuits are loosely coupled, whereas iron-core transformers used in power systems are tightly coupled. The linear transformers discussed in Section 18.4 are mostly air-core; the ideal transformers discussed in Sections 18.5 and 18.6 are principally iron-core.

Example 18.3

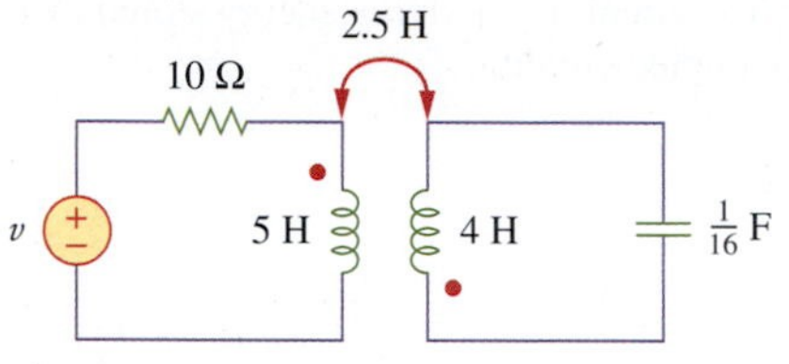

Figure 18.16
For Example 18.3.

Consider the circuit in Fig. 18.16. Determine the coupling coefficient. Calculate the energy stored in the coupled inductors at time $t = 1$ s if $v = 60\cos(4t + 30°)$ V.

Solution:
The coupling coefficient is

$$k = \frac{M}{\sqrt{L_1L_2}} = \frac{2.5}{\sqrt{20}} = 0.56$$

indicating that the inductors are tightly coupled. To find the currents required in calculating the energy stored, we need to obtain the frequency-domain equivalent of the circuit.

$$\begin{aligned} 60\cos(4t + 30°) \quad &\Rightarrow \quad 60\underline{/30°}, \quad \omega = 4 \text{ rad/s} \\ 5 \text{ H} \quad &\Rightarrow \quad j\omega L_1 = j20\ \Omega \\ 2.5 \text{ H} \quad &\Rightarrow \quad j\omega M = j10\ \Omega \\ 4 \text{ H} \quad &\Rightarrow \quad j\omega L_2 = j16\ \Omega \\ \frac{1}{16}\text{ F} \quad &\Rightarrow \quad \frac{1}{j\omega C} = -j4\ \Omega \end{aligned}$$

The frequency-domain equivalent is shown in Fig. 18.17. We now apply mesh analysis. For mesh 1,

$$(10 + j20)\mathbf{I}_1 + j10\mathbf{I}_2 = 60\underline{/30°} \tag{18.3.1}$$

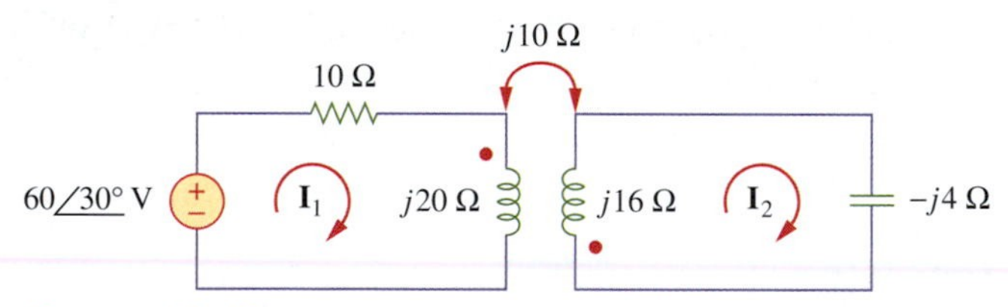

Figure 18.17
Frequency-domain equivalent of the circuit in Fig. 18.16.

For mesh 2,

$$j10\mathbf{I}_1 + (j16 - j4)\mathbf{I}_2 = 0$$

or

$$\mathbf{I}_1 = -1.2\mathbf{I}_2 \tag{18.3.2}$$

Substituting this into Eq. (18.3.1) yields

$$\mathbf{I}_2(-12 - j14) = 60\underline{/30°} \quad \Rightarrow \quad \mathbf{I}_2 = 3.254\underline{/-199.4°}$$

and

$$\mathbf{I}_1 = -1.2\mathbf{I}_2 = 3.905\underline{/-19.4°} \text{ A}$$

In the time-domain,

$$i_1(t) = 3.905\cos(4t - 19.4°), \quad i_2(t) = 3.254\cos(4t - 199.4°)$$

At time $t = 1$ s, $4t = 4$ rad $= 229.2°$, and

$$i_1 = 3.905\cos(229.2° - 19.4°) = -3.389 \text{ A}$$
$$i_2 = 3.254\cos(229.2° - 199.4°) = 2.824 \text{ A}$$

The total energy stored in the coupled inductors is

$$w = \frac{1}{2}L_1 i_1^2 + \frac{1}{2}L_2 i_2^2 + M i_1 i_2$$
$$= \frac{1}{2}(5)(-3.389)^2 + \frac{1}{2}(4)(2.824)^2 + 2.5(-3.389)(2.824) = 20.73 \text{ J}$$

Practice Problem 18.3

For the circuit in Fig. 18.18, determine the coupling coefficient and the energy stored in the coupled inductors at $t = 1.5$ s.

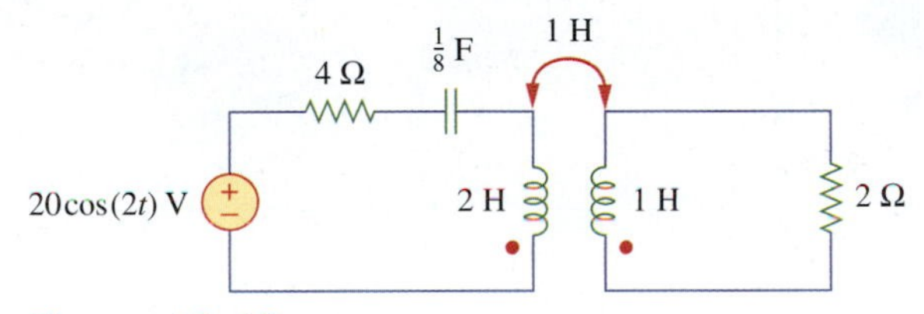

Figure 18.18
For Practice Problem 18.3.

Answer: 0.7071; 9.85 J

18.4 Linear Transformers

As shown in Fig. 18.19, the coil that is directly connected to the voltage source is called the *primary winding*. The coil connected to the load is called the *secondary winding*. The resistances R_1 and R_2 are included to account for the losses (power dissipation) in the coils. The transformer is said to be *linear* if the coils are wound on a magnetically linear material—a material for which the magnetic permeability is constant. Such materials include air, plastic, bakelite, and wood. In fact, most materials are magnetically linear. (Nonlinear materials include iron and steel.) Linear transformers are sometimes called *air-core transformers*, although not all of them are necessarily air-core. They find applications in radio and TV sets. Figure 18.20 portrays different types of transformers.

> A **transformer** is generally a four-terminal device comprising two (or more) magnetically coupled coils used to step up or step down voltages.

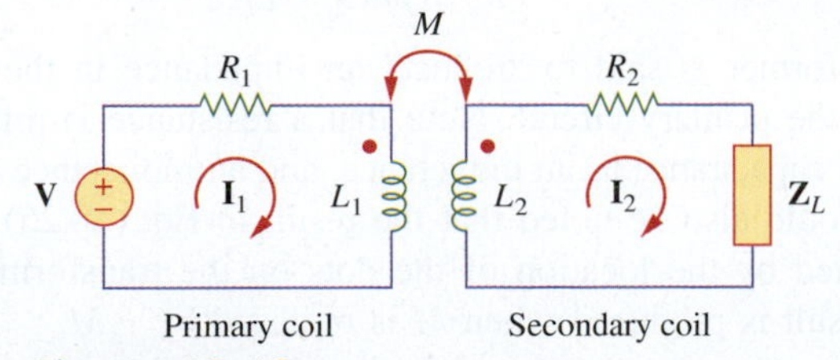

Figure 18.19
A linear transformer.

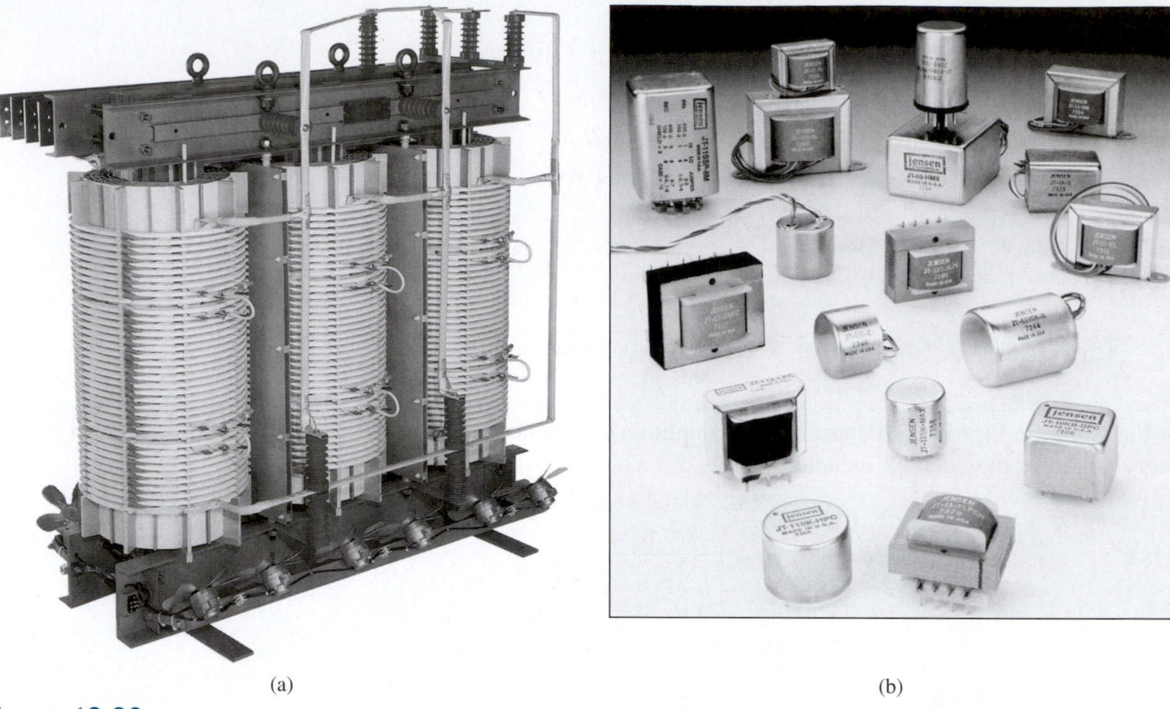

(a) (b)

Figure 18.20
Different types of transformers: (a) copper wound dry power transformers; (b) audio transformers.
(a) © The Electric Service Company (b) © Jensen Transformers, Inc.

We would like to obtain the input impedance $\mathbf{Z}_{\text{in}}$ as seen by the source because $\mathbf{Z}_{\text{in}}$ governs the behavior of the primary circuit. Applying KVL to the two meshes in Fig. 18.19, we obtain

$$\mathbf{V} = (R_1 + j\omega L_1)\mathbf{I}_1 - j\omega M\mathbf{I}_2 \tag{18.24a}$$

$$0 = -j\omega M\mathbf{I}_1 + (R_2 + j\omega L_2 + \mathbf{Z}_L)\mathbf{I}_2 \tag{18.24b}$$

In Eq. (18.24b), we express $\mathbf{I}_2$ in terms of $\mathbf{I}_1$ and substitute it into Eq. (18.24a). We get the input impedance as

$$\mathbf{Z}_{\text{in}} = \mathbf{V}/\mathbf{I}_1 = R_1 + j\omega L_1 + \frac{\omega^2 M^2}{R_2 + j\omega L_2 + Z_L} \tag{18.25}$$

We notice that the input impedance comprises two terms. The first term $(R_1 + j\omega L_1)$ is the primary impedance. The second term is due to the coupling between the primary and secondary windings. It is as though this impedance is reflected to the primary. Thus, it is known as the *reflected impedance*[1] $\mathbf{Z}_R$; that is,

$$\mathbf{Z}_R = \frac{\omega^2 M^2}{R_2 + j\omega L_2 + Z_L} \tag{18.26}$$

A transformer is said to "reflect" an impedance in the secondary circuit into the primary circuit. Note that a resistance is reflected as a resistance, a capacitance as an inductance, and an inductance as a capacitance. It should also be noted that the result in Eq. (18.25) or (18.26) is not affected by the location of the dots on the transformer because the same result is produced when M is replaced by $-M$.

[1] Note: Some authors call it *coupled impedance*.

The little experience gained in Sections 18.2 and 18.3 in analyzing magnetically coupled circuits is enough to convince anyone that analyzing such circuits is not easy. For this reason, it is sometimes convenient to replace a magnetically coupled circuit by an equivalent circuit that does not involve magnetic coupling. We want to replace the linear transformer in Fig. 18.21 by an equivalent T or Π circuit, a circuit that would have no mutual inductance.

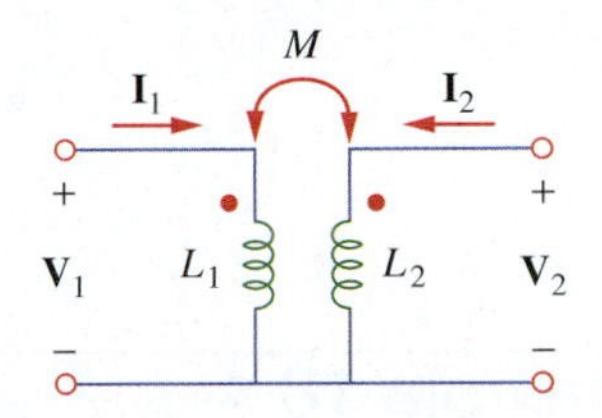

Figure 18.21
Determining the equivalent circuit of a linear transformer.

The voltage–current relationships for the primary and secondary coils give the matrix equation

$$\begin{bmatrix} V_1 \\ V_2 \end{bmatrix} = \begin{bmatrix} j\omega L_1 & j\omega M \\ j\omega M & j\omega L_2 \end{bmatrix} \begin{bmatrix} I_1 \\ I_2 \end{bmatrix} \tag{18.27}$$

By matrix inversion (see Appendix A), Eq. (18.27) can be written as

$$\begin{bmatrix} I_1 \\ I_2 \end{bmatrix} = \begin{bmatrix} \dfrac{L_2}{j\omega(L_1L_2 - M^2)} & \dfrac{-M}{j\omega(L_1L_2 - M^2)} \\ \dfrac{-M}{j\omega(L_1L_2 - M^2)} & \dfrac{L_1}{j\omega(L_1L_2 - M^2)} \end{bmatrix} \begin{bmatrix} V_1 \\ V_2 \end{bmatrix} \tag{18.28}$$

Our goal is to match Eqs. (18.27) and (18.28) with the corresponding equations for the T and Π networks.

For the T (or Y) network of Fig. 18.22, mesh analysis provides the terminal equations as

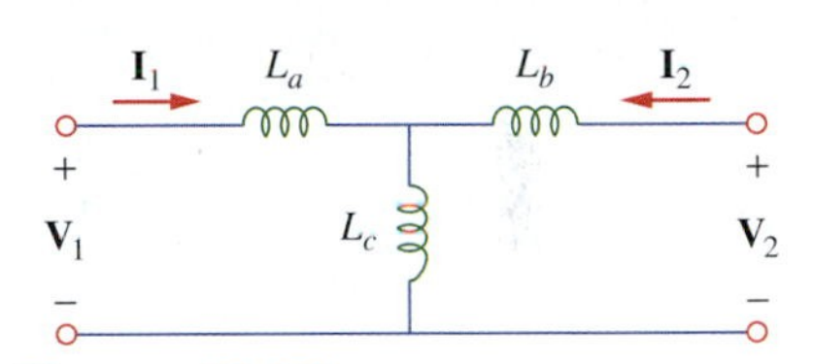

Figure 18.22
An equivalent T circuit.

$$\begin{bmatrix} V_1 \\ V_2 \end{bmatrix} = \begin{bmatrix} j\omega(L_a + L_c) & j\omega L_c \\ j\omega L_c & j\omega(L_b + L_c) \end{bmatrix} \begin{bmatrix} I_1 \\ I_2 \end{bmatrix} \tag{18.29}$$

If the circuits in Figs. 18.21 and 18.22 are equivalents, Eqs. (18.27) and (18.29) must be identical. Equating terms in the impedance matrices of Eqs. (18.27) and (18.29) leads to

$$L_a = L_1 - M, \qquad L_b = L_2 - M, \qquad L_c = M \tag{18.30}$$

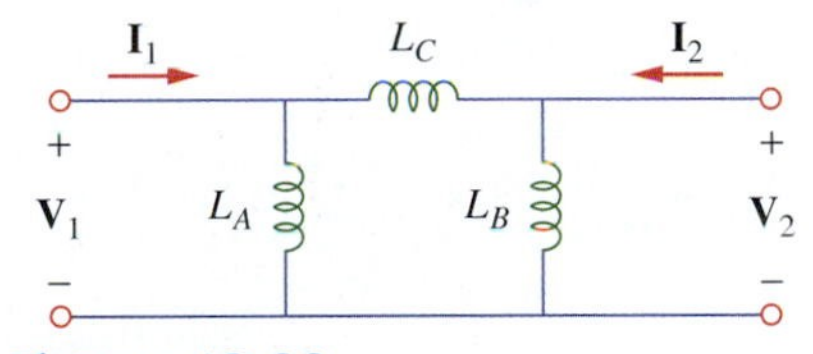

Figure 18.23
An equivalent Π circuit.

For the Π (or Δ) network in Fig. 18.23, nodal analysis gives the terminal equations as

$$\begin{bmatrix} I_1 \\ I_2 \end{bmatrix} = \begin{bmatrix} \dfrac{1}{j\omega L_A} + \dfrac{1}{j\omega L_C} & -\dfrac{1}{j\omega L_C} \\ -\dfrac{1}{j\omega L_C} & \dfrac{1}{j\omega L_B} + \dfrac{1}{j\omega L_C} \end{bmatrix} \begin{bmatrix} V_1 \\ V_2 \end{bmatrix} \tag{18.31}$$

Equating terms in admittance matrices of Eqs. (18.28) and (18.31), we obtain

$$L_A = \frac{L_1L_2 - M^2}{L_2 - M}, \qquad L_B = \frac{L_1L_2 - M^2}{L_1 - M}, \qquad L_C = \frac{L_1L_2 - M^2}{M} \tag{18.32}$$

Note that in Figs. 18.22 and 18.23, the inductors are not magnetically coupled. It should also be noted that changing the location of either of the dots in Fig. 18.21 can cause M to become $-M$. As Example 18.6 illustrates, negative value of M is physically unrealizable, but the equivalent model is still mathematically valid.

Example 18.4

In the circuit of Fig. 18.24, calculate the input impedance and current $\mathbf{I}_1$. Take $\mathbf{Z}_1 = 60 - j100\ \Omega$, $\mathbf{Z}_2 = 30 + j40\ \Omega$, and $\mathbf{Z}_L = 80 + j60\ \Omega$.

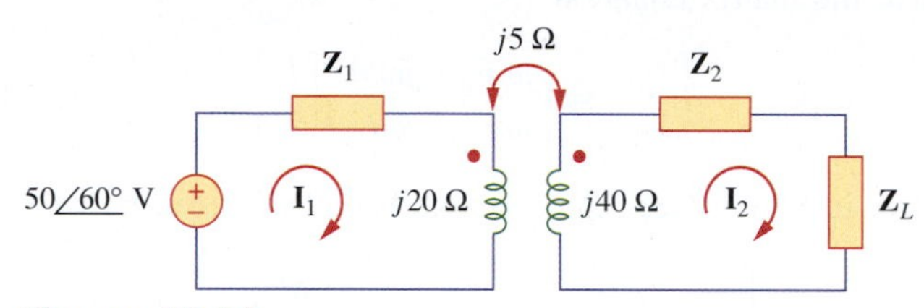

Figure 18.24
For Example 18.4.

Solution:
From Eq. (18.25),

$$\begin{aligned}
\mathbf{Z}_{\text{in}} &= \mathbf{Z}_1 + j20 + \frac{(5)^2}{j40 + Z_2 + Z_L} \\
&= 60 - j100 + j20 + \frac{25}{110 + j140} \\
&= 60 - j80 + 0.14\angle{-51.84^\circ} = 60.09 - j80.11 \\
&= 100.14\angle{-53.1^\circ}\ \Omega
\end{aligned}$$

Thus,

$$\mathbf{I}_1 = \mathbf{V}/\mathbf{Z}_{\text{in}} = \frac{50\angle{60^\circ}}{100.14\angle{-53.1^\circ}} = 0.5\angle{113.1^\circ}\text{ A}$$

Practice Problem 18.4

Find the input impedance of the circuit of Fig. 18.25 and the current from the voltage source.

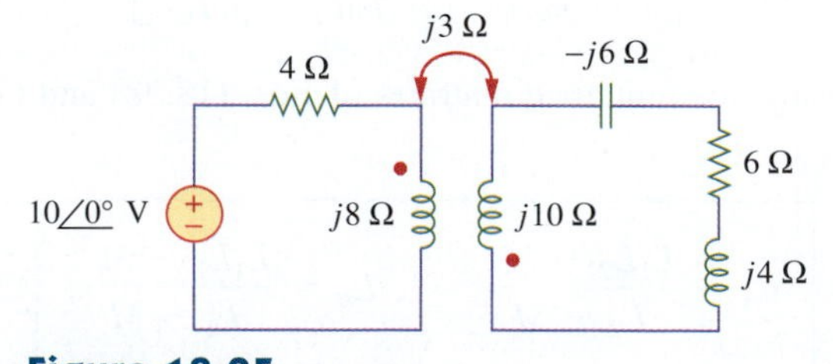

Figure 18.25
For Practice Problem 18.4.

Answer: $8.58\angle{58.05^\circ}\ \Omega$; $1.165\angle{-58.05^\circ}$ A

Example 18.5

Determine the T-equivalent circuit of the linear transformer in Fig. 18.26(a).

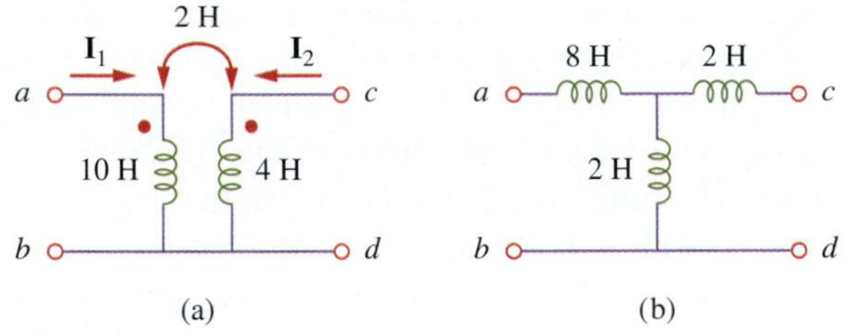

Figure 18.26
For Example 18.5 and Practice Problem 18.5: (a) a linear transformer, (b) its T-equivalent circuit.

Solution:
Given that $L_1 = 10$ H, $L_2 = 4$ H, and $M = 2$ H, the T-equivalent network has the following parameters

$$L_a = L_1 - M = 10 - 2 = 8 \text{ H}$$
$$L_b = L_2 - \text{M} = 4 - 2 = 2 \text{ H}$$
$$L_c = M = 2 \text{ H}$$

Hence, the T-equivalent circuit is shown in Fig. 18.26(b). We have assumed that reference directions for currents and voltage polarities in the primary and secondary windings conform to those in Fig. 18.21. Otherwise, we may need to replace M with $-M$, as Example 18.6 illustrates later.

Practice Problem 18.5

For the linear transformer in Fig. 18.26(a), find the Π equivalent network.

Answer: $L_A = 18$ H; $L_B = 4.5$ H; $L_C = 18$ H

Example 18.6

Solve for $\mathbf{I}_1$, $\mathbf{I}_2$, and $\mathbf{V}_o$ in Fig. 18.27 (same circuit for Practice Problem 18.1) using the T-equivalent circuit for the linear transformer.

Solution:
Notice that the circuit in Fig. 18.27 is the same as that in Fig. 18.10, except that the reference direction for current $\mathbf{I}_2$ has been reversed, just to make the reference directions for the currents for the magnetically coupled coils conform with those in Fig. 18.21.

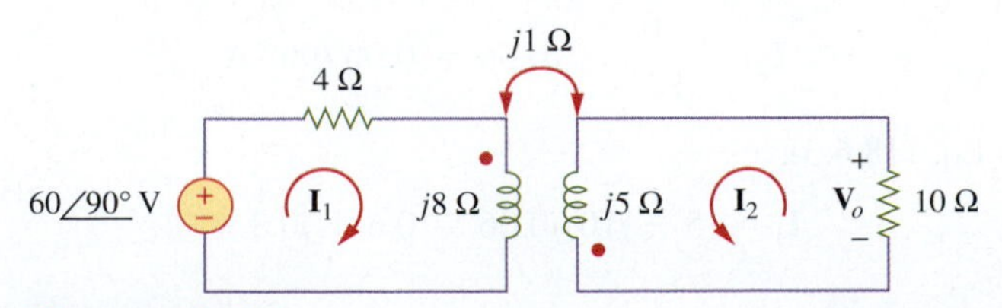

Figure 18.27
For Example 18.6.

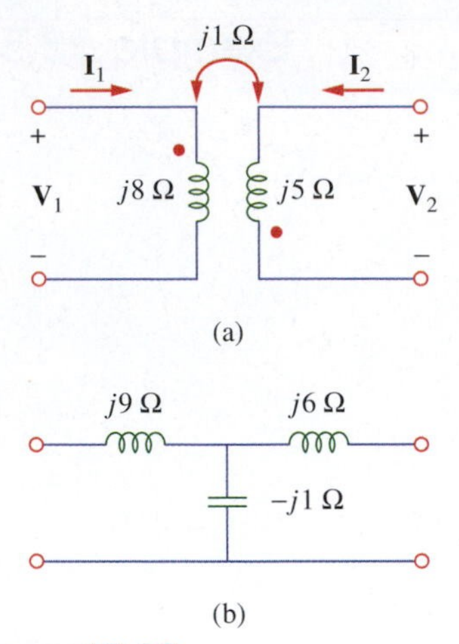

Figure 18.28
For Example 18.6: (a) circuit for coupled coils of Fig. 18.27; (b) *T*-equivalent circuit.

We need to replace the magnetically coupled coils with the *T*-equivalent circuit. The relevant portion of the circuit in Fig. 18.27 is shown in Fig. 18.28(a). Comparing Fig. 18.28(a) with Fig. 18.21 shows that there are two differences. First, due to the current reference directions and voltage polarities, we need to replace M with $-M$ to make Fig. 18.28(a) conform to Fig. 18.21. Second, the circuit in Fig. 18.21 is in the time-domain, whereas the circuit in Fig. 18.28(a) is in the frequency domain. The difference is the factor $j\omega$; that is, L in Fig. 18.21 has been replaced with $j\omega L$ and M with $j\omega M$. Because ω is not specified, we can assume $\omega = 1$ or any other value; it really does not matter. With these two differences in mind,

$$L_a = L_1 - (-M) = 8 + 1 = 9 \text{ H}$$
$$L_b = L_2 - (-M) = 5 + 1 = 6 \text{ H}$$
$$L_c = -M = -1 \text{ H}$$

Thus, the *T*-equivalent circuit for the coupled coils is shown in Fig. 18.28(b). Inserting the *T*-equivalent circuit in Fig. 18.28(b) to replace the two coils in Fig. 18.27 gives the equivalent circuit in Fig. 18.29, which can be solved using nodal or mesh analysis. If we apply mesh analysis, we obtain

$$j6 = \mathbf{I}_1(4 + j9 - j1) + \mathbf{I}_2(-j1) \qquad \textbf{(18.6.1)}$$

and

$$0 = \mathbf{I}_1(-j1) + \mathbf{I}_2(10 + j6 - j1) \qquad \textbf{(18.6.2)}$$

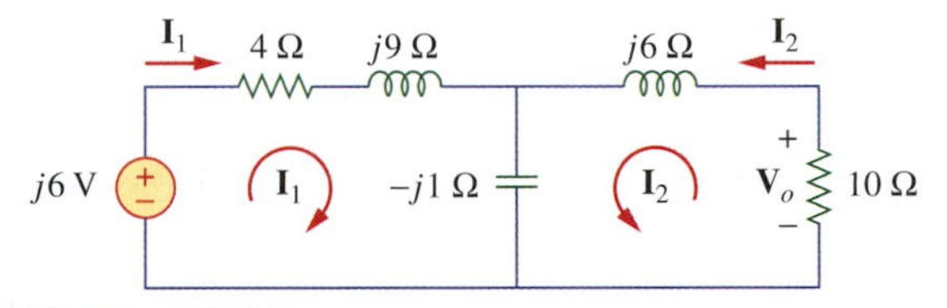

Figure 18.29
For Example 18.6.

From Eq. (18.6.2),

$$\mathbf{I}_1 = \frac{(10 + j5)}{j}\mathbf{I}_2 = (5 - j10)\mathbf{I}_2 \qquad \textbf{(18.6.3)}$$

Substituting Eq. (18.6.3) into Eq. (18.6.1) gives

$$\begin{aligned} j6 &= (4 + j8)(5 - j10)\mathbf{I}_2 - j\mathbf{I}_2 \\ &= (100 - j)\mathbf{I}_2 \cong 100\mathbf{I}_2 \end{aligned}$$

Because 100 is very large compared to 1, the imaginary part of $(100 - j)$ can be ignored so that $100 - j \cong 100$. Hence,

$$\mathbf{I}_2 = \frac{j6}{100} = j0.06 = 0.06\underline{/90^\circ} \text{ A}$$

From Eq. (18.6.3),

$$\mathbf{I}_1 = (5 - j10)j0.06 = 0.6 + j0.3 \text{ A}$$

and

$$\mathbf{V}_o = -10\mathbf{I}_2 = -j0.6 = 0.6\underline{/-90^\circ} \text{ V}$$

This agrees with the answer to Practice Problem 18.1. Of course, the direction of $\mathbf{I}_2$ in Fig. 18.10 is opposite to that in Fig. 18.27. This will not affect $\mathbf{V}_o$, but the value of $\mathbf{I}_2$ in this example is negative that of $\mathbf{I}_2$ in Practice Problem 18.1. The advantage of using the T-equivalent model for the magnetically coupled coils is that in Fig. 18.29, we do not need to bother with the dot on the coupled coils.

Practice Problem 18.6

Solve the problem in Example 18.1 (see Fig. 18.9) using the T-equivalent model for the magnetically coupled coils.

Answer: $13\underline{/-49.4^\circ}$ A; $2.91\underline{/14.04^\circ}$ A

18.5 Ideal Transformers

A transformer is said to be ideal if it has the following properties:

1. Coils have very large reactances—that is, $L_1, L_2, M \rightarrow \infty$
2. Coupling coefficient is equal to unity—that is, $k = 1$
3. Primary and secondary coils are lossless—that is, $R_1 = 0 = R_2$

> An **ideal transformer** is a unity-coupled, lossless transformer in which the primary and secondary coils have infinite self-inductances.

An ideal transformer is one with perfect coupling ($k = 1$). It consists of two (or more) coils with a large number of turns wound on a common core of high permeability. Because of the high permeability of the core, the flux links all the turns of both coils, thereby resulting in a perfect coupling.

To see how an ideal transformer is the limiting case of two coupled inductors, where the inductances approach infinity and the coupling is perfect, let us reexamine the circuit in Fig. 18.14. In the frequency domain,

$$\mathbf{V}_1 = j\omega L_1\mathbf{I}_1 + j\omega M\mathbf{I}_2 \tag{18.33a}$$

$$\mathbf{V}_2 = j\omega M\mathbf{I}_1 + j\omega L_2\mathbf{I}_2 \tag{18.33b}$$

From Eq. (18.33a), $\mathbf{I}_1 = (\mathbf{V}_1 - j\omega M\mathbf{I}_2)/j\omega L_1$. Substituting this in Eq. (18.33b) gives

$$\mathbf{V}_2 = j\omega L_2\mathbf{I}_2 + \frac{M\mathbf{V}_1}{L_1} - \frac{j\omega M^2\mathbf{I}_2}{L_1} \tag{18.34}$$

But $M = \sqrt{L_1L_2}$ for perfect coupling ($k = 1$). Hence,

$$\mathbf{V}_2 = j\omega L_2\mathbf{I}_2 + \frac{\sqrt{L_1L_2}\mathbf{V}_1}{L_1} - \frac{j\omega L_1L_2\mathbf{I}_2}{L_1} = \sqrt{\frac{L_2}{L_1}}\mathbf{V}_1 = n\mathbf{V}_1 \tag{18.35}$$

where $n = \sqrt{L_2/L_1}$ and is called the *turns ratio*. As $L_1, L_2, M \rightarrow \infty$ such that n remains the same, the coupled coils become an ideal transformer. The ideal transformer neglects losses to resistive heating in the primary coil and assumes ideal coupling to the secondary coil (i.e., no magnetic losses).

Iron-core transformers are close approximations to ideal transformers. These transformers are used mainly in power systems and also in electronics.

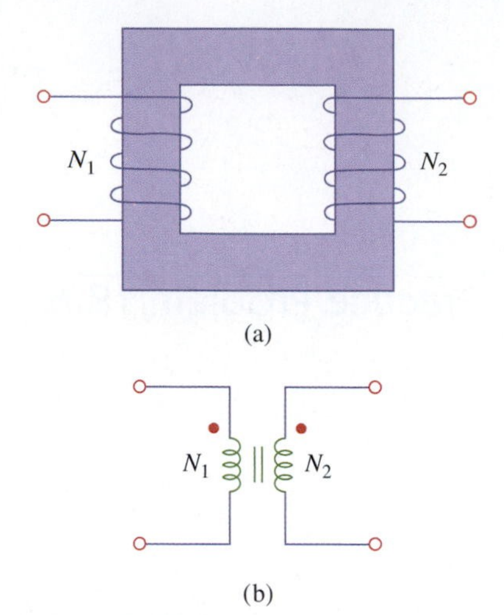

Figure 18.30
(a) Ideal transformer; (b) circuit symbol for ideal transformers.

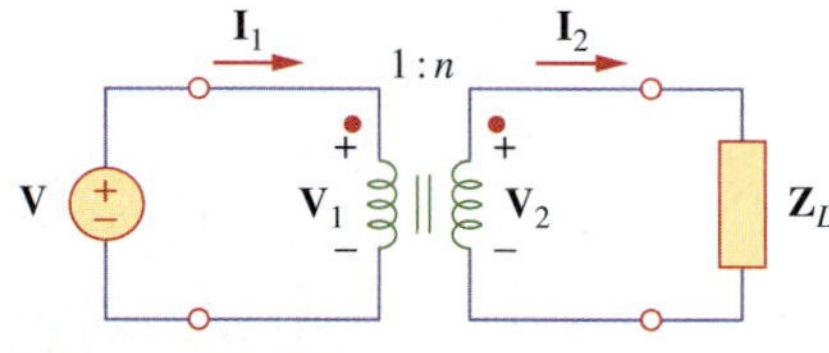

Figure 18.31
Relating primary and secondary quantities in an ideal transformer.

Figure 18.30(a) shows a typical ideal transformer; the circuit symbol for an ideal transformer is seen in Fig. 18.30(b). The vertical lines between the coils indicate iron-core, as opposed to the air-core used in linear transformers. The primary winding has N_1 turns, while the secondary winding has N_2 turns.

When a sinusoidal voltage is applied to the primary winding as shown in Fig. 18.31, the same magnetic flux ϕ goes through both windings. According to Faraday's law, the voltage across the primary winding is

$$v_1 = N_1 \frac{d\phi}{dt} \tag{18.36}$$

while that across the secondary winding is

$$v_2 = N_2 \frac{d\phi}{dt} \tag{18.37}$$

Dividing Eq. (18.37) by Eq. (18.36), we get

$$\frac{v_2}{v_1} = \frac{N_2}{N_1} = n \tag{18.38}$$

where n is the *turns ratio* or *transformation ratio*. We can use the phasor voltages $\mathbf{V}_1$ and $\mathbf{V}_2$ rather than the instantaneous values v_1 and v_2. Thus, Eq. (18.38) may be written as

$$\boxed{\frac{\mathbf{V}_2}{\mathbf{V}_1} = \frac{N_2}{N_1} = n} \tag{18.39}$$

For reasons of power conservation, the energy supplied to the primary circuit must equal the energy absorbed by the secondary line because there are no losses in an ideal transformer. This implies that

$$v_1 i_1 = v_2 i_2 \tag{18.40}$$

In phasor form, Eq. (18.40) in conjunction with Eq. (18.39) becomes

$$\frac{\mathbf{I}_1}{\mathbf{I}_2} = \frac{\mathbf{V}_2}{\mathbf{V}_1} = n \tag{18.41}$$

showing that the primary and secondary currents are related to the turns ratio in the inverse manner as the voltages. Thus,

$$\boxed{\frac{\mathbf{I}_2}{\mathbf{I}_1} = \frac{N_1}{N_2} = \frac{1}{n}} \tag{18.42}$$

When $n = 1$, we generally call the transformer an *isolation transformer*. The reason will become obvious in Section 18.9.1. If $n > 1$, we have a *step-up* transformer because the voltage is increased from primary to secondary ($\mathbf{V}_2 > \mathbf{V}_1$). On the other hand, if $n < 1$, the transformer is a *step-down transformer* because the voltage is decreased from primary to secondary ($\mathbf{V}_2 < \mathbf{V}_1$). Thus,

> A **step-down transformer** is one whose secondary voltage is less than its primary voltage.

> A **step-up transformer** is one whose secondary voltage is greater than its primary voltage.

The ratings of transformers are usually specified as V_1/V_2. A transformer with rating 2400/120 V should have 2400 V on the primary and 120 in the secondary—that is, a step-down transformer. Keep in mind that the voltage ratings are in rms.

Power companies often generate electricity at some convenient voltage and use a step-up transformer to increase the voltage so that the power can be transmitted at very high voltage and low current over transmission lines, resulting in significant cost savings. The cost savings come from lower wasted power dissipated as ohmnic loss in the transmission lines. Near residential customer premises, step-down transformers are used to bring the voltage down to 120 V. Section 18.8.3 will elaborate on this.

It is important that we know how to get the proper polarity of the voltages and the direction of the currents for the transformer in Fig. 18.31. If the polarity of $\mathbf{V}_1$ or $\mathbf{V}_2$ or the direction of $\mathbf{I}_1$ or $\mathbf{I}_2$ is changed, n in Eqs. (18.38) through (18.42) may need to be replaced by $-n$. The two simple rules to follow are:

1. If $\mathbf{V}_1$ and $\mathbf{V}_2$ are *both* positive or both negative at the dotted terminals, use $+n$ in Eq. (18.39). Otherwise, use $-n$.
2. If $\mathbf{I}_1$ and $\mathbf{I}_2$ *both* enter into or both leave the dotted terminals, use $-n$ in Eq. (18.42). Otherwise, use $+n$.

The rules are demonstrated with the four circuits in Fig. 18.32.

Using Eqs. (18.39) and (18.42), we can always express $\mathbf{V}_1$ in terms of $\mathbf{V}_2$ and $\mathbf{I}_1$ in terms of $\mathbf{I}_2$ or vice versa; that is,

$$\mathbf{V}_1 = \mathbf{V}_2/n \quad \text{or} \quad \mathbf{V}_2 = n\mathbf{V}_1 \tag{18.43}$$

$$\mathbf{I}_1 = n\mathbf{I}_2 \quad \text{or} \quad \mathbf{I}_2 = \mathbf{I}_1/n \tag{18.44}$$

The complex power in the primary winding is

$$\mathbf{S}_1 = \mathbf{V}_1\mathbf{I}_1^* = (\mathbf{V}_2/n)(n\mathbf{I}_2)^* = \mathbf{V}_2\mathbf{I}_2^* = \mathbf{S}_2 \tag{18.45}$$

showing that the complex power supplied to the primary is delivered to the secondary without loss. The transformer absorbs no power. Of course, we should expect this because the ideal transformer is lossless. The input impedance as seen by the source in Fig. 18.31 is found from Eqs. (18.43) and (18.44) as

$$\mathbf{Z}_{\text{in}} = \frac{\mathbf{V}_1}{\mathbf{I}_1} = \frac{1}{n^2}\frac{\mathbf{V}_2}{\mathbf{I}_2} \tag{18.46}$$

It is evident from Fig. 18.31 that $\mathbf{V}_2/\mathbf{I}_2 = \mathbf{Z}_L$, so that

$$\mathbf{Z}_{\text{in}} = \frac{\mathbf{Z}_L}{n^2} \tag{18.47}$$

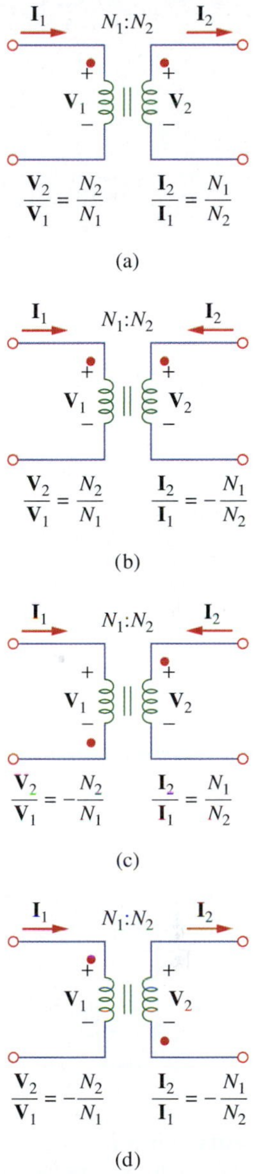

(a) $\frac{\mathbf{V}_2}{\mathbf{V}_1} = \frac{N_2}{N_1}$, $\frac{\mathbf{I}_2}{\mathbf{I}_1} = \frac{N_1}{N_2}$

(b) $\frac{\mathbf{V}_2}{\mathbf{V}_1} = \frac{N_2}{N_1}$, $\frac{\mathbf{I}_2}{\mathbf{I}_1} = -\frac{N_1}{N_2}$

(c) $\frac{\mathbf{V}_2}{\mathbf{V}_1} = -\frac{N_2}{N_1}$, $\frac{\mathbf{I}_2}{\mathbf{I}_1} = \frac{N_1}{N_2}$

(d) $\frac{\mathbf{V}_2}{\mathbf{V}_1} = -\frac{N_2}{N_1}$, $\frac{\mathbf{I}_2}{\mathbf{I}_1} = -\frac{N_1}{N_2}$

Figure 18.32
Typical circuits illustrating proper voltage polarities and current directions in an ideal transformer.

The input impedance is also called the *reflected impedance* because it appears as if the load impedance is reflected to the primary side.[2] This ability of the transformer to transform a given impedance into another impedance provides us a means of *impedance matching* to ensure maximum power transfer. The idea of impedance matching is very useful in practice and will be discussed in more detail in Section 18.8.2.

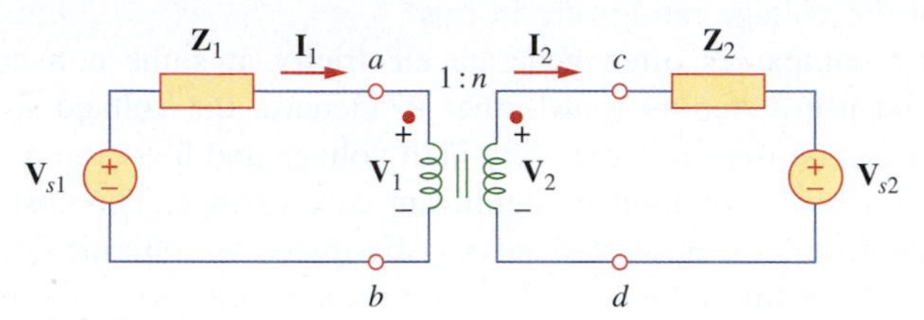

Figure 18.33
Ideal transformer circuit whose equivalent circuits are to be found.

In analyzing a circuit containing an ideal transformer, it is common practice to eliminate the transformer by reflecting impedances and sources from one side of the transformer to the other. In the circuit of Fig. 18.33, suppose we want to reflect the secondary side of the circuit to the primary side. We find the Thevenin equivalent of the circuit to the right of the terminals *a-b*. We obtain V_{Th} as the open-circuit voltage at terminals *a-b*, as shown in Fig. 18.34(a). Because terminals *a-b* are open, $\mathbf{I}_1 = 0 = \mathbf{I}_2$ so that $\mathbf{V}_2 = \mathbf{V}_{s2}$. Hence, from Eq. (18.43),

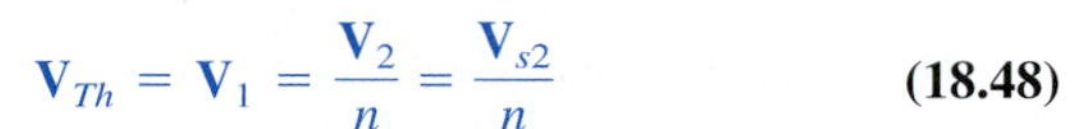

$$\mathbf{V}_{Th} = \mathbf{V}_1 = \frac{\mathbf{V}_2}{n} = \frac{\mathbf{V}_{s2}}{n} \qquad \textbf{(18.48)}$$

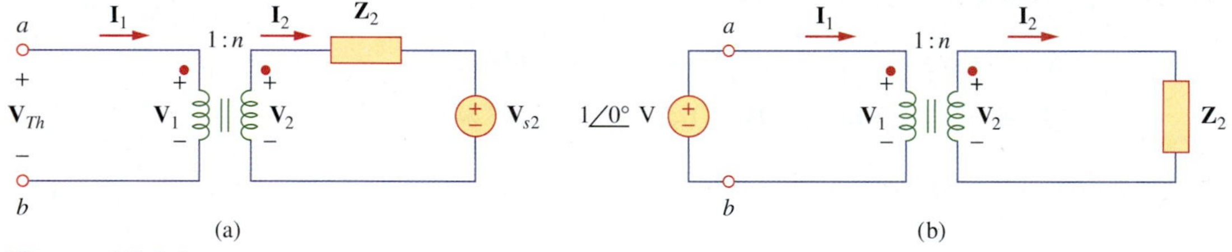

Figure 18.34
(a) Obtaining $\mathbf{V}_{Th}$ for the circuit in Fig. 18.33, (b) obtaining $\mathbf{Z}_{Th}$ for the circuit in Fig. 18.33.

To get $\mathbf{Z}_{Th}$, we remove the voltage source in the primary winding and insert a unit voltage source at terminals *a-b*, as in Fig. 18.34(b). From Eq. (18.47), $\mathbf{I}_1 = n\mathbf{I}_2$ and $\mathbf{V}_1 = \mathbf{V}_2/n$ so that

$$\mathbf{Z}_{Th} = \mathbf{V}_1/\mathbf{I}_1 = \frac{\mathbf{V}_2/n}{nI_2} = \frac{\mathbf{Z}_2}{n^2}, \qquad \text{where} \qquad \mathbf{V}_2 = \mathbf{Z}_2\mathbf{I}_2 \qquad \textbf{(18.49)}$$

which is what we should have expected from Eq. (18.47). Once $\mathbf{V}_{Th}$ and $\mathbf{Z}_{Th}$ are obtained, we add the Thevenin equivalent to the part of the circuit in Fig. 18.33 to the right of terminals *a-b*. The result is depicted in Fig. 18.35. We can also reflect the primary side of the circuit in Fig. 18.33 to the secondary side. The equivalent circuit is shown in Fig. 18.36.

The general rule for eliminating the transformer and reflecting the secondary circuit to the primary side is: divide the secondary impedance by n^2, divide the secondary voltage by n, and multiply the secondary current by n.

[2] Note: Notice that an ideal transformer reflects impedance as the square of the turns ratio.

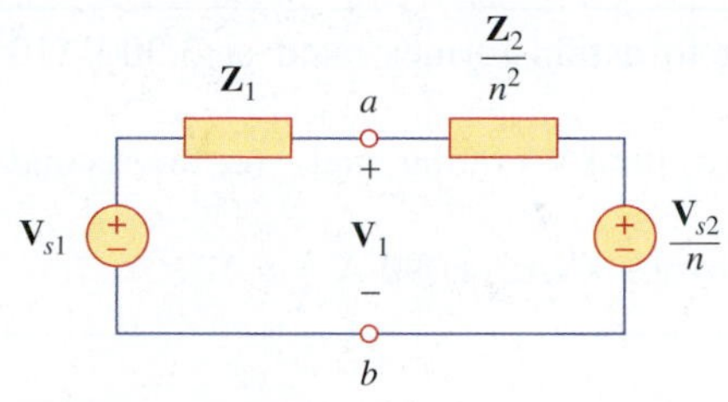

Figure 18.35
Equivalent circuit for Fig. 18.33 obtained by reflecting the secondary circuit to the primary side.

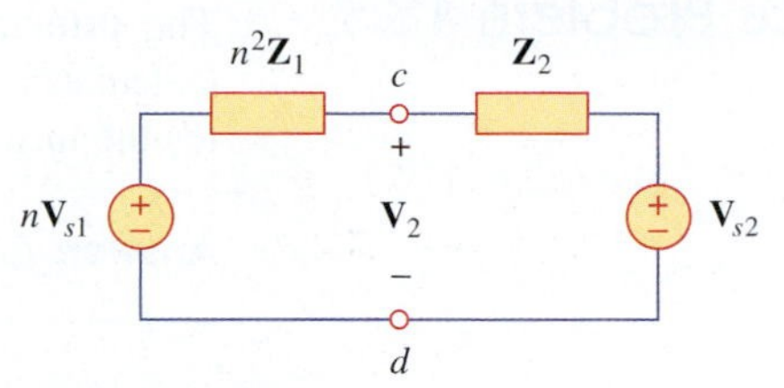

Figure 18.36
Equivalent circuit for Fig. 18.33 obtained by reflecting the primary circuit to the secondary side.

> The rule for eliminating the transformer and reflecting the primary circuit to the secondary side is multiply the primary impedance by n^2, multiply the primary voltage by n, and divide the primary current by n.

According to Eq. (18.45), powers remain the same, whether calculated on the primary or secondary side. It should be pointed out that this reflection approach only applies if there are no external connections between the primary and secondary windings. When we have external connections between the primary and secondary windings, we simply use regular mesh and nodal analyses. Examples of circuits where there are external connections between the primary and secondary windings are in Figs. 18.39 and 18.40. It should also be noted that if the locations of the dots in Fig. 18.33 are changed, we may have to replace n by $-n$ in order to obey the dot rule, as stated earlier and illustrated in Fig. 18.32.

Example 18.7

A transformer is rated at 2,400/120 V, 9.6 kVA, and has 50 turns on the secondary side. Calculate:
(a) the turns ratio,
(b) the number of turns on the primary side,
(c) the current ratings for the primary and secondary.

Solution:

(a) This is a step-down transformer because $V_1 = 2{,}400 \text{ V} > V_2 = 120 \text{ V}$.

$$n = \frac{V_2}{V_1} = \frac{120}{2{,}400} = 0.05$$

(b)

$$n = \frac{N_2}{N_1} \quad \Rightarrow \quad 0.05 = \frac{50}{N_1}$$

or

$$N_1 = \frac{50}{0.05} = 1{,}000 \text{ turns}$$

(c)

$$S = V_1 I_1 = V_2 I_2 = 9.6 \text{ kVA}$$

Hence,

$$I_1 = \frac{9{,}600}{V_1} = \frac{9{,}600}{2{,}400} = 4 \text{ A}$$

$$I_2 = \frac{9{,}600}{V_2} = \frac{9{,}600}{120} = 80 \text{ A}$$

or

$$I_2 = \frac{I_1}{n} = \frac{4}{0.05} = 80 \text{ A}$$

Practice Problem 18.7

The primary current to a transformer rated at 3,300/110 V is 3 A. Calculate:
(a) the turns ratio, (b) the kVA rating, and (c) the secondary current.

Answer: (a) 1/30; (b) 9.9 kVA; (c) 90 A

Example 18.8

For the ideal transformer circuit of Fig. 18.37, find:
(a) the source current $\mathbf{I}_1$,
(b) the output voltage $\mathbf{V}_o$, and
(c) the complex power supplied by the source.

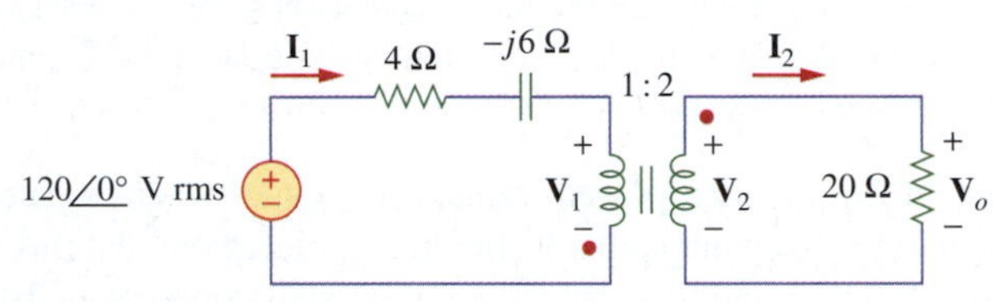

Figure 18.37
For Example 18.8.

Solution:
(a) The 20-Ω impedance can be reflected to the primary side and we get

$$\mathbf{Z}_R = \frac{20}{n^2} = \frac{20}{4} = 5\ \Omega$$

Thus,

$$\mathbf{Z}_{\text{in}} = 4 - j6 + \mathbf{Z}_R = 9 - j6 = 10.82\underline{/-33.69^\circ}\ \Omega$$

$$\mathbf{I}_1 = \frac{120\underline{/0^\circ}}{\mathbf{Z}_{\text{in}}} = \frac{120\underline{/0^\circ}}{10.82\underline{/-33.69^\circ}} = 11.09\underline{/33.69^\circ}\ \text{A}$$

(b) Because both $\mathbf{I}_1$ and $\mathbf{I}_2$ leave the dotted terminals,

$$\mathbf{I}_2 = -\frac{1}{n}\mathbf{I}_1 = -5.545\underline{/33.69^\circ}\ \text{A}$$

$$\mathbf{V}_o = 20\mathbf{I}_2 = 110.9\underline{/213.69^\circ}$$

(c) The complex power supplied is

$$\mathbf{S} = \mathbf{V}_s\mathbf{I}_1^* = (120\underline{/0^\circ})(11.09\underline{/-33.69^\circ}) = 1{,}330.8\underline{/-33.69^\circ}\ \text{VA}$$

Practice Problem 18.8

In the ideal transformer circuit of Fig. 18.38, find $\mathbf{V}_o$ and the complex power supplied by the source.

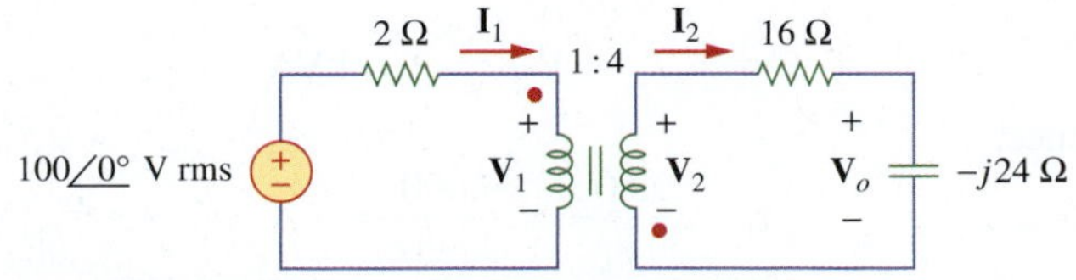

Figure 18.38
For Practice Problem 18.8.

Answer: $178.9\underline{/116.56^\circ}$ V; $2{,}981.5\underline{/-26.56^\circ}$ VA

Example 18.9

Calculate the power supplied to the 10-Ω resistor in the ideal transformer circuit of Fig. 18.39.

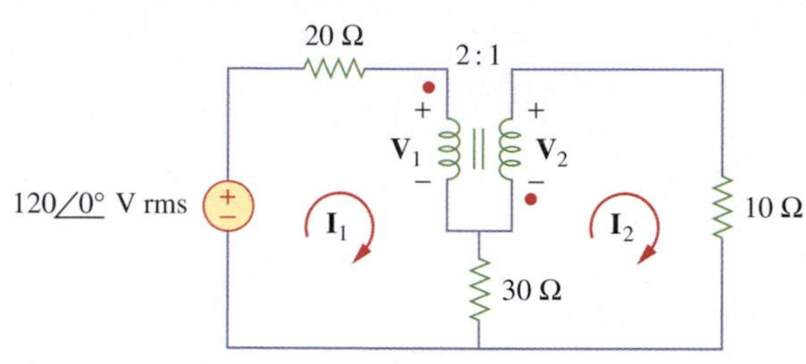

Figure 18.39
For Example 18.9.

Solution:

Reflection to the secondary or primary side cannot be done with this circuit because there is direct connection between the primary and secondary sides due to the 30-Ω resistor. We apply mesh analysis. For mesh 1,

$$-120 + (20 + 30)\mathbf{I}_1 - 30\mathbf{I}_2 + \mathbf{V}_1 = 0$$

or

$$50\mathbf{I}_1 - 30\mathbf{I}_2 + \mathbf{V}_1 = 120 \qquad \textbf{(18.9.1)}$$

For mesh 2,

$$-\mathbf{V}_2 + (10 + 30)\mathbf{I}_2 - 30\mathbf{I}_1 = 0$$

or

$$-30\mathbf{I}_1 + 40\mathbf{I}_2 - \mathbf{V}_2 = 0 \qquad \textbf{(18.9.2)}$$

At the transformer terminals,

$$\mathbf{V}_2 = -\frac{1}{2}\mathbf{V}_1 \qquad \textbf{(18.9.3)}$$

$$\mathbf{I}_2 = -2\mathbf{I}_1 \qquad \textbf{(18.9.4)}$$

(Note that $n = 1/2$). We now have four equations and four unknowns, but our goal is to get $\mathbf{I}_2$. So we substitute for $\mathbf{V}_1$ and $\mathbf{I}_1$ in terms of $\mathbf{V}_2$ and $\mathbf{I}_2$ in Eqs. (18.9.1) and (18.9.2). Equation (18.9.1) becomes

$$-55\mathbf{I}_2 - 2\mathbf{V}_2 = 120 \qquad \textbf{(18.9.5)}$$

and Eq. (18.9.2) becomes

$$15\mathbf{I}_2 + 40\mathbf{I}_2 - \mathbf{V}_2 = 0 \quad \Rightarrow \quad \mathbf{V}_2 = 55\mathbf{I}_2 \qquad \textbf{(18.9.6)}$$

Substituting Eq. (18.9.6) in Eq. (18.9.5),

$$-165\mathbf{I}_2 = 120 \quad \Rightarrow \quad \mathbf{I}_2 = -\frac{120}{165} = -0.7272 \text{ A}$$

The power absorbed by the 10-Ω resistor is

$$P = (-0.7272)^2(10) = 5.3 \text{ W}$$

Practice Problem 18.9

Find $\mathbf{V}_o$ in the circuit of Fig. 18.40.

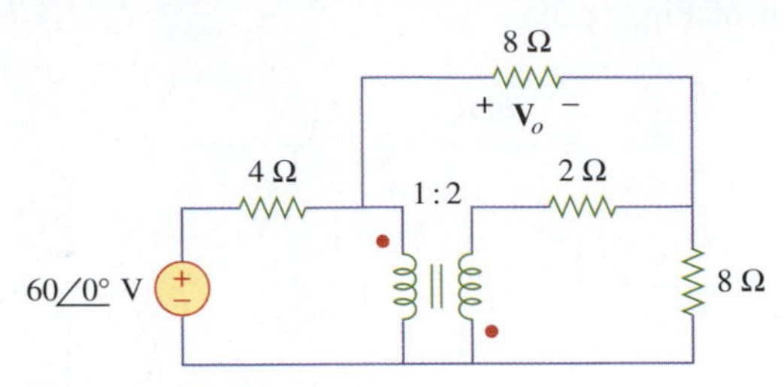

Figure 18.40
For Practice Problem 18.9.

Answer: 24 V

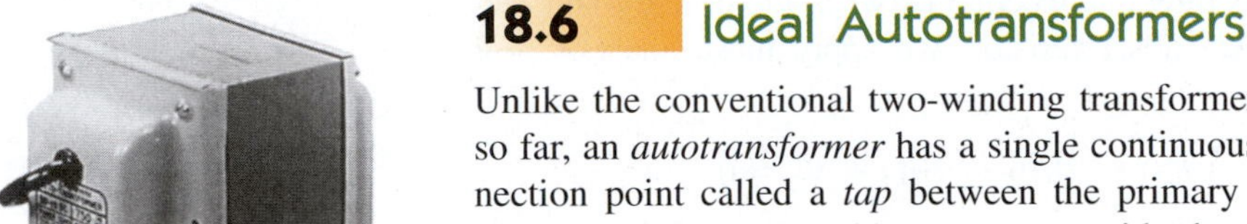

18.6 Ideal Autotransformers

Unlike the conventional two-winding transformer we have considered so far, an *autotransformer* has a single continuous winding with a connection point called a *tap* between the primary and secondary sides. The tap is often adjustable so as to provide the desired turns ratio for stepping up or stepping down the voltage. This way, a variable voltage is provided to the load connected to the autotransformer. An autotransformer is used to make an ac source adjustable. Thus,

> An **autotransformer** is a transformer in which both the primary and the secondary windings are in a single winding.

Figure 18.41
A typical autotransformer.
© Todd Systems, Inc., Yonkers, NY

A typical autotransformer is shown in Fig. 18.41. As shown in Fig. 18.42, the autotransformer can operate in the step-down or step-up mode. The autotransformer is a type of power transformer. Its major advantage over the two-winding transformer is its ability to transfer larger apparent power. This will be demonstrated in Example 18.10. Another advantage is that an autotransformer is smaller and lighter than an equivalent two-circuit transformer. However, because both the primary and secondary windings are one winding, *electrical isolation* (no direct electrical connection) is lost. (We will see how the property of electrical isolation in the conventional transformer is practically employed in Section 18.8.1.) The lack of electrical isolation between the primary and secondary windings is a major disadvantage of the autotransformer.

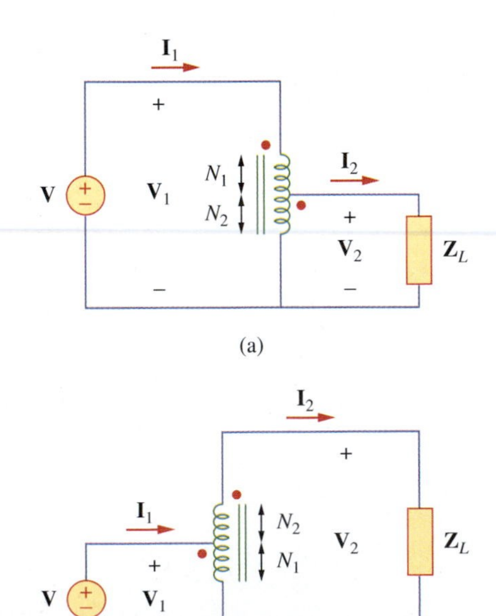

Figure 18.42
(a) Step-down autotransformer, (b) step-up autotransformer.

Some of the formulas we derived for ideal transformers apply to ideal autotransformers as well. For the step-down autotransformer circuit in Fig. 18.42(a), Eq. (18.39) gives

$$\frac{\mathbf{V}_1}{\mathbf{V}_2} = \frac{N_1 + N_2}{N_2} = 1 + \frac{N_1}{N_2} \tag{18.50}$$

As an ideal autotransformer, there are no losses so that the complex power remains the same in the primary and secondary windings:

$$\mathbf{S}_1 = \mathbf{V}_1\mathbf{I}_1^* = \mathbf{S}_2 = \mathbf{V}_2\mathbf{I}_2^* \tag{18.51}$$

Equation (18.51) can also be expressed with rms values as

$$\mathbf{V}_1\mathbf{I}_1 = \mathbf{V}_2\mathbf{I}_2$$

or

$$\frac{\mathbf{V}_2}{\mathbf{V}_1} = \frac{\mathbf{I}_1}{\mathbf{I}_2} \tag{18.52}$$

Thus, the current relationship is

$$\frac{\mathbf{I}_1}{\mathbf{I}_2} = \frac{N_2}{N_1 + N_2} \tag{18.53}$$

For the step-up autotransformer circuit of Fig. 18.42(b),

$$\frac{\mathbf{V}_1}{N_1} = \frac{\mathbf{V}_2}{N_1 + N_2}$$

or

$$\boxed{\frac{\mathbf{V}_1}{\mathbf{V}_2} = \frac{N_1}{N_1 + N_2}} \tag{18.54}$$

The complex power given by Eq. (18.51) also applies to the step-up autotransformer so that Eq. (18.52) again applies. Hence, the current relationship is

$$\frac{\mathbf{I}_1}{\mathbf{I}_2} = \frac{N_1 + N_2}{N_1} = 1 + \frac{N_2}{N_1} \tag{18.55}$$

A major difference between conventional transformers and autotransformers is that the primary and secondary sides of the autotransformer are not only coupled magnetically but also coupled conductively. The autotransformer can be used in place of a conventional transformer provided that electrical isolation is not required.

The larger the turns ratio, the less economical an autotransformer becomes. As a result, autotransformers with turns ratios greater than 2 are seldom used. Autotransformers of large sizes are used for interconnecting high-voltage power systems. They are used in small sizes for starting of motors. They are also used in the transmission of voice or sound signals.

Example 18.10

Compare the power ratings of the two-winding transformer in Fig. 18.43(a) and the autotransformer in Fig. 18.43(b).

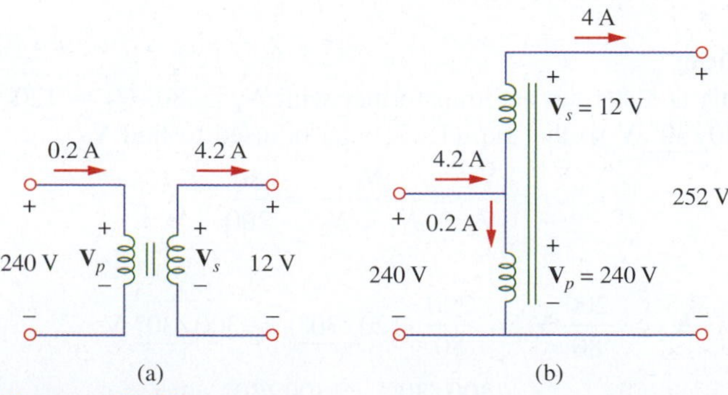

Figure 18.43
For Example 18.10.

Solution:

Although the primary and secondary windings of the autotransformer are together as a continuous winding, they are separated in Fig. 18.43(b) for clarity. We note that the current and voltage of each winding of the autotransformer in Fig. 18.43(b) are the same as those for the two-winding transformer in Fig. 18.43(a). This is the basis of comparing their power ratings.

For the two-winding transformer, the power rating is

$$S_1 = 0.2(240) = 48 \text{ VA}$$

or

$$S_2 = 4(12) = 48 \text{ VA}$$

For the autotransformer, the power rating is

$$S_1 = 4.2(240) = 1{,}008 \text{ VA}$$

or

$$S_2 = 4(252) = 1{,}008 \text{ VA}$$

which is 21 times the power rating of the two-winding transformer.

Practice Problem 18.10

Refer to Fig. 18.43. If the two-winding transformer is a 60-VA, 120/10-V transformer, what is the power rating of the autotransformer?

Answer: 780 VA

Example 18.11

Refer to the autotransformer circuit in Fig. 18.44. Calculate:
(a) $\mathbf{I}_1$, $\mathbf{I}_2$, and $\mathbf{I}_o$ if $\mathbf{Z}_L = 8 + j6\ \Omega$; and
(b) the complex power supplied to the load.

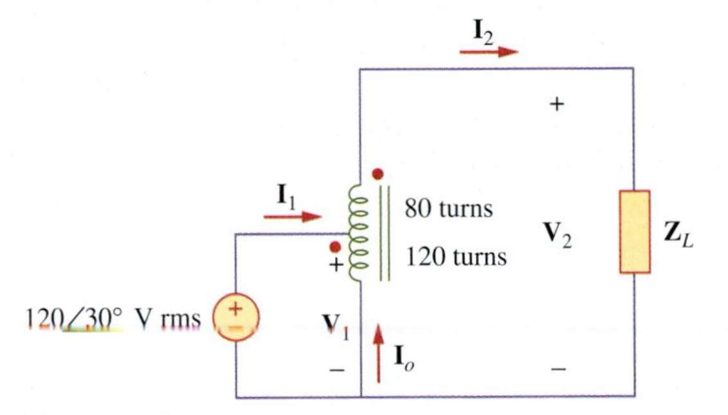

Figure 18.44
For Example 18.11.

Solution:

(a) This is a step-up autotransformer with $N_1 = 80$, $N_2 = 120$, $\mathbf{V}_1 = 120\underline{/30^\circ}$ V so that Eq. (18.54) can be used to find $\mathbf{V}_2$.

$$\frac{\mathbf{V}_1}{\mathbf{V}_2} = \frac{N_1}{N_1 + N_2} = \frac{80}{200}$$

or

$$\mathbf{V}_2 = \frac{200}{80}\mathbf{V}_1 = \frac{200}{80}(120\underline{/30^\circ}) = 300\underline{/30^\circ} \text{ V}$$

$$\mathbf{I}_2 = \mathbf{V}_2/\mathbf{Z}_L = \frac{300\underline{/30^\circ}}{8 + j6} = \frac{300\underline{/30^\circ}}{10\underline{/36.87^\circ}} = 30\underline{/-6.87^\circ} \text{ A}$$

But

$$\mathbf{I}_1/\mathbf{I}_2 = \frac{N_1 + N_2}{N_1} = \frac{200}{80}$$

or

$$\mathbf{I}_1 = \frac{200}{80}\mathbf{I}_2 = \frac{200}{80}(30\angle -6.87^\circ) = 75\angle -6.87^\circ \text{ V}$$

At the tap, KCL gives

$$\mathbf{I}_1 + \mathbf{I}_o = \mathbf{I}_2$$

or

$$\mathbf{I}_o = \mathbf{I}_2 - \mathbf{I}_1 = 30\angle -6.87^\circ - 75\angle -6.87^\circ = 45\angle 173.13^\circ \text{ A}$$

(b) The complex power supplied to the load is

$$\mathbf{S}_2 = \mathbf{V}_2\mathbf{I}_2^* = |\mathbf{I}_2|^2\mathbf{Z}_L = (30)^2(10\angle 36.87^\circ) = 9\angle 36.87^\circ \text{ kVA}$$

Practice Problem 18.11

In the autotransformer circuit in Fig. 18.45, find currents $\mathbf{I}_1$, $\mathbf{I}_2$, and $\mathbf{I}_o$. Take $\mathbf{V}_1 = 1{,}250$ V and $\mathbf{V}_2 = 800$ V.

Answer: 12.8 A; 20 A; 7.2 A

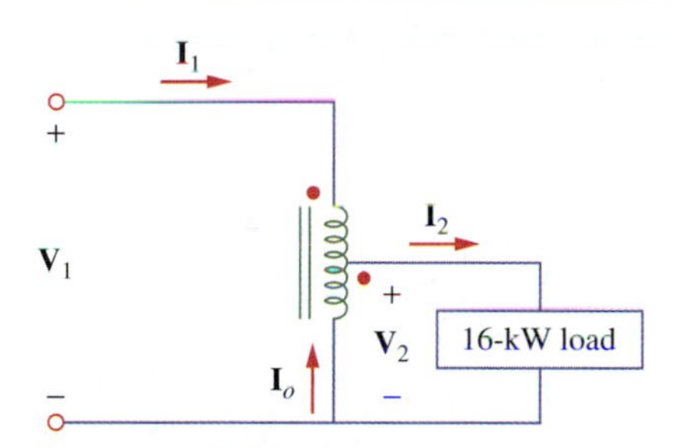

Figure 18.45
For Practice Problem 18.11.

18.7 Computer Analysis

PSpice analyzes magnetically coupled circuits just like inductor circuits except that the dot convention must be followed. In PSpice Schematic, the dot (not shown) is always next to pin 1, which is the left-hand terminal of the inductor when the inductor with partname L is placed (horizontally) without rotation on a schematic. Thus, the dot or pin 1 will be at the bottom after one 90° counterclockwise rotation because rotation is always around pin 1. Once the magnetically coupled inductors are arranged with the dot convention in mind and their value attributes are set in henrys, we use the coupling symbol KBREAK to define the coupling. For each pair of coupled inductors, take the following steps:

1. Select **Place/Part** and type K_LINEAR.
2. **DCLICK** and place K_LINEAR symbol on the schematic, as shown in Fig. 18.46. (Notice that K_LINEAR is not a component and therefore has no pins.)
3. **DCLICKL** on COUPLING and set the value of the coupling coefficient k.
4. **DCLICKL** on the boxed K (the coupling symbol) and enter the reference designator names for the coupled inductors as values of Li, $i = 1, 2, \ldots, 6$. For example, if inductors L20 and L23 are coupled, we set L1 = L20 and L2 = L23. L1 and at least one other Li must be assigned values; other Li values may be left blank.

```
[K] K1
K_Linear
 COUPLING=1
```

Figure 18.46
K_Linear for defining coupling.

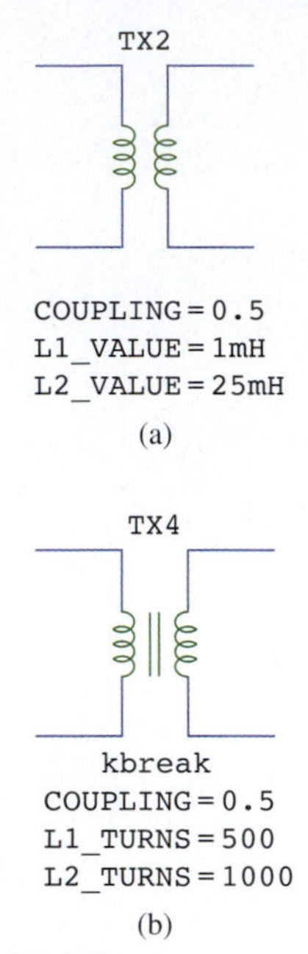

Figure 18.47
(a) Linear transformer XFRM_LINEAR; (b) ideal transformer XFRM_NONLINEAR.

In step 4, up to six coupled inductors with equal coupling can be specified.

For the air-core transformer, the partname is XFRM_LINEAR. It can be inserted in a circuit by selecting **Place/Part** and then typing in the part name or by selecting the partname from the analog.slb library. As shown typically in Fig. 18.47(a), the main attributes of the linear transformer are the coupling coefficient k and the inductance values L1 and L2 in henrys. If the mutual inductance M is specified, its value must be used along with L1 and L2 to calculate k. Keep in mind that the value of k should lie between 0 and 1.

For the ideal transformer, the part name is XFRM_NONLINEAR and is located in the ANALOG library. It can be selected by clicking **Place/Part** and then typing in the part name. Its attributes are the coupling coefficient and the numbers of turns associated with L1 and L2, as illustrated typically in Fig. 18.47(b). The value of the coefficient of mutual coupling must lie between 0 and 1. Although it was mentioned in Section 18.5 that $k = 1$ for an ideal transform, PSpice allows k to vary between 0 and 1.

PSpice has some additional transformer configurations that will not be discussed here.

Example 18.12

Use PSpice to find i_1, i_2, and i_3 in the circuit displayed in Fig. 18.48.

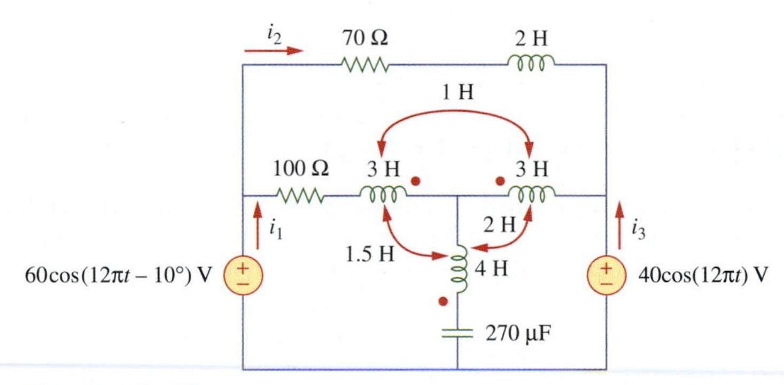

Figure 18.48
For Example 18.12.

Solution:

The coupling coefficients of the three coupled inductors are determined as follows.

$$k_{12} = \frac{M_{12}}{\sqrt{L_1 L_2}} = \frac{1}{\sqrt{3 \times 3}} = 0.3333$$

$$k_{13} = \frac{M_{13}}{\sqrt{L_1 L_3}} = \frac{1.5}{\sqrt{3 \times 4}} = 0.433$$

$$k_{23} = \frac{M_{23}}{\sqrt{L_2 L_3}} = \frac{2}{\sqrt{3 \times 4}} = 0.5774$$

The operating frequency f is obtained from Fig. 18.48 as

$$\omega = 12\pi = 2\pi f \qquad \Rightarrow \qquad f = 6 \text{ Hz.}$$

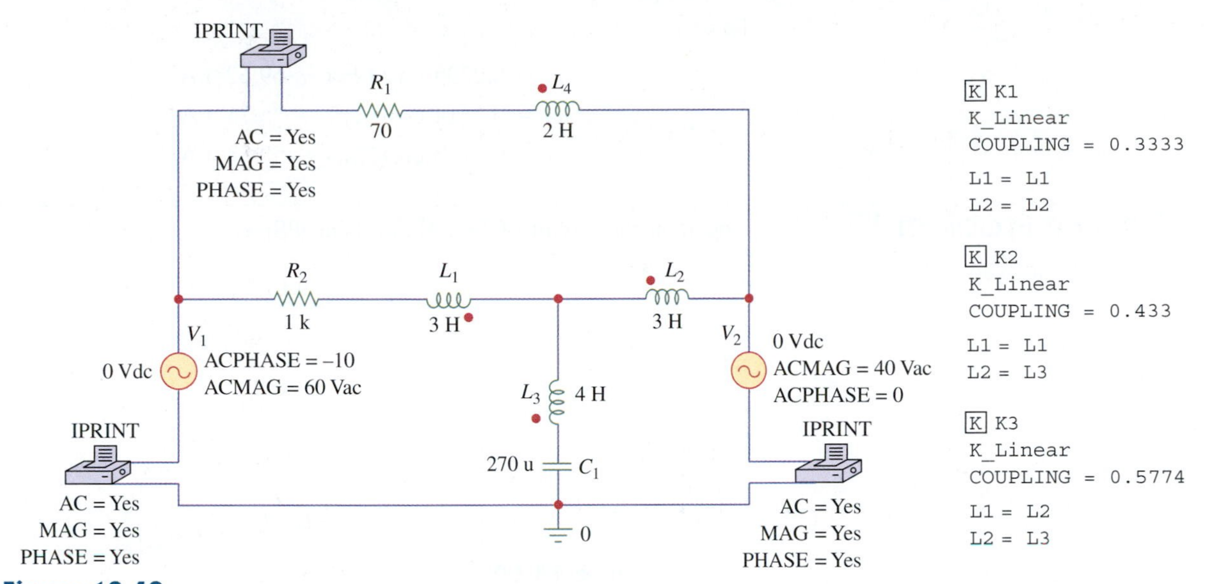

Figure 18.49
Schematic of the circuit in Fig. 18.48.

The schematic of the circuit is portrayed in Fig. 18.49. Three IPRINT pseudocomponents are inserted in the appropriate branches to obtain the required currents i_1, i_2, and i_3. Notice how the dot convention is adhered to. For L2, the dot is on pin 1 (the left-hand terminal) and is therefore placed without rotation. For L1, in order for the dot to be on the right-hand side of the inductor, the inductor must be rotated through 180°. For L3, the inductor must be rotated through 90° so that the dot will be at the bottom. Note that the 2-H inductor (L4) is not coupled.

After drawing the circuit as in Fig. 18.49 and saving as exam1812.dsn, we select **PSpice/New Simulation Profile**. This leads to the New Simulation dialog box. Type "exam1812" as the name of the file and click Create. This leads to the Simulation Settings dialog box. In the Simulation Settings dialog box, select AC Sweep/Noise under *Analysis Type*, select Linear under *AC Sweep Type*. We type 6 as the *Start Freq*, 6 as the *Final Freq*, and 1 as *Total Points*.

We simulate the circuit by selecting **PSpice/Run.** We obtain the output file by selecting **PSpice/View Output file**. The output file includes the following:

```
FREQ          IM(V_PRINT3)     IP(V_PRINT3)
6.000E+00     2.335E-01        -6.962E+01
FREQ          IM(V_PRINT1)     IP(V_PRINT1)
6.000E+00     2.114E-01        -7.575E+01
FREQ          IM(V_PRINT2)     IP(V_PRINT2)
6.000E+00     1.143E-01        -5.058E+01
```

From this we obtain

$$\mathbf{I}_1 = 0.2335\underline{/-69.62^\circ}$$

$$\mathbf{I}_2 = 0.2114\underline{/-75.75^\circ}$$

$$\mathbf{I}_3 = 0.1143\underline{/-50.58^\circ}$$

Thus,

$$i_1 = 0.2335 \cos(12\pi t - 69.62°) \text{ A}$$
$$i_2 = 0.2114 \cos(12\pi t - 75.75°) \text{ A}$$
$$i_3 = 0.1143 \cos(12\pi t - 50.58°) \text{ A}$$

Practice Problem 18.12

Find i_o in the circuit of Fig. 18.50 using PSpice.

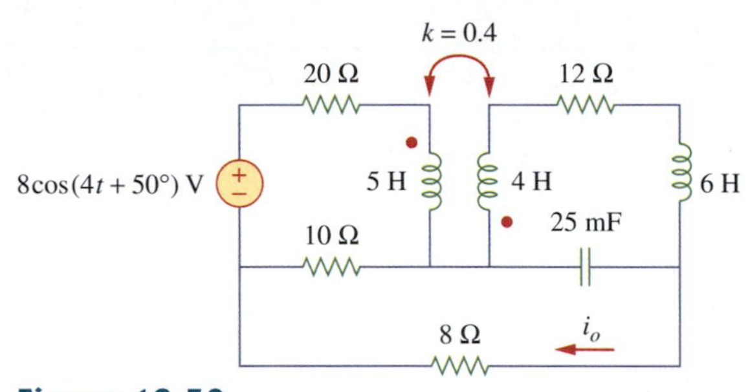

Figure 18.50
For Practice Problem 18.12.

Answer: $0.1006 \cos(4t + 68.52°)$ A

Example 18.13

Find $\mathbf{V}_1$ and $\mathbf{V}_2$ in the ideal transformer circuit of Fig. 18.51 using PSpice.

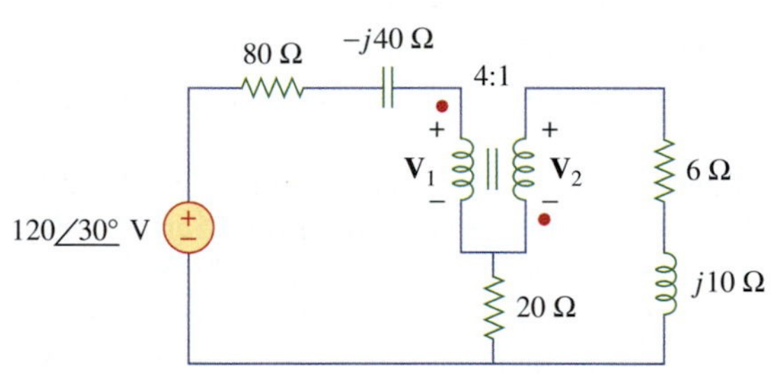

Figure 18.51
For Example 18.13.

Solution:
As usual, we assume $\omega = 1$ and find the corresponding values of capacitance and inductance of the elements.

$$j10 = j\omega L \quad \Rightarrow \quad L = 10 \text{ H}$$
$$-j40 = \frac{1}{j\omega C} \quad \Rightarrow \quad C = 25 \text{ mF}$$

Figure 18.52 shows the schematic. For the ideal transformer, we set the coupling factor to 0.999 and the numbers of turns to 400,000 and 100,000.[3] The two VPRINT2 pseudocomponents are connected across the transformer terminals to obtain $\mathbf{V}_1$ and $\mathbf{V}_2$.

[3] Reminder: Keep in mind that for an ideal transformer, the inductances of both the primary and secondary windings are infinitely large.

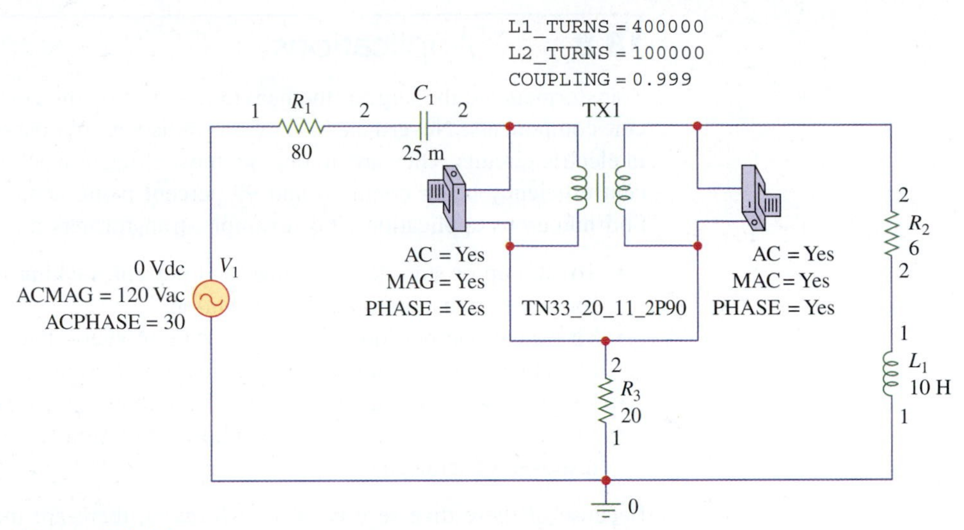

Figure 18.52
Schematic for the circuit in Fig. 18.51.

After drawing the circuit as in Fig. 18.52 and saving as exam1813.dsn, we select **PSpice/New Simulation Profile**. This leads to the New Simulation dialog box. Type "exam1813" as the name of the file and click Create. This leads to the Simulation Settings dialog box. In the Simulation Settings dialog box, select AC Sweep/Noise under *Analysis Type*, select Linear under *AC Sweep Type*. We type 0.1592 as the *Start Freq*, 0.1592 as the *Final Freq*, and 1 as *Total Points.*

We simulate the circuit by selecting **PSpice/Run**. We obtain the output file by selecting **PSpice/View Output file**. The output file includes the following:

```
FREQ          VM(C,A)       VP(C,A)
1.592E-01     1.212E+02     -1.435E+02
FREQ          VM(B,C)       VP(B,C)
1.592E-01     2.775E+02     2.789E+01
```

From this we obtain

$$\mathbf{V}_1 = -V(C, A) = 121.2\underline{/36.5^\circ}\text{ V}$$
$$\mathbf{V}_2 = V(B, C) = 277.5\underline{/27.89^\circ}\text{ V}$$

Practice Problem 18.13

Obtain $\mathbf{V}_1$ and $\mathbf{V}_2$ in the circuit of Fig. 18.53 using PSpice.

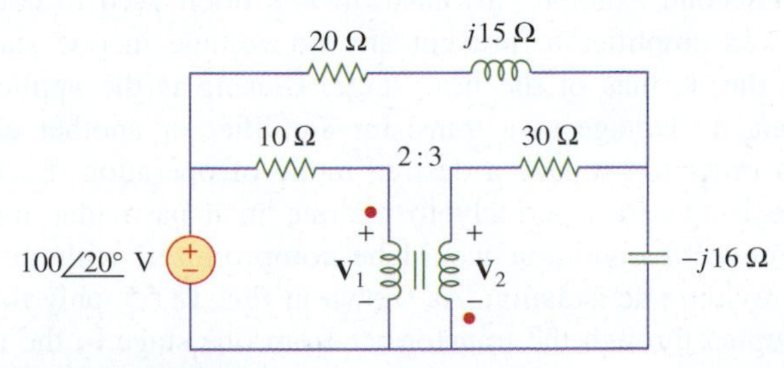

Figure 18.53
For Practice Problem 18.13.

Answer: $63.1\underline{/28.65^\circ}$ V; $94.64\underline{/-151.4^\circ}$ V

18.8 †Applications

Transformers are the largest, the heaviest, and often the costliest of circuit components. Nevertheless, they are indispensable passive devices in electric circuits. They are among the most efficient machines, 95 percent efficiency being common and 99 percent being achievable. They find numerous applications. For example, transformers are used:

- To step up or step down voltage and current, making them useful for power transmission and distribution.
- To isolate one portion of a circuit from another—that is, to transfer power without any wired electrical connection.
- As an impedance-matching device for maximum power transfer.
- In frequency selective circuits whose operation depends on the response of inductances.

Because of these diverse uses of transformers, there are many special designs for transformers—voltage transformers, current transformers, power transformers, distribution transformers, impedance-matching transformers, audio transformers, single-phase transformers, three-phase transformers, rectifier transformers, inverter transformers, and the like—only some of which are discussed in this chapter.[4] In this section, we will consider three important applications of transformers: as an isolation device, as a matching device, and as a power distribution system.

18.8.1 Transformers as Isolation Devices

Electrical isolation is said to exist between two devices when there is no wired electrical connection between them. In a transformer, energy is transferred by magnetic coupling, without an electrical connection between the primary circuit and secondary circuit. We now consider three simple practical examples of how we take advantage of this property.

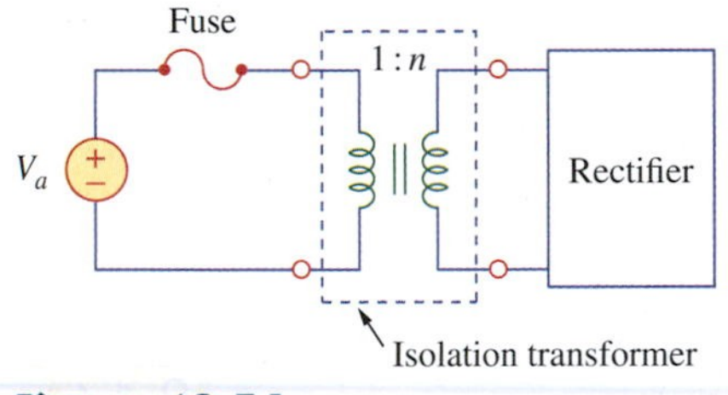

Figure 18.54
A transformer used to isolate an ac supply from a rectifier.

Consider the circuit in Fig. 18.54. A rectifier is an electronic circuit that converts an ac voltage to a dc voltage. A transformer is often used to couple the ac voltage to the rectifier. The transformer serves two purposes. First, it steps up or steps down the voltage. Second, it provides electrical isolation between the ac power supply and the rectifier, thereby reducing the risk of shock in handling the electronic device.

As a second example, a transformer is often used to couple two stages of an amplifier to prevent any dc voltage in one stage from affecting the dc bias of the next stage. Biasing is the application of appropriate dc voltage to a transistor amplifier or another electronic device in order to produce a desired mode of operation. Each amplifier stage is biased separately to operate in a particular mode; the desired mode of operation would be compromised without a transformer providing dc isolation. As shown in Fig. 18.55, only the ac signal is coupled through the transformer from one stage to the next. We

[4] Note: For more information on the many kinds of transformers, a good text is W. M. Flanagan, *Handbook of Transformer Design and Applications,* 2nd ed., New York: McGraw-Hill, 1993.

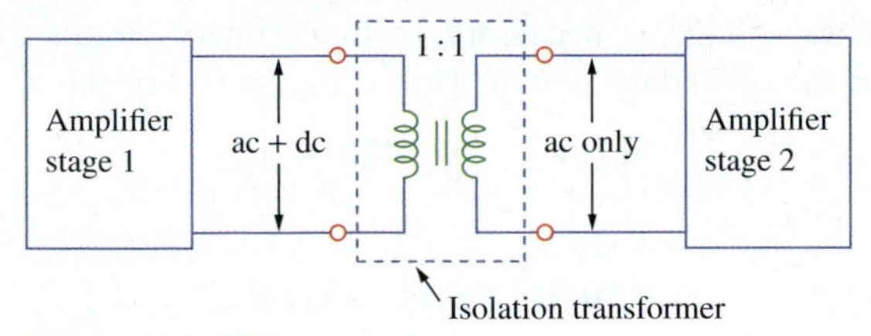

Figure 18.55
A transformer providing dc isolation between two amplifier stages.

recall that magnetic coupling does not exist with a dc voltage source. Transformers are used in radio and TV receivers to couple stages of high-frequency amplifiers. When the sole purpose of a transformer is to provide isolation, its turns ratio n is made unity. Thus, an isolation transformer has $n = 1$.

As a third example, consider measuring the voltage across 13.2-kV lines. It is obviously not safe to connect a voltmeter directly to such high-voltage lines. A transformer can be used both to electrically isolate the line power from the voltmeter and to step down the voltage to a safe level, as shown in Fig. 18.56. Once the voltmeter is used to measure the secondary voltage, the turns ratio is used to determine the line voltage on the primary side.

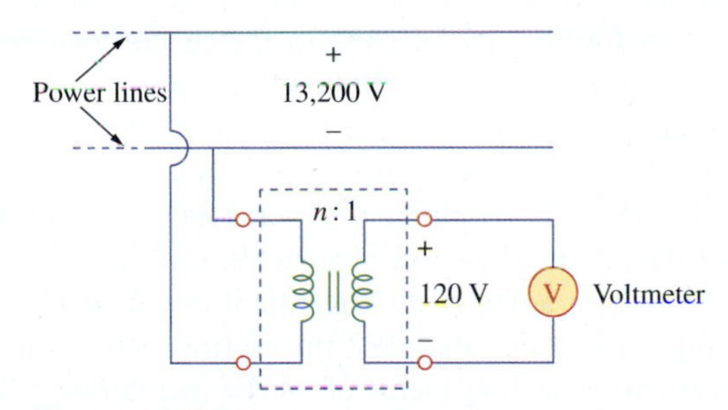

Figure 18.56
A transformer providing isolation between the power lines and the voltmeter.

Example 18.14

Determine the voltage across the load in Fig. 18.57.

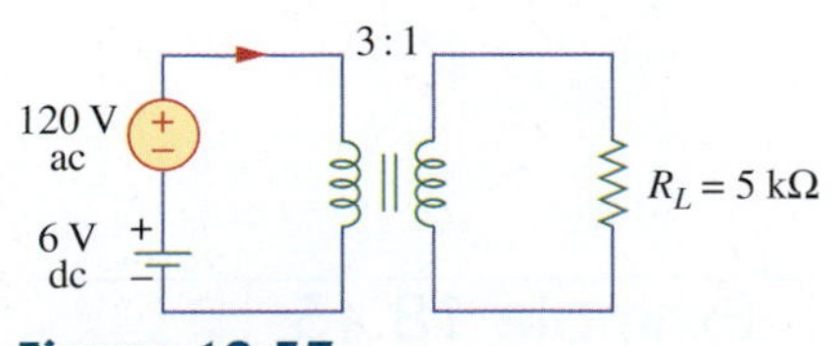

Figure 18.57
For Example 18.14.

Solution:
We can apply the superposition principle to find the load voltage. Let $v_L = v_{L1} + v_{L2}$, where v_{L1} is due to the dc source and v_{L2} is due to the ac source. We consider the dc and ac sources separately, as shown in Fig. 18.58. The load voltage due to the dc source is zero because a

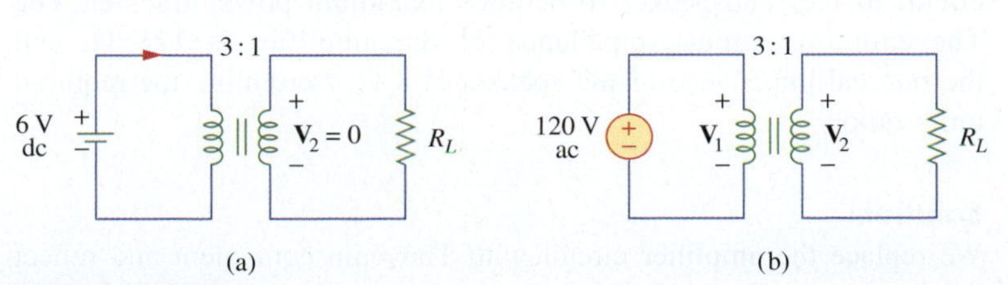

Figure 18.58
For Example 18.14: (a) dc source; (b) ac source.

time-varying voltage is necessary in the primary circuit to induce a voltage in the secondary circuit. Hence $v_{L1} = 0$. For the ac source,

$$\mathbf{V}_2/\mathbf{V}_1 = \mathbf{V}_2/120 = 1/3$$

or

$$\mathbf{V}_2 = 120/3 = 40 \text{ V}$$

Hence, the load voltage due to the ac source is $\mathbf{V}_{L2} = 40$ Vac or $v_{L2} = 40 \cos(\omega t)$; that is, only the ac voltage is passed to the load by the transformer. This example shows how the transformer provides dc isolation.

Practice Problem 18.14

Refer to Fig. 18.56. Calculate the turns ratio required to step down the 18.2-kV line voltage to a safe level of 120 V.

Answer: 110

18.8.2 Transformers as Matching Devices

We recall that for maximum power transfer, the load resistance R_L must be matched with the source resistance R_s. In most cases, the two resistances are not matched; both are fixed and cannot be altered. However, an iron-core transformer can be used to match the load resistance to the source resistance. This is called *impedance matching*. For example, to connect a loudspeaker to an audio power amplifier requires a transformer because the speaker's resistance is only a few ohms, while the internal resistance of the amplifier is several thousand ohms.

Consider the circuit shown in Fig. 18.59. (In Fig. 18.59, V_s and R_s may be replaced with a Thevenin equivalent circuit if necessary.) We recall from Eq. (18.47) that the ideal transformer reflects its load back to the primary with a scaling factor of n^2. To match this reflected load R_L/n^2 with the source resistance R_s, we set them equal; that is,

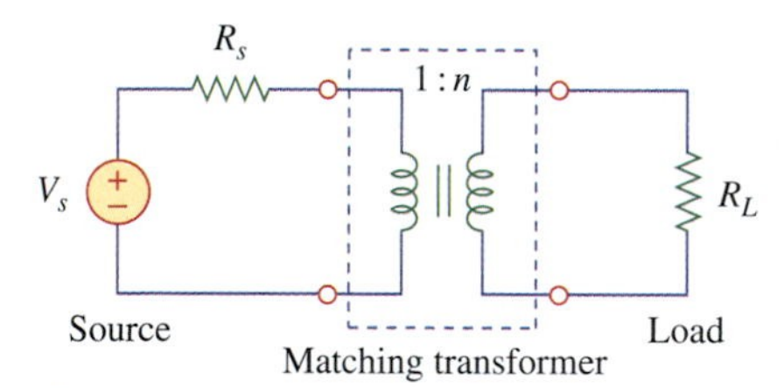

Figure 18.59
Transformer used as a matching device.

$$R_s = \frac{R_L}{n^2} \tag{18.56}$$

Equation (18.56) can be satisfied by proper selection of the turns ratio n. From Eq. (18.56), we notice that a step-down transformer ($n < 1$) is needed as the matching device when $R_s > R_L$ and a step-up ($n > 1$) is required when $R_s < R_L$.

Example 18.15

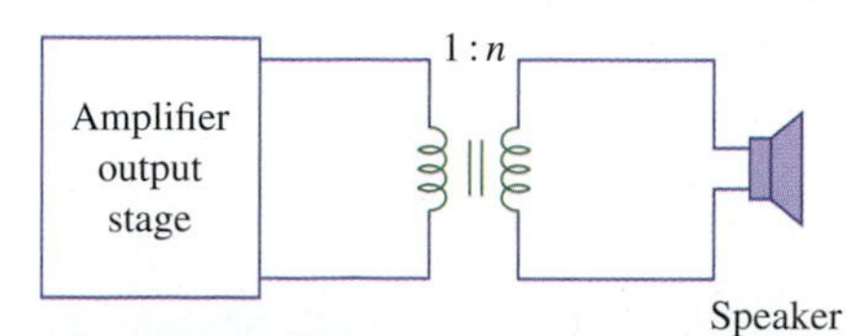

Figure 18.60
Using an ideal transformer to match the speaker to the amplifier; for Example 18.15.

The ideal transformer in Fig. 18.60 is used to match the amplifier circuit to the loudspeaker to achieve maximum power transfer. The Thevenin (or output) impedance of the amplifier is 128 Ω, and the internal impedance of the speaker is 8 Ω. Determine the required turns ratio.

Solution:

We replace the amplifier circuit with Thevenin equivalent and reflect the impedance $Z_L = 12\ \Omega$ of the speaker to the primary side of the ideal transformer. The result is shown in Fig. 18.61. For maximum power transfer,

$$Z_{Th} = \frac{Z_L}{n^2}$$

or

$$n^2 = \frac{Z_L}{Z_{Th}} = \frac{8}{128} = \frac{1}{16}$$

Thus the turns ratio is $n = 1/4 = 0.25$.

Using $P = I^2R$, we can show that indeed the power delivered to the speaker is much larger than without the ideal transformer. Without the ideal transformer, the amplifier is directly connected to the speaker. The power delivered to the speaker is

$$P_L = \left(\frac{V_{Th}}{Z_{Th} + Z_L}\right)^2 Z_L = 432.5V_{Th}^2\ \mu\text{W}$$

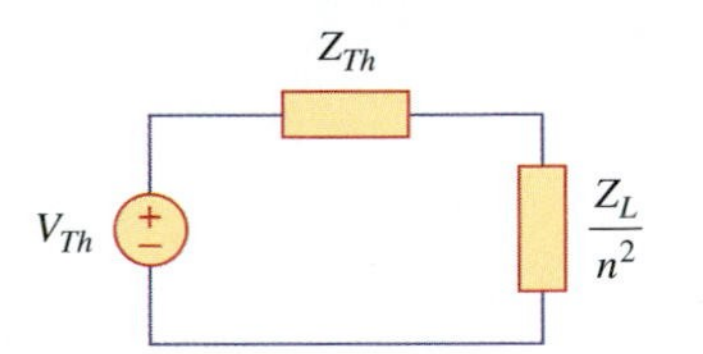

Figure 18.61
Equivalent circuit of the circuit in Fig. 18.60; for Example 18.15.

With the transformer in place, the primary and secondary currents are

$$I_p = \frac{V_{Th}}{Z_{Th} + Z_L/n^2}, \qquad I_s = \frac{I_p}{n}$$

Hence,

$$P_L = I_s^2 Z_L = \left(\frac{V_{Th}/n}{Z_{Th} + Z_L/n^2}\right)^2 Z_L = \left(\frac{nV_{Th}}{n^2 Z_{Th} + Z_L}\right)^2 Z_L$$

$$= 1{,}953V_{Th}^2\ \mu\text{W}$$

confirming what was said earlier.

Practice Problem 18.15

Calculate the turns ratio of an ideal transformer required to match a 1-kΩ load to a source with internal impedance of 25 kΩ. Find the load voltage when the source voltage is 30 V.

Answer: 0.2; 3 V

18.8.3 Power Distribution

A power system basically consists of three parts: generation, transmission, and distribution. The electric company operates a plant that generates several hundreds of megavolt-amperes (MVA) at typically about 18 kV. As illustrated in Fig. 18.62, three-phase step-up transformers are used to feed the generated power to the transmission line.

Why do we need the transformer? Suppose we need to transmit 100,000 VA over a distance of 50 km. Because $S = VI$, using a line voltage of 1,000 V implies that the transmission line must carry 100 A, and this requires a transmission line of a large diameter. If, on the other hand, we use a line voltage of 10,000 V, the current would be only 10 A. A smaller current reduces the required conductor size, producing considerable savings as well as minimizing I^2R losses. To minimize losses requires a step-up transformer.[5] Without the transformer, the majority

[5]Note: One may ask, "How would increasing the voltage not increase the current, thereby increasing I^2R losses?" Keep in mind that $IV_\ell = V_\ell^2/R$, where V_ℓ is the potential difference between the sending and receiving ends of the line. The voltage that is stepped-up is the sending end voltage V, not V_ℓ. If the receiving end is V_R, then $V_\ell = V - V_R$. Because V and V_R are close to each other, V_ℓ is small even when V is stepped up.

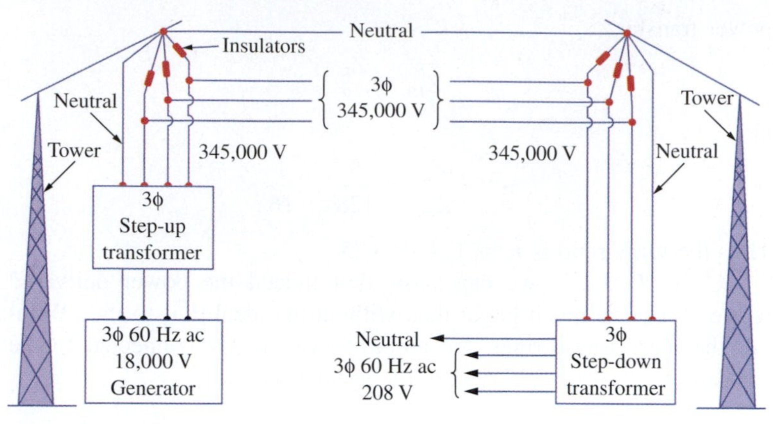

Figure 18.62
A typical power distribution system.
A. Marcus and C. M. Thomson, *Electricity for Technicians,* 2nd edition, © 1975, p. 337. Reprinted by permission of Pearson Education, Inc., Upper Saddle River, NJ.

of the power generated would be lost on the transmission line. The ability of the transformer to step up or step down voltage and distribute power economically is one of the major reasons for generating ac rather than dc. Thus, for a given power, the larger the voltage, the better. Today, 1 MV is the largest voltage in use; the level may increase as a result of research and experiments.

Beyond the generation plant, the power is transmitted for hundreds of miles through an electric network called the *power grid*. The three-phase power in the power grid is conveyed by transmission lines hung overhead from steel towers, which come in a variety of sizes and shapes. The (aluminum-conductor, steel-reinforced) lines typically have overall diameters up to about 40 mm and can carry current of up to 1,380 A.

At the substations, distribution transformers are used to step down the voltage. The step-down process is usually carried out in stages. The power may be distributed throughout a locality by means of either overhead or underground cables. The substations distribute the power to residential, commercial, and industrial customers. At the receiving end, a residential customer is eventually supplied with 120/240 V, while industrial or commercial customers are fed with higher voltages such as 460/208 V. Residential customers are usually supplied by distribution transformers, often mounted on the poles of the electric utility company or underground. When direct current is needed, the alternating current is converted to dc electronically.

Example 18.16

A distribution transformer is used to supply a household as in Fig. 18.63. The load consists of eight 100-W bulbs, a 350-W TV, and a 15-kW kitchen range. If the secondary side of the transformer has 72 turns, calculate:
(a) the number of turns of the primary winding and
(b) the current I_p in the primary winding.

Solution:

(a) The dot locations on the winding are not important because we are only interested in the magnitudes of the variables involved. Because

$$\frac{N_p}{N_s} = \frac{V_p}{V_s}$$

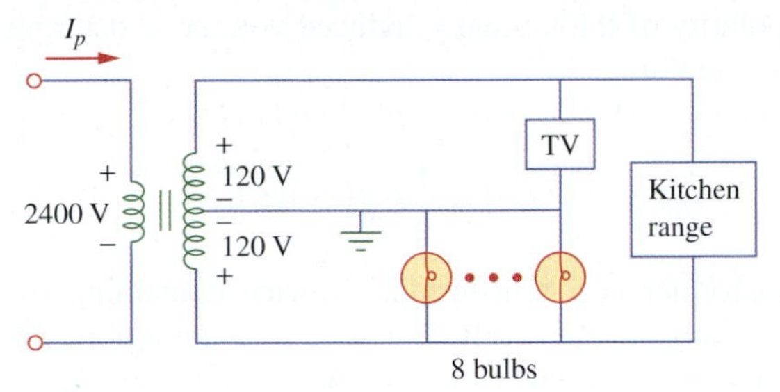

Figure 18.63
For Example 18.16 and Practice Problem 18.16.

we get

$$N_p = N_s \frac{V_p}{V_s} = 72 \times \frac{2{,}400}{240} = 720 \text{ turns}$$

(b) The total power absorbed by the load is

$$S = 8 \times 100 + 350 + 15{,}000 = 16.15 \text{ kW}$$

But

$$S = V_p I_p = V_s I_s$$

so that

$$I_p = \frac{S}{V_p} = \frac{16{,}150}{2{,}400} = 6.729 \text{ A}$$

Practice Problem 18.16

In Example 18.16, if the eight 100-W bulbs are replaced by twelve 60-W bulbs and the kitchen range is replaced by a 4.5-kW air-conditioner, find: (a) the total power supplied and (b) the current I_p in the primary winding.

Answer: (a) 5.57 kW; (b) 2.321 A

18.9 Summary

1. Two coils are said to be mutually coupled if the magnetic flux emanating from one passes through the other. The mutual inductance between the two coils is given by

$$M = k\sqrt{L_1 L_2}$$

where k is the coupling coefficient, $0 < k < 1$.

2. If v_1 and i_1 are the voltage and current in coil 1, while v_2 and i_2 are the voltage and current in coil 2, then

$$v_1 = L_1 \frac{di_1}{dt} + M \frac{di_2}{dt}$$

and

$$v_2 = L_2 \frac{di_2}{dt} + M \frac{di_1}{dt}$$

Thus, the voltage induced in a coupled coil consists of self-induced voltage and mutual voltage.

3. The polarity of the mutually induced voltage is determined by the dot convention.
4. The energy stored in two coupled coils is

$$\frac{1}{2}L_1 i_1^2 + \frac{1}{2}L_2 i_2^2 \pm M i_1 i_2$$

5. A transformer is a four-terminal device containing two or more magnetically coupled coils. It is used in changing the current, voltage, or impedance level in a circuit.
6. A linear (or loosely coupled) transformer has its coils wound on a magnetically linear material. It can be replaced by an equivalent T or Π network for the purposes of analysis.
7. An ideal (or iron-core) transformer is a lossless ($R_1 = R_2 = 0$) transformer with unity coupling coefficient ($k = 1$) and infinite inductances (L_1, L_2, $M \rightarrow \infty$).
8. For an ideal transformer,

$$\mathbf{V}_2 = n\mathbf{V}_1, \qquad \mathbf{I}_2 = \mathbf{I}_1/n, \qquad \mathbf{S}_1 = \mathbf{S}_2, \ \mathbf{Z}_R = \frac{Z_L}{n^2}$$

where $n = N_2/N_1$ is the turns ratio, N_1 is the number of turns of the primary winding, and N_2 is the number of turns of the secondary winding. The transformer steps up the primary voltage when $n > 1$, steps it down when $n < 1$, or serves as an isolation device when $n = 1$.
9. An autotransformer is a transformer with a single winding common to both the primary and the secondary circuits.
10. PSpice is a useful tool for analyzing magnetically coupled circuits.
11. Transformers are necessary in all stages of power distribution systems.
12. Important uses of transformers in electronics applications are as electrical isolation devices and as impedance matching devices.

Review Questions

18.1 Refer to the two magnetically coupled coils of Fig. 18.64(a). The polarity of the mutual voltage is:

(a) Positive (b) Negative

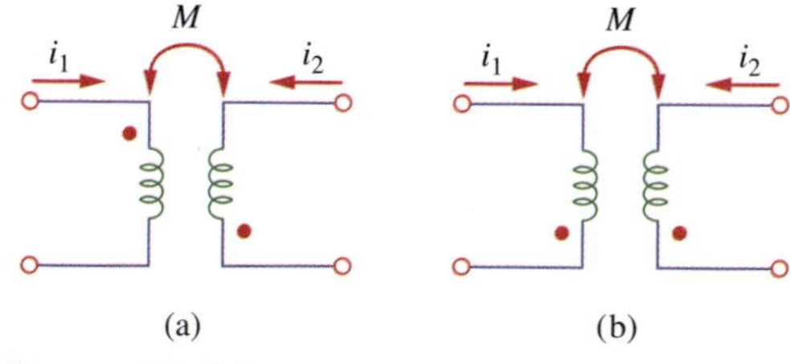

Figure 18.64
For Review Questions 18.1 and 18.2.

18.2 For the two magnetically coupled coils of Fig. 18.64(b), the polarity of the mutual voltage is:

(a) Positive (b) Negative

18.3 The coefficient of coupling for two coils having $L_1 = 2$ H, $L_2 = 8$ H, and $M = 3$ H is:

(a) 0.1875 (b) 0.75 (c) 1.333 (d) 5.333

18.4 A transformer is used in stepping down or stepping up:

(a) dc voltages (b) ac voltages
(c) both dc and ac voltages

18.5 If a 10-kΩ resistor is connected to the secondary side of a transformer with turns ratio of 10, the source "sees" a reflected load of:

(a) 10 kΩ (b) 1 kΩ (c) 100 Ω (d) 10 Ω

18.6 The ideal transformer in Fig. 18.65(a) has $N_2/N_1 = 10$. The ratio V_2/V_1 is:

(a) 10 (b) 0.1 (c) −0.1 (d) −10

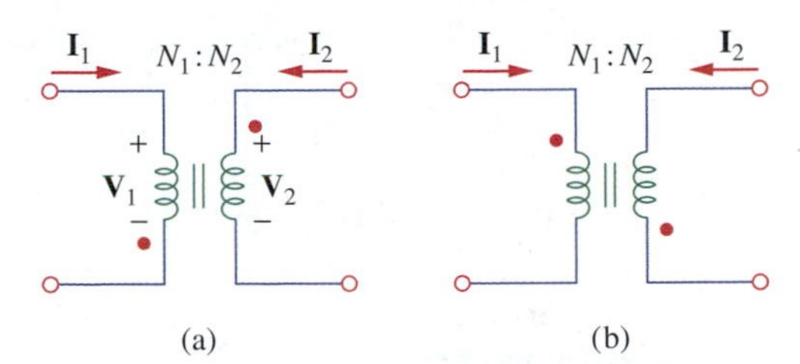

Figure 18.65
For Review Questions 18.6 and 18.7.

18.7 For the ideal transformer in Fig. 18.65(b), $N_2/N_1 = 10$. The ratio I_2/I_1 is:

(a) 10 (b) 0.1
(c) −0.1 (d) −10

18.8 An ideal transformer is rated 25000/240 V. What kind of transformer is this?

(a) step-up (b) step-down (c) isolation

18.9 In order to match a source with internal impedance of 500 Ω to a 15-Ω load, what is needed is:

(a) a step-up linear transformer
(b) a step-down linear transformer
(c) a step-up ideal transformer
(d) a step-down ideal transformer
(e) an autotransformer

18.10 Which of these transformers can be used as an isolation device?

(a) a linear transformer
(b) an ideal transformer
(c) an autotransformer
(d) all of the above

Answers: 18.1b, 18.2a, 18.3b, 18.4b, 18.5c, 18.6d, 18.7b, 18.8b, 18.9d, 18.10b

Problems

Section 18. 2 Mutual Inductance

18.1 For the three coupled coils in Fig. 18.66, calculate the total inductance.

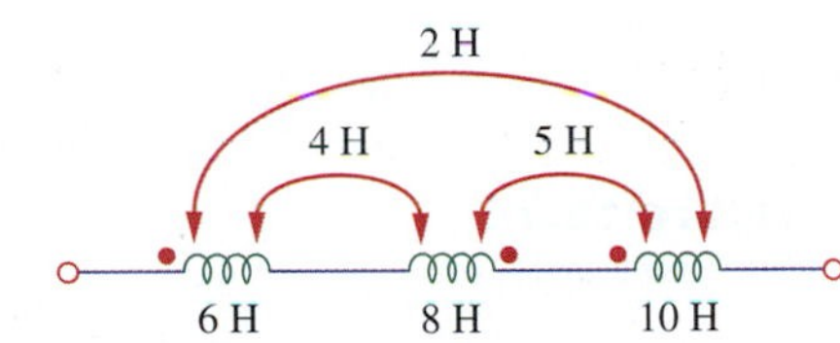

Figure 18.66
For Problem 18.1.

18.2 Determine the total inductance of the three series-connected inductors in Fig. 18.67.

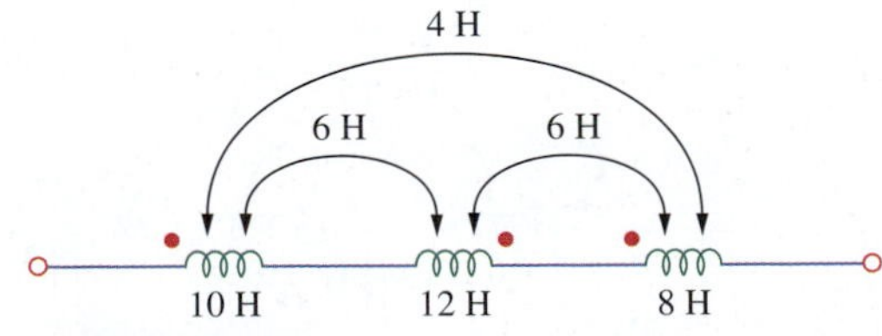

Figure 18.67
For Problem 18.2.

18.3 Two coils connected in series-aiding fashion have a total inductance of 250 mH. When connected in a series-opposing configuration, the coils have a total inductance of 150 mH. If the inductance of one coil (L_1) is three times the other, find L_1, L_2, and M. What is the coupling coefficient?

18.4 Two coils are mutually coupled, with $L_1 = 25$ mH, $L_2 = 60$ mH, and $k = 0.5$. Calculate the maximum possible equivalent inductance if:

(a) the two coils are connected in series
(b) the two coils are connected in parallel

18.5 Determine $\mathbf{V}_1$ and $\mathbf{V}_2$ in terms of $\mathbf{I}_1$ and $\mathbf{I}_2$ in the circuit in Fig. 18.68.

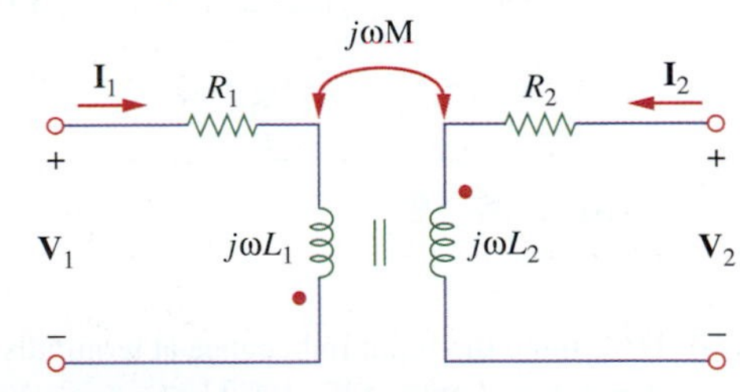

Figure 18.68
For Problem 18.5.

18.6 Find $\mathbf{V}_o$ in the circuit in Fig. 18.69.

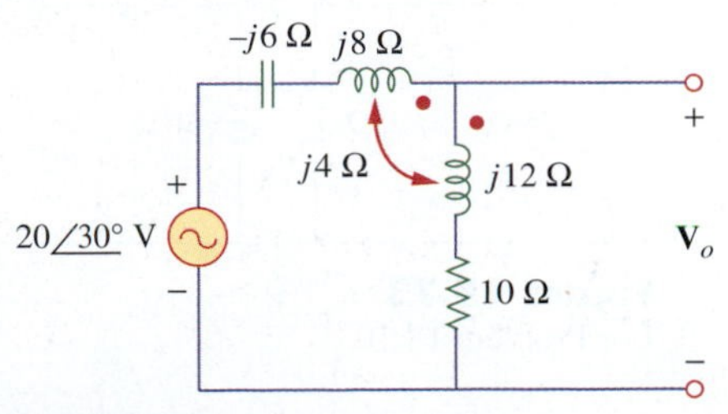

Figure 18.69
For Problem 18.6.

18.7 Obtain $\mathbf{V}_o$ in the circuit in Fig. 18.70.

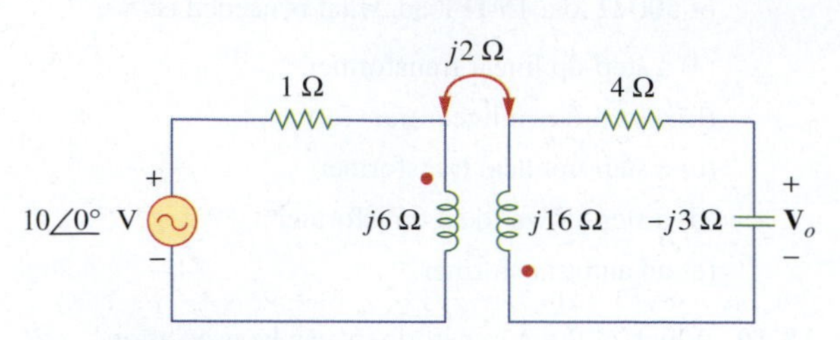

Figure 18.70
For Problem 18.7.

18.8 Find $\mathbf{V}_x$ in the network shown in Fig. 18.71.

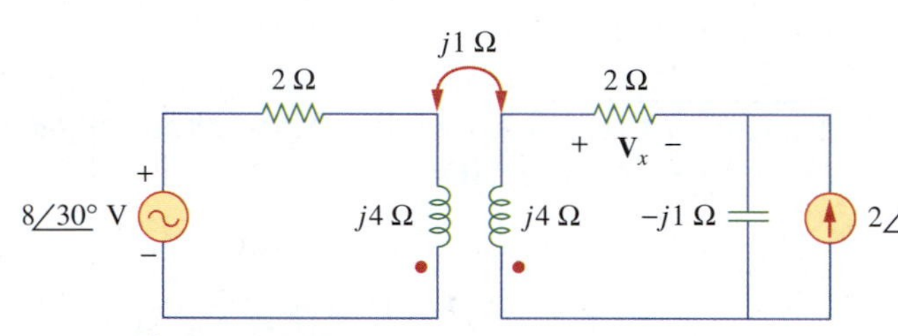

Figure 18.71
For Problem 18.8.

18.9 Calculate the equivalent impedance in the circuit in Fig. 18.72.

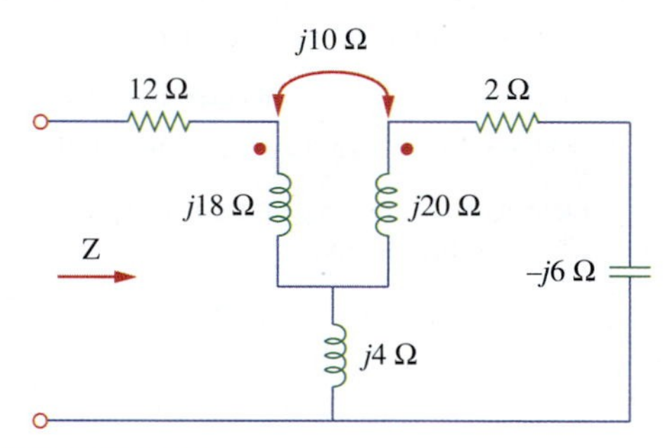

Figure 18.72
For Problem 18.9.

18.10 Determine the input impedance at terminals *a*-*b* of the circuit shown in Fig. 18.73.

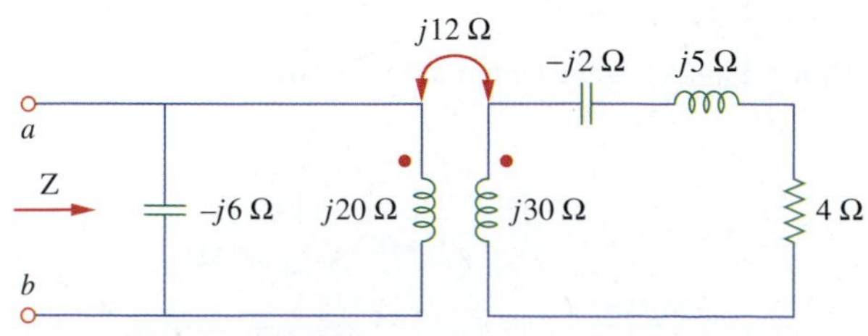

Figure 18.73
For Problem 18.10.

18.11 Determine an equivalent *T*-section that can be used to replace the transformer in Fig. 18.74.

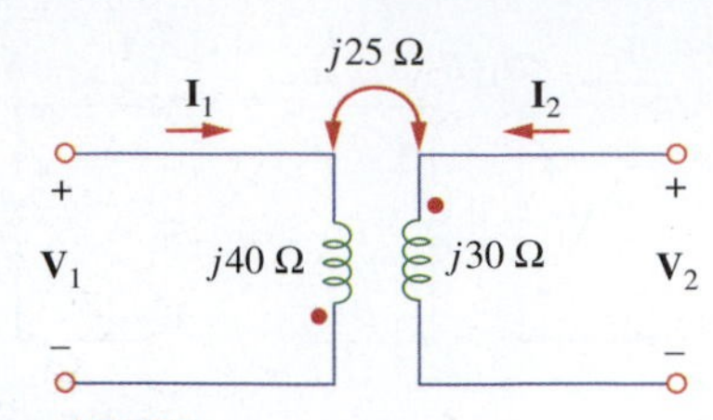

Figure 18.74
For Problem 18.11.

18.12 Use mesh analysis to obtain $\mathbf{I}_1$ and $\mathbf{I}_2$ in the circuit in Fig. 18.75.

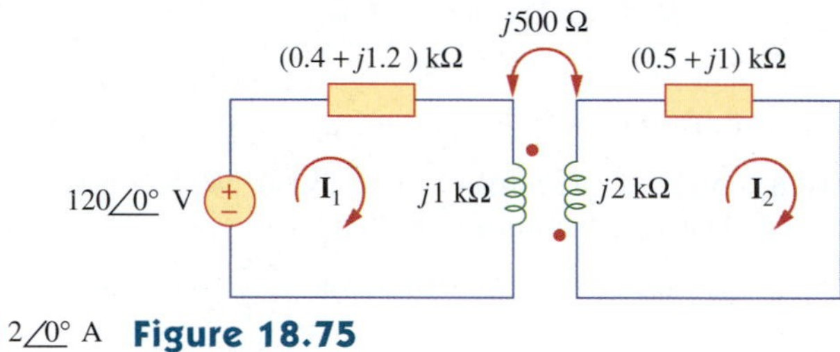

Figure 18.75
For Problem 18.12.

18.13 Find $\mathbf{I}_1$, $\mathbf{I}_2$, and $\mathbf{V}_o$ in the circuit of Fig. 18.76.

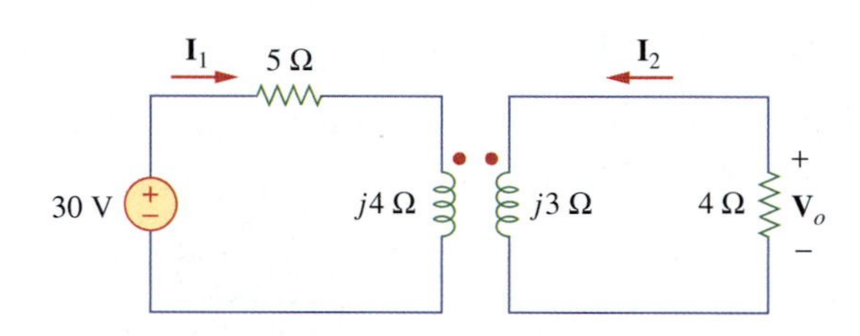

Figure 18.76
For Problem 18.13.

Section 18.3 Energy in a Coupled Circuit

18.14 Determine currents $\mathbf{I}_1$, $\mathbf{I}_2$, and $\mathbf{I}_3$ in the circuit in Fig. 18.77. Find the energy stored in the coupled coils at $t = 2$ ms. Take $\omega = 1{,}000$ rad/s.

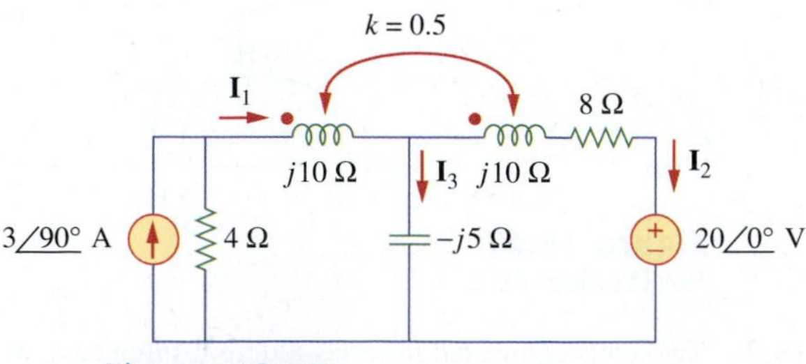

Figure 18.77
For Problem 18.14.

18.15 Find $\mathbf{I}_1$ and $\mathbf{I}_2$ in the circuit in Fig. 18.78. Calculate the power absorbed by the 4-Ω resistor.

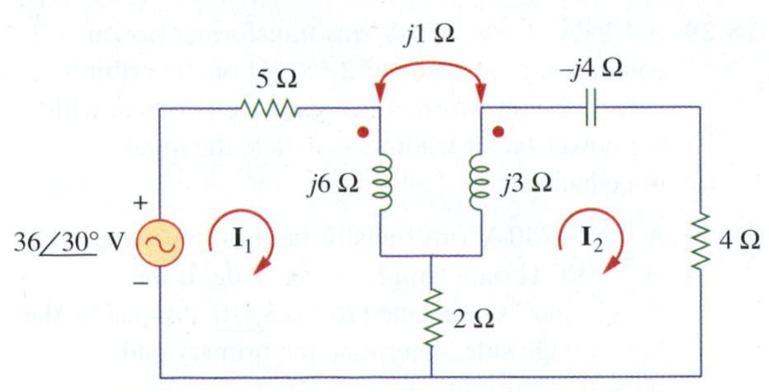

Figure 18.78
For Problems 18.15 and 18.49.

18.16 If $M = 0.2$ H and $v_s = 12\cos(10t)$ V in the circuit in Fig. 18.79, find i_1 and i_2. Calculate the energy stored in the coupled coils at $t = 15$ ms.

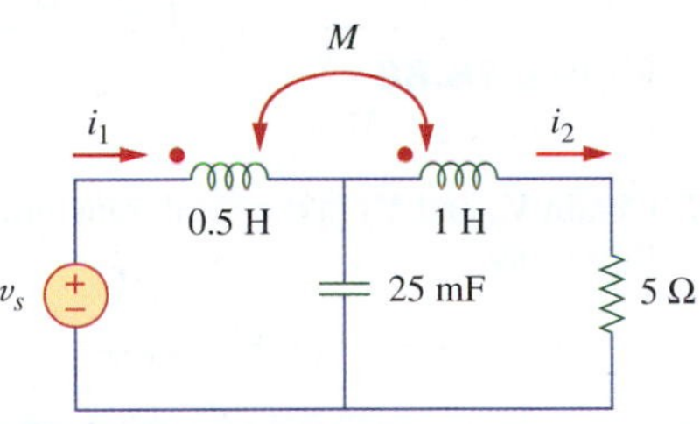

Figure 18.79
For Problem 18.16.

18.17 In the circuit in Fig. 18.80,

(a) find the coupling coefficient

(b) calculate v_o

(c) determine the energy stored in the coupled inductors at $t = 2$s

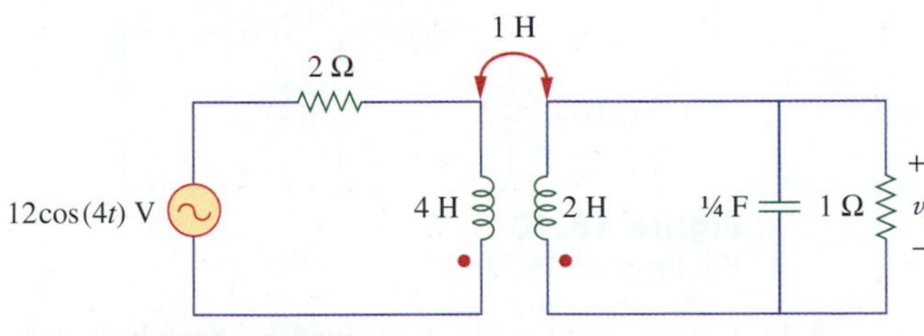

Figure 18.80
For Problem 18.17.

18.18 Find $\mathbf{I}_o$ in the circuit in Fig. 18.81. Switch the dot on the winding on the right and calculate $\mathbf{I}_o$ again.

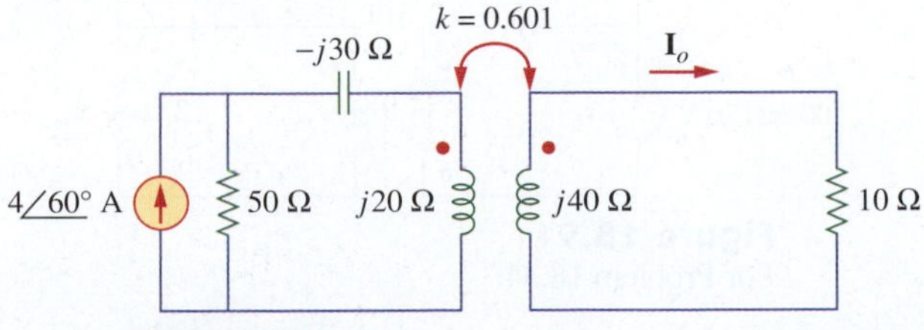

Figure 18.81
For Problem 18.18.

18.19 Determine the coupling coefficient for two coils with inductances of $L_1 = 5$ mH, $L_2 = 2.4$ mH, and $M = 3.2$ mH.

Section 18.4 Linear Transformers

18.20 Rework Example 18.1 using the concept of reflected impedance.

18.21 In the circuit in Fig. 18.82, find the value of the coupling coefficient k that will cause the 10-Ω resistor to dissipate 320 W. For this value of k, find the energy stored in the coupled coil at $t = 1.5$ ms.

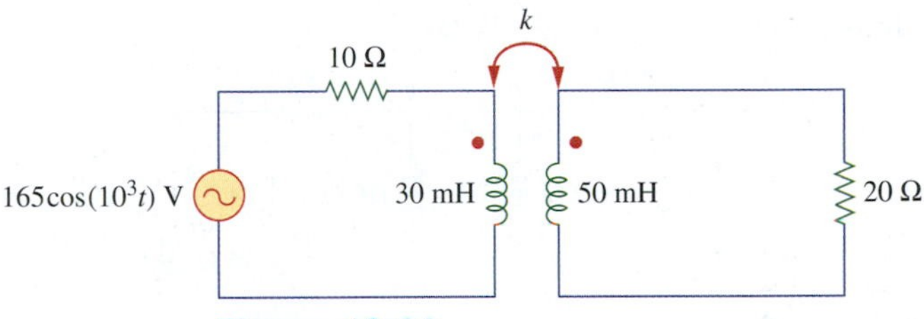

Figure 18.82
For Problem 18.21.

18.22 (a) Find the input impedance of the circuit in Fig. 18.83 using the concept of reflected impedance.

(b) Obtain the input impedance by replacing the linear transformer by its T equivalent.

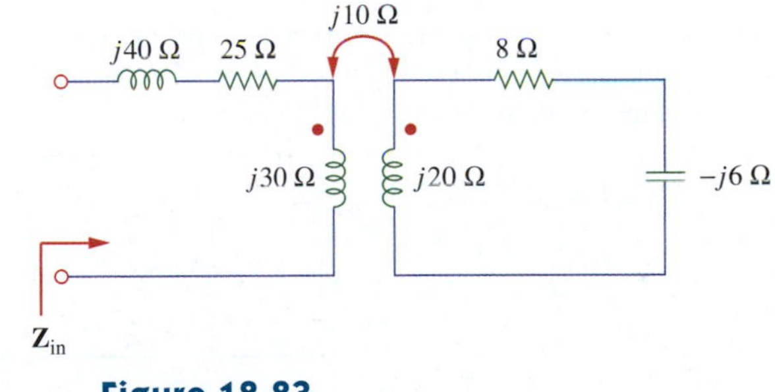

Figure 18.83
For Problem 18.22.

18.23 For the circuit in Fig. 18.84, find:

(a) the T-equivalent circuit

(b) the Π-equivalent circuit

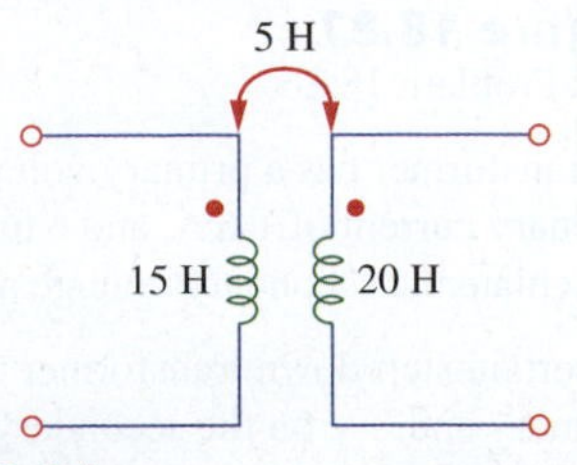

Figure 18.84
For Problem 18.23.

18.24 Determine the input impedance of the air-core transformer circuit in Fig. 18.85.

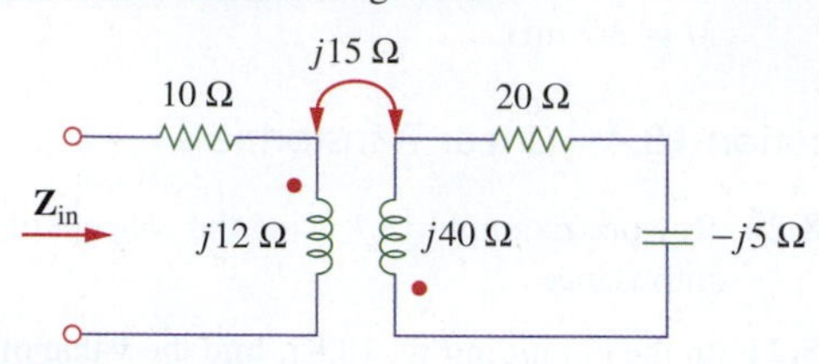

Figure 18.85
For Problem 18.24.

18.25 Find the input impedance of the circuit in Fig. 18.86.

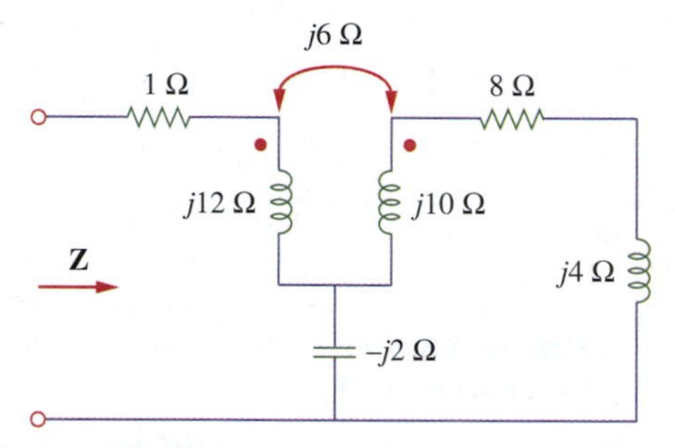

Figure 18.86
For Problem 18.25.

Section 18.5 Ideal Transformers

18.26 Give the relationships between terminal voltages and currents for each of the ideal transformers in Fig. 18.87, as shown in Fig. 18.32.

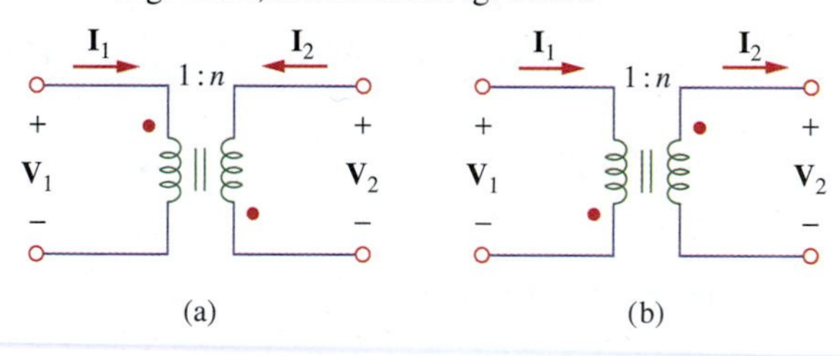

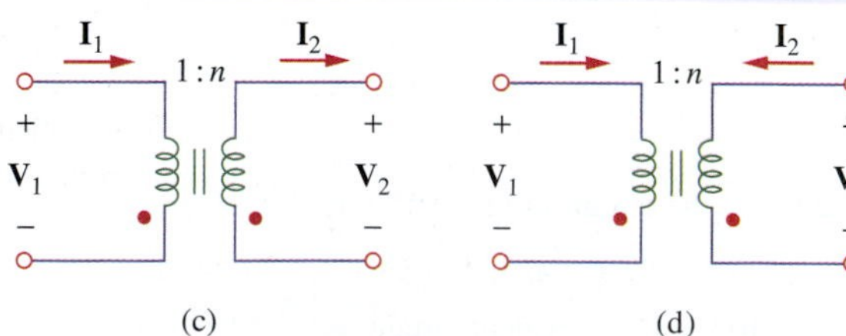

Figure 18.87
For Problem 18.26.

18.27 A transformer has a primary voltage of 210 V, a primary current of 1.6 A, and a turns ratio of 2. Calculate the secondary voltage and current.

18.28 A certain step-down transformer uses 120 V on the primary and 8 V on the secondary. If the secondary is rated for a maximum of 2 A, determine the rating of the primary side.

18.29 A 4-kVA, 2,300/230-V rms transformer has an equivalent impedance of $2\underline{/10^\circ}\ \Omega$ on the primary side. If the transformer is connected to a load with 0.6 power factor leading, calculate the input impedance.

18.30 A 1,200/240-V rms transformer has impedance $60\underline{/-30^\circ}\ \Omega$ on the high-voltage side. If the transformer is connected to a $0.8\underline{/10^\circ}$-Ω load on the low-voltage side, determine the primary and secondary currents.

18.31 Determine $\mathbf{I}_1$ and $\mathbf{I}_2$ in the circuit of Fig. 18.88.

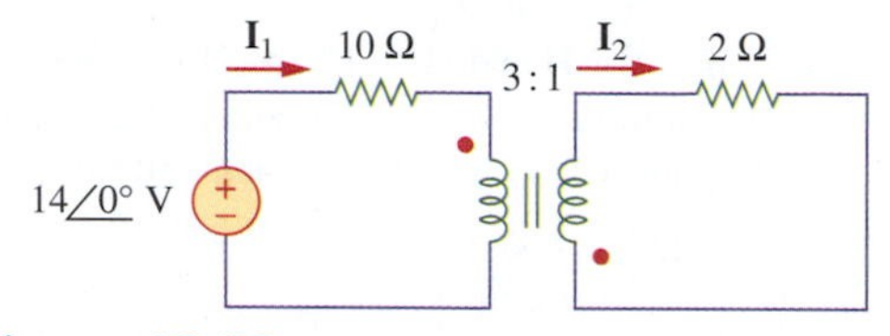

Figure 18.88
For Problem 18.31.

18.32 Obtain $\mathbf{V}_1$ and $\mathbf{V}_2$ in the ideal transformer circuit of Fig. 18.89.

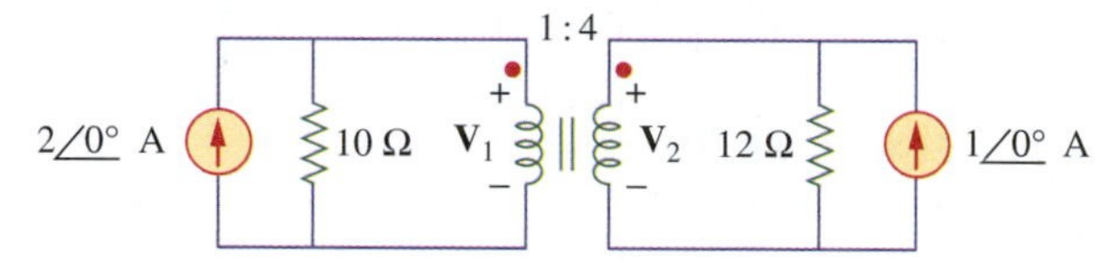

Figure 18.89
For Problem 18.32.

18.33 For the circuit in Fig. 18.90, find the value of the power absorbed by the 8-Ω resistor.

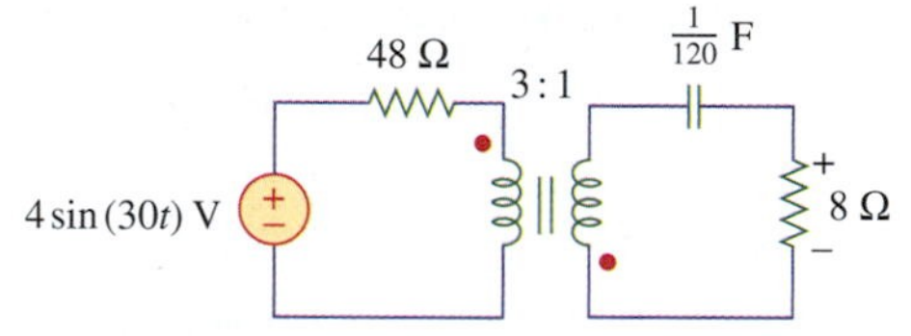

Figure 18.90
For Problem 18.33.

18.34 For the circuit in Fig. 18.91, find $\mathbf{V}_o$. Switch the dot on the secondary side and find $\mathbf{V}_o$ again.

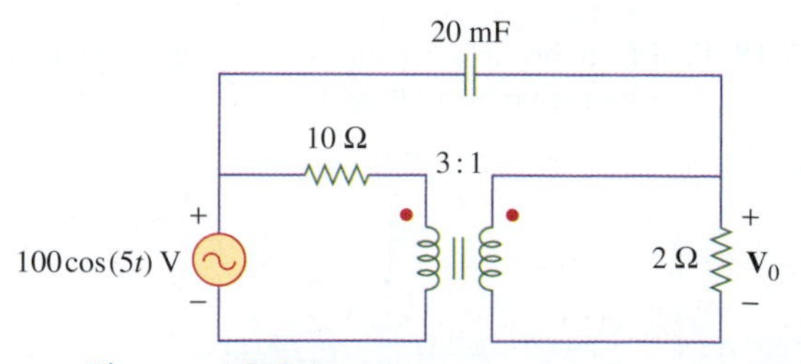

Figure 18.91
For Problem 18.34.

18.35 Find $\mathbf{I}_x$ in the ideal transformer circuit in Fig. 18.92.

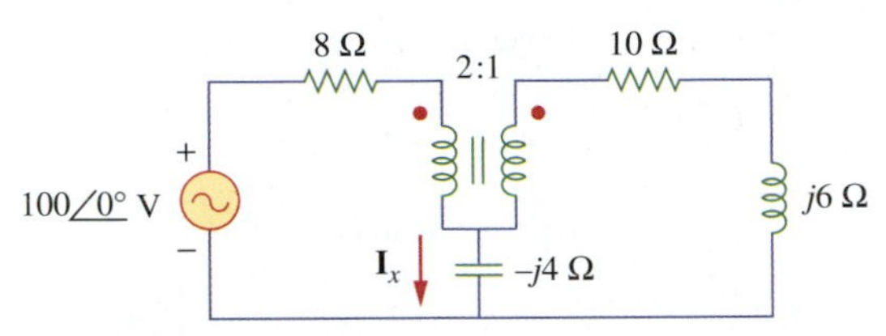

Figure 18.92
For Problem 18.35.

18.36 For the circuit in Fig. 18.93, determine the turns ratio that will cause maximum average power transfer to the load. Calculate that maximum average power.

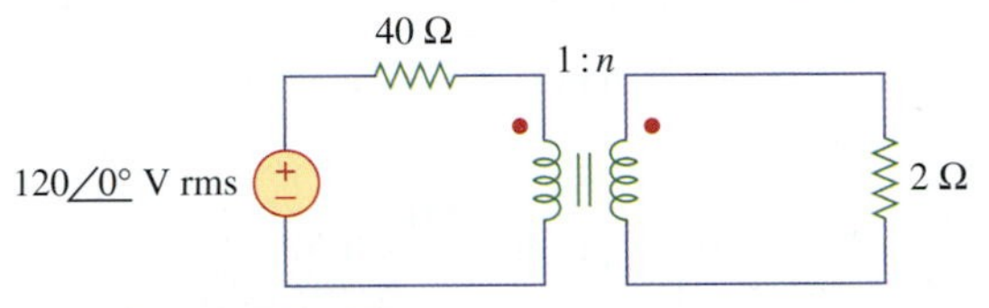

Figure 18.93
For Problem 18.36.

18.37 Refer to the network in Fig. 18.94. (a) Find n for maximum power supplied to the 200-Ω load. (b) Determine the power in the 200-Ω load if $n = 10$.

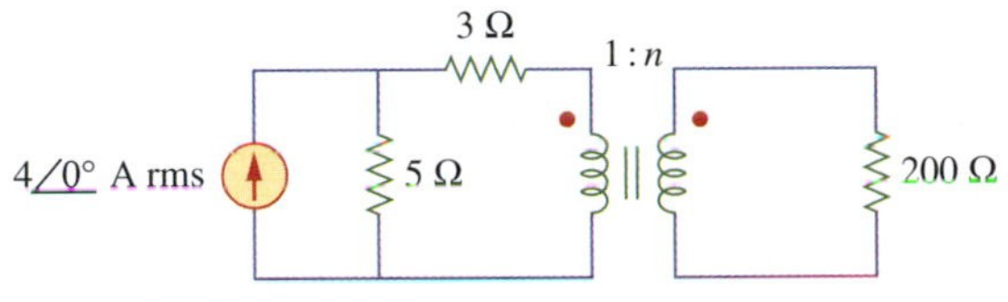

Figure 18.94
For Problem 18.37.

18.38 A transformer is used to match an amplifier with an 8-Ω load as shown in Fig. 18.95. The Thevenin equivalent of the amplifier is: $\mathbf{V}_{Th} = 10$ V, $\mathbf{Z}_{Th} = 128\ \Omega$.

(a) Find the required turns ratio for maximum energy power transfer.

(b) Determine the primary and secondary currents.

(c) Calculate the primary and secondary voltages.

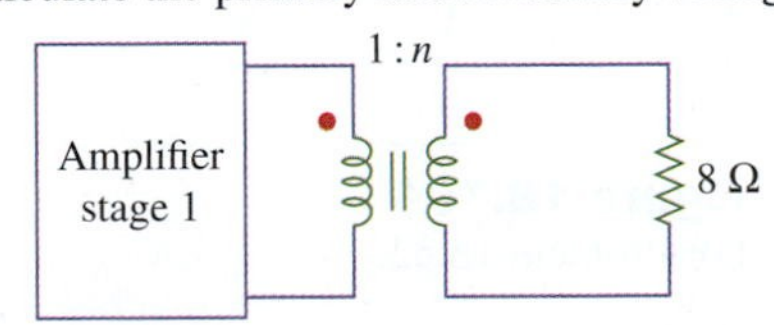

Figure 18.95
For Problem 18.38.

18.39 In Fig. 18.96, determine the average power delivered to $\mathbf{Z}_s$.

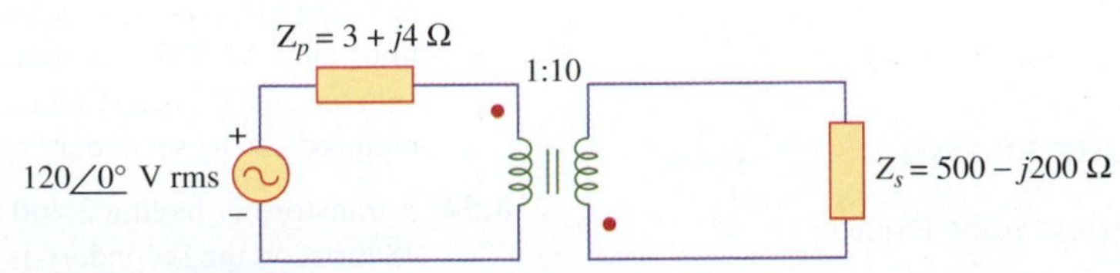

Figure 18.96
For Problem 18.39.

18.40 Find the power absorbed by the 10-Ω resistor in the ideal transformer circuit of Fig. 18.97.

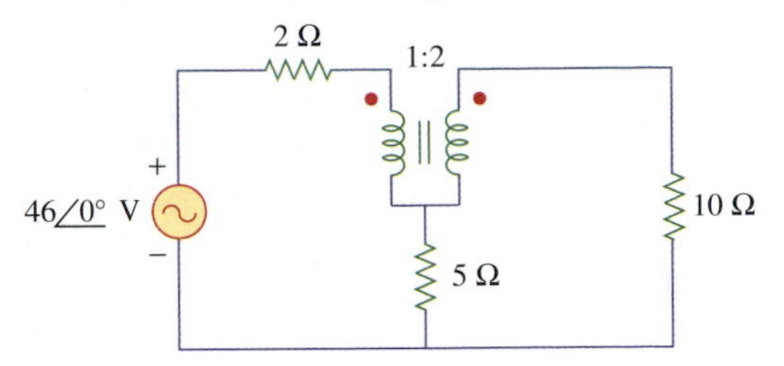

Figure 18.97
For Problem 18.40.

18.41 An ideal transformer is used in matching a 8-Ω load resistance to a source with an internal resistance of 96 Ω. (a) Determine the turns ratio. (b) What is the load voltage when the source voltage is 6 V?

18.42 A bell transformer has 200 turns on the primary side and 35 turns on the secondary. If the bell draws 0.2 A from a 120-V line, determine: (a) the secondary current, (b) the secondary voltage, and (c) the power delivered by the secondary.

18.43 Find the load current I in the circuit of Fig. 18.98.

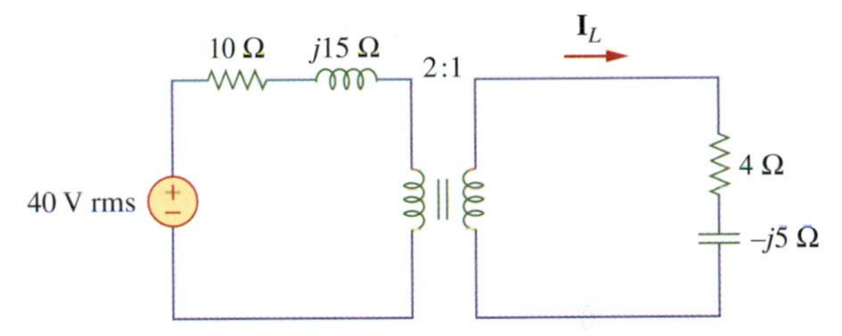

Figure 18.98
For Problem 18.43.

Section 18.6 Ideal Autotransformers

18.44 An ideal autotransformer with a 1:4 step-up turns ratio has its secondary connected to a 120-Ω load and the primary to a 420-V source. Determine the primary current.

18.45 An autotransformer with a 40 percent tap is supplied by a 400-V, 60-Hz source and is used for step-down operation. A 5-kVA load operating at unity power factor is connected to the secondary terminals. Find: (a) the secondary voltage, (b) the secondary current, and (c) the primary current.

18.46 In the ideal transformer in Fig. 18.99, calculate $\mathbf{I}_1$, $\mathbf{I}_2$, and $\mathbf{I}_o$. Find the average power delivered to the load.

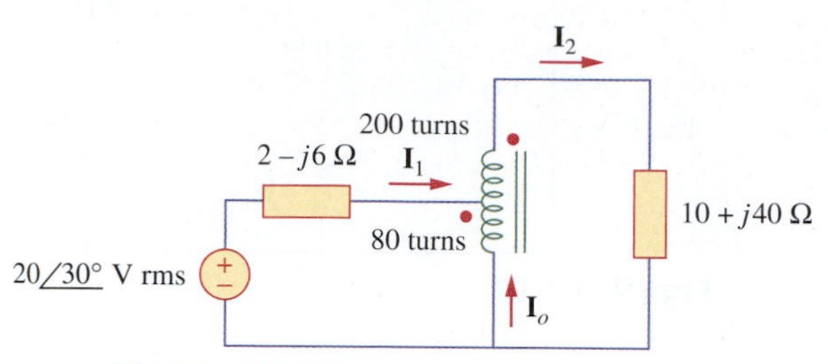

Figure 18.99
For Problem 18.46.

***18.47** In the circuit of Fig. 18.100, $\mathbf{Z}_L$ is adjusted until maximum average power is delivered to $\mathbf{Z}_L$. Find $\mathbf{Z}_L$ and the maximum average power transferred to it. Take $N_1 = 600$ turns and $N_2 = 200$ turns.

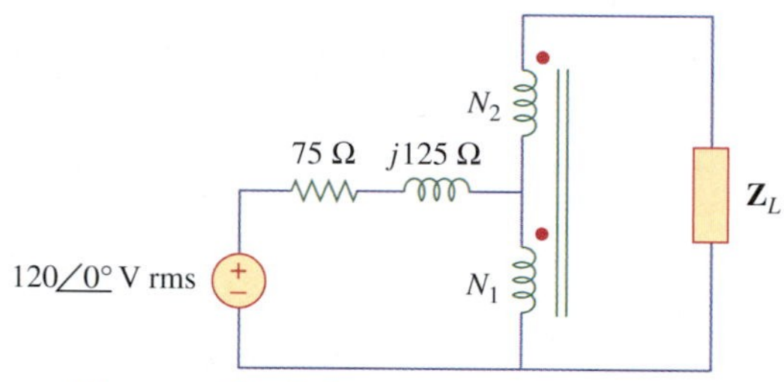

Figure 18.100
For Problem 18.47.

18.48 In the ideal transformer circuit shown in Fig. 18.101, determine the average power delivered to the load.

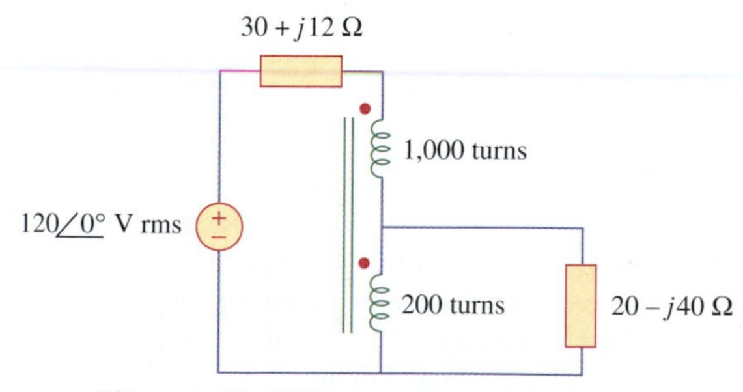

Figure 18.101
For Problem 18.48.

Section 18.7 Computer Analysis

18.49 Rework Problem 18.15 using PSpice.

18.50 Use PSpice to find $\mathbf{I}_1$, $\mathbf{I}_2$, and $\mathbf{I}_3$ in the circuit of Fig. 18.102.

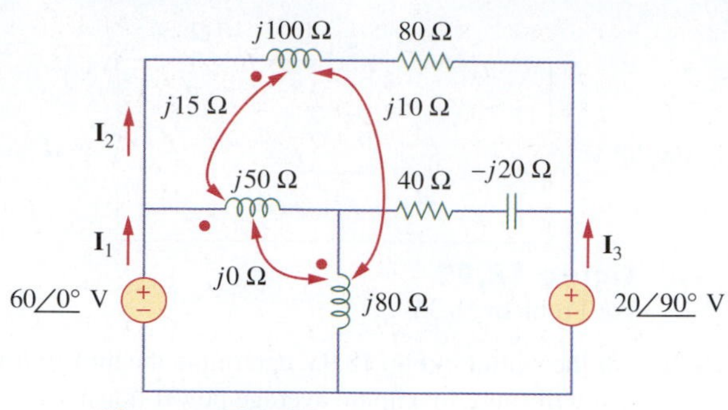

Figure 18.102
For Problem 18.50.

18.51 Use PSpice to find $\mathbf{I}_1$, $\mathbf{I}_2$, and $\mathbf{I}_3$ in the circuit of Fig. 18.103.

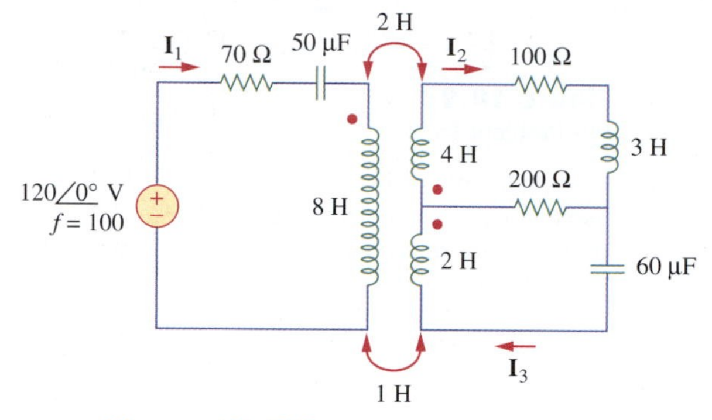

Figure 18.103
For Problem 18.51.

18.52 Use PSpice to find $\mathbf{V}_1$, $\mathbf{V}_2$, and $\mathbf{I}_o$ in the circuit of Fig. 18.104.

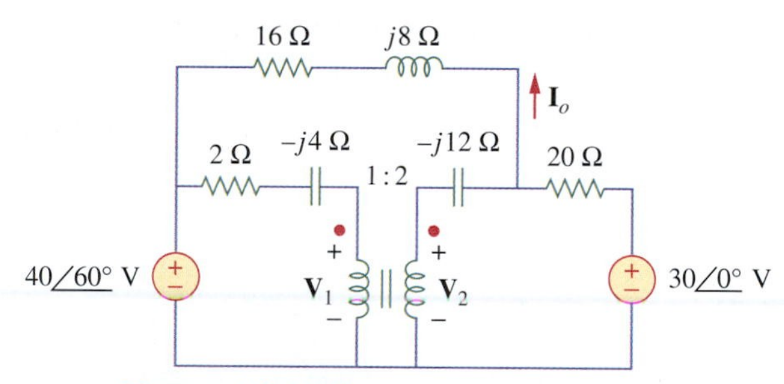

Figure 18.104
For Problem 18.52.

Section 18.8 Applications

18.53 A stereo amplifier circuit with an output impedance of 7.2 kΩ is to be matched to a speaker with input impedance of 8 Ω by a transformer whose primary side has 3,000 turns. Calculate the number of turns required on the secondary side.

18.54 A transformer having 2,400 turns on the primary and 48 turns on the secondary is used as an impedance-matching device. What is the reflected value of a 3-Ω load connected to the secondary?

18.55 A radio receiver has an input resistance of 300 Ω. When it is connected directly to an antenna system with characteristic impedance of 75 Ω, an impedance mismatch occurs. By inserting an impedance-matching transformer ahead of the receiver, maximum power can be realized. Calculate the required turns ratio.

18.56 A step-down power transformer with a turns ratio of $n = 0.1$ supplies 12.6 V rms to a resistive load. If the primary current is 2.5 A rms, how much power is delivered to the load?

18.57 A 240/120-V rms power transformer is rated at 10 kVA. Determine the turns ratio, the primary current, and the secondary current.

18.58 A 4-kVA, 2400/240-V rms transformer has 250 turns on the primary side. Calculate: (a) the turns ratio, (b) the number of turns on the secondary side, and (c) the primary and secondary currents.

18.59 A 25000/240-V rms distribution transformer has a primary current rating of 75 A. (a) Find the transformer kVA rating and (b) calculate the secondary current.

18.60 A 4,800-V rms transmission line feeds a distribution transformer with 1,200 turns on the primary and 28 turns on the secondary. When a 10-Ω load is connected across the secondary, find: (a) the secondary voltage, (b) the primary and secondary currents, and (c) the power supplied to the load.

18.61 Ten bulbs in parallel are supplied by a 7200/120-V transformer as shown in Fig. 18.105, where the bulbs are modeled by the 144-Ω resistors. Find: (a) turns ratio n and (b) the current through the primary winding.

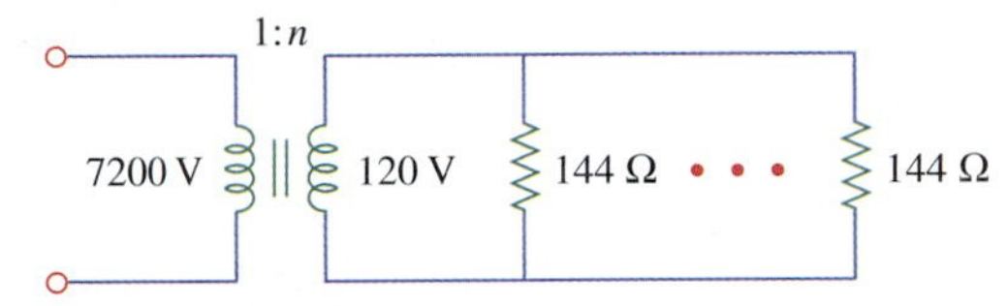

Figure 18.105
For Problem 18.61.

18.62 A 5-kΩ output impedance of an electronic amplifier is to be matched to a 8-Ω load. If the secondary winding has 80 turns, how many turns will the primary have?

18.63 A European coffee maker operates from a 240-V (rms) line to obtain 940 W. (a) How would you operate the coffee maker in America with 120 V (rms)? (b) Find the current it will draw from the 120-V line.

chapter

19

Two-Port Networks

Education is what survives when what has been learned has been forgotten.

—Burrhus F. Skinner

Enhancing Your Career

Career in Communications Systems

Communications systems apply the principles of circuit analysis. A communication system is designed to convey information from a source (the transmitter) to a destination (the receiver) via a channel (the propagation medium). Communication technologists design systems for transmitting and receiving information. The information can be in the form of voice, data, or video.

Photo by James Watson
© Chuck Alexander

We live in the information age—news, weather, sports, shopping, financial, business inventory, and other sources make information available to us almost instantly via communication systems. Some obvious examples of communications systems are the Internet, telephone network, mobile cellular telephones, radio, cable TV, satellite TV, fax, and radar. Mobile radio, used by police and fire departments, aircraft, and various businesses is another example.

The field of communications is perhaps the fastest growing area in electrical engineering technology. The merging of the communications field with computer engineering technology in recent years has led to digital data communications networks such as local area networks, metropolitan area networks, and broadband integrated services digital networks. For example, the Internet (the "information superhighway") allows educators, businesspeople, and others to send electronic mail from their computers worldwide, log onto remote databases, and transfer files. The Internet has hit the world like a tidal wave and is drastically changing the way people do business, communicate, and get information. This trend will continue.

A communications technologist designs and maintains systems that provide high-quality information services. The systems include hardware for generating, transmitting, and receiving information signals. Communications technologists are employed in numerous communications industries and places where communication systems are routinely used. More and more government agencies, academic departments, and businesses are demanding faster and more accurate transmission of information. To meet these needs, communications technologists are in high demand.

19.1 Introduction

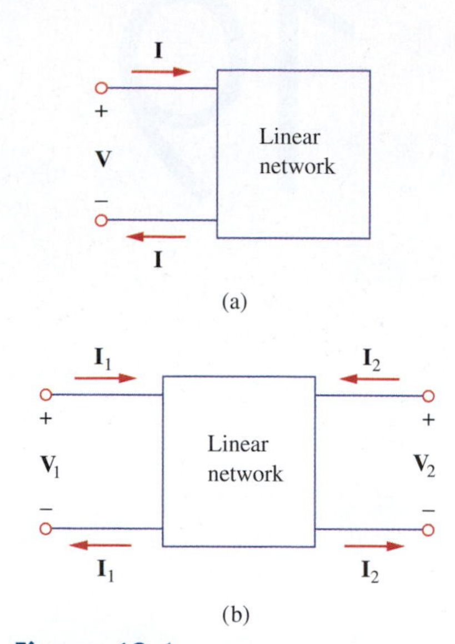

Figure 19.1
(a) A one-port network; (b) a two-port network.

A pair of terminals through which a current may enter or leave a network is known as a *port*. Two-terminal devices or elements (such as resistors, capacitors, and inductors) result in one-port networks. Most of the circuits we have dealt with so far are two-terminal or one-port circuits, represented in Fig. 19.1(a). They are passive linear networks absorbing power at an input or linear networks containing independent sources and supplying power at an output. We have considered the voltage across or current through a single pair of terminals—such as the two terminals of a resistor, a capacitor, or an inductor. We have also studied four-terminal or two-port circuits involving transistors and transformers, as shown in Fig. 19.1(b). In general, a network may have *n* ports. A port provides an access to the network and consists of a pair of terminals; the current entering one terminal leaves through the other terminal so that the net current entering the port equals zero.

In this chapter, we are mainly concerned with *two-port* networks or simply *two-ports.* Thus, a two-port network has two terminal pairs acting as access points. As shown in Fig. 19.1(b), the current entering one terminal of a pair leaves the other terminal in the pair.[1] Three-terminal devices such as transistors can be configured into two-port networks.

> A **two-port network** is an electrical network with two separate ports for input and output.

Our study of two-port networks is important for at least two reasons. First, such networks are useful in communications, control systems, power systems, and electronics. For example, they are used in electronics to model transistors and to facilitate cascaded design. They are also used in designing filters. Second, knowing the parameters of a two-port network enables us to treat it as a "black box" when embedded within a larger network.

To characterize a two-port network requires that we relate the terminal quantities $\mathbf{V}_1$, $\mathbf{V}_2$, $\mathbf{I}_1$, and $\mathbf{I}_2$ in Fig. 19.1(b), of which two are independent. The various terms that relate these voltages and currents are called *parameters.* Our goal in this chapter is to derive three sets of these parameters. We will show the relationship between these parameters and how two-port networks can be connected in series, parallel, or cascade. Finally, we will apply some of the concepts developed in this chapter to the analysis of transistor circuits.

19.2 Impedance Parameters

Impedance and admittance parameters are commonly used in the synthesis of filters. They are also useful in the design and analysis of impedance-matching networks and power distribution networks. We discuss impedance parameters in this section; admittance parameters will be discussed in the next section.

A two-port network may be voltage-driven as in Fig. 19.2(a) or current-driven as in Fig. 19.2(b). From either Fig. 19.2(a) or (b), the

[1] Note: A two-port network has one important characteristic: the current into the port (input or output) equals the current out of the port.

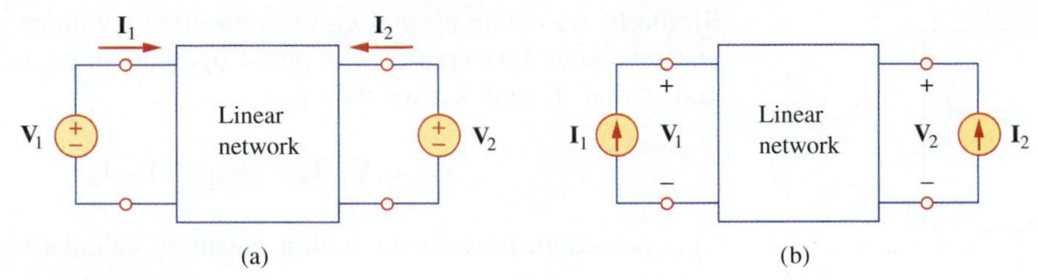

Figure 19.2
The linear two-port network: (a) driven by voltage sources; (b) driven by current sources.

terminal voltages can be related to the terminal currents as

$$\mathbf{V}_1 = \mathbf{z}_{11}\mathbf{I}_1 + \mathbf{z}_{12}\mathbf{I}_2$$
$$\mathbf{V}_2 = \mathbf{z}_{21}\mathbf{I}_1 + \mathbf{z}_{22}\mathbf{I}_2 \tag{19.1}$$

or in matrix form as[2]

$$\begin{bmatrix} \mathbf{V}_1 \\ \mathbf{V}_2 \end{bmatrix} = \begin{bmatrix} \mathbf{z}_{11} & \mathbf{z}_{12} \\ \mathbf{z}_{21} & \mathbf{z}_{22} \end{bmatrix} \begin{bmatrix} \mathbf{I}_1 \\ \mathbf{I}_2 \end{bmatrix} = [\mathbf{z}] \begin{bmatrix} \mathbf{I}_1 \\ \mathbf{I}_2 \end{bmatrix} \tag{19.2}$$

where the **z** terms are called the *impedance parameters* or simply *z parameters* and have units of ohms.

The values of the parameters can be evaluated by setting $\mathbf{I}_1 = 0$ (input port open-circuited) or $\mathbf{I}_2 = 0$ (output port open-circuited). Thus,

$$\mathbf{z}_{11} = \left.\frac{\mathbf{V}_1}{\mathbf{I}_1}\right|_{\mathbf{I}_2=0} \tag{19.3a}$$

$$\mathbf{z}_{12} = \left.\frac{\mathbf{V}_1}{\mathbf{I}_2}\right|_{\mathbf{I}_1=0} \tag{19.3b}$$

$$\mathbf{z}_{21} = \left.\frac{\mathbf{V}_2}{\mathbf{I}_1}\right|_{\mathbf{I}_2=0} \tag{19.3c}$$

$$\mathbf{z}_{22} = \left.\frac{\mathbf{V}_2}{\mathbf{I}_2}\right|_{\mathbf{I}_1=0} \tag{19.3d}$$

Because the z parameters are obtained by open-circuiting the input or output port, they are also called the *open-circuit impedance parameters*. Specifically,

$\mathbf{z}_{11}$ = open-circuit input impedance (19.4a)

$\mathbf{z}_{12}$ = open-circuit transfer impedance (19.4b)

$\mathbf{z}_{21}$ = open-circuit transfer impedance (19.4c)

$\mathbf{z}_{22}$ = open-circuit output impedance (19.4d)

According to Eq. (19.3), we obtain $\mathbf{z}_{11}$ and $\mathbf{z}_{21}$ by connecting a voltage source $\mathbf{V}_1$ (or a current source $\mathbf{I}_1$) to port 1 with port 2 open-circuited as in Fig. 19.3(a) and finding $\mathbf{I}_1$ and $\mathbf{V}_2$; we then get

$$\mathbf{z}_{11} = \mathbf{V}_1/\mathbf{I}_1, \qquad \mathbf{z}_{21} = \mathbf{V}_2/\mathbf{I}_1 \tag{19.5}$$

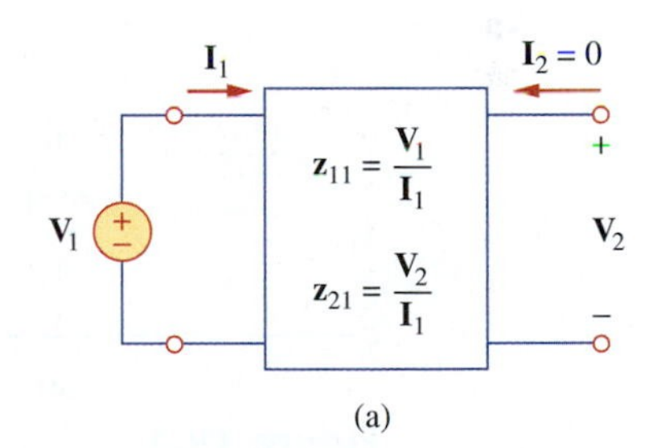

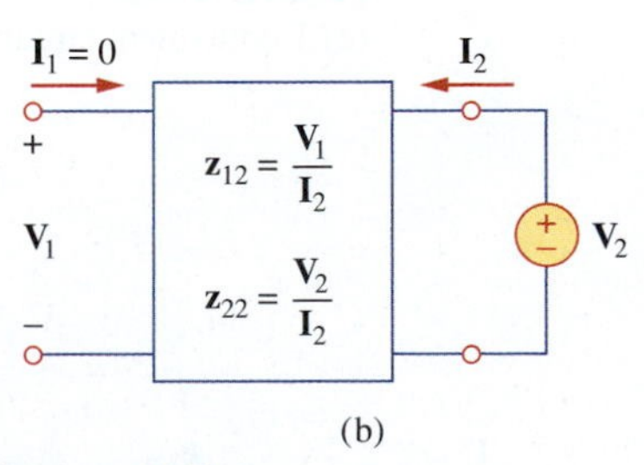

Figure 19.3
Determination of the z parameters: (a) finding $\mathbf{z}_{11}$ and $\mathbf{z}_{21}$; (b) finding $\mathbf{z}_{12}$ and $\mathbf{z}_{22}$.

[2] Reminder: Only two of the four variables ($\mathbf{V}_1$, $\mathbf{V}_2$, $\mathbf{I}_1$, and $\mathbf{I}_2$) are independent. The other two can be found using Eq. (19.1).

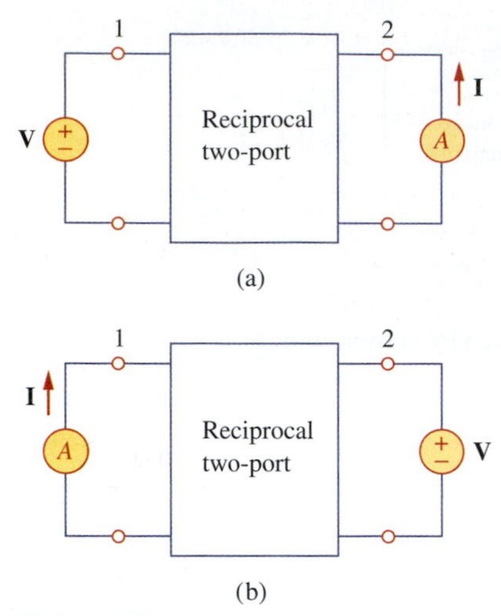

Figure 19.4
Interchanging a voltage source of one port with an ideal ammeter at the other port produces the same reading in a reciprocal two-port network.

Similarly, we obtain $\mathbf{z}_{12}$ and $\mathbf{z}_{22}$ by connecting a voltage source $\mathbf{V}_2$ (or a current source $\mathbf{I}_2$) to port 2 with port 1 open-circuited as in Fig. 19.3(b) and finding $\mathbf{I}_2$ and $\mathbf{V}_1$; we then get

$$\mathbf{z}_{12} = \mathbf{V}_1/\mathbf{I}_2, \qquad \mathbf{z}_{22} = \mathbf{V}_2/\mathbf{I}_2 \tag{19.6}$$

This procedure provides us with a means of calculating or measuring the z parameters.

When $\mathbf{z}_{11} = \mathbf{z}_{22}$, the two-port network is said to be *symmetrical*. This implies that the network has mirrorlike symmetry about some centerline; that is, a line can be found that divides the network into two similar halves.

When the two-port network is linear and has no dependent sources, the transfer impedances are equal ($\mathbf{z}_{12} = \mathbf{z}_{21}$) and the two-port network is said to be *reciprocal*. This means that if the points of excitation and response are interchanged, the transfer impedances remain the same. As illustrated in Fig. 19.4, a two-port network is reciprocal if interchanging an ideal voltage source at one port with an ideal ammeter at the other port gives the same ammeter reading. The reciprocal network yields $\mathbf{V} = \mathbf{z}_{12}\mathbf{I}$ according to Eq. (19.1) when connected as in Fig. 19.4(a) but yields $\mathbf{V} = \mathbf{z}_{21}\mathbf{I}$ when connected as in Fig. 19.4(b). This is possible only if $\mathbf{z}_{12} = \mathbf{z}_{21}$. Any two-port network that is made entirely of resistors, capacitors, inductors, and transformers must be reciprocal. For a reciprocal network, the T-equivalent circuit in Fig. 19.5(a) can be used. If the network is not reciprocal, a more general equivalent network is shown in Fig. 19.5(b). Notice that Fig. 19.5(b) follows directly from Eq. (19.1).

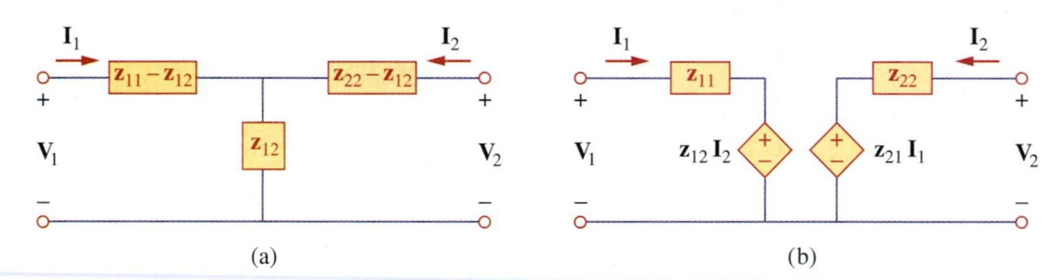

Figure 19.5
(a) T-equivalent circuit (for reciprocal case only); (b) general equivalent circuit.

It should be mentioned that for some two-port networks, the z parameters do not exist because they cannot be described by Eq. (19.1). As an example, consider the ideal transformer of Fig. 19.6. The defining equations for the two-port network are:

$$\mathbf{V}_1 = \frac{1}{n}\mathbf{V}_2, \qquad \mathbf{I}_1 = -n\mathbf{I}_2 \tag{19.7}$$

We observe that it is impossible to express the voltages in terms of the currents and vice versa as Eq. (19.1) requires. Thus, the ideal transformer has no z parameters. However, it does have hybrid parameters, as we shall see in Section 19.4.

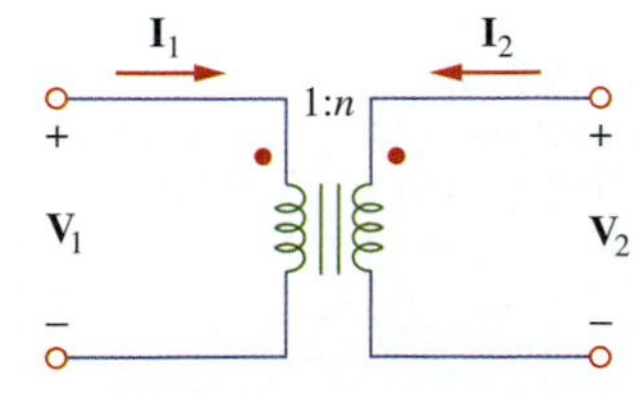

Figure 19.6
An ideal transformer has no z parameters.

Example 19.1

Determine the z parameters for the circuit in Fig. 19.7.

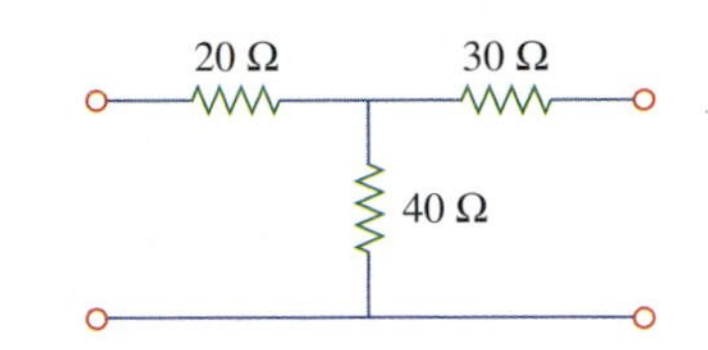

Figure 19.7
For Example 19.1.

Solution:

■ METHOD 1 To determine $\mathbf{z}_{11}$ and $\mathbf{z}_{21}$, we apply a voltage source $\mathbf{V}_1$ to the input port and leave the output port open as in Fig. 19.8(a).

$$\mathbf{z}_{11} = \frac{\mathbf{V}_1}{\mathbf{I}_1} = \frac{(20 + 40)\mathbf{I}_1}{\mathbf{I}_1} = 60\ \Omega$$

that is, $\mathbf{z}_{11}$ is the input impedance at port 1.

$$\mathbf{z}_{21} = \frac{\mathbf{V}_2}{\mathbf{I}_1} = \frac{40\mathbf{I}_1}{\mathbf{I}_1} = 40\ \Omega$$

To find $\mathbf{z}_{12}$ and $\mathbf{z}_{22}$, we apply a voltage source $\mathbf{V}_2$ to the output port and leave the input port open as in Fig. 19.8(b).

$$\mathbf{z}_{12} = \frac{\mathbf{V}_1}{\mathbf{I}_2} = \frac{40\mathbf{I}_2}{\mathbf{I}_2} = 40\ \Omega, \qquad \mathbf{z}_{22} = \frac{\mathbf{V}_2}{\mathbf{I}_2} = \frac{(30 + 40)\mathbf{I}_2}{\mathbf{I}_2} = 70\ \Omega$$

that is, $\mathbf{z}_{22}$ is the output impedance at port 2. Thus,

$$[\mathbf{z}] = \begin{bmatrix} 60\ \Omega & 40\ \Omega \\ 40\ \Omega & 70\ \Omega \end{bmatrix}$$

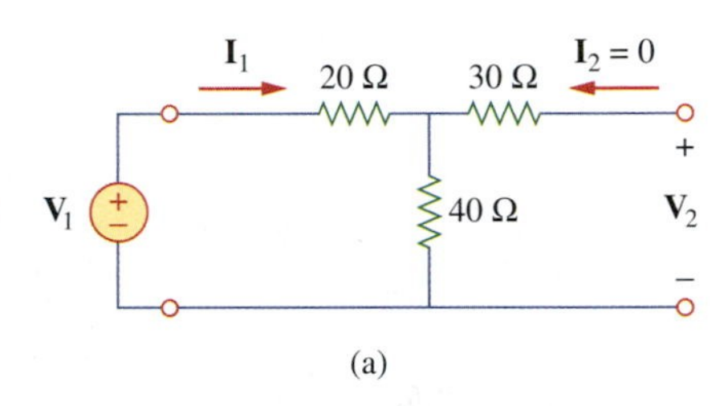

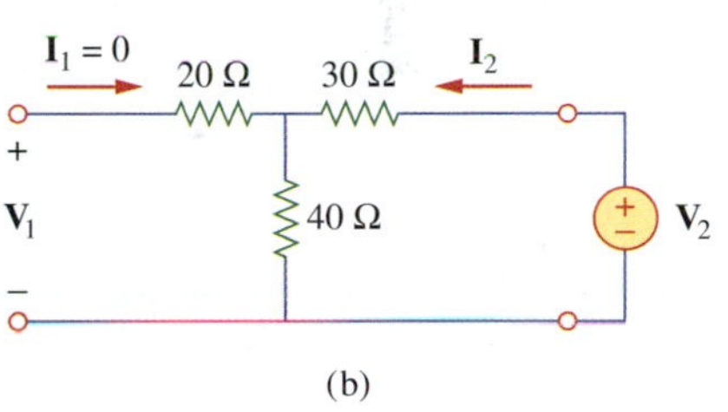

Figure 19.8
For Example 19.1: (a) finding $\mathbf{z}_{11}$ and $\mathbf{z}_{21}$; (b) finding $\mathbf{z}_{12}$ and $\mathbf{z}_{22}$.

■ METHOD 2 Alternatively, because there is no dependent source in the given circuit, $\mathbf{z}_{12} = \mathbf{z}_{21}$ and we can use Fig. 19.5(a). Comparing Fig. 19.7 with Fig. 19.5(a), we get

$$\begin{aligned} \mathbf{z}_{12} &= 40\ \Omega = \mathbf{z}_{21} \\ \mathbf{z}_{11} - \mathbf{z}_{12} &= 20 \quad \Rightarrow \quad \mathbf{z}_{11} = 20 + \mathbf{z}_{12} = 60\ \Omega \\ \mathbf{z}_{22} - \mathbf{z}_{12} &= 30 \quad \Rightarrow \quad \mathbf{z}_{22} = 30 + \mathbf{z}_{12} = 70\ \Omega \end{aligned}$$

Practice Problem 19.1

Find the z parameters of the two-port network in Fig. 19.9.

Answer: $\mathbf{z}_{11} = 14\ \Omega$; $\mathbf{z}_{12} = \mathbf{z}_{21} = \mathbf{z}_{22} = 6\ \Omega$.

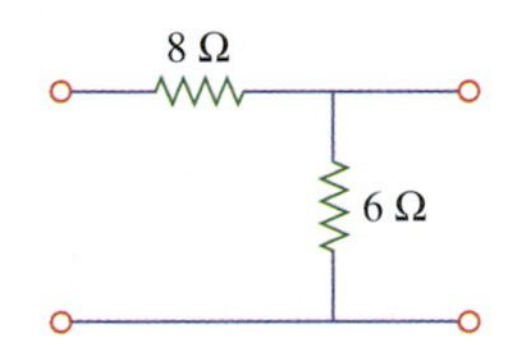

Figure 19.9
For Practice Problem 19.1.

Example 19.2

Find $\mathbf{I}_1$ and $\mathbf{I}_2$ in the circuit of Fig. 19.10.

$\mathbf{I}_1$, $\mathbf{I}_2$; $100\underline{/0^\circ}$ V; $\mathbf{V}_1$; $\mathbf{V}_2$; 10 Ω

$\mathbf{z}_{11} = 40\ \Omega$
$\mathbf{z}_{12} = j20\ \Omega$
$\mathbf{z}_{21} = j30\ \Omega$
$\mathbf{z}_{22} = 50\ \Omega$

Figure 19.10
For Example 19.2.

Solution:

This is not a reciprocal network since $\mathbf{z}_{12}$ does not equal $\mathbf{z}_{21}$. We may use the equivalent circuit in Fig. 19.5(b) but we can also use Eq. (19.1) directly. Substituting the given z parameters into Eq. (19.1), we obtain

$$\mathbf{V}_1 = 40\mathbf{I}_1 + j20\mathbf{I}_2 \quad \textbf{(19.2.1)}$$

$$\mathbf{V}_2 = j30\mathbf{I}_1 + 50\mathbf{I}_2 \quad \textbf{(19.2.2)}$$

Because we are looking for $\mathbf{I}_1$ and $\mathbf{I}_2$, we substitute

$$\mathbf{V}_1 = 100\underline{/0^\circ}, \qquad \mathbf{V}_2 = -10\mathbf{I}_2$$

into Eqs. (19.2.1) and (19.2.2), which become

$$100 = 40\mathbf{I}_1 + j20\mathbf{I}_2 \quad \textbf{(19.2.3)}$$

$$-10\,\mathbf{I}_2 = j30\mathbf{I}_1 + 50\mathbf{I}_2 \qquad \Rightarrow \qquad \mathbf{I}_1 = j2\mathbf{I}_2 \quad \textbf{(19.2.4)}$$

Substituting Eq. (19.2.4) into Eq. (19.2.3) gives

$$100 = j80\mathbf{I}_2 + j20\mathbf{I}_2 \qquad \Rightarrow \qquad \mathbf{I}_2 = \frac{100}{j100} = -j$$

From Eq. (19.2.4),

$$\mathbf{I}_1 = j2(-j) = 2$$

Thus,

$$\mathbf{I}_1 = 2\underline{/0^\circ}\text{ A}, \qquad \mathbf{I}_2 = 1\underline{/-90^\circ}\text{ A}$$

Practice Problem 19.2

Calculate $\mathbf{I}_1$ and $\mathbf{I}_2$ in the two-port network of Fig. 19.11.

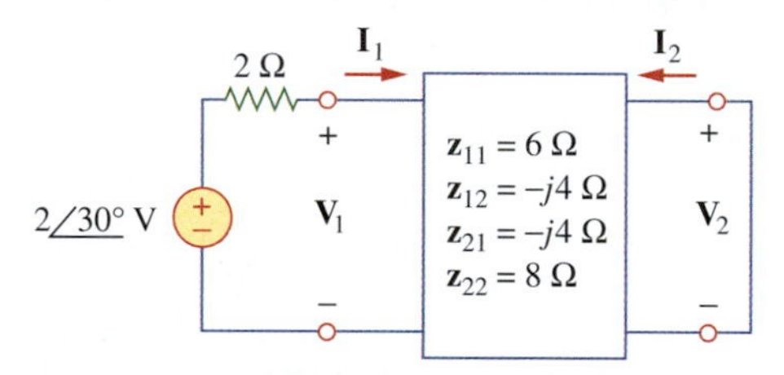

Figure 19.11
For Practice Problem 19.2.

Answer: $2\underline{/20^\circ}$ A; $1\underline{/-60^\circ}$ A.

19.3 Admittance Parameters

The second set of parameters is obtained by expressing the terminal currents in terms of the terminal voltages. In either Fig. 19.2(a) or (b), the terminal currents can be expressed in terms of the terminal voltages as

$$\boxed{\begin{aligned}\mathbf{I}_1 &= \mathbf{y}_{11}\mathbf{V}_1 + \mathbf{y}_{12}\mathbf{V}_2 \\ \mathbf{I}_2 &= \mathbf{y}_{21}\mathbf{V}_1 + \mathbf{y}_{22}\mathbf{V}_2\end{aligned}} \quad \textbf{(19.8)}$$

or in matrix form as

$$\begin{bmatrix}\mathbf{I}_1 \\ \mathbf{I}_2\end{bmatrix} = \begin{bmatrix}\mathbf{y}_{11} & \mathbf{y}_{12} \\ \mathbf{y}_{21} & \mathbf{y}_{22}\end{bmatrix}\begin{bmatrix}\mathbf{V}_1 \\ \mathbf{V}_2\end{bmatrix} = [\mathbf{y}]\begin{bmatrix}\mathbf{V}_1 \\ \mathbf{V}_2\end{bmatrix} \quad \textbf{(19.9)}$$

The $\mathbf{y}$ terms are known as the *admittance parameters* or simply *y parameters* and have units of siemens.

The values of the parameters can be determined by setting $\mathbf{V}_1 = 0$ (input port short-circuited) or $\mathbf{V}_2 = 0$ (output port short-circuited). Thus,

$$\mathbf{y}_{11} = \left.\frac{\mathbf{I}_1}{\mathbf{V}_1}\right|_{\mathbf{V}_2=0}, \qquad \mathbf{y}_{12} = \left.\frac{\mathbf{I}_1}{\mathbf{V}_2}\right|_{\mathbf{V}_1=0}$$
$$\mathbf{y}_{21} = \left.\frac{\mathbf{I}_2}{\mathbf{V}_1}\right|_{\mathbf{V}_2=0}, \qquad \mathbf{y}_{22} = \left.\frac{\mathbf{I}_2}{\mathbf{V}_2}\right|_{\mathbf{V}_1=0} \tag{19.10}$$

Because the y parameters are obtained by short-circuiting the input or output port, they are also called the *short-circuit admittance parameters*. Thus,

$\mathbf{y}_{11}$ = short-circuit input admittance **(19.11a)**

$\mathbf{y}_{12}$ = short-circuit transfer admittance from port 2 to port 1 **(19.11b)**

$\mathbf{y}_{21}$ = short-circuit transfer admittance from port 1 to port 2 **(19.11c)**

$\mathbf{y}_{22}$ = short-circuit output admittance **(19.11d)**

Following Eq. (19.10), we obtain $\mathbf{y}_{11}$ and $\mathbf{y}_{21}$ by connecting a current source $\mathbf{I}_1$ to port 1 and short-circuiting port 2 as in Fig. 19.12(a), finding $\mathbf{V}_1$ and $\mathbf{I}_2$, and then calculating

$$\mathbf{y}_{11} = \mathbf{I}_1/\mathbf{V}_1, \qquad \mathbf{y}_{21} = \mathbf{I}_2/\mathbf{V}_1 \tag{19.12}$$

Similarly, we obtain $\mathbf{y}_{12}$ and $\mathbf{y}_{22}$ by connecting a current source $\mathbf{I}_2$ to port 2 and short-circuiting port 1, as in Fig. 19.12(b), finding $\mathbf{I}_1$ and $\mathbf{V}_2$, and then getting

$$\mathbf{y}_{12} = \mathbf{I}_1/\mathbf{V}_2, \qquad \mathbf{y}_{22} = \mathbf{I}_2/\mathbf{V}_2 \tag{19.13}$$

This procedure provides us with a means of calculating or measuring the y parameters. The impedance and admittance parameters are collectively referred to as *immittance* parameters.

For a two-port network that is linear and has no dependent sources, the transfer admittances are equal ($\mathbf{y}_{12} = \mathbf{y}_{21}$). This can be proved in the same way as for the z parameters. A reciprocal network ($\mathbf{y}_{12} = \mathbf{y}_{21}$) can be modeled by the Π equivalent circuit in Fig. 19.13(a). If the network is not reciprocal, a more general equivalent network is shown in Fig. 19.13(b).

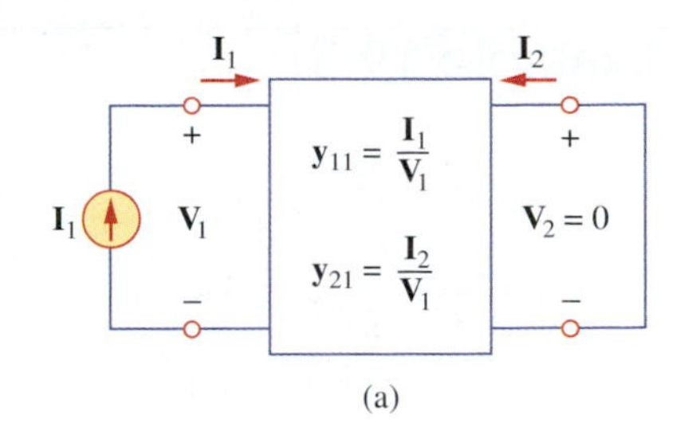

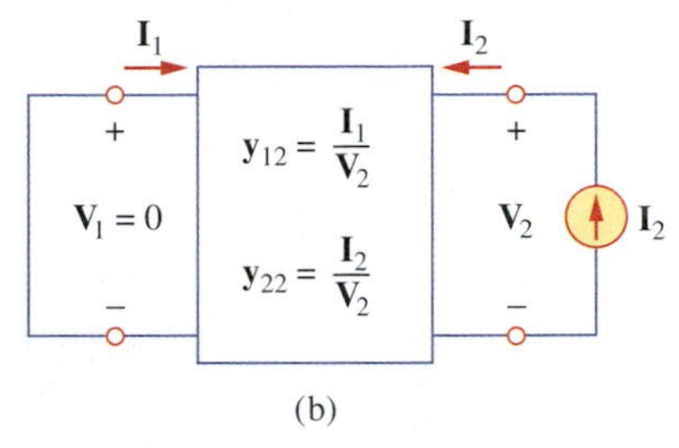

Figure 19.12
Determination of the y parameters: (a) finding $\mathbf{y}_{11}$ and $\mathbf{y}_{21}$; (b) finding $\mathbf{y}_{12}$ and $\mathbf{y}_{22}$.

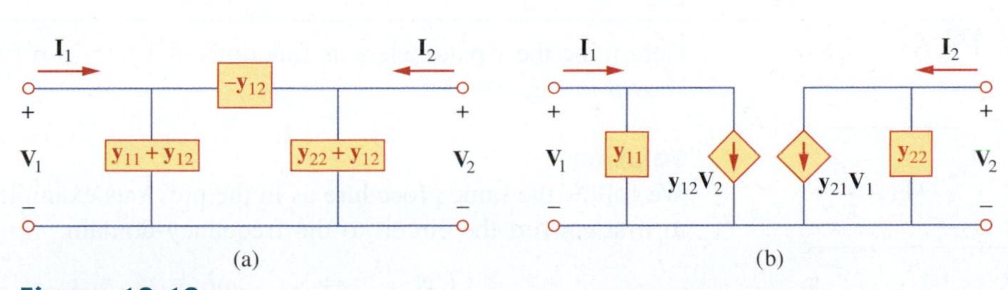

Figure 19.13
(a) Π-equivalent circuit (for reciprocal case only); (b) general equivalent circuit.

Example 19.3

Obtain the y parameters for the Π network shown in Fig. 19.14.

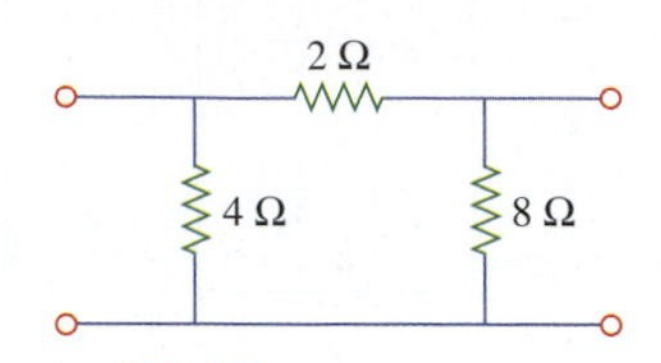

Figure 19.14
For Example 19.3.

Solution:

■ **METHOD 1** To find $\mathbf{y}_{11}$ and $\mathbf{y}_{21}$, short-circuit the output port and connect a current source $\mathbf{I}_1$ to the input port as in Fig. 19.15(a). Because the 8-Ω resistor is short-circuited, the 2-Ω resistor is in parallel with the 4-Ω resistor. Hence,

$$\mathbf{V}_1 = \mathbf{I}_1(4 \parallel 2) = \frac{4}{3}\mathbf{I}_1, \qquad \mathbf{y}_{11} = \frac{\mathbf{I}_1}{\mathbf{V}_1} = \frac{\mathbf{I}_1}{\frac{4}{3}\mathbf{I}_1} = 0.75 \text{ S}$$

By the current divider rule,

$$-\mathbf{I}_2 = \frac{4}{4+2}\mathbf{I}_1 = \frac{2}{3}\mathbf{I}_1, \qquad \mathbf{y}_{21} = \frac{\mathbf{I}_2}{\mathbf{V}_1} = \frac{-\frac{2}{3}\mathbf{I}_1}{\frac{4}{3}\mathbf{I}_1} = -0.5 \text{ S}$$

To get $\mathbf{y}_{12}$ and $\mathbf{y}_{22}$, short-circuit the input port and connect a current source $\mathbf{I}_2$ to the output port as in Fig. 19.15(b). The 4-Ω resistor is short-circuited so that 2-Ω and 8-Ω resistors are in parallel.

$$\mathbf{V}_2 = \mathbf{I}_2(8 \parallel 2) = \frac{8}{5}\mathbf{I}_2, \qquad \mathbf{y}_{22} = \frac{\mathbf{I}_2}{\mathbf{V}_2} = \frac{\mathbf{I}_2}{\frac{8}{5}\mathbf{I}_2} = \frac{5}{8} = 0.625 \text{ S}$$

By the current divider rule,

$$-\mathbf{I}_1 = \frac{8}{8+2}\mathbf{I}_2 = \frac{4}{5}\mathbf{I}_2, \qquad \mathbf{y}_{12} = \frac{\mathbf{I}_1}{\mathbf{V}_2} = \frac{-\frac{4}{5}\mathbf{I}_2}{\frac{8}{5}\mathbf{I}_2} = -0.5 \text{ S}$$

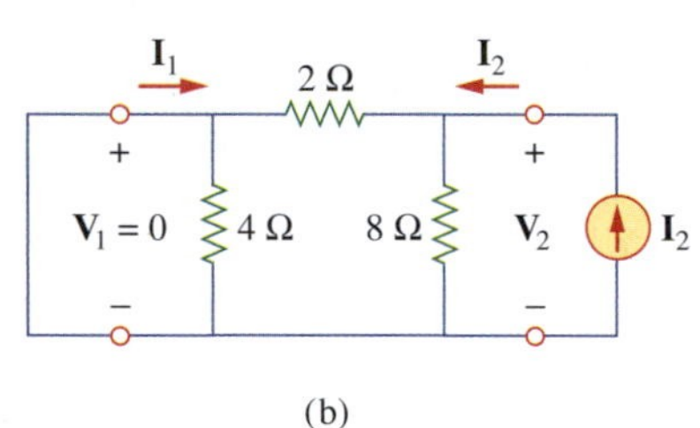

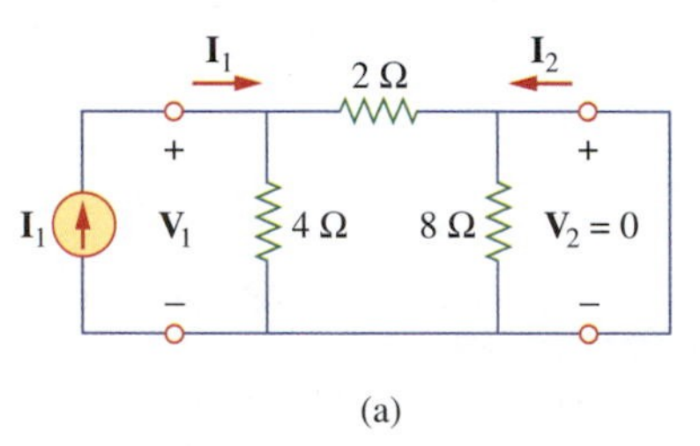

Figure 19.15
For Example 19.3: (a) finding $\mathbf{y}_{11}$ and $\mathbf{y}_{21}$; (b) finding $\mathbf{y}_{12}$ and $\mathbf{y}_{22}$.

■ **METHOD 2** Alternatively, comparing Fig. 19.14 with Fig. 19.13(a),

$$\mathbf{y}_{12} = -\frac{1}{2}\text{ S} = \mathbf{y}_{21}$$

$$\mathbf{y}_{11} + \mathbf{y}_{12} = \frac{1}{4} \qquad \Rightarrow \qquad \mathbf{y}_{11} = 0.5 - \mathbf{y}_{12} = 0.75 \text{ S}$$

$$\mathbf{y}_{22} + \mathbf{y}_{12} = \frac{1}{8} \qquad \Rightarrow \qquad \mathbf{y}_{22} = \frac{1}{8} - \mathbf{y}_{12} = 0.625 \text{ S}$$

as obtained previously.

Practice Problem 19.3

Obtain the y parameters for the T network shown in Fig. 19.16.

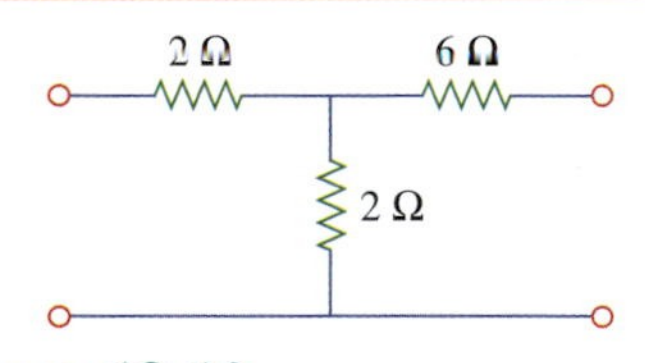

Figure 19.16
For Practice Problem 19.3.

Answer: $\mathbf{y}_{11} = 0.2273$ S; $\mathbf{y}_{12} = \mathbf{y}_{21} = -0.0909$ S; $\mathbf{y}_{22} = 0.1364$ S

Example 19.4

Determine the y parameters as functions of s ($s = j\omega$) for the two-port shown in Fig. 19.17.

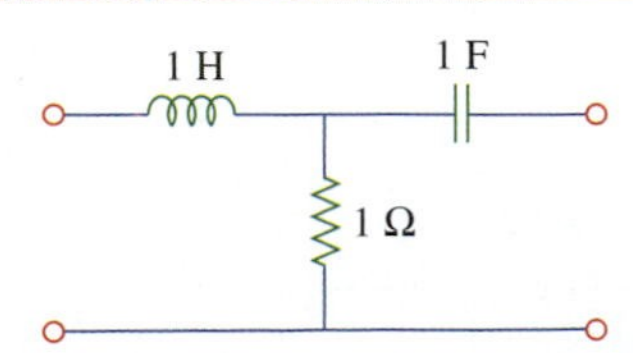

Figure 19.17
For Example 19.4.

Solution:
We follow the same procedure as in the previous example. But we need to first convert the circuit to the frequency-domain.

$$1 \text{ H} \qquad \Rightarrow \qquad j\omega L = sL = s$$

$$1 \text{ F} \qquad \Rightarrow \qquad \frac{1}{j\omega C} = \frac{1}{sC} = \frac{1}{s}$$

To get $\mathbf{y}_{11}$ and $\mathbf{y}_{21}$, we use the circuit in Fig. 19.18(a), in which port 2 is short-circuited and a current source is applied to port 1.

$$\mathbf{V}_1 = \mathbf{I}_1\left(s + 1 \parallel \frac{1}{s}\right) = \mathbf{I}_1\left(s + \frac{1}{s+1}\right) = \mathbf{I}_1\left(\frac{s^2+s+1}{s+1}\right)$$

$$\mathbf{y}_{11} = \frac{\mathbf{I}_1}{\mathbf{V}_1} = \frac{s+1}{s^2+s+1}$$

By the current divider rule,

$$\mathbf{I}_2 = -\frac{1}{1+1/s}\mathbf{I}_1 = -\frac{s}{s+1}\mathbf{I}_1$$

$$\mathbf{y}_{21} = \frac{\mathbf{I}_2}{\mathbf{V}_1} = -\frac{s}{s^2+s+1}$$

Similarly, we get $\mathbf{y}_{12}$ and $\mathbf{y}_{22}$ using Fig. 19.18(b):

$$\mathbf{V}_2 = \mathbf{I}_2\left(\frac{1}{s} + 1 \parallel s\right) = I_2\left(\frac{1}{s} + \frac{s}{s+1}\right) = \frac{s^2+s+1}{s(s+1)}\mathbf{I}_2$$

$$\mathbf{y}_{22} = \frac{\mathbf{I}_2}{\mathbf{V}_2} = \frac{s(s+1)}{s^2+s+1}$$

By the current divider rule,

$$\mathbf{I}_1 = -\mathbf{I}_2\frac{1}{s+1}$$

$$\mathbf{y}_{12} = \frac{\mathbf{I}_1}{\mathbf{V}_2} = -\frac{1}{s+1}\,\frac{s(s+1)}{s^2+s+1} = -\frac{s}{s^2+s+1}$$

Thus,

$$[\mathbf{y}] = \begin{bmatrix} \dfrac{s+1}{s^2+s+1} & -\dfrac{s}{s^2+s+1} \\ -\dfrac{s}{s^2+s+1} & \dfrac{s(s+1)}{s^2+s+1} \end{bmatrix}$$

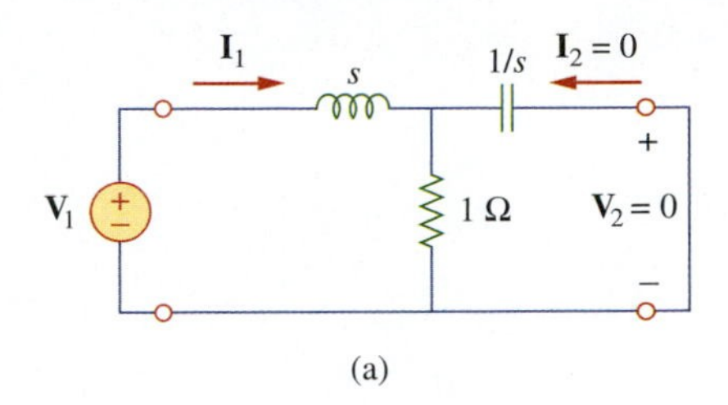

(a)

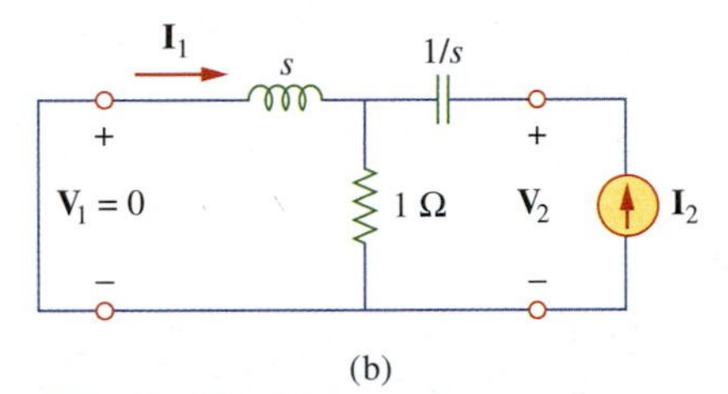

(b)

Figure 19.18
Solution of Example 19.4: (a) finding $\mathbf{y}_{11}$ and $\mathbf{y}_{21}$; (b) finding $\mathbf{y}_{12}$ and $\mathbf{y}_{22}$.

Practice Problem 19.4

Obtain the y parameters as functions of s for the ladder network in Fig. 19.19.

Answer:

$$[\mathbf{y}] = \begin{bmatrix} \dfrac{s+1}{s(s+2)} & \dfrac{-1}{s(s+2)} \\ \dfrac{-1}{s(s+2)} & \dfrac{s^2+3s+1}{s(s+2)} \end{bmatrix}$$

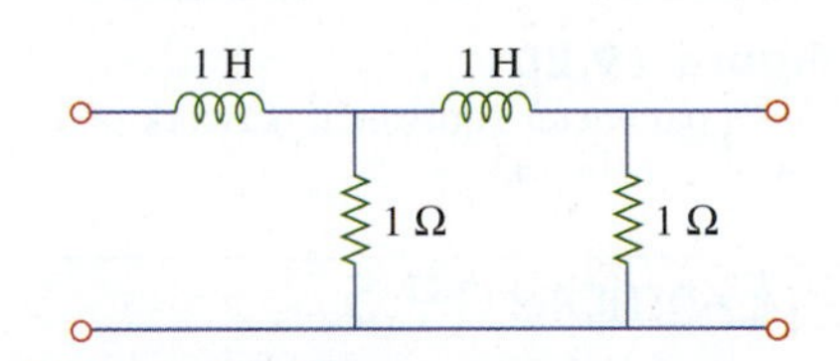

Figure 19.19
For Practice Problem 19.4.

19.4 Hybrid Parameters

The third set of parameters is based on making $\mathbf{V}_1$ and $\mathbf{I}_2$ the dependent variables. Thus, we obtain

$$\begin{aligned} \mathbf{V}_1 &= \mathbf{h}_{11}\mathbf{I}_1 + \mathbf{h}_{12}\mathbf{V}_2 \\ \mathbf{I}_2 &= \mathbf{h}_{21}\mathbf{I}_1 + \mathbf{h}_{22}\mathbf{V}_2 \end{aligned} \tag{19.14}$$

or in matrix form

$$\begin{bmatrix} \mathbf{V}_1 \\ \mathbf{I}_2 \end{bmatrix} = \begin{bmatrix} \mathbf{h}_{11} & \mathbf{h}_{12} \\ \mathbf{h}_{21} & \mathbf{h}_{22} \end{bmatrix} \begin{bmatrix} \mathbf{I}_1 \\ \mathbf{V}_2 \end{bmatrix} = [\mathbf{h}] \begin{bmatrix} \mathbf{I}_1 \\ \mathbf{V}_2 \end{bmatrix} \tag{19.15}$$

The **h** terms are known as the *hybrid parameters* or simply *h parameters* because they are a hybrid combination of ratios. They are very useful for describing electronic devices such as transistors (see Section 19.8) because it is much easier to experimentally measure the h parameters of such devices than to measure their z or y parameters. In fact, we have seen that the ideal transformer in Fig. 19.6, described by Eq. (19.7), does not have z parameters. The ideal transformer can be described by the hybrid parameters because Eq. (19.7) conforms with Eq. (19.14).

The values of the parameters are determined as:

$$\begin{aligned} \mathbf{h}_{11} &= \left.\frac{\mathbf{V}_1}{\mathbf{I}_1}\right|_{\mathbf{V}_2=0}, & \mathbf{h}_{12} &= \left.\frac{\mathbf{V}_1}{\mathbf{V}_2}\right|_{\mathbf{I}_1=0} \\ \mathbf{h}_{21} &= \left.\frac{\mathbf{I}_2}{\mathbf{I}_1}\right|_{\mathbf{V}_2=0}, & \mathbf{h}_{22} &= \left.\frac{\mathbf{I}_2}{\mathbf{V}_2}\right|_{\mathbf{I}_1=0} \end{aligned} \tag{19.16}$$

It is evident from Eq. (19.16) that the parameters $\mathbf{h}_{11}$, $\mathbf{h}_{12}$, $\mathbf{h}_{21}$, and $\mathbf{h}_{22}$ represent an impedance, a voltage gain, a current gain, and an admittance, respectively. This is why they are called the *hybrid parameters*. To be specific,

$$\begin{aligned} \mathbf{h}_{11} &= \text{short-circuit input impedance} \\ \mathbf{h}_{12} &= \text{open-circuit reverse voltage gain} \\ \mathbf{h}_{21} &= \text{short-circuit forward current gain} \\ \mathbf{h}_{22} &= \text{open-circuit output admittance} \end{aligned} \tag{19.17}$$

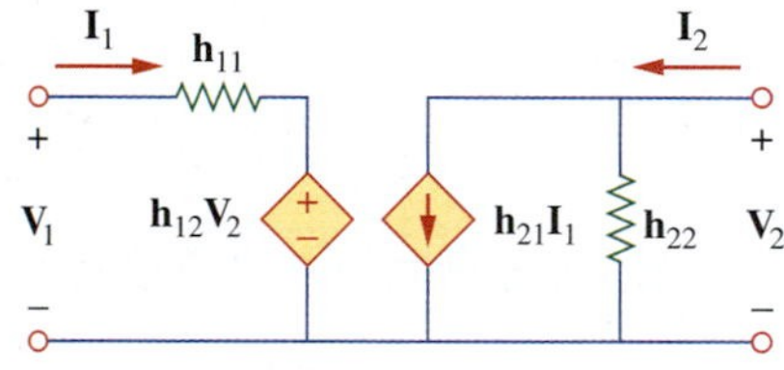

Figure 19.20
The h parameter equivalent network of a two-port network.

The procedure for calculating the h parameters is similar to that used for the z or y parameters. We apply a voltage or current source to the appropriate port; short-circuit or open-circuit the other port, depending on the parameter of interest; and perform regular circuit analysis. For reciprocal networks, $\mathbf{h}_{12} = -\mathbf{h}_{21}$. This can be proved in the same way as we proved that $\mathbf{z}_{12} = \mathbf{z}_{21}$. The hybrid model of a two-port network is shown in Fig. 19.20.

Example 19.5

Find the hybrid parameters for the two-port network of Fig. 19.21.

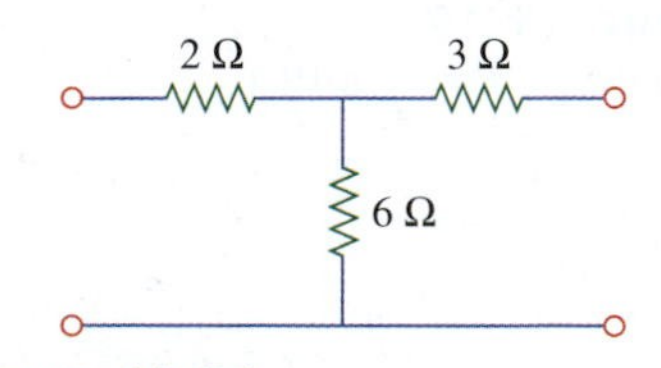

Figure 19.21
For Example 19.5.

Solution:

To find $\mathbf{h}_{11}$ and $\mathbf{h}_{21}$, we short-circuit the output port and connect a current source $\mathbf{I}_1$ to the input port as shown in Fig. 19.22(a). From Fig. 19.22(a),

$$\mathbf{V}_1 = \mathbf{I}_1 (2 + 3 \parallel 6) = 4\mathbf{I}_1$$

Hence,

$$\mathbf{h}_{11} = \mathbf{V}_1/\mathbf{I}_1 = 4\ \Omega$$

Also, from Fig. 19.22(a), we obtain by the current divider rule,

$$-\mathbf{I}_2 = \frac{6}{6+3}\mathbf{I}_1 = \frac{2}{3}\mathbf{I}_1$$

Hence,

$$\mathbf{h}_{21} = \frac{\mathbf{I}_2}{\mathbf{I}_1} = -\frac{2}{3}$$

To obtain $\mathbf{h}_{12}$ and $\mathbf{h}_{22}$, we open-circuit the input port and connect a voltage source $\mathbf{V}_2$ to the output port as in Fig. 19.22(b). By the voltage divider rule,

$$\mathbf{V}_1 = \frac{6}{6+3}\mathbf{V}_2 = \frac{2}{3}\mathbf{V}_2$$

Hence,

$$\mathbf{h}_{12} = \mathbf{V}_1/\mathbf{V}_2 = \frac{2}{3}$$

Also,

$$\mathbf{V}_2 = (3+6)\mathbf{I}_2 = 9\mathbf{I}_2$$

Thus,

$$\mathbf{h}_{22} = \frac{\mathbf{I}_2}{\mathbf{V}_2} = \frac{1}{9}\text{ S}$$

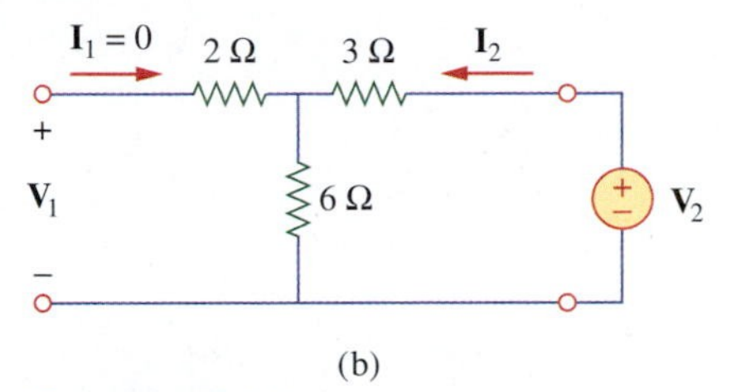

Figure 19.22
For Example 19.5: (a) computing $\mathbf{h}_{11}$ and $\mathbf{h}_{21}$; (b) computing $\mathbf{h}_{12}$ and $\mathbf{h}_{22}$.

Practice Problem 19.5

Determine the h parameters for the circuit in Fig. 19.23.

Answer: $\mathbf{h}_{11} = 1.2\ \Omega$; $\mathbf{h}_{12} = 0.4$; $\mathbf{h}_{21} = -0.4$; $\mathbf{h}_{22} = 0.4$ S

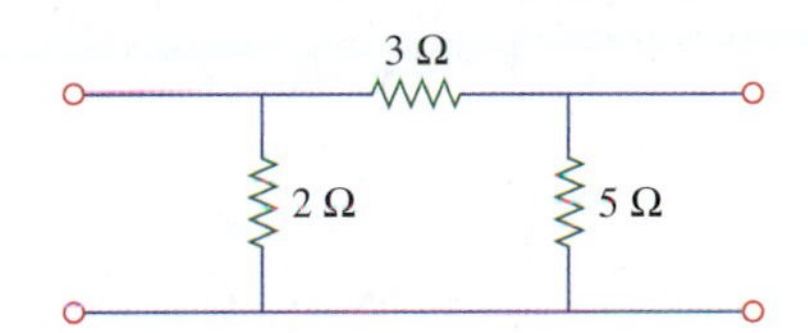

Figure 19.23
For Practice Problem 19.5.

Example 19.6

Determine the Thevenin equivalent at the output port of the circuit in Fig. 19.24.

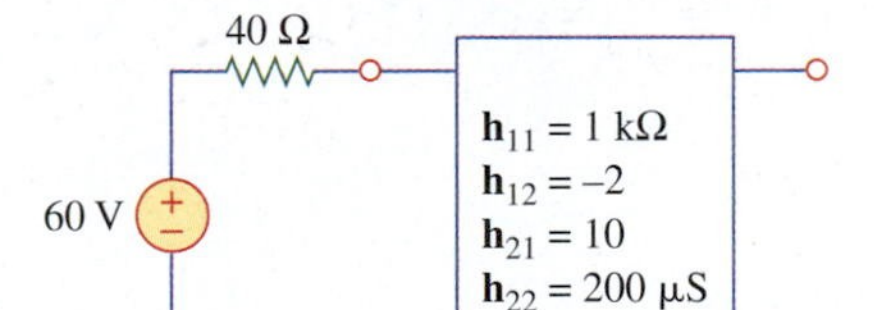

Figure 19.24
For Example 19.6.

Solution:
To find $\mathbf{Z}_{Th}$ and $\mathbf{V}_{Th}$, we apply the normal procedure, keeping in mind the formulas relating the input and output ports of the h model. To obtain $\mathbf{Z}_{Th}$, remove the 60-V voltage source at the input port and apply a 1-V voltage source at the output port, as shown in Fig. 19.25(a). From Eq. (19.14),

$$\mathbf{V}_1 = \mathbf{h}_{11}\mathbf{I}_1 + \mathbf{h}_{12}\mathbf{V}_2 \quad (19.6.1)$$
$$\mathbf{I}_2 = \mathbf{h}_{21}\mathbf{I}_1 + \mathbf{h}_{22}\mathbf{V}_2 \quad (19.6.2)$$

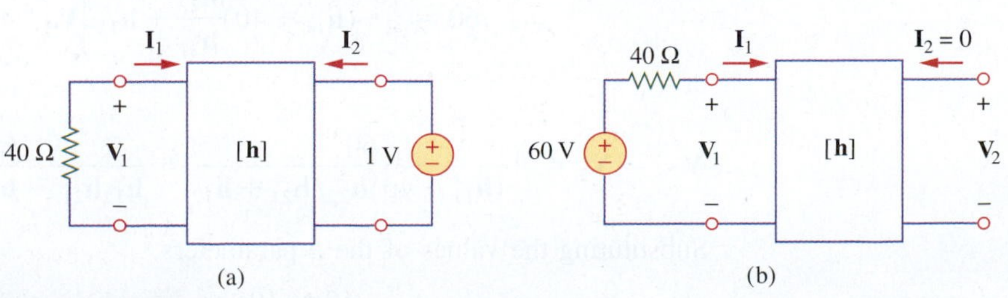

Figure 19.25
For Example 19.6: (a) finding $\mathbf{Z}_{Th}$; (b) finding $\mathbf{V}_{Th}$.

But

$$\mathbf{V}_2 = 1, \qquad \mathbf{V}_1 = -40\,\mathbf{I}_1$$

Substituting these into Eqs. (19.6.1) and (19.6.2), we get

$$-40\mathbf{I}_1 = \mathbf{h}_{11}\mathbf{I}_1 + \mathbf{h}_{12} \quad \Rightarrow \quad \mathbf{I}_1 = -\frac{\mathbf{h}_{12}}{40 + \mathbf{h}_{11}} \tag{19.6.3}$$

$$\mathbf{I}_2 = \mathbf{h}_{21}\mathbf{I}_1 + \mathbf{h}_{22} \tag{19.6.4}$$

Substituting Eq. (19.6.3) into Eq. (19.6.4) gives

$$\mathbf{I}_2 = \mathbf{h}_{22} - \frac{\mathbf{h}_{21}\mathbf{h}_{12}}{\mathbf{h}_{11} + 40} = \frac{\mathbf{h}_{11}\mathbf{h}_{22} - \mathbf{h}_{21}\mathbf{h}_{12} + \mathbf{h}_{22}40}{\mathbf{h}_{11} + 40}$$

Therefore,

$$\mathbf{Z}_{Th} = \mathbf{V}_2/\mathbf{I}_2 = 1/\mathbf{I}_2 = \frac{\mathbf{h}_{11} + 40}{\mathbf{h}_{11}\mathbf{h}_{22} - \mathbf{h}_{21}\mathbf{h}_{12} + \mathbf{h}_{22}40}$$

Substituting the values of the h parameters,

$$\mathbf{Z}_{Th} = \frac{1{,}000 + 40}{10^3 \times 200 \times 10^{-6} + 20 + 40 \times 200 \times 10^{-6}} = \frac{1{,}040}{20.21}$$
$$= 51.46\ \Omega$$

To get $\mathbf{V}_{Th}$, we find the open-circuit voltage $\mathbf{V}_2$ in Fig. 19.25(b). At the input port,

$$-60 + 40\mathbf{I}_1 + \mathbf{V}_1 = 0 \quad \Rightarrow \quad \mathbf{V}_1 = 60 - 40\mathbf{I}_1 \tag{19.6.5}$$

At the output,

$$\mathbf{I}_2 = 0 \tag{19.6.6}$$

Substituting Eqs. (19.6.5) and (19.6.6) into Eqs. (19.6.1) and (19.6.2), we obtain

$$60 - 40\mathbf{I}_1 = \mathbf{h}_{11}\mathbf{I}_1 + \mathbf{h}_{12}\mathbf{V}_2$$

or

$$60 = (\mathbf{h}_{11} + 40)\mathbf{I}_1 + \mathbf{h}_{12}\mathbf{V}_2 \tag{19.6.7}$$

and

$$0 = \mathbf{h}_{21}\mathbf{I}_1 + \mathbf{h}_{22}\mathbf{V}_2 \quad \Rightarrow \quad \mathbf{I}_1 = -\frac{\mathbf{h}_{22}}{\mathbf{h}_{21}}\mathbf{V}_2 \tag{19.6.8}$$

Substituting Eq. (19.6.8) into Eq. (19.6.7) gives

$$60 = \left[-(\mathbf{h}_{11} + 40)\frac{\mathbf{h}_{22}}{\mathbf{h}_{21}} + \mathbf{h}_{12}\right]\mathbf{V}_2$$

or

$$\mathbf{V}_{Th} = \mathbf{V}_2 = \frac{60}{-(\mathbf{h}_{11} + 40)\mathbf{h}_{22}/\mathbf{h}_{21} + \mathbf{h}_{12}} = \frac{60\mathbf{h}_{21}}{\mathbf{h}_{12}\mathbf{h}_{21} - \mathbf{h}_{11}\mathbf{h}_{22} - 40\mathbf{h}_{22}}$$

Substituting the values of the h parameters,

$$\mathbf{V}_{Th} = \frac{60 \times 10}{-20.21} = -29.69\ \text{V}$$

Practice Problem 19.6

Find the impedance at the input port of the circuit in Fig. 19.26.

Answer: 1667 Ω

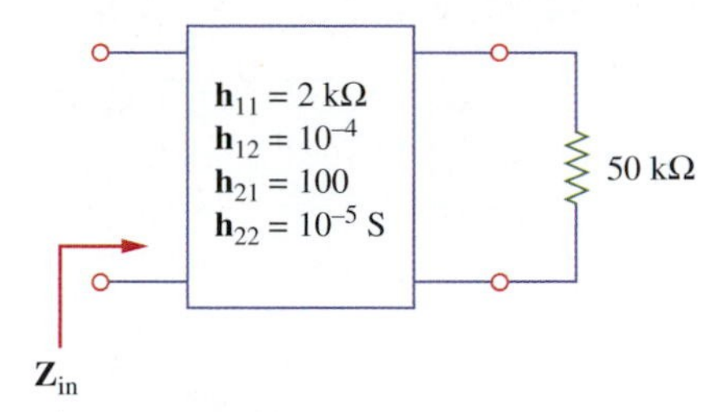

Figure 19.26
For Practice Problem 19.6.

19.5 †Relationships Between Parameters

Because the three sets of parameters relate the same input and output terminal variables of the same two-port network, they should be interrelated. If two sets of parameters exist, we can relate one set to the other set. Let us demonstrate the process with two examples.

Given the z parameters, let us obtain the y parameters. From Eq. (19.2),

$$\begin{bmatrix} \mathbf{V}_1 \\ \mathbf{V}_2 \end{bmatrix} = \begin{bmatrix} \mathbf{z}_{11} & \mathbf{z}_{12} \\ \mathbf{z}_{21} & \mathbf{z}_{22} \end{bmatrix} \begin{bmatrix} \mathbf{I}_1 \\ \mathbf{I}_2 \end{bmatrix} = [\mathbf{z}] \begin{bmatrix} \mathbf{I}_1 \\ \mathbf{I}_2 \end{bmatrix} \tag{19.18}$$

or

$$\begin{bmatrix} \mathbf{I}_1 \\ \mathbf{I}_2 \end{bmatrix} = [\mathbf{z}]^{-1} \begin{bmatrix} \mathbf{V}_1 \\ \mathbf{V}_2 \end{bmatrix} \tag{19.19}$$

Also, from Eq. (19.9),

$$\begin{bmatrix} \mathbf{I}_1 \\ \mathbf{I}_2 \end{bmatrix} = \begin{bmatrix} \mathbf{y}_{11} & \mathbf{y}_{12} \\ \mathbf{y}_{21} & \mathbf{y}_{22} \end{bmatrix} \begin{bmatrix} \mathbf{V}_1 \\ \mathbf{V}_2 \end{bmatrix} = [\mathbf{y}] \begin{bmatrix} \mathbf{V}_1 \\ \mathbf{V}_2 \end{bmatrix} \tag{19.20}$$

Comparing Eqs. (19.19) and (19.20), we see that

$$[\mathbf{y}] = [\mathbf{z}]^{-1} \tag{19.21}$$

The adjoint of the $[\mathbf{z}]$ matrix is

$$\begin{bmatrix} \mathbf{z}_{22} & -\mathbf{z}_{12} \\ -\mathbf{z}_{21} & \mathbf{z}_{11} \end{bmatrix} \tag{19.22}$$

and its determinant is

$$\Delta_z = \mathbf{z}_{11}\mathbf{z}_{22} - \mathbf{z}_{12}\mathbf{z}_{21}$$

Substituting these into Eq. (19.21), we get

$$\begin{bmatrix} \mathbf{y}_{11} & \mathbf{y}_{12} \\ \mathbf{y}_{21} & \mathbf{y}_{22} \end{bmatrix} = \frac{\begin{bmatrix} \mathbf{z}_{22} & -\mathbf{z}_{12} \\ -\mathbf{z}_{21} & \mathbf{z}_{11} \end{bmatrix}}{\Delta_z} \tag{19.23}$$

Equating terms yields

$$\mathbf{y}_{11} = \frac{\mathbf{z}_{22}}{\Delta_z}, \qquad \mathbf{y}_{12} = -\frac{\mathbf{z}_{12}}{\Delta_z}, \qquad \mathbf{y}_{21} = -\frac{\mathbf{z}_{21}}{\Delta_z}, \qquad \mathbf{y}_{22} = \frac{\mathbf{z}_{11}}{\Delta_z} \tag{19.24}$$

As a second example, let us determine the h parameters from the z parameters. From Eq. (19.1),

$$\mathbf{V}_1 = \mathbf{z}_{11}\mathbf{I}_1 + \mathbf{z}_{12}\mathbf{I}_2 \tag{19.25a}$$
$$\mathbf{V}_2 = \mathbf{z}_{21}\mathbf{I}_1 + \mathbf{z}_{22}\mathbf{I}_2 \tag{19.25b}$$

Making $\mathbf{I}_2$ the subject of Eq. (19.25b),

$$\mathbf{I}_2 = -\frac{\mathbf{z}_{21}}{\mathbf{z}_{22}}\mathbf{I}_1 + \frac{1}{\mathbf{z}_{22}}\mathbf{V}_2 \tag{19.26}$$

Substituting this into Eq. (19.25a),

$$\mathbf{V}_1 = \frac{\mathbf{z}_{11}\mathbf{z}_{22} - \mathbf{z}_{12}\mathbf{z}_{21}}{\mathbf{z}_{22}}\mathbf{I}_1 + \frac{\mathbf{z}_{12}}{\mathbf{z}_{22}}\mathbf{V}_2 \tag{19.27}$$

Putting Eqs. (19.26) and (19.27) in matrix form,

$$\begin{bmatrix} \mathbf{V}_1 \\ \mathbf{I}_2 \end{bmatrix} = \begin{bmatrix} \dfrac{\Delta_Z}{\mathbf{z}_{22}} & \dfrac{\mathbf{z}_{12}}{\mathbf{z}_{22}} \\ -\dfrac{\mathbf{z}_{21}}{\mathbf{z}_{22}} & \dfrac{1}{\mathbf{z}_{22}} \end{bmatrix} \begin{bmatrix} \mathbf{I}_1 \\ \mathbf{V}_2 \end{bmatrix} \tag{19.28}$$

From Eq. (19.15),

$$\begin{bmatrix} \mathbf{V}_1 \\ \mathbf{I}_2 \end{bmatrix} = \begin{bmatrix} \mathbf{h}_{11} & \mathbf{h}_{12} \\ \mathbf{h}_{21} & \mathbf{h}_{22} \end{bmatrix} \begin{bmatrix} \mathbf{I}_1 \\ \mathbf{V}_2 \end{bmatrix} = [\mathbf{h}] \begin{bmatrix} \mathbf{I}_1 \\ \mathbf{V}_2 \end{bmatrix} \tag{19.29}$$

Comparing Eqs. (19.28) and (19.29), we obtain

$$\mathbf{h}_{11} = \frac{\Delta_z}{\mathbf{z}_{22}}, \qquad \mathbf{h}_{12} = \frac{\mathbf{z}_{12}}{\mathbf{z}_{22}}, \qquad \mathbf{h}_{21} = -\frac{\mathbf{z}_{21}}{\mathbf{z}_{22}}, \qquad \mathbf{h}_{22} = \frac{1}{\mathbf{z}_{22}} \tag{19.30}$$

Table 19.1 provides the conversion formulas for the three sets of two-port parameters. Given one set of parameters, Table 19.1 can be used to find other two parameter sets. For example, given the h parameters, we find the corresponding y parameters in the second row of the third column. Also, given that $\mathbf{z}_{21} = \mathbf{z}_{12}$ for a reciprocal network, the table can be used to express this condition in terms of other parameters.

TABLE 19.1

Conversion of two-port parameters.

	z		y		h	
z	$\mathbf{z}_{11}$	$\mathbf{z}_{12}$	$\dfrac{\mathbf{y}_{22}}{\Delta_y}$	$-\dfrac{\mathbf{y}_{12}}{\Delta_y}$	$\dfrac{\Delta_h}{\mathbf{h}_{22}}$	$\dfrac{\mathbf{h}_{12}}{\mathbf{h}_{22}}$
	$\mathbf{z}_{21}$	$\mathbf{z}_{22}$	$-\dfrac{\mathbf{y}_{21}}{\Delta_y}$	$\dfrac{\mathbf{y}_{11}}{\Delta_y}$	$-\dfrac{\mathbf{h}_{21}}{\mathbf{h}_{22}}$	$\dfrac{1}{\mathbf{h}_{22}}$
y	$\dfrac{\mathbf{z}_{22}}{\Delta_z}$	$-\dfrac{\mathbf{z}_{12}}{\Delta_z}$	$\mathbf{y}_{11}$	$\mathbf{y}_{12}$	$\dfrac{1}{\mathbf{h}_{11}}$	$-\dfrac{\mathbf{h}_{12}}{\mathbf{h}_{11}}$
	$-\dfrac{\mathbf{z}_{21}}{\Delta_z}$	$\dfrac{\mathbf{z}_{11}}{\Delta_z}$	$\mathbf{y}_{21}$	$\mathbf{y}_{22}$	$\dfrac{\mathbf{h}_{21}}{\mathbf{h}_{11}}$	$\dfrac{\Delta_h}{\mathbf{h}_{11}}$
h	$\dfrac{\Delta_z}{\mathbf{z}_{22}}$	$\dfrac{\mathbf{z}_{12}}{\mathbf{z}_{22}}$	$\dfrac{1}{\mathbf{y}_{11}}$	$-\dfrac{\mathbf{y}_{12}}{\mathbf{y}_{11}}$	$\mathbf{h}_{11}$	$\mathbf{h}_{12}$
	$-\dfrac{\mathbf{z}_{21}}{\mathbf{z}_{22}}$	$\dfrac{1}{\mathbf{z}_{22}}$	$\dfrac{\mathbf{y}_{21}}{\mathbf{y}_{11}}$	$\dfrac{\Delta_y}{\mathbf{y}_{11}}$	$\mathbf{h}_{21}$	$\mathbf{h}_{22}$

where $\Delta_z = \mathbf{z}_{11}\mathbf{z}_{22} - \mathbf{z}_{12}\mathbf{z}_{21}$

$\Delta_y = \mathbf{y}_{11}\mathbf{y}_{22} - \mathbf{y}_{12}\mathbf{y}_{21}$

$\Delta_h = \mathbf{h}_{11}\mathbf{h}_{22} - \mathbf{h}_{12}\mathbf{h}_{21}$

Example 19.7

Find [**z**] of a two-port network if $[\mathbf{h}] = \begin{bmatrix} 20\ \Omega & 3 \\ -2 & 0.01\ S \end{bmatrix}$

Solution:

From [**h**], we obtain the determinant as

$$\Delta_h = 20 \times 0.01 - (3)(-2) = 6.2$$

Thus, from Table 9.1,

$$[\mathbf{z}] = \begin{bmatrix} \frac{\Delta_h}{\mathbf{h}_{22}} & \frac{\mathbf{h}_{12}}{\mathbf{h}_{22}} \\ -\frac{\mathbf{h}_{21}}{\mathbf{h}_{22}} & \frac{1}{\mathbf{h}_{22}} \end{bmatrix} = \begin{bmatrix} \frac{6.2}{0.01} & \frac{3}{0.01} \\ \frac{2}{0.01} & \frac{1}{0.01} \end{bmatrix} = \begin{bmatrix} 620 & 300 \\ 200 & 100 \end{bmatrix}$$

Thus,

$$\mathbf{z}_{11} = 620, \qquad \mathbf{z}_{12} = 300, \qquad \mathbf{z}_{21} = 200, \qquad \mathbf{z}_{22} = 100\ \Omega$$

Practice Problem 19.7

Determine $[\mathbf{y}]$ of a two-port network whose z parameters are

$$[\mathbf{z}] = \begin{bmatrix} 6 & 4 \\ 4 & 6 \end{bmatrix} \Omega$$

Answer $[\mathbf{y}] = \begin{bmatrix} 0.3 & -0.2 \\ -0.2 & 0.3 \end{bmatrix}$ S

19.6 Interconnection of Networks

A large, complex network may be divided into subnetworks for the purpose of analysis and design. The subnetworks are modeled as two-port networks and are interconnected to form the original network. The two-port networks may therefore be regarded as building blocks, which can be interconnected to form a complex network. The interconnection can be in series, in parallel, or in cascade. Although the interconnected network can be described by any of the three parameter sets mentioned in this chapter, a certain set of parameters may have a definite advantage. For example, when the networks are in series, their individual z parameters add up to give the z parameters of the larger network.

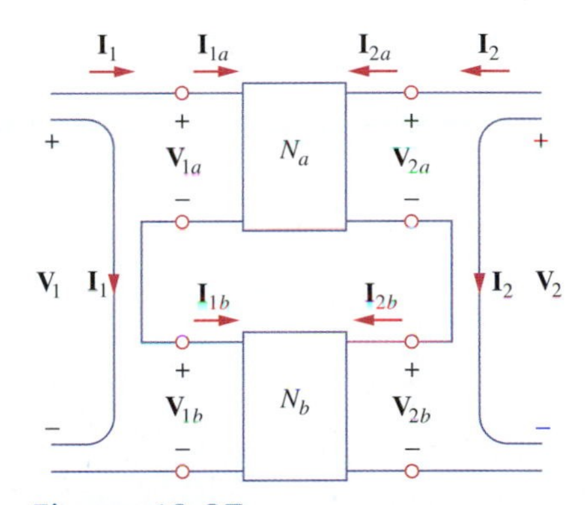

Figure 19.27
Series connection of two two-port networks.

Consider the series connection of two two-port networks shown in Fig. 19.27. The networks are regarded as being in series because their input currents are the same and their voltages add. For network N_a,

$$\begin{aligned} \mathbf{V}_{1a} &= \mathbf{z}_{11a}\mathbf{I}_{1a} + \mathbf{z}_{12a}\mathbf{I}_{2a} \\ \mathbf{V}_{2a} &= \mathbf{z}_{21a}\mathbf{I}_{1a} + \mathbf{z}_{22a}\mathbf{I}_{2a} \end{aligned} \tag{19.31}$$

and for network N_b,

$$\begin{aligned} \mathbf{V}_{1b} &= \mathbf{z}_{11b}\mathbf{I}_{1b} + \mathbf{z}_{12b}\mathbf{I}_{2b} \\ \mathbf{V}_{2b} &= \mathbf{z}_{21b}\mathbf{I}_{1b} + \mathbf{z}_{22b}\mathbf{I}_{2b} \end{aligned} \tag{19.32}$$

We notice from Fig. 19.27 that

$$\mathbf{I}_1 = \mathbf{I}_{1a} = \mathbf{I}_{1b} \tag{19.33a}$$

$$\mathbf{I}_2 = \mathbf{I}_{2a} = \mathbf{I}_{2b} \tag{19.33b}$$

and that

$$\begin{aligned} \mathbf{V}_1 &= \mathbf{V}_{1a} + \mathbf{V}_{1b} = (\mathbf{z}_{11a} + \mathbf{z}_{11b})\mathbf{I}_1 + (\mathbf{z}_{12a} + \mathbf{z}_{12b})\mathbf{I}_2 \\ \mathbf{V}_2 &= \mathbf{V}_{2a} + \mathbf{V}_{2b} = (\mathbf{z}_{21a} + \mathbf{z}_{21b})\mathbf{I}_1 + (\mathbf{z}_{22a} + \mathbf{z}_{22b})\mathbf{I}_2 \end{aligned} \tag{19.34}$$

Thus, the z parameters for the overall network are

$$\begin{bmatrix} \mathbf{z}_{11} & \mathbf{z}_{12} \\ \mathbf{z}_{21} & \mathbf{z}_{22} \end{bmatrix} = \begin{bmatrix} \mathbf{z}_{11a} + \mathbf{z}_{11b} & \mathbf{z}_{12a} + \mathbf{z}_{12b} \\ \mathbf{z}_{21a} + \mathbf{z}_{21b} & \mathbf{z}_{22a} + \mathbf{z}_{22b} \end{bmatrix} \tag{19.35}$$

or

$$[\mathbf{z}] = [\mathbf{z}_a] + [\mathbf{z}_b] \tag{19.36}$$

showing that the z parameters for the overall network are the sum of the z parameters for the individual networks. This can be extended to n two-port networks in series. If two two-port networks in the [**h**] model, for example, are connected in series, we use Table 19.1 to convert the **h** to **z** and then apply Eq. (19.36). We finally convert the result back to **h** using Table 19.1.

Two two-port networks are in parallel when their port voltages are equal and the port currents of the larger network are the sums of the individual port currents. In addition, each circuit must have a common reference, and when the networks are connected together, they must all have their common references tied together. The parallel connection of two two-port networks is shown in Fig. 19.28. As we did for the series connection, we can show that the y parameters for the overall network are

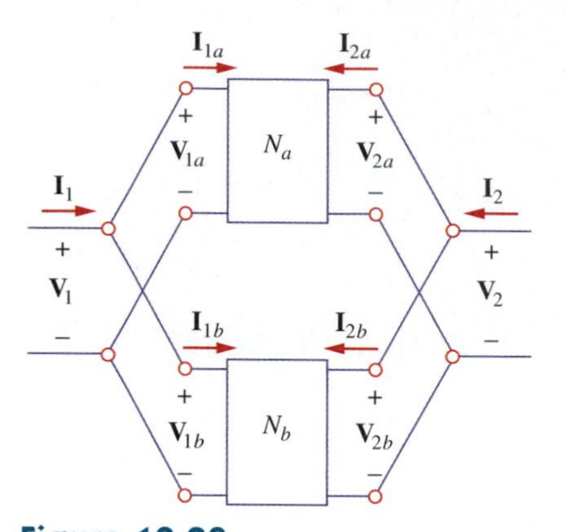

Figure 19.28
Parallel connection of two two-port networks.

$$\begin{bmatrix} \mathbf{y}_{11} & \mathbf{y}_{12} \\ \mathbf{y}_{21} & \mathbf{y}_{22} \end{bmatrix} = \begin{bmatrix} \mathbf{y}_{11a} + \mathbf{y}_{11b} & \mathbf{y}_{12a} + \mathbf{y}_{12b} \\ \mathbf{y}_{21a} + \mathbf{y}_{21b} & \mathbf{y}_{22a} + \mathbf{y}_{22b} \end{bmatrix} \tag{19.37}$$

or

$$[\mathbf{y}] = [\mathbf{y}_a] + [\mathbf{y}_b] \tag{19.38}$$

showing that the y parameters of the overall network are the sum of the y parameters of the individual networks. The result can be extended to n two-port networks in parallel.

Two networks are said to be *cascaded* when the output of one is the input of the other. The connection of two two-port networks in cascade is shown in Fig. 19.29. Analysis of the cascaded connection will involve using transmission parameters, which are beyond the scope of this textbook.

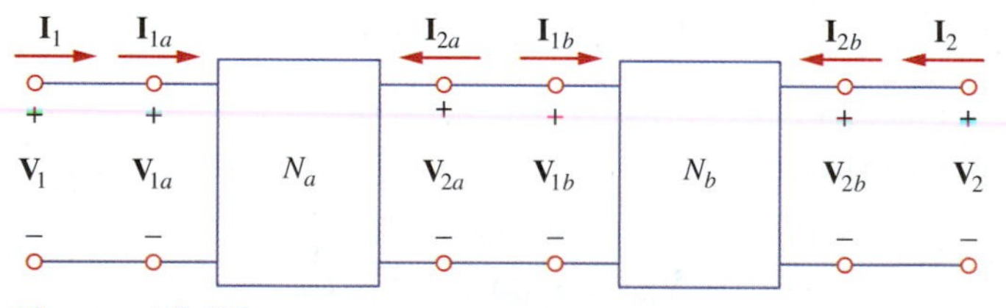

Figure 19.29
Cascade connection of two two-port networks.

Example 19.8

Evaluate $\mathbf{V}_2/\mathbf{V}_s$ in the circuit in Fig. 19.30.

Solution:

This may be regarded as two two-ports in series. For N_b,

$$\mathbf{z}_{12b} = \mathbf{z}_{21b} = 10 = \mathbf{z}_{11b} = \mathbf{z}_{22b}$$

Thus,

$$[\mathbf{z}] = [\mathbf{z}_a] + [\mathbf{z}_b] = \begin{bmatrix} 12 & 8 \\ 8 & 20 \end{bmatrix} + \begin{bmatrix} 10 & 10 \\ 10 & 10 \end{bmatrix} = \begin{bmatrix} 22 & 18 \\ 18 & 30 \end{bmatrix}$$

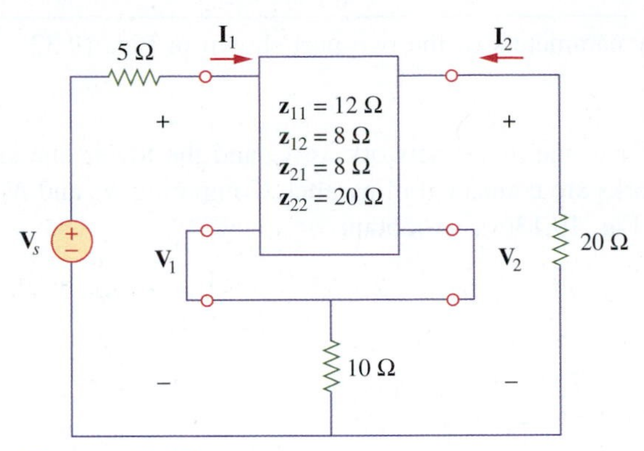

Figure 19.30
For Example 19.8.

But

$$\mathbf{V}_1 = \mathbf{z}_{11}\mathbf{I}_1 + \mathbf{z}_{12}\mathbf{I}_2 = 22\mathbf{I}_1 + 18\mathbf{I}_2 \quad (19.8.1)$$

$$\mathbf{V}_2 = \mathbf{z}_{21}\mathbf{I}_1 + \mathbf{z}_{22}\mathbf{I}_2 = 18\mathbf{I}_1 + 30\mathbf{I}_2 \quad (19.8.2)$$

Also at the input port

$$\mathbf{V}_1 = \mathbf{V}_s - 5\mathbf{I}_1 \quad (19.8.3)$$

and at the output port

$$\mathbf{V}_2 = -20\mathbf{I}_2 \quad \Rightarrow \quad \mathbf{I}_2 = -\mathbf{V}_2/20 \quad (19.8.4)$$

Substituting Eqs. (19.8.3) and (19.8.4) into Eq. (19.8.1) gives

$$\mathbf{V}_s - 5\mathbf{I}_1 = 22\mathbf{I}_1 - \frac{18}{20}\mathbf{V}_2 \quad \Rightarrow \quad \mathbf{V}_s = 27\mathbf{I}_1 - 0.9\,\mathbf{V}_2 \quad (19.8.5)$$

while substituting Eq. (19.8.4) into Eq. (19.8.2) yields

$$\mathbf{V}_2 = 18\mathbf{I}_1 - \frac{30}{20}\mathbf{V}_2 \quad \Rightarrow \quad \mathbf{I}_1 = \frac{2.5}{18}\mathbf{V}_2 \quad (19.8.6)$$

Substituting Eq. (19.8.6) into Eq. (19.8.5), we get

$$\mathbf{V}_s = 27 \times \frac{2.5}{18}\mathbf{V}_2 - 0.9\mathbf{V}_2 = 2.85\mathbf{V}_2$$

Thus,

$$\frac{\mathbf{V}_2}{\mathbf{V}_s} = \frac{1}{2.85} = 0.3509$$

Practice Problem 19.8

Find $\mathbf{V}_2/\mathbf{V}_s$ in the circuit of Fig. 19.31.

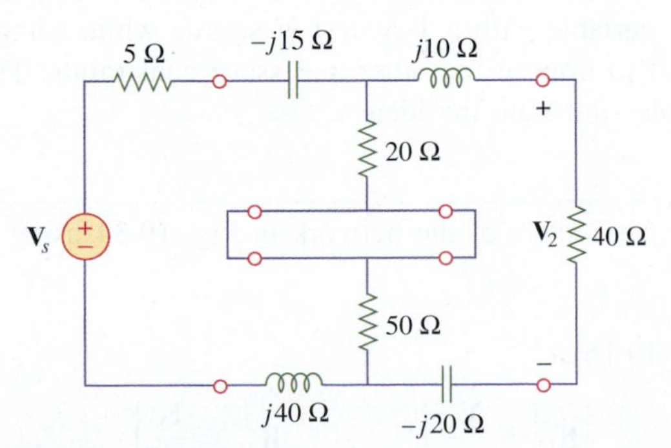

Figure 19.31
For Practice Problem 19.8.

Answer: $0.58\,\underline{/40^\circ}$

Example 19.9

Find the y parameters of the two-port shown in Fig. 19.32.

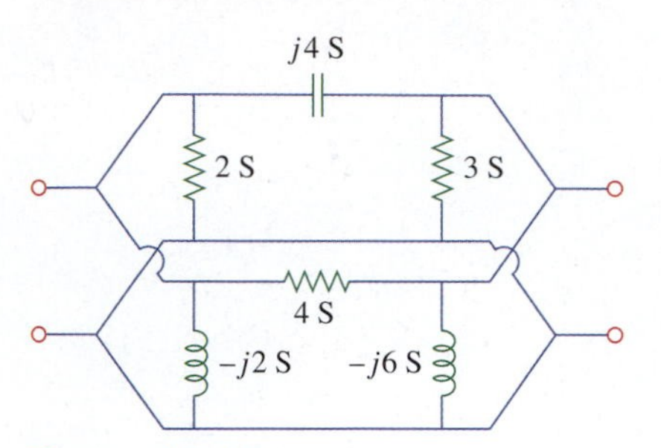

Figure 19.32
For Example 19.9.

Solution:
Let us refer to the upper network as N_a and the lower one as N_b. The two networks are connected in parallel. Comparing N_a and N_b with the circuit in Fig. 19.13(a), we obtain

$$\mathbf{y}_{12a} = -j4 = \mathbf{y}_{21a}, \qquad \mathbf{y}_{11a} = 2 + j4, \qquad \mathbf{y}_{22a} = 3 + j4$$

or

$$[\mathbf{y}_a] = \begin{bmatrix} 2+j4 & -j4 \\ -j4 & 3+j4 \end{bmatrix} \text{S}$$

$$\mathbf{y}_{12b} = -4 = \mathbf{y}_{21b}, \qquad \mathbf{y}_{11b} = 4 - j2, \qquad \mathbf{y}_{22b} = 4 - j6$$

or

$$[\mathbf{y}_b] = \begin{bmatrix} 4-j2 & -4 \\ -4 & 4-j6 \end{bmatrix} \text{S}$$

The overall y parameters are

$$[\mathbf{y}] = [\mathbf{y}_a] + [\mathbf{y}_b] = \begin{bmatrix} 6+j2 & -4-j4 \\ -4-j4 & 7-j2 \end{bmatrix} \text{S}$$

Practice Problem 19.9

Obtain the y parameters for the network in Fig. 19.33.

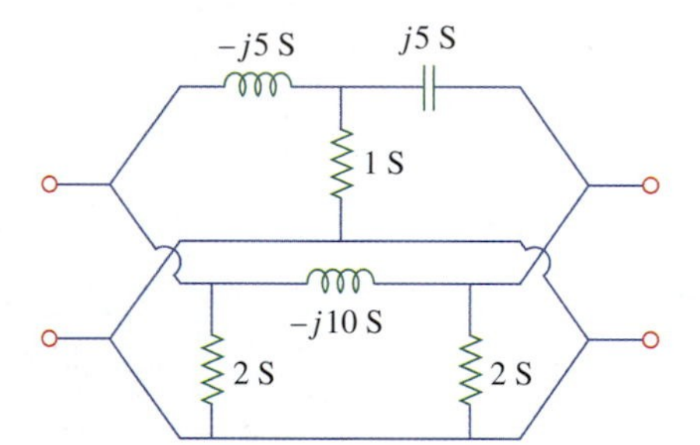

Figure 19.33
For Practice Problem 19.9.

Answer: $\begin{bmatrix} 27-j15 & -25+j10 \\ -25+j10 & 27-j5 \end{bmatrix} \text{S}$

19.7 Computer Analysis

Hand calculation of the two-port parameters may become difficult when the two-port is complicated. We resort to PSpice in such situations. If the circuit is purely resistive, PSpice dc analysis may be used; otherwise, PSpice ac analysis is required at a specific frequency. The key to using PSpice in computing a particular two-port parameter is to remember how that parameter is defined and to constrain the appropriate port variable with a 1-A or 1-V source while using an open or short-circuit to impose the other necessary constraints. The following two examples illustrate the idea.

Example 19.10

Find the h parameters of the network in Fig. 19.34 using PSpice.

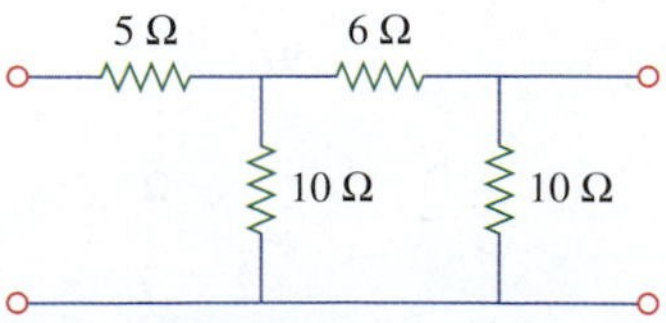

Figure 19.34
For Example 19.10.

Solution:
From Eq. (19.16),

$$\mathbf{h}_{11} = \left.\frac{\mathbf{V}_1}{\mathbf{I}_1}\right|_{\mathbf{V}_2=0}, \qquad \mathbf{h}_{21} = \left.\frac{\mathbf{I}_2}{\mathbf{I}_1}\right|_{\mathbf{V}_2=0}$$

showing that $\mathbf{h}_{11}$ and $\mathbf{h}_{21}$ can be found by setting $\mathbf{V}_2 = 0$. Also by setting $\mathbf{I}_1 = 1$, A, $\mathbf{h}_{11}$ becomes $\mathbf{V}_1/1$, while $\mathbf{h}_{21}$ becomes $\mathbf{I}_2/1$. With this in mind, we draw the schematic in Fig. 19.35(a). We insert a 1-A dc current source IDC to take care of $\mathbf{I}_1 = 1$ A. Because the output is supposed to be short-circuited, we can use a tiny resistance $\mathbf{R}_5$ as the short-circuit.

After drawing the circuit as in Fig. 19.35(a) and saving as exam1910a.dsn, we select **PSpice/New Simulation Profile**. This leads to the New Simulation dialog box. Type "exam1910a" as the name of the file and click Create. This leads to the Simulation Settings dialog box. In the Simulation Settings dialog box, select Bias Point as *Analysis Type* and click OK. We simulate the circuit by selecting **PSpice/Run**. We can show the voltage on the nodes or currents through the branches by pressing on icon **I** or **V** (below Help). By pressing **V**, we see on display the voltage across $\mathbf{R}_3$ as 3.75 so that the $\mathbf{V}_1$ = voltage across $\mathbf{I}_1$ is $3.75 + 5(1) = 8.75$ V. By pressing **I**, we see on display the current through $\mathbf{R}_5$ as 624.9 mA.

$$\mathbf{h}_{11} = \frac{\mathbf{V}_1}{1} = 8.75\ \Omega, \qquad \mathbf{h}_{21} = \frac{\mathbf{I}_2}{1} = 0.625$$

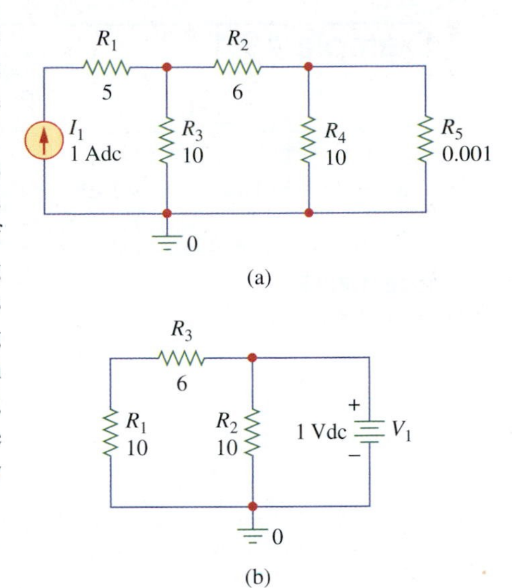

Figure 19.35
Schmematics for Example 19.10: (a) computing $\mathbf{h}_{11}$ and $\mathbf{h}_{21}$; (b) computing $\mathbf{h}_{12}$ and $\mathbf{h}_{22}$.

Similarly, from Eq. (19.16),

$$\mathbf{h}_{12} = \left.\frac{\mathbf{V}_1}{\mathbf{V}_2}\right|_{\mathbf{I}_1=0}, \qquad \mathbf{h}_{22} = \left.\frac{\mathbf{I}_2}{\mathbf{V}_2}\right|_{\mathbf{I}_1=0}$$

indicating that we obtain $\mathbf{h}_{12}$ and $\mathbf{h}_{22}$ by open-circuiting the input port ($\mathbf{I}_1 = 0$). By making $\mathbf{V}_2 = 1$ V, $\mathbf{h}_{12}$ becomes $\mathbf{V}_1/1$, while $\mathbf{h}_{22}$ becomes $\mathbf{I}_2/1$. Thus, we use the schematic in Fig. 19.35(b) with a 1-V dc voltage source VDC inserted at the output terminal to take care of $\mathbf{V}_2 = 1$ V. [Notice that in Fig. 19.35(b), the 5-Ω resistor is ignored because the input port is open-circuited and PSpice will not allow the 5-ohm resistor to be left hanging. We may include the 5-Ω resistor if we replace the open circuit with a very large resistor, say, 10 MΩ.]

After drawing the circuit as in Fig. 19.35(b) and saving as exam1910b.dsn, we select **PSpice/New Simulation Profile**. This leads to the New Simulation dialog box. Type "exam1910b" as the name of the file and click Create. This leads to the Simulation Settings dialog box. In the Simulation Settings dialog box, select Bias Point as *Analysis Type* and click OK. We simulate the circuit by selecting **PSpice/Run**. We can show the voltage on the nodes or currents through the branches by pressing on icon **I** or **V** (below Help). By pressing **V**, we see on display the voltage across $\mathbf{R}_1$ as 625 mV so that $\mathbf{V}_1 = 0.625$ V. By pressing **I**, we see on display the current through $\mathbf{V}_1$ as $\mathbf{I}_2 = 162.5$ mA.

$$\mathbf{h}_{12} = \frac{\mathbf{V}_1}{1} = 0.625, \qquad \mathbf{h}_{22} = \frac{\mathbf{I}_2}{1} = 0.1625\ \text{S}$$

Practice Problem 19.10

Obtain the h parameters for the network in Fig. 19.36 using PSpice.

Answer: $[\mathbf{h}] = \begin{bmatrix} 5.899\ \Omega & 0.1111 \\ 0.1111 & 0.1111\ \text{S} \end{bmatrix}$

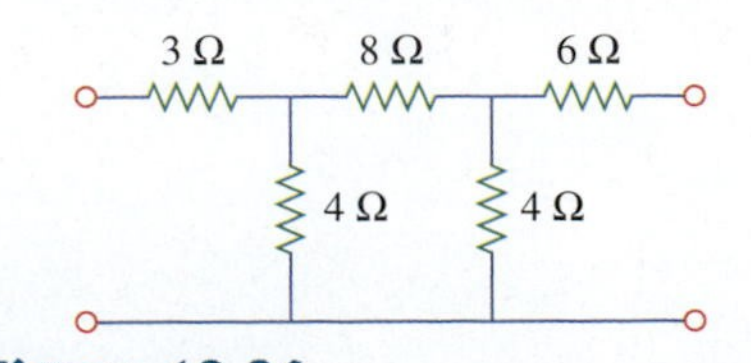

Figure 19.36
For Practice Problem 19.10.

Example 19.11

Find the z parameters for the circuit in Fig. 19.37 at $\omega = 10^6$ rad/s.

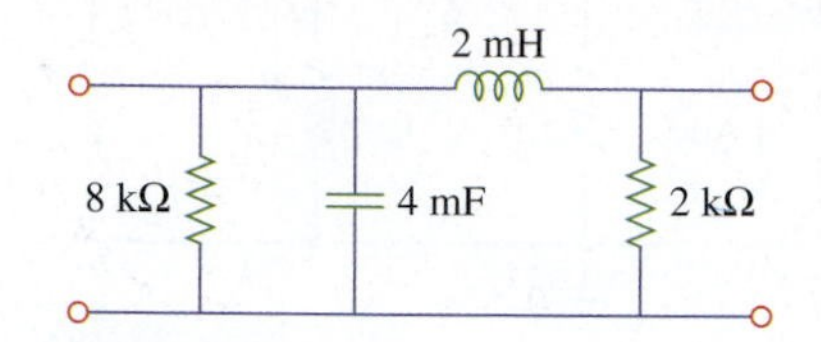

Figure 19.37
For Example 19.11.

Solution:

Notice that dc analysis was used in Example 19.10 because the circuit in Fig. 19.34 is purely resistive. Here, ac analysis at $f = \omega/2\pi = 0.15915$ MHz is used because L and C are frequency dependent.

In Eq. (19.3), we defined the z parameters as

$$\mathbf{z}_{11} = \frac{\mathbf{V}_1}{\mathbf{I}_1}\bigg|_{\mathbf{I}_2=0}, \qquad \mathbf{z}_{21} = \frac{\mathbf{V}_2}{\mathbf{I}_1}\bigg|_{\mathbf{I}_2=0}$$

This suggests that if we let $\mathbf{I}_1 = 1$ A and open-circuit the output port so that $\mathbf{I}_2 = 0$, then we obtain

$$\mathbf{z}_{11} = \frac{\mathbf{V}_1}{1}, \qquad \mathbf{z}_{21} = \frac{\mathbf{V}_2}{1}$$

We realize this with the schematic in Fig. 19.38(a). We insert a 1-A ac current source IAC at the input terminal of the circuit and two VPRINT1 pseudocomponents to obtain $\mathbf{V}_1$ and $\mathbf{V}_2$. The attributes of each VPRINT1 are set as AC = yes, MAG = yes, and PHASE = yes to print the magnitude and phase values of the voltages.

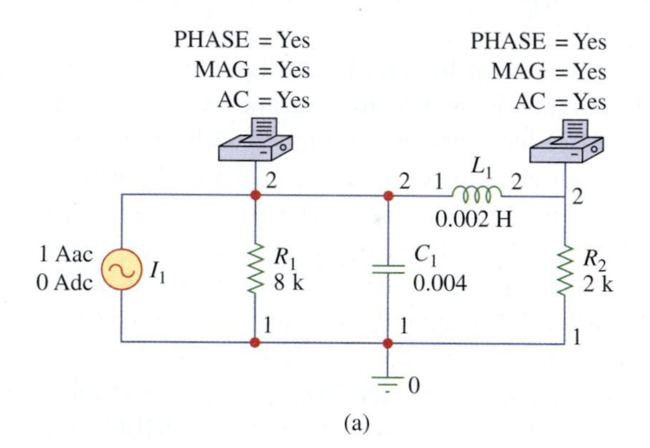

(a)

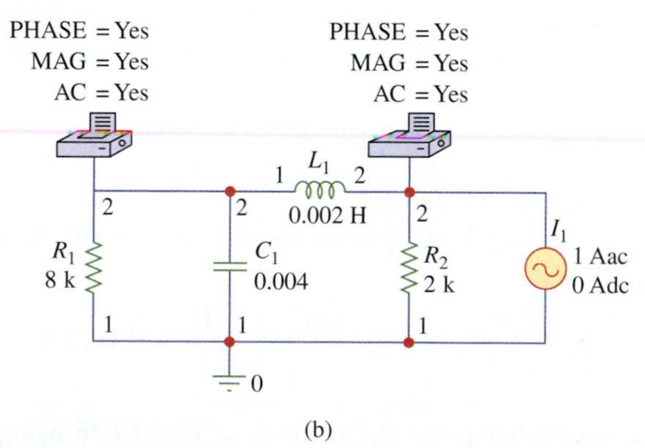

(b)

Figure 19.38
Schematics for Example 19.11: (a) computing $\mathbf{z}_{11}$ and $\mathbf{z}_{21}$; (b) computing $\mathbf{z}_{12}$ and $\mathbf{z}_{22}$.

After drawing the circuit as in Fig. 19.38(a) and saving as exam1911a.dsn, we select **PSpice/New Simulation Profile**. This leads to the New Simulation dialog box. Type "exam1911a" as the name of the file and click Create. This leads to the Simulation Settings dialog box. In the Simulation Settings dialog box, select AC Sweep/Noise under *Analysis Type*, select Linear under *AC Sweep Type.* We type 0.1519MEG as the *Start Freq,* 0.1519MEG as the *Final Freq*, and 1 as *Total Points.*

We simulate the circuit by selecting **PSpice/Run**. We obtain the output file by selecting **PSpice/View Output file**. The output file includes the following:

```
FREQ            VM(N00139)      VP(N00139)
1.519E+05       2.619E-04       9.000E+01
FREQ            VM(N00159)      VP(N00159)
1.519E+05       1.895E-04       4.634E+01
```

From this, we obtain $\mathbf{V}_1$ and $\mathbf{V}_2$. Thus,

$$\mathbf{z}_{11} = \frac{\mathbf{V}_1}{1} = 261.9\underline{/90^\circ}\ \mu\Omega$$

$$\mathbf{z}_{21} = \frac{\mathbf{V}_2}{1} = 189.5\underline{/46.34^\circ}\ \mu\Omega$$

In a similar manner, from Eq. (19.3),

$$\mathbf{z}_{12} = \left.\frac{\mathbf{V}_1}{\mathbf{I}_2}\right|_{\mathbf{I}_1=0}, \qquad \mathbf{z}_{22} = \left.\frac{\mathbf{V}_2}{\mathbf{I}_2}\right|_{\mathbf{I}_1=0}$$

suggesting that if we let $\mathbf{I}_2 = 1$ A and open-circuit the input port, $\mathbf{z}_{12} = \mathbf{V}_1/1$ and $\mathbf{z}_{22} = \mathbf{V}_2/1$. This leads to the schematic in Fig. 19.38(b). The only difference between this schematic and the one in Fig. 19.38(a) is that the 1-A ac current source IAC is now at the output terminal. We run the schematic in Fig. 19.38(b) and obtain the output file, which includes:

```
FREQ            VM(N00163)      VP(N00163)
1.519E+05       1.895E-04       4.634E+01
FREQ            VM(N00139)      VP(N00139)
1.519E+05       1.381E+03       -1.337E+02
```

Thus,

$$\mathbf{z}_{12} = \frac{\mathbf{V}_1}{1} = 189.5\underline{/46.34^\circ}\ \mu\Omega$$

$$\mathbf{z}_{22} = \frac{\mathbf{V}_2}{1} = 1381\underline{/-133.7^\circ}\ \Omega$$

Practice Problem 19.11

Obtain the z parameters of the circuit in Fig. 19.39 at $f = 60$ Hz using PSpice.

Answer: $[\mathbf{z}] = \begin{bmatrix} 11.89\underline{/-177.2^\circ} & 0.2629\underline{/99.07^\circ} \\ 0.2629\underline{/99.07^\circ} & 0.2662\underline{/90.03^\circ} \end{bmatrix}\ \Omega$

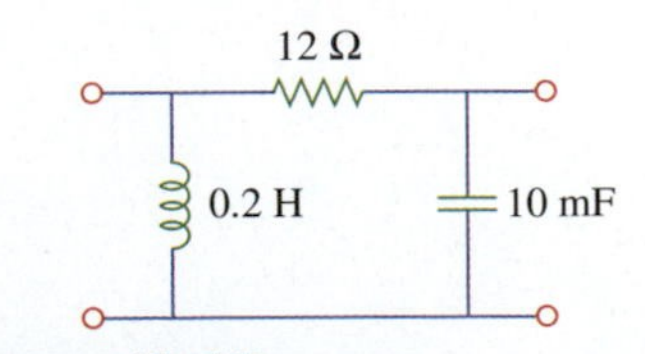

Figure 19.39
For Practice Problem 19.11.

19.8 †Applications

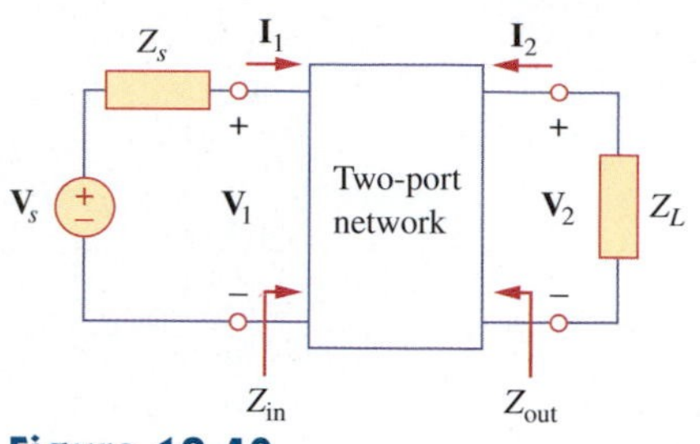

Figure 19.40
Two-port network isolating source and load.

We have seen how the three sets of network parameters can be used to characterize a wide range of two-port networks. Depending on how two-ports are interconnected to form a larger network, a particular set of parameters may have advantages over others, as we noticed in Section 19.5. In this section, we will consider an important application area of two-port parameters: transistor circuits.

The two-port network is often used to isolate a load from the excitation of a circuit. For example, the two-port in Fig. 19.40 may represent an amplifier, a filter, or some other network. When the two-port represents an amplifier, expressions for the voltage gain A_v, the current gain A_i, the input impedance Z_{in}, and the output impedance Z_{out} can be derived with ease. They are defined as follows:

$$A_v = \frac{\mathbf{V}_2(s)}{\mathbf{V}_1(s)} \tag{19.39}$$

$$A_i = \frac{\mathbf{I}_2(s)}{\mathbf{I}_1(s)} \tag{19.40}$$

$$Z_{in} = \frac{\mathbf{V}_1(s)}{\mathbf{I}_1(s)} \tag{19.41}$$

$$Z_{out} = \left.\frac{\mathbf{V}_2(s)}{\mathbf{I}_2(s)}\right|_{V_s=0} \tag{19.42}$$

Any of the three sets of two-port parameters can be used to derive the expressions in Eqs. (19.39) through (19.42). Here, we will specifically use the hybrid parameters to obtain them for transistor amplifiers.

The hybrid (h) parameters are the most useful for transistors because they are easily measured and are often provided in the manufacturer's data or spec sheets for transistors. The h parameters provide a quick estimate of the performance of transistor circuits. They are used for finding the exact voltage gain, input impedance, and output impedance of a transistor.

The h parameters for transistors have specific meanings expressed by their subscripts. They are listed by the first subscript and related to the general h parameters as follows:

$$h_i = h_{11}, \qquad h_r = h_{12}, \qquad h_f = h_{21}, \qquad h_o = h_{22} \tag{19.43}$$

where the subscripts i, r, f, and o stand for input, reverse, forward, and output, respectively. The second subscript specifies the type of transistor connection used: e for common emitter (CE), c for common collector (CC), and b for common base (CB). Here, we are mainly concerned with the common emitter connection. Thus, the four h parameters for the common-emitter amplifier are:

$$\begin{aligned} h_{ie} &= \text{base input impedance} \\ h_{re} &= \text{reverse voltage feedback ratio} \\ h_{fe} &= \text{base-collector current gain} \\ h_{oe} &= \text{output admittace} \end{aligned} \tag{19.44}$$

These are calculated or measured in the same way as the general h parameters. Typical values are $h_{ie} = 6\ \text{k}\Omega$, $h_{re} = 1.5 \times 10^{-4}$, $h_{fe} = 200$,

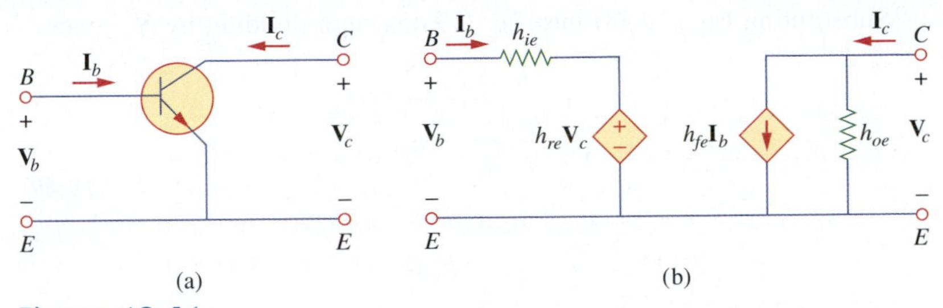

Figure 19.41
Common emitter amplifier: (a) circuit schematic, (b) hybrid model.

and $h_{oe} = 8\ \mu$S. We must keep in mind that these values represent ac characteristics of the transistor, measured under specific circumstances.

The circuit schematic for the common emitter amplifier and the equivalent hybrid model are shown in Fig. 19.41. From Fig. 19.41, we see that

$$\mathbf{V}_b = h_{ie}\mathbf{I}_b + h_{re}\mathbf{V}_c \tag{19.45a}$$

$$\mathbf{I}_c = h_{fe}\mathbf{I}_b + h_{oe}\mathbf{V}_c \tag{19.45b}$$

Consider the transistor amplifier connected to an ac source and a load as shown in Fig. 19.42. This is an example of a two-port network embedded within a larger network. We can analyze the hybrid equivalent circuit as usual with Eq. (19.45) in mind. Recognizing from Fig. 19.42 that $\mathbf{V}_c = -R_L\mathbf{I}_c$ and substituting this into Eq. (19.45b) gives

$$\mathbf{I}_c = h_{fe}\mathbf{I}_b - h_{oe}R_L\mathbf{I}_c$$

or

$$(1 + h_{oe}R_L)\mathbf{I}_c = h_{fe}\mathbf{I}_b \tag{19.46}$$

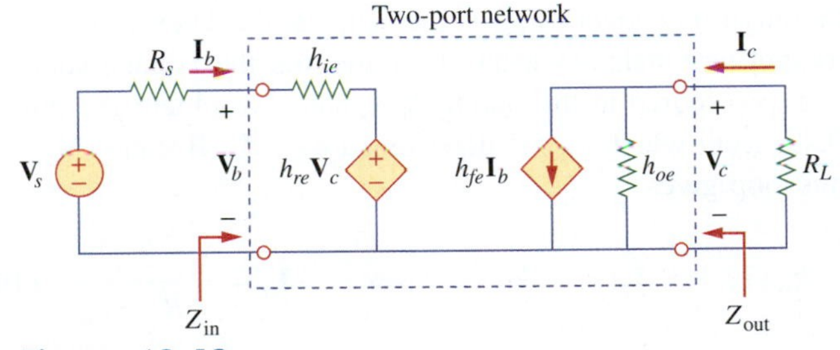

Figure 19.42
Transistor amplifier with source and load resistance.

From this, we obtain the current gain as

$$\boxed{A_i = \frac{\mathbf{I}_c}{\mathbf{I}_b} = \frac{h_{fe}}{1 + h_{oe}R_L}} \tag{19.47}$$

From Eqs. (19.45b) and (19.46), we can express $\mathbf{I}_b$ in terms of $\mathbf{V}_c$:

$$\mathbf{I}_c = \frac{h_{fe}}{1 + h_{oe}R_L}\mathbf{I}_b = h_{fe}\mathbf{I}_b + h_{oe}\mathbf{V}_c$$

or

$$\mathbf{I}_b = \frac{h_{oe}}{\dfrac{h_{fe}}{1 + h_{oe}R_L} - h_{fe}}\mathbf{V}_c \tag{19.48}$$

Substituting Eq. (19.48) into Eq. (19.45a) and dividing by $\mathbf{V}_c$ gives

$$\frac{\mathbf{V}_b}{\mathbf{V}_c} = \frac{h_{oe}h_{ie}}{\dfrac{h_{fe}}{1 + h_{oe}R_L} - h_{fe}} + h_{re} = \frac{h_{ie} + h_{ie}h_{oe}R_L - h_{re}h_{fe}R_L}{-h_{fe}R_L} \tag{19.49}$$

Thus, the voltage gain is

$$A_v = \frac{\mathbf{V}_c}{\mathbf{V}_b} = \frac{-h_{fe}R_L}{h_{ie} + (h_{ie}h_{oe} - h_{re}h_{fe})R_L} \tag{19.50}$$

Substituting $\mathbf{V}_c = -R_L\mathbf{I}_c$ into Eq. (19.45a) gives

$$\mathbf{V}_b = h_{ie}\mathbf{I}_b - h_{re}R_L\mathbf{I}_c$$

or

$$\frac{\mathbf{V}_b}{\mathbf{I}_b} = h_{ie} - h_{re}R_L\frac{\mathbf{I}_c}{\mathbf{I}_b} \tag{19.51}$$

Replacing $\mathbf{I}_c/\mathbf{I}_b$ by the current gain in Eq. (19.47) yields the input impedance as

$$Z_{\text{in}} = \frac{\mathbf{V}_b}{\mathbf{I}_b} = h_{ie} - \frac{h_{re}h_{fe}R_L}{1 + h_{oe}R_L} \tag{19.52}$$

The output impedance Z_{out} is the same as the Thevenin equivalent at the output terminals. As usual, by removing the voltage source and placing a 1-V source at the output terminals, we obtain the circuit in Fig. 19.43, from which Z_{out} is determined as $1/\mathbf{I}_c$. Because $\mathbf{V}_c = 1$ V, the input loop gives

$$h_{re}(1) = -\mathbf{I}_b(R_s + h_{ie}) \qquad \Rightarrow \qquad \mathbf{I}_b = -\frac{h_{re}}{R_s + h_{ie}} \tag{19.53}$$

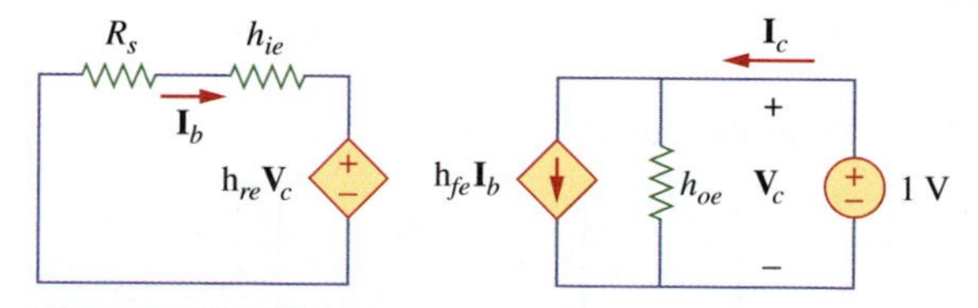

Figure 19.43
Finding the output impedance of the amplifier circuit in Fig. 19.42.

For the output loop,

$$\mathbf{I}_c = h_{oe}(1) + h_{fe}\mathbf{I}_b \tag{19.54}$$

Substituting Eq. (19.53) into Eq. (19.54) gives

$$\mathbf{I}_c = \frac{(R_s + h_{ie})h_{oe} - h_{re}h_{fe}}{R_s + h_{ie}} \tag{19.55}$$

From this, we obtain the output impedance Z_{out} as $1/\mathbf{I}_c$, that is,

$$Z_{out} = \frac{R_s + h_{ie}}{(R_s + h_{ie})h_{oe} - h_{re}h_{fe}} \tag{19.56}$$

Example 19.12

Consider the common-emitter amplifier circuit of Fig. 19.44.
(a) Determine the voltage gain, current gain, input impedance, and output impedance using these h parameters:

$$h_{ie} = 1\ \text{k}\Omega, \qquad h_{re} = 2.5 \times 10^{-4}, \qquad h_{fe} = 50, \qquad h_{oe} = 20\ \mu\text{S}$$

(b) Find the output voltage $\mathbf{V}_o$.

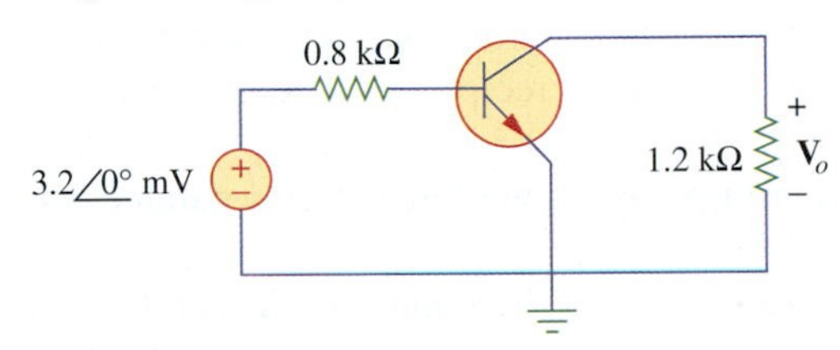

Figure 19.44
For Example 19.12.

Solution:
(a) We note that $R_s = 0.8\ \text{k}\Omega$ and $R_L = 1.2\ \text{k}\Omega$. We treat the transistor of Fig. 19.44 as a two-port network and apply Eqs. (19.47) through (19.56).

$$h_{ie}h_{oe} - h_{re}h_{fe} = 10^3 \times 20 \times 10^{-6} - 2.5 \times 10^{-4} \times 50 = 7.5 \times 10^{-3}$$

$$A_v = \frac{-h_{fe}R_L}{h_{ie} + (h_{ie}h_{oe} - h_{re}h_{fe})R_L} = \frac{-50 \times 1{,}200}{1{,}000 + 7.5 \times 10^{-3} \times 1{,}200} = -59.46$$

$$A_i = \frac{h_{fe}}{1 + h_{oe}R_L} = \frac{50}{1 + 20 \times 10^{-6} \times 1{,}200} = 48.83$$

$$Z_{in} = h_{ie} - \frac{h_{re}h_{fe}R_L}{1 + h_{oe}R_L} = 1{,}000 - \frac{2.5 \times 10^{-4} \times 50 \times 1{,}200}{1 + 20 \times 10^{-6} \times 1{,}200} = 985.4\ \Omega$$

$$(R_s + h_{ie})h_{oe} - h_{re}h_{fe} = (800 + 1{,}000) \times 20 \times 10^{-6} - 2.5 \times 10^{-4} \times 50 = 23.5 \times 10^{-3}$$

$$Z_{out} = \frac{R_s + h_{ie}}{(R_s + h_{ie})h_{oe} - h_{re}h_{fe}} = \frac{800 + 1{,}000}{23.5 \times 10^{-3}} = 76.6\ \text{k}\Omega$$

(b) The output voltage is

$$\mathbf{V}_o = A_V\mathbf{V}_s = -59.46(3.2\angle 0^\circ)\ \text{mV} = 0.19\angle 180^\circ\ \text{V}$$

Practice Problem 19.12

For the transistor amplifier of Fig. 19.45, find the voltage gain, current gain, input impedance, and output impedance. Assume that:

$$h_{ie} = 6\ \text{k}\Omega, \qquad h_{re} = 1.5 \times 10^{-4}, \qquad h_{fe} = 200, \qquad h_{oe} = 8\ \mu\text{S}$$

Answer: -123.61,194.17; 6 kΩ; 128.08 kΩ

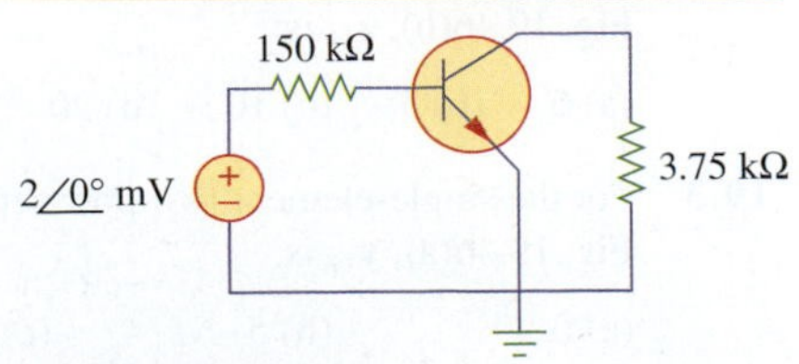

Figure 19.45
For Practice Problem 19.12.

19.9 Summary

1. A two-port network is one with two ports (or two pairs of access terminals) known as input and output ports.
2. The three parameters used to model a two-port network are the impedance [z], admittance [y], and hybrid [h].
3. The parameters relate the input and output port variables as:

$$\begin{bmatrix} \mathbf{V}_1 \\ \mathbf{V}_2 \end{bmatrix} = [\mathbf{z}] \begin{bmatrix} \mathbf{I}_1 \\ \mathbf{I}_2 \end{bmatrix}, \quad \begin{bmatrix} \mathbf{I}_1 \\ \mathbf{I}_2 \end{bmatrix} = [\mathbf{y}] \begin{bmatrix} \mathbf{V}_1 \\ \mathbf{V}_2 \end{bmatrix}, \quad \begin{bmatrix} \mathbf{V}_1 \\ \mathbf{I}_2 \end{bmatrix} = [\mathbf{h}] \begin{bmatrix} \mathbf{I}_1 \\ \mathbf{V}_2 \end{bmatrix}$$

4. The parameters can be calculated or measured by short-circuiting or open-circuiting the appropriate input or output port.
5. A two-port network is reciprocal if $\mathbf{z}_{12} = \mathbf{z}_{21}$, $\mathbf{y}_{12} = \mathbf{y}_{21}$, and $\mathbf{h}_{12} = -\mathbf{h}_{21}$.
6. The relationships among the three sets of parameters are provided in Table 19.1.
7. Two-port networks may be connected in series or in parallel. In the series connection, the z parameters are added; in the parallel connection the y parameters are added.
8. PSpice can be used to compute the two-port parameters by constraining the appropriate port variables with a 1-A or 1-V source while using an open- or short-circuit to impose the other necessary constraints.
9. The hybrid (h) parameters are specifically applied in the analysis of transistor circuits.

Review Questions

19.1 For the single-element two-port network in Fig. 19.46(a), $\mathbf{z}_{11}$ is:

(a) 0 (b) 5 (c) 10 (d) 20 (e) nonexistent.

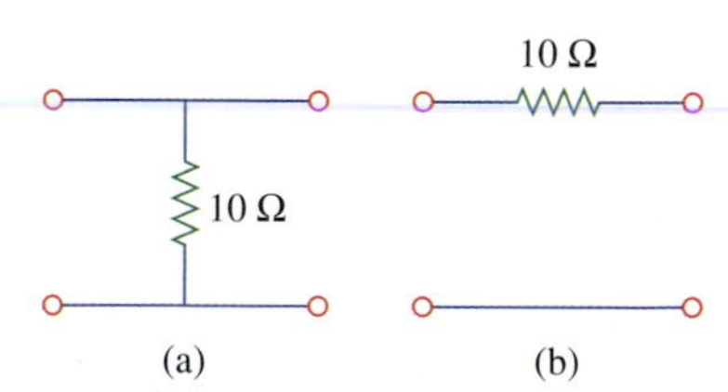

Figure 19.46
For Review Questions.

19.2 For the single-element two-port network in Fig. 19.46(b), $\mathbf{z}_{11}$ is:

(a) 0 (b) 5 (c) 10 (d) 20 (e) nonexistent.

19.3 For the single-element two-port network in Fig. 19.46(a), $\mathbf{y}_{11}$ is:

(a) 0 (b) 5 (c) 10
(d) 20 (e) nonexistent.

19.4 For the single-element two-port network in Fig. 19.46(b), $\mathbf{h}_{21}$ is:

(a) −0.1 (b) −1 (c) 0
(d) 10 (e) nonexistent.

19.5 When port 1 of a two-port circuit is short-circuited, $\mathbf{I}_1 = 4\mathbf{I}_2$ and $\mathbf{V}_2 = 0.25\mathbf{I}_2$. Which of the following is true?

(a) $\mathbf{y}_{11} = 4$ (b) $\mathbf{y}_{12} = 16$
(c) $\mathbf{y}_{21} = 16$ (d) $\mathbf{y}_{22} = 0.25$

19.6 Which of the following is *not* true? A two-port is described by the following equations:

$$\mathbf{V}_1 = 50\mathbf{I}_1 + 10\mathbf{I}_2$$
$$\mathbf{V}_2 = 30\mathbf{I}_1 + 20\mathbf{I}_2$$

(a) $\mathbf{z}_{12} = 10$ (b) $\mathbf{y}_{12} = -0.0143$
(c) $\mathbf{h}_{12} = 0.5$ (d) $\mathbf{h}_{21} = 0.5$

19.7 If a two-port is reciprocal, which of the following is *not* true?

(a) $\mathbf{z}_{21} = \mathbf{z}_{12}$ (b) $\mathbf{y}_{21} = \mathbf{y}_{12}$
(c) $\mathbf{h}_{21} = \mathbf{h}_{12}$ (d) $\mathbf{h}_{21} = -\mathbf{h}_{12}$

19.8 Given the following z parameters:

$$[\mathbf{z}] = \begin{bmatrix} 4 & 3 \\ 2 & 5 \end{bmatrix} \Omega$$

the corresponding $\mathbf{y}_{11}$ is:

(a) $-3/14$ (b) $-2/14$ (c) $4/14$ (d) $5/14$

19.9 For the z parameters in the previous problem, the corresponding $\mathbf{h}_{11}$ is:

(a) $14/5$ (b) $4/5$ (c) $3/5$ (d) $1/5$

19.10 If the two single-element two-port networks in Fig. 19.46 are connected in series, then $\mathbf{z}_{11}$ is:

(a) 0 (b) 0.1 (c) 2
(d) 10 (e) nonexistent.

Answers: 19.1c, 19.2e, 19.3e, 19.4b, 19.5b, 19.6d, 19.7c, 19.8d, 19.9a, 19.10e

Problems

Section 19.2 Impedance Parameters

19.1 Obtain the z parameters for the network in Fig. 19.47.

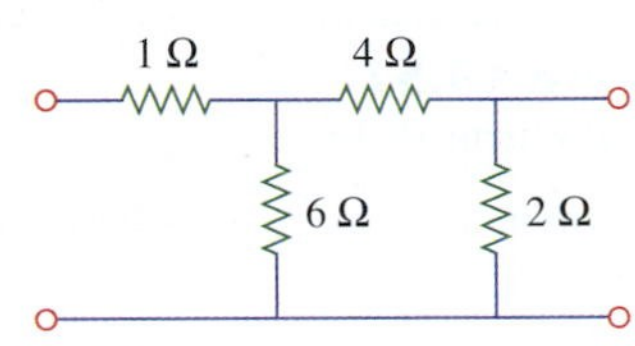

Figure 19.47
For Problem 19.1.

19.2 Determine the z parameters of the two-port network shown in Fig. 19.48.

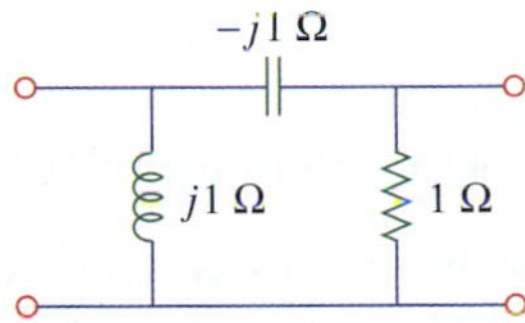

Figure 19.48
For Problem 19.2.

19.3 Calculate the z parameters for the circuit in Fig. 19.49.

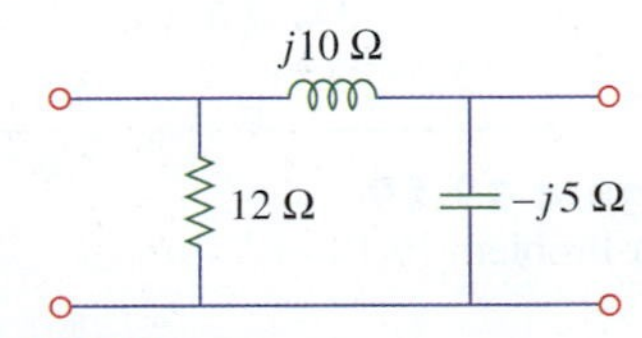

Figure 19.49
For Problem 19.3.

19.4 Obtain the z parameters for the network in Fig. 19.50 as functions of s.

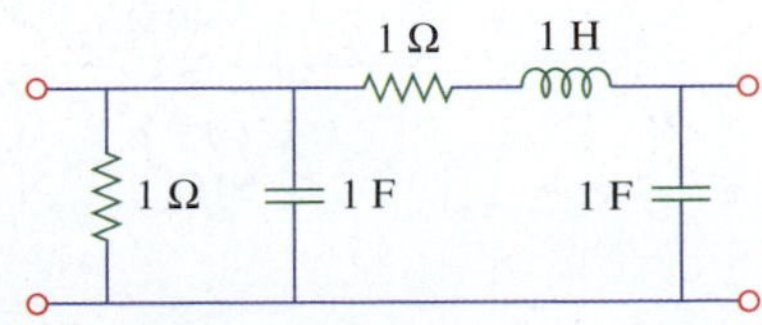

Figure 19.50
For Problem 19.4.

19.5 Design a two-port network that realizes the following z parameters:

$$[\mathbf{z}] = \begin{bmatrix} 10 & 4 \\ 4 & 6 \end{bmatrix} \Omega$$

19.6 Determine the two-port network that is represented by the following z parameters,

$$[\mathbf{z}] = \begin{bmatrix} 6 + j3 & 5 - j2 \\ 5 - j2 & 8 - j \end{bmatrix} \Omega$$

19.7 For the two-port network shown in Fig. 19.51,

$$[\mathbf{z}] = \begin{bmatrix} 40 & 60 \\ 80 & 120 \end{bmatrix} \Omega$$

(a) Find $\mathbf{Z}_L$ for the maximum power transfer.

(b) Calculate the maximum power delivered to the load.

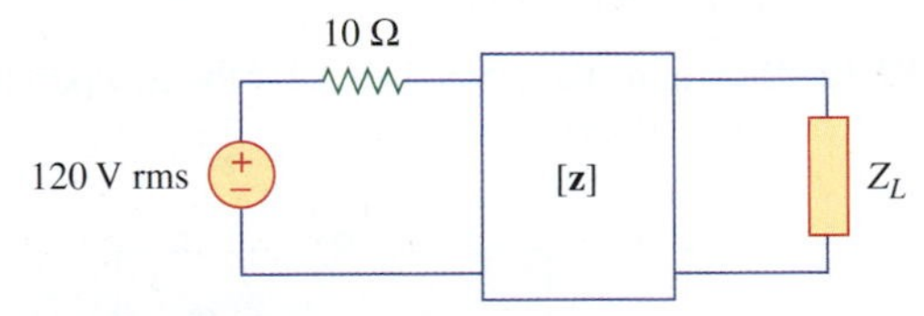

Figure 19.51
For Problem 19.7.

19.8 Find the z parameters for the circuit in Fig. 19.52.

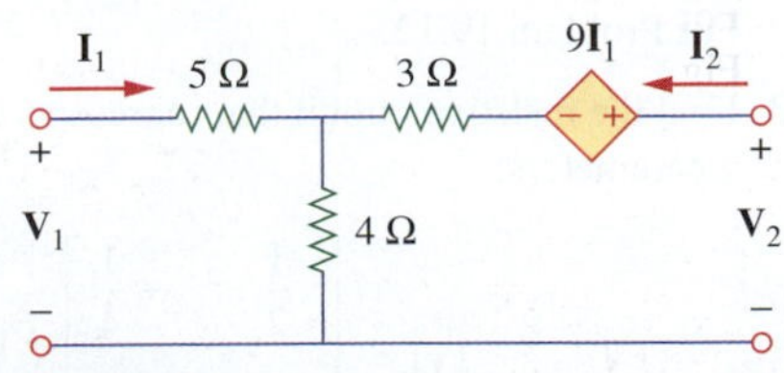

Figure 19.52
For Problem 19.8.

19.9 Determine the z parameter for the two-port circuit in Fig. 19.53.

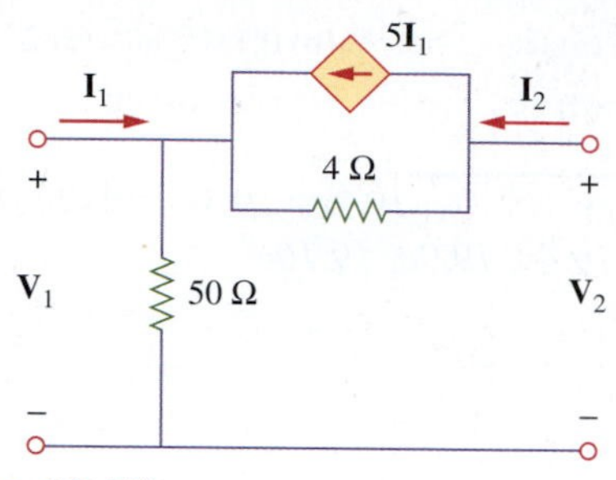

Figure 19.53
For Problem 19.9.

Section 19.3 Admittance Parameters

19.10 Calculate the y parameters for the two-port network in Fig. 19.54.

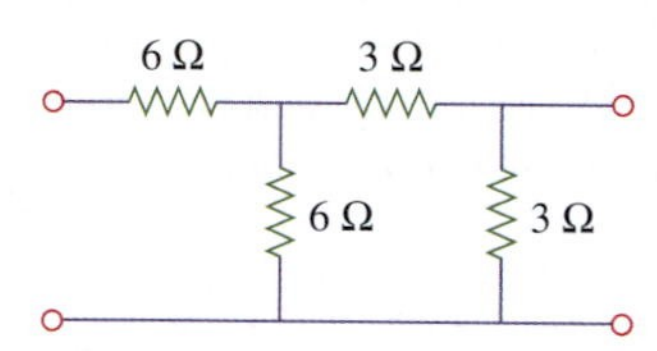

Figure 19.54
For Problem 19.10.

19.11 Find the y parameters of the two-port network in Fig. 19.55 in terms of s.

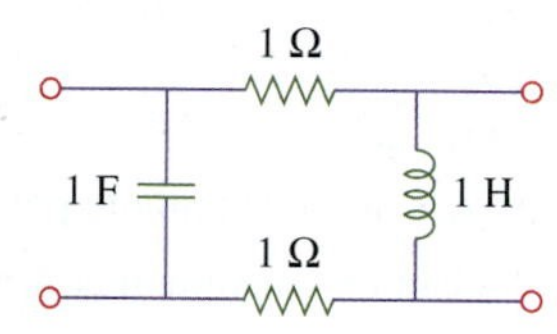

Figure 19.55
For Problem 19.11.

19.12 Determine the y parameters for the two-port network in Fig. 19.56.

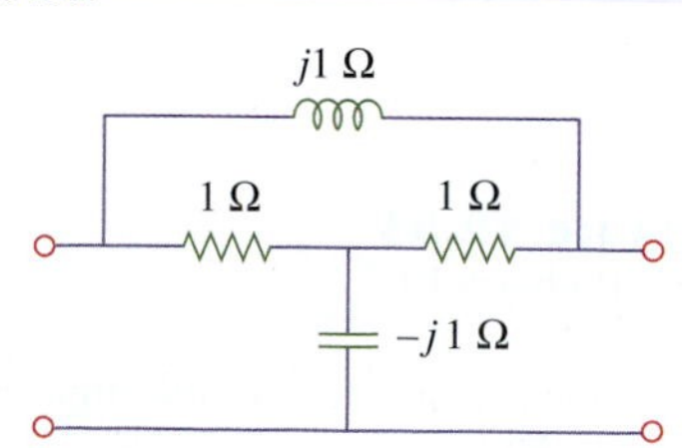

Figure 19.56
For Problem 19.12.

19.13 Find the resistive circuit that represents these y parameters:

$$[\mathbf{y}] = \begin{bmatrix} \frac{1}{2} & -\frac{1}{4} \\ -\frac{1}{4} & \frac{3}{8} \end{bmatrix} \text{S}$$

19.14 Draw the two-port network that has the following y parameters:

$$[\mathbf{y}] = \begin{bmatrix} 1 & -0.5 \\ -0.5 & 1.5 \end{bmatrix} \text{S}$$

19.15 In the bridge circuit of Fig. 19.57, $\mathbf{I}_1 = 10$ A and $\mathbf{I}_2 = -4$ A.

(a) Find $\mathbf{V}_1$ and $\mathbf{V}_2$ using y parameters.

(b) Confirm the results in part (a) by direct circuit analysis.

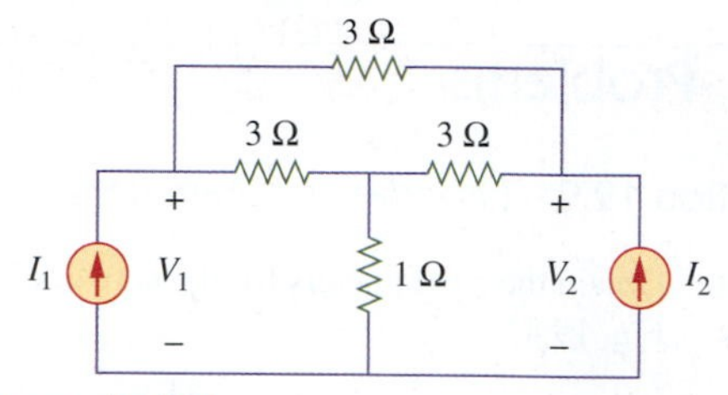

Figure 19.57
For Problem 19.15.

19.16 Determine the y parameters for the two-port network of Fig. 19.58.

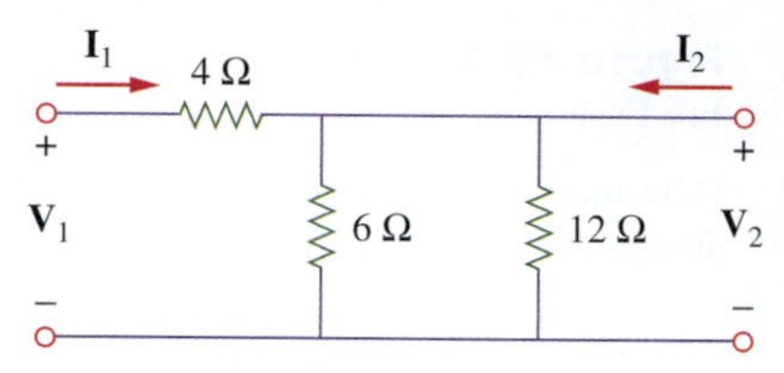

Figure 19.58
For Problems 19.16 and 19.24.

19.17 Obtain the y parameters for the circuit shown in Fig. 19.59.

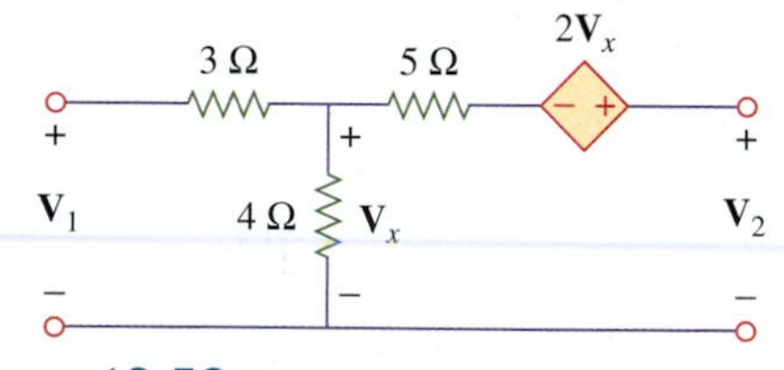

Figure 19.59
For Problem 19.17.

Section 19.4 Hybrid Parameters

19.18 Find the h parameters for the network in Fig. 19.60.

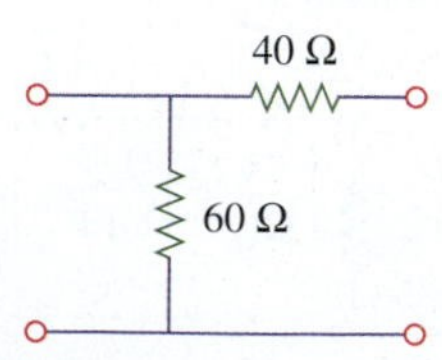

Figure 19.60
For Problem 19.18.

19.19 Determine the h parameters for the networks in Fig. 19.61.

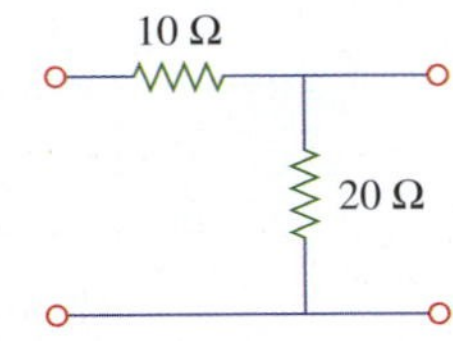

Figure 19.61
For Problem 19.19.

19.20 Find the h parameters of the two-port network in Fig. 19.62 as functions of s.

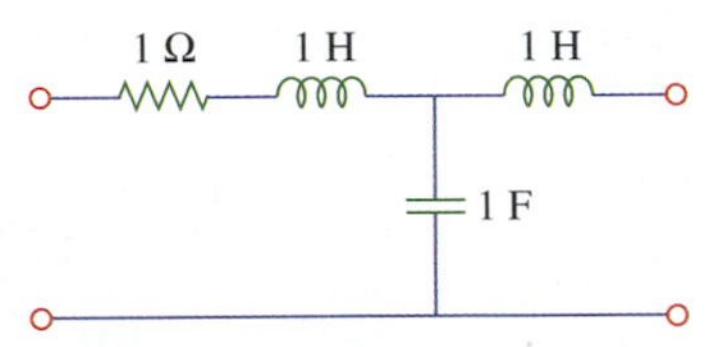

Figure 19.62
For Problem 19.20.

19.21 Obtain the h parameters of the two-port network in Fig. 19.63.

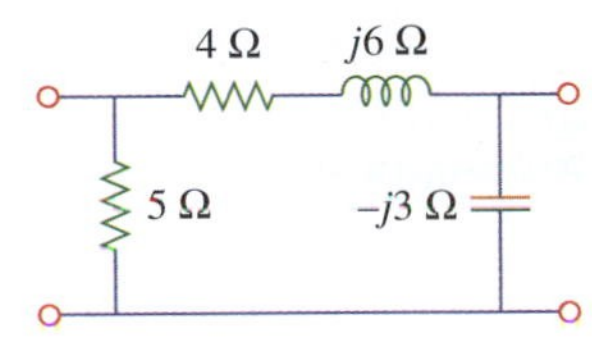

Figure 19.63
For Problem 19.21.

19.22 Determine the h parameters for the two-port network in Fig. 19.64.

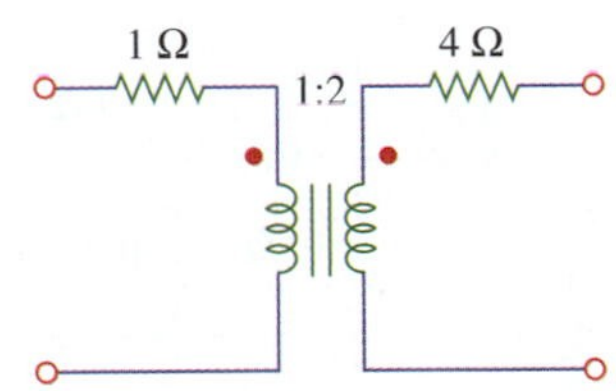

Figure 19.64
For Problem 19.22.

19.23 For the two-port network in Fig. 19.65,

$$[\mathbf{h}] = \begin{bmatrix} 16\ \Omega & 3 \\ -2 & 0.01\ \text{S} \end{bmatrix}$$

Find: (a) $\mathbf{V}_2/\mathbf{V}_1$, (b) $\mathbf{I}_2/\mathbf{I}_1$, (c) $\mathbf{I}_1/\mathbf{V}_1$, (d) $\mathbf{V}_2/\mathbf{I}_1$.

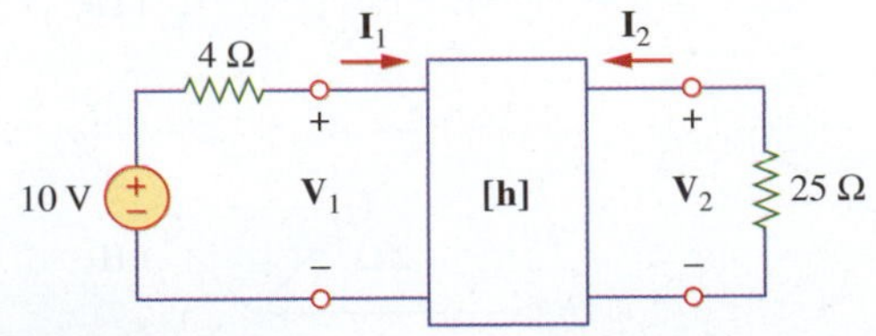

Figure 19.65
For Problem 19.23.

19.24 For the network in Fig. 19.58 (Problem 19.16), find the h parameters.

19.25 For the T network in Fig. 19.66, show that the h parameters are:

$$\mathbf{h}_{11} = \mathbf{R}_1 + \frac{\mathbf{R}_2\mathbf{R}_3}{\mathbf{R}_2 + \mathbf{R}_3}, \qquad \mathbf{h}_{12} = \frac{\mathbf{R}_2}{\mathbf{R}_2 + \mathbf{R}_3},$$

$$\mathbf{h}_{21} = -\frac{\mathbf{R}_2}{\mathbf{R}_2 + \mathbf{R}_3}, \qquad \mathbf{h}_{22} = \frac{1}{\mathbf{R}_2 + \mathbf{R}_3}.$$

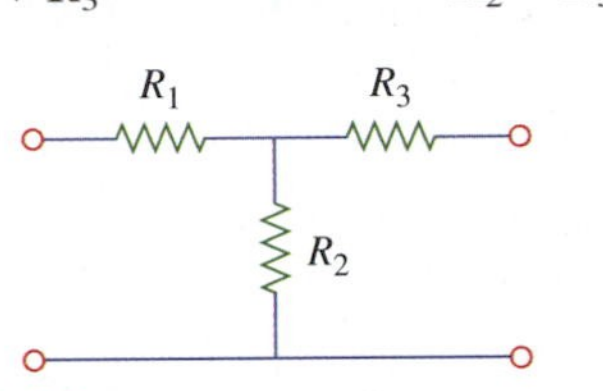

Figure 19.66
For Problem 19.25.

19.26 Obtain the h parameters for the circuit in Fig. 19.67.

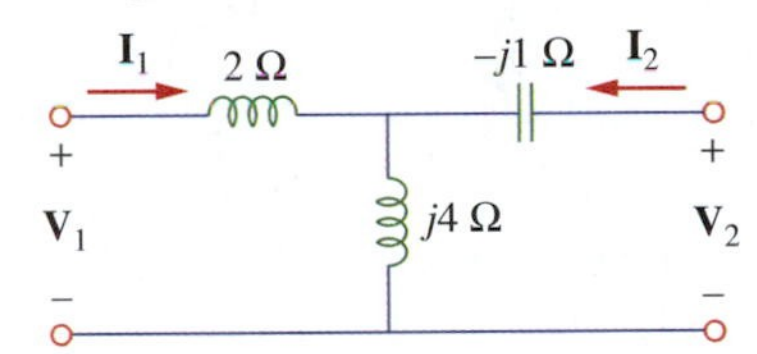

Figure 19.67
For Problem 19.26.

Section 19.5 Relationships between Parameters

19.27 If $[\mathbf{z}] = \begin{bmatrix} 2+j & j4 \\ -1 & 3-j \end{bmatrix} \Omega$, find: (a) $[\mathbf{y}]$, (b) $[\mathbf{h}]$.

19.28 For the bridge circuit of Fig. 19.68, obtain: (a) the z parameters, (b) the h parameters.

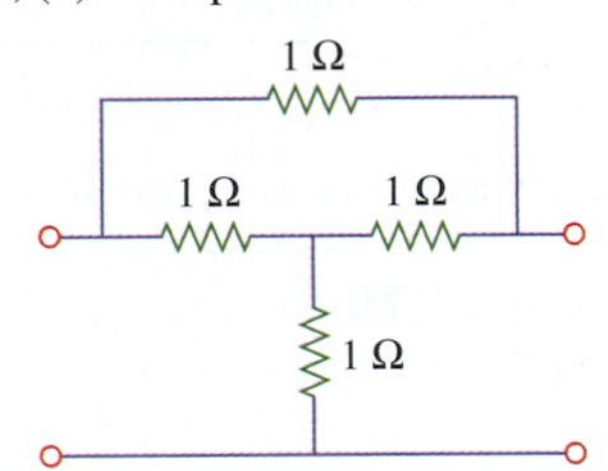

Figure 19.68
For Problem 19.28.

19.29 A two-port is described by

$$\mathbf{V}_1 = \mathbf{I}_1 + 2\mathbf{V}_2$$

$$\mathbf{I}_2 = -2\mathbf{I}_1 + 0.4\mathbf{V}_2$$

Find: (a) the z parameters, (b) the y parameters.

19.30 Let

$$[\mathbf{y}] = \begin{bmatrix} 0.6 & -0.2 \\ -0.1 & 0.5 \end{bmatrix} \text{S}$$

Find: (a) $[\mathbf{z}]$ (b) $[\mathbf{h}]$

19.31 For a two-port network, the z parameters are $\mathbf{z}_{11} = 5\ \text{k}\Omega$, $\mathbf{z}_{12} = \mathbf{z}_{21} = 2\ \text{k}\Omega$, and $\mathbf{z}_{22} = 10\ \text{k}\Omega$.

(a) Find the y parameters.

(b) Determine the h parameters.

19.32 Given that the h parameters of a two-port network are:

$$\begin{bmatrix} \mathbf{h}_{11} & \mathbf{h}_{12} \\ \mathbf{h}_{21} & \mathbf{h}_{22} \end{bmatrix} = \begin{bmatrix} 10\ \Omega & 1 \\ 12 & 0.4\ \text{S} \end{bmatrix}$$

Find the z parameters.

19.33 For the circuit shown in Fig. 19.69, find the z, y, and h parameters.

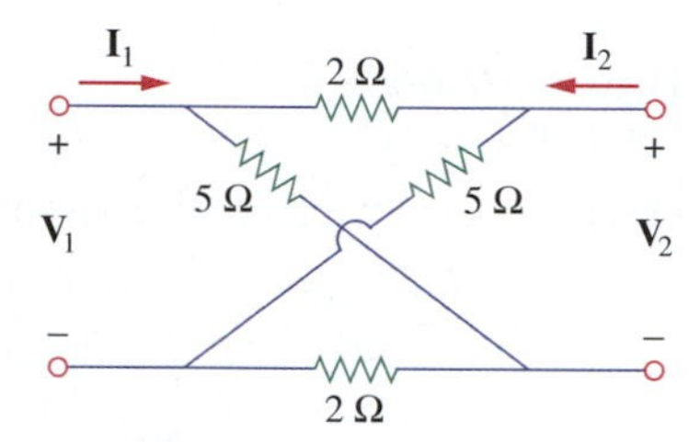

Figure 19.69
For Problem 19.33.

Section 19.6 Interconnection of Networks

19.34 What is the y parameter presentation of the circuit in Fig. 19.70?

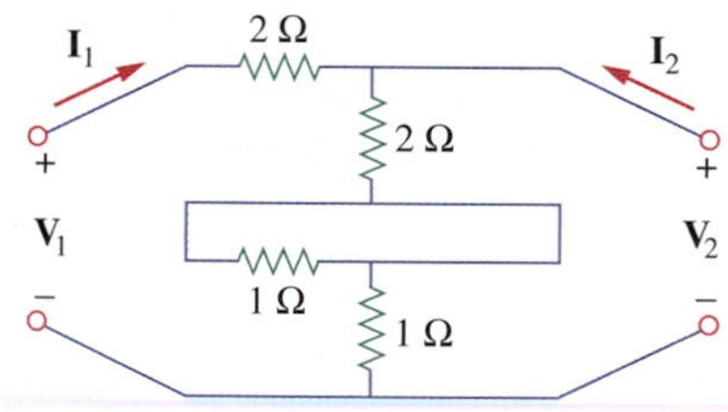

Figure 19.70
For Problem 19.34.

19.35 For the two-port network of Fig. 19.71, let $\mathbf{y}_{12} = \mathbf{y}_{21} = 0$, $\mathbf{y}_{11} = 2$ mS, and $\mathbf{y}_{22} = 10$ mS. Find $\mathbf{V}_o/\mathbf{V}_s$.

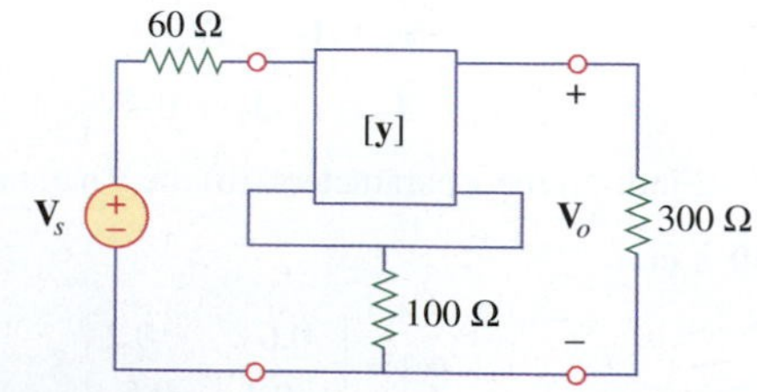

Figure 19.71
For Problem 19.35.

19.36 Figure 19.72 shows two two-ports in series. Find the z parameters.

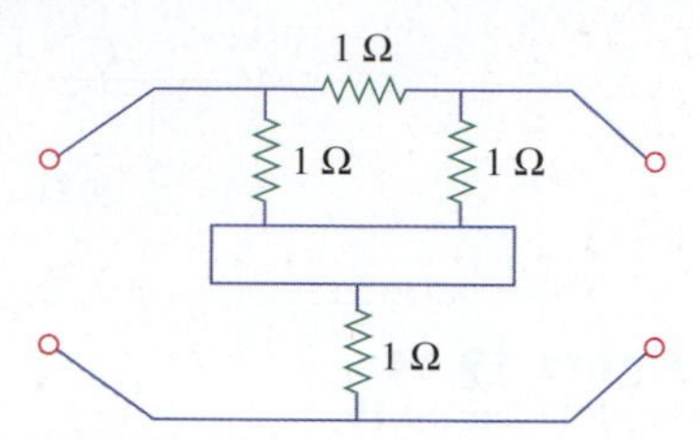

Figure 19.72
For Problem 19.36.

19.37 Obtain the h parameters for the network in Fig. 19.73.

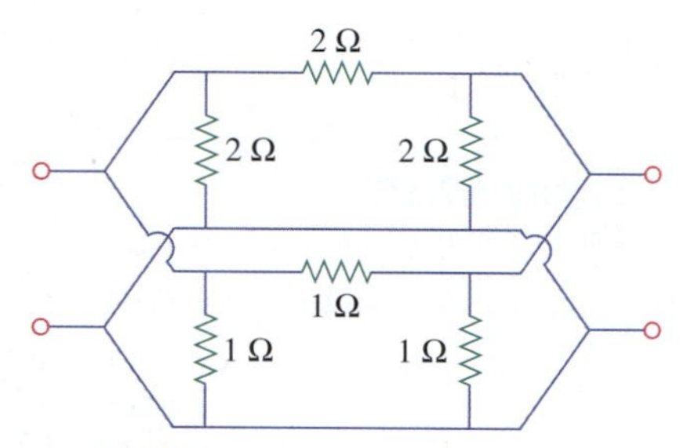

Figure 19.73
For Problem 19.37.

Section 19.7 Computer Analysis

19.38 Use PSpice to compute the z parameters for the circuit in Fig. 19.74.

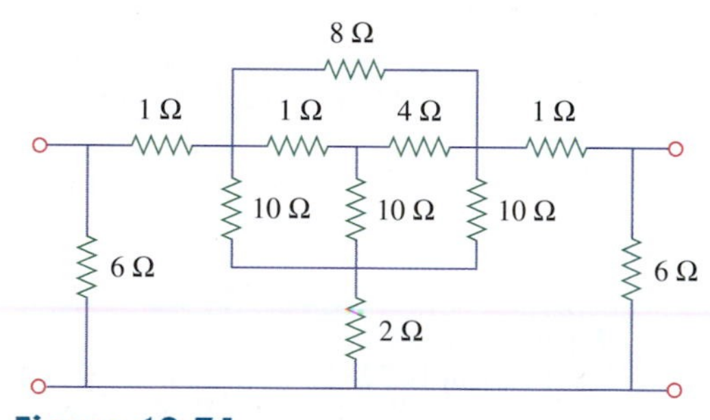

Figure 19.74
For Problem 19.38.

19.39 Using PSpice, find the h parameters of the network in Fig. 19.75. Let $\omega = 1$ rad/s.

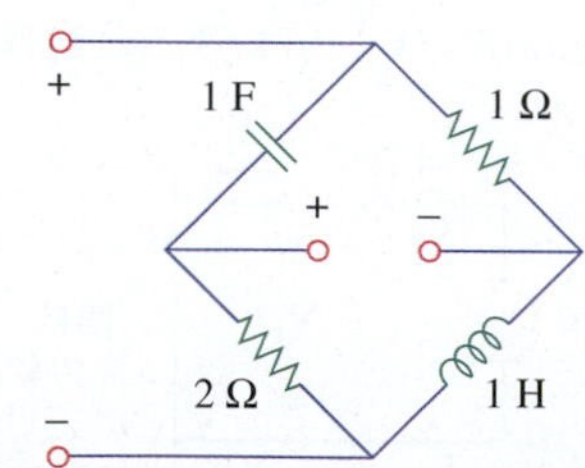

Figure 19.75
For Problem 19.39.

19.40 Obtain the h parameters at $\omega = 4$ rad/s for the circuit in Fig. 19.76 using PSpice.

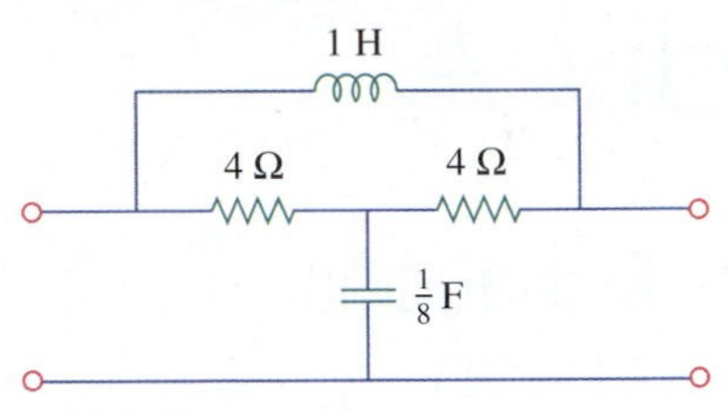

Figure 19.76
For Problem 19.40.

19.41 Use PSpice to determine the z parameters of the circuit in Fig. 19.77. Take $\omega = 2$ rad/s.

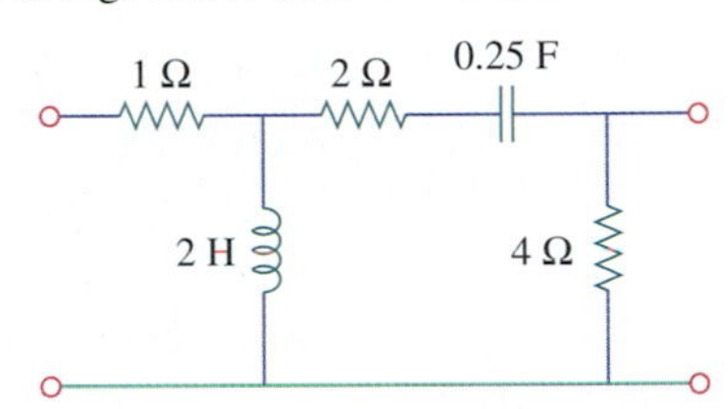

Figure 19.77
For Problem 19.41.

19.42 At $\omega = 1$ rad/s, find the z parameters of the network in Fig. 19.78 using PSpice.

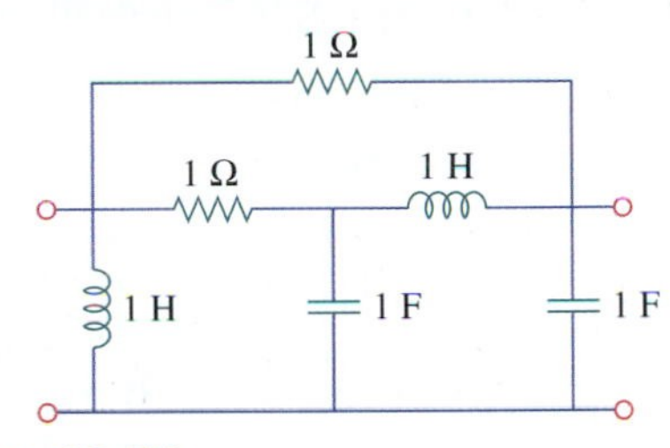

Figure 19.78
For Problem 19.42.

Section 19.8 Applications

19.43 A transistor has the following parameters in a common-emitter circuit:

$$h_{ie} = 2640\ \Omega,\ h_{re} = 2.6 \times 10^{-4},\ h_{fe} = 72,$$
$$h_{oe} = 16\ \mu\text{S},\ R_L = 100\ \text{k}\Omega$$

What is the voltage amplification of the transistor? How many decibels gain is this?

19.44 A transistor with

$$h_{fe} = 120,\ h_{ie} = 2\ \text{k}\Omega,\ h_{re} = 10^{-4},\ h_{oe} = 20\ \mu\text{S}$$

is used for a CE amplifier to provide an input resistance of 1.5 kΩ.

(a) Determine the necessary load resistance R_L.

(b) Calculate A_v, A_i, and Z_{out} if the amplifier is driven by a 4-mV source having an internal resistance of 600 Ω.

(c) Find the voltage across the load.

***19.45** For the transistor network of Fig. 19.79,

$$h_{fe} = 80,\ h_{ie} = 1.2\ \text{k}\Omega,\ h_{re} = 1.5 \times 10^{-4},\ h_{oe} = 20\ \mu\text{S}$$

Determine the following:

(a) voltage gain $A_v = V_o/V_s$

(b) current gain $A_i = I_o/I_i$

(c) input impedance Z_{in}

(d) output impedance Z_{out}

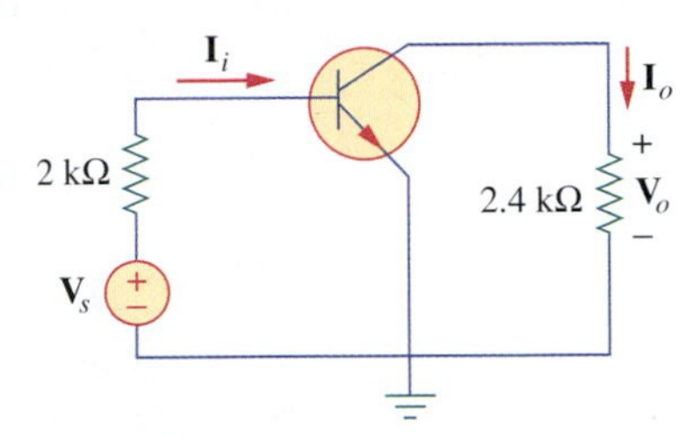

Figure 19.79
For Problem 19.45.

***19.46** Determine A_v, A_i, Z_{in}, and Z_{out} for the amplifier shown in Fig. 19.80. Assume that

$$h_{ie} = 4\ \text{k}\Omega,\ h_{re} = 10^{-4},\ h_{fe} = 100,\ h_{oe} = 30\ \mu\text{S}$$

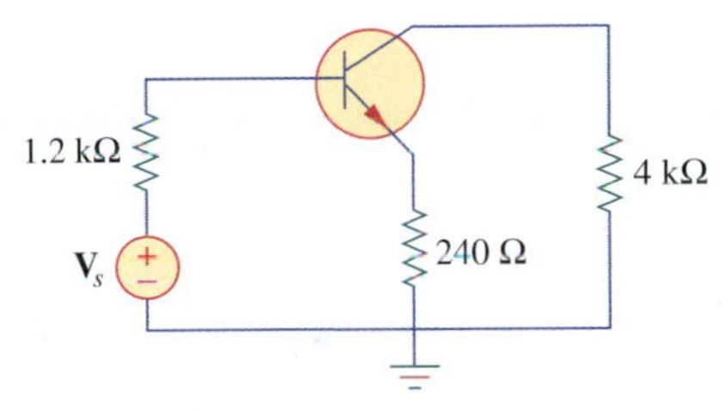

Figure 19.80
For Problem 19.46.

***19.47** Calculate A_v, A_i, Z_{in}, and Z_{out} for the transistor network in Fig. 19.81. Assume that

$$h_{ie} = 2\ \text{k}\Omega,\ h_{re} = 2.5 \times 10^{-4},\ h_{fe} = 150,\ h_{oe} = 10\ \mu\text{S}$$

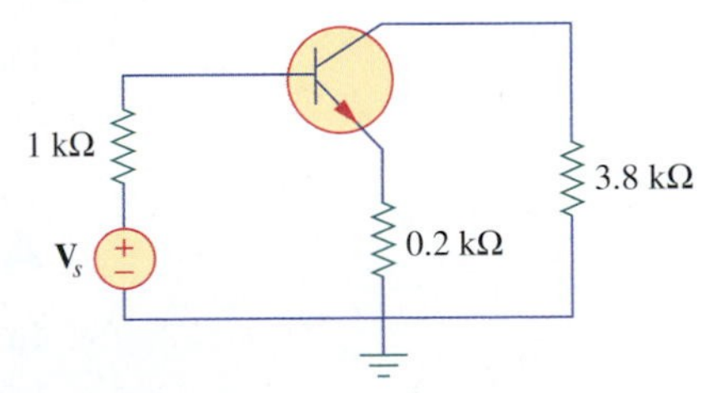

Figure 19.81
For Problem 19.47.

Appendix A

Simultaneous Equations and Matrix Inversion

In circuit analysis, we often encounter a set of simultaneous equations having the form

$$\begin{aligned} a_{11}x_1 + a_{12}x_2 + \cdots + a_{1n}x_n &= b_1 \\ a_{21}x_1 + a_{22}x_2 + \cdots + a_{2n}x_n &= b_2 \\ \vdots \qquad \vdots \qquad & \vdots \\ a_{n1}x_1 + a_{n2}x_2 + \cdots + a_{nn}x_n &= b_n \end{aligned} \tag{A.1}$$

where there are n unknown $x_1, x_2, \ldots, x_n$ to be determined. Equation (A.1) can be written in matrix form as

$$\begin{bmatrix} a_{11} & a_{12} & \cdots & a_{1n} \\ a_{21} & a_{22} & \cdots & a_{2n} \\ \vdots & \vdots & \cdots & \vdots \\ a_{n1} & a_{n2} & \cdots & a_{nn} \end{bmatrix} \begin{bmatrix} x_1 \\ x_2 \\ \vdots \\ x_n \end{bmatrix} = \begin{bmatrix} b_2 \\ b_2 \\ \vdots \\ b_n \end{bmatrix} \tag{A.2}$$

This matrix equation can be put in a compact form as

$$\mathbf{AX} = \mathbf{B} \tag{A.3}$$

where

$$\mathbf{A} = \begin{bmatrix} a_{11} & a_{12} & \cdots & a_{1n} \\ a_{21} & a_{22} & \cdots & a_{2n} \\ \vdots & \vdots & \cdots & \vdots \\ a_{n1} & a_{n2} & \cdots & a_{nn} \end{bmatrix}, \qquad \mathbf{X} = \begin{bmatrix} x_1 \\ x_2 \\ \vdots \\ x_n \end{bmatrix}, \qquad \mathbf{B} = \begin{bmatrix} b_1 \\ b_2 \\ \vdots \\ b_n \end{bmatrix} \tag{A.4}$$

A is a square $(n \times n)$ matrix while **X** and **B** are column matrices.

There are several methods for solving Eq. (A.1) or (A.3). These include substitution, Gaussian elimination, Cramer's rule, matrix inversion, and numerical analysis.

A.1 Cramer's Rule

In many cases, Cramer's rule can be used to solve the simultaneous equations we encounter in circuit analysis. Cramer's rule states that the solution to Eq. (A.1) or (A.3) is

$$\boxed{\begin{aligned} x_1 &= \frac{\Delta_1}{\Delta} \\ x_2 &= \frac{\Delta_2}{\Delta} \\ &\vdots \\ x_n &= \frac{\Delta_n}{\Delta} \end{aligned}} \tag{A.5}$$

where the Δ's are the determinants given by

$$\Delta = \begin{vmatrix} a_{11} & a_{12} & \cdots & a_{1n} \\ a_{21} & a_{22} & \cdots & a_{2n} \\ \vdots & \vdots & \cdots & \vdots \\ a_{n1} & a_{n2} & \cdots & a_{nn} \end{vmatrix}, \qquad \Delta_1 = \begin{vmatrix} b_1 & a_{12} & \cdots & a_{1n} \\ b_2 & a_{22} & \cdots & a_{2n} \\ \vdots & \vdots & \cdots & \vdots \\ b_n & a_{n2} & \cdots & a_{nn} \end{vmatrix}$$

$$\vdots \qquad\qquad \vdots$$

$$\Delta_2 = \begin{vmatrix} a_{11} & b_1 & \cdots & a_{1n} \\ a_{21} & b_2 & \cdots & a_{2n} \\ \vdots & \vdots & \cdots & \vdots \\ a_{n1} & b_n & \cdots & a_{nn} \end{vmatrix}, \ldots, \Delta_n = \begin{vmatrix} a_{11} & a_{12} & \cdots & b_1 \\ a_{21} & a_{22} & \cdots & b_2 \\ \vdots & \vdots & \cdots & \vdots \\ a_{n1} & a_{n2} & \cdots & b_n \end{vmatrix} \qquad \textbf{(A.6)}$$

Notice that Δ is the determinant of matrix **A** and Δ_k is the determinant of the matrix formed by replacing the kth column of **A** by **B**. It is evident from Eq. (A.5) that Cramer's rule applies only when $\Delta \neq 0$. When $\Delta = 0$, the set of equations has no unique solution, because the equations are linearly dependent.

The value of the determinant Δ, for example, can be obtained by expanding along the first row:

$$\Delta = \begin{vmatrix} a_{11} & a_{12} & a_{13} & \cdots & a_{1n} \\ a_{21} & a_{22} & a_{23} & \cdots & a_{2n} \\ a_{31} & a_{32} & a_{33} & \cdots & a_{3n} \\ \vdots & \vdots & \vdots & \cdots & \vdots \\ a_{n1} & a_{n2} & a_{n3} & \cdots & a_{nn} \end{vmatrix} \qquad \textbf{(A.7)}$$

$$= a_{11}M_{11} - a_{12}M_{12} + a_{13}M_{13} + \cdots + (-1)^{1+n}a_{1n}M_{1n}$$

where the minor M_{ij} is an $(n-1) \times (n-1)$ determinant of the matrix formed by striking out the ith row and jth column. The value of Δ may also be obtained by expanding along the first column:

$$\Delta = a_{11}M_{11} - a_{21}M_{21} + a_{31}M_{31} + \cdots + (-1)^{n+1}a_{n1}M_{n1} \qquad \textbf{(A.8)}$$

We now specifically develop the formulas for calculating the determinants of 2 × 2 and 3 × 3 matrices, because of their frequent occurrence in this text. For a 2 × 2 matrix,

$$\Delta = \begin{vmatrix} a_{11} & a_{12} \\ a_{21} & a_{22} \end{vmatrix} = a_{11}a_{22} - a_{12}a_{21} \qquad \textbf{(A.9)}$$

For a 3 × 3 matrix,

$$\Delta = \begin{vmatrix} a_{11} & a_{12} & a_{13} \\ a_{21} & a_{22} & a_{23} \\ a_{31} & a_{32} & a_{33} \end{vmatrix} = a_{11}(-1)^2\begin{vmatrix} a_{22} & a_{23} \\ a_{32} & a_{33} \end{vmatrix} + a_{21}(-1)^3\begin{vmatrix} a_{12} & a_{13} \\ a_{32} & a_{33} \end{vmatrix}$$

$$+ a_{31}(-1)^4\begin{vmatrix} a_{12} & a_{13} \\ a_{22} & a_{23} \end{vmatrix}$$

$$= a_{11}(a_{22}a_{33} - a_{32}a_{23}) - a_{21}(a_{12}a_{33} - a_{32}a_{13})$$
$$+ a_{31}(a_{12}a_{23} - a_{22}a_{13}) \qquad \textbf{(A.10)}$$

An alternative method of obtaining the determinant of a 3×3 matrix is by repeating the first two rows and multiplying the terms diagonally as follows.

$$\Delta = \begin{vmatrix} a_{11} & a_{12} & a_{13} \\ a_{21} & a_{22} & a_{23} \\ a_{31} & a_{32} & a_{33} \\ a_{11} & a_{12} & a_{13} \\ a_{21} & a_{22} & a_{23} \end{vmatrix}$$

$$= a_{11}a_{22}a_{33} + a_{21}a_{32}a_{13} + a_{31}a_{12}a_{23} - a_{13}a_{22}a_{31} - a_{23}a_{32}a_{11} - a_{33}a_{12}a_{21} \qquad \textbf{(A.11)}$$

In summary:

The solution of linear simultaneous equations by Cramer's rule boils down to finding

$$x_k = \frac{\Delta_k}{\Delta}, \qquad k = 1, 2, \dots, n \qquad \textbf{(A.12)}$$

where Δ is the determinant of matrix $\mathbf{A}$ and Δ_k is the determinant of the matrix formed by replacing the kth column of $\mathbf{A}$ by $\mathbf{B}$.

One may use other methods, such as matrix inversion and elimination. Only Cramer's method is covered here, because of its simplicity and also because of the availability of powerful calculators.

You may not find much need to use Cramer's method described in this appendix, in view of the availability of calculators, computers, and software packages such as *MATLAB*, which can be used easily to solve a set of linear equations. But in case you need to solve the equations by hand, the material covered in this appendix becomes useful. At any rate, it is important to know the mathematical basis of those calculators and software packages.

Example A.1

Solve the simultaneous equations

$$4x_1 - 3x_2 = 17, \qquad -3x_1 + 5x_2 = -21$$

Solution:

The given set of equations is cast in matrix form as

$$\begin{bmatrix} 4 & -3 \\ -3 & 5 \end{bmatrix} \begin{bmatrix} x_1 \\ x_2 \end{bmatrix} = \begin{bmatrix} 17 \\ -21 \end{bmatrix}$$

The determinants are evaluated as

$$\Delta = \begin{vmatrix} 4 & -3 \\ -3 & 5 \end{vmatrix} = 4 \times 5 - (-3)(-3) = 11$$

$$\Delta_1 = \begin{vmatrix} 17 & -3 \\ -21 & 5 \end{vmatrix} = 17 \times 5 - (-3)(-21) = 22$$

$$\Delta_2 = \begin{vmatrix} 4 & 17 \\ -3 & -21 \end{vmatrix} = 4 \times (-21) - 17 \times (-3) = -33$$

Hence,

$$x_1 = \frac{\Delta_1}{\Delta} = \frac{22}{11} = 2, \qquad x_2 = \frac{\Delta_2}{\Delta} = \frac{-33}{11} = -3$$

Practice Problem A.1

Find the solution to the following simultaneous equations:

$$3x_1 - x_2 = 4, \qquad -6x_1 + 18x_2 = 16$$

Answer: $x_1 = 1.833$; $x_2 = 1.5$

Example A.2

Determine x_1, x_2, and x_3 for this set of simultaneous equations:

$$\begin{aligned} 25x_1 - 5x_2 - 20x_3 &= 50 \\ -5x_1 + 10x_2 - 4x_3 &= 0 \\ -5x_1 - 4x_2 + 9x_3 &= 0 \end{aligned}$$

Solution:

In matrix form, the given set of equations becomes

$$\begin{bmatrix} 25 & -5 & -20 \\ -5 & 10 & -4 \\ -5 & -4 & 9 \end{bmatrix} \begin{bmatrix} x_1 \\ x_2 \\ x_3 \end{bmatrix} = \begin{bmatrix} 50 \\ 0 \\ 0 \end{bmatrix}$$

We apply Eq. (A.11) to find the determinants. This requires that we repeat the first two rows of the matrix. Thus,

$$\Delta = \begin{vmatrix} 25 & -5 & -20 \\ -5 & 10 & -4 \\ -5 & -4 & 9 \end{vmatrix} = \begin{vmatrix} 25 & -5 & -20 \\ -5 & 10 & -4 \\ -5 & -4 & 9 \\ 25 & -5 & -20 \\ -5 & 10 & -4 \end{vmatrix} \begin{matrix} \\ \\ \\ - \quad + \\ - \quad + \\ - \quad + \end{matrix}$$

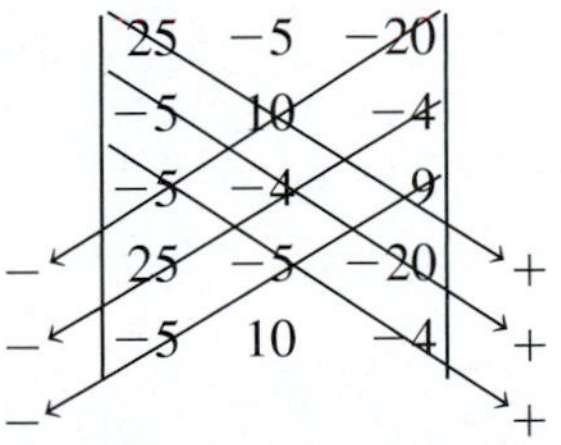

$$\begin{aligned} &= 25(10)9 + (-5)(-4)(-20) + (-5)(-5)(-4) \\ &\quad - (-20)(10)(-5) - (-4)(-4)25 - 9(-5)(-5) \\ &= 2{,}250 - 400 - 100 - 1{,}000 - 400 - 225 = 125 \end{aligned}$$

Similarly,

$$\Delta_1 = \begin{vmatrix} 50 & -5 & -20 \\ 0 & 10 & -4 \\ 0 & -4 & 9 \end{vmatrix} = \begin{vmatrix} 50 & -5 & -20 \\ 0 & 10 & -4 \\ 0 & -4 & 9 \\ 50 & -5 & -20 \\ 0 & 10 & -4 \end{vmatrix} \begin{matrix} \\ \\ \\ - \quad + \\ - \quad + \\ - \quad + \end{matrix}$$

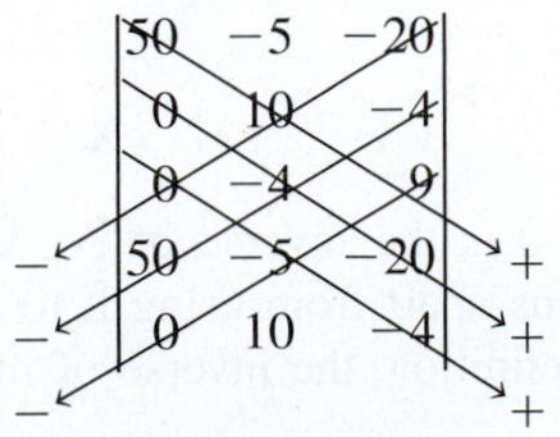

$$= 4{,}500 + 0 + 0 - 0 - 800 - 0 = 3{,}700$$

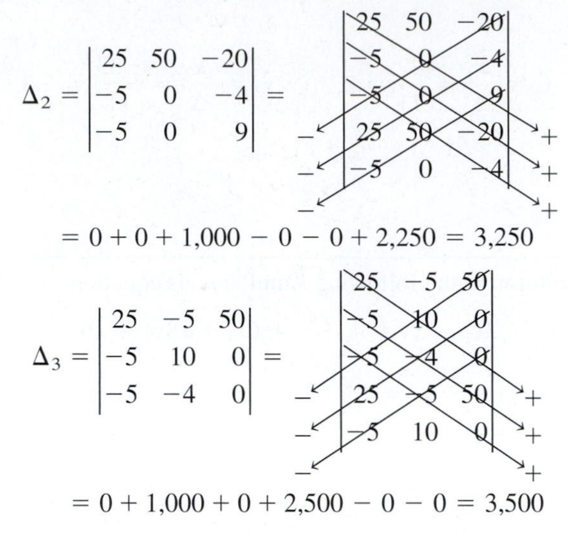

$$\Delta_2 = \begin{vmatrix} 25 & 50 & -20 \\ -5 & 0 & -4 \\ -5 & 0 & 9 \end{vmatrix}$$

$$= 0 + 0 + 1{,}000 - 0 - 0 + 2{,}250 = 3{,}250$$

$$\Delta_3 = \begin{vmatrix} 25 & -5 & 50 \\ -5 & 10 & 0 \\ -5 & -4 & 0 \end{vmatrix}$$

$$= 0 + 1{,}000 + 0 + 2{,}500 - 0 - 0 = 3{,}500$$

Hence, we now find

$$x_1 = \frac{\Delta_1}{\Delta} = \frac{3{,}700}{125} = 29.6$$

$$x_2 = \frac{\Delta_2}{\Delta} = \frac{3{,}250}{125} = 26$$

$$x_3 = \frac{\Delta_2}{\Delta} = \frac{3{,}500}{125} = 28$$

Practice Problem A.2

Obtain the solution of this set of simultaneous equations:

$$3x_1 - x_2 - 2x_3 = 1$$
$$-x_1 + 6x_2 - 3x_3 = 0$$
$$-2x_1 - 3x_2 + 6x_3 = 6$$

Answer: $x_1 = 3 = x_3; x_2 = 2$

A.2 Matrix Inversion

The linear system of equations in Eq. (A.3) can be solved by matrix inversion. In the matrix equation $\mathbf{AX} = \mathbf{B}$, we may invert $\mathbf{A}$ to get $\mathbf{X}$, i.e.,

$$\mathbf{X} = \mathbf{A}^{-1}\mathbf{B} \tag{A.13}$$

where $\mathbf{A}^{-1}$ is the inverse of $\mathbf{A}$. Matrix inversion is needed in other applications apart from using it to solve a set of equations.

By definition, the inverse of matrix $\mathbf{A}$ satisfies

$$\mathbf{A}^{-1}\mathbf{A} = \mathbf{A}\mathbf{A}^{-1} = \mathbf{I} \tag{A.14}$$

where $\mathbf{I}$ is an identity matrix. $\mathbf{A}^{-1}$ is given by

$$\mathbf{A}^{-1} = \frac{\text{adj } \mathbf{A}}{\det \mathbf{A}} \tag{A.15}$$

where adj $\mathbf{A}$ is the adjoint of $\mathbf{A}$ and $\det \mathbf{A} = |\mathbf{A}|$ is the determinant of $\mathbf{A}$. The adjoint of $\mathbf{A}$ is the transpose of the cofactors of $\mathbf{A}$. Suppose we are given an $n \times n$ matrix $\mathbf{A}$ as

$$\mathbf{A} = \begin{bmatrix} a_{11} & a_{12} & \cdots & a_{1n} \\ a_{21} & a_{22} & \cdots & a_{2n} \\ \vdots & & & \\ a_{n1} & a_{n2} & \cdots & a_{nn} \end{bmatrix} \tag{A.16}$$

The cofactors of $\mathbf{A}$ are defined as

$$\mathbf{C} = \text{cof}\,(\mathbf{A}) = \begin{bmatrix} c_{11} & c_{12} & \cdots & c_{1n} \\ c_{21} & c_{22} & \cdots & c_{2n} \\ \vdots & & & \\ c_{n1} & c_{n2} & \cdots & c_{nn} \end{bmatrix} \tag{A.17}$$

where the cofactor c_{ij} is the product of $(-1)^{i+j}$ and the determinant of the $(n-1) \times (n-1)$ submatrix is obtained by deleting the ith row and jth column from $\mathbf{A}$. For example, by deleting the first row and the first column of $\mathbf{A}$ in Eq. (A.16), we obtain the cofactor c_{11} as

$$c_{11} = (-1)^2 \begin{vmatrix} a_{22} & a_{23} & \cdots & a_{2n} \\ a_{32} & a_{33} & \cdots & a_{3n} \\ \vdots & & & \\ a_{n2} & a_{n3} & \cdots & a_{nn} \end{vmatrix} \tag{A.18}$$

Once the cofactors are found, the adjoint of $\mathbf{A}$ is obtained as

$$\text{adj}\,(\mathbf{A}) = \begin{bmatrix} c_{11} & c_{12} & \cdots & c_{1n} \\ c_{21} & c_{22} & \cdots & c_{2n} \\ \vdots & & & \\ c_{n1} & c_{n2} & \cdots & c_{nn} \end{bmatrix}^T = \mathbf{C}^T \tag{A.19}$$

where T denotes transpose.

In addition to using the cofactors to find the adjoint of $\mathbf{A}$, they are also used in finding the determinant of $\mathbf{A}$ which is given by

$$|\mathbf{A}| = \sum_{j=1}^{n} a_{ij} c_{ij} \tag{A.20}$$

where i is any value from 1 to n. By substituting Eqs. (A.19) and (A.20) into Eq. (A.15), we obtain the inverse of $\mathbf{A}$ as

$$\boxed{\mathbf{A}^{-1} = \frac{\mathbf{C}^T}{|\mathbf{A}|}} \tag{A.21}$$

For a 2×2 matrix, if

$$\mathbf{A} = \begin{bmatrix} a & b \\ c & d \end{bmatrix} \tag{A.22}$$

its inverse is

$$\mathbf{A}^{-1} = \frac{1}{|\mathbf{A}|}\begin{bmatrix} d & -b \\ -c & a \end{bmatrix} = \frac{1}{ad - bc}\begin{bmatrix} d & -b \\ -c & a \end{bmatrix} \qquad \textbf{(A.23)}$$

For a 3×3 matrix, if

$$\mathbf{A} = \begin{bmatrix} a_{11} & a_{12} & a_{13} \\ a_{21} & a_{22} & a_{23} \\ a_{31} & a_{32} & a_{33} \end{bmatrix} \qquad \textbf{(A.24)}$$

we first obtain the cofactors as

$$\mathbf{C} = \begin{bmatrix} c_{11} & c_{12} & c_{13} \\ c_{21} & c_{22} & c_{23} \\ c_{31} & c_{32} & c_{33} \end{bmatrix} \qquad \textbf{(A.25)}$$

where

$$c_{11} = \begin{vmatrix} a_{22} & a_{23} \\ a_{32} & a_{33} \end{vmatrix}, \quad c_{12} = -\begin{vmatrix} a_{21} & a_{23} \\ a_{31} & a_{33} \end{vmatrix}, \quad c_{13} = \begin{vmatrix} a_{21} & a_{22} \\ a_{31} & a_{32} \end{vmatrix},$$
$$c_{21} = -\begin{vmatrix} a_{12} & a_{13} \\ a_{32} & a_{33} \end{vmatrix}, \quad c_{22} = \begin{vmatrix} a_{11} & a_{13} \\ a_{31} & a_{33} \end{vmatrix}, \quad c_{23} = -\begin{vmatrix} a_{11} & a_{12} \\ a_{31} & a_{32} \end{vmatrix},$$
$$c_{31} = \begin{vmatrix} a_{12} & a_{13} \\ a_{22} & a_{23} \end{vmatrix}, \quad c_{32} = -\begin{vmatrix} a_{11} & a_{13} \\ a_{21} & a_{23} \end{vmatrix}, \quad c_{33} = \begin{vmatrix} a_{11} & a_{12} \\ a_{21} & a_{22} \end{vmatrix} \qquad \textbf{(A.26)}$$

The determinant of the 3×3 matrix can be found using Eq. (A.11). Here, we want to use Eq. (A.20), i.e.,

$$|\mathbf{A}| = a_{11}c_{11} + a_{12}c_{12} + a_{13}c_{13} \qquad \textbf{(A.27)}$$

The idea can be extended $n > 3$, but we deal mainly with 2×2 and 3×3 matrices in this book.

Example A.3

Use matrix inversion to solve the simultaneous equations

$$2x_1 + 10x_2 = 2, \qquad -x_1 + 3x_2 = 7$$

Solution:

We first express the two equations in matrix form as

$$\begin{bmatrix} 2 & 10 \\ -1 & 3 \end{bmatrix}\begin{bmatrix} x_1 \\ x_2 \end{bmatrix} = \begin{bmatrix} 2 \\ 7 \end{bmatrix}$$

or

$$\mathbf{AX} = \mathbf{B} \longrightarrow \mathbf{X} = \mathbf{A}^{-1}\mathbf{B}$$

where

$$\mathbf{A} = \begin{bmatrix} 2 & 10 \\ -1 & 3 \end{bmatrix}, \qquad \mathbf{X} = \begin{bmatrix} x_1 \\ x_2 \end{bmatrix}, \qquad \mathbf{B} = \begin{bmatrix} 2 \\ 7 \end{bmatrix}$$

The determinant of $\mathbf{A}$ is $|\mathbf{A}| = 2 \times 3 - 10(-1) = 16$, so the inverse of $\mathbf{A}$ is

$$\mathbf{A}^{-1} = \frac{1}{16}\begin{bmatrix} 3 & -10 \\ 1 & 2 \end{bmatrix}$$

Hence,

$$\mathbf{X} = \mathbf{A}^{-1}\mathbf{B} = \frac{1}{16}\begin{bmatrix} 3 & -10 \\ 1 & 2 \end{bmatrix}\begin{bmatrix} 2 \\ 7 \end{bmatrix} = \frac{1}{16}\begin{bmatrix} -64 \\ 16 \end{bmatrix} = \begin{bmatrix} -4 \\ 1 \end{bmatrix}$$

that is, $x_1 = -4$ and $x_2 = 1$.

Practice Problem A.3

Solve the following two equations by matrix inversion.

$$2y_1 - y_2 = 4, \quad y_1 + 3y_2 = 9$$

Answer: $y_1 = 3; y_2 = 2$

Example A.4

Determine x_1, x_2, and x_3 for the following simultaneous equations using matrix inversion.

$$\begin{aligned} x_1 + x_2 + x_3 &= 5 \\ -x_1 + 2x_2 &= 9 \\ 4x_1 + x_2 - x_3 &= -2 \end{aligned}$$

Solution:

In matrix form, the equations become

$$\begin{bmatrix} 1 & 1 & 1 \\ -1 & 2 & 0 \\ 4 & 1 & -1 \end{bmatrix}\begin{bmatrix} x_1 \\ x_2 \\ x_3 \end{bmatrix} = \begin{bmatrix} 5 \\ 9 \\ -2 \end{bmatrix}$$

or

$$\mathbf{AX} = \mathbf{B} \longrightarrow \mathbf{X} = \mathbf{A}^{-1}\mathbf{B}$$

where

$$\mathbf{A} = \begin{bmatrix} 1 & 1 & 1 \\ -1 & 2 & 0 \\ 4 & 1 & -1 \end{bmatrix}, \quad \mathbf{X} = \begin{bmatrix} x_1 \\ x_2 \\ x_3 \end{bmatrix}, \quad \mathbf{B} = \begin{bmatrix} 5 \\ 9 \\ -2 \end{bmatrix}$$

We now find the cofactors

$$c_{11} = \begin{vmatrix} 2 & 0 \\ 1 & -1 \end{vmatrix} = -2, \quad c_{12} = -\begin{vmatrix} -1 & 0 \\ 4 & -1 \end{vmatrix} = -1, \; c_{13} = \begin{vmatrix} -1 & 2 \\ 4 & 1 \end{vmatrix} = -9$$

$$c_{21} = -\begin{vmatrix} 1 & 1 \\ 1 & -1 \end{vmatrix} = 2, \; c_{22} = \begin{vmatrix} 1 & 1 \\ 4 & -1 \end{vmatrix} = -5, \quad c_{23} = -\begin{vmatrix} 1 & 1 \\ 4 & 1 \end{vmatrix} = 3$$

$$c_{31} = \begin{vmatrix} 1 & 1 \\ 2 & 0 \end{vmatrix} = -2, \quad c_{32} = -\begin{vmatrix} 1 & 1 \\ -1 & 0 \end{vmatrix} = -1, \quad c_{33} = \begin{vmatrix} 1 & 1 \\ -1 & 2 \end{vmatrix} = 3$$

The adjoint of matrix **A** is

$$\text{adj}\,\mathbf{A} = \begin{bmatrix} -2 & -1 & -9 \\ 2 & -5 & 3 \\ -2 & -1 & 3 \end{bmatrix}^T = \begin{bmatrix} -2 & 2 & -2 \\ -1 & -5 & -1 \\ -9 & 3 & 3 \end{bmatrix}$$

We can find the determinant of **A** using any row or column of **A**. Because one element of the second row is 0, we can take advantage of this to find the determinant as

$$|\mathbf{A}| = -1c_{21} + 2c_{22} + (0)c_{23} = -1(2) + 2(-5) = -12$$

Hence, the inverse of **A** is

$$\mathbf{A}^{-1} = \frac{1}{-12}\begin{bmatrix} -2 & 2 & -2 \\ -1 & -5 & -1 \\ -9 & 3 & 3 \end{bmatrix}$$

$$\mathbf{X} = \mathbf{A}^{-1}\mathbf{B} = \frac{1}{-12}\begin{bmatrix} -2 & 2 & -2 \\ -1 & -5 & -1 \\ -9 & 3 & 3 \end{bmatrix}\begin{bmatrix} 5 \\ 9 \\ -2 \end{bmatrix} = \begin{bmatrix} -1 \\ 4 \\ 2 \end{bmatrix}$$

that is, $x_1 = -1, x_2 = 4, x_3 = 2$.

Practice Problem A.4

Solve the following equations using matrix inversion.

$$\begin{aligned} y_1 - y_3 &= 1 \\ 2y_1 + 3y_2 - y_3 &= 1 \\ y_1 - y_2 - y_3 &= 3 \end{aligned}$$

Answer: $y_1 = 6; y_2 = -2; y_3 = 5$

Appendix B

Complex Numbers

The ability to manipulate complex numbers is very handy in circuit analysis and in electrical engineering technology in general. Complex numbers are particularly useful in the analysis of ac circuits. Again, although calculators and computer software packages are now available to manipulate complex numbers, it is still advisable for a student to be familiar with how to handle them by hand.

B.1 Representations of Complex Numbers

A complex number z may be written in *rectangular form* as

$$z = x + jy \tag{B.1}$$

where $j = \sqrt{-1}$; x is the *real part* of z while y is the *imaginary part* of z; that is,

$$x = \text{Re}(z), \qquad y = \text{Im}(z) \tag{B.2}$$

The complex plane looks like the two-dimensional curvilinear coordinate space, but it is not.

The complex number z is shown plotted in the complex plane in Fig. B.1. Because $j = \sqrt{-1}$,

$$\begin{aligned} \frac{1}{j} &= -j \\ j^2 &= -1 \\ j^3 &= j \cdot j^2 = -j \\ j^4 &= j^2 \cdot j^2 = 1 \\ j^5 &= j \cdot j^4 = j \\ &\vdots \\ j^{n+4} &= j^n \end{aligned} \tag{B.3}$$

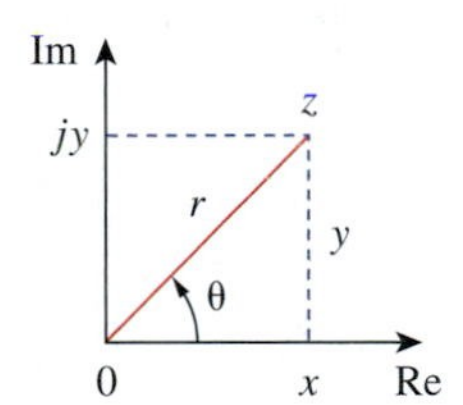

Figure B.1
Graphical representation of a complex number.

A second way of representing the complex number z is by specifying its magnitude r and the angle θ it makes with the real axis, as Fig. B.1 shows. This is known as the *polar form.* It is given by

$$z = |z|\underline{/\theta} = r\underline{/\theta} \tag{B.4}$$

where

$$r = \sqrt{x^2 + y^2}, \qquad \theta = \tan^{-1}\frac{y}{x} \tag{B.5a}$$

or

$$x = r\cos\theta, \qquad y = r\sin\theta \tag{B.5b}$$

that is,

$$z = x + jy = r\underline{/\theta} = r\cos\theta + jr\sin\theta \tag{B.6}$$

In converting from rectangular to polar form using Eq. (B.5), we must exercise care in determining the correct value of θ. These are the four possibilities:

$$\begin{aligned}
z &= x + jy, & \theta &= \tan^{-1}\frac{y}{x} & &\text{(1st quadrant)}\\
z &= -x + jy, & \theta &= 180^\circ - \tan^{-1}\frac{y}{x} & &\text{(2nd quadrant)}\\
z &= -x - jy, & \theta &= 180^\circ + \tan^{-1}\frac{y}{x} & &\text{(3rd quadrant)}\\
z &= x - jy, & \theta &= 360^\circ - \tan^{-1}\frac{y}{x} & &\text{(4th quadrant)}
\end{aligned} \tag{B.7}$$

assuming that x and y are positive.

In the exponential form, $z = re^{j\theta}$ so that $dz/d\theta = jre^{j\theta} = jz$.

The third way of representing the complex z is the *exponential form*:

$$z = re^{j\theta} \tag{B.8}$$

This is almost the same as the polar form, because we use the same magnitude r and the angle θ.

The three forms of representing a complex number are summarized as follows.

$$\begin{aligned}
z &= x + jy, & &(x = r\cos\theta,\ y = r\sin\theta) & &\text{Rectangular form}\\
z &= r\underline{/\theta}, & &\left(r = \sqrt{x^2 + y^2},\ \theta = \tan^{-1}\frac{y}{x}\right) & &\text{Polar form}\\
z &= re^{j\theta}, & &\left(r = \sqrt{x^2 + y^2},\ \theta = \tan^{-1}\frac{y}{x}\right) & &\text{Exponential form}
\end{aligned} \tag{B.9}$$

The first two forms are related by Eqs. (B.5) and (B.6). In Section B.3 we will derive Euler's formula, which proves that the third form is also equivalent to the first two.

Example B.1

Express the following complex numbers in polar and exponential form: (a) $z_1 = 6 + j8$, (b) $z_2 = 6 - j8$, (c) $z_3 = -6 + j8$, (d) $z_4 = -6 - j8$.

Solution:

Notice that we have deliberately chosen these complex numbers to fall in the four quadrants, as shown in Fig. B.2.

(a) For $z_1 = 6 + j8$ (1st quadrant),

$$r_1 = \sqrt{6^2 + 8^2} = 10, \qquad \theta_1 = \tan^{-1}\frac{8}{6} = 53.13^\circ$$

Hence, the polar form is $10\underline{/53.13^\circ}$ and the exponential form is $10e^{j53.13^\circ}$.

(b) For $z_2 = 6 - j8$ (4th quadrant),

$$r_2 = \sqrt{6^2 + (-8)^2} = 10, \qquad \theta_2 = 360^\circ - \tan^{-1}\frac{8}{6} = 306.87^\circ$$

so that the polar form is $10\angle 306.87^\circ$ and the exponential form is $10e^{j306.87^\circ}$. The angle θ_2 may also be taken as -53.13°, as shown in Fig. B.2, so that the polar form becomes $10\angle -53.13^\circ$ and the exponential form becomes $10e^{-j53.13^\circ}$.

(c) For $z_3 = -6 + j8$ (2nd quadrant),

$$r_3 = \sqrt{(-6)^2 + 8^2} = 10, \qquad \theta_3 = 180^\circ - \tan^{-1}\frac{8}{6} = 126.87^\circ$$

Hence, the polar form is $10\angle 126.87^\circ$ and the exponential form is $10e^{j126.87^\circ}$.

(d) For $z_4 = -6 - j8$ (3rd quadrant),

$$r_4 = \sqrt{(-6)^2 + (-8)^2} = 10, \qquad \theta_4 = 180^\circ + \tan^{-1}\frac{8}{6} = 233.13^\circ$$

so that the polar form is $10\angle 233.13^\circ$ and the exponential form is $10e^{j233.13^\circ}$.

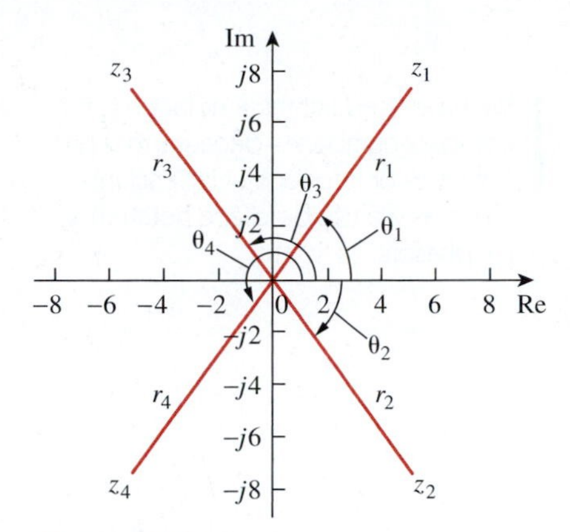

Figure B.2
For Example B.1.

Practice Problem B.1

Convert the following complex numbers to polar and exponential forms: (a) $z_1 = 3 - j4$, (b) $z_2 = 5 + j12$, (c) $z_3 = -3 - j9$, (d) $z_4 = -7 + j$.

Answer: (a) $5\angle 306.9^\circ$, $5e^{j306.9^\circ}$; (b) $13\angle 67.38^\circ$, $13e^{j67.38^\circ}$; (c) $9.487\angle 251.6^\circ$, $9.487e^{j251.6^\circ}$; (d) $7.071\angle 171.9^\circ$, $7.071e^{j171.9^\circ}$

Example B.2

Convert the following complex numbers into rectangular form: (a) $12\angle -60^\circ$, (b) $-50\angle 285^\circ$, (c) $8e^{j10^\circ}$, (d) $20e^{-j\pi/3}$.

Solution:

(a) Using Eq. (B.6),

$$12\angle -60^\circ = 12\cos(-60^\circ) + j12\sin(-60^\circ) = 6 - j10.39$$

Note that $\theta = -60^\circ$ is the same as $\theta = 360^\circ - 60^\circ = 300^\circ$.

(b) We can write

$$-50\angle 285^\circ = -50\cos 285^\circ - j50\sin 285^\circ = -12.94 + j48.3$$

(c) Similarly,

$$8e^{j10^\circ} = 8\cos 10^\circ + j8\sin 10^\circ = 7.878 + j1.389$$

(d) Finally,

$$20e^{-j\pi/3} = 20\cos(-\pi/3) + j20\sin(-\pi/3) = 10 - j17.32$$

Practice Problem B.2

Find the rectangular form of the following complex numbers: (a) $-8\angle 210^\circ$, (b) $40\angle 305^\circ$, (c) $10e^{-j30^\circ}$, (d) $50e^{j\pi/2}$.

Answer: (a) $6.928 + j4$; (b) $22.94 - j32.77$; (c) $8.66 - j5$; (d) $j50$

B.2 Mathematical Operations

We have used lightface notation for complex numbers—because they are not time- or frequency-dependent—whereas we use boldface notation for phasors.

Two complex numbers $z_1 = x_1 + jy_1$ and $z_2 = x_2 + jy_2$ are equal if and only if their real parts are equal and their imaginary parts are equal.

$$x_1 = x_2, \qquad y_1 = y_2 \tag{B.10}$$

The *complex conjugate* of the complex number $z = x + jy$ is

$$z^* = x - jy = r\underline{/-\theta} = re^{-j\theta} \tag{B.11}$$

Thus, the complex conjugate of a complex number is found by replacing every j by $-j$.

Given two complex numbers $z_1 = x_1 + jy_1 = r_1\underline{/\theta_1}$ and $z_2 = x_2 + jy_2 = r_2\underline{/\theta_2}$, their sum is

$$z_1 + z_2 = (x_1 + x_2) + j(y_1 + y_2) \tag{B.12}$$

and their difference is

$$z_1 - z_2 = (x_1 - x_2) + j(y_1 - y_2) \tag{B.13}$$

While it is more convenient to perform addition and subtraction of complex numbers in rectangular form, the product and quotient of the two complex numbers are best done in polar or exponential form. For their product,

$$z_1 z_2 = r_1 r_2\underline{/\theta_1 + \theta_2} \tag{B.14}$$

Alternatively, using the rectangular form,

$$\begin{aligned} z_1 z_2 &= (x_1 + jy_1)(x_2 + jy_2) \\ &= (x_1x_2 - y_1y_2) + j(x_1y_2 + x_2y_1) \end{aligned} \tag{B.15}$$

For their quotient,

$$\frac{z_1}{z_2} = \frac{r_1}{r_2}\underline{/\theta_1 - \theta_2} \tag{B.16}$$

Alternatively, using the rectangular form,

$$\frac{z_1}{z_2} = \frac{x_1 + jy_1}{x_2 + jy_2} \tag{B.17}$$

We rationalize the denominator by multiplying both the numerator and denominator by z_2^*.

$$\frac{z_1}{z_2} = \frac{(x_1 + jy_1)(x_2 - jy_2)}{(x_2 + jy_2)(x_2 - jy_2)} = \frac{x_1x_2 + y_1y_2}{x_2^2 + y_2^2} + j\frac{x_2y_1 - x_1y_2}{x_2^2 + y_2^2} \tag{B.18}$$

Example B.3

If $A = 2 + j5$, $B = 4 - j6$, find: (a) $A^*(A + B)$, (b) $(A + B)/(A - B)$.

Solution:

(a) If $A = 2 + j5$, then $A^* = 2 - j5$ and

$$A + B = (2 + 4) + j(5 - 6) = 6 - j$$

so that

$$A^*(A + B) = (2 - j5)(6 - j) = 12 - j2 - j30 - 5 = 7 - j32$$

(b) Similarly,

$$A - B = (2 - 4) + j(5 - -6) = -2 + j11$$

Hence,

$$\frac{A+B}{A-B} = \frac{6-j}{-2+j11} = \frac{(6-j)(-2-j11)}{(-2+j11)(-2-j11)}$$
$$= \frac{-12 - j66 + j2 - 11}{(-2)^2 + 11^2} = \frac{-23 - j64}{125} = -0.184 - j0.512$$

Practice Problem B.3

Given that $C = -3 + j7$ and $D = 8 + j$, calculate:
(a) $(C - D^*)(C + D^*)$, (b) D^2/C^*, (c) $2CD/(C + D)$.

Answer: (a) $-103 - j26$; (b) $-5.19 + j6.776$; (c) $6.045 + j11.53$

Example B.4

Evaluate:

(a) $\dfrac{(2 + j5)(8e^{j10^\circ})}{2 + j4 + 2\underline{/-40^\circ}}$ (b) $\dfrac{j(3 - j4)^*}{(-1 + j6)(2 + j)^2}$

Solution:

(a) Because there are terms in polar and exponential forms, it may be best to express all terms in polar form:

$$2 + j5 = \sqrt{2^2 + 5^2}\underline{/\tan^{-1}5/2} = 5.385\underline{/68.2^\circ}$$
$$(2 + j5)(8e^{j10^\circ}) = (5.385\underline{/68.2^\circ})(8\underline{/10^\circ}) = 43.08\underline{/78.2^\circ}$$
$$2 + j4 + 2\underline{/-40^\circ} = 2 + j4 + 2\cos(-40^\circ) + j2\sin(-40^\circ)$$
$$= 3.532 + j2.714 = 4.454\underline{/37.54^\circ}$$

Thus,

$$\frac{(2 + j5)(8e^{j10^\circ})}{2 + j4 + 2\underline{/-40^\circ}} = \frac{43.08\underline{/78.2^\circ}}{4.454\underline{/37.54^\circ}} = 9.672\underline{/40.66^\circ}$$

(b) We can evaluate this in rectangular form because all terms are in that form. But

$$j(3 - j4)^* = j(3 + j4) = -4 + j3$$
$$(2 + j)^2 = 4 + j4 - 1 = 3 + j4$$
$$(-1 + j6)(2 + j)^2 = (-1 + j6)(3 + j4) = -3 - 4j + j18 - 24$$
$$= -27 + j14$$

Hence,

$$\frac{j(3 - j4)^*}{(-1 + j6)(2 + j)^2} = \frac{-4 + j3}{-27 + j14} = \frac{(-4 + j3)(-27 - j14)}{27^2 + 14^2}$$
$$= \frac{108 + j56 - j81 + 42}{925} = 0.1622 - j0.027$$

Practice Problem B.4

Evaluate these complex fractions:

(a) $\dfrac{6\angle 30^\circ + j5 - 3}{-1 + j + 2e^{j45^\circ}}$ (b) $\left[\dfrac{(15 - j7)(3 + j2)^*}{(4 + j6)^*(3\angle 70^\circ)}\right]^*$

Answer: (a) $3.387\angle -5.615^\circ$; (b) $2.759\angle -287.6^\circ$

B.3 Euler's Formula

Euler's formula is an important result in complex variables. We derive it from the series expansion of e^x, $\cos\theta$, and $\sin\theta$. We know that

$$e^x = 1 + x + \frac{x^2}{2!} + \frac{x^3}{3!} + \frac{x^4}{4!} + \cdots \tag{B.19}$$

Replacing x by $j\theta$ gives

$$e^{j\theta} = 1 + j\theta - \frac{\theta^2}{2!} - j\frac{\theta^3}{3!} + \frac{\theta^4}{4!} + \cdots \tag{B.20}$$

Also,

$$\begin{aligned} \cos\theta &= 1 - \frac{\theta^2}{2!} + \frac{\theta^4}{4!} - \frac{\theta^6}{6!} + \cdots \\ \sin\theta &= \theta - \frac{\theta^3}{3!} + \frac{\theta^5}{5!} - \frac{\theta^7}{7!} + \cdots \end{aligned} \tag{B.21}$$

so that

$$\cos\theta + j\sin\theta = 1 + j\theta - \frac{\theta^2}{2!} - j\frac{\theta^3}{3!} + \frac{\theta^4}{4!} + j\frac{\theta^5}{5!} - \cdots \tag{B.22}$$

Comparing Eqs. (B.20) and (B.22), we conclude that

$$\boxed{e^{j\theta} = \cos\theta + j\sin\theta} \tag{B.23}$$

This is known as *Euler's formula.* The exponential form of representing a complex number as in Eq. (B.8) is based on Euler's formula. From Eq. (B.23), notice that

$$\boxed{\cos\theta = \text{Re}(e^{j\theta}), \qquad \sin\theta = \text{Im}(e^{j\theta})} \tag{B.24}$$

and that

$$|e^{j\theta}| = \sqrt{\cos^2\theta + \sin^2\theta} = 1$$

Replacing θ by $-\theta$ in Eq. (B.23) gives

$$e^{-j\theta} = \cos\theta - j\sin\theta \tag{B.25}$$

Adding Eqs. (B.23) and (B.25) yields

$$\boxed{\cos\theta = \frac{1}{2}(e^{j\theta} + e^{-j\theta})} \tag{B.26}$$

Subtracting Eq. (B.25) from Eq. (B.23) yields

$$\sin\theta = \frac{1}{2j}(e^{j\theta} - e^{-j\theta}) \tag{B.27}$$

Useful Identities

The following identities are useful in dealing with complex numbers. If $z = x + jy = r\underline{/\theta}$, then

$$zz^* = x^2 + y^2 = r^2 \tag{B.28}$$

$$\sqrt{z} = \sqrt{x + jy} = \sqrt{r}e^{j\theta/2} = \sqrt{r}\underline{/\theta/2} \tag{B.29}$$

$$z^n = (x + jy)^n = r^n\underline{/n\theta} = r^n e^{jn\theta} = r^n(\cos n\theta + j\sin n\theta) \tag{B.30}$$

$$\begin{aligned} z^{1/n} &= (x + jy)^{1/n} = r^{1/n}\underline{/\theta/n + 2\pi k/n} \\ k &= 0, 1, 2, \dots, n - 1 \end{aligned} \tag{B.31}$$

$$\begin{aligned} &\ln(re^{j\theta}) = \ln r + \ln e^{j\theta} = \ln r + j\theta + j2k\pi \\ &(k = \text{integer}) \end{aligned} \tag{B.32}$$

$$\begin{aligned} \frac{1}{j} &= -j \\ e^{\pm j\pi} &= -1 \\ e^{\pm j2\pi} &= 1 \\ e^{j\pi/2} &= j \\ e^{-j\pi/2} &= -j \end{aligned} \tag{B.33}$$

$$\begin{aligned} \text{Re}(e^{(\alpha+j\omega)t}) &= \text{Re}(e^{\alpha t}e^{j\omega t}) = e^{\alpha t}\cos\omega t \\ \text{Im}(e^{(\alpha+j\omega)t}) &= \text{Im}(e^{\alpha t}e^{j\omega t}) = e^{\alpha t}\sin\omega t \end{aligned} \tag{B.34}$$

Example B.5

If $A = 6 + j8$, find: (a) $\sqrt{A}$, (b) A^4.

Solution:

(a) First, convert A to polar form:

$$r = \sqrt{6^2 + 8^2} = 10, \qquad \theta = \tan^{-1}\frac{8}{6} = 53.13^\circ, \qquad A = 10\underline{/53.13^\circ}$$

Then

$$\sqrt{A} = \sqrt{10}\underline{/53.13^\circ/2} = 3.162\underline{/26.56^\circ}$$

(b) Because $A = 10\underline{/53.13^\circ}$,

$$A^4 = r^4\underline{/4\theta} = 10^4\underline{/4 \times 53.13^\circ} = 10{,}000\underline{/212.52^\circ}$$

Practice Problem B.5

If $A = 3 - j4$, find: (a) $A^{1/3}$ (3 roots), and (b) ln A.

Answer: (a) $1.71\underline{/102.3^\circ}$, $1.71\underline{/222.3^\circ}$, $1.71\underline{/342.3^\circ}$;
(b) $1.609 + j5.356 + j2n\pi$ $(n = 0, 1, 2, \dots)$

Appendix C

PSpice for Windows

There are several computer software packages such as Spice, Multisim, Mathcad, Quattro, MATLAB, and Maple that can be used for circuit analysis. The most popular is Spice, which stands for Simulation Program with Integrated-Circuit Emphasis. Spice was developed at the Department of Electrical and Computer Engineering at the University of California at Berkeley in the 1970s for mainframe computers. Since then, more than 20 versions have been developed. PSpice, a version of Spice for personal computers, was developed by MicroSim Corporation in California and made available in 1984 and later by OrCAD and Cadence. PSpice has been made available in different operating systems: DOS, Windows, and Unix, to name a few. A free copy of PSpice can be downloaded from the Internet.

Because the Windows version of PSpice is becoming more and more popular, it is this version that is used in book. Specifically, version 16.3 of PSpice/OrCAD from Cadence is used throughout this book.

The objective of this appendix is to provide a short tutorial on using the Windows-based PSpice on an IBM PC or equivalent.

PSpice can analyze up to roughly 130 elements and 100 nodes. It is capable of performing three major types of circuit analysis: (1) dc analysis, (2) transient analysis and, (3) ac analysis. In addition, it can also perform transfer function analysis, Fourier analysis, and operating point analysis. The circuit can contain resistors, inductors, capacitors, independent and dependent voltage and current sources, op amps, transformers, transmission lines, and semiconductor devices.

We will assume that you are familiar with using the Microsoft Windows operating system and that PSpice for Windows is already installed in your computer.

C.1 Design Center for Windows

In earlier versions of PSpice, PSpice for Windows is formally known as the MicroSim Design Center, which is a computer environment for simulating electric circuits. The Design Center for Windows includes the following programs:

- *Orcad Capture*: This program (used to be known as Schematics) is a graphical editor used to draw the circuit to be simulated on the screen. It allows the user to enter the components, wire the components together to form the circuit, and specify the type of analysis to be performed.

- *PSpice*: This program simulates the circuit created using Orcad Capture. By simulation, we mean a method of analysis described in a program by which a circuit is represented by mathematical models of the components comprising the circuit.
- *Probe*: This program provides a graphic display of the output generated by the PSpice program. It can be used to observe any voltage or current in the circuit.

One may think of *Orcad Capture* as the computer breadboard for setting up the circuit topology, *PSpice* as the simulator (performing the computation), and *Probe* as the oscilloscope. Using the *Orcad Capture* is perhaps the hardest part of circuit simulation using PSpice. The next section covers the essential skills needed to operate the *Orcad Capture*.

C.2 Creating a Circuit

For a circuit to be analyzed by PSpice, we must take three steps: (1) create the circuit, (2) simulate it, and (3) print or plot the results. In this section, we learn how to create the circuit using the Schematics program.

Before we discuss how to use the Schematics capture, we need to know how to use the mouse to select an object and perform an action. The mouse is used in Schematics in conjunction with the keyboard to carry out various instructions. Throughout this text, we will use the following terms to represent actions to be performed by the mouse:

- **CLICKL**: click the left button once to select an item.
- **CLICKR**: click the right button once to abort a mode.
- **DCLICKL**: double-click the left button to edit a selection or end a mode.
- **DCLICKR**: double-click the right button to repeat an action.
- **CLICKLH**: click the left button, hold down, and move the mouse to drag a selected item. Release the left button after the item has been placed.
- **DRAG**: Drag the mouse (without clicking) to move an item.

When the term "click" is used, we mean that you quickly press and release the **left** mouse button. To select an item requires **CLICKL**, while to perform an action requires **DCLICKL**. Also, to avoid writing "click" several times, the menu to be clicked will be highlighted in bold. For example, click **Place**, click **Wire** will be written as **Place/Wire**. Of course, we can always press the <Esc> key to abort any action.

Assuming that you are using Windows XP, you can access PSpice by clicking the Start icon on the left-hand corner of your PC, drag the cursor to All Programs, to Cadence, to Orcad 16.3 Demo, and to Orcad Capture CIS Demo, as shown in Fig. C.1. This will lead you to a screen or window. You select **File/New/Project**. (We use "/" throughout to indicate levels of a hierarchical menu.) It will ask for the name of the new project. For example, you may type "example" and click OK. This will lead you to a dialog box. Select "Create a blank project" and click OK. The blank screen, with the Main Menu bar (from **File** to **Help**) at the top, will appear as shown in Fig. C.2.

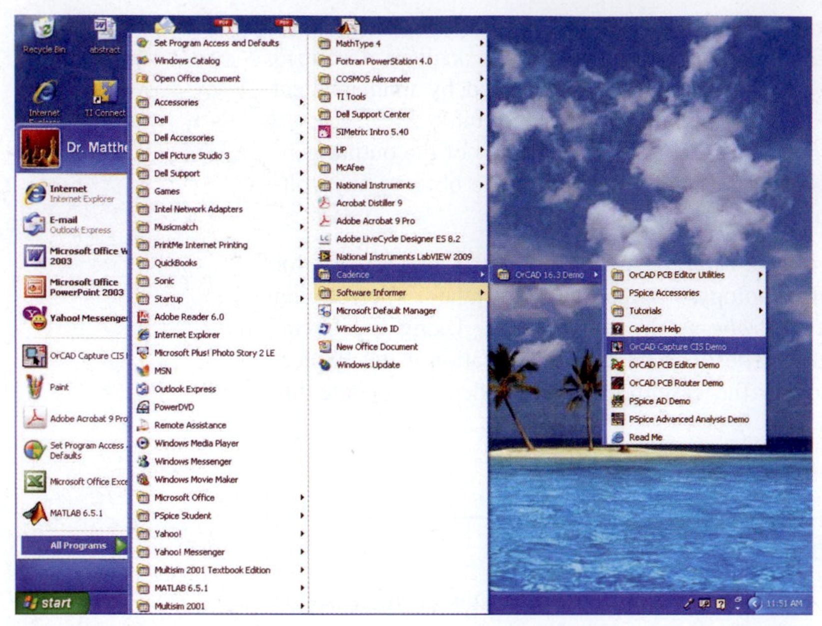

Figure C.1
Accessing PSpice in Windows.

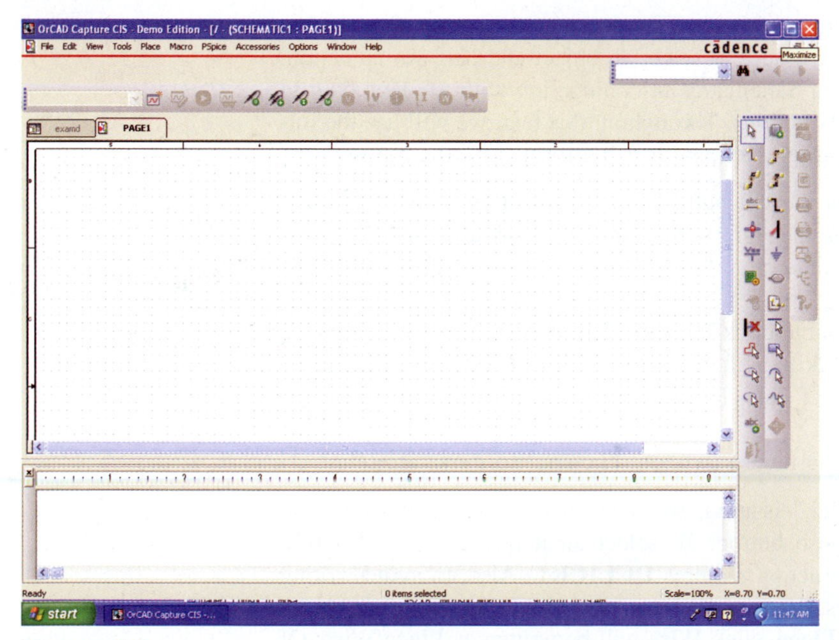

Figure C.2
Window for Orcad Capture.

To create a circuit using Schematics requires three steps: (1) placing the parts or components of the circuit, (2) wiring the parts together to form the circuit, and (3) changing attributes of the parts.

Step 1: Placing the Parts

Each circuit part is retrieved by following this procedure:

- Select **Place/Part** to pop up the Place Part dialog box shown in Fig. C.3. For resistor, for example, type r in the box. (You can also

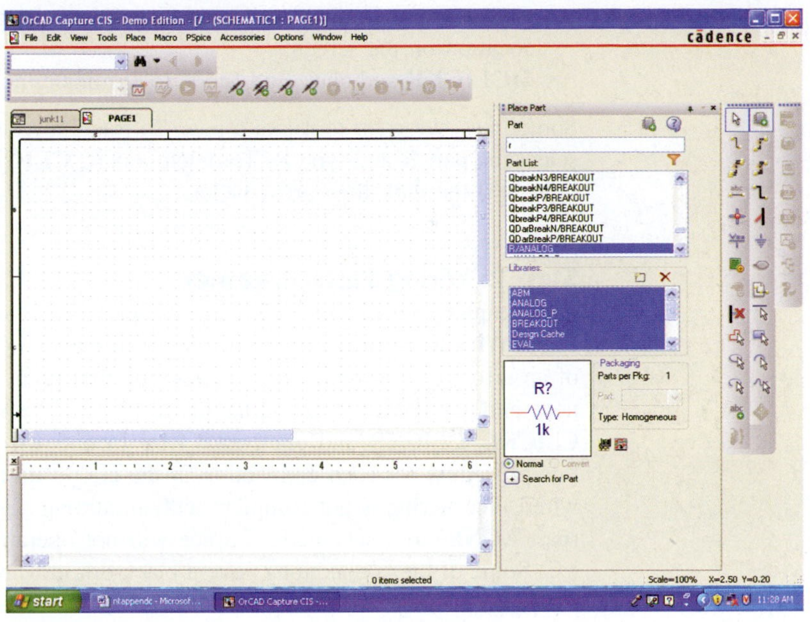

Figure C.3
Place Part dialog box on the right-hand side.

type R, because PSpice is not case sensitive.) Table C.1 shows some part names and symbols for circuit elements and independent voltage and current sources and the PSpice libraries where the parts are located.

TABLE C.1

PSpice parts and their libraries.

Symbol	Description/PSpice name	Library
R_1 1 k	Resistor/R	ANALOG
L_1 10 uH	Inductor/L	ANALOG
C_1 1 n	Capacitor/C	ANALOG
0 Vdc + − V_1	DC voltage source/VDC	SOURCE
0 Adc I_1	DC current source/IDC	SOURCE
0	Ground/0	CAPSYM

- **DCLICK** on R/ANALOG in blue and drag the part to the desired location on the screen.
- **DCLICKR** and click End Mode to terminate the placement mode.

Sometimes, we want to rotate a part. To rotate a resistor, for example, select the part R and type R. To delete a part, **CLICK** to select (highlight red) the part, then press Delete.

Step 2: Wiring Parts Together

We complete the circuit by wiring the parts together. We first select **Place/Wire** to be in wiring mode. A pencil cursor will appear in place of an arrow cursor. Drag the cursor to the first point you want to connect and **CLICKL**. Next, drag the cursor to the second point and CLICKL. CLICKR and click *End Wire* to end wiring mode. Repeat this procedure for each connection in the circuit until all the parts are wired. The wiring is not complete without adding a ground connection (part AGND) to a schematic; PSpice will not operate without it.

Some of the connections have a black dot indicating a connection. Although it is not necessary to have a dot where a wire joins a pin, having the dot shows the presence of a connection. To be sure a dot appears, make sure the wire overlaps the pin.

Step 3: Changing Attributes of Parts

As shown in Table C.1, each component has an attribute in addition to its symbol. Attributes are the labels for parts. (A component may have several attributes, of which some are displayed by default. If need be, we may add more attributes for display. However, we should hide unimportant attributes to avoid clutter.) Each attribute consists of a *name* and its designated *value*. For example, R and VDC are the names of the resistor and dc voltage source, respectively, while 2k and 10 V are the designated values of the resistor and voltage source, respectively.

As parts are placed on the screen, they are automatically assigned names by successive numbers (e.g., R1, R2, R3, etc.). Also, some parts are assigned some predetermined values. For example, all resistors are placed horizontally and assigned a value of 1 kΩ. We may need to change the attributes (names and values) of a part. Although there are several ways of changing the attributes, one simple way will be shown here.

To change the name R3 to RX, for example, **DCLICKL** on the text R3 to bring up the Display Properties dialog box. Type the new name RX and click the OK button to accept the change. The same procedure can be used to change VDC to V1 or whatever.

To change the value 1k to 10Meg, for example, **DCLICKL** on the 1k attribute (not the symbol) to open up the Display Properties dialog box. Type the new value 10Meg (no space between 1 and Meg) and click the OK button to accept the change. Similarly, to change the default value 0V to 15kV for voltage source VDC, **DCLICKL** the symbol for 0V to bring up the Display Properties dialog box. Type 15kV in the value box and click OK. For convenience, numbers can be expressed with the scale factors in Table C.2. For example 6.6×10^{-8} can be written as 66N or 0.066U.

TABLE C.2

Scale factors.

Symbol	Value	Name of suffix
T	10^{12}	Tera
G	10^{9}	Giga
MEG	10^{6}	Mega
K	10^{3}	Kilo
M	10^{-3}	Milli
U	10^{-6}	Micro
N	10^{-9}	Nano
P	10^{-12}	Pico
F	10^{-15}	Femto

Example C.1

Draw the circuit in Fig. C.4 using PSpice.

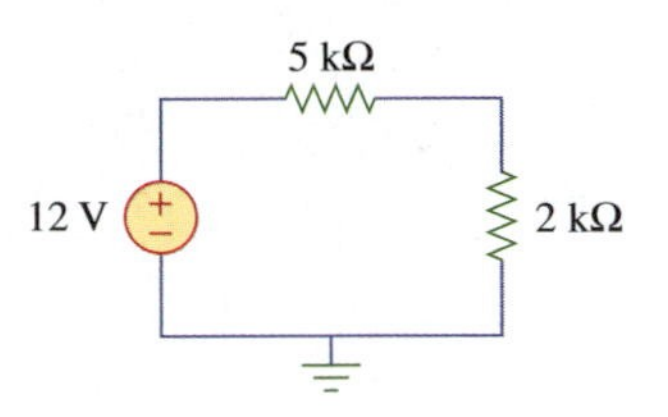

Figure C.4
For Example C.1.

Solution:

We assume that you are already on the Schematic screen shown in Fig. C.2 or Fig. C.3. So we have a blank screen as worksheet to draw the circuit on. We now take the following steps to create the circuit in Fig. C.4.

To place the voltage source, we need to:

1. Select **Place/Part**.
2. Type VDC in the Part box.
3. **DCLICKL** on "VDC/Design Cashe" in blue.
4. Drag the part to the desired location on the screen.
5. **CLICKL** to place VDC and **CLICKR** to terminate placement mode.

At this point, only the voltage source V1 in Fig. C.5(a) is shown on the screen, highlighted red. To place the resistors, we need to:

1. Select **Place/Part**. (This step is unnecessary if you are already there.)
2. Type R the Part box.
3. **DCLICKL** on "R/ANALOG" in blue.
4. Drag the part to the desired location on the screen.
5. Press R to rotate R2 and drag it to its desired location on the screen.
6. **CLICKR** to terminate placement mode.

At this point, the three parts have been created as shown in Fig. C.5(a). The next step is to connect the parts by wiring. To do this requires that we:

1. Select **Place/Wire** to be in wiring mode indicated by the plus sign (+) cursor.
2. **DRAG** the cursor to the top of V1 CLICKL to join the wire to the top of V1.
3. **DRAG** the cursor to the top corner and then to the left of R1.
4. **CLICKR** to end placement mode.

Follow the preceding steps to connect R1 with R2 and V1 with R2. At this point, we have the circuit in Fig. C.5(b), except that the ground symbol is missing. We insert the ground by taking the following steps:

1. Select Place/Ground.
2. **DCLICKL** on 0/CAPSYM.
3. **DRAG** the part to the desired location on the screen.
4. **CLICKR** to terminate placement mode.

The last thing to be done is to change or assign values to the attributes. To assign the attribute 12 V to V1, we take these steps:

1. **DCLICKL** on 0Vdc to open up the *Display Properties* dialog box.
2. Type 12Vdc in the *Value* box.
3. Click OK.

To assign 5k to R1, we follow these steps:

1. **DCLICKL** on 1k attribute of R1 to bring up the *Display Properties* dialog box.
2. Type 5k in the *Value* box.
3. Click OK.

The same procedure is used in assigning value 2k to R2. The final circuit is as shown in Fig. C.5(c).

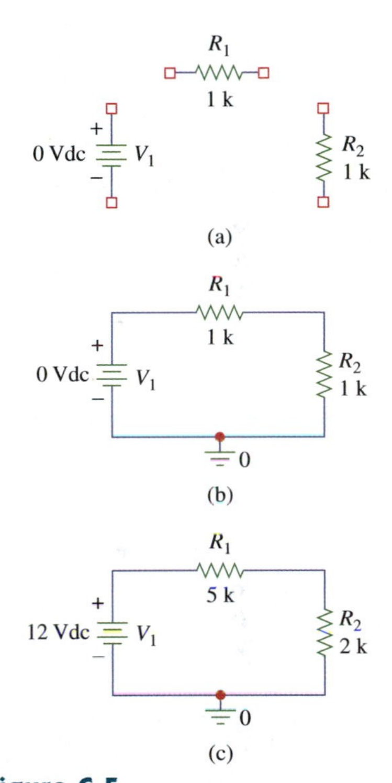

Figure C.5
Creating the circuit in Fig. C.4:
(a) placing the parts; (b) wiring the parts together; (c) changing attributes.

Practice Problem C.1

Construct the circuit in Fig. C.6 with PSpice Schematics.

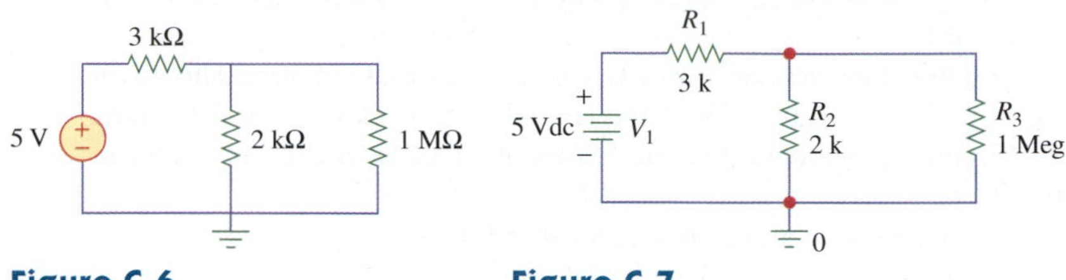

Figure C.6
For Practice Problem C.1.

Figure C.7
For Practice Problem C.1.

Answer: See the schematic in Fig. C.7.

C.3 DC Analysis

DC analysis is one of the standard analyses that can be performed using PSpice. Other standard analyses include transient, ac, and Fourier. Under dc analysis, there are two kinds of simulation that can be executed using PSpice: (1) dc nodal analysis and (2) DC Sweep.

1. DC Nodal Analysis

PSpice allows dc nodal analysis to be performed on dc sources and provides the dc voltage at each node of the circuit and dc branch currents if required. For illustration, we may use the schematic in Fig. C.5. At this point, the schematic must be saved because PSpice will not run without first saving the schematic to be simulated. Before learning how to run PSpice, the following points should be noted:

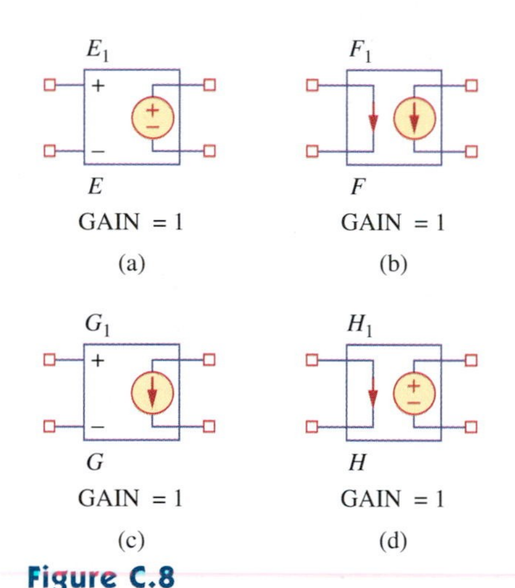

Figure C.8
(a) Voltage-controlled voltage source (VCVS); (b) current-controlled current source (CCCS); (c) voltage-controlled current source (VCCS); (d) current-controlled voltage source (CCVS).

1. It is important that there is a reference node or ground connection in the schematic. Any node can be used as a ground, and the voltages at other nodes will be oriented with respect to the selected ground.
2. Dependent sources are found in the ANALOG library. They can be obtained by selecting **Place/Part** and typing the part name. The part name for each type with the gain is shown in Fig. C.8, where E is a voltage-controlled voltage source with gain E; F is a current-controlled current source with gain F; G is a voltage-controlled current source with a transconductance gain G; and H is a current-controlled voltage source with transresistance gain H.
3. By convention, it is assumed in dc analysis that all capacitors are open circuits and all inductors are short circuits.

We run PSpice by selecting **PSpice/Run**. This invokes the *electric rule check* (ERC), which generates the *netlist*. The ERC performs a connectivity check on the schematic before creating the netlist. The netlist is a list describing the operational behavior of each component in the circuit and its connections. Each line in the netlist represents a single component of the circuit. The netlist can be examined by selecting **PSpice/View Netlist**. If errors are found in the schematic, an *error* window will appear. There are two kinds of common errors in PSpice: (1) errors involving wiring of the circuit and (2) errors that occur during simulation. Once the errors are noted, exit from the Error List and go back to Schematics to correct the errors. If no errors are found, the system automatically enters PSpice and performs the simulation (nodal analysis). When the analysis is complete, the nodal voltages are automatically displayed on the circuit and the result/output file is created. To examine the output file, click **PSpice/View Output File** from Schematics. The values displayed on the circuit should be the same as those in the output file.

2. DC Sweep

DC nodal analysis allows simulation for dc sources with fixed voltages or currents. DC Sweep provides more flexibility in that it allows the calculation of node voltages and branch currents of a circuit when a source is swept over a range of values. For example, suppose we desire to perform a DC Sweep of voltage source V1 in Fig. C.5(c) from 0 to 20 V in 1-V increments. We proceed as follows:

1. Select **PSpice/New Simulation Profile**. This leads to Simulation Settings dialog box, shown in Fig. C.9.
2. Under *Analysis type*, select DC Sweep.
3. On *Name*, type V1.
4. Under *Sweep type*, select Linear.
5. In *Start Value* box, type 0.
6. In *End Value* box, type 20.
7. In *Increment* box, type 1.
8. Click OK to save parameters.

Notice that in Fig. C.9, the default setting is *Voltage Source* for the *Sweep Variable*, while it is *Linear* for *Sweep Type*. If needed, other options can be selected by clicking the appropriate buttons.

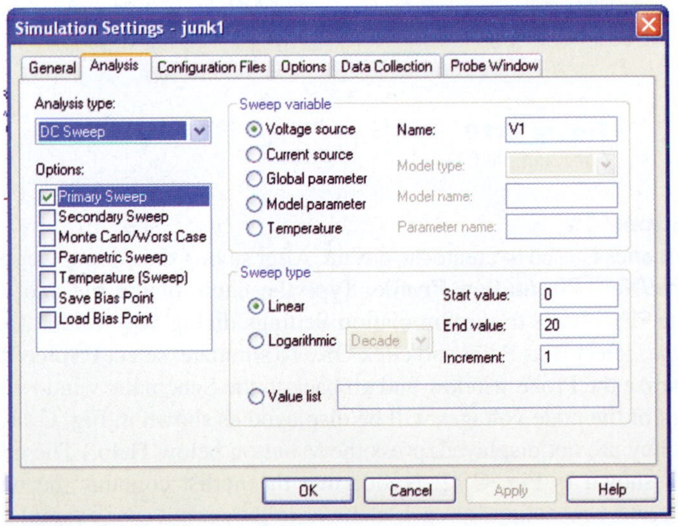

Figure C.9
Simulation Settings dialog box.

To run DC sweep analysis, select **PSpice/Run**. Schematics will create a netlist and then run PSpice if no errors are found.

If no errors are found, the data generated by PSpice are passed to Probe. The Probe window will appear, displaying a graph in which the *x*-axis is by default set to the DC sweep variable and range and the *y*-axis is blank for now. To display some specific plots, select **Trace/Add Trace}** in the Probe menu to open the *Add Traces* dialog box. The box contains traces that are the output variables (node voltages and branch currents) in the data file available for display. Select the traces to be displayed by clicking or typing them and click OK. The selected traces will be plotted and displayed on the screen. As many traces as you want may be added to the same plot or on different windows. To delete a trace, click the trace name in the legend of the plot to highlight it and press <Delete>.

It is important to understand how to interpret the traces. We must interpret the voltage and current variables according to the passive sign convention. As parts are initially placed horizontally in a schematic, the left-hand terminal is named pin 1 while the right-hand terminal is pin 2. When a component (say, R1) is rotated counterclockwise once, pin 2 would be on the top because rotation is about pin 1. Therefore, if current enters through pin 2, the current I(R1) through R1 would be negative. In other words, positive current implies that the current enters through pin 1, while negative current means that the current enters through pin 2. As for voltage variables, they are always with respect to the ground. For example, V(R1:2) is the voltage (with respect to the ground) at pin 2 of resistor R1; V(V1:+) is the voltage (with respect to the ground) at the positive terminal of voltage source V1; and V(E2:1) is the voltage at pin 1 of component E2 with respect to ground, regardless of the polarity.

Example C.2

For the circuit in Fig. C.10, find the dc node voltages and the current i_o.

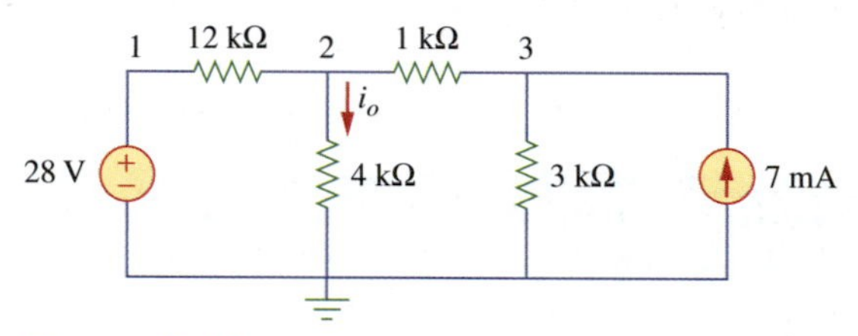

Figure C.10
For Example C.2.

Solution:

Schematics is used to create the circuit. After saving the circuit, we select **PSpice/New Simulation Profile**. Type the name of the file and click Create. This leads to the Simulation Settings dialog box. Under *Analysis type*, select Bias Point and click OK. To simulate, select **PSpice/Run**. Minimize the Probe window and go back to the Schematic window. The values of the node voltages will be displayed as shown in Fig. C.11. (In case they are not displayed, press the V button below Help.) The netlist file is shown in Fig. C.12. Notice that the netlist contains the name, value, and connection for each element in the circuit. For example, the

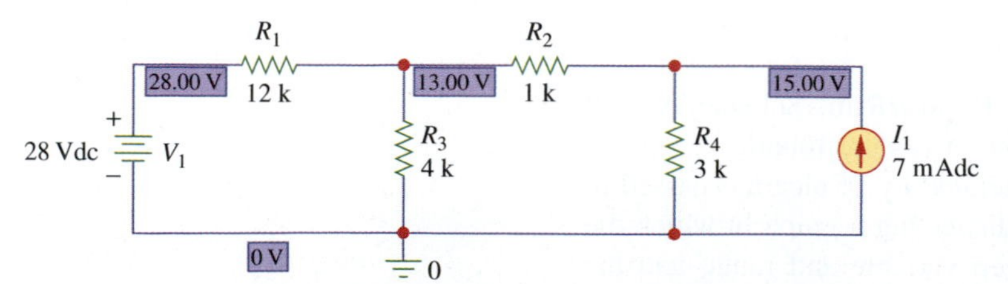

Figure C.11
For Example C.2; the schematic for the circuit in Fig. C.10.

```
R_R2   N00567 N00579 1k TC=0,0
I_I1   0 N00579 DC 7mAdc
R_R4   0 N00579 3k TC=0,0
V_V1   N00543 0 28Vdc
R_R1   N00543 N00567 12k TC=0,0
R_R3   0 N00567 4k TC=0,0
```

Figure C.12
The Netlist file for Example C.2.

first line shows that resistor R2 is connected between nodes N00567 and N00579 and its value is 1k. The edited version of the output file is shown in Fig. C.13. The output file also contains the Netlist file but this was removed from Fig. C.13. From the output file, we obtain i_o as 3.25 mA. Alternatively, because we know the voltage across R3 to be 13 V from Fig. C.11, $i_o = 13/4\text{k} = 3.25$ mA.

```
**** 09/21/10 14:29:04 ******** PSpice Lite (June 2009) ******* ID# 10813 ****

** Profile: "SCHEMATIC1-examc2a" [
C:\OrCAD\OrCAD_16.3_Demo\tools\examc2a-PSpiceFiles\SCHEMATIC1\examc2a.sim ]

**** CIRCUIT DESCRIPTION

******************************************************************************

** Creating circuit file "examc2a.cir"
** WARNING: THIS AUTOMATICALLY GENERATED FILE MAY BE OVERWRITTEN BY SUBSEQUENT SIMULATIONS

* Libraries:
* Profile Libraries :
* Local Libraries :
* From [PSPICE NETLIST] section of
C:\OrCAD\OrCAD_16.3_Demo\tools\PSpice\PSpice.ini file:.lib "nomd.lib"

*Analysis directives:
.PROBE V(alias(*)) I(alias(*)) W(alias(*)) D(alias(*)) NOISE(alias(*))
.INC "..\SCHEMATIC1.net"

**** INCLUDING SCHEMATIC1.net ****
* source EXAMC2A
R_R2       N00567 N00579 1k TC=0,0
I_I1       0 N00579 DC 7mAdc
R_R4       0 N00579 3k TC=0,0
V_V1       N00543 0 28Vdc
R_R1       N00543 N00567 12k TC=0,0
R_R3       0 N00567 4k TC=0,0

**** RESUMING examc2a.cir ****
.END

**** 09/21/10 14:29:04 ******** PSpice Lite (June 2009) ******* ID# 10813 ****

** Profile: "SCHEMATIC1-examc2"
[ C:\OrCAD\OrCAD_16.3_Demo\tools\examc2-PSpiceFiles\SCHEMATIC1\examc2.sim ]

**** SMALL SIGNAL BIAS SOLUTION TEMPERATURE= 27.000 DEG C

******************************************************************************
NODE VOLTAGE          NODE VOLTAGE          NODE VOLTAGE          NODE VOLTAGE

(N00543) 28.0000      (N00567) 13.0000      (N00579) 15.0000

VOLTAGE SOURCE CURRENTS
NAME        CURRENT

V_V1        -1.250E-03
V_V2        3.250E-03

TOTAL POWER DISSIPATION 3.50E-02 WATTS
```

Figure C.13
Output file (edited version) for Example C.2.

Practice Problem C.2

Use PSpice to determine the node voltages and the current i_x in the circuit of Fig. C.14.

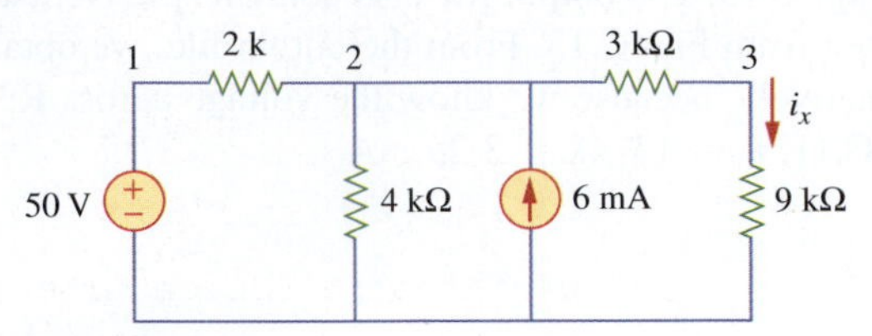

Figure C.14
For Practice Problem C.2.

Answer: $V_1 = 50$; $V_2 = 37.2$; $V_3 = 27.9$; $i_x = 3.1$ mA

Example C.3

Plot I_1 and I_2 if the dc voltage source in Fig. C.15 is swept from 2 to 10 V.

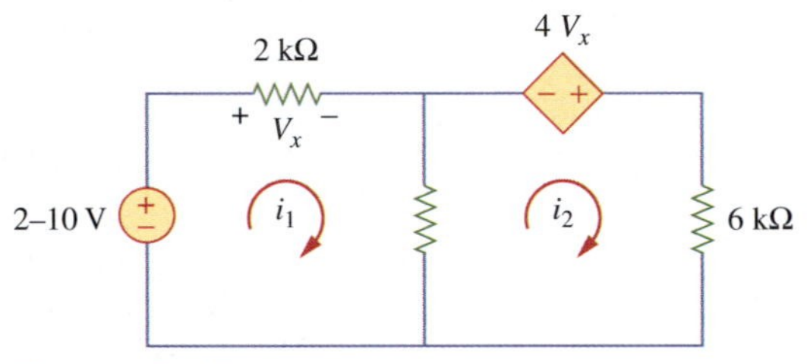

Figure C.15
For Example C.3.

Solution:
We draw the schematic of the circuit and set the attributes as shown in Fig. C.16. Notice how the voltage-controlled voltage source E1 is connected. Once the schematic is completed and saved, we select **PSpice/New Simulation Profile**. In the New Simulation dialog box, type the filename and click Create. This will lead to the Simulation Settings dialog box. Under Analysis type, select *DC Sweep*. Under Sweep Variable, select *Voltage source*. Under Name, type V1. Type 2, 10, and 0.5 for *Start Value*, *End Value*, and *Increment*, respectively. Click OK. Select **PSpice/Run**. This leads to Probe window. Select **Trace/Add Trace**. Click I(R1) and –I(R3) to be displayed. (The negative sign is needed to make the current through R3 positive.) Fig. C.17 shows the result.

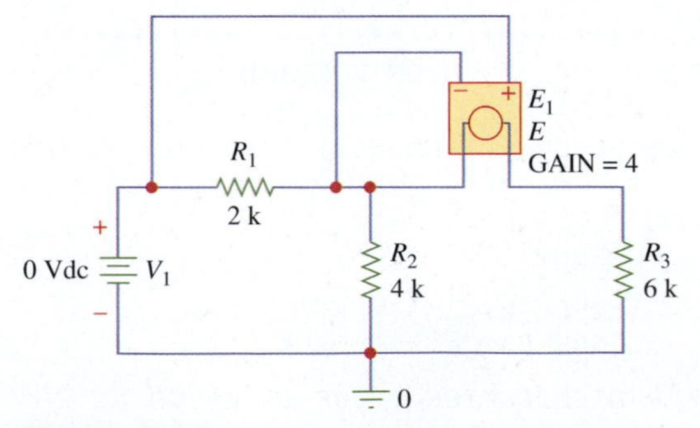

Figure C.16
The schematic for the circuit in Fig. C.15.

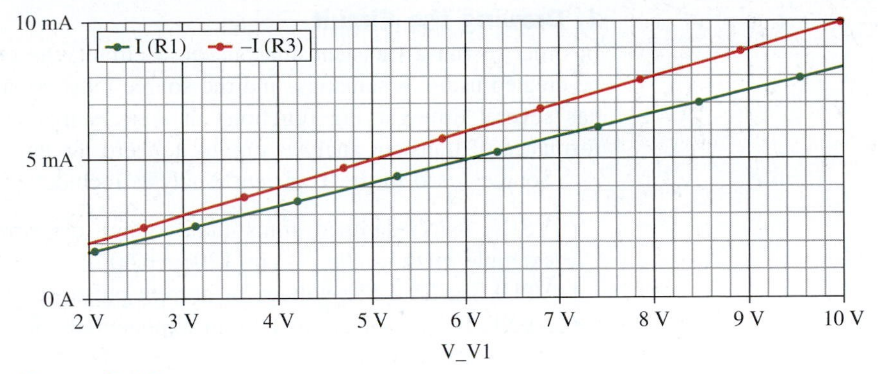

Figure C.17
Plots of I_1 and I_2 against V1.

Practice Problem C.3

Use PSpice to obtain the plots of i_x and i_o if the dc voltage source in Fig. C.18 is swept from 2 to 10 V.

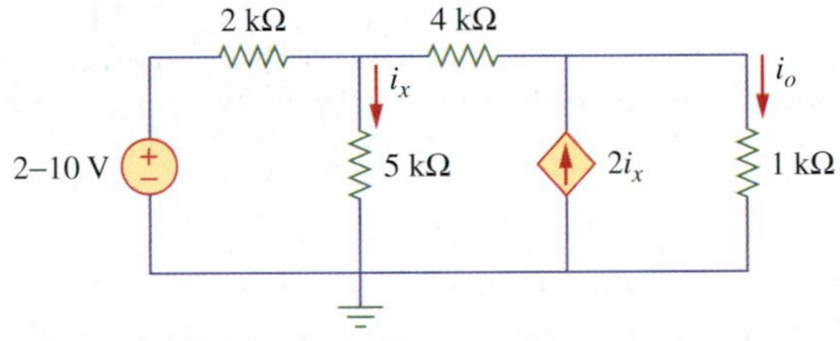

Figure C.18
For Practice Problem C.3.

Answer: The plots of i_x and i_o are displayed in Fig. C.19.

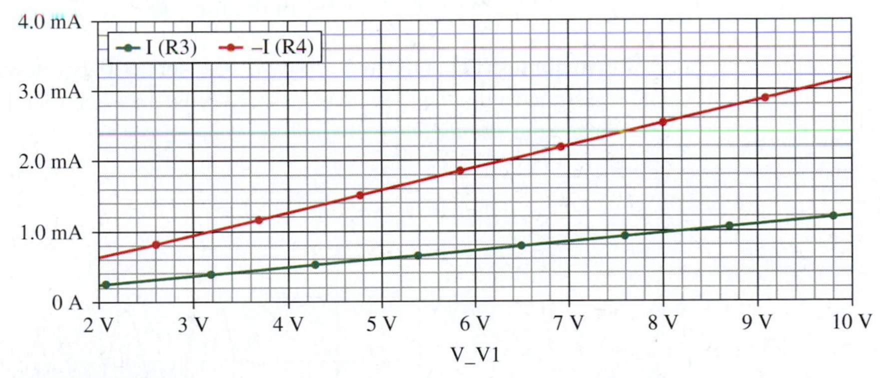

Figure C.19
Plots of i_x and i_o versus V1.

C.4 Transient Analysis

In PSpice, Transient Analysis is generally used to examine the behavior of a waveform (voltage or current) as time varies. (Transient Analysis is used to view the transient response of inductors and capacitors.) Transient Analysis solves some differential equations describing a circuit and obtains voltages and currents versus time. Transient Analysis is also used to obtain Fourier analysis. To perform transient analysis on a circuit using PSpice usually involves these steps: (1) drawing the circuit, (2) providing specifications, and (3) simulating the circuit.

1. Drawing the Circuit

In order to run a transient analysis on a circuit, the circuit must first be created using Schematics, and the source must be specified. PSpice has several time-varying functions or sources that enhance the performance of transient analysis due to nonzero inputs.

Sources used in the Transient Analysis include:

- VSIN, ISIN: damped sinusoidal voltage or current source, for example, $v(t) = 10e^{-0.2t}\sin(120\pi - 60°)$.
- VPULSE, IPULSE: voltage or current pulse.
- VEXP, IEXP: voltage or current exponential source, for example, $i(t) = 6[1 - \exp(0.5t)]$.
- VPWL, IPWL: piecewise linear voltage or current function which can be used to create an arbitrary waveform.

It is expedient that we take a close look at these functions.

VSIN is the exponentially damped sinusoidal voltage source; that is,

$$v(t) = V_o + V_m e^{-\alpha(t - t_d)}\sin[2\pi f(t - t_d) + \phi] \tag{C.1}$$

The VSIN source has the following attributes, which are illustrated in Fig. C.20 and compared with Eq. (C.1).

$$\begin{aligned}
\text{VOFF} &= \text{Offset voltage, } V_o \\
\text{VAMPL} &= \text{Amplitude, } V_m \\
\text{TD} &= \text{Time delay in seconds, } t_d \\
\text{FREQ} &= \text{Frequency in Hz, } f \\
\text{DF} &= \text{Damping factor (dimensionless), } \alpha \\
\text{PHASE} &= \text{Phase in degrees, } \phi
\end{aligned} \tag{C.2}$$

Attributes TD, DF, and PHASE are set to 0 by default but can be assigned other values if necessary. What has been said about VSIN is also true for ISIN.

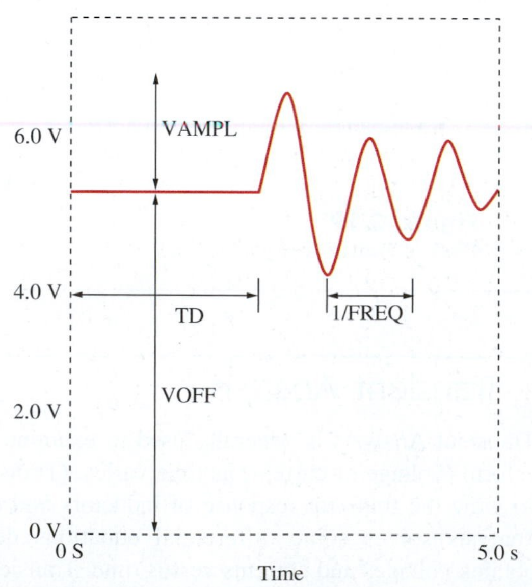

Figure C.20
Sinusoidal voltage source VSIN.

The VPULSE source has the following attributes, which are portrayed in Fig. C.21.

$$\begin{aligned}
\text{V1} &= \text{Low voltage} \\
\text{V2} &= \text{High voltage} \\
\text{TD} &= \text{Initial time delay in seconds} \\
\text{TR} &= \text{Rise time in seconds} \\
\text{TF} &= \text{Fall time in seconds} \\
\text{PW} &= \text{Pulse width in seconds} \\
\text{PER} &= \text{Period in seconds}
\end{aligned} \tag{C.3}$$

Attributes V1 and V2 must be assigned values. By default, attribute TD is assigned 0; TR and TF are assigned the *print step* value; and PW and PER are assigned the *final time* value. The values of the *print*

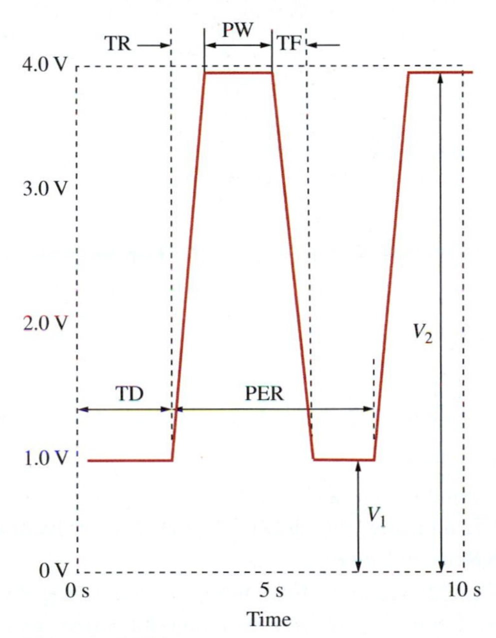

Figure C.21
Pulse voltage source VPULSE.

time and *final time* are obtained as default values from the specifications provided by the User in preparation for transient analysis.

The exponential voltage source VEXP has the following attributes, which are illustrated in Fig. C.22.

$$\begin{aligned}
\text{V1} &= \text{Initial voltage} \\
\text{V2} &= \text{Final voltage} \\
\text{TD1} &= \text{Rise delay in seconds} \\
\text{TC1} &= \text{Rise time constant in seconds} \\
\text{TD2} &= \text{Fall delay in seconds} \\
\text{TC2} &= \text{Fall time in seconds}
\end{aligned} \tag{C.4}$$

The piecewise linear voltage source VPWL, such as shown in Fig. C.23, requires specifying pairs of TN, VN, where VN is the

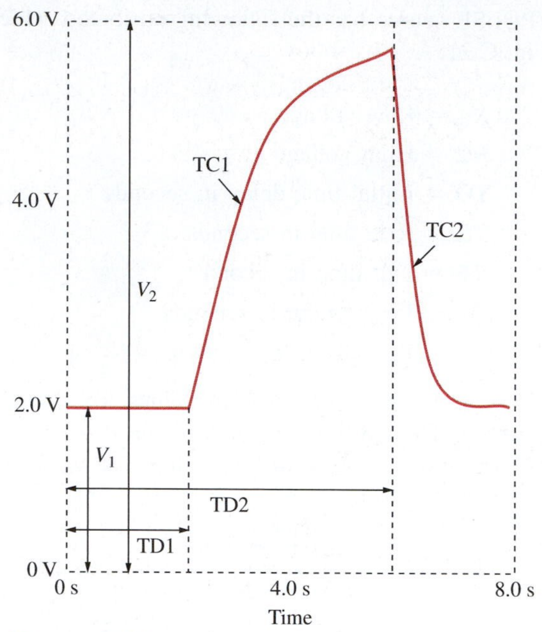

Figure C.22
Exponential voltage source VEXP.

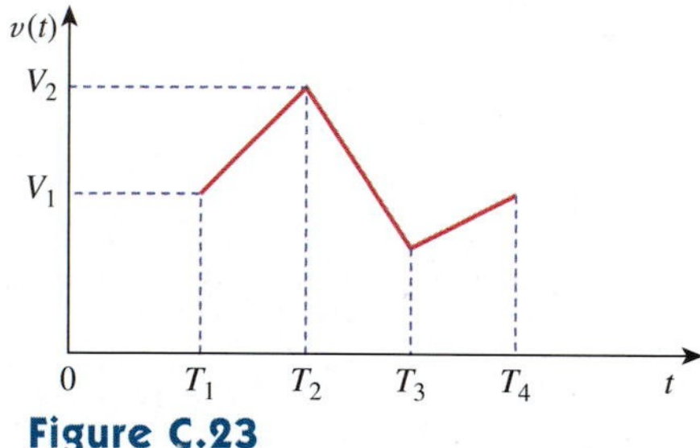

Figure C.23
Piecewise linear voltage source VPWL.

voltage at time TN for $N = 1, 2, \ldots, 10$. For example, for the function in Fig. C.24, we will need to specify the attributes

T1 = 0, V1 = 0, T2 = 2, V2 = 4, T3 = 6, V3 = 4, T4 = 8, V4 = −2

A source is added to the schematic by taking the following steps:

1. Select **Place/Part**.
2. Type the name of the source.
3. **DCLICKL** and drag the symbol to the desired location.
4. **DCLICKR** to end mode.
5. **DCLICKL** the symbol of the source to open up the Property Editor dialog box that will allow you to change the attributes of the source.

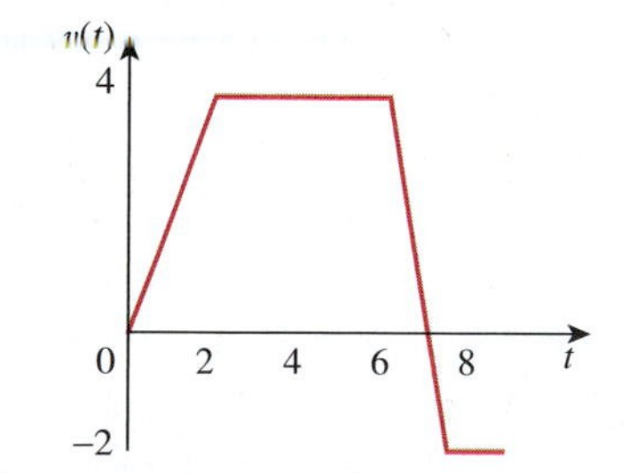

Figure C.24
An example of a piecewise linear voltage source VPWL.

In step 5, the attributes may not be shown on the schematic after entering their values. To display an attribute, **DCLICKL** the symbol of the source to open up the Property Editor dialog again. Highlight the attribute that you want to display and click Display. In the Display Properties dialog box, select Name and Value.

In addition to specifying the source to be used in transient analysis, there may be a need to set initial conditions on capacitors and inductors in the circuit. To do so, **DCLICKL** the part symbol to bring up the Property Editor dialog box, click IC, and type in the initial condition. Or, while in the Property Editor dialog box, click New Column, type IC as Name, and type the initial value as Value. The IC attribute allows for setting the initial conditions on a capacitor or inductor. The default value of IC is 0. The attributes of open/close switches (with part names Sw-tClose and Sw-tOpen) can be changed in a similar manner.

2. Providing Specifications

After the circuit is drawn and the source is specified with its attributes, we need to add some specifications for the transient analysis. For example, suppose we want the analysis to run from 0 to 10 ms; we enter these specifications as follows:

1. Select **PSpice/New Simulation Profile** to open up the *Simulation Settings* dialog box.
2. Type 10 ms as the Run to Time (or TSTOP).
3. Click OK to accept specifications.

3. Simulating the Circuit

Once the circuit is drawn, the specifications for the transient analysis are given, and the circuit is saved, we are ready to simulate it. To perform transient analysis, we select **PSpice/Run**. If there are no errors, the Probe window will automatically appear. As usual, the time axis (or x-axis) is drawn but no curves are drawn yet. Select **Trace/Add Trace** and click on the variables to be displayed.

An alternative way of displaying the results is to use *markers*. Although there are many types of markers, we will discuss only voltage and current markers. A voltage marker is used to display voltage at a node relative to ground, while a current marker is for displaying current through a component pin. Markers are placed at the appropriate place in the schematic. This will cause two things to happen immediately. The voltage marker becomes part of the circuit, and the appropriate node voltage is displayed by Probe. To go back to the schematic, press <Alt-Esc>. It is important that the current marker be placed at the pin of the component; otherwise, the system would reject the marker. As many voltage and current markers can be placed on a circuit.

Example C.4

Assuming that $i(0) = 10$ A, plot the response $i(t)$ in the circuit of Fig. C.25 for $0 < t < 4$ s using PSpice.

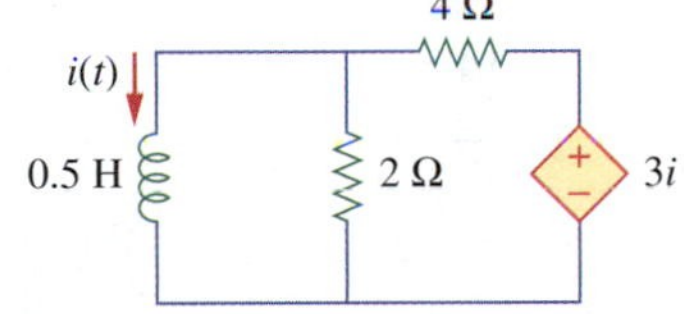

Figure C.25
For Example C.4.

Solution:

For PSpice analysis, the schematic is given in Fig. C.26, where the current-controlled voltage source H1 has been wired to agree with the circuit in Fig. C.25. The voltage of H1 is three times the current through inductor L1. Therefore, for H1, we set GAIN = 3 and for the inductor L1, we **DCLICKL** on the symbol and set the initial condition IC = 10. After drawing the circuit, select **PSpice/New Simulation Profile**. In the New Simulation dialog box, type the filename and click Create. This will lead to the Simulation Settings dialog box. Under *Analysis type*, select Time Domain (Transient). Type 4s as *Run to time*. Click OK. Select **PSpice/Run**. This will lead to Probe Window. Select **Trace/Add Trace** and click on I(L1), which is shown in Fig. C.27.

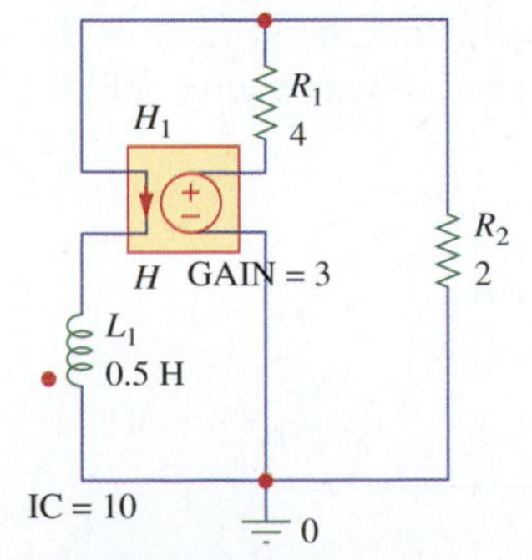

Figure C.26
The schematic of the circuit in Fig. C.25.

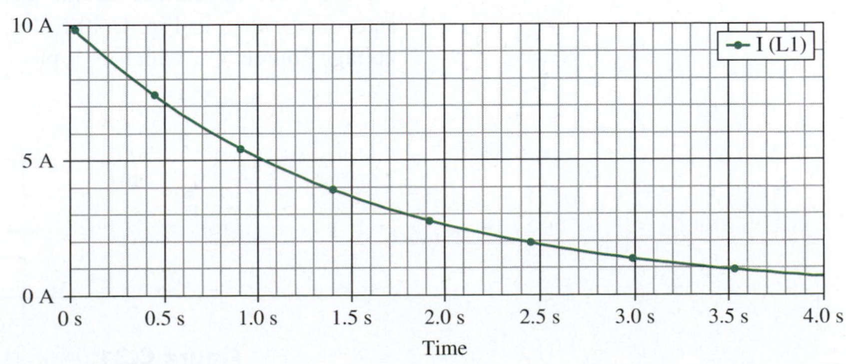

Figure C.27
Output plot for Example C.4.

Practice Problem C.4

Using PSpice, plot the source-free response $v(t)$ in the circuit of Fig. C.28 for $0 < t < 500$ ms. Assume that $v(0) = 10$ V.

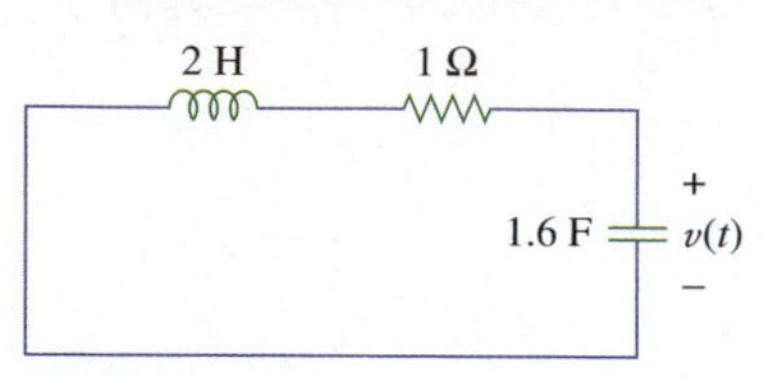

Figure C.28
For Practice Problem C.4.

Answer: The plot is shown in Fig. C.29. Note that $v(t) = 10e^{-0.25t}\cos 0.5t + 5e^{-0.25t}\sin 0.5t$ V.

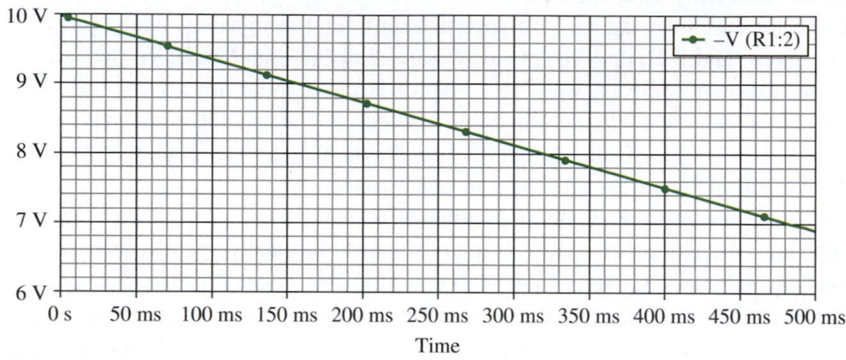

Figure C.29
Output plot for Practice Problem C.4.

Example C.5

Plot the forced response $v_o(t)$ in the circuit of Fig. C.30(a) for $0 < t < 5$ s if the source voltage is shown in Fig. C.30(b).

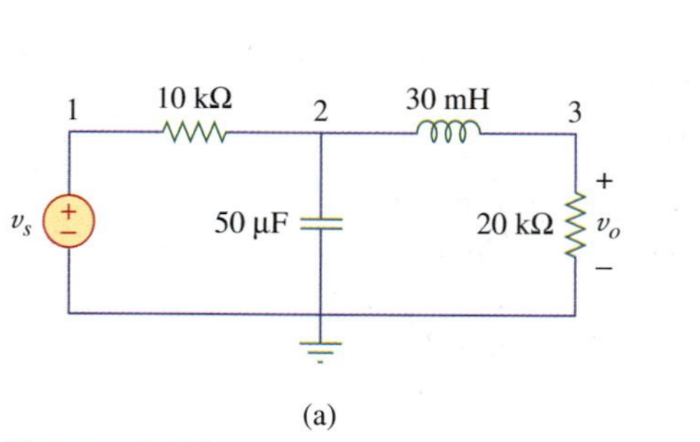

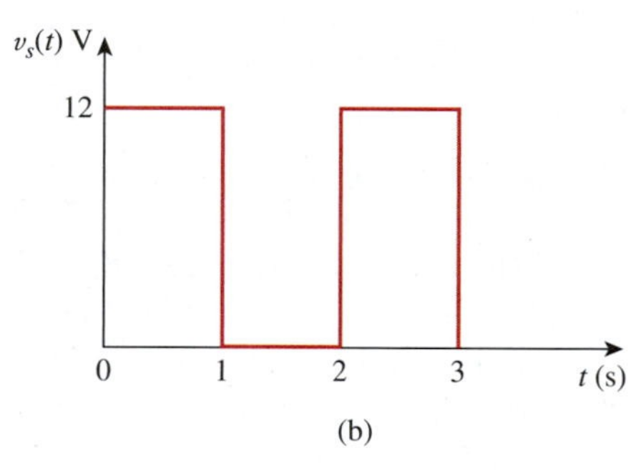

Figure C.30
For Example C.5.

Solution:
We draw the circuit and set the attributes as shown in Fig. C.31. We enter in the data in Fig. C.30(b) by double-clicking the symbol of the voltage source V1, which is a piecewise linear voltage source VPWL

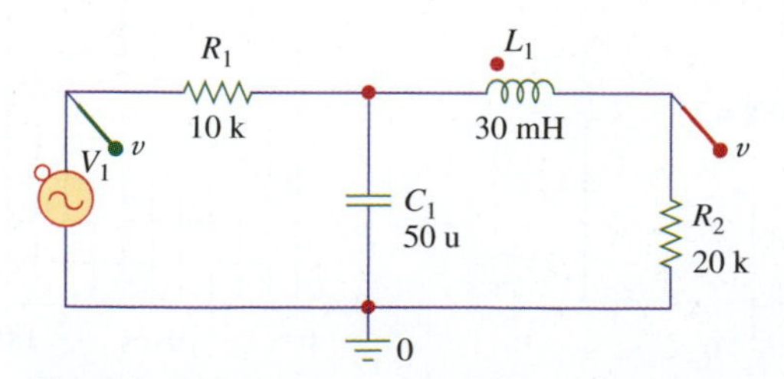

Figure C.31
The schematic of the circuit in Fig. C.30.

and typing in T1=0, V1=0, T2=1ns, V2=12, T3=1s, V3=12, T4=1.001s, V4=0, T5=2s, V5=0, T6=2.001s, V6=12, T7=3s, V7=12, T8=3.001s, V8=0. When you double click on the VPWL symbol, you get into Property Editor and click on New Column. In the Add New Column dialog box (shown in Fig. C.32), type the name (e.g., T1) and value (e.g., 1) of each entry at a time and press Apply. This way, you enter all names and values of V and T one at a time.

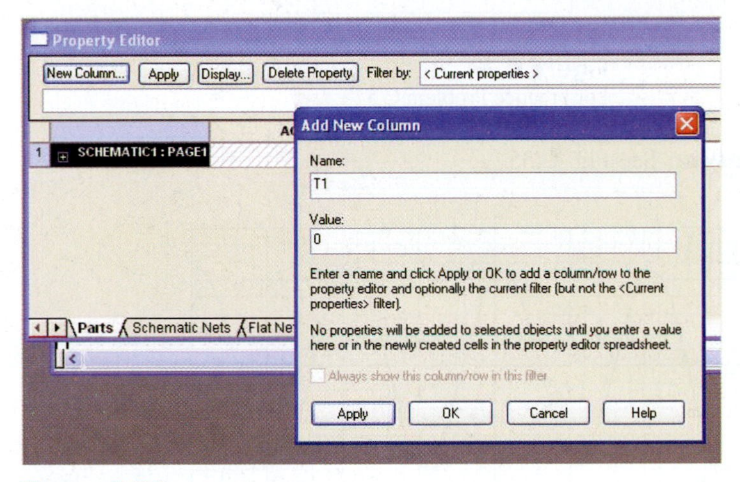

Figure C.32
Add New Column dialog box.

After drawing the circuit, select **PSpice/New Simulation Profile**. In the New Simulation dialog box, type the filename and click Create. This will lead to the Simulation Settings dialog box. Under *Analysis type*, select Time Domain (Transient). Type 5 s as *Run to time*. Click OK. Select **PSpice/Run**. This will lead to the Probe Window. Select **Trace/Add Trace** and click on V(V1:+) and V(L2:2), which are shown in Fig. C.32. Alternatively, we press <Alt-Esc> to get out of the Probe window and go back to Schematic window. We insert two voltage markers as shown in Fig. C.31 to get the plots of input VS and output VO automatically when we select **PSpice/Run**. Either way, we obtain the plots shown in Fig. C.33.

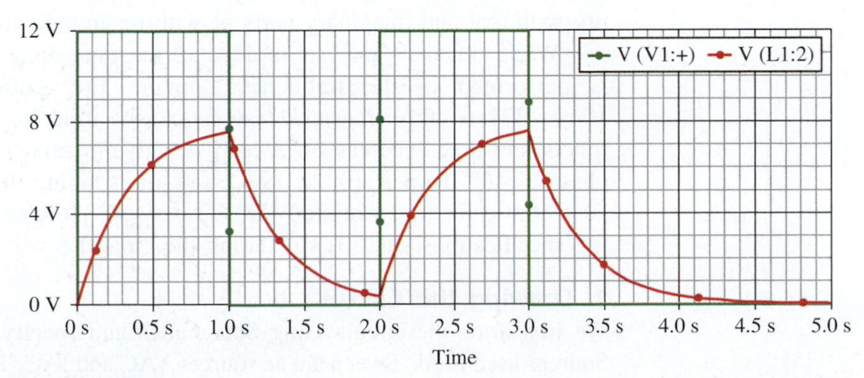

Figure C.33
Output plot for Example C.5.

Practice Problem C.5

Obtain the plot of $v(t)$ in the circuit in Fig. C.34 for $0 < t < 0.5$ s if

$$i_s(t) = 2e^{-t}\sin 2\pi(5)t \text{ A}$$

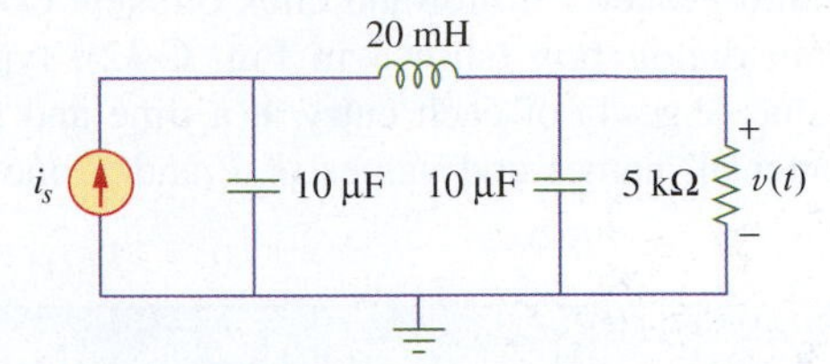

Figure C.34
For Practice Problem C.5.

Answer: See Fig. C.35

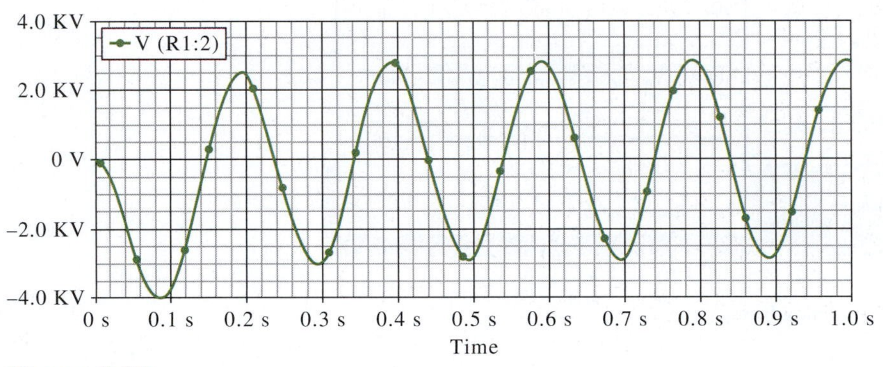

Figure C.35
Output plot for Practice Problem C.5.

C.5 AC Analysis/Frequency Response

Using AC Sweep, PSpice can perform ac analysis of a circuit for a single frequency or over a range of frequencies in increments that can vary linearly, by decade, or by octave. In AC Sweep, one or more sources are swept over a range of frequencies, while the voltages and currents of the circuit are calculated. Thus, AC Sweep is used both for phasor analysis and for frequency response analysis; it will output Bode gain and phase plots. (Keep in mind that a phasor is a complex quantity with real and imaginary parts or with magnitude and phase.)

While transient analysis is done in the time domain, ac analysis is performed in the frequency domain. For example, if $v_s = 10\cos(377t + 40°)$, transient analysis can be used to display v_s as a function of time, whereas AC Sweep will give the magnitude as 10 and phase as 40°. To perform AC Sweep requires taking three steps similar to those for transient analysis: (1) drawing the circuit, (2) providing specifications, and (3) simulating the circuit.

1. Drawing the Circuit

We first draw the circuit using Schematics and specify the source(s). Sources used in AC Sweep are ac sources VAC and IAC. The sources and attributes are entered into the Schematics as stated in the previous section. For each independent source, we must specify its magnitude and phase.

2. Providing Specifications

Before simulating the circuit, we need to add some specifications for AC Sweep. Select **PSpice/New Simulation Profile**. In the New Simulation dialog box, type the filename and click Create. This will lead to the Simulation Settings dialog box. For example, suppose we want a linear sweep at frequencies 50, 100, and 150 Hz. The parameters are entered as follows:

1. Under *Analysis type*, select AC Sweep/Noise.
2. Under AC Sweep Type, select Linear for the *x*-axis to have a linear scale.
3. Type 50 in the *Start Freq* box.
4. Type 150 in the *End Freq* box.
5. Type 100 in the *Total Points* box.
6. **CLICKL** OK to accept specifications.

A linear sweep implies that simulation points are spread uniformly between the starting and ending frequencies. Note that the *Start Freq* cannot be zero because 0 Hz corresponds to dc analysis. If we want the simulation to be done at a single frequency, we enter 1 in step 3 and the same frequency in steps 4 and 5. If we want the AC Sweep to simulate the circuit from 1 Hz to 10 MHz at 10 points per decade, we **CLICKL** on *Decade* in step 2 to make the *x*-axis logarithmic, enter 10 in the *Total Points* box in step 3, enter 1 in the *Start Freq* box, and enter 10Meg in the *End Freq* box. Keep in mind that a decade is a factor of 10. In this case, a decade is from 1 to 10 Hz, from 10 to 100 Hz, from 100 Hz to 1 kHz, and so on.

3. Simulating the Circuit

After providing the necessary specifications and saving the circuit, we perform AC Sweep by selecting **PSpice/Run**. If no errors are encountered, the circuit is simulated. At the end of the simulation, the system creates an output file. Also, the Probe program will automatically run if there are no errors. The frequency axis (or *x*-axis) is drawn, but no curves are shown yet. Select **Trace/Add Trace** from the Probe menu bar and click on the variables to be displayed. We may also use current or voltage markers to display the traces as explained in the previous section.

In case the resolution of the trace is not good enough, we may need to check the data points to see if they are enough. To do so, select **Tools/Options/Mark Data Points/OK** in the Probe menu and the data points will be displayed. If necessary, we can improve the resolution by increasing the value of the entry in the *Total Points* box in the Simulation Settings dialog box.

To generate Bode plots involves using the AC Sweep and the dB command in Probe. Bode plots are separate plots of magnitude and phase versus frequency. To obtain Bode plots, it is common to use an ac source—say, V1—with 1-V magnitude and zero phase. After we have selected **PSpice/Run** and have the Probe program running, we can display the magnitude and phase plots as mentioned earlier. Suppose we want to display a Bode magnitude plot of V(Vo). We select **Trace/Add Trace** and type dB(V(Vo)) in *Trace Command* box; dB(V(Vo)) is equivalent to 20log(V(Vo)) and because the magnitude of V1 or V(R1:1) is unity, and dB(V(Vo)) actually corresponds to dB(V(Vo)/V(R1:1)), which is the gain. Adding the trace dB(V(Vo)) will give a Bode magnitude/gain plot with the y-axis in dB.

As an alternative approach, we can avoid running the Probe program by using *pseudocomponents* to send results to the output file. Pseudocomponents are like parts that can be inserted into a schematic as if they were circuit elements, but they do not correspond to circuit elements. They can be added to the circuit for specifying initial conditions or for output control. Important pseudocomponents and their usage are shown in Fig. C.36 and listed in Table C.3. The pseudocomponents are added to the schematic. They are available in the SPECIAL library. To add a pseudocomponent, select **Place/Parts** in the Schematics window, place it at the desired location, and add the appropriate attributes as usual.

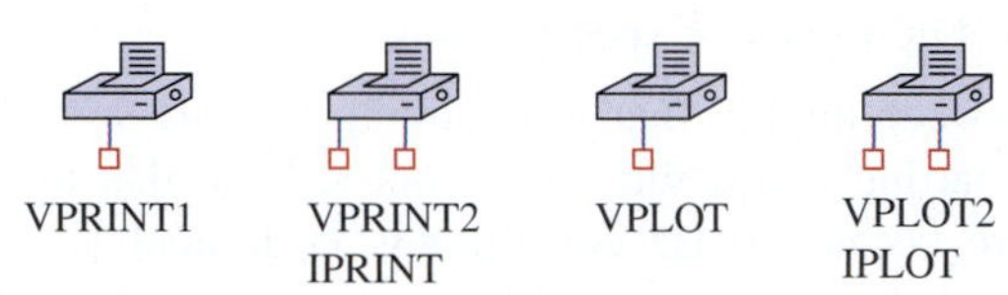

Figure C.36
Print and plot pseudocomponents.

TABLE C.3

Print and Plot pseudocomponents.

Symbol	Description
IPLOT	Plot showing branch current; symbol must be placed in series
IPRINT	Table showing branch current; symbol must be placed in series
VPLOT1	Plot showing voltage at the node to which the symbol is connected
VPLOT2	Plot showing voltage differentials between two points to which the symbol is connected
VPRINT1	Table showing voltage at the node to which the symbol is connected
VPRINT2	Table showing voltage differentials between two points to which the symbol is connected

Example C.6

Find current i in the circuit in Fig. C.37.

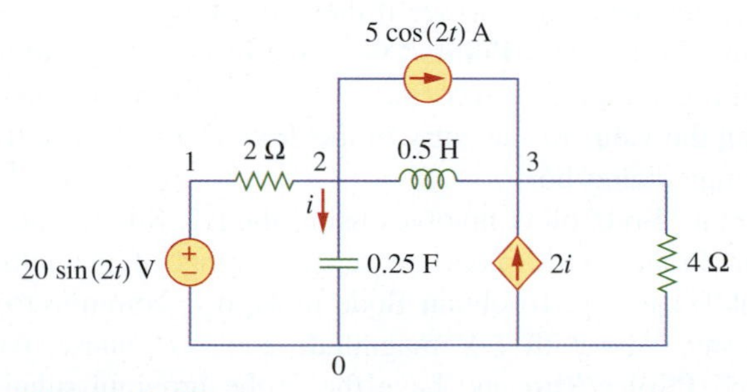

Figure C.37
For Example C.6.

Solution:
Recall that $20 \sin 2t = 20 \cos(2t - 90°)$ and that $f = \omega/2\pi = 2/2\pi = 0.31831$. The schematic is shown in Fig. C.38. The attributes of V1 or

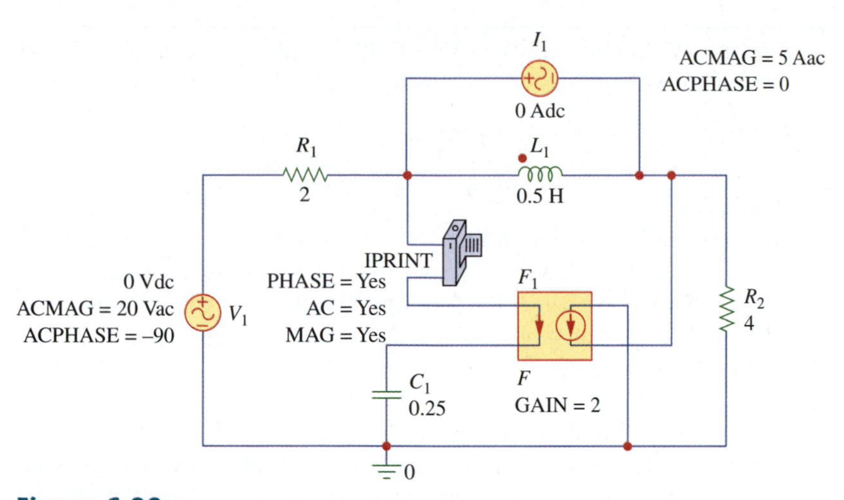

Figure C.38
The schematic of the circuit in Fig. C.37.

VAC are set as *ACMAG* = 20, *ACPHASE* = –90; while the attributes of IAC are set as ACMAG = 5 and ACPHASE = 0. The current controlled current source F1 is connected in such a way as to conform with the original circuit in Fig. C.37; its gain is set equal to 2. The attributes of the pseudocomponent IPRINT are set as *AC* = yes, *MAG* = yes, *PHASE* = ok. Because this is a single-frequency ac analysis, we select **PSpice/New Simulation Profile**. In the Simulation Settings dialog box, select AC Sweep/Noise under *Analysis type* and Linear under *AC Sweep Type*. Type Start Frequency = 0.31831, End Frequency = 0.31831, and Total Points = 1. Click OK. Select **PSpice/Run**. This leads to the Probe Window. You go back to Schematic window by pressing Alt-Esc. Select **PSpice/View Output File**. This will give the output file, which includes:

```
FREQ          IM(V_PRINT1)      IP(V_PRINT3)
3.183E-01     7.906E+00         4.349E+01
```

From the output file, we obtain

$$I = 7.906\underline{/43.49^\circ} \text{ A} \quad \text{or} \quad i(t) = 7.906 \cos(2t + 43.49^\circ) \text{ A}$$

This example is for a single-frequency ac analysis; Example C.7 is for AC Sweep over a range of frequencies.

Practice Problem C.6

Find $i_x(t)$ in the circuit in Fig. C.39.

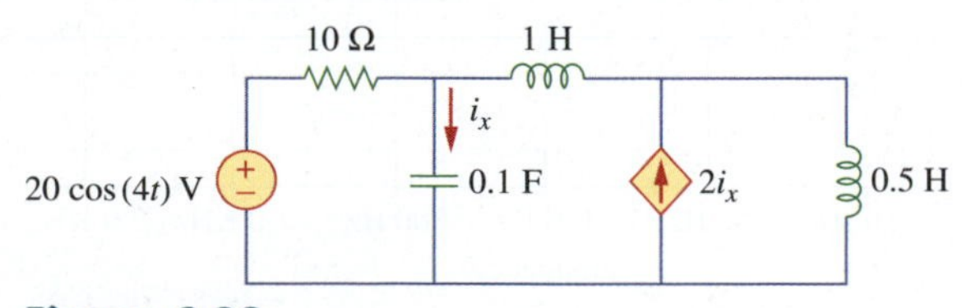

Figure C.39
For Practice Problem C.6.

Answer: From the output file,

$$i_x(t) = 7.59 \cos(4t + 108.43^\circ) \text{ A}.$$

Example C.7

For the *RC* circuit shown in Fig. C.40, obtain the magnitude plot of the output voltage v_o for frequencies from 1 Hz to 10 kHz. Let $R = 1\ \text{k}\Omega$ and $C = 4\ \mu\text{F}$.

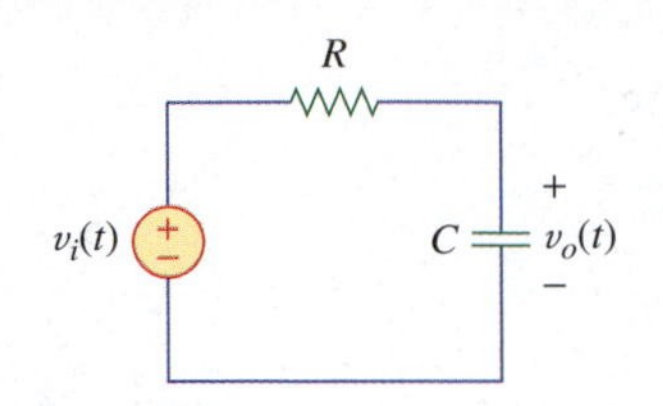

Figure C.40
For Example C.7.

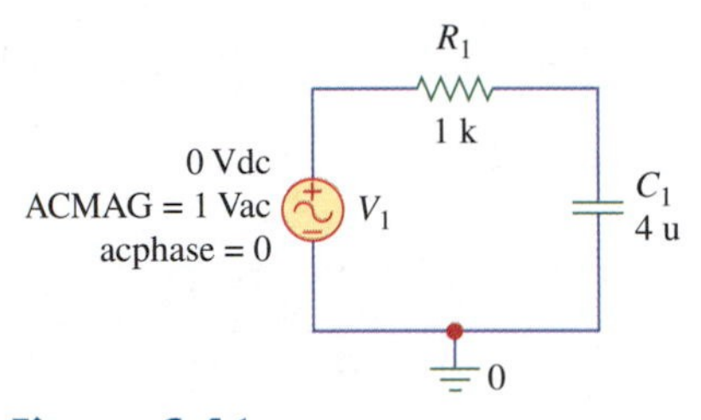

Figure C.41
The schematic of the circuit in Fig. C.40.

Solution:
The schematic is shown in Fig. C.41. It is assumed that the magnitude of V1 is 1 and its phase is zero; we enter these as the attributes of V1. For AC Sweep, we select **PSpice/New Simulation Profile**. In the Simulation Settings dialog box, select AC Sweep/Noise under *Analysis type* and Logarithmic under *AC Sweep Type*. Type Start Frequency = 1, End Frequency = 10k, and Total Points = 10. Click OK. After saving the circuit, select **PSpice/Run**. This leads to the Probe Window. We obtain the plot in Fig. C.42(a) by selecting **Trace/Add Trace** and clicking V(R1:2). Also, by selecting **Trace/Add Trace** and typing dB(V(R1:2)) in the *Trace Command* box, we obtain the Bode plot in Fig. C.42(b). The two plots in Fig. C.42 indicate that the circuit is a lowpass filter because low frequencies are passed while high frequencies are blocked by the circuit.

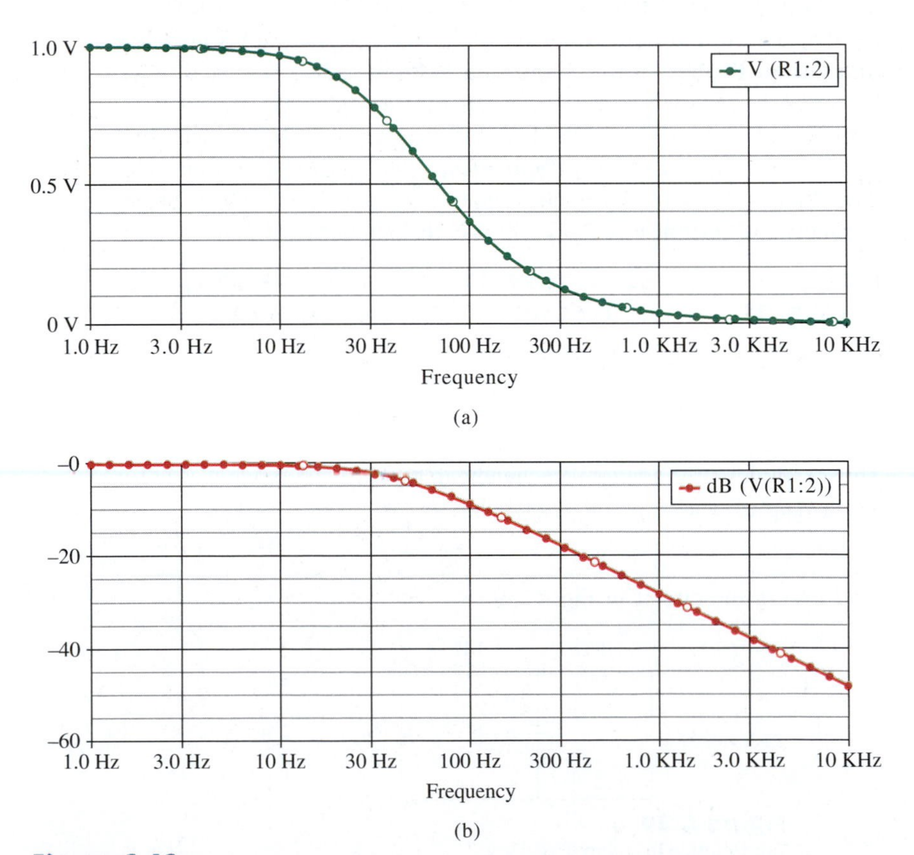

Figure C.42
Result of Example C.7: (a) linear plot; (b) Bode plot.

Practice Problem C.7

For the circuit in Fig. C.40, replace the capacitor C with an inductor $L = 4$ mH and obtain the magnitude plot (both linear and Bode) for v_o for $10 < f < 100$ MHz.

Answer: See the plots in Fig. C.43

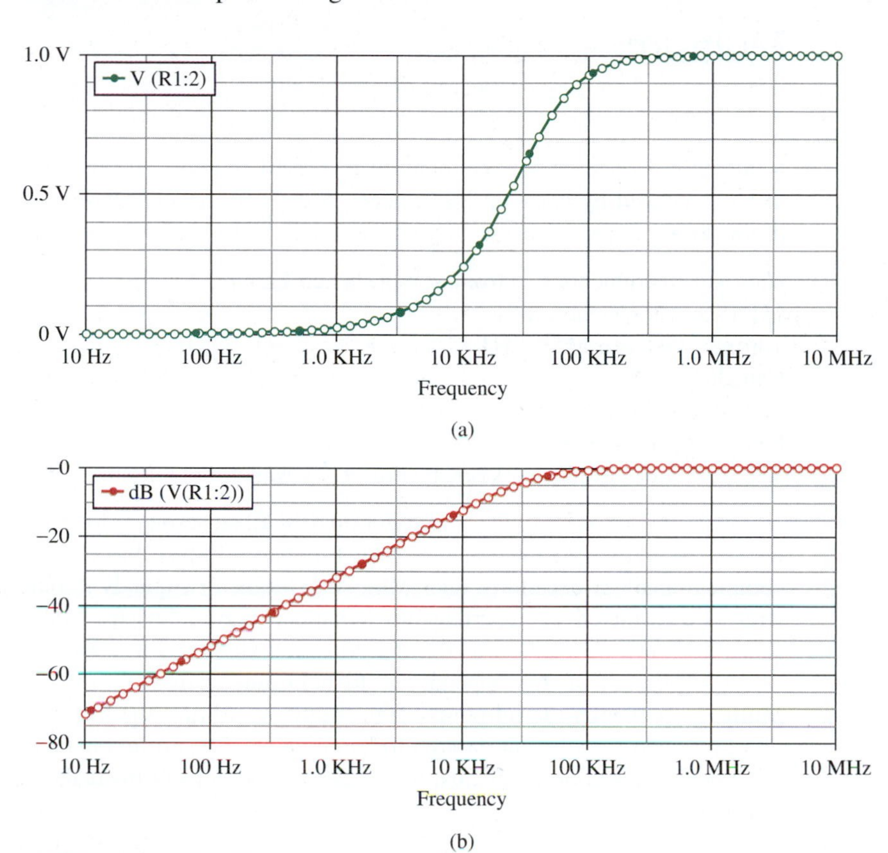

Figure C.43
Result of Practice Problem C.7: (a) linear plot; (b) Bode plot.

Appendix D

Multisim

Multisim (also known as Electronics Workbench) is a software package that is produced by National Instruments. A copy of the student version of the package can be obtained by calling 1-800-263-5552 or by writing:

National Instruments Electronics Workbench Group
111 Peter Street, Suite 801
Toronto, Ontario, M5V 2H1
Canada
Tel: (416) 977-5550 or 1-800-263-5552
Fax: (416) 977-1818
Web: www.electronicsworkbench.com or www.ni.com/multisim

Both PSpice and Multisim are built on SPICE, an acronym for Simulation Program with Integrated Circuit Emphasis. PSpice has a more complete analysis capability than Multisim. Multisim expands PSpice with its own mixed-signal extensions to support digital devices. Multisim allows you to work on a circuit just as you would do in a lab.

The purpose of this appendix is to help you learn the ABCs of Multisim. Although the introduction to Multisim is brief, the information presented is sufficient enough to handle the Multisim problems in this book. For additional details, the reader should consult manuals on Multisim.

D.1 Multisim Screen

Assuming that you have already loaded Multisim into your computer system, you open the Multism program by clicking **Start > Program > Multisim**. A screen similar to the one in Fig. D.1 will appear. The screen is so large it will not fit on one page. The left side of the screen is shown in Fig. D.1(b). You will find menu items and icons along the top and sides of the screen.

Pulldown menus contain commands for all functions. They include familiar Windows items such as File, Edit, Place, MCU, Simulate, Transfer, Tools, Options, Windows, and Help.

The **Component Toolbar** contains buttons that enable you select components. Multisim contains everything necessary to build circuits, from ac voltage sources to zener diodes. This toolbar is the most important toolbar you will working with; it is shown in Fig. D.2.

Place source allows you to place power sources, current/voltages source, and ground. **Place basic** permits you to place basic components such as resistor, capacitor, inductor, switch, relay, 555 timer, and transformer. The place indicators allow you to place ammeter, voltmeter, and probe. Other icons in the component toolbar are not important or useful at this level.

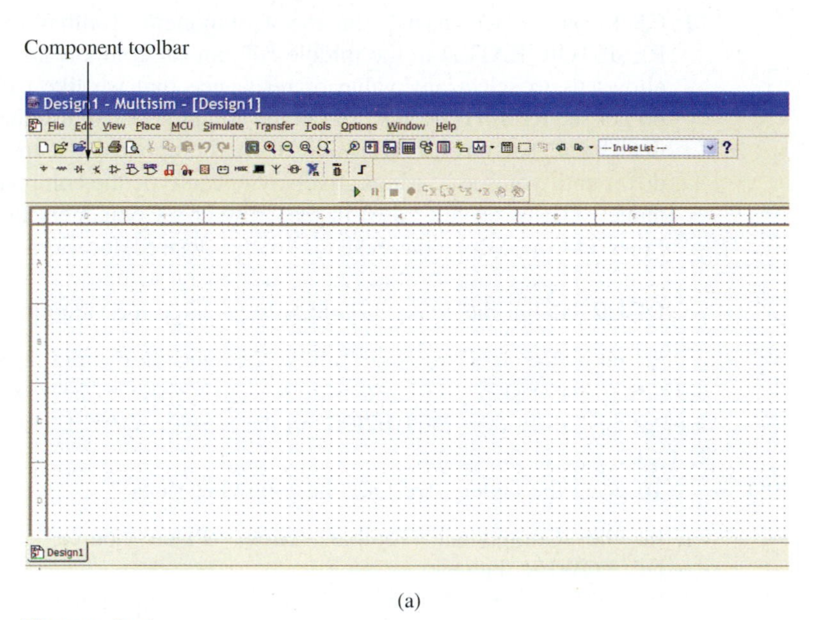

(a)

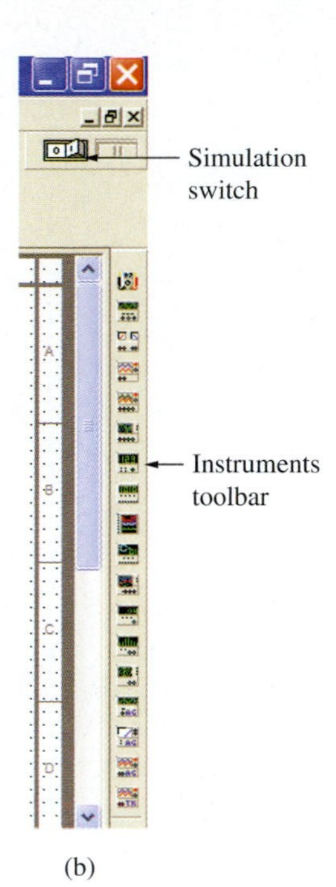

(b)

Figure D.1
Multisim user interface screen.

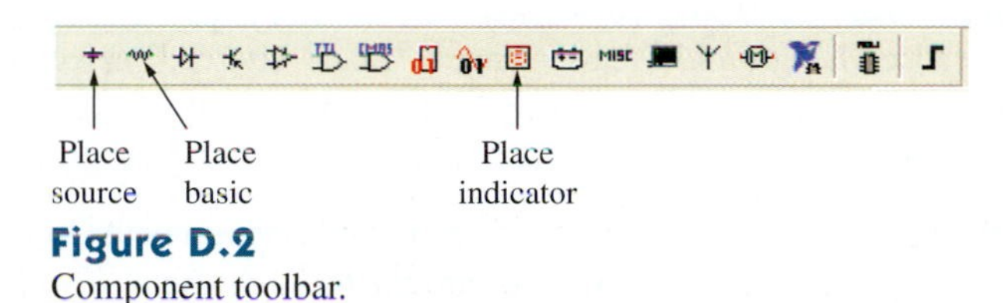

Figure D.2
Component toolbar.

The **Instruments Toolbar** contains buttons for different instruments such as multimeter, function generator, wattmeter, oscilloscope, Bode plotter, and logic analyzer.

D.2 Creating a Circuit

Having become familiar with the Multisim screen, we are now ready to create and simulate circuits. We will begin with a simple circuit that has few components. The steps involved in creating a circuit include choosing the components, wiring them together, saving the circuit, and simulating it. Throughout the book, we will use the following terms to represent actions to be performed by the mouse:

- **CLICKL**: click the left button once to select an item.
- **CLICKR**: click the right button once to abort a mode.
- **DCLICKL**: double-click the left button to edit a selection.
- **DCLICKR**: double-click the right button to repeat an action.
- **CLICKLH**: click the left button, hold down, and move the mouse to drag a selected item. Release the left button after placing the item.

For concreteness, let us consider the simple circuit in Fig. D.3. Our goal is to use Multisim to measure the voltage across the resistors and the current through them.

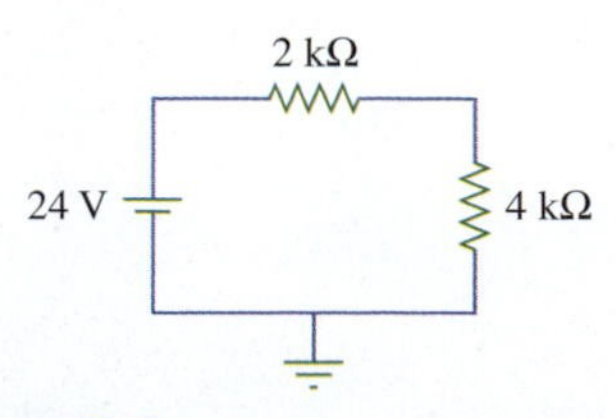

Figure D.3
Circuit to be created and simulated.

To place the 2-kΩ resistor, we take the following steps:

1. Click on "Place basic" in the Component Toolbar. Select RESISTOR_RATED in the middle column for components. (This allows us to select any value of resistance that we like.) If you do not see RESISTOR_RATED in the component list, use the vertical scroll bar on the right side of the component list to scroll down until you see it. Alternatively, you can type the component's name (resistor, in this case) in the browser's Component field.
2. Press OK and drag the resistor to the desired location on the screen. Press Close to close the "Place basic" window.
3. **DCLICKL** on the resistor symbol and change the value to 2 kΩ.

To place the 4-kΩ resistor, follow the preceding three steps. In addition to these steps, we need to rotate the resistor to have vertical orientation. To do this, **DCLICKR** on the resistor symbol and select **90 Clockwise** in the menu.

To place the battery, we take the following steps:

1. In the Component Toolbar, click "Place source." Select DC_POWER (battery).
2. Press OK and drag the battery to the desired location on the screen.
3. The default value of the battery is 12 V. To change this, **DCLICKL** on the battery symbol. A dialog box appears. Change 12 V to 24 V in the dialog box and click OK.
4. Because the ground is also under "Place source," select GDGN and place it on the screen. Press Close to close the "Place source" window.

We now need to place two voltmeters and one ammeter to measure the voltage across the resistors and current through them. To place the voltmeters, we take the following steps.

1. In the Component Toolbar, click "Place indicator."
2. Select VOLTMETER_V for the 4-kΩ resistor and VOLTMETER_H for the 2-kΩ resistor.

We follow the same steps to place the ammeter. At this point, the circuit looks like that shown in Fig. D.4.

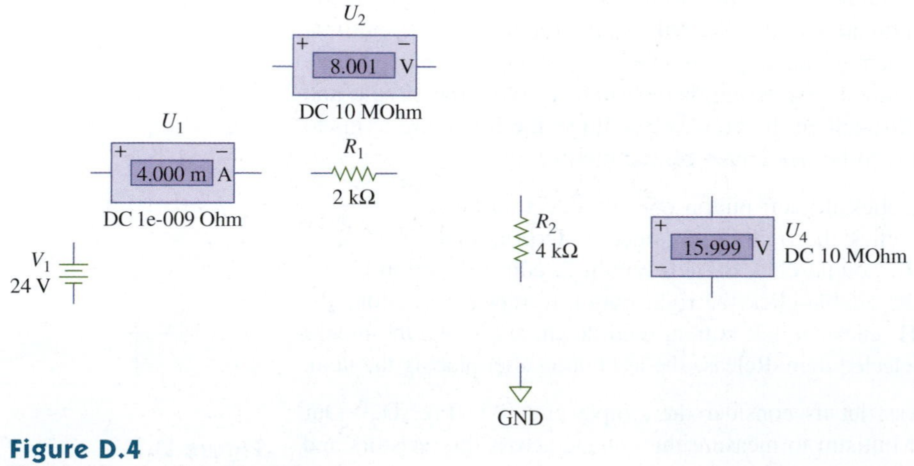

Figure D.4
Creating the circuit in Fig. D.3.

The next step is to wire the components together. All components have pins that you can use to wire them to other components. For example, to connect the ammeter to the 2-kΩ resistor, place the cursor on the right pin of the ammeter. A small dot and a crosshair will appear. **CLICKL** and move the cursor to the left pin of the resistor. Release the mouse to terminate the wiring. We repeat this process to connect other components as shown in Fig. D.5.

Finally, we need to save the circuit by selecting **File/Save As** as is done in Windows. Once the circuit is saved, we are ready for simulation. There are two ways to simulate the circuit. One way is by selecting **Simulate/Run**. The other way is by pressing the power switch at the top right-hand corner of the screen. Either way, we obtain the results displayed in the readings of the voltmeters and ammeter as shown in Fig. D.5.

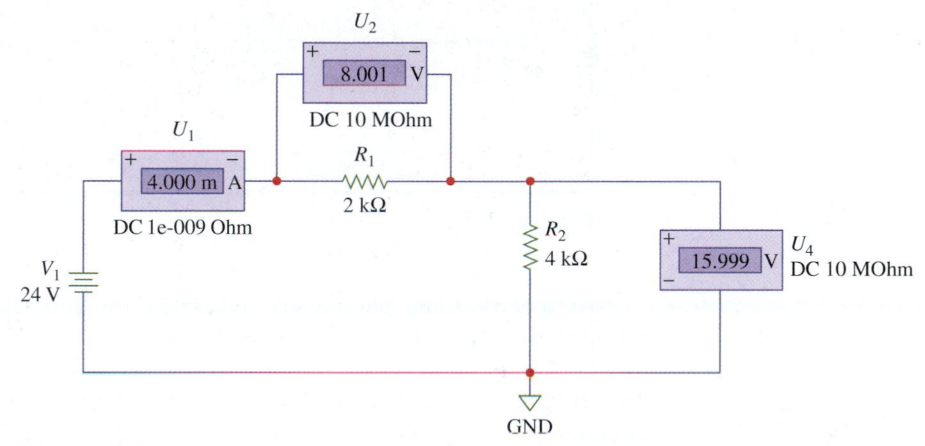

Figure D.5
Simulation of the circuit in Fig. D.3.

D.3 Transient Analysis

Multisim offers us many analyses—dc analysis, transient analysis, ac analysis, Fourier analysis, and so on. The analysis carried out in the previous section is dc analysis. In this section, we examine how to use Multisim for the transient analysis of a circuit. While dc and ac analyses are essentially steady (one-moment-in-time) analyses, transient analysis adds a time factor to your circuit's readings. Transient analysis can be used for almost any circuit for which you need to view voltage or current over time. To perform transient analysis on a circuit using Multisum usually require three steps: (1) drawing the circuit, (2) providing specifications, and (3) simulating the circuit and displaying results.

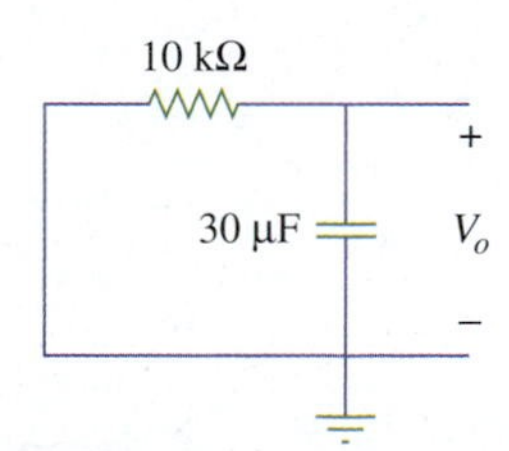

Figure D.6
Circuit for transient analysis.

For the purpose of illustration, suppose we want to perform transient analysis on the circuit shown in Fig. D.6, where the initial voltage on the capacitor is 12 V. This is basically a discharging circuit. First, we draw the circuit using Multisim as shown in Fig. D.7. **DCLICKL** the capacitor symbol and set the **Initial Condition** equal to 12 V. Multisim automatically labels or numbers the nodes, but it does not show the numbers automatically. To show the node number, select **Options/Sheet Properties**. Under *Net Names*, select "Show all." This will number the nodes (1 and GND in this case) as shown in Fig. D.7.

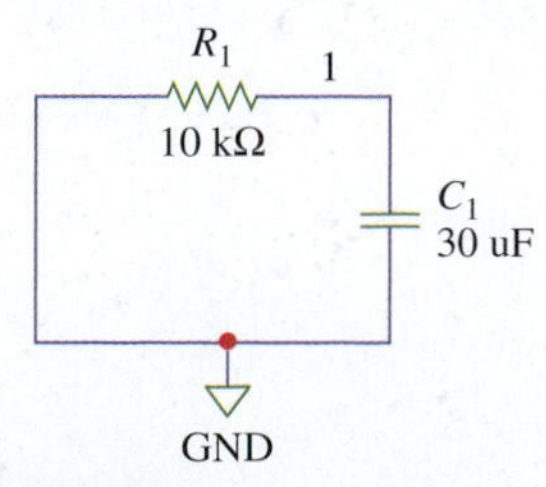

Figure D.7
Multisim simulation of the circuit in Fig. D.6.

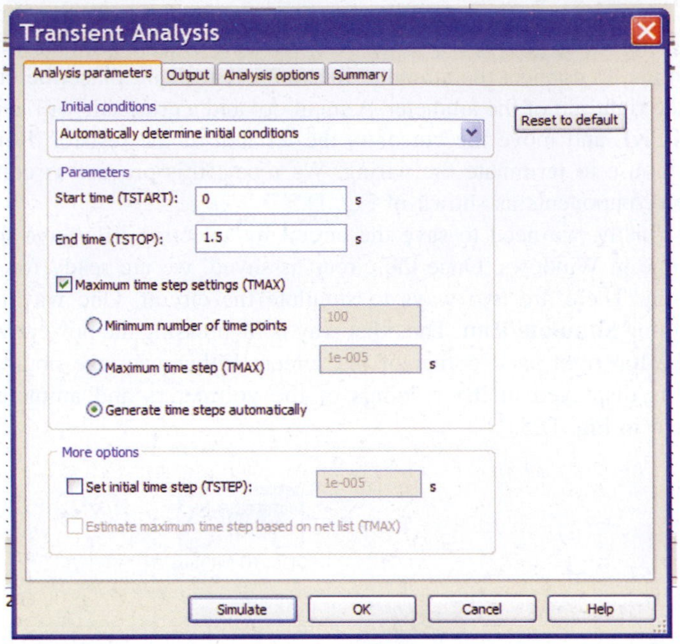

Transient Analysis

Analysis parameters | Output | Analysis options | Summary

Initial conditions

Automatically determine initial conditions

Reset to default

Parameters

Start time (TSTART): 0 s

End time (TSTOP): 1.5 s

Maximum time step settings (TMAX)

Minimum number of time points 100

Maximum time step (TMAX) 1e-005 s

Generate time steps automatically

More options

Set initial time step (TSTEP): 1e-005 s

Estimate maximum time step based on net list (TMAX)

Simulate | OK | Cancel | Help

Figure D.8
Transient Analysis dialog box.

To specify some parameters necessary for simulation, select **Simulate/Analyses/Transient Analysis**. The Transient Analysis dialog box comes up as shown in Fig. D.8. We select TSTART as 0 and TSTOP as 1.5 s, which is at least five times the time constant of the circuit,

$$\tau = RC = 10 \times 10^3 \times 30 \times 10^{-6} = 0.3s, \qquad 5\tau = 1.5s$$

In a situation where it is not easy to determine the time constant, the value of TSTOP can be selected by trial and error. Under **Initial Conditions**, select *Set to zero*. We now select **Output**. On the left-hand side, we have a list of variables that appear in the circuit. On the right-hand side, we have a list of selected variables for analysis. In between the two lists, we have **Remove/Add**. To move a variable from the left list to the right list, we select it from the left list and select **Add**. We can also move a variable from the right list to the left list by selecting it and then selecting **Remove**. For our purposes, we are interested in the voltage at node 1. So we move V(1) from the left list to the right list. We finally select **Simulate**, and the result of the simulation will automatically be displayed as shown in Fig. D.9. To stop the simulation at any time, we switch off the power switch at the top right-hand corner of the screen.

As another example, let us consider the circuit whose schematic is shown in Fig. D.10. The source is a 15-V pulse voltage source generating square waves that cycles between 0 and 15 V with a frequency of 400 kHz. The period is $T = 1/f = 1/400 \times 10^3 = 2.5\mu s$, which is comparable to

$$5\tau\left(\tau = \frac{L}{R} = \frac{10^{-3}}{2 \times 10^3} = 0.5\mu s, \qquad 5\tau = 2.5\mu s\right)$$

To simulate the circuit, select **Simulate/Analyses/Transient**. In the Transient Analysis dialog box, we change the end time TSTOP = 15 μs and

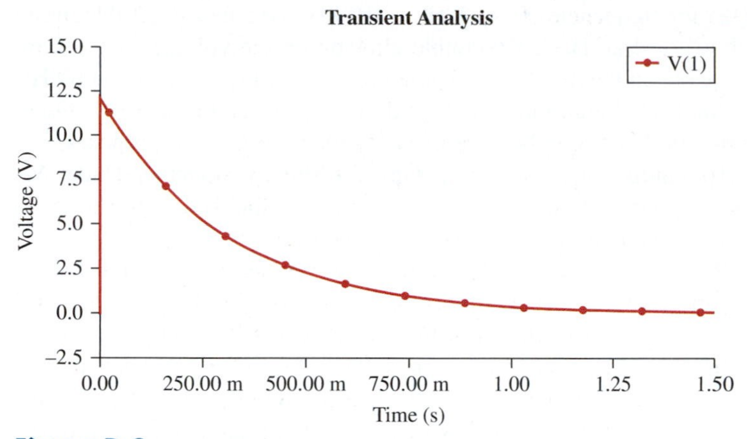

Figure D.9
Transient analysis of the circuit in Fig. D.6.

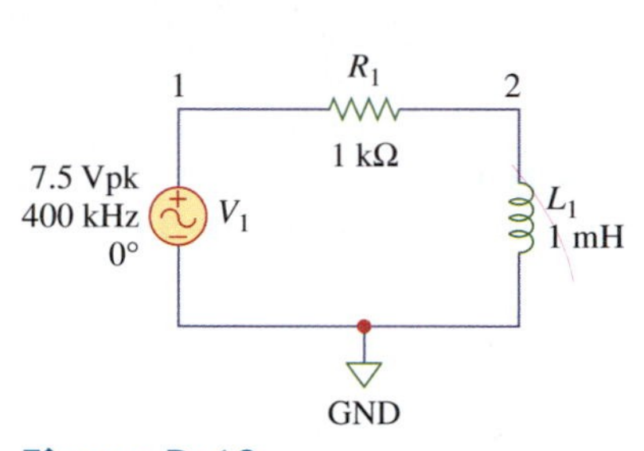

Figure D.10
Using a square wave generator as a source.

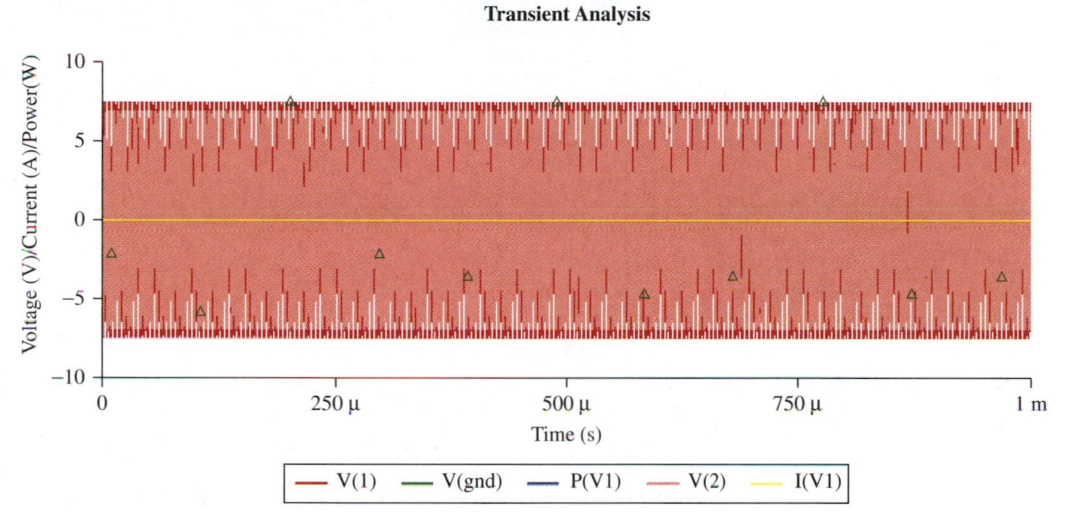

Figure D.11
Transient analysis of the circuit in Fig. D.10.

select **Simulate**. The result in Fig. D.11 is displayed. The signal in blue is the input wave, and the signal in red is the output wave, which is the voltage across the inductor.

D.4 AC Analysis

AC analysis is used to determine the frequency response of circuits. In ac analysis, dc sources are assigned zero values, while ac sources, capacitors, and inductors are represented by their ac models. In fact, ac sources do not matter because Multisim inputs its own sinusoidal waveform and determines the ac circuit response as a function of frequency. To perform an ac sweep on a circuit requires taking three steps similar to those for transient analysis: (1) drawing the circuit, (2) providing specification, and (3) simulating the circuit.

For the purpose of illustration, consider the circuit in Fig. D.12. Suppose we want to obtain the frequency response (magnitude and

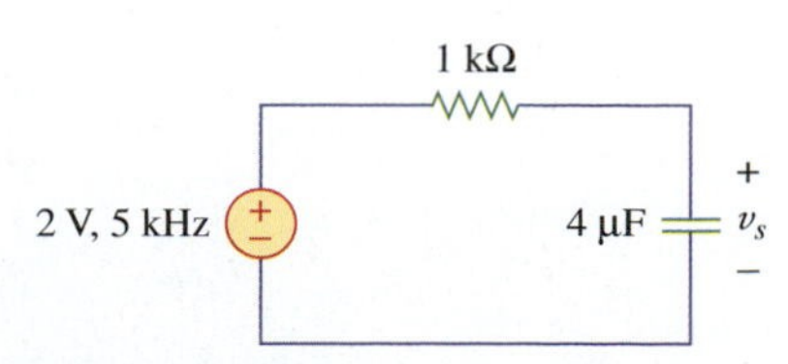

Figure D.12
Circuit for ac analysis.

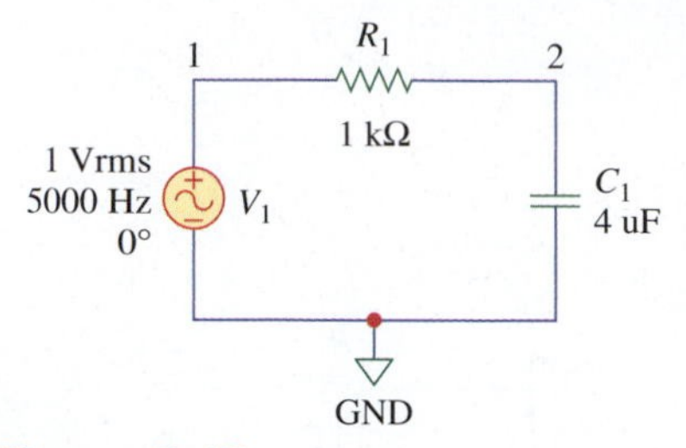

Figure D.13
Simulation of the circuit in Fig. D.12.

phase) for frequencies from 1 Hz to 10 kHz. We first draw the circuit as shown in Fig. D.13. By double-clicking on the voltage source symbol, we obtain the AC_Power dialog box. We set Frequency to 5,000 Hz, AC Analysis Magnitude to 2, and AC Analysis Phase to 0. Other settings in this dialog box are used for other simulation purposes.

To number the nodes, select **Options/Sheet Properties**. Under *Net Names*, select "Show all." This will number the nodes as shown in Fig. D.13. Our response is taken from node 2.

We now need to specify the range of frequencies required for the simulation. We select **Simulate/Analyses/AC Analysis**. In the AC Analysis dialog box (shown in Fig. D.14), we select *FSTART* as 1 Hz, *FSTOP* as 10 kHz, and *Swept type* as Decade. In the AC Analysis dialog box, select Output and move V(2) from the left list to the right list so that the voltage at node 2 is displayed. We finally select **Simulate**. The plot of the magnitude and phase of v_o in the frequency domain is shown in Fig. D.15.

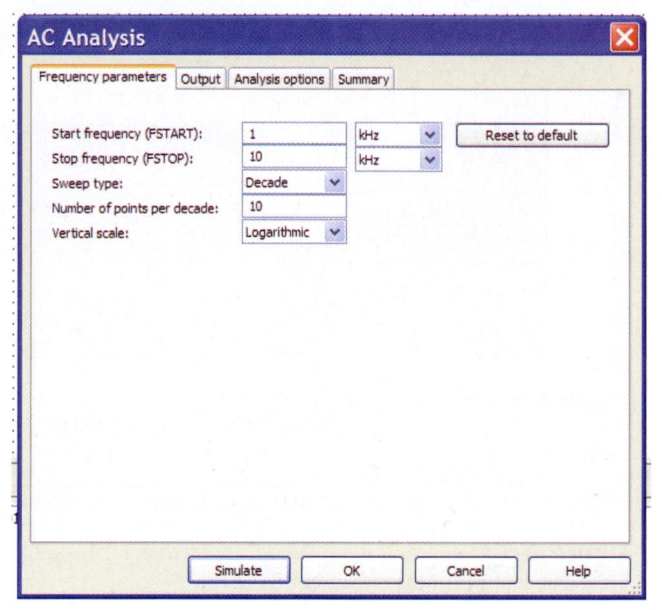

Figure D.14
AC Analysis dialog box.

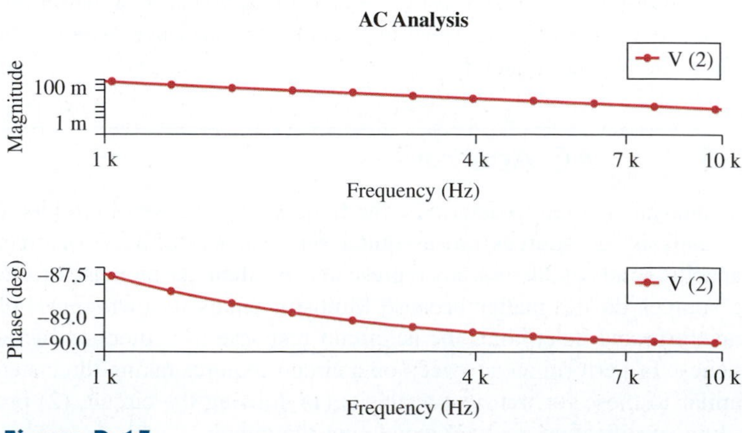

Figure D.15
AC sweep of the circuit in Fig. D.12.

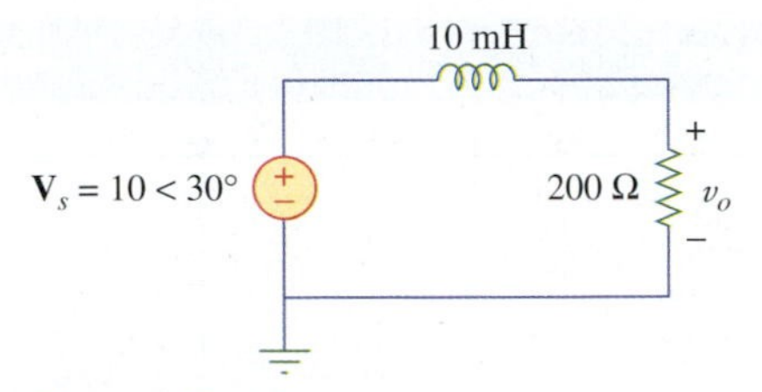

Figure D.16
A typical ac circuit to be simulated.

Another way to observe waveforms is to use Multisim's oscilloscope. Consider the circuit shown in Fig. D.16, where v_s is the time-domain version of $V_s = 10 < 30°$. Let v_s be an ac voltage source with $\omega = 2\pi \times 10^3$ rad/s so that $f = \omega/2\pi = 1{,}000$ Hz. Our goal is to observe both v_s and v_o on the oscilloscope. We first draw the circuit as shown in Fig. D.17. The ac voltage is selected from the Component Toolbar. We double-click the voltage symbol, and the AC_Power dialog box comes up. We set Frequency to 1,000 Hz, AC Analysis Magnitude to 10, and AC Analysis Phase to 0.

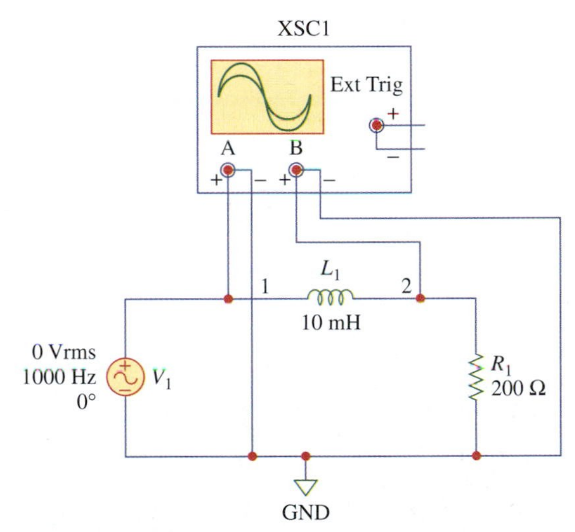

Figure D.17
Simulation of the circuit in Fig. D.16.

For the oscilloscope (XSC1), we select it from the Instrument Toolbar. It is a dual-channel oscilloscope with all the major controls of a typical laboratory oscilloscope. We connect channel A to measure v_s and channel B to measure v_o. It is not necessary to ground the oscilloscope if the circuit to which it is attached contains a ground. To change the settings of the oscilloscope, we double-click on the oscilloscope symbol. We change Timebase scale to 1μs/Div, Channel A scale to 2V/Div, and Channel B scale to 2V/Div. We save the circuit and simulate it by selecting **Simulate/Run**. To display the result, we double the oscilloscope symbol. The result is shown in Fig. D.18. The red curve is for channel A (or v_s), while the blue curve is for channel B (or v_o). The color of the curve is dictated by the wire color. For example, the wire to channel A is red so that the corresponding curve is red.

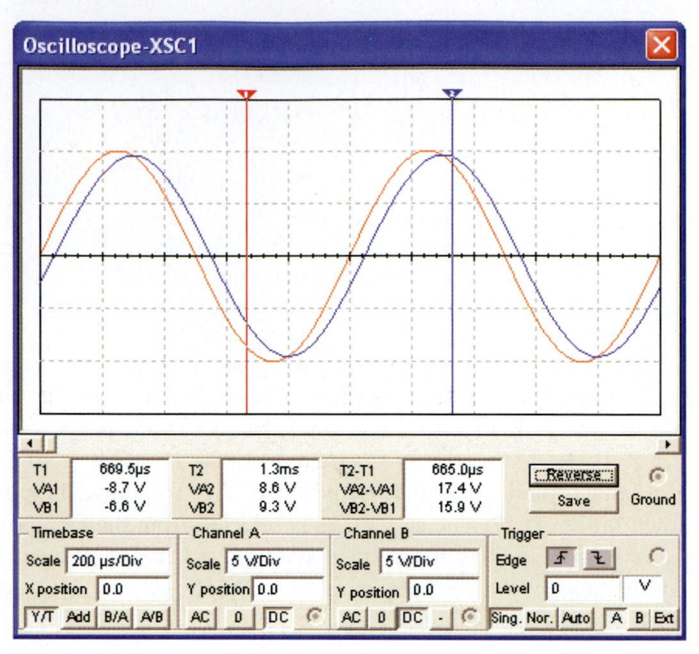

Figure D.18
Display of v_s and v_o on the oscilloscope.

Appendix E

MATLAB

MATLAB has become a powerful tool of technical professionals worldwide. The term MATLAB is an abbreviation for MATrix LABoratory, implying that MATLAB is a computational tool that uses matrices and vectors (or arrays) to carry out numerical analysis, signal processing, and scientific visualization tasks. Because MATLAB uses matrices as its fundamental building blocks, one can write mathematical expressions involving matrices just as easily as one would on paper. MATLAB is available for Macintosh, Unix, and Windows operating systems. A student version of MATLAB is available for personal computers (PCs). A copy of MATLAB can be obtained from

The Mathworks, Inc.
3 Apple Hill Drive
Natick, MA 01760-2098
Phone: (508) 647-7000
Website: http://www.mathworks.com

A brief introduction to MATLAB is presented in this appendix. What is presented is sufficient for solving problems in this book. More about MATLAB can be found in MATLAB books and from online help. The best way to learn MATLAB is to work with it after having learned the basics.

E.1 MATLAB Fundamentals

The Command window is the primary area where you interact with MATLAB. A little later, we will learn how to use the text editor to create M-files, which allow for execution of sequences of commands. For now, we focus on how to work in the Command window. We will first learn how to use MATLAB as a calculator.

Using MATLAB as a Calculator

The following are algebraic operators used in MATLAB:

+	Addition
-	Subtraction
*	Multiplication
^	Exponentiation
/	Right division (`a/b` means $a \div b$)
\	Left division (`a\b` means $b \div a$)

To begin to use MATLAB, we use these operators. Type commands to the MATLAB prompt ">>" in the Command window (correct any mistakes by backspacing) and press the Enter key. For example,

```
>>  a = 2; b = 4;c = -6;
>>  dat = b^2 - 4*a*c
dat =
      64
>>  e = sqrt(dat)/10
e =
      0.8000
```

The first command assigns the values 2, 4, and −6 to the variables a, b, and c, respectively. MATLAB does not respond because this line ends with a colon. The second command sets dat to $b^2 - 4ac$ and MATLAB returns the answer as 64. Finally, the third line sets e equal to the square root of dat and divides by 10. MATLAB prints the answer as 0.8. Other mathematical functions, listed in Table E.1, can be used similarly to how the function sqrt is used here. Table E.1 provides just a tiny sample of MATLAB functions. Others can be obtained from the on-line help. To get help, type

```
>> help
```

A long list of topics will come up. For a specific topic, type the command name. For example, to get help on "log to base 2," type

```
>> help log2
```

A help message on the log function will be displayed. Note that MATLAB is case sensitive, so sin(a) is not the same as sin(A).

TABLE E.1

Typical elementary math functions.

Function	Remark
abs(x)	Absolute value or complex magnitude of x
acos, acosh(x)	Inverse cosine and inverse hyperbolic cosine of x in radians
acot, acoth(x)	Inverse cotangent and inverse hyperbolic cotangent of x in radians
angle(x)	Phase angle (in radian) of a complex number x
asin, asinh(x)	Inverse sine and inverse hyperbolic sine of x in radians
atan, atanh(x)	Inverse tangent and inverse hyperbolic tangent of x in radians
conj(x)	Complex conjugate of x
cos, cosh(x)	Cosine and hyperbolic cosine of x in radians
cot, coth(x)	Cotangent and hyperbolic cotangent of x in radians
exp(x)	Exponential of x
fix	Round toward zero
imag(x)	Imaginary part of a complex number x
log(x)	Natural logarithm of x
log2(x)	Logarithm of x to base 2
log10(x)	Common logarithms (base 10) of x
real(x)	Real part of a complex number x
sin, sinh(x)	Sine and hyperbolic sine of x in radians
sqrt(x)	Square root of x
tan, tanh	Tangent and hyperbolic tangent of x in radians

Try the following examples:

```
>> 3^(log10(25.6))
>> y = 2* sin(pi/3)
>> exp(y+4-1)
```

In addition to operating on mathematical functions, MATLAB allows one to work easily with vectors and matrices. A vector (or array) is a special matrix with one row or one column. For example,

```
>> a = [1 -3 6 10 -8 11 14];
```

is a row vector. Defining a matrix is similar to defining a vector. For example, a 3 × 3 matrix can be entered as

```
>> A = [1 2 3; 4 5 6; 7 8 9]
```

or as

```
>> A = [ 1 2 3
         4 5 6
         7 8 9]
```

In addition to the arithmetic operations that can be performed on a matrix, the operations in Table E.2 can be implemented.

Using the operations in Table E.2, we can manipulate matrices as follows:

```
>> B = A'
B =
      1   4   7
      2   5   8
      3   6   9
>> C = A + B
C =
      2   6  10
      6  10  14
     10  14  18
>> D = A^3 - B*C
D =
    372   432   492
    948  1131  1314
   1524  1830  2136
>> e = [1 2; 3 4]
e =
   1 2
   3 4
>> f = det(e)
f =
   -2
>> g = inv(e)
g =
   -2.0000   1.0000
    1.5000  -0.5000
```

TABLE E.2

Matrix operations.

Operation	Remark
`A'`	Finds the transpose of matrix A
`det(A)`	Evaluates the determinant of matrix A
`inv(A)`	Calculates the inverse of matrix A
`eig(A)`	Determines the eigenvalues of matrix A
`diag(A)`	Finds the diagonal elements of matrix A

TABLE E.3

Special matrices, variables, and constants.

Matrix, Variable, Constant	Remark
`eye`	Identity matrix
`ones`	An array of 1s
`zeros`	An array of 0s
`i or j`	Imaginary unit or `sqrt(-1)`
`pi`	3.142
`NaN`	Not a number
`inf`	Infinity
`eps`	A very small number, `2.2e - 16`
`rand`	Random element

```
>> H = eig(g)
H =
   -2.6861
    0.1861
```

Note that not all matrices can be inverted. A matrix can be inverted if and only if its determinant is nonzero. Special matrices, variables, and constants are listed in Table E.3. For example, type

```
>> eye(3)
ans =
      1  0  0
      0  1  0
      0  0  1
```

to get a 3×3 identity matrix.

Plotting

To plot using MATLAB is easy. For a two-dimensional plot, use the plot command with two arguments as follows:

```
>> plot(xdata,ydata)
```

where `xdata` and `ydata` are vectors of the same length containing the data to be plotted.

For example, suppose we want to plot `y = 10*sin(2*pi*x)` from `0` to `5*pi`. We will proceed with the following commands:

```
>> x = 0:pi/100:5*pi;     % x is a vector, 0 <= x <=
                          5*pi, increments of pi/100
>> y = 10*sin(2*pi*x);    % creates a vector y
>> plot(x,y);             % creates the plot
```

With this, MATLAB responds with the plot in Fig. E.1.

MATLAB will let you graph multiple plots together and distinguish them with different colors. This is obtained with the format `plot(xdata, ydata, 'color')`, where the color is indicated by using a character string from the options listed in Table E.4.

For example,

```
>> plot (x1,y1, 'r', x2,y2, 'b', x3,y3, '--');
```

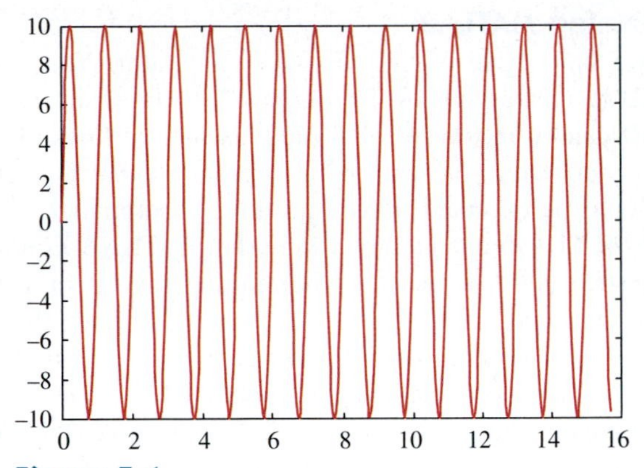

Figure E.1
MATLAB plot of `y = 10*sin(2*pi*x)`.

TABLE E.4

Various color and line types.

y	Yellow	.	Point
m	Magenta	o	Circle
c	Cyan	x	x mark
r	Red	+	Plus
g	Green	-	Solid
b	Blue	*	Star
w	White	:	Dotted
k	Black	-.	Dashdot
		--	Dashed

will graph data `(x1, y1)` in red, data `(x2, y2)` in blue, and data `(x3, y3)` in dashed line all on the same plot.

MATLAB also allows for logarithm scaling. Rather than using the plot command, we use

`loglog log(y)` versus `log(x)`

`semilogx y` versus `log(x)`

`semilogy log(y)` versus `x`

Three-dimensional plots are drawn using the functions mesh and meshdom (mesh domain). For example, to draw the graph of `z = x*exp( - x^2 - y^2)` over the domain `-1 < x, y < 1`, we type the following commands:

```
>> xx = -1:.1:1;
>> yy = xx;
>> [x,y] = meshgrid(xx,yy);
>> z = x.*exp(-x.^2 -y.^2);
>> mesh(z);
```

(The dot symbol used in `x.` and `y.` allows element-by-element multiplication.) The result is shown in Fig. E.2.

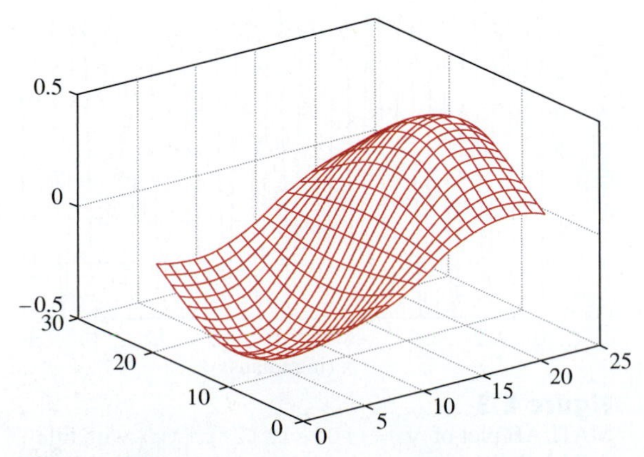

Figure E.2
A three-dimensional plot.

Programming MATLAB

So far we have used MATLAB as a calculator. You can also use MATLAB to create your own program. The command line editing in MATLAB can be inconvenient if one has several lines to execute. To avoid this problem, you can create a program that is a sequence of statements to be executed. If you are in the Command window, click **File/New/M**-files to open a new file in the MATLAB Editor/Debugger or simple text editor. Type the program and save it in a file with an extension `.m`, say `filename.m`; it is for this reason that it is called an M-file. Once the program is saved as an M-file, exit the Debugger window. You are now back in the Command window. Type the file without the extension `.m` to get results. For example, the plot that was made in Fig. E.2 can be improved by adding title and labels and being typed as an M-file called `example1.m`.

```
x = 0:pi/100:5*pi;              % x is a vector, 0 <= x <= 5*pi, increments of pi/100
y = 10*sin(2*pi*x);             % creates a vector y
plot(x,y);                      % create the plot
xlabel('x (in radians)');       % label the x axis
ylabel('10*sin(2*pi*x)');       % label the y axis
title('A sine functions');      % title the plot
grid                            % add grid
```

Once the file is saved as `example1.m` and you exit the text editor, type

```
>> example1
```

in the Command window and hit **Enter** to obtain the result shown in Fig. E.3.

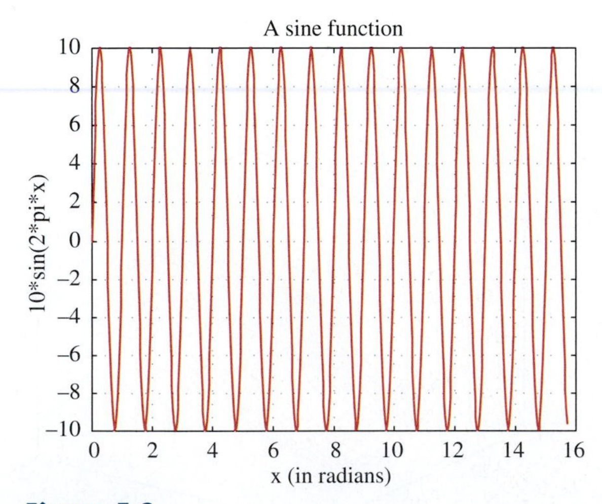

Figure E.3
MATLAB plot of `y = 10*sin(2*pi*x)` with title and labels.

To allow flow control in a program, certain relational and logical operators are necessary. They are shown in Table E.5. Perhaps the most commonly used flow control statements are `for` and `if`. The `for` statement is used to create a loop or a repetitive procedure and has the general form

```
for x = array
   [commands]
end
```

The `if` statement is used when certain conditions need to be met before an expression is executed. It has the general form

```
if expression
   [commands if expression is True]
else
   [commands if expression is False]
end
```

For example, suppose we have an array `y(x)` and we want to determine the minimum value of `y` and its corresponding index `x`. This can be done by creating an M-file as shown here.

```
% example2.m
% This program finds the minimum y value and
  its corresponding x index
x=[1 2 3 4 5 6 7 8 9 10]; %the nth term in y
y=[3 9 15 8 1 0 -2 4 12 5];
min1=y(1); for k = 1:10
   min2 = y(k);
   if(min2 < min1)
      min1 = min2;
      xo = x(k);
   else
      min1 = min1;
   end
end
diary
min1, xo
diary off
```

Note the use of the `for` and `if` statements. When this program is saved as `example2.m`, we execute it in the Command window and obtain the minimum value of `y` as `-2` and the corresponding value of `x` as `7`, as expected.

```
>> example2
min1 =
  -2
xo =
   7
```

TABLE E.5

Relational and logical operators.

Operator	Remark
<	Less than
<=	Less than or equal
>	Greater than
>=	Greater than or equal
==	Equal
~=	Not equal
&	And
|	Or
~	Not

If we are not interested in the corresponding index, we could do the same thing using the command

```
>> min(y)
```

The following tips are helpful in working effectively with MATLAB:

- Comment your M-file by adding lines beginning with a % character.
- To suppress output, end each command with a semicolon (;); you may remove the semicolon when debugging the file.
- Press the up and down arrow keys to retrieve previously executed commands.
- If your expression does not fit on one line, use an ellipse (. . .) at the end of the line and continue on the next line. For example, MATLAB considers

  ```
  y = sin(x + log10(2x + 3)) + cos(x + ...
  log10(2x + 3));
  ```

  as one line of expression.
- Keep in mind that variable and function names are case sensitive.

Solving Equations

Consider the general system of n simultaneous equations:

$$\begin{aligned} a_{11}x_1 + a_{12}x_2 + \cdots + a_{1n}x_n &= b_1 \\ a_{21}x_1 + a_{22}x_2 + \cdots + a_{2n}x_n &= b_2 \\ &\vdots \\ a_{n1}x_1 + a_{n2}x_2 + \cdots + a_{nn}x_n &= b_n \end{aligned}$$

or in matrix form

$$\mathbf{AX} = \mathbf{B}$$

where

$$\mathbf{A} = \begin{bmatrix} a_{11} & a_{12} & \cdots & a_{1n} \\ a_{21} & a_{22} & \cdots & a_{2n} \\ \cdots & \cdots & \cdots & \cdots \\ a_{n1} & a_{n2} & a_{n3} & a_{n4} \end{bmatrix} \qquad \mathbf{X} = \begin{bmatrix} x_1 \\ x_2 \\ \cdots \\ x_n \end{bmatrix} \qquad \mathbf{B} = \begin{bmatrix} b_1 \\ b_2 \\ \cdots \\ b_n \end{bmatrix}$$

A is a square matrix and is known as the coefficient matrix, while **X** and **B** are vectors. **X** is the solution vector we are seeking to get. There are two ways to solve for **X** in MATLAB. First, we can use the backslash operator(\) so that

```
X = A\B
```

Second, we can solve for **X** as

$$X = A^{-1}B$$

which in MATLAB is the same as

```
X = inv(A)*B
```

Example E.1

Use MATLAB to solve Example A.2 in Appendix A.

Solution:
From Example A.2, we obtain matrix **A** and vector **B** and enter them in MATLAB as follows.

```
>> A = [25 -5 -20; -5 10 -4; -5 -4 9]
A =
    25  -5  -20
    -5  10   -4
    -5  -4    9
>> B = [50 0 0]'
B =
    50
     0
     0
>> X = inv(A)*B
X =
    29.6000
    26.0000
    28.0000
>> X = A\B
X =
    29.6000
    26.0000
    28.0000
```

Thus, $x_1 = 29.6$, $x_2 = 26$, and $x_3 = 28$.

Practice Problem E.1

Solve the problem in Practice Problem A.2 using MATLAB.

Answer: $x_1 = 3 = x_3$; $x_2 = 2$

E.2 DC Circuit Analysis

There is nothing special in applying MATLAB to resistive dc circuits. We apply mesh and nodal analysis as usual and solve the resulting simultaneous equations using MATLAB as is described in Section E.1. Examples E.2 through E.5 illustrate.

Example E.2

Use nodal analysis to solve for the nodal voltages in the circuit of Fig. E.4.

Solution:
At node 1,

$$2 = \frac{V_1 - V_2}{4} + \frac{V_1 - 0}{8} \rightarrow 16 = 3V_1 - 2V_2 \qquad \textbf{(E.2.1)}$$

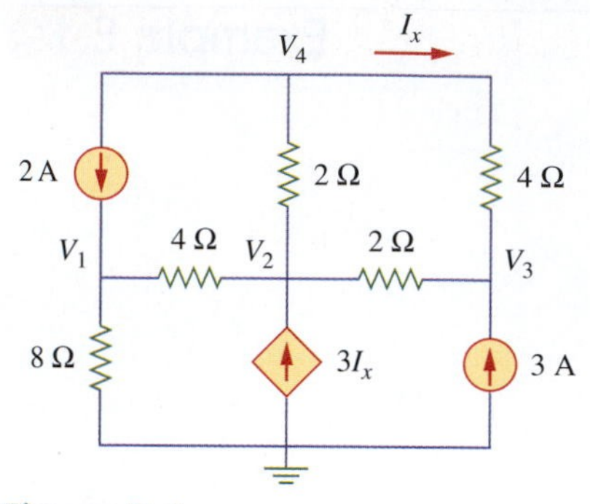

Figure E.4
For Example E.2.

At node 2,

$$3I_x = \frac{V_2 - V_1}{4} + \frac{V_2 - V_3}{2} + \frac{V_2 - V_4}{2}$$

But

$$I_x = \frac{V_4 - V_3}{4}$$

so that

$$3\left(\frac{V_4 - V_3}{4}\right) = \frac{V_2 - V_1}{4} + \frac{V_2 - V_3}{2} + \frac{V_2 - V_4}{2} \rightarrow \tag{E.2.2}$$
$$0 = -V_1 + 5V_2 + V_3 - 5V_4$$

At node 3,

$$3 = \frac{V_3 - V_2}{2} + \frac{V_3 - V_4}{4} \rightarrow 12 = -2V_2 + 3V_3 - V_4 \tag{E.2.3}$$

At node 4,

$$0 = 2 + \frac{V_4 - V_2}{2} + \frac{V_4 - V_3}{4} \rightarrow -8 = -2V_2 - V_3 + 3V_4 \tag{E.2.4}$$

Combining Eqs. (E.2.1) through (E.2.4) gives

$$\begin{bmatrix} 3 & -2 & 0 & 0 \\ -1 & 5 & 1 & -5 \\ 0 & -2 & 3 & -1 \\ 0 & -2 & -1 & 3 \end{bmatrix} \begin{bmatrix} V_1 \\ V_2 \\ V_3 \\ V_4 \end{bmatrix} = \begin{bmatrix} 16 \\ 0 \\ 12 \\ -8 \end{bmatrix}$$

or

$$\mathbf{AV} = \mathbf{B}$$

We now use MATLAB to determine the nodal voltages contained in vector **V**.

```
>> A = [ 3 -2   0   0;
        -1  5   1  -5;
         0 -2   3  -1;
         0 -2  -1   3];
>> B = [16  0  12  -8]';
>> V = inv(A)*B
V =
    -6.0000
   -17.0000
   -13.5000
   -18.5000
```

Hence, $V_1 = -6.0$, $V_2 = -17$, $V_3 = -13.5$, and $V_4 = -18.5$ V.

Practice Problem E.2

Find the nodal voltages in the circuit in Fig. E.5 using MATLAB.

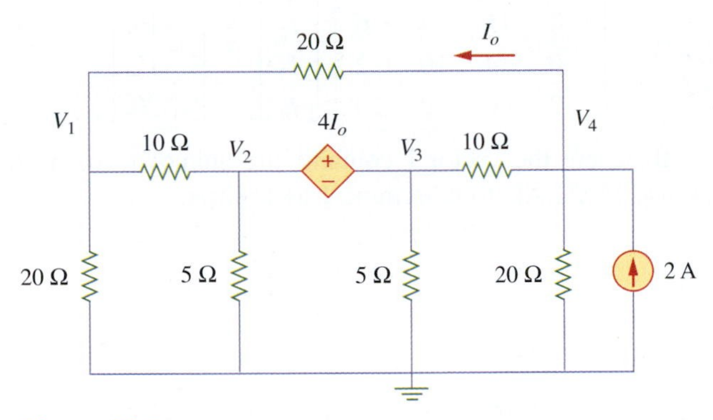

Figure E.5
For Practice Problem E.2.

Answer:
$V_1 = 14.55$; $V_2 = 38.18$; $V_3 = -34.55$; $V_4 = -3.636$ V

Example E.3

Use MATLAB to solve for the mesh currents in the circuit in Fig. E.6.

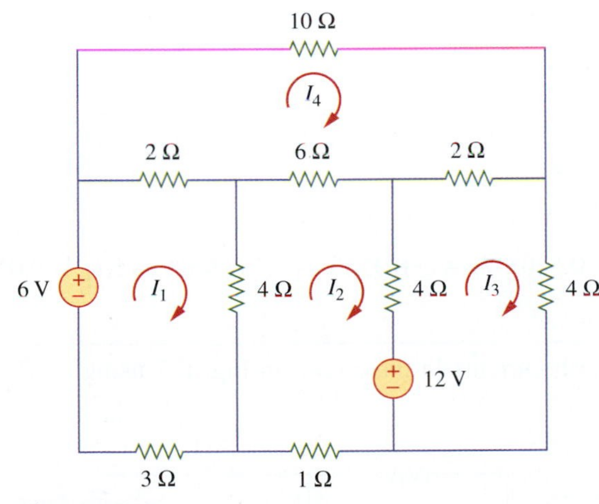

Figure E.6
For Example E.3.

Solution:
For the four meshes,

$$-6 + 9I_1 - 4I_2 - 2I_4 = 0 \longrightarrow 6 = 9I_1 - 4I_2 - 2I_4 \quad \textbf{(E.3.1)}$$

$$12 + 15I_2 - 4I_1 - 4I_3 - 6I_4 = 0 \longrightarrow$$
$$-12 = -4I_1 + 15I_2 - 4I_3 - 6I_4 \quad \textbf{(E.3.2)}$$

$$-12 + 10I_3 - 4I_2 - 2I_4 = 0 \longrightarrow 12 = -4I_2 + 10I_3 - 2I_4 \quad \textbf{(E.3.3)}$$

$$20I_4 - 2I_1 - 6I_2 - 2I_3 = 0 \longrightarrow 0 = -2I_1 - 6I_2 - 2I_3 + 20I_4 \quad \textbf{(E.3.4)}$$

Putting Eqs. (E.3.1) through (E.3.4) together in matrix form, we have

$$\begin{bmatrix} 9 & -4 & 0 & -2 \\ -4 & 15 & -4 & -6 \\ 0 & -4 & 10 & -2 \\ -2 & -6 & -2 & 20 \end{bmatrix} \begin{bmatrix} I_1 \\ I_2 \\ I_3 \\ I_4 \end{bmatrix} = \begin{bmatrix} 6 \\ -12 \\ 12 \\ 0 \end{bmatrix}$$

or $\mathbf{AI} = \mathbf{B}$, where the vector **I** contains the unknown mesh currents. We now use MATLAB to determine **I** as follows:

```
>> A = [9 -4  0 -2; -4 15 -4 -6;
        0 -4 10 -2;  -2 -6 -2 20]

A =

   9  -4   0  -2
  -4  15  -4  -6
   0  -4  10  -2
  -2  -6  -2  20

>> B = [6 -12 12 0]'

B =

    6
  -12
   12
    0

>> I = inv(A)*B

I =

   0.5203
  -0.3555
   1.0682
   0.0522
```

Thus, $I_1 = 0.5203$, $I_2 = -0.3555$, $I_3 = 1.0682$, and $I_4 = 0.0522$ A.

Practice Problem E.3

Find the mesh currents in the circuit in Fig. E.7 using MATLAB.

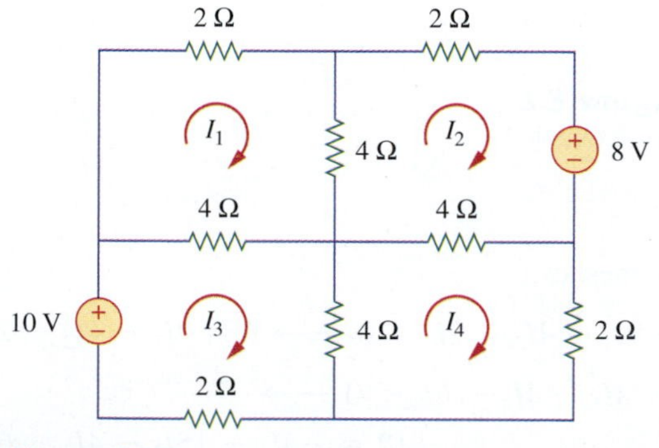

Figure E.7
For Practice Problem E.3.

Answer: $I_1 = 0.2222$; $I_2 = -0.6222$; $I_3 = 1.1778$; $I_4 = 0.2222$ A

E.3 AC Circuit Analysis

Using MATLAB in ac circuit analysis is similar to how MATLAB is used for dc circuit analysis. We must first apply nodal or mesh analysis to the circuit and then use MATLAB to solve the resulting system of equations. However, the circuit is in the frequency domain, and we are dealing with phasors or complex numbers. So in addition to what we learned in Section E.2, we need to understand how MATLAB handles complex numbers.

MATLAB expresses complex numbers in the usual manner, except that the imaginary part can be either j or i representing $\sqrt{-1}$. Thus, $3 - j4$ can be written in MATLAB as `3 - j4`, `3 - j*4`, `3 - i4`, or `3 - i*4`. Here are the other complex functions:

`abs(A)`	Absolute value of magnitude of `A`
`angle(A)`	Angle of `A` in radians
`conj(A)`	Complex conjugate of `A`
`imag(A)`	Imaginary part of `A`
`real(A)`	Real part of `A`

Keep in mind that an angle in radians must be multiplied by $180/\pi$ to convert it to degrees, and vice versa. Also, the transpose operator (`'`) gives the complex conjugate transpose, whereas the dot-transpose (`.'`) transposes an array without conjugating it.

Example E.4

In the circuit of Fig. E.8, let $v = 4\cos(5t - 30°)$ V and $i = 0.8\cos(5t)$ A. Find v_1 and v_2.

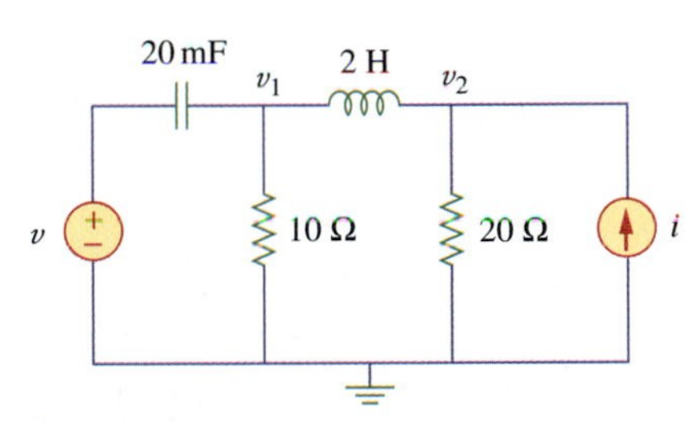

Figure E.8
For Example E.4.

Solution:
As usual, we convert the circuit in the time-domain to its frequency-domain equivalent.

$$v = 4\cos(5t - 30°) \longrightarrow \mathbf{V} = 4\underline{/-30°}, \quad \omega = 5$$
$$i = 0.8\cos(5t) \longrightarrow \mathbf{I} = 8\underline{/0°}$$
$$2\text{ H} \longrightarrow j\omega L = j5 \times 2 = j10$$
$$20\text{ mF} \longrightarrow \frac{1}{j\omega C} = \frac{1}{\text{j}10\ \Omega \times 10^{-3}} = -j10$$

Thus, the frequency-domain equivalent circuit is shown in Fig. E.9. We now apply nodal analysis to this.

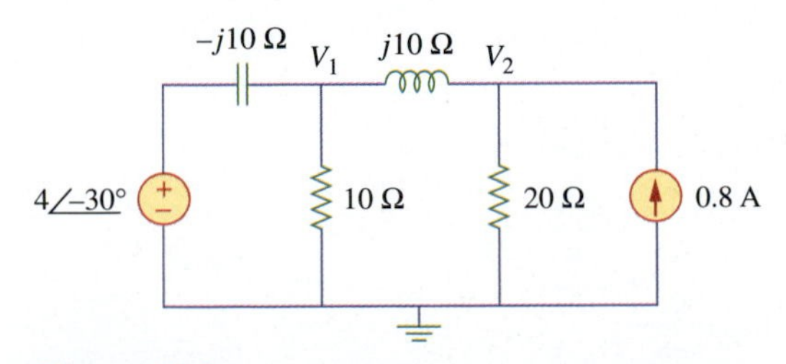

Figure E.9
The frequency-domain equivalent circuit of the circuit in Fig. E.8.

At node 1,

$$\frac{4\angle -30^\circ - V_1}{-j10} = \frac{V_1}{10} + \frac{V_1 - V_2}{j10} \longrightarrow 4\angle -30^\circ = 3.468 - j2 = -jV_1 + V_2 \qquad \textbf{(E.4.1)}$$

At node 2,

$$0.8 = \frac{V_2}{20} + \frac{V_2 - V_1}{j10} \longrightarrow j16 = -2V_1 + (2 + j)V_2 \qquad \textbf{(E.4.2)}$$

Equations (E.4.1) and (E.4.2) can be cast in matrix form as

$$\begin{bmatrix} -j & 1 \\ -2 & (2 + j) \end{bmatrix} \begin{bmatrix} V_1 \\ V_2 \end{bmatrix} = \begin{bmatrix} 3.468 - j2 \\ j16 \end{bmatrix}$$

or **AV** = **B**. We use MATLAB to invert **A** and multiply the inverse by **B** to get **V**.

```
>> A = [-j 1; -2 (2+j)]
A =
   0 - 1.0000i 1.000
   -2.0000 2.0000 + 1.000 i
>> B = [(3.468 - 2j) 16j].' %note the dot-transpose
B =
   3.4680 - 2.0000i
   0 + 16.0000i
>> V = inv(A)*B
V =
   4.6055 - 2.4403i
   5.9083 + 2.6055i
>> abs(V(1))
ans =
   5.2121
>> angle(V(1))*180/pi %converts angle from
   radians to degrees
ans =
   -27.9175
>> abs(V(2))
ans =
   6.4573
>> angle(V(2))*180/pi
ans =
   23.7973
```

Thus,

$$V_1 = 4.6055 - j2.4403 = 5.212\angle -27.92^\circ$$
$$V_2 = 5.908 + j2.605 = 6.457\angle 23.8^\circ$$

In the time domain,

$$v_1 = 4.605 \cos(5t - 27.92°) \text{ V}, \qquad v_2 = 6.457 \cos(5t + 23.8°) \text{ V}$$

Practice Problem E.4

Calculate v_1 and v_2 in the circuit in Fig. E.10 given $i = 4\cos(10t + 40°)$ A and $v = 12\cos(10t)$ V.

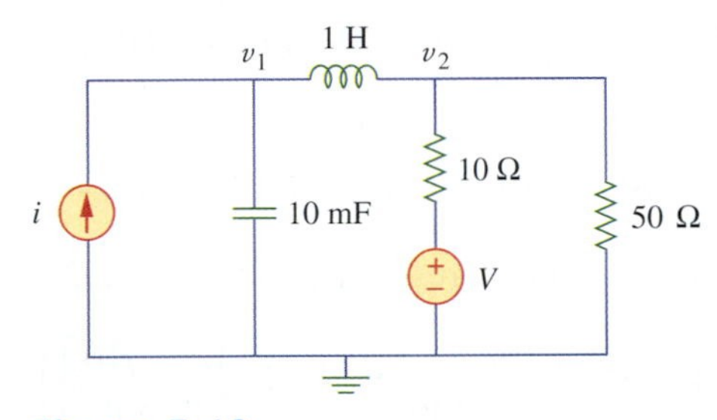

Figure E.10
For Practice Problem E.4.

Answer: $63.58\cos(10t - 10.68°)$ V; $40\cos(10t - 50°)$ V

Example E.5

In the unbalanced three-phase system shown in Fig. E.11, find currents I_1, I_2, I_3, and I_{Bb}. Let

$$Z_A = 12 + j10\ \Omega, \qquad Z_B = 10 - j8\ \Omega, \qquad Z_C = 15 + j6\ \Omega$$

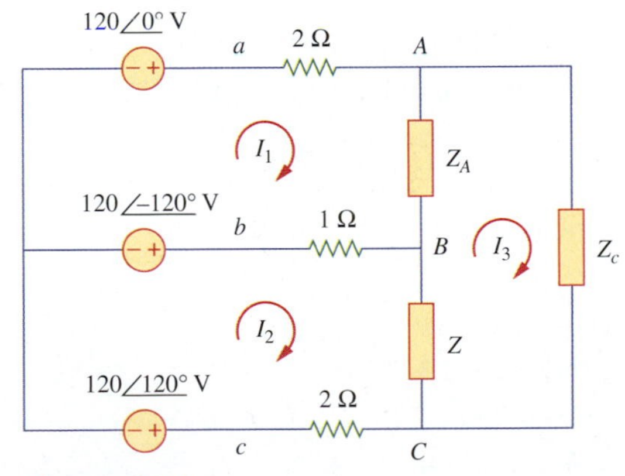

Figure E.11
For Example E.5.

Solution:
For mesh 1,

$$120\angle{-120°} - 120\angle{0°} + I_1(2 + 1 + 12 + j10) - I_2 - I_3(12 + j10) = 0$$

or

$$I_1(15 + j10) - I_2 - I_3(12 + j10) = 120\angle{0°} - 120\angle{-120°} \qquad \textbf{(E.5.1)}$$

For mesh 2,

$$120\angle 120^\circ - 120\angle -120^\circ + I_2(2 + 1 + 10 - j8) - I_1 - I_3(10 - j8) = 0$$

or

$$-I_1 + I_2(13 - j8) - I_3(10 - j8) = 120\angle -120^\circ - 120\angle 120^\circ \qquad \textbf{(E.5.2)}$$

For mesh 3,

$$I_3(12 + j10 + 10 - j8 + 15 + j6) - I_1(12 + j10) - I_2(10 - j8) = 0$$

or

$$-I_1(12 + j10) - I_2(10 - j8) - I_3(37 + j8) = 0 \qquad \textbf{(E.5.3)}$$

In matrix form, we can express Eqs. (E.5.1), (E.5.2), and (E.5.3) as

$$\begin{bmatrix} 15 + j10 & -1 & -12 - j10 \\ -1 & 13 - j8 & -10 + j8 \\ -12 - j10 & -10 + j8 & 37 + j8 \end{bmatrix} \begin{bmatrix} I_1 \\ I_2 \\ I_3 \end{bmatrix} = \begin{bmatrix} 120\angle 0^\circ - 120\angle -120^\circ \\ 120\angle -120^\circ - 120\angle 120^\circ \\ 0 \end{bmatrix}$$

or

$$\mathbf{ZI} = \mathbf{V}$$

We input matrices **Z** and **V** into MATLAB to get I.

```
>> z = [(15 + 10j) -1 (-12 - 10j);
        -1 (13 - 8j) (-10 + 8j);
        (-12 - 10j) (-10 + 8j) (37 + 8j)];
>> c1=120*exp(j*pi*(-120)/180);
>> c2=120*exp(j*pi*(120)/180);
>> a1=120 - c1; a2=c1 - c2;
>> V = [a1; a2; 0]
>> I = inv(z)*V
I=
   16.9910 - 6.5953i
   12.4023 - 16.9993i
    5.6621 - 6.0471i
>> IbB = I(2) - I(1)
IbB =
   -4.5887 - 10.4039i
>> abs(I(1))
ans =
    18.2261
>> angle(I(1))*180/pi
ans =
   -21.2146
```

```
>> abs (I(2))
ans =
    21.0426
>> angle(I(2))*180/pi
ans =
  -53.8864
>> abs(I(3))
ans =
    8.2841
>> angle(I(3))*180/pi
ans =
  -46.8833
>> abs(IbB)
ans =
   11.3709
>> angle(IbB)*180/pi
ans =
  -113.8001
```

Thus, $I_1 = 18.23\underline{/-21.21^\circ}$, $I_2 = 21.04\underline{/-58.89^\circ}$, $I_3 = 8.284\underline{/-46.88^\circ}$, and $I_{bB} = 11.37\underline{/-113.8^\circ}$A.

Practice Problem E.5

In the unbalanced wye-wye three-phase system in Fig. E.12, find the line currents I_1, I_2, and I_3 and the phase voltage V_{CN}.

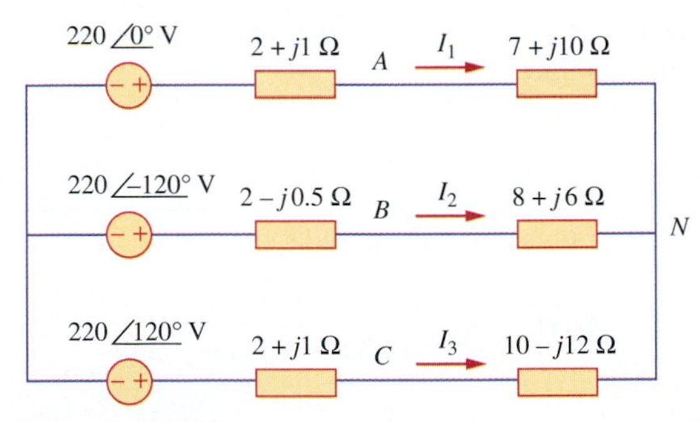

Figure E.12
For Practice Problem E.5.

Answer: $22.66\underline{/-26.54^\circ}$ A; $6.036\underline{/-150.48^\circ}$ A; $19.93\underline{/138.9^\circ}$ A; $94.29\underline{/159.3^\circ}$ V

E.4 Frequency Response

Frequency response involves plotting the magnitude and phase of the transfer function $H(s) = D(s)/N(s)$ or obtaining the Bode magnitude and phase plots of $H(s)$. One hard way to obtain the plots is to generate data using the `for` loop for each value of $s = j\omega$ for a given range of ω and then plot the data as we did in Section E.1. However, there is an easy way that allows us to use one of two MATLAB commands: `freqs` and `bode`. For each command, we must first specify `H(s)` as `num` and `den`, where `num` and `den` are the vectors of coefficients of the numerator `N(s)` and denominator `D(s)` in descending powers of `s`, i.e., from the highest power to the constant term. The general form of the `bode` function is

```
bode(num, den, range);
```

where `range` is a specified frequency interval for the plot. If `range` is omitted, MATLAB automatically selects the frequency range. The `range` could be linear or logarithmic. For example, for $1 < \omega < 1{,}000$ rad/s with 50 plot points, we can specify a linear `range` as

```
range = linspace(1,1000,50);
```

For a logarithmic `range` with $10^{-2} < \omega < 10^{4}$ rad/s and 100 plot points in between, we specify `range` as

```
range = logspace(-2,4,100);
```

For the `freqs` function, the general form is

```
hs = freqs(num, den, range);
```

where `hs` is the frequency response (generally complex). We still need to calculate the magnitude in decibels as

```
mag = 20*log 10(abs(hs))
```

and phase in degrees as

```
phase = angle(hs)*180/pi
```

and plot them, whereas the `bode` function does it all at once. We illustrate with an example.

Example E.6

Use MATLAB to obtain the Bode plots of

$$G(s) = \frac{s^3}{s^3 + 14.8s^2 + 38.1s + 2554}$$

Solution:

With the explanation previously given, we develop the MATLAB code as shown here.

```
% for example e.6
num=[1 0 0 0];
den = [1 14.8 38.1 2554];
w = logspace(-1,3);
bode(num, den, w);
title('Bode plot for a highpass filter')
```

Running the program produces the Bode plots in Fig. E.13. It is evident from the magnitude plot that $G(s)$ represents a highpass filter.

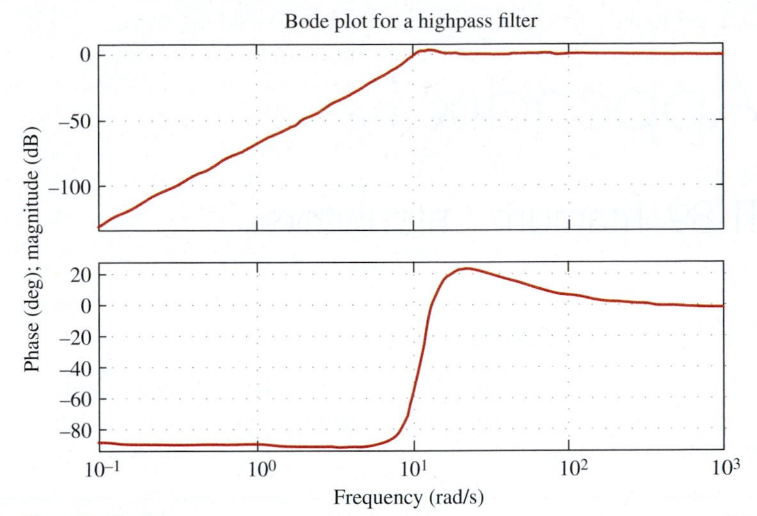

Figure E.13
For Example E.6.

Practice Problem E.6

Use MATLAB to determine the frequency response of

$$H(s) = \frac{10(s + 1)}{s^2 + 6s + 100}$$

Answer: See Fig. E.14

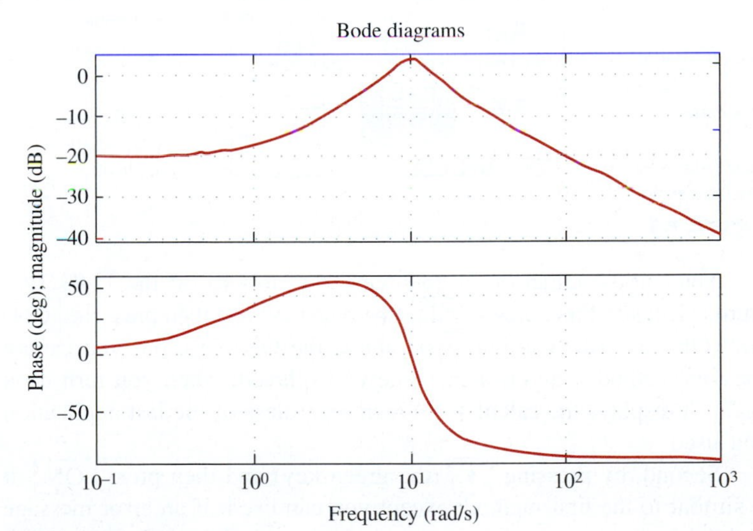

Figure E.14
For Practice Problem E.6.

Appendix F

TI-89 Titanium Calculators

F.1 Batteries

Use four AAA batteries for the TI-89 Titanium calculator, and make sure you arrange the batteries in the battery compartment according to the polarity (+ and −).

F.2 Power

To turn on the power for the TI-89 Titanium calculator, simply press [ON] which is located in the bottom-left corner on the keyboard. Figure F.1 shows the screen when it is starting.

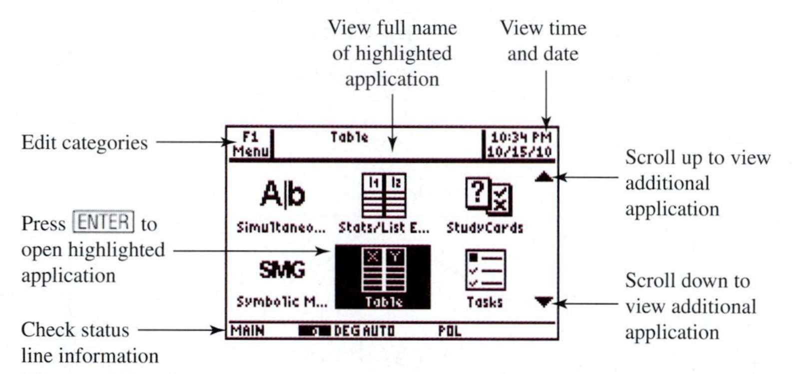

Figure F.1

You can use either of the following keys to turn off the TI-89 Titanium calculator. First, press [2ND] (the blue key) and then press the [ON] key (OFF is written in blue type above the ON key), but you cannot use this method if an error message is displayed; when you turn it on again, it displays the calculator screen regardless of the last application you used.

Second, by pressing [♦] (the green key) and then press [ON], it is similar to the first method, except you can use it if an error message is displayed; when you turn it on again, it will be exactly as you left it. If no keys are pressed for several minutes, the calculator will shut off automatically.

F.3 Screen Contrast

To darken the display, press and hold [♦] and tap [+].

To lighten the display, press and hold [♦] and tap [−].

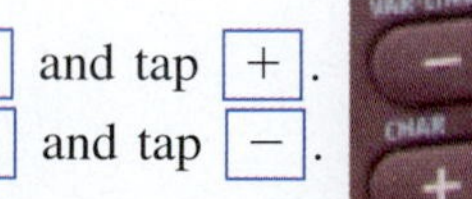

View on the keyboard of the calculator

F.4 TI-89 Titanium Keys

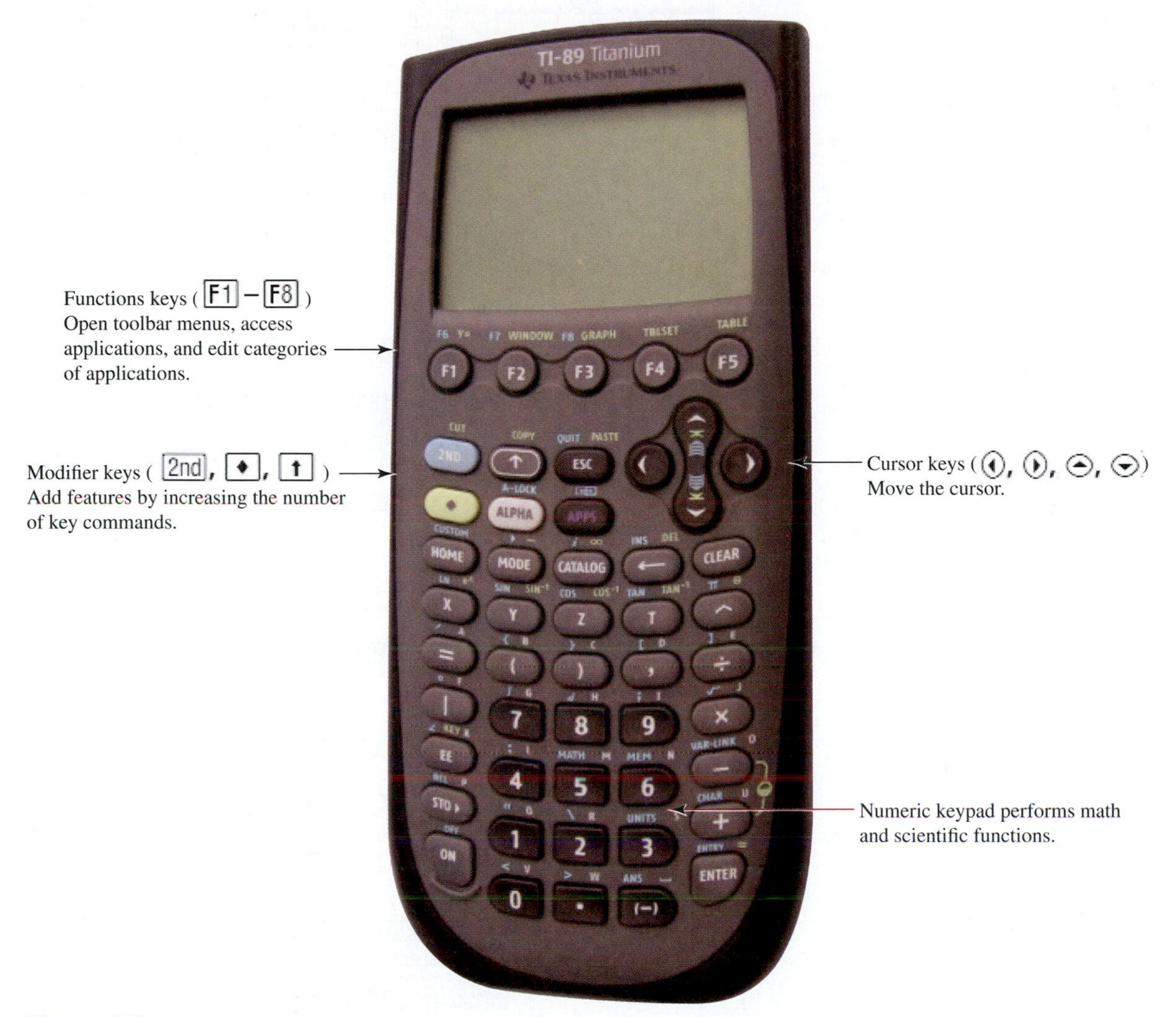

Figure F.2
© Sarhan M. Musa

Mode Settings

Modes control how the calculator displays and interprets information. To view the calculator mode settings:

1. Press MODE. Page 1 of the MODE dialog box appears.

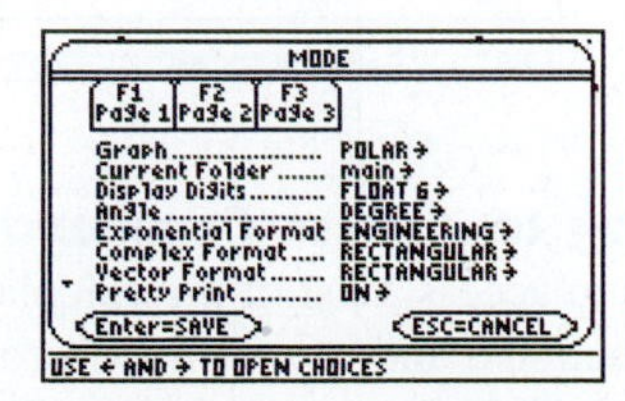

2. Press F2 or F3 to display the modes listed on Page 2 or Page 3.

Pressing F2

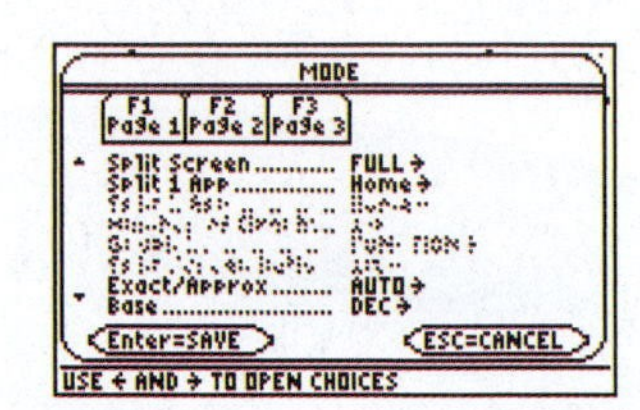

Pressing F3

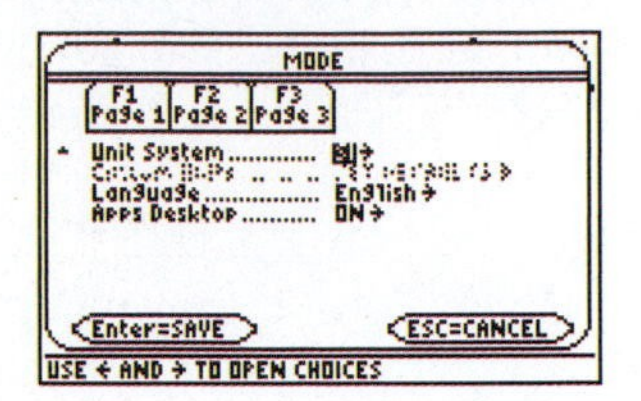

Modes that are grayed out are available only if other required mode settings are selected.

Example of Changing Mode settings:

Press MODE

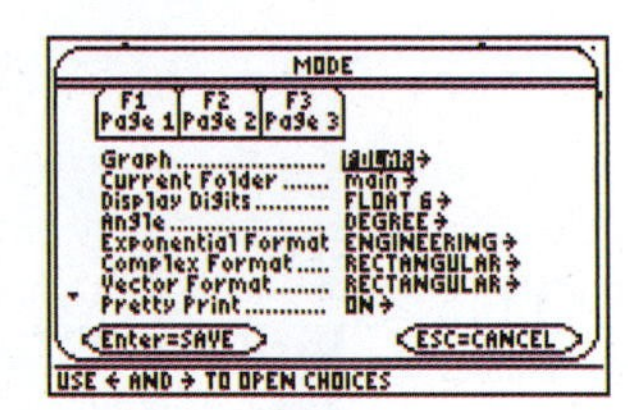

Press F3

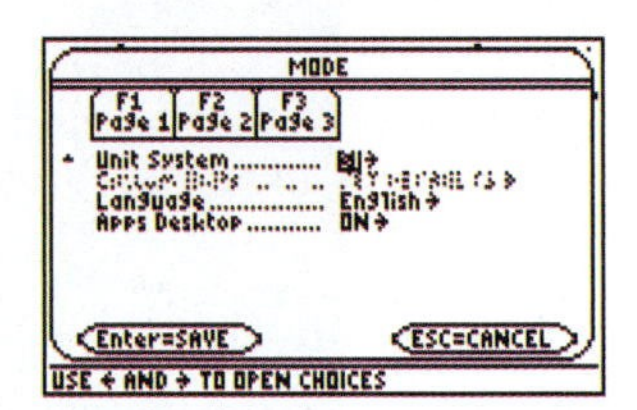

Scroll down to the language field (▼), to select the language and then Press (▶), to choose the language such as English, and then press ENTER to save it.

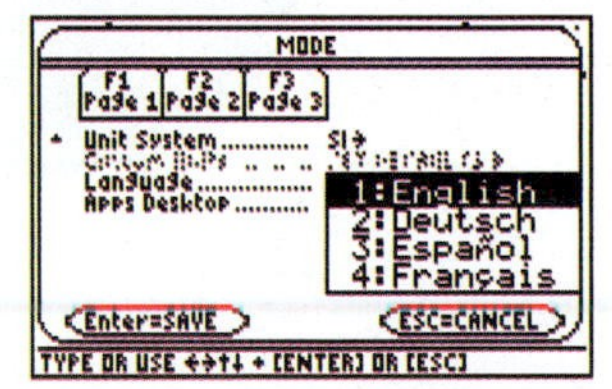

Press ENTER.

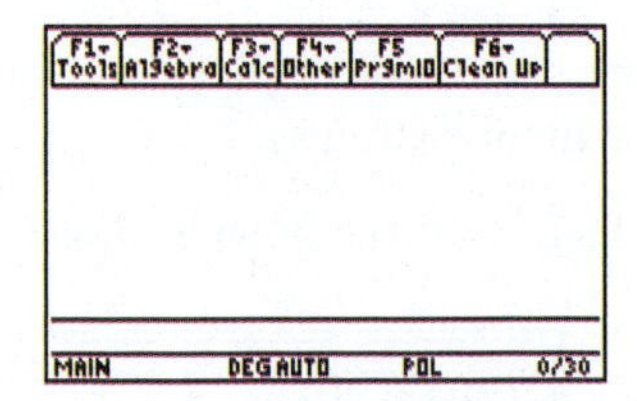

Using the Catalog to Access Commands

The Catalog is used to access a list of the calculator commands, including functions, instructions, and user-defined programs. Commands are listed alphabetically. The Catalog Help application includes details about each command.

Using Menus

To select menus, press the function keys corresponding to the toolbars at the top of the calculator Home screen and most application screens. Other menus can be selected by using key commands as in the following table.

Key	Description
2nd [CHAR]	To display CHAR menu. Lists characters not available on the keyboard; characters are organized by category (Greek, math, punctuation, special, and international).
2nd [MATH]	To display MATH menu. Lists math operations by category.
APPS	To display APPLICATIONS menu. Lists the installed applications. (Menu is available only when the Applications desktop is turned off; applications are normally accessed from the Applications desktop.)
♦ APPS	To display FLASH APPLICATIONS menu. Lists the installed Flash applications. (Menu is available only when Applications desktop is turned off; Flash applications are normally accessed from the Applications desktop.)

Special Characters

To enter special characters, use CHAR Menu and key commands. The CHAR menu lets you access Greek, math, international, and other special characters.

To select characters from the CHAR menu:

1. Press 2nd [CHAR]. The CHAR menu appears.
2. Use the cursor keys to select a category where a submenu lists the characters in the category.
3. Use the cursor keys to select a character, and press ENTER.

To open the keyboard map, press ♦ [KEY].

To type most characters, press ♦ and the corresponding key. Press ESC to close the map.

Modifiers Keys

These keys add features by increasing the number of keyboard operators at your fingertips. To access a modifier function, press a modifier key and then press the key for the corresponding operation. Table F.1 shows the most important keys.

TABLE F.1

Important Keys

Keys	Description
2nd (Second)	To access applications, menu options, and other operations. Second functions are printed above their corresponding keys in the same color as the 2nd key.
♦ (Diamond)	To access applications, menu options, and other operations. Diamond functions are printed above their corresponding keys in the same color as the ♦ key.

TABLE F.1 Continued

Keys	Description
[alpha] (Alpha)	To type alphabetic characters without a QWERTY keypad. Alpha characters are printed above their corresponding keys in the same color as the [alpha] key.
[↑] (Shift)	To type an uppercase character for the next letter key you press it. Also, used with ◀ and ▶ to highlight characters when editing.
[ESC]	It is equivalent of the Escape key on a computer. It gets you out of whatever you are doing.
[ENTER]	To execute commands and to evaluate expressions.
[CLEAR]	This is the erase key. If you are entering something into the calculator and you change your mind, press it two times.
[♦] [Y=]	To display the Y= Editor.
[♦] [WINDOW]	To display the Window Editor.
[♦] [GRAPH]	To display the graph screen.
[♦] [TBLSET]	To set parameters for the Table screen.
[♦] [TABLE]	To display the table screen.
[♦] [CUT] [♦] [COPY] [♦] [PASTE]	All these keys used to edit entered information by performing a cut, copy, or paste operation, respectively.
[APPS]	To display the Applications desktop.
[♦] [APPS]	With the Applications desktop off, to display the FLASH APPLICATIONS menu.
[2nd] [[↔]]	To switch between the last two chosen applications.
[2nd] [CUSTOM]	To turn the custom menu on and off.
[2nd] [▶]	To convert measurement units.
[♦] [_]	To designate a measurement unit.
[←]	To delete the character to the left of the cursor (backspace).
[♦] [DEL]	To delete the character to the right of the cursor.
[2nd] [INS]	To switch between insert and overwrite modes.
[2nd] [MEM]	To display the MEMORY screen.
[CATALOG]	To display a list of commands.
[2nd] [RCL]	To recall the contents of a variable.
[STO▶]	To store a value to a variable.
[2nd] [CHAR]	To display the CHAR menu, which lets you select Greek letters, international accented characters, and other special characters.
[2nd] [QUIT]	• In full-screen mode, to display the Applications desktop. • In split-screen mode, to display the full-screen view of the active application. • With the Applications desktop off, to display the calculator Home screen.

Numeric Keypad

We use the numeric keypad to enter positive and negative numbers. To enter a negative number, Press (−) before typing the number. Do not use the subtraction key − to represent the negative.

Do the following to enter a number in scientific notation:

1. Type the numbers that precede the exponent.
2. Press EE. The exponent symbol (E) follows the numbers you entered.
3. Type the exponent as an integer with up to three digits.

Enter 0.00786 using scientific notation.

Example F.1

Solution:

1. Press 7 . 8 6
2. Press EE. Then (−) 3.

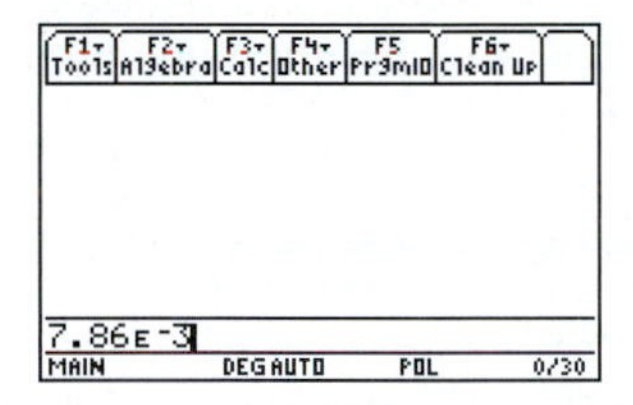

3. Press ENTER.

F.5 Calculator Home Screen

The calculator Home screen is the starting point for math operations, including executing instructions, evaluating expressions, and viewing results. You can display the calculator Home screen by two ways:

1. Press HOME.
2. From the Applications desktop by highlighting the Home icon and pressing ENTER.

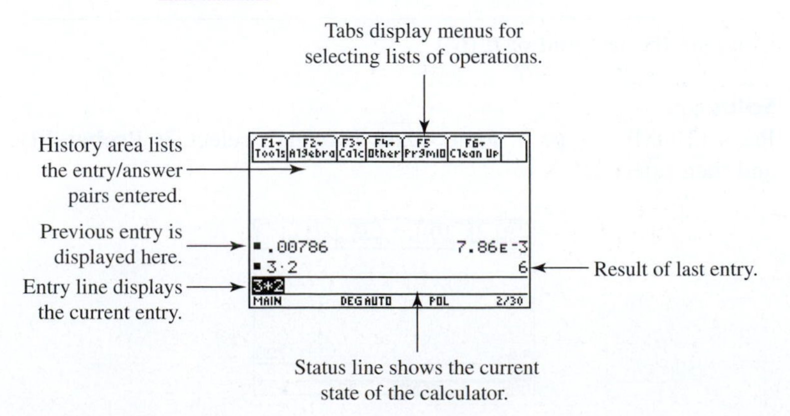

To return to the Applications desktop from the calculator Home screen, press [APPS].

F.6 Performing Computations

Before performing any activity, always clear the history area in each screen by pressing [F1] and selecting **8: Clear Home**.

Example F.2

Compute $\cos(\pi/3)$ and display the result in symbolic and numeric format.

Solution:
First make sure you choose radian for angle by pressing [MODE] and selecting radian for angle.

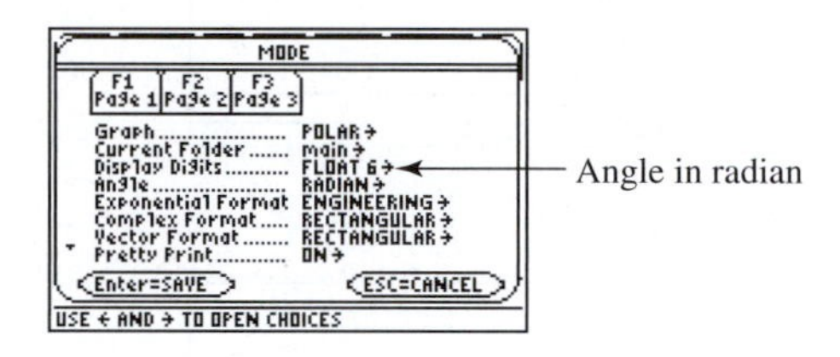

Angle in radian

Then, press [HOME], then

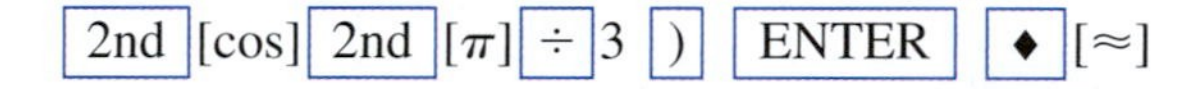

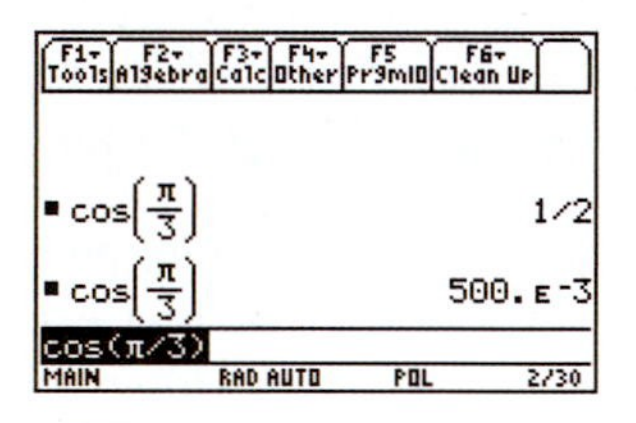

Example F.3

Compute the factorial of 6 (6!).

Solution:
Press [HOME], type 6, Press [2nd] [MATH], select **7: Probability**, and then select **1:!**.

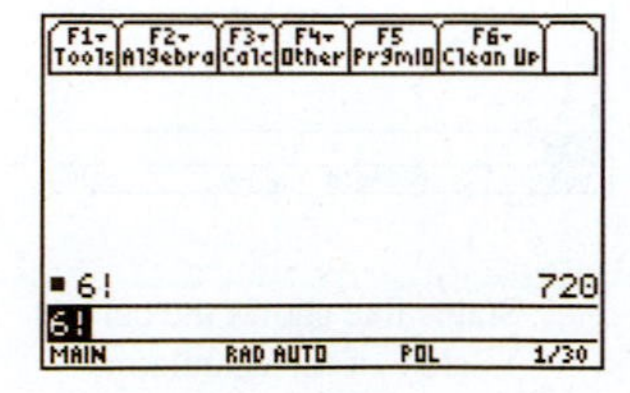

Example F.4

Compute the complex number $(5 + 7i)^3$.

Solution:

Press (5 + 7 2nd [i]) ^ **3** ENTER

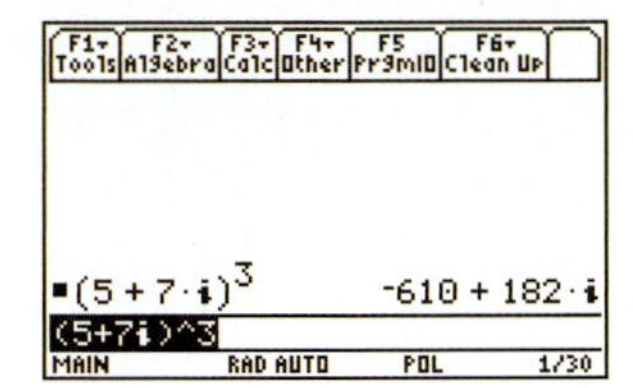

Example F.5

Compute the square root of 20 ($\sqrt{20}$).

Solution:

Press 2nd and then √; enter 20) and press ENTER.

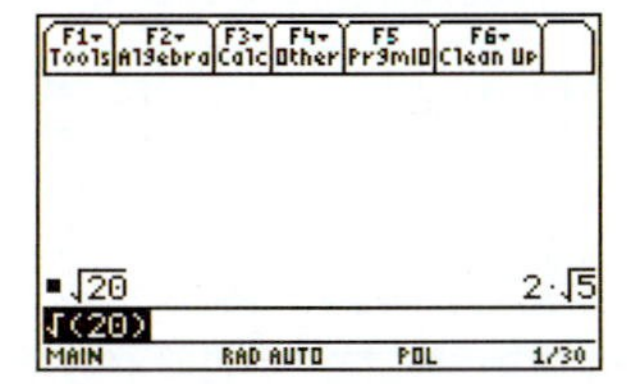

Example F.6

Store the variable **a** as 15, then compute 5**a**.

Solution:

Press HOME, type 15, press STO▶, press alpha, then **a**, then ENTER, type 5 ×, STO▶, press alpha, **a**, ENTER.

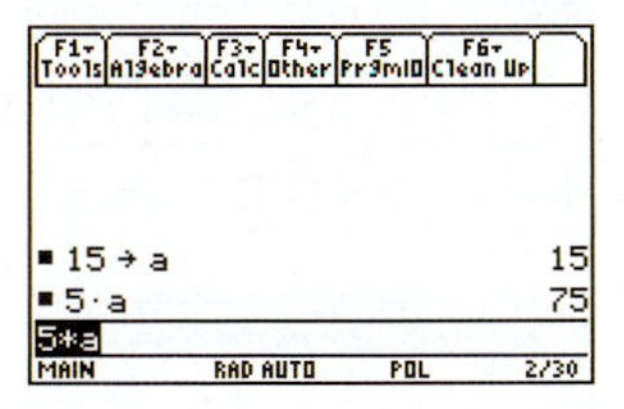

Use Press 2nd F1, **1: Clear a-z**, to delete the variable.

F.7 Matrices

A matrix is a rectangular array of elements arranged in rows and columns. The calculator can define a matrix in two ways; on the Home screen or in the Data/Matrix editor.

Define a Matrix on the Home Screen

You need to do the following:

- Enclose the elements of the matrix in square bracket, press 2nd , and 2nd ÷.

- The elements in the rows of the matrix are separated by commas, [,].
- The rows are delineated by semicolons, [2nd] [9].

Example F.7

Let $a = \begin{bmatrix} 2 & 4 \\ 6 & 8 \end{bmatrix}$, $b = \begin{bmatrix} 0 & 3 \\ 5 & 1 \end{bmatrix}$ and $c = \begin{bmatrix} 1 & 3 & 5 \\ 7 & 9 & 11 \end{bmatrix}$

Evaluate the following:

1. $2a$
2. a − b
3. $a \times c$
4. determinant of a

Solution:

1. Press [2nd] [,] 2 [,] 4 [2nd] [9] 6 [,] 8 [2nd] [÷] [STO▶] [alpha] a [ENTER]. Then 2 [×] a.

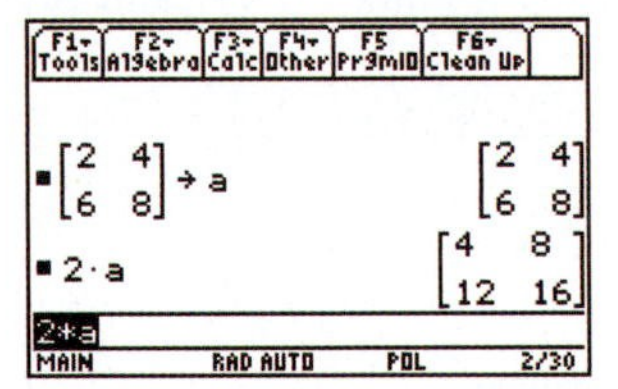

2. Press [2nd] [,] 0 [,] 3 [2nd] [9] 5 [,] 1 [2nd] [÷] [STO▶] [alpha] b [ENTER]. Then a [−] b.

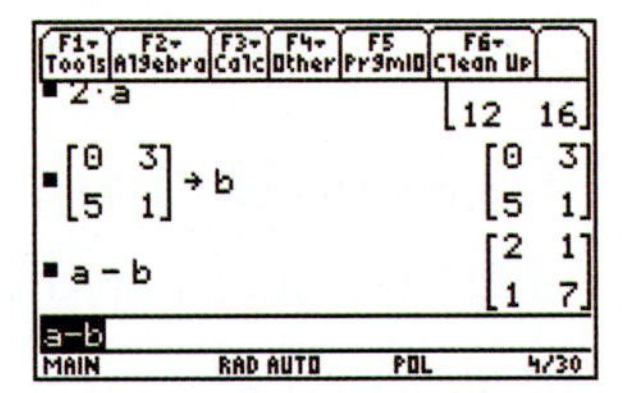

3. Press [2nd] [,] 1 [,] 3 [,] 5 [2nd] [9] 7 [,] 9 [,] 11 [2nd] [÷] [STO▶] [alpha] c [ENTER]. Then a [×] c.

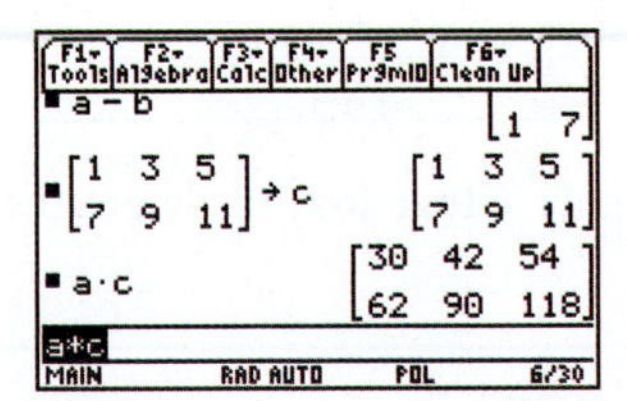

4. Press [2nd] 5 **4: Matrix** (▶), **2: det(** [ENTER] [alpha] a [ )] [ENTER]

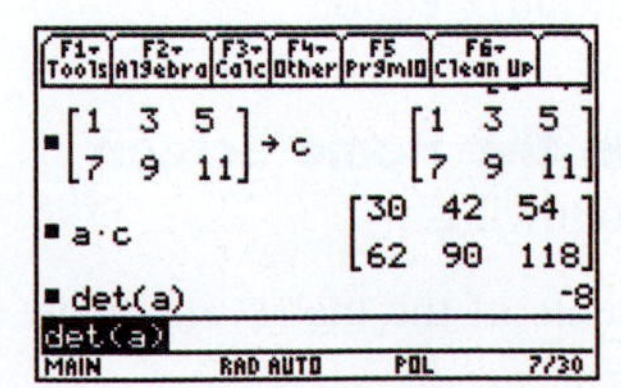

Define and Edit a Matrix in the Data/Matrix Editor

The Data/Matrix editor is a useful place to define and edit matrices, especially for several large matrices.

Example F.8

Define $d = \begin{bmatrix} 1 & 2 \\ 3 & 4 \end{bmatrix}$ in Data/Matrix editor, and then find $5d$.

Solution:

Press APPS, then select Data/Matrix

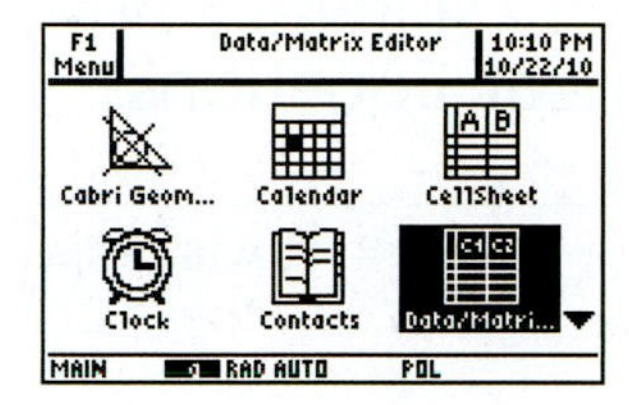

Press ENTER

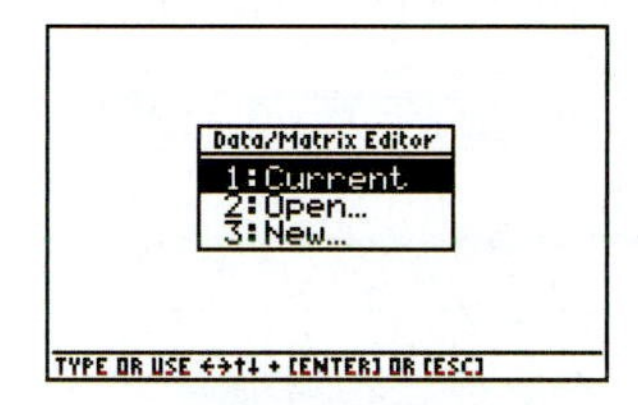

Press ENTER, and select **2: Matrix** in type.

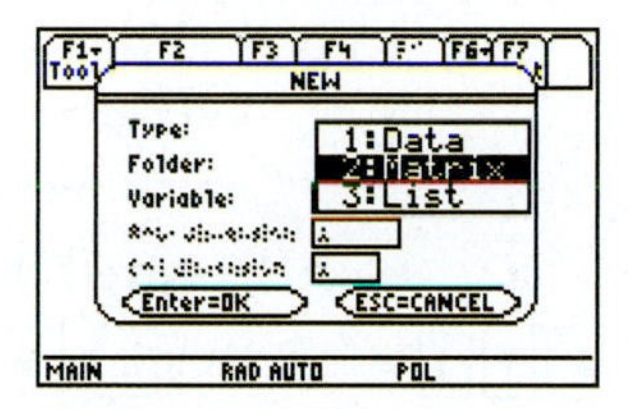

Type in variable d, row dimension 2, col dimension 2.

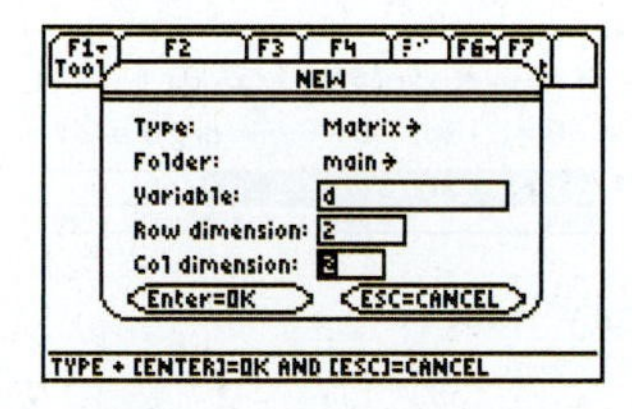

Press ENTER, and type the elements.

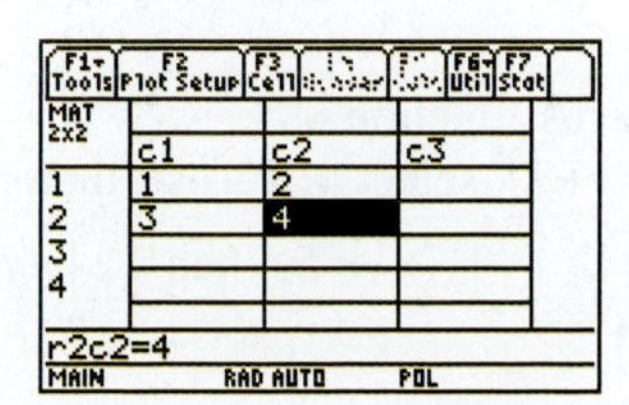

Press [HOME] and type 5 [×] d.

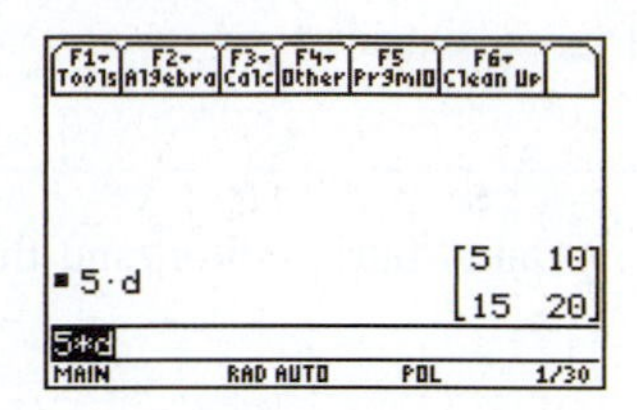

F.8 Solving Equations and Systems of Linear Equations

Example F.9

Solve the equation $x^2 - 4x + 2 = 0$, with respect to x.

Solution:

Press [HOME], press [F2], select **1: solve(,** then [ENTER], type 1 × [^] 2 [−] 4 × [+] 2 [=] 0 [,] × [)] [ENTER]

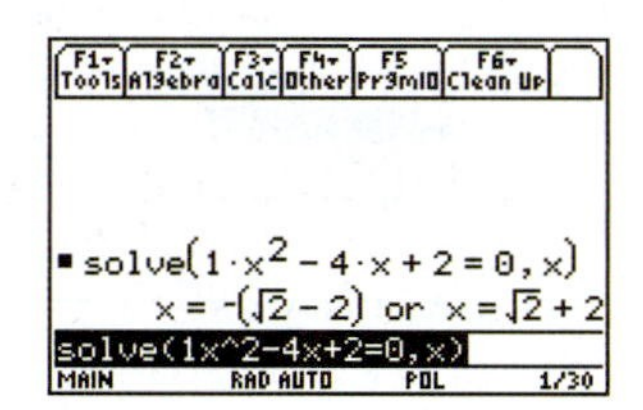

Example F.10

Solve $\sin(\theta) = 1/2$.

Solution:

Press [HOME], press [F2], select **1: solve(,** then [ENTER], press [2nd] y, then press [♦] [^] [)] = 1/2 then [,] [♦] [^] [)] [ENTER]

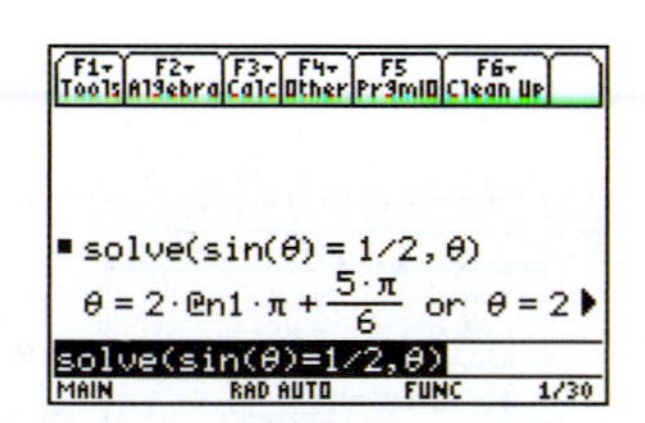

where n1 is an arbitrary integer.

Example F.11

Find the zeros of the equation $2x^3 - 3x^2 - 7x = -5$

Solution:

The solution using **zeros** function as:
Press [HOME], press [F2], select **4: zeros(,** then press [ENTER] type 2x [^] 3 [−] 3x [^] 2 [−] 7x + 5, then [,] type x, then press [)] [ENTER]

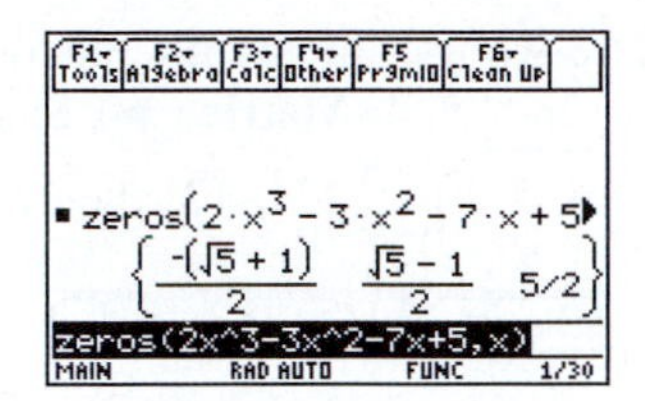

So the solution is $\left\{\dfrac{-(\sqrt{5}+1)}{2}, \dfrac{\sqrt{5}-1}{2}, \dfrac{5}{2}\right\}$

Example F.12

Find the zeros of the equation $\sin(2x) - 5\cos(x) = 0$, over the interval $0 \le x \le 2\pi$.

Solution:

The solution using **zeros** function as:

Press HOME, press F2, select **4: zeros(,** then press ENTER type $\sin(2x) - 5\cos(x)$ press , x) press | type 0 then press 2nd 5 8 to get ≤, press 2nd 8 8 (to get and), press 2nd 5 8 to get ≤, then type 2 2nd ^ ENTER

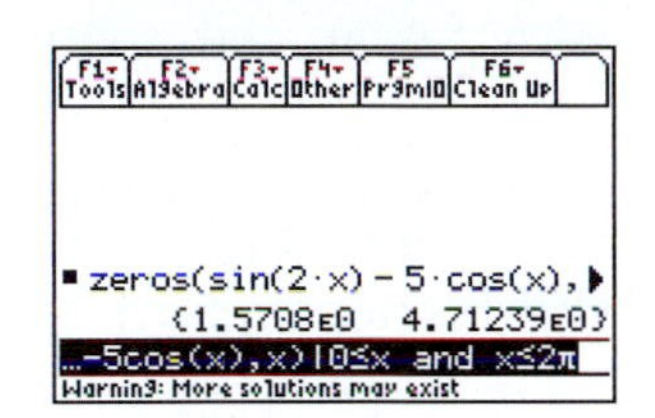

Example F.13

Solve the system of equations $\begin{cases} 3x - 6y = 24 \\ 5x + 4y = 12 \end{cases}$.

Solution:

■ **METHOD 1** Here we use the reduced row echelon form using the **rref** function.

Press HOME, press 2nd 5, **4: Matrix** ▶, **4: rref(**, press ENTER, type [3, −6, 24; 5, 4, 12], then press) ENTER

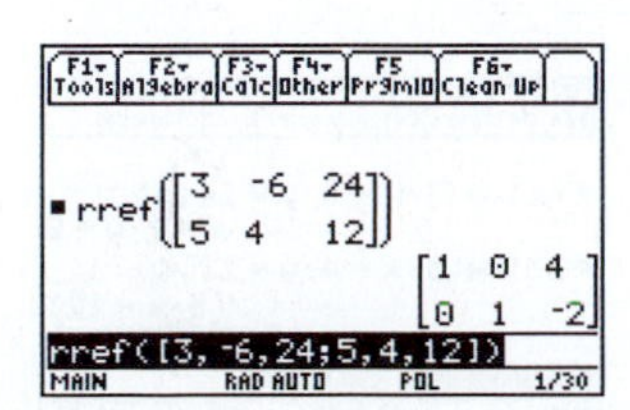

Reduced row echelon form $= \begin{bmatrix} 1 & 0 & 4 \\ 0 & 1 & -2 \end{bmatrix}$, the solution of the system $x = 4$, $y = -2$.

METHOD 2 Here we use the **simult** function.
Press HOME, press 2nd 5, **4: Matrix** ▶, **5: simult(**, press ENTER, type [3, −6; 5, 4] , [24; 12] then press) ENTER

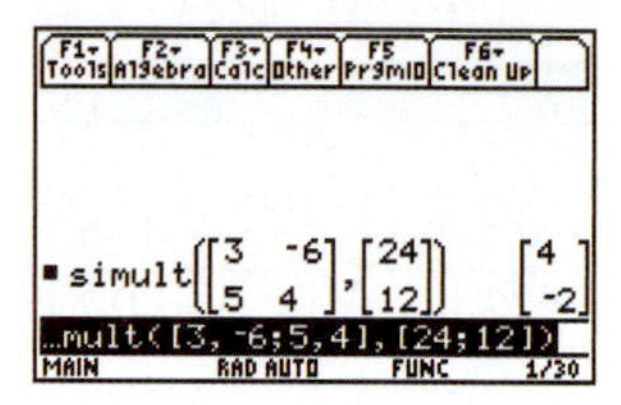

The solution of the system $x = 4$, $y = -2$.

METHOD 3 You can find the solution using the **solve** function.
Press HOME, press F2, select **1: solve(**, then type $3x$ − $6y = 24$, then press 2nd [MATH] 8 8 (to make and) type $5x + 4y = 12$, then , 2nd [{] × , Y 2nd [}]) ENTER

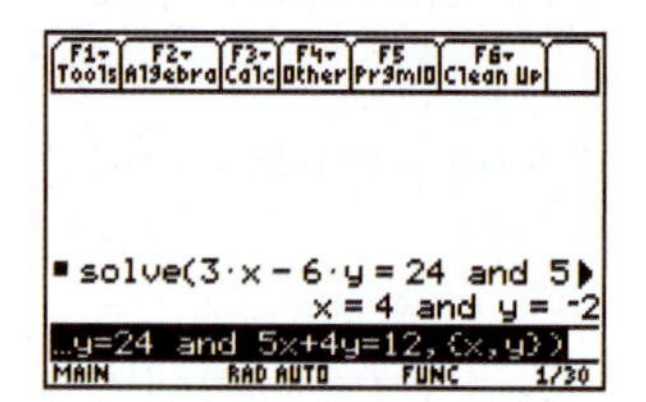

METHOD 4 Here, we use the substitution method:
Solve the following equations $\begin{cases} 3x - 6y = 24 \\ 5x + 4y = 12 \end{cases}$.

Solution:

Press HOME, Press F2, select **1: Solve(**, press ENTER, type $3x - 6y = 24$, x then press) ENTER.

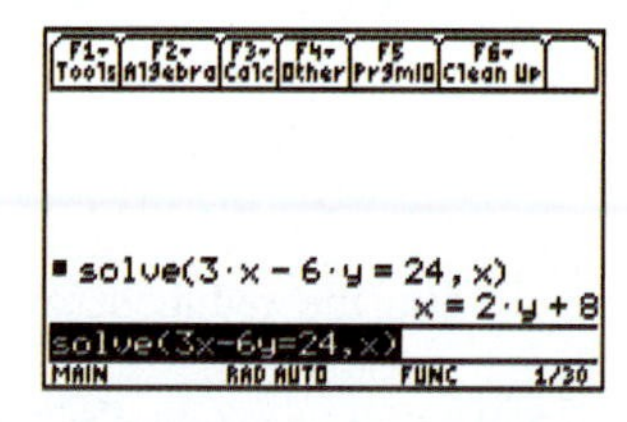

Press F2, select **1: Solve(** , press ENTER, type $5x + 4y = 12$, y then press) ENTER.

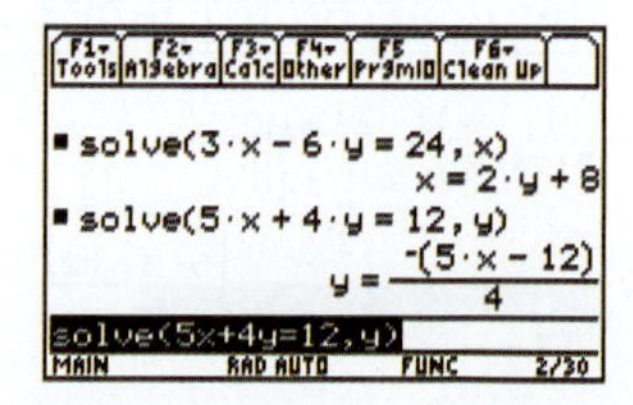

Press ▶ to remove the highlight on Solve, then press | to substitute the expression for x by copying the $x = 2y + 8$ using the up arrow key (▲) to highlight $x = 5y + 2$, then press ENTER to paste it after |, then press again ENTER, now we have the value of $y = -2$.

Now, press the up arrow key (▲), to highlight the $x = 2y + 8$, and press ENTER, then press | again and press the up arrow key (▲), to highlight the $y = 6/17$, press ENTER to paste it after |, then press again ENTER, this will give the value of $x = 4$.

Example F.14

Solve the following equations $\begin{cases} x + y - z = 6 \\ 3x - 2y + z = -5. \\ x + 3y - 2z = 14 \end{cases}$

Solution:

Press HOME, Press 2nd 5, **4: Matrix** (▶), **5: simult(** , press ENTER, type [1, 1, −1; 3, −2, 1; 1, 3, −2] , [6; −5; 14] then press) ENTER the solution of the system $x = 1$, $y = 3$, $z = -2$.

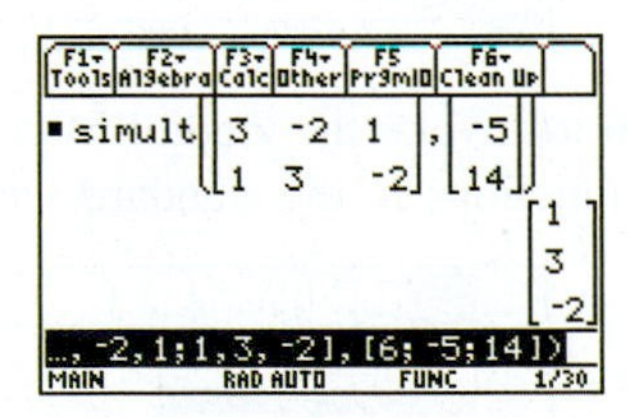

F.9 Plotting

To graph a function in the calculator you need to perform the following:

1. Press MODE and select **function** for graph, then press ENTER. Put **Radian** for angle when plotting Trig functions.

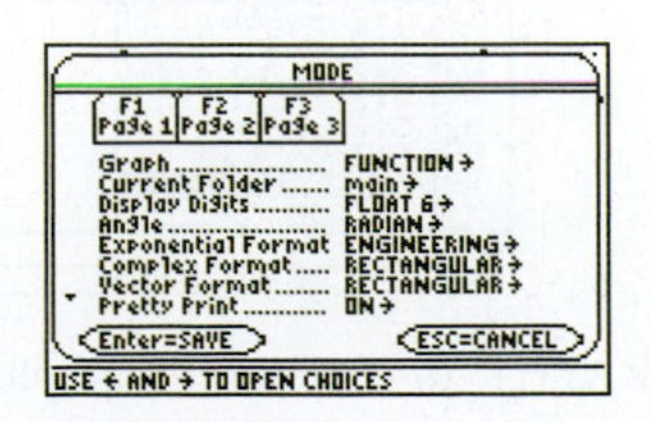

2. Press [♦] [F1] to access the Y = editor and to enter the functions.

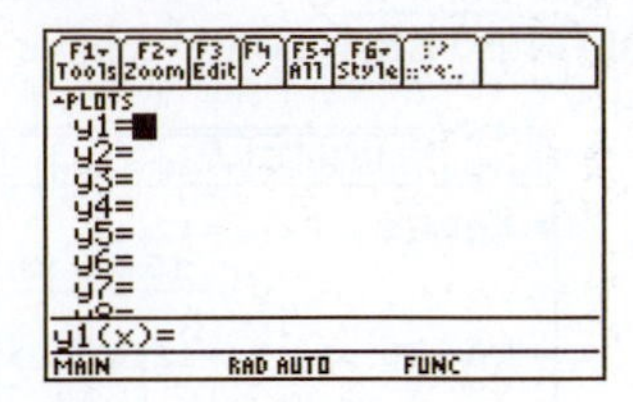

3. Press [F1] 9 to access the graph formats menu.

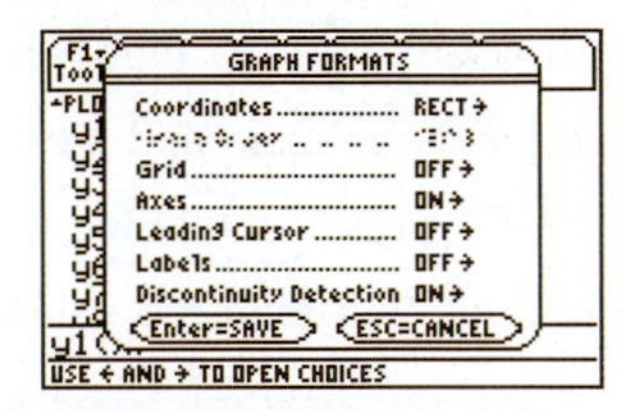

4. Press [♦] [F2] to access the Window editor.

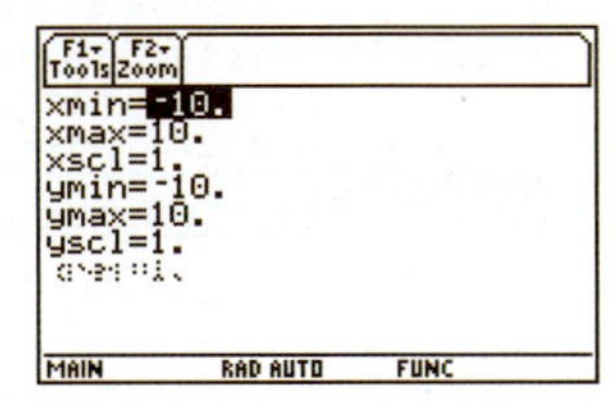

5. Press [F2] then **A** to invoke the **ZoomFit** command; it uses the **xmin** and **xmax** setting to determine the appropriate setting for **ymin** and **ymax** and then automatically draws the graph.

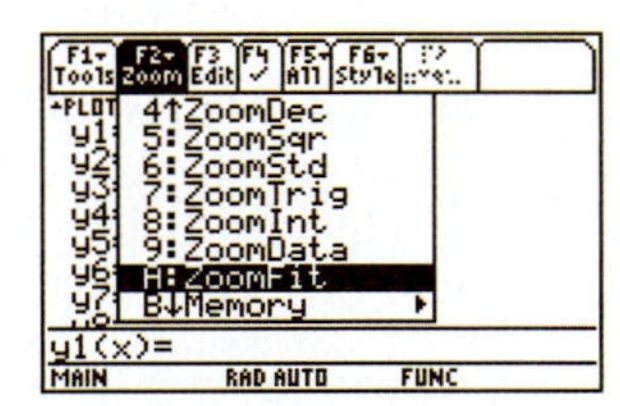

6. Press [F2] then **6** to invoke the **ZoomStd** command; it automatically graphs the functions in the standard viewing window.

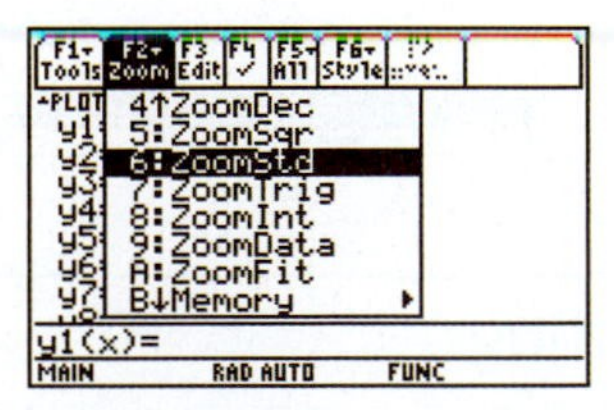

7. Press [F2] **7** to invoke **ZoomTrig** command to graph the Trig function.

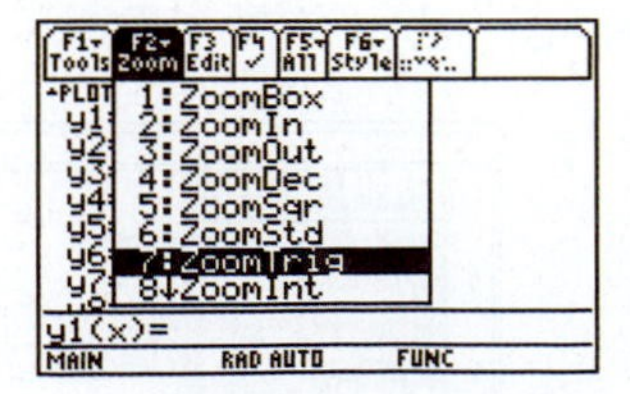

8. Press [♦] [GRAPH] to execute the graph.

Example F.15

Graph the function $y = 5x - x^3$.

Solution:

Press ♦ F1 to access the Y = editor and type in $y_1 = 5x - x^3$, press ENTER

Press F2 then **A: ZoomFit** command, press ENTER.

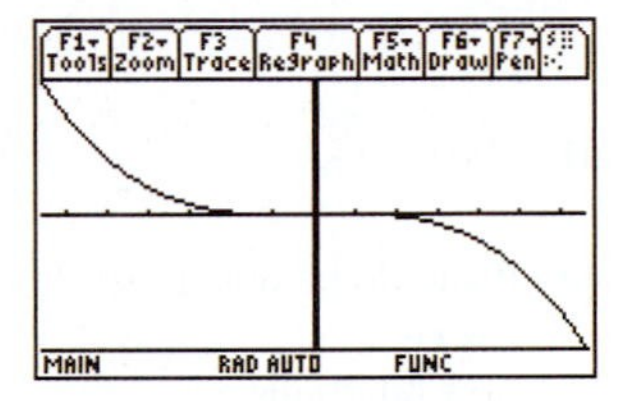

Example F.16

Graph the function $y = -3\sin(x)$.

Solution:

Press ♦ F1, $-3\sin(x)$, then press ENTER.

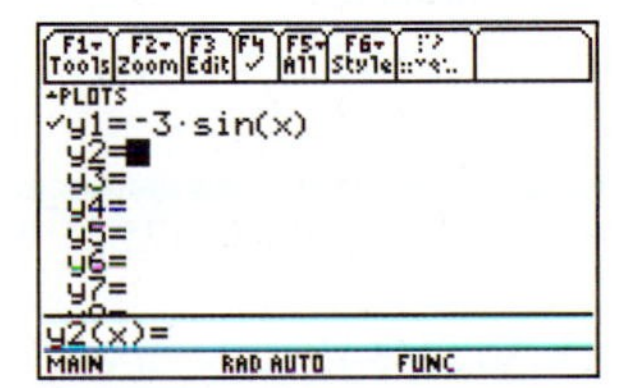

Press F2 then **7: ZoomTrig** command,

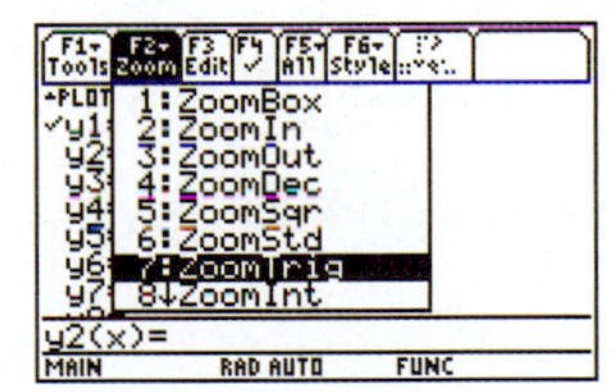

Press ENTER.

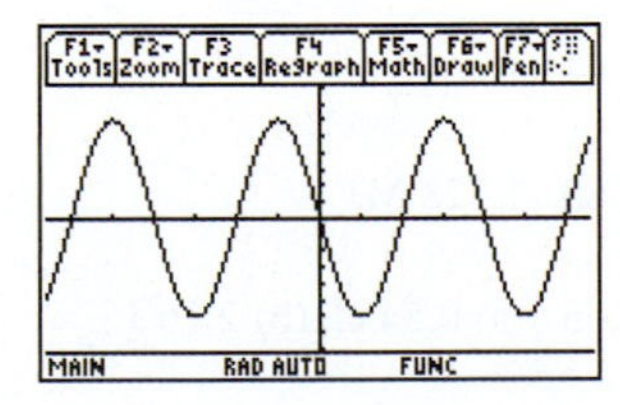

Appendix G

Answers to Odd-Numbered Problems

Chapter 1

1.1 (a) 13.716 m, (b) 3.658 m, (c) 5148.8 m, (d) 0.010668 m

1.3 23.872 kW

1.5 (a) 4.5×10^{-3}, (b) 9.26×10^{-3}, (c) 7.421×10^{3}, (d) 26.356×10^{6}

1.7 (a) 1.26×10^{-4}, (b) 9.8×10^{4}, (c) 5.0×10^{-7}

1.9 (a) 1.2×10^{10}, (b) 2.24×10^{-8}, (c) 5.625×10^{11}, (d) 1.25×10^{5}

1.11 (a) 24, (b) 0.8411 rad or 48.19°

1.13 (a) 0.99, (b) 9666.8

1.15 (a) −0.10384 C, (b) −0.19865 C, (c) −3.941 C, (d) −26.08 C

1.17 16.02 C

1.19 3.6 A

1.21 2.0833 ms

1.23 162.5 ms

1.25 $V_{ab} = -3$ V
$V_{bc} = 8$ V
$V_{ac} = 5$ V
$V_{ba} = 3$ V

1.27 0.333 A

1.29 10.8 V

1.31 0.333 A

1.33 0.375 A

1.35 (a) 15 V, (b) −5 V

1.37 4.56 kW

1.39 3 mC

1.41 500 C, 6 kJ

1.43 It should be noted that these are only typical answers.

(a) Light bulb	60 W, 100 W
(b) Radio set	4 W
(c) TV set	110 W
(d) Refrigerator	700 W
(e) PC	120 W
(f) PC printer	18 W
(g) Microwave oven	1000 W
(h) Blender	350 W

1.45 21.6 cents

1.47 (a) 0.1667 A, (b) 175.2 kWh, $21.02

1.49 $0.1355

1.51 0.96 kWh, 8.16 cents

1.53 14,000 ft-lb/s, 25.45 hp

1.55 (a) 14.11 MJ, (b) $1.18

1.57 6 C

1.59 −961.2 J

1.61 16.667 A

1.63 1.728 MJ

1.65 (a) 0.54 C, (b) 2.16 J

Chapter 2

2.1 1.131 Ω

2.3 8.13 $\mu\Omega$

2.5 3.427 m

2.7 6.6×10^{-6} Ωm

2.9 If we shorten the length of the conductor, its resistance decreases due to the linear relationship between resistance and length.

2.11 0.61

2.13 The graph in (c) represents Ohm's law.

2.15 3.2 mA

2.17 40 A

2.19 162 V

2.21 428.57 Ω

2.23 For $V = 10$, 4 A
For $V = 20$, 16 A
For $V = 50$, 100 A

2.25 (a) 40 V, the top terminal of the resistor is positive.
(b) 0.2 V, the bottom terminal of the resistor is positive.
(c) 12 mV, the top terminal of the resistor is positive.

2.27 (a) 0.4 S, (b) 25 μS, (c) 83.33 nS

2.29 20.83 μS

2.31 0.8 V

2.33 AWG # 1 will be appropriate.

2.35 (a) 144 CM,
(b) 78,540 CM

2.37 (a) 0.62 MΩ ± 10%,
(b) 50 kΩ ± 5%

2.39 (a) Green, red, black,
(b) Orange, red, brown,
(c) Blue, gray, red,
(d) Orange, red, green

2.41 (a) maximum value of 0.682 MΩ and minimum value of 0.558 MΩ.
(b) maximum value of 52.5 kΩ and minimum value of 47.5 kΩ.

2.43 0.25 V

2.45 You connect the light bulb terminals to the ohmmeter. If the ohmmeter reads infinity, it means there is an open circuit and the bulb is burnt out.

2.47 The voltmeter is connected across R_1 as in Figure G.1.

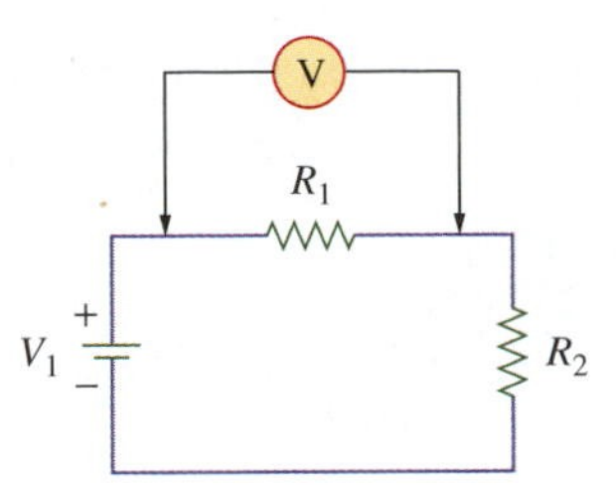

Figure G.1
For Problem 2.47.

2.49 The ohmmeter is connected as in Figure G.2.

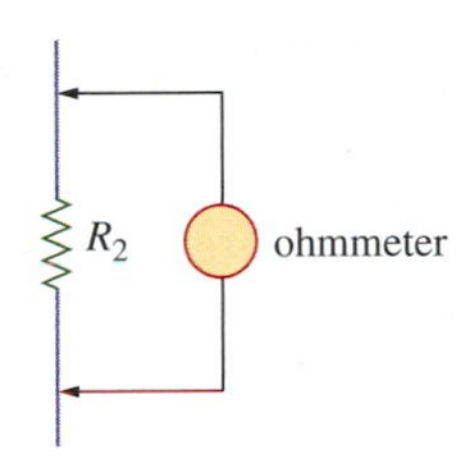

Figure G.2
For Problem 2.49.

2.51 Electric shock is caused by an electrical current passing through a body.

Chapter 3

3.1 0.3611 W

3.3 40 s

3.5 60.48 kWh

3.7 (a) 0.0031 kWh, (b) 0.36 kWh, (c) 160 kWh

3.9 18.75 s

3.11 1.682 A

3.13 116.071 V

3.15 86.4 Ah

3.17 (a) 30 A, (b) 3.6 kW, (c) 204.84 Btu/min

3.19 10 W

3.21 1.55 A

3.23 $P_{10} = 40$ W
$P_{12} = 48$ W
$P_{20} = 80$ W

3.25 2.828 mA, 141.42 V

3.27 5.801 MJ

3.29 The power has increased to 8 times the original value.

3.31 For R_1, 162 W
For R_2, 259.2 W

3.33 (a) 120 W, (b) −60 W

3.35 22.36 mA

3.37 (a) 14.23 W—can be damaged,
(b) 8 μW—fine,
(c) 32 W—can be damaged

3.39 105.83 V

3.41 $38.25

3.43 93.83%

3.45 56.82%

3.47 1.802 hp

3.49 13.33%

3.51 $36.86

3.53 56%

3.55 (a) 80%,
(b) 6 kJ,
(c) Most of it is converted to heat energy.

3.57 (a) 52.22%, (b) 1,433.4 W

3.59 (a) 2400 W, (b) 1119 W, (c) 46.6%, (d) 1281 W,
(e) 72/89 Btu/min

Chapter 4

4.1 6 branches, 6 nodes, and 1 loop.

4.3 9 nodes, 7 loops, and 14 branches.

4.5 350 Ω

4.7 2.2 kΩ

4.9 235 kΩ

4.11 7.136 MΩ

4.13 $R_1 = 5.333\ \Omega$
$R_2 = 2.667\ \Omega$

4.15 4 Ω or 16 Ω

4.17 15 mA

4.19 $I_x = 0.5$ mA
$P_8 = 2$ mW
$P_{10} = 2.5$ mW
$P_{12} = 3$ mW

4.21 20 V

4.23 20 V, 20 Ω

4.25 0.1143 mA, 2.286 V

4.27 0.2286 A

4.29 1.8 mA, 7 V

4.31 (a) 1.667 A, (b) 1.63 V, (c) *y*

4.33 −10 V

4.35 8

4.37 The simplified circuit is shown in Figure G.3.

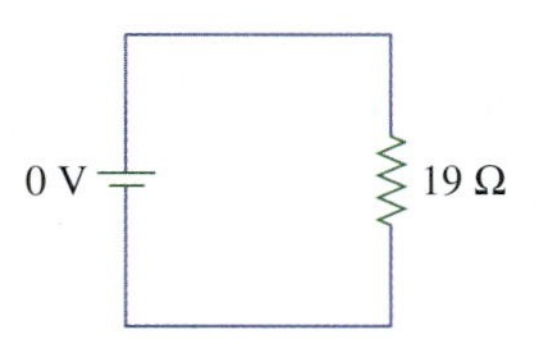

Figure G.3
For Problem 4.37.

4.39 4 V

4.41 9.333 kΩ

4.43 (a) For minimum value, 41.74 V
For maximum value, 67.83 V,
(b) The minimum value is 0 V
The maximum value is 40 V

4.45 9.202 V

4.47 $V_1 = 16.32$, $V_2 = 12.48$, $V_3 = 0.98$ V

4.49 $V_1 = 1.74$ V, $V_2 = 21.74$ V

4.51 62.442 V

4.53 4 V

4.55 1.2 A

4.57 10 Ω

4.59 At position 1, 3 mA
At position 2, 3.593 mA
At position 3, 5 mA

4.61 (a) 0.2 A, (b) 112 V, (c) 22.4 W

4.63 (a) 1 A, (b) 0.3 V, (c) 11.7 V, (d) 0.3 W, (e) 11.7 W, (f) 97.5%

Chapter 5

5.1 4 A

5.3 $I_1 = 6$ A, $I_2 = 14$ A, $I_3 = 4$ A

5.5 $I_1 = 7$ A, $I_2 = -5$ A

5.7 3 A

5.9 8 mA

5.11 10 Ω

5.13 (a) 33.275 Ω, (b) 8.0385 kΩ, (c) 0.4028 MΩ

5.15 2.12 S

5.17 31.58 Ω

5.19 8 Ω

5.21 5 kΩ

5.23 100 Ω

5.25 A1 reads: 138 mA
A2 reads: 130 mA

5.27 For R_1, 1.915 mA, 17.235 mW
For R_2, 4.091 mA, 36.82 mW
For R_3, 5 mA, 45 mW
For R_4, 2.727 mA, 24.545 mW

5.29 (a) 2.1 A, (b) 120 V, (c) 252 W

5.31 (a) 6 V, (b) 3 W, (c) 3 W

5.33 4.033 Ω, 12.1 V

5.35 $I_1 = 7.5$ A
$I_2 = 2.5$ A

5.37 $I_1 = 1.0527$ A
$I_2 = 1.3157$ A
$I_3 = 2.6314$ A

5.39 $I_{100} = 25$ mA
$I_{300} = 75$ mA
$I_{600} = 150$ mA

5.41 (a) 55.1 mS,
(b) 18.15 Ω, 6.614 A, 300 W

5.43 $I_1 = 5.555$ mA
$I_2 = 8.333$ mA
$I_3 = 11.111$ mA

5.45 (a) 4.396 Ω,
(b) 0.2275 S,
(c) $I_T = 27.3$ A, $I_x = 15$ A

5.47 $I_1 = 6.857$ A
$I_2 = 3.428$ A
$I_3 = 1.714$ A

5.49 8.4 A

5.51 $I_1 = 5$ A, and $I_2 = 4$ A.

5.53 $I_{12} = 4$ A, $I_8 = 6$ A, $I_4 = 12$ A, $I_2 = 24$ A

5.55 $V_1 = 20$ V
$V_2 = 30$ V
$V_3 = 0$ V
$V_4 = 10$ V

5.57 (a) R_2 is open circuited.
(b) R_1 is open circuited.
(c) R_3 is open circuited.

5.59 Under normal operation, 0.2857 A
Under abnormal condition, 0 A

5.61 195 W

5.63 $I_1 = 1.29$ A
$I_2 = 1.5$ A
$I_3 = 1.8$ A
$I_T = 4.6$ A

5.65 $P_{max} = 2.469$ kW
$P_{min} = 600$ W

Chapter 6

6.1 R_1 and R_2 are in parallel.
R_3 and R_4 are in series.
R_5 and R_6 are in parallel.

6.3 34.5 Ω

6.5 50 Ω

6.7 20 Ω

6.9 13.94 Ω

6.11 0.1 A

6.13 3 A

6.15 8 V

6.17 $V_2 = 4$ V
$V_{10} = 20$ V
$V_4 = 6$ V
$V_7 = 3.5$ V
$V_5 = 2.5$ V

6.19 $V_{10} = 20$ V
$V_3 = 4$ V
$V_{12} = 16$ V

6.21 0.133 mA

6.23 (a) 0 Ω, (b) 1.429 Ω

6.25 4 V

6.27 52.976 kΩ

6.29 (a) 14.17 Ω,
(b) 16.94 A

6.31 $I_7 = I_{50} = 0.15$ A
$I_{46} = 0.075$ A
$I_{15} = 0.075$ A
$I_{20} = 0.06$ A
$I_{30} = I_{10} = I_{40} = 0.015$ A

6.33 0.5 V

6.35 $I_1 = I_2 = \frac{1}{2}I_0$

$I_3 = I_4 = \frac{1}{4}I_0$

$I_6 = I_5 = \frac{1}{8}I_0$

6.37 2 V, $\frac{1}{3}$ Ω

6.39 −4.8 A

6.41 6 mV

6.43 (a) 24 V, (b) 18 V

6.45 (a) 10.4 V, (b) 12 V, (c) 13.08%

6.47 68 V

6.49 $V_1 = 25.71$ V, $V_2 = 5.153$ V, $V_3 = 3.857$ V

6.51 990.8 mA

6.53 14.47 V

6.55 14.472 V

6.57 (a) 4.949 kΩ,
(b) $I_1 = 5.365$ mA, $I_2 = 1.247$ mA.

6.59 2 kΩ

6.61 6.667 V

6.63 At no load, 83.33 Ω
Under load, 83.54 Ω

Chapter 7

7.1 (a) 62, (b) −306

7.3 $I_1 = 1.2143$ A, $I_2 = -1.5714$ A

7.5 1.5 A

7.7 $i_1 = -0.4286$ A, $i_2 = 0.4286$ A

7.9 1.188 A

7.11 $i_1 = 5.25$ mA, $i_2 = 8.5$ mA, $i_3 = 10.25$ mA

7.13 $I_a = 530$ mA
$I_b = 90$ mA
$I_c = 0.170$ mA

7.15 $I_1 = 2.462$ A, $I_2 = 0.1538$ A

7.17 $29I_1 - 8I_2 - 7I_3 = 10$
$-4I_2 + 10I_4 = -12$
$-3I_2 - I_3 + 4I_5 = 12$

7.19 33.78 V, 10.667 A

7.21 $i_1 = 3.5$ A, $i_2 = -0.5$ A, $i_3 = 2.5$ A

7.23 −2.286 V

7.25 $V_1 = 24$, $V_2 = 8$ V, $V_3 = 0$ V

7.27 20 V

7.29 571 mA

7.31 $V_1 = 18$ V, $V_2 = 26$ V

7.33 37 V

7.35 $V_1 = 5.9296$ V, $V_2 = 7.5377$ V

7.37 $V_1 = 10$ V
$V_4 = 5$ V
$V_3 = 4.884$ V
$V_2 = 6.628$ V

7.39 $V = 2.004$ V
$I = 3.6673$ A

7.41 $V_1 = 9.143$ V, $V_2 = -10.286$ V
$P_{8\Omega} = 10.45$ W
$P_{4\Omega} = 94.37$ W
$P_{2\Omega} = 52.9$ W

7.43 39.67 mA

7.45
$$\begin{bmatrix} 6 & -2 & 0 \\ -2 & 12 & -2 \\ 0 & -2 & 7 \end{bmatrix} \begin{bmatrix} i_1 \\ i_2 \\ i_3 \end{bmatrix} = \begin{bmatrix} 12 \\ 8 \\ -20 \end{bmatrix}$$
$P_{8\Omega} = 2.73$ W

7.47
$$\begin{bmatrix} 1050 & -150 & -800 \\ -150 & 990 & -600 \\ -800 & -600 & 2150 \end{bmatrix} \begin{bmatrix} i_1 \\ i_2 \\ i_3 \end{bmatrix} = \begin{bmatrix} 24 \\ -10 \\ 10 \end{bmatrix}$$

7.49 $V_1 = 2.333$ volts, $V_2 = 3.275$ volts
$V_3 = 2.745$ volts

7.51
$$\begin{bmatrix} 0.6 & -0.5 & 0 & 0 \\ -0.5 & 1 & -0.25 & -0.25 \\ 0 & -0.25 & 0.25 & 0 \\ 0 & -0.25 & 0 & 1.25 \end{bmatrix} \begin{bmatrix} V_1 \\ V_2 \\ V_3 \\ V_4 \end{bmatrix} = \begin{bmatrix} 3 \\ 2 \\ -1 \\ -5 \end{bmatrix}$$

7.53 (a) $R_1 = R_2 = R_3 = 4\ \Omega$,
(b) $R_1 = 18\ \Omega$, $R_2 = 6\ \Omega$, $R_3 = 3\ \Omega$

7.55 (a) 142.32 Ω, (b) 33.33 Ω

7.57 0.9974 A

7.59 −0.8095 A

7.61 $i_1 = 99.61$ mA, $i_2 = 31.84$ mA, $i_3 = 42.3$ mA

7.63 $V_1 = 12.35$ V, $V_2 = 8.824$ V, $V_3 = 4.824$
and $V_4 = -2.235$ V.

7.65 20 V

7.67 10.69 V

7.69 1.5 V

7.71 $I_B = 0.61\ \mu$A
$V_0 = 49$ mV
$V_{CE} = 8.341$ V

Chapter 8

8.1 0.1 A, 1 A

8.3 1.5 A

8.5 5 A

8.7 3 A

8.9 5.8876 mA

8.11 3.875 A, 45.05 W

8.13 −8.48 V

8.15 15.2 V

8.17 3 A

8.19 555.5 mA

8.21 −125 mV

8.23 3 V

8.25 1.6 A

8.27 28 ohms, 92 V

8.29 500 mA

8.31 $R_{Th} = 28$ ohms, $V_{Th} = -160$ volts, $V_x = -48$ V

8.33 $R_{Th} = 10$ ohms, $V_{Th} = 0$ volts, $I_x = 0$ A

8.35 $R_N = 3$ ohms, $I_N = 2$ A

8.37 $R_N = 10$ ohms, $I_N = -0.4$ A, $I = 2.4$ A

8.39 $R_{Th} = 21.667\ \Omega$
$V_{Th} = 1.5$ V
$R_N = 21.667\ \Omega$
$I_N = 69.23$ mA

8.41 (a) $R_{Th} = 3.636\ \Omega$
$V_{Th} = 60$ V
$I_N = 16.501$ A
$R_N = 3.636\ \Omega$,
(b) $R_{Th} = 195.35\ \Omega$
$V_{Th} = 398.15$ V
$R_N = R_{Th} = 195.35\ \Omega$
$I_N = 2.04$ A

8.43 V_L ranges from 1.445 V to 2.399 V

8.45 $V_{Th} = 9.45$ V
$R_{Th} = R_{\text{norton}} = 1.31\ \Omega$
$I_{\text{norton}} = 7.21$ A

8.47 $R_{Th} = 2$ k ohms
$V_{Th} = -160$ V

8.49 $R_{Th} = 10$ ohms
$V_{Th} = 166.67$ V
$I_N = 16.667$ A
$R_N = 10$ ohms

8.51 625 mW

8.53 (a) $R_{Th} = 12$ ohms
$V_{Th} = 40$ V,
(b) 2 A,
(c) 12 ohms,
(d) 33.33 watts

8.55 10 V

8.57 6.469 V

8.59 0.1793 A

8.61 10.4 mA

8.63 0.2 A

8.65 0.1667 A

8.67 (a) 2.4 A, (b) 2.4 A, (c) Yes

8.69 $V_{Th} = 6$V, $R_{Th} = 13\ \Omega$
$I_N = 68$ mA, $G_N = 44$ mS

8.71 (a) $V = 60$ V,
(b) $V = 60$ V, confirming the reciprocity theorem

8.73 $V_{Th} = 8$V, $R_{Th} = 9.5\ \Omega$

8.75 $R_{Th} = 28\ \Omega$
$V_{Th} = 92$ V

8.77 $R_{Th} = R_N = 13.333\ \Omega$
$V_{Th} = 34$ V
$I_N = 2.55$ A

8.79 $V_{Th} = 12$ V, $R_{Th} = 8$ ohms

8.81 (a) $V_{Th} = 24$ V, $R_{Th} = 30$ kΩ
(b) 9.6 V

Chapter 9

9.1 (a) 0.000268 μF, (b) 45,000 pF, (c) 2.4 pF

9.3 22.5 V

9.5 15 μF

9.7 The second capacitor (C_2) has 25 times the stored energy of the first capacitor (C_1).

9.9 106.1 nF

9.11 (a) 88.42 pF, (b) 12 kV/m, (c) 10.61 nC

9.13 375 kV/m

9.15 3

9.17 $C_2 = 6C_1$

9.19 (a) 120 mF, (b) 7.5 mF

9.21 20 μF

9.23 10 μF

9.25 4 mF

9.27 (a) 35 μF,
(b) $Q_1 = 0.75$ mC
$Q_2 = 1.5$ mC
$Q_3 = 3$ mC,
(c) 393.8 mJ

9.29 (a) 200 μF, (b) 10 mC, (c) 200 V

9.31 $C_1 = 5.52$ nF, $C_2 = 2.76$ nF, $C_3 = 55.2$ nF

9.33 5.455 μF

9.35 1.15 MF

9.37 480 mA

9.39 $I_1 = 5/3$ A
$I_2 = 10/3$ A

9.41 $i(t) = \begin{cases} 32\text{ A}, & 0 < t < 1\text{ ms} \\ 0, & 1 < t < 3\text{ ms} \\ -32\text{ A}, & 3 < t < 4\text{ ms} \end{cases}$

9.43 $i(t) = \begin{cases} 800\ \mu\text{A}, & 0 < t < 1 \\ 0, & 1 < t < 2 \\ -800\ \mu\text{A}, & 2 < t < 3 \end{cases}$

9.45 (a) 10 ms, (b) 6 s

9.47 $4e^{-12.5t}$ V

9.49 $20e^{-t/2}$ V

9.51 $12 - 12e^{-t/0.6}$ V, $0.3e^{-t/0.6}$ mA

9.53 $v_c(t) = 15(1 - e^{-t/\tau})$ V, $t = 5.333$ ms

9.55 5.545 s

9.57 10.75 ms

9.59 (a) 1.2 μF, (b) $V_1 = 7.2$ V, $V_2 = 4.8$ V, (c) 0.83 μs

9.61 200 ms

9.63 The voltage $v(t)$ across the resistor $R1$ as in Figure G.4.

9.65 The voltage $v(t)$ across of the capacitor is shown in Figure G.5.

9.67 The voltage $v(t)$ across the capacitor is shown in Figure G.6.

9.69 The $v(t) = $ V(2) is plotted in Figure G.7.

9.71 The capacitor is short-circuited.

9.73 8 groups in parallel with each group made up of 2 capacitors in series.

9.75 (a) 1250 μF, (b) 400 J

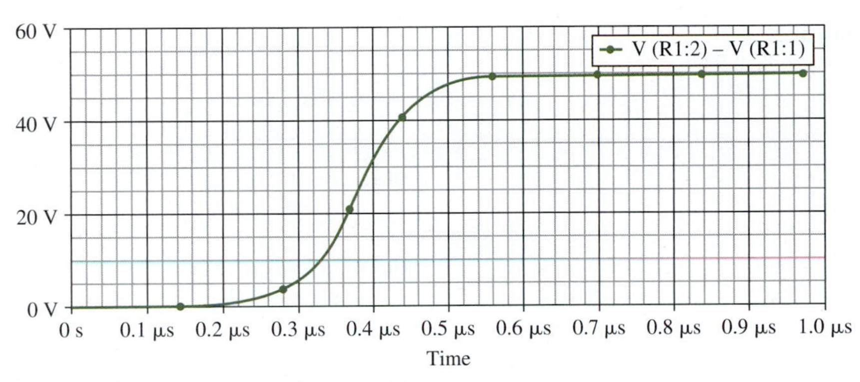

Figure G.4
For Problem 9.63.

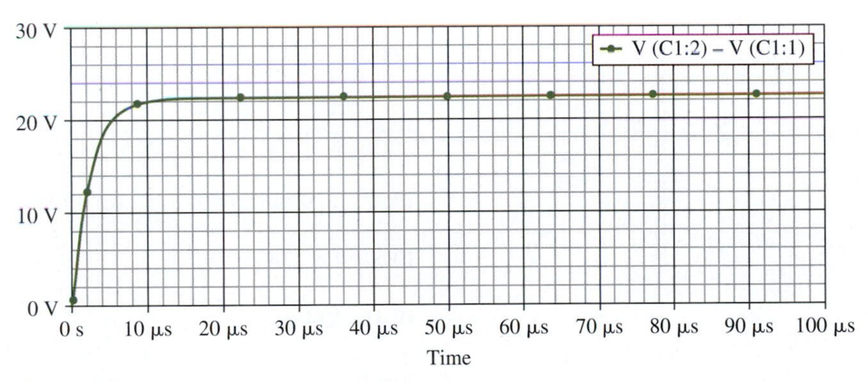

Figure G.5
For Problem 9.65.

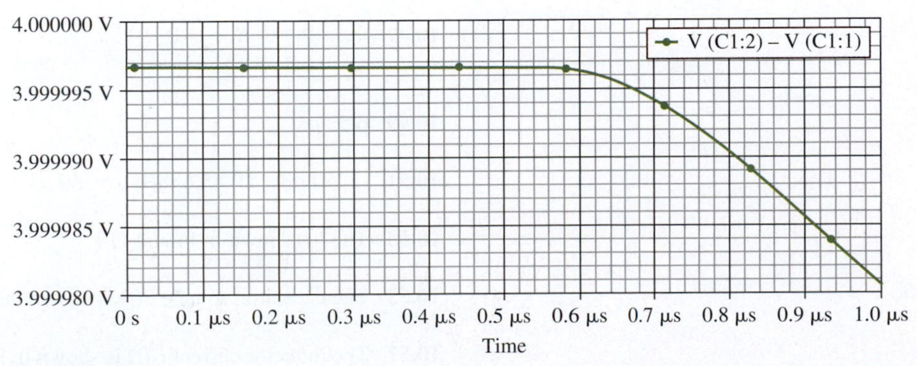

Figure G.6
For Problem 9.67.

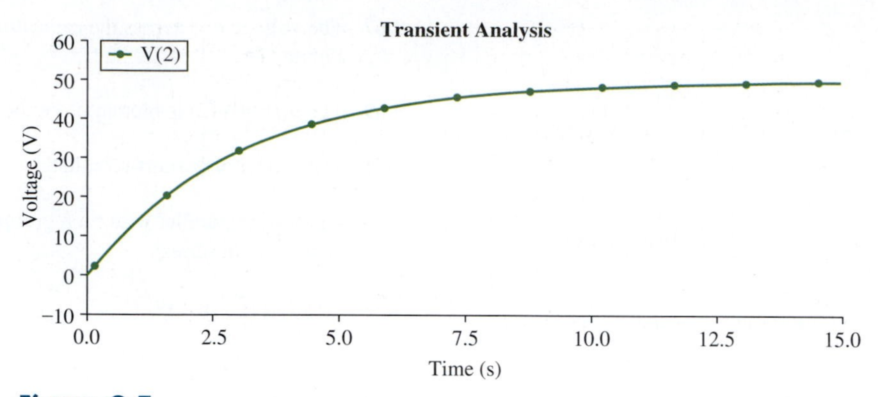

Figure G.7
For Problem 9.69.

Chapter 10

10.1 30 mV

10.3 50 turns

10.5 2.4 V

10.7 Plotted as in Figure G.8.

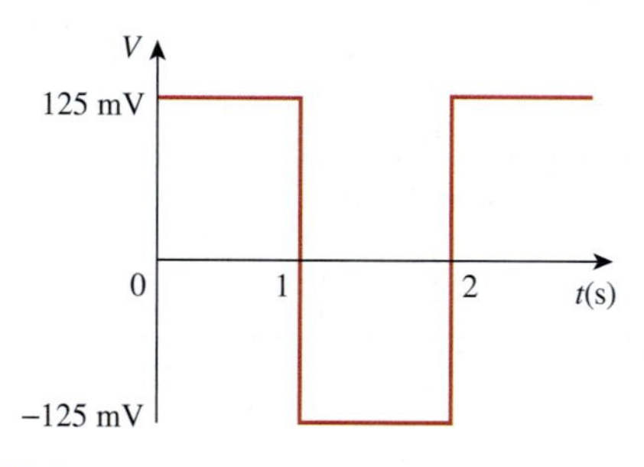

Figure G.8
For Problem 10.7.

10.9 160 mH

10.11 30 H

10.13 48 mV

10.15 3.25 mH

10.17 142

10.19 903

10.21 0.625 A/s

10.23 1.118 A

10.25 $v = 6$ V
$i_1 = i_2 = 2$ A

10.27 $i_{L_1} = i_{L_2} = 3$ A
$v_{C_1} = 18$ V
$v_{C_2} = 0$ V

10.29 160 μH

10.31 12.5 mH

10.33 0.8 H

10.35 $\frac{5}{8}L$

10.37 (a) 0.8187, (b) 0.8187, (c) 0.1355

10.39 $4\,e^{-2t}$ A

10.41 3.14 ms

10.43 $-2\,e^{16t}$ V

10.45 (a) 20 ms, (b) 80 mH, 4 Ω

10.47 $v_o = 2\,e^{-4t}$ V, $t > 0$
$v_x = 0.5\,e^{-4t}$ V, $t > 0$

10.49 6.52 μs

10.51 (a) 50 mA, (b) 45.6 mH, (c) 380 μs

10.53 The $i(t)$ shown in Figure G.9.

10.55 The capacitor voltage $v(t)$ is shown in Figure G.10.

10.57 The inductor current $i_o(t)$ is shown in Figure G.11.

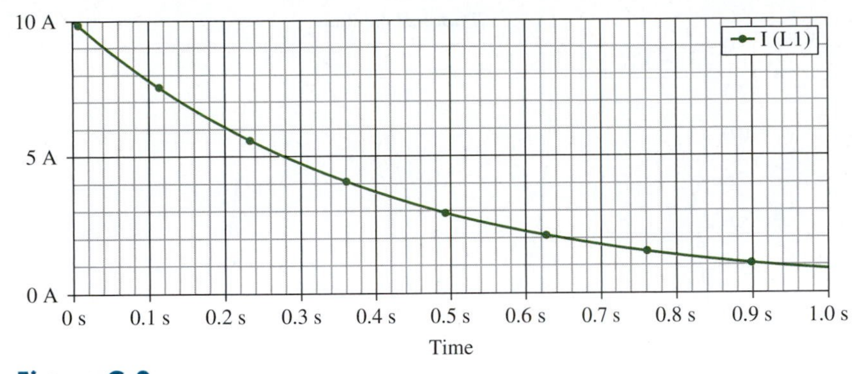

Figure G.9
For Problem 10.53.

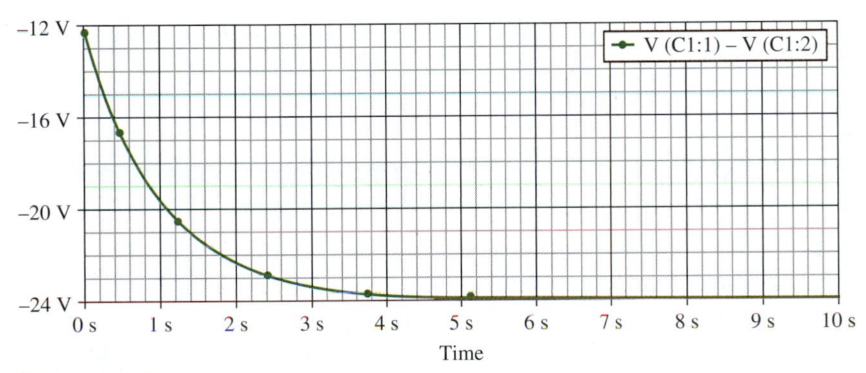

Figure G.10
For Problem 10.55.

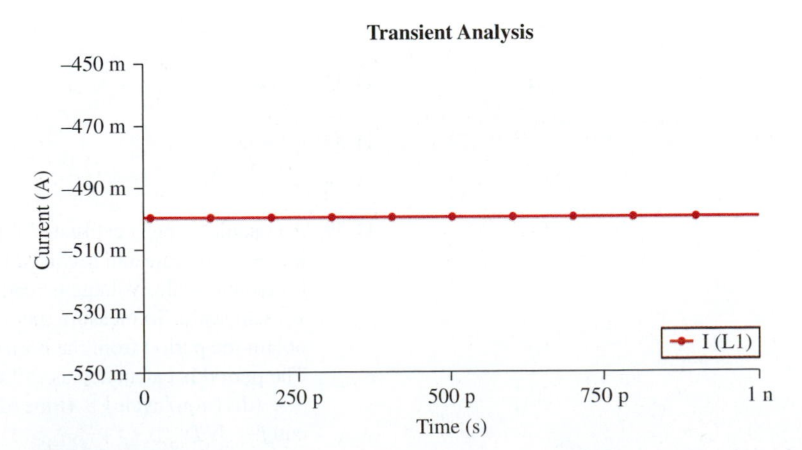

Figure G.11
For Problem 10.57.

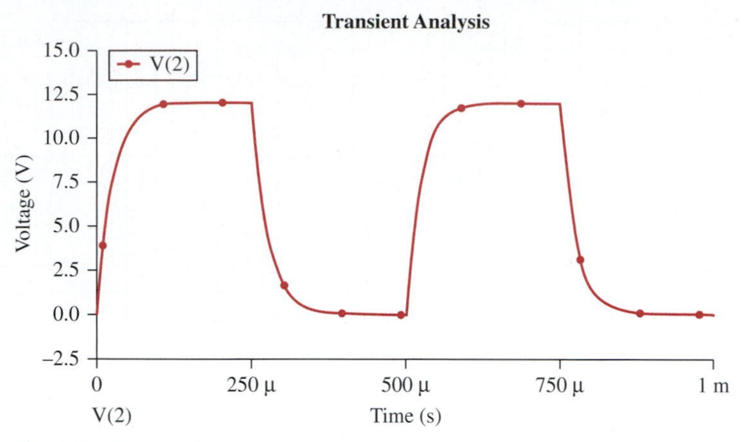

Figure G.12
For Problem 10.59.

10.59 The $v_o(t)$ is shown in Figure G.12.

10.61 0.441 A

10.63 1.271 Ω

Chapter 11

11.1 2.4 V

11.3 $i(0) = 24$ mA
$i(10 \text{ ms}) = -19.415$ mA
$i(40 \text{ ms}) = -19.421$ mA

11.5 At $t = 2$ ms, 6.846 V
At $t = 14.5$ ms, -7.289 V
At $t = 25.2$ ms, -0.7515 V

11.7 (a) 36°,
(b) 154.29°,
(c) 135.68°,
(d) 257.83°

11.9 (a) 2.5 Hz,
(b) 500 Hz,
(c) 33.33 kHz

11.11 (a) 10 s,
(b) 0.1 Hz,
(c) 3

11.13 50 Hz

11.15 12.5 Hz

11.17 1570.8 rad/s

11.19 (a) 10^3 rad/s, (b) 159.2 Hz, (c) 6.283 ms,
(d) 2.65 V

11.21 (a) $4 \cos(\omega t - 120°)$,
(b) $-2 \cos(6t + 90°)$,
(c) $-10 \cos(\omega t + 110°)$

11.23 (a) $8 \sin(7t + 105°)$,
(b) $10 \cos(3t + 5°)$

11.25 (a) 200,
(b) 82.92 Hz,
(c) 12.06 ms,
(d) 25°

11.27 (a) $i(t)$ leads $v(t)$ by 20°.
(b) $v_2(t)$ leads $v_1(t)$ by 170°.
(c) $y(t)$ leads $x(t)$ by 9.24°.

11.29 3.535 V

11.31 (a) 7.071 V,
(b) 1.414 mA

11.33 $V_{ave} = 8.333$ V
$V_{rms} = 9.574$ V

11.35 3.266 V

11.37 (a) 9 Ω,
(b) 13.3 A

11.39 An oscilloscope is calibrated for volts/division on the vertical scale and seconds/division on the horizontal scale. Voltage is read directly from the vertical scale. To measure the frequency, we first obtain the period from the horizontal scale.
The period is calculated as follows.
$T = (\text{division/cycle}) \times (\text{time/division})$
and $f = 1/T$

Chapter 12

12.1 (a) $3 + j9$,
(b) $39 + j43$,
(c) $10.250 + j2.929$,
(d) $2.759 - j1.038$

12.3 (a) $-0.4243 + j4.97$,
(b) $0.4151 - j0.6281$,
(c) $109.25 - j31.07$

12.5 (a) $-56 + j33$,
(b) $-0.3314 + j0.1953$,
(c) $-0.6372 - j0.5575$

12.7 (a) $-1.2749 + j0.1520$,
(b) -2.0833,
(c) $35 + j14$

12.9 (a) $40 \cos(\omega t - 60°)$,
(b) $38.36 \cos(\omega t + 96.8°)$,
(c) $6 \cos(\omega t + 80°)$,
(d) $11.5 \cos(\omega t - 52.06°)$

12.11 (a) $20\underline{/-60°}$,
(b) $5\underline{/340°}$

12.13 $13.75 \cos(377t)$ A

12.15 15.08 Ω, 7.295 A

12.17 $v(t) = 2 \sin(10^6 t - 65°)$ V

12.19 The element is a resistor with $R = 6.5$ Ω.

12.21 69.82 V

12.23 $i(t) = 2.1466 \sin(1000t + 66.56°)$ A

12.25 $i(t) = 4.472 \cos(3t - 18.43°)$ A,
$v(t) = 17.89 \cos(3t - 18.43°)$ V

12.27 $i_o(t) = 3.328 \cos(2t + 33.69°)$ A

12.29 $i_s(t) = 25 \cos(2t - 53.13°)$ A

12.31 263.9 mA

12.33 $54.18\underline{/-67.19°}$ Ω

12.35 $0.3171 - j0.1463$ S

12.37 $1 + j0.5$ Ω

12.39 $19 - j5$ Ω, $1.527\underline{/104.7°}$ A

12.41 $0.4724 + j0.219$ S

12.43 $7.567 + j0.5946$ Ω

12.45 $8.137\underline{/-44.69°}$ Ω

12.47 (a) $14.19\underline{/-5.95°}$ Ω,
(b) $7.05\underline{/5.95°}$ A

12.49 The current $i(t)$ is shown in Figure G.13.

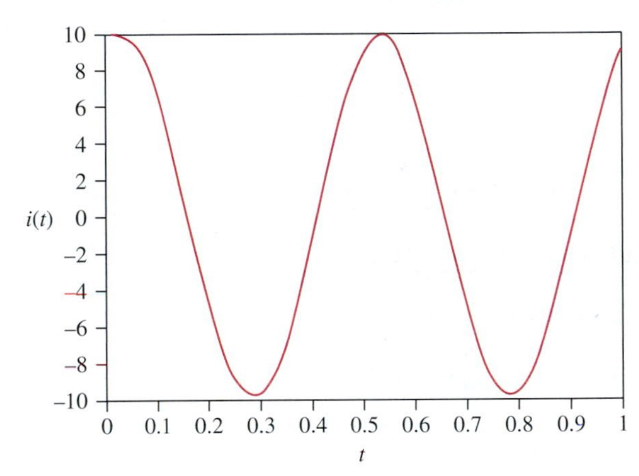

Figure G.13
For Problem 12.49.

12.51 $-1.2832 + 8.0232i$

12.53 $v_o(t) = -5 \cos(2t)$ V

12.55 $i_x(t) = 2.12 \sin(5t + 32°)$ A

12.57 This is achieved by the RL circuit shown in Figure G.14.

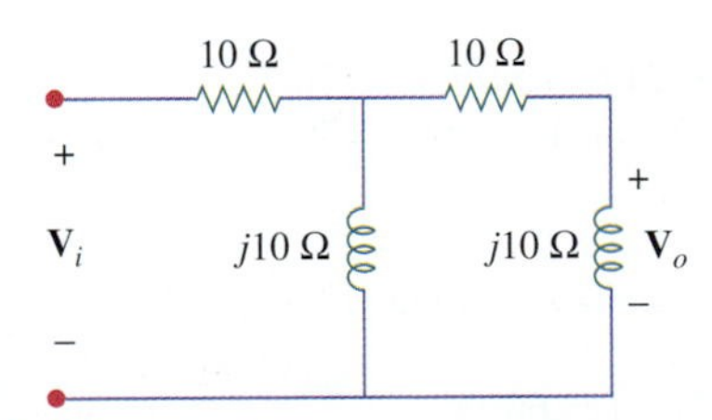

Figure G.14
For Problem 12.57.

12.59 (a) The phase shift is 51.49° lagging,
(b) 1.5915 MHz

12.61 0.1 μF

12.63 104.17 mH

12.65 (a) $471.4\underline{/13.5°}$ Ω,
(b) $0.2121\underline{/61.5°}$

Chapter 13

13.1 $i_o(t) = 1.414\cos(2t + 45°)$ A

13.3 $\mathbf{I}_1 = 4.67\angle{-20.17°}$ A
$\mathbf{I}_2 = 1.79\angle{37.35°}$ A

13.5 $I_1 = 0.3814\angle{109.6°}$ A
$I_2 = 0.3443\angle{124.4°}$ A
$I_3 = 0.1455\angle{-60.42°}$ A
$I_x = 0.1005\angle{48.5°}$ A

13.7 $39.5\cos(10^3 t - 18.43°)$ mA

13.9 $I_1 = 4.547\angle{-50.06°}$ A
$I_2 = 2.206\angle{-36.03°}$ A

13.11 $14.377\angle{-57.77°}$ A

13.13 $124.08\angle{-154°}$ V

13.15 $29.36\angle{62.88°}$ A

13.17 $\mathbf{V}_1 = 22.87\angle{132.27°}$ V
$\mathbf{V}_2 = 27.87\angle{140.6°}$ V

13.19 $13.775\angle{99.94°}$ V

13.21 $3.632\angle{77.45°}$ A

13.23 $V_0 = 27.27\angle{2.882°}$ V
$I_1 = 695\angle{-27.88°}$ mA
$I_2 = 33.3\angle{36.38°}$ mA

13.25 $8 + 1.5811\cos(4t - 71.57°)$ A

13.27 $9.902\cos(2t - 129.17°)$ A

13.29 $10 + 21.47\sin(2t + 26.56°) +$
$10.73\cos(3t - 26.56°)$ V

13.31 $0.7911\cos(10t + 21.47°) +$
$0.2995\sin(4t + 176.6°)$ A

13.33 $9.37\angle{-128.66°}$ V

13.35 $3.615\cos(10^5 t - 40.6°)$ V

13.37 $(3.529 - j5.883)$ V

13.39 $4.98\cos(4t - 175.25°)$ V

13.41 $\mathbf{Z}_N = \mathbf{Z}_{Th} = 5.423\angle{-77.47°}\ \Omega$
$\mathbf{I}_N = 3.578\angle{18.43°}$ A
$\mathbf{V}_{Th} = 19.4\angle{-59°}$ V

13.43 $\mathbf{Z}_{Th} = 11.18\angle{26.56°}\ \Omega$
$\mathbf{V}_{Th} = 55.9\angle{71.56°}$ V

13.45 $\mathbf{Z}_N = 44.72\angle{63.43°}\ \Omega$
$\mathbf{I}_N = 3\angle{60°}$ A

13.47 $Z_N = Z_{Th} = 11.243 + j1.079\ \Omega$
$V_{Th} = 4.945\angle{-69.76°}$ V,
$I_N = 0.4378\angle{-75.24°}$ A

13.49 $\mathbf{Z}_N = \mathbf{Z}_{Th} = 0.67\angle{129.56°}\ \Omega$
$\mathbf{V}_{Th} = 29.79\angle{-3.6°}$ V
$\mathbf{I}_N = 44.46\angle{-133.16°}$ A

13.51 $9.8686\angle{-19.84°}$ V

13.53 $\mathbf{V}_1 = 28.91\angle{135.4°}$ V
$\mathbf{V}_2 = 49.18\angle{124.1°}$ V

13.55 $6.639\cos(10^3 t - 160°)$ V

13.57 $\mathbf{V}_1 = 15.91\angle{169.6°}$ V, $\mathbf{V}_2 = 5.172\angle{-138.6°}$ V,
$\mathbf{V}_3 = 2.27\angle{-152.4°}$ V

Chapter 14

14.1 $p(t) = 800 + 1600\cos(100t + 60°)$ W, $P = 800$ W

14.3 The average power absorbed by the source = −7.5 W.
For the 4-Ω resistor, the average power absorbed is 5 W.
The average power absorbed by the inductor = 0 W.
For the 2-Ω resistor, the average power absorbed is 2.5 W.
The average power absorbed by the capacitor = 0 W.

14.5 $P_{1\Omega} = 1.4159$ W
$P_{3\text{H}} = P_{0.25\text{F}} = 0$
$P_{2\Omega} = 5.097$ W

14.7 (a) $12.798 + j49.6\ \Omega$,
(b) 90.08 W

14.9 $Z_L = 20\ \Omega$, $P_{max} = 31.25$ W

14.11 Apparent power = 275.6 VA
pf = 0.1876 (lagging)

14.13 (a) $P_{j30\Omega} = 0 = P_{-j40\Omega}$
$P_{10\Omega} = 8.665$ W
$P_{50\Omega} = 46.03$ W
$P_{20\Omega} = 87.86$ W,
(b) 177.8 VA,
(c) pf = 0.8015 (leading)

14.15 Real power = 1320 W
Reactive power = 900 VAR
Apparent power = 1597.6 VA

14.17 $8.158\angle{78.23°}$ VA

14.19 (a) 3600 VA, (b) 0.8233, (c) 2353 VAR

14.21 Average power = 32.5 kW,
Reactive power = 38 kVAR

14.23 (a) Complex power = 112 + j194 VA
Average power = 112 W
Reactive power = 194 VAR,
(b) Complex power = 226.3 − j226.3
Average power = 226.3 W
Reactive power = −226.3 VAR,
(c) Complex power = 110.85 + j64
Average power = 110.85 W
Reactive power = 64 VAR,
(d) Complex power = 7.071 + j7.071 kVA
Average power = 7.071 kW
Reactive power = 7.071 kVAR

14.25 (a) 30.98 − j23.23 Ω, (b) 10.42 + j13.89 Ω, (c) 0.8 + j1.386 Ω

14.27 (a) 0.9845 leading,
(b) $35.55\underline{/55.11^\circ}$ A.

14.29 (a) 1600 VA,
(b) 1.1314 KW,
(c) 1.1314 KVAR,
(d) The power triangle is shown in Figure G.15.

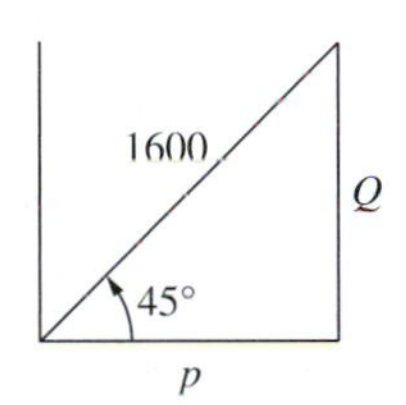

Figure G.15
For Problem 14.29.

14.31 6.167 + j3.561 Ω

14.33 (a) 0.4429 + j0 VA, (b) 0.98 W, (c) 2000 − j5 Ω

14.35 For the 40-V source, 140 − j20 VA
For the capacitor, −j250 VA
For the resistor, 290 VA
For the inductor, j130 VA
For the j50-V source, −150 + j100 VA

14.37 $7.098\underline{/32.29^\circ}$, 0.8454 (lagging)

14.39 (a) 0.6402 (lagging), (b) 295.1 W, (c) 130.4 μF

14.41 10.33 μF

14.43 (a) 12.21 kVA,
(b) $50.86\underline{/-35^\circ}$ A,
(c) 4.083 kVAR, 188.03 μF,
(d) $43.4\underline{/-16.26^\circ}$ A

14.45 172.8 W

14.47 $76.26

14.49 (a) $120,000, (b) $0.10 per kWh

14.51 (a) Apparent power = 12 kVA
Complex power = 9.36 + j7.51 kVA,
(b) 2.866 + j2.3 Ω

14.53 547.3 W

Chapter 15

15.1 $\mathbf{Z}(\omega_0) = 2\ \text{k}\Omega$
$\mathbf{Z}(\omega_0/4) = 2 - j0.75\ \text{k}\Omega$
$\mathbf{Z}(\omega_0/2) = 2 - j0.3\ \text{k}\Omega$
$\mathbf{Z}(2\omega_0) = 2 + j0.3\ \text{k}\Omega$
$\mathbf{Z}(4\omega_0) = 2 + j0.75\ \text{k}\Omega$

15.3 $R = 10\ \Omega$, $L = 16$ H, $C = 25\ \mu$F, $BW = 0.625$ rad/s

15.5 0.7861 rad/s

15.7 (a) 2.513 krad/s,
(b) $\omega_1 = 48$ krad/s
$\omega_2 = 51.51$ krad/s,
(c) 200 Ω

15.9 251.3 rad/s, 8.66

15.11 78.18 nF

15.13 (a) 25 μF, (b) 1, (c) 2 krad/s

15.15 8.796×10^6 rad/s

15.17 56.84 pF, 14.21 μH

15.19 $R = 40\ \Omega$
$C = 10\ \mu$F
$L = 2.5\ \mu$H
$BW = 2.5$ krad/s
$\omega_1 = 198.75$ krad/s
$\omega_2 = 201.25$ krad/s

15.21 4841 rad/s

15.23 (a) $\omega_0 = 1.5811$ rad/s
$Q = 0.1976$
$BW = 8$ rad/s,
(b) $\omega_0 = 5$ krad/s
$Q = 20$
$BW = 250$ rad/s

15.25 298.4 μH, 21.22 pF

15.27 8.443 mH

15.29 The schematic is shown in Figure G.16.

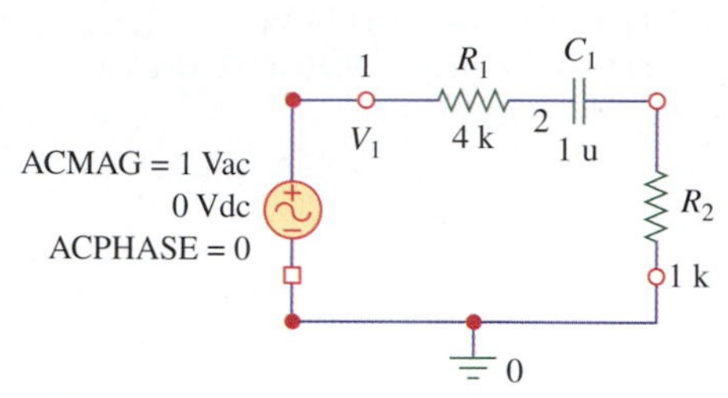

Figure G.16
For Problem 15.29.

15.31 The schematic is shown in Figure G.17.

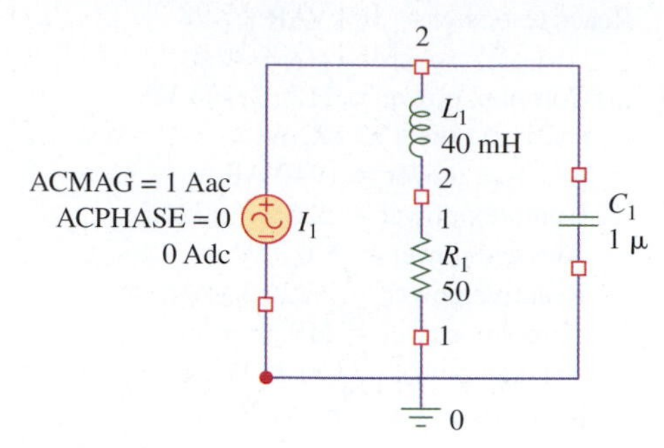

Figure G.17
For Problem 15.31.

15.33 The frequency response (magnitude and phase) is shown in Figure G.18.

15.35 The magnitude and phase plots are shown in Figure G.19.

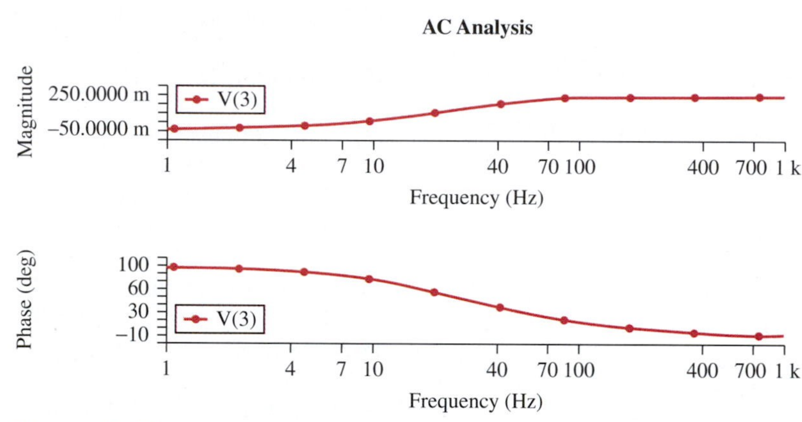

Figure G.18
For Problem 15.33.

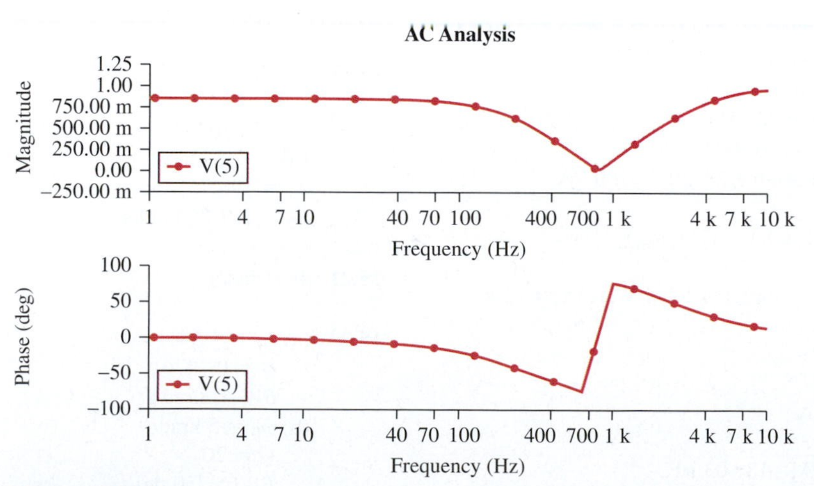

Figure G.19
For Problem 15.35.

15.37 (a) 0.541 MHz $< f_0 <$ 1.624 MHz,
(b) At $f_0 = 0.541$ MHz, $Q = 67.98$
At $f_0 = 1.624$ MHz, $Q = 204.1$

15.39 8.165 MHz, 4.188×10^6 rad/s

Chapter 16

16.1 (a) −4, (b) 4.663, (c) 8, (d) −0.6749

16.3 (a) 12.218 dB, (b) 1,000 dB, (c) 37.781 dB, (d) 88.52 dB

16.5 (a) 1.005773, (b) 0.4898, (c) 1.718×10^5

16.7 43.8 dB

16.9 0.3162 mW, 3.981 mW, 10 W

16.11 11.4×10^{-9} nW, 29.4 dB

16.13 −13.98 dB

16.15 $\mathbf{H}(\omega) = \dfrac{j\omega/\omega_0}{1 + j\omega/\omega_0}$, where $\omega_0 = \dfrac{1}{RC}$

16.17 $\mathbf{H}(s) = \dfrac{1}{s^2R^2C^2 + 3sRC + 1}$

16.19 $\mathbf{H}(\omega) = \dfrac{-\omega^2LC + j\omega RC}{1 - \omega^2LC + j\omega RC}$

16.21 $\mathbf{H}(\omega) = \dfrac{j\omega L - \omega^2RLC}{R + j\omega L - \omega^2RLC}$

16.23 The magnitude and phase plots are shown in Figure G.20.

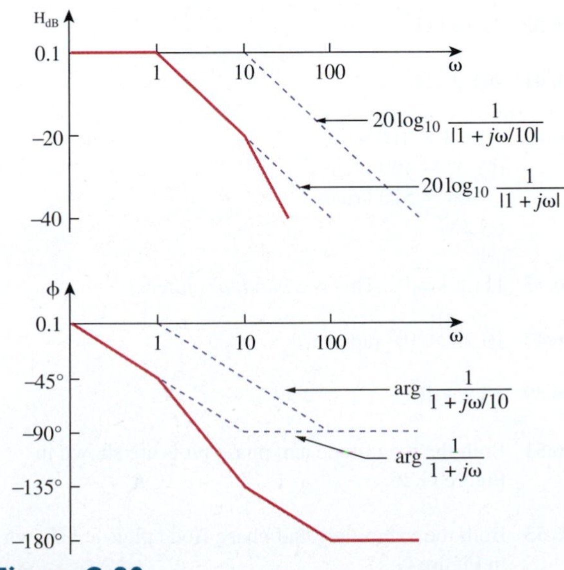

Figure G.20
For Problem 16.23.

16.25 The magnitude and phase plots are shown in Figure G.21.

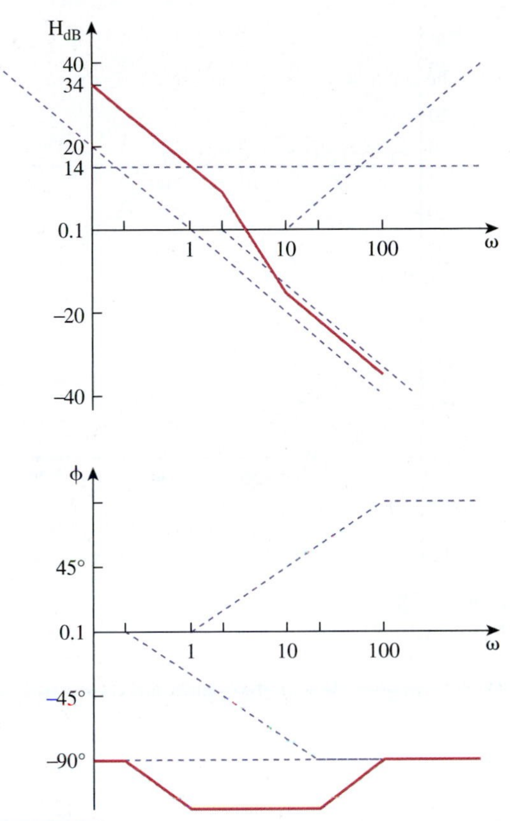

Figure G.21
For Problem 16.25.

16.27 The magnitude and phase plots are shown in Figure G.22.

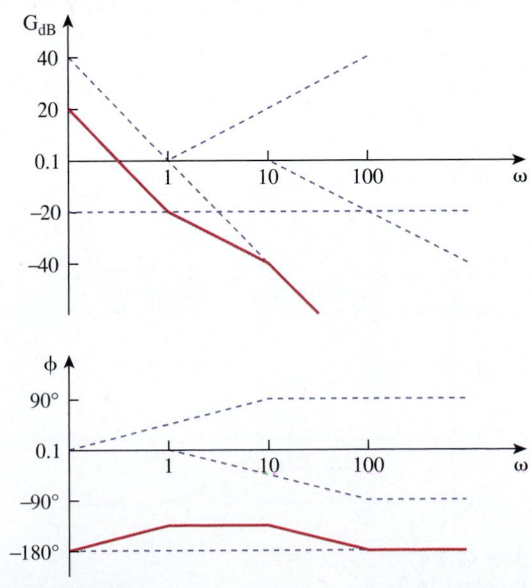

Figure G.22
For Problem 16.27.

16.29 The magnitude and phase plots are shown in Figure G.23.

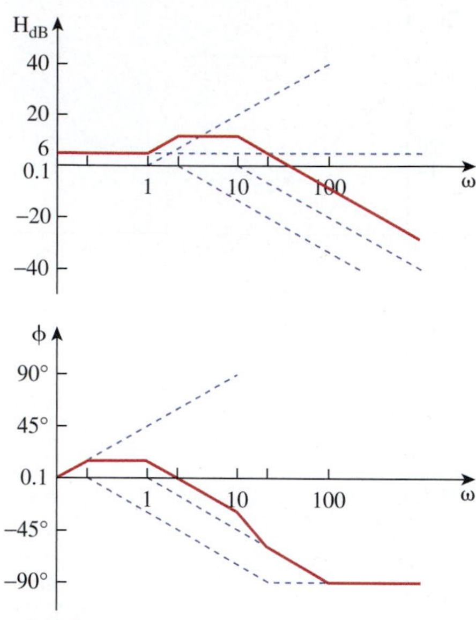

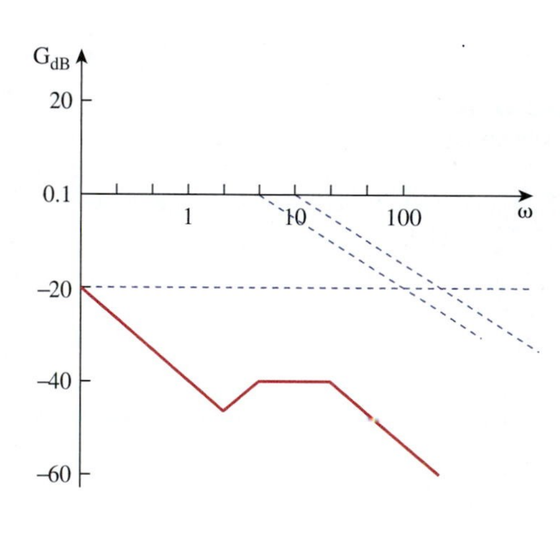

Figure G.23
For Problem 16.29.

16.31 The magnitude and phase plots are shown in Figure G.24.

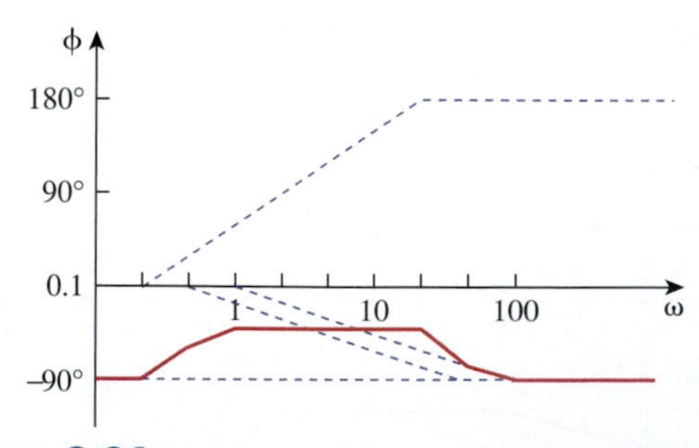

Figure G.24
For Problem 16.31.

16.33 (a) The Bode plot is shown in Figure G.25 (a):

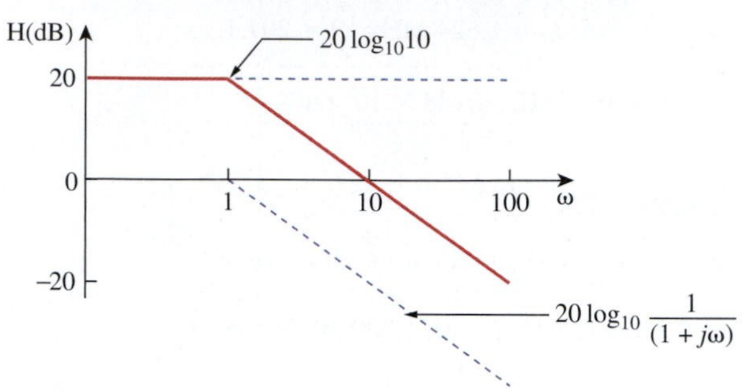

Figure G.25(a)
For Problem 16.33 (a).

(b) The Bode plot is shown in Figure G.25 (b):

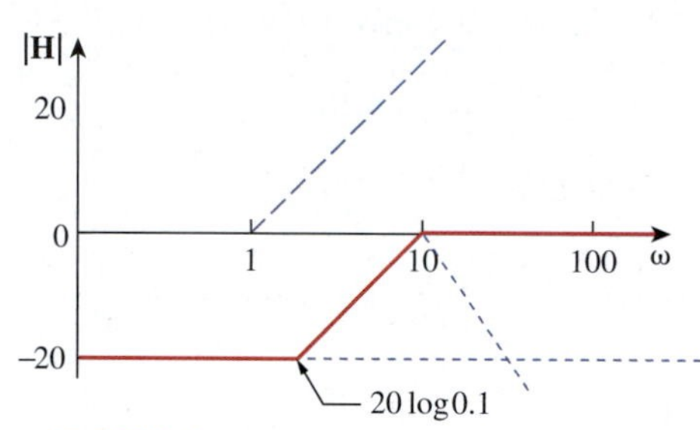

Figure G.25(b)
For Problem 16.33 (b).

16.35 $\mathbf{H}(\omega) = \dfrac{\mathbf{V}_o}{\mathbf{V}_i} = \dfrac{R}{R + j\omega L - \omega^2 RLC}$

H(0) = 1 and H(∞) = 0 showing that this circuit is a lowpass filter

16.37 This circuit is a highpass filter, 318.3 Hz

16.39 25.13 kΩ

16.41 265.3 kΩ

16.43 (a) 0.5×10^6 rad/s,
(b) $\omega_1 = 490$ krad/s,
$\omega_2 = 510$ krad/s,
(c) 25

16.45 111.8 krad/s. This is a *bandpass filter.*

16.47 10.22×10^6 rad/s

16.49 −28.0 dB

16.51 Both the magnitude and phase plots are shown in Figure G.26.

16.53 Both the magnitude and phase Bode plots are shown in Figure G.27.

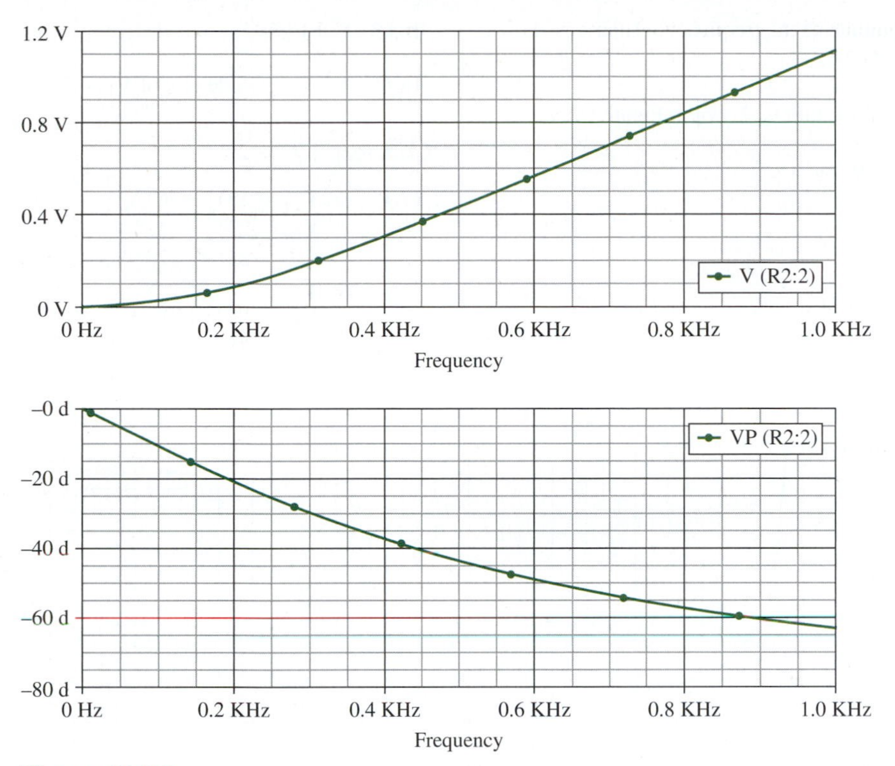

Figure G.26
For Problem 16.51.

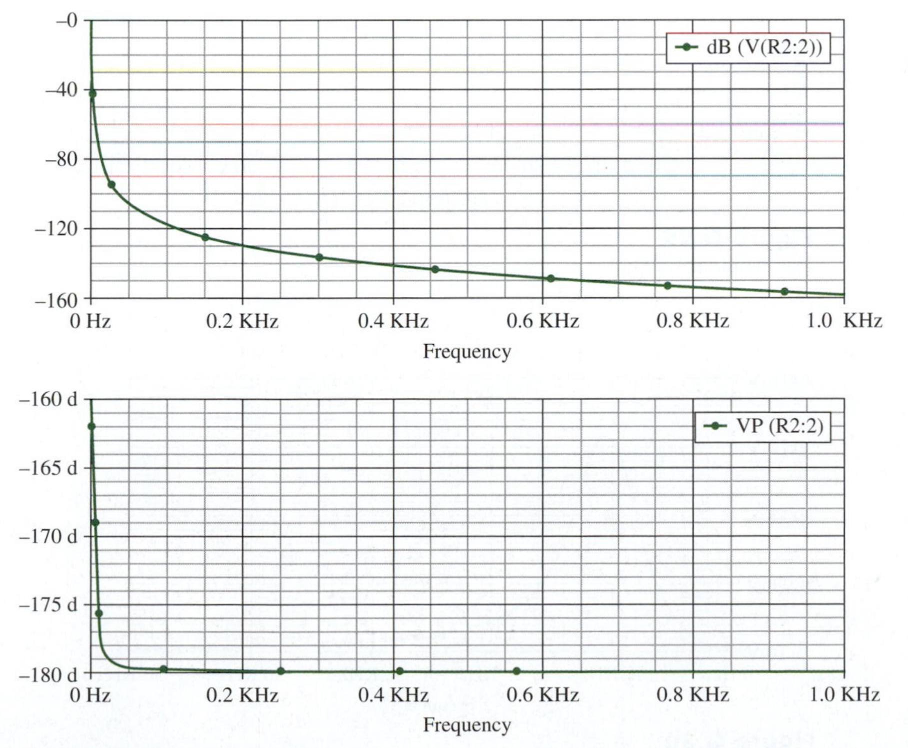

Figure G.27
For Problem 16.53.

16.55 The magnitude of the frequency response is shown in Figure G.28.

16.57 The magnitude plot is shown in Figure G.29.

16.59 114.55×10^6 rad/s

16.61

$$\mathbf{H}(\omega) = \frac{s^3 L R_L C_1 C_2}{(sR_i C_1 + 1)(s^2 L C_2 + sR_L C_2 + 1) + s^2 L C_1 (sR_L C_2 + 1)}$$

where $s = j\omega$.

16.63 1.061 kΩ

16.65 The magnitude plot of V_o is shown in Figure G.30.

Chapter 17

17.1 (a) $\mathbf{V}_{an} = 231\underline{/-30^\circ}$ V
$\mathbf{V}_{bn} = 231\underline{/-150^\circ}$ V
$\mathbf{V}_{cn} = 231\underline{/-270^\circ}$ V,
(b) $\mathbf{V}_{an} = 231\underline{/30^\circ}$ V
$\mathbf{V}_{bn} = 231\underline{/150^\circ}$ V
$\mathbf{V}_{cn} = 231\underline{/-90^\circ}$ V

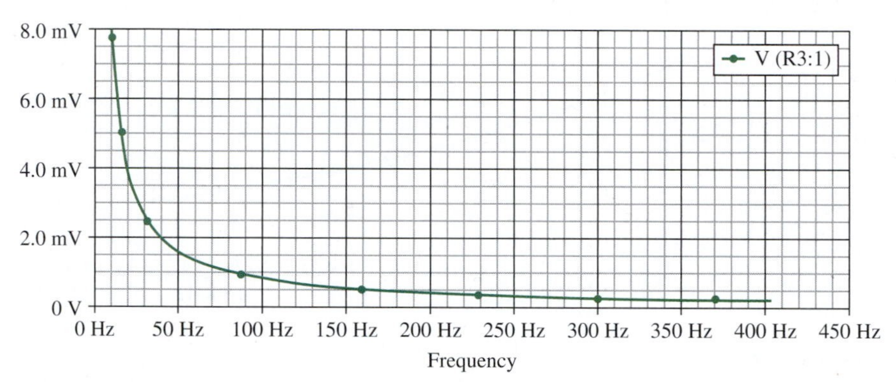

Figure G.28
For Problem 16.55.

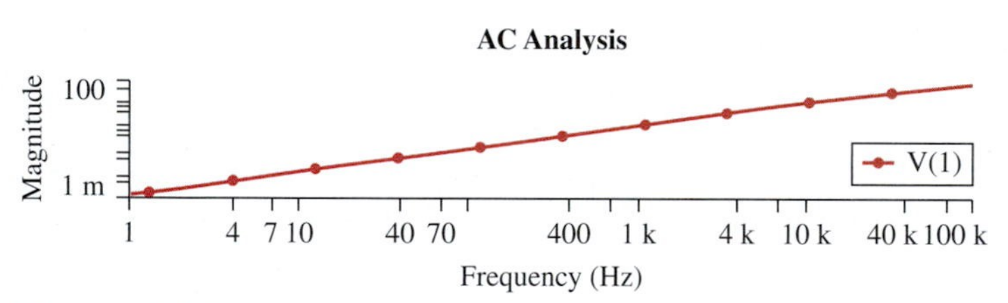

Figure G.29
For Problem 16.57.

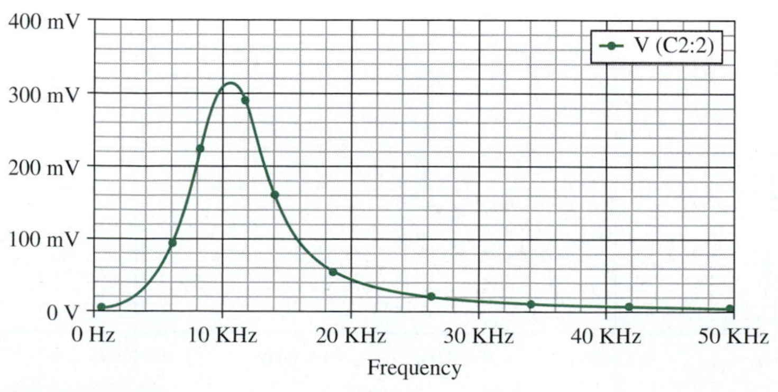

Figure G.30
For Problem 16.65.

17.3 An abc sequence, $208\angle 250^\circ$ V

17.5 $440\angle 210^\circ$ V

17.7 254.03 V

17.9 (a) $\mathbf{V}_{BN} = 230\angle -105^\circ$ V
$\mathbf{V}_{CN} = 230\angle 135^\circ$ V,
(b) $\mathbf{V}_{AB} = 398.37\angle 45^\circ$ V
$\mathbf{V}_{BC} = 398.37\angle -75^\circ$ V
$\mathbf{V}_{CA} = 398.37\angle 165^\circ$ V

17.11 $\mathbf{I}_a = 44\angle 53.13^\circ$ A
$\mathbf{I}_b = 44\angle -66.87^\circ$ A
$\mathbf{I}_c = 44\angle 173.13^\circ$ A

17.13 $\mathbf{I}_a = 4.8\angle -36.87^\circ$ A
$\mathbf{I}_b = 4.8\angle -156.87^\circ$ A
$\mathbf{I}_c = 4.8\angle 83.13^\circ$ A
$\mathbf{I}_n = 0$ A

17.15 $\mathbf{V}_{an} = 127\angle 100^\circ$ V
$\mathbf{V}_{AB} = 220\angle 130^\circ$ V
$\mathbf{I}_{AC} = 17.32\angle 150^\circ$ A
$\mathbf{Z} = 12.7\angle -80^\circ\ \Omega$

17.17 13.66 A

17.19 (a) 75 A(rms), (b) 43.3 A, (c) $5.54\angle 36.9^\circ\ \Omega$

17.21 $\mathbf{I}_{AB} = 14\angle -36.87^\circ$ A
$\mathbf{I}_{BC} = 14\angle -156.87^\circ$ A
$\mathbf{I}_{CA} = 14\angle 83.13^\circ$ A
$\mathbf{I}_a = 24.25\angle -66.87^\circ$ A
$\mathbf{I}_b = 24.25\angle -186.87^\circ$ A
$\mathbf{I}_c = 24.25\angle 53.13^\circ$ A

17.23 $\mathbf{I}_a = 11.24\angle -31^\circ$ A
$\mathbf{I}_b = 11.24\angle -151^\circ$ A
$\mathbf{I}_c = 11.24\angle 89^\circ$ A
$\mathbf{I}_{AB} = 6.489\angle -1^\circ$ A
$\mathbf{I}_{BC} = 6.489\angle -121^\circ$ A
$\mathbf{I}_{CA} = 6.489\angle 119^\circ$ A

17.25 $\mathbf{I}_a = 17.74\angle 4.78^\circ$ A
$\mathbf{I}_b = 17.74\angle -115.22^\circ$ A
$\mathbf{I}_c = 17.74\angle 124.78^\circ$ A

17.27 $\mathbf{I}_a = 5.081\angle -46.87^\circ$ A
$\mathbf{I}_b = 5.081\angle -166.87^\circ$ A
$\mathbf{I}_c = 5.081\angle 73.13^\circ$ A

17.29 13.856 A

17.31 12 kW

17.33 65.61 A, $2.323 + j3.098\ \Omega$

17.35 6.732 A, $1359.2 - j2175$ VA

17.37 39.19 A, 0.9982 (lagging)

17.39 $\mathbf{I}_{AB} = 3.464\angle 30^\circ$ A
$\mathbf{I}_{BC} = 3.464\angle 0^\circ$ A
$\mathbf{I}_{CA} = 3.464\angle 60^\circ$ A

17.41 $I_a = 1.9585\angle -18.1^\circ$ A, $I_b = 1.4656\angle -130.55^\circ$ A,
$I_c = 1.947\angle 117.8^\circ$ A

17.43 $I_{aA} = 1.9\angle -71.6^\circ$ A rms
$I_{bB} = 4\angle -210^\circ$ A rms
$I_{cC} = 1.5\angle 106^\circ$ A rms
The total complex power $= -111.5 + j563.2$ VA

17.45 $\mathbf{V}_{AN} = 150\angle 50^\circ$, $\mathbf{V}_{CN} = 54.33\angle -148.6^\circ$,
$\mathbf{V}_{BN} = 206.8\angle 77.04^\circ$

17.47 $\mathbf{I}_{ab} = 3.667 \times 10^7\angle 60^\circ$ A, $\mathbf{I}_{bB} = 8.822\angle -67.61^\circ$ A

17.49 0.4472 (leading), $40\angle -63.43^\circ\ \Omega$

17.51 206.06 W, 371.65 W

17.53 (a) 192.45 A, (b) 128 kW, (c) 96 kVAR

Chapter 18

18.1 10 H

18.3 $L_1 = 150$ mH
$L_2 = 50$ mH
$M = 25$ mH
$k = 0.2887$

18.5 $\mathbf{V}_1 = (R_1 + j\omega L_1)I_1 - j\omega MI_2$
$\mathbf{V}_2 = -j\omega MI_1 + (R_2 + j\omega L_2)I_2$

18.7 $2.392\angle 94.57^\circ$

18.9 $13.195 + j11.244\ \Omega$

18.11 The T-section is shown in Figure G. 31.

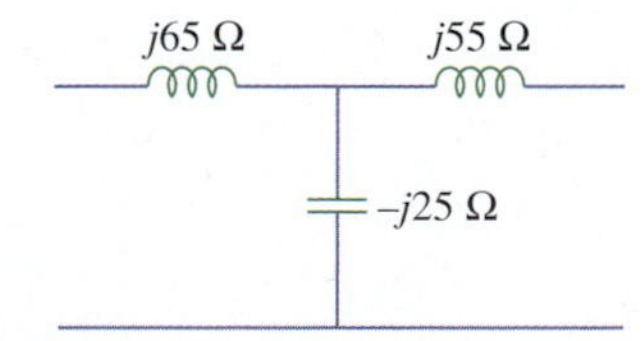

Figure G.31
For Problem 18.11.

18.13 $\mathbf{I}_1 = 4.512\angle -31.97^\circ$ A
$\mathbf{I}_2 = 1.805\angle 21.16^\circ$ A
$\mathbf{V}_o = 7.22\angle 201.16^\circ$ V

18.15 $I_1 = 4.254\angle -8.51^\circ$ A, $I_2 = 1.5637\angle 27.52^\circ$ A
Power absorbed by the 4-ohm resistor is 4.89 watts

18.17 (a) 0.3535, (b) $321.7 \cos(4t + 57.6°)$ mV, (c) 1.168 J

18.19 0.9238

18.21 0.984, 112.35 mJ

18.23 (a) $L_a = 10$ H
$L_b = 15$ H
$L_c = 5$ H,
(b) $L_A = 18.33$ H
$L_B = 27.5$ H
$L_C = 55$ H

18.25 $9.219\underline{/79.91°}\ \Omega$

18.27 420 V, 0.8 A

18.29 $1.324\underline{/-53.05°}$ k ohms

18.31 $I_1 = 0.5$ A and $I_2 = -1.5$ A

18.33 36.71 mW

18.35 $1.923\underline{/157.4°}$ A

18.37 (a) 5, (b) 8 watts

18.39 1059 watts

18.41 (a) 0.2887, (b) 1.732 V

18.43 $4.28\underline{/10.89°}$ A

18.45 (a) 160 V, (b) 31.25 A, (c) 12.5 A

18.47 $(1.2 - j2)$ kΩ, 5.333 watts

18.49 $\mathbf{I}_1 = 4.253\underline{/-8.53°}$, $\mathbf{I}_2 = 1.564\underline{/27.49°}$
The power absorbed by the 4-ohm resistor = 4.892 W

18.51 $\mathbf{I}_1 = 6\underline{/89.8}$ mA, $\mathbf{I}_2 = 4.003 \times 10^{-5}\underline{/-1.42°}$ A,
$\mathbf{I}_3 = 1.204 \times 10^{-7}\underline{/-95.78°}$ A

18.53 100 turns

18.55 0.5

18.57 $n = 0.5$
The primary current = 41.67 A
The secondary current = 83.33 A

18.59 (a) 1875 kVA, (b) 7812 A

18.61 (a) 1/60, (b) 139 mA

18.63 (a) 2, (b) $I_1 = 7.83$ A
The secondary current = $I_2 = 3.92$ A

Chapter 19

19.1 $[\mathbf{z}] = \begin{bmatrix} 4 & 1 \\ 1 & 1.667 \end{bmatrix} \Omega$

19.3 $[\mathbf{z}] = \begin{bmatrix} 1.775 + j4.26 & -1.775 - j4.26 \\ -1.775 - j4.26 & 1.775 - j5.739 \end{bmatrix} \Omega$

19.5 A T network is appropriate for realizing the z parameters.
$R_1 = 6\ \Omega$
$R_2 = 2\ \Omega$
$R_3 = 4\ \Omega$

19.7 (a) 24 Ω, (b) 192 W

19.9 $[z] = \begin{bmatrix} 50 & 50 \\ 30 & 54 \end{bmatrix} \Omega$

19.11 $[\mathbf{y}] = \begin{bmatrix} s + 0.5 & -0.5 \\ -0.5 & 0.5 + 1/s \end{bmatrix}$ S

19.13 A network is appropriate, as shown below.
$\mathbf{Z}_1 = 4\ \Omega$, $\mathbf{Z}_2 = 4\ \Omega, 8\ \Omega$

19.15 (a) $\mathbf{V}_1 = 22$ V
$\mathbf{V}_2 = 8$ V,
(b) $\mathbf{V}_1 = 22$ V
$\mathbf{V}_2 = 8$ V

19.17 $[y] = \begin{bmatrix} 0.24 & -0.17 \\ -0.6 & 1.05 \end{bmatrix}$ S

19.19 $[\mathbf{h}] = \begin{bmatrix} 10\ \Omega & 1 \\ -1 & 0.05\ \text{S} \end{bmatrix}$

19.21 $[h] = \begin{bmatrix} 3.0769 + j1.2821 & 0.3846 - j0.2564 \\ -0.3846 + j0.2564 & 0.0769 + j0.2821 \end{bmatrix}$

19.23 (a) 0.2941, (b) −1.6, (c) 7.353 mS, (d) 40 Ω

19.25 $[\mathbf{z}] = \begin{bmatrix} R_1 + R_2 & R_2 \\ R_2 & R_2 + R_3 \end{bmatrix}$

$$\Delta_z = (R_1 + R_2)(R_2 + R_3) - R_2^2$$
$$= R_1R_2 + R_2R_3 + R_3R_1$$

$$[\mathbf{h}] = \begin{bmatrix} \dfrac{\Delta_z}{\mathbf{z}_{22}} & \dfrac{\mathbf{z}_{12}}{\mathbf{z}_{22}} \\ \dfrac{-\mathbf{z}_{21}}{\mathbf{z}_{22}} & \dfrac{1}{\mathbf{z}_{22}} \end{bmatrix}$$

$$= \begin{bmatrix} \dfrac{R_1R_2 + R_2R_3 + R_3R_1}{R_2 + R_3} & \dfrac{R_2}{R_2 + R_3} \\ \dfrac{-R_2}{R_2 + R_3} & \dfrac{1}{R_2 + R_3} \end{bmatrix}$$

Thus,

$$h_{11} = R_1 + \frac{R_2R_3}{R_2 + R_3}, \qquad h_{12} = \frac{R_2}{R_2 + R_3} = h_{21},$$

$$h_{22} = \frac{1}{R_2 + R_3}$$

as required.

19.27 (a) $[y] = \begin{bmatrix} (0.4 + j0.2) & (0.4 - j0.4) \\ (0.1 + j0.1) & (0.1 + j0.3) \end{bmatrix}$,

(b) $[h] = \begin{bmatrix} (2 - j) & (-0.4 + j1.2) \\ (0.3 + j0.1) & (0.3 + j0.1) \end{bmatrix}$

19.29 (a) $[z] = \begin{bmatrix} 11 & 5 \\ -5 & 2.5 \end{bmatrix} \Omega$,

(b) $[\mathbf{y}] = \begin{bmatrix} 1 & -2 \\ -2 & 4.4 \end{bmatrix} \text{S}$

19.31 (a) $[y] = \begin{bmatrix} \frac{10}{46} & \frac{-2}{46} \\ \frac{-2}{46} & \frac{5}{46} \end{bmatrix}$,

(b) $[h] = \begin{bmatrix} \frac{46}{10} & \frac{2}{10} \\ \frac{-2}{10} & \frac{1}{10} \end{bmatrix}$

19.33 $[z] = \begin{bmatrix} 3.5 & 1.5 \\ 1.5 & 3.5 \end{bmatrix} \Omega$

$[y] = \begin{bmatrix} 0.35 & -0.15 \\ -0.15 & 0.35 \end{bmatrix} \text{S}$

$[h] = \begin{bmatrix} 2.86 & 0.43 \\ -0.43 & 0.29 \end{bmatrix}$

19.35 0.09375

19.37 $[\mathbf{h}] = \begin{bmatrix} \frac{1}{6}\Omega & \frac{1}{2} \\ -\frac{1}{2} & \frac{9}{2}\text{S} \end{bmatrix}$

19.39 $[h] = \begin{bmatrix} 0.9488\angle{-161.6^\circ} & 0.3163\angle{18.42^\circ} \\ 0.3163\angle{-161.6^\circ} & 0.9488\angle{-161.6^\circ} \end{bmatrix}$

19.41 $[z] = \begin{bmatrix} 4.669\angle{-136.7^\circ} & 2.53\angle{-108.4^\circ} \\ 2.53\angle{-108.4^\circ} & 1.789\angle{-153.4^\circ} \end{bmatrix}$

19.43 −1442, 63.18 dB

19.45 (a) −156.2, (b) 76.3, (c) ≅1.2 kΩ, (d) 114.3 kΩ

19.47 $A_v = -17.74$
$A_i = 144.5$
$Z_{in} = 31.17$ kΩ
$Z_{out} = -6.148$ MΩ

Index

E

F

G

H

I

J

K

L

M

T

U

V

W

Y

Z